Human Physiology

The Basis of Medicine

Third Edition

Gillian Pocock
Senior Lecturer in Clinical Science
Canterbury Christ Church University College
Canterbury, UK

and

Christopher D. Richards
Professor of Experimental Physiology
Department of Physiology
University College London

OXFORD
UNIVERSITY PRESS

OXFORD

UNIVERSITY PRESS

Great Clarendon Street, Oxford OX2 6DP

Oxford University Press is a department of the University of Oxford.
It furthers the University's objective of excellence in research, scholarship,
and education by publishing worldwide in

Oxford New York

Auckland Cape Town Dar es Salaam Hong Kong Karachi
Kuala Lumpur Madrid Melbourne Mexico City Nairobi
New Delhi Shanghai Taipei Toronto

With offices in

Argentina Austria Brazil Chile Czech Republic France Greece
Guatemala Hungary Italy Japan Poland Portugal Singapore
South Korea Switzerland Thailand Turkey Ukraine Vietnam

Oxford is a registered trade mark of Oxford University Press
in the UK and in certain other countries

Published in the United States
by Oxford University Press Inc., New York

First edition published 1999
Reprinted (with corrections) 2001
Second edition published 2004
Third edition published 2006

British Library Cataloguing in Publication Data

Data available

Library of Congress Cataloging in Publication Data

Data available

Typeset by EXPO Holdings, Malaysia
Printed in China
on acid-free paper through Asia Pacific Offset

ISBN 0-19-856878-9 978-0-19-856878-0

5 7 9 10 8 6 4

OXFORD MEDICAL PUBLICATIONS

Human Physiology:
The Basis of Medicine

Oxford Core Texts

To: Chris, David P., David R., James, Sue, and Rebecca and to Joan *in memoriam*,

and to

Michael de Burgh Daly (1922–2002)

Preface to the first edition

The idea for this book grew out of regular discussions between the authors when we were both on the staff of the Department of Physiology, Royal Free Hospital School of Medicine in London. We felt that there was a need for a modern, concise textbook of physiology which covered all aspects of the preclinical course in physiology. The text is written primarily for students of medicine and related subjects, so that the clinical implications of the subject are deliberately emphasized. Nevertheless, we hope that the book will also prove useful as core material for first- and second-year science students. We have assumed a knowledge of chemistry and biology similar to that expected from British students with 'AS' levels in these subjects. Our intention has been to provide clear explanations of the basic principles that govern the physiological processes of the human body and to show how these principles can be applied to the understanding of disease processes.

The book begins with cell physiology (including some elementary biochemistry), and proceeds to consider how cells interact both by direct contact and by longer-distance signaling. The nervous system and endocrine system are dealt with at this point. The physiology of the main body systems is then discussed. These extensive chapters are followed by a series of shorter chapters describing integrated physiological responses including the control of growth, the regulation of body temperature, the physiology of exercise, and the regulation of body fluid volume. The final chapters are mainly concerned with the clinical applications of physiology, including acid–base balance, heart failure, hypertension, liver failure, and renal failure. This structure is not a reflection of the organization of a particular course but is intended to show how, by understanding the way in which cells work and how their activity is integrated, one can arrive at a satisfying explanation of body function.

In providing straightforward accounts of specific topics, it has occasionally been necessary to omit some details or alternative explanations. Although this approach occasionally presents a picture that is more clear cut than the evidence warrants, we believe that this is justified in the interests of clarity. Key points are illustrated by simple line drawings as we have found that they are a useful aid to students in understanding and remembering important concepts. We have not included extensive accounts of the experimental techniques of physiology but have tried to make clear the importance of experimental evidence in elucidating underlying mechanisms. Normal values have been given throughout the text in SI units but important physiological variables have also been given in traditional units (e.g. mmHg for pressure measurements).

Each chapter is organized in the same way. In answer to the frequently heard plea 'what do I need to know?' we have set out the key learning objectives for each chapter. This is followed, where appropriate, by a brief account of the physical and chemical principles required to understand the physiological processes under discussion. The essential anatomy and histology are then discussed, as a proper appreciation of any physiological process must be grounded on a knowledge of the main anatomical features of the organs involved. Detailed discussion of the main physiological topics then follows.

To aid student learning, short numbered summaries are given after each major section. From time to time we have set out important biological questions or major statements as section headings. We hope that this will help students to identify more clearly why a particular topic is being discussed. The reading material given at the end of each chapter is intended both to provide links with other subjects commonly studied as part of the medical curriculum and to provide sources from which more detailed information can be obtained. Self-testing is encouraged by the provision of multiple-choice questions or quantitative problems (or both) at the end of each chapter. Annotated answers to the questions are given. Some numerical problems have also been given which are intended to familiarize students with the key formulae and to encourage them to think in quantitative terms.

We are deeply indebted to Professor Michael de Burgh Daly and Dr Ted Debnam, who not only advised us on their specialist topics but also read through and constructively criticized the entire manuscript. Any remaining obscurities or errors are entirely our responsibility. Finally, we wish to thank the staff of Oxford University Press for their belief in the project, their forbearance when writing was slow, and their help in the realization of the final product.

London   G.P.
February 1999   C.D.R.

Preface to the third edition

Our aim in the third edition of this book remains that of the first edition: we have attempted to provide clear explanations of the basic principles that govern the physiological processes of the human body and to show how these principles can be applied to the practice of medicine.

In this new edition, we have taken the opportunity of eliminating errors, clarifying presentation, and expanding our treatment of many topics. To this end we have included around 30 new figures and redrawn many others. New material includes a more detailed discussion of glaucoma and its treatment; the clinical significance of eye movement defects; cerebellar ataxia; autonomic failure; an updated discussion of iron metabolism; more detailed discussion of the MHC complex and complement; antibody presentation and the role of dendritic cells. The discussion of immunological disorders now includes deficiencies in different complement factors. Carbohydrate absorption has been revised and updated. Finally, Chapter 31 (Clinical physiology) now includes new sections on clinical aspects of heart sounds and electrocardiography.

We are grateful for the comments we have received from our readers, and for the helpful and detailed advice we have received from our colleagues on a number of topics. Finally, we wish to thank the staff of Oxford University Press for their help and encouragement.

London   G.P.
August 2005   C.D.R.

Contents

Acknowledgments

We wish to acknowledge the help of many colleagues who have helped us to clarify our thinking on a wide variety of topics. Those to whom especial thanks are due for detailed criticisms of particular chapters are listed here:

Professor J.F. Ashmore F.R.S., Department of Physiology, University College London, UK

Professor S. Bevan, The Novartis Institute for Medical Sciences, 5 Gower Place, London, UK

Dr T.V.P. Bliss, F.R.S., Division of Neurophysiology, The National Institute for Medical Research, Mill Hill, London, UK

Professor M. de Burgh Daly, Department of Physiology, University College London, UK

Dr E.S. Debnam, Department of Physiology, University College London, UK

Professor D.A. Eisner, Department of Preclinical Veterinary Sciences, University of Liverpool, Liverpool, UK

Dr B.D. Higgs, Division of Anaesthesia, Royal Free Hospital, London, UK

Professor R. Levick, Department of Physiology, St George's Hospital Medical School, London, UK.

Dr A. Mathie, Department of Biology, Imperial College, London, UK

Dr D.A. Richards, Department of Anatomy, Cell Biology and Neuroscience, University of Cincinnati, Ohio, USA

Professor I.C.A.F. Robinson, Laboratory of Endocrine Physiology, The National Institute for Medical Research, Mill Hill, London, UK

Dr A.H. Short, Department of Physiology and Pharmacology, Medical School, Queen's Medical Centre, Nottingham, UK

The authors wish to thank all those people who have granted permission to reproduce figures from books and original articles either unmodified or in a modified form. The original sources are listed here:

Figures 3.1 and 14.11 are from Figures 1.6 and 8.8 of J.M. Austyn and K.J. Wood (1993) *Principles of Cellular and Molecular Immunology*, Oxford University Press, Oxford, UK. Figure 3.5 is based on Figure 2-18 of B. Alberts, D. Bray, J. Lewis, M. Raff, K. Roberts, and J.D. Watson (1989) *Molecular Biology of the Cell* 2nd edition, Garland, New York. Figures 3.6 and 3.7 are from Figures 8.7 and 8.13 of W.H. Elliott and D.C. Elliott (1997) *Biochemistry and Cell Biology*, Oxford University Press, Oxford, UK. Figure 4.3 is based on Figure 13 of P.C. Caldwell *et al.* (1960) *Journal of Physiology* vol. 152, pp 561–590. Figure 4.7 is from an original figure of Dr E.S. Debnam. Figure 4.10 is from an original figure of Dr P. Charlesworth. Figures 5.3 and 15.35 are based on Figures 2.3 and 13.6 of H.P. Rang, M.M. Dale, and J.M Ritter (1995) *Pharmacology* 3rd edition, Churchill-Livingstone, Edinburgh. Figures 6.1, 6.4, 6.6, 8.5, 8.11, 8.12, 8.37, 8.51, 9.10, 9.15, 9.18, 9.19, 9.20, 15.39 and 15.40 are based on Figures 2.1, 2.6, 2.7, 2.28, 4.1, 14.13, 14.14, 13.24, 13.26, 13.12, 9.1, 11.2, 11.13, 10.2, 10.3, 1.15, 2.38 and 2.39 of P. Brodal (1992) *The Central Nervous System. Structure and Function*, Oxford University Press, New York. Figure 6.10 is based on Figure 17 of A.L. Hodgkin and B. Katz (1949) *Journal of Physiology* vol. 108, pp 37–77. Figure 6.11 is based on Figure 17 of A.L. Hodgkin and A.F. Huxley (1952) *Journal of Physiology* vol. 117, pp 500–540. Figure 6.12 is based on Figure 1 of P. Fatt and B. Katz (1952) *Journal of Physiology* vol. 117, pp 109–128. Figure 7.1 is modified after Figure 11.19 of W. Bloom and D.W. Fawcett (1975) *Textbook of Histology*, W.B. Saunders & Co. Figure 7.2 is based on Figure 10.18 in *Human Physiology 3E* (illustrator – Sylvia Colard Keene), reproduced by permission of Edward Arnold. Figure 7.11 is based on Figures 12 and 14 of A.M. Gordon, A.F. Huxley, and F.J. Julian (1966) *Journal of Physiology* vol. 184, pp 170–192. Figures 8.7, 8.9, 8.50, 8.52 and 8.53 are based on Figures 16.3, 16.6, 16.10, 17.2 and 17.6 of H.B. Barlow and J.D. Mollon (eds) (1982) *The Senses*, Cambridge University Press, Cambridge. Figure 8.10 is based on Figure 30.6 of D. Ottoson (1984) *The Physiology of the Nervous System*, Macmillan Press, London. Figures 8.18, 8.26 and 11.2 are based on Figures 7.3, 7.18 and 13.4 of R.H.S. Carpenter (1996) *Neurophysiology* 3rd edition, Edward Arnold, London. Figure 8.20 is based on Figure 20.6 of E.R. Kandel, J.H. Schwartz, and T.M. Jessell (Eds) (1991) *Principles of Neuroscience* 3rd edition, Elsevier Science, New York. Figure 8.23 is from Figure 4.5 of R.F. Schmidt (Ed) (1986) *Fundamentals of Sensory Physiology* 3rd edition, Springer-Verlag, Berlin. Figures 8.20 and 8.22, are based on Figures 7.4 and 7.2 in vol. 3 of P.C.B MacKinnon and J.F. Morris *Oxford Textbook of Functional Anatomy*, Oxford University Press, Oxford. Figure 8.30 is from P.H. Schiller (1992) *Trends in Neurosciences* vol. 15, p 87 with permission of Elsevier Science. Figure 8.31 is derived from data of D.H. Hubel and T.N. Wiesel. Figure 8.41 is based on Figures 2.4 and 2.5 of J.O. Pickles (1982) *An Introduction to the Physiology of Hearing*, Academic Press, London.

Figure 8.44 is partly based on Figure 5 of I.J. Russell and P.M. Sellick (1978) *Journal of Physiology*, vol. 284, pp 261–290. Figure 8.45 is partly based on J.O. Pickles and D.P. Corey (1992) *Trends in Neurosciences* vol. 15 p. 255, with permission of Elsevier Science. Figure 8.46 is based on data of E.F. Evans, with permission. Figure 8.47 is based on Figure 14.14a from *The Senses* edited by H.B. Barlow and J.D. Mollon, published by Cambridge University Press, with permission from Professor Evans, University of Keele. Figure 8.49 is adapted from Lindemann (1969), *Ergebnisse der Anatomie* vol. 42 pp 1–113. Figure 10.2 is adapted from Figure 18.2 of G.M. Shepherd *Neurobiology*, Oxford University Press, New York. Figure 11.3 is modified from *Left Brain, Right Brain*, 3/e by Sally P. Springer and Georg Deutsch. © 1981, 1985, 1989 by Sally P. Springer and Georg Deutsch. Reprinted by permission of W.H. Freeman and Company. Figure 11.12 is from original data of Dr D.A. Richards. Figures 12.9, 12.13, 12.21 and 12.24 are adapted from Figures 3.4, 5.2, 4.1 and 4.12 of C. Brook and N. Marshall (1996) *Essential Endocrinology*, Blackwell Science, Oxford. Figures 12.11, 12.18 and 12.19(b) are from plates 4.1, 4.2, 9.1, 9.2 and 9.3 of J. Laycock and P. Wise (1996) *Essential Endocrinology* 3rd edition, Oxford Medical Publications, Oxford. Figures 12.17, 12.26, 12.27, 19.3, 19.4, 19.6 and 23.15 are printed courtesy of the Wellcome Library. Figure 12.19(a) is reproduced from Braverman & Utiger, The Thyroid – A Fundamental and Clinical Text 9E, © 2004, by permission of Lippincott, Williams & Wilkins. Figure 13.4 is from an original figure of Dr E.S. Debnam. Figures 14.1, 14.7 and 14.10 are based on Figures 9.1, 9.2, 13.1 and 13.2 of J.H. Playfair (1995) *Infection and Immunity*, Oxford University Press, Oxford. Figures 15.1, 15.19, 15.29 and 15.40 are based on Figures 1.4, 1.6, 9.2 and 13.6 of J.R. Levick (1995) *An Introduction to Cardiovascular Physiology* 2nd edition, Butterworth-Heinemann, Oxford. Figures 15.2 and 15.3 by courtesy of S. Ruehm. Figures 15.4, 15.10, 15.34, 16.1, 16.5 are based on Figures 5.4.10, 5.4.6, 5.4.7, 5.3.3 and 5.3.4 in vol. 2 of P.C.B MacKinnon and J.F. Morris *Oxford Textbook of Functional Anatomy*, Oxford University Press, Oxford. Figure 15.15 is based on Figures 4 and 10 of O.F. Hutter and W. Trautwein (1956) *Journal of General Physiology*, vol. 39, pp 715–733 by permission of the Rockefeller University Press. Figure 15.24 is based on Figure 19.14 of A.C. Guyton (1986) *Textbook of Medical Physiology* 7th edition, W.B. Saunders & Co, Philadelphia. Figure 15.25 is based on Figure 7 of A.E. Pollack and E.H. Wood (1949) *Journal of Applied Physiology* vol. 1 pp 649–662. Figure 15.28 is based on Figure 5 of L.H. Smaje et al. (1970) *Microvascular Research* vol. 2, pp 96–110. Figure 15.34 is based on Figure 5.9 of R.F. Rushmer (1976) *Cardiovascular Dynamics*, W.B. Saunders & Co, Philadelphia. Figure 15.39 is based on data of R.M. Berne and R. Rubio (1979) *Coronary Circulation*, in *American Handbook of Physiology, Section 2, The Cardiovascular System*, Oxford University Press, New York. Figures 16.2, 16.12, 16.21 and 16.30 are based on Figures 1.5, 2.4, 5.4 and 8.5 of J. Widdicombe and A. Davies (1991) *Respiratory Physiology*, Edward Arnold, London. Figure 16.4 is based on Figure 11.1 of E.R. Weibel (1984) *The Pathway for Oxygen. Structure and Function of the Mammalian Respiratory System*, Harvard University Press, Boston. Figures 16.8, 16.9, 16.10 and 24.3 are based on Figures III.4, III.3, III.11, III.80 and III.81 of C.A. Keele, E. Neil, and N. Joels *Samson Wright's Applied Physiology* 13th edition, Oxford University Press, Oxford. Figure 16.18 is adapted from Figure 3.8 of M.G. Levitsky (1991) *Pulmonary Physiology* 3rd edn, McGraw-Hill, New York. Figure 16.19 is based on Figure 8.1 of J.F. Nunn (1993), *Nunn's Applied Respiratory Physiology* 4th edition, Butterworth-Heinemann, Oxford. Figures 16.20 and 25.3 are from original figures of Professor M. de Burgh Daly. Figure 16.22 is based on Figure 19 of J.B. West *Ventilation/Blood Flow and Gas Exchange*, Blackwell Science, Oxford. Figure 16.25 is based on data of W. Barron and J.H. Coote (1973) *Journal of Physiology* vol. 235, pp 423–436. Figure 16.28 is based on Figure 39 of P. Dejours (1966) *Respiration*, Oxford University Press, Oxford. Figures 17.2 and 17.5 are based on Figures 1 and 2 of W. Kritz and L. Bankir (1988) *American Journal of Physiology* vol. 254 pp. F1–F8. Figure 17.4 is based on Figure 2.3 of B.M. Keoppen and B.A. Stanton (1992) *Renal Physiology*, Mosby, St Louis. Figure 18.3 is based on Figure 21.13 of D.F. Moffett, S.B. Moffett, and C.L. Schauf (1993) *Human Physiology*, McGraw-Hill, New York. Figures 18.9, 18.12, 18.18, 18.19, 18.25 are based on Figures 6.6.2, 6.7.1 and 7.4 in vol. 2 and Figures 7.7 and 7.8 of vol. 3 of P. C. B MacKinnon and J.F. Morris, *Oxford Textbook of Functional Anatomy*, Oxford University Press, Oxford. Figure 18.17 is based on Figure 13 of J.H. Szurszewiski, *Journal of Physiology* vol. 252 pp 335–361. Figure 18.35 is based on Figures 6.1 and 6.2 of P.A. Sanford (1992) *Digestive System Physiology* 2nd edition, Edward Arnold. London. Figures 19.1 and 19.2 are from Figures 17.1 and 17.2 of J.A. Mann and S. Truswell (1998) *Essentials of Human Nutrition*, Oxford Medical Publications, Oxford. Figures 19.5 and 28.9 are by courtesy of WHO. Figure 20.7 is based on Figure 6.10.3 in vol. 2 of P.C.B MacKinnon and J.F. Morris *Oxford Textbook of Functional Anatomy*, Oxford University Press, Oxford. Figure 21.12 is based on data of F. Hytten and G. Chamberlain in *Clinical Physiology in Obstetrics*, Blackwell Scientific Publications, Boston. Figure 21.13 is based on data of R.M. Pitkin (1976) *Clinical Obstetrics and Gynecology*, vol 19, pp 489–513. Figure 22.5 is adapted from Figure 13.1 of D. J. Begley, J.A. Firth, and J.R.S. Hoult (1980) *Human Reproduction and Developmental Biology*, Macmillan Press, London. Figures 22.10 and 22.11 are based on Figures 7.3 and 7.2 of N.E. Griffin and S.R. Ojeda (1995) *Textbook of Endocrine Physiology* 2nd edition, Oxford University Press, Oxford. Figure 23.8 is reproduced by permission of Edward Arnold from Don W. Fawcett and Ronald P. Jensh, *Concise Histology*, 2/e (Arnold 2002), © D.W. Fawcett and R.P. Jensh. Figure 23.1 is from Figure 16 of J.M. Tanner (1989) *Foetus into Man* 2nd edition, Castlemead, London. Figure 23.9 and Figure 2 of Box 23.1 were kindly provided by Ruth Denton. Figures 25.1 and 25.2 are from Figures 7.3 and 7.10 of P.-O. Astrand and K. Rohdal (1986) *Textbook of Work Physiology. Physiological Basis of Exercise* 3rd edition, McGraw-Hill, New York. Figure 26.3 is from J. Werner (1977) *Pflugers Archiv* vol. 367 pp 291–294. Figure 28.8 was kindly provided by Professor R. Levick. Figure 28.10 is from plates 9 and 10 of R.A. Hope, J.M. Longmore, S.K. McManus, and C.A. Wood-Allum (1998) *Oxford Handbook of Clinical Medicine* 4th edition, Oxford University Press, Oxford. Figure 30.2 is from Figure 1 of N. Pace, B. Meyer and B.E. Vaughan (1956) *Journal of Applied Physiology* vol. 9 pp 141–144 by permission of the American Physiological Society. Figure 31.2 is based on data of S. Landhal *et al.* (1986) *Hypertension* vol. 8, pp 1044–9. Figure 31.3 is based on data in Table 11.3 of J.F. Nunn (1993), *Nunn's Applied Respiratory Physiology* 4th edition, Butterworth-Heinemann, Oxford. Figure 31.4 is based on data of U. Klotz *et al.* (1975) *Journal of Clinical Investigation* vol. 55 pp 347–359.

List of abbreviations

ACh	acetylcholine
ACTH	adrenocorticotropic hormone (corticotropin)
ADH	antidiuretic hormone (vasopressin)
ADP	adenosine diphosphate
AIDS	acquired immunodeficiency syndrome
AMP	adenosine monophosphate
ANP	atrial natriuretic peptide
ARDS	adult respiratory distress syndrome
ATP	adenosine triphosphate
ATPS	ambient temperature and pressure saturated with water vapor (with reference to respiratory gas)
AV	atrioventricular, arteriovenous
BER	basic electrical rhythm
BMI	body mass index
BMR	basal metabolic rate
BP	blood pressure
2,3-BPG	2,3-bisphosphoglycerate
b.p.m.	beats per minute
BTPS	body temperature and pressure saturated (with water vapor)
C	gas content of blood (e.g. C_{vO_2})
CCK	cholecystokinin
CJD	Creutzfeld–Jakob disease
CLIP	corticotropin-like peptide
CN	cranial nerve
CNS	central nervous system
CO	cardiac output
CoA	coenzyme A
CRH	corticotropin-releasing hormone
CSF	cerebrospinal fluid
CVP	central venous pressure
DAG	diacylglycerol
dB	decibel
DIT	di-iodotyrosine
DMT1	divalent metal ion transporter 1
DNA	deoxyribonucleic acid
ECF	extracellular fluid
ECG	electrocardiogram

ECV	effective circulating volume
EDRF	endothelium-derived relaxing factor
EDTA	ethylenediaminetetraacetic acid
EDV	end-diastolic volume
EEG	electroencephalogram
ENS	enteric nervous system
epp	end-plate potential
epsp	excitatory postsynaptic potential
ER	endoplasmic reticulum
ESV	end-systolic volume
FAD	flavine adenine dinucleotide
$FADH_2$	reduced flavine adenine dinucleotide
FEV_1	forced expiratory volume at 1 second
FRC	functional residual volume
FSH	follicle-stimulating hormone
FVC	forced vital capacity
GABA	γ-aminobutyric acid
GALT	gut-associated lymphoid tissue
GDP	guanosine diphosphate
GFR	glomerular filtration rate
GH	growth hormone (somatotropin)
GHIH	growth hormone-inhibiting hormone (somato-statin)
GHRH	growth hormone-releasing hormone
GI	gastrointestinal
GIP	gastric inhibitory peptide
GLUT2	glucose transporter family member 2
GLUT5	glucose transporter family member 5
GMP	guanosine monophosphate
GnRH	gonadotropin-releasing hormone
G protein	heterotrimeric GTP-binding protein
GTP	guanosine triphosphate
Hb	hemoglobin
HbF	fetal hemoglobin
HbS	sickle cell hemoglobin
hCG	human chorionic gonadotropin
hGH	human growth hormone
HIV	human immunodeficiency virus
HLA	human leukocyte antigen

hPL	human placental lactogen
HPNS	high-pressure nervous syndrome
HRT	hormone replacement therapy
HSL	hormone-sensitive lipase
5-HT	5-hydroxytryptamine
ICF	intracellular fluid
ICSH	interstitial cell stimulating hormone (identical with luteinizing hormone)
IgA	immunoglobin A
IgE	immunoglobin E
IgG	immunoglobin G
IgM	immunoglobin M
IGF-1, IGF-2	insulin-like growth factor 1, insulin-like growth factor 2
IP_3	inositol trisphosphate
ipsp	inhibitory postsynaptic potential
kph	kilometers per hour
LDL	low-density lipoprotein
LH	luteinizing hormone
LHRH	luteinizing hormone releasing hormone
LTB_4	leukotriene B_4
LTP	long-term potentiation
MALT	mucosa-associated lymphoid tissue
MAP	mean arterial pressure
mepp	miniature end-plate potential
MHC	major histocompatibility complex
MIH	Mullerian inhibiting hormone
MIT	mono-iodotyrosine
MMC	migrating motility complex
MODS	multiple organ system dysfunction syndrome
M_r	relative molecular mass
mRNA	messenger RNA
MSH	melanophore stimulating hormone
MVV	maximum ventilatory volume
NAD	nicotinamide adenine dinucleotide
NADH	reduced nicotinamide adenine dinucleotide
NTS	nucleus of the tractus solitarius
P	pressure (see Chapter 16, Box 16.1, for explanation of symbols)
P_{50}	pressure for half saturation
PAH	p-aminohippurate
PDGF	platelet-derived growth factor
PGE_2	prostaglandin E_2
$PGF_{2\alpha}$	prostaglandin $F_{2\alpha}$
PGI_2	prostacyclin
PIH	prolactin inhibitory hormone (dopamine)
PRL	prolactin
PTH	parathyroid hormone
RBF	renal blood flow
RDA	recommended daily amount
REM	rapid eye movement
Rh	rhesus factor (D antigen)
RNA	ribonucleic acid
RPF	renal plasma flow
RQ	respiratory quotient (also known as the respiratory exchange ratio)
RV	residual volume
SA	sinoatrial
SGLT1	sodium linked glucose transporter 1
SPL	sound pressure level
STP	standard temperature and pressure
STPD	standard temperature and pressure dry
SV	stroke volume
SWS	slow-wave sleep
T	absolute temperature
T_3	tri-iodothyronine
T_4	thyroxine
TENS	transcutaneous electrical nerve stimulation
TGF-α	transforming growth factor-α
TGF-β	transforming growth factor-β
TH	thyroid hormone
T_m	transport maximum
TPA	tissue plasminogen activator
TPR	total peripheral resistance
TRH	thyrotropin-releasing hormone
tRNA	transfer RNA
TSH	thyroid-stimulating hormone
TXA_2	thromboxane A_2
V	volume (usually of gas, see Chapter 16, Box 16.1, for explanation of subscripts)
$\dot{V}$	flow rate
$\dot{V}/\dot{Q}$	ventilation /perfusion ratio (in lungs)
VC	vital capacity
VIP	vasoactive intestinal polypeptide

A note to the reader

The chapters in this book cover the physiological material normally taught in the first and second years of the medical curriculum and degree courses in physiology. While each chapter can be read on its own, the book has been laid out logically in six parts. Section 1 is a broad introduction to the subject. Section 2 (Chapters 2–4) presents basic information on the properties of cells and how they communicate. Section 3 (Chapter 5) deals with the mechanisms by which the body is able to coordinate and regulate the activities of its various parts. Section 4 (Chapters 6–18) includes much of the core material of traditional courses in physiology and discusses the functioning of the principal organ systems. Section 5 (Chapters 19–22) is concerned with the physiology of reproduction and that of the neonate. Section 6 (Chapters 23–30) is concerned with the interactions between different organ systems with an emphasis on the clinical applications of physiology. In an effort to bridge the gap between the basic science taught in the preclinical years and clinical practice, the final chapter is devoted to clinical physiology.

Each chapter begins with a list of learning objectives which set out the principal points that we think you, the reader, should try to assimilate. We have assumed a basic knowledge of chemistry and biology, but important physical topics are briefly discussed where necessary. Key terms and definitions are given in *italics* where they first occur. The contents of each chapter are arranged in numbered sections in the same order as the learning objectives, and each major section ends with a summary of the main points. We have tried to avoid repetition as far as possible by cross-referencing. Many chapters have boxes which contain material that is more advanced or deal with numerical examples. It is not necessary to read these boxes to understand the core material.

At the end of each chapter there is a reading list which is intended to link physiology with other key subjects in the medical curriculum (particularly with anatomy, biochemistry, and pharmacology). These have been chosen for their clarity of exposition but many other good sources are available.

For those who wish to study a particular physiological topic in greater depth, we have also included in our reading lists some monographs which we have found helpful in preparing this book. We have also suggested specific chapters in more advanced texts. In addition to the sources listed, the *American Handbook of Physiology* has more detailed articles on specific topics of physiological interest. These sources will provide you with a guide to the primary source literature which, like other areas of biomedical science, is still advancing rapidly. Articles in mainstream review journals such as the *Annual Review of Physiology* and *Physiological Reviews* will provide an introduction in the most recent developments in particular fields. Many specialist journals now also regularly carry review articles relating to their areas of interest.

Most chapters end with a set of problems. These are mainly in the form of multiple-choice questions but some numerical problems are also included. Do make the effort to test your knowledge—it will help to lodge the key information in your mind. Answers (with explanations where appropriate) are given at the end of each chapter. When you find that a particular topic is difficult to understand, break it down into its components to identify where your difficulties lie. This is the first step towards resolving them. If, after further study, you still have difficulty, seek help from your tutor or lecturer.

What a piece of work is a man! How noble in reason! How infinite in faculty! In form, in moving, how express and admirable! … The paragon of animals!

William Shakespeare, *Hamlet*, Act 2

What is physiology?

After reading this chapter you should understand:

♦ The subject matter of physiology
♦ The hierarchical organization of the body
♦ The concept of homeostasis

1.1 Introduction

Physiology is the study of the functions of living matter. It is concerned with *how* an organism performs its varied activities: how it feeds, how it moves, how it adapts to changing circumstances, how it spawns new generations. The subject is vast and embraces the whole of life. The success of physiology in explaining how organisms perform their daily tasks is based on the notion that they are intricate and exquisite machines whose operation is governed by the laws of physics and chemistry. Although some processes are similar across the whole spectrum of biology—the replication of the genetic code for example—many are specific to particular groups of organisms. For this reason it is necessary to divide the subject into various parts such as bacterial physiology, plant physiology, and animal physiology. The focus of this book is the physiology of mammals, particularly that of humans.

To study how an animal works it is first necessary to know how it is built. A full appreciation of the physiology of an organism must therefore be based on a sound knowledge of its anatomy. Experiments can then be carried out to establish how particular parts perform their functions. Although there have been many important physiological investigations on human volunteers, the need for precise control over the experimental conditions has meant that much of our present physiological knowledge has been derived from studies on other animals such as frogs, rabbits, cats, and dogs. When it is clear that a specific physiological process has a common basis in a wide variety of animal species, it is reasonable to assume that the same principles will apply to humans. The knowledge gained from this approach has given us a great insight into human physiology and endowed us with a solid foundation for the effective treatment of many diseases.

1.2 The organization of the body

The building blocks of the body are the *cells*, which are grouped together to form *tissues*. The principal types of tissue are epithelial, connective (including blood and lymphoid tissue), nervous, and muscular, each with its own characteristics. Many connective tissues have relatively few cells but have an extensive extracellular matrix. In contrast, smooth muscle consists of densely packed layers of muscle cells linked together via specific cell junctions. *Organs* such as the brain, the heart, the lungs, the intestines, and the liver are formed by the aggregation of different kinds of tissue. The organs are themselves parts of distinct physiological *systems*. The heart and blood vessels form the cardiovascular system; the lungs, trachea, and bronchi together with the chest wall and diaphragm form the respiratory system; the skeleton and skeletal muscles form the musculoskeletal system; the brain, spinal cord, autonomic nerves and ganglia, and peripheral somatic nerves form the nervous system, and so on.

Cells differ widely in form and function but they all have certain common characteristics. First, they are bounded by a limiting membrane, the plasma membrane. Secondly, they have the ability to break down large molecules to smaller ones to liberate energy for their activities. Thirdly, at some point in their life history, they possess a nucleus which contains genetic information in the form of deoxyribonucleic acid (DNA). Further details of the fine structure of cells will be considered in Chapter 3.

Living cells continually transform materials. They break down glucose and fats to provide energy for other activities such as motility and the synthesis of proteins for growth and repair. These chemical changes are collectively called *metabolism*. The breakdown of large molecules to smaller ones is called *catabolism* and the synthesis of large molecules from smaller ones *anabolism*.

In the course of evolution, cells began to differentiate to serve different functions. Some developed the ability to contract (muscle cells), others to conduct electrical signals (nerve cells). A further group developed the ability to secrete different substances such as hormones (endocrine cells) or enzymes (e.g. the acinar cells of the salivary glands). During embryological development, this process of *differentiation* is re-enacted as many different types of cell are formed from the fertilized egg.

Most tissues contain a mixture of cell types. For example, blood consists of red cells, white cells, and platelets. Red cells transport oxygen around the body. The white cells play an important role in defense against infection and the platelets are vital components in the process of blood clotting. There are a number of different types of connective tissue but all are characterized by having cells distributed within an extensive noncellular matrix. Nerve tissue contains nerve cells (of which there are many different kinds) and glial cells.

The principal organ systems
The cardiovascular system

The cells of large multicellular animals cannot derive the oxygen and nutrients they need directly from the external environment. These must be transported to the cells. This is one of the principal functions of the blood, which circulates within blood vessels by virtue of the pumping action of the heart. The heart, blood vessels, and associated tissues form the cardiovascular system.

The heart consists of four chambers, two atria and two ventricles, which form a pair of pumps arranged side by side. The right ventricle pumps deoxygenated blood to the lungs where it absorbs oxygen from the air, while the left ventricle pumps oxygenated blood returning from the lungs to the rest of body to supply the tissues. Physiologists are concerned with establishing the factors responsible for the heartbeat, how the heart pumps the blood around the circulation, and how it is distributed to perfuse the tissues according to their needs. Fluid exchanged between the blood plasma and the tissues passes into the *lymphatic system*, which eventually drains back into the blood.

The respiratory system

The energy required for performing the various activities of the body is ultimately derived from respiration. This process involves the oxidation of foodstuffs (principally sugars and fats) to release the energy they contain. The oxygen needed for this process is absorbed from the air in the lungs and carried to the tissues by the blood. The carbon dioxide produced by the respiratory activity of the tissues is carried to the lungs by the blood in the pulmonary artery where it is excreted in the expired air. The basic questions to be answered include the following: How is the air moved in and out of the lungs? How is the volume of air breathed adjusted to meet the requirements of the body? What limits the rate of oxygen uptake in the lungs?

The digestive system

The nutrients needed by the body are derived from the diet. Food is taken in by the mouth and broken down into its component parts by enzymes in the gastrointestinal tract (or gut). The digestive products are then absorbed into the blood across the wall of the intestine and pass to the liver via the portal vein. The liver makes nutrients available to the tissues both for their growth and repair and for the production of energy. In the case of the digestive system, key physiological questions are: How is food ingested? How is it broken down and digested? How are the individual nutrients absorbed? How is the food moved through the gut? How are the indigestible remains eliminated from the body?

The kidneys and urinary tract

The chief function of the kidneys is to control the composition of the extracellular fluid (the fluid which bathes the cells). In the course of this process, they also eliminate non-volatile waste products from the blood. To perform these functions, the kidneys produce urine of variable composition which is temporarily stored in the bladder before voiding. The key physiological questions in this case are: how do the kidneys regulate the composition of the blood? How do they eliminate toxic waste? How do they respond to stresses such as dehydration? What mechanisms allow the storage and elimination of the urine?

The reproductive system

Reproduction is one of the fundamental characteristics of living organisms. The gonads (the testes in the male and the ovaries in the female) produce specialized sex cells known as gametes. At the core of sexual reproduction is the creation and fusion of the male and female gametes, the sperm and ova (eggs), with the result that the genetic characteristics of two separate individuals are mixed to produce offspring that differ genetically from their parents. Key questions are: How are the sperm and eggs produced? What is the mechanism of fertilization? How does the embryo grow and develop? How is it delivered and nourished until it can fend for itself?

The musculoskeletal system

This consists of the bones of the skeleton, skeletal muscles, joints, and their associated tissues. Its primary function is to provide a means of movement, which is required for locomotion, for the maintenance of posture, and for breathing. It also provides physical support for the internal organs. Here the mechanism of muscle contraction is a central issue.

The endocrine and nervous systems

The activities of the different organ systems need to be coordinated and regulated so that they act together to meet the needs of the body. Two coordinating systems have evolved: the nervous system and the endocrine system. The nervous system uses electrical signals to transmit information very rapidly to specific cells. Thus the nerves pass electrical signals to the skeletal muscles to control their contraction. The endocrine system secretes chemical agents, *hormones*, which travel in the bloodstream to the cells upon which they exert a regulatory effect. Hormones play a major role in the regulation of many different organs and are particularly important in the regulation of the menstrual cycle and other aspects of reproduction.

The *immune system* provides the body's defenses against infection both by killing invading organisms and by eliminating diseased or damaged cells.

Although it is helpful to study how each organ performs its functions, it is essential to recognize that the activity of the body as a whole is dependent on the intricate interactions between the various organ systems. If one part fails, the consequences are found in other organ systems throughout the whole body. For example, if the kidneys begin to fail, the regulation of the internal environment is impaired which in turn leads to disorders of function elsewhere.

1.3 Homeostasis

Complex mechanisms are at work to regulate the composition of the extracellular fluid and individual cells have their own mechanisms for regulating their internal composition. The regulatory mechanisms stabilize the internal environment despite variations in both the external world and the activity of the animal. The process of stabilization of the internal environment is called *homeostasis* and is essential if the cells of the body are to function normally.

To take one example, the beating of the heart depends on the rhythmical contractions of cardiac muscle cells. This activity depends on electrical signals which, in turn, depend on the concentration of sodium and potassium ions in the extracellular and intracellular fluids. If there is an excess of potassium in the extracellular fluid, the cardiac muscle cells become too excitable and may contract at inappropriate times rather than in a coordinated manner. Consequently, the concentration of potassium in the extracellular fluid must be kept within a narrow range if the heart is to beat normally.

How does the body regulate its own composition?
The concept of balance

In the course of a day, an adult consumes approximately 1 kg of food and drinks 2–3 liters of fluid. In a month, this is equivalent to around 30 kg of food and 60–90 liters of fluid. Yet, in general, body weight remains remarkably constant. Such individuals are said to be *in balance*; the intake of food and drink matches the amounts used to generate energy for normal bodily activities plus the losses in urine and feces. In some circumstances, such as starvation, intake does not match the needs of the body and muscle tissue is broken down to provide glucose for the generation of energy. Here, the intake of protein is less than the rate of breakdown and the individual is said to have a *negative nitrogen balance* (nitrogen is a characteristic component of the amino acids that make up the protein—see Chapter 2). Equally, if the body tissues are being built up, as is the case for growing children, pregnant women and athletes in the early stages of training, the daily intake of protein is greater than the normal body turnover and the individual is in *positive nitrogen balance*.

This concept of balance can be applied to any of the body constituents including water and salt (NaCl) and is important in considering how the body regulates its own composition. Intake must match requirements and any excess must be excreted for balance to be maintained. Additionally, for each chemical constituent of the body there is a desirable concentration range, which the control mechanisms are adapted to maintain. For example, the concentration of glucose in the plasma (the fluid part of the blood) is about 4–5 mmol.l^{-1} between meals. Shortly after a meal, plasma glucose rises above this level and this stimulates the secretion of the hormone insulin by the pancreas, which acts to bring the concentration down. As the concentration of glucose falls, so does the secretion of insulin. In each case, the changes in the circulating level of insulin act (together with other mechanisms) to maintain the plasma glucose at an appropriate level. This type of regulation is known as *negative feedback*. During the period of insulin secretion, the glucose is being stored as either glycogen (mainly in the liver and muscles) or fat (in specialized fat cells in adipose tissue).

A *negative feedback loop* is a control system that acts to maintain the level of some variable within a given range following a disturbance. Although the example given above refers to plasma glucose, the basic principle can be applied to other physiological variables such as body temperature, blood pressure, and the osmolality of the plasma.

A negative feedback loop requires a *sensor* of some kind that responds to the variable in question but not to other physiological variables. Thus an osmoreceptor should respond to changes in osmolality of the body fluids but not to changes in body temperature or blood pressure. The information from the sensor must be compared in some way with the desired level (known as the 'set point' of the system) by some form of *comparator*. If the two do not match, an error signal is transmitted to an *effector*, a system that can act to restore the variable to its desired level. The basic features of a negative feedback loop are summarized

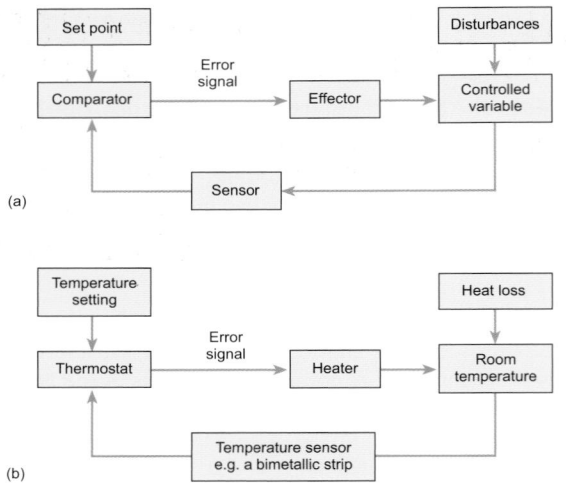

Fig. 1.1 Schematic drawings of a negative feedback control loop (a) compared with a simple heating system (b).

in Fig. 1.1. These features of negative feedback can be appreciated by examining a simple heating system. The controlled variable is room temperature, which is sensed by a thermostat. The effector is a heater of some kind. When the room temperature falls below the set point, the temperature difference is detected by the thermostat which switches on the heater. This heats the room until the temperature reaches the pre-set level whereupon the heater is switched off.

Although negative feedback is the principal mechanism for maintaining a constant internal environment, it does have certain disadvantages. First, negative feedback control can only be exerted after the controlled variable has been disturbed. Secondly, the correction to be applied can only be assessed by the magnitude of the error signal (the difference between the desired value and the displaced value of the variable in question). In practice, this means that negative feedback systems provide incomplete correction. Thirdly, overcorrection has the potential for causing oscillations in the controlled variable. These disadvantages are largely overcome in physiological systems by means of multiple regulatory processes. In the example above, blood glucose is maintained within a narrow range by two mechanisms that act in opposition (push–pull). Insulin acts to lower plasma glucose while another pancreatic hormone, glucagon, acts to mobilize glucose from the body's stores.

While it is difficult to overemphasize the importance of negative feedback control loops in homeostatic mechanisms, they are frequently reset or overridden in stresses of various kinds. For example, arterial blood pressure is monitored by receptors, known as baroreceptors, which are found in the walls of the aortic arch and carotid sinus. These receptors are the sensors for a negative feedback loop that maintains the arterial blood pressure within close limits. If the blood pressure rises, compensatory changes occur that tend to restore it to normal. In exercise, however, this mechanism is reset. Indeed, if it were not, the amount of exercise we could undertake would be very limited.

Negative feedback loops operate to maintain a particular variable within a specific range. They are a stabilizing force in the economy of the body. However, in some circumstances *positive feedback* occurs. In this case, the feedback loop is inherently unstable as the error signal acts to increase the initial deviation. An example from everyday life is the howling that occurs when a microphone is placed near one of the loudspeakers of a public address system. The microphone picks up the initial sound and this is amplified by the electronic circuitry. This drives the loudspeaker to emit a louder sound, which is again picked up by the microphone and amplified so that the loudspeaker makes an even louder sound, and so on until the amplifying circuitry reaches the limit of its power—and the hearers run for cover!

An example of the interaction between negative and positive feedback mechanisms is the hormonal regulation of the menstrual cycle. Cyclical alterations in the plasma levels of two hormones from the pituitary gland known as follicle-stimulating hormone (FSH) and luteinizing hormone (LH) are involved in the regulation of fertility. Steroid hormones from the ovaries can exert both negative and positive feedback control on the output of FSH and LH, depending upon the concentration of hormone present. Low or moderate levels of a hormone called estradiol-17β, tend to inhibit secretion of FSH and LH (negative feedback). However, if estradiol-17β is present in high concentrations for several days, it stimulates the secretion of FSH and LH (positive feedback). As a result, there is a sharp increase in the output of both FSH and LH just before midcycle. This rise is responsible for ovulation. Once ovulation has taken place, estrogen levels fall sharply and the output of FSH and LH drops as negative feedback reasserts control.

Recommended reading

Houk, J.C. (1980). Homeostasis and control principles. In *Medical physiology* (14th edn) (ed. V.B. Mountcastle), Chapter 8, pp. 246–267. Mosby, St Louis, MO.

Paton, W.D.M. (1993). *Man and mouse. Animals in medical research* (2nd edn). Oxford University Press, Oxford.

The chemical constitution of the body

After reading this chapter you should understand:

◆ The chemical composition of the body
◆ The properties of water as a biological solvent: polar and non-polar compounds
◆ The osmotic pressure and tonicity of aqueous solutions
◆ The structure and functions of the carbohydrates
◆ The chemical nature and functions of lipids
◆ The structure of the amino acids and proteins
◆ The structure of the nucleotides and the nucleic acids

2.1 Introduction

The human body consists largely of four elements: oxygen, carbon, hydrogen, and nitrogen. These are combined in many different ways to make a huge variety of chemical compounds. About 70 per cent of the lean body tissues is water, the remaining 30 per cent being made up of organic (i.e. carbon-containing) molecules and minerals. The principal organic constituents of mammalian cells are the *carbohydrates*, *fats*, *proteins*, and *nucleic acids*, which are built from small molecules belonging to four classes of chemical compounds: the sugars, the fatty acids, the amino acids, and the nucleotides. Of the minerals found in the tissues, the most abundant are calcium, iron, magnesium, phosphorous, potassium, and sodium.

The chemical composition of the body given in Table 2.1 is an approximate average of all the tissues of an adult. The proportions of the various constituents vary between tissues and change during development (see Chapter 28).

2.2 Body water

As mentioned above, water is the principal constituent of the human body and is essential for life. It is the chief solvent in living cells. Molecules of biological interest can be divided into those that dissolve readily in water and those that do not. Substances that dissolve readily in water are called *polar* or *hydrophilic*, while those that are insoluble in water are called *non-polar* or *hydrophobic*. Examples of polar substances are sodium chloride, glucose, and ethanol, while examples of non-polar

TABLE 2.1 The approximate chemical constitution of the body (all values are expressed as percentage body weight)	
Oxygen	65
Carbon	18
Hydrogen	10
Nitrogen	3.4
Minerals total	3.6
Na$^+$	0.17
K$^+$	0.28
Cl$^-$	0.16
Mg^{2+}	0.05
Ca^{2+}	1.5
Phosphorus	1.2
Sulfur	0.25
Fe^{2+}/$^{3+}$	0.007
Zn^{2+}	0.002

The body contains trace amounts of other elements in addition to those listed above.

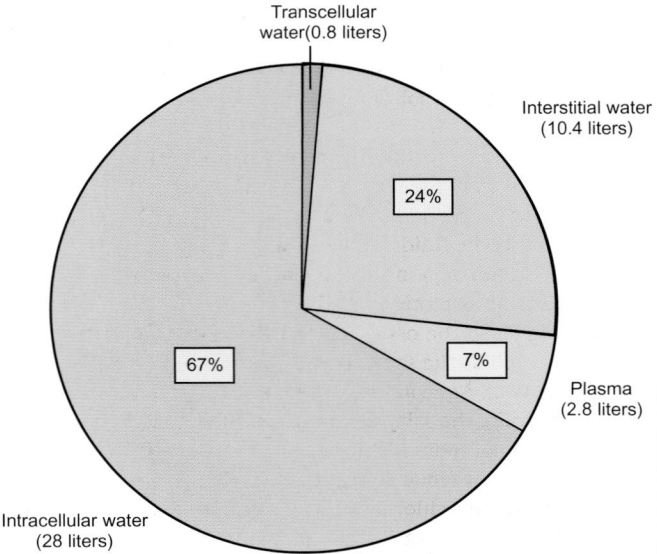

Fig. 2.1 The distribution of body water between the various compartments for a 70 kg man.

materials are fats and cholesterol. Many molecules of biological interest have mixed properties, so that one part is polar while another part is non-polar. These are known as *amphiphilic* substances. Examples of amphiphilic substances are the phospholipids and the bile salts.

The intracellular and extracellular fluids

Body water can be divided into that within the cells, the *intracellular water*, and that which lies outside the cells, the *extracellular water*. As the body water contains many different substances in solution, the liquid portions (i.e. the water plus the dissolved materials) of cells and tissues are known as fluids. The fluid of the space outside the cells is called the *extracellular fluid*, while that inside the cells is the *intracellular fluid*. The extracellular fluid in the serosal spaces such as the ventricles of the brain, the peritoneal cavity, the joint capsules, and the ocular fluids is called *transcellular fluid* (see Fig. 2.1). The extracellular fluid is further subdivided into the plasma and the interstitial fluid. The plasma is the liquid fraction of the blood while the interstitial fluid lies outside the blood vessels and bathes the cells. The distribution of water between the different body compartments and the mechanisms that regulate body water balance are considered in Chapter 28 (pp. 547–549).

The intracellular fluid is separated from the extracellular fluid by the plasma membrane of the individual cells which is mainly composed of lipids (fats) and has a non-polar core (see Chapter 3). Consequently, polar molecules cannot readily cross from the extracellular fluid to the intracellular fluid. Indeed, this barrier is used to create concentration gradients that the cells exploit to perform various functions.

Diffusion

When a substance (the *solute*) is dissolved in a *solvent* to form a solution, the individual solute molecules become dispersed within the solvent and are free to move in a random way. Thus,

in an aqueous solution, the molecules of both water and solute are in continuous random motion with frequent collisions between them. This process leads to *diffusion*, the random dispersion of molecules in solution. When a drop of a concentrated solution (e.g. 5 per cent w/v glucose) is added to a volume of pure water, the random motion of the glucose molecules results in their slow dispersion throughout the whole volume. If the drop of 5 per cent solution had been added to a 1 per cent solution of glucose, the same process of dispersion of the glucose molecules would occur until the whole solution was of uniform concentration. There is a tendency for the glucose (or any other solute) to diffuse from a region of high concentration to one of a lower concentration (i.e. down its concentration gradient).

The rate of diffusion in a solvent depends on temperature (it is faster at higher temperatures), the magnitude of the concentration gradient, and the area over which diffusion can occur. The molecular characteristics of the solute and solvent also affect the rate of diffusion. These characteristics are reflected in a physical constant known as the *diffusion coefficient*. The role of these different factors is expressed in Fick's law of diffusion which is discussed briefly in Box 15.6 (p. 294). In general, large molecules diffuse more slowly than small ones. Note that diffusion is not confined to the fluids of the body but also occurs through cell membranes, which are largely made of lipids (see pp. 20–22).

The osmotic pressure of the body fluids

When an aqueous solution is separated from pure water by a membrane that is permeable to water but not to the solute, water moves across the membrane into the solution by a process known as *osmosis*. This movement can be opposed by applying a hydrostatic pressure to the solution. The pressure that is just sufficient to prevent the uptake of water is known as the *osmotic pressure* (π) of the solution. The osmotic pressure of a solution of known molar composition (*M*) can be calculated from the following simple equation:

$$\pi = MRT$$

where R is the universal gas constant (8.31 J K^{-1} mol^{-1}) and T is the absolute temperature (310K at normal body temperature). The osmotic pressure is thus directly related to the number of particles present in a solution, and is independent of their chemical nature.

Rather than measuring osmotic pressure directly, it is more convenient to state the *osmolarity* (moles per liter of solution) or *osmolality* (moles per kg of water). In clinical medicine, osmotic pressure of body fluids is generally expressed as *osmolality*. One gram mole of a non-dissociating substance in 1 kg of water exerts an osmotic pressure of 1 Osmole (abbreviated as 1 Osmol kg^{-1}). So the osmotic pressure exerted by a mmole of glucose (M_r 180) is the same as that exerted by a mmole of albumin (M_r 69 000). Aqueous salt solutions are an important exception to this rule: the salts separate into their constituent ions so that a solution of sodium chloride will exert an osmotic pressure double that of its molal concentration. Hence a 100 mmol kg^{-1} solution of sodium chloride in water will have an osmotic pressure of 200 mOsmol kg^{-1}, of which a half is due to the sodium ions and half to the chloride ions.

The total osmolality of a solution is the sum of the osmolality due to each of the constituents. The extracellular fluid and plasma have an osmolality of around 0.3 Osmol kg^{-1} (300 mOsmol kg^{-1}). The principal ions (Na$^+$, K$^+$, Cl$^-$, etc.) contribute about 290 mOsmol kg^{-1} (about 96 per cent) while glucose, amino acids, and other small non-ionic substances contribute approximately 10 mOsmol kg^{-1}. Proteins contribute only around 0.5 per cent to the total osmolality of plasma and still less to the osmolality of the extracellular fluid (which has little plasma protein). This is made clear by the following calculations: A liter of blood plasma is 95 per cent water and contains about 6.42 g of sodium chloride, and 45 g of albumin. These figures correspond to 6.76 g of sodium chloride and 47.4 g of albumin per kilogram of plasma water. The osmolality of a solution of 6.76 g of sodium chloride (M_r 58.4) in 1 kg water is:

$$(2 \times 6.76) \div 58.4 = 0.231 \text{ Osmol kg}^{-1} \text{ or } 231 \text{ mOsmol kg}^{-1}$$

The osmotic pressure exerted by 47.4 g of albumin is:

$$47.4 \div 69\,000 = 6.87 \times 10^{-4} \text{ Osmol kg}^{-1} \text{ or } 0.687 \text{ mOsmol kg}^{-1}$$

Thus the osmotic pressure exerted by 47 g of albumin is only about 0.3 per cent that of 6.76 g of sodium chloride. This makes clear that **the osmotic pressure exerted by proteins is far less than that exerted by the principal ions of the biological fluids.** Nevertheless, the small osmotic pressure that the proteins do exert (known as the *colloid osmotic pressure* or *oncotic pressure*) plays an important role in the exchange of fluids between body compartments.

Although lipid membranes are hydrophobic, they are more permeable to water than they are to ions so that the osmolality of the intracellular fluid is the same as that of the extracellular fluid (i.e. the two fluids have an osmolality of about 300 mOsmol kg^{-1} and are *iso-osmotic*). If the osmotic pressure in one compartment is higher than the other, water will move from the region of low osmotic pressure to that of the higher osmotic pressure until the two become equalized.

The tonicity of solutions

The tonicity of a solution refers to the influence of its osmolality on the volume of cells. For example, red blood cells placed in a solution of 0.9 per cent sodium chloride in water (i.e. 0.9 g sodium chloride in 100 ml of water) neither swell nor shrink. This concentration has an osmolality ≈ 310 mOsmol kg^{-1} and is said to be *isotonic* with the cells. (This solution is sometimes referred to as 'normal saline' but would be better called isotonic saline). If the same cells are added to a solution of sodium chloride with an osmolality of 260 mOsmol kg^{-1}, they will swell as they take up water to equalize the osmotic pressure across their cell membranes (see Fig. 2.2). This concentration of sodium chloride is said to be *hypotonic* with respect to the cells. Solutions that have a very low osmolality cause cells placed in them to swell so much that they burst, a process called *lysis*. Conversely, red blood cells placed in a solution of sodium chloride that has an osmolality of 360 mOsmol kg^{-1} will shrink as water is drawn from the cells. In this case the fluid is *hypertonic* and the surface of the cells becomes irregularly folded. Cells with such an appearance are said to be *crenated*.

Not all solutions that are iso-osmotic with respect to the intracellular fluid are isotonic with cells. A solution containing 310 mOsmol kg^{-1} of urea is iso-osmotic with both normal saline and the intracellular fluid, but it is not isotonic, as cells placed in such a solution would swell and burst (or lyse). This behavior is explained by the fact that urea can penetrate the cell membrane

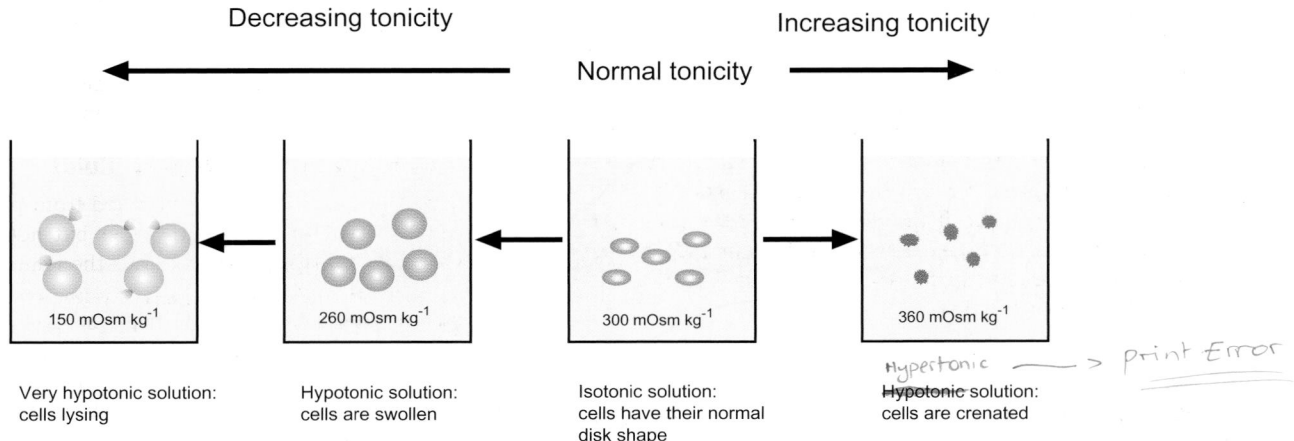

Decreasing tonicity ←   → Increasing tonicity

Normal tonicity →

150 mOsm kg^{-1} — Very hypotonic solution: cells lysing

260 mOsm kg^{-1} — Hypotonic solution: cells are swollen

300 mOsm kg^{-1} — Isotonic solution: cells have their normal disk shape

360 mOsm kg^{-1} — Hypotonic solution: cells are crenated

Fig. 2.2 The changes in cell volume that occur when cells are placed in solutions of different osmolality. The changes illustrated are for red blood cells.

Summary

1. Water is the chief solvent of the body and accounts for about 50–60 per cent of body mass. Substances that dissolve readily in water are said to be polar (or hydrophilic) while those that are insoluble in water are non-polar (or hydrophobic).

2. Body water can be divided into intracellular water (that within the cells) and extracellular water. The solutes and water of the space inside the cells is called the intracellular fluid, while that outside the cells is the extracellular fluid.

3. When a substance dissolves in water it exerts an osmotic pressure that is related to its molal concentration. The osmotic pressure of a solution is expressed as its osmolality, which is related to the number of particles present per kilogram of solvent, independent of their chemical nature. The total osmolality of a solution is the sum of the osmolality due to each of the constituents.

4. The osmolality of the intracellular fluid is the same as that of the extracellular fluid (i.e. the two fluids are iso-osmotic).

relatively freely. When it does so, it diffuses down its concentration gradient and water will follow (otherwise the osmolality of the intracellular fluid would increase and become *hyper-osmotic*). Since there is an excess of urea outside the cells, it will continue to diffuse into the cells, attracting water via osmosis, and the cells will progressively swell until they burst.

Filtration

When a fluid passes through a permeable membrane, it leaves behind those particles that are larger in diameter than the pores of the membrane. This process is known as *filtration* and is driven by the pressure gradient between the two sides of the membrane. When filtration separates large solutes, such as proteins, from small ones, such as glucose and inorganic ions (Na^+, K^+, Cl^-, etc.), the process is called *ultrafiltration*.

The walls of the capillaries are not normally permeable to plasma proteins (e.g. albumin) but are permeable to small solutes. The pumping action of the heart causes a pressure gradient across the walls of the capillaries, which tends to force fluid from the capillaries into the interstitial space. This process occurs in all vascular beds but is particularly important in the glomerular capillaries of the kidney, which filter large volumes of plasma each day.

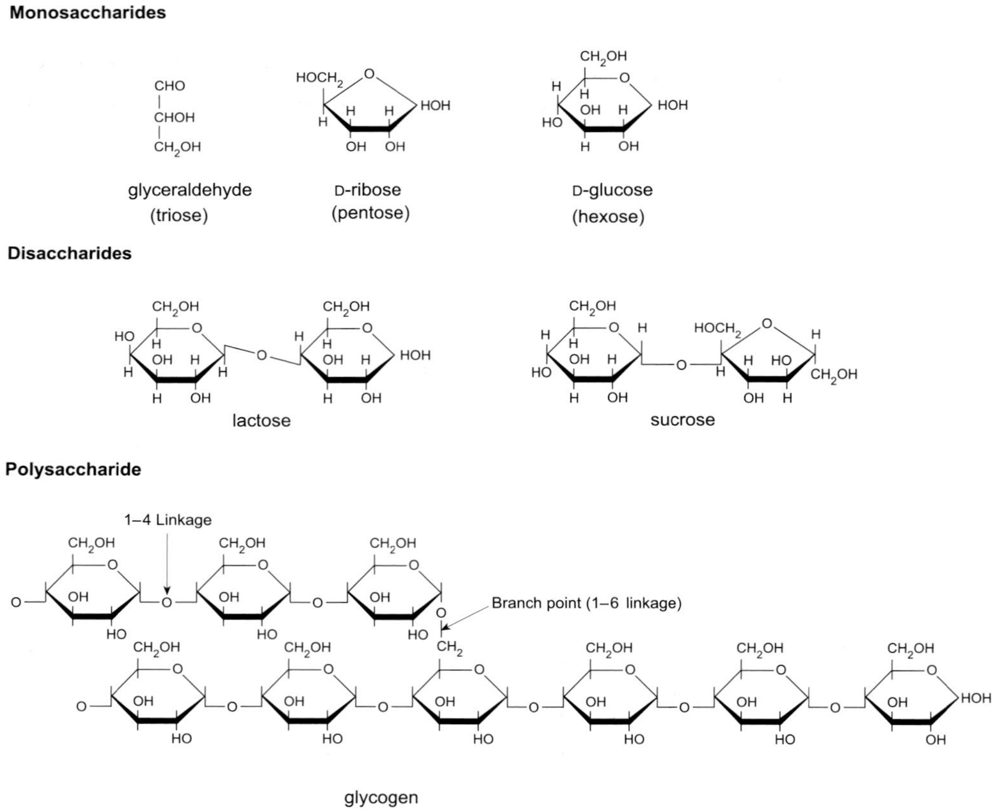

Fig. 2.3 The structure of representative members of the carbohydrates. The polysaccharide glycogen consists of many glucose molecules joined together by 1–4 linkages known as glycosidic bonds to form a long chain. A number of glucose chains are joined together by 1–6 linkages to form a single glycogen molecule.

2.3 The sugars

The sugars are the principal source of energy for cellular reactions. They have the general formula $C_n(H_2O)_m$ and some examples are shown in Fig. 2.3. Sugars containing three carbon atoms are known as *trioses*, those with five carbons are *pentoses*, and those containing six are *hexoses*. Examples are glyceraldehyde (a triose), ribose (a pentose), and fructose and glucose (both hexoses). When two sugar molecules are joined together with the elimination of one molecule of water, they form a *disaccharide*. Fructose and glucose combine to form sucrose while glucose and galactose (another hexose) form lactose, the principal sugar of milk. When many sugar molecules are joined together they form a *polysaccharide*. Examples of polysaccharides are starch, which is an important constituent of the diet, and *glycogen*, which is the main store of carbohydrate within the muscles and liver.

Although sugars are the major source of energy for cells, they are also constituents of a number of important molecules. The nucleic acids DNA and RNA contain the pentose sugars 2-deoxyribose and ribose. Ribose is also one of the components of the purine nucleotides which play a central role in cellular metabolism. (The structure of the nucleotides is given below in Section 2.6.)

Some hexoses have an amino group in place of one of the hydroxyl groups. These are known as the *amino sugars* or *hexosamines*. The amino sugars are found in the *glycoproteins* (= sugar + protein) and the *glycolipids* (= sugar + lipid). In the glycoproteins, a polysaccharide chain is linked to a protein by a covalent bond. The glycoproteins are important constituents of bone and connective tissue. The glycolipids consist of a polysaccharide chain linked to the glycerol residue of a sphingosine lipid (see below).

Summary

1. The carbohydrates, especially glucose (a hexose sugar), are broken down to provide energy for cellular reactions. The body stores carbohydrate for energy metabolism as glycogen, which is a polysaccharide.

2. While sugars are the major source of energy for cells, they are also constituents of a large number of molecules of biological importance such as the purine nucleotides and the nucleic acids.

Glycolipids are found in the cell membranes, particularly those of the white matter of the brain and spinal cord.

2.4 The lipids

The lipids are a chemically diverse group of substances that share the property of being insoluble in water but soluble in organic solvents such as ether and chloroform. They serve a wide variety of functions.

◆ They are the main structural element of cell membranes (Chapter 3).

◆ They are an important reserve of energy.

◆ Some act as chemical signals (e.g. the steroid hormones and prostaglandins).

◆ They provide a layer of heat insulation beneath the skin.

◆ Some provide electrical insulation for the conduction of nerve impulses.

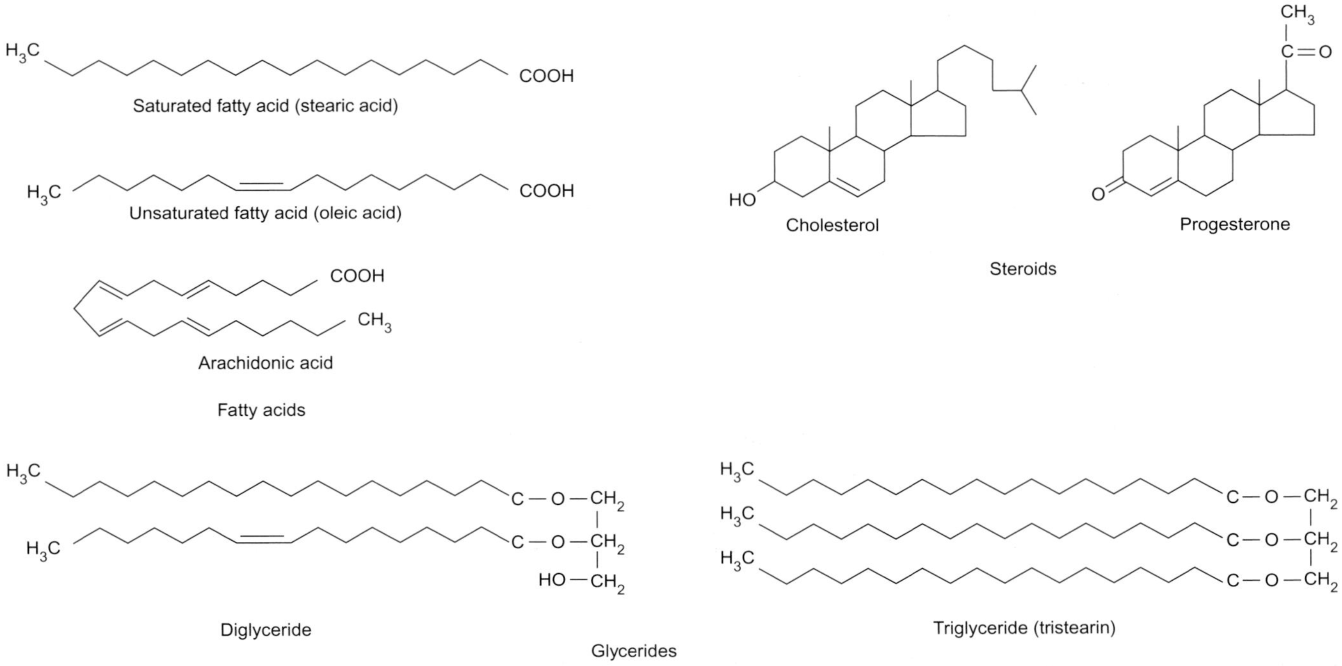

Fig. 2.4 The chemical structures of the fatty acids, glycerides, and steroids.

The *triglycerides* or *triacylglycerols* are the body's main store of energy and can be laid down in adipose tissue in virtually unlimited amounts. They consist of three fatty acids joined by ester linkages to glycerol as shown in Fig. 2.4. *Diglycerides* have two fatty acids linked to glycerol while *monoglycerides* have only one. The fatty acids have the general formula $CH_3(CH_2)_nCOOH$. Typical fatty acids are acetic acid (with two carbon atoms), butyric acid (with four carbon atoms), palmitic acid (with 16 carbon atoms), and stearic acid (with 18 carbon atoms). Triglycerides generally contain fatty acids with many carbon atoms, e.g. palmitic and stearic acids, and the middle fatty-acid chain frequently has an unsaturated fatty acid such as linoleic acid (18 carbon atoms with two double bonds) and *arachidonic acid* (20 carbon atoms with four double bonds). Although mammals, including humans, are unable to synthesize these unsaturated fatty acids, they play an important role in cellular metabolism. Consequently they must be provided by the diet and are known as the *essential fatty acids*. The essential fatty acids are precursors for an important group of lipids known as the prostaglandins (see below).

The *structural lipids* are the main component of the cell membranes. They fall into three main groups: *phospholipids*, *glycolipids*, and *cholesterol*. The basic chemical structures of these key constituents can be seen in Fig. 2.5. The phospholipids fall into two groups: those based on glycerol and those based on sphingosine. The glycerophospholipids are the most abundant in the mammalian plasma membranes and are classified on the basis of the polar groups attached to the phosphate. Phosphatidylcholine, phosphatidylserine, phosphatidylethanolamine, and phosphatidylinositol are examples of glycerophospholipids. The glycerophosphate head groups are linked to long-chain fatty-acid residues via ester linkages. However, there is another class of phospholipid, the plasmalogens, in which one hydrocarbon chain is linked to the glycerol of the head group via an ether linkage. The fatty-acid residues vary in chain length from 14 to 24 carbons and may contain one or more double bonds. Commonly, one fatty-acid chain is fully saturated and one possesses a *cis* double bond; for example, *oleic acid* has a double bond between carbons 9 and 10.

The *glycolipids* are based on sphingosine which is linked to a fatty acid to form ceramide. There are two classes of glycolipid: the cerebrosides, in which the ceramide is linked to a monosaccharide such as galactose (see Fig. 2.5), and the gangliosides in which it is linked to an oligosaccharide.

The *steroids* are lipids with a structure based on four carbon rings known as the steroid nucleus. The most abundant steroid is cholesterol (see Fig. 2.4) which is a major constituent of cell membranes and which acts as the precursor for the synthesis of many steroid hormones. The *prostaglandins* are lipids that are derived from the unsaturated fatty acid arachidonic acid (Fig. 2.6). Their biosynthesis and physiological roles are discussed in Chapter 5.

The long-chain fatty acids and steroids are insoluble in water but they naturally form micelles in which the polar head groups face outwards towards the water (the aqueous phase) and the long hydrophobic chains associate together. They are transported in the blood and body fluids in association with proteins as *lipoprotein* particles. Each particle consists of a lipid micelle protected by a coat of proteins known as apoproteins.

In cell membranes, the lipids form bilayers which are arranged so that their polar headgroups are oriented towards the aqueous phase while the hydrophobic fatty acid chains face inwards to form a central hydrophobic region. This provides a barrier to the diffusion of polar molecules (e.g. glucose) and ions but not to small non-polar molecules such as urea. The cell membranes divide the cell into discrete compartments that provide the means of storage of various materials and permit the

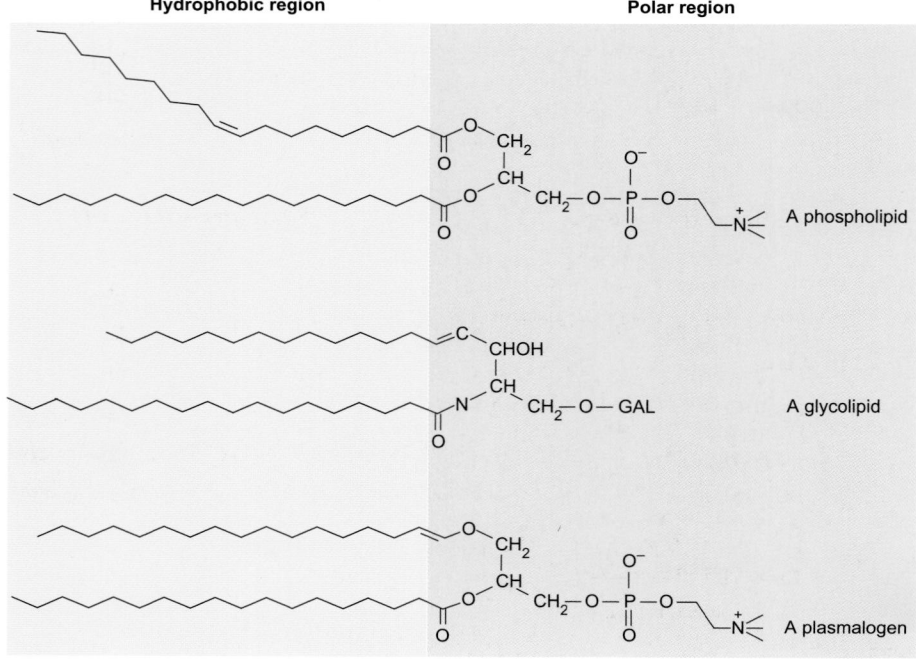

Fig. 2.5 The structure of some of the structural lipids (lipids that form the cell membranes). Note that they have a polar head group region and a long hydrophobic tail.

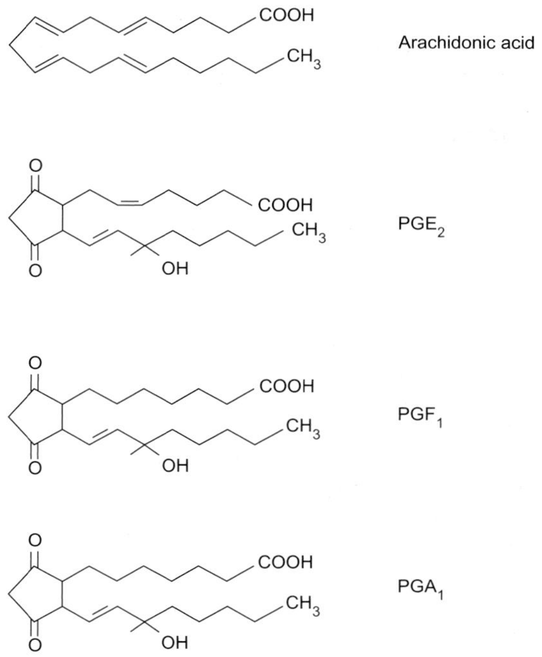

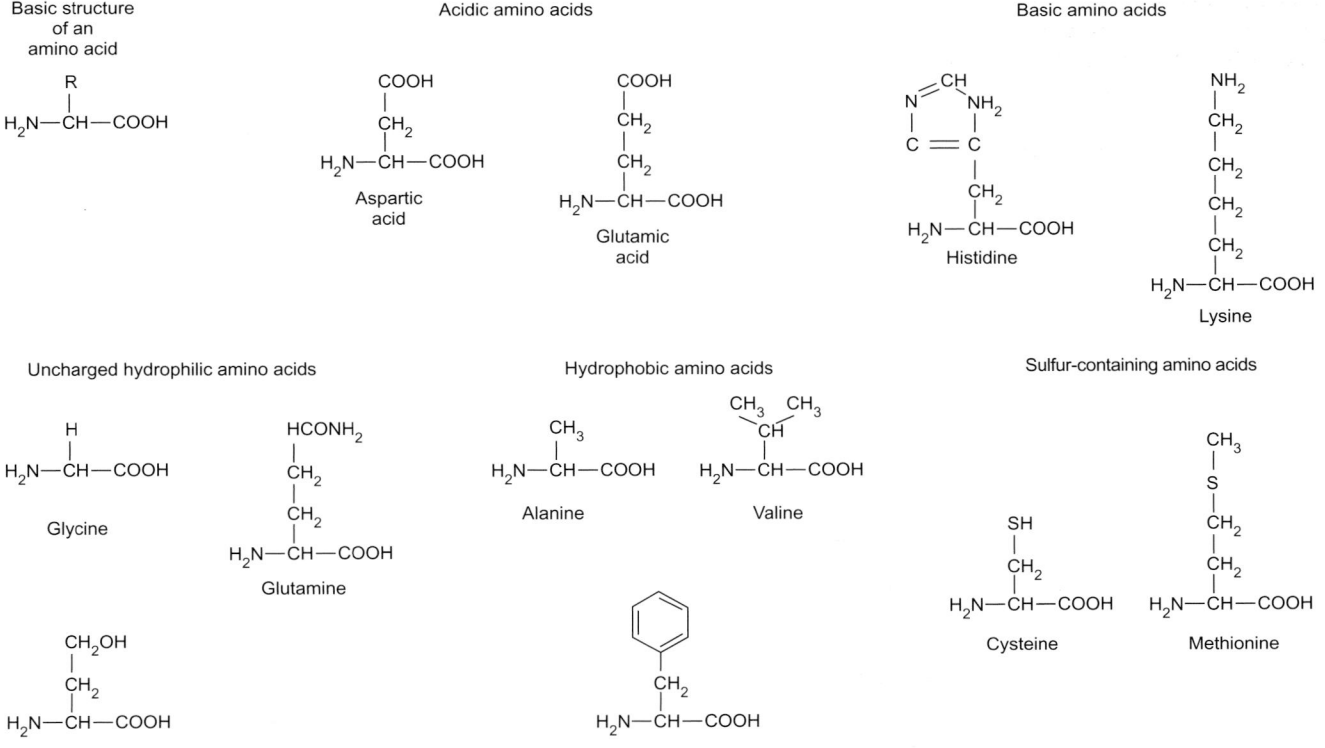

Fig. 2.6 The chemical structures of some of the prostaglandins.

segregation of different metabolic processes. This compartment-alization of cells by lipid membranes is discussed in greater detail in the next chapter.

Summary

The lipids are a chemically diverse group of substances that are insoluble in water but are soluble in organic solvents such as ether and chloroform. They serve a wide variety of functions: the phospholipids form the main structural element of cell membranes, the triglycerides are an important reserve of energy, and many steroids and prostaglandins act as chemical signals.

2.5 The amino acids and proteins

Proteins serve an extraordinarily wide variety of functions in the body.

- They form the enzymes that catalyze the chemical reactions of living things.
- They are involved in the transport of molecules and ions around the body.
- Proteins bind ions and small molecules for storage inside cells.
- They are responsible for the transport of molecules and ions across cell membranes.
- Proteins such as tubulin form the cytoskeleton that provides the structural strength of cells.
- They form the motile components of muscle and of cilia.
- They form the connective tissues that bind cells together and transmit the force of muscle contraction to the skeleton.

Fig. 2.7 The chemical structures of representative α-amino acids.

◆ Proteins known as the immunoglobulins play an important part in the body's defense against infection.

◆ As if all this were not enough, some proteins act as signaling molecules—the hormone insulin is one example of this type of protein.

Proteins are assembled from a set of twenty α-amino acids

The basic structural units of proteins are the α-amino acids. An α-amino acid is a carboxylic acid that has an amine group and a side-chain attached to the carbon atom next to the carboxyl group (the α-carbon atom) as shown in Fig. 2.7. Thus the α-carbon atom is attached to four different groups and exhibits optical asymmetry with an L form and a D form. All naturally occurring amino acids belong to the L series.

Proteins are built from 20 different L α-amino acids, which can be grouped into five different classes:

(1) acidic amino acids (aspartic acid and glutamic acid);

(2) basic amino acids (arginine, histidine, and lysine);

(3) uncharged hydrophilic amino acids (asparagine, glycine, glutamine, serine and threonine);

(4) hydrophobic amino acids (alanine, leucine, isoleucine, phenylalanine, proline, tyrosine, tryptophan and valine);

(5) sulfur-containing amino acids (cysteine and methionine).

Amino acids can be combined together by linking the amine group of one with the carboxyl group of another and eliminating water to form a *dipeptide* as shown in Fig. 2.8. The linkage between two amino acids joined in this way is known as a *peptide bond*. The addition of a third amino acid would give a tripeptide, a fourth a tetrapeptide, and so on. Peptides with large numbers of amino acids linked together are known as *polypeptides*. Proteins are large polypeptides. By convention, the structure of a peptide begins at the end with the free amine group (the amino terminus) on the left and ends with the free carboxyl group on the right, and the order in which the amino

acids are arranged is known as the peptide sequence. Since proteins and most peptides are large structures, the sequence of amino acids would be tedious to write out in full and so a single-letter or three-letter code is used as shown in Table 2.2.

Since proteins are made from 20 L-amino acids and there is no specific limit to the number of amino acids that can be linked

TABLE 2.2   The α-amino acids of proteins and their customary abbreviations

Name	Three-letter code	Single-letter code
Alanine	Ala	A
Cysteine	Cys	C
Aspartic acid	Asp	D
Glutamic acid	Glu	E
Phenylalanine	Phe	F
Glycine	Gly	G
Histidine	His	H
Isoleucine	Ile	I
Lysine	Lys	K
Leucine	Leu	L
Methionine	Met	M
Asparagine	Asn	N
Proline	Pro	P
Glutamine	Gln	Q
Arginine	Arg	R
Serine	Ser	S
Threonine	Thr	T
Valine	Val	V
Tryptophan	Trp	W
Tyrosine	Tyr	Y

The amino acids are arranged in alphabetical order of their single-letter codes. Asparagine and glutamine are amides of aspartic and glutamic acids.

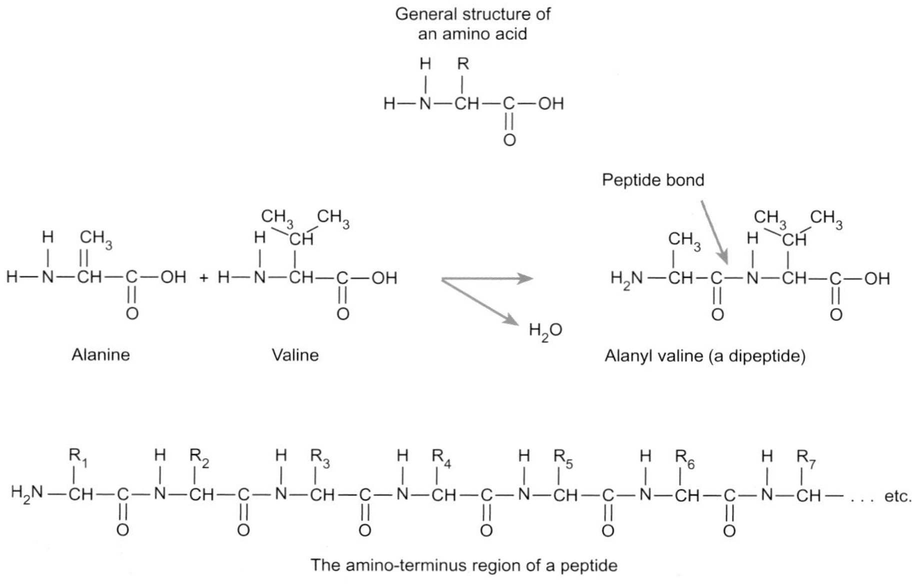

Fig. 2.8 The formation of a peptide bond and the structure of the amino terminus region of a polypeptide showing the peptide bonds. R_1, R_2, etc. represent different amino acid side-chains.

together, the number of possible protein structures is essentially infinite. It is this that makes them so versatile. Different proteins have different shapes and different physical properties. The fact that some amino acid side-chains are hydrophilic while others are hydrophobic results in different proteins having differing degrees of hydrophobicity. As a result, some are soluble in water while others are not. Proteins with large hydrophobic regions are associated with the lipid membranes of cells.

Many cellular structures consist of protein assemblies, i.e. units made up of several different kinds of protein. Examples are the myofilaments of the skeletal muscle fibers. These contain the proteins actin, myosin, troponin, and tropomyosin. Actin molecules also assemble together to form microfibrils. Enzymes are frequently arranged so that the product of one enzyme can be passed directly to another and so on. These multi-enzyme assemblies increase the efficiency of cell metabolism.

Some important amino acids are not found in proteins

Some amino acids of physiological importance are not found in proteins but have other important functions. Coenzyme A contains an isomer of alanine called β-alanine. The amino acid γ-aminobutyric acid (GABA) plays a major role as a neurotransmitter in the brain and spinal cord. Creatine is phosphorylated in muscle to form *creatine phosphate*, which is an important source of energy in muscle contraction. Ornithine is an intermediate in the urea cycle.

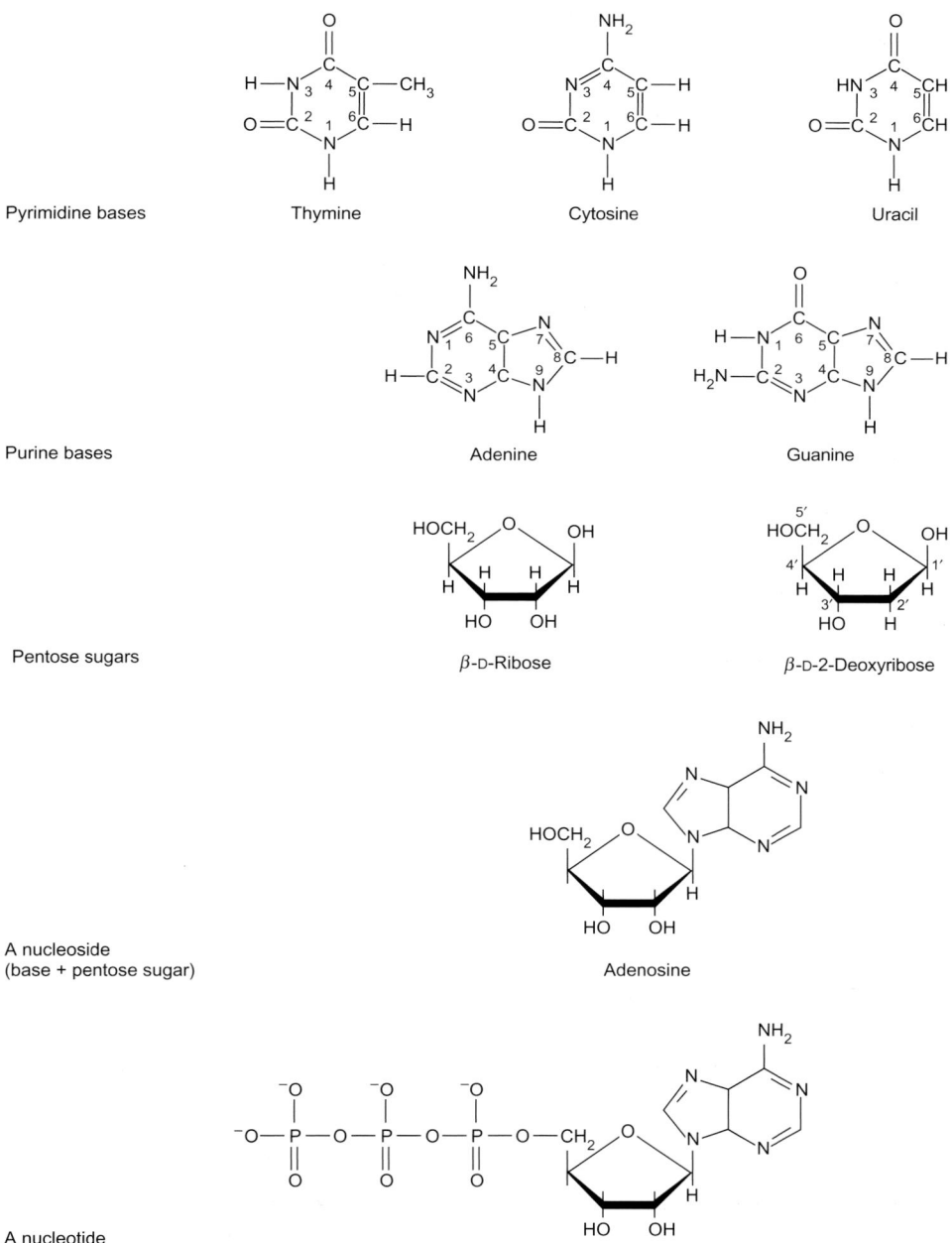

Fig. 2.9 The structural components of the nucleotides and nucleic acids.

Summary

Proteins are assembled from a set of 20 α-amino acids which are linked together by peptide bonds. Proteins serve a wide variety of functions in the body. They form the enzymes, which catalyze the chemical reactions of living things. They are involved in the transport of molecules and ions around the body and across cell membranes. They form the cytoskeleton, which provides the structural strength of cells and the motile components of muscle and cilia. The immunoglobulins are proteins that play an important part in the body's defense against infection. Some proteins such as growth hormone and insulin act as signaling molecules.

2.6 The nucleosides, nucleotides, and nucleic acids

The genetic information of the body resides in its DNA (deoxyribonucleic acid), which is stored in the chromosomes of the nucleus. DNA is made by assembling smaller components known as *nucleotides* into a long chain. Ribonucleic acid (RNA) has a similar primary structure. Each nucleotide consists of a base linked to a pentose sugar, which is in turn linked to a phosphate group as shown in Fig. 2.9. The bases in the nucleic acids are cytosine, thymine, and uracil, which are based on the structure of pyrimidine (the *pyrimidine bases*), and adenine and guanine, which are based on the structure of purine (the *purine bases*). Other bases of importance are nicotinamide and dimethylisoalloxazine. These form the nicotinamide and flavine nucleotides, which play an important role in cellular metabolism (see Chapter 3).

Nucleosides

When a base combines with a pentose sugar it forms a nucleoside. Thus the combination of adenine and ribose forms adenosine, the combination of thymine with ribose forms thymidine, and so on.

Nucleotides

A nucleotide is formed when a nucleoside becomes linked to one or more phosphate groups. Thus adenosine may become linked to one phosphate to form adenosine monophosphate, uridine will form uridine monophosphate, and so on. The nucleotides are the basic building blocks of the nucleic acids.

The nucleotide coenzymes

Nucleotides can be combined together or with other molecules to form coenzymes. Adenosine monophosphate (AMP) may become linked to a further phosphate group to form adenosine diphosphate (ADP) or to two further phosphate groups to form adenosine triphosphate (ATP) (see Fig. 2.9). Similarly, guanosine can form guanosine mono-, di-, and triphosphate, and uridine can form uridine mono-, di-, and triphosphate. The higher phosphates of the nucleotides play a vital role in cellular energy metabolism.

The phosphate group of nucleotides is attached to the 5′ position of the ribose residue and has two negative charges. It can link with the hydroxyl of the 3′ position to form 3′, 5′ cyclic

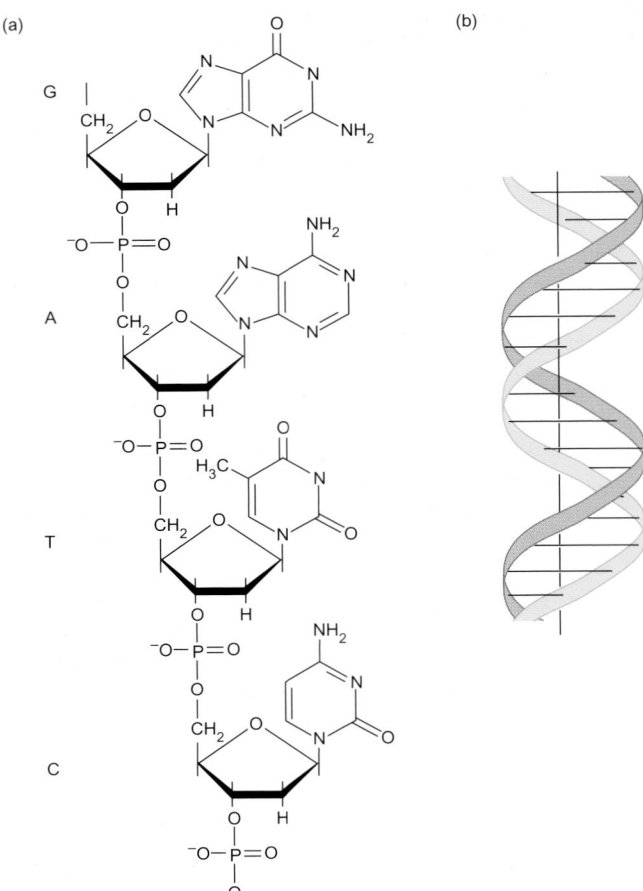

Fig. 2.10 The structure of DNA: (a) represents a short length of one of the strands of DNA; (b) is a diagrammatic representation of the two complementary strands. Note that the sequence of one strand runs in the opposite sense to the other.

adenosine monophosphate or *cyclic AMP*, which plays an important role as an intracellular messenger.

The nucleic acids

In nature there are two main types of nucleic acid: DNA and RNA. In DNA the sugar of the nucleotides is deoxyribose and the bases are adenine, guanine, cytosine, and thymine (abbreviated A, G, C, and T). In RNA the sugar is ribose and the bases are adenine, guanine, cytosine and uracil (A, G, C and U). In both DNA and RNA the nucleotides are joined by phosphate linkages between the 5′ position of one nucleotide and the 3′ position of the next as shown in Fig. 2.10(a).

A molecule of DNA consists of a pair of nucleotide chains linked together by hydrogen bonds in such a way that adenine links with thymine and guanine links with cytosine. The hydrogen bonding between the two chains is so precise that the sequence of bases on one chain automatically determines that of the second. The pair of chains is twisted to form a double helix in which the complementary strands run in opposite directions (Fig. 2.10(b)). The discovery of this base pairing was crucial to the understanding of the three-dimensional structure of DNA and to the subsequent unraveling of the genetic code. For their work in

BOX 2.1 THE HUMAN GENOME PROJECT

During the closing years of the last century, a huge effort was put into determining the sequence of all the DNA in human cells. This was known as the Human Genome Project, and it succeeded in determining the full sequence of the entire DNA that goes to make a human being. There are some 3.2 billion (3 200 000 000) base pairs in human DNA and the DNA sequence of different individuals differs by only one base in every 1000 or so. From this we can say that 99.9 per cent of the human DNA sequence is shared between all members of the human population. The 0.1 per cent difference between individuals corresponds to about three million variations in sequence. As each of these variations is independent and arises from random changes, each person is genetically unique (except identical twins who share exactly the same DNA sequence). Most of the DNA (around 97 per cent) has no known function but the remaining 3 per cent or so codes for all the genes (around 35 000 in all) that are required to make a human being.

Each amino acid is coded by a sequence of three bases (a sequence triplet or codon) and errors occur during the DNA replication that accompanies cell division. The substitution of one base for another in a sequence triplet will result in the substitution of one amino acid for another in the amino acid sequence of the protein encoded by the gene. Depending on the position of the base substitution, the effect of the mutation on the function of the protein may be either trivial or severe. Examples of genetic disorders that arise from the substitution of one base for another (single nucleotide polymorphism—a DNA base is equivalent to a nucleotide) are sickle cell disease and hemophilia B.

It is hoped that the knowledge of the human genome and its variation between different individuals will help in the diagnosis and treatment of a range of genetic disorders. For diseases such as sickle cell disease and cystic fibrosis, where a single gene is at fault, there is the distant possibility of gene therapy to provide a cure. However, for many other diseases it is the balance of activity between different proteins that matters. A good example is the regulation of lipid transport by the blood. A protein known as the low-density lipoprotein (LDL) receptor is responsible for the removal of some lipoprotein complexes (including those containing cholesterol) from the blood. A single error in the sequence causes the substitution of the amino acid serine in place of asparagine. This mutation is associated with increased cholesterol levels in the blood and a greater risk of coronary heart disease. In this case, genetic profiling of an individual can be of help in deciding whether a modification to the diet will be beneficial in reducing a major risk factor.

There is also the prospect that knowledge of the genetic profile of an individual will be of benefit in the selection of drugs for the treatment of particular non-genetic disorders and the avoidance of undesirable side-effects. For example, some liver enzymes are responsible for breaking down certain drugs before they are excreted. One of these enzymes plays an important role in the breakdown of the drug debrisoquine, which is used in the treatment of high blood pressure. A single mutation results in the production of a defective version of the enzyme. Individuals who have this form of the enzyme are unable to break down debrisoquine (as well as some other drugs which are broken down by the same enzyme). If they are given the debrisoquine for their high blood pressure, the drug accumulates in the body and causes the same symptoms as a debrisoquine overdose. Prior knowledge of the mutation would allow a better choice of drug for the particular patient or permit a more suitable drug dosage.

---

this area, J.D. Watson, F. Crick, and M. Wilkins were awarded the Nobel Prize in 1962. Unlike DNA, each RNA molecule has only one polynucleotide chain and this property is exploited during protein synthesis (see below).

Gene transcription and translation

A sequence of DNA bases that codes for a specific protein is known as a *gene* and the totality of all the genes present in an animal is called its *genome*. The genome contains all the information required for a fertilized egg to make another individual of the same species. During the closing years of the last century, a huge international effort was put into determining the sequence of the entire DNA in human cells. This project, which was known as the *Human Genome Project*, has discovered that there are about 35 000 genes in the human genome (see Box 2.1).

For a gene to perform its task, it must instruct a cell to make a specific protein. Protein synthesis takes place in the cytoplasm of a cell in association with small subcellular particles called *ribosomes*. However, a strand of DNA is too large to leave the nucleus. To overcome this, the genetic information encoded by the DNA sequence is passed to the cellular machinery responsible for making proteins by messenger RNA (mRNA). The conveying of the gene sequence from DNA to mRNA is called *transcription*.

The mRNA molecule coding for a specific gene is synthesized in the nucleus and migrates to the cytoplasm where it provides a template for the synthesis of a specific protein (Fig. 2.11). For protein synthesis to take place, the mRNA strand must bind to a ribosome, which is made of another form of RNA, ribosomal RNA, and certain proteins. A new protein is synthesized by progressive elongation of a peptide chain. For this to occur, each amino acid must be arranged in the correct order. The position of each amino acid is coded by a sequence of three bases (a *codon*) on the mRNA. To form a polypeptide chain, the amino acids are assembled in the correct order by a ribosome–mRNA complex as described below. The synthesis of protein from mRNA is called *translation*.

Before it can form a peptide, an amino acid must first be bound to a specific kind of RNA molecule known as transfer RNA (tRNA). Transfer RNA molecules exist in many forms but a particular amino acid will only bind to one specific form of tRNA. For example, the tRNA for alanine will only bind alanine,

Summary

1. Nucleotides are made of a base, a pentose sugar, and a phosphate residue. They can be combined with other molecules to form coenzymes such as the nicotinamide and flavine nucleotides. The nucleotide adenosine triphosphate (ATP) is the most important carrier of chemical energy in cells and the metabolic breakdown of glucose and fatty acids is directed to the formation of ATP.

2. The DNA (deoxyribonucleic acid) of a cell contains the genetic information for making proteins. DNA is made by assembling nucleotides into a long chain which has a specific sequence. Each DNA molecule consists of two complementary helical strands linked together by hydrogen bonds.

3. Ribonucleic acid (RNA) has a similar primary structure to a single DNA strand and exists in three different forms known as messenger RNA, transfer RNA, and ribosomal RNA. The various forms of RNA play a central role in the synthesis of proteins. In short, DNA makes RNA make proteins.

that for glutamate will only bind glutamate, and so on. Each tRNA–amino acid complex (amino acyl tRNA) has a specific coding triplet of bases which matches its complementary sequence on the mRNA (an anticodon). To assemble the peptide, the ribo-

some matches the anticodon on the tRNA with codon on the mRNA. The ribosome assembles the peptide by first allowing one amino acyl tRNA to bond with the mRNA strand. Then the next amino acyl tRNA is allowed to bind to the mRNA. The ribosome then catalyzes the formation of a peptide bond between the two amino acids and moves one codon along the mRNA strand and binds the next amino acyl tRNA molecule to extend the peptide chain. This process continues until the ribosome reaches a signal telling it to stop adding amino acids (a stop codon). At this point the protein is complete and is released into the cytosol. The ribosome is then able to catalyze the synthesis of another peptide chain. The main stages of protein synthesis are summarized in Fig. 2.11.

Recommended reading

Alberts, B., Johnson, A., Lewis, J., Raff, M., Roberts, K., and Walter, P. (2002). *Molecular biology of the cell* (4th edn), Chapters 1–7. Garland, New York.

Berg, J.M., Tymoczko, J.L., and Stryer, L. (2002). *Biochemistry* (5th edn), Chapters 1–5 and 27–29. Freeman, New York.

Elliott, W.H, and Elliott, D.C. (2005). *Biochemistry and molecular biology*, Chapters 1 and 22–25. Oxford University Press, Oxford

Numerical problems

Here are some simple calculations of molarities and osmolarities. Their purpose is to familiarize you with some basic chemical concepts that are important in physiology.

1. Calculate the molarity (moles per liter) of the following.
   a. 58.4 g NaCl dissolved in water to make a liter of solution (M_r of NaCl is 58.4).
   b. 8.77 g NaCl dissolved in water to make a liter of solution.
   c. 18.03 g urea dissolved in water to make a liter of solution (M_r of urea is 60.1).
   d. 1 mg/ml of glucose in water (M_r of glucose is 180.2).
   e. 45 g of albumin dissolved to make a liter of solution (M_r of albumin is 69 000).

2. Calculate the osmolarity (in Osmoles per liter) of the solutions given in question 1.

3. a. If 29.2 g of NaCl were present in 1 kg of NaCl solution what would be its osmolality?
   b. Calculate the osmolality of the same quantity of NaCl dissolved in water to make a liter of solution (the specific gravity of solid NaCl is approximately 2.2).
   c. If, in a sample of blood, there are approximately 6.15 g of NaCl per liter of plasma and solids make up 5.5 per cent of plasma by weight, what is the osmolality contributed by the NaCl?
   d. What is the osmotic pressure (in kPa) exerted by a solution of 9 g of NaCl in a liter of water at body temperature (310K)? (Hint: RT = 2577 J mole^{-1} and 1 liter = 0.001 m³).
   e. What is the osmotic pressure of a solution of 50 g of albumin in a liter of water at body temperature (310K)?

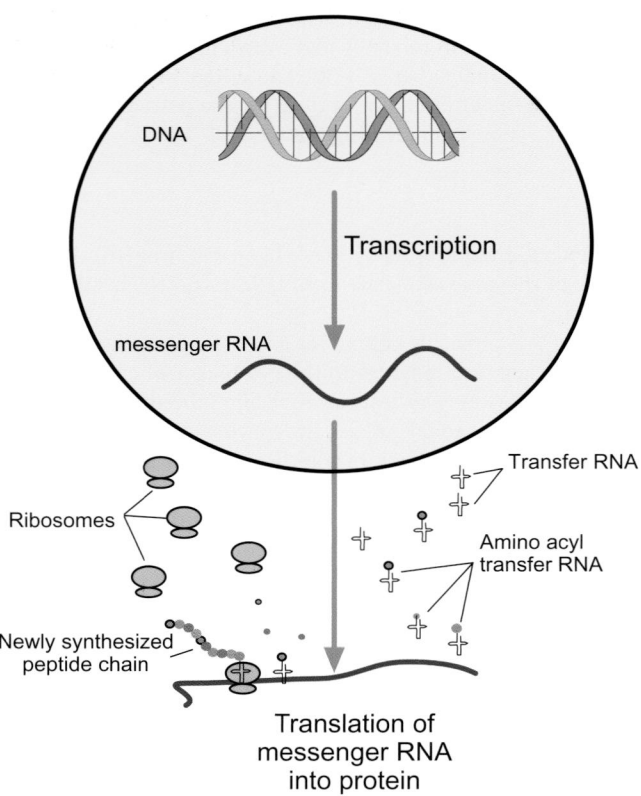

Fig. 2.11 The principal steps in the conversion of the genetic information encoded by DNA to the synthesis of specific proteins.

4. If a small number of red cells were to be isolated from the blood and placed in solutions of the composition given in (a)–(d), would they swell, shrink, or stay approximately the same size?

   a. A solution of 0.9 g NaCl per 100 ml (0.9 per cent saline solution or normal saline).

   b. A solution of 7.5 g NaCl per liter.

   c. A solution of 10.5 g NaCl per liter.

   d. A solution of 18 g of urea per liter.

Answers

1. a. 1 molar.

   b. 0.15 molar or 150 mmol l^{-1} (150 mmolar).

   c. 0.3 molar or 300 mmol l^{-1} (300 mmolar).

   d. 0.0055 molar or 5.55 mmol l^{-1} (5.55 mmolar).

   e. 0.00067 molar or 670 micromol l^{-1} (0.67 mmolar).

2. a. 2 Osmol l^{-1} (2 Osmolar) (NaCl dissociates into its constituent ions so the osmolarity of a solution of NaCl is twice the molar concentration).

   b. 300 mOsmol l^{-1} (300 mOsm l^{-1} or 300 mOsmolar).

   c. 300 mOsmol l^{-1} (urea does not dissociate so mol l^{-1} = Osmol l^{-1}).

   d. 0.0055 Osmol l^{-1} or 5.55 mOsm l^{-1}.

   e. 0.00067 Osmoles l^{-1} or 0.67 mOsmoles l^{-1}.

Note the low osmolarity of the albumin solution for each gram dissolved compared with that calculated for NaCl. This quantity of albumin corresponds to the normal level in plasma.

3. a. The solution would be 1 Osmol kg^{-1} or an osmolal solution.

   b. The second would be 1.01 Osmolar. (The difference is between Osmolar (Osmoles per liter of solution) and osmolal (Osmoles per kilogram of solvent).

   c. 6.15 g NaCl in a liter of plasma is equivalent to 6.15 g in 945 ml of water or 6.50 g per kilogram of water. This is 223 mOsm kg^{-1}. This accounts for about two thirds of the sodium in plasma; the rest is present mainly as sodium bicarbonate.

   d. 793 kPa (equivalent to 7.85 × atmopheric pressure).

   e. 1.87 kPa (equivalent to 14 mmHg).

Osmolal solutions become important in physiology when there are many constituents in the solution and not all are fully defined, e.g. blood where water makes up about 95 per cent of the plasma by weight. In lean tissues, water only makes up about 70 per cent of the total by weight so that the osmolality of the tissue constituents is usually expressed per kilogram of cell water.

4. a. The osmolarity of the solution is 308 mOsm l^{-1}, approximately isotonic with the blood so that the cells will neither swell nor shrink.

   b. The osmolarity of the solution is 256 mOsm l^{-1}, hypotonic to the blood so that the cells will swell and may burst (lyse).

   c. The osmolarity of the solution is 360 mOsm l^{-1}, hypertonic to the blood so that the cells will shrink.

   d. The osmolarity of the solution is 299 mOsm l^{-1}, approximately iso-osmotic with the blood, but the solution is not isotonic and so the cells will swell and ultimately burst.

Introducing cells

After reading this chapter you should understand:

+ The structural organization of cells
+ The functions of the different cellular organelles
+ Cell division
+ The organization of epithelia
+ The principles of cellular energy metabolism

3.1 Introduction

The cells are the building blocks of the body. There are many different types of cell, each with its own characteristic size and shape. Some cells are very large. For example, the cells of skeletal muscle (skeletal muscle fibers) may extend for up to 30 cm along the length of a muscle and be up to 100 μm in diameter. Others are very small. For example, the red cells of the blood are small biconcave disks with diameters in the region of 7 μm. Skeletal muscle fibers and red cells represent some of the more striking variations in cell morphology, but all cells have certain characteristics, some of which may only be evident during differentiation. The structure of a typical mammalian cell is illustrated in Fig. 3.1, which shows it to be bounded by a *cell membrane* also called the *plasma membrane* or *plasmalemma*. The cell membrane is a continuous sheet which separates the watery phase inside the cell, the *cytoplasm*, from that outside the cell, the extracellular fluid. The shape of an individual cell is maintained by an array of protein filaments known as the *cytoskeleton*.

At some stage of their life cycle, all cells possess a prominent structure called the *nucleus*, which contains the hereditary material DNA. Most cells have just one nucleus but skeletal muscle cells have many nuclei, reflecting their embryological origin from the fusion of large numbers of progenitor cells known as myoblasts. In contrast, the red cells of the blood lose their nucleus as they mature. Cells possess other structures which perform specific functions such as energy production, protein synthesis, and the secretion of various materials. The internal structures of a cell are collectively known as *organelles* and include the nucleus, the mitochondria, the Golgi apparatus, the endoplasmic reticulum, and various membrane-bound vesicles (see below).

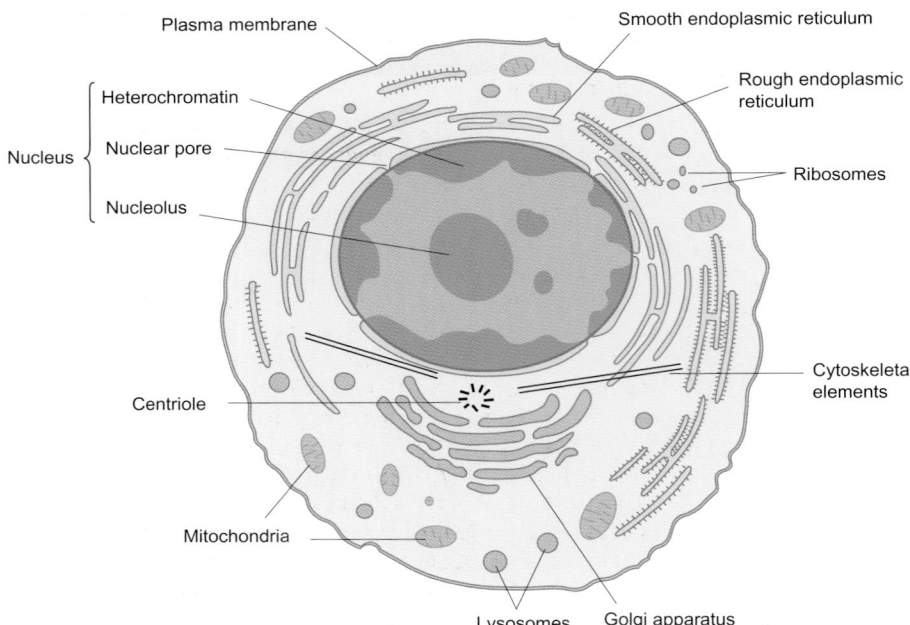

Fig. 3.1 Diagram of a typical mammalian cell, showing the major organelles.

3.2 The structure and functions of the cellular organelles

The plasma membrane

The plasma or cell membrane regulates the movement of substances into and out of a cell. It is also responsible for regulating a cell's response to a variety of signals such as hormones and neurotransmitters. Therefore, an intact plasma membrane is essential for the proper function of a cell. When viewed at high power in an electron microscope, the plasma membrane appears as a sandwich-like structure 5–10 nm thick (i.e. $5–10 \times 10^{-9}$ m). A layer of fine filaments, which form the glycocalyx or cell coat, covers the outer surface. The membranes of the intracellular organelles (e.g. endoplasmic reticulum, Golgi apparatus, lysosomes, and mitochondria) have a three-layered structure similar to that of the plasma membrane.

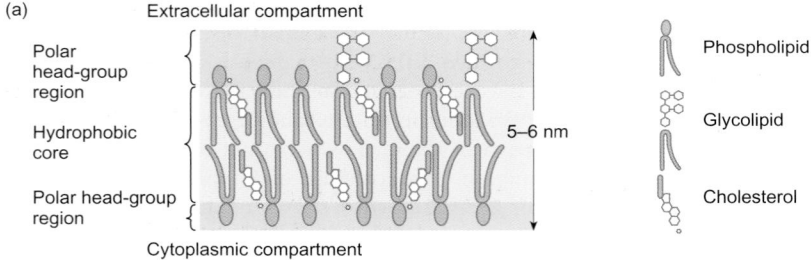

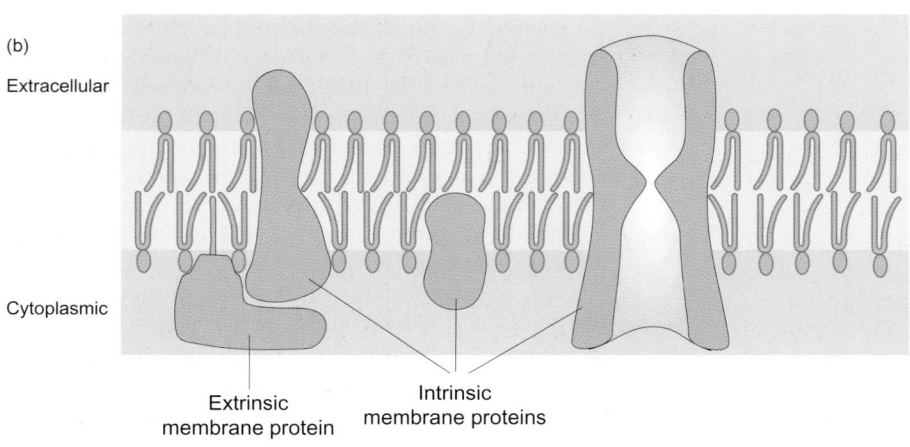

Fig. 3.2 The structure of the plasma membrane. (a) The basic arrangement of the lipid bilayer. Note the presence of glycolipid only in the outer leaflet of the bilayer. (b) A simplified model of the plasma membrane showing the arrangement of some of the membrane proteins.

The plasma membrane is a lipid bilayer containing proteins

Chemical analysis shows that the plasma membrane is made of lipid and protein in approximately equal amounts by weight. The lipids are arranged so that their polar head groups are oriented towards the aqueous phase and the hydrophobic fatty-acid chains face inwards to form a central hydrophobic region as shown in Fig. 3.2. The membrane proteins either span the lipid bilayer or are anchored to it in various ways.

The principal lipids of the plasma membrane belong to one of three classes: phospholipids, glycolipids, and cholesterol. There is now good evidence that the composition of the outer and inner layers of the plasma membrane differs significantly. The outer leaflet consists of glycolipids, phosphatidylcholine, and sphingomyelin. The inner leaflet is richer in negatively charged phospholipids such as phosphatidylinositol. Cholesterol is present in both leaflets of the bilayer. The presence of phosphatidylinositol in the inner leaflet of the bilayer is important, as inositol phosphates play a major role in the transmission of certain signals from the cell membrane to the interior of the cell (see Chapter 5). Each phospholipid molecule is able to diffuse freely in one plane of the bilayer, but phospholipids rarely flip from one leaf of the bilayer to the other. This indicates that the lipid bilayer is an inherently stable structure.

Artificial lipid membranes are not very permeable to ions or polar molecules

Artificial lipid membranes are relatively permeable to carbon dioxide, oxygen and lipid-soluble molecules, but they are almost impermeable to polar molecules and ions such as glucose and sodium. Moreover, such membranes are relatively impermeable to water. However, natural cell membranes are permeable to a wide range of polar materials which cross the hydrophobic barrier formed by the lipid bilayer via specific protein molecules—the aquaporins, ion channels, and carrier proteins.

Membrane proteins

The membrane proteins can be divided into two broad groups: *intrinsic*—those which are embedded in the bilayer itself—and *extrinsic*—those which are external to the bilayer but linked to it in some way. The proteins that facilitate the movement of ions and other polar materials across the plasma membrane are all intrinsic membrane proteins. Some extrinsic proteins link the cell to its surroundings (the extracellular matrix) or to neighboring cells. Others play a role in the transmission of signals from the plasma membrane to the interior of the cell. The physio-logical roles of the membrane proteins are discussed in greater detail in subsequent chapters.

The nucleus

The nucleus is separated from the rest of the cytoplasm by the *nuclear membrane* (also known as the *nuclear envelope*), which consists of two lipid bilayers separated by a narrow space. The nuclear membrane is furnished with small holes known as *nuclear pores* that provide a means of communication between the nucleus and the cytoplasm.

The nucleus contains the DNA of a cell, which associates with proteins called histones to form chromatin fibers. The chromatin may be either very condensed (*heterochromatin*) or relatively dispersed (*euchromatin*). When a cell synthesizes a protein it first needs to read the base sequence from its DNA (a process known as *transcription*). Heterochromatin is chromatin that is not taking part in transcription while euchromatin is chromatin that is doing so. From this it is evident that a cell with a small densely staining nucleus is transcribing very little of its DNA and therefore is not using many of its genes. Conversely, a cell with a large pale nucleus is involved in wholesale gene transcription. During cell division (mitosis), the chromatin becomes distributed into pairs of *chromosomes* which attach to a structure known as the mitotic spindle before they separate as the cell divides (see below).

A structure called the *nucleolus* is the most prominent feature visible within the nucleus. It is concerned with the manufacture of the *ribosomes*. Within the nucleolus are one or more weakly staining regions formed from a type of DNA called nucleolar organizer DNA which codes for ribosomal RNA. Associated with these regions are densely packed fibers, which are formed from the primary transcripts of the ribosomal RNA and protein. A further region known as the pars granulosa consists of maturing ribosomes, which are subsequently released into the cytoplasm where they play an important part in the synthesis of new protein molecules (see Chapter 2). Prominent nucleoli are seen in cells that are synthesizing large amounts of protein, such as embryonic cells, secretory cells, and those of rapidly growing malignant tumors.

The organelles of the cytoplasm

The cytoplasm contains many different organelles which perform specific tasks within the cell. Some of these organelles are separated from the rest of the cytoplasm by membranes, so that the cell's internal membranes divide the cytoplasm into various compartments. Examples are the mitochondria, the endoplasmic reticulum, the Golgi apparatus, and vesicles such as lysosomes.

Mitochondria

As discussed in Chapter 1, living cells continually transform materials. They oxidize glucose and fats to provide energy for other activities such as motility and the synthesis of proteins for growth and repair. The energy is provided as ATP. In most cells, the bulk of ATP synthesis occurs in the mitochondria by a process known as oxidative phosphorylation (see below).

The mitochondria are 2–6 μm long and about 0.2 μm in diameter. They have two distinct membranes, an outer membrane which is smooth and regular in appearance and an inner membrane which is thrown into a number of folds known as *cristae*.

Summary

The plasma membrane consists of roughly equal amounts by weight of protein and lipid. The lipid is arranged as a bilayer whose inner and outer leaflets have a different composition. The lipid bilayer forms a barrier to the passage of polar materials so that these substances must enter or leave a cell via specialized transport proteins. The combination of lipid bilayer and transport proteins allows cells to maintain an internal composition that is very different from that of the extracellular fluid.

It is here, on the inner membrane, that the synthesis of ATP takes place via the tricarboxylic acid cycle and the electron transport chain. Cells that have a high demand for ATP (e.g. the muscle cells of the heart) have many mitochondria, which are often located close to the site of ATP utilization. In the case of heart muscle, the mitochondria lie close to the contractile elements of the cells, the myofibrils. In addition to their role in energy production, mitochondria can also accumulate significant amounts of calcium. Thus they play a significant role in regulating the concentration of ionized calcium within cells.

The numbers of mitochondria in a cell are regulated according to the metabolic requirements. They may increase substantially if required, for example in skeletal muscle cells subject to prolonged periods of contractile activity. Mitochondria are not assembled from scratch from their molecular constituents but increase in number by the division of existing mitochondria. This division takes place in the interphase of the cell cycle (the interval between cell divisions). When cells divide, the new mitochondria are shared out between the daughter cells.

Endoplasmic reticulum

The endoplasmic reticulum is a system of membranes that extends throughout the cytoplasm. These membranes are continuous with the nuclear membrane and enclose a significant space within the cell. The endoplasmic reticulum is classified as rough or smooth according to its appearance under the electron microscope. The rough endoplasmic reticulum has polyribosomes attached to its cytoplasmic surface and plays an important role in the synthesis of certain proteins. It is also important for the addition of carbohydrates to membrane proteins (glycosylation), which takes place on its inner surface. The smooth endoplasmic reticulum is involved in the synthesis of lipids and other aspects of metabolism. It also accumulates calcium to provide an intracellular store for calcium signaling which is regulated in many cells by the inositol lipids of the inner leaflet of the plasma membrane (see Chapter 5). In muscle cells the endoplasmic reticulum is called the sarcoplasmic reticulum and plays an important role in the initiation of muscle contraction.

Golgi apparatus

The Golgi apparatus (also called the Golgi complex or the Golgi membranes) is a system of flattened membranous sacs that are involved in modifying and packaging proteins for secretion. In cells adapted for secretion as their main function (e.g. the acinar cells of the pancreas, mucus-secreting goblet cells) the Golgi apparatus lies between the nucleus and the apical surface (where the secretion takes place). It has a characteristic appearance in which one face (the *cis* face) appears convex while the other (the *trans* face) is concave. Whilst the Golgi apparatus is effectively an extension of the endoplasmic reticulum, it is not in direct continuity with it. Rather, small vesicles known as transport vesicles pinch off from the endoplasmic reticulum and migrate to the Golgi apparatus with which they fuse. The Golgi apparatus itself gives rise to specific secretory vesicles that migrate to the plasma membrane. The *cis* Golgi is the site that receives transport vesicles from the endoplasmic reticulum and the *trans* face is the site from which secretory vesicles bud off. During their passage from the *cis* to the *trans* Golgi, membrane proteins are modified in various ways including the addition of oligosaccharide chains (glycosylation), the addition of phosphate groups (phosphorylation), and/or the addition of sulfate groups (sulfation). The chemical changes to newly synthesized proteins are called post-translational modification. Further details of the mechanism of secretion are discussed in Chapter 4.

Membrane-bound vesicles

Cells contain a variety of membrane-bound vesicles that are integral to their function. Secretory vesicles are formed by the Golgi apparatus. Under the electron microscope, some vesicles appear as simple round profiles while others have an electron-dense core. After they have discharged their contents, the secretory vesicles are retrieved to form endocytotic vesicles (see Chapter 4). Other cytoplasmic vesicles include the lysosomes and the peroxisomes. Peroxisomes contain enzymes which can synthesize and destroy hydrogen peroxide. The lysosomes contain hydrolytic enzymes which allow cells to digest materials that they are recycling (e.g. plasma membrane components) or have taken up during endocytosis.

The importance of the lysosomes to the economy of a cell is strikingly illustrated by Tay–Sachs disease in which gangliosides taken up into the lysosomes are not degraded. The sufferers lack an enzyme known as β-hexosaminidase A which is responsible for the breakdown of gangliosides derived from recycled plasma membrane. Consequently, the lysosomes accumulate lipid and become swollen. Although other cells are affected, gangliosides are especially abundant in nerve cells which show severe pathological changes as their lysosomes become swollen with undegraded lipid. This leads to the premature death of neurons, resulting in muscular weakness and retarded development. The condition is fatal and those with the disease usually die before they are 3 years old.

Ribosomes

Ribosomes consist of proteins and RNA. They are formed in the nucleolus and migrate to the cytoplasm where they may occur free or in groups called polyribosomes. Ribosomes play an important role in the synthesis of new proteins as outlined in Chapter 2. Some ribosomes become attached to the outer membrane of the endoplasmic reticulum to form the rough endoplasmic reticulum, which is the site of synthesis of membrane proteins.

The cytoskeleton

A cell is not just a bag of enzymes and isolated organelles. Different cell types each have a distinctive and stable morphology that is maintained by an internal array of protein filaments known as the cytoskeleton. The protein filaments are of three main kinds: actin filaments, intermediate filaments, and microtubules.

Actin filaments

Actin filaments play an important role in cell movement such as the contraction of skeletal muscle. They also help to maintain cell shape in non-motile cells.

Intermediate filaments

Intermediate filaments were originally so called because their diameter, estimated from electron micrographs, lay between that of the thin actin filaments and the thick myosin filaments of skeletal muscle. They play an important role in the mechanical stability of cells. Those cells that are subject to a large amount of mechanical stress (such as epithelia and smooth muscle) are particularly rich in intermediate filaments which link cells together via specialized junctions.

Microtubules

As their name suggests, microtubules are hollow tubes which have an external diameter of about 25 nm and a wall thickness of 5–7 nm. They are formed from a protein called tubulin. Microtubules play an important role in moving organelles (e.g. secretory vesicles) through the cytoplasm. This is particularly important in nerve cells where they are responsible for axoplasmic flow (see Chapter 6). They also play a major role in the movement of cilia and flagella. Microtubules originate from a complex structure known as a *centrosome*. Between cell divisions, the centrosome is located at the center of a cell near the nucleus. Embedded in the centrosome are two *centrioles*, which are cylindrical structures arranged at right angles to each other. At the beginning of cell division the centrosome divides and the two daughter centrosomes move to opposite poles of the nucleus to form the mitotic spindle (see below).

Summary

1. Cells are the building blocks from which all animals are made. Each cell is bounded by a plasma membrane, which separates the intracellular compartment from the extracellular compartment. The plasma membrane also plays an important role in cell signaling. Within the cells are structures known as organelles that carry out specific cellular functions.

2. The nucleus is the most prominent feature of most cells. It is bounded by the nuclear membrane, which is pierced by pores that permit the passage of proteins and RNA between the nucleus and the cytoplasm. The nucleolus is the site of assembly of the ribosomes, which migrate to the cytoplasm where they play an essential part in protein synthesis.

3. Membrane-bound organelles divide the cytoplasm into a number of separate compartments: the endoplasmic reticulum, the Golgi apparatus, the lysosomes, and the mitochondria.

4. The endoplasmic reticulum plays an important role in protein synthesis while the Golgi apparatus is involved in the post-translational modification of proteins and the formation of secretory vesicles.

5. The mitochondria provide ATP for the energy requirements of the cell by oxidizing carbohydrates and fatty acids.

6. Cell shape is maintained by protein filaments within the cell that form the cytoskeleton. Protein filaments also form the internal structure of the cilia and flagella of motile cells

Cilia

Cilia are very small hair-like projections from certain cells which have a characteristic array of microtubules at their core. In mammals, cilia beat in an orderly wave-like motion to propel material over the surface of an epithelial layer such as the lining of the upper respiratory tract. In this case, the cilia beat in such a way as to move the layer of mucus covering the epithelium towards the mouth. The mucus traps dust, cell debris, and invading organisms, and the action of the cilia moves this material away from the respiratory surface towards the throat where it can be coughed up. Flagella are similar in structure to cilia but are much longer. While they are common in single-cell organisms, flagella are only found in mammals as the motile part of the sperm (see Chapter 20).

3.3 Cell division

During life animals grow by two processes:

(1) the addition of new material to pre-existing cells;

(2) increasing the number of cells by division.

Cell division occurs by one of two processes: *mitosis*, in which each daughter cell has the same number of genetically identical chromosomes as the parent cell; *meiosis*, in which each daughter cell has half the number of chromosomes as its parent. Most cells that divide do so by mitosis. Meiosis occurs only in the germinal cells during the formation of the eggs and sperm.

Mitosis

The process of mitosis and DNA replication can be divided into six phases (see Fig. 3.3).

Prophase

During the early part of prophase the nuclear chromatin condenses into well-defined chromosomes, each of which consists of two identical *chromatids* (sister chromosomes) linked by a specific sequence of DNA known as a *centromere*. As prophase proceeds, the cytoplasmic microtubules disassemble and the mitotic spindle begins to form outside the nucleus between a pair of separating centrioles.

Prometaphase

This begins with the dissolution of the nuclear membrane. It is followed by the movement of the microtubules of the mitotic spindle into the nuclear region. The chromosomes then become attached to the mitotic spindle by their centromeres.

Metaphase

During metaphase the chromosomes become aligned along the central region of the mitotic spindle.

Anaphase

This begins with the separation of the two chromatids to form the chromosomes of the daughter cells. The poles of the mitotic spindle move further apart.

Telophase

In telophase, the separated chromosomes reach the poles of the mitotic spindle which begins to disappear. A new nuclear mem-

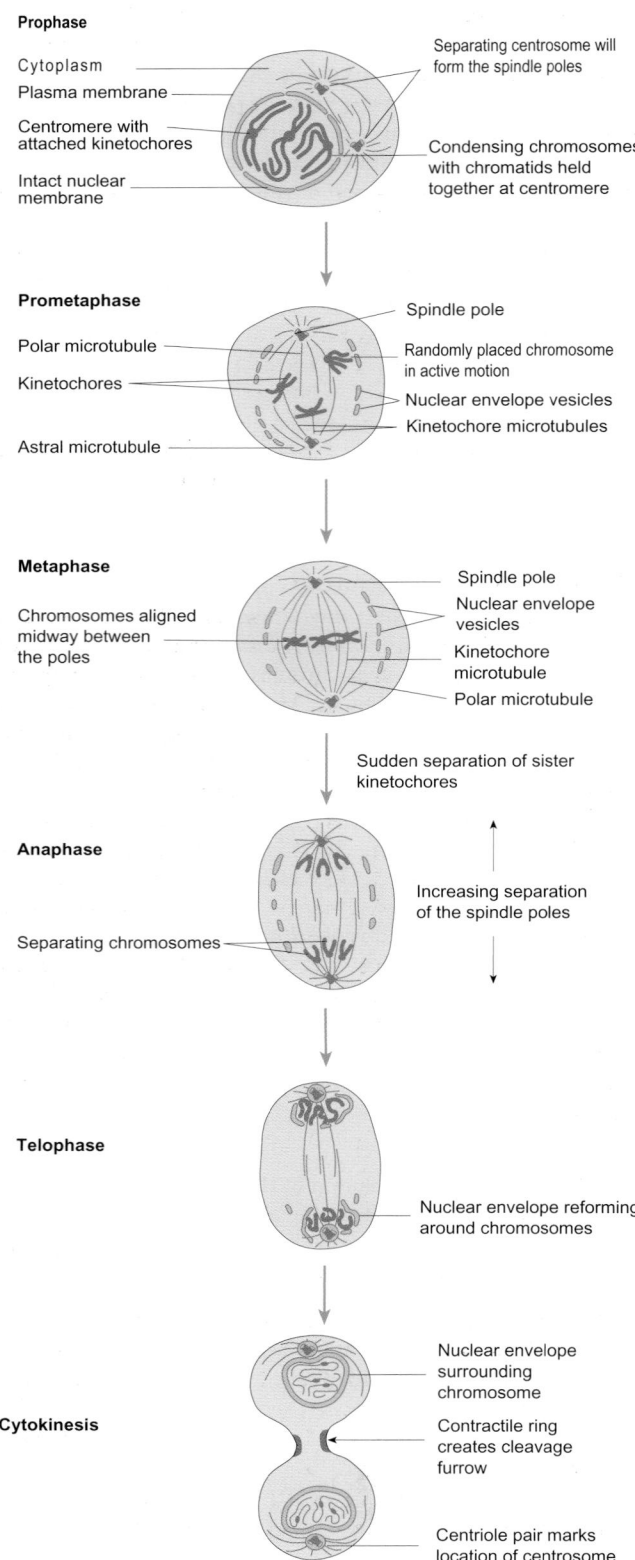

Prophase

Cytoplasm

Plasma membrane

Centromere with attached kinetochores

Intact nuclear membrane

Separating centrosome will form the spindle poles

Condensing chromosomes with chromatids held together at centromere

Prometaphase

Polar microtubule

Kinetochores

Astral microtubule

Spindle pole

Randomly placed chromosome in active motion

Nuclear envelope vesicles

Kinetochore microtubules

Metaphase

Chromosomes aligned midway between the poles

Spindle pole

Nuclear envelope vesicles

Kinetochore microtubule

Polar microtubule

Sudden separation of sister kinetochores

Anaphase

Separating chromosomes

Increasing separation of the spindle poles

Telophase

Nuclear envelope reforming around chromosomes

Cytokinesis

Nuclear envelope surrounding chromosome

Contractile ring creates cleavage furrow

Centriole pair marks location of centrosome

Fig. 3.3 The principal stages of cell division by mitosis.

brane is formed around each daughter set of chromosomes and mitosis proper is complete.

Cytokinesis

This is the division of the cytoplasm between the two daughter cells. It begins during anaphase but is not completed until after the end of telophase. The cell membrane invaginates in the center of the cell at right angles to the long axis of the mitotic spindle to form a cleavage furrow. The furrow deepens until only a narrow neck of cytoplasm joins the two daughter cells. This finally breaks, separating the two daughter cells.

Meiosis

In humans, somatic cells (i.e. those of the body) are *diploid* with 46 chromosomes (23 pairs). The gametes (eggs and sperm) have 23 (unpaired) chromosomes. They are *haploid* cells. In diploid cells half the chromosomes originated from the father (paternal chromosomes) and half from the mother (maternal chromosomes). The maternal and paternal forms of a specific chromosome are known as *homologs*. With the exception of the X and Y chromosomes (the chromosomes that determine sex), homologous chromosomes carry identical sets of genes arranged in precisely the same order along their length. This arrangement is crucial for the genetic recombination that takes place during meiosis.

During meiosis the chromosome number is halved to form the gametes. This process consists of two successive cell divisions. As for mitosis, meiosis begins with DNA replication. The first stage (*prophase I*) of the first cell division is divided into five substages : leptotene, zygotene, pachytene, diplotene, and diakinesis.

During leptotene the chromosomes condense in a similar manner to that seen in mitosis. The homologous pairs of chromosomes then become aligned along their length so that the order of their genes exactly matches (zygotene). Each pair of chromosomes is bound together by proteins to form a *bivalent* (i.e. a complex with four chromosomes; see Fig. 3.4). When all 23 pairs of chromosomes have become aligned, new combinations of maternal and paternal genes occur by the crossing-over of chromosomal segments (pachytene). The pachytene ends with the disassembly of the protein linkage between the homologous pairs of chromosomes which begin to separate. However, they do remain joined by two DNA linkages. It is at this stage (diplotene) that the eggs (ova) arrest their meiotic division and start to accumulate material. (For this reason the diplotene is also called the synthesis stage.) The final stage of prophase I is similar to mitosis with spindle formation and repulsion of the chromosome pairs (diakinesis).

The dissolution of the nuclear envelope marks the start of *metaphase I*. The bivalents align themselves so that each pair of homologs faces opposite poles of the spindle. This process is random so that the homologs facing any one pole of the spindle are a mixture of maternal and paternal chromosomes. At *anaphase I* the chromosome pairs separate. As sister chromatids remain joined, they move towards the spindle poles as a unit so that the daughter cells each receive two copies of one of the two homologs.

Formation of the gametes now proceeds by a second cell division (cell division II) which resembles a normal mitotic cell division. It differs in that (a) it occurs without further DNA repli-

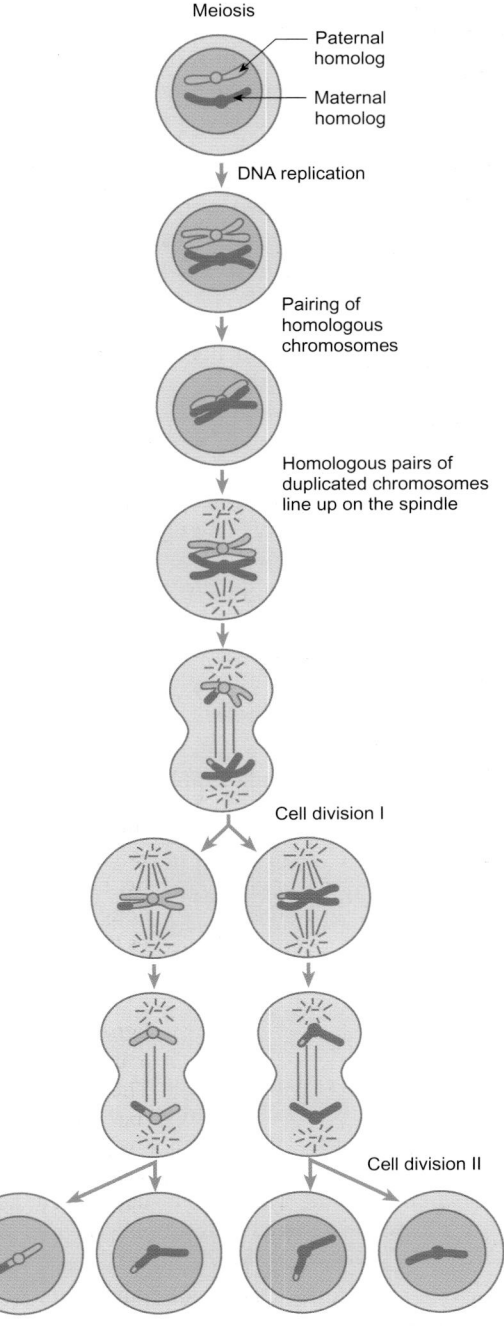

Meiosis

Paternal homolog

Maternal homolog

DNA replication

Pairing of homologous chromosomes

Homologous pairs of duplicated chromosomes line up on the spindle

Cell division I

Cell division II

Fig. 3.4 The main stages of gamete formation by meiosis.

Summary

Cell division occurs by either mitosis or meiosis. Most cells divide by mitosis and each daughter cell is genetically identical to its parent. Meiosis occurs only in the germinal cells during the formation of the gametes (eggs and sperm). Meiosis permits genetic recombination to occur so that the offspring possess genes from both parents.

vidual with one X and one Y chromosome is male. How do X and Y chromosomes form homologous pairs during the formation of the sperm? The answer resides in a small region of homology shared by both X and Y chromosomes. Thus they are able to pair and go through meiotic cell division I in a similar manner to other chromosomes. The daughter cells will then either have two X chromosomes or two Y chromosomes and will give rise to gametes with a single X or a single Y chromosome. Non-disjunction of the X or Y chromosomes gives rise to abnormal sexual differentiation (see Chapter 22, Box 22.1).

3.4 Epithelia

The interior of the body is physically separated from the outside world by the skin, which forms a continuous sheet of cells known as an *epithelium*. Epithelia also line the hollow organs of the body such as the gut, lungs, and urinary tract, as well as the fluid-filled spaces such as the peritoneal cavity. Any continuous cell layer that separates an internal space from the rest of the body is an epithelium. Note, however, that the cell layer that lines the blood vessels is generally called an *endothelium*, as is the lining of the fluid-filled spaces of the brain, while the epithelial coverings of the pericardium, pleura, and peritoneal cavity are known as *mesothelium*, reflecting their origins in the mesoderm.

Epithelia have three main functions: protective, secretory, and absorptive. Their structure reflects their differing functional requirements. For example, the epithelium of the skin is thick and tough to resist abrasion and to prevent the loss of water from the body. In contrast, the epithelial lining of the alveoli of the lungs is very delicate and thin to permit free exchange of the respiratory gases.

Despite such differences in form and function, all epithelia share certain features:

◆ They are composed entirely of cells.

◆ Their cells are tightly joined together via specialized cell–cell junctions to form a continuous sheet.

◆ Except for the ependymal lining of the cerebral ventricles, epithelia lie on a matrix of connective tissue fibers called the *basement membrane*, which is 0.1–2.0 μm thick, depending on tissue type. The basement membrane consists of a *basal lamina*, 30–70 nm thick, overlying a matrix of collagen fibrils. The basement membrane provides physical support and separates the epithelium from the underlying vascular connective tissue, which is known as the *lamina propria*.

◆ To replace damaged and dead cells, all epithelia undergo continuous cell replacement. The natural loss of dead epithelial cells is known as *desquamation*. The rate of replacement

cation and (b) the daughter cells have half the number of chromosomes of the parent. Occasionally, some chromosomes do not separate properly and the daughter cells will either lack one or more chromosomes or have a greater number than normal. This is known as *non-disjunction*. Non-disjunction causes a number of genetic diseases of which Down's syndrome is perhaps the best known. In this disease, the sufferer has inherited an extra copy of chromosome 21.

The X and Y chromosomes determine the sex of an individual. An individual with two X chromosomes is female while an indi-

depends on the physiological role of the epithelium and is highest in the skin and gut, both of which are continually subject to abrasive forces.

◆ Finally, unlike cells scattered throughout a tissue, the arrangement of cells into epithelial sheets permits the directional transport of materials either into or out of a compartment. In a word, epithelia show polarity. In the gut, kidney, and many glandular tissues this feature of epithelia is of great functional significance. The surface of an epithelial layer that is oriented towards the central space of a gland or hollow organ is known as the *apical surface*. The surface that is orientated towards the basement membrane and the interior of the body is called the *basolateral surface*.

Close to the apical surface of an epithelial cell lies a characteristic structure known as the *junctional complex* (Fig. 3.5). This consists of three structural components: the *tight junction* (also known as the zonula occludens), the *adherens junction* (or zonula adherens), and the *desmosomes*. Within the junctional complexes specialized regions of contact, called *gap junctions*, are found. Gap junctions allow small molecules to diffuse between adjacent cells. In this way they play a role in communication between neighboring cells (see Chapter 5, p. 59).

The tight junction is a continuous region of contact between the membranes of adjacent cells. It links an epithelial cell to each of the surrounding cells to seal off the space above the apical surface from that surrounding the basolateral surface. In a tight junction, the plasma membranes of the adjacent cells are held so closely apposed that the extracellular space is eliminated. This prevents ions and molecules from leaking between the cells. The proteins responsible for linking the epithelial cells together are transmembrane proteins called *claudins*.

The tight junctions have little structural rigidity. This property is supplied by the adherens junctions and desmosomes. The adherens junctions form a continuous band around each cell. On the cytoplasmic side, intracellular anchor proteins form a distinct zone of attachment for actin and intermediate filaments of the cytoskeleton. This is known as an attachment plaque and appears as a densely staining band in electron micrographs. The anchor proteins are connected to transmembrane adhesion proteins (*cadherins*) that bind neighboring cells together.

The desmosomes are points of contact between the plasma membranes of adjacent cells. They are distributed in clusters along the lateral surfaces of the epithelial cells. The attachment plaques consist of various anchoring proteins that link intermediate filaments of the cytoskeleton to the cadherins. The cadherins of one cell bind to those of its neighbors to form focal attachments of great mechanical strength.

Hemidesmosomes, as their name implies, have a similar appearance to half a desomosome, as seen in an electron micrograph. However, they are formed by different anchoring proteins that bind cytoskeletal intermediate filaments to transmembrane adhesion proteins, known as *integrins*. The integrins fix the epithelial cells to the basal lamina, so linking the cell layer to the underlying connective tissue.

The classification of epithelia

Epithelia vary considerably in their morphology. The main types are known as *simple*, *stratified*, and *pseudostratified* epithelia. Simple epithelia consist of a single cell layer and are classified according to the shape of the constituent cell type. Stratified epithelia have more than one cell layer, while pseudostratified epithelia consist of a single layer of cells in contact with the basement membrane, but the varying height and shape of the constituent cells gives the appearance of more than one cell layer.

The morphological characteristics of the main types of epithelium are shown in Fig. 3.6 and are summarized below:

◆ Simple squamous epithelium (squamous – flattened) consists of thin and flattened cells, as shown in Fig. 3.6 (i) and (ii). These epithelia are adapted for the exchange of small molecules between the separated compartments. Squamous epithelium forms the walls of the alveoli of the lungs, lines the abdominal cavity, and forms the endothelium of the blood vessels.

◆ Simple cuboidal epithelium, as the name implies, consists of a single layer of cuboidal cells, of approximately equal width and height. Simple cuboidal epithelium forms the walls of the small collecting ducts of the kidneys.

◆ Simple columnar epithelium is adapted to perform secretory or absorptive functions. In this form of epithelium, the height of the cells is much greater than their width, as shown in Fig. 3.6 (iv). It occurs in the large-diameter collecting ducts of the kidneys. It is also found lining the small intestine, where the apical surface is covered with microvilli to increase the surface area available for absorption.

◆ Ciliated epithelium consists of cells that may be cuboidal or columnar in shape. The apical surface is ciliated, although

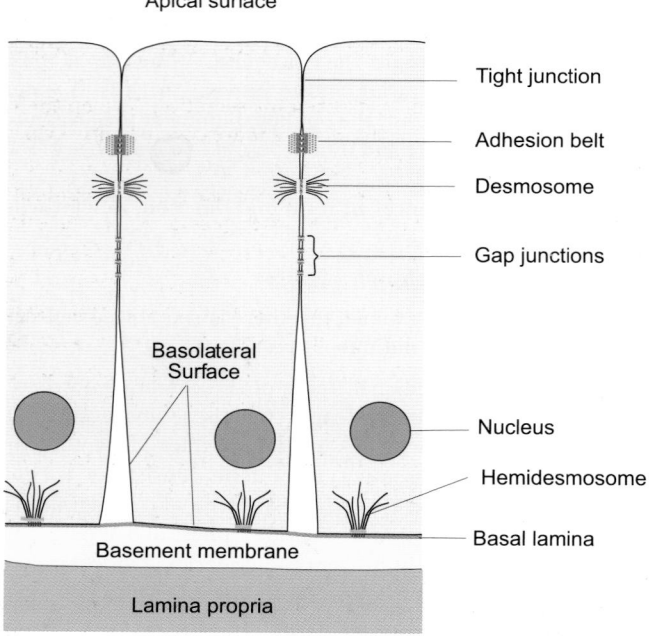

Apical surface

Tight junction

Adhesion belt

Desmosome

Gap junctions

Basolateral Surface

Nucleus

Hemidesmosome

Basal lamina

Basement membrane

Lamina propria

Fig. 3.5 A diagrammatic representation of the main features of the junctional complexes of epithelia.

a) Simple epithelia

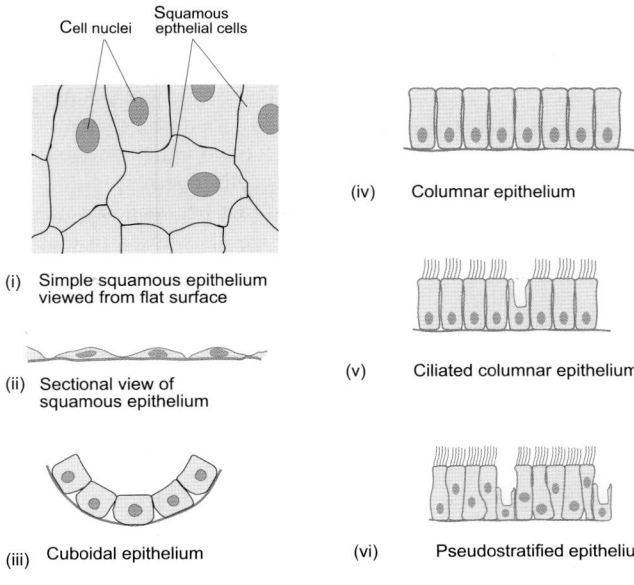

(i) Simple squamous epithelium viewed from flat surface

(ii) Sectional view of squamous epithelium

(iii) Cuboidal epithelium

(iv) Columnar epithelium

(v) Ciliated columnar epithelium

(vi) Pseudostratified epithelium

b) Stratified epithelia

(vii) Stratified squamous epithelium

(viii) Transitional epithelium

c) Glandular epithelia

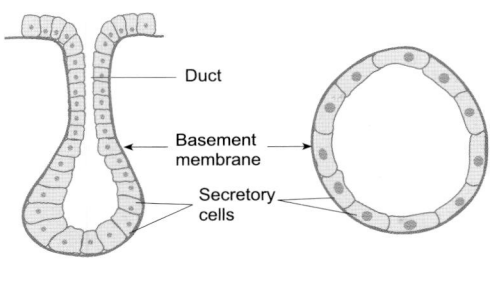

Duct

Basement membrane

Secretory cells

(ix) Exocrine

(x) Endocrine

Fig. 3.6 Diagrammatic representation of the principal types of epithelium. Panel (a) shows the general structure of the common types of simple epithelia. Note that when viewed from the apical surface there are no gaps between the cells (i). Pseudostratified epithelia (vi) have cells of differing shape and size, so giving a false appearance of multiple cell layers. The nuclei are found at many levels within the epithelium. Panels (b) and (c) show the general structure of stratified, transitional, and glandular epithelia. See text for further details.

non-ciliated cells are also interspersed between the ciliated cells, as illustrated in Fig. 3.6 (v). The non-ciliated cells may have a secretory role. Examples of ciliated epithelia are those

of the fallopian tubes and the ependymal lining of the cerebral ventricles.

◆ Pseudostratified columnar ciliated epithelium consists of cells of differing shapes and height, as shown in Fig. 3.6 (vi). This type of epithelium predominates in the upper airways (trachea and bronchi). Epithelia of this type combine a mechanical and secretory function: the goblet cells secrete mucus to trap airborne particles while the columnar ciliated cells move the mucus film towards the mouth.

◆ Stratified squamous epithelium is adapted to withstand chemical and physical stresses. The best known stratified epithelium is the epidermis of the skin. In this case, the flattened epithelial cells form many layers, only the lowest layer being in direct contact with the basement membrane (see Fig. 3.6 (viii)). The more superficial cells are filled with a special protein called keratin which renders the skin almost impervious to water and provides an effective barrier against invading organisms, such as bacteria. If the outer layers of the skin are damaged, for example by burns, there is a loss of fluid and a risk of infection. Moreover, if the area involved is very large, the loss of fluid may become life threatening.

◆ Transitional epithelium is found in the bladder and ureters. It is similar in structure to stratified squamous epithelium except that the superficial cells are larger and rounded. This adaptation allows stretching of the epithelial layer as the bladder fills.

Serous membranes enclose a fluid-filled space such as the peritoneal cavity. The peritoneal membranes consist of a layer of epithelial cells (called the mesothelium) overlying a thin basement membrane, beneath which is a narrow band of connective tissue. The fluid that separates the two epithelial surfaces is formed from the plasma by the same mechanism as the interstitial fluid (see p. 295). The pleural membranes that separate the lungs from the wall of the thoracic cavity are another example of serous epithelia.

Glandular epithelia are specialized for secretion. The epithelia that line the airways and part of the gut are covered with a thick secretion called mucus that is provided by specialized secretory cells known as goblet cells, which discharge their contents directly onto an epithelial surface. Other glands secrete material on to an epithelial surface via a specialized duct, as shown in Fig. 3.6 (ix). These are known as *exocrine glands*. Examples are the pancreas, salivary glands, and sweat glands. An exocrine gland may be a simple coiled tube (as in the case of the sweat glands), or it may consist of a complex set of branching ducts linking groups of cells together to form assemblies called *acini* (singular *acinus*). The acinar cells of an individual acinus are held in a tight ball by a capsule of connective tissue. This type of organization is found in the pancreas and salivary glands. Cells that produce a copious watery secretion when they are stimulated are called *serous cells*; examples are the acinar cells of the pancreas and salivary glands. Other glandular epithelia lack a duct and secrete material across their basolateral surfaces into the blood. These are the *endocrine glands*. Examples include the thyroid gland and the irregular clusters of epithelial cells that constitute the islets of Langerhans of the pancreas.

Summary

1. Epithelia act both to separate the interior of the body from the external environment and to separate one compartment of the body from another.

2. Individual epithelia are formed entirely from a sheet of cells and consist of one or more cell layers. An epithelium consisting of a single cell layer is known as a simple epithelium, while those with more than one layer are called stratified epithelia. Some have the appearance of being stratified because they contain cells of differing height and shape. These are known as pseudostratified epithelia.

3. Glandular epithelia are specialized for secretion. If their secretion is via a duct, they form part of an exocrine gland. If their secretion passes directly into the blood, they form part of an endocrine gland.

3.5 Energy metabolism in cells

Animals take in food as carbohydrates, fats, and protein. These complex molecules are broken down in the gut to simple molecules which are then absorbed. The carbohydrates of the diet are mainly starch which is broken down to glucose. The fats are broken down to fatty acids and glycerol while the dietary pro-

tein is broken down to its constituent amino acids. These breakdown products are then used by the cells of the body to make ATP, which provides a convenient way of harnessing chemical energy.

ATP can be synthesized in two ways: first, by the glycolytic breakdown of glucose to pyruvate, and secondly by the oxidative metabolism of pyruvate and acetate via the tricarboxylic acid cycle. (The tricarboxylic acid cycle is also known as the citric acid cycle or Krebs cycle.) The utilization of glucose, fatty acids, and amino acids for the synthesis of ATP is summarized in Fig. 3.7. In each case the synthesis of ATP is accompanied by the production of carbon dioxide and water.

The generation of ATP by glycolysis

Of the simple sugars, glucose is the most important in the synthesis of ATP. It is transported into cells where it is phosphoryl-

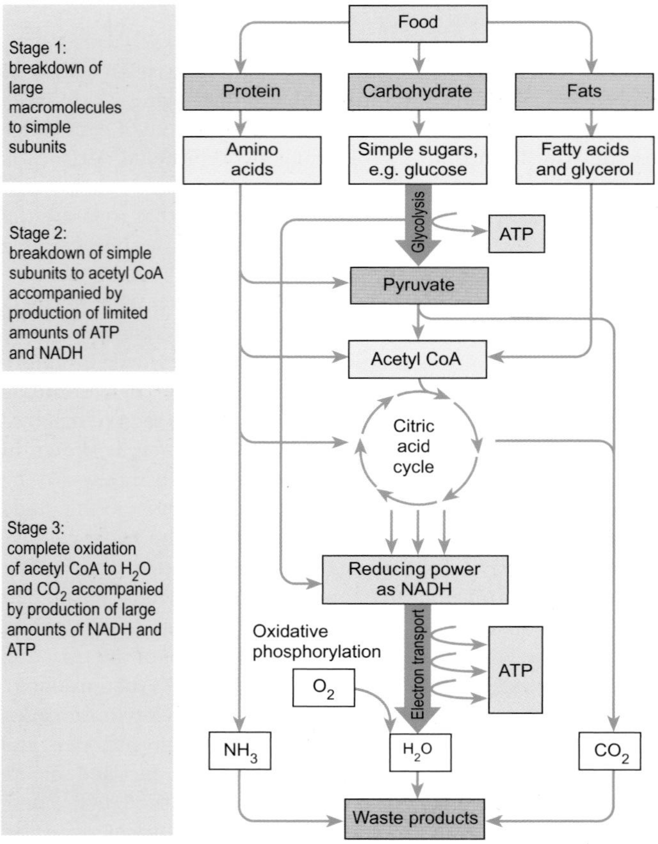

Fig. 3.7 A schematic diagram showing the three main stages by which foodstuffs are broken down to yield ATP for cellular metabolism.

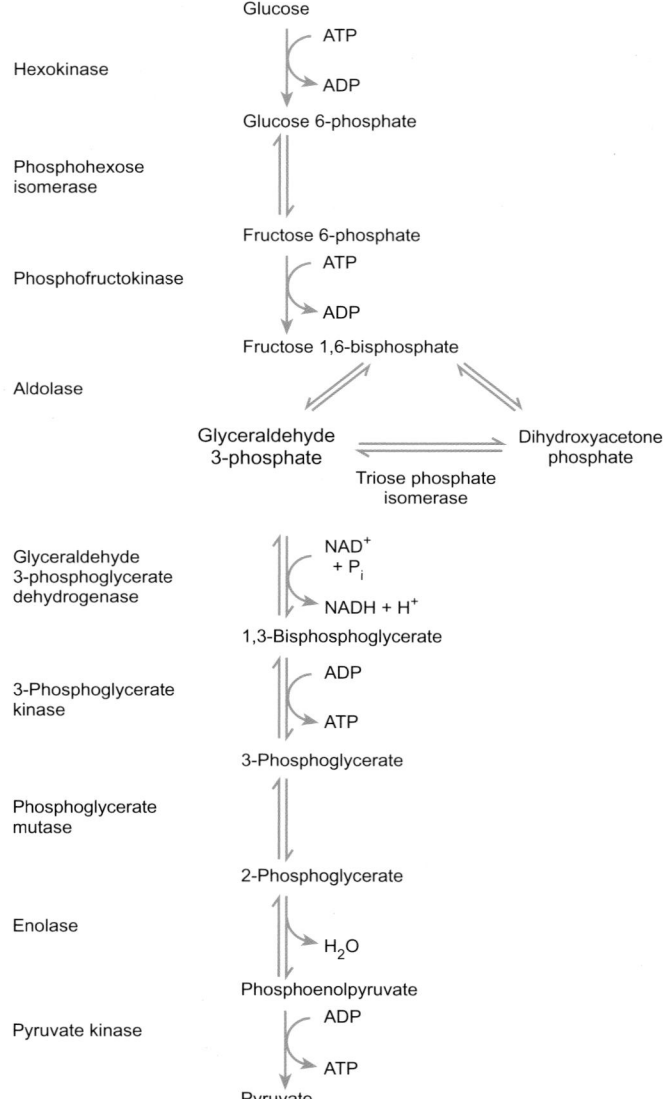

Fig. 3.8 A schematic chart showing the principal steps in the glycolytic breakdown of glucose.

ated to form glucose 6-phosphate. This is the first stage of its breakdown by the process of glycolysis in which glucose is broken down to form pyruvate. Glycolysis takes place in the cytoplasm of the cell outside the mitochondria, and does not require the presence of oxygen. For this reason, the glycolytic breakdown of glucose is said to be *anaerobic*. The glycolytic pathway is summarized in Fig. 3.8. Further details can be found in any standard textbook of biochemistry.

The breakdown of a molecule of glucose by glycolysis yields two molecules of pyruvate and two molecules of ATP. In addition, two molecules of reduced nicotinamide adenine dinucleotide (NADH) are produced. When oxygen is present, the NADH generated by glycolysis is oxidized by the mitochondria via the electron transport chain resulting in the synthesis of about three molecules of ATP and the regeneration of two molecules of nicotinamide adenine dinucleotide (NAD).

In normal circumstances, the pyruvate that is generated during glycolysis combines with coenzyme A (CoA) to form acetyl CoA which is oxidized via the tricarboxylic acid cycle to yield ATP. In the absence of sufficient oxygen, however, some of the pyruvate is reduced by NADH in the cytosol to generate lactate. This step regenerates the NAD used in the early stages of glycolysis so that it can participate in the breakdown of a further molecule of glucose. Thus *glycolysis can generate ATP even in the absence of oxygen* and becomes an important source of ATP for skeletal muscle during heavy exercise (Chapter 25).

The breakdown of pyruvate by the tricarboxylic acid cycle

The two molecules of pyruvate formed by the glycolytic breakdown of glucose combine with CoA to form acetyl CoA before they enter the tricarboxylic acid cycle. This step occurs in the mitochondria and results in the formation of two molecules of carbon dioxide and two molecules of NADH. Each molecule of acetyl CoA combines with a molecule of oxaloacetate to form the tricarboxylic acid citrate, which then undergoes a series of reactions that result in the complete oxidation of the acetyl CoA to carbon dioxide and water. These reactions are summarized in Fig. 3.9.

During each turn of the tricarboxylic acid cycle three molecules of NADH, one molecule of reduced flavine adenine dinucleotide ($FADH_2$), and one molecule of guanosine triphosphate (GTP) are generated. The mitochondria utilize the NADH and $FADH_2$ to generate ATP via an enzyme complex known as the electron transport chain. This process requires molecular oxygen and is called *aerobic metabolism*. The aerobic metabolism is very efficient and yields about 10 molecules of ATP for each molecule of acetyl CoA that enters the tricarboxylic acid cycle. Overall, the complete oxidation of one molecule of glucose to carbon dioxide and water yields about 30 molecules of ATP. This should be compared with the generation of ATP by anaerobic metabolism where only two molecules of ATP are generated for each molecule of glucose used.

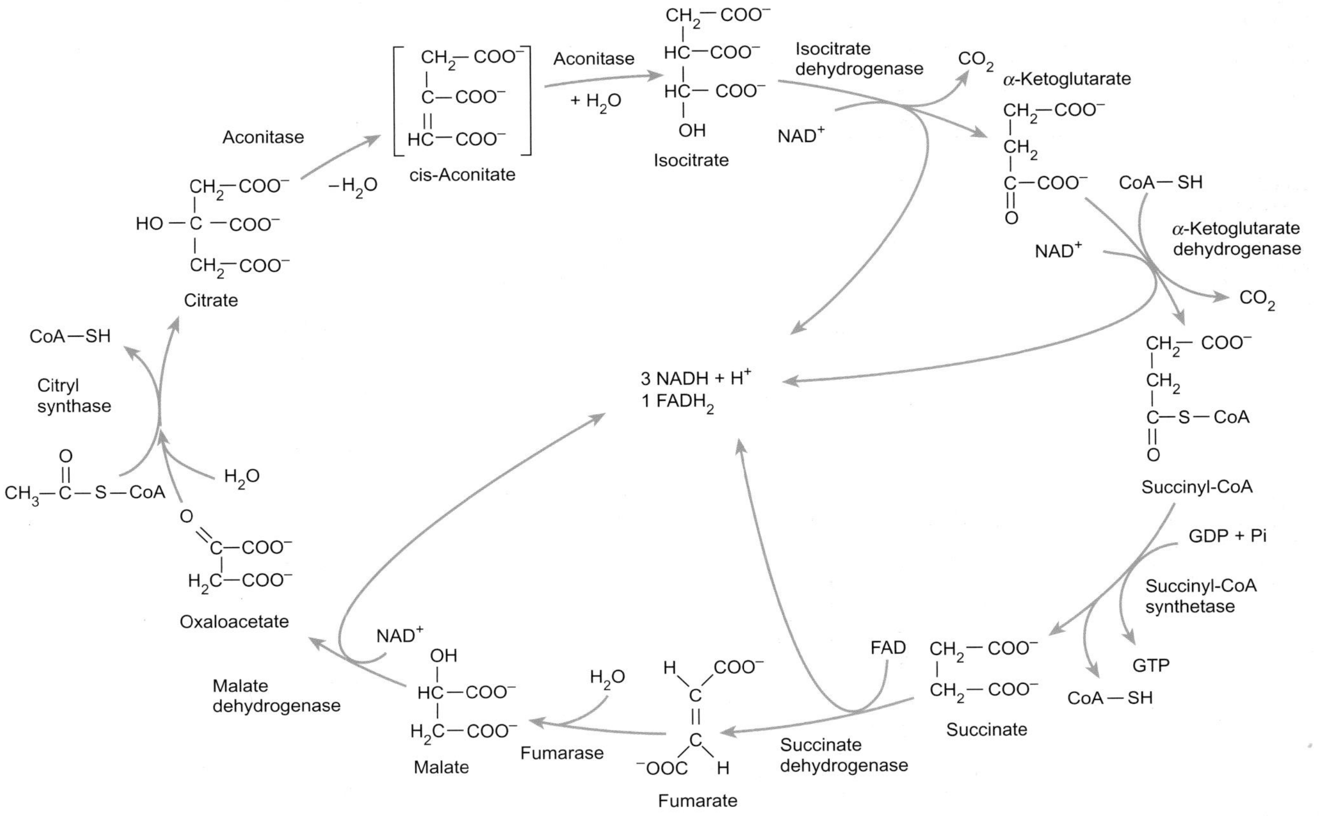

Fig. 3.9 The tricarboxylic acid cycle.

Fatty acids are the body's largest store of food energy. They are stored in fat cells (adipocytes) as triglycerols. Fat cells are widespread but are most abundant in the adipose tissues. Stored fat is broken down to fatty acids and glycerol by lipases in a process known as lipolysis. This takes place in the cytoplasm of the cell. The glycolytic pathway metabolizes the glycerol while the fatty acids combine with CoA to form acyl CoA before being broken down by a process known as β-oxidation. The breakdown of fatty acids to provide energy takes place within the mitochondria. The first step is the formation of acyl CoA followed by its transport from outside the mitochondria to the mitochondrial matrix where it undergoes a series of metabolic steps that result in the formation of acetyl CoA, NADH, and $FADH_2$ as shown in Fig. 3.10. The acetyl CoA is oxidized via the tricarboxylic acid cycle and the NADH and $FADH_2$ are oxidized via the electron transport chain to provide ATP. For each two-carbon unit metabolized, 13 molecules of ATP and one molecule of GTP are produced. Overall, the complete oxidation of a molecule of palmitic acid (which has 16 carbon atoms) yields over 100 molecules of ATP.

Although animals can synthesize fats from carbohydrates via acetyl CoA, they cannot synthesize carbohydrates from fatty acids. When glucose reserves are low, many tissues preferentially utilize fatty acids liberated from the fat reserves. Under these circumstances, the liver relies on the oxidation of fatty acids for energy production and may produce more acetyl CoA than it can utilize in the tricarboxylic acid cycle. Under such conditions, it synthesizes acetoacetate and D-3-hydroxybutyrate (also called β-hydroxybutyrate). These compounds are known as *ketone bodies* and are produced in large amounts when the utilization of glucose by the tissues is severely restricted as in starvation or poorly controlled diabetes mellitus (see Chapter 27). Acetoacetate and β-hydroxybutyrate are not waste products but can be utilized for ATP production by the heart and the kidney. In severe uncontrolled diabetes mellitus, significant amounts of acetone form spontaneously from acetoacetate, and this gives the breath a characteristic sweet smell which can be a useful aid in diagnosis.

Proteins are the main structural components of cells and the principal use for the protein of the diet is the synthesis of new

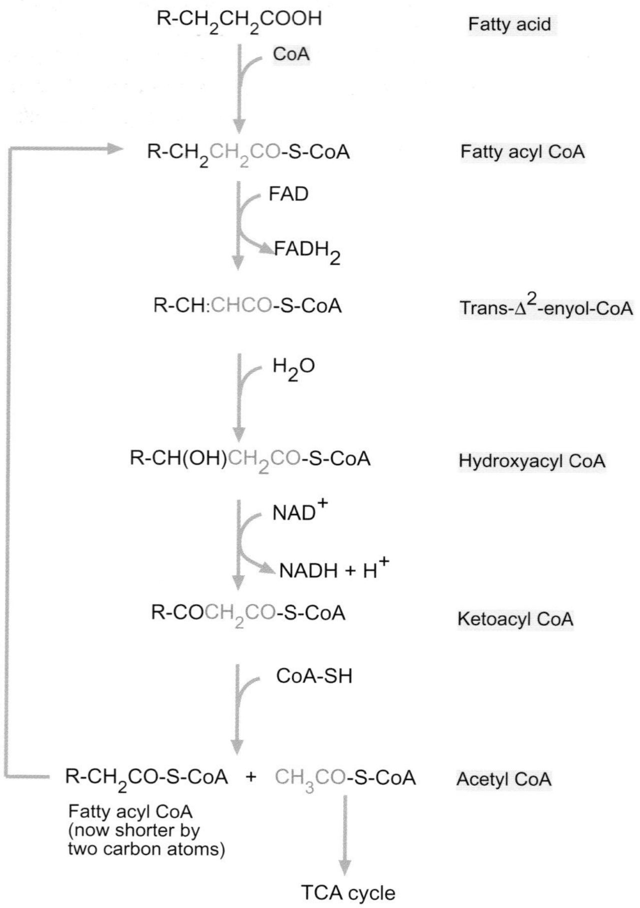

R-CH$_2$CH$_2$COOH   Fatty acid

CoA

R-CH$_2$CH$_2$CO-S-CoA   Fatty acyl CoA

FAD

FADH$_2$

R-CH:CHCO-S-CoA   Trans-Δ^2-enyol-CoA

H$_2$O

R-CH(OH)CH$_2$CO-S-CoA   Hydroxyacyl CoA

NAD$^+$

NADH + H$^+$

R-COCH$_2$CO-S-CoA   Ketoacyl CoA

CoA-SH

R-CH$_2$CO-S-CoA + CH$_3$CO-S-CoA   Acetyl CoA

Fatty acyl CoA
(now shorter by
two carbon atoms)

TCA cycle

Fig. 3.10 The pathway by which fatty acids are broken down to form acetyl CoA which can be metabolized via the tricarboxylic acid cycle to produce ATP. In these reactions each two-carbon unit generates one molecule of NADH and one molecule of $FADH_2$ in addition to that generated by the tricarboxylic acid cycle.

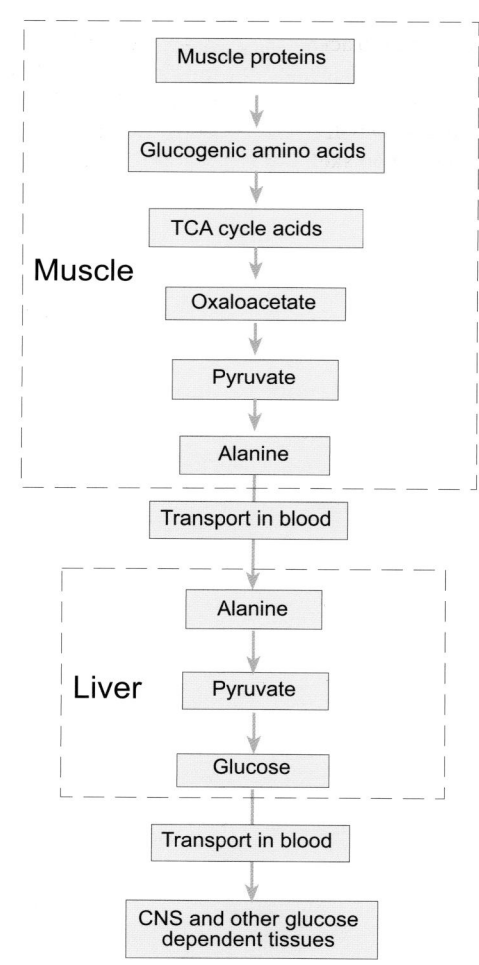

Fig. 3.11 An outline of the pathways by which amino acids are broken down for energy (ATP) production and the formation of glucose (gluconeogenesis). TCA, tricarboxylic acid; CNS, central nervous system.

Summary

1. Animals generate the energy they require for movement and growth of body tissues by the progressive breakdown of foodstuffs.

2. Carbohydrates are first broken down by the glycolytic pathway to form pyruvate. Fats are broken down by β-oxidation to form acetyl CoA. Proteins are first broken down to amino acids which may then generate pyruvate (glucogenic amino acids) or acetyl CoA (ketogenic amino acids).

3. Pyruvate and acetyl CoA are utilized by the tricarboxylic acid cycle and electron transport chain of the mitochondria to synthesize ATP.

4. Although animals can synthesize fats from carbohydrates via acetyl CoA, they cannot synthesize carbohydrates from fatty acids. When glucose reserves are low, many tissues preferentially utilize fatty acids.

5. During starvation, the glucose required by the brain is synthesized in the liver and kidneys by gluconeogenesis.

protein. Those amino acids that are not required for protein synthesis are deaminated. This process results in the replacement of the amino group ($-NH_2$) by a keto group ($>C=O$) and the liberation of ammonia, which is subsequently metabolized to urea by the liver. The carbon skeleton of most amino acids can be used to synthesize glucose (a process known as *gluconeogenesis*), and those that can be so utilized are known as the glucogenic amino acids (see Fig. 3.11). Of the 20 amino acids found in proteins, only leucine and lysine cannot be used for gluconeogenesis and are oxidized via the same pathway as fats. Such amino acids are called ketogenic. The glucogenic amino acids ultimately form pyruvate or oxaloacetate, which can be oxidized via the tricarboxylic acid cycle to produce ATP or to generate glucose. The carbon skeleton of the ketogenic amino acids is oxidized by β-oxidation to form acetyl CoA which, like the fatty acids, is metabolized via the tricarboxylic acid cycle to form ATP. Although the amino acids are generally considered to be either glucogenic or ketogenic, several, including the aromatic amino acids phenylalanine, tryptophan and tyrosine, can be either glucogenic or ketogenic depending on the metabolic pathway by which they are broken down.

Recommended reading

Alberts, B., Johnson, A., Lewis, J., Raff, M., Roberts, K., and Walter, P. (2002). *Molecular biology of the cell* (4th edn), Chapters 1, 10, 16, and 18. Garland, New York.

Berg, J.M., Tymoczko, J.L., and Stryer, L. (2002). *Biochemistry* (5th edn), Chapters 16–18, 21, and 22. Freeman, New York.

Elliott, W.H., and Elliott, D.C. (2005). *Biochemistry and molecular biology*, Chapters 3, 7, 11–15. Oxford University Press, Oxford.

Junquieira, L.C., and Carneiro, J. (2003). *Basic histology* (10th edn), Chapters 2–4. McGraw-Hill, New York.

CHAPTER 4

The transport functions of the plasma membrane

After reading this chapter you should understand:

- The principal types of proteins involved in membrane transport
- The difference between passive and active transport
- How polar molecules and ions cross the plasma membrane
- The role of metabolic pumps in generating and maintaining ionic gradients
- How the sodium pump generates the Na^+ and K^+ gradients across the plasma membrane
- The importance of the sodium pump in maintaining constant cell volume
- How cells regulate their intracellular free Ca^{2+}
- How cells make use of the Na^+ gradient to regulate intracellular $[H^+]$
- How cells exploit the ionic gradients established by the sodium pump to permit the secondary active transport of glucose and amino acids across epithelial layers
- The origin of the membrane potential
- The nature of ion channels: ligand- and voltage-gated channels
- The mechanism of secretion: constitutive and regulated secretion by exocytosis
- Endocytosis and the retrieval of membrane constituents
- The mechanism by which certain cells ingest cell debris and foreign material—phagocytosis

4.1 Introduction

As explained in Chapter 3, each cell is bounded by a plasma membrane. Thus the region outside the cell (the *extracellular compartment*) is separated from the inside of the cell (the *intracellular compartment*). This physical separation allows each cell to regulate its internal composition independently of other cells. Chemical analysis has shown that the composition of the intracellular fluid is very different from that of the extracellular fluid (Table 4.1). It is rich in potassium ions (K^+) but relatively poor in both sodium ions (Na^+) and chloride ions (Cl^-). It is also rich in proteins (enzymes and structural proteins) and the small organic molecules that are involved in metabolism and signaling (amino acids, ATP, fatty acids, etc.). The first part of this chapter is concerned with the mechanisms responsible for establishing and

TABLE 4.1 The approximate ionic composition of the intracellular and extracellular fluid of mammalian muscle

Ionic species	Extracellular fluid (mmol l−1)	Intracellular fluid (mmol l−1)	Nernst equilibrium potential (mV)
Na+	145	20	+53
K+	4	150	−97
Ca2+	1.8	c. 2 × 10−4	+120
Cl−	114	3	−97
HCO3−	31	10	−30

The equilibrium potentials were calculated from the Nernst equation (see Box 4.1). Note that the resting membrane potential is about −90 mV, close to the equilibrium potentials for potassium (the principal intracellular cation) and Cl−.

maintaining the difference in ionic composition between the intracellular and the extracellular compartments. The ways in which cells utilize ionic gradients to perform their essential physiological roles are then discussed and the chapter concludes with a discussion of the mechanisms by which proteins and other large molecules cross the cell membrane—secretion and endocytosis.

4.2 The permeability of cell membranes to ions and uncharged molecules

The plasma membrane consists of a lipid bilayer in which many different proteins are embedded (see Chapter 3). Both natural membranes and artificial lipid bilayers are permeable to gases and lipid-soluble molecules (*hydrophobic molecules*). However, compared with artificial lipid bilayers, natural membranes have a high permeability to water and water-soluble molecules (*hydrophilic* or *polar molecules*) such as glucose, and to ions such as sodium, potassium, and chloride. The relatively high permeability of natural membranes to ions and polar molecules can be ascribed to the presence of two classes of integral membrane proteins: the *channel proteins* and the *carrier proteins*.

What determines the direction in which molecules and ions move across the cell membrane?

For uncharged molecules such as carbon dioxide, oxygen, and urea, the direction of movement across the plasma membrane is simply determined by the prevailing concentration gradient. However, the situation is more complicated for charged molecules and ions. Measurements have shown that mammalian cells have an electrical potential across their plasma membrane called the *membrane potential*, which is discussed in greater detail on p. 40. Although the magnitude of the membrane potential varies from one type of cell to another (from about −35 to −90 mV), the inside of a cell is always negative with respect to the outside. The existence of the membrane potential influences the diffusion of charged molecules and ions. Positively charged chemical species will tend to be attracted into the cell while negatively charged species will tend to be repelled. Overall, the direction in which ions and charged molecules move across the cell membrane is determined by three factors:

THE NERNST EQUATION AND THE RESTING MEMBRANE POTENTIAL

The flow of any ion across the membrane via an ion channel is governed by its electrochemical gradient, which reflects the charge and the concentration gradient for the ion in question together with the membrane potential. The potential at which the tendency of the ion to move down its concentration gradient is exactly balanced by the membrane potential is known as the equilibrium potential for that ion, so that at the equilibrium potential the rate at which ions enter the cell is exactly balanced by the rate at which they leave. The equilibrium potential can be calculated from the Nernst equation:

$$E = \frac{RT}{zF} \ln \frac{[C]_o}{[C]_i}$$

where E is the equilibrium potential, ln is the natural logarithm ($\log_e$), $[C]_o$ and $[C]_i$ are the extracellular (outside) and intracellular (inside) concentrations of the ion in question, R is the gas constant (8.31 J K−1 mol−1), T is the absolute temperature, F is the Faraday constant (96 487 C mol−1), and z is the charge on the ion (+1 for Na+, +2 for Ca2+, −1 for Cl−, etc.). At 37°C, $RT/F = 26.7$ mV.

The equilibrium potential for K+, using the data shown in Table 4.1, can be calculated as:

$$E_K = \frac{RT}{zF} \ln \frac{[K^+]_o}{[K^+]_i}$$

$$= 26.7 \times \ln \frac{4}{150} = 26.7 \times (-3.62).$$

Hence:

$$E_K = -96.8 \text{ mV}.$$

(1) the concentration gradient;

(2) the charge of the molecule or ion;

(3) the membrane potential.

These factors combine to give rise to the *electrochemical gradient*, which can be calculated from the difference between the *equilibrium potential* for the ion in question and the membrane potential. If the intracellular and extracellular concentrations for a particular ion are known, the equilibrium potential can be calculated using the Nernst equation (Box 4.1).

When molecules and ions diffuse across the plasma membrane down their electrochemical gradients they do so by *passive transport*. As polar materials cross the plasma membrane much more readily than artificial bilayers, passive transport of these substances is sometimes called *facilitated diffusion*. Cells can also transport molecules and ions against their prevailing electrochemical gradients. This uphill transport requires a cell to expend metabolic energy either directly or indirectly and is called *active transport*.

While the *direction* of movement of a particular substance is determined by its electrochemical gradient, the number and properties of the channel and/or carrier proteins present determine the *rate* at which it can cross the plasma membrane.

If there are few channels or carriers for a particular molecule or ion, the permeability of the membrane to that substance will be low: the more channels or carriers that are available, the greater is the permeability of the membrane to that substance.

Channel proteins and carrier proteins permit polar molecules to cross the plasma membrane

Water channels

The relatively high permeability of natural membranes to water can be attributed to the presence of specialized water channels known as *aquaporins*. These are proteins possessing pores that allow water to pass from one side of the membrane to the other according to the prevailing osmotic gradient. Currently, six different types of aquaporin are known in mammals, several of which are found in tissues which transport large volumes of water in the course of a day; examples of such tissues are the tubules of the kidney and the secretory glands of the gut. Aquaporins 2 and 3 play a central role in regulating water reabsorption by the collecting ducts of the kidney and aquaporin 5 plays a significant role in the production of salivary secretions.

Ion channels

The permeability of natural membranes to ions is due to specific membrane proteins called *ion channels* which have a pore that spans the membrane to provide a route for a particular ion to diffuse down its electrochemical gradient (Fig 4.1(a)). Such channels exist in one of two states: they are either open and allow the passage of the appropriate ion from one side of the membrane to the other, or they are closed and prevent such movements. Many ion channels have a very high capacity for transport. For example, some potassium channels permit as many as 10^8 ions to cross the membrane each second. Despite this, ion channels are able to discriminate one type of ion from another, so that they show selectivity with respect to the kind of ion they allow through their pore. They are named after the principal ion to which they are permeable. For example, sodium channels allow sodium ions to cross the membrane but not other cations such as potassium and calcium. Potassium channels are permeable to potassium ions but not to sodium or calcium ions. Chloride ions cross the membrane via chloride channels, and so on. The permeability of a cell membrane to a particular ion will depend on how many channels of the appropriate type are open and the number of ions that can pass through each channel in a given period of time.

Some ion channels allow a variety of positively charged ions to cross the plasma membrane but do not allow the passage of negatively charged ions. Such channels are known as *non-selective cation channels*. An important example of this type of channel is the acetylcholine receptor of the neuromuscular junction. Many channel proteins have now been identified and their peptide sequence determined with the aid of the techniques of molecular biology. This has revealed that they can be grouped into large families according to certain features of their structure. For example, the peptide chains of many types of potassium channel have a similar sequence of amino acids in certain regions and make the same number of loops across the plasma membrane.

Carrier proteins

Carrier proteins (carriers) bind specific substances—usually small organic molecules such as glucose or inorganic ions (Na^+, K^+, etc.)—and then undergo a change of shape (known as a *conformational change*) to move the solute from one side of the membrane to the other (Fig 4.1(b)). The capacity of a cell to transport a molecule is limited by both the number of carrier molecules and the number of molecules each carrier is able to translocate in a given period of time (the 'turnover number'). Carriers tend to transport fewer ions or molecules than channels. The fastest carriers transport about 10^4 molecules each second but more typical values are between 10^2 and 10^3 molecules a second. Carriers are selective for a particular type of molecule and can even

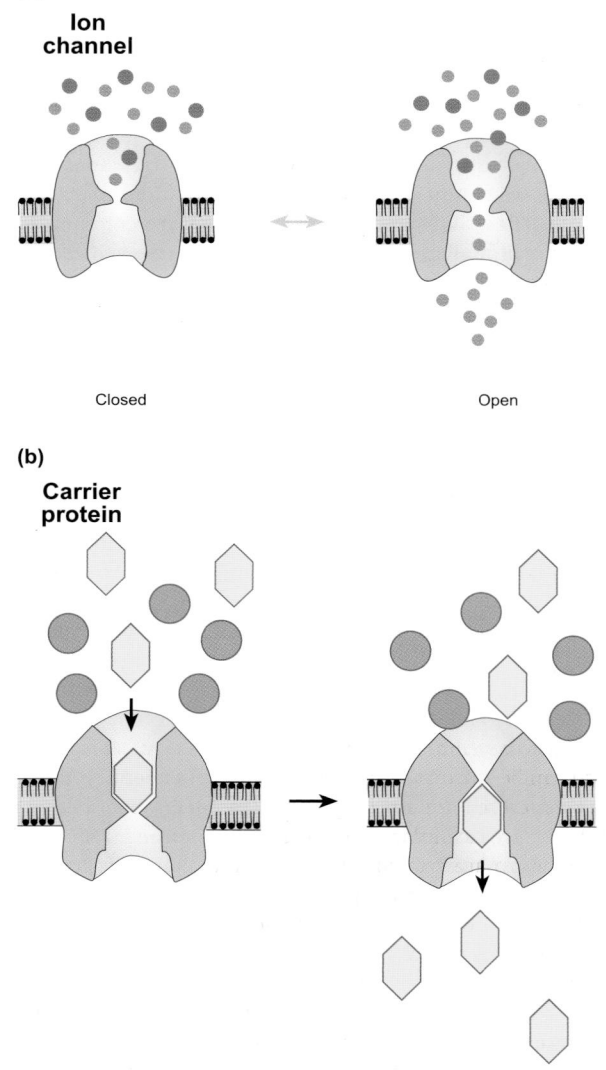

(a)

Ion channel

Closed   Open

(b)

Carrier protein

Fig. 4.1 Schematic drawings illustrating the differing modes of action of (a) an ion channel and (b) a membrane carrier. Molecules that are transported across biological membranes by carrier proteins first bind to a specific site. The carrier then undergoes a conformational change and the bound molecule is able to leave the carrier on the other side of the membrane. Ion channels operate quite differently. When they are activated, a pore is opened and ions are able to diffuse from one side of the membrane to the other in a continuous stream. Both show selectivity for a particular ion or molecule.

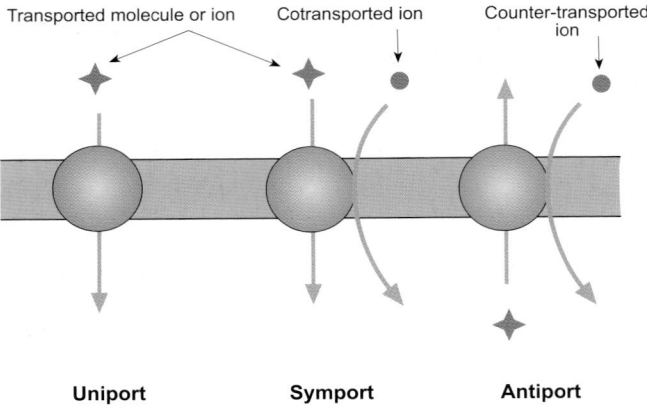

Fig. 4.2 A diagrammatic representation of the main types of carrier proteins employed by mammalian cells. Unidirectional transport and counter transport may be linked to hydrolysis of ATP (and thus play a role in active transport). Symports exploit existing ionic gradients for secondary active transport.

discriminate between optical isomers. For example, the natural form of glucose (D-glucose) is readily transported by specific carrier proteins, but the synthetic L-isomer is not transported. Although both isomers have the same chemical constitution, the D-and L-isomers of glucose are mirror images, like our left and right hands. This proves that the carrier can distinguish between optical isomers solely by their shape. This property is known as *stereoselectivity*.

Carrier proteins (also called *transporters*) are subdivided into three main groups according to the way in which they permit molecules to cross the plasma membrane. *Uniports* bind a specific molecule on one face of the membrane and then transfer it to the other side as shown in Fig 4.1(b) and 4.2. However, many substances are transported across the membrane only in association with a second molecule or ion. Carriers of this type are called cotransporters and the transport itself is called *cotransport* or *coupled transport*. When both molecular entities move in the same direction across the membrane, the carrier is called a *symport*; when the movement of an ion or molecule into a cell is coupled to the movement of a second ion or molecule out of the cell, the carrier is called an *antiport*. Figure 4.2 shows schematic representations of the different kinds of transport proteins.

Summary

1. While lipid-soluble molecules can cross pure lipid membranes relatively easily, water-soluble molecules cross only with difficulty. Two groups of membrane proteins facilitate the movement of water-soluble molecules from one side of the membrane to the other: the ion channels and the carrier proteins.

2. Ion channels permit the passage of ions from one side of the membrane to the other via a pore.

3. Carrier molecules translocate a molecule from one side of the membrane to the other by binding the molecule on one face of the membrane and undergoing a conformational change to release it on the other.

4.3 The active transport of ions and other molecules across cell membranes

Active transport requires a cell to expend metabolic energy either directly or indirectly and involves carrier proteins. In many cases, the activity of a carrier protein is directly dependent on metabolic energy derived from the hydrolysis of ATP (e.g. the sodium pump discussed below). In other cases, the transport of a substance (e.g. glucose) can occur against its concentration gradient by coupling its 'uphill' movement to the 'downhill' movement of Na^+ into the cell. This type of active transport is known as *secondary active transport*. It depends on the ability of the sodium pump to keep the intracellular concentration of Na^+ significantly lower than that of the extracellular fluid.

The sodium pump is present in all cells and exchanges intracellular sodium for extracellular potassium

As shown in Table 4.1, the intracellular concentration of sodium, calcium, and chloride in muscle cells is much lower than the extracellular concentration. Conversely, the intracellular concentration of potassium is much greater than that of the extracellular fluid. These differences in composition are common to all healthy mammalian cells, although the precise values for the concentrations of intracellular ions vary from one kind of cell to another. What mechanisms are responsible for these differences in ionic composition?

The first clues about the mechanisms by which cells regulate their intracellular sodium came from the problems associated with the storage of blood for transfusion. Like other cells, the red cells of human blood have a high intracellular potassium concentration and a low intracellular sodium concentration. When blood is stored at a low temperature in a blood bank, the red cells lose potassium and gain sodium over a period of time—a trend that can be reversed by warming the blood to body temperature (37°C). If red cells that have lost their potassium during storage are incubated at 37°C in an artificial solution similar in ionic composition to that of plasma, they only re-accumulate potassium if glucose is present. This glucose-dependent uptake of potassium and extrusion of sodium by the red cells occurs against the concentration gradients for these ions. It is therefore clear that the movement of these ions is dependent on the activity of a membrane pump driven by the energy liberated by the metabolic breakdown of glucose, in this case the *sodium pump*.

The sodium pump is found in all mammalian cells and plays a central role in regulating the intracellular environment. How does it work? Important clues were provided by experiments on the giant axon of the squid. This preparation was chosen for these experiments because it has a large diameter for a single cell (1–2 mm) and this property made direct measurements of ionic movements across the plasma membrane possible. By injecting radioactive sodium ($^{22}Na^+$) into the axon, the rate at which the axon pumped out sodium ions could be followed by measuring the appearance of radioactive sodium in the bathing solution. When the metabolic inhibitor cyanide was used to block ATP generation, the rate of pumping declined. When ATP was subsequently injected into an axon that had been poisoned by cyanide, the sodium pumping was restored to almost normal levels (Fig. 4.3). This experiment showed that ATP was required

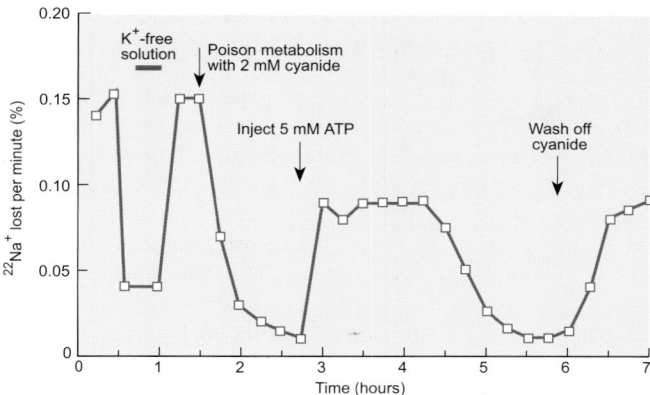

Fig. 4.3 The sodium pump is driven by intracellular ATP. In this experiment, which was conducted on a large axon isolated from a squid, normal oxidative metabolism was blocked by cyanide. As ATP levels fell, the axon became unable to pump sodium ions from its cytoplasm until ATP was injected. Removal of the cyanide gradually restored the ability of the axon to pump sodium ions. The first part of the experiment, shown on the left, demonstrates that sodium can only be actively pumped out of the axon if it can be exchanged for extracellular potassium.

for sodium to be pumped out of the axon against its electro-chemical gradient. In the same series of experiments it was found that the sodium efflux was also inhibited if potassium was removed from the extracellular fluid, thus demonstrating that the uptake of potassium is closely coupled to the efflux of sodium.

Subsequent studies on red blood cells showed that the hydro-lysis of ATP is tightly coupled to the efflux of sodium and to the influx of potassium in such a way that, for each ATP molecule hydrolysed, a cell pumps out three sodium ions in exchange for two potassium ions. For this reason the sodium pump is also called the Na^+, K^+ ATPase. The sodium pump can be inhibited by a glycoside called ouabain that binds to the extracellular face of the protein. A schematic diagram of the operation of the sodium pump is given in Fig. 4.4.

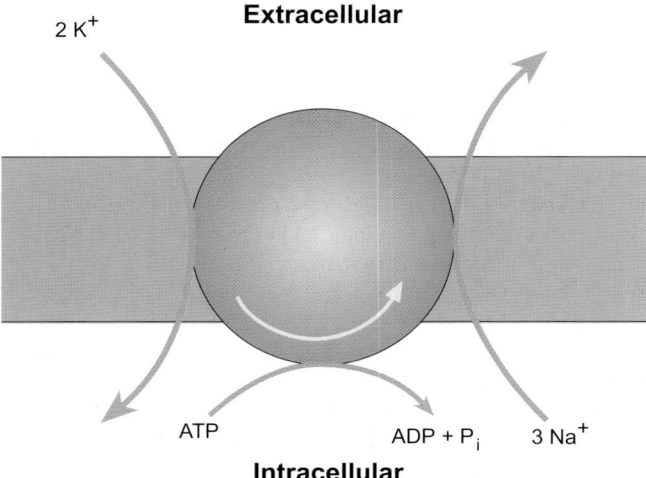

Fig. 4.4 Schematic diagram of the sodium pump. Note that for every molecule of ATP hydrolysed, three sodium ions are pumped out of the cell and two potassium ions are pumped into the cell.

The sodium pump stabilizes cell volume by maintaining a low intracellular sodium concentration

The total number of particles present in a solution determines its osmotic pressure. *Outside the cell* the osmolality is due chiefly to the large number of small inorganic ions present in the extra-cellular fluid (Na^+, K^+, Cl^-, etc.). *Inside the cell* the osmolality is due to inorganic ions and to a large number of membrane-imperme-ant molecules such as ATP, amino acids, and proteins. Over time, there is a propensity for the intracellular concentration of ions to become equal to that of the extracellular fluid as they diffuse down their electrochemical gradients. If this tendency were not countered, the total osmolality of the intracellular fluid would tend to increase because the large impermeant molecules can-not pass out of the cell to compensate for the inward movement of small ions. The increase in osmolality would cause the cells to take up water and swell. By keeping the intracellular sodium ion concentration low, the sodium pump maintains the total osmo-lality of the intracellular compartment equal to that of the extra-cellular fluid. As a result, cell volume is kept relatively constant.

Cells use a number of transport proteins to regulate the intracellular hydrogen ion concentration

All respiring cells continually produce metabolic acids (e.g. car-bon dioxide and carboxylic acids) which are capable of altering the intracellular concentration of hydrogen ions [H^+]. Measure-ments have shown that cells maintain the hydrogen ion concen-tration of their cytoplasm close to 10^{-7} mol l^{-1}. Although this is higher than that of the extracellular fluid (which is usually about 4×10^{-8} mol l^{-1}), it is about a tenth of that expected if hydrogen ions were simply at electrochemical equilibrium. Therefore cells must actively regulate their intracellular hydro-gen ion concentration.

The hydrogen ion concentration of a solution is often expressed on the pH scale, on which a low pH value corresponds to a high value for hydrogen ion concentration and a high pH value corresponds to a low hydrogen ion concentration (see Chapter 29, Section 29.2, for further details). An extracellular hydrogen ion concentration of 4×10^{-8} mol l^{-1} corresponds to a pH value of 7.4, and an intracellular hydrogen ion concentration of 10^{-7} mols l^{-1} corresponds to a pH value of 7.0.

Since hydrogen ions are very reactive and bind readily to a wide variety of proteins, changes in the intracellular concentra-tion of hydrogen ions can have major consequences for cellular activity. For example, many enzymes work best at a particular hydrogen ion concentration (their pH optimum). Changes in intracellular hydrogen ion concentration also influence the function of other proteins such as ion channels, and the con-tractile proteins actin and myosin. When the hydrogen ion concentration rises during cellular activity, many hydrogen ions will bind to intracellular molecules, thus limiting the rise. This phenomenon is known as *buffering* (see Chapter 29, Section 29.3, for further details). However, buffering can only limit the change in hydrogen ion concentration. To restore its original hydrogen ion concentration a cell must pump out the excess hydrogen ions. To enable them to do so, cells have evolved a number of regulatory mechanisms, three of which are discussed here:

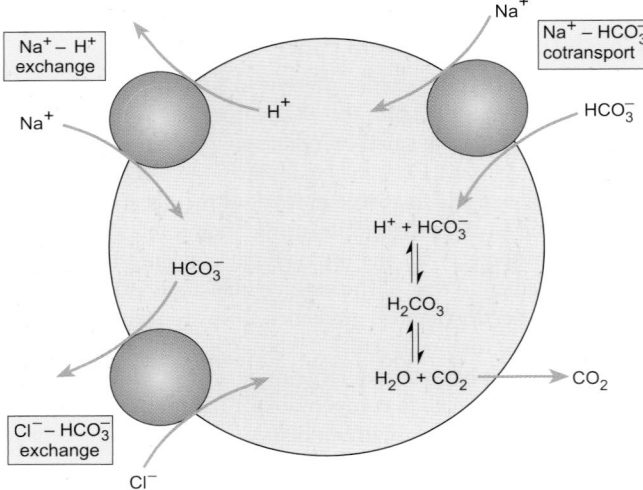

Fig. 4.5 Cells regulate their internal H^+ concentration by three separate mechanisms: they exchange internal H^+ for external Na^+; they import HCO_3^- ions to combine with H^+ to form carbonic acid and ultimately CO_2 which is able to cross the cell membrane; when the cell is alkaline (deficient in H^+), HCO_3^- can be exchanged for Cl^-.

(1) sodium–hydrogen ion exchange;

(2) cotransport of Na^+ and HCO_3^- into cells to increase intracellular HCO_3^-;

(3) chloride–bicarbonate exchange.

Sodium–hydrogen ion exchange occurs via an antiport that couples the outward movement of hydrogen ions against their electrochemical gradient to the inward movement of sodium ions down their electrochemical gradient (Fig. 4.5). Thus it is an example of secondary active transport, the driving force for the transport of hydrogen ions out of the cell being provided by the sodium gradient established by the sodium pump. This carrier is found in a wide variety of different cell types and is particularly important in the epithelial cells of the kidney.

Many cells also possess a symporter that transports both sodium and bicarbonate ions into the cell. By linking the influx of bicarbonate to that of sodium ions, the bicarbonate ions are transported into a cell against their electrochemical gradient. Once inside, these ions bind excess hydrogen ions to form carbonic acid, which is in equilibrium with dissolved carbon dioxide that can cross the cell membrane by diffusion. For each molecule of carbon dioxide that leaves the cell as a result of this transport, one hydrogen ion is used up to form one molecule of water (Fig. 4.5).

Under most circumstances the cells are acting to prevent an increase in their hydrogen ion concentration. At high altitude, however, the extra breathing required to keep the tissues supplied with oxygen leads to a fall in the carbon dioxide concentration in the blood and tissues. This makes the cells more alkaline than they should be (i.e. they have a hydrogen ion deficit). To maintain the intracellular hydrogen ion concentration within the normal range, intracellular bicarbonate is exchanged for extracellular chloride. This mechanism (known as chloride–bicarbonate exchange) provides a means of defense

against a fall in the intracellular hydrogen ion concentration. Chloride–bicarbonate exchange is freely reversible and plays an important role in the carriage of carbon dioxide by the red cells (see Chapter 13).

Many processes regulate intracellular calcium

The intracellular fluid of mammalian cells has a very low concentration of calcium ions—typical values for a resting cell are about 10^{-7} mol l^{-1} while that of the extracellular fluid is 1–2×10^{-3} mol l^{-1}. Therefore, there is a very steep concentration gradient of calcium ions across the plasma membrane and this is exploited by a wide variety of cells to provide a means of transmitting signals to the cell interior. A rise in intracellular Ca^{2+} is the trigger for many processes including the contraction of muscle and the initiation of secretion. Therefore it is essential that resting cells maintain a low resting level of intracellular Ca^{2+}. As shown in Fig. 4.6 this is accomplished in a rather interesting and intricate way.

1. Calcium ions can be pumped from the inside of the cell across the plasma membrane via a Ca^{2+}-ATPase. This is known as the *calcium pump*, and, like the sodium pump, it uses energy derived from ATP to pump calcium against its concentration gradient.

2. Intracellular calcium can be exchanged for extracellular sodium (the Na^+–Ca^{2+} exchanger). In this case the inward movement of sodium ions down their electrochemical gradient provides the energy for the uphill movement of calcium ions from the inside to the outside of the cell against their electrochemical gradient (another example of secondary active transport—in this case via an antiport).

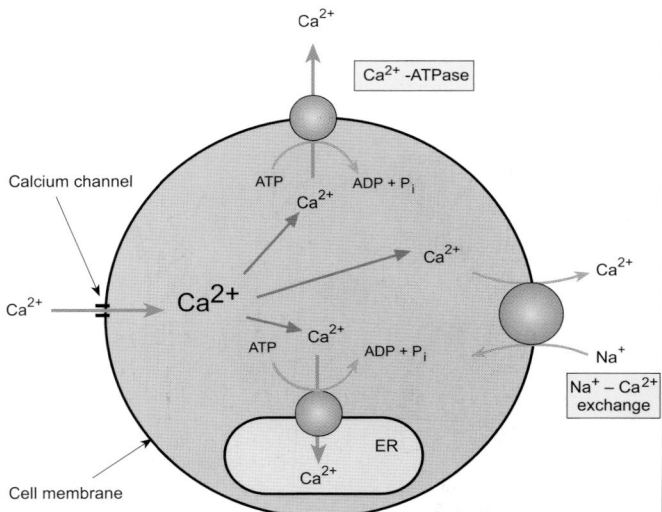

Fig. 4.6 As the level of intracellular Ca^{2+} is widely employed by cells as a signal to initiate a wide variety of physiological responses, it is essential that intracellular Ca^{2+} is closely regulated. The figure is a schematic drawing illustrating the main ways in which this is achieved. Calcium ions enter cells by way of calcium channels. Calcium can be stored within the cell either in the endoplasmic reticulum (sarcoplasmic reticulum in muscle) or in the mitochondria (not shown). It can be removed from the cell either by the calcium pump (Ca^{2+}-ATPase) or by Na^+–Ca^{2+} exchange.

3. Another type of calcium pump is used to pump calcium into the endoplasmic reticulum to provide a store of calcium within the cell itself while keeping the calcium concentration of the cytosol very low. (In muscle the endoplasmic reticulum is called the sarcoplasmic reticulum.) This store of calcium can be released in response to signals from the plasma membrane (see Chapter 5 for further details). In addition, the mitochondria can also take up calcium from the cytosol.

The transepithelial transport of glucose and amino acids is linked to the activity of the sodium pump

Glucose and amino acids are required to generate energy and for cell growth. Both are obtained by the digestion of food so that the cells that line the small intestine (the enterocytes) must be able to transport glucose and amino acids from the central cavity of the gut (the *lumen*) to the blood. To take full advantage of all the available food, the enterocytes must be able to continue this transport even when the concentration of these substances in the lumen has fallen below that of the blood. This requires active transport across the intestinal epithelium. How is this transepithelial transport achieved?

Like all epithelia, those of the gut are polarized. Their apical surfaces face the lumen and their basolateral surfaces are oriented towards the blood stream. The two regions are separated by the zonula occludens, which is made up of tight junctions (see Chapter 3 and Fig. 4.7) that are relatively impermeable to glucose and other small solutes. Therefore, to be absorbed, these substances must pass through the cells. This is called *transcellular absorption* to distinguish it from absorption across the tight junctions, which is called *paracellular absorption*.

The plasma membranes of the apical and basolateral regions of the enterocytes possess different carriers. The *apical membrane* contains a symport, called SGLT1, that binds both sodium and glucose. As the concentration of sodium in the enterocytes is about 20 mmol l⁻¹, the movement of sodium from the lumen

(where the sodium concentration is about 150 mmol l⁻¹) into these cells is favored by the concentration gradient. The carrier links the inward movement of sodium down its electrochemical gradient to the uptake of glucose against its concentration gradient and this enables the enterocytes to accumulate glucose. The *basolateral membrane* has a different type of glucose carrier—a uniport, called GLUT2—which permits the movement of glucose from the cell interior down its concentration gradient into the space surrounding the basolateral surface. This is an example of facilitated diffusion. As this carrier does not link the movement of glucose to that of sodium, the reverse transport of glucose from the blood into the lumen is effectively prevented. The sodium absorbed with the glucose is removed from the enterocyte by the sodium pump of the basolateral membrane. Thus the energy for the uptake of glucose is provided by the sodium gradient established by the sodium pump, and the asymmetric arrangement of the glucose carriers on the cell permits the *secondary active transport* of glucose across the wall of the intestine from where it can be absorbed into the blood (Fig. 4.7). Similar mechanisms exist for the transport of the amino acids.

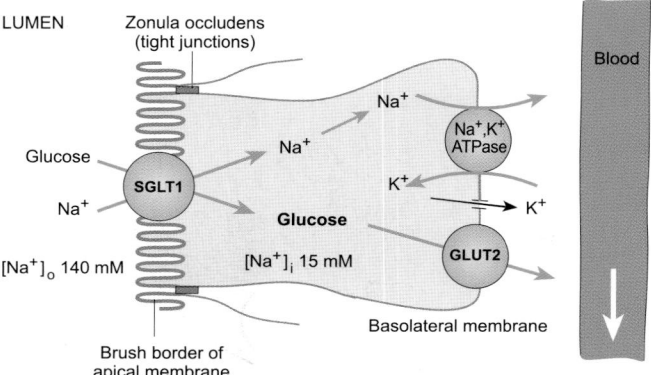

Fig. 4.7 The transport of glucose across the epithelium of the small intestine occurs in two stages. First, glucose entry into the enterocyte is coupled to the movement of sodium ions down their concentration gradient. This coupled transport allows the cell to accumulate glucose until its concentration inside the cell exceeds that bathing the basolateral surface of the cell. Glucose crosses the basolateral membrane down its concentration gradient via another carrier protein that is not dependent on extracellular sodium. The sodium pump removes the sodium ions accumulated during glucose uptake.

Summary

1. Molecules and ions that cross the plasma membrane by diffusion down their electrochemical gradients are said to undergo passive transport. Passive transport may be mediated by ion channels or by carrier molecules. When a carrier molecule transports an ion or molecule against the prevailing electrochemical gradient the cell must directly or indirectly expend metabolic energy. This is known as active transport.

2. The net movement of ions across the membrane via ion channels is always down an electrochemical gradient via a pore. Ionic movements through channels do not require the expenditure of metabolic energy and are examples of passive transport.

3. Carriers work by binding the transported molecule on one face of the membrane, undergoing a conformational change, and exposing the transported molecule to the other face of the membrane. The activity of some carrier molecules is directly linked to the hydrolysis of ATP to permit the active transport of a substance uphill against its electrochemical gradient. The sodium pump is the prime example of active transport.

4. Certain carriers use the ionic gradient established by the sodium pump to provide the energy to move another molecule (e.g. glucose) against its concentration gradient. This process is known as secondary active transport.

5. Cells use carriers in combination to regulate the intracellular concentration of certain ions. Thus the intracellular concentration of calcium ions is regulated by Na⁺–Ca²⁺ exchange and by a Ca²⁺ pump. Intracellular H⁺ is regulated by Na⁺–H⁺ exchange, Na⁺–HCO₃⁻ cotransport, and Cl⁻–HCO₃⁻ exchange.

4.4   The resting membrane potential of cells is determined by the K⁺ gradient across the plasma membrane

The activity of the sodium pump leads to an accumulation of potassium ions inside the cell. However, the plasma membrane, is not totally impermeable to potassium ions so that some are able to diffuse out of the cell down their concentration gradient via potassium channels. The membrane is much less permeable to sodium ions (the membrane at rest is between 10 and 100 times more permeable to potassium ions than it is to sodium ions) and so the lost potassium ions cannot readily be replaced by sodium ions. The leakage of potassium ions from the cell leads to the build up of negative charge on the inside of the membrane. This negative charge gives rise to a potential difference across the membrane known as the *membrane potential*.

The membrane potential is a physiological variable that is used by many cells to control various aspects of their activity. The resting membrane potential is the membrane potential of cells that are not engaged in a major physiological response that involves the plasma membrane, such as contraction or secretion. In many cases physiological responses are triggered by a fall in the membrane potential (i.e. the membrane potential becomes less negative). This fall is known as a *depolarization*. If the membrane potential becomes more negative, the change is called a *hyperpolarization*.

The negative value of the membrane potential tends to attract (positively charged) potassium ions into the cell. Thus, on one hand, potassium ions tend to diffuse out of the cell down their concentration gradient and, on the other, the negative charge on the inside of the membrane tends to attract potassium ions from the external medium into the cell. The potential at which these two opposing tendencies are exactly balanced is known as the *potassium equilibrium potential* and is very close to the *resting membrane potential* of many cells. Indeed, if the intracellular and extracellular concentrations of potassium ions are known, the approximate value of the resting membrane potential can be calculated using the Nernst equation (see Box 4.1). As the resting membrane potential is negative, the inward movement of positively charged ions such as sodium and calcium is favored and they are able to diffuse down their respective concentration gradients into the cell. In contrast, the negative value of the membrane potential opposes the inward movement of negatively charged ions such as chloride even though their concentration gradient favors net inward movement (Table 4.1).

The membrane potential of cells can be measured with fine glass electrodes (microelectrodes) that can puncture the cell membrane without destroying the cell. The magnitude of the resting membrane potential varies from one type of cell to another but is a few tens of millivolts (1 mV = 1/1000 V). It is greatest in nerve and muscle cells (excitable cells) where it is generally –70 to –90 mV (the minus sign indicates that the inside of the cell is negative with respect to the outside). In non-excitable cells the membrane potential may be significantly lower. For example, the membrane potential of the hepatocytes of the liver is about –35 mV.

The importance of the sodium pump in establishing the membrane potential is illustrated during embryonic development. Early in development, the sodium pump is not very active and embryonic cells have low membrane potentials. As they develop, sodium pump activity increases, the potassium gradient becomes established, and membrane potentials reach the levels seen in mature tissues.

The distribution of chloride across the plasma membrane is determined largely by the potassium ion gradient

Intracellular chloride is much lower than extracellular chloride (see Table 4.1), yet few cells are able to pump chloride ions across their plasma membrane against their electrochemical gradient. What factors regulate intracellular chloride? Intracellular fluid contains molecules such as proteins, amino acids, and ATP that are negatively charged. These molecules cannot cross the cell membrane so that the inside of the cell contains a fixed quantity of negative charge. Moreover, to keep the cell volume constant, the osmolality of the cell must be kept close to that of the extracellular environment. As the sum of the charges of all the ionized groups inside the cell must be zero (i.e. the cell interior must be electroneutral), the negative charges of the fixed anions largely balance the positive charge due to the cations, principally potassium. The difference between the total negative charge due to the intracellular fixed anions (ATP, proteins, etc.) and the positive charge due to potassium ions is made up of small diffusible anions (chiefly chloride and bicarbonate). If the potassium gradient is altered, chloride (the principal extracellular anion) is redistributed to maintain electroneutrality.

Ion-channel activity can be regulated by membrane voltage or by chemical signals

Some ion channels open when they bind a specific chemical agent known as an *agonist* or *ligand* (Fig. 4.8). This kind of channel is known as a *ligand-gated ion channel*. (A ligand is a molecule that binds to another; an agonist is a molecule that *both binds to and activates* a physiological system.) Other ion channels open when the membrane potential changes, usually when it becomes depolarized. These are known as *voltage-gated ion channels*. Both kinds of channel are widely distributed throughout the cells of the body, but a particular type of cell will possess a specific set of channels that are appropriate to its function.

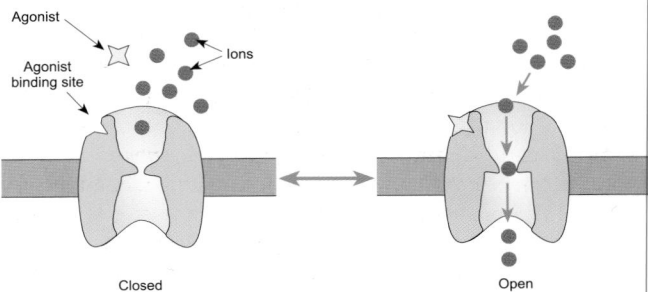

Fig. 4.8 Agonists bind to specific sites on the channels that they activate. Once they are bound they cause the channel to open and ions can pass through the pore. When the agonist dissociates from its binding site the channel is able to revert to the closed state. This type of channel is called a ligand-gated channel.

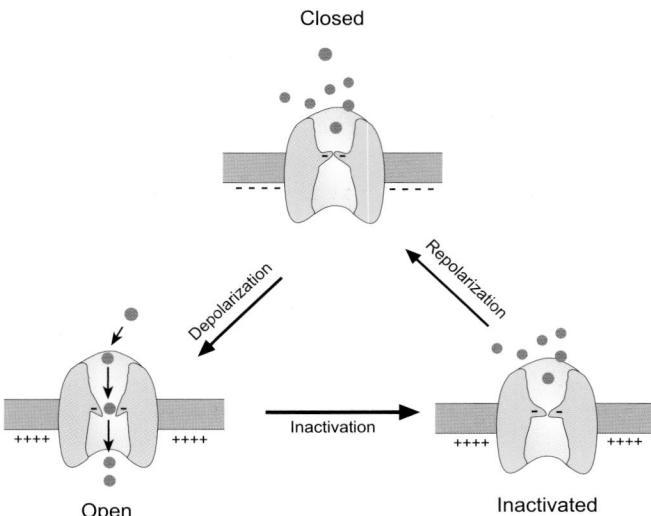

Fig. 4.9 Voltage-gated ion channels are opened by changes in membrane potential (usually by a depolarization) and ions are then able to pass through the pore down their concentration gradient. The open state of most voltage-gated channels is unstable and the channels close spontaneously even though the membrane remains depolarized. This is known as channel inactivation. When the membrane is repolarized the channel returns to its normal closed state from which it can again be opened by membrane depolarization.

Some ion channels are regulated by both a ligand and changes in membrane potential.

Ligand-gated ion channels are widely employed by cells to send signals from one to another. The plasma membrane of a cell has many different kinds of ion channel. As each type of channel exhibits different properties, one cell can respond in a variety of ways to different agonists. Often, however, a cell is specialized to perform a specific function (e.g. contraction or secretion) and responds to a specific agonist. To take one important example, the activity of skeletal muscles is controlled by acetylcholine released from the endings of motor nerves. The acetylcholine activates non-selective cation channels, which leads to depolarization of the muscle membrane and muscle contraction (see Chapter 6).

Voltage-gated ion channels generally open in response to depolarization of the membrane (i.e. they open when the membrane potential becomes less negative). When they are open they allow certain ions to cross the membrane. Quite commonly, the channels spontaneously close after a brief period of time even though the membrane remains depolarized—a property known as *inactivation*. Thus this type of channel can exist in three distinct states: it can be primed ready to open (the closed or resting state), it can be open and allow ions to cross the membrane, or it can be in an inactivated state from which it must first return to the closed state before it can reopen (Fig. 4.9). Examples are voltage-gated sodium channels, which are employed by nerve and muscle cells to generate action potentials, and voltage-gated calcium channels, which are utilized by cells to control a variety of cell functions including secretion.

As the passage of ions represents the movement of charge from one side of the membrane to the other, ions moving

---

Recent developments have enabled physiologists to study how ions move through channels. A small glass pipette filled with a solution containing electrolytes (e.g. NaCl) is pressed against the surface of a cell. By applying a little suction, the electrode becomes sealed against the cell membrane and a small patch of membrane is isolated from the remainder of the cell by the electrode. The movement of ions through channels in the isolated patch of membrane can be detected as weak electrical signals using a suitable amplifier. This is called patch–clamp recording.

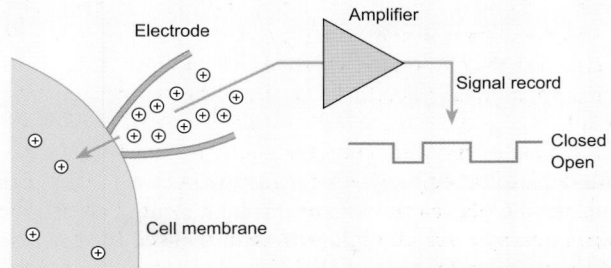

A schematic drawing of a typical patch–clamp recording circuit is shown. The cell (bottom left) has an electrode pressed against its surface to isolate a small patch of membrane. The currents passing through the electrode are measured with a special amplifier. When single ion channels open and close, small step-like currents can be detected.

The recordings made with this method unequivocally showed that channels open in discrete steps. Thus they either open fully (and can pass ions) or they are closed. If the patch of membrane under the electrode is ruptured, it is then possible to record all the current that passes across the cell membrane. This is called *whole-cell recording* and has proved to be of great value in understanding how many different types of cell perform their specific functions. Other variants of the method exist in which the patches of membrane are pulled from the cell.

---

through a channel generate a small electrical current, which, for a single ion channel, is of the order of a picoampere (10^{-12} A). These minute currents can be detected with modern patch–clamp recording techniques (see Box 4.2). It is now clear that ion channels are either in a state that permits specific ions to cross the membrane (the open or conducting state) or the path for ion movement is blocked (the closed or inactivated state) (see Fig. 4.10).

How do changes in ionic permeability alter the membrane potential?

As mentioned earlier, the membrane potential of a cell is an important physiological variable that is used to control many aspects of cell behavior. Since the equilibrium potentials of the various ions present on either side of the plasma membrane differ (see Table 4.1), the electrochemical gradients for ion move-

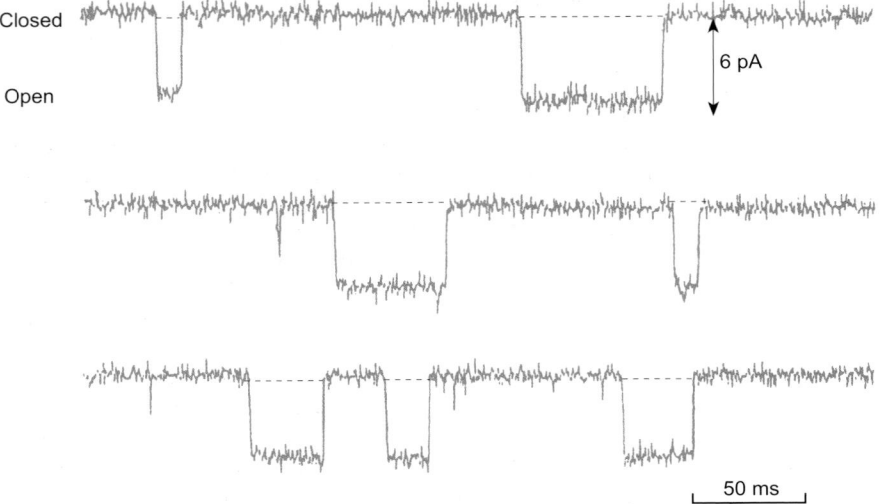

Closed

Open

6 pA

50 ms

Fig. 4.10 Examples of single-channel currents recorded by the patch–clamp technique. In this example, the ion channel opens (downward deflection) when it has bound acetylcholine. The record is of the activity of a nicotinic receptor channel in a chromaffin cell.

ments also differ. At the resting membrane potential, the potassium ion distribution is close to the equilibrium potential and the tendency of potassium ions to diffuse out of the cell down their concentration gradient is balanced by their inward movement due to the membrane potential. For sodium ions the situation is very different. At the resting membrane potential, the electrochemical gradient strongly favors sodium influx but few sodium channels are open, so the permeability of the membrane to sodium is low. When sodium channels open, the permeability of the membrane to sodium increases and the membrane potential adopts a value nearer to the sodium equilibrium potential

(i.e. the membrane becomes more depolarized). This shows how the membrane potential depends on both the ionic gradients across the cell membrane and the permeability of the membrane to the different kinds of ion present. The precise relationship between the membrane potential, the ionic gradients, and the permeability of the membrane to specific ions is given by the Goldman equation (see Box 4.3 for further information).

As an example of the way in which ionic permeabilities control cell activity, consider how a secretory cell responds to stimulation by an agonist. When the agonist activates the cell, the membrane depolarizes and this depolarization triggers secretion

---

BOX 4.3 THE GOLDMAN EQUATION EXPLAINS MEMBRANE POTENTIAL CHANGES DURING CELL ACTIVITY

If the membrane potential was determined solely by the distribution of potassium across the membrane it should be exactly equal to the potassium equilibrium potential. In practice it is found that while the resting membrane potential is close to the potassium equilibrium potential, it is seldom equal to it. Instead, it is usually somewhat less negative. Moreover, during periods of activity the membrane potential may be very different to the potassium equilibrium potential. For example, during the secretory response of an exocrine cell the membrane potential may be very low (i.e. it will be *depolarized*) and during the peak of a nerve action potential it is actually positive.

A modified form of the Nernst equation known as the Goldman constant-field equation (or, more simply, the Goldman equation) can explain these dynamic aspects of membrane potential behavior. The equation takes into account not only the ionic gradients that exist across the membrane but the permeability of the membrane to the different ions. The equation is

$$E = \frac{RT}{F} \ln \frac{P_{Na}[\text{Na}^+]_o + P_K[\text{K}^+]_o + P_{Cl}[\text{Cl}^-]_i}{P_{Na}[\text{Na}^+]_i + P_K[\text{K}^+]_i + P_{Cl}[\text{Cl}^-]_o}$$

where R, T, and F are physical constants (the gas constant, the absolute temperature, and the Faraday constant), and P_{Na} P_K, and P_{Cl} are the permeability coefficients of the membrane to Na$^+$, K$^+$, and Cl$^-$ respectively. [Na$^+$]$_o$, [Na$^+$]$_i$, etc., are the extracellular and intracellular concentrations of Na$^+$, K$^+$, and Cl$^-$. Note that [Cl$^-$]$_i$ and [Cl$^-$]$_o$ are reversed compared with [Na$^+$]$_i$, [Na$^+$]$_o$ and [K$^+$]$_i$, [K$^+$]$_o$ as Cl$^-$ has a negative charge.

As the resting membrane is much more permeable to K$^+$ than to Na$^+$ ($P_{Na}/P_K \approx 0.01$), the resting membrane potential lies close to the potassium equilibrium potential. (Remember that the Cl$^-$ gradient is determined by the K$^+$ gradient (see p. 40)). During an action potential, the membrane becomes much more permeable to Na$^+$ than to K$^+$ (at the peak of the action potential $P_{Na}/P_K \approx 20$) and the membrane potential is closer to the sodium equilibrium potential which is positive (+53 mV). Thus the Goldman equation shows how the membrane potential of a cell can be altered by changes in the relative permeability of its membrane to Na$^+$, K$^+$, Cl$^-$, and other ions without any change in the ionic gradients themselves.

Summary

1. As potassium ions are able to diffuse out of the cell via potassium channels, the potassium gradient generated by the activity of the sodium pump gives rise to a membrane potential in which the cell interior is negative with respect to the outside.

2. The exact value of the membrane potential is determined by the ion gradients across the plasma membrane and by the number and kind of ion channels that are open.

3. The membrane potential of a resting cell varies with the cell type but generally lies between –40 and –90 mV. The activity of many cells is governed by changes in their membrane potential.

4. Ion channels may be opened as a result of either binding a chemical (ligand-gated channels) or depolarization of the cell membrane (voltage-gated channels).

(see next section). At rest, the membrane potential is close to the potassium equilibrium potential (about –90 mV) because there are more open potassium channels than there are open sodium channels. Consequently, the permeability of the membrane to potassium is much greater than its permeability to sodium. When the cell is stimulated, more sodium channels open, and the permeability of the membrane to sodium increases relative to that of potassium. As the sodium equilibrium potential is positive (about +50 mV), the membrane potential becomes depolarized and this triggers the opening of voltage-gated calcium channels. The calcium concentration within the cell rises and this initiates the secretory response.

4.5 Secretion, exocytosis, and endocytosis

Many cells release molecules that they have synthesized into the extracellular environment. This is known as *secretion*. In some cases secretion occurs by simple diffusion through the plasma membrane—the secretion of steroid hormones by the cells of the adrenal cortex occurs in this way. (A hormone is a signaling molecule carried in the blood—see Chapter 5). However, this method of secretion is limited to those molecules that can penetrate the lipid barrier of the cell membrane. Polar molecules (e.g. digestive enzymes) are packaged in membrane-bound vesicles which can fuse with the plasma membrane to release their contents into the extracellular space in a process called *exocytosis*.

Exocytosis occurs by two pathways: constitutive and regulated

Exocytosis is a complex process so that it is not surprising that many details remain to be worked out. Key questions that need to be answered are:

1. What are the mechanisms by which the vesicles are formed?

2. How are they filled with the appropriate molecules?

3. How are they moved into place next to the plasma membrane?

4. What causes them to fuse with the plasma membrane?

5. What happens to the vesicles when they have emptied their contents?

The formation of vesicles

The proteins associated with the secretory vesicles are synthesized on the rough endoplasmic reticulum and the proteins required for export become internalized within the space enclosed by the endoplasmic reticulum itself. Others become embedded in the membrane of the endoplasmic reticulum. Part of the endoplasmic reticulum buds off to form specialized transport vesicles which carry the newly synthesized proteins to the Golgi apparatus where they may undergo further modification. The secretory vesicles bud off from the Golgi apparatus and migrate towards the plasma membrane where they form a secretory store that can be rapidly released in response to an appropriate stimulus. (See Chapter 3 for details of the Golgi apparatus and other aspects of cell structure.)

How are the vesicles filled with the appropriate molecules?

Small molecules that are secreted, such as histamine or acetylcholine, are synthesized by enzymes in the cytosol and are actively pumped into preformed vesicles by specific transport proteins. Large proteins and peptide hormones are synthesized as precursor molecules by the ribosomes and are then transported to the Golgi apparatus via specific transport vesicles as described above. The active form of many peptide hormones is stored in association with a specific binding protein—for example, the hormone vasopressin is stored in this way. Digestive enzymes are stored as inactive precursors.

Less is known about the details of vesicle mobilization and fusion. The vesicles must first move into close apposition to the plasma membrane in a process known as docking. This is followed by the fusion of the two membranes, and during this event any proteins that were embedded in the vesicle membrane become incorporated into the plasma membrane where their orientation is preserved. The contents of the vesicles are released into the extracellular space where they may influence neighboring cells.

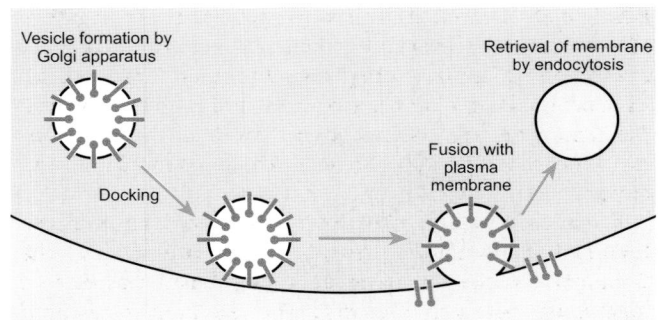

Fig. 4.11 The principal phases of constitutive secretion. The Golgi apparatus forms vesicles. The fully formed vesicles move towards the plasma membrane, become closely apposed to the plasma membrane ('docking'), and finally fuse with it. After fusion of the two membranes, the proteins embedded in the vesicular membrane are able to diffuse into the plasma membrane itself. Note that the orientation of the proteins is preserved so that the protein on the inside of the vesicle becomes exposed to the extracellular space.

Secretion occurs via two pathways: *constitutive secretion*, which is continuous, and *regulated secretion*, which, as the name implies, is secretion in response to a specific signal. Constitutive secretion is common to all cells and is the mechanism by which cells are able to insert newly synthesized lipids and proteins such as carriers and ion channels into their cell membranes (Fig. 4.11).

Regulated secretion provides cells with a mechanism for precisely timed release of molecules into the extracellular space

All exocrine and endocrine cells use regulated secretion to control the timing and rate of release of their vesicles into the extracellular space. The signal may be precisely timed or it may vary continuously. Nerve activity provides a means of achieving precise timing of secretion. The acinar cells of the salivary glands provide one example of the nervous control of secretion. These cells secrete the enzyme amylase in response to nerve impulses arising from the salivary nerves. In other cases secretion is stimulated by the release of a hormone into the blood. Many hormonal secretions are regulated by the concentration of a substance that is continuously circulating in the plasma. This may be another hormone or some other chemical constituent of the blood. For example, the rate at which the β cells of the pancreatic islets of Langerhans secrete the hormone insulin is regulated by the glucose concentration in the plasma.

To control the secretory event, the extracellular signal must be translated into an intracellular signal that can regulate the rate at which the preformed secretory vesicles will fuse with the plasma membrane. In many types of cell, regulated exocytosis is triggered by an increase in the concentration of ionized calcium within the cytosol. This occurs via two processes:

(1) entry of calcium through calcium channels in the plasma membrane;

(2) release of calcium from intracellular stores (mainly the endoplasmic reticulum).

The entry of calcium ions into the cytosol usually occurs through voltage-gated calcium channels which open following depolarization of the plasma membrane in response to a chemical signal. Calcium is then able to diffuse into the cell down its electrochemical gradient. It may also be mobilized from internal stores following activation of G protein-coupled receptor systems (see Chapter 5, pp. 54–56). In both cases the intracellular free calcium is increased and this triggers the fusion of docked secretory vesicles with the plasma membrane. As fusion proceeds, a pore is formed that connects the extracellular space with the interior of the vesicle and this pore provides a pathway for the contents of the vesicle to diffuse into the extracellular space (Fig. 4.12).

Many secretory cells are polarized so that one part of their membrane is specialized to receive a signal (e.g. a hormone) while another region is adapted to permit the fusion of secretory vesicles. The acinar cells of exocrine glands (e.g. those of the salivary glands) provide a clear example of this zonation. The basolateral surface receives chemical signals from circulating hormones or from nerve endings. These signals activate receptors that control the secretory response (see Chapter 5). Exocytosis occurs at the apical surface of the cells which is in direct communication with the secretory duct.

Endocytosis is used by cells to retrieve components of the plasma membrane and to take up macromolecules from the extracellular space

When cells undergo exocytosis their cell membrane increases in area as the vesicular membrane fuses with the plasma membrane. This increase in area is offset by membrane retrieval known as *endocytosis* in which small areas of the plasma membrane are pinched off to form endocytotic vesicles. The vesicles formed are generally small (less than 150 nm in diameter) but may contain macromolecules derived from the extracellular space. Almost all the cells of the body continuously undergo membrane retrieval via endocytosis. Since the formation of the vesicles traps some of the extracellular fluid, this process is also known as pinocytosis ('cell drinking'). Pinocytosis and endocytosis are often employed as interchangeable terms. Some proteins and other macromolecules are absorbed with the extracellular fluid (*fluid-phase endocytosis*). In other cases proteins bind to specific surface receptors. For example, the cholesterol required for the formation of new membranes is absorbed by cells via the binding of specific carrier protein–cholesterol complexes (low-density lipoproteins) to a surface receptor. This is called *receptor-mediated endocytosis*.

The endocytotic vesicles fuse with larger vesicles known as *endosomes* which are located either just beneath the plasma membrane (the peripheral endosomes) or near the cell nucleus (the perinuclear endosomes). The perinuclear endosomes fuse with vesicles containing lysosomal enzymes derived from the Golgi apparatus to form endolysosomes. Although the exact processes involved are not yet clear, the endolysosomes become transformed into lysosomes. The interior of the lysosomes is acidified by the activity of an ATP-driven proton pump. This acidic environment allows the lysosomal enzymes to break down the macromolecules trapped during endocytosis into their constituent amino acids and sugars. These small molecules are then transported out of the lysosomes to the cytosol where they may be reincorporated into newly synthesized proteins.

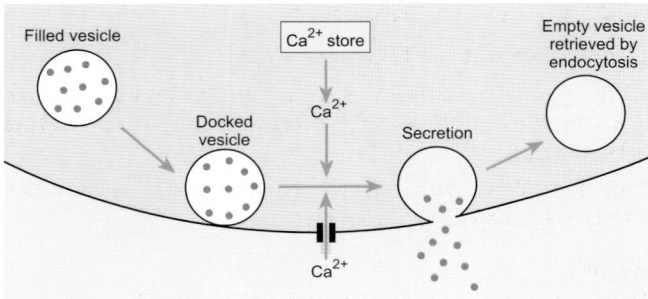

Fig. 4.12 A rise in intracellular Ca²⁺ can trigger regulated secretion. In exocrine cells and many endocrine cells secretion is dependent on extracellular Ca²⁺. Following stimulation the cell depolarizes and this opens Ca²⁺ channels. The Ca²⁺ ions enter and cause docked vesicles to fuse with the plasma membrane and release their contents. Calcium ions can also be released into the cytoplasm from intracellular stores to initiate secretion—this calcium is derived mainly from the endoplasmic reticulum.

Not all endocytosed molecules are broken down by the lysosomes. Some elements of the plasma membrane that are internalized during endocytosis are eventually returned to the plasma membrane by transport vesicles. This is often the case with membrane lipids, carrier proteins, and receptors. Indeed, in many cases cells regulate the number of active carriers or receptors by adjustment of the rate at which these proteins are reinserted into the plasma membrane. For example, when the body needs to conserve water, the cells of the collecting duct of the kidney increase their permeability to water, by increasing the rate at which water channels (aquaporins) are inserted into the apical membrane by transport vesicles (see Chapter 17).

Phagocytosis is important both for defense against infection and for the retrieval of material from dead and dying cells

Phagocytosis ('cell eating') is a specialized form of endocytosis in which large particles (e.g. bacteria or cell debris) are ingested by cells. Within the mammalian body this activity is confined to specialized cells (*phagocytes*). The principal cells that perform this function in humans and other mammals are the *neutrophils* of the blood and the *macrophages*, which are widely distributed throughout the body (see Chapters 13 and 14). Unlike endocytosis, phagocytosis is triggered only when receptors on the surface of the cell bind to the particle to be engulfed. Both macrophages and neutrophils have a rich repertoire of receptors, but any particle coated with an antibody is avidly ingested. Phagocytosis occurs by the extension of processes known as pseudopodia around the particle. The extension of the pseudopodia is guided by contact between the particle and the cell surface. When the pseudopodia have engulfed the particle a membrane-bound *phagosome* is formed around it, the size of which is determined by that of the ingested particle. The phagosomes fuse with lysosomes to form phagolysosomes, which are able to digest the ingested material (Fig. 4.13) and transport the metabolically useful components to the cytosol. The material that cannot be digested remains in the phagocyte as a membrane-bound particle called a residual body.

Not all the surface protein of the foreign particle is digested. Some is combined with cell-surface glycoproteins and inserted into the plasma membrane of the phagocyte. This exposes the foreign protein to scrutiny by the T lymphocytes, which are

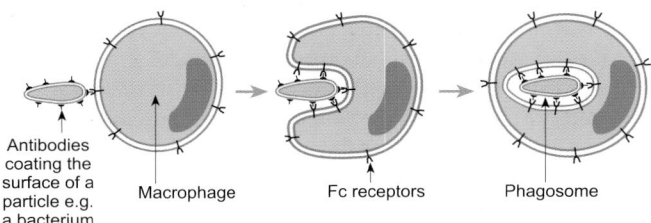

Antibodies coating the surface of a particle e.g. a bacterium   Macrophage   Fc receptors   Phagosome

Fig. 4.13 The stages of phagocytosis. A bacterium or other particle must first adhere to the surface of a phagocyte before it can be engulfed. This is achieved by linking surface receptors on the phagocyte with complementary binding sites on the microbe (e.g. the Fc region of antibodies manufactured by the host). The microbe is then engulfed by a zipper-like action until it becomes fully enclosed in a vacuole.

Summary

1. Cells release materials that they have synthesized into the extracellular space by means of secretion.

2. Secretion of lipid-soluble materials such as steroid hormones is by diffusion across the cell membrane.

3. Small water-soluble molecules and macromolecules are secreted by exocytosis—a process by which vesicles containing the secreted material fuse with the cell membrane to release their contents. Two pathways exist: constitutive secretion, which operates continuously, and regulated secretion, which occurs in response to specific signals. The principal signal for regulated secretion is a rise in intracellular Ca^{2+}.

4. The vesicular membrane that became incorporated into the plasma membrane during fusion is later retrieved by endocytosis.

5. Specialized cells engulf foreign particles and cell debris by a process called phagocytosis, which is triggered when receptors on the cell surface recognize specific proteins on the surface of a foreign particle. The ingested material is digested in vacuoles called phagosomes.

then able to signal the immune system to increase the production of the appropriate antibody (see Chapter 14).

Although phagocytes play an important role in protecting the body from infection by ingesting bacteria, this activity is dwarfed by their role in the scavenging of dying cells and cellular debris. For example about 1 per cent of red cells die and are replaced each day. This requires the phagocytes of the reticuloendothelial system to ingest about 10^{11} red cells each day. The signals that trigger this phagocytic response are not known.

Recommended reading

Biochemistry and cell biology

Alberts, B., Johnson, A., Lewis, J., Raff, M., Roberts, K., and Walter, P. (2002). *Molecular biology of the cell* (4th edn), Chapters 11 and 13. Garland, New York.

Berg, J.M., Tymoczko, J.L., and Stryer, L. (2002). *Biochemistry* (5th edn), Chapter 12. Freeman, New York.

Biophysics

Aidley, D.J., and Stanfield, P.R. (1996). *Ion channels: molecules in action*. Cambridge University Press, Cambridge.

Hille, B. (2001). *Ionic channels of excitable membrane* (3rd edn). Sinauer, Sunderland, MA.

Nicholls, J.G., Martin, A.R., Wallace, B.G., and Fuchs, P.A. (2001). *From neuron to brain* (4th edn), Chapters 2–5. Sinauer, Sunderland, MA.

Multiple choice questions

The following questions are designed to help you test your knowledge of the material discussed in this chapter. Each statement is either true or false. Answers are given below.

1. Gases such as oxygen and carbon dioxide cross the plasma membrane by:
   a. Active transport.
   b. Passive diffusion through the lipid bilayer.
   c. A specific carrier protein.

2. Ions can cross the plasma membrane by:
   a. Diffusion through the lipid bilayer.
   b. Diffusion through channel proteins.
   c. Binding to specific carrier proteins.

3. A substance can be accumulated against its electrochemical gradient by:
   a. Active transport.
   b. Facilitated diffusion.
   c. Ion channels.
   d. A symport.

4. The principal intracellular cation is:
   a. Na^+
   b. Cl^-
   c. Ca^{2+}
   d. K^+

5. The following are examples of active transport:
   a. The sodium pump.
   b. Cl^-–HCO_3^- exchange.
   c. Na^+–Ca^{2+} exchange.
   d. Na^+-linked glucose uptake by the enterocytes.
   e. Na^+–H^+ exchange.

6. The sodium pump:
   a. Exchanges intracellular Na^+ for extracellular K^+.
   b. Requires ATP.
   c. Directly links Na^+ efflux with K^+ influx.
   d. Is an ion channel.
   e. Can be inhibited by metabolic poisons.
   f. Is important for maintaining a constant cell volume.

7. The resting membrane potential of a muscle fiber is close to:
   a. 0 mV
   b. –90 mV
   c. +50 mV
   d. The K^+ equilibrium potential

8. The resting membrane potential is mainly determined by:
   a. The K^+ gradient.
   b. The Na^+ gradient.
   c. The Ca^{2+} gradient.

9. Secretion:
   a. Always involves membrane vesicles.
   b. May be triggered by a rise in intracellular Ca^{2+}.
   c. Provides a means of inserting proteins into the plasma membrane.

10. Endocytosis is used by cells to:
    a. Ingest bacteria and cell debris.
    b. Retrieve elements of the plasma membrane after exocytosis.
    c. Take up large molecules from the extracellular space.

Quantitative problems

Answers are given below.

1. Use the Nernst equation to calculate the equilibrium potential for the following distribution of ions measured in mammalian skeletal muscle (values are in mmoles per liter). $RT/F = 26.7$ mV at 37°C.

Ionic species	Extracellular fluid	Intracellular fluid
Na^+	145	20
K^+	4	150
Ca^{2+}	1	1×10^{-4}
Cl^-	114	3
HCO_3^-	31	10

2. Using the data from question 1 calculate the electrochemical gradients for Na^+, K^+, Ca^{2+}, Cl^-, and HCO_3^- when the membrane potential is –90 mV, –30 mV, +50 mV, and +100 mV.

3. If the membrane potential is –90 mV at rest and +30 mV at the peak of the action potential what changes in the relative permeability of Na^+ and K^+ take place? (use the data from question 1). Hint: Use the Goldman equation to solve this but ignore the term for Cl^- as it is passively distributed according to the prevailing K^+ gradient (you could do the experiment in a chloride-free medium).

Answers to the multiple choice questions

1. Gases are very lipid soluble and readily pass through the lipid bilayer. They diffuse down their concentration gradient by *passive transport*.

a. False

b. True

c. False

2. Unlike gases, ions and other polar molecules cannot diffuse through the lipid bilayer but cross the membrane via channel proteins or are transported from one side of the membrane to the other by carrier proteins.

a. False

b. True

c. True

3. For any substance to be accumulated against its electrochemical gradient energy must be expended. In active transport this is provided either by the hydrolysis of ATP (e.g. the sodium pump) or by coupling the movement of one substance against its electrochemical gradient to the movement of another down its electrochemical gradient. This is secondary active transport. The Na^+-dependent uptake of glucose by the enterocytes is one example.

a. True

b. False

c. False

d. True

4. Na^+ is the *principal extracellular cation* and K^+ is the *principal intracellular cation*. Cl^- is the *principal extracellular anion*.

a. False

b. False

c. False

d. True

5. The sodium pump is an example of ATP driven active transport. Na^+–Ca^{2+} exchange and Na^+–H^+ exchange are examples of secondary active transport. In these cases the energy is provided by the Na^+ gradient generated by the sodium pump.

a. True

b. False

c. True

d. True

e. True

6. The sodium pump is a carrier protein not an ion channel.

a. True

b. True

c. True

d. False

e. True

f. True

7. a. False

b. True

c. False

d. True

8. The resting membrane potential is determined by the K^+ gradient because there are many more open K^+ channels than Na^+ channels. Consequently the resting membrane potential is close to the K^+ equilibrium potential.

a. True

b. False

c. False

9. Lipid-soluble molecules like the steroids and prostaglandins are secreted as they are synthesized and pass across the lipid bilayer. Vesicle-mediated secretion is called exocytosis.

a. False

b. True

c. True

10. Cells take up large proteins from the extracellular fluid by *receptor mediated endocytosis*. Bacteria and cell debris are taken up by *phagocytosis*.

a. False

b. True

c. True

Answers to the quantitative problems

1. By entering the values from the table into the Nernst Equation and taking $RT/F=26.7$ mV the following calculated values are obtained.

Ionic species	Nernst equilibrium potential (mV)
Na^+	+53
K^+	−97
Ca^{2+}	+184
Cl^-	−97
HCO_3^-	−30

2. The electrochemical gradient for a particular ion is given by the difference between the membrane potential (MP) and the equilibrium potential. For the data given in question 1 the electrochemical gradients are shown in the table.

Ionic species	Electrochemical gradient (mV)			
	MP = −90 mV	MP = −30 mV	MP = +50 mV	MP = +100 mV
Na^+	−143	−83	−3	47
K^+	−7	67	147	197
Ca^{2+}	−274	−214	−134	−84
Cl^-	−7	67	147	197
HCO_3^-	−60	0	80	130

By convention, inward currents are represented by a negative sign, so for all those values where the electrochemical gradient is negative, the inward movement of cations (influx) is favored. For positive values outward movement of cations (efflux) is favored. A negative electrochemical gradient would give rise to an efflux of anions and a positive value would give rise to an influx of anions.

3. At the resting membrane potential the Na^+/K^+ permeability ratio is 0.008 and during the action potential it is 5.46, an increase in the permeability of the membrane to Na^+ by a factor of about 700.

Working:

Since $E = -90$ mV when $[K^+]_o = 4$ mM and $[K^+]_i = 150$ mM, we can write

$$-90 = 26.7 \, \ln \frac{a[145]+[4]}{a[20]+[150]}$$

so that

$$-3.37 = \ln \frac{a[145]+[4]}{a[20]+[150]}$$

and

$$\frac{a[145]+[4]}{a[20]+[150]} = 0.0343.$$

Hence

$$a = 0.00789 \approx 0.008.$$

A similar calculation is made for the peak of the action potential when $E = +30$ mV. It is worth stressing that this approach provides an understanding of the way the membrane potential can change without changes in the ionic gradients themselves. Selective alterations in membrane permeability are the key to understanding why synaptic potentials may either be depolarizing or hyperpolarizing (see Chapter 6). They also explain the changes in membrane potential seen in non-excitable cells in response to agonists.

CHAPTER 5

Principles of cell signaling

Chapter contents

After reading this chapter you should understand:

♦ The need for cell signaling

♦ The differing roles of paracrine, endocrine, and synaptic signaling

♦ How receptors in the plasma membrane regulate the activity of target cells

♦ The roles of cyclic AMP and inositol trisphosphate as second messengers

♦ How steroid and thyroid hormones control gene expression via intracellular receptors

♦ The role of cell-surface proteins in cell–cell adhesion and cell recognition

♦ The functions of gap junctions between cells

5.1 Introduction

Individual cells are specialized to carry out a specific physiological role such as secretion or contraction. In order to coordinate their activities they need to receive and transmit signals of various kinds. The ways in which cells make use of such signals form the substance of this chapter. In essence, cells communicate with each other in three different ways:

(1) by diffusible chemical signals;

(2) by direct contact between the plasma membranes of adjacent cells;

(3) by direct cytoplasmic contact via gap junctions.

Diffusible chemical signals allow cells to communicate at a distance, while direct contact between cells is particularly important in cell–cell recognition during development and during the passage of lymphocytes through the tissues where they scan cells for the presence of foreign antigens (see Chapter 14). Direct cytoplasmic contact between neighboring cells via gap junctions permits the electrical coupling of cells and plays an important role in the spread of excitation between adjacent cardiac muscle cells. It also allows the direct exchange of chemical signals between adjacent cells.

5.2 Cells use diffusible chemical signals for paracrine, endocrine, and synaptic signaling

Cells release a variety of chemical signals. Some are local mediators that act on neighboring cells, reaching their targets by diffusion over relatively short distances (up to a few millimeters). This is known as *paracrine signaling*. When the secreted chemical also acts on the cells that secreted it, the signal is said to be an *autocrine signal*. Frequently, substances are secreted into the blood by specialized glands—the *endocrine glands*—to act on various tissues around the body. The secreted chemicals themselves are called *hormones*. Finally, nerve cells release chemicals at their endings to affect the cells they contact. This is known as *synaptic signaling*.

Local chemical signals—paracrine and autocrine secretions

Paracrine secretions are derived from individual cells rather than from a collection of similar cells or a specific endocrine gland. They act on cells close to the point of secretion and have a local effect (Fig. 5.1(a)). The signaling molecules are rapidly destroyed by extracellular enzymes or by uptake into the target cells. Consequently, very little of the secreted material enters the blood.

One example of paracrine signaling is provided by mast cells. These cells are found in connective tissues all over the body and have large secretory granules that contain histamine, which is secreted in response to injury or infection. The secreted histamine dilates the local arterioles and this results in an increase in the local blood flow. In addition, the histamine increases the permeability of the nearby capillaries to proteins such as immunoglobulins. However, this increase in capillary permeability must be local and not widespread, otherwise there would be a significant loss of protein from the plasma to the interstitial fluid, which could result in circulatory failure (see Chapter 28, Section 28.5). The mast cells also secrete small peptides that are released with the histamine to stimulate the invasion of the affected tissue by white blood cells—phagocytes and eosinophils. These actions form part of the inflammatory response and play an important role in halting the spread of infection (see Chapter 14).

The inflammatory response is also associated with an increase in the synthesis and secretion of a group of local chemical mediators called the prostaglandins (see pp. 56–57). The prostaglandins secreted by a cell act on neighboring cells to stimulate them to produce more prostaglandins. This is another example of paracrine signaling. In addition, the secreted prostaglandin stimulates further prostaglandin production by the cell that initiated the response. Here the prostaglandin is acting as an *autocrine signal*. This autocrine action amplifies the initial signal and may help it to spread throughout a population of cells, thus ensuring a rapid mobilization of the body's defenses in response to injury or infection.

Hormonal secretions provide a means of diffuse, long-distance signaling to regulate the activity of distant tissues

Hormones play an extensive and vital role in regulating many physiological processes and their physiology will be discussed at

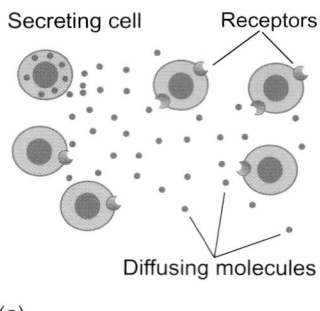

Secreting cell   Receptors

Diffusing molecules

(a)

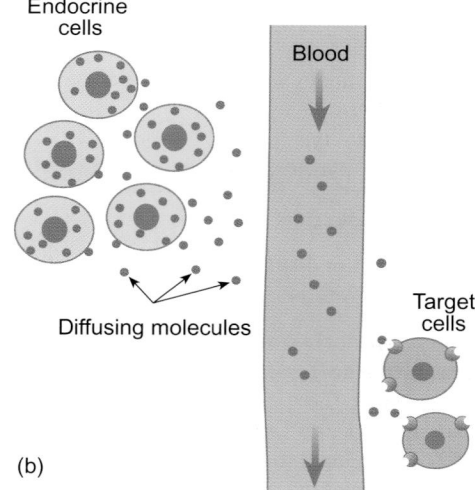

Endocrine cells

Blood

Diffusing molecules

Target cells

(b)

Fig. 5.1 A comparison between paracrine and endocrine signaling. (a) Schematic diagram illustrating paracrine secretion. The secreting cell releases a signaling molecule (e.g. a prostaglandin) into the extracellular fluid from where it reaches the receptors of neighboring cells by diffusion. (b) The chief features of endocrine signaling. Endocrine cells secrete a hormone into the extracellular space in sufficient quantities for it to enter the bloodstream from where it can be distributed to other tissues. The target tissues may be at a considerable distance from the secreting cells.

length in Chapter 12 and subsequent chapters. Here their role as cell signals will be discussed only in general terms.

Endocrine cells synthesize hormones and secrete them into the extracellular space from where they are able to diffuse into the blood. Once in the bloodstream a hormone will be distributed throughout the body and is then able to influence the activity of tissues remote from the gland that secreted it. In this important respect hormones differ from local chemical mediators. The secretion and distribution of hormones is shown schematically in Fig. 5.1(b).

The endocrine glands secrete hormones in response to a variety of signals.

1. They may respond to the level of some constituent of the blood. For example, insulin secretion from the β cells of the islets of Langerhans is regulated by the blood glucose concentration.

2. The circulating levels of other hormones may closely regulate their activity. This is the case for the secretion of the sex hormones from the ovaries (estrogens) and testes (testosterone), which are secreted in response to hormonal signals from the anterior pituitary gland.

3. They may be directly regulated by the activity of nerves. This is how oxytocin secretion from the posterior pituitary gland is controlled during lactation.

Since hormones are distributed throughout the body via the circulation, they are capable of affecting widely dispersed populations of cells. The releasing hormones of the hypothalamus (a small region in the base of the brain) are an important exception to this rule. These hormones are secreted into the portal blood vessels in minute quantities and travel a few millimeters to the anterior pituitary where they control the secretion of the anterior pituitary hormones (see Chapter 12). While cells are exposed to almost all the hormones secreted into the bloodstream, a particular cell will only respond to a hormone if it possesses receptors of the appropriate type (see Section 5.3). Thus the ability of a cell to respond to a particular hormone depends on whether it has the right kind of receptor. This diffuse signaling is beautifully adapted to regulate a wide variety of cellular activities in different tissues.

As the endocrine glands secrete their products into the blood, the concentrations of the various hormones in the blood are generally very low indeed—typically about 10^{-9} mol l^{-1}. This means that the receptors must bind hormones very effectively. To put it another way, the individual receptors must have a high affinity for their particular hormone.

Compared with the signals mediated by nerve cells (synaptic signaling), the effects of hormones are usually relatively slow in onset (ranging from seconds to hours). Nevertheless, the effects of hormones can be very long lasting, as in the control of growth (Chapter 23). Perhaps the most striking example of a permanent change triggered by a hormone is the effect of testosterone on the development of the male reproductive system. In its absence a genetically male fetus will fail to develop male genitalia (see Chapter 22, Section 22.8).

The difference between endocrine and paracrine signaling fundamentally depends on the quantity of the chemical signal secreted. If sufficient quantities of a signaling molecule are secreted for it to enter the blood, it is being employed as a hormone. However, if the amounts secreted are only sufficient to affect neighboring cells, the chemical signal is acting as a paracrine signal. Consequently, a substance can be a hormone in one situation and a paracrine signal in another. For example, the peptide somatostatin is found in the hypothalamus. It is secreted into the portal blood vessels, which carry it to the anterior pituitary gland where it acts to inhibit the release of growth hormone (see Chapter 12). As somatostatin has entered the blood to be carried to its target tissue, it is acting as a hormone. Somatostatin is also found in the D cells of the gastric mucosa. It is secreted when the [H+] in the stomach rises and it inhibits the secretion of gastrin by the adjacent G cells. (Gastrin is the hormone that stimulates acid secretion by the parietal cells of the gastric mucosa—see Chapter 18.) In this situation somatostatin acts on neighboring cells as a paracrine signal rather than as a hormone.

Fast signaling over long distances is accomplished by nerve cells

Chemical signals that are released into the extracellular space and need to diffuse some distance to reach their target cells have two disadvantages:

(1) their effect cannot be restricted to an individual cell;

(2) their speed of signaling is relatively slow, particularly if there is any great distance between the secreting cell and its target.

For many purposes these factors are not important, but there are circumstances where the signaling needs to be both rapid (occurring within a few milliseconds) and discrete. For example, during locomotion different sets of muscles are called into play at different times to provide coordinated movement of the limbs. This type of rapid signaling is performed by nerve cells. The contact between the nerve ending and the target cell is called a *synapse* and the overall process is known as synaptic signaling.

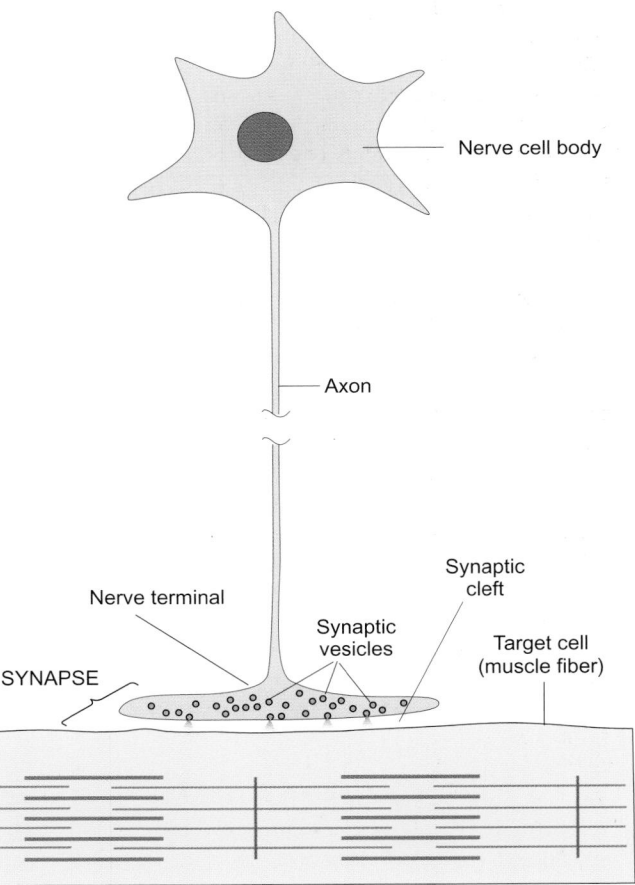

Fig. 5.2 Synaptic signaling is performed by nerve cells. Electrical signals (action potentials) originating in the cell body pass along the axon and trigger the secretion of a signaling molecule by the nerve terminal. As the nerve terminal is closely apposed to the target cell (in this case a skeletal muscle fiber), the signal is highly localized. The junction between the nerve cell and its target is called a synapse.

To perform their role, nerve cells need to make direct contact with their target cells. They do this via long, thin, hair-like extensions of the cell called *axons*. Each nerve cell gives rise to a single axon, which may branch to contact a number of different targets. As axons may extend over considerable distances (in some cases up to 1 m), nerve cells need to be able to transmit their signals at relatively high speeds. This is achieved by means of an electrical signal (an *action potential*) that passes along the length of the axon from the cell body to its terminal (the *nerve terminal*). When an action potential reaches a nerve terminal it triggers the release of a small quantity of a chemical (a *neurotransmitter*), which then acts on the target cell. The nerve terminal is usually very closely apposed to its target and so the neurotransmitter released by the nerve terminal has to diffuse a very short distance (about 20 nm) to reach its point of action. Since the receptors for the neurotransmitter are located directly under the nerve terminal, only very small quantities of neurotransmitter are required to activate the target cell and neighboring cells will not be affected. Therefore the combination of electrical signaling and very short diffusion time permits both rapid and discrete activation of the target. The essential features of synaptic signaling are summarized in Fig. 5.2. The organization and properties of nerve cells and synapses will be discussed more fully in the next chapter (Chapter 6).

Signaling molecules are very diverse in structure

The chemical signals employed by cells are very diverse. Cells use both water-soluble and hydrophobic molecules for signaling. Some, such as nitric oxide and glycine, have a small molecular mass (< 100) while others, such as growth hormone (191 amino acids linked together and a molecular mass of 21 500), are very large molecules. Many of the water-soluble signaling molecules are derived from amino acids—examples are peptides, proteins, and the biological amines. Hydrophobic signaling molecules include the prostaglandins (see below), the steroid hormones such as testosterone, and thyroid hormone.

Many chemical signaling molecules are stored in membrane-bound vesicles prior to secretion by exocytosis (see Chapter 4). This is the case with many hormones (e.g. epinephrine (adrenaline) and vasopressin) and neurotransmitters (e.g. acetylcholine). Other signaling molecules are so lipid-soluble that they cannot be stored in vesicles on their own but must be stored bound to a specific storage protein. This is the case with thyroid hormone. Finally, some chemical mediators are secreted as they are formed. This happens with the steroid hormones and the prostaglandins. As lipid-soluble hormones (thyroid hormone and the steroids) are not very soluble in water, they are carried to their target tissues bound to plasma proteins. The principal families of signaling molecule and their chief modes of action are listed in Table 5.1.

TABLE 5.1	Examples of signaling molecules used in cell–cell communication	
Class of molecule	**Specific example**	**Physiological role and mode of action**
(a) Signaling molecules secreted by fusion of membrane bound vesicles		
Ester	Acetylcholine	Synaptic signaling molecule, opens ligand-gated ion channels and also activates G protein-linked receptors
Amino acid	Glycine	Synaptic signaling molecule, opens a specific type of ligand-gated ion channel
	Glutamate	Synaptic signaling molecule, affects specific ligand-gated ion channels and also activates G protein-linked receptors
Amine (bioamine)	Epinephrine (adrenaline)	Hormone, wide variety of effects, acts via G protein-linked receptors
	5-Hydroxytryptamine	Local mediator and synaptic signaling molecule, acts via G protein-linked receptors (5-HT, serotonin)
	Histamine	Local mediator, acts via G protein-linked receptors
Peptide	Somatostatin	Hormone and local mediator, inhibits secretion of growth hormone by anterior pituitary via G protein-linked receptor
	Vasopressin (antidiuretic hormone (ADH))	Hormone, increases water reabsorption by collecting tubule of the kidney via G protein-linked receptor
Protein	Insulin	Hormone, activates a catalytic receptor in plasma membrane; increases uptake of glucose by liver, fat, and muscle cells
	Growth hormone	Hormone, activates a non-receptor tyrosine kinase in target cell
(b) Signaling molecules that can diffuse through the plasma membrane		
Steroid	Estradiol-17β	Hormone, binds to cytosolic receptor; hormone–receptor complex regulates gene expression
Thyroid hormone	Tri-iodothyronine (T_3)	Hormone, binds to nuclear receptor; hormone–receptor complex regulates gene expression
Eicosanoid	Prostaglandin E_2 (PGE_2)	Local mediator, diverse actions on many tissues; activates G protein-linked receptors in plasma membrane
Inorganic gas	Nitric oxide	Local mediator; acts by binding to guanylyl cyclase in target cell

Summary

1. In order to coordinate their activities, cells need to send and receive signals of various kinds. They do this by the secretion of specific chemical signals (diffusible signals), by direct cell–cell contact, and by gap junctions.

2. Cells use diffusible chemical signals in three ways: as local signals (paracrine signaling), as diffuse signals that reach their target tissues via the bloodstream (endocrine signaling), and as rapid discrete signals (synaptic signaling).

3. A wide variety of molecules are employed as chemical signals, ranging in size from small highly diffusible molecules such as nitric oxide to large proteins. Signaling molecules may be secreted by exocytosis (e.g. acetylcholine) or they may be secreted by diffusion across the plasma membrane (e.g. testosterone).

How do receptors control the activity of the target cells?

Cells respond to chemical signals by initiating an appropriate physiological response. This process is called *transduction*. There are four basic ways in which activation of a receptor can alter the activity of a cell (Fig. 5.3):

(1) it may open an ion channel and so modulate the membrane potential;

(2) it may directly activate a membrane-bound enzyme;

(3) it may activate a G protein-linked receptor which may modulate an ion channel or change the intracellular concentration of a specific chemical called a *second messenger*—the original signaling molecule is the first messenger;

(4) it may act on an intracellular receptor to modulate the transcription of specific genes.

5.3 Chemical signals are detected by specific receptor molecules

Cells are exposed to a wide variety of chemical signals and need to have a means of detecting those signals that are intended for them. They do so by means of molecules known as *receptors*. Receptors are specific for particular chemical signals so that, for example, an acetylcholine receptor binds acetylcholine but does not bind epinephrine or nitric oxide. When a receptor has detected a chemical signal, it initiates an appropriate cellular response. The link between the detection of the signal and the response is called *transduction*. Substances that bind to and activate a particular receptor are called *agonists*, while those drugs that block the effect of an agonist are called *antagonists*.

All receptors are proteins and many are located in the plasma membrane where they are able to bind the water-soluble signaling molecules that are present in the extracellular fluid. Hydrophobic signaling molecules such as the steroid hormones can cross the plasma membrane, and these bind to cytoplasmic and nuclear receptors. Finally, some intracellular organelles possess receptors for molecules that are generated within the cell (second messengers—see below).

In recent years it has become clear that an individual cell may possess many different types of receptor so that it is able to respond to a variety of extracellular signals. The response of a cell to a specific signal depends on which receptors are activated. Consequently, a particular chemical mediator can produce different responses in different cell types. For example, acetylcholine released from the nerve terminals of motor nerve fibers onto skeletal muscle causes the muscle to contract. When it is released from the endings of the vagus nerve it slows the rate at which the heart beats (the heart rate). The same mediator has different effects in these two tissues because it acts on different receptors. The acetylcholine receptors of skeletal muscle are known as *nicotinic receptors* because the alkaloid nicotine can also activate them. Those of the heart have a different structure and are called *muscarinic receptors* as they can be activated by another chemical, *muscarine*.

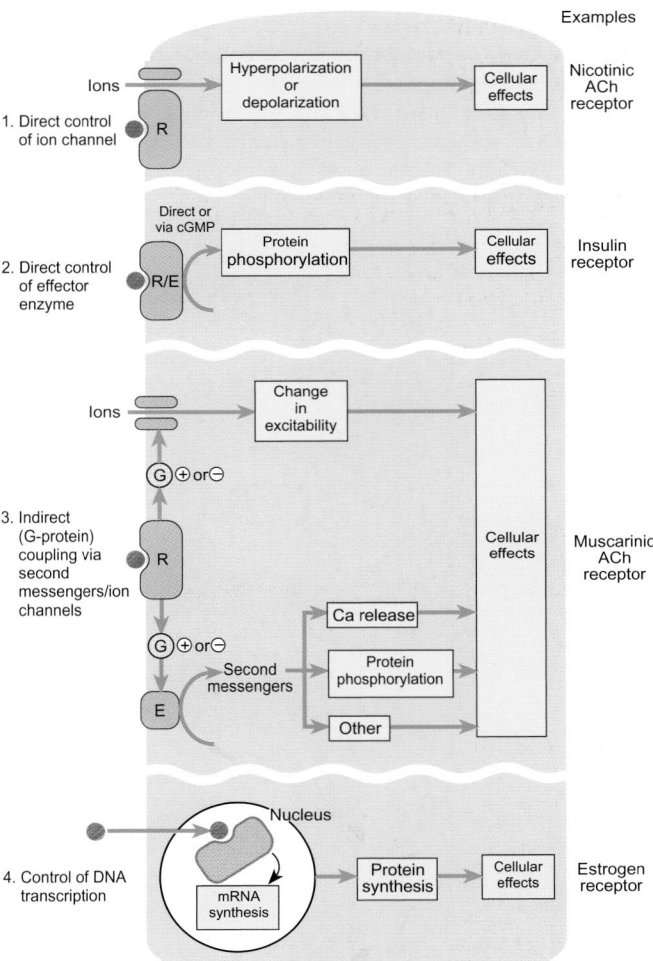

Fig. 5.3 Schematic drawing to show the principal ways in which chemical signals affect their target cells. Examples of each type of coupling are shown. R, receptor; E, enzyme; G, G protein; ⊕, increased activity; ⊖, decreased activity.

Ion-channel modulation

Many receptors are directly coupled to ion channels. These receptor–channel complexes are called *ligand-gated ion channels* (see Chapter 4, p. 41) and they are employed by cells to regulate a variety of functions. In general, ligand-gated channels open for a short period of time following the binding of their specific agonist and this transiently alters the membrane potential of the target cell and thereby modulates its physiological activity. In some instances, control of the membrane potential is the final response. This is the situation when one nerve cell inhibits the activity of another (see Chapter 6, Section 6.4). More often the change in membrane potential triggers some further event. Thus, in many cases, activation of ligand-gated channels causes the target cell to depolarize. This depolarization then activates voltage-gated ion channels that trigger the appropriate cellular response. In this way the activation of a ligand-gated ion channel can be used to control the activity of even the largest of cells.

This pattern of events is illustrated by the stimulatory effect of acetylcholine on epinephrine secretion by the chromaffin cells of the adrenal gland. Acetylcholine released by the terminals of the splanchnic nerve binds to nicotinic receptors on the plasma membrane and this increases the permeability of the membrane to Na^+ so that the membrane depolarizes. The depolarization results in the opening of voltage-gated calcium channels and calcium ions flow down their concentration gradient into the cell via these channels. The intracellular Ca^{2+} concentration rises and triggers the secretion of epinephrine. This complex sequence of events is summarized in Fig. 5.4.

Activation of catalytic receptors

Catalytic receptors are *membrane-bound protein kinases* that become activated when they bind their specific ligand. (A kinase is an enzyme that adds a phosphate group to its substrate, which can be another enzyme). A typical example of a catalytic receptor is the insulin receptor that is found in liver, muscle, and fat cells. This receptor is activated when it binds insulin and in turn it activates other enzymes by adding a phosphate group to tyrosine residues. This results in an increase in the activity of the affected enzymes which culminates in an increase in the rate of glucose uptake. Many peptide hormone and growth factor receptors are tyrosine-specific kinases.

G proteins link receptor activation directly to the control of a second messenger

Second messengers are synthesized in response to the activation of specific receptors to transmit the signal from the plasma membrane to particular enzymes (e.g. cyclic AMP) or intracellular receptors (e.g. inositol trisphosphate (IP_3)). The series of events linking the change in the level of the second messenger to the final response is called a *signaling cascade*.

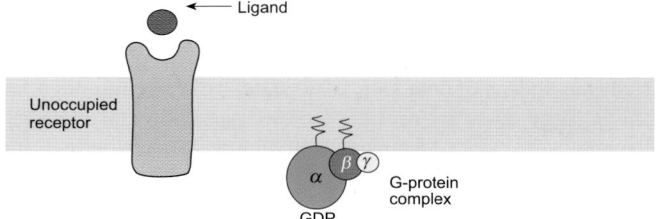

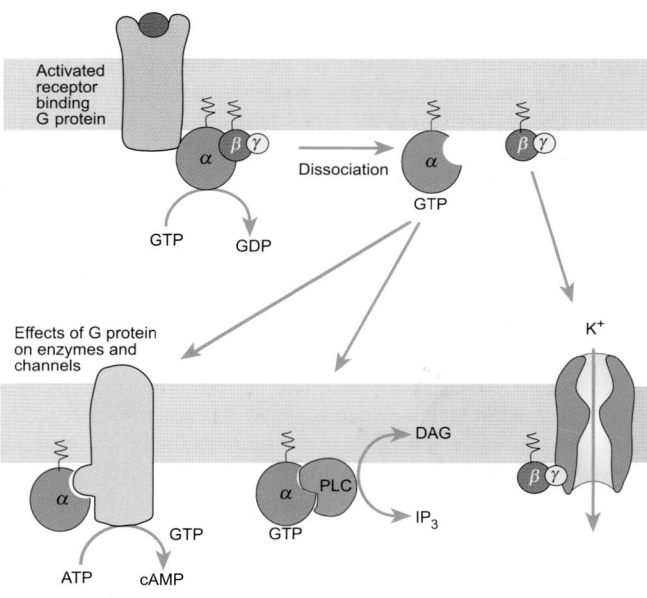

Fig. 5.5 Receptor activation of heterotrimeric G proteins leads to activation of enzymes and ion channels. The upper panels show a schematic representation of a receptor–G protein interaction. The ligand binds to its receptor which is then able to associate with a G protein. When this occurs, the α subunit exchanges bound GDP for GTP and dissociates from the β and γ subunits (center). The free α subunit can then either interact with adenylyl cyclase (bottom left) or phospholipase C (bottom center) or directly open a channel (bottom right).

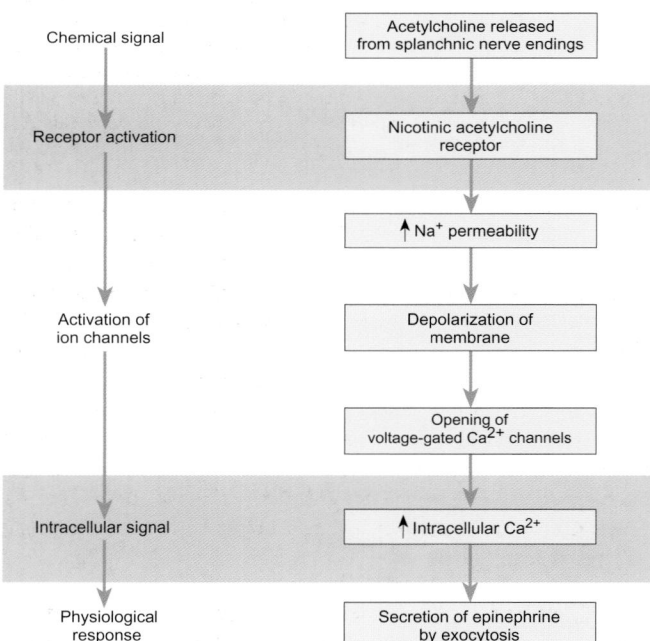

Fig. 5.4 The sequence of events that links activation of a nicotinic receptor to the secretion of the hormone epinephrine by adrenal chromaffin cells. Acetylcholine binds to a nicotinic receptor on the plasma membrane of the chromaffin cell and opens an ion channel. This increases the permeability of the membrane to Na^+ and the membrane depolarizes. The depolarization activates voltage-gated Ca^{2+} channels, and Ca^{2+} ions enter the cell to trigger the secretion of epinephrine by exocytosis.

GTP-binding regulatory proteins or G proteins are a specific class of membrane-bound regulatory proteins that are activated when a receptor binds its specific ligand. The receptor-linked G proteins have three subunits (α, β and γ), each with a different amino acid composition. Therefore they are called *heterotrimeric G proteins*. When the largest subunit (the α subunit) binds GDP the three subunits associate together. Activation of a G protein-linked receptor results in the G protein exchanging bound GDP for GTP, and this causes it to dissociate into two parts, the α subunit and the βγ subunit complex. The α and βγ subunits can then migrate laterally in the plasma membrane to modulate the activity of ion channels or membrane-bound enzymes (Fig. 5.5). There are many different kinds of G protein but they either act to open an ion channel or they alter the rate of production of a second messenger (e.g. cyclic AMP or IP₃). In turn, the second messengers regulate a variety of intracellular events. Changes in the level of cyclic AMP alter the activity of a variety of enzymes while IP₃ mainly acts by releasing Ca^{2+} from intracellular stores.

There are other GTP binding proteins which consist of a single polypeptide chain. These are called monomeric G proteins and they play an important role in the control of cell growth. In this book any reference to G proteins will mean the heterotrimeric G proteins.

5.4 G protein activation of signaling cascades

Adenylyl cyclase and phosphodiesterase regulate cyclic AMP concentrations inside cells

Cyclic AMP is generated when adenylyl cyclase (often incorrectly called adenylate cyclase) is activated by binding the α subunit of a G protein called G_s. The cyclic AMP formed as a result of receptor activation then binds to other proteins (enzymes and ion channels) within the cell and thereby alters their activity. The exact response elicited by cyclic AMP in a particular type of cell will depend on which enzymes are expressed by that cell. Only one molecule of hormone or other chemical mediator is required to activate the membrane receptor, and once adenylyl cyclase is activated it can produce many molecules of cyclic AMP with the result that activation of adenylyl cyclase allows a cell to amplify the initial signal many times. The signal is terminated by conversion of cyclic AMP to AMP by enzymes known as phosphodiesterases. Consequently, the balance of activity between adenylyl cyclase and phosphodiesterase controls the intracellular concentration of cyclic AMP. The principal features of G protein activation of adenylyl cyclase are summarized in Fig. 5.5.

The action of epinephrine (adrenaline) on skeletal muscle highlights the role of G proteins in the regulation of cyclic AMP. Skeletal muscle stores glucose as glycogen which is a large polysaccharide (Chapter 2). ATP is required during exercise to fuel muscle contraction and this necessitates the breakdown of glycogen to glucose. This change in metabolism is triggered by the hormone epinephrine that is secreted into the blood from the adrenal medulla. Increased levels of circulating epinephrine activate a particular kind of adrenergic receptor on the muscle membrane called a β-*adrenergic receptor* or β-*adrenoceptor*. These receptors are linked to G_s, and when the α subunit of G_s dissociates it activates adenylyl cyclase. The activation of adenylyl

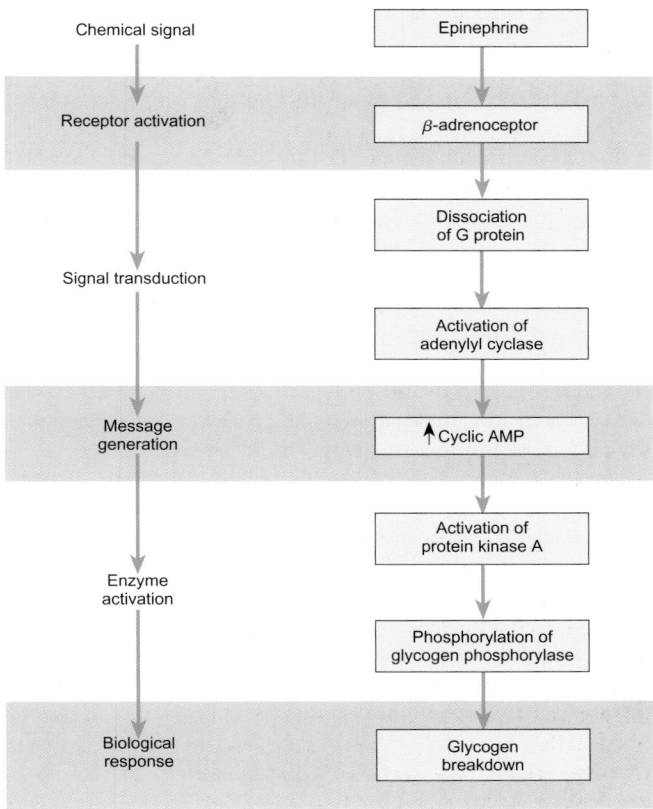

Fig. 5.6 A simplified diagram to show the transduction pathway for the action of epinephrine (adrenaline) on the glycogen stores of skeletal muscle. On the left is a key to the main stages in the signaling pathway. The details of the individual steps are shown on the right. Epinephrine binds to β-adrenoceptors which are linked to a specific type of G protein (G_s) that can activate adenylyl cyclase. This enzyme increases the intracellular concentration of the second messenger, cyclic AMP, which in turn leads to the activation of enzymes that break glycogen down to glucose.

cyclase leads to an increase in the intracellular concentration of cyclic AMP. In turn, cyclic AMP activates another enzyme called protein kinase A which, in its turn, activates another enzyme called glycogen phosphorylase which breaks glycogen down to glucose. The individual steps progressively amplify the initial signal and lead to the rapid mobilization of glucose. The main features of this cascade are summarized in Fig. 5.6.

While G_s activates adenylyl cyclase and so stimulates the production of cyclic AMP, another G protein (G_i) inhibits adenylyl cyclase. Activation of receptors coupled to G_i causes the intracellular level of cyclic AMP to fall. This, for example, is how somatostatin inhibits the release of gastrin by the G cells of the gastric mucosa.

Inositol trisphosphate and diacylglycerol are formed by the enzymatic breakdown of inositol lipids. Both act as second messengers

The inner leaf of the plasma membrane contains a small quantity of a phospholipid called phosphatidyl inositol 4, 5-bisphosphate. This phospholipid is the starting point for another important second-messenger cascade. Certain G protein-linked receptors such as the muscarinic receptor (Section 5.3) activate an enzyme known as phospholipase C. This enzyme hydrolyses phosphatidyl inositol 4,5-bisphosphate to produce diacylglycerol and inositol 1,4,5-trisphosphate (IP_3), both of which act as intracellular mediators.

IP_3 is a water-soluble molecule that binds to a specific receptor (the IP_3 receptor) to mobilize Ca^{2+} from the store within the endoplasmic reticulum. Therefore IP_3 generation is able to couple activation of a receptor in the plasma membrane to the release of Ca^{2+} from an intracellular store (Fig. 5.7). Many cellular responses depend on this pathway. Examples are enzyme secretion by the pancreatic acinar cells and smooth muscle contraction. A number of cytosolic proteins bind Ca^{2+} and one, calmodulin, is an activator of a specific set of protein kinases. In this way activation of the IP_3 signaling pathway can also regulate the pattern of enzymatic activity within the cell.

The diacylglycerol that is generated by hydrolysis of phosphatidyl inositol is a hydrophobic molecule. Consequently it is retained in the membrane when IP_3 is formed. Like other membrane lipids, it is able to diffuse in the plane of the membrane where it is able to interact with and activate another enzyme called protein kinase C. This enzyme activates other enzymes in its turn and thereby regulates a variety of cellular responses (Fig. 5.7) including DNA transcription. Diacylglycerol can also be metabolized to form *arachidonic acid* (see below).

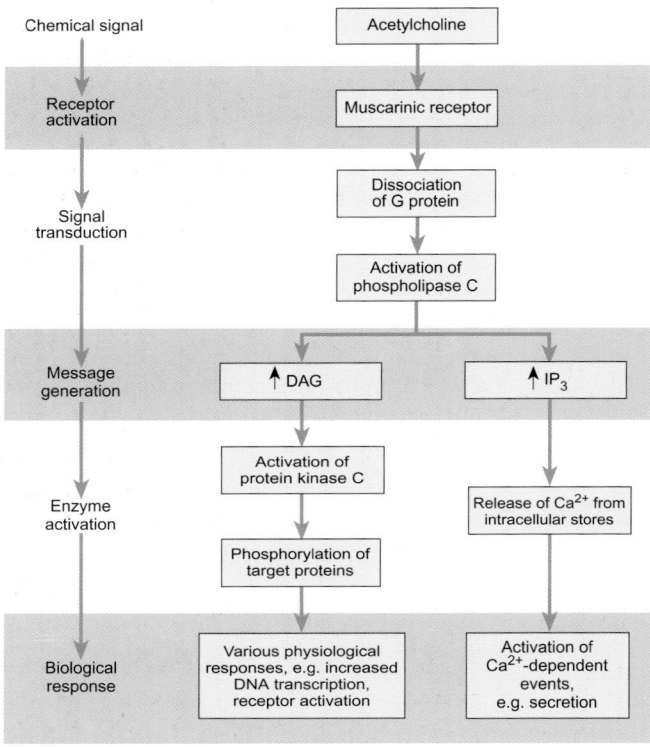

Fig. 5.7 A simplified diagram to show the transduction pathway for the formation of inositol trisphosphate (IP_3) and diacylglycerol (DAG). Both IP_3 and DAG act as second messengers. The main stages in the signal transduction pathway are shown on the left of the figure while the detailed steps are shown on the right. In this example, acetylcholine acts on muscarinic receptors that are linked to a G protein that can activate phospholipase C. This enzyme breaks down membrane phosphoinositides to form DAG and IP_3.

Summary

1. Chemical signals are detected by specific receptor molecules. When a receptor has bound a signaling molecule it must have a means of altering the behavior of the target cell. It does this in one of two ways: by opening an ion channel, or by activating a membrane bound enzyme.

2. Some receptors are membrane-bound protein kinases that become activated when they bind a ligand. Others activate a G protein, which alters the level of a second messenger (e.g. cyclic AMP) that can diffuse through the cytosol of the target cell to activate intracellular enzymes.

5.5 Some local mediators are synthesized as they are needed

Certain chemical signals are very lipid soluble and, unlike peptides and amino acids, cannot be stored in vesicles. Instead, the cells synthesize them as they are required. Important examples are the eicosanoids and *nitric oxide*, which regulate a wide variety of physiological processes.

Diacylglycerol is formed from membrane phospholipids by the action of phospholipases A_2, C, and D. It is subsequently metabolized to arachidonic acid which itself gives rise to a large number of important metabolites by one of two pathways: via cyclo-oxygenase to generate prostaglandins, thromboxanes, and prostacylin, or via lipoxygenases to give rise to leukotrienes and lipoxins (Fig. 5.8). These metabolites form a group of 20-carbon compounds known as the *eicosanoids*. The secretion of the eicosanoids is continuously regulated by increasing or decreasing their rate of synthesis from membrane phospholipids. Once formed, the eicosanoids are rapidly degraded by enzyme activity.

Eicosanoid synthesis is initiated in response to stimuli that are specific for a particular type of cell and different cell types produce different kinds of eicosanoid— over 16 different kinds are known. As local chemical mediators, the eicosanoids have many effects throughout the body (Table 5.2) and the specific effect exerted by a particular eicosanoid depends on the individual tissue. For example, prostaglandins PGE_1 and PGE_2 relax vascular

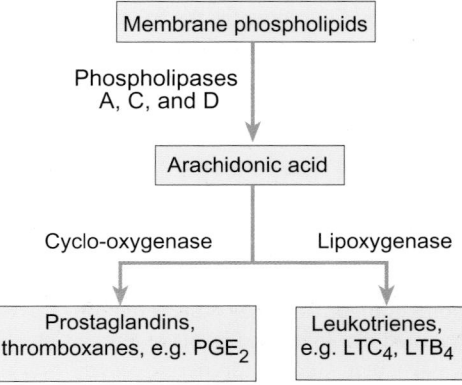

Fig. 5.8 A simple schematic diagram to illustrate the pathways leading to the synthesis of prostaglandins and leukotrienes from membrane phospholipids. LTB_4, leukotriene B_4; LTC_4, leukotriene C_4; PGE_2, prostaglandin E_2.

TABLE 5.2 Some actions of eicosanoids

Eicosanoid	Effect on blood vessels	Effect on platelets	Effects on the lung
Prostaglandin E_1 (PGE$_1$)	Vasodilatation	Inhibition of aggregation	Bronchodilatation
Prostaglandin E_2 (PGE$_2$)	Vasodilatation	Variable effects	Bronchodilatation
Prostacyclin (PGI$_2$)	Vasodilatation	Inhibition of both aggregation and adhesion	Bronchodilatation
Thromboxane A_2 (TXA$_2$)	Vasoconstriction	Aggregation	Bronchoconstriction
Leukotriene C_4 (LTC$_4$)	Vasoconstriction	–	Bronchoconstriction, increased secretion of mucus

smooth muscle and are powerful vasodilators. However, in the gut and uterus, they cause contraction of the smooth muscle. The diversity of effects in response to a particular prostaglandin is explained by the presence of different prostaglandin receptors in different tissues. These receptors are located in the plasma membrane of the target cells and are linked to second-messenger cascades via G proteins.

Thromboxane A_2 (TXA$_2$) plays an important role in hemostasis (blood clotting—see Chapter 13, pp. 239–243) by causing platelets to aggregate (i.e. to stick together). It is produced by platelets in response to the blood clotting factor thrombin (which is formed in response to tissue damage). Thrombin acts on a receptor in the cell membrane which activates phospholipase C. In turn the phospholipase C liberates diacylglycerol from which TXA$_2$ is formed from arachidonic acid. The TXA$_2$ then acts on cell surface receptors via an autocoid action to give rise to more TXA$_2$ (a positive feedback effect). It also diffuses to neighboring platelets inducing them to generate more TXA$_2$. The TXA$_2$ also activates a protein called β_3 integrin (see p. 59), which enables platelets to stick to each other and to the blood clotting protein fibrin. In this way, tissue damage leads to the formation of a blood clot. This process is normally held in check by prostacyclin (PGI$_2$), which is secreted by the endothelial cells that line the blood vessels.

Both prostaglandins and leukotrienes play a complex role in regulating the inflammatory response to injury and infection. When the tissues become inflamed, the affected region reddens, becomes swollen, and feels hot and painful. These effects are, in part, the result of the actions of prostaglandins and leukotrienes, which cause vasodilatation in the affected region. These substances also increase the permeability of the capillary walls to immunoglobulins and this leads to local accumulation of tissue fluid and swelling. Leukotriene B_4 (LTB$_4$) also attracts phagocytes. Non-steroidal anti-inflammatory drugs such as aspirin are used when the inflammatory response becomes excessively painful or persistent (as in arthritis). They act as inhibitors of cyclo-oxygenase to prevent prostaglandin synthesis. The inflammatory response is described in greater detail in Chapter 14.

Nitric oxide dilates blood vessels by increasing the production of cyclic GMP in smooth muscle

Acetylcholine is able to relax the smooth muscle of the walls of certain blood vessels and this leads to vasodilatation. If the vascular endothelium (the layer of cells lining the blood vessels) is first removed, acetylcholine causes contraction of the smooth muscle rather than relaxation. This experiment indicates that acetylcholine must release another substance in intact blood vessels—endothelium-derived relaxing factor (EDRF)—which is now known to be the highly reactive gas, nitric oxide (NO). Many other vasoactive materials including adenine nucleotides, bradykinin and histamine also act by releasing NO. It is now thought that the vasodilatation that occurs when the walls of blood vessels are subjected to stretch is also attributable to the release of NO by the endothelial cells and that this may play an important role in the local regulation of blood flow (see Chapter 15).

How do the endothelial cells form NO and how does NO cause the smooth muscle to relax? NO is derived from the amino acid arginine by an enzyme called nitric oxide synthase. This enzyme is activated when the intracellular free Ca^{2+} concentration in the endothelial cells is increased by various ligands (acetylcholine, bradykinin, etc.) or by the opening of stress-activated ion channels (i.e. ion channels that are activated by stretching of the plasma membrane). As it is a gas, the newly synthesized NO readily diffuses across the plasma membrane of the endothelial cell and into the neighboring smooth muscle cells. In the smooth muscle cells NO binds to and activates an enzyme called guanylyl cyclase (sometimes wrongly called guanylate cyclase). This enzyme converts GTP into cyclic GMP. Thus stimulation of the endothelial cells leads to an increase in cyclic GMP within the smooth muscle and this in turn activates other enzymes to bring about muscle relaxation. This sequence of events is summarized in Fig. 5.9.

The NO formed by endothelial cells depends on the presence of the enzyme nitric oxide synthase, which is regulated by intracellular Ca^{2+}. Nitric oxide synthase is not normally present in macrophages, but when these cells are exposed to bacterial toxins the gene controlling the synthesis of this enzyme is switched on (a process known as induction) and the cells begin to make NO. In this case the NO is not used as a signaling molecule but as a lethal agent to kill invading organisms.

Organic nitrites and nitrates such as amyl nitrite and nitroglycerine have been used for over 100 years to treat the pain that occurs when the blood flow to the heart muscle is insufficient. (This pain is called angina pectoris or *angina*.) These compounds promote the relaxation of the smooth muscle in the walls of blood vessels. Detailed investigation has revealed that this effect can be attributed to the formation of NO by enzymatic conversion of nitrite ions derived from the organic nitrates. This exogenous NO then acts in a similar way to that derived from normal metabolism.

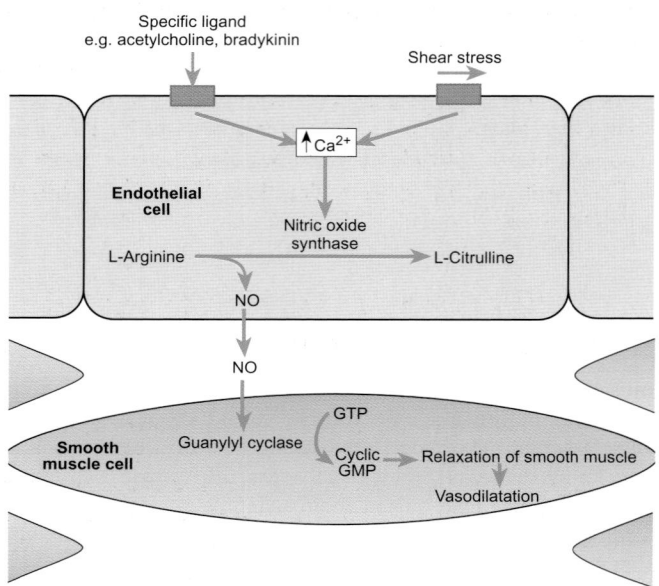

Fig. 5.9 Diagram to show the synthesis of NO by endothelial cells and its action on vascular smooth muscle. The trigger for increased synthesis of NO is a rise in Ca^{2+} in the endothelial cell. This can occur as a result of stimulation by chemical signals (acetylcholine, bradykinin, ADP, etc.) acting on receptors in the plasma membrane or as a result of the opening of ion channels by stretching of the plasma membrane (shear stress). The NO diffuses across the plasma membrane of the endothelial cell into the neighboring smooth muscle cells and converts guanylyl cyclase into its active form. The increased production of cyclic GMP leads to relaxation of the smooth muscle.

5.6 Steroid and thyroid hormones bind to intracellular receptors to regulate gene transcription

The steroid hormones are themselves lipids, so they are able to pass through the plasma membrane freely, unlike more polar, water-soluble signaling molecules such as peptide hormones. Thus, steroid hormones are not only able to bind to specific receptors in the plasma membrane of the target cell, but they can also bind to receptors within its cytoplasm and nucleus. The existence of cytoplasmic receptors for steroid hormones was first shown for estradiol. This hormone is accumulated by its specific target tissues (the uterus and vagina) but not by other tissues. It was found that the target tissues possess a cytoplasmic receptor protein for estradiol which, when it has bound the hormone, increases the synthesis of specific proteins.

The full sequence of events for the action of estradiol can be summarized as follows: the hormone first crosses the plasma membrane by diffusing through the lipid bilayer and binds to its cytoplasmic receptor. The receptor–hormone complex then migrates to the nucleus where it increases the transcription of DNA into the appropriate mRNA. The new mRNA is then used as a template for protein synthesis (see Chapter 2). Other steroid hormones, such as the glucocorticoids and aldosterone, are now known to act in a similar way. This scheme is outlined in Fig. 5.10. Thyroid hormones regulate gene expression in a simi-

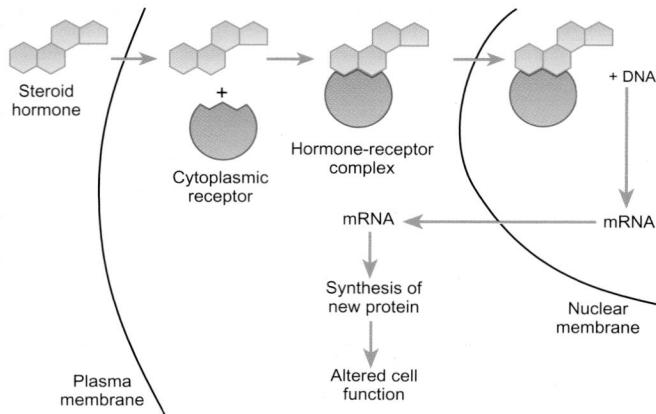

Fig. 5.10 A simplified diagram to show how steroid hormones regulate gene transcription in their target cells. Steroid hormones are lipophilic and pass through the plasma membrane to bind to specific receptor proteins in the cytoplasm of the target cells. The hormone–receptor complex diffuses to the cell nucleus where it binds to a specific region of DNA to regulate gene transcription.

lar manner to steroid hormones, but gain entry to their target tissues via specific transport proteins.

The receptors for the steroid and thyroid hormones are members of a large group of proteins involved in regulating gene expression, the *nuclear receptor superfamily*. However, not all nuclear receptors are located in the cytoplasm prior to binding their hormone, some are bound to DNA in the nucleus, even in the absence of their normal ligand. For example, the thyroid hormone receptor is bound to DNA in the nucleus, even in the absence of thyroid hormone.

In the absence of their specific ligand, all members of the nuclear receptor superfamily are bound to other regulatory proteins to form inactive complexes. When they bind their ligand, the receptors undergo a conformational change that leads to the dissociation of the regulatory proteins from the receptor. The activated receptor protein then regulates the transcription of a specific gene, or set of genes. This occurs in successive steps: initially, the hormone–receptor complex activates a small set of genes relatively quickly. This is called the primary response, and occurs in less than an hour. The proteins synthesized in this early phase then activate other genes. This is the secondary response. Since many proteins are involved in gene activation, the initial hormonal signal is able to initiate very complex changes in the pattern of protein synthesis within its target cells.

The response to thyroid hormone, the steroid hormones, and other ligands that bind to members of the nuclear receptor superfamily is determined by the nature of the target cells themselves. It depends on the presence of the appropriate intracellular receptor and the particular set of regulatory proteins present, both of which are specific to certain cell populations. The final physiological effects are slow in onset and long lasting. For example, sodium retention by the kidney occurs after a delay of 2–4 hours following the administration of aldosterone (see Fig. 17.20) and may last for many hours.

Summary

Steroid and thyroid hormones are very hydrophobic and are carried in the blood bound to specific carrier proteins. They enter cells by diffusing across the plasma membrane and bind to cytoplasmic and nuclear receptors to modulate the transcription of specific genes. This leads to the increased synthesis of specific proteins.

5.7 Cells use specific cell surface molecules to assemble into tissues

To form complex tissues, different cell types must aggregate together. Therefore some cells need to migrate from their point of origin to another part of the body during development. When they arrive at the appropriate region they must recognize their target cells and participate in the differentiation of the tissue. To do so, they must attach to other cells and to the extracellular matrix. What signals are employed by developing cells to establish their correct positions and why do they cease their migrations when they have found their correct target?

Unlike adult cells, embryonic cells do not form strong attachments to each other. Instead, when they interact, their cell membranes become closely apposed to each other leaving a very small gap of only 10–20 nm. Exactly how cells are able to recognize their correct associations is not known, but it is likely that each type of cell has a specific marker on its surface. When cell membranes touch each other the marker proteins on the surfaces of the cells can interact. If the cells have complementary proteins they are then able to cross-link and the cells will adhere. This must be an early step in tissue formation. It has been shown that cells will associate only if they recognize the correct surface markers. Thus, if differentiated embryonic liver cells are dispersed by treatment with enzymes and grown in culture with cells from the retina, the two cell types aggregate with others of the same kind. Thus liver cells aggregate together and exclude the retinal cells and vice versa.

The various kinds of cell–cell and cell–matrix junctions have already been discussed in Chapter 3 and many of the proteins involved have been characterized. These can be grouped into several families including the cadherins that form desmosomes, the connexins that form gap junctions (see below), the immunoglobulin-like (Ig-like) cell adhesion molecules (e.g. N-CAM), and the integrins that form hemidesmosomes that attach cells to the extracellular matrix. The integrins also play an important role in development and wound repair. The cadherins, integrins, and Ig-like cell adhesion molecules are also involved in non-junctional cell–cell adhesion, which must play an important role in the formation of integrated tissues.

While the integrins are important in maintaining attachments between the majority of cells, those of platelets are not normally adhesive. If they were, blood clots would form spontaneously with disastrous consequences. However, during hemostasis (blood clotting) platelets adhere to fibrin and to the damaged wall of the blood vessel (see pp. 239–243). This adhesion results from a change in the properties of the platelets when a non-adhering integrin precursor present in the platelet membrane is transformed into an adhesive protein. This transformation is triggered by factors released from the walls of damaged blood vessels that activate second-messenger cascades within the platelets which, in turn, trigger modification of the structure of preformed integrins so that they act as receptors for extracellular molecules including fibrin. The final result is an increase in platelet adhesion and clotting of the blood.

5.8 Gap junctions permit the exchange of small molecules and ions between neighboring cells

Some cells are joined together by a specific type of junction known as a *gap junction*. Specific membrane proteins associate to form doughnut-shaped structures known as connexons. When the connexons of two adjacent cells are aligned, the cells become joined by a water-filled pore. As the connexons jut out above the surface of the plasma membrane, the cell membranes of the two cells forming the junction are separated by a small gap, hence the name gap junction (see Chapter 3).

Unlike ion channels, the pores of gap junctions are kept open most of the time so that small molecules (less than 1500 Da) and inorganic ions can readily pass from one cell to another, as shown in Fig. 5.11. Consequently gap junctions form a low-resistance pathway between the cells and electrical current can spread from one cell to another. The cells are thus *electrically coupled*. This property is exploited by the myocytes of the heart, which are connected by gap junctions. Since the cells are electrically coupled, depolarization of one myocyte causes current to pass between it and its immediate neighbors, which become depolarized. In turn, these cells cause the depolarization of their neighbors and so on. Consequently, current from a single point of excitation spreads across the whole of the heart via the gap junctions and the heart muscle behaves as a *syncytium* (a collection of cells fused together). By this means the electrical and contractile activity of individual myocytes is coordinated. This allows the muscle of the heart to provide the wave of contrac-

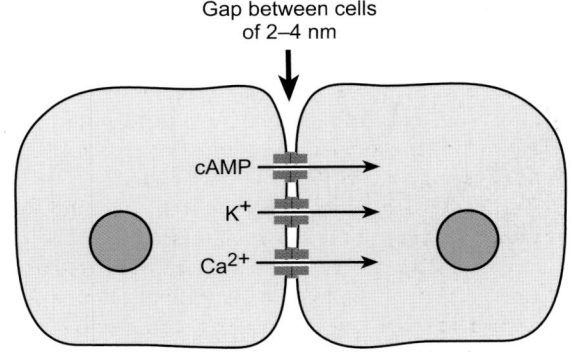

Fig. 5.11 A simple diagram showing two cells linked by gap junctions. Ions can pass between the neighboring cells so the cells are electrically coupled. In addition, small organic molecules (M_r < 1500) can also pass through gap junctions, allowing for the spread of second messenger molecules (e.g. cyclic AMP) between adjacent cells.

Summary

1. For cells to assemble into tissues they need to adhere to other cells of the correct type. This recognition requires tissue-specific cell-surface marker molecules.

2. Cell–cell adhesion and cell–matrix interaction play an important role in tissue maintenance and development. Several different families of proteins are involved in these processes.

3. Gap junctions between adjacent cells permit the diffusion of small molecules and ions directly from the cytoplasm of one cell to its neighbor. The direct contact between the cytoplasm of adjacent cells allows electrical current to flow from one cell to another. Therefore gap junctions, allow the electrical coupling of cells.

tion (the heart beat) that propels the blood around the body (see Chapter 15).

In the liver, the role of gap junctions between adjacent liver cells (hepatocytes) is quite different. The gap junctions allow the exchange of intracellular signals (second messengers) between cells. For example, the hormone glucagon stimulates the break-down of glycogen to glucose by increasing the level of cyclic AMP. The cyclic AMP diffuses through the water-filled pores of the gap junctions from one cell to another. Thus cells not directly activated by glucagon can be stimulated to initiate glycogen breakdown. The gap junctions provide a means of spreading the initial stimulus from one cell to another.

Recommended reading

Biochemistry and cell biology

Alberts, B., Johnson, A., Lewis, J., Raff, M., Roberts, K., and Walter, P. (2002). *The molecular biology of the cell* (4th edn), Chapters 15 and 19. Garland, New York.

Barritt, G.J. (1992). *Communication within animal cells*, Chapters 1, 3–7 and 10. Oxford University Press, Oxford.

Berg, J.M., Tymoczko, J.L., and Stryer, L. (2002). *Biochemistry* (5th edn), Chapter 15. Freeman, New York.

Gomperts, B.D., Tatham, P.E.R., and Kramer, I.M. (2002). *Signal transduction*. Academic Press, San Diego, CA.

Pharmacology

Rang, H.P., Dale, M.M., Ritter, J.M., and Moore, P. (2003). *Pharmacology* (5th edn), Chapters 2, 10, and 11. Churchill-Livingstone, Edinburgh.

Multiple choice questions

Each statement is either true or false. Answers are given below.

1. Hormones:
   a. Are chemical signals that are secreted into the blood.
   b. Can influence the behavior of many different cell types.
   c. Act only on neighboring cells.
   d. Are secreted by specialized glands.

2. The following are released by one type of cell specifically to regulate the activity of others:
   a. NO
   b. Prostaglandins
   c. Insulin
   d. Epinephrine
   e. Ca^{2+}
   f. Glucose
   g. Cyclic AMP

3. Receptors:
   a. Are always proteins.
   b. Are always located in the plasma membrane.
   c. May be membrane-bound enzymes.
   d. Can activate second-messenger cascades via G proteins.

4. Prostaglandins:
   a. Are secreted by exocytosis.
   b. Are synthesized from phospholipids.
   c. Are hormones.
   d. Act on G protein-linked receptors.

5. Steroid hormones:
   a. Are lipid soluble.
   b. Directly activate ion channels.
   c. Can alter gene transcription.
   d. Are secreted as they are synthesized.

Answers

1. Hormones are secreted by endocrine glands into the blood stream to act on cells at a distance. Paracrine signals act locally.
   a. True
   b. True
   c. False
   d. True

2. NO and prostaglandins are paracrine signaling molecules. Insulin and adrenaline are hormones. Ca^{2+} and cyclic AMP are both intracellular mediators but cyclic AMP can spread from cell to cell via gap junctions. Glucose is a substrate for metabolism.
   a. True
   b. True
   c. True
   d. True
   e. False
   f. False
   g. True

3. Many receptors are membrane proteins but some are intracellular proteins (e.g. steroid hormone receptors). Receptors may directly activate ion channels but some receptors are protein kinases that are activated when they bind their ligand. Many receptors activate G proteins and thereby modulate the levels of second messengers.
   a. True
   b. False
   c. True
   d. True

4. Prostaglandins are metabolites of arachidonic acid, which is derived from membrane phospholipids by the action of phospholipases. They are lipid-soluble molecules that are secreted as they are formed. They are paracrine and autocrine signaling molecules which activate G protein-linked receptors in the plasma membrane.
   a. False
   b. True
   c. False
   d. True

5. Steroid hormones are lipid signaling molecules that act mainly by altering gene transcription. As they are lipids they cannot be stored in membrane-bound vesicles and therefore are secreted as they are synthesized.
   a. True
   b. False
   c. True
   d. True

Nerve cells and their connections

After reading this chapter you should understand:

- The basic anatomical organization of the nervous system
- The structure of nerve cells
- The ionic basis of the action potential in neurons and axons
- The principal features of synaptic transmission between nerve cells—the synaptic basis of excitation and inhibition
- Neuromuscular transmission: the neuromuscular junction as a model chemical synapse
- The effects of denervation of skeletal muscle
- Axonal transport

6.1 Introduction

The nervous system is adapted to provide rapid and discrete signaling over long distances (from millimeters to a meter or more), and this chapter is chiefly concerned with the underlying mechanisms. It begins with a simple outline of the structure of the central nervous system and peripheral nerves and goes on to discuss the key cellular events involved in neuronal signaling: action potential generation and synaptic transmission.

The nervous system can be divided into five main parts:

(1) the brain;

(2) the spinal cord;

(3) the peripheral nerves;

(4) the autonomic nervous system;

(5) the enteric nervous system.

The brain and spinal cord constitute the *central nervous system* (CNS), while the peripheral nerves, autonomic nervous system, and enteric nervous system make up the *peripheral nervous system*. The *autonomic nervous system* is the part of the nervous system that is concerned with the innervation of the blood vessels and the internal organs. It includes the autonomic ganglia that run parallel to the spinal column (the paravertebral ganglia) and their associated nerves. The *enteric nervous system* controls the activity of the gut. The organization and functions of the autonomic and enteric divisions of the nervous system will be discussed in Chapters 10 and 18 respectively.

6.2 The organization of the brain and spinal cord

As Figure 6.1 shows, the brain and spinal cord lie within a bony case formed by the skull and vertebral canal of the spinal column and are covered by three membranes called the *meninges*. Immediately beneath the skull lies a tough membrane of dense connective tissue called the *dura mater* (or *dura*) that envelops the whole brain and extends in the form of a tube over the spinal cord. Attached to the inner face of the dura is the *arachnoid membrane*. Beneath the arachnoid lies the highly vascularized *pia mater* (or *pia*) which is attached to the surface of the brain and spinal cord, following every contour. The narrow space between the pia and arachnoid membranes is filled with a clear fluid called the *cerebrospinal fluid* (CSF) that is actively secreted by the choroid plexuses. These are vascular structures situated in fluid-filled spaces within the brain (the *cerebral ventricles*). The CSF plays an important role in regulating the extracellular environment of nerve cells, and its composition, formation, and circulation are discussed in Chapter 15, p. 307.

Since the space between the skull and the brain is filled with CSF, the brain floats in a fluid-filled container. Moreover, deep infoldings of the dura divide the fluid-filled space between the skull and the brain into smaller compartments. This arrangement restricts the displacement of the brain within the skull during movements of the head and limits the stresses on the blood vessels and the cranial nerves. The spinal cord is attached to the meninges by thin bands of connective tissue known as the dentate ligaments.

The surface of the human brain has many folds called *sulci* (singular *sulcus*) and the smooth regions of the brain surface that lie between the folds are known as *gyri* (singular *gyrus*). Viewing the brain from the dorsal surface reveals a deep cleft known as the longitudinal cerebral fissure that divides the brain into two *cerebral hemispheres*. Each hemisphere can be broadly divided into four lobes: the frontal, parietal, occipital, and temporal. Posterior to the cerebral hemispheres lies a smaller highly convoluted structure known as the *cerebellum* (Fig. 6.2).

If the brain is cut in half along the midline (a midsagittal section) between the cerebral hemispheres, some details of its internal organization can be seen (Fig. 6.3). On the medial surface is a broad white band known as the *corpus callosum* which interconnects the two hemispheres. Immediately below the corpus callosum is a membranous structure called the septum pellucidum which separates two internal spaces known as the lateral cerebral ventricles, which are filled with CSF.

Beneath the septum pellucidum and lateral ventricles is the *thalamus*, a major site for the processing of information from sense organs. Lying just in front of and below the thalamus is the *hypothalamus* which plays a vital role in the regulation of the endocrine system via its control of the pituitary gland (see Chapter 12). Posterior and ventral to the thalamus lies the midbrain (Fig. 6.3), which merges into the pons, a large swelling containing fibers interconnecting the two halves of the cerebellum.

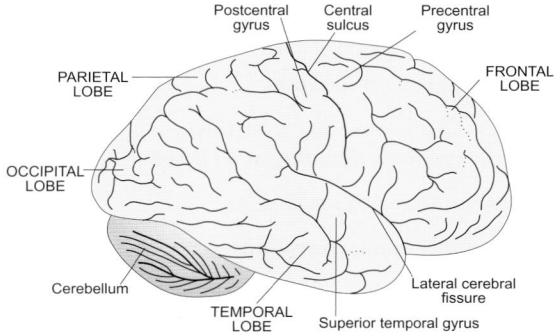

Fig. 6.2 A side view of the human brain showing the cerebellum and the principal divisions of the right cerebral hemisphere.

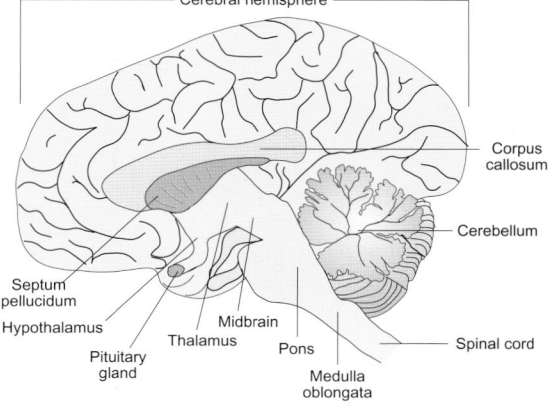

Fig. 6.3 A midsagittal view of the right side of the human brain to show the relationships of the main structures.

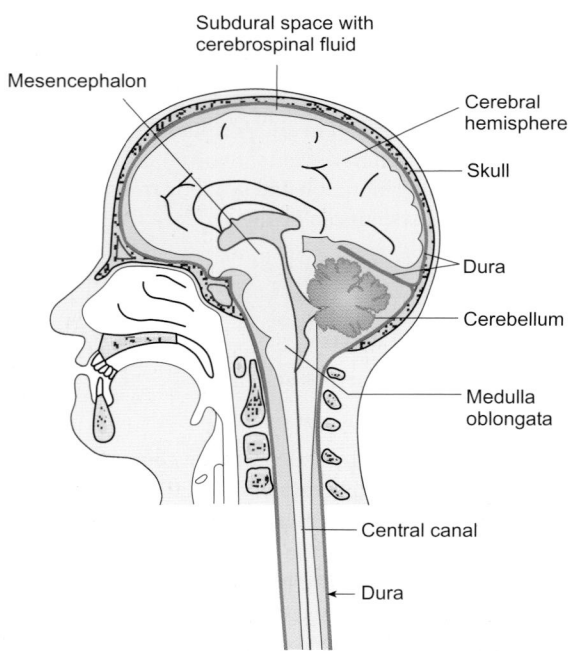

Fig. 6.1 A diagrammatic representation of a midsagittal view of the CNS.

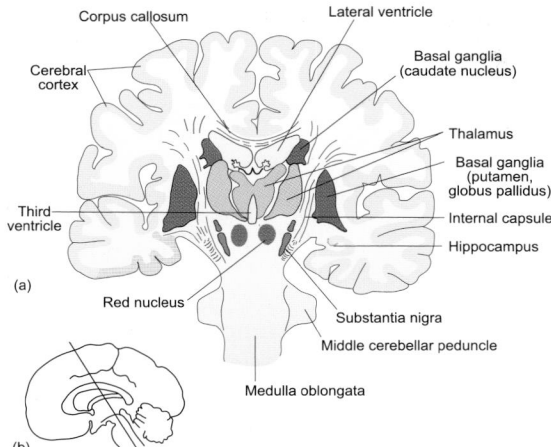

Fig. 6.4 An oblique cross-section through the human brain showing the spatial relationship of the cerebral cortex, basal ganglia, and thalamus. The small inset diagram (b) shows the plane of section.

Below and behind the pons lies the *medulla oblongata* (or *medulla*) which, in turn, connects with the spinal cord that runs through the spinal canal of the vertebral column. As it passes down the spinal column, the spinal cord gives rise to a series of paired nerves that connect the CNS to peripheral organs (see below).

If the brain is cut at right angles to the midline, a coronal section is obtained which reveals the internal structure of the brain. The outer part or *cerebral cortex* is grayish in appearance. The cortex and other parts of the brain that have a similar appearance are *gray matter,* which contains large numbers of nerve cell bodies. Inside the gray matter of the cerebral cortex is the *white matter,* which is composed of bundles of nerve fibers such as those of the corpus callosum and the internal capsule.

An oblique section through the brain reveals a number of other important structures which are shown diagrammatically in Fig. 6.4. The caudate nucleus, the putamen, and the globus pallidus together form the *corpus striatum.* Between the caudate nucleus and putamen runs the *internal capsule* which contains

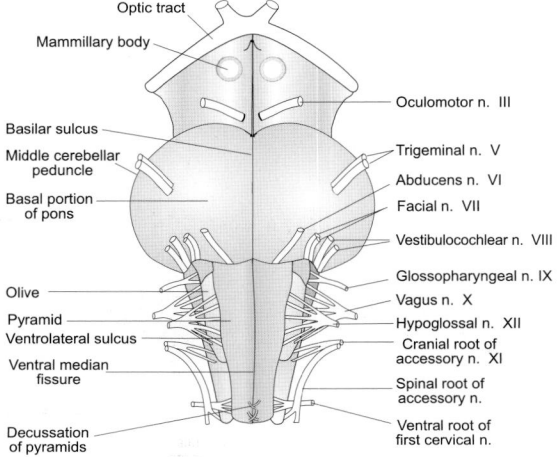

Fig. 6.5 A ventral view of the human brain showing the cranial nerves, the optic chiasm, the pons and the decussation of the pyramids (crossing over of the nerve fibers of the pyramidal tracts).

nerve fibers connecting the cerebral cortex to the spinal cord. A small region known as the *substantia nigra* lies beneath the thalamus. (The substantia nigra is so named because it contains a characteristic black pigment). All these structures play an important role in the control of movement (see Chapter 9 for further details).

The cranial nerves

On the base of the brain there are 12 pairs of nerves that serve the motor and sensory functions of the head (Fig. 6.5). These are the *cranial nerves*, which are numbered I–XII. Some contribute to the parasympathetic division of the autonomic nervous system (Chapter 10). These are the oculomotor (III), the facial (VII), the glossopharyngeal (IX) and the vagus (X). The names and the main functions of all the cranial nerves are given in Table 6.1.

The organization of the spinal cord

While the human brain is a large structure, the spinal cord is delicate—scarcely as thick as a pencil for much of its length. A cross-section of the spinal cord shows that it has a central region of gray matter surrounded by white matter (Fig. 6.6). The white matter of the spinal cord is arranged in columns that contain the nerve

TABLE 6.1	The functions of the cranial nerves	
Number	**Name**	**Chief functions**
I	Olfactory	Sensory nerve subserving the sense of smell
II	Optic	Sensory nerve subserving vision (output of the retina)
III	Oculomotor	Chiefly motor control of the extrinsic muscles of the eye and the parasympathetic supply for the intrinsic muscles of the iris and ciliary body
IV	Trochlear	Chiefly motor control of the extrinsic muscles of the eye
V	Trigeminal	Sensory and motor—motor control of the jaw and facial sensation
VI	Abducens	Chiefly motor control of the extrinsic muscles of the eye
VII	Facial	Sensory and motor—motor control of the facial muscles and parasympathetic supply to the salivary glands.
		Subserves the sense of taste via the chorda tympani
VIII	Vestibulo-cochlear	Sensory—hearing and balance
IX	Glossopharyngeal	Sensory and motor—control of swallowing and parasympathetic supply to the salivary glands
		Subserves the sense of taste from the back of the tongue (bitter sensations)
X	Vagus	Major parasympathetic outflow to the chest and abdomen
		Afferent inputs from the viscera
XI	Spinal accessory	Motor—control of neck muscles and larynx
XII	Hypoglossal	Motor control of the tongue

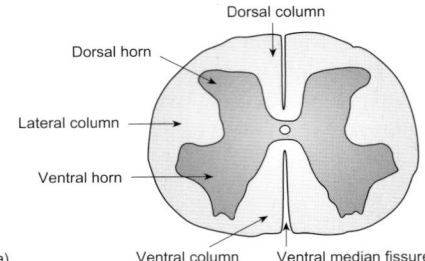

(a)

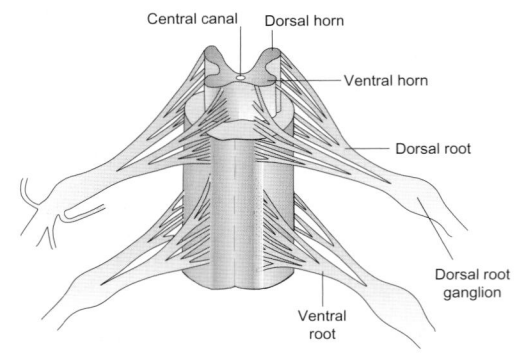

(b)

Fig. 6.6 Diagrams illustrating (a) the structure of the spinal cord at the level of the lumbar enlargement and (b) the arrangement of the spinal roots. In (b) part of the white matter of the spinal cord is cut away to show the direct entry of the spinal roots into the central gray matter.

fibers connecting the brain and spinal cord. The gray matter is roughly shaped in the form of a butterfly around a central canal, although the exact appearance depends on whether the spinal cord has been cut across at the cervical, thoracic, lumbar, or sacral level. This gray matter can be broadly divided into two *dorsal horns* and two *ventral horns*. (The dorsal and ventral horns are also known as the posterior and anterior horns, respectively.)

At each segmental level the spinal cord gives rise to a pair of spinal nerves each of which is formed by the fusion of nerve segments known as the *dorsal and ventral roots* as shown in Fig. 6.6(b). In most individuals, the spinal cord has 31 pairs of spinal nerves. Each dorsal root has an enlargement known as a *dorsal root ganglion* which contains the cell bodies of the nerve fibers making up the dorsal root. The fibers of the ventral root originate from nerve cells in the ventral horn of the spinal gray matter. To leave the spinal canal, the spinal nerves pierce the dura mater between the vertebrae. Thereafter, they form the peripheral nerve trunks that innervate the muscles and organs of the body.

Sensory information enters the spinal cord via the dorsal root ganglia. Since the sensory fibers carry information from sense organs to the spinal cord, they are known as *afferent nerve fibers*. The ventral root fibers are known as *efferent nerve fibers*. They carry motor information from the spinal cord to the muscles and secretory glands (the *effectors*). The nerves that leave the spinal cord to supply the skeletal muscles are known as somatic nerves, while those that supply the blood vessels and viscera are sympathetic efferent fibers (see Chapter 10 for further details of the organization and function of the sympathetic fibers).

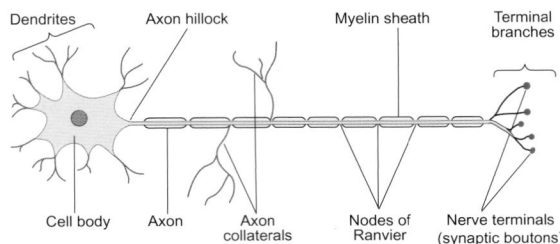

Fig. 6.7 A diagrammatic representation of a CNS neuron.

The neuron is the principal functional unit of the nervous system

The CNS is made up of two main types of cell: *nerve cells* or *neurons*, and *glial cells* or *neuroglia*. The cell bodies of the neurons are found throughout the gray matter of the brain and spinal cord.

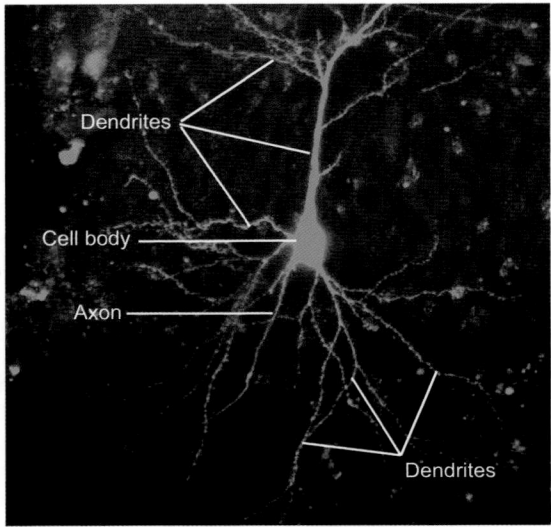

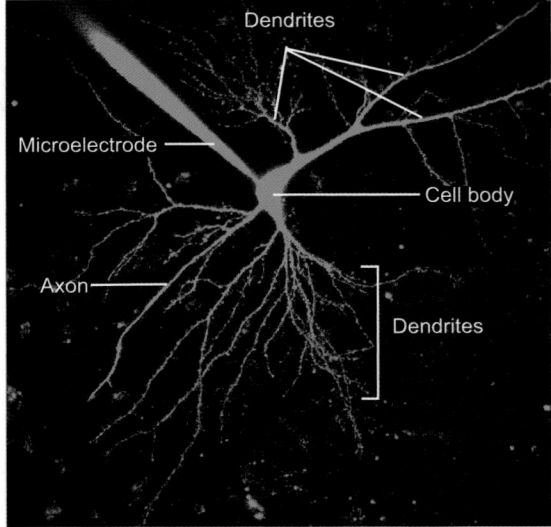

Fig. 6.8 Two neurons stained with a green fluorescent dye (Oregon green BAPTA 1). The lower panel shows the microelectrode used to introduce the dye into the cell. Note the extensive branching of the dendrites and the single axon.

They are very varied in both size and shape but all stain strongly with basic dyes. The stained material is called Nissl substance. It contains a high proportion of RNA. Neurons have extensive branches called *dendrites* which receive information from other neurons. They transmit information to their targets (which may be other neurons) via a threadlike extension of the cell body called an *axon* (Fig. 6.7). Axons and dendrites are collectively called neuronal processes. The space between the cell bodies is known as the neuropil and contains the cytoplasmic extensions of both neurons and glia.

The dendrites of neurons within the CNS generally branch extensively as shown in Fig. 6.8. They are, therefore, able to receive information from many different sources. Each nerve cell gives rise to a single axon, which subsequently branches to contact a number of different target cells. The axon branches are called collaterals. Each branch ends in a small swelling—an axon terminal (also called a *nerve terminal*). The contact between an axon terminal and its target is called a *synapse*. Synapses may be made between nerve cells or between an axon and a non-neuronal cell such as a muscle fiber.

Crushing or cutting an axon prevents a neuron from controlling its target cell. In the case of nerves supplying a skeletal muscle, such injuries result in paralysis of the muscle even though the nerve sheath may remain intact. Control of movement is regained only if the appropriate axons regenerate. Similarly, the death of neurons within the brain following a stroke can cause paralysis even though the peripheral nerves remain intact. (A stroke is a loss of cerebral function due to a cerebral thrombosis, i.e. a blockage of a cerebral blood vessel caused by a blood clot, or, more rarely, due to a cerebral hemorrhage caused by rupture of a blood vessel.) These observations demonstrate that the neurons are the principal functional units of the nervous system. Their role is to transmit signals to each other and to target cells outside the nervous system.

The non-neuronal cells of the CNS

Four main classes of non-neuronal cell occur in the brain and spinal cord.

1. *Astroglia or astrocytes.* These are cells with long processes that make firm attachments to blood vessels. The ends of the astrocytic processes seal closely together and form an additional barrier between the blood and the extracellular fluid of the brain and spinal cord (see Chapter 15, Fig. 15.44). This barrier is known as the blood–brain barrier and it serves to prevent changes in the composition of the blood influencing the activity of the nerve cells within the CNS.

2. *Oligodendroglia or oligodendrocytes.* Oligodendrocytes account for about 75 per cent of all glial cells in white matter where they form the myelin sheaths of axons. In the peripheral nervous system, the myelin sheaths are formed by Schwann cells (see below).

3. *Microglia.* These are scattered throughout the gray and white matter. They are phagocytes and rapidly converge on a site of injury or infection within the CNS.

4. *Ependymal cells.* These are ciliated cells that line the central fluid-filled spaces of the brain (the cerebral ventricles) and the central canal of the spinal cord. They form a cuboidal-columnar epithelium called the ependyma.

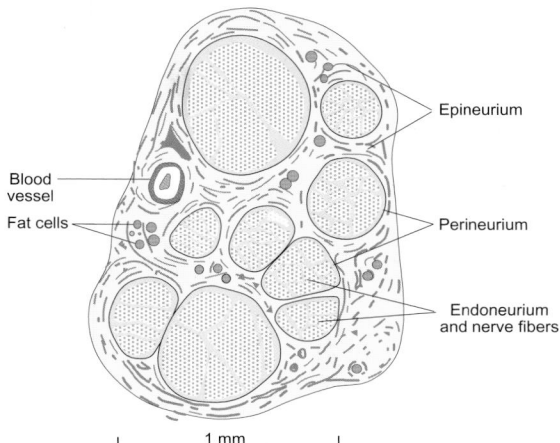

Fig. 6.9 A cross-section through a small peripheral nerve to show the relationship between the nerve fibers and the surrounding layers of connective tissue (the epineurium, perineurium, and endoneurium). Note that at this magnification individual nerve fibers are not clearly resolved.

The structure of peripheral nerve trunks

Axons are delicate structures which may traverse considerable distances to reach their target organs. Outside the CNS they run in peripheral nerve trunks alongside the major blood vessels where they are protected from damage by layers of connective tissue (Fig. 6.9). The outermost layer of a peripheral nerve is a loose aggregate of connective tissue called the *epineurium*, which serves to anchor the nerve trunk to the adjacent tissue. Within the epineurium, axons run in bundles called fascicles and each bundle is surrounded by a tough layer of connective tissue called the *perineurium*. Within the perineural sheath, individual nerve fibers are protected by a further layer of connective tissue called the *endoneurium*. Individual axons are covered by specialized cells called Schwann cells.

Some nerve trunks transmit information from specific sensory end organs to the CNS (*sensory nerves*) while others transmit signals from the CNS to specific effectors (*motor nerves*). Nerve trunks that contain both sensory and motor fibers are called *mixed nerves*. Peripheral nerves also contain sympathetic postganglionic fibers, which innervate the blood vessels and sweat glands (see Chapter 10).

Axons may be either myelinated or unmyelinated. *Myelinated axons* are covered by a thick layer of fatty material called myelin that is formed by layers of plasma membrane derived from specific satellite cells. In peripheral nerves the myelin is derived from Schwann cells (Fig. 6.10), while in the CNS the myelin is formed by the oligodendrocytes. Although the myelin sheath extends along the length of an axon, it is interrupted at regular intervals by gaps known as the *nodes of Ranvier*. At the nodes of Ranvier the axon membrane is not covered by myelin but is in direct contact with the extracellular fluid. The distance between adjacent nodes varies with the axon diameter, with larger fibers having a greater internodal distance. In peripheral nerves, *unmyelinated axons* are also covered by Schwann cells, but in this case there is no layer of myelin and a number of nerve fibers are covered by a single Schwann cell. Unmyelinated axons are in

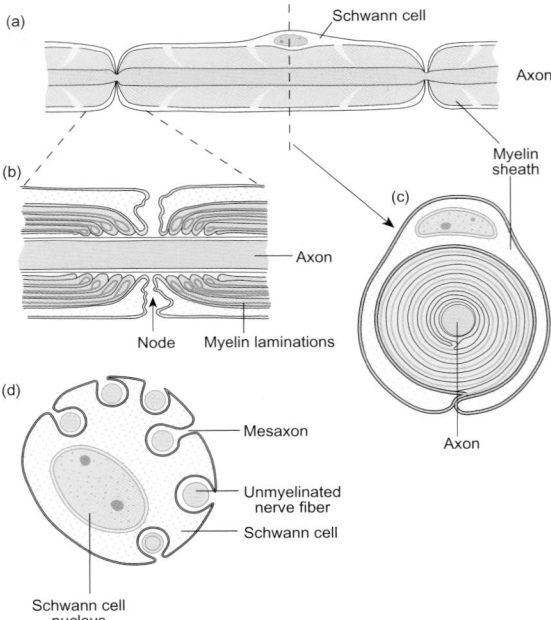

Fig. 6.10 Drawing to show myelinated and unmyelinated nerve fibers and their relationship to the surrounding Schwann cells. (a) A short length of myelinated nerve. (b) Illustration of the detailed structure of a single node of Ranvier. (c) Diagrammatic view of a cross-section of a myelinated nerve fiber showing how the Schwann cell forms the myelin layers. (d) The arrangement of a number of unmyelinated nerve fibers within a Schwann cell.

direct communication with the extracellular fluid via a longitudinal cleft in the Schwann cell called a mesaxon (Fig. 6.10).

6.3 The primary function of an axon is to transmit information coded as a sequence of action potentials

In the late eighteenth century, Galvani showed that electrical stimulation of the nerve in a frog's leg caused the muscles to twitch. This key observation led to the discovery that the excitation of nerves was accompanied by an electrical wave that passed along the nerve. This wave of excitation is now called the *nerve impulse* or *action potential*. To generate an action potential, an axon requires a stimulus of a certain minimum strength known as the threshold stimulus or *threshold*. An electrical stimulus that is below the threshold (a *subthreshold* stimulus) will not elicit an action potential while a stimulus that is above the threshold (a *suprathreshold* stimulus) will do so. With stimuli above the threshold, each action potential has approximately the same magnitude and duration irrespective of the strength of the stimulus. This is known as the 'all-or-none' law of action potential transmission.

In a mammalian axon, each action potential lasts about 0.5–1 ms. If a stimulus is given immediately after an action potential has been elicited, a second action potential is not generated until a certain minimum time has elapsed. The interval during which it is impossible to elicit a second action potential is known as the *absolute refractory period*. This determines the upper limit to the number of action potentials a particular axon can

Summary

1. The nervous system can be broadly divided into five main parts:
    * the brain
    * the spinal cord
    * the autonomic nervous system
    * the enteric nervous system
    * the peripheral nerves.

2. Within the CNS two distinct types of tissue can be discerned on the basis of their appearance: gray matter and white matter. Gray matter contains the cell bodies of the neurons. White matter mainly consists of myelinated nerve axons and oligodendrocytes.

3. The neuron is the principal functional unit of the nervous system. The cell bodies of neurons give rise to two types of process: dendrites and axons. The dendrites are highly branched and receive information from many other nerve cells. Each cell body gives rise to a single axon, which subsequently branches to make contact with a number of other cells.

4. Individual axons may be either myelinated or unmyelinated. Myelin is formed by oligodendrocytes in the CNS and by Schwann cells in the periphery. The axon of a neuron transmits information to other neurons or to non-neuronal cells such as muscles.

5. After leaving the CNS, axons run in peripheral nerve trunks which provide structural support.

transmit in a given period. Following the absolute refractory period (which in mammals is usually 0.5–1 ms in duration) a stronger stimulus than normal is required to elicit an action potential. This phase of reduced excitability lasts for about 5 ms and is called the *relative refractory period*.

What mechanisms are responsible for the generation of the action potential? The answer to this important question was provided by a brilliant series of experiments carried out on the giant axon of the squid between 1939 and 1950 by K.C. Cole in the United States and by A.L. Hodgkin and A.F. Huxley in England. Like mammalian nerve cells, the resting membrane potential of the squid axon is negative (about –70 mV), close to the equilibrium potential for potassium ions. Hodgkin and Huxley found that during the action potential the membrane potential briefly reversed in polarity, reaching a peak value of +40 to +50 mV before falling back to its resting level of about –70 mV (Fig. 6.11). This discovery provided an important clue to the underlying mechanism. At the peak of the action potential, the membrane potential is close to the equilibrium potential for sodium ions, unlike the resting membrane potential which lies close the equilibrium potential for potassium ions. This suggested that the action potential resulted from a large increase in the permeability of the axon membrane to sodium ions. This was confirmed when it was shown that removal of sodium ions from the extracellular solution prevented the axon from generating an action potential. In addition, if the sodium concentration in

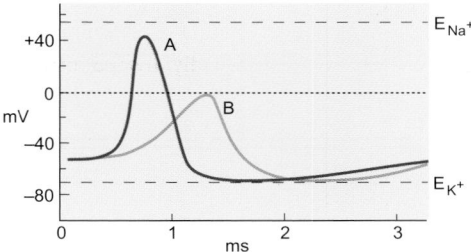

Fig. 6.11 The membrane potential changes that occur during the action potential of the giant axon of the squid in normal seawater (curve A) and when external sodium is reduced by two-thirds (curve B). Note that, at the peak of the action potential, the membrane potential is positive and close to the equilibrium potential for sodium (E_{Na}). Reduction of extracellular sodium both reduces the maximum amplitude of the action potential and slows its time course. The effect is reversible.

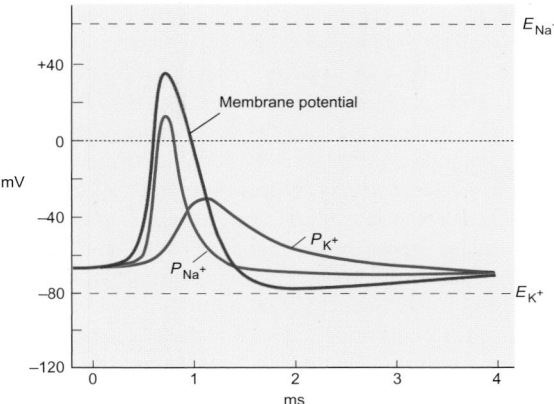

Fig. 6.12 The permeability changes that are responsible for the action potential of the squid giant axon. During the upstroke of the action potential there is a very abrupt but short-lived rise in the membrane's permeability to sodium (P_{Na}), and this is followed by a rise in the membrane permeability to potassium (P_k) that returns to normal as the membrane potential returns to its resting level.

the bathing medium was reduced by two-thirds, the action potential was slower and smaller than normal (Fig. 6.11).

What underlies the change in the permeability of the axon membrane to sodium ions during an action potential? The axonal membrane contains ion channels of a specific type called voltage-gated sodium channels. Following a stimulus, the membrane depolarizes and this causes some of the sodium channels to open, permitting sodium ions to move into the axon down their electrochemical gradient. This inward movement of sodium depolarizes the membrane further, thus leading to the opening of more sodium channels. This results in a greater influx of sodium, which causes a greater depolarization. This process continues until, at the peak of the action potential, the highest proportion of available sodium channels is open and the membrane is, for a brief time, highly permeable to sodium ions. Indeed, at the peak of the action potential, the membrane is more permeable to sodium ions than to potassium ions, and this explains why the membrane potential is positive at the peak of the action potential (see Chapter 4, pp. 41–43).

Why does the membrane repolarize after the action potential? The open state of the sodium channels is unstable and the channels show a time-dependent inactivation. Immediately after the action potential has peaked, the sodium channels begin to inactivate (see Chapter 4) and the permeability of the axon membrane to sodium begins to fall. At the same time, voltage-activated potassium channels begin to open in response to the depolarization and potassium ions leave the axon down their electrochemical gradient. This increases the permeability of the membrane to potassium at the same time as the permeability of the membrane to sodium begins to fall. The decrease in sodium permeability and the increase in potassium permeability combine to drive the membrane potential from its positive value at the peak of the action potential towards the equilibrium potential for potassium ions. As the membrane potential approaches its resting level, the voltage-activated potassium channels close and the membrane potential assumes its resting level. At the resting membrane potential, the inactivated sodium channels revert to their closed state and the axon is primed to generate a fresh action potential. The time course of the changes in the ionic permeability of the axon membrane during the action potential is shown in Fig. 6.12.

This sequence of events explains both the threshold and the refractory period. When a weak stimulus is given, the axon does not depolarize sufficiently to allow enough voltage-gated sodium channels to open to depolarize the membrane further. Such a stimulus is subthreshold. At the threshold, sufficient sodium channels open to permit further depolarization of the membrane as a result of the increased permeability to sodium. The resulting depolarization causes the opening of more sodium channels and the process continues until all available sodium channels are open. Immediately after the action potential, most sodium channels are in their inactivated state and cannot reopen until they have returned to their closed state. To do so, they must spend a brief period at the resting membrane potential. This accounts for the absolute refractory period. One plausible explanation of the relative refractory period of a single nerve fiber is that an action potential can be supported when enough sodium channels have returned to the closed state provided that they are all activated together. This will require a stronger stimulus than normal to ensure that all the sodium channels open at the same time.

The effect of the passage of an action potential along an axon is to leave the nerve with a little more sodium and a little less potassium than it had before. Nevertheless, *the quantities of ions exchanged during a single action potential are very small and are not sufficient to alter the ionic gradients across the membrane*. In a healthy axon, the ionic gradients are maintained by the continuous activity of the sodium pump. If the sodium pump is blocked by a metabolic poison, a nerve fiber is able to conduct impulses only until its ionic gradients are dissipated. Thus the sodium gradient established by the sodium pump indirectly powers the action potential.

How does an action potential start?

While it is convenient to use electrical stimuli to provide a precise and controlled way of exciting an isolated nerve, action potentials in living animals do not arise in this way. So how do they start? In many cases, they arise as a result of specific stimuli arriving at the sense organs. The sense organs convert the external stimulus into a sequence of action potentials that is transmitted to the CNS. The process that activates a particular receptor is known as *sensory transduction*. The details differ from one kind of receptor to another (see Chapter 8). Within the CNS, the action potentials in the sensory nerves activate the neurons involved in processing the appropriate sensory information. This kind of direct activation of neurons occurs continuously as the sense organs report the current state of the environment. As the brain is constantly processing this information, it is perhaps not surprising to discover that the membrane potential of most neurons in the CNS is not stable but fluctuates according to the level of activity in the excitatory and inhibitory nerves impinging on the cell. When all the influences on a particular neuron cause its membrane potential to fall below threshold, an action potential results. There is also some evidence that certain nerve cells have an intrinsic pacemaker activity so that they generate action potentials spontaneously.

How is the action potential propagated along an axon?

When an action potential has been initiated, it is transmitted rapidly along the entire length of an axon. How does this happen? At rest the membrane potential is about –70 mV, while during the peak of the action potential the membrane potential is positive (about +50 mV). Thus, when an action potential is being propagated along an axon, the active zone and the resting membrane will be at different potentials and a small electrical current will flow between the two regions. This forms a local circuit that links the active zone to the neighboring resting membrane (Fig. 6.13), which then depolarizes. This causes sodium channels to open and, when sufficient channels have opened, the action potential invades this part of the membrane. This in turn spreads the excitation farther along the axon where the process is repeated until the action potential has traversed the length of the axon.

In the intact nervous system, an action potential is propagated from its point of origin in one direction only along an axon

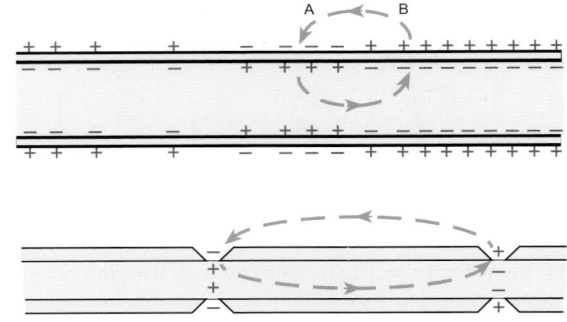

Fig. 6.13 A simplified diagram illustrating the local circuit theory of action potential propagation in unmyelinated and myelinated axons. In unmyelinated axons (top) the local circuit current passes from the active region (point A) to the neighboring resting membrane (point B). The current flow progresses smoothly along the axon membrane without discontinuous jumps. In myelinated fibers (bottom) the current can only cross the axon membrane at the nodes of Ranvier where there are breaks in the insulating layer of myelin. As a result, the action potential is propagated in a series of jumps (known as saltatory conduction).

(orthodromic propagation). Why is this the case? What prevents the retrograde (or antidromic) propagation of an action potential? When an action potential has invaded one part of the membrane and has moved on, it leaves the sodium channels in an inactivated state from which they cannot be reactivated until they have passed through their closed (or resting) state. The time taken for this transition is the absolute refractory period of the axon, during which the membrane cannot support another action potential (see above). By the time the sodium channels have reverted to their closed state, the peak depolarization has passed further along the axon. It is then no longer capable of depolarizing the regions through which it has passed.

The precise way in which an action potential is propagated depends on whether the axon is myelinated or unmyelinated. In unmyelinated axons, the axonal membrane is in direct electrical contact with the extracellular fluid for the whole of its length via

TABLE 6.2 Classification of peripheral nerve fibers according to their function and conduction velocity

Diameter (µm)	Conduction velocity (m s⁻¹)	Fiber classification	Cutaneous nerve classification	Function	
15–20	70–120	Aα	–	Motor	Control of skeletal muscles
15–20	70–120	Aα	Ia	Sensory	Innervation of primary muscle spindles
			Ib		Innervation of Golgi tendon organs
5–10	30–70	Aβ	II	Sensory	Cutaneous senses; secondary innervation of muscle spindles
3–6	15–30	Aγ	–	Motor	Control of intrafusal muscle fibers
2–5	12–30	Aδ	III	Sensory	Cutaneous senses, especially pain and temperature
≈3	3–15	B	–	Motor	Autonomic preganglionic fibers
0.5–1.0	0.5–2.0	C	–	Motor	Autonomic postganglionic nerve fibers
0.5–1.0	0.5–2.0	C	IV	Sensory	Cutaneous senses—pain and temperature

As the table indicates, two separate classifications have been used for axons. That shown in the third column was first introduced by Erlanger and Gasser to label the peaks seen in the compound action potential of peripheral nerves and so classifies axons solely on the basis of their conduction velocity. The classification shown in the fourth column was introduced by Lloyd and Chang. It applies to afferent (sensory) fibers only and is intended to indicate the position and innervation of specific types of receptor. Both systems of classification are in current use.

the mesaxon (see above). Each region depolarized by the action potential results in the spread of current to the immediately adjacent membrane, which then becomes depolarized. This ensures that the action potential propagates smoothly along the axon in a continuous wave. In myelinated axons, the axon membrane is insulated from the extracellular fluid by the layers of myelin, except at the nodes of Ranvier where the axon membrane is in contact with the extracellular fluid. In this case, an action potential at one node of Ranvier completes its local circuit via the next node. This enables the action potential to jump from one node to the next—a process called *saltatory conduction*. This adaptation serves to increase the rate at which the action potential is propagated so that, size for size, myelinated nerve fibers have a much higher conduction velocity than unmyelinated fibers.

What factors determine the conduction velocity of axons?

The speed with which axons conduct action potentials (i.e. their *conduction velocity*) varies from less than $0.5\,\mathrm{m\,s^{-1}}$ for small unmyelinated fibers to over $100\,\mathrm{m\,s^{-1}}$ for large myelinated fibers (see Table 6.2). Why are there such wide variations in conduction velocity and what factors determine the speed of conduction?

The spread of current along an axon from an active region to an inactive region mainly depends on three factors:

(1) the resistance of the axon to the flow of electrical current along its length (its internal electrical resistance);

(2) the electrical resistance of the axon membrane;

(3) the electrical capacitance of the axon membrane.

In unmyelinated fibers, the electrical resistance and capacitance are the same for each unit area of membrane regardless of the diameter of the axon. In these fibers, the spread of current from the active to the neighboring inactive region is determined mainly by the internal electrical resistance of the axon, which decreases as the axon diameter increases. Thus large unmyelinated fibers have a lower internal electrical resistance and conduct action potentials faster than small unmyelinated axons because the spread of current from the active region is greater.

In myelinated fibers, the situation is a little more complicated as the axon membrane is electrically insulated by the layers of myelin. This has two effects:

(1) the electrical resistance of the axon membrane is higher than that of unmyelinated fibers;

(2) the membrane capacitance is lower than that of unmyelinated fibers.

These two factors combine to allow the depolarizing influence of an action potential to spread much farther from the active region. To take advantage of the greater current spread, the axon membrane is only exposed at the nodes of Ranvier. Thus the action potential in a myelinated axon must jump from node to node as described above. The thickness of the myelin and the internodal distance are directly related to fiber size. Large axons have the thickest myelin and greatest internodal distance. As a result, large myelinated axons have the highest conduction velocity.

If myelination of axons confers such advantages in terms of conduction velocity, why are some axons unmyelinated? Myelin is largely made of fats and represents a considerable investment in terms of metabolic energy so that a balance needs to be struck between the requirements for rapid signaling and the cost of maintaining such a sophisticated structure. Not all information needs to be transmitted rapidly and some axons are very short (e.g. those that remain wholly within a particular region of the brain or spinal cord). In these cases, the speed advantages conferred by myelination are not great. Where the speed of conduction is not of paramount importance, axons have either a thin layer of myelin (Aδ fibers) or they are completely unmyelinated (C-fibers) (Table 6.2). Some invertebrates, such as the squid, have very large unmyelinated nerve fibers that can conduct as rapidly as group I mammalian nerve fibers. These fibers are 0.5–1 mm in diameter compared with about 18 μm for a group I fiber with a similar conduction velocity. Therefore a mammalian nerve trunk of the same cross-sectional area as one squid giant axon can contain more than 400 nerve fibers. Thus the evolution of myelinated nerve fibers has permitted many more rapidly conducting fibers to be packed into the amount of space that would be occupied by one giant axon.

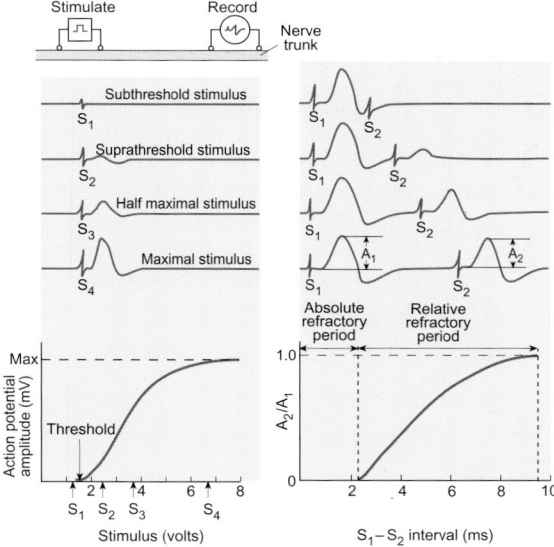

Fig. 6.14 Some characteristic properties of compound action potentials recorded from nerve trunks. Top left shows a simple diagram of the arrangement of the stimulating and recording electrodes. The left panel shows how increasing the strength of the electrical stimulus applied to the nerve gives rise to an action potential that becomes progressively larger in amplitude until it reaches a maximum value. S_1 is a very weak shock that does not elicit an action potential (a *subthreshold* stimulus). S_2–S_4 elicit action potentials of progressively larger amplitude until a maximum is reached (S_4 is called a *maximal* stimulus). The progressive increase in the amplitude of the compound action potential with increasing stimulus strength is shown at the bottom of the panel. The right panel shows the change in excitability that follows the passage of an action potential. Two shocks (S_1 and S_2) of equal intensity are given at various intervals. If S_2 follows S_1 within 2 ms, no fibers are excited (top record—this is the absolute refractory period). As the interval between S_1 and S_2 is increased, more and more fibers are excited by S_2 (middle two records) until all are excited (bottom record). The change in the amplitude of the compound action potential with the interval between the stimuli is shown at the bottom of the panel.

Action potentials can be recorded from intact nerves

While it is possible to record action potentials from individual axons, for diagnostic purposes it is more useful to stimulate a nerve trunk through the skin and record the summed action potentials of all the different fibers present. This summed signal is called a *compound action potential*. Unlike individual nerve fibers, which show an all-or-none response, the action potential of a nerve trunk is graded with stimulus strength (left panel of Fig. 6.14). For weak stimuli below the threshold, no action potential is elicited. Above the threshold, those axons with the lowest threshold are excited first. As the stimulus intensity is increased, more axons are excited and the action potential becomes larger. In this way, the compound action potential grows with increasing stimulus strength until all the available nerve fibers have been excited.

Compound action potentials exhibit both absolute and relative refractory periods. Just as with a single axon, a second stimulus applied within 1–2 ms of the first will fail to elicit an action potential (absolute refractory period). As the interval between successive stimuli is increased, a second action potential is generated which progressively grows in amplitude until it reaches the amplitude of the first (right panel of Fig. 6.14). During the relative refractory period, more and more of the fibers in the nerve trunk recover their excitability as the interval between the two stimuli increases.

When a long length of nerve is stimulated and the compound action potential is recorded well away from the stimulating electrode, a number of peaks become visible in the record (Fig. 6.15). These different peaks reflect differences in the conduction velocity of the different axons present within the nerve trunk. Axons can be classified on the basis of their conduction velocity and physiological role (Table 6.2). The axons with the highest conduction velocity are motor fibers and the sensory

fibers involved in motor control (see Chapter 9). Unmyelinated fibers (sometimes called C-fibers) have the slowest conduction velocity and serve to transmit sensory information (e.g. pain and temperature) to the CNS.

Some diseases such as multiple sclerosis are characterized by a loss of myelin from axons. These diseases, which are known as *demyelinating diseases*, can be diagnosed by measuring the conduction velocity of peripheral nerves. Affected nerves have an abnormally slow conduction velocity. The loss of myelin affects conduction velocity because the normal pattern of current spread is disrupted. Action potentials are propagated through the affected region by continuous conduction similar to that seen in small unmyelinated fibers rather than by saltatory conduction as seen in healthy fibers. In severe cases, there may even be a total failure of conduction. In either eventuality, the function of the affected pathways is severely impaired.

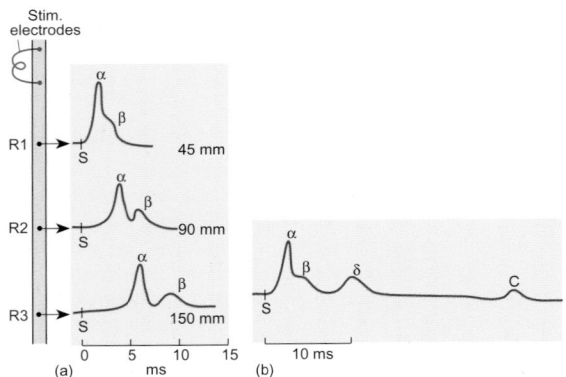

Fig. 6.15 (a) A compound action potential recorded at different points along an intact nerve (shown as R1–R3 in the figure). Note that as the action potential travels along the nerve it breaks up into two distinct components that propagate at different velocities (marked α and β—the slower components are omitted for clarity). (b) All the components of the compound action potential recorded at position R1. Each wave reflects the activity of a group of fibers with a similar conduction velocity. The fastest are the α-fibers, which have a maximum conduction velocity of about 100 m s⁻¹, and the slowest are the C-fibers which have a conduction velocity of about 1.3 m s⁻¹.

Summary

1. Nerve cells transmit information along their axons by means of action potentials. This enables them to transmit signals rapidly over considerable distances.

2. Action potentials are generated when a neuron is activated by a stimulus of a certain minimum strength known as the threshold. With stimuli above threshold, each action potential has approximately the same magnitude and duration. This is known as the 'all-or-none' law.

3. The action potential is caused by a large, short-lived increase in the permeability of the plasma membrane to sodium ions. The increase in sodium permeability is caused by the opening of voltage-gated sodium channels. Sodium channels spontaneously inactivate a short time after they have opened and this limits the duration of the action potential to about 1 ms. Voltage-gated sodium channels cannot reopen until they have been reprimed by spending a period at the resting membrane potential.

4. After the passage of an action potential, the axon cannot propagate another until sufficient sodium channels have returned to their resting state. This period of inexcitability is known as the absolute refractory period. The axon is less excitable than normal for a few milliseconds after the absolute refractory period has ended. This period is known as the relative refractory period.

5. Large-diameter axons conduct faster than small-diameter axons, and myelinated axons conduct impulses faster than unmyelinated axons. Myelinated axons conduct impulses by saltatory conduction.

6. An electrical stimulus applied to a nerve trunk elicits a compound action potential, which is the summed activity of all the nerve fibers present in the nerve trunk. Compound action potentials grow in amplitude as the stimulus strength is increased above the threshold until all the axons in the nerve are recruited.

6.4 Chemical synapses

When an axon reaches its target cell, it forms a specialized junction known as a *synapse*. The nerve cell that gave rise to the axon is called the *presynaptic neuron* and the target cell is called the *postsynaptic cell*. The target cell may be another neuron, a muscle cell, or a gland cell. A synapse has the function of transmitting the information coded in a sequence of action potentials to the postsynaptic cell so that it responds in an appropriate way. This section is concerned with the fundamental properties of chemical synaptic transmission, both between nerve cells and from nerve cells to effector organs such as skeletal muscle and secretory glands.

Synaptic transmission is a one-way signaling mechanism—information flows from the presynaptic to the postsynaptic cell and not in the other direction. When the activity of the postsynaptic cell is increased, the synapse is called an *excitatory synapse*. Conversely, when activity in the presynaptic neuron leads to a fall in activity of the postsynaptic cell, the synapse is called an *inhibitory synapse*. In mammals, including man, most synapses operate by the secretion of a small quantity of a chemical (a neurotransmitter) from the nerve terminal. This type of synapse is called a *chemical synapse*.

In some instances, a synapse operates by transmitting the electrical current generated by the action potential to the postsynaptic cell via gap junctions. This type of synapse is called an *electrical synapse*. Electrical synapses are found in the retina and in some invertebrates such as the crayfish.

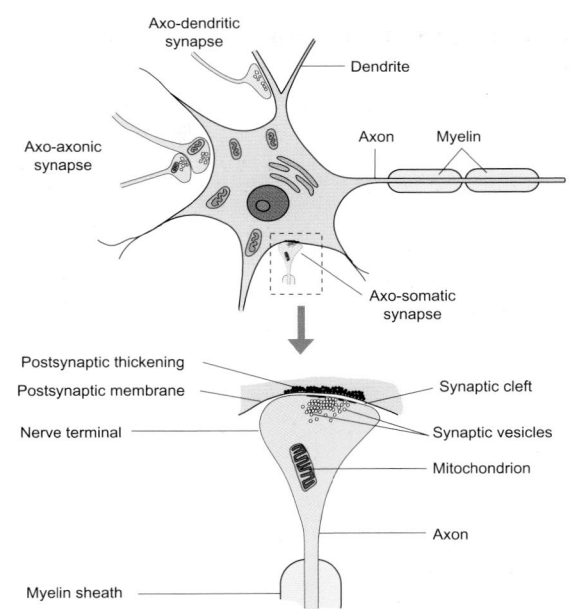

Fig. 6.16 A diagrammatic view of the principal types of synaptic contact in the CNS and the detailed structure of a typical CNS chemical synapse.

TABLE 6.3	Some neurotransmitters and their receptors			
Class of compound	**Specific example**	**Receptor types**	**Physiological role**	**Mechanism of action**
Ester	Acetylcholine	Nicotinic	Fast excitatory synaptic transmission especially at neuromuscular junction	Activates ion channels
		Muscarinic	Both excitatory and inhibitory slow synaptic transmission depending on tissue, e.g. slowing of heart; contraction of visceral smooth muscle	Acts via G protein
Monoamine	Norepinephrine	Various α and β adrenoceptors	Slow synaptic transmission in CNS and smooth muscle	Acts via G protein
	Serotonin (5-HT)	Various 5HT receptors (e.g. $5HT_{1A}$, $5HT_{2A}$, etc)	Slow synaptic transmission in CNS and periphery (smooth muscle and gut)	Acts via G protein
		$5HT_3$	Fast excitatory synaptic transmission	Activates ion channels
	Dopamine	D_1, D_2 receptors	Slow synaptic transmission in CNS and periphery (blood vessels and gut)	Acts via G protein
Amino acid	Glutamate	AMPA	Fast excitatory synaptic transmission in CNS	Activates ion channels
		NMDA	Slow excitatory synaptic transmission in CNS	Activates ion channels
		Metabotropic	Neuromodulation	Acts via G protein
	GABA	$GABA_A$	Fast inhibitory synaptic transmission in CNS	Activates ion channels
		$GABA_B$	Slow inhibitory synaptic transmission in CNS	Acts via G protein
Peptide	Substance P	NK_1	Slow excitation of smooth muscle and neurons in CNS	Acts via G protein
	Enkephalins	μ/δ-Opioid	Slow synaptic signaling (reduction in excitability)	Act via G protein
			Decrease in gut motility	
			Cause analgesia	
	β-Endorphin	κ-Opioid	Slow synaptic signaling	Acts via G protein
			Analgesia	

The structure of chemical synapses

When an axon reaches its target cell, it loses its myelin sheath and ends in a small swelling known as a *nerve terminal* or synaptic bouton (Fig. 6.16). A nerve terminal together with the underlying membrane on the target cell constitutes a synapse. The nerve terminals contain mitochondria and a large number of small vesicles known as synaptic vesicles. The membrane immediately under the nerve terminal is called the *postsynaptic membrane*. It contains electron-dense material which makes it appear thicker than that of the plasma membrane outside the synaptic region. This is known as postsynaptic thickening (Fig. 6.16). The postsynaptic membrane contains specific receptor molecules for the neurotransmitter released by the nerve terminal. There is a small gap of about 20 nm, which is known as the *synaptic cleft*, between a nerve terminal and the postsynaptic membrane.

Nerve cells employ a wide variety of signaling molecules as neurotransmitters, including acetylcholine, norepinephrine (noradrenaline), glutamate, γ-amino butyric acid (GABA), serotonin (5-HT), and many peptides such as substance P and enkephalins (Table 6.3). Chemical analysis of nerve terminals isolated from the brain has shown that they contain high concentrations of neurotransmitters. By careful subcellular fractionation of nerve terminals, the synaptic vesicles can be separated from the other intracellular organelles. Analysis of the contents of the isolated vesicles has shown that they contain almost all of the neurotransmitter present in the terminal.

How does a chemical synapse work?

The transmission of information across a synapse occurs when an action potential reaches the presynaptic nerve terminal. The nerve

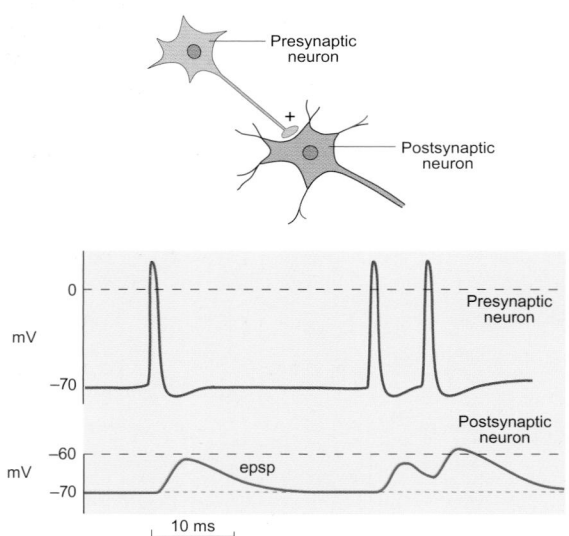

Fig. 6.17 Diagram of the relationship between an action potential in a presynaptic neuron and the generation of an excitatory postsynaptic potential (epsp) in the postsynaptic neuron. Note that if two epsps occur in rapid succession they summate and the final degree of depolarization is greater than either epsp could achieve on its own (temporal summation—see also Fig. 6.19). If the depolarization reaches the threshold, an action potential will occur in the postsynaptic neuron.

terminal becomes depolarized and this depolarization causes voltage-gated calcium channels in the presynaptic membrane to open. Calcium ions flow into the nerve terminal down their electrochemical gradient and the consequential rise in free calcium triggers the fusion of one or more synaptic vesicles with the presynaptic membrane, resulting in the secretion of neurotransmitter into the synaptic cleft. In nerve terminals, this secretory process is extremely rapid and occurs within 0.5 ms of the arrival of the action potential. The secreted transmitter diffuses across the synaptic cleft and binds to receptors on the postsynaptic membrane. Subsequent events depend on the kind of receptor present. If the transmitter activates a ligand-gated ion channel, synaptic transmission is usually both rapid and short lived. This type of transmission is called fast synaptic transmission and is typified by the action of acetylcholine on the neuromuscular junction (see Section 6.5). If the neurotransmitter activates a G protein-linked receptor, the change in the postsynaptic cell is much slower in onset and lasts much longer. This type of synaptic transmission is called slow synaptic transmission. A typical example is the excitatory action of norepinephrine on α_1-adrenoceptors in the peripheral blood vessels.

Fast excitatory synaptic transmission occurs when a neurotransmitter (e.g. acetylcholine or glutamate) is released from the presynaptic nerve ending and is able to bind to and open nonselective cation channels. The opening of these channels causes a brief *depolarization* of the postsynaptic cell. This shifts the membrane potential closer to the threshold for action potential generation and so renders the postsynaptic cell more excitable. When the postsynaptic cell is a neuron, the depolarization is called an *excitatory post-synaptic potential* or *epsp*. A single epsp occurring at a fast synapse reaches its peak value within 1–5 ms of the arrival of the action potential in the nerve terminal and decays to nothing over the ensuing 20–50 ms (Fig. 6.17).

How does activation of a non-selective cation channel lead to depolarization of the postsynaptic membrane? The membrane potential is determined by the distribution of ions across the plasma membrane and the permeability of the membrane to those ions (see Chapter 4). At rest, the membrane is much more permeable to potassium ions than it is to sodium ions. Therefore the membrane potential (about –70 mV) is close to the equilibrium potential for potassium (about –80 mV). However, if the membrane were equally permeable to sodium and potassium ions, the membrane potential would be close to zero (i.e. the membrane would be depolarized). Consequently, when a neurotransmitter such as acetylcholine opens a non-selective cation channel in the postsynaptic membrane, the membrane depolarizes at the point of excitation. The exact value of the depolarization will depend on how many channels have been opened, as this will determine how far the membrane's permeability to sodium ions has increased relative to that of potassium. In neurons, single epsps rarely exceed a few millivolts, while at the neuromuscular junction the synaptic potential (known as the *end-plate potential*—see p. 78) has an amplitude of about 40 mV.

Fast inhibitory synaptic transmission occurs when a neurotransmitter such as GABA or glycine is released from a presynaptic nerve ending and is able to activate chloride channels in the postsynaptic membrane. The opening of these channels causes the postsynaptic cell to become *hyperpolarized* for a brief

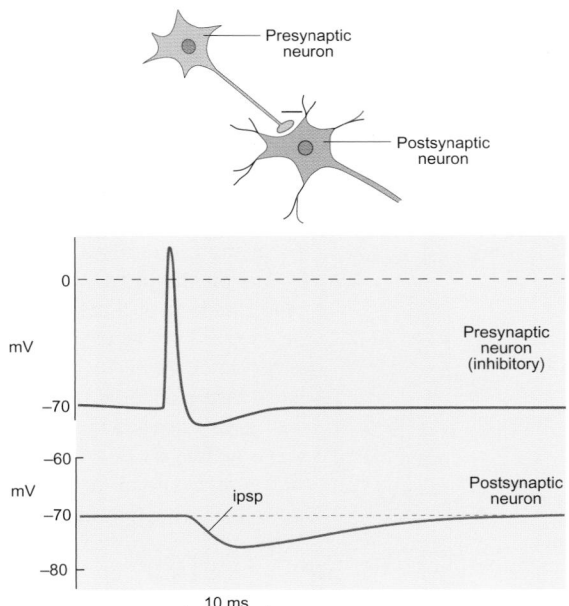

Fig. 6.18 Diagram to show the relationship between an action potential in a presynaptic neuron and the generation of an inhibitory postsynaptic potential (ipsp) in the postsynaptic neuron. Note that in this case the membrane potential becomes more negative (it hyperpolarizes) and thus moves further away from the threshold for action potential generation. This leads to a decrease in the excitability of the postsynaptic neuron.

period. This negative shift in membrane potential is called an *inhibitory postsynaptic potential* or *ipsp* as the membrane potential is moved further away from threshold. A single ipsp occurring at a fast synapse reaches its peak value within 1–5 ms of the arrival of the action potential in the nerve terminal and decays to nothing within a few tens of milliseconds (Fig. 6.18).

Why does the membrane hyperpolarize during an ipsp? The resting membrane potential is less (at about –70 mV) than the equilibrium potential for potassium ions (which is about –80 mV) as the membrane has a small permeability to sodium ions. However, the distribution of chloride ions across the plasma membrane mirrors that of potassium (see Chapter 4) so that the chloride equilibrium potential lies close to that of potassium. When an inhibitory neurotransmitter such as GABA opens chloride channels, this increases the permeability of the membrane to chloride ions which flow into the cell down their electrochemical gradient. This increase in chloride permeability shifts the membrane potential towards a more negative value that is closer to the equilibrium potential for chloride ions. The extent of the hyperpolarization will depend on how much the chloride permeability has increased relative to that of both potassium and sodium. Ipsps generally have a small amplitude, about 1–5 mV. Nevertheless, they tend to last for tens of milliseconds and play an important role in determining the membrane potential of neurons.

Epsps and ipsps are not all-or-none events but are graded with the intensity of activation. Consequently, they can be superimposed on each other (a process called *summation*). If a synapse is activated repeatedly, the resulting synaptic potentials become superimposed (Fig. 6.19ii, iii). This phenomenon is called *tempo-*

ral summation. As individual neurons receive very many contacts, it is possible for two synapses on different parts of the cell to be activated at the same time. The resulting synaptic potentials also summate but, because they were activated at different points on the cell, this type of summation is called *spatial summation* (Fig 6.19iv, v).

The changes in the membrane potential of a neuron due to the synaptic activity occurring all over its surface determine its ability to generate an action potential. Within the CNS and autonomic ganglia, the epsps that occur as a result of activation of individual synapses are small—often only a few millivolts in amplitude. If the cell is already relatively depolarized by the activity of various excitatory synapses, activation of a further set of excitatory synapses may depolarize the cell sufficiently for the membrane potential to reach the threshold for action potential generation. The cell will be excited and will transmit an action potential to its target cells. If the cell is already hyperpolarized by ipsps following the activation of inhibitory synapses, the same stimulus may not be sufficient to trigger an action potential.

Presynaptic inhibition

As Fig. 6.16 shows, synapses may occur between a nerve terminal and the cell body of the postsynaptic cell (axo-somatic synapses), between a nerve terminal and a dendrite on the postsynaptic cell (axo-dendritic synapses), and between a nerve terminal and the terminal region of another axon (axo-axonic synapses). Axo-axonic synapses are generally believed to be inhibitory, and activation of an axo-axonic synapse prevents an action potential invading the nerve terminal of the postsynaptic axon. This leads to a blockade of synaptic transmission at the affected synapse. This blockade is called *presynaptic inhibition* to distinguish it from the inhibition that results from an ipsp occurring in a postsynaptic neuron (*postsynaptic inhibition*). Unlike postsynaptic inhibition, which changes the excitability of the postsynaptic cell, presynaptic inhibition permits the selective blockade of a specific synaptic connection without altering the excitability of the postsynaptic neuron.

Neurotransmitters may directly activate ion channels or act via second-messenger systems

As with other kinds of chemical signaling, synaptic signaling is mediated by a wide variety of substances. They can be grouped into six main classes:

(1) esters, such as acetylcholine (ACh);

(2) monoamines, such as norepinephrene, dopamine, and serotonin;

(3) amino acids, such as glutamate and GABA;

(4) purines, such as adenosine and ATP;

(5) peptides, such as the enkephalins, substance P, and vasoactive intestinal polypeptide (VIP);

(6) inorganic gases, such as NO.

Most of these transmitters activate both ion channels and G protein-linked receptors as shown in Table 6.3, which also includes a brief outline of the role of various types of receptor.

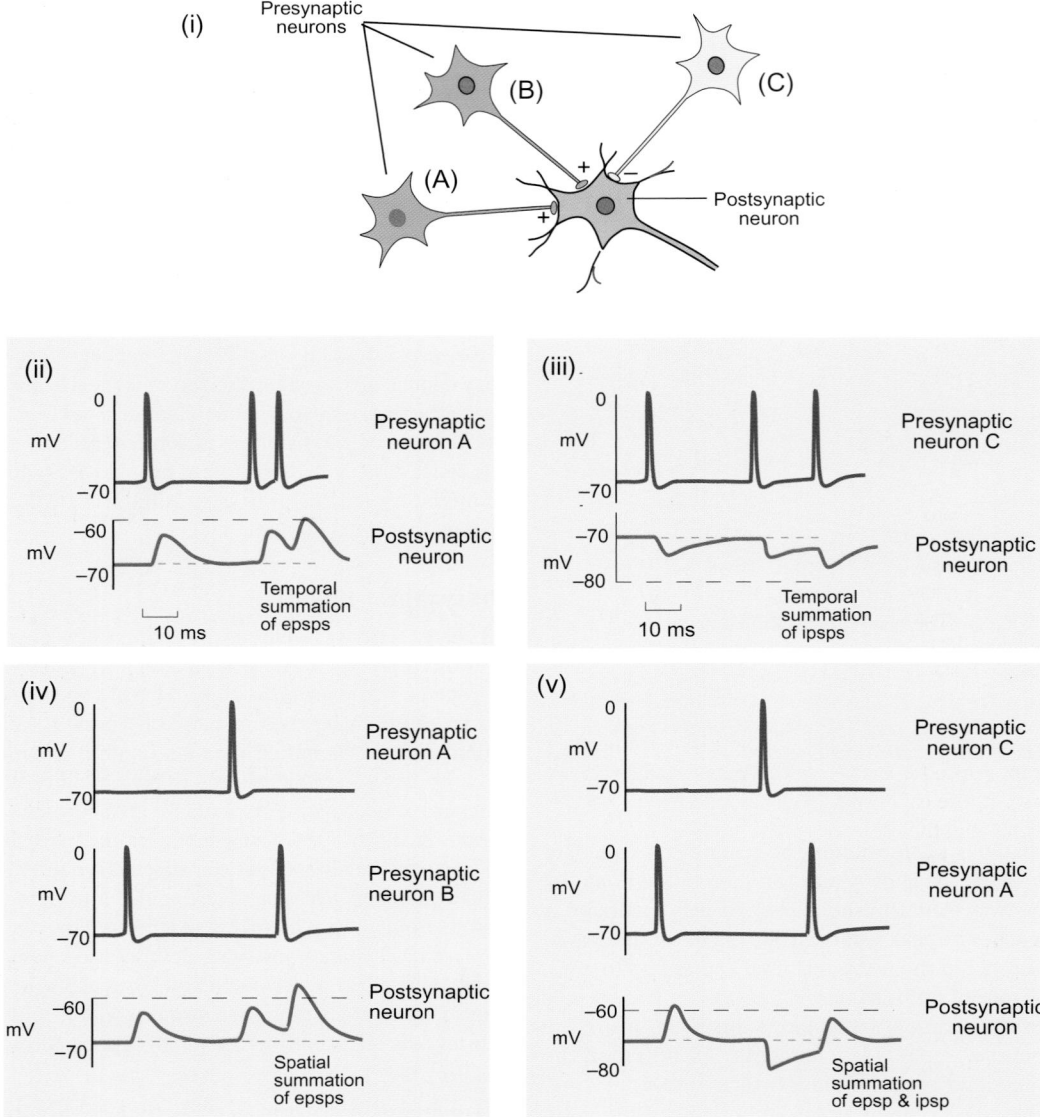

Fig. 6.19 A simple diagram to illustrate the temporal and spatial summation of synaptic potentials. Consider a group of four neurons such as those at the top of the figure. Action potentials in the presynaptic neurons (A–C) elicit epsps and ipsps in the postsynaptic neuron. If two epsps occur in quick succession at a particular synaptic contact, they summate as shown in (ii) (temporal summation). Ipsps also show temporal summation (iii). If epsps occur at different synaptic contacts within a short time of one another, they exhibit spatial summation (iv). Epsps and ipsps can also undergo spatial summation as shown in panel (v).

Slow synaptic transmission plays a major role in regulating the internal environment

Like fast synaptic transmission, slow synaptic excitation and inhibition result from changes in the relative permeability of the membrane to the principal extracellular ions. They differ in their time course because the channels that are responsible for slow synaptic transmission are regulated over a longer timescale by both second messengers such as cyclic AMP and G protein subunits (see Chapter 5).

Fast and slow synaptic transmission serve different functions. In those synapses that support fast synaptic excitation or inhibition, the ligand-gated channels lie under the nerve terminal and are rapidly activated by high concentrations of neurotransmitter. The

high concentration is achieved by the release of a small quantity of neurotransmitter into the narrow synaptic cleft. As the transmitter diffuses away from the synaptic cleft, its concentration falls rapidly and is too low to affect neighboring cells. Therefore this type of transmission is highly specific for the contact between a presynaptic neuron and its target cell and is well adapted to serve a role in the rapid processing of sensory information and the control of locomotion.

In slow synaptic transmission the neurotransmitters are usually not secreted onto a specific point on the postsynaptic cell. Instead, the nerve fibers have a number of swellings (varicosities) along their length that secrete neurotransmitter into the extracellular fluid close to a number of cells. The synapses of the autonomic

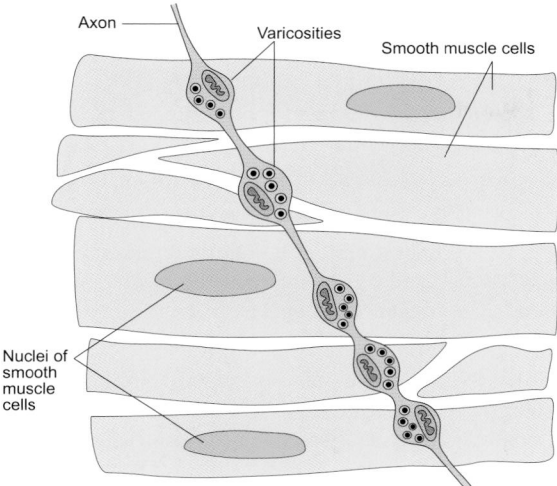

Fig. 6.20 A diagram of the axonal varicosities of adrenergic autonomic nerve fibers. Once an axon has reached its target, it courses through the tissue (in this case smooth muscle) and forms a series of varicosities, each of which contains neurotransmitter (norepinephrine) packaged in vesicles. Unlike most CNS synapses or the skeletal neuromuscular junction, the autonomic fibers release their neurotransmitter into the extracellular space in a diffuse manner where it affects a number of muscle cells.

Summary

1. At a chemical synapse the axon terminal is separated from the postsynaptic membrane by the synaptic cleft. In response to an action potential the nerve terminal secretes a neurotransmitter into the synaptic cleft. The secretion of neurotransmitter occurs by calcium-dependent exocytosis of synaptic vesicles.

2. The neurotransmitter rapidly diffuses across the synaptic cleft and binds to specific receptors to cause a short-lived change in the membrane potential of the postsynaptic cell.

3. Activation of an excitatory synapse causes depolarization of the membrane potential which is called an excitatory postsynaptic potential or epsp. Activation of an inhibitory synapse leads to hyperpolarization of the membrane potential, which is called an inhibitory postsynaptic potential or ipsp.

4. Epsps and ipsps are graded in intensity. They greatly outlast the action potentials that initiated them. As a result, synaptic potentials can be superimposed on each other, leading to temporal and spatial summation.

5. Many different kinds of chemical can serve as neurotransmitters. Examples are acetylcholine, norepinephrine and peptides such as VIP. Some nerve terminals secrete more than one kind of neurotransmitter, and some neurotransmitters activate more than one kind of receptor at the same synapse.

nervous system are of this kind (Fig. 6.20). Slow synaptic transmission is of great importance for the control of such varied functions as the cardiac output, the caliber of blood vessels, and the secretion of hormones. Within the CNS slow synaptic transmission may underlie changes of mood and the control of appetites, e.g. hunger and thirst.

Dual transmission

Some nerve terminals are known to contain two different kinds of neurotransmitter. When such a nerve ending is activated, both neurotransmitters may be released. This is called cotransmission. As an example, the parasympathetic nerves of the salivary glands release both acetylcholine and VIP when they are activated. In this case, the acetylcholine acts on the acinar cells to increase secretion and the VIP acts on the smooth muscle of the arterioles to increase the local blood flow.

What limits the duration of action of a neurotransmitter?

Neurotransmitters are highly potent chemical signals that are secreted in response to very specific stimuli. If the effect of a particular neurotransmitter is to be restricted to a particular synapse at a given time, there needs to be some means of terminating its action. This can be achieved in three different ways:

(1) by rapid enzymatic destruction;

(2) by uptake either into the secreting nerve terminals or into neighboring cells;

(3) by diffusion away from the synapse followed by enzymatic destruction, uptake, or both.

Of the currently known neurotransmitters, only acetylcholine is inactivated by rapid enzymatic destruction. It is hydrolyzed to acetate and choline by the enzyme acetylcholinesterase. The acetate and choline are not effective in stimulating the cholinergic receptors but they may be taken up by the nerve terminal and resynthesized into new acetylcholine. The role of acetylcholine as a fast neurotransmitter at the neuromuscular junction will be discussed below in more detail.

The monoamines (such as norepinephrine) are inactivated by uptake into the nerve terminals where they may be re-incorporated into synaptic vesicles for subsequent release. Any monoamine that is not removed by uptake into a nerve terminal is metabolized either by monoamine oxidase or by catechol-O-methyl transferase. These enzymes are present in nerve terminals. They are also found in other tissues such as the liver.

Peptide neurotransmitters become diluted in the extracellular fluid as they diffuse away from their site of action. They are subsequently destroyed by extracellular peptidases. The amino acids released in the process are taken up by the surrounding cells where they enter the normal metabolic pathways.

6.5 Neuromuscular transmission is an example of fast synaptic signaling at a chemical synapse

The neurons that are directly responsible for controlling the activity of muscle fibers are known as *motoneurons*. The nerves that transmit signals from the CNS to the skeletal muscles are known as *motor nerves* and the process of transmitting a signal from a motor nerve to a skeletal muscle to cause it to contract is called *neuromuscular transmission*. The individual axons of the motor nerves that supply skeletal muscle are myelinated and branch as they enter a muscle, so that each individual moto-neuron controls the activity of a number of muscle fibers. The motoneuron and the muscle fibers it controls form a *motor unit* and an action potential in the motoneuron will cause the contraction of all the muscle fibers to which it is connected.

The region of contact between a motor axon and a muscle fiber is called the *neuromuscular junction* or *motor end-plate*. As an axon approaches its terminal, it loses its myelin sheath and runs along grooves in the surface of the muscle membrane. The terminals of motor axons contain mitochondria and large numbers of synaptic vesicles, which contain the neurotransmitter acetylcholine (ACh). Beneath the axon terminal, the muscle membrane is thrown into elaborate folds known as junctional folds. This is the postsynaptic region of the muscle fiber membrane and it contains the nicotinic receptors that bind the acetylcholine. Finally, as with other chemical synapses, there is a small gap of about 20 nm, the synaptic cleft, separating the nerve membrane from that of the muscle fiber. The synaptic cleft contains acetylcholinesterase, which rapidly inactivates acetylcholine by breaking it down to acetate and choline. A diagram of the neuromuscular junction is shown in Figure 6.21.

The neuromuscular junction operates in a similar way to other chemical synapses:

• First, an action potential in the motor axon invades the nerve terminal and depolarizes it.

• This depolarization opens voltage-gated calcium channels in the terminal membrane and calcium ions flow into the nerve terminal down their electrochemical gradient.

• This leads to a local rise in free calcium within the terminal, which triggers the fusion of docked synaptic vesicles with the plasma membrane.

• The acetylcholine contained within the synaptic vesicles is released into the synaptic cleft.

• Acetylcholine molecules diffuse across the cleft and bind to the nicotinic receptors on the post-synaptic membrane.

• When the receptors bind acetylcholine, non-selective cation channels open and this depolarizes the muscle membrane in the end-plate region. This depolarization is called the *end-plate potential* or epp.

• Finally, when the epp has reached threshold, the muscle membrane generates an action potential that propagates along the length of the fiber. This action potential triggers the contraction of the muscle fiber—a process called excitation–contraction coupling (see Chapter 7 p. 86).

The epp is confined to the end-plate region of the muscle fiber (Fig. 6.22). If the electrical activity of the muscle membrane at the end-plate is examined closely, small spontaneous depolarizations of about 1 mV are observed even in the absence of action potentials in the motor nerve. These small depolarizations are similar in time course to the epp and are called *miniature end-plate potentials* or *mepps*. Each mepp is thought to reflect the release of the acetylcholine contained within a single synaptic vesicle. The mepps occur only in the junctional region and are both random and relatively infrequent in the resting muscle membrane. However, if the nerve terminal is artificially depolarized, the frequency of mepps increases. In 1952, this led B. Katz and P. Falt to suggest that depolarization of the motor nerve terminal triggers the simultaneous release of many synaptic vesicles to give rise to the epp. This is known as the vesicular hypothesis of transmitter release. Subsequent work has general-

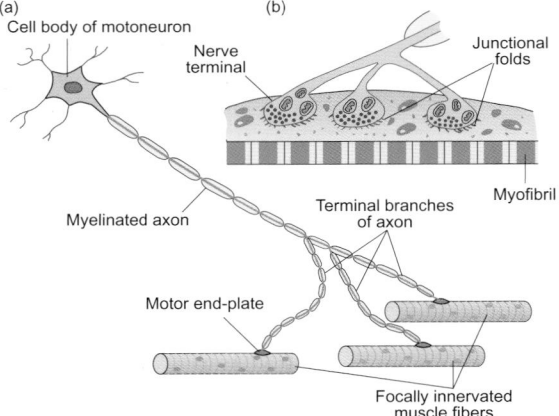

Fig. 6.21 A diagram of the innervation of mammalian skeletal muscle and the structure of the neuromuscular junction. (a) The general organization of a motor unit. A motoneuron in the CNS gives rise to a motor axon, which branches to supply a number of muscle fibers. The motoneuron and the muscle fibers it supplies form a single motor unit. (b) A schematic drawing showing a neuromuscular junction. Note the folds under each nerve ending (the junctional folds) and the large number of synaptic vesicles in the nerve endings.

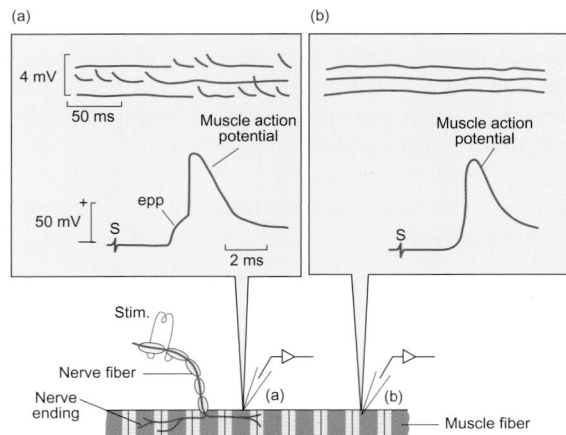

Fig. 6.22 A schematic drawing illustrating the electrical activity in the membrane of frog skeletal muscle fibers. Stimulating electrodes are applied to the motor nerve fiber and intracellular recording electrodes record the changes in the membrane potential of the muscle at the neuromuscular junction (a) and at a point remote from the junctional region (b). Note that small spontaneous randomly occurring mepps occur in the junctional region but not elsewhere and that the muscle action potential is preceded by the end-plate potential (epp) only in the junctional region.

ly supported this idea. Indeed, it has proved possible to capture a picture of the fusion of synaptic vesicles with the plasma membrane of the nerve terminal by rapidly freezing the neuromuscular junction at the moment of transmitter release.

After the acetylcholine has activated the nicotinic receptors it is rapidly broken down by acetylcholinesterase to acetate and choline. This enzymatic breakdown limits the action of acetylcholine to a few milliseconds. The acetate and choline can be taken up by the nerve terminal and resynthesized into acetylcholine for recycling as a neurotransmitter. If acetylcholinesterase is inhibited by a specific blocker such as *eserine*, activation of the motor nerve leads to a maintained depolarization of the muscle and neuromuscular transmission is blocked.

The nicotinic acetylcholine receptors of the neuromuscular junction are amongst the most well characterized of all receptors. It is now clear that the binding site for acetylcholine is part of the same molecular complex as the ion channel. Thus the binding of acetylcholine to the receptor directly opens the ion channel without an intervening step—an ideal adaptation for fast synaptic transmission. The receptors can be blocked by a number of specific drugs and poisons (called neuromuscular blockers), the best known of which is curare—an arrow poison used by South American Indians for hunting. Neuromuscular blockers such as D-tubocurarine and succinylcholine are now routinely used in conjunction with general anesthetics to prevent muscle contractions and so provide complete muscular relaxation during complex surgery. Since the respiratory muscles also rely on acetylcholine to trigger their contractions, this way of providing muscle relaxation is only possible while the patient's breathing is being supported by artificial respiration.

Denervation of skeletal muscle leads to atrophy and supersensitivity of the membrane to acetylcholine

If the motor nerve to a muscle is cut or crushed, the muscle becomes paralyzed. After a time the nerve terminal degenerates and disappears and the end-plate region becomes less sensitive to acetylcholine. The muscle becomes weak and atrophies.

Except at the end-plate region, the muscle membrane is not normally sensitive to acetylcholine, but following denervation and degeneration of the end-plate, the non-junctional region becomes sensitive to acetylcholine. This is known as *denervation supersensitivity* and it shows that the motor nerve exerts an influence on the muscle in addition to signaling. If the nerve regenerates and the end-plate is re-formed, the supersensitivity disappears and the muscle progressively regains its original strength. This type of influence of a nerve on its target cell is known as a *trophic action*, but the factors that are responsible for this type of interaction are, at present, largely unknown.

Functional denervation occurs in a disease of the neuromuscular junction called *myasthenia gravis*. Patients with this disease make antibodies against their own nicotinic receptors. The disease itself is characterized by progressive muscular weakness, especially of the cranial muscles, and drooping of the eyelids is a characteristic early sign. The circulating antibodies bind to the nicotinic receptors of the neuromuscular junction and so reduce the number available for neuromuscular transmission. The decline in the number of active receptors leads to a progressive failure of neuromuscular transmission and progressive paralysis

of the affected muscles. The symptoms of the disease can be ameliorated by giving the sufferer a low dose of an inhibitor of acetylcholinesterase (an anticholinesterase), such as neostigmine, to prolong the activity of the acetylcholine.

Fast and slow axonal transport are used to provide the distal parts of the axon with newly synthesized proteins and organelles

In an adult man or woman there is often a great distance between a motor end-plate and the cell body of the motoneuron that gives rise to the axon. This raises an important issue. How do proteins and other membrane constituents that are synthesized in the cell body reach the nerve terminal? The first experiments to answer this question were performed by P. Weiss in 1948 when he tied a ligature around the sciatic nerve of a rat so that it occluded the nerve without totally crushing it. After a few weeks, the axons had become swollen on the proximal side of the ligature (the side nearest the cell body), indicating that material had accumulated at the constriction (Fig. 6.23). When the ligature was removed, the accumulated material was found to progress towards the axon terminal at 1–2 mm a day. Weiss concluded that the cell body exported material to the distal parts of the axon. Subsequent studies have shown that axons transport materials in three ways:

(1) by fast anterograde axonal transport, i.e. from the cell body towards the nerve terminal (this occurs at up to 400 mm a day);

(2) by slow anterograde axonal transport (0.2–4 mm a day);

(3) by fast retrograde axonal transport, i.e. from the terminal to the cell body (this takes place at a rate of about 200–300 mm a day).

All newly synthesized organelles are exported to the axon and the distal dendrites by fast anterograde transport. This includes the synaptic vesicles and precursor molecules for peptide neurotransmission. Slow anterograde axonal transport is used to export both soluble cytoplasmic and cytoskeletal proteins. Both slow and fast anterograde axonal transport are too rapid to occur by diffusion along the length of the axon and it is known that axonal transport occurs along the microtubules. Drugs such as colchicine that disrupt the microtubules also inhibit axonal transport.

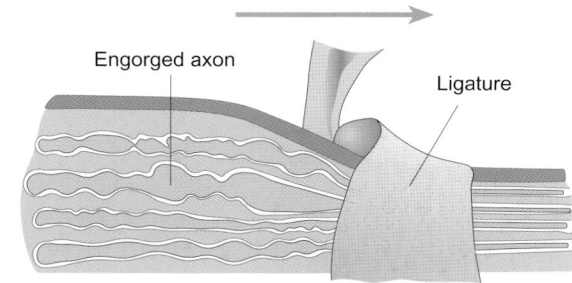

Fig. 6.23 The effect of compression of a localized region of the sciatic nerve of a rat caused by a ligature. Note the engorged and distorted appearance of the axons proximal to the ligature. The arrow indicates the direction of axoplasmic flow.

Retrograde axonal transport from the nerve terminal to the cell body enables neurons to recycle materials used by the nerve terminal during synaptic transmission. While retrograde axonal transport allows the recycling of structural components, it also allows the transport of growth factors (e.g. nerve growth factor) from the nerve terminal to the cell body. This permits a two-way traffic between nerves and their target tissues and may play an important role during the development of the nervous system. Retrograde transport is exploited by some viruses to gain entry to the CNS (e.g. herpes simplex, polio, and rabies). Tetanus toxin also gains entry to the CNS via retrograde transport.

Summary

1. The neuromuscular junction of skeletal muscle is a classic example of a fast chemical synapse. The motor nerve endings release acetylcholine, which activates nicotinic cholinergic receptors to depolarize the muscle membrane. The depolarization is known as an end-plate potential or epp.

2. The effect of acetylcholine is terminated by the enzyme acetylcholinesterase. If this enzyme is inhibited, the muscle membrane becomes depolarized and neuromuscular transmission is blocked.

3. Neuromuscular transmission can also be blocked by drugs such as curare that compete with acetylcholine for binding sites on the nicotinic receptors.

4. The epp triggers an action potential in the muscle membrane that leads to contraction of the muscle. This is called excitation–contraction coupling.

5. Axons transport materials required for the normal function of the nerve terminal by anterograde axoplasmic flow and recover expended materials by retrograde transport.

Recommended reading

Anatomy and histology of the nervous system

Brodal, P. (2003). *The central nervous system. Structure and function* (3rd edn). Oxford University Press, Oxford.

Kiernan, J.A. (2004). *Barr's The human nervous system: an anatomical viewpoint* (8th edn). Lippincott–Williams & Wilkins, Baltimore, MD.

Pharmacology of synaptic transmission

Rang, H.P., Dale, M.M., Ritter, J.M., and Moore, P. (2005). *Pharmacology* (5th edn), Chapters 31–33. Churchill-Livingstone, Edinburgh.

Physiology

Alberts, B., Johnson, A., Lewis, J., Raff, M., Roberts, K., and Walters, P. (2002). *Molecular biology of the cell.* (4th edn), pp. 637–57. Garland, New York.

Katz, B. (1966). *Nerve, muscle, and synapse.* McGraw-Hill, New York. [A classic account of the physiology of nerves and muscle by one of the great figures in the field.]

Nicholls, J.G., Martin, A.R., Wallace, B.G., and Fuchs, P.A. (2001). *From neuron to brain* (4th edn), Chapters 6–14. Sinauer, Sunderland, MA.

Shepherd, G.M. (1994). *Neurobiology* (3rd edn), Chapters 4–8. Oxford University Press, Oxford.

Multiple choice questions

Each statement is either true or false. Answers are given below.

1. In the CNS:
   a. White matter contains large numbers of nerve cell bodies.
   b. The myelin of axons is formed by oligodendrocytes.
   c. The end-feet of astrocytes cover the capillaries of the brain.
   d. The extracellular fluid surrounding nerve cells is insulated from changes in the composition of the plasma.

2. The axons of peripheral nerves:
   a. Are protected by three layers of connective tissue.
   b. Are always associated with Schwann cells.
   c. Are always covered by a layer of myelin.
   d. Only conduct action potentials from the CNS to a target organ.

3. The action potential of a single nerve fiber:
   a. Is caused by a large change in the permeability of the membrane to sodium.
   b. Is terminated when the sodium channels have inactivated.
   c. Can summate with an earlier action potential.
   d. Becomes larger as the stimulus is increased above threshold.

4. The velocity of action potential propagation:
   a. Is faster size for size in myelinated axons than in unmyelinated axons.
   b. Is faster the larger the diameter of the nerve fiber.
   c. Is independent of the thickness of the myelin.
   d. Will fall if an axon loses its myelin sheath (demyelination).

5. If a motor nerve has a conduction velocity of 50 m s^{-1}, how long will it take an action potential to reach a muscle that is 0.5 m from the cell body?
   a. 15 ms
   b. 10 ms
   c. 2 ms
   d. 20 ms

6. The following statements relate to excitatory and inhibitory synaptic transmission:
   a. During an epsp the membrane potential of the postsynaptic neuron always depolarizes.
   b. Unlike action potentials, epsps can summate to produce larger depolarizations of the postsynaptic membrane.
   c. The effects of a synaptic transmitter are always terminated by enzymatic destruction.
   d. Both epsps and ipsps can result from activation of second-messenger systems.
   e. During an ipsp the postsynaptic membrane hyperpolarizes.
   f. Ipsps alter the threshold for action potential generation.
   g. Both excitatory and inhibitory synaptic transmission is triggered by an influx of calcium into the presynaptic nerve terminal.

7. The following questions relate to neuromuscular transmission at the motor end-plate:
   a. Excitation of a motor nerve fiber leads to the contraction of all the muscle fibers innervated by its branches.
   b. The epp is the result of a summation of many mepps.
   c. The epp is the result of an increase in the permeability of the junctional membrane to chloride.
   d. At the neuromuscular junction the cholinergic receptors are muscarinic.
   e. Neuromuscular transmission can be blocked by curare.
   f. End-plate potentials can be prolonged by drugs that inhibit acetylcholinesterase.

Answers

1. The white matter contains large numbers of myelinated nerve fibers. The gray matter contains large numbers of nerve cell bodies. The end-feet of the astrocytes form part of the blood–brain barrier that isolates the extracellular fluid around the neurons from changes in the composition of the blood.
   a. False
   b. True
   c. True
   d. True

2. Peripheral axons are protected by the epineurium, the perineurium, and the endoneurium. All peripheral axons are associated with Schwann cells but only the larger fibers are covered by layers of myelin. Efferent (motor) fibers convey action potentials from the CNS to target organs while afferent (sensory) fibers convey action potentials from sense organs and other receptors to the CNS.
   a. True
   b. True
   c. False
   d. False

3. Once the stimulus intensity has reached threshold, an action potential is elicited but the amplitude of the action potential does not increase further with increasing stimulus strength. The action potential is all or none. When one action potential has passed along an axon the nerve passes through its refractory period and a second action potential cannot occur until the membrane regains its excitability. Action potentials cannot summate.
   a. True
   b. True
   c. False
   d. False

4. Myelin thickness and internodal distance increase with increasing fiber diameter. Thus large myelinated nerve fibers conduct faster than small myelinated (or unmyelinated) fibers. If the myelin sheath is lost, saltatory conduction cannot occur and the speed of conduction falls dramatically.

   a. True
   b. True
   c. False
   d. True

5. 10 ms

6. The actions of synaptic transmitters are terminated by enzymatic destruction, uptake, or diffusion away from the synaptic region. Ipsps hyperpolarize the membrane potential and this makes the postsynaptic cell less excitable. They do not alter the threshold for action potential generation. The release of a synaptic transmitter in response to a nerve impulse occurs by exocytosis and this is triggered by the entry of calcium into the nerve terminal.
   a. True
   b. True
   c. False
   d. True
   e. True
   f. False
   g. True

7. A motor nerve together with the muscle fibers it innervates is called a motor unit. The end-plate potential is the result of an increase the permeability of the junctional membrane to sodium and potassium leading to depolarization. The junctional receptors are nicotinic cholinergic receptors and these receptors can be blocked by curare. If acetylcholinesterase is inhibited, the acetylcholine released by the motor nerve endings is not destroyed and can continue to stimulate the nicotinic receptors. Very prolonged action of acetylcholine ultimately leads to depolarization of the muscle and block of neuromuscular transmission.
   a. True
   b. True
   c. False
   d. False
   e. True
   f. True

Muscle

After reading this chapter you should understand:

- The different morphological characteristics of the three principal types of muscle
- The detailed structure of skeletal muscle
- Excitation–contraction coupling in skeletal muscle
- The mechanism of contraction in skeletal muscle
- The mechanical properties of skeletal muscle
- The cardiac action potential and pacemaker activity
- The mechanical properties of cardiac muscle
- Excitation–contraction coupling in cardiac muscle
- The role of smooth muscle
- Excitation–contraction coupling in smooth muscle: single- and multi-unit smooth muscles
- The mechanical and electrical properties of smooth muscle

7.1 Introduction

One of the distinguishing characteristics of animals is their ability to use coordinated movement to explore their environment. For large multicellular animals, this movement is achieved by the use of muscles, which consist of cells that can change their length by a specific contractile process. In vertebrates, including man, three types of muscle can be identified on the basis of their structure and function. They are skeletal muscle, cardiac muscle, and smooth muscle. As its name implies, *skeletal muscle* is the muscle directly attached to the bones of the skeleton and its role is both to maintain posture and to move the limbs by contracting. *Cardiac muscle* is the muscle of the heart and *smooth muscle* is the muscle that lines the blood vessels and the hollow organs of the body. Together the three kinds of muscle account for nearly half of body weight, the bulk of which is contributed by skeletal muscle (about 40 per cent of total body weight).

The cells of skeletal, cardiac, and smooth muscle are known as *myocytes*. When the cells of skeletal and cardiac muscle are viewed down a microscope they are seen to have characteristic

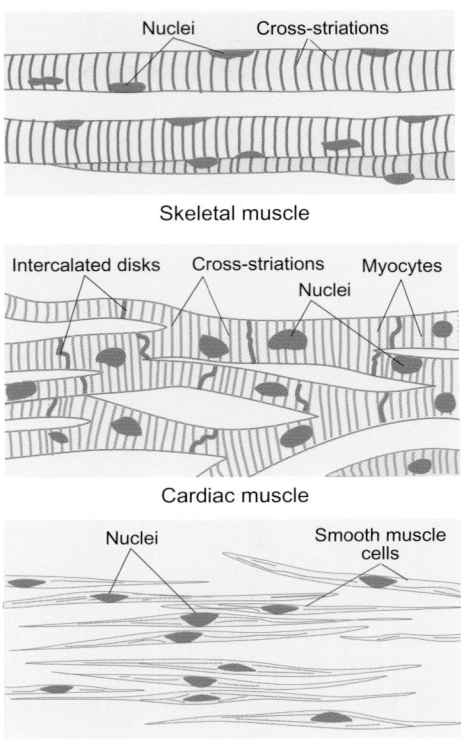

Fig. 7.1 Simplified drawings illustrating the microscopical appearance of skeletal, cardiac, and smooth muscle at similar magnification.

striations—small regular stripes running across the individual muscle cells. For this reason, skeletal and cardiac muscles are sometimes called striated muscles. Smooth muscle lacks striations and consists of sheets of spindle-shaped cells. The microscopical appearance of the different kinds of muscle is shown in Fig. 7.1. Despite these differences in structure, the molecular basis of the contractile process is very similar for all types of muscle.

This chapter is chiefly concerned with the physiological properties of the different kinds of muscle. It will also outline the cellular and molecular basis of the contractile process itself.

7.2 The structure of skeletal and cardiac muscle

Skeletal muscle

Each skeletal muscle is made up of a large number of skeletal *muscle fibers*, which are long thin cylindrical cells that contain many nuclei. The length of individual muscle fibers varies according to the length of the muscle and ranges from a few millimeters to 10 cm or more. Their diameter also depends on the size of the individual muscle and ranges from about 50 to 100 μm. Despite their great length, few muscle fibers extend for the full length of a muscle. Individual muscle fibers are embedded in connective tissue called the *endomysium*, and groups of muscle fibers are bound together by connective tissue called the

perimysium to form bundles called muscle *fascicules*. Surrounding the whole muscle is a coat of connective tissue called the *epimysium* or fascia which binds the individual fascicules together. The connective tissue matrix of the muscle is secreted by fibroblasts that lie between the individual muscle fibers. It contains collagen and elastic fibers that merge with the connective tissue of the tendons where it serves to transmit the mechanical force generated by the muscle to the skeleton. Finally, within the body of a skeletal muscle there are specialized sense organs known as muscle spindles that play an important role in the regulation of muscle length (see Chapter 9).

The individual muscle fibers are made up of filamentous bundles that run along the length of the fiber. These bundles are called *myofibrils* and have a diameter of about 1 μm. Each myofibril consists of a repeating unit known as a *sarcomere*. The alignment of the sarcomeres between adjacent myofibrils gives rise to the characteristic striations of skeletal muscle. The sarcomere is the fundamental contractile unit within skeletal and cardiac muscle. Each sarcomere is only about 2 μm long so that each myofibril is made up of many sarcomeres placed end to end.

When a muscle fiber is viewed by polarized light the sarcomeres are seen as alternating dark and light zones. The regions that appear dark do so because they refract the polarized light. This property is called *anisotropy* and the corresponding band is known as an *A band*. The light regions do not refract polarized light and are said to be *isotropic*. These regions are called *I bands*. Each I band is divided by a characteristic line known as a Z line and the unit between successive Z lines is a sarcomere. At high magnification in the electron microscope the A bands are seen to be composed of thick filaments arranged in a regular order. The I bands consist of thin filaments. At the normal resting length of a muscle, a pale area can be seen in the center of the A band. This is known as the H zone, and it corresponds to the region where the thick and thin filaments do not overlap. In the center of each H zone is the M line at which links are formed between adjacent thick filaments. The principal protein of the A bands is myosin, while that of the I bands is actin. The interaction between these proteins is fundamental to the contractile process (see Section 7.4). The actin filaments of the I band are made by joining many G (globular) actin subunits together by polymerization to form F (filamentous) actin which is stabilized by binding to the Z line. The thick filaments are made by assembling myosin molecules together. Each myosin molecule consists of two heavy chains, each of which has two light chains associated with their globular head region. The junction between the head region and the long tail contains the hinge that allows the myosin to generate the force required for muscle contraction. The tail regions of the myosin molecules associate together to form the thick filaments. Each thick filament consists of several hundred myosin molecules. This elaborate cellular architecture is maintained by a number of structural proteins, one of which, titin, is a long molecule that links the myosin filaments to the Z line. The Z line itself contains a number of proteins including α-actinin, which binds the actin filaments, and desmin, which links adjacent Z lines and serves to keep them in register. The main features of the structure of skeletal muscle are summarized in Fig. 7.2.

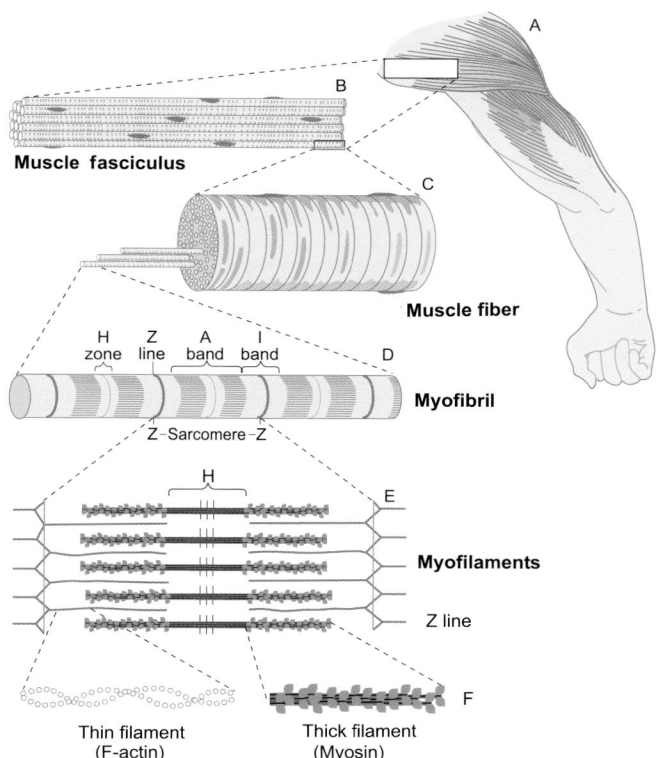

Fig. 7.2 The organization of skeletal muscle at various degrees of magnification: A, the appearance of the whole muscle; B, the appearance of a muscle fascicule; C, the appearance of a single muscle fiber; D, the structure of individual myofibrils; E, the appearance of the protein filaments that make up the individual sarcomeres; F, the organization of actin and myosin in the thin and thick filaments.

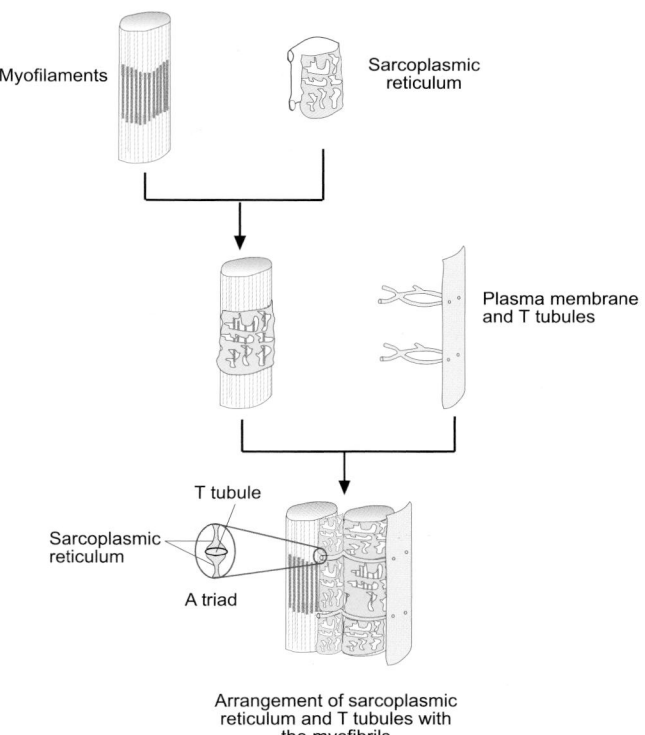

Fig. 7.3 The detailed organization of the T system and sarcoplasmic reticulum of skeletal muscle. The sarcoplasmic reticulum is a membranous sac that envelops the myofibrils. It stores and releases calcium to regulate muscle contraction. The T tubules have their origin at the sarcolemma (plasma membrane) and form a network that crosses the muscle fiber. At the junction of the A and I bands the T tubules come into close apposition with the terminal cisternae of the sarcoplasmic reticulum to form the triads.

Like all cells, skeletal muscle fibers are bounded by a cell membrane which, in the case of muscle, is known as the *sarcolemma*. At its normal resting length, the sarcolemma is folded, forming small indentations known as caveolae. These are probably important in permitting the muscle fiber to be stretched without causing damage to the sarcolemma. Beneath the sarcolemma lie the nuclei and many mitochondria. In mammalian muscle, narrow tubules run from the sarcolemma transversely across the fiber at the junction of the A and I bands. These are known as *T tubules*. Each myofibril is surrounded by the *sarcoplasmic reticulum*, which is a membranous structure homologous with the endoplasmic reticulum of other cell types. Where the T tubules and the sarcoplasmic reticulum come into contact, the sarcoplasmic reticulum is enlarged to form the *terminal cisternae*. Each T tubule is in close contact with the cisternae of two regions of sarcoplasmic reticulum and the whole complex is called a *triad* (Fig. 7.3). The T tubules and triads play an important role in excitation-contraction coupling (see Section 7.4).

Cardiac muscle

Cardiac muscle consists of individual cells linked together by junctions called intercalated disks. Characteristically, the intercalated disks cross the muscle in irregular lines. Individual cardiac muscle cells are aligned so that they run in arrays that often branch to link adjacent groups of fibers together. The individual

cells are about 15 μm in diameter and up to 100 μm in length, and adjacent cells are coupled electrically by gap junctions. This arrangement permits activity to spread from one cell to another. Cardiac muscle cells (or *cardiac myocytes*) usually have only one

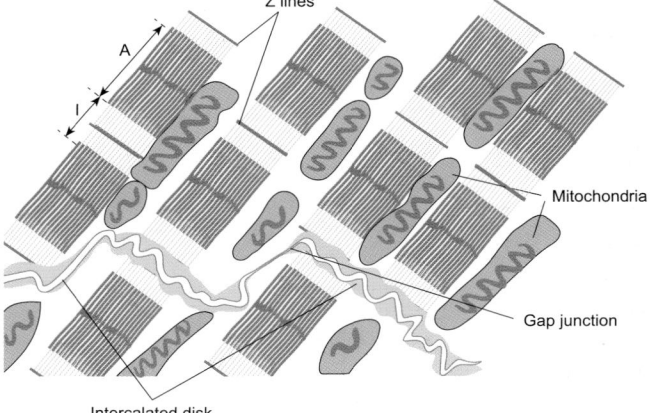

Fig. 7.4 The appearance of cardiac muscle at high magnification. The field of view shows two myocytes joined by an intercalated disk which courses irregularly across the lower part of the field. Note the close apposition of the plasma membrane at the gap junctions, the large numbers of mitochondria, and the irregular arrangement of the Z lines.

Summary

1. Skeletal muscle is made up of long cylindrical multinucleated cells called muscle fibers. Muscle fibers contain myofibrils that are made up of repeating units called sarcomeres, which are the fundamental contractile units. Each sarcomere is separated from its neighbors by Z lines and consists of two half I bands, one at each end, separated by a central A band. The myofibrils are surrounded by a membranous structure known as the sarcoplasmic reticulum.

2. Cardiac muscle is made up of many individual cells (cardiac myocytes) linked together via intercalated disks. Unlike skeletal muscle fibers, cardiac myocytes generally have a single nucleus. The striated appearance of cardiac myocytes is due to the presence of an orderly array of sarcomeres similar to that seen in skeletal muscle.

3. The principal contractile proteins of the sarcomeres of both skeletal and cardiac muscle are actin and myosin.

centrally located nucleus and their mitochondria are distributed throughout the cytoplasm (Fig. 7.4). The structure of the contractile elements of cardiac myocytes is very similar to that of skeletal muscle but the individual sarcomeres are not as regularly arranged. The myocytes also contain T tubules and sarcoplasmic reticulum but the arrangement is less ordered than that seen in skeletal muscle. Usually one T tubule is associated with one of the cisternae to form a diad.

7.3 How does a skeletal muscle contract?

Skeletal muscle, like nerve, is an excitable tissue, and stimulation of a muscle fiber at one point will rapidly lead to excitation of the whole cell. In the body, each skeletal muscle fiber is innervated by one motoneuron. As it enters a muscle, the axon branches and each branch supplies a single muscle fiber. An action potential in the motor neuron leads to the generation of a motor end-plate potential (epp) in each of the muscle fibers to which it is connected as described in Chapter 6. The epp depolarizes the muscle fiber membrane in the region adjacent to the end-plate and, in its turn, triggers an action potential that propagates away from the end-plate along the whole length of the muscle fiber. The passage of the muscle action potential is followed by contraction of the muscle fiber and the development of tension. The process by which a muscle action potential triggers a contraction is known as *excitation–contraction coupling*.

What steps link the muscle action potential to the contractile response? It has been known for a long time that injection of Ca^{2+} into a muscle fiber causes it to contract. It was later discovered that significant amounts of Ca^{2+} are stored in the sarcoplasmic reticulum and that much of this is released during contraction. It is now thought that the depolarization of the plasma membrane during the muscle action potential spreads along the T tubules where it causes Ca^{2+} channels in the sarco-

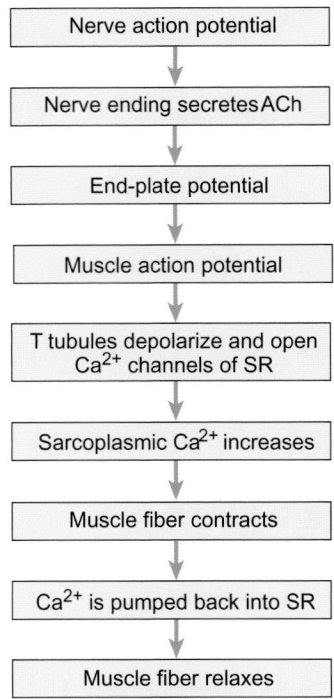

Fig. 7.5 Flow diagram to illustrate the sequence of events leading to the contraction and subsequent relaxation of a skeletal muscle fiber. SR = sarcoplasmic reticulum

plasmic reticulum to open. As a result, Ca^{2+} stored in the sarcoplasmic reticulum is released and the level of Ca^{2+} in the sarcoplasm rises. This rise in Ca^{2+} triggers the contraction of the muscle fiber. Relaxation of the muscle occurs as the Ca^{2+} in the sarcoplasm is pumped back into the sarcoplasmic reticulum by a Ca^{2+} pump of the kind described in Chapter 4. These events are summarized in Fig. 7.5.

How is tension generated and how does a rise in Ca^{2+} lead to the contractile response? All muscles contain two proteins, *actin* and *myosin*. In skeletal and cardiac muscle, the thick filaments are chiefly composed of myosin while the thin filaments contain actin (the principal protein) and lesser quantities of two other proteins known as troponin and tropomyosin. Muscle contraction is now known to occur through interactions between actin and myosin, which cause the thick and thin filaments to slide past each other. This is known as the *sliding-filament theory* of muscle contraction.

The individual myosin molecules of the thick filaments have two globular heads and a long thin tail region. They are arranged such that the thin tail regions associate together to form the backbone of the thick filaments while the thicker head regions project outwards to form cross-bridges with the neighboring thin filaments (Fig. 7.2, parts E and F). The actin molecules link together to form a long polymer chain (actin F). Each actin molecule in the chain is able to bind one myosin head region. Actin and myosin molecules dissociate when a molecule of ATP is bound by the myosin. The breakdown of ATP and the subsequent release of inorganic phosphate cause a change in the angle of the head region of the myosin molecule, enabling it

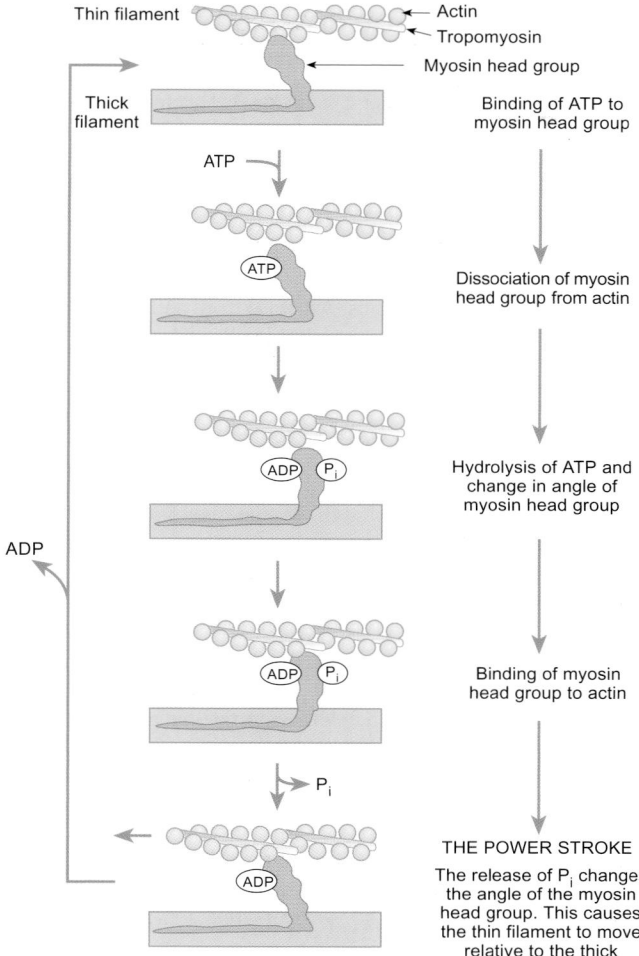

Fig. 7.6 A schematic representation of the molecular events responsible for the relative movement of the thin and thick filaments of a striated muscle.

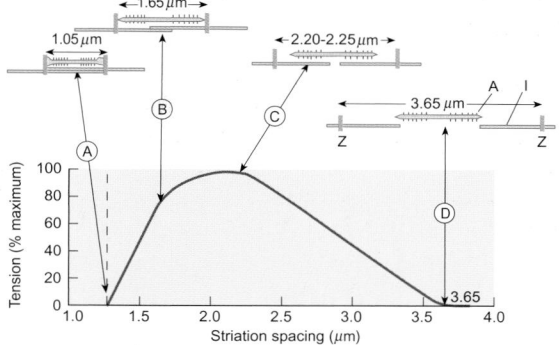

Fig. 7.7 The sliding-filament theory of muscle contraction and the length–tension relationship of a skeletal muscle fiber. Sequence A–D represents the force developed as the resting length of a single skeletal muscle fiber is increased. When the thin and thick filaments fully overlap and the A band is compressed against the Z line (point A) the muscle is unable to develop tension. As the fiber is stretched so that the thin and thick filaments overlap without compressing the A band, active tension is generated when the muscle is stimulated (point B). Further stretching provides optimal overlap between the thin and thick filaments, leading to maximal development of tension (point C). If the fiber is stretched to such a degree that the thin and thick filaments no longer overlap, no tension is developed (point D).

to move relative to the thin filament (Fig. 7.6). Once again, ATP causes the dissociation between the actin and myosin and the cycle is repeated. This process is known as *cross-bridge cycling* and results in the thick and thin filaments sliding past each other, thus shortening the fiber. The formation of cross-bridges between the actin and myosin head groups is not synchronized across the myofibril. While some myosin head groups are dissociating from the actin, others are binding or developing their power stroke. Consequently, tension is developed smoothly

The role of calcium ions in muscle contraction

If purified actin and myosin are mixed in a test tube, they form a gel (a jelly-like mixture). If ATP is then added to this mixture it contracts and the ATP is hydrolyzed. What prevents actin and myosin continuously hydrolyzing ATP in an intact muscle fiber? The answer to this problem came when it was discovered that muscle contained two other proteins called troponin and tropomyosin. These proteins form a complex with actin that prevents it binding to the myosin head groups. When Ca^{2+} is released from the sarcoplasmic reticulum, the concentration of free Ca^{2+}

in the sarcoplasm is transiently raised from the low levels found in resting muscle (0.1–0.2 μM) to a peak value of about 10 μM at the beginning of a contraction. The troponin complex binds the released Ca^{2+} and, in doing so, it changes its position on the actin molecule so that actin and the myosin head groups can interact as described above.

The sliding-filament theory of muscle contraction provides a clear explanation for the length–tension relationship of skeletal muscle (Fig. 7.7). When the muscle is at its natural resting length, the thin and thick filaments overlap optimally and form the maximum number of cross-bridges. When the muscle is stretched, the degree of overlap between the thin and thick filaments is reduced and the number of cross-bridges falls. This leads to a decline in the ability of the muscle to generate tension. When the muscle is shorter than its natural resting length, the thin filaments already fully overlap the thick filaments, but the filaments from each end of the sarcomere touch in the center of the A band and each interferes with the motion of the other. As a result, tension development declines. When the thin and thick filaments fully overlap, the A bands abut the Z lines and tension development is no longer possible.

The role of ATP and creatine phosphate

At rest, a skeletal muscle is plastic and can readily be stretched. In this state, although the myosin head groups have bound ATP, no cross-bridges are being formed between the thick and thin filaments because the troponin complex prevents the interaction between actin and myosin. When ATP levels fall to zero after death, the cross-bridges between the actin and myosin do not dissociate. The muscles lose their plasticity and become stiff—a state known as *rigor mortis*.

The energy for contraction is derived from the hydrolysis of ATP. As with other tissues, the ATP is derived from the oxidative metabolism of glucose and fats (see Chapter 3). However, it is important for the levels of ATP to be maintained during the contractile cycle. Since the blood flow through a muscle during contraction may be intermittent, a store of high-potential phosphate is needed to maintain contraction, as the available ATP (about 3 mM) would all be hydrolyzed within a few seconds. This need is met by *creatine phosphate* (also known as *phosphocreatine*), which is present in muscle at high concentrations (about 15–20 $mmol\,l^{-1}$) and has a phosphate group that is readily transferred to ATP. This reaction is catalyzed by the enzyme *creatine kinase*:

$$\text{creatine phosphate} + \text{ADP} + \text{H}^+ \rightleftharpoons \text{creatine} + \text{ATP}.$$

During heavy exercise insufficient oxygen for oxidative metabolism may be delivered to the exercising muscles. In this situation, the generation of ATP from glucose and fats via the tricarboxylic acid cycle is compromised and ATP is generated from glucose via the glycolytic pathway instead. This anaerobic phase of muscle contraction is much less efficient in generating ATP (see Chapter 3, p. 28), and hydrogen ions, lactate and phosphate ions are produced in increasing quantities. As these ions accumulate, muscle pH falls, the muscular effort becomes progressively weaker and the muscle relaxes more slowly. This is known as *fatigue*. During muscle fatigue, the muscle fibers remain able to propagate action potentials but their ability to develop tension is impaired.

Summary

1. The link between the electrical activity of a muscle and the contractile response is called excitation–contraction coupling. The action potential of the sarcolemma depolarizes the T tubules and this causes Ca^{2+} channels in the sarcoplasmic reticulum to open. This leads to an increase in Ca^{2+} around the myofibrils and the development of tension.

2. The A band of each sarcomere consists mainly of myosin molecules arranged in thick filaments and the I band consists mainly of actin, tropomyosin, and troponin arranged in thin filaments. Tension development occurs when actin and myosin interact. This is a calcium-dependent process in which actin and myosin form a series of cross-bridges that break and re-form as ATP is hydrolyzed. As a result, the thick and thin filaments slide past each other and force is generated.

3. The energy for muscle contraction is provided by the hydrolysis of ATP. As ATP is used up, it is rapidly replenished from the reserves of creatine phosphate. In prolonged exercise, the ATP for muscle contraction is derived from either glycogen breakdown to lactate (anaerobic activity) or the oxidative metabolism of glucose and fats (aerobic activity). During phases of anaerobic activity, muscles accumulate hydrogen ions, lactate, and phosphate ions. The increased concentration of these metabolites causes a decline in the development of tension known as muscular fatigue.

7.4 The activation and mechanical properties of skeletal muscle

Under normal circumstances, a skeletal muscle will only contract when the motor nerve to the muscle is activated. Therefore the signal to cause contraction of a skeletal muscle originates in the CNS and the resulting contraction is said to be *neurogenic* in origin. In contrast, provided that it is placed in a suitable nutrient medium, the heart will continue to beat spontaneously even when it is isolated from the body. Some smooth muscles behave in the same way. Contractions that arise from activity within the muscle itself are said to be *myogenic* in origin.

The innervation of skeletal muscles

The nerves supplying a mammalian skeletal muscle are myelinated. As a motor nerve enters the muscle, it branches to supply different groups of muscle fibers. Within the motor nerve bundle, individual axons also branch, so that one motor fiber makes contact with a number of muscle fibers. When a motoneuron is activated, all the muscle fibers it supplies contract in an all-or-none fashion so that they act as a single unit—a *motor unit*. The size of motor units varies from muscle to muscle according to the degree of control required. Where fine control of a movement is not required the motor units are large. Thus, in the gastrocnemius muscle of the calf, a motor unit may contain up to 2000 muscle fibers. In contrast, the motor units of the extraocular muscles (which control the direction of the gaze) are much smaller (as few as 6–10 muscle fibers are supplied by a single motoneuron).

The detailed mechanism by which a motoneuron activates skeletal muscle (neuromuscular transmission) has already been considered in Chapter 6.

The mechanical properties of skeletal muscle

Like nerve, skeletal muscle is an excitable tissue that can be activated by direct electrical stimulation. When a muscle is activated, it shortens and, in doing so, exerts a force on the tendons to which it is attached. The amount of force exerted depends on many factors:

(1) the number of active muscle fibers;

(2) the frequency of stimulation;

(3) the rate at which the muscle shortens;

(4) the initial resting length of the muscle;

(5) the cross-sectional area of the muscle.

Consider first the situation where the muscle is being artificially activated by a single electrical shock to its motor nerve. The response of a muscle to a single stimulus is called a *muscle twitch* and the force developed is called the *twitch tension*. If the electrical shock is weak, only a small proportion of the nerve fibers will be activated. Consequently, only a small number of the muscle fibers will contract and the amount of force generated will be weak. If the intensity of the stimulus is increased, more motor fibers will be recruited and there will be a corresponding increase in both the number of active muscle fibers and the total tension developed. When all the muscle fibers are activated, the total tension developed will reach its maximum. A stimulus that is sufficient to activate all the motor fibers in a muscle is called a *maximal stimulus*.

Muscles do not elongate when they relax unless they are stretched. In the body, this stretching is achieved for skeletal muscles by the arrangement of muscle pairs acting on particular joints. These pairs of muscles are known as *antagonists*. In cardiac muscle the blood returning to the heart performs the stretching action.

The force developed by a muscle increases during repetitive activation

When a muscle fiber is activated, the process of contraction begins with an action potential passing along its length. This is followed by the contractile response, which consists of an initial phase of acceleration during which the tension in the muscle increases until it equals that of the load to which it is attached. The fiber then begins to shorten. At first, the fiber shortens at a relatively constant rate but, as the muscle continues to contract, the rate of shortening progressively falls—the phase of deceleration. Finally, the muscle relaxes and the tension it exerts declines to zero. The whole cycle of shortening and relaxation takes several tens of milliseconds, while the muscle action potential is over in a matter of two or three milliseconds. Consequently, the mechanical response greatly outlasts the electrical signal that initiated it (Fig. 7.8).

If a muscle is activated by a pair of stimuli that are so close together that the second stimulus arrives before the muscle has fully relaxed after the first, the total tension developed following the second stimulus is greater than that developed in response to a single stimulus. This increase in developed tension is called *summation* and is illustrated in Fig. 7.9(b). The degree of summation is at its maximum with short intervals and declines as the interstimulus interval increases. If a number of stimuli are given in quick succession, the tension developed progressively summates and the response is called a *tetanus*. At low frequencies, the tension oscillates at the frequency of stimulation as shown in Fig. 7.9(c) but, as the frequency of stimulation rises, the development of tension proceeds more and more smoothly. When tension develops smoothly (Fig. 7.9(d)), the contraction is known as a *fused tetanus*.

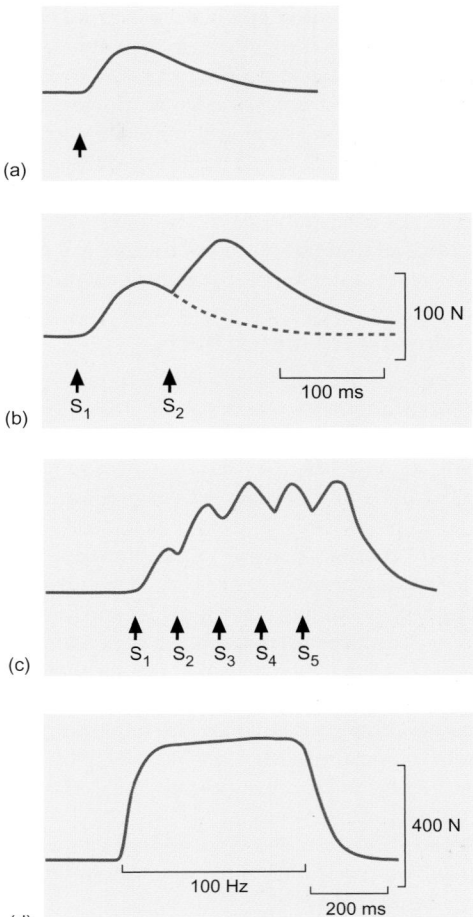

Fig. 7.9 The summation and fusion of the mechanical response of a muscle in response to repetitive stimulation: (a) the response to a single electrical stimulus (arrow); (b) the summation of tension when a second stimulus (S_2) is given before the mechanical response to the first has decayed to zero; (c) the summation of tension during a brief train of stimuli; (d) the smooth development of tension during a high-frequency train of stimuli (giving a fused tetanus). Note the different time and tension calibrations for (a) and (b) compared with (c) and (d).

Why is a muscle able to develop more tension when two or more stimuli are given in rapid succession? Consider first how tension develops during a single twitch. When the muscle is activated, it must transmit the tension developed to the load. To do this the tension of the tendons and of the connective tissue of the muscle itself must be raised to that of the load. As the tendons are to some degree elastic, they need to be stretched a little until the tension that they exert on the load is equal to that of the muscle. This takes a short, but finite, amount of time during which the response of the tension generating machinery begins to decline. The transmission of force to the load is accordingly reduced in efficiency. This also accounts for the initial acceleration in the rate of contraction described above. However, during a train of impulses the contractile machinery is activated repeatedly and the tension in the tendons has little time to decay between successive contractions. Transmission of tension to the load is more efficient and a greater tension is developed.

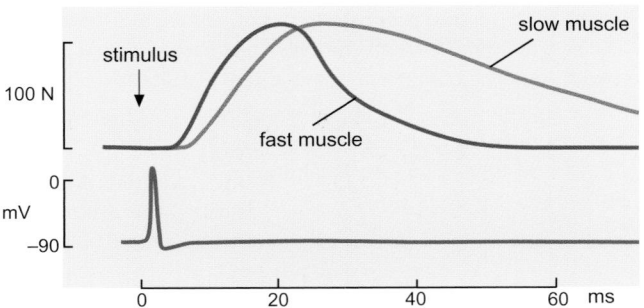

Fig. 7.8 The time course of the contractile response (a twitch) in response to a single action potential for a fast and a slow skeletal muscle. Note that the contractile response begins after the action potential and lasts much longer. Force is expressed in newtons (N).

Furthermore, the concentration of Ca^{2+} in the sarcoplasm that is attained in response to a single action potential is insufficient for maximal troponin binding. Consequently, not all actin molecules can interact with the myosin heads. However, when stimuli are given in rapid succession, the contractile machinery will be primed because not all the Ca^{2+} released during one stimulus will have been pumped back into the sarcoplasmic reticulum before the arrival of the next. As a result, more of the troponin binds Ca^{2+}; this permits the formation of a greater number of cross-bridges and the development of greater tension.

Fast- and slow-twitch muscle fibers

Different types of muscle show different rates of contraction and susceptibility to fatigue. Broadly speaking, skeletal muscle fibers can be classified as slow (type 1) or fast (type 2) according to their rate of contraction. Most muscles contain a mixture of both kinds of fiber but some muscles have a predominance of one type.

Slow muscles contract at about 15 mm s^{-1} and relax relatively slowly. They are composed of thin fibers, rich in both mitochondria and the oxygen-binding protein myoglobin (see Chapter 13, Section 13.6). This gives them a reddish appearance. They rely mainly on the oxidative metabolism of fats for their energy supply and have a rich blood supply, which makes them very resistant to fatigue. Slow muscles play an important role in the maintenance of posture. They are activated by their motoneurons at a continuous steady rate, which enables them to maintain a steady muscle tone.

Fast muscles shorten at about 40–45 mm s^{-1} and relax relatively quickly. Two types of fast fiber (type 2A and type 2B) exist. Type 2A are fast oxidative–glycolytic fibers which have a high myosin ATPase activity compared with the slow fibers discussed above. The type 2A fibers are relatively thin, have a good blood supply, and, like the type 1 fibers, are rich in both mitochondria and myoglobin. They rely on oxidative metabolism for their energy requirements and utilize either glucose or fats as their source of energy. They are relatively resistant to fatigue. Type 2B fibers are fast glycolytic fibers with a large diameter. They have a high myosin ATPase activity, contain large quantities of glycogen, and have high concentrations of glycolytic enzymes. This enables type 2B fibers to develop great tension rapidly. However, as they have a limited blood supply, few mitochondria and little myoglobin, they are easily fatigued. Therefore they are well adapted to providing short periods of high tension development during anaerobic exercise. Unsurprisingly, these fibers are recruited for short periods of intense muscular activity (e.g. during sprinting). Their relative lack of mitochondria and myoglobin gives them a pale appearance.

Smooth tension development in intact muscle is due to asynchronous activation of motor units

In the absence of organic disease, the development of tension during normal movement is smooth and progressive. This arises because the CNS recruits motoneurons progressively. Consequently, the motor units comprising a muscle are activated at different times—very unlike the experimental situations described in the previous sections. Individual motor units may be activated by relatively low frequencies of stimulation but the maintenance of a steady tension ensures the efficient transmission of force to the load. The ability of a muscle to maintain a smooth contraction is further enhanced by the presence of both fast and slow muscle fibers in most muscles.

The power of a muscle depends on the rate at which it shortens

The rate at which a muscle can shorten depends on the load against which it acts. If there is no external load, a muscle shortens at its maximum rate. With progressively greater loads, the rate of shortening decreases until the load is too great for the muscle to move. The relationship between the load imposed on a

---

BOX 7.1 THE EFFICIENCY AND POWER OF MUSCLES

It is a matter of common experience that it is more difficult to move heavy objects than light ones, but how efficient are muscles in converting chemical energy into useful work?

The *force* exerted on a given load is defined as:

$$\text{force} = \text{mass} \times \text{acceleration} \qquad (1)$$

and is given in *newtons* (N): 1N is the force that will give a mass of 1 kg an acceleration of 1 m s^{-1}.

The work performed on a load is the product of the load and the distance through which the load is moved. Thus

$$\text{work} = \text{force} \times \text{distance}. \qquad (2)$$

Therefore the unit for work is N. m and one N. m is a joule (J).

Power is defined as the capacity to do work or the work per unit time and is expressed in J s^{-1} or *watts* (W):

$$\text{power} = \frac{\text{work}}{\text{time}}$$
$$= \text{force} \times \frac{\text{distance}}{\text{time}}$$
$$= \text{force} \times \text{velocity}. \qquad (3)$$

The key to understanding the power and efficiency of a muscle is its force–velocity curve. For an isometric contraction, maximum force is exerted but the load is not moved, so that no work is done and the power is also zero. Similarly. for an isotonic contraction against zero load no useful work is done. Between these two extremes the work is given by equation (2) and the power by Equation (3). The power is usually at a maximum when the muscle is shortening at about one-third of the maximum possible rate (Fig. 7.10).

The mechanical efficiency of muscular activity or work is expressed as the percentage of work done relative to the increase in metabolic rate attributable to the activity of the muscles employed in the task:

$$\text{efficiency} = \frac{\text{work done}}{\text{energy expended in task}} \times 100$$

In our examples above, both isometric contraction and contraction with no external load have zero efficiency. When a muscle does external work (e.g. walking up stairs or cycling) its efficiency is about 20–25 per cent.

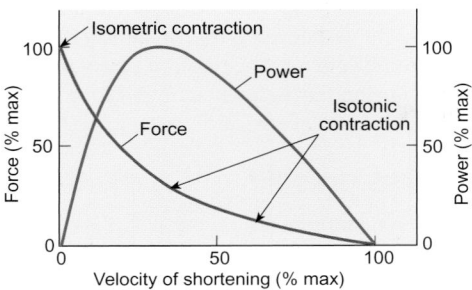

Fig. 7.10 The force–velocity relation for a skeletal muscle. Note that maximum force is developed during an isometric contraction but maximum speed of shortening occurs in an unloaded muscle. Maximum power is developed when the muscle shortens at about one-third of maximum velocity.

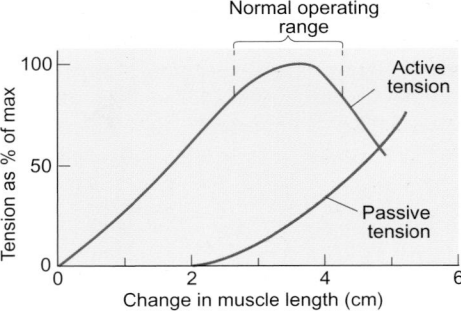

Fig. 7.11 Isometric force–tension relationship at different muscle lengths. The data are for human triceps muscle, which is about 20 cm in length. As the muscle is stretched, passive tension increases. Active tension increases from zero to a maximum and then declines with further stretching. The total tension developed during a contraction is the sum of the active and passive tensions.

muscle and its rate of shortening is known as the *force–velocity curve* (Fig. 7.10). If a muscle contracts against a load which prevents shortening, the muscle is said to undergo an *isometric contraction*, while if it shortens against a constant load it is said to undergo an *isotonic contraction*. Thus isometric contraction and isotonic contraction with no external load represent the extreme positions of the relationship between the force developed by a muscle and the rate at which it shortens.

The work of a muscle is determined by the distance it is able to move a given load and the power of a muscle is the rate at which it performs work (see Box 7.1). Thus

$$power = force \times velocity.$$

From the curve relating the velocity of shortening to the power developed (Fig. 7.10), it is clear that the power developed by a muscle passes through a definite maximum. When a muscle shortens isometrically, it does no work as the load is not moved through a distance. Consequently, no power is developed. Equally, if the muscle contracts while it is not acting on an external load, no work is done and no power is developed. Between these two extremes, the muscle performs useful work and develops power. In general, the greatest power is developed when the muscle is shortening at about one-third of its maximum rate.

The effect of muscle length on the development of tension

If the force generated by a muscle during isometric contraction is measured for different initial resting lengths a characteristic relationship is found. In the absence of stimulation, the tension increases progressively as the muscle is stretched beyond its normal resting length. This is known as *passive tension* and is due to the stretching of the muscle fibers themselves and of the connective tissue of the muscle and tendons. The extra tension developed as a result of stimulation (called the *active tension*) is at its maximum when the muscle is close to its resting length (i.e. the length that it would have had in its resting state in the body). If the muscle is stimulated when it is shorter than normal, it develops less tension, and if it is stretched beyond normal resting length, the tension developed during contraction is also less than normal. Overall, the relationship between the initial length of a muscle and the active tension is described by a bell-shaped curve (Fig. 7.11) that is very similar to the length–tension rela-

tionship seen for individual skeletal muscle fibers (Fig. 7.7). In the body the range over which a muscle can shorten is determined by the anatomical arrangement of the joint on which it acts. The lengths of muscles attached to the skeleton range from 0.7 to 1.2 times their equilibrium length.

Effect of cross-sectional area on the power of muscles

The force generated by a single skeletal muscle fiber does not depend on its length but on its cross-sectional area (which determines the number of myofibrils that can act in parallel). This can be seen if we consider the force generated by a myofibril consisting of two sarcomeres. The forces generated by the two central half-sarcomeres cancel out as one pulls to the left and the other to the right. Consequently, it is the two end half-sarcomeres that generate useful force. This remains true whether the myofibril has 100 or 1000 sarcomeres arranged in series. Thicker fibers have more myofibrils arranged in parallel. Therefore, they develop more tension. When muscles hypertrophy (i.e. enlarge) in response to training, the number of muscle fibers does not increase. Rather, there is an increase in the number of myofibrils in the individual fibers and this leads to an increase in their cross-sectional area. For both men and women between the ages of 12 and 20 the maximal isotonic force in flexor muscles is approximately 60 N cm^{-2}. The difference in strength between individuals is due to the difference in cross-sectional area of the individual muscles.

As long muscle fibers have more sarcomeres than short ones, they can shorten to a greater degree. In addition, they shorten faster. Consider a muscle 1 cm long in which each fiber runs the complete length of the muscle. Each fiber will have approximately 4000 sarcomeres arranged end to end (in series). If each sarcomere were to shorten from 2.5 to 2.0 μm, the muscle would shorten by 2.0 mm (20 per cent of the resting length). If this shortening occurred in 100 ms, the rate of shortening would be 0.2 ÷ 0.10 = 2 cm/s^{-1} since each sarcomere shortens at approximately the same rate. Following the same line of argument, a muscle 30 cm long can shorten at 60 cms^{-1}, i.e. 30 times faster.

The power of a muscle is equal to the force generated multiplied by the rate of contraction (see above). As the force of contraction depends on the cross-sectional area of the muscle and the rate of contraction depends on the length of the muscle, the

Summary

1. The nerves supplying a skeletal muscle are known as motor nerves. They are myelinated and individual axons branch to make contact with a number of muscle fibers. A motor neuron and its associated muscle fibers are called a *motor unit*.

2. A skeletal muscle contracts in response to an action potential in its motor nerve. Therefore its activity is said to be neurogenic in origin. A single action potential gives rise to a contractile response called a twitch. During repeated activation of a muscle, the tension summates and the muscle is said to undergo tetanic contraction. Maximum tension is developed during a fused tetanus.

3. There are two principal types of skeletal muscle fiber: fast or twitch fibers which develop tension and relax rapidly, and slow fibers which develop tension much more slowly but are able to maintain tension for long periods.

4. The force developed by a muscle depends on the number of active motor units, its cross-sectional area, and the frequency of stimulation.

power of a muscle is proportional to its volume. Therefore a short thick muscle will develop the same power as a long thin muscle of the same volume. The thick muscle will develop more force but will shorten more slowly than the long thin one.

7.5 Cardiac muscle

In the body, skeletal muscle only contracts in response to activity in the appropriate motor nerve. Denervated skeletal muscle does not contract. In contrast, denervated cardiac muscle continues to contract rhythmically. It is this intrinsic or *myogenic* activity that is responsible for the steady beating of the heart and enables the organ to be transplanted. This intrinsic rhythm has its origin in the cardiac myocytes found at the junction between the great veins and the right atrium. This region is called the sinoatrial (SA) node and the cells of the SA node are known as *pacemaker cells* as it is their activity that sets the basic heart rate. Action potentials spread from the SA node across the whole of the heart so that cardiac muscle behaves as a functional syncytium. In disease states, myocytes in other parts of the heart can show pacemaker activity which causes irregular beating of the heart known as *arrhythmia*.

BOX 7.2 DISEASES OF SKELETAL MUSCLE

Like any other complex tissue, skeletal muscle is subject to many disorders. Muscle diseases (or *myopathies*) usually become apparent either because of evident weakness or because there is inappropriate contractile activity. They can be broadly divided into myopathies that have been acquired and those that are due to abnormal genes. In the case of the genetic disorders, the specific gene loci responsible are now known, offering the distant prospect of a cure.

The acquired myopathies can be divided into:

(a) Inflammatory myopathies such as polymyositis (inflammation of muscle tissue), parasitic infection, and malignancy.

(b) Disorders of the neuromuscular junction such as myasthenia gravis, in which antibodies to the muscle nicotinic receptor are present in the blood resulting in impaired neuromuscular transmission, and the Lambert–Eaton syndrome, in which there is defective release of acetylcholine from the motor nerve endings.

(c) Acquired metabolic and endocrine myopathies arise from a number of causes such as Cushing's syndrome, in which there are high concentrations of corticosteroids circulating in the blood and thyrotoxicosis (high circulating levels of thyroid hormone). Myopathy of the proximal muscles occurs in hypocalcaemia, rickets, and osteomalacia.

The genetic myopathies can be divided into four main groups:

(a) Muscular dystrophies in which there is muscle destruction such as Duchenne muscular dystrophy where the gene for the protein dystrophin is faulty or even missing.

(b) The myotonias in which there is sustained contraction with slow relaxation. Myotonic dystrophy is associated with weakness of the distal muscles and is caused by an abnormal trinucleotide repeat in a protein kinase.

(c) The channelopathies in which the function of ion channels is impaired. These are associated with intermittent loss of muscle tone. Though rare, two principal types are known: hypokalaemic periodic paralysis in which plasma potassium falls below 3 mmol l^{-1} and hyperkalaemic periodic paralysis in which serum potassium is elevated above its normal level of 5 mmol l^{-1}. In hypokalaemic periodic paralysis there are a number of possible genetic defects, the most common of which is a defect in a type of calcium channel called the dihydropyridine-sensitive (or L-type) calcium channel. In the case of the hyperkalaemic form of the disease there is commonly a mutation in the voltage-gated sodium channel.

(d) Specific metabolic disorders of muscle are due to defects of specific enzymes. The glycogen storage disease McArdle disease is perhaps the best known. In this case there is a defect in muscle phosphorylase, which is responsible for mobilization of glucose from muscle glycogen. Another disorder, which is of considerable practical importance, is malignant hyperpyrexia in which there is a generalized muscle rigidity and elevation of body temperature. This is provoked by administration of general anesthetics and is caused by a defect in the ryanodine receptor of skeletal muscle which is responsible for regulating the release of calcium from the sarcoplasmic reticulum.

The action potential of cardiac myocytes is of long duration and is maintained by a prolonged inward movement of calcium ions

Like skeletal muscle, the contractile response of a cardiac muscle fiber is associated with an action potential. The characteristics of the action potential depend on the position of the myocyte within the heart. Cardiac action potentials are between 150 and 300 ms in duration. Thus they are of far longer duration than those of nerve and skeletal muscle (which last for 1–2 ms) and this has important consequences for the contractile response of cardiac muscle (see below). Cardiac myocytes may have either rapidly activating action potentials, as shown for atrial fibers and ventricular fibers in Fig. 7.12, or slowly activating action potentials, as shown for the myocytes of the SA node. The action potentials of the Purkinje fibers, which form the conducting system of the ventricles, have a similar appearance to those of ventricular fibers, while those of the atrioventricular (AV) node have an appearance similar to those of the SA node (see Fig. 15.7).

As for other cells, the resting membrane potential of atrial and ventricular myocytes is determined mainly by the permeability of the membrane to potassium ions. The upstroke of the action potential is due to a rapid increase in the permeability of the membrane to sodium ions similar to that seen in nerve axons and skeletal muscle fibers. The initial rapid phase of repolarization is due to inactivation of these sodium channels and the transient opening of potassium channels. The long phase of depolarization (called the plateau phase) is due to slowly activating calcium channels, which increase the permeability of the membrane to calcium ions. As these channels progressively inactivate, the membrane repolarizes and assumes a value close to the potassium equilibrium potential.

The myocytes of the sinoatrial node develop a pacemaker potential

The membrane potential of the cells of the SA node fluctuates spontaneously (Fig. 7.12). It is at its most negative (about –60 mV)

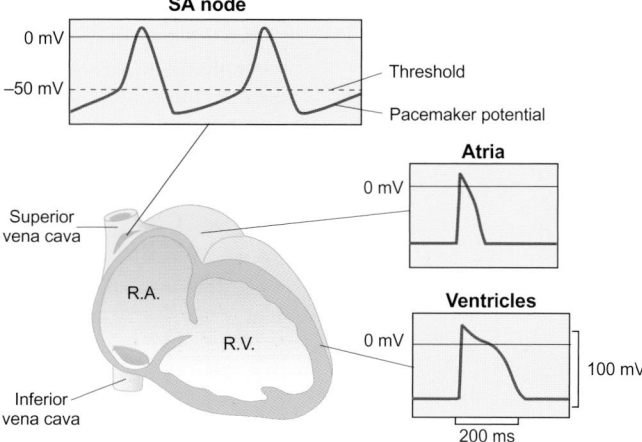

Fig. 7.12 The characteristics of the cardiac action potential in the myocytes of the SA node, atria, and ventricles. Note the very different appearance of the action potential in the different cell types and the presence of the pacemaker potential in the cells of the SA node. R.A., right atrium; R.V., right ventricle

immediately after the action potential and slowly falls (i.e. becomes less negative) until it reaches a value of about –50 mV which is the threshold for action potential generation. The action potential of the SA node cells has a slow rise time (time to peak is about 50 ms) and the whole action potential lasts for 150–200 ms. The slow depolarization that precedes the action potential is known as the *pacemaker potential* and the rate at which it falls towards threshold (i.e. its slope) is an important factor in setting the heart rate. Thus nerves and hormones may alter the heart rate as follows:

- they may hyperpolarize the membrane of the SA node cells (e.g. inhibition following brief stimulation of the vagus nerve);
- they may change the slope of the pacemaker potential (e.g. stimulation of the cardiac sympathetic nerves);
- they may both hyperpolarize the membrane and change the slope of the pacemaker potential (as seen following strong vagal stimulation—see Chapter 15, Section 15.7).

The pacemaker potential of the SA node cells arises because a sodium current is slowly activated when the membrane potential becomes more negative than about –50 mV during repolarization. The ion channels responsible for this current are quite distinct from the rapidly activating sodium channels responsible for the upstroke of the action potential. The slowly activating sodium current opposes the potassium current responsible for repolarization of the SA node cells and the membrane potential progressively becomes less negative. The resulting slow depolarization activates a calcium current which sums with the sodium current to accelerate the rate of depolarization until an action potential is triggered. The upstroke of the action potential of SA node cells is caused by a large increase in the permeability of the membrane to calcium ions (not sodium ions as is the case for the myocytes of the atria and ventricles). Since the equilibrium potential for calcium ions is positive (just as it is for sodium ions), the rise in calcium permeability leads to a reversal of the membrane potential. Repolarization occurs as the permeability of the membrane to potassium increases while the permeability to calcium falls. The membrane potential becomes more hyperpolarized; this activates the slow sodium current and the cycle is repeated.

The contractile response of cardiac muscle

The long duration of the cardiac action potential has the important consequence that the mechanical response of the muscle occurs while the muscle membrane remains depolarized (Fig. 7.13). Since a second action potential cannot occur until the first has ended, and since the mechanical response of the cardiac muscle largely coincides with the action potential, cardiac muscle cannot be tetanized. This is an important adaptation, as the heart needs to relax fully between beats if it is to allow time for filling. Fibrillation (rapid and irregular contractions) may occur if the duration of the cardiac action potential (and therefore that of the refractory period) is substantially decreased.

Calcium activates the contractile machinery of cardiac muscle in much the same way as it does in skeletal muscle, but there is one important difference. If heart muscle is placed in a physio-

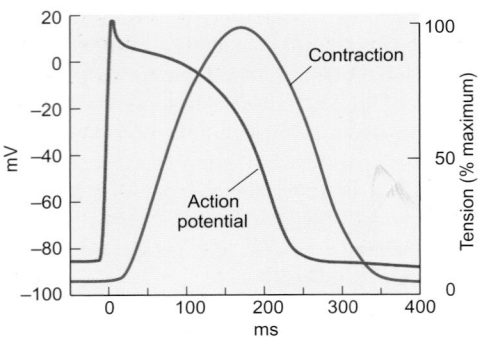

Fig. 7.13 The relationship between the action potential and the development of isometric tension in ventricular muscle.

logical solution lacking calcium, it quickly stops contracting, unlike skeletal muscle which will continue to contract each time it is stimulated. In cardiac muscle, the rise in Ca^{2+} within the myocyte during the plateau phase of the action potential is derived both from the calcium store of the sarcoplasmic reticulum, and from calcium influx through voltage-gated ion channels in the plasma membrane. Indeed, this calcium influx is the trigger for the release of the stored calcium. This is known as *Ca^{2+}-induced Ca^{2+} release* (CICR). Relaxation of cardiac muscle occurs as calcium ions are pumped from the sarcoplasm either into the sarcoplasmic reticulum or out of the cells.

The force of contraction in cardiac muscle is even more closely linked to the initial length of the sarcomeres than that in skeletal muscle. As cardiac muscle is stretched beyond its normal resting length more force is generated until it is about 40 per cent longer than normal. Further stretching of the muscle then leads to a decline in tension development. These characteristics are summarized in Fig. 7.14.

In the normal course of events, the degree to which the muscle fibers of the heart are stretched is determined by the amount

of blood returning to the heart (the venous return). If the venous return is increased, the ventricular muscle will be stretched to a greater degree as the ventricle fills with blood and will respond with a more forceful contraction (Fig. 7.15(a)). Similarly, if the work of the left ventricle is increased by a rise in blood pressure while the venous return remains constant, the muscle will again be stretched to a greater degree as the ventricle fills. The ventricle will respond with a more forceful contraction. This principle is enshrined in Starling's law of the heart (see Chapter 15, p. 280). This property of cardiac muscle ensures that, in normal circumstances, the heart will pump out all the blood it receives.

The force with which the heart contracts varies according to the needs of the circulation. During exercise, the heart beats more strongly and frequently. These changes are mediated by the sympathetic nerves that innervate the SA node and the ventricles, and by circulating epinephrine secreted by the adrenal medulla. The change in rate is called a *chronotropic* effect while the change in the force of contraction (or contractility) is called an *inotropic* effect. The intrinsic contractility of the heart (its *inotropic state*) determines its efficiency as a pump. The increased contractility of the heart seen following stimulation of the sympathetic nerves, or with increased circulating epinephrine, is called a positive inotropic effect, while a decrease in contractility is a negative inotropic effect.

The positive inotropic effect occurs without any change in the length of the cardiac muscle fibers (Fig. 7.15(b)). There is no similar effect in skeletal muscle. The increase in contractility is

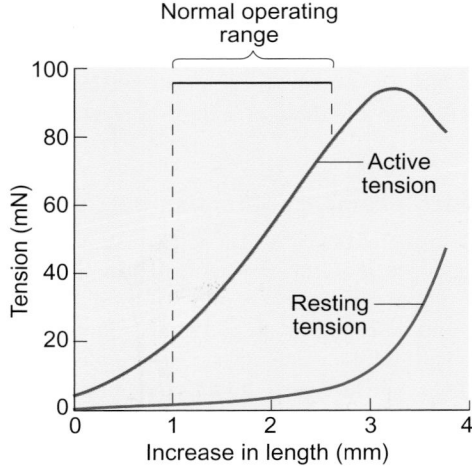

Fig. 7.14 The relationship between the resting length of a papillary muscle from the ventricle of the heart and the maximum isometric force generated in response to stimulation. The passive tension increases steeply as the muscle is stretched. Note that cardiac muscle normally operates on the ascending phase of the active tension curve, unlike skeletal muscle which operates around the peak (see Fig. 7.8 for data for a skeletal muscle).

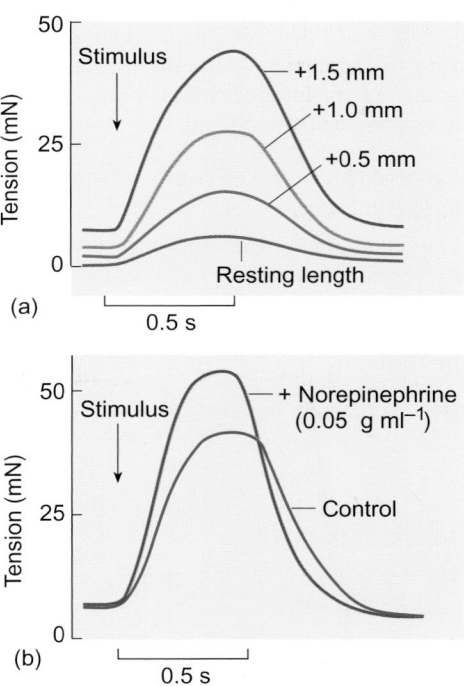

Fig. 7.15 Comparison of the intrinsic and extrinsic regulation of the force of contraction in cardiac muscle. (a) The effect of increasing the initial length on the force of contraction is shown. Note the increase in resting tension with each 0.5 mm increment of the muscle above its resting length. As the muscle is stretched, the force developed in response to stimulation increases. (b) The effect of norepinephrine on the force of contraction. In this case the increased force of contraction occurs for the same initial resting length.

Summary

1. The beating of the heart is due to an intrinsic or myogenic rhythm. Pacemaker cells in the SA node set the rate at which the heart beats. In these cells the action potential is preceded by a pacemaker potential, which sets the frequency of action potentials and thus the intrinsic rhythm of the normal heart.

2. In the atria and ventricles, the upstroke of the action potential is due to a rapid increase in the permeability of the membrane to sodium ions, while the long plateau phase is maintained by an influx of calcium ions. Repolarization is due to an increase in the permeability of the membrane to potassium ions.

3. The action potential of cardiac muscle varies according to the position of the cells in the heart but it is always of long duration (150–300 ms). The contractile response of cardiac muscle largely overlaps the action potential. Consequently, for much of the contractile response, cardiac muscle cannot be re-excited and this prevents it from undergoing tetanic contractions.

4. As in skeletal muscle, the contractile response of cardiac muscle is triggered by a rise in intracellular free Ca^{2+}. Shortening occurs by the relative movement of actin and myosin filaments by the same mechanism as that seen in skeletal muscle. The long duration of the contraction is sustained by a steady influx of Ca^{2+} during the plateau phase of the action potential.

5. The force of contraction of cardiac muscle is determined by two factors: the initial length of the cardiac fibers (the Frank–Starling law) and the inotropic effect whereby the fibers can develop increased force for the same initial fiber length in response to circulating epinephrine or stimulation of the sympathetic nerves.

caused by an increase in calcium influx following activation of β-adrenergic receptors and the subsequent generation of cyclic AMP. The cyclic AMP activates protein kinase A, which phosphorylates the calcium channels of the plasma membrane. The phosphorylated channels remain open for longer following depolarization and this, in turn, leads to an increased calcium influx that results in an increase in the force of contraction.

7.6 Smooth muscle

Smooth muscle is the muscle of the internal organs such as the gut, blood vessels, bladder, and uterus. It forms a heterogeneous group with a range of physiological properties. Each type of smooth muscle is adapted to serve a particular function. In some cases the muscle must maintain a steady contraction for long periods of time and then rapidly relax (as in the case of the sphincter muscles controlling emptying of the bladder and rectum). In others the muscles are constantly active (such as those of the stomach and small intestine). A muscle may express different properties at different times, as in the case of uterine muscle which must be quiescent during pregnancy but contract

forcefully during labor. Therefore it is not surprising that the smooth muscle serving a specific function will have distinct properties. For this reason, rather than classifying smooth muscles into particular types, it is more useful to determine how the properties of a specific muscle are adapted to serve its particular function.

Each smooth muscle consists of sheets of many small spindle-shaped cells (Fig. 7.1) linked together by two types of junctional contact as shown in Fig. 7.16. These are mechanical attachments between neighboring cells and gap junctions, which provide electrical continuity between cells and thus provide a pathway for the passage of electrical signals between cells. Each smooth muscle cell has a single nucleus and is about 2–5 μm in diameter at its widest point and about 50–200 μm in length. In some tissues, such as the alveoli of the mammary gland and in some small blood vessels, the smooth muscle cells are arranged in a single layer known as myoepithelium. Myoepithelial cells have broadly similar physiological properties to other smooth muscle cells.

In smooth muscle, no cross-striations are visible under the microscope (hence its name) but, like skeletal and cardiac muscle, smooth muscle contains actin and myosin filaments. These are not arranged in a regular manner like those of skeletal and cardiac muscle but are arranged in a loose lattice with the filaments running obliquely across the smooth muscle cells as

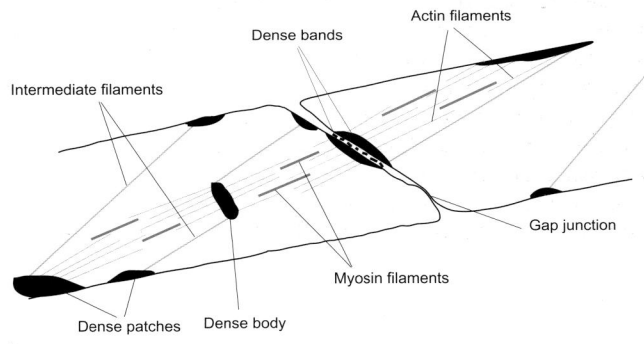

a)

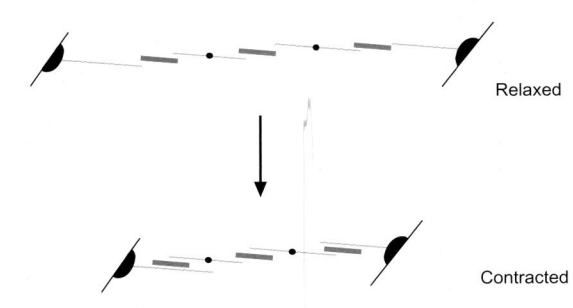

(b)

Fig. 7.16 The organization of the contractile elements of smooth muscle fibers. (a) Note the points of close contact for mechanical coupling (dense bands) and the gap junction for electrical signaling between cells. (b) A simple model of the contraction of smooth muscle. As the obliquely running contractile elements contract, the muscle shortens.

shown in Fig 7.16(a). The ratio of actin filaments to myosin filaments is much higher in smooth muscle than in skeletal muscle. (The ratio of thin to thick filaments is around 10:1 in smooth muscle compared with 2:1 in skeletal muscle.) In addition to actin and myosin filaments, smooth muscle cells contain cytoskeletal intermediate filaments that assist in transmission of the force generated during contraction to the neighboring smooth muscle cells and connective tissue. While there are no Z lines in smooth muscle, they have a functional counterpart in *dense bodies* that are distributed throughout the cytoplasm and which serve as attachments for both the thin and intermediate filaments. The filaments of contractile proteins are attached to the plasma membrane at the junctional complexes between neighboring cells, and this is illustrated in Fig. 7.16. Smooth muscle cells do not have a T system and the sarcoplasmic reticulum is not as extensive as that found in other types of muscle. However, they do have a large number of small membrane infoldings known as caveolae, which may serve to increase their surface area for ion fluxes. Some of the properties of smooth, cardiac, and skeletal muscle are compared in Table 7.1.

Smooth muscle is innervated by fibers of the autonomic nervous system which have varicosities along their length. These varicosities correspond to the nerve endings of the motor axons of the neuromuscular junction (see Chapter 6). In some tissues each varicosity is closely associated with an individual muscle cell (e.g. the piloerector muscles of the hairs), while in others the axon varicosities remain in small bundles within the bulk of the muscle and are not closely associated with individual fibers (e.g. the smooth muscle of the gut). The varicosities release their neurotransmitter into the space surrounding the muscle fibers rather than onto a clearly defined synaptic region as is the case at the neuromuscular junction of skeletal muscle. The neurotransmitter receptors are distributed over the surface of the cells instead of being concentrated at one region of the membrane as they are at the motor end-plate.

In many tissues, particularly those of the viscera, the individual smooth muscle cells are grouped loosely into clusters that extend in three dimensions. Gap junctions connect the cells so that the whole muscle behaves as a functional syncytium. In this type of muscle, activity originating in one part spreads throughout the remainder. This is known as *single-unit smooth muscle*. The smooth muscle of the gut, uterus and bladder are good examples of single-unit smooth muscle. In some tissues, such as the gut, there are regular spontaneous contractions (*myogenic contractures*) that originate in specific pacemaker areas (Fig. 7.17).

The activity of many single-unit muscles is strongly influenced by hormones circulating in the bloodstream as well as by the activity of autonomic nerves. For example, during pregnancy

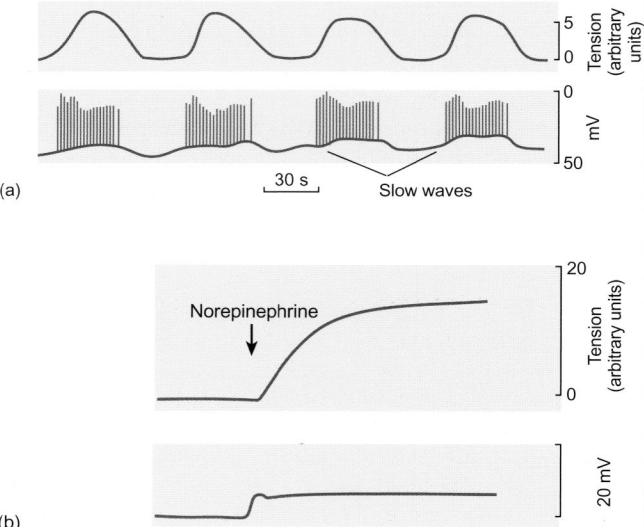

Fig. 7.17 Patterns of electrical activity recorded from different kinds of smooth muscle. (a) The electrical and mechanical response of uterine smooth muscle isolated from a rat giving birth. Note the slow waves leading to a burst of action potentials and the slow and sustained development of tension. (b) The development of tension in a sheep carotid artery following application of norepinephrine. Note that in this case the development of tension is not preceded by an action potential.

TABLE 7.1	Comparison of the properties of cardiac, skeletal, and smooth muscle		
Property	**Skeletal muscle**	**Cardiac muscle**	**Smooth muscle**
Cell characteristics	Very long cylindrical cells with many nuclei	Irregular rod-shaped cells usually with a single nucleus	Spindle-shaped cells with a single nucleus
Maximum cell size (length × diameter)	30 cm × 100 μm	100 μm × 15 μ	200 μm × 5 μm
Visible striations	Yes	Yes	No
Myogenic activity	No	Yes	Yes
Motor innervation	Somatic	Autonomic (sympathetic and parasympathetic)	Autonomic (sympathetic and parasympathetic
Type of contracture	Phasic	Rhythmic	Mostly tonic, some phasic
Basis of muscle tone	Neural activity	None	Intrinsic and extrinsic factors
Cells electrically coupled	No	Yes	Yes
T system	Yes	Only in ventricular muscle	No
Mechanism of excitation–contraction coupling	Action potential and T system	Action potential and T system	Action potential, Ca^{2+} channels, and second messengers
Force of contraction regulated by hormones	No	Yes	Yes

the motor activity of the uterine muscle (the myometrium) is much reduced due to the presence of high circulating levels of the hormone progesterone. In this instance, the progesterone decreases the expression of certain proteins involved in the formation of gap junctions (e.g. connexin 43) and this reduces the excitability of the myometrium. Another steroid, estriol, antagonizes this effect and increases the expression of gap junction proteins, thus increasing the excitability of the muscle. The increase in excitability is important during parturition.

Certain smooth muscles do not contract spontaneously and are normally activated by motor nerves. The muscles themselves are organized into motor units similar to those of skeletal muscle except that the motor units are more diffuse. These muscles are known as *multi-unit smooth muscle*. The intrinsic muscles of the eye (e.g. the smooth muscle of the iris), the piloerector muscles of the skin, and the smooth muscle of the larger blood vessels are all examples of the multi-unit type. Nevertheless, the distinction between the two types of muscle is not rigid as, for example, the smooth muscle of certain arteries and veins shows spontaneous activity but also responds to stimulation of the appropriate sympathetic nerves.

Excitation–contraction coupling in smooth muscle

The membrane potential of smooth muscle is often quite low, typically about –50 to –60 mV, which is some 30 mV more positive than the potassium equilibrium potential. This low value of the resting membrane potential arises because the sodium ion permeability of the cell membrane is about one-fifth that of potassium (compared with a Na^+/K^+ permeability ratio of about 1:100 for skeletal muscle). When a smooth muscle fiber generates an action potential, the depolarization depends on an influx of both sodium and calcium ions, although the exact contribution of each ion depends on the individual muscle. For example, in the smooth muscle of the vas deferens and the gut, the action potential appears to be mainly dependent on an influx of calcium ions. In contrast, the action potentials of the smooth muscle of the bladder and ureters depends on an influx of sodium ions in just the same way as the action potential of skeletal muscle. However, unlike skeletal muscle, the action potential of this smooth muscle lasts 10–50 ms (i.e. 5 to 10 times as long). In addition, in some smooth muscles the action potential may develop a prolonged plateau phase similar in appearance to that seen in cardiac muscle.

In single-unit smooth muscle, certain cells act as pacemaker cells and these show spontaneous fluctuations of the membrane potential known as slow waves. During an excitatory phase, slow-wave activity builds up progressively until the membrane potential falls below about –35 mV when a series of action potentials is generated. These are propagated through many cells via gap junctions and the muscle slowly contracts. This pattern of electrical activity and force generation is seen in the gut during peristalsis and in uterine muscle during parturition (see Fig. 7.17(a)).

In many smooth muscles the pacemaker activity is regulated by the activity of the sympathetic and parasympathetic nerves. In the intestine, the release of acetylcholine from the parasympathetic nerve varicosities causes a depolarization that results in the slow waves occurring at a more depolarized membrane potential. Consequently, the membrane potential exceeds the threshold for action potential generation for a greater period of time during the slow-wave cycle, so making the muscle more active. Conversely, activity in the sympathetic nerves results in membrane hyperpolarization, thus maintaining the membrane potential below threshold for a longer period of time and inhibiting contractile activity. Neither acetylcholine nor norepinephrine appear to have a direct action on the pacemaker activity, which is intrinsic to the muscle itself. In other smooth muscles, the role of the sympathetic and parasympathetic innervation is reversed, with sympathetic activity resulting in excitation and parasympathetic activity resulting in inhibition (see Chapter 10 for further details).

Like skeletal and cardiac muscle, smooth muscle contracts when intracellular Ca^{2+} rises. As smooth muscle does not possess a T system, the rise in intracellular Ca^{2+} can either occur as a result of calcium influx through calcium channels in the plasma membrane or by the release of calcium from the sarcoplasmic reticulum following the activation of receptors that increase the formation of IP_3 (see Chapter 5). The contractile response is slower and much longer lasting than that of skeletal and cardiac muscle (Fig. 7.17). Furthermore, not all smooth muscles require an action potential to occur before they contract. In some large blood vessels (e.g. the carotid and pulmonary arteries) norepinephrine causes a strong contraction but only a small change in membrane potential (Fig. 7.17(b)). In this case, the contractile response is initiated by a rise in intracellular Ca^{2+} in response to the generation of IP_3 following the activation of α-adrenoceptors as described in Chapter 5.

The excitation–contraction coupling of smooth muscle is regulated in a different manner to that of cardiac and skeletal muscle. The thin filaments of smooth muscle do not possess the regulatory protein troponin. Other proteins regulate the cross-bridge cycling of actin and myosin. In most smooth muscles the

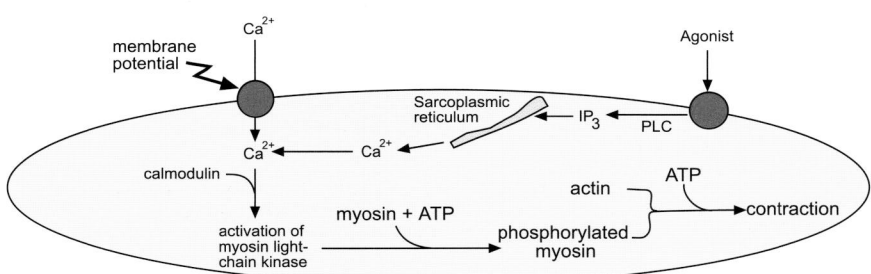

Fig. 7.18 Excitation–contraction coupling in smooth muscle. As in cardiac and skeletal muscle, the development of tension is regulated by calcium which can enter the sarcoplasm via either voltage-gated calcium channels in the plasma membrane or from the intracellular calcium stores in the sarcoplasmic reticulum. Calcium binds to calmodulin, which regulates the interaction between actin and myosin via myosin light-chain kinase.

regulation is performed by a calcium-binding protein called calmodulin which combines with calcium to form a calcium–calmodulin complex which activates an enzyme called myosin light-chain kinase (Fig. 7.18). This enzyme phosphorylates the regulatory region on the myosin light chains. When the myosin light chains are phosphorylated, the myosin head groups can bind to actin and undergo cross-bridge cycling. In other smooth muscles, the phosphorylation of the myosin light chains is regulated in a more complex manner by other proteins (e.g. caldesmon) that are associated with the thin filaments.

The slow rate at which smooth muscle is able to hydrolyze ATP explains the slowness of the contractile process in smooth muscle compared with that seen in cardiac and skeletal muscle. Smooth muscle relaxation occurs when intracellular free calcium falls. This leads to dephosphorylation of myosin light-chain kinase by a specific phosphatase. The direct dependence of the activity of the myosin light-chain kinase on the availability of the calcium–calmodulin complex permits the contractile response to be smoothly graded. Moreover, the slow and steady generation of tension enables smooth muscle to generate and maintain tension with relatively little expenditure of energy.

Smooth muscle is able to maintain a steady level of tension called tone

The smooth muscle of the hollow organs maintains a steady level of contraction that is known as *tone* or tonus. Tone is important in maintaining the capacity of the hollow organs. For example, the flow of blood through a particular tissue depends on the caliber of the arterioles and this, in turn, is determined by the tone in the smooth muscle of the vessel wall (i.e. by the degree of contraction of the smooth muscle). Smooth muscle tone depends on many factors, which may be either extrinsic or intrinsic to the muscle. Extrinsic factors include activity in the autonomic nerves and circulating hormones, while intrinsic factors include the slow rate of cross-bridge cycling, the response to stretch, local metabolites, locally secreted chemical agents (e.g. NO in the blood vessels) and temperature. Thus, smooth muscle tone does not depend solely on activity in the autonomic nerves or on circulating hormones.

Length–tension relationships in single-unit smooth muscle

If a smooth muscle is stretched, there is a corresponding increase in tension immediately following the stretch. This is followed by a progressive relaxation of the tension towards its initial value. This property is unique to smooth muscle and is called stress relaxation or plasticity. The converse happens if the tension on smooth muscle is decreased (e.g. by voiding the contents of a hollow organ such as the gut or bladder). In this case the tension initially falls but returns to its original level in a short period of time. This is called reverse stress relaxation. By adjusting its tension in this way smooth muscle is able to maintain a level of tone in the wall of a hollow organ. This permits the internal diameter of the organ to be adjusted to suit the volume of material it contains

Compared with skeletal or cardiac muscle, smooth muscle can shorten to a far greater degree. A stretched striated muscle can shorten by as much as a third of its resting length while a normal resting muscle would shorten by perhaps a fifth, perfectly adequate for it to perform its normal physiological role. In contrast, a smooth muscle may be able to shorten by more than two-thirds of its initial length. This unusual property is conferred by the loose arrangement of the thick and thin myofilaments in smooth muscle cells. This is a crucial adaptation as the volume contained by an organ such as the bladder depends on the cube of the length of the individual muscle fibers. Thus the ability of smooth muscle to change its length to such a large degree permits the hollow organs to adjust to much wider variations in the volume of their contents than would be possible for skeletal muscle. A simple calculation shows the advantage conferred by this property of smooth muscle. When the bladder is full it contains about 400 ml of urine. Almost all the urine is expelled when the smooth muscle in the bladder wall contracts. If the bladder were a simple sphere, its circumference would be about 30 cm when it contained 400 ml of urine but would be only about 6 cm if 4.0 ml of urine were left after it had emptied (i.e. if it contained only 1 per cent of the original volume). This corresponds to a change in muscle length of about 80 per cent. However, if the bladder were made of skeletal muscle, the maximum length change would be only about 30 per cent and the bladder would only be able to void about 70 per cent of its contents, leaving behind about 120 ml of urine.

Summary

1. Smooth muscle consists of sheets containing many small spindle-shaped cells linked together at specific junctions. Smooth muscle cells contain actin and myosin, but these proteins are not arranged in regular sarcomeres. Instead, each smooth muscle cell has a loose matrix of contractile proteins that is attached to the plasma membrane at the junctional complexes between neighboring cells.

2. Smooth muscle is of two types: single-unit (or visceral) and multi-unit. Single-unit smooth muscle shows myogenic activity and behaves as a syncytium. Multi-unit smooth muscle has little spontaneous activity and is activated by impulses in specific motor nerves. Fibers from the autonomic nervous system innervate both types of smooth muscle.

3. The contractile response of smooth muscle is slow and is regulated by the activity of myosin light-chain kinase, which in turn depends on the level of free calcium in the sarcoplasm.

4. Smooth muscle maintains a steady level of tension known as tone. The tone exhibited by a particular muscle may be increased or decreased by circulating hormones, by local factors, or by activity in autonomic nerves.

5. Smooth muscle is much more plastic in its properties than other types of muscle and is able to adjust its length over a much wider range than skeletal or cardiac muscle.

Recommended reading

Histology

Junquiera, L.C. and Carneiro, J. (2003). *Basic histology*, (10th edition), Chapter 10. McGraw-Hill, New York.

Biochemistry of muscle contraction

Alberts, B., Johnson, A., Lewis, J., Raff, M., Roberts, K., and Walter, P. (2002). *Molecular biology of the cell* (4th edn), pp. 949–68. Garland, New York.

Berg, J.M., Tymoczko, J.L., and Stryer, L. (2002). *Biochemistry* (5th edn), pp. 956–62. Freeman, New York.

Physiology

Aidley, D.J. (1998). *The physiology of excitable cells* (4th edn), Chapters 18–21. Cambridge University Press, Cambridge.

Åstrand, P.-O., Rodahl, K., Dahl, H., and Stromme, S. (2003). *Textbook of work physiology* (4th edn), Chapter 3. Human Kinetics, Champaign, IL.

Jones, D.A., Round, J.M., and de Haan, A. (2004). *Skeletal muscle from molecules to movement*. Churchill-Livingstone, Edinburgh.

Levick, J.R. (2002). *An introduction to cardiovascular physiology* (5th edn), Chapter 11. Hodder Arnold, London.

Medicine

Ledingham, J.G.G., and Warrell, D.A. (eds.) (2000). *Concise Oxford textbook of medicine*, Chapters 13.32–13.40. Oxford University Press, Oxford.

Multiple choice questions

Each statement is either true or false. Answers are given below.

1. The following statements relate to the structure of muscle tissue:
   a. All muscle cells contain actin and myosin.
   b. Skeletal muscle has the same structure as cardiac muscle.
   c. In skeletal and cardiac muscle, actin is the major protein of the thin filaments.
   d. The myofibrils of skeletal muscle are surrounded by the sarcoplasmic reticulum.
   e. Skeletal muscle has a system of T tubules while smooth muscle does not.

2. The following statements relate to the role of Ca^{2+} in muscle contraction:
   a. The sarcoplasmic reticulum acts as a store of Ca^{2+} for the contractile process.
   b. Ca^{2+} entry across the plasma membrane is important in sustaining the contraction of cardiac muscle.
   c. A muscle will relax when intracellular Ca^{2+} is raised.
   d. The tension of a smooth muscle fiber is partly regulated by second messengers.
   e. A rise in intracellular Ca^{2+} allows actin to interact with myosin.

3. In skeletal muscle:
   a. A motor unit consists of a single motor neuron and the muscle fibers it innervates.
   b. The action potential propagates from the neuromuscular junction to both ends of the muscle fiber.
   c. The muscle action potential is an essential step in excitation–contraction coupling.
   d. The muscle fibers are electrically coupled so that one nerve fiber can control the activity of several muscle fibers.
   e. The energy for muscle contraction comes from the hydrolysis of ATP.
   f. The muscle can contract by more than two-thirds of its resting length.

4. In cardiac muscle:
   a. The cardiac action potential is initiated by action potentials in sympathetic nerve fibers.
   b. The action potential is much longer than that of skeletal muscle.
   c. All the Ca^{2+} required for contraction enters a myocyte during the plateau phase of the action potential.
   d. Only the SA node cells show pacemaker activity.
   e. The action potential spreads from cell to cell via gap junctions.
   f. The force of contraction is increased by circulating epinephrine.

5. In smooth muscle:
   a. Autonomic nerves never innervate individual muscle fibers.
   b. In some muscles the action potential is due to pacemaker activity.
   c. All smooth muscles behave as a single motor unit.
   d. In many types of smooth muscle the action potential results from an increase in the permeability of the sarcolemma to both Ca^{2+} and Na^+.
   e. Some smooth muscles contract without an action potential.

Answers

1. Skeletal and cardiac muscle are both striated but the organization of each differs considerably. Skeletal muscle fibers are
   large multinucleated cells with a well-developed T system.
   Unlike cardiac muscle, individual fibers are not coupled via
   gap junctions.
   a. True
   b. False
   c. True
   d. True
   e. True

2. Muscles relax when Ca^{2+} is pumped out of the sarcoplasm
   (i.e. when intracellular Ca^{2+} falls).
   a. True
   b. True
   c. False
   d. True
   e. True

3. Skeletal muscle fibers act independently of each other, unlike
   cardiac and smooth muscle cells which are linked by gap
   junctions. Skeletal muscle contracts by about one-fifth of its
   resting length. Smooth muscle can contract by more than
   two-thirds of its resting length.
   a. True
   b. True
   c. True
   d. False

   e. True
   f. False

4. Although some of the Ca^{2+} required for contraction does
   come from influx across the sarcolemma, the majority comes
   from the sarcoplasmic reticulum. The myocytes of the SA
   node generate spontaneous action potentials that set the
   heart rate. The atrial and ventricular myocytes do not normally generate pacemaker potentials. The heart rate is
   increased by stimulation of the cardiac sympathetic nerves
   and inhibited by stimulation of the vagus nerve.
   a. False
   b. True
   c. False
   d. True
   e. True
   f. True

5. Autonomic nerves have varicosities in the terminal regions
   which usually secrete neurotransmitter in a diffuse manner
   close to several fibers, but in the piloerector muscles each
   varicosity is closely associated with a single fiber. Smooth
   muscles can be broadly considered to act either as a single
   unit (e.g. the bladder) or as a multi-unit (e.g. the intrinsic
   muscles of the eye).
   a. False
   b. True
   c. False
   d. True
   e. True

Sensory systems

After reading this chapter you should understand:

- The principles by which sensory receptors derive information about the environment
- The physiological basis of somatic sensation: the senses of touch, pressure, vibration, and temperature
- The pathophysiology of pain and itch
- The physiology of the eye and the visual pathways
- The physiology of the ear and the auditory pathways
- The organization of the vestibular system and its role in the sense of balance
- The physiological basis of the senses of smell (olfaction) and taste

8.1 Introduction

We smell the air, taste our food, feel the earth under our feet, hear and see what is around us. To do all this, and more, we must have some means of converting the physical and chemical properties of the environment into nerve impulses, which are used for signaling between the neurons of the nervous system. The process by which specific properties of the external (or internal) environment become encoded as nerve impulses is called *sensory transduction*. It is carried out by specialized structures called *sensory receptors*, often simply called receptors.

Sensory receptors can be classified in several ways, of which two will be considered here (Table 8.1). They can be classified on the basis of the specific environmental qualities to which they are sensitive—chemoreceptors, mechanoreceptors, nociceptors, photoreceptors, and thermoreceptors. Alternatively, they can be classified according to the source of the quality that they sense. Thus there are receptors that sense events that originate at some distance from the body—the eye, ear, and nose—which are sometimes called *teleceptors*. Others sense changes occurring in the immediate external environment—touch, pressure, and temperature. These are called *exteroceptors*. Then there are receptors that signal changes in the internal environment—the *interoceptors*. These sense blood pressure (baroreceptors), the oxygen

TABLE 8.1   Classification of sensory receptors

Receptor type		Other classification	Example
Mechanoreceptors	Special senses (ear)	Teleceptor	Cochlear hair cells
		Interoceptor	The hair cells of the vestibular system
	Muscle and joints	Proprioceptor	Muscle spindles Golgi tendon organs
	Skin and viscera	Exteroceptors	Pacinian corpuscle Ruffini ending Meissner's corpuscle Bare nerve endings
	Cardiovascular	Interoceptor	Arterial baroreceptors (sense high pressures) Atrial volume receptors (low-pressure receptors)
Chemoreceptors	Special senses	Teleceptor	Olfactory receptors
		Exteroceptors	Taste receptors
	Skin and viscera	Exteroceptors	Nociceptors
		Interoceptors	Nociceptors Glomus cells (carotid body, sense arterial P_{O_2}) Hypothalamic osmoreceptors and glucose receptors
Photoreceptors	Special senses	Teleceptor	Retinal rods and cones
Thermoreceptors	Skin	Exteroceptor	Warm and cold receptors
	CNS	Interoceptor	Temperature-sensing hypothalmic neurons

and carbon dioxide levels of the blood (chemoreceptors), and substances released following tissue damage (nociceptors). Other receptors provide information about our position in space and the disposition of our limbs—the gravitational receptors and proprioceptors.

There is usually one kind of stimulus to which a particular receptor will be especially sensitive. This is known as the *adequate stimulus*. Thus vibration is the adequate stimulus for the Pacinian corpuscles of the skin, small changes in temperature excite specific cutaneous thermoreceptors, light is the adequate stimulus for the photoreceptors of the eye, and so on.

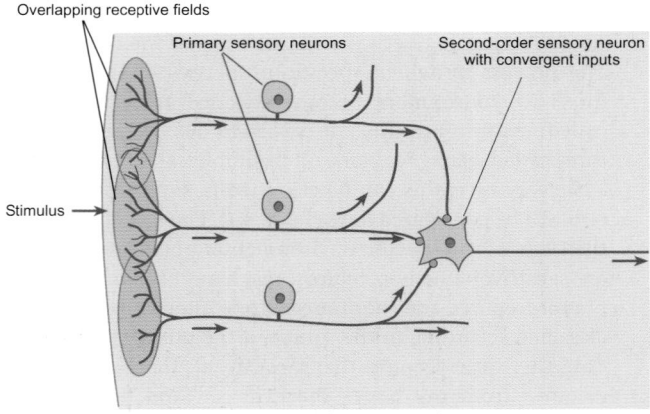

Fig. 8.1 A simple diagram illustrating overlap between receptive fields of neighboring primary afferents and their convergence onto a second-order sensory neuron. The primary afferents also send axon branches to other spinal neurons as indicated by the arrows.

The organization of sensory pathways in the CNS

Receptors send their information to the CNS (the brain and spinal cord) via *afferent nerve fibers* often called primary afferents. A given afferent nerve fiber often serves a number of receptors of the same kind. It will respond to a stimulus over a certain area of space and intensity that is called its *receptive field*. The receptive fields of neighboring afferents often overlap as shown in Fig. 8.1.

Individual nerve cells in the CNS may receive inputs from many primary afferent fibers so that the receptive field of a particular sensory neuron in the spinal cord is usually larger than that of the afferents to which it is connected. This is called *convergence* and is illustrated in Figs 8.1 and 8.2. The second-order sensory nerve cells then make contact with many other nerve cells as information is processed by the CNS. Thus a specific piece of sensory information tends to be spread amongst more and more nerve cells. This is called *divergence* (Fig. 8.2). Convergence and divergence are an essential part of the processing of sensory information. Ultimately, this information is incorporated into an internal representation of the world that the brain can use to determine appropriate patterns of behavior.

To avoid the excessive spread of excitation in response to a single stimulus, nerve cells in the CNS receive inhibitory connections from neighboring cells via small neurons known as interneurons as shown in Fig. 8.3. In this way a strongly excited cell will exert a powerful inhibitory effect on those of its neighbors that were less intensely excited. This is known as *lateral inhibition* or surround inhibition which features prominently in the processing of sensory information at all levels of the CNS. At the highest levels, in the cerebral cortex for example, this type of neural interaction allows the brain to extract informa-

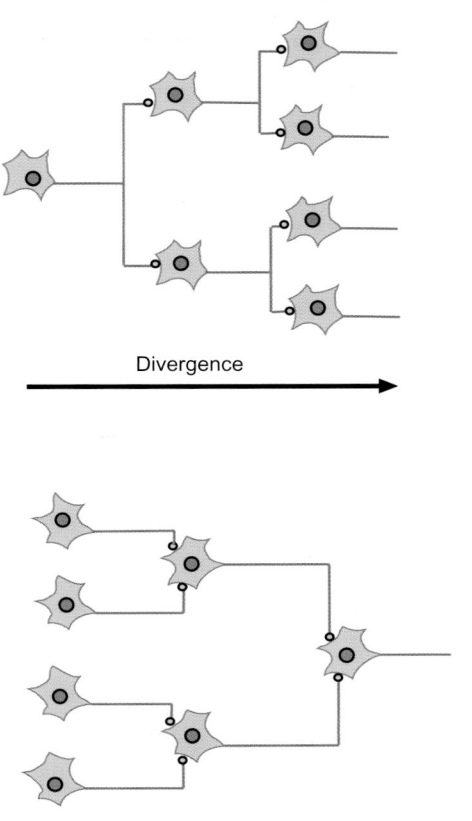

Fig. 8.2 A simple diagram to illustrate the principles of neural divergence (top) and convergence (bottom). See text for further details.

tion about the specific features of a stimulus. For example, the visual cortex is able to discriminate the position of an object in space, its illumination, and its relation to other nearby objects. All of this complex processing is achieved by the interplay of excitatory and inhibitory synaptic connections of the kind illustrated in Fig. 8.3. The synaptic basis of excitation and inhibition was described in Chapter 6.

Principles of transduction

Although different receptors respond to environmental stimuli in different ways, in all cases the adequate stimulus ultimately leads to a change in membrane potential called a *receptor potential*. For most sensory receptors, stimulation causes cation-permeable ion channels to open which leads to the depolarization of the fiber and the generation of a receptor potential. When the threshold is reached, an action potential is generated. The magnitude and duration of a receptor potential governs the number and frequency of action potentials transmitted by the afferent nerve

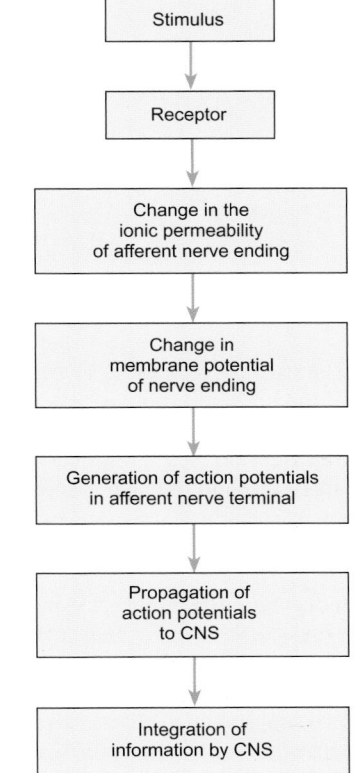

Fig. 8.4 The main steps in sensory transduction for cutaneous receptors.

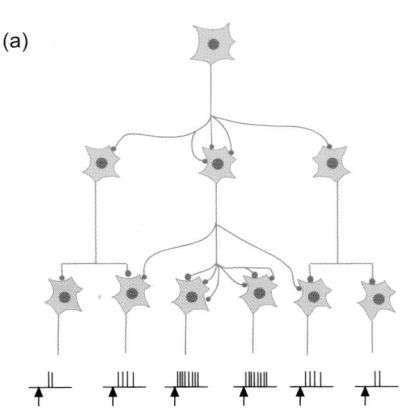

Wide spatial spread of activity

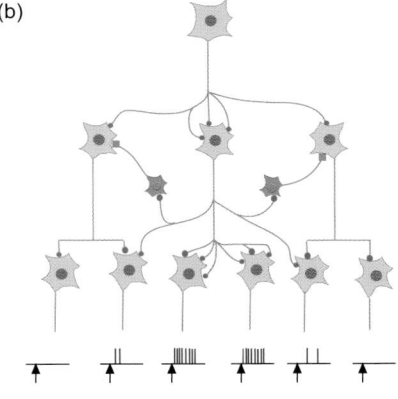

Narrow spread of activity due to lateral inhibition by interneurons

Fig. 8.3 Lateral inhibition limits the spread of excitation within neural networks.
(a) This diagram shows how excitation is spread across a population as a consequence of divergent neural connections.
(b) This spread can be limited by the activity of interneurons shown in red. The arrows indicate the timing of the stimulus.

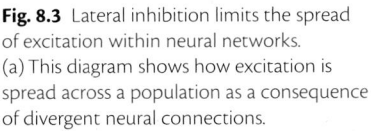

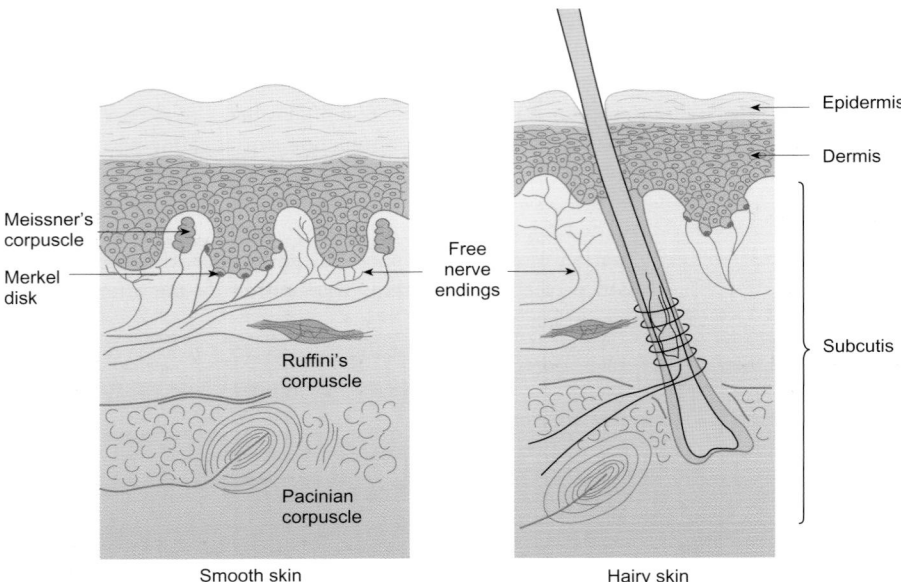

Fig. 8.5 The disposition of the various types of sensory receptor in smooth (glabrous) skin and hairy skin. Note the thickness of the epidermis in glabrous skin and the location of Meissner's corpuscles between the dermal ridges.

fibers to the CNS. The basic steps in sensory transduction are illustrated in Fig. 8.4.

Receptors may be bare nerve endings or they may consist of specialized cells in close association with a nerve fiber. Bare nerve endings are commonly found in the cornea and the superficial region of the skin. Some respond to touch and temperature while others respond to chemicals released into their local environment. For example, the sensory endings of pain fibers can be excited by a variety of substances including hydrogen ions.

In the skin, encapsulated receptors are mechanoreceptors; examples are Merkel's disks and Pacinian corpuscles (see also Fig. 8.5). The intricate structure of these receptors permits the nervous system to discriminate the specific features of a stimulus. The role of the encapsulating cells in determining the characteristics of a particular kind of receptor is illustrated by the Pacinian corpuscle which is surrounded by many layers of flattened fibroblasts. An intact Pacinian corpuscle responds to deformation by a brief depolarization when the stimulus is first

applied and when it is removed ('on' and 'off' responses) as shown in Fig. 8.6. If the associated fibroblasts are removed following treatment with enzymes, the naked nerve ending remains depolarized for as long as the mechanical stimulus is applied. Thus the fibroblasts allow the intact corpuscle to respond to rapid tissue movement (such as vibration) rather than maintained pressure.

The coding of stimulus intensity and duration

The nervous system needs to establish the location, physical nature, and intensity of all kinds of stimuli. Since most receptors respond to very specific stimuli, the primary afferent fibers to which they are connected can be considered as 'labeled lines'. For example, the skin has receptors that respond selectively to touch and others that respond to a small fall or a small rise in temperature. Therefore the activation of a specific population of receptors will inform the CNS of the nature and location of the stimulus. The intensity of the stimulus is coded by both the number of active receptors and the number of action potentials

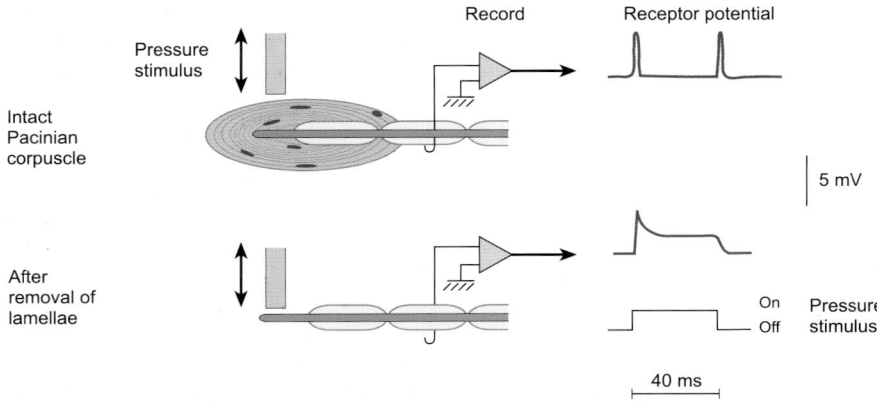

Fig. 8.6 The role of the fibroblast lamellae in shaping the response of the Pacinian corpuscle to a pressure stimulus. The experimental arrangement is shown on the left and the receptor potentials are shown on the right. Note that the intact corpuscle signals the onset and offset of the stimulus while the receptor potential is maintained for the duration of the stimulus in the desheathed corpuscle.

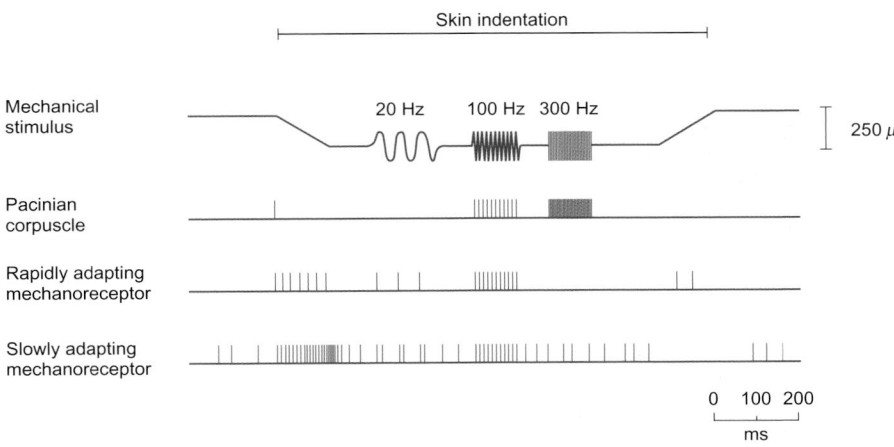

Fig. 8.7 The responses of three different types of mechanoreceptor to pressure stimuli applied to the skin. The top record shows the characteristics of the mechanical stimulus. The lower three records show the responses of a Pacinian corpuscle, a rapidly adapting mechanoreceptor, and a slowly adapting mechanoreceptor. Each vertical spike represents an action potential. Note that the Pacinian corpuscle responds to maintained skin indentation with a single action potential while the slowly adapting mechanoreceptor continues to generate action potentials. Only the Pacinian corpuscle is able to respond to the 300 Hz vibration of the skin.

that each receptor elicits; the timing of the sequence of action potentials signals its onset and duration.

Many receptors generate action potentials when they are first stimulated, but the action potential frequency then falls with time even though the intensity of the stimulus is unchanged. This property is known as *adaptation*. Some receptors respond to the onset of a stimulus with a few action potentials and then become quiescent. This type of receptor is called a *rapidly adapting* receptor. Other receptors maintain a steady flow of action potentials for as long as the stimulus is maintained. These are known as *slowly adapting* or *non-adapting* receptors. These different types of response are illustrated in Fig. 8.7.

Summary

1. The external and internal environment is continuously monitored by the sensory receptors. Each kind of receptor is excited most effectively by a specific type of stimulus known as its adequate stimulus. The nerve fibers that convey information from the sensory receptors to the CNS are known as afferent or sensory nerve fibers.

2. The process by which an environmental stimulus becomes encoded as a sequence of nerve impulses in an afferent nerve fiber is called sensory transduction. Different kinds of receptor are activated in different ways but the first stage in sensory transduction is the generation of a receptor potential.

3. Different types of receptor show differing degrees of adaptation. Some (the rapidly adapting receptors) signal the onset and offset of a stimulus, while others (the slowly adapting receptors) continuously signal the intensity of the stimulus.

8.2 The somatosensory system

The skin is the interface between the body and the outside world. It is richly endowed with receptors that sense pressure, touch, temperature, vibration, pain, and itch. The muscles and joints also possess sensory receptors that provide information concern-

ing the disposition and movement of the limbs. All this information is relayed by the afferent nerves of the somatosensory system to the brain and spinal cord. Careful exploration of the skin has shown that specific points are sensitive to touch, while others are sensitive to cooling or warming. There is little overlap between the different modalities of cutaneous sensation.

The various kinds of receptors present in the skin are illustrated in Fig. 8.5. Both bare nerve endings and encapsulated receptors are present. Each of the specific kinds of receptor subserves a specific submodality of cutaneous sensation (Table 8.2). By identifying a point sensitive to touch, for example, and then excising it and subjecting it to histological examination, it has been possible to associate particular types of receptor with specific modalities of sensation. Each kind of receptor is innervated by a particular type of nerve fiber. Pacinian corpuscles and Merkel's disks are innervated by relatively large Aβ myelinated nerve fibers while the bare nerve endings that subserve temperature and pain sensations are derived either from small Aδ myelinated fibers or from slowly conducting unmyelinated C-fibers (Table 8.2). (For the classification of nerve fibers, see Chapter 6.)

In general, the receptive fields of touch receptors overlap considerably as illustrated in Fig. 8.1. The finer the discrimination required, the higher is the density of receptors, the smaller their receptive fields, and the greater the degree of overlap. The receptive fields for touch are particularly small at the tips of the fingers and tongue (about 1 mm²) where fine tactile discrimination is required. In other areas, such as the small of the back, the buttocks, and the calf, the receptive fields are about 100 times larger.

The distance between two points on the skin which can just be detected as separate stimuli is closely allied to the density of touch receptors and the size of their receptive fields. This is known as the two-point discrimination threshold. Not surprisingly, the greatest discrimination is at the tips of the fingers, the tip of the tongue, and the lips. It is least precise for the skin of the back (Fig. 8.8). A loss of precision in two-point discrimination can be used to localize specific neurological lesions.

The sensitivity of the skin to touch can be assessed by measuring the smallest indentation that can be detected, either subject-

TABLE 8.2 The receptor types and modalities of the somatosensory system

Modality	Receptor type	Afferent nerve fiber type and conduction velocity
Touch	Rapidly adapting mechanoreceptor, e.g. hair follicle receptors, bare nerve endings, Pacinian corpuscles	Aβ 6–12 μm diameter 33–75 m s⁻¹
Touch and pressure	Slowly adapting mechanoreceptors, e.g. Merkel's cells, Ruffini end-organs	Aβ 6–12 μm diameter 33–75 m s⁻¹
	Bare nerve endings	Aδ 1–5 μm diameter 5–30 m s⁻¹
Vibration	Meissner's corpuscles Pacinian corpuscles	Aβ 6–12 μm diameter 33–75 m s⁻¹
Temperature	Cold receptors	Aδ 1–5 μm diameter 5–30 m s⁻¹
	Warm receptors	C-fibers 0.2–1.5 μm diameter 0.5–2.0 m s⁻¹
Pain	Bare nerve endings—fast 'pricking' pain	Aδ 1–5 μm diameter 5–30 m s⁻¹
	Bare nerve endings—slow burning pain, itch	C-fibers 0.2–1.5 μm diameter 0.5–2.0 m s⁻¹

ively or by recording the action potentials in a single afferent fiber. For the fingertips, an indentation of as little as 6–7 μm can be detected which corresponds to the diameter of a single red cell. Elsewhere on the hand, the skin is less sensitive to deformation so that, for example, an indentation of about 20 μm is

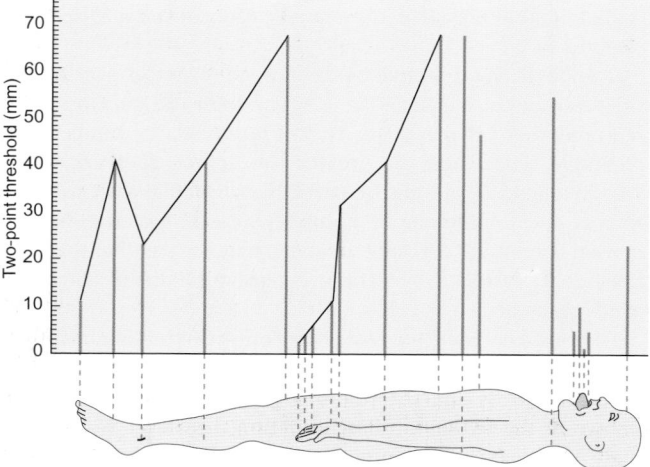

Fig. 8.8 The variation of two-point discrimination across the body surface. Each vertical line represents the minimum distance that two points can be distinguished as being separate when they are simultaneously stimulated. Note the fine discrimination achieved by the fingertips, lips, and tongue and compare this with the poor discrimination on the thigh, chest, and neck.

required to evoke an action potential in an afferent fiber serving the sense of touch on the palm. The skin of the back or that of the soles of the feet is even less sensitive to touch.

Thermoreceptors

The skin has two kinds of thermoreceptor. One type specifically responds to cooling of the skin and another responds to warming. Their receptive fields are small (about 1 mm²) and they do not overlap. Thus, exploration of the skin with a small temperature probe reveals specific points sensitive to cold or warmth. Histological examination of a cold- or warm-sensitive point shows only bare nerve endings. This suggests that the cutaneous thermoreceptors are probably a specific subset of bare nerve endings. The cold receptors are innervated by Aδ myelinated afferents while the warm receptors are innervated by C-fibers.

Cutaneous thermoreceptors are generally insensitive to mechanical and chemical stimuli and maintain a constant rate of discharge for a particular skin temperature. They respond to a change in temperature with an increase or decrease in firing rate, the cold receptors showing a maximal rate of discharge around 25–30 °C while the warm receptors have a maximal rate of discharge around 40 °C (Fig. 8.9). Thus, a given frequency of discharge from the cold receptors may reflect a temperature that is either above or below the maximum firing rate for a particular receptor. This ambiguity may explain the well known paradoxical sense of cooling when cold hands are being rapidly warmed by immersion in hot water.

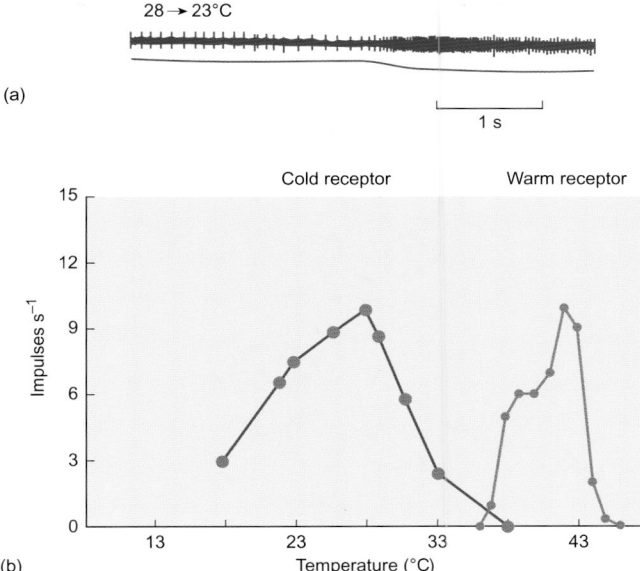

(a)

28 → 23°C

1 s

(b)

Cold receptor   Warm receptor

Fig. 8.9 Typical response patterns of a warm and a cold skin thermoreceptor. (a) The action potential discharge of a cold receptor increases as the skin is cooled from 28 to 23 °C. (b) The rate of action potential discharge as a function of temperature for a single cold and a single warm thermoreceptor.

Kinesthesia and haptic touch

People move through their environment—they lift objects, move them around, and feel their texture. The constant stimulation of different receptors prevents them adapting. As a result, the brain is provided with more information about an object than would be possible with a single contact. The active exploration of an object to determine its shape and texture is known as *haptic touch*, which is used to great effect by the blind.

We know the position of our limbs even when we are blindfold. This sense is called *kinesthesia*. Two sources of information provide the brain with information about the disposition of the limbs. These are the corollary discharge of the motor efferents to the sensory cortex (see Chapter 9), which provides information about the intended movement, and the sensory feedback that directly informs the sensory cortex of the actual progress of the movement. As the load on the muscles moving the limbs cannot

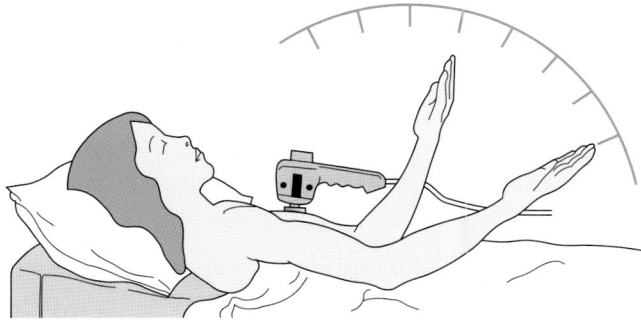

Fig. 8.10 An illustration of a limb-position mismatch induced by vibratory stimulation. The blindfold subject was asked to align her forearms while her left biceps was being vibrated. The mismatch illustrates the illusion in limb position sense elicited by the vibration.

be known by the brain in advance, both these sources of information are required (see Chapter 9 for further information).

The importance of the muscle spindles in kinesthesia is revealed by the vibration illusion. A normal subject can accurately replicate the position of one arm by moving the other to the same angle. The application of vibration to a muscle such as the biceps can give the illusion that the arm has been moved. If both arms are initially placed at 90° to the horizontal and the left biceps is then vibrated, a blindfold subject will have the illusion that the arm has moved and will alter the position of the right arm to report the perceived (but incorrect) position of the left arm (Fig. 8.10). This occurs because the vibration stimulates the stretch receptors. The CNS interprets the increased afferent discharge as indicating that the muscle is longer than it actually is.

Visceral receptors

The internal organs are much less well innervated than the skin. Nevertheless, all the internal organs have an afferent innervation although the activity of these afferents rarely reaches consciousness except as a vague sense of 'fullness' or as pain (see Section 8.3). The afferent fibers reach the spinal cord by way of the visceral nerves which also carry the sympathetic and parasympathetic fibers that provide motor innervation to the viscera (see Chapter 10).

The visceral receptors include both rapidly adapting and slowly adapting mechanoreceptors and chemoreceptors. Many of these afferents are an essential component of visceral reflexes which control vital body functions. Examples are the baroreceptors of the aortic arch and carotid sinus, which monitor arterial blood pressure, and the chemoreceptors of the carotid bodies that detect the P_{O_2}, P_{CO_2}, and pH of the arterial blood.

Afferent information from the somatosensory receptors reaches the brain via the dorsal columns and the spinothalamic tracts

The cutaneous and visceral afferent fibers enter the spinal cord via the dorsal (posterior) roots. The large-diameter afferents branch after they have entered the spinal cord and travel in the dorsal columns to synapse in the dorsal column nuclei (the cuneate and gracile nuclei) of the medulla oblongata. The second-order fibers leave the dorsal column nuclei as a discrete fiber bundle called the medial lemniscus. The fibers first run anteriorly and then cross the midline before reaching the ventral thalamus. From the thalamus they project to the somatosensory regions of the cerebral cortex (Fig. 8.11).

The small afferent fibers join a bundle of fibers at the dorsolateral margin of the spinal cord known as Lissauer's tract (Fig. 8.12). These thin afferent fibers only travel for a few spinal segments at most before entering the gray matter of the spinal cord where they synapse on spinal interneurons in the substantia gelatinosa. These interneurons then synapse on the neurons whose axons form the spinothalamic tract. The axons from these neurons then cross the midline and project to the thalamus via the spinothalamic tract, which runs in the anterolateral quadrant of the spinal cord. From the thalamus, the sensory projections reach the somatosensory regions of the cerebral cortex as shown in Fig. 8.13. Finally, mention must be made of the

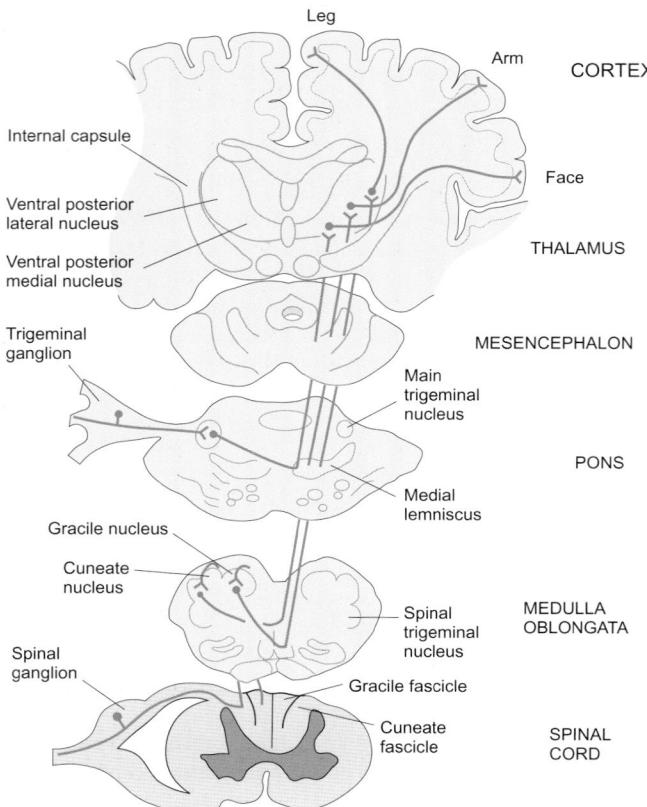

Fig. 8.11 The dorsal-column lemniscal pathway for cutaneous sensation.

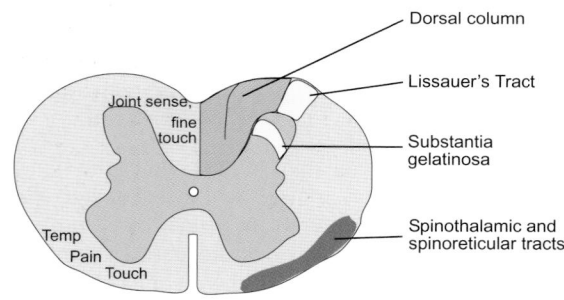

Fig. 8.12 Diagrammatic representation of the spinal cord showing the positions of the main sensory tracts and the substantia gelatinosa of the dorsal horn.

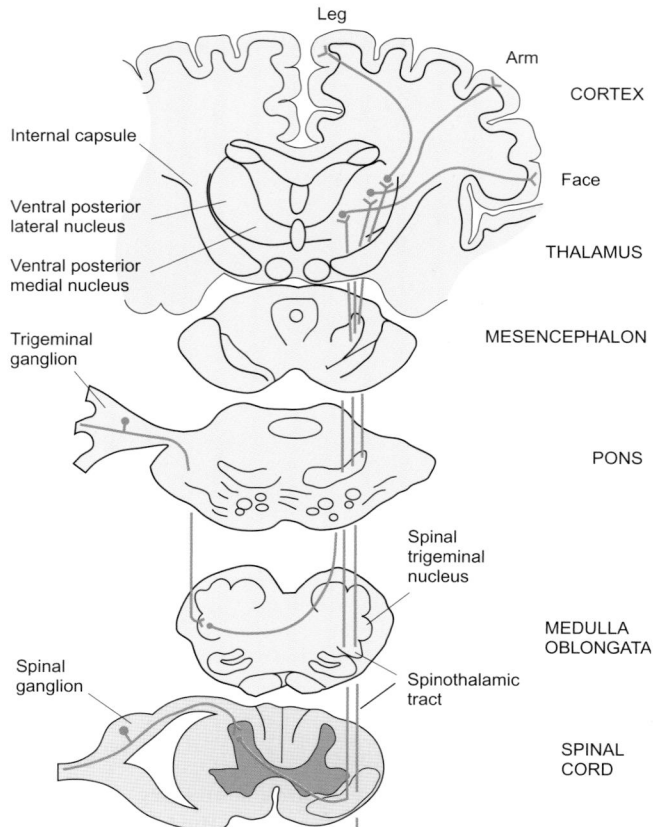

Fig. 8.13 The spinothalamic tract and its projection to the cerebral cortex.

All the afferents of a particular type that enter one dorsal root tend to run together in the lower regions of the spinal cord. Initially this segmental organization is preserved but, as they ascend, the fibers from the different segments become rearranged so that those from the leg run together, as do those of the trunk, hand, and so on. Thus, for example, the afferents from the hand project to cells in a particular part of the dorsal column nuclei and thalamus while those from the forearm project to an adjacent group of cells. This orderly arrangement provides topographical maps of the body in those parts of the brain that are responsible for integrating information from the different sensory receptors.

Although the cells of the dorsal column nuclei project chiefly to the specific sensory nuclei in the ventrobasal region of the thalamus, those from the spinothalamic tract have a wider distribution and project to the intralaminar nuclei and the posterior thalamus. The sensory cells of the thalamus project to two specific regions of the cerebral cortex, the primary and secondary sensory cortex.

Sensory information is represented topographically on the postcentral gyrus of the cerebral cortex

The exploration of the human somatosensory cortex by Wilder Penfield and his colleagues is one of the most remarkable investigations in neurology. While treating patients for epilepsy, Penfield attempted to localize the site of the lesions responsible

spinoreticular tract, which receives afferents from the sensory nerves and projects mainly to the reticular formation of the brainstem on the same side. These fibers ascend in the anterolateral spinal cord. This tract consists of a chain of short fibers that synapse many times as they ascend the spinal cord.

The dorsal columns of the spinal cord contain large-diameter afferent fibers that are mainly concerned with touch and proprioception. They relay information that is concerned with fine discriminatory touch, vibration, and position sense (kinesthetic information). The spinothalamic and spinoreticular tracts receive information from the smaller afferent fibers and transmit information concerning crude touch, temperature, and pain to the brain (see Table 8.2).

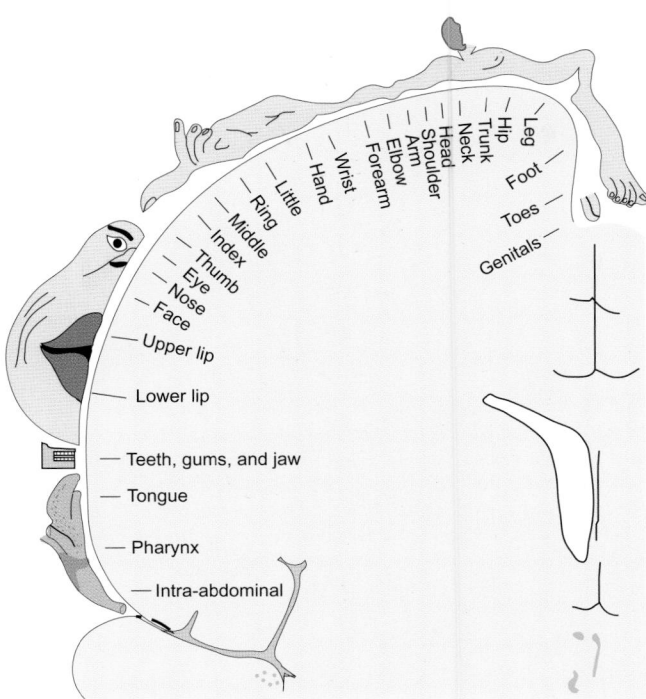

Fig. 8.14 The representation of the body surface on the postcentral gyrus revealed by electrical stimulation of the cerebral cortex of conscious subjects.

Summary

1. The somatosensory system is concerned with stimuli arising from the skin, joints, muscles, and viscera. It provides the CNS with information concerning touch, peripheral temperature, limb position (kinesthesia), and tissue damage (nociception).

2. Individual skin receptors respond to stimuli applied to specific areas of the body surface. The area within which a stimulus will excite a specific afferent fiber is known as the receptive field of that fiber. The receptive fields of adjacent fibers often overlap. The size of the receptive fields in different regions of the skin and the degree of overlap between adjacent receptive fields both play an important role in the spatial discrimination of a stimulus.

3. Information from the somatosensory receptors reaches the cerebral cortex by way of the dorsal column–medial lemniscal pathway and by the spinothalamic tract. The dorsal column pathway is primarily concerned with fine discriminatory touch and position sense while the spinothalamic tract is concerned with crude touch, temperature, and nociception. The postcentral gyrus of the cerebral cortex possesses a topographical map of the contralateral surface of the body.

for the condition by electrically stimulating the cerebral cortex. During this procedure, the patients were conscious but the cut edges of the scalp, skull, and dura were infiltrated with local anesthetic. Since the brain has no nociceptive fibers, this procedure did not elicit pain but it did elicit specific movements or specific sensations depending on which area of the cortex was stimulated.

Systematic exploration of the postcentral gyrus revealed that it was organized somatotopically with the different regions of the opposite side of the body represented as shown in Fig. 8.14. The genitalia and feet are mapped on to an area adjacent to the central fissure, while the face, tongue, and lips are mapped on the lateral aspect of the postcentral gyrus. Although the area of representation is disproportionate with respect to the body surface, it is appropriate to the degree of importance of the different areas in sensation. Thus the hands, lips, and tongue all have a relatively large area of cortex devoted to them compared with the areas devoted to the legs, upper arm, and back. The body is also mapped on to the superior wall of the Sylvian gyrus. This area is called the SII region. Unlike the postcentral gyrus (the SI region), the SII region has a representation of both sides of the body surface.

The trigeminal system

The sensory inputs from the face are relayed to the brain via cranial nerve (CN) V, the trigeminal nerve. The trigeminal nerve is mixed, having both somatic afferent and efferent fibers, although the afferent innervation dominates. It arises in the pontine region of the brain stem. Shortly after its origin, it expands to form the semilunar ganglion which contains the primary sensory neurons which are analogous to the dorsal root ganglion

neurons. Three large nerves leave the ganglion to innervate the face. These are the opthalmic, maxillary, and mandibular nerves. These nerves relay information to the brain concerning touch, temperature and pain from the face. They also relay information from the mucous membranes and the teeth. The large afferent fibers transmit information from the mechanoreceptors to the thalamus via the medial lemniscus (Fig. 8.11) while the Aδ and C-fiber afferents (which are mainly concerned with temperature and nociception) join the spinothalamic tract as shown in Fig. 8.13. From the thalamus, information from both large and small afferents is transmitted to the face region of the primary sensory cortex.

8.3 Pain

Pain is an unpleasant experience associated with acute tissue damage. It is the main sensation experienced by most people following injury. Pain also accompanies certain organic diseases such as advanced cancer. The weight of evidence now suggests that pain is conveyed by specific sets of afferent nerve fibers and is not simply the result of a massive stimulation of afferent fibers. Nevertheless, pain may arise spontaneously without an obvious organic cause or in response to an earlier injury, long since healed. This kind of pain often has its origin within the CNS itself. Although it is not obviously associated with tissue damage, pain of central origin is no less real to the sufferer.

Unlike most other sensory modalities, pain is almost invariably accompanied by an emotional reaction of some kind such as fear or anxiety. If it is intense, pain elicits autonomic responses such as sweating and an increase in blood pressure and heart rate. Pain may be short lasting and directly related to the injury that elicited it (*acute pain*) or it may persist for many days or even months (*chronic pain*).

Broadly, pain can be classed under one of three headings.

1. *Pricking pain* is rapidly appreciated and accurately localized but elicits little by way of autonomic responses. This kind of pain is usually transient and has a sharp pricking quality. It is sometimes called 'first' or 'fast' pain and is transmitted to the CNS via small myelinated Aδ fibers. In normal physiology, it serves an important protective function as activation of these fibers triggers the reflex withdrawal of the affected region of the body from the source of injury.

2. *Burning pain* is more intense and less easy to endure than pricking pain. It has a diffuse quality and is difficult to localize. Burning pain readily evokes autonomic responses including an increased heart rate, elevated blood pressure, dilation of the pupils, and sweating (see Chapter 10). The pattern of breathing may also be altered, with rapid shallow breaths interrupted with periods of apnea (breath-holding) during severe episodes. Burning pain is of slower onset and greater persistence than fast pain. It reaches the CNS via non-myelinated C-fibers and is sometimes called 'second' or 'slow' pain. The difference in quality between fast pricking pain and slow burning pain will be known to anyone who has inadvertently stepped into an excessively hot bath. The initial response is to withdraw the affected foot quickly, but an intense pain slowly develops which is also slow to subside.

3. *Deep pain* arises when deep structures such as muscles or visceral organs are diseased or injured. Deep pain has an aching quality, sometimes with the additional feeling of burning. It is usually difficult to localize and, when it arises from visceral organs, it may be felt at a site other than that at which it originates. This is known as *referred pain* (see below).

Nociceptors are activated by specific substances released from damaged tissue

As for the sense of touch and temperature, the distribution of the pain receptors in the skin is punctate. Histological examination of a pain spot reveals a dense innervation with bare nerve endings, which are believed to be the nociceptors. The adequate stimulus for the nociceptors is not known with certainty, but the application of pain-provoking stimuli such as radiant heat elicits reddening of the skin and other inflammatory changes. It is probable that a number of chemical agents called pain-producing substances (*algogens*) are released following injury to the skin and cause the pain endings to discharge. These agents include ATP, bradykinin, histamine, serotonin (5-HT), hydrogen ions, and a number of inflammatory mediators such as prostaglandins.

The triple response

If a small area of skin is injured, for example by a burn, there is a local vasodilatation which elicits a reddening of the skin. This is followed by a swelling (a weal or wheal) that is localized to the site of the injury and its immediate surroundings. The original site of injury is then surrounded by a much wider area of less intense vasodilatation known as the 'flare'. The local reddening

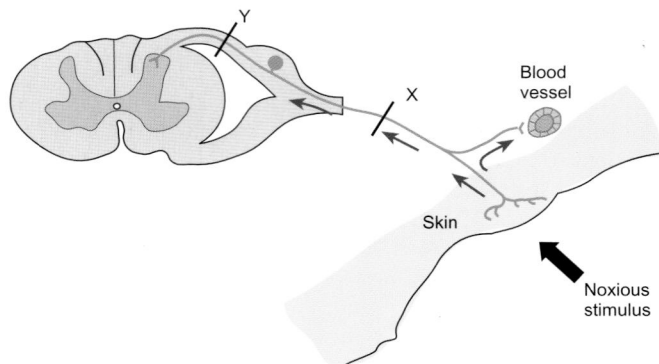

Fig. 8.15 A schematic diagram of the axon reflex that gives rise to the flare of the triple response. Nociceptive fibers from the skin branch and send collateral fibers to nearby blood vessels. If the skin is injured, action potentials pass to the spinal cord via the dorsal root and to the axon collaterals where they cause a vasodilatation that gives rise to the flare. If the nerve trunk is cut at point X and sufficient time allowed for the nerve to degenerate, the flare reaction is abolished. If the cut is made beyond the dorsal root ganglion at point Y, the segmental spinal reflexes are abolished but not the flare reaction.

('red reaction'), flare, and weal formation comprise the *triple response* that was first described by Thomas Lewis. The weal is a local edema caused by the accumulation of fluid in the damaged area. The red-reaction is due to arteriolar dilatation in response to vasodilator substances released from the damaged skin, and the flare is due to dilatation of arterioles in the area surrounding the site of injury.

Within the injured area and across the surrounding weal, the sensitivity to mildly painful stimuli such as a pinprick is much greater than before the injury. This is known as *primary hyperalgesia*, and may persist for many days. In the region covered by the flare, outside the area of tissue damage, there is also an increased sensitivity to pain which may last for some hours. This is known as *secondary hyperalgesia*.

Unlike the weal and local red reaction, the flare is abolished by infiltration of the skin with local anesthetic. However, it is not blocked if the nerve trunk supplying the affected region is anesthetized. Moreover, if the nerve trunk is cut (e.g. at point X in Fig. 8.15) and allowed to degenerate before eliciting the triple response, only the local reddening and weal formation occur—the flare is absent. The triple response can be elicited in animals that have had complete removal of their sympathetic innervation. Furthermore, cutting the nerve supply to the dorsal root of the appropriate segment (point Y in Fig. 8.15) does not block the flare reaction. These experiments show that the flare is a local *axon reflex* rather than a reflex vasodilatation involving the spinal cord or brainstem.

CNS pathways in pain perception

Nociceptive fibers are a specific set of small-diameter dorsal root afferents that subserve pain sensation in a particular region. These fibers enter Lissauer's tract and synapse in the substantia gelatinosa close to their site of entry into the spinal cord (see Fig. 8.12). The second-order fibers cross the midline and ascend to the brainstem reticular formation and thalamus via the spinoreticular and spinothalamic tracts. The fibers of the spinoreticular tract may be concerned with cortical arousal mechanisms and with eliciting the defense reaction (see p. 302).

Although the neurons of the ventrobasal thalamus project to the primary sensory cortex, there is no compelling evidence that noxious stimuli evoke neural responses in this region. However, electrical stimulation of the posterior thalamic nuclei in conscious human subjects does elicit pain. Moreover, degenerative lesions of the posterior thalamic nuclei can give rise to a severe and intractable pain of central origin known as *thalamic pain*. It is known that the neurons of the posterior thalamus project to the secondary sensory cortex on the upper wall of the lateral fissure. Electrical stimulation of the white matter immediately beneath the secondary somatosensory cortex (SII) also elicits the sensation of pain. This suggests that pain sensations are relayed to the SII region. Other areas that appear to be involved in the whole pain experience are the reticular formation, the structures of the limbic system such as the amygdala, and the frontal cortex.

The perception of pain may be greatly modified by circumstances. It is widely known that the pain from a bruise can be relieved by vigorous rubbing of the skin in the affected area. Pain can also be relieved by the electrical stimulation of the peripheral nerves. In both cases, the large-diameter afferent fibers are activated. This activation appears to inhibit the transmission of pain signals by the small unmyelinated nociceptive afferents. This inhibition occurs at the segmental level in the local networks of the spinal cord and prevents the onward transmission of pain signals to the brain. Transcutaneous electrical nerve stimulation (TENS) is now frequently used to control pain during childbirth and for some other conditions where prolonged use of powerful pain-suppressing drugs is undesirable.

Stressful situations can also produce a profound loss of sensation to pain (analgesia). It has been known for many years that soldiers with very severe battlefield injuries are often surprisingly free of pain. This suggests that the brain is able to control the level of pain in some way. Evidence for this view has come from experiments in which specific brain regions of conscious animals were stimulated electrically. When the central mass of gray matter surrounding the aqueduct was stimulated, the animals were almost immune to pain. Fibers from this region (known as the periaqueductal gray matter) project to the spinal cord via a series of small nuclei in the brainstem called the raphe nuclei. It was subsequently found that stimulation of the largest of these, the raphe magnus nucleus, also inhibited the transmission of pain to the brain. The analgesic ('pain-suppressing') effect of stimulation of these brain regions has subsequently been demonstrated in man. Axons from the raphe magnus nucleus terminate on neurons in the dorsal horn of the spinal cord as shown in Fig. 8.16.

The brain and spinal cord possess a number of peptides which have actions similar to morphine and other opiate drugs that have a powerful analgesic effect. These are the *enkephalins*, the *dynorphins*, and the *endorphins*. When these peptides are injected into certain areas of the brain, such as the periaqueductal gray matter or the spinal cord, they have powerful analgesic effects. β-endorphin is also secreted by the anterior hypothalamus and by the adrenal medulla, along with epinephrine and norepinephrine, as part of the body's overall response to stress.

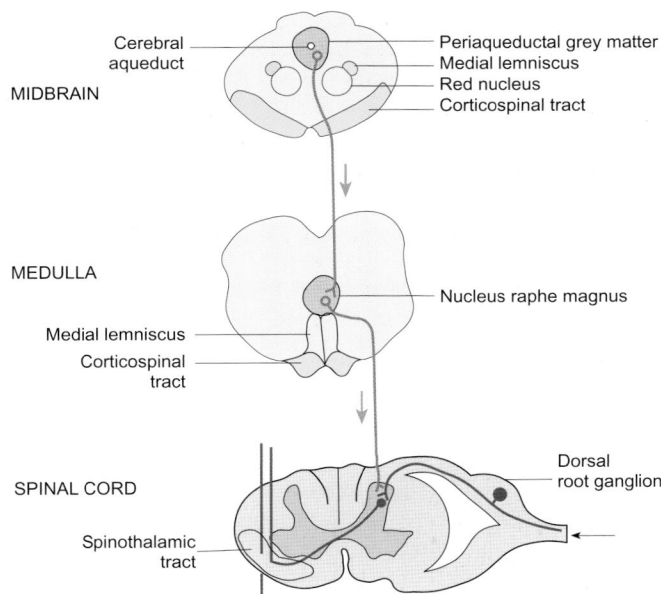

Fig. 8.16 A schematic diagram of the descending pathways that inhibit the activity of nociceptive neurons in the dorsal horn. The different sections are not drawn to scale. Descending fibers are shown in red and ascending fibers in blue.

Visceral pain

Although the viscera are not as densely innervated with nociceptive afferents as the skin, they are nevertheless capable of transmitting pain signals. The majority of the visceral afferent fibers that signal pain have their cell bodies in the dorsal root ganglia of the thoracic and upper two or three lumbar segments. Their axons may run in the sympathetic trunk for several segments before leaving to innervate the viscera via the cardiac, pulmonary and splanchnic nerves. The nociceptors of the fundus of the uterus and the bladder run in the sympathetic nerves of the hypogastric plexus. As a general rule, the fibers that convey information from the nociceptors of the viscera follow the course of the sympathetic nerves. The nociceptive afferents of the neck of the bladder, the prostate, the cervix, and the rectum are an exception to this general rule. They reach the spinal cord via the pelvic parasympathetic nerves.

The nociceptive visceral afferents enter the spinal cord via the dorsal roots. They synapse in the substantia gelatinosa of the dorsal horn and the second-order axons project to the brain, mainly via the spinothalamic tract.

The nociceptive nerve endings are stimulated by a variety of stimuli. When the walls of hollow structures such as the bile duct, intestine, and ureter are stretched, the nociceptive afferents are activated. Thus, when a gallstone is being passed along the bile duct from the gall bladder it causes an obstruction that the smooth muscle of the bile duct tries to overcome by a forceful contraction. The nerve endings are stretched and the nociceptive afferents are activated. Each time the smooth muscle contracts the sufferer will experience a bout of severe pain known as *colic* until the stone is passed into the intestine. Similar bouts of spasmodic pain accompany the passage of a kidney stone along a ureter. Ischemia also causes intense pain. The classic example is the pain felt following narrowing or occlusion of one of the coronary arteries. Chemical irritants are also a potent source of pain originating in the viscera. This is commonly caused by gastric acid coming into direct contact with the gastric or esophageal mucosa. This gives rise to a sensation known as heartburn see p. 387.

Referred pain

The pain arising from the viscera is often scarcely felt at the site of origin but may be felt as a diffuse pain at the body surface in the region innervated by the same spinal segments. This separation between the site of origin of the pain and the site at which it is apparently felt is known as referral and the pain is called *referred pain*. The site from which the pain is reported, together with its characteristics (sharp, dull, throbbing, etc.), provide valuable information for the diagnosis of organic disease. For example, the passage of gallstones along the bile duct causes an intermittent but intense pain that appears to travel along the base of the right shoulder. The pain arising from cardiac ischemia is strikingly referred to the base of the neck and radiates down the left arm (Fig. 8.17). It may also be felt directly under the sternum.

The mechanism of pain referral is unknown. One hypothesis proposes that the pain afferents from the viscera converge on the same set of spinal neurons as the somatic afferents arising from the same segment. Activity in these neurons is normally associated with somatic (i.e. cutaneous) sensation and, as the brain does not normally receive much information from the visceral nociceptive afferents, it interprets the afferent discharge arising from the visceral nociceptors as arising from that region of the body surface innervated by the same spinal segment.

When a nerve trunk is stimulated somewhere along its length rather than at its termination, there is an unpleasant confused sensation known as *paresthesia*. This presumably arises because the CNS attempts to interpret the sequence of action potentials in terms of its normal physiology. The somatic nerve fibers are specific labeled lines. Each afferent is connected to a particular set of sensory receptors. The abnormal pattern of activation is then projected to the area of the body served by the nerve in question. Sometimes this sensation is one of explicit pain and as this is projected to the body surface, it is called *projected pain*. A well known example is the pain felt when the ulnar nerve is knocked, giving rise to a tingling sensation in the third and fourth fingers of the affected hand.

Causalgia, neuralgia, tic douloureux, and phantom limb pain

Gunshot wounds and other traumatic injuries which damage, but do not sever, major peripheral nerves such as the sciatic nerve can give rise to a severe burning pain in the area served by the nerve. This is known as *causalgia*, which fortunately occurs in less than 5 per cent of patients receiving such an injury. During an acute episode, the affected limb is initially warm and dry but later it becomes cool and the affected area sweats profusely. The pain is so great that the patient cannot bear the mildest of stimuli to the affected area. Even a puff of air elicits unbearable pain. Treatment is often ineffective but injection of local anesthetic into the sympathetic ganglia supplying the affected region may provide relief, as may electrical stimulation of the large myelinated afferents using a TENS stimulator (see above).

Other kinds of damage to peripheral nerves may also cause severe and unremitting pain that is resistant to treatment. This kind of pain is known as *neuralgia*. The causes include viral infections, neural degeneration, and nerve damage from poisons. In *trigeminal neuralgia*, which is sometimes called *tic douloureux*, pain of a lacerating quality is felt on one side of the face within the distribution of the trigeminal nerve. Between attacks, the patient is free of pain and no abnormalities of function can be detected. The attacks are not provoked by thermal or nociceptive stimuli but may be triggered by light mechanical stimulation of the face or lip. An attack may also be triggered during eating by food coming into contact with a trigger point on the mucous membranes of mouth or the throat. In this case, the pain may be relieved by section of the appropriate branch of the trigeminal nerve, although this will leave the affected area without normal sensation.

After limb amputations, the nerve fibers in the stump begin to sprout and eventually form a tangle of fibers, fibroblasts, and Schwann cells called a *neuroma*. The nerve fibers in a neuroma often become very sensitive to mechanical stimulation that may give rise to severe shooting pains, which may be blocked by injection of local anesthetic into the mass of the neuroma. Some patients develop a strong illusion that the amputated limb is still present. The phantom limb may appear to be held in a very uncomfortable or painful position. One patient described a phantom hand as being held as a tightly clenched fist.

Although the pain evoked by noxious stimuli serves an obvious protective function by warning that a specific action can lead to tissue damage, much pain is of pathological origin. It is this pathological pain that drives many patients to seek medical help. Although the characteristics of the pain may be helpful in

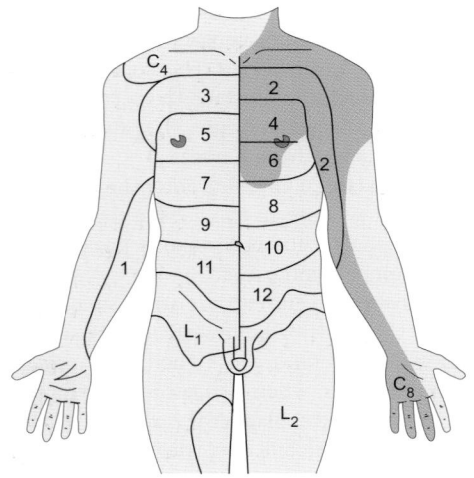

Fig. 8.17 Diagram to show the overlap between the segmental innervation of the trunk and the distribution of referred pain from the heart. The left-hand side shows those regions innervated by nerves from C4, T1 (inner arm), T3, T5, T7, T9, T11, and L_1, while the right-hand side illustrates the innervation for T2, T4, T6, T8, T10, and T12. L_2 innervates the thigh. The dark shading shows the distribution of the referred pain.

Summary

1. Pain is the sensation we experience when we injure ourselves or when we have some organic disease. It is an unpleasant experience that is generally associated with actual or impending tissue damage.

2. The nociceptors appear to be a specific set of Aδ and C-fiber afferents that are triggered by strong thermal, mechanical, or chemical stimuli. Their afferents synapse extensively within the spinal cord at a segmental level and project to the CNS by way of the spinothalamic and spinoreticular tracts.

3. Pain arising from the body can be classified as follows.

   (a) Pricking pain, which is accurately localized but elicits little by way of autonomic or emotional responses.

   (b) Burning pain which arises following activation of C-fiber nociceptors. This type of pain is difficult to endure and readily elicits both autonomic and emotional reactions.

   (c) Deep pain which, like burning pain, elicits both autonomic and emotional reactions. It is difficult to localize accurately and is often referred to a site distant to its site of origin (referred pain).

4. Many pains are of central origin and reflect hyperexcitability in the nociceptive pathways, which may follow a severe injury. This is the case with phantom limb pain. Other pains may be elicited by stimulation of specific trigger points (e.g. trigeminal neuralgia).

reaching a diagnosis, it is essential to understand that severe unremitting pain will color every aspect of the sufferer's life, and prompt and effective treatment is therefore required. This terrifying aspect of chronic pain was very well put in 1872 by S.W. Mitchell who wrote:

> Perhaps few persons who are not physicians can realize the influence which long-continued and unendurable pain may have upon the mind and body. Under such torments the temper changes, the most amiable grow irritable, the soldier becomes a coward and the strongest man is scarcely less nervous than the most hysterical girl.

8.4 Itch

Itch or *pruritus* is a well known sensation that is associated with the desire to scratch. There are many causes of itch including skin parasites such as scabies and ringworm, insect bites, chemical irritants derived from plants (e.g. 'itching powder'), and contact with rough cloth or with inorganic materials such as fiber glass. A variety of common skin diseases such as eczema also give rise to the sensation of itch, as do obstructive jaundice and renal failure. The intensity of the sensation is highly variable; it may be so mild as to be scarcely noticeable or so severe as to be as unendurable as chronic pain.

What is the physiological basis of this sensation? Exploration of the skin with itch-provoking stimuli has shown that there are specific itch points with a punctate distribution. If one of these points is cut out and examined histologically, no specific sensory structures are found but, as for pain points, there is an increased density of bare nerve endings. These are thought to be the receptors that give rise to the sensation of itch.

Several pieces of evidence suggest that itch is transmitted to the CNS via a subset of peripheral C-fibers. The sensation disappears along with burning pain following block of cutaneous C-fibers by local anesthetics. It persists when a nerve trunk is blocked by pressure at a time when only C-fibers remain active. Finally, the reaction time to pruritic stimuli is characteristically slow, being around 1–3 s after the stimulus. In these respects, there is a similarity to cutaneous pain. However, itch differs from superficial pain in a number of important respects: it can only be elicited from the most superficial layers of the skin, the mucous membranes and the cornea. Increasing the frequency of electrical stimulation of a cutaneous itch spot increases the intensity of the itch without eliciting the sensation of pain. Moreover, skin stripped of the epidermis is exquisitely sensitive to pain but is insensitive to itch-provoking stimuli. Finally, the reflexes evoked by pruritic (itch-provoking) stimuli and nociceptive stimuli are different. Pruritic stimuli elicit the well-known scratch reflex, while nociceptive stimuli elicit withdrawal and guarding reflexes. On these grounds, it is reasonable to conclude that itch is a distinct modality of cutaneous sensation.

Exactly how the afferent fibers subserving the sensation of itch are excited remains unknown. However, a number of naturally occurring substances give rise to itch including certain proteolytic enzymes and histamine. Histamine is of interest because it is released by mast cells when they become activated by antigens. When histamine is applied to the superficial regions of the skin, it produces a distinct sensation of itch that is graded with its local concentration. Furthermore, there is good evidence that histamine can excite the unmyelinated sensory fibers that subserve the sense of itch. In a recent study of human volunteers, a small subset of unmyelinated fibers were found to respond with sustained discharge to localized injections of histamine into the superficial regions of the skin. Moreover, the time course of the discharge corresponded to the time course of the itch sensations. However, as antihistamine drugs are not always effective in controlling itch, other chemical mediators are likely to be involved.

Pruritic afferents travel to the spinal cord in nerve trunks along with other sensory fibers. They enter the cord via the dorsal roots in a similar manner to those fibers that subserve pain sensation in the same region and synapse close to their site of entry into the spinal cord. The second-order fibers cross the midline and ascend to the brainstem reticular formation and thalamus via the spinothalamic tract. In addition to projecting centrally, activity in pruritic afferents elicits the scratch reflex, which is a polysynaptic spinal reflex that is clearly directed towards the removal of the itch-provoking stimulus. This reflex has been thoroughly examined in dogs but the principle also applies to man. In the first stage of the reflex, a limb (in man usually the hand) is positioned above the irritated area. This is followed by the rhythmic alternation of extension and flexion of the limb such that the extremity repeatedly passes over the affected area in a manner calculated to dislodge the offending item.

Summary

Itch is a specific cutaneous sensation chiefly arising in the superficial regions of the skin. The end-organs responsible appear to be bare nerve endings that are excited by histamine and other chemical mediators. Itch is transmitted to the CNS via specific afferents where it may elicit a scratch reflex. Fibers conveying information about pruritic stimuli travel to the brain via the spinothalamic tract.

8.5 The physiology of the eye and visual pathways

The properties of light and the sensitivity of the eye

Man is pre-eminently a visual animal. Those blessed with normal vision largely react to the world as they see it, rather than to its feel or its sounds or smells. Objects are judged at a distance from the light they emit or reflect into the eye. Light itself is a component of the electromagnetic spectrum and the human eye responds to radiation with wavelengths between about 380 nm (deep violet) and 800 nm (deep red). This is a relatively narrow part of the total spectrum and other animals may have a slightly different range of sensitivities. Like all electromagnetic radiation, light has the properties of both a wave and a particle, but for many purposes it can be considered to travel in straight lines.

Objects either emit light or reflect it to varying degrees. Light emitted from a source is measured in candelas per square meter ($cd\ m^{-2}$) while reflected light is measured in lux. Figure 8.18 illustrates the range of light intensities that are normally encoun-

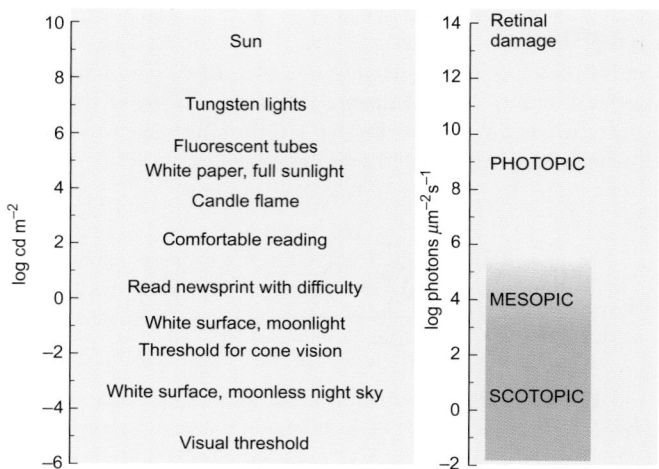

Fig. 8.18 Illustration of the range of luminances (left) and retinal illumination (right) found in the environment. Note that the logarithmic scales represent a range of 15 orders of magnitude. Photopic, mesopic, and scotopic refer to vision in bright, low-level, and dark conditions.

tered in daily life. Note that the eye is able to respond to light of intensities over 15 orders of magnitude.

The anatomy of the eye

The eyes are protected by their location in the bony cavities of the orbits. Only about a third of the eyeball is unprotected by bone. The eyeball itself its roughly spherical and its wall consists of three layers (Fig. 8.19): a tough outer coat, the *sclera*, which is white in appearance, a pigmented layer called the *choroid*, which

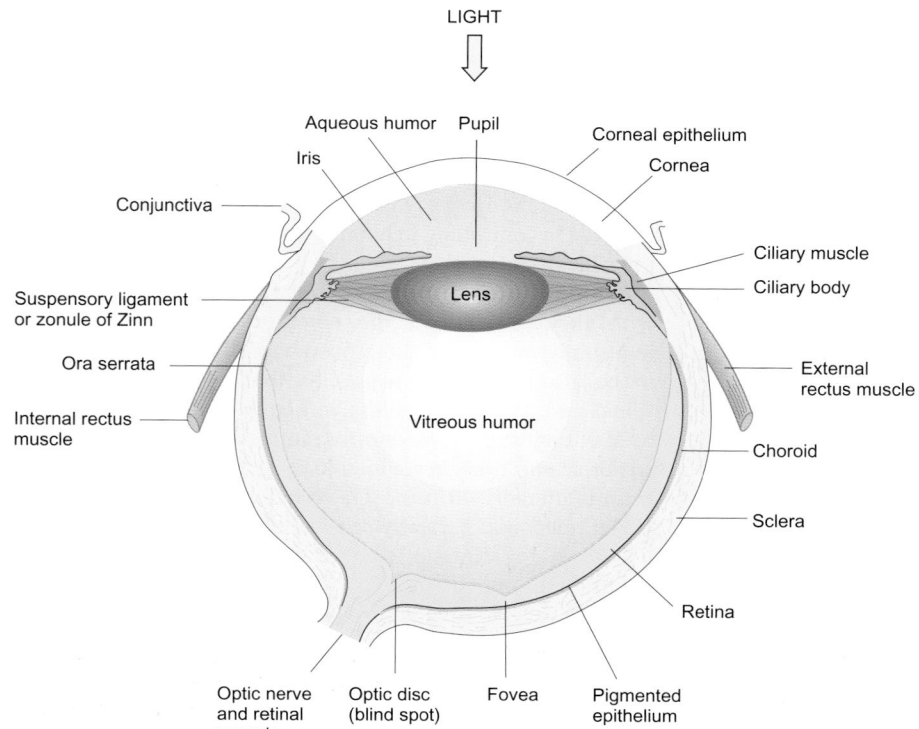

Fig. 8.19 A cross-sectional view of the eye to show the principal structures.

is highly vascular, and the *retina*, which contains the photo-receptors (rods and cones) together with an extensive network of nerve cells. The retinal ganglion cells are the output cells of the retina and they send their axons to the brain via the optic nerves.

At the front of the eye, the sclera gives way to the transparent cornea which consists of a special kind of connective tissue that lacks blood vessels. The health and transparency of the cornea are maintained by the tear fluid secreted by the lacrimal glands and by the aqueous humor that is secreted by the ciliary body within the eye itself. The pigmented iris covers much of the transparent opening of the eye formed by the cornea, leaving a central opening, the pupil, to admit light to the photoreceptors of the retina.

The pupil diameter is controlled by two muscles, the circular sphincter pupillae and the radial dilator pupillae of the iris, both of which are innervated by the autonomic nervous system. The sphincter pupillae receives parasympathetic innervation via the ciliary ganglion while the dilator pupillae receives sympathetic innervation via the superior cervical ganglion.

Behind the iris lies the ciliary body, which contains smooth muscle fibers. The lens of the eye is attached to the ciliary body by a circular array of fibers called the zonule of Zinn or the suspensory ligament. The lens is formed as a series of cell layers, which arise from the cuboidal epithelial cells that cover its anterior surface. The cells of the lens synthesize proteins known as crystallins that are important for maintaining its transparency. Like the cornea, the lens has no blood vessels and depends on the diffusion of nutrients from the aqueous humor for its nourishment. The lens itself is elastic and can change its shape according to the tension placed on it by the zonal fibers. The ability of the lens to change its shape is an essential part of the mechanism by which the eye can bring images into focus on the retina. This process is controlled by the ciliary muscles and is called *accommodation*.

The organization of the retina

The retina is the sensory region of the eye. It consists of eight layers. Starting from the vascular choroid layer, the first, most outward, component of the retina is the *pigmented epithelium*. The next three layers contain the photoreceptors—the *rods* and *cones* and the terminal regions of the photoreceptors where they make synaptic contact with other retinal cells. Above this are two layers that consist of the cell bodies of the bipolar cells, horizontal cells, and amacrine cells and their processes. The final two layers contain the output cells of the retina, the *ganglion cells* and their axons. Individual photoreceptors consist of an outer segment, which contains the photosensitive pigment, an inner segment, where the cell nucleus is located, and a rod pedicle, which is the site at which the photoreceptors make synaptic contact with the bipolar and amacrine cells of the retina.

A highly schematic diagram of the organization of the retina is shown in Fig. 8.20. Note that light passes through the cell layers to reach the photoreceptors, which are located next to the pigmented epithelium. Rods and cones are distributed throughout the retina but, in the central region known as the *fovea centralis*, the retina is very thin and consists of a densely packed layer of cones. In the surrounding region, the parafoveal region, both rods and cones are present in abundance together with the bipolar, amacrine, and horizontal cells connected to the cones of the fovea. In Fig. 8.20 one cone is shown connected to one ganglion cell. This is only the case for the central region of the retina; elsewhere the signals from a number of photoreceptors converge on a single ganglion cell. In the extreme periphery, as many as 100 rods are connected to a single ganglion cell. The region where the ganglion cell axons pass out of the eye to form the optic nerve (the papilla or *optic disk*) is devoid of photoreceptors (the blind spot).

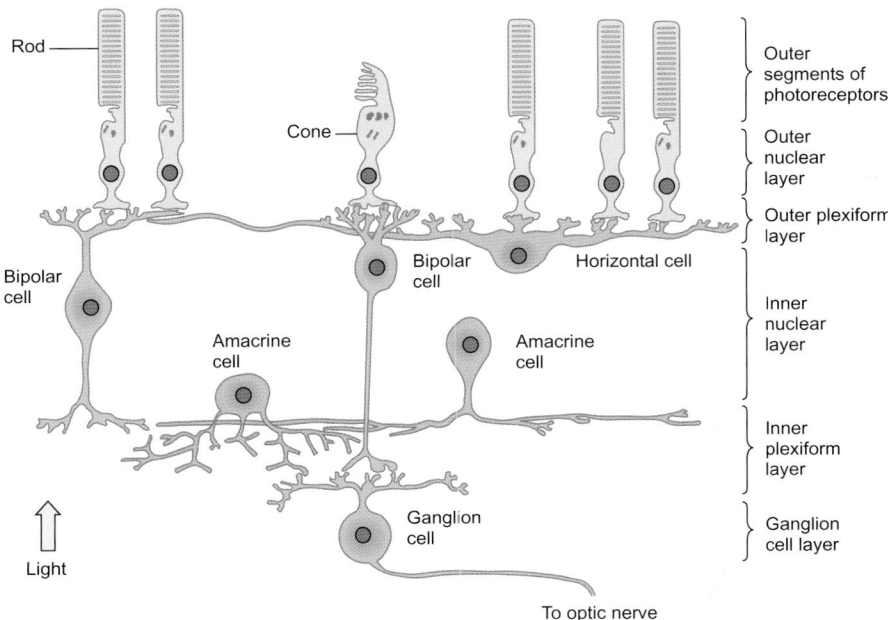

Fig. 8.20 A simple diagram to show the cellular organization of the retina. Note that light has to pass through the innermost layers of the retina to reach the photoreceptors.

The general physiology of the eye

The lacrimal glands and tear fluid

Each orbit is endowed with lacrimal glands which provide a constant secretion of tear fluid that serves to lubricate the movement of the eyelids and keep the outer surface of the cornea moist, thus providing a good optical surface. The lachrimal glands are innervated by the parasympathetic outflow of the facial nerve (CN VII). Tear fluid has a pH similar to that of plasma (7.4) and is isotonic with blood. It possesses a mucolytic enzyme called lysozyme that has a bactericidal action. Under normal circumstances about 1 ml of tear fluid is produced each day, most of which is lost by evaporation; the remainder is drained into the nasal cavity via the tear duct. Irritation of the corneal surface (the conjunctiva) increases the production of tear fluid and this helps to flush away noxious agents.

The intraocular pressure is maintained by the balance between the rate of production and absorption of the aqueous humor

The space behind the cornea and surrounding the lens is filled with a clear fluid called the *aqueous humor*, which supplies the lens and cornea with nutrients. Although the aqueous humor is isotonic with the blood, it is not a simple ultrafiltrate of plasma as it is relatively enriched in bicarbonate. Moreover, inhibitors of the enzyme carbonic anhydrase decrease the rate of aqueous humor production (see below). The aqueous humor is secreted by the processes of the ciliary body that lie just behind the iris, and it flows into the anterior chamber from which it drains into the canal of Schlemm though a fibrous mesh (the trabecular meshwork) situated at the junction between the cornea and the sclera (see Fig. 8.21). From the canal of Schlemm the fluid is returned to the venous blood.

The constant production of aqueous humor generates a pressure within the eye known as the *intraocular pressure*. In normal individuals the intraocular pressure is about 2 kPa (15 mmHg) and serves to maintain the rigidity of the eye, which is essential for clear image formation. This pressure is sustained by keeping the balance between production and drainage of aqueous humor constant. If the drainage is obstructed, the intraocular pressure rises and a condition known as *glaucoma* results.

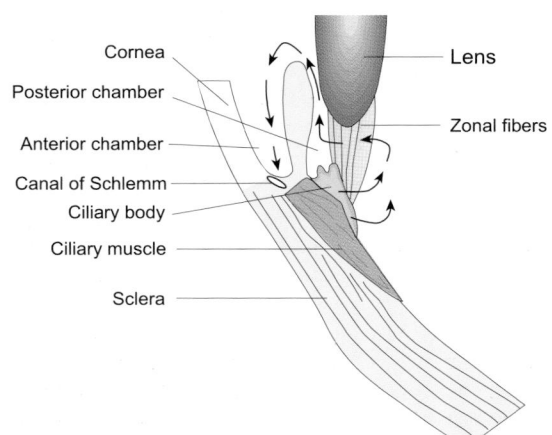

Fig. 8.21 Diagrammatic representation of the circulation of the aqueous humor from the ciliary body to the canal of Schlemm.

Obstructions to the drainage of aqueous humor arise in a variety of ways:

- In *open-angle glaucoma*, the obstruction lies in the canal of Schlemm or the trabecular meshwork itself.

- In *closed-angle glaucoma* there is forward displacement of the iris towards the cornea, such that the angle between the iris and the cornea (the iridocorneal angle) is narrowed and the trabecular meshwork becomes covered by the root of the iris. Consequently, the outflow of fluid from the anterior chamber is impeded. This form of obstruction is occasionally seen following a blunt injury to the eye. It may develop slowly and progressively over many years, or it may be sudden in onset, with acute closure of the iridocorneal angle. In such cases there is an abrupt rise in intraocular pressure, pain, and visual disturbances which, if left untreated, may lead to blindness within days or even hours.

- Occasionally, in old age, the characteristic thickening of the lens is sufficient to put pressure on the canal of Schlemm and impede drainage of aqueous humor from the anterior chamber.

- Glaucoma may be the result of abnormal blood vessel formations within the eye (these are often genetic in origin), or it may result from degenerative changes to the drainage area or swelling of the iris (uveitis).

In all cases of glaucoma, peripheral vision is normally lost first, followed by central vision impairment. Total loss of vision may occur if the condition remains undiagnosed and untreated for many years. Loss of visual acuity is the result of pressure on the optic nerve. Over time, this results in disruption of the supply of nutrients to the retinal neurons and peripheral optic fibers entering the brain, and widespread cell death.

Available treatments for glaucoma focus on trying to reduce the rate of production of aqueous humor, or to increase its rate of drainage. Drug treatments, surgical procedures, or a combination of both are used. These may slow or even halt degeneration of vision, but they cannot restore vision that has already been lost.

Drugs that block adrenergic β-receptors (beta-antagonists) reduce the rate of fluid production by the eye, but these are not suitable for people with heart conditions as they can alter cardiac and lung function. Other drugs with similar actions are prostaglandin analogues and alpha-2 agonists (see Chapter 10, p. 170). Acetylcholine agonists such as pilocarpine are also used to treat patients with glaucoma. These drugs cause pupillary constriction through contraction of the sphincter pupillae muscle of the iris, which increases the ability of aqueous humor to drain from the anterior chamber and thus lowers intraocular pressure.

Surgical treatment options for glaucoma include trabeculoplasty and trabeculectomy. In the former procedure a highly focused laser beam is used to create a series of tiny holes (usually between 40 and 50) in the trabecular meshwork to enhance drainage. Trabeculectomy may be performed in cases of advanced glaucoma. In this procedure an artificial opening is made in the sclera (sclerostomy), through which excess fluid is allowed to escape. A tiny piece of the iris may also be removed so that fluid can flow back into the eye.

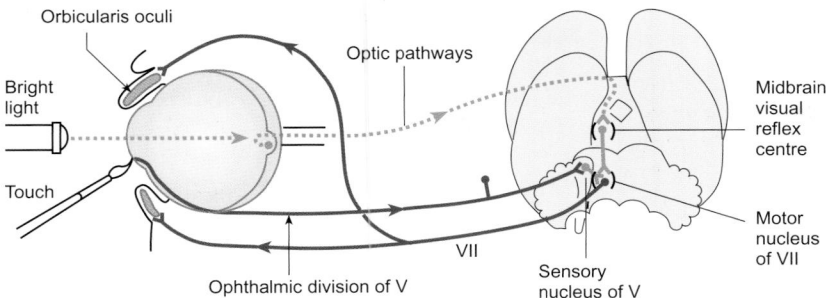

Fig. 8.22 The neural pathways that subserve the blink and dazzle reflexes.

If the glaucoma is unresponsive to other forms of treatment, an artificial device may be implanted in the eye to facilitate the drainage of aqueous humor from the anterior chamber. A thin silicon tube is inserted into the anterior chamber and fluid drains on to a tiny plate sewn to the side of the eye. The fluid is eventually absorbed from the plate by the tissues of the eye. In the case of an acute closed-angle glaucoma that threatens to destroy vision within hours, emergency treatment is required and, in such cases, surgical removal of the iris (*iridectomy*) may be performed. Very occasionally, a procedure known as *cyclophotocoagulation* may be used to reduce the rate at which fluid is secreted by the ciliary body. Essentially the ciliary body is destroyed using a laser beam, so this procedure is normally reserved for end-stage glaucoma, or for patients whose disease has failed to respond satisfactorily to any of the treatments described above.

The blink reflex and the dazzle reflex

The eyelids are closed by relaxation of the levator palpabrae muscles, which are supplied by the oculomotor nerve (CN III) coupled with contraction of the orbicularis oculi muscles, which are supplied by the facial nerve (CN VII). The reflex closure of the eyelids can result from corneal irritation due to specks of dust and other debris. In this case, the afferent fibers run in the trigeminal nerve (CN V). This is known as the *blink reflex* or corneal reflex.

Very bright light shone directly into the eyes also elicits closure of the eyelids. Lid closure will cut off more than 99 per cent of incident light. In this case, the afferent arm of the reflex originates in the retina and collateral fibers pass to the oculomotor nuclei. This is known as the *dazzle reflex* and a similar response can be elicited by an object that rapidly approaches the eye—the 'menace' or 'threat' reflex. The neural pathways involved in these reflexes are shown in Fig. 8.22.

Pupillary reflexes

The pupils are the central dark regions of the eyes. They define the area over which light can pass to reach the retina. If a light is shone directly into one eye, its pupil constricts. This response is known as the *direct pupillary response*. The pupil of the other eye also constricts, and this is known as the *consensual response*. These reflexes have a reaction time of about 0.2 s. As Fig. 8.23 shows, the afferent arm is a collateral projection from the optic nerve to the oculomotor nucleus from where the parasympathetic fibers pass to the ciliary ganglion and the short postganglionic fibers then travel to the circular muscles of the iris (the sphincter pupillae) to cause them to contract and narrow the pupils. The pupils also constrict when the eye is focused on a close object. This is known as the *accommodation reflex*, and has the effect of improving the depth of focus.

Dilatation of the pupils is brought about by increased activity in the sympathetic nerve supply, which causes contraction of the radial dilator muscles. Thus, the iris has a dual antagonistic innervation by the two divisions of the autonomic nervous system (see Chapter 10 for further details of the autonomic nervous system).

The muscles of the iris can only exert a slight degree of control over the amount of light admitted to the eye. Under steady state conditions, the pupil is about 4–5 mm in diameter. In extremely bright light it may constrict to 2 mm, and in very dim illumination it may assume a diameter of about 8 mm. Thus the maximal change in light transmission that can be controlled by the pupil is only about 16-fold—far less than the total range of

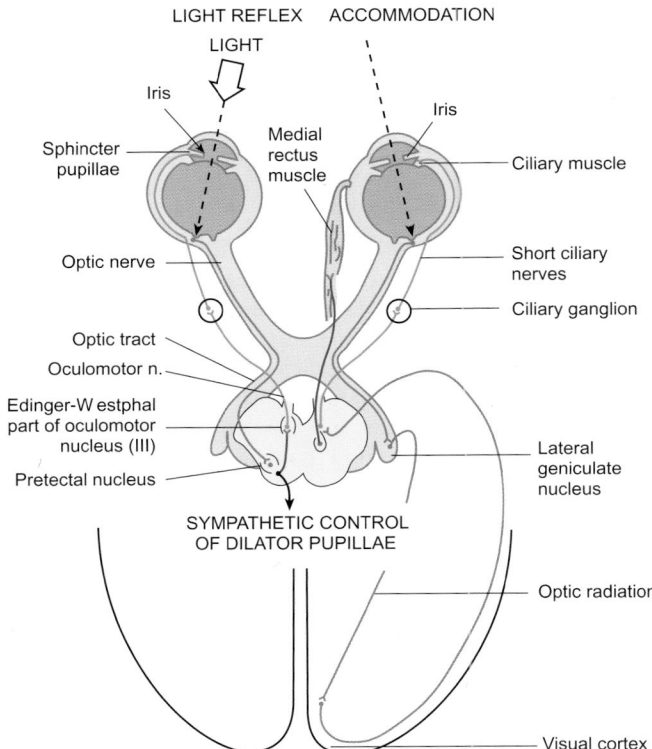

Fig. 8.23 The neural pathways of the pupillary reflexes.

Summary

1. The eye consists of an outer coat of connective tissue (the sclera), a highly vascular layer (the choroid), and a photo-receptive layer (the retina). Light enters the eye via a clear zone (the cornea) and is focused on the retina by the lens.

2. The cornea is protected from damage by the blink reflex, which is elicited by irritants on the corneal surface. The eyelids also close in response to excessively bright light. This is known as the dazzle reflex.

3. The amount of light falling on the retina is controlled by the pupil. If light is shone into one eye, the pupil constricts (the direct pupillary response). The pupil of the other eye also constricts (the consensual pupillary response).

4. The diameter of the pupils is controlled by the autonomic nervous system. Sympathetic stimulation leading to dilation of the pupils is a result of contraction of the radial smooth muscle of the iris. Parasympathetic stimulation elicits pupillary constriction as a result of contraction of the circular smooth muscle of the iris.

BOX 8.1 CALCULATION OF THE POWER OF LENSES

The refractive power of a lens is measured in *diopters*. The dioptric power (D) of a lens is the reciprocal of the focal length measured in meters:

$$\text{dioptric power} = \frac{1}{\text{focal length}}.$$

A converging lens with a focal length of 1 m has a power of 1 diopter. A lens with a focal length of 10 cm will have a power of 10 D and one with a focal length of 17 mm (the approximate focal length of the lens system of the human eye) has a dioptric power of

$$\frac{1}{17 \times 10^{-3}} = 58.8 \text{ D}$$

Expressing the power of lenses in this way has the advantage that the focal length of lenses in combination is given by adding together the dioptric power of each lens. For diverging lenses, the dioptric power is negative, so that a diverging lens with a focal length of 10 cm has a dioptric power of –10 D.

For example, a patient has an eye with a focal length of 59 D but the retina is only 16 mm behind the lens instead of the usual 17 mm. (The patient is affected by hyperopia, which is also called hypermetropia.) What power of spectacle lens is required for correction? To bring parallel light into focus in 16 mm requires a power of approximately 62.5 D. Thus, in this case, a converging lens of 62.5 – 58.8 = 3.7 D is needed.

If the retina were 18 mm behind the lens, the lens system would need to be 55.5 D to bring parallel light into focus on the retina. Thus a lens of 55.5 – 58.8 = –3.3 D would be needed for correction (i.e. a diverging lens of 3.3 D).

sensitivity of the eye. This indicates that the wide range of light intensity to which the eye can respond is due to the adaptation of the photoreceptors themselves. However, the changes in pupil diameter do help the eye to adjust to sudden changes in illumination.

While the benefit of the corneal, dazzle, and pupillary reflexes in protecting the eye from damage is obvious, they are also of value in the assessment of the physiological state of the brain-stem nuclei and in diagnosing brain death following traumatic head injury. In a condition known as *Argyll Robertson pupil*, the pupils constrict during accommodation but the direct and consensual pupillary responses to light are absent. It is an important clinical sign indicating CNS disease such as tertiary syphilis.

The optics of the eye—image formation

The eye behaves much as a pinhole camera. The image is inverted so that light falling on the retina nearest the nose (the *nasal retina*) comes from the lateral part of the visual field (the *temporal field*), while the more lateral part of the retina (the *temporal retina*) receives light from the central region of the visual field (the *nasal field*).

Since objects may lie at different distances from the eye, the optical elements of the eye must be able to vary its focus in order to form a clear image on the retina. The eyes of a young adult with normal vision are able to focus objects as distant as the stars and as close as 25 cm (the *near point*). The optical power of the eye is at its minimum when distant objects are brought into focus and at its maximum when it is focused on the near point. This variation in optical power is achieved by the lens.

As for any lens system, the optical power of the eye is measured in *diopters* (see Box 8.1 for further details). The total optical power of a normal relaxed eye is about 58 diopters. Most of the optical power of the eye is provided by the refractive power of the cornea (about 43 diopters) while the lens contributes a fur-

ther 15 diopters when the eye is focused on a distant object, and about 30 diopters when it is focused on the near point. When the lens loses its elasticity, which commonly occurs with advancing age, the power of accommodation is reduced and the nearest point of focus recedes. This condition is known as *presbyopia*.

The ciliary muscles control the focusing of the eye

The lens is suspended from the ciliary muscle by the zonal fibers as shown in Fig. 8.24. When the ciliary muscles are relaxed there is a constant tension on the zonal fibers exerted by the effect of the intraocular pressure on the sclera. This tension stretches the lens and minimizes its curvature. When the eye switches its focus from a distant to a near object, the ciliary muscle contracts and this opposes the tension in the sclera. As a result, the tension on the zonal fibers decreases and the lens is able to assume a more spheroidal shape owing to its inherent elasticity: the more spheroidal the lens, the greater is its optical power. The ciliary muscle is innervated by parasympathetic fibers from the ciliary ganglion.

Refractive errors—myopia, hyperopia, and astigmatism

When the ciliary muscle of a normal eye is relaxed, the eye itself is focused on infinity so that the parallel rays of light from a distant object will be brought into sharp focus on the retina. This is called *emmetropia*.

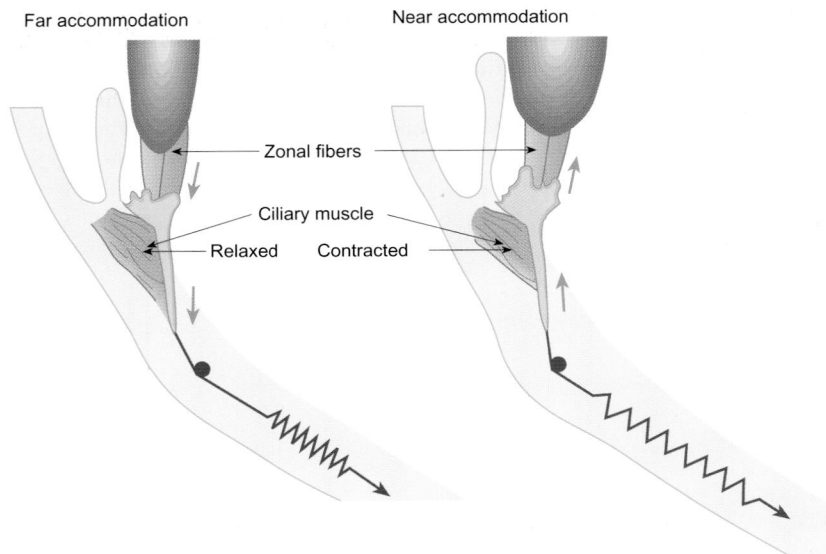

Far accommodation   Near accommodation

Zonal fibers

Ciliary muscle

Relaxed   Contracted

Fig. 8.24 A diagram to illustrate the mechanism of accommodation. When the ciliary muscles are relaxed, the tension in the wall of the eye is transmitted to the lens via the zonal fibers. The lens is stretched and adapted for distant accommodation (left). When the ciliary muscles contract, they relieve the tension on the zonal fibers and the lens assumes a more rounded shape suited to near accommodation.

If the eyeball is too long, the parallel rays of light from a distant object will be brought into focus in front of the retina and vision will be blurred. This situation can also arise if the lens system is too powerful. In both cases, the eye is able to focus objects much nearer than normal. This optical condition is known as *myopia*. It can be corrected with a diverging (or concave) lens as shown in Fig. 8.25.

If the eye is too short, parallel light from a distant object is brought into focus behind the retina. This can also arise if the lens system is insufficiently powerful. People with this optical condition have difficulty in bringing near objects into focus and are said to suffer from *hyperopia* or *hypermetropia*. It can be corrected with a converging (or convex) lens as shown in Fig. 8.25.

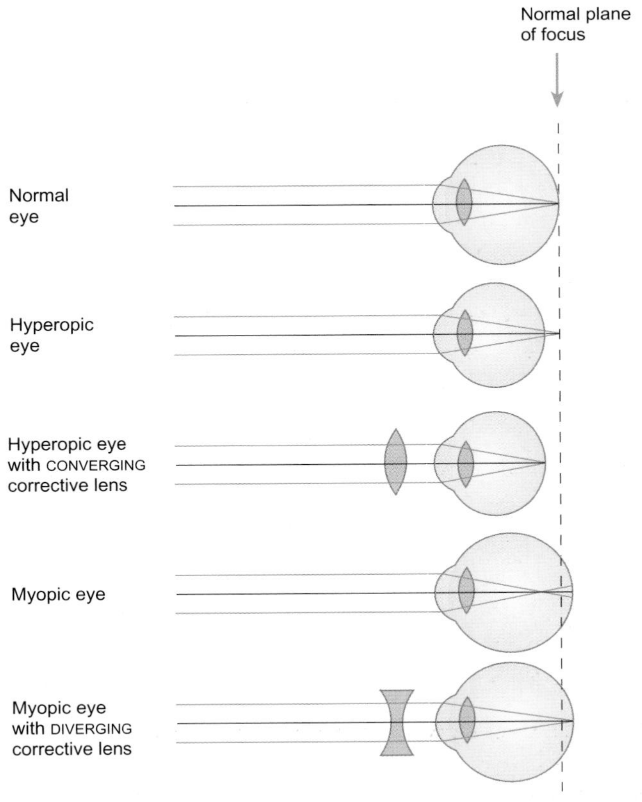

Normal plane of focus

Normal eye

Hyperopic eye

Hyperopic eye with CONVERGING corrective lens

Myopic eye

Myopic eye with DIVERGING corrective lens

Fig. 8.25 A simple diagram illustrating the principal refractive errors of the eye and their correction by external lenses.

Summary

1. The optics of the eye are very like those of a camera. The image of the world is brought into focus on the retina by the action of the lens system of the eye. The total optical power of the eye when it is fixated on a distant object is about 58 diopters (D). Of this, the refractive power of the cornea accounts for some 43 D and that of the lens accounts for about 15 D. The lens is able to alter its refractive power by a further 14 or 15 D to bring near objects into focus.

2. The focusing of the lens is controlled by the ciliary muscle which modulates the tension in the zonal fibers. Near accommodation occurs by contraction of the ciliary muscle which relieves the tension in the zonal fibers. The reduction in tension allows the lens to become more rounded, so increasing its optical power.

3. The major problems in image formation are due to malformation of the eyeball. The most common defects are myopia, hypermetropia (or hyperopia), and astigmatism. They can be corrected with an external lens of appropriate power.

4. The capacity of the eye to resolve the detail of an object is its visual acuity. Under photopic conditions, visual acuity is best in the central region of the visual field but, under scotopic conditions, it is best in the area surrounding the central region. This difference reflects the distribution of cones and rods in the retina and their differing roles in photopic and scotopic vision.

BOX 8.2 THE EFFECT OF LESIONS IN DIFFERENT PARTS OF THE VISUAL PATHWAYS ON THE VISUAL FIELD

The visual field of an eye covers that part of the external world that can be seen without changing the point of fixation. It is measured by *perimetry*, which is a useful clinical tool for determining the health of the eye and visual pathways. The extent of the visual field is measured with an instrument called a perimeter. The visual field of each eye is measured in turn by moving a small spot of light along a meridian until the subject signals that they have seen it. The point is marked and another meridian tested until the full 360° has been tested.

If the eye were not protected within the orbit, the visual field would be a circle. In reality, the features of the face—the nose, eyebrows, and cheekbones—limit the peripheral extent of the visual field, as shown in Fig. 1. The visual fields of the two eyes overlap extensively in the central region.

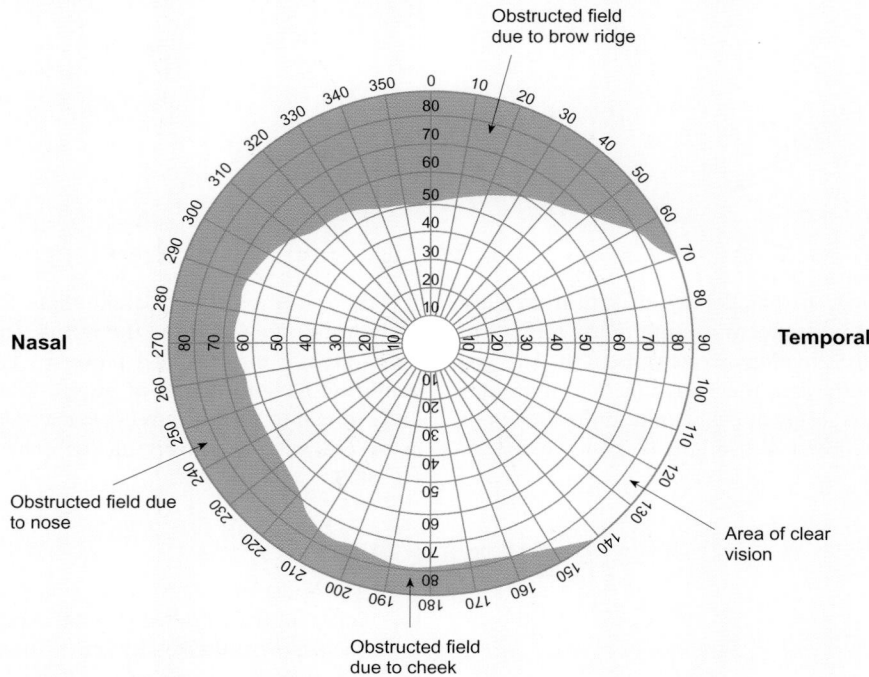

Fig. 1 A perimeter chart to show the field of vision for the right eye. The nasal field (left) is obstructed by the nose while the temporal field extends a full 90°. Note the obstruction produced by the brow ridge.

Since the visual system is organized in a highly characteristic way, defects in specific parts of the visual pathways give rise to characteristic regions of blindness in the visual field (Fig. 2). If one optic nerve is cut or compressed, as in panel (a), the eye served by that nerve is blind. In contrast, if the optic fibers are damaged after the optic chiasm, then the patient will be blind in the visual field on the opposite side to the lesion (*homonymous hemianopia*; panel (b)). If the optic chiasm is compressed there will be a loss of the crossing fibers, resulting in *bitemporal hemianopia*, or *tunnel vision* as shown in panel (c). This commonly occurs as a result of the growth of a pituitary tumor which initially presses on the middle part of the optic chiasm. Damage to the optic radiation produces an area of blindness (a *scotoma*) in the corresponding region of the contralateral visual field, as shown in panel (d).

In some cases, the curvature of the cornea or lens is not uniform in all directions. As a result the power of the optical system of the eye is different in different planes. This is known as *astigmatism* and it can be corrected by a cylindrical lens placed so that the refraction of light is the same in all planes.

The visual field

The visual field of an eye is that area of space which can be seen at any instant of time. It is measured in a routine clinical procedure called *perimetry*. The eye to be tested is kept focused ('fixated') on a central point and a test light is gradually moved from the periphery towards the center and the subject asked to signal when they see it. The result shows that the field of view is restricted by the nose and by the roof of the orbit so that the visual field is at its maximum laterally and inferiorly. The visual fields of the two eyes overlap extensively in the nasal region. This is the basis of binocular vision which confers the ability to judge distances precisely. Defects in the visual field can be used to diagnose damage to different parts of the optical pathways as described in Box 8.2.

Photopic and scotopic vision

The cones are the main photoreceptors used during the day when the ambient light levels are high. This is called *photopic*

BOX 8.2 *(CONTINUED)*

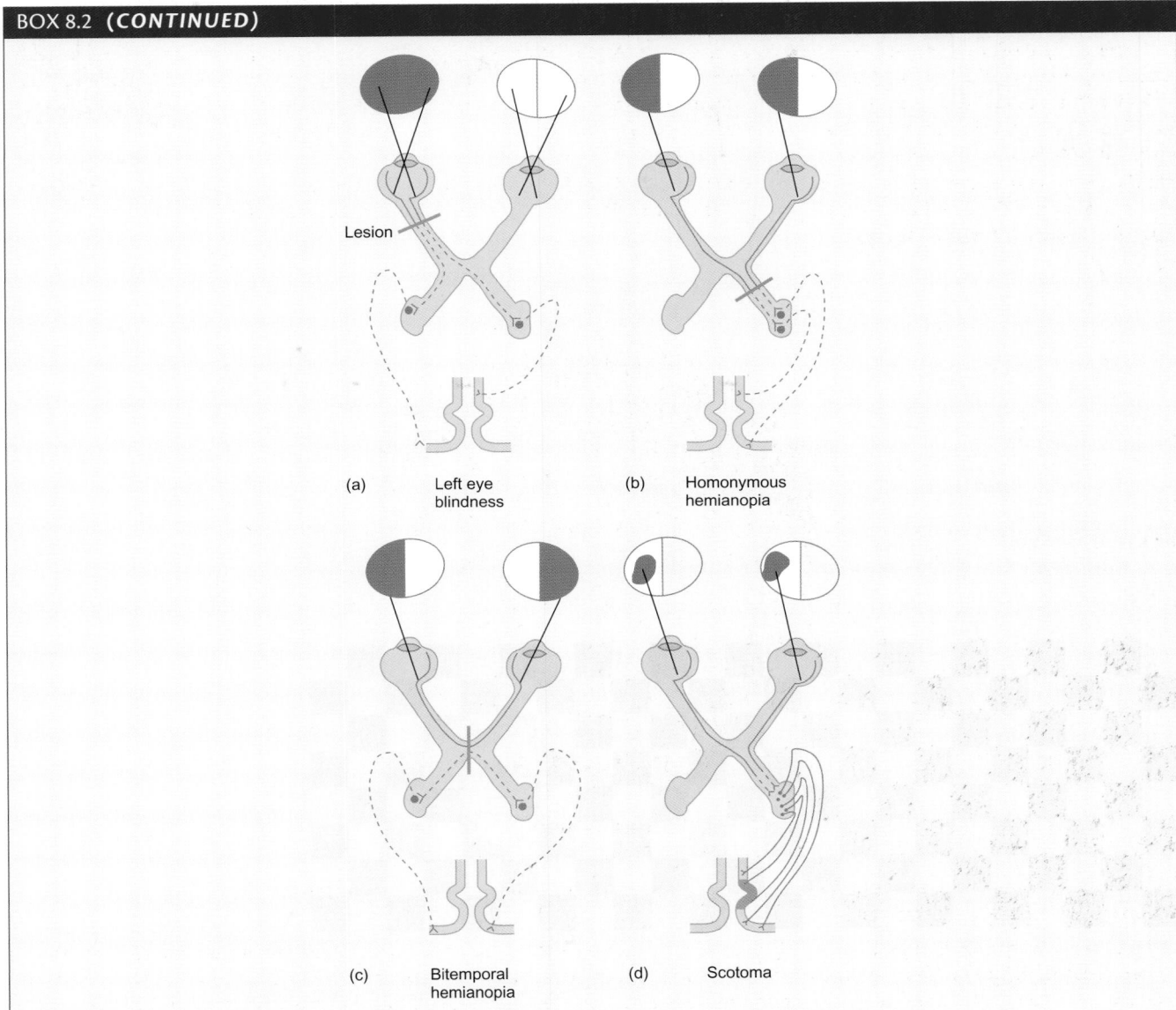

Fig. 2 The effect of lesions in (a) the left optic nerve; (b) the right optic tract; (c) the optic chiasm and (d) the visual cortex.

It is important to note that patients suffering from partial loss of vision do not experience 'darkness', they have a loss of vision in the affected part of the visual field. Often the first signs of a visual deficit are an apparent clumsiness on the part of patients. They fail to notice things in a particular region of the visual field. Therefore proper testing of the visual field should be carried out on people who have suffered a traumatic head injury or if there are signs of abnormal pituitary function.

vision. In photopic conditions, visual acuity (see below) is high and there is color vision. At night, when light levels are low, the rods are the main photoreceptors. This is known as *scotopic vision* and is characterized by high sensitivity but poor acuity and a lack of color vision. During twilight, both rods and cones are used and this is called *mesopic vision.*

Visual acuity varies across the retina

Objects fixated at the center of the visual field are seen in great detail, while those in the peripheral regions of the field are seen less distinctly. The capacity of the eye to resolve the detail of an object is its *visual acuity*, which is measured in terms of the angle subtended by two points that can just be distinguished as being separate entities. Visual acuity is best in bright light. If visual acuity is plotted against the distance from the main optical axis of the eye, it is found to be greatest at the central region and progressively falls towards the periphery under photopic conditions. In dim light (i.e. under scotopic conditions), the cones are not stimulated and visual acuity is zero at the center of the visual field and at its (modest) maximum in the parafoveal region (Fig. 8.26).

Light from objects in the center of the visual field falls onto the fovea centralis, which has the highest density of cones and

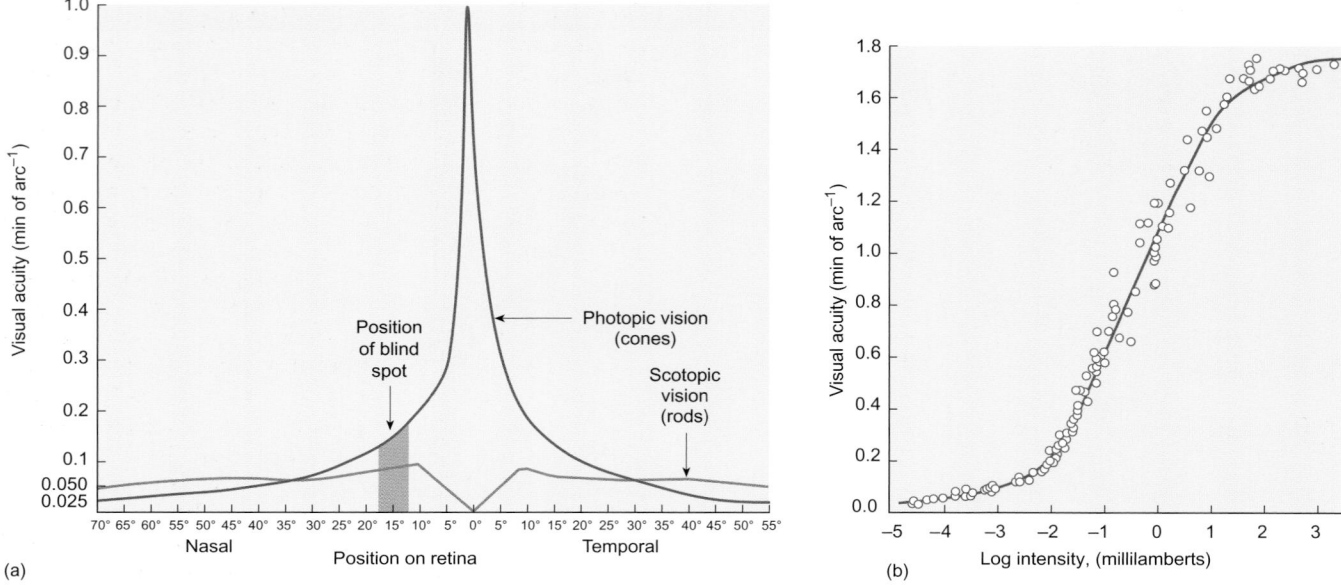

(a)

(b)

Fig. 8.26 The variation of visual acuity (a) across the retina and (b) with the intensity of light falling on the fovea. In (a) note that, while visual acuity is greatest for an image falling on the fovea, in the dark-adapted eye it is least at the fovea and highest in the parafoveal region.

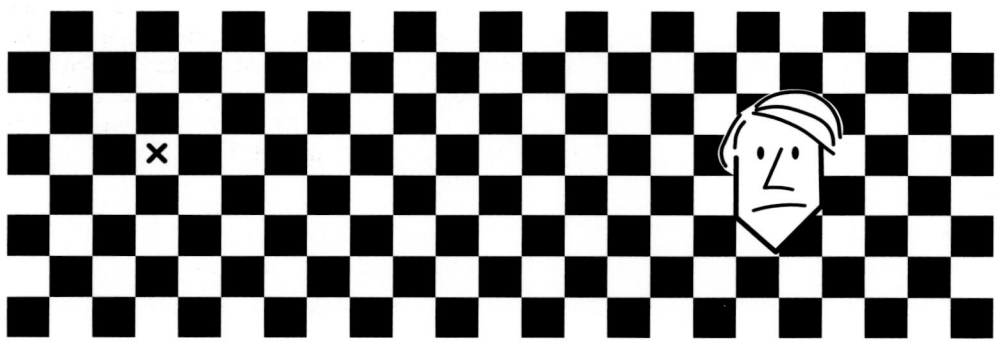

Fig. 8.27 Demonstration of the blind spot. Close the left eye and focus on the cross with the right eye. Move the page so that is about 25 cm from you and the face will disappear with no discontinuity in the background.

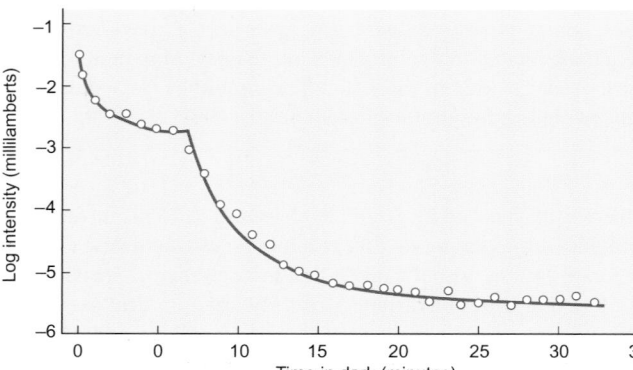

Fig. 8.28 The dark adaptation curve obtained by plotting the visual threshold as a function of time spent in the dark following a period spent in bright light. The first segment of the curve is due to the adaptation of the cones, which is complete in a little over 5 min. The adaptation of the rods takes a further 15–20 min before maximum sensitivity is achieved.

the smallest degree of convergence between adjacent photo-receptors. Outside the fovea, the degree of convergence increases and visual acuity falls. The rods are specialized for the detection of low levels of light and have a high degree of convergence. Thus they provide the eye with a high sensitivity at the expense of visual acuity.

Light falling on the area of the optic disk (the point where the optic nerve leaves the eye) will not be detected, as there are no rods or cones in this region. This is known as the *blind spot*, and it can be demonstrated as described in the legend to Fig. 8.27.

Dark adaptation

When one first moves from a well-lit room into a garden lit only by the stars, it is difficult to see anything. After a short time, the surrounding shrubs and trees become increasingly visible. The process that occurs during this change in the sensitivity of the eye is called *dark adaptation* and its time course can be followed by measuring the threshold intensity of light that can be detected at different times following the switch from photopic conditions to scotopic conditions. The dark adaptation curve

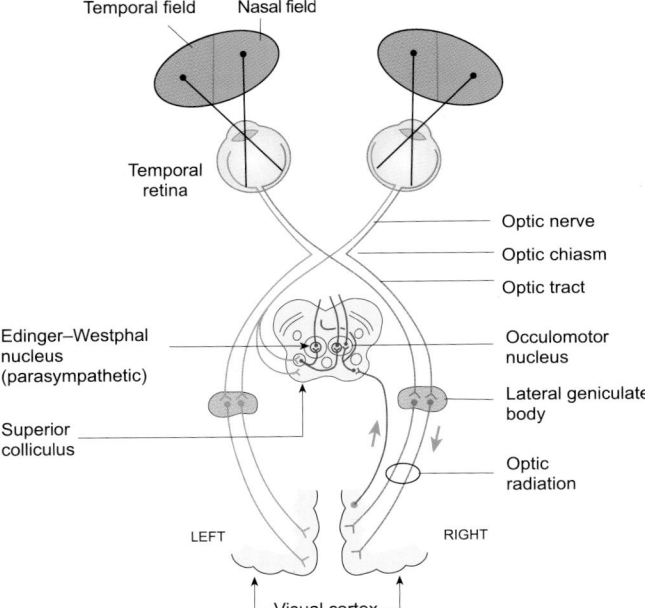

Fig. 8.29 A schematic drawing of the visual pathways. Note that the nasal field projects to the temporal retina and that fibers from this region do not cross. Thus each half of the visual field projects to the opposite hemisphere. The connections between the two hemispheres are not shown. The projections to the brainstem mediate the corneal and pupillary reflexes.

obtained in this way shows that the greatest sensitivity is attained after about 20–30 min. in the dark (Fig. 8.28). The adaptation curve shows two distinct phases. The first shows a relatively rapid change in threshold, which appears to reach its limit after about 5–10 min. This is attributed to adaptation of the cones. The second phase is much slower, taking a further 10–20 minutes to reach its limit. This phase is attributed to adaptation of the rods.

The change from scotopic vision to photopic vision is called *light adaptation*. This process is faster than dark adaptation and is readily appreciated on moving from a darkened area into bright light. The initial sense of being blinded by the light rapidly wears off.

The neurophysiology of vision

The organization of the visual pathways

From the retina, the axons of the ganglion cells pass out of the eye in the optic nerve. At the optic chiasm, the medial bundle of fibers (which carry information from the nasal side of the retina) crosses to the other side of the brain. This partial decussation of the optic nerve allows all the information arriving in one field of vision to project to the opposite side of the brain. Thus the left visual field projects to the right visual cortex and the right visual field projects to the left visual cortex (Fig. 8.29). From the optic chiasm, the fibers pass to the lateral geniculate bodies giving off collateral fibers that pass to the superior colliculi and oculomotor nuclei in the brain stem. From the lateral geniculate bodies the main optic radiation passes to the primary visual cortex located in the occipital lobe. Nerve fibers connecting the two hemispheres pass through the corpus callosum.

The role of the retina in visual processing

The receptors of the retina are the rods and cones. Both contain photosensitive pigments. The pigment in the rods is known as rhodopsin and is made up of the aldehyde of vitamin A (11-*cis*-retinal) and a protein called opsin. The cones also have photosensitive pigments containing 11-*cis*-retinal but, in their case, the aldehyde is conjugated to different photoreceptor proteins that make the different types of cone sensitive to light of different wavelengths.

In both rods and cones, the visual pigment is located in the membranous disks of the outer segment. When a photon of light is captured by one of the visual pigment molecules, the shape of the retinal changes and this triggers a change in the properties of the photoreceptor protein to which it is bound. The activated photoreceptor protein then activates a G protein called transducin. Transducin activates a phosphodiesterase that breaks down cyclic GMP to 5′-GMP. In the dark, the levels of cyclic GMP in the photoreceptors are high. The cyclic GMP binds to the internal surface of ion channels permeable to Na^+ and causes them to open. As a result, the membrane potential of the photoreceptors is low (about –40 mV). When the rhodopsin is activated by light, the levels of cyclic GMP fall and so fewer of the Na^+-permeable ion channels are open. The fall in Na^+ permeability results in a hyperpolarization of the photoreceptor. When the photoreceptors become hyperpolarized following the absorption of light, the secretion of neurotransmitter (glutamate) by the photoreceptors is decreased and the bipolar cells respond with a change in their membrane potential. The sequence of events that immediately follows the absorption of light is shown schematically in Figure 8.30.

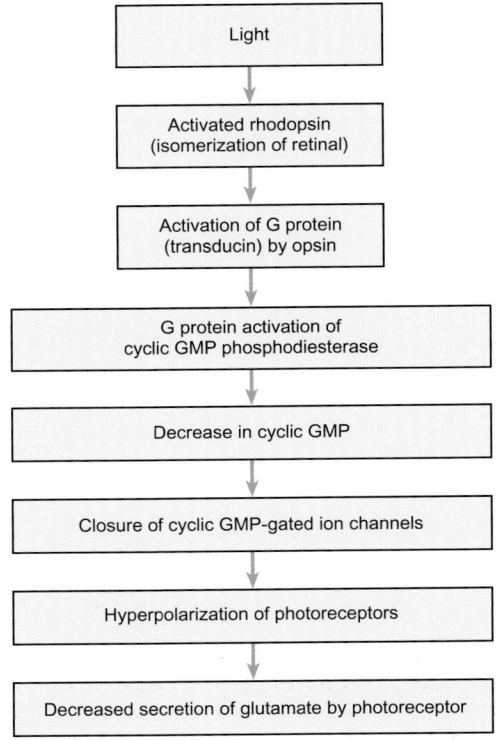

Fig. 8.30 The principal steps in phototransduction.

Within the retina there is a considerable degree of neural processing. There are two types of bipolar cell. One type depolarizes in response to light and the other type hyperpolarizes. Moreover, the horizontal cells and amacrine cells permit the spread of excitation and inhibition across the retina (see Fig. 8.20). As a result, the responses of an individual retinal ganglion cell will depend on the properties of the cells to which they are connected. In general, retinal ganglion cells have receptive fields with a center-surround organization of the kind shown in Fig. 8.31. Two types are generally recognized: 'on-center off-surround' and the 'off-center on-surround'. In the first type, the ganglion cell discharge is increased when a spot of light is shone

in the center of its receptive field and decreased when a light is shone in the surround. The second type has the reverse arrangement: the action potential frequency falls when light is shone in the central region of the field but rises when light falls on the surround. The receptive fields of the neurons of the lateral geniculate body to which the ganglion cells project have a similar organization.

The neurons of the primary visual cortex respond to specific features of an image

Many neurons in the visual cortex respond to stimuli presented to both eyes. Moreover, as the visual cortices of the two hemispheres are interconnected via the corpus callosum, some cortical neurons receive information from both halves of the visual field and can detect small differences between the images in each retina. This is the basis of stereoscopic vision, which provides cues about the distance of objects.

Unlike the receptive fields of cells in the early part of the visual pathway, neurons in the visual cortex often respond to

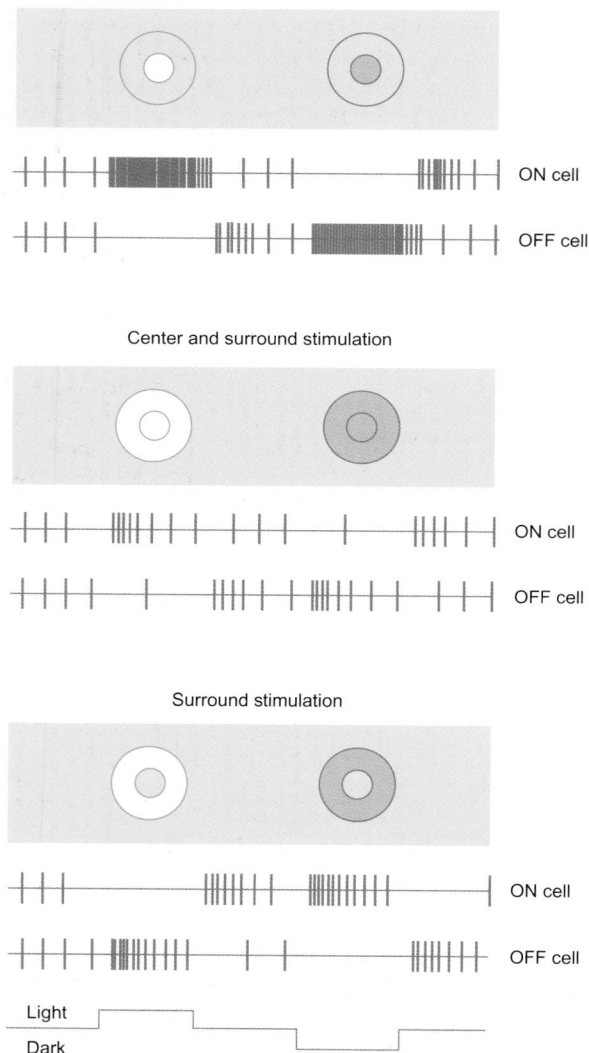

Fig. 8.31 The functional organization of the receptive fields of two types of retinal ganglion cells. For 'on-center' cells, a spot of light shone into the center of the receptive field excites the ganglion cell. A similar spot of light shone into the surround region inhibits ganglion cell discharge. For 'off-center' ganglion cells the organization is reversed. For both kinds, diffuse illumination of the whole of the receptive field leads to weak excitation.

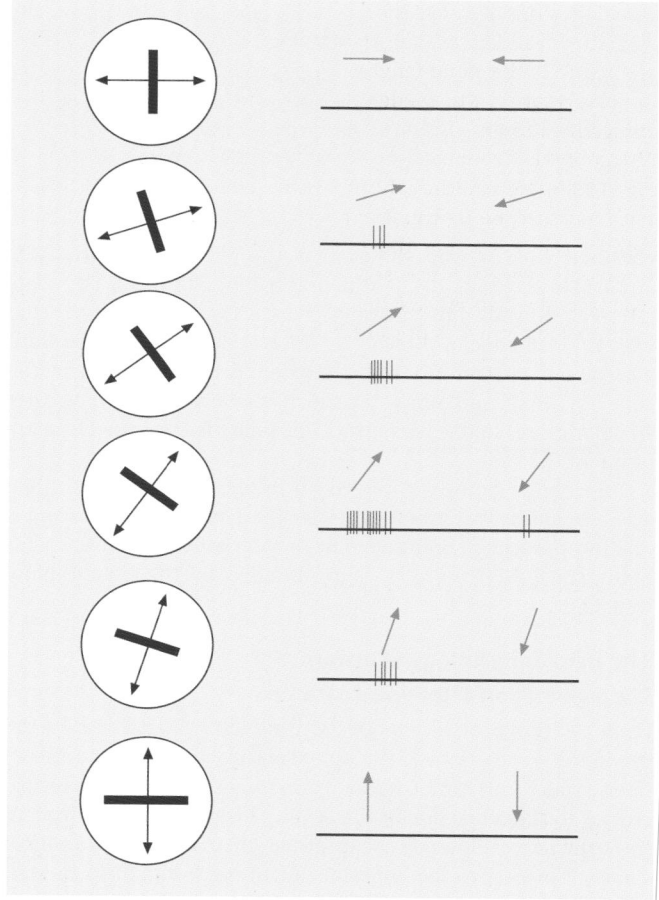

Fig. 8.32 The response of a cell in the visual cortex to a black bar passing across an illuminated region of the visual field (shown by the white circles). The response of the neuron is shown in the right-hand panel by a series of action potential records. The cortical neuron shows no response to a vertically or horizontally oriented bar but gives a strong response to a bar placed at 36° to the horizontal. In this example, the cell shows a distinct directional preference, firing strongly to a rightward movement but weakly or not at all to a leftward movement.

particular features of objects in the visual field. Some respond optimally to a bright or dark bar passing across the retina at a specific angle (Fig. 8.32). Others respond to a light–dark border and so on. In many cases the direction of movement is also critical. In this way, the brain is able to identify the specific features of an object, which it then uses to form its internal representation of that object in the visual field. Such representations are crucial to our ability to interpret the visual aspects of the world around us.

Vision is an active process. We learn to interpret the images we see. We associate particular properties with specific visual objects and have a remarkable capacity to identify objects, particularly other people, from quite subtle visual cues. We are able to recognize a person we know at a distance. Yet we do not notice that they appear larger as they approach us; rather, we assume that they have a particular size and use that fact as a means of judging how far away they are. This phenomenon is known as *size constancy*.

The importance of the way the brain uses visual cues to form its internal representation of an object is illustrated by many visual illusions. Perhaps the best known is the Müller–Leyer illusion shown in Fig. 8.33(a). Here the central line in each half of the figure is the same length but the line with the arrowheads appears shorter. Knowledge of this fact does not diminish the subjective illusion. The fact that the brain judges the size of an object by its context is revealed by the illusion shown in Fig. 8.33(b). Here, although the central circle is the same size in both parts of the figure, it appears larger when surrounded by small circles. Another illusion that illustrates how the brain interprets what the eye sees is *form vision*. In the case of the example given in Fig. 8.33(c) a white square is 'seen' when, in fact, no such image is present. It appears that the brain uses the

information from the surrounding black circles to complete the square. In Fig. 8.33(d) the displacement of alternate rows of tiles gives rise to a gross distortion that creates the impression of a series of long wedges. In fact, the black and white squares are all exactly the same size. Illusions are not just related to the shape of an object; the apparent lightness or darkness of an object is judged by its context as Fig. 8.34 shows.

The brain processes color, motion, and shape by parallel pathways.

As the illusions mentioned above show, the brain processes visual information in a very different manner to a camera system. Detailed neurophysiological studies have shown that different aspects of a visual image are processed by different parts of the brain. The shape of an object is processed by one pathway while another processes its color. Movement is processed by yet another pathway. This notion is supported by deficits in visual function that arise from damage to different parts of the brain. The failure to name and recognize a common object is known as *object agnosia* (agnosia means not knowing). This occurs when there is damage to certain visual association areas in the left hemisphere. Damage to nearby areas can lead to difficulty in distinguishing colors (a condition known as *achromatopsia*). Failure to discern the movement of an object is associated with bilateral lesions at the border of the occipital and temporal lobes. Exactly how the different aspects of an object are subsequently resynthesized into a unified visual image is not known.

Color vision

In full daylight, the cones are the principal photoreceptors. Most people experience color vision—individual objects have their own intrinsic color. An unripe apple appears to us as green while a ripe tomato appears red. This has obvious advantages in distin-

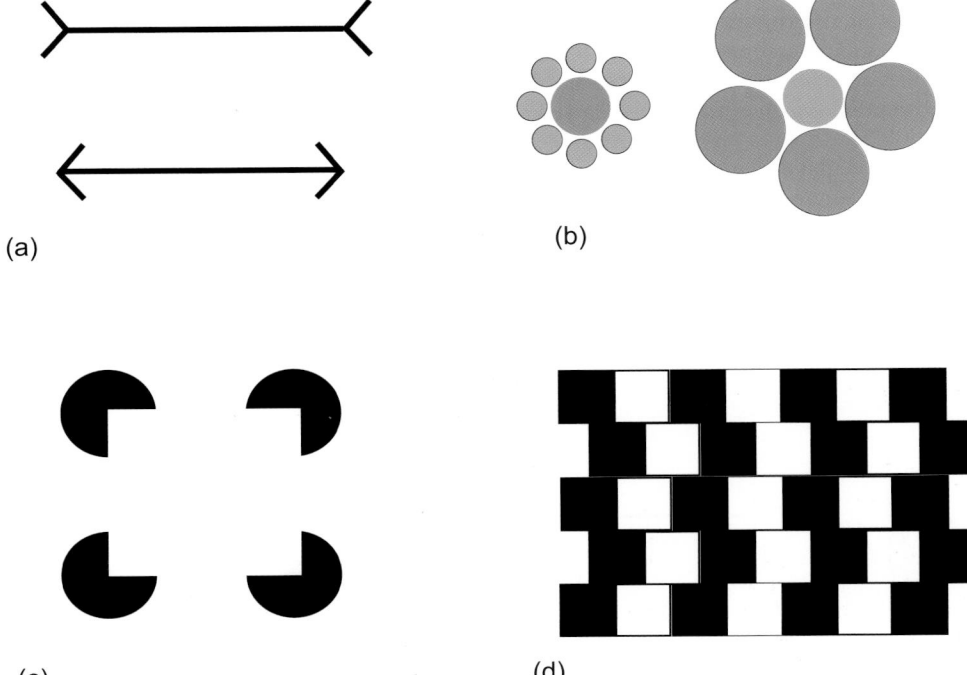

(a)

(b)

(c)

(d)

Fig. 8.33 Some visual illusions. (a) The Müller–Leyer illusion. Here the central line appears longer in the figure with outwardly directed fins although it is the same length as that with inwardly directed fins. The length of the lines can be confirmed by measurement with a ruler, yet the illusion does not disappear. (b) Errors of size perception—the Tichener illusion. Here the central circle is the same diameter in both parts of the figure, although it looks much larger when surrounded by small circles. (c) Form perception in a Kanizsa figure. Here the brain completes the outline of a square, although no square is present. (d) The café-wall illusion. The displacement of the alternate rows of tiles creates a false impression of a wedge. In fact, the black and white tiles are all exactly the same size.

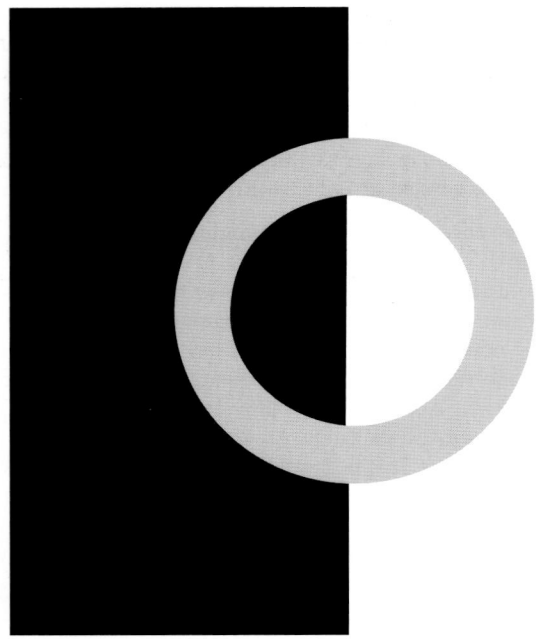

Fig. 8.34 Simultaneous contrast. The half of the grey circle abutting the black rectangle appears lighter that that over the white background. The effect is even more marked if a black thread is placed across the circle at the border with the black area.

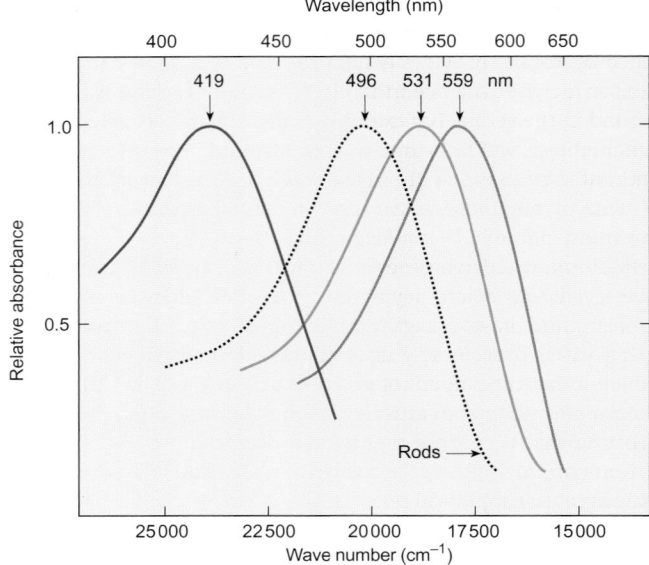

Fig. 8.35 The absorbance spectra of the pigments in the three types of human cones and that of rhodopsin (which is found in the rods).

guishing the different objects in the environment, but how is this remarkable feat achieved? People with normal color vision can match the color of an object by mixing varying amounts of just three colored lights—blue, green, and red. This suggests that there should be three different cone pigments—one sensitive to blue light, one to green light, and one to red light. This supposition has now been validated by direct measurements of the absorption of pigments found in individual human cones. The blue-sensitive cones show an absorption maximum at 420 nm, the green-sensitive cones have a maximum absorption at about 530 nm, and the red-sensitive cones absorb maximally at 560 nm. The rod pigments have an absorption maximum at 496 nm (Fig. 8.35).

The absorption spectra measure the likelihood that a pigment will absorb a photon of light at a given wavelength. Thus, at

least two different pigments are required for any color vision and the brain must be able to compare the intensity of the signals emanating from different cones. For normal human color vision, a green light is seen when the green-sensitive cones are more strongly stimulated than the red- and blue-sensitive cones and so on. White light reflects equal intensity of stimulation of all three types of cone. This is the basis of the trichromatic theory of color vision proposed by Thomas Young at the beginning of the nineteenth century.

Useful as this theory is, it fails to explain some well known observations. First, certain color combinations, such as a reddish green or a bluish yellow, do not occur. Yet it is possible to see a reddish yellow (orange) or a bluish green (cyan). Second, if one stares at a blue spot for a short time and then looks at a white page, a yellow after-image is seen. Similarly, a green after-image will be seen after looking at a red spot (Fig. 8.36). To answer these difficulties, E. Hering proposed the existence of neural processes in which blue and yellow were considered opponent colors, as

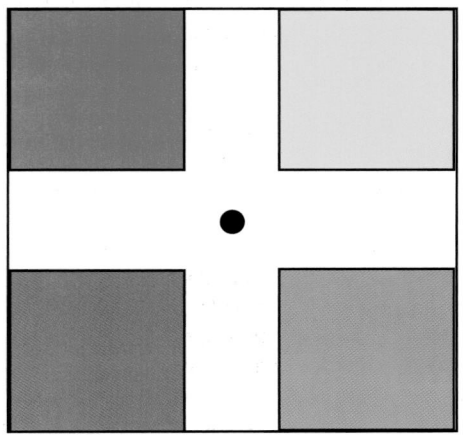

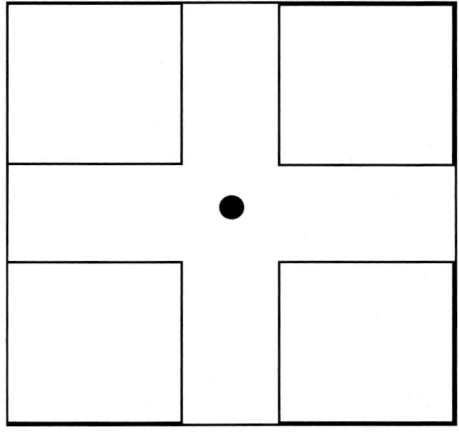

Fig. 8.36 Color after-images show complementary colors. Close one eye and focus on the black spot in the center of the color panel on the left for about a minute. Then switch your eye to the black spot of the right-hand figure. A yellow after-image will be seen in the top left square, a blue after-image in the top right, a green after-image in the bottom left, and a red after-image in the bottom right.

were green and red. This color-opponent theory (with some later modifications), together with the trichromatic principle enunciated by Young, provides a basis for understanding color vision. Experimental evidence in favor of the color-opponent theory is found in the retina. For example, some retinal ganglion cells are excited by a red light in the center of their receptive field but inhibited by a green light in the surround.

One of the more remarkable aspects of color vision is the phenomenon of *color constancy*. An apple will appear green and a ripe tomato red whatever the time of day. This distinction is observed even when they are illuminated with artificial light. As the mixture of wavelengths making up the light (its spectral content) varies considerably under these differing circumstances, it follows that the amount of green, blue, and red light reflected by the apple and tomato must be different under these different circumstances. How does the brain compensate for the changing illumination? If the apple and tomato are illuminated with (say) red light, the apple will reflect less light that the tomato and so it will appear darker than the tomato. If the illuminating light is green, the situation will be reversed; the apple will appear bright and the tomato dark. Thus the comparative intensity of the light detected by the red and green cones provides information about the color of the two fruits regardless of the spectral content of the illumination. Thus the brain is able to compensate for the effect of changes in the quality of the illuminating light.

Color blindness

More than 90 per cent of the human race can match a given color by mixing the appropriate proportions of red, green, and blue light. However, some people (mostly males) can match any color with only two colors. These people are *dichromats* (rather than the *trichromats* of the general mass of the population).

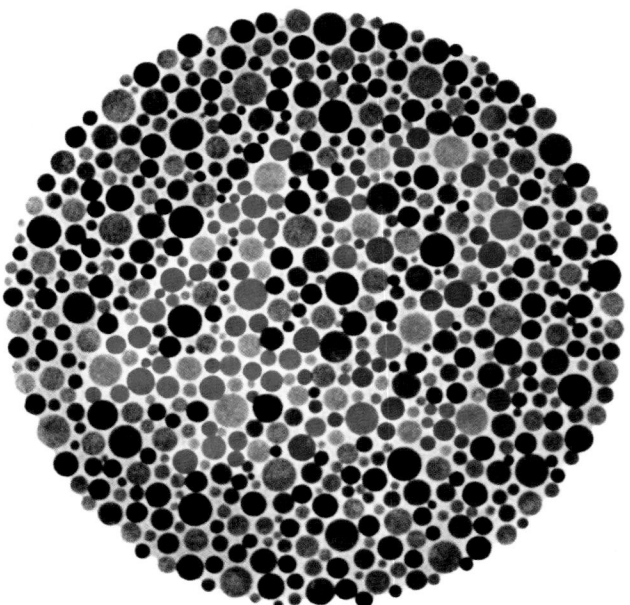

Fig. 8.37 An Ishihara card. People with normal color vision see '42', a red-blind person (a protanope) will only see the 2, while a green-blind person (a deuteranope) will only see the 4. A complete test uses many such cards; different cards use different combinations of colors.

Although they are generally considered to be color blind, they do distinguish different colors. It is just that their color matching is abnormal. About 6 per cent of trichromats match colors with abnormal proportions of one or other of the primary colors. These are called *anomalous trichromats*.

Dichromats are classified as *protanopes* (a relative insensitivity to red), *deuteranopes* (a relative insensitivity to green), or *tritanopes* (a relative insensitivity to blue). Protanopes and deuteranopes each constitute about 1 per cent of the male population. Tritanopes are exceedingly rare, as are those who have no cones at all (rod monochromats).

Apart from the rod monochromats, those suffering from color blindness are not debilitated, but in certain industries where color codes are important it is necessary to be aware of any potential color confusions. This diagnosis is generally done with the Ishihara card test (Figure 8.37).

Eye movements

Our eyes constantly scan the world around us. An image appearing in the peripheral field of vision is rapidly centered onto the fovea by a jerky movement of the eyes. These rapid eye move-

Summary

1. The photoreceptors contain photosensitive pigments made up of vitamin A aldehyde (11-*cis*-retinal) and a photoreceptor protein. In the case of the rods, the photoreceptor protein is called opsin. When light is absorbed by a rod, the retinal isomerizes and dissociates from the opsin. This activates a cascade of reactions that leads to the hyperpolarization of the photoreceptor, which is the first step in a sequence of events that leads to the generation of action potentials by the retinal ganglion cells.

2. The visual pathways are arranged so that each half of the visual field is represented in the visual cortex of the contralateral hemisphere. To achieve this, the fibers arising from the ganglion cells of the nasal retina cross to the other side while those of the temporal retina do not. Therefore the right visual field is represented in the left hemisphere and vice versa.

3. The axons of the ganglion cells reach the lateral geniculate bodies via the optic tracts. The axons of the lateral geniculate neurons project in their turn to the visual cortex where the specific features of objects in the visual field are resolved.

4. There are three kinds of cone, each of which is sensitive to light of different wavelengths. This confers the ability to discriminate between different colors. Some people (most of whom are male) have defective cone pigments and are said to be color blind. The most common are the anomalous trichromats who match colors with unusual proportions of red and green. True color-blind people lack one or other of the cone pigments. The most common are those who lack either the red pigment (protanopes) or the green pigment (deuteranopes).

ments are called *saccades*. During a saccade, the eyes move at angular velocities of between 200 and 600 deg s^{-1}. In contrast, when watching a race or when playing a ball game, the eye follows the object of interest, keeping its position on the retina fairly constant. This smooth tracking of an object is called a *pursuit movement*, and it can have a maximum angular velocity of about 50 deg s^{-1}. These two types of eye movement can be combined, for example when looking out of a moving vehicle. One object is first fixated and followed until the eyes reach the limit of their travel. The eyes then flick to fixate and follow another object and so on. The continuous switching of the point of fixation is seen as a pursuit movement followed by a saccade and then another pursuit movement. This pattern is known as *optokinetic nystagmus*. A nystagmus is a rapid involuntary movement of the eyes.

If an object such as a pencil is fixated and then moved around the visual field both eyes track the object. These are known as *conjugate eye movements*. If the pencil is moved first away from the face and then towards it, the two eyes move in mirror-image fashion to keep the image in focus on each retina. When the object approaches the eyes, the visual axes converge and as the object moves away they progressively diverge until they are parallel with each other. These eye movements are called *vergence movements*. If the movements of the two eyes are not properly coordinated, double vision (*diplopia*) results.

The position of each eye is controlled by six extraocular muscles which are innervated by CN III, CN IV, and CN VI (the oculomotor, trochlear, and abducent nerves). The lateral and medial rectus muscles control the sideways movements, the superior and inferior oblique muscles control diagonal movements, and the superior and inferior rectus muscles control the up and down movements, as shown in Fig. 8.38.

The eye movements require full coordination of all of the extraocular muscles with activation of synergists and an appropriate degree of inhibition of antagonists. If we look to one side, the right for example, the right lateral rectus and the left medial rectus are both activated while the right medial rectus and left lateral rectus are inhibited as shown in Fig. 8.39. Diagonal movements involve more muscles. Consequently they require an even greater complexity of control. Therefore, it is not surprising that about 10 per cent of all motoneurons are employed in serving the extraocular eye muscles. These muscles have the smallest motor units in the body, with only 5–10 muscle fibers per motoneuron. This permits a high degree of precision in the control of eye position. The role of the vestibular system in the control of the eye movements is discussed on p. 136.

Defective control of the eye movements is often revealed by a squint, in which the eyes fail to look in the same direction and the sufferer experiences double vision. As the majority of the extraocular muscles are innervated by the oculomotor nerve (CN III), damage to this nerve is the most common source of such problems. There are a number of causes, including meningitis and compression of the nerve by a tumor. A less common problem is loss of ocular abduction, in which the affected eye cannot traverse in the temporal direction. This results from a

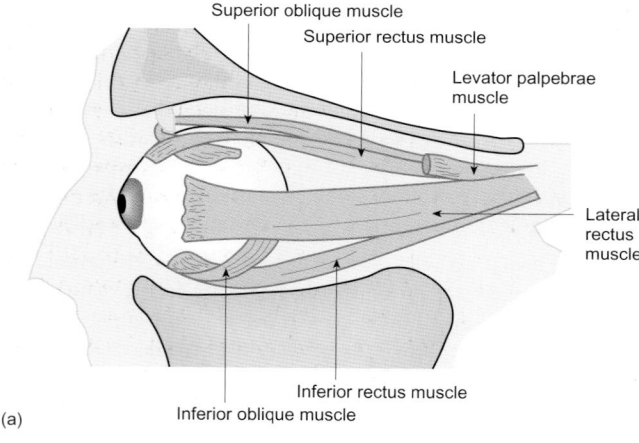

(a)

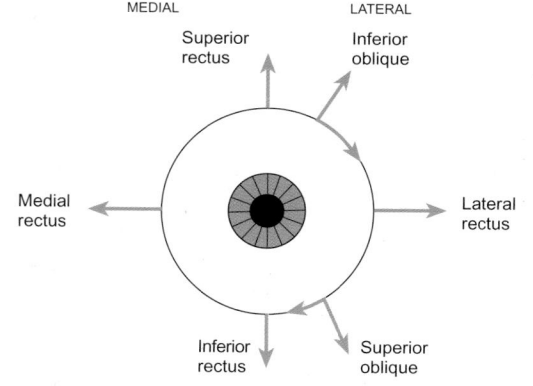

(b)

Fig. 8.38 (a) The extraocular muscles of the human eye and (b) the direction of eye movement controlled by each.

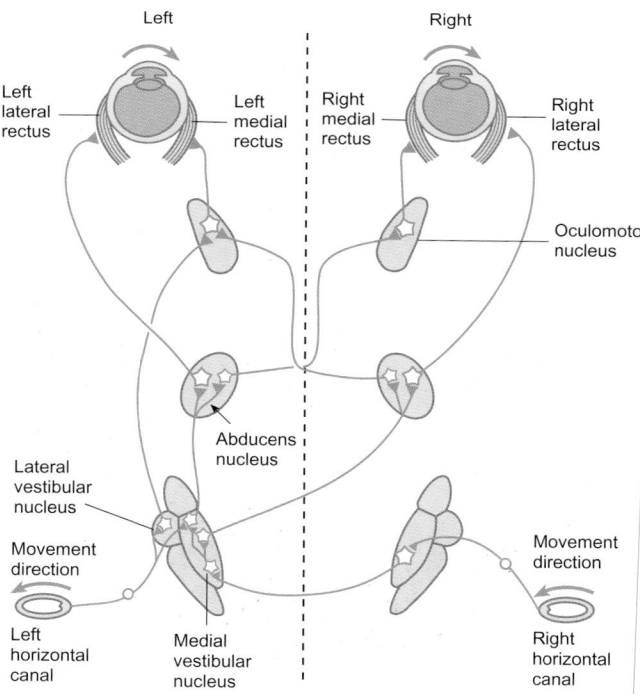

Fig. 8.39 The semicircular canals regulate the activity of the extraocular muscles. This diagram shows the neural pathways that control the medial and lateral rectus muscles of the eyes.

Summary

1. The eye movements are controlled by six extraocular eye muscles. These muscles are innervated by CN III, CN IV, and CN VI. The movements of the eyes are driven both by visual information and by information arising from the vestibular system. Their role is to keep objects of importance centered on the central region of the retina.

2. The eye movements are classified as saccades, in which the eyes move at very high angular velocities, smooth pursuit movements, and vergence movements. In saccades and pursuit movements the eyes move together and these are known as conjugate eye movements. Nystagmus occurs when saccadic movements are repeatedly followed by a smooth pursuit movement. In vergence movements, the two eyes move in mirror-image fashion so that their optical axes converge.

lesion of the abducent nerve (CN VI). Very rarely, head injury can result in damage to the trochlear nerve (CN IV), and the resulting paralysis affects the downward and inward movement of the affected eye.

8.6 The physiology of the ear—hearing and balance

The physical nature of sound

Sound consists of pressure variations in the air or some other medium. These pressure variations originate at a point in space and radiate outwards as a series of waves. Subjectively, sounds are characterized by their *loudness*, their *pitch* (i.e. how low or high they seem), and their specific tonal quality or *timbre*. To be able to extract the maximum information from the sounds in the environment, the auditory system must be able to determine their origin and analyze their specific qualities.

When a sound is propagated, the air pressure alternately increases and decreases above the mean atmospheric pressure. The larger the pressure variation, the higher the energy of the wave and the louder the sound will appear to be. While it is perfectly possible to express loudness in terms of the peak pressure change, it is much more convenient to express the intensity of sounds in relation to an arbitrary standard using the *decibel scale* (abbreviated as dB). This scale has the advantage that equal increments in sound intensity expressed as dB approximately correspond to equal increments in loudness:

$$dB = 10 \times \log_{10} \frac{\text{sound intensity}}{\text{reference intensity}}.$$

Since the energy of a sound wave depends on the square of the pressure change, the scale is more often written

$$dB = 20 \times \log_{10} \frac{\text{sound pressure}}{\text{reference pressure}}.$$

The reference pressure used in auditory physiology is 20 μPa which is close to the average threshold for human hearing at 1 kHz. Because the dB scale is expressed relative to an arbitrary standard, it is sometimes expressed as dB SPL (sound pressure level). If a sound pressure is 10 times that of the reference pressure, this will correspond to 20 dB SPL. A sound pressure 1000 times that of the reference pressure will be 60 dB SPL and so on.

A *pure tone* is a sound consisting of just one frequency. By playing a series of pure tones of varying intensity, it is possible to determine the threshold of hearing of a given subject for different frequencies. The results can be plotted as intensity (in dB SPL) against the logarithm of the frequency to produce a graphical representation of the auditory threshold known as an *audiogram* (Fig. 8.40). As the figure shows, the threshold of hearing varies with frequency. The ear is very sensitive to sounds around 1–3 kHz but its sensitivity declines at lower and higher frequencies. For healthy young people hearing extends from about 20 Hz to 20 kHz. In the most sensitive range, the threshold of hearing is close to 20 μPa or 0 dB SPL. The loudest sounds to which the ear can be exposed without irreversible damage are about 140 dB above threshold.

The range of frequencies and intensities of sounds encountered in speech are also indicated in Fig. 8.40. Note that the ear is most sensitive to the frequencies of normal speech. In the most sensitive part of the auditory spectrum, the ear is able to distinguish between frequencies that differ by only 0.3 per cent, i.e. by 3 cycles in 1 kHz. It is also able to distinguish between sounds of the same frequency that differ in intensity by only 1–2 dB.

Individual tones gives rise to distinct sensations of pitch—a high-frequency pure tone will subjectively appear as high pitched while a low frequency tone will have a low pitch. Pitch is a feature of continuous sounds and is especially clear in music and speech despite the fact that these sounds are made of a mixture of frequencies. The pitch of such a sound is related to its fundamental frequency—the lowest number of vibrations or pressure waves that occur each second. If the frequency increases, the pitch of the sound will be higher and vice versa. If a particular note is sung by different people, or the same note is played by two different instruments, a flute and a violin for example, it is easy to distinguish between them because they sound different despite having the same pitch. They have a different mixture of frequencies, which gives the sounds a distinctive quality or timbre.

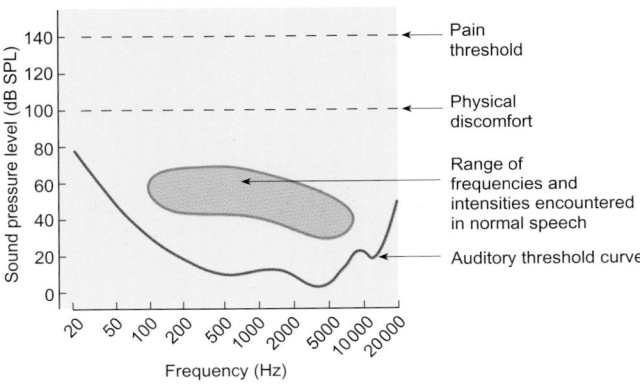

Fig. 8.40 A plot of the auditory threshold for human hearing. Note that greatest sensitivity is found in the 2–3 kHz region.

Our ability to distinguish between sounds depends partly on the mixture of frequencies present and their relative intensity and partly on the way in which a sound begins and ends. A percussive sound such as that arising from the snapping of a twig begins abruptly and dies away quickly. In contrast, speech and music are characterized by a continuous production of sound that is modulated (i.e. varied) in both pitch and intensity to provide specific meaning.

Structure of the auditory system

The auditory system can be arbitrarily divided into the *peripheral auditory system*, which consists of the ear, the auditory nerve, and the neurons of the spiral ganglia, and the *central auditory system*, which consists of the neural pathways concerned with the analysis of sound from the cochlear nuclei to the auditory cortex. The ear itself is conventionally divided into the outer ear (the pinna and external auditory meatus), the middle ear (the tympanic membrane, ossicles, and associated muscles), and the inner ear (the cochlea and auditory nerve). These structures are illustrated in Fig. 8.41.

The *outer ear* consists of the *pinna* (or auricle) which is highly convoluted. These convolutions play an important role in helping to determine the direction of a sound. The pinna leads to the *external auditory meatus* or *auditory canal*, which is about 25 mm in length and about 7 mm in diameter. The head, pinna, and auditory canal act together to form an acoustic resonator that increases the intensity of sound waves with frequencies in the range 2–5 kHz. In this way they effectively increase sound pressure at the tympanic membrane for those frequencies that are important in speech.

The *middle ear* is an air-filled space bounded laterally by the *tympanic membrane* or *ear drum* and medially by the *oval window* of the cochlea. These two membranes are coupled via three small bones, the *ossicles*. The middle ear is connected to the pharynx via the *Eustachian tube*. As long as the Eustachian tube is not blocked, the middle ear will be maintained at atmospheric pressure and the pressures on either side of the tympanic membrane will be equalized. The tympanic membrane is roughly cone-shaped and it is connected to the first of the ossicles, the *malleus* (or hammer), via an elongated extension known as the manubrium. The malleus is connected to the second of the ossicles, the *incus* (or anvil), which is itself connected to the *stapes* (or stirrup). The footplate of the stapes connects directly with the oval window of the cochlea (see Fig. 8.42).

The ossicles help to ensure the efficient transmission of airborne sounds to the cochlea by a process known as *impedance matching*. The matching of the acoustic impedances is achieved largely by the difference in area between the tympanic membrane and the footplate of the stapes. Since force is equal to pressure times area, the pressure collected over the tympanic membrane is applied to the much smaller area of the stapes footplate and a 17-fold net pressure gain is achieved (Fig. 8.42). This is further assisted by the lever action of the ossicles, which provides a mechanical advantage of 1:1.3. Overall, the efficient operation of the middle ear permits about 60 per cent of the energy of an airborne sound to be transmitted to the cochlea. If the function of the middle ear is impaired, for example by fluid accumulation or as a result of fixation of the footplate of the stapes to the bone surrounding the oval window, the threshold of hearing will be elevated. This type of deafness is known as *conductive hearing loss* and will be discussed further below.

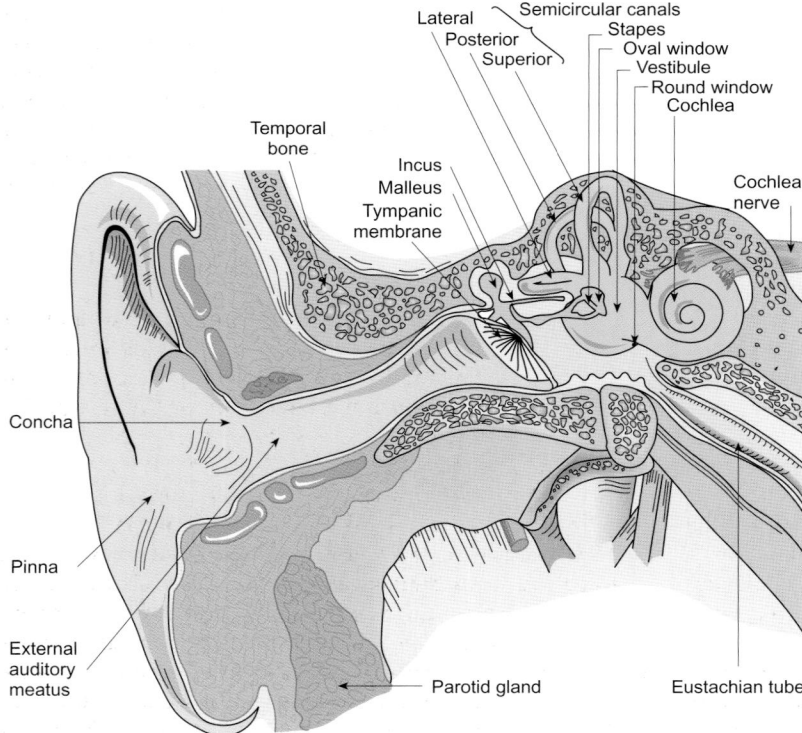

Fig. 8.41 Diagrammatic representation of the structure of the human ear.

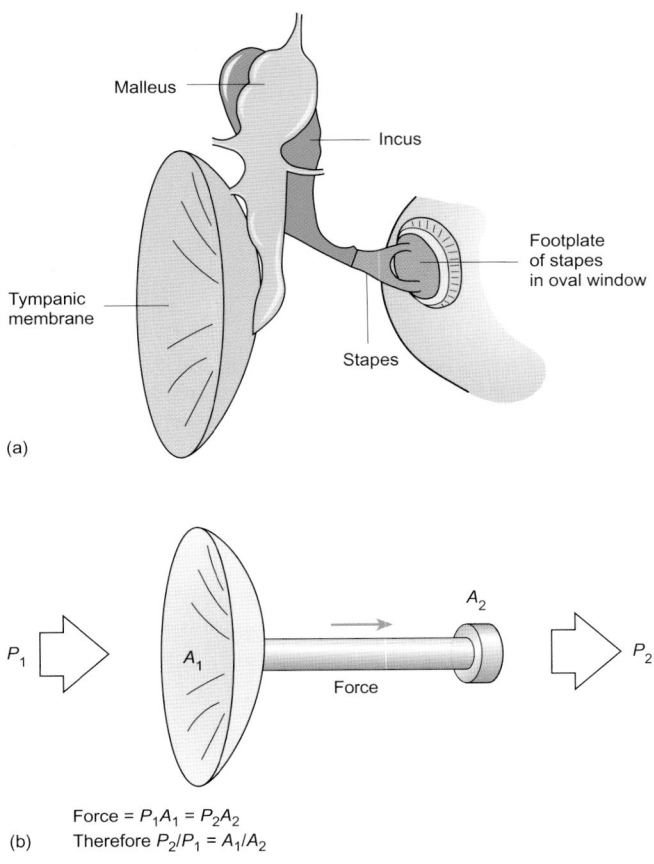

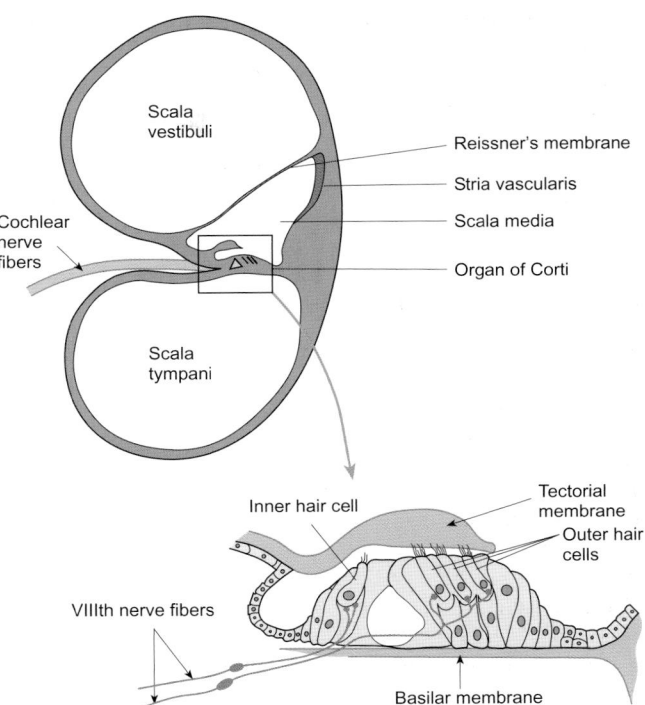

Fig. 8.43 A schematic cross-sectional drawing of the structure of the cochlea illustrating the position and structure of the organ of Corti.

Fig. 8.42 (a) A simple diagram to illustrate the transmission of sound energy from the tympanic membrane to the oval window via the ossicles. The pressure waves are collected over the whole of the area of the tympanic membrane and become concentrated over the much smaller area of the footplate of the stapes to provide a higher pressure at the oval window as shown in (b).

The transmission of energy through the middle ear can be affected by the muscles of the middle ear, the *tensor tympani* and *stapedius* muscles. These small muscles act to reduce the efficiency with which the ossicles transmit energy to the cochlea and so attenuate the perceived sound. They contract reflexly in response to loud sounds and during speech. While the latency of the reflex (*c.* 150 ms) is too great for protection against sudden loud sounds such as gunfire, the reflex is able to protect the ear from damage caused by the continuous loud sounds of noisy environments.

The *inner ear* consists of the organ of hearing, the *cochlea*, and the organ of balance—the *semicircular canals*, *saccule*, and *utricle* (see below). It is located in the temporal bone and consists of a series of passages called the *bony labyrinth* containing a further series of sacs and tubes, the *membranous labyrinth*. The fluid of the bony labyrinth is called *perilymph* and is similar in composition to cerebrospinal fluid, having a high sodium concentration and a low potassium concentration. The fluid within the membranous labyrinth is called *endolymph* and has a high potassium concentration and a low sodium concentration.

As Figure 8.41 shows, the cochlea is a coiled structure which, in man, is about 1 cm in diameter at the base and 5 mm in height. In cross-section (Fig. 8.43) it is seen to consist of three tubes which spiral together for about $2\frac{1}{2}$ turns. These are the *scala vestibuli* (filled with perilymph), the *scala media* (filled with endolymph), and the *scala tympani* (also filled with perilymph). The scala vestibuli and scala tympani communicate at the apex of the cochlear spiral at the *helicotrema*. The scala media is separated from the scala vestibuli by *Reissner's membrane* and from the scala tympani by the *basilar membrane* on which sits the organ of Corti, the true organ of hearing. The basilar membrane is about 34 mm long and varies in width from about 100 μm at the base to 500 μm at the helicotrema. The footplate of the stapes is connected with the fluid of the scala vestibuli via the *oval window*. The flexible membrane of the *round window* separates the fluid of the scala tympani from the cavity of the middle ear.

Mechanism of sound transduction

The movement of the stapes in response to sound results in pressure changes within the cochlea. If the basilar membrane were rigid, the pressure waves would be transmitted along the scala vestibuli to the helicotrema and down the scala tympani to be dissipated at the round window. However, the basilar membrane is not rigid but flexible so that pressure waves arriving at the oval window set up a series of traveling waves in the basilar membrane itself. These waves begin at the basal end by the oval window and progressively grow in amplitude as they proceed along the basilar membrane until they reach a peak, whereupon they rapidly decline as shown in Fig. 8.44(b). The position of this peak depends on the frequency of sound wave impinging on the ear. The peak amplitude for high frequencies is near to the base of the cochlea while low-frequency sounds elicit the largest motion nearer to the helicotrema (Fig. 8.44(c)). The frequency

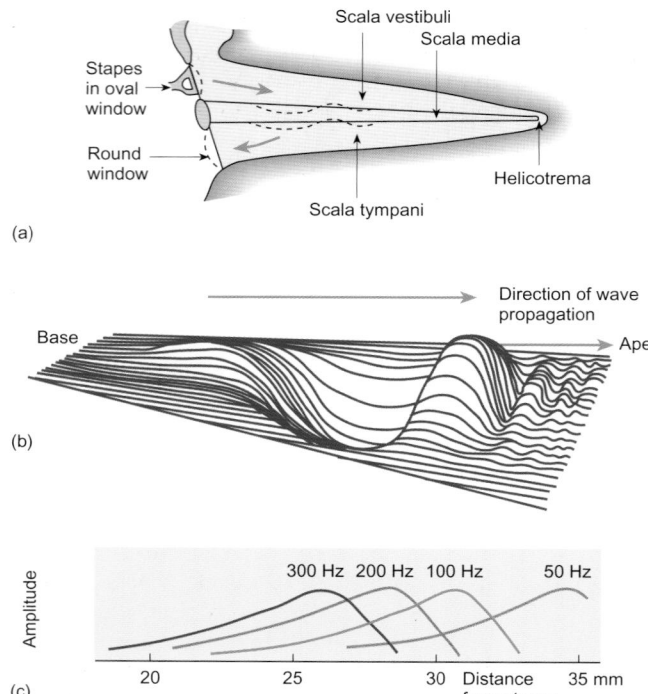

(a)

(b)

(c)

Fig. 8.44 (a) A diagrammatic representation of the way in which sound energy elicits the traveling wave in the basilar membrane. Panel (b) shows how the traveling wave progressively reaches a maximum as it passes along the basilar membrane before abruptly dying out. (c) The peak amplitude of the traveling wave is reached at different places along the basilar membrane for different frequencies of sound.

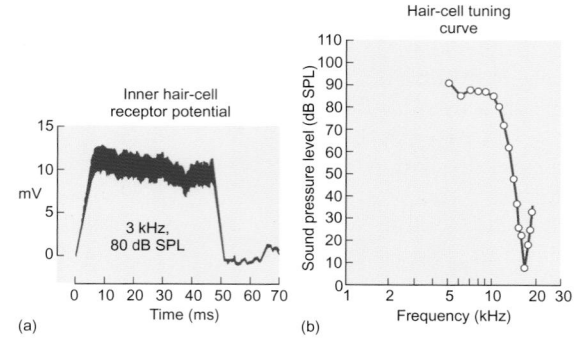

(a)

(b)

Fig. 8.45 (a) A receptor potential observed in a mammalian inner hair cell for a 3 kHz tone of 80 dB SPL. (b) The variation in sound intensity needed to elicit a receptor potential of a fixed amplitude for different frequencies. In this case the cell was most sensitive to a pure tone at 18 kHz.

also elicit a receptor potential, and the range of frequencies and intensities that excite a particular hair cell constitute its receptive field.

The hair cells are stimulated by the bending of the stereocilia that results from the motion of the basilar membrane. The

of a sound wave is thus mapped onto the basilar membrane by the point of maximum displacement and the constituent frequencies of complex sounds are mapped onto different regions of the basilar membrane.

How is the displacement of the basilar membrane by the traveling wave converted into nerve impulses? The cells responsible for this process are the *hair cells* of the organ of Corti. If these are absent (as in some genetic defects) or have been destroyed by ototoxic antibiotics such as kanamycin, the affected person will be deaf. The hair cells are arranged in two groups, the outer hair cells that are arranged in three rows and a single row of inner hair cells (Fig. 8.43). *Stereocilia* project from the upper surface of the hair cells. The stereocilia of the inner hair cells are arranged in a rough line, at the middle of which there is a basal body lying in a gap in the cuticular plate. The stereocilia of the outer hair cells are arranged in a rough V-shape with the basal body at the apex. In immature cochlear hair cells and in mature vestibular hair cells, a *kinocilium* emerges from the basal body.

In the absence of stimulation, the hair cells have a membrane potential of about –60 mV. When the stereocilia are moved towards the basal body the hair cells become depolarized, and when they are moved away from the basal body the hair cells become hyperpolarized. Thus the displacement of the stereocilia elicits a *receptor potential* in the hair cells. For any particular hair cell, the specific frequency that most easily elicits a receptor potential is known as the characteristic frequency for that cell (Fig. 8.45). More intense sounds of neighboring frequencies will

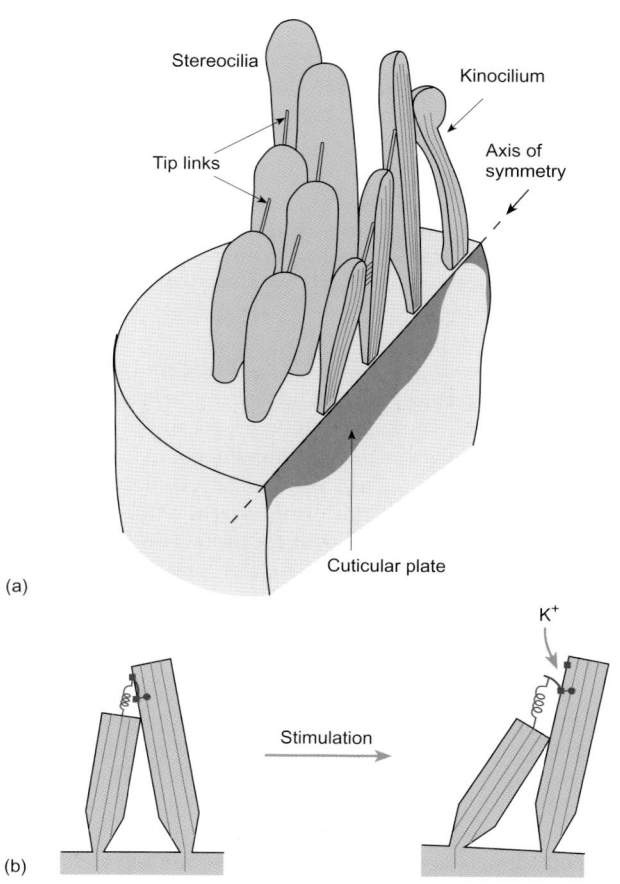

(a)

(b)

Fig. 8.46 (a) The organization of the stereocilia on hair cells and (b) the probable mechanism by which the fine tip links activate ion channels to depolarize the hair cells in response to sound waves. Although the kinocilium is present on vestibular hair cells, it is normally absent in mature cochlear hair cells.

stereocilia are linked to each other by fine threads of protein known as tip links (Fig. 8.46), and it is now believed that these links control the opening of specific ion channels in the stereocilia known as transduction channels which are very permeable to potassium. Bending the stereocilia towards the basal body generates tension in the links and this increases the chances of a transduction channel being open. In this way the movement of the stereocilia controls the permeability of the hair cells to potassium ions. When these channels are opened, they *depolarize* the hair cells. The more the stereocilia are bent in this direction, the more channels will be opened, the greater the permeability to potassium, and the greater the depolarization of the hair cell. Conversely, when the stereocilia are bent away from the basal body, the tension in the links is reduced and more of the transduction channels become closed.

Why do the hair cells depolarize when the transduction channels are open? The basolateral surface of the hair cells is exposed to perilymph (which has a low potassium ion concentration) so that there is a substantial potassium gradient across this part of the plasma membrane. At rest, the stereocilia and the apical surface of the hair cells have a very low permeability to potassium so that the resting membrane potential of the hair cells is determined mainly by the potassium gradient across the basolateral surface. When the hair cells are activated, the transduction channels open and the permeability of the stereocilia to potassium increases; this results in a depolarization of the hair cells because the stereocilia are not bathed in perilymph but in endolymph which is rich in potassium. This depolarization will result in the opening of voltage-gated calcium channels in the basolateral surface and the secretion of neurotransmitter onto the afferent nerve endings of the cochlear nerve, thus exciting them.

The afferent fibers have their cell bodies in the spiral ganglion and a specific set of afferent fibers innervate a particular region of the basilar membrane. Since the basilar membrane is tuned so that a sound of a given frequency will elicit maximum motion in a particular region, specific sounds will excite a specific set of afferent fibers. This is known as tonotopic mapping. For a given cochlear nerve fiber, there is a specific frequency to which it is most sensitive. This is known as the characteristic frequency of that fiber (cf. characteristic frequency of a hair cell) as shown in Fig. 8.47.

The nerve fibers of the cochlear nerve show spontaneous activity (i.e. action potentials in the absence of specific stimulation). When they are stimulated by a specific sound, the frequency of action potential discharge is increased and the increased discharge frequency is related to the intensity of the sound. When a sound ends, the action potential frequency falls below its natural spontaneous rate for a short period and this may be important in signaling the timing of specific sounds.

Central auditory processing

The organization of the auditory pathway is shown in Fig. 8.48. The cochlear nerve divides as it enters the cochlear nucleus and sends branches to the three main subdivisions: the antero- and posteroventral divisions and the dorsal division. Since the arrangement of the cochlear nerve fibers reflects their origin in the cochlea, this organization is preserved in the cochlear nucleus and the neurons of each subdivision are arranged so that they respond to different frequencies in a tonotopic order.

The fibers of the cochlear nuclei pass to the olivary nuclei, which is the first part of the auditory system to receive inputs from both ears. From each olivary nucleus, fibers project to the ipsilateral inferior colliculus via the lateral lemniscus and thence to the medial geniculate body of the thalamus. From the

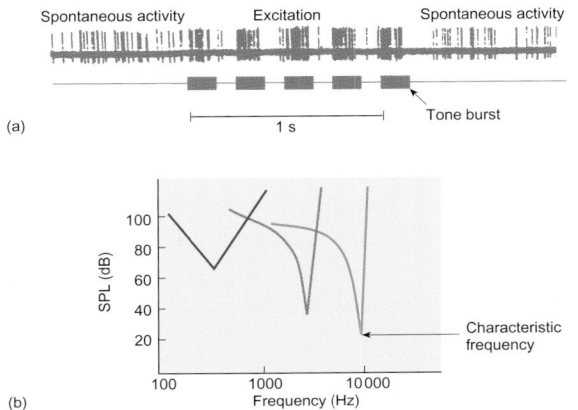

Fig. 8.47 (a) The characteristics of action potential discharge in the auditory nerve. Note the spontaneous firing of the auditory nerve fiber and the increase in action potential discharge during each tone burst. (b) Tuning curves for various auditory nerve fibers. The lines represent the threshold intensity required to elicit an increase in action potential discharge. Note the overlap between the receptive fields of the different fibers.

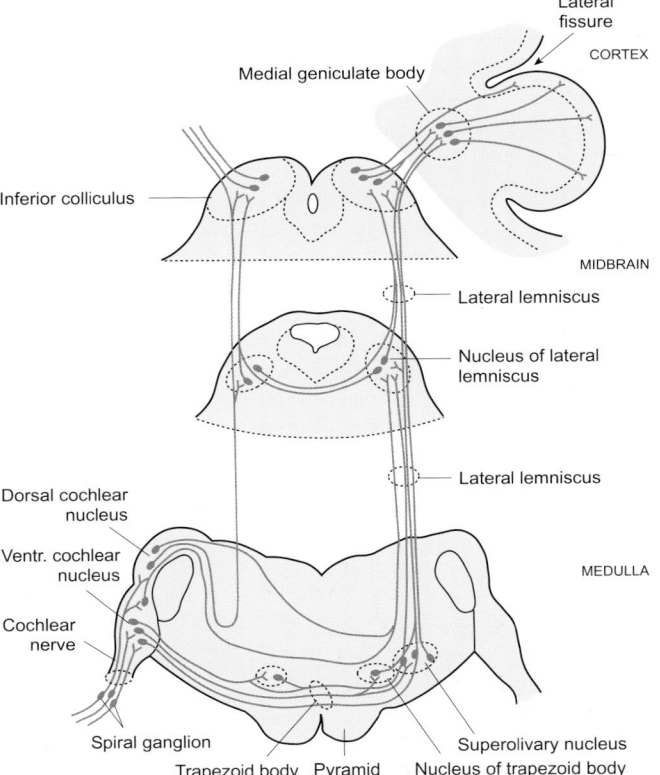

Fig. 8.48 The organization of the ascending pathways of the auditory system.

medial geniculate body, auditory fibers project to the primary auditory cortex, which is located on the upper aspect of the lateral (or Sylvian) fissure. The area of cortex devoted to hearing in the left hemisphere is greater than that in the right and is closely associated with Wernicke's area—a region of the brain specifically concerned with speech (see Chapter 11). Note that at all levels above the cochlear nuclei, there is strong bilateral representation of auditory information.

The localization of sound

One important function of the auditory system is to localize the source of a sound. It has been shown that humans, in common with many other animals, are able to do so with considerable accuracy. The fact that the two ears are located at different points in space is of considerable significance in this task. As sound travels through the air at a finite speed (c. 360 m s⁻¹), it reaches one ear before the other unless the source is directly in front of, or behind, the head. Therefore, the brain is able to use the time delay as one cue to localizing a sound. A sound originating on the left side has to travel an extra 15 cm to reach the right ear. Thus the sound will reach the right ear approximately 0.4 ms after it has reached the left. Under optimum conditions, the brain can detect time delays as brief as 30 μsc for sounds arriving at the two ears.

The head also casts a sound shadow which gives rise to an intensity difference for the sounds arriving at the two ears, at least for those sounds whose wavelength is similar to, or smaller than, the dimensions of the head. The sound shadow cast by the pinnae provides a means of distinguishing sounds that originate in front, from those that originate behind the head. Finally, the convolutions of the pinnae of the ears provide information regarding the localization of sounds in the vertical plane

Hearing deficits and their clinical evaluation

Hearing is a complex physiological process that can be disrupted in a variety of different ways. Broadly, hearing deficits are classified as conductive deafness, sensorineural deafness, and central deafness according to the site of the primary lesion.

Hearing loss

People who are partially deaf have a raised threshold to hearing so that a higher pressure change is required for the ear to detect sound. This represents a loss of sensitivity compared with normal subjects and is called hearing loss. It is expressed as the difference in dB SPL between the threshold of hearing for the person concerned and that of the average sensitivity of normal healthy subjects. If the threshold intensity of a particular tone were to be 20 dB higher than that of normal subjects, this would be indicated as a 20 dB hearing loss.

Conductive deafness

This hearing loss results from a defect in the middle or outer ear. Essentially the cause is a failure of the outer and middle ear to transmit sound efficiently to the inner ear. This type of hearing loss is readily distinguished from sensorineural and central deafness by *Rinne's test* which compares the ability of a patient to hear airborne sounds with their ability to detect those reaching

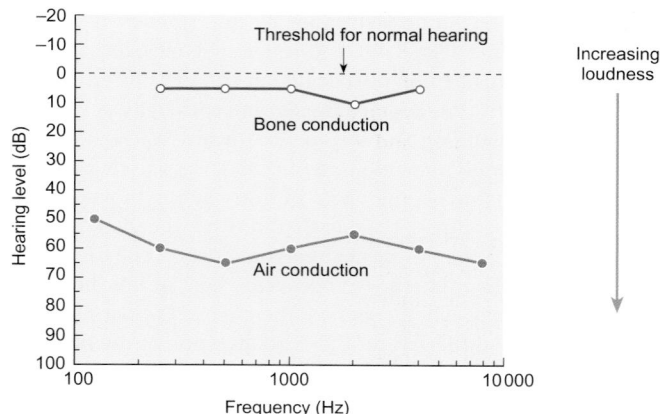

Fig. 8.49 An audiogram for a patient with conduction deafness in one ear. Note that while the auditory threshold is elevated for airborne sounds, bone conduction is nearly normal.

the cochlea via vibrations of the skull ('bone conduction'). A vibrating tuning fork is moved close to the ear and, when its tone is no longer heard, the base of the fork is applied to the mastoid process behind the ear. If the patient can hear the tone once more, then the middle ear is affected; if not, then the deafness is likely to be due to sensorineural hearing loss. A more precise diagnosis can be obtained by *pure tone audiometry*. This is used to determine the auditory threshold of subjects by asking them to respond to a series of tones of progressively lower intensity. Two tests are performed on each ear: one for air conduction and one for bone conduction. The auditory threshold curve is plotted and compared with the average threshold for healthy subjects. A typical audiogram for conductive hearing loss is shown in Fig. 8.49.

Conductive hearing loss can result from various causes including middle-ear infections (which can lead to a condition known as 'glue ear'), serous otitis media in which fluid accumulates in the middle ear, and otosclerosis in which the movement of the footplate of the stapes is impeded by the growth of bone around the oval window.

Sensorineural hearing loss

Sensorineural hearing loss results when some part of the cochlea or auditory nerve is damaged. It is quite commonly the result of traumatic damage to the cochlea by high-intensity sounds caused by industrial processes such as the riveting of steel plates ('boilermakers' disease') or by sounds resulting from overamplified music. Age-related hearing loss (known as *presbyacusis*) is a type of sensorineural deafness that specifically affects the high frequencies. Loss of hair cells can be caused by ototoxic drugs such as the aminoglycoside antibiotics (e.g. streptomycin and neomycin) and certain diuretics (e.g. furosemide). The excessive growth of Schwann cells in the cochlear nerve can result in compression of the nerve, and lead to deafness. A further very distressing but common cause of sensorineural hearing loss is *tinnitus*—the unremitting sensation of sound generated within the ear itself. The characteristics of tinnitus vary from subject to subject and include high-pitched continuous notes and buzzing pulsing sounds. All tend to mask the natural sounds reaching the ear, thus impairing the sufferer's hearing.

Summary

1. Sounds consist of pressure variations in the air. Subjectively, they are characterized by their pitch, timbre (specific qualities), and loudness. The human ear is most sensitive to frequencies between 1 and 3 kHz although it can detect sounds ranging from 20 Hz to 20 kHz.

2. The auditory system consists of the ear and the auditory pathways. The ear is divided into the outer ear, the middle ear, and the inner ear. The inner ear consists of the cochlea, which is the organ of hearing, and the vestibular system, which is concerned with balance.

3. The outer and middle ear serve to collect sound energy and focus it on to the oval window. This allows the efficient transfer of sound energy to the cochlea. Incident sound waves cause pressure waves to be set up in the fluids of the cochlea. These pressure waves evoke a traveling wave in the basilar membrane that activates the hair cells, which are responsible for sound transduction.

4. The basilar membrane is tuned so that the constituent frequencies of a sound are represented as a map on the basilar membrane. High frequencies are represented near the oval window while low frequencies are represented near to the apex of the cochlea. Since individual auditory fibers terminate in specific regions of the basilar membrane, activation of a specific set of fibers codes for the frequency components of a sound wave. The number of action potentials in a specific fiber codes for the intensity of the sound.

5. Hearing deficits are classified as conductive deafness, sensorineural deafness, or central deafness. Conductive and sensorineural deafness can be distinguished by pure tone audiometry. In conductive deafness, the auditory threshold for bone conduction is normal while that for air conduction is elevated. In sensorineural hearing loss, the thresholds of both air and bone conduction are elevated.

Central deafness

Because the auditory pathways are extensively crossed at all levels above the cochlear nuclei, the hearing deficits resulting from damage to the pathways subserving the sense of hearing are usually very subtle. Generally speaking, unilateral damage to the auditory cortex does not result in deafness, although the affected person may have difficulty in localizing sounds. Extensive damage to the auditory cortex of the dominant hemisphere (usually the left hemisphere), will lead to difficulties in the recognition of speech, while extensive damage to the auditory cortex of the minor (right) hemisphere affects recognition of timbre and the interpretation of temporal sequences of sound, both of which are important in music and speech.

The vestibular system and the sense of balance

The sense of balance plays an important role in the maintenance of normal posture and in stabilizing the retinal image, particularly during locomotion. Indeed, people who have lost the function of their vestibular system have difficulty in walking over an irregular or compliant surface when they are deprived of visual cues by a blindfold. Such people also have difficulty with their vision during walking, as the visual world appears to move up and down rather than remaining stable as it does for a normal individual.

Structure of the vestibular system

The vestibular portion of the membranous labyrinth is the organ of balance. The whole structure is about 1 cm in diameter and, as illustrated in Fig. 8.50, it consists of two chambers, the utricle and saccule, and three semicircular canals. The nerve supply is via the vestibular branch of CN VIII and Scarpa's ganglion.

The utricle and saccule are arranged horizontally and vertically, respectively, while the three semicircular canals are arranged at right angles to each other with the lateral canal inclined at an angle of about 30° from the horizontal. The other two canals lie in vertical planes as shown in Fig. 8.51(a). Near the utricle, each canal has an enlargement known as the ampulla. The fluid within the semicircular canals, utricle, and saccule is

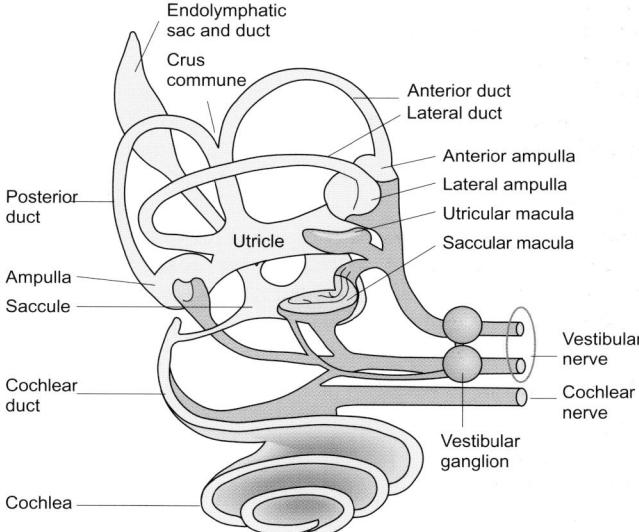

Fig. 8.50 The principal structures of the membranous labyrinth and their neural connections (shown in blue).

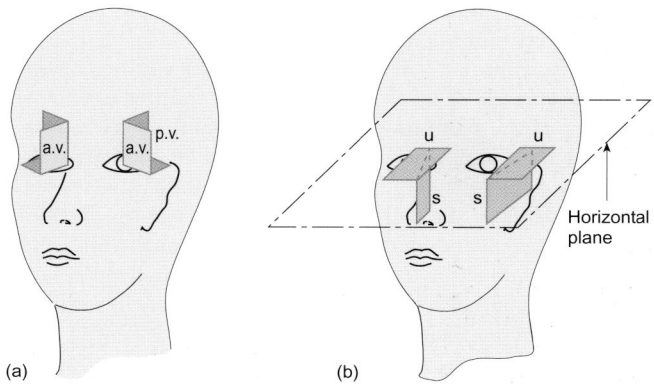

Fig. 8.51 The principal planes of the three semicircular canals are shown in (a), those of the utricle (u), and the saccule (s) (b).

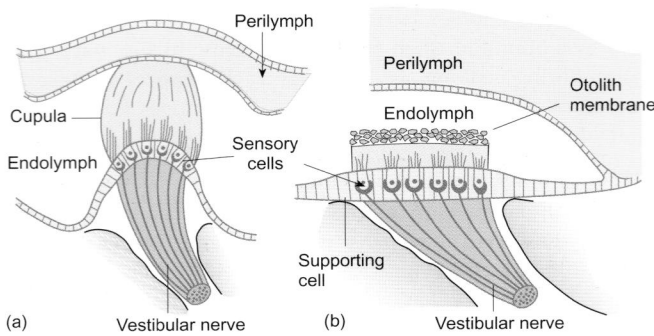

Fig. 8.52 The structure of (a) the ampulla of the semicircular canals and (b) the saccule. Note that the sensory hairs of the crista ampullaris project into a gelatinous mass called the cupula that forms a barrier across the ampulla, preventing the endolymph from circulating. The hair cells of the saccule project into a gelatinous sheet covered with crystals of calcium carbonate.

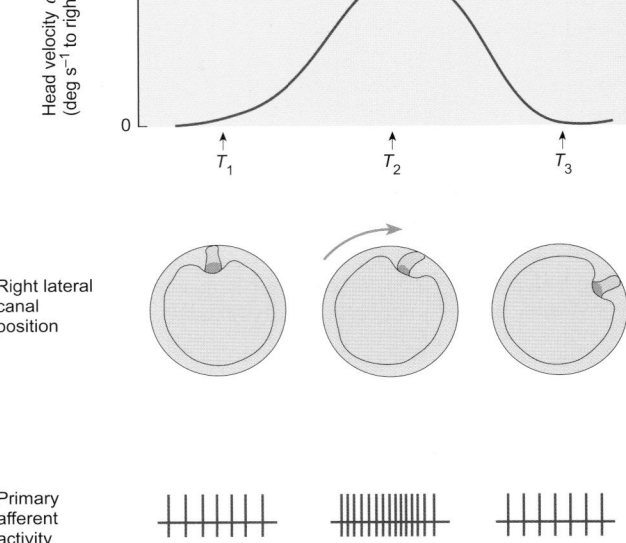

Fig. 8.53 The response of vestibular afferent fibers following displacement of the head to the left. Immediately following the movement the cupula is displaced and the hair cells excite the afferent nerve endings. Once the movement ceases, the cupula resumes its normal position and the action potential discharge returns to its resting level.

endolymph, and the whole structure floats in the perilymph that is contained within the bony labyrinth.

The sensory portion of the semicircular canals is located within the ampulla. A cross-section of the ampulla reveals that the wall of the canal projects inward to form the *crista ampullaris* (Fig. 8.52a). The hair cells are located on the epithelial layer that covers the crista and the hairs that project from their upper surface are embedded into a large gelatinous mass, the *cupula*, which is in loose contact with the wall of the ampulla at its free end. As a result, it forms a compliant seal that closes the lumen of the canal, preventing free circulation of the endolymph. The sensory epithelium of the utricle is in the horizontal plane while that of the saccule lies in the vertical plane as shown in Fig. 8.51(b). As in the crista ampullaris, the sensory cells are hair cells. Afferent axons from the vestibular nerve terminate on the hair cells via synaptic contacts similar to those seen in the cochlea.

The cilia of the hair cells consist of a single large hair known as the *kinocilium* and several rows of smaller stereocilia. The orientation of the hair cells is rather specific for different parts of the vestibular apparatus. In the crista ampullaris, the hair cells are oriented so that they all have their kinocilium pointing in the same direction. In the utricle and saccule the orientation of the hair cells is rather more complex. This morphological polarization coupled with the specific orientation of the semicircular canals, utricle, and saccule enables the vestibular system to interpret any movement of the head.

The mechanism of hair cell excitation in the vestibular system is essentially the same as that seen in the cochlea (see above). Bending of the hairs towards the kinocilium opens transduction channels and, since the upper surface of the hair cells is bathed in a high-potassium medium (endolymph), this results in depolarization of the hair cells and excitation of the vestibular afferent fibers. Bending of the hairs away from the kinocilium leads to closure of the transduction channels and hyperpolarization of the hair cells, resulting in a reduction of the vestibular afferent fiber discharge.

Operation of the crista ampullaris

Although the cupula of the crista effectively seals the semicircular canal at the ampulla, it is able to hinge about the axis of the crista. When the head is turned, the walls of the labyrinth must move as they are attached to the skull but, as the endolymph in the semicircular canals is less constrained, it tends to lag behind by virtue of its inertia. The effect is to deflect the cupula away from the direction of movement as shown in Fig. 8.53. For a simple turning movement to the left in a horizontal plane the hair cells in the left horizontal canal excite their afferent fibers while those in the right canal inhibit theirs. The outputs of the two cristae act in a push–pull manner to signal the angular movement of the head. Similar coupled reactions will occur in the other pairs of canals in response to movements in other planes, and the disposition of the semicircular canals in space is such that, whatever the angular movement, at least one pair of the semicircular canals will be stimulated.

Signals from the semicircular canals control eye movements

The information derived from the semicircular canals is used to control eye movements (see above). Direct stimulation of the ampullary nerves elicits specific movements of the eyes. Stimulation of the afferent fibers that serve the left horizontal canal results in the eyes turning to the right as shown in Fig. 8.39. This is one of a group of *vestibulo-ocular reflexes*. The eye movements elicited by activation of the ampullary receptors are specifically adapted to permit the gaze to remain steady during movement of the head. Such compensatory movements can also be seen in human subjects seated in a rotating chair. Movement of the chair to the left causes the eyes to move to the right to preserve the direction of the gaze. If the subject's eyes are first defocused using a pair of strong lenses, the eyes move in the direction opposite to the induced rotation until they reach their limit of travel when they flick back to their center position and the process of drift recurs. This is known as *vestibular nystagmus*. If the chair is rotated at a steady speed, the nystagmus decreases and is finally lost after about 20 s. Sudden stopping of the chair then leads to nystagmus in the direction opposite to that of the imposed rotation (post-rotatory nystagmus). Such procedures are sometimes used clinically to assess the function of the vestibular system.

The receptors of the utricle and saccule signal linear accelerations such as gravity

The sensory epithelium of the utricle and saccule consists of a layer of hair cells covered by a gelatinous membrane containing small crystals of calcium carbonate called *otoliths* (see Fig. 8.51b). This membrane (the otolith membrane) is able to move relative to the hair cells as the orientation of the head is changed. The movement of the membrane leads to deflection of the stereocilia and excitation of the vestibular afferents which, in turn, provide the brain with sufficient information to define the position of the head with respect to gravity. The principle is illustrated in Fig. 8.54.

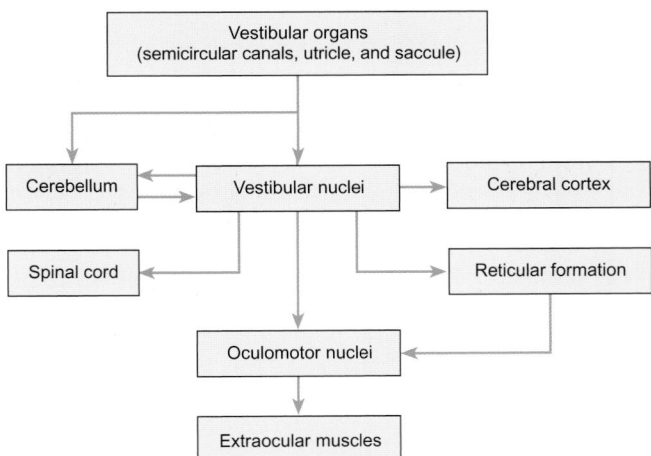

Fig. 8.55 The principal neural connections of the vestibular system.

Information from the utricle and saccule is important in the control of posture. Unlike those of the semicircular canals, the afferents of the utricle and saccule continuously relay information to the brain about the position of the head. The role of these signals can be studied with a tilting table. If a subject is blindfolded and positioned over the pivot, the arms are seen to extend and flex as the table is tilted in an attempt to keep the head level and maintain a stable center of gravity.

The nervous connections that permit the postural and oculomotor adjustments are complex and are shown in outline only in Fig. 8.55. The afferents from the hair cells send their axons to the vestibular nuclear complex in the brainstem. These afferents give off collaterals to the cerebellar nuclei. The vestibular nuclei send fibers to the spinal cord via the lateral vestibulospinal tracts and proprioceptive information reaches the vestibular nuclear complex via the spinovestibular tract. Eye movements are controlled via an anterior projection to the pons, which ultimately relays information to the oculomotor nuclei.

Disorders of the vestibular system

From the previous discussion, it should be clear that the vestibular system is concerned with maintaining our posture and stabilizing the visual field on the retina. People with unilateral damage to the vestibular system have a sense of turning and vertigo and may have abnormal eye movements (see below). (Vertigo is an illusion in which affected subjects feel that either they or their surroundings are moving even though they are, in fact, stationary. For example, they may feel as though they are being pulled to one side by an invisible force or the floor may appear to tilt or sink.) Initially they have difficulty in maintaining their balance although they eventually learn to compensate. Remarkably, if the vestibular system is damaged on both sides (e.g. as a result of the ototoxic effects of some drugs) affected subjects are essentially unaware that they have a sensory deficit. They are able to stand, walk, and run in an apparently normal fashion. However, if they are blindfolded they become unsure of their posture and will fall over if they are asked to walk over a compliant surface such as a mattress. A normal subject would have no such difficulty.

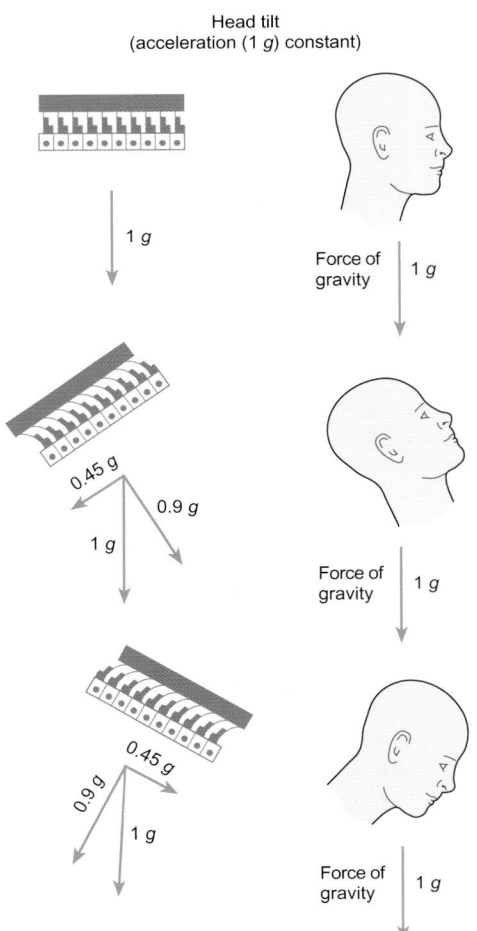

Fig. 8.54 Diagrammatic representation of the displacement of the otolith membrane of the utricle to show how the hair cell stereocilia are displaced during tilting movements of the head.

The vestibular nerve has a basal level of activity which signals that the hair cells of the semicircular canals are in their resting position. Head rotation elicits signals that are dependent on the angle and direction of movement, and the eyes move to stabilize the direction of gaze. However, if there is some abnormality of the semicircular canals, the signals from the various receptors will be out of balance and this will result in abnormal eye movements. Indeed, vestibular damage is characterized by a persistent nystagmus that is evident even when the head is still.

The vestibulo-ocular reflexes have proved valuable in the diagnosis of vestibular disease. Thermal stimulation of the horizontal canal is generally used. The head is tilted backwards by 60° so that the horizontal canal is in the vertical plane. Then the auditory canal is rinsed with warm water. The heat from the auditory canal is conducted to the horizontal semicircular canal and this elicits convective currents that cause deflection of the cupula resulting in a nystagmus. The test can easily be repeated on the other side to establish whether one canal is defective. This test is also one of several that are used to establish whether a comatose patient is brain dead.

Overproduction of endolymph may result in a condition known as *Menière's disease* which affects both hearing and balance. There is a loss of sensitivity to low-frequency sounds accompanied by attacks of dizziness or vertigo that may be so severe that the sufferer is unable to stand. These attacks are frequently accompanied by nausea and vomiting. Since the labyrinth is affected, there is also a pathological vestibular nystagmus.

Motion sickness does not reflect damage to the vestibular system; rather, it is caused by a conflict between the information arising from the vestibular system and that provided from other sensory systems such as vision and proprioception. Indeed, subjects with bilateral damage to the vestibular system do not appear to suffer from motion sickness.

Summary

1. The vestibular system on each side consists of three semicircular canals and two chambers, the utricle and the saccule. The three semicircular canals are arranged at right angles to each other while the utricle and saccule are arranged horizontally and vertically respectively. The receptors are hair cells, which have specific orientations in each of the vestibular organs. The nerve supply is via the vestibular branch of CN VIII (auditory nerve).

2. The semicircular canals are arranged to signal angular accelerations of the head, while the utricle and saccule signal linear accelerations such as gravity. The semicircular canals act as paired detectors to signal the angular movement of the head.

3. The information derived from the semicircular canals is used to control eye movements via the vestibulo-ocular reflexes whose role is to stabilize the visual field on the retina. Information from the utricle and saccule plays an important role in the maintenance of posture.

8.7 The chemical senses—smell and taste

The chemical senses, smell (or olfaction) and taste (or gustation), are amongst the most basic responses of living organisms to their environment. Although man makes comparatively little use of his sense of smell, it does play a significant part in social interactions as revealed by the widespread use of perfumes and aftershave. The sense of taste plays a vital role in the selection of foods and avoidance of poisons, and both smell and taste play a significant role in the enjoyment of food. Indeed, the taste of food is profoundly disturbed when the sense of smell is temporarily impaired during a severe head cold.

The sense of smell (olfaction)

Although their sense of smell is not highly developed, humans are able to discriminate between the odors of many thousands of different substances and some odors, such as methyl mercaptan, can be detected at astonishingly low concentrations. (The olfactory threshold for this substance in air is less than 1 part per billion). As few as one or two molecules are believed to be sufficient to excite an individual olfactory receptor.

The sensory organ for the sense of smell is the olfactory epithelium which lies high in the nasal cavity above the turbinate bones (Fig. 8.56). It is about 2–3 cm² in area on each

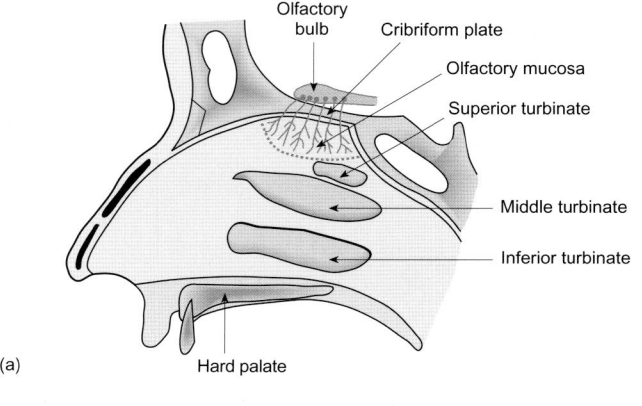

Olfactory bulb
Cribriform plate
Olfactory mucosa
Superior turbinate
Middle turbinate
Inferior turbinate
(a)
Hard palate

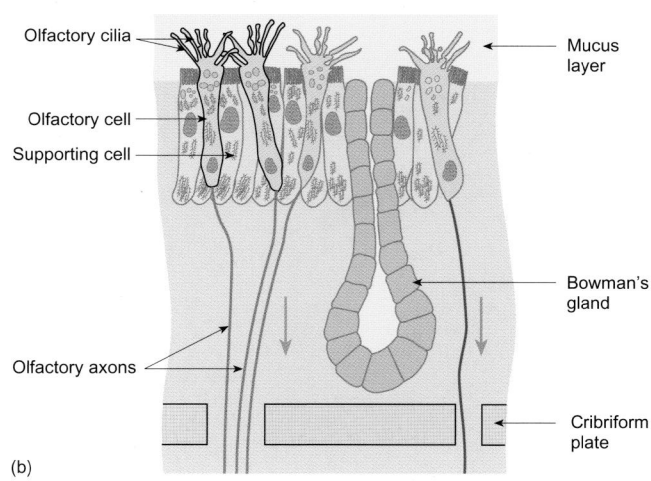

Olfactory cilia
Mucus layer
Olfactory cell
Supporting cell
Bowman's gland
Olfactory axons
Cribriform plate
(b)

Fig. 8.56 A sectional view of the nose to show (a) the location and (b) the detailed arrangement of the olfactory epithelium.

side and consists of ciliated receptor cells and supporting cells. It is covered by a layer of mucus secreted by Bowman's glands which lie beneath the epithelial layer. The axons from the olfactory receptor cells pass through the cribriform plate to make their synaptic contacts in the glomeruli of the olfactory bulbs.

There is no generally agreed classification of odors, but some authorities group odiferous substances into seven classes: flowery, ethereal, musky, camphorous, sweaty, rotten, and pungent. However, there is no simple relationship between chemical structure and odor; thus acetic acid (the principal ingredient of vinegar) is pungent, while butyric acid has an unpleasant sweaty odor. The smell of everyday objects and food is a complex mixture of many different odors.

Olfactory transduction depends on activation of specific G protein-linked receptors

To excite an olfactory receptor, a substance must be both volatile and able to dissolve in the layer of mucus that covers the olfactory epithelium. The olfactory receptor molecules are located on the cilia of the olfactory cells; there are as many as 1000 different odorant-binding proteins. Each olfactory receptor protein is coupled with a G protein that activates adenylyl cyclase. Thus, when an odorant molecule is bound by an appropriate receptor molecule, the intracellular concentration of cyclic AMP rises in the receptor cell. This increase in cyclic AMP opens a cation-selective channel, which leads to the depolarization of the olfactory receptor. If the depolarization reaches the threshold for action potential generation, an action potential will be propagated to the olfactory bulb.

Individual olfactory receptors respond to more than one odiferous substance, although a receptor is usually excited best by one odor. Therefore it is likely that olfactory information is coded in the pattern of incoming information which the brain learns to interpret.

The central connections of the olfactory system

The bipolar olfactory cells of the olfactory epithelium send their axons through the cribriform plate to the olfactory bulbs via a short olfactory nerve. The olfactory bulbs have a complex organization but they project to the ipsilateral olfactory cortex via the lateral olfactory tract and to the contralateral olfactory cortex via the anterior commisure. The fibers of the lateral olfactory tract also project to the hypothalamus where they play an important role in triggering sexual behavior in animals, though probably not in man. Olfactory projections reach the hippocampus, amygdala, and other structures of the limbic system. Olfactory information reaches the frontal lobe via the thalamus.

The sense of taste (gustation)

The tongue is the principal organ of taste. All over its surface there are small projections called *papillae* that give the tongue its roughness. Four different types can be identified: filiform, folate, fungiform, and vallate. The organs of taste are the *taste buds* that are found on the folate, fungiform, and vallate papillae only. The taste buds are located just below the surface epithelium and communicate with the surface via a small opening called a taste pore (Fig. 8.57). Each taste bud consists of a group of taste cells and supporting cells. The taste cells are innervated by fibers from the facial and glossopharyngeal nerves. A few taste buds are scattered around the oral cavity and on the soft palate.

There are four basic modalities of taste

There are four distinct modalities of taste: salty, sour, sweet, and bitter. The complexity of the taste of foodstuffs arises partly from the mixed sensations arising from stimulation of the different modalities of taste but chiefly from the additional stimulation of the olfactory receptors. Different parts of the tongue show different sensitivities to these taste modalities. As shown in Fig. 8.57, the base of the tongue is most sensitive to bitter, the sides to sour and salt, while the tip is most sensitive to sweet. Despite these regional variations in sensitivity, it is important to remember that all modalities may be detected by the taste buds. Only the dorsum of the tongue is insensitive to specific taste sensations.

Mechanisms of transduction

Solutions that are salty or sour activate taste cells by opening a specific ion channel that has a high permeability to sodium ions. This channel is inhibited by a substance called amiloride. The opening of this ion channel depolarizes the taste cell, which leads to excitation of the afferent taste fibers to which it is connected. Sour solutions are always of low pH and the increased concentration of hydrogen ions leads to closure of a specific potassium channel in addition to the opening of the sodium channel. Again, activation of the taste receptor leads to depolarization of the taste cell. The depolarization opens voltage-gated Ca^{2+} channels and this triggers the exocytosis of neurotransmitter by the taste cells onto the appropriate taste afferent nerve fibers, thus exciting them.

Sweet and bitter substances bind to specific receptors that are coupled to G proteins. Sweet substances activate adenylyl cyclase and increase the intracellular concentration of cyclic AMP. The cyclic AMP then closes a potassium channel and this results in depolarization of the taste cell and excitation of the appropriate

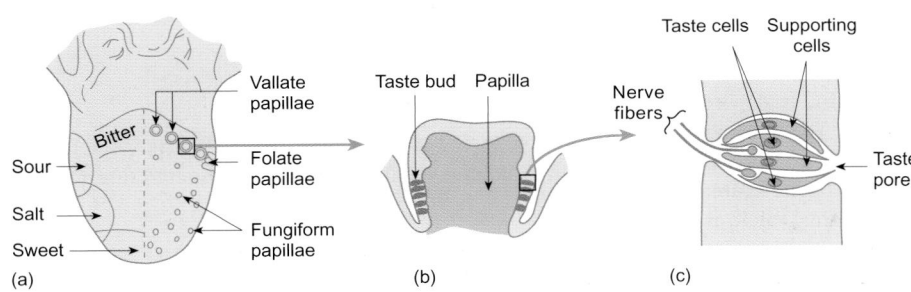

(a)   (b)   (c)

Fig. 8.57 (a) The regions of the tongue that show greatest sensitivity to the four modalities of taste. (b) A cross-sectional view of a vallate papilla showing the location of the taste buds. (c) Detailed structure of a single taste bud.

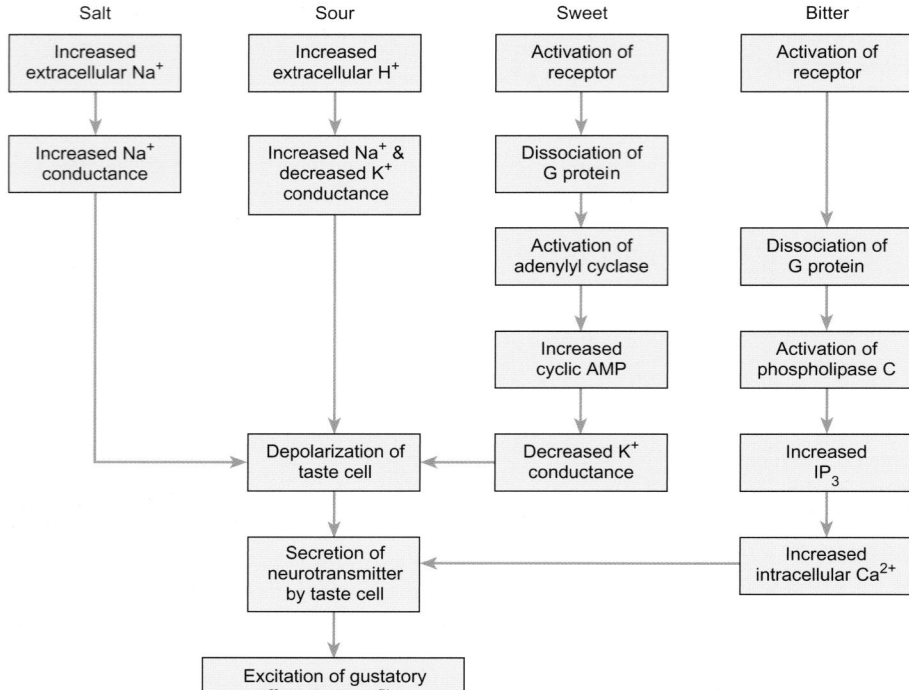

Salt	Sour	Sweet	Bitter

Fig. 8.58 The cellular mechanisms that allow the detection of the principal modalities of taste.

afferents. Bitter substances activate phospholipase C and the resultant increase in intracellular calcium causes the release of a transmitter onto the taste afferents to excite them. The transduction mechanisms for each modality are illustrated in Fig. 8.58.

Individual taste cells respond to more than one modality of taste. Nevertheless, they respond to one particular modality more strongly than to others (cf. olfactory coding discussed earlier). In view of this, the brain must interpret the pattern of activity from the active population of taste cells to discriminate the taste of a substance that is present in the mouth.

Taste sensations reach the cerebral cortex primarily via the facial and glossopharyngeal nerves. The facial nerve supplies the front two-thirds of the tongue while the glossopharyngeal nerve innervates the vallate papillae at the back of the tongue. The taste buds on the walls of the mouth and soft palate are innervated by the vagus nerves. The nerves subserving taste project to the nucleus of the tractus solitarius (NTS) in the medulla from where the taste sensations project to the thalamus and the lateral region of the postcentral gyrus adjacent to the somatic area for the tongue. Taste afferents from the NTS also play a role in various visceral reflexes (e.g. the secretion of gastric juices).

The sense of taste plays an important role in regulating some gastrointestinal secretions and in selecting foodstuffs. In general, sweet substances are nutritious while bitter substances are often poisonous. Therefore it is not surprising to find that sweet foods are sought out while bitter ones are avoided. The ingestion of sweet substances will lead to the secretion of insulin from the pancreas. Sour and salty tasting substances can be tolerated, even approved of, provided that the sensations they elicit are not too intense, in which case they are likely to provoke adverse reactions and will be avoided.

Summary

1. The chemical senses are olfaction (sense of smell) and gustation (sense of taste). The olfactory receptors are located in the epithelium above the third turbinate in the nose. The olfactory receptor proteins are located on the cilia of the olfactory neurons. Individual olfactory receptor cells respond to more than one odor and so specific odors are encoded in the pattern of information that reaches the brain. In many animals, olfactory stimuli play an important role in sexual behavior although this is not strikingly evident in humans.

2. The taste receptors are primarily found on the tongue. There are four modalities of taste: salty, sour, sweet, and bitter. Although individual taste cells respond preferentially to one modality of taste, other modalities will excite them so that, like the olfactory system, the specific sense of taste is coded in the pattern of information reaching the brain. The sense of taste is important in discriminating between different foodstuffs and in regulating some gastrointestinal secretions.

Recommended reading

Anatomy of sensory systems

Brodal, P. (2003). *The central nervous system. Structure and function* (3rd edn), Chapters 5–10, 15. Oxford University Press, Oxford.

Physiology

Barlow, H.B., and Mollon, J.D. (eds.) (1982). *The senses.* Cambridge University Press, Cambridge.

Berg, J.M., Tymoczko, J.L., and Stryer, L. (2002). *Biochemistry* (5th edn), Chapter 32. Freeman, New York.

Carpenter, R.H.S. (2002). *Neurophysiology* (4th edn), Chapters 4–8. Hodder Arnold, London.

Gomperts, B.D., Tatham, P.E.R., and Kramer, I.J. (2002). *Signal transduction*, Chapter 6. Academic Press, San Diego, CA.

Gregory, R.L. (1998). *Eye and brain: the psychology of seeing* (5th edn). Oxford University Press, Oxford.

Kandel, E.R., Schwartz, J.H., and Jessell, T.M. (eds.) (1991). *Principles of neural science* (3rd edn), Chapters 23–34. Prentice-Hall International, London.

Melzack, R., and Wall, P.D. (1996). *The challenge of pain* (2nd edn). Penguin, London.

Nicholls, J.G., Martin, A.R., Wallace, B.G., and Fuchs, P.A. (2001). *From neuron to brain* (4th edn), Chapters 17–21. Sinauer, Sunderland, MA.

Pickles, J.O. (1988). *An introduction to the physiology of hearing* (2nd edn). Academic Press, London.

Shepherd, G.M. (1994). *Neurobiology* (3rd edn), Chapters 10–16. Oxford University Press, Oxford.

Zeki, S. (1993). *A vision of the brain*. Blackwell Science, Oxford.

Zigmond, M.J., Bloom, F.E., Landis, S.C., Roberts, J.L., and Squire, L.R. (1999). *Fundamental neuroscience*, Chapters 23–28. Academic Press, San Diego, CA.

Medicine

Donaghy, M. (2005). *Neurology* (2nd edn), Chapters 6, 14–17. Oxford University Press, Oxford.

Multiple choice questions

Each statement is either true or false. Answers are given below.

1. a. Particular types of receptor respond most easily to a specific quality of a stimulus.
   b. The frequency of a train of action potentials in an afferent fiber reflects the intensity of the stimulus given to its receptor.
   c. A constant stimulus given to a touch receptor will result in a constant rate of discharge of its axon for as long as the receptor is stimulated.
   d. The first step in sensory transduction is the generation of a receptor potential.
   e. The receptive fields of adjacent touch receptors overlap.

2. a. All touch receptors are encapsulated receptors.
   b. The sizes of the receptive fields of touch receptors are uniform over the skin.
   c. Skin thermoreceptors are bare nerve endings.
   d. Sensory information from the touch receptors of the skin reaches the brain via the dorsal column pathway.
   e. The spinothalamic tract relays information from the skin thermoreceptors.

3. a. Signals from the somatosensory system reach consciousness in the precentral gyrus.
   b. Pain sensations can be elicited by stimulating the primary somatosensory cortex.
   c. Pain receptors are bare nerve endings.
   d. Pain receptors are activated following the release of bradykinin.
   e. Pain arising from the visceral organs is unpleasant in quality and usually poorly localized.

4. a. The pressure within the eye (the intraocular pressure) is about 15 mm Hg (2 kPa).
   b. The aqueous humor is an ultrafiltrate of plasma.
   c. The pupils constrict when the eye is focused on a near object.
   d. When light is shone in one eye, both pupils constrict.
   e. The diameter of the pupils is controlled exclusively by the sympathetic innervation.

5. a. The lens is the chief refractive element of the eye.
   b. When the eye focuses on a distant object, the ciliary muscle contracts.
   c. The focus of the eye is controlled exclusively by parasympathetic innervation of the ciliary body.
   d. In myopia, either the lens system of the eye is too strong or the eyeball is too long.
   e. Hypermetropia (hyperopia) can be corrected by a diverging lens.

6. a. In good light, the visual acuity falls towards the edge of the visual field.
   b. Full dark adaptation takes nearly 30 min.
   c. In a dark-adapted eye, the visual acuity is best at the center of the visual field.
   d. Full color vision is possible in dim light.
   e. Protanopes cannot distinguish between red and green because they lack the pigment for detecting green light.

7. a. The range of human hearing is from 20 Hz to 20 kHz.
   b. The ear is most sensitive to frequencies between about 1 kHz and 3 kHz.
   c. The ossicles of the middle ear are essential for the efficient transmission of airborne sounds to the cochlea.
   d. The endolymph of the scala media is similar in composition to plasma.
   e. Conductive hearing loss would be evident if a patient had a similar degree of hearing loss for air conduction and bone conduction.

8. a. The semicircular canals play an essential role in the maintenance of posture.
   b. The utricle will respond to the deceleration of a moving vehicle.

c. The semicircular canals contain perilymph.

d. When a subject is rotated in a swiveling chair, his eyes will exhibit an optokinetic nystagmus.

e. Thermal stimulation of the horizontal semicircular canal on one side elicits a nystagmus.

Answers

1. The stimulus to which a receptor is most sensitive is called its adequate stimulus. When a receptor is activated, the rate of discharge of the afferent fiber will tend to fall with time. This is known as adaptation. The initial step in sensory transduction varies with receptor type. There is then a change in the number of open ion channels which results in the generation of the receptor potential.
   a. True
   b. True
   c. False
   d. False
   e. True

2. Touch is subserved by a number of different types of receptor including bare nerve endings. The size of the receptive field of touch receptors varies. It is smallest in the fingertips, lips, and tip of the tongue, and largest on the back.
   a. False
   b. False
   c. True
   d. True
   e. True

3. Signals from the somatosensory system reach consciousness in the postcentral gyrus, which does not appear to receive any signals from the nociceptors.
   a. False
   b. False
   c. True
   d. True
   e. True

4. The intraocular pressure is the result of the balance between the secretion of aqueous humor and its drainage. The secretion is active and can be reduced by drugs that inhibit the enzyme carbonic anhydrase. The diameter of the pupils is controlled by sympathetic innervation from the superior cervical ganglion (m. dilator pupillae) and by parasympathetic nerves from the ciliary ganglion (m. sphincter pupillae).
   a. True
   b. False
   c. True
   d. True
   e. False

5. The main refractive element of the eye is the cornea; the lens is the focusing element. When the eye is focused on a near object the ciliary muscle contracts; it relaxes when the eye focuses on a distant object. Hyperopia (hypermetropia) is corrected with a converging lens.
   a. False
   b. False
   c. True
   d. True
   e. False

6. In the dark-adapted eye, visual acuity is best in the parafoveal region. The fovea has no rods, only cones. Color vision requires stimulation of the cones, which are active in bright light. Protanopes lack the pigment for detecting red light. Deuteranopes lack the pigment for detecting green light.
   a. True
   b. True
   c. False
   d. False
   e. False

7. The endolymph has a high potassium content and a low sodium content, unlike plasma which has a high sodium and low potassium content. In conductive hearing loss, bone conduction would be near normal while air conduction would be impaired.
   a. True
   b. True
   c. True
   d. False
   e. False

8. The semicircular canals are important in the control of eye movements. The utricle and saccule are important in the control of posture. The semicircular canals and the chambers of the utricle and saccule contain endolymph as does the scala media of the cochlea. When a subject is rotated in a swiveling chair, the eyes tend to fixate on one object until they reach the limit of their movement whereupon they fixate on another. The alternation between smooth pursuit movement and the subsequent saccade is optokinetic nystagmus. However, if the eyes are defocused, they cannot fixate and a vestibular nystagmus becomes evident.
   a. False
   b. True
   c. False
   d. True
   e. True

The physiology of motor systems

After reading this chapter you should understand:

♦ The hierarchical nature of control within the motor systems of the body

♦ The organization of the spinal cord and its role in reflex activity

♦ The contribution of spinal reflexes to postural control

♦ The descending pathways that modify the output of the spinal cord

♦ The role of the motor cortical regions in the programming and execution of voluntary activity

♦ The organization of the cerebellum and its role in refined coordinated movements

♦ The role of the basal ganglia in the planning and execution of defined motor patterns

♦ The effects of lesions at various levels of the motor hierarchy

9.1 Introduction

Intrauterine movement is detectable by ultrasound from a very early stage of gestation and is felt by the mother for the first time between 16 and 20 weeks ('quickening' of the fetus). By the time of birth, a baby is capable of some coordinated movement. During the first 2 years of life, as the brain and spinal cord continue to develop and mature, the child learns to defy gravity, first by sitting, then standing, and later walking, running, jumping, and climbing. At the same time, the capacity to perform the precise movements needed for complex manipulations and speech is acquired. In short, coordinated purposeful movement is a fundamental aspect of human existence.

The simplest form of motor act controlled by the nervous system is called a *reflex*. This is a rapidly executed, automatic, and stereotyped response to a given stimulus. As reflexes are not under the direct control of the brain, they are described as involuntary motor acts. Nevertheless, most reflexes involve coordination between groups of muscles and this is achieved by interconnections between various groups of neurons. The neurons forming the pathway taken by the nerve impulses responsible for a reflex make up a *reflex arc*. Many voluntary motor acts are guided by the intrinsic properties of reflex arcs but are modified by commands from higher centers in the brain and from sensory inputs.

Two kinds of voluntary motor function can be distinguished: (i) the maintenance of position (posture), and (ii) goal-directed movements. They are inextricably linked in practice. A goal-directed movement will only be performed successfully if the moving limb is first correctly positioned. Similarly, a posture can only be maintained if appropriate compensatory movements are made to counteract any force tending to oppose that posture.

Physiological understanding is still insufficient to allow a full explanation of the events occurring within the central nervous system during the execution of a voluntary movement. There is no doubt that voluntary movement is impaired whenever there is an interruption of the afferent pathways to the brain arising from sense organs such as the labyrinth, eyes, proprioceptors, and mechanoreceptors. However, what happens between the arrival of afferent information and the execution of an appropriate movement is unclear. Similarly, the processes of motor learning and the relationship between the 'will' to carry out a movement and the movement itself remain poorly understood.

Investigation of the ways in which the CNS controls movement is far more difficult than the study of sensory systems. There are a number of reasons for this. Firstly, movements themselves are difficult to describe in a precise quantitative manner. Secondly, experimentation on anesthetized animals can give little useful information regarding what is, by its nature, a voluntary process. Thirdly, the many complex motor pathways operate in parallel, making it hard to assign particular roles to each. Furthermore, every motor action results in sensory feedback that may modify it further.

Despite these difficulties, it is possible to define some key questions which must be addressed if we are to make progress towards a greater understanding of motor systems.

1. What structural components of the central and peripheral nervous systems participate in the maintenance of posture and movement of the head, trunk, and limbs?

2. How are reflexes organized within the spinal cord?

3. How do 'higher centers' influence the fundamental motor patterns contained within the spinal cord?

4. How is information from the peripheral sense organs used to plan and refine both postural mechanisms and voluntary movements?

9.2 The hierarchical nature of motor control systems

The term *motor system* refers to the neural pathways that control the sequence and pattern of muscle contractions. The structures responsible for the neural control of posture and movement are distributed throughout the brain and spinal cord, as indicated in Fig. 9.1.

As skeletal muscles can only contract in response to excitation of the motoneurons that supply them, all motor acts depend on

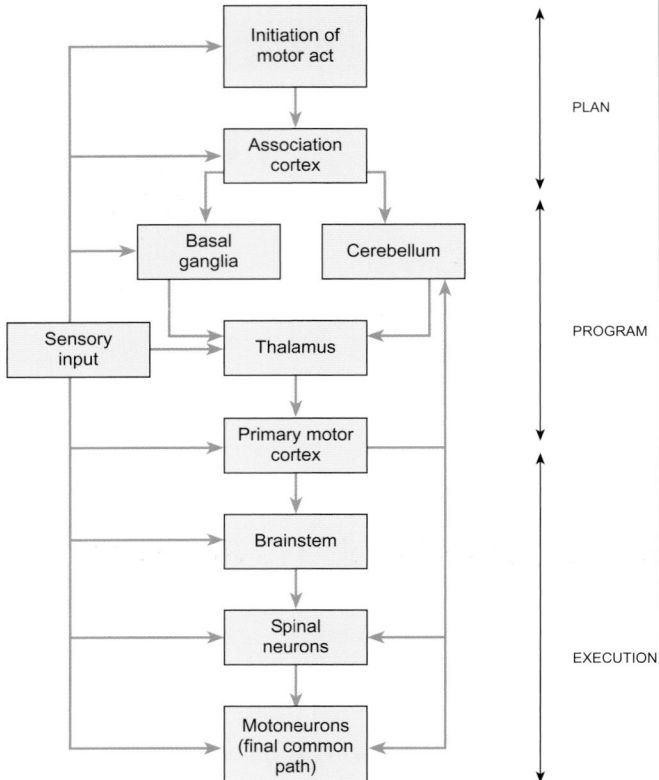

Fig. 9.1 A diagrammatic representation of the motor systems of the body showing possible interactions between them.

neural circuits that eventually impinge on the α-motoneurons that form the output of the motor system. As discussed in Chapter 7, each motoneuron supplies a number of skeletal muscle fibers, and an α-motoneuron together with the skeletal muscle fibers it innervates constitute a *motor unit*, which is the basic element of motor control. For this reason α-motoneurons are often referred to as the final common pathway of the motor system. The processes that underlie neuromuscular transmission have been discussed in Chapter 6 pp. 77–79.

The motoneurons are found in the brainstem and spinal cord and their excitability is influenced by neural pathways that may form local circuits or may arise in a variety of brain areas. Thus there is a kind of hierarchical arrangement of so-called 'motor centers' from the spinal cord up to the cerebral cortex. A vast array of reflexes are controlled by neural circuits within the spinal cord, and these reflex circuits form a system that organizes the basic motor patterns of posture and movement. Superimposed on these local circuits are influences from higher centers in the brain. Postural control is exerted largely at the level of the brainstem, while goal-directed movements require the participation of the cerebral cortex. The basal ganglia and cerebellum both play an important role in motor control though neither is directly connected with the spinal motoneurons. Instead, they influence the motor cortex by way of the thalamic nuclei (see Fig. 9.1).

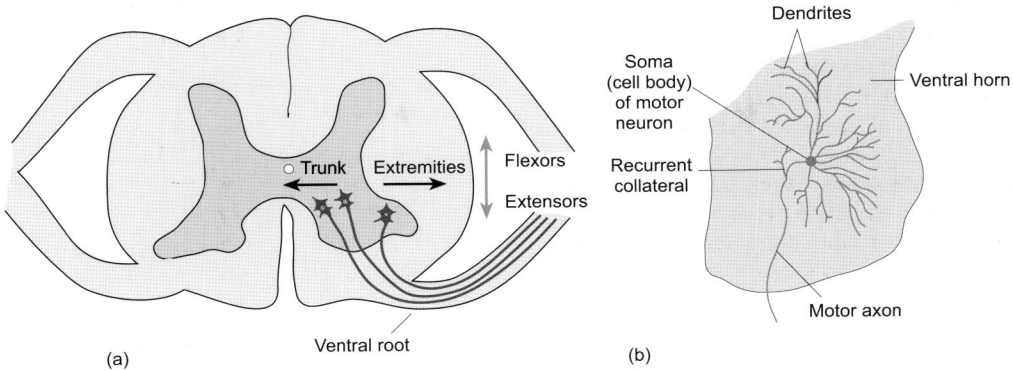

Fig. 9.2 (a) A transverse section of the spinal cord. The localization of motor neurons corresponding to various groups of muscles are indicated with flexors represented more dorsally while extensors are represented ventrally within the cord. The muscles of the trunk are represented medially and the extremities are represented laterally. (b) An α-motoneuron in the ventral horn of the spinal cord, illustrating the elaborate dendritic tree. Although not shown in this diagram, numerous synaptic connections are made with these dendrites.

9.3 Organization of the spinal cord

The α-motoneurons (or somatic motor neurons) are large neurons whose cell bodies lie in clumps (sometimes called motor nuclei) within the ventral horn of the spinal cord and in the brainstem. Each motoneuron innervates a motor unit that may consist of anything between 6 and 1500 skeletal muscle fibers. α-Motoneurons have long dendrites and receive many synaptic connections (Fig. 9.2(b)). These include afferents from interneurons and proprioceptors as well as descending fibers from higher levels of the CNS. The axons of α-motoneurons collect in bundles that leave the ventral horn and pass through the ventral white matter of the spinal cord before entering the ventral root. Some axons send off branches that turn back into the cord and make excitatory synaptic contact with small interneurons called *Renshaw cells*. These cells in turn have short axons that synapse with the pool of motor neurons by which they are stimulated. These synapses are inhibitory and bring about recurrent, or feedback, inhibition.

The ventral horn motor neurons show an orderly topographical arrangement within the spinal cord, as illustrated in Fig. 9.2(a). Motor neurons supplying the muscles of the trunk are situated in the medial ventral horn while those supplying more distal muscle groups tend to be situated laterally. Muscles that flex the limbs (flexors) are under the control of neurons that lie dorsal to those that control the muscles that extend the limbs (extensors).

The spinal cord receives afferent input from proprioceptors in muscles and joints

For movements to be carried out in a functionally appropriate way, it is essential for sensory and motor information to be integrated. All the neural structures involved in the execution of movements are continually informed of the position of the body and of the progress of the movement by sensory receptors within the muscles and joints. These provide information regarding the position of our limbs and their movements relative to each other and to our surroundings. These receptors are called *proprioceptors* and the information they provide is used to control muscle length and posture (see Section 9.8).

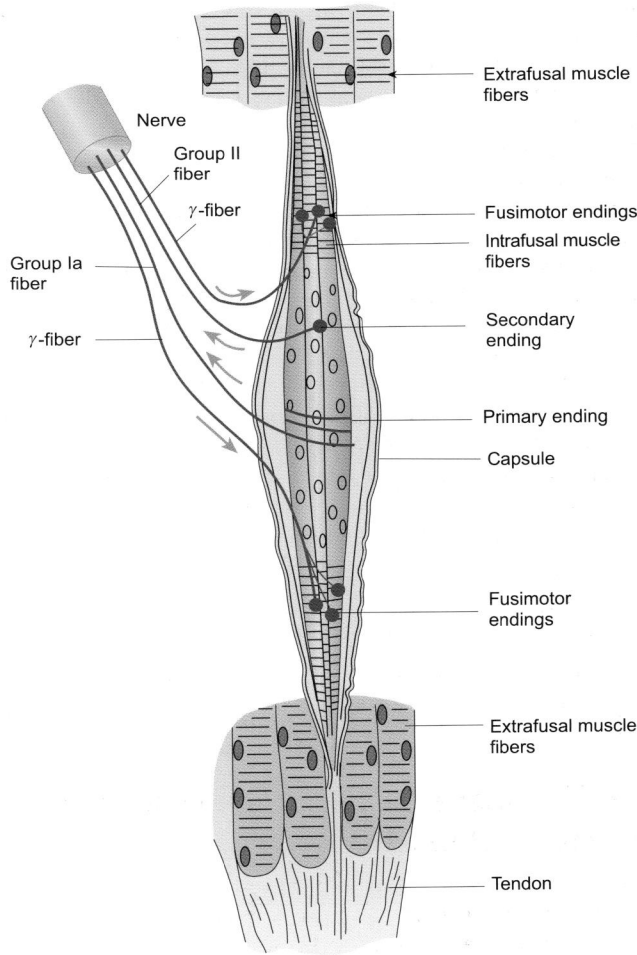

Fig. 9.3 The basic organization of a muscle spindle. These sensory organs lie in parallel with the extrafusal muscle fibers and therefore are adapted to monitor muscle length. Note that the muscle spindle is innervated by both motor and sensory nerve fibers.

The main proprioceptors are the *muscle spindles* and *Golgi tendon organs*. Both provide information about the state of the musculature. Muscle spindles lie within the muscles between and in parallel with the skeletal muscle fibers. Therefore they can respond to muscle length and its rate of change. Golgi tendon organs lie within the tendons and are in series with the contractile elements of the muscle. They are sensitive to the force generated within that muscle during contraction.

Muscle spindles

Although muscle spindles are found in most skeletal muscles, they are particularly numerous in those muscles that are responsible for fine motor control, such as those of the eyes, neck, and hands. The basic organization of a muscle spindle is illustrated in Fig. 9.3. The muscle spindle consists of a small bundle of modified muscle fibers innervated by both sensory and motor neurons. The muscle fibers of the muscle spindles are called the *intrafusal fibers* while those of the main body of the muscle are the *extrafusal fibers*. The nerve endings and intrafusal fibers of the muscle spindles are enclosed within a capsule.

There are two different types of intrafusal fibers—the nuclear bag fibers and nuclear chain fibers, so called because of the arrangement of their nuclei. They are shown diagrammatically in Fig. 9.4. Nuclear bag fibers have a cluster of nuclei near their midpoint. Nuclear chain fibers are smaller, and have a single row of nuclei near their midpoint. The central regions of both bag and chain fibers contain no myofibrils and are the most elastic parts of the fibers, so that this central region stretches preferentially when the muscle spindle is stretched.

The afferent fibers of muscle spindles are of two types. Each spindle receives fibers from a primary afferent (group 1a fibers). These wind around the middle section of both bag and chain fibers forming so-called *annulospiral endings* (or primary sensory endings). Many muscle spindles are also innervated by one or more afferent fibers of group II (see Chapter 6 for the classification of nerve fibers). They terminate more peripherally than the primary endings, almost exclusively on the nuclear chain fibers. They are known as secondary sensory endings (or flower-spray endings because of their multiple branched nature).

The two types of afferent nerve endings respond to muscle stretch in different ways. The rate of firing of both kinds is proportional to the degree of stretch of the muscle spindle at any

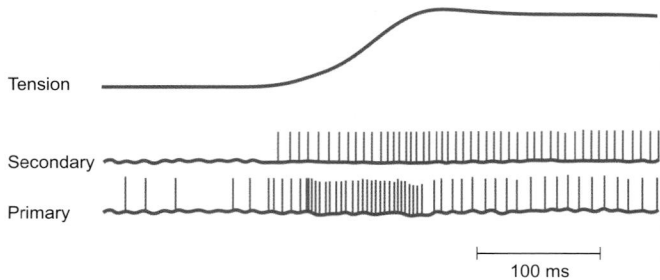

Fig. 9.5 Responses of primary (Ia) and secondary (II) muscle spindle afferent fibers to muscle stretch. Note the very intense period of activity of the primary ending during the stretch.

moment. Although the primary (Ia) fibers respond in proportion to the degree of stretch, they are much more sensitive to rapid changes in muscle length and for this reason they are classed as rapidly adapting or dynamic endings. The secondary endings are non-adapting and are said to be static endings. The differing nature of the responses to stretch of the primary and secondary endings is illustrated in Fig. 9.5.

The motor neurons innervating the intrafusal fibers are known as γ-*motoneurons* to distinguish them from the large α-motoneurons which innervate the extrafusal fibers. The cell bodies of the γ-motoneurons lie in the ventral horn of the spinal cord and their axons, which are also known as *fusimotor fibers*, leave the spinal cord via the ventral roots. The fusimotor fibers have diameters in the range 3–6 μm and conduction velocities of 15–30 m s^{-1}. The α-motoneurons have large-diameter fibers that range in size between 15 and 20 μm and have conduction velocities of 70–120 m s^{-1} (see Table 6.2).

Within the muscle, the fusimotor fibers branch to supply several muscle spindles and, within these, branch further to supply several intrafusal fibers. The γ-motoneurons innervate both the nuclear chain and nuclear bag fibers and bring about contraction of the peripheral regions of the muscle spindles. Note, however, that contraction of the intrafusal fibers is too weak to effect movements of the muscle as a whole; rather, the tension in the intrafusal fibers regulates the sensitivity of the spindles. When the intrafusal fibers contract in response to fusimotor stimulation, their sensory endings are stimulated by the stretch. The consequent excitation of the Ia fibers is superimposed on the firing that results from the degree of stretch imposed by the extrafusal muscle fibers.

To sum up, muscle spindles can be stimulated in two ways:

(1) by stretching the entire skeletal muscle;

(2) by causing the intrafusal fibers to contract while the extrafusal muscle fibers remain at the same length.

In either case, stretching a muscle spindle will increase the rate of discharge of the group Ia and group II afferent nerve fibers to which it is connected.

Golgi tendon organs

The Golgi tendon organs are mechanoreceptors that lie within the tendons of muscles immediately beyond their attachments to the muscle fibers (Fig. 9.6). Around 10 or 15 muscle fibers are

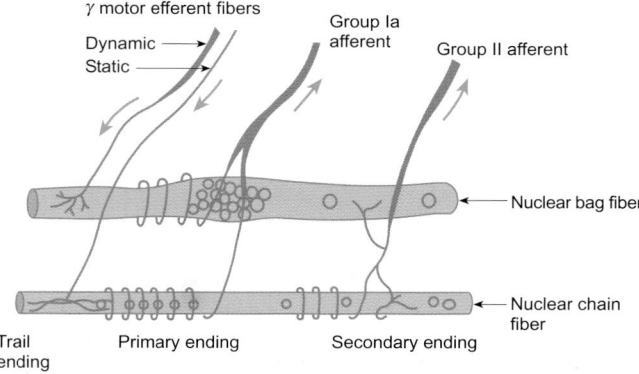

Fig. 9.4 A diagram showing nuclear bag and chain fibers of a muscle spindle, with their sensory and motor nerve supplies.

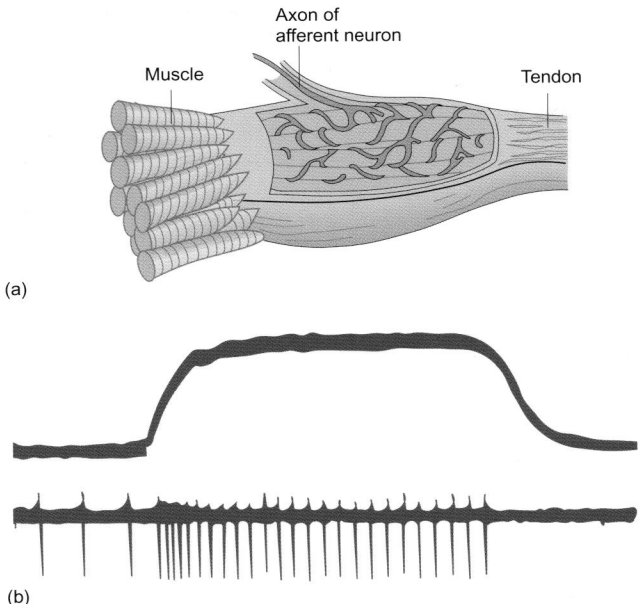

Fig. 9.6 The Golgi tendon organ. (a) The basic organization of a Golgi tendon organ. (b) The response of a Golgi tendon organ to tension in the muscle. The upper trace shows the tension in the muscle to which the tendon organ is attached and the lower trace shows the firing pattern in the Ib fiber supplying the receptor. Note that the Golgi tendon organ lies in series with the muscle and so is adapted to monitor muscle tension.

usually connected to each Golgi tendon organ, which is then stimulated by the tension produced by this bundle of fibers. Impulses are carried from the tendon organs to the CNS (particularly the spinal cord and the cerebellum) by group Ib afferent fibers.

Summary

1. The basic element of motor control is the motor unit. The cell bodies of α-motoneurons are topographically arranged within the ventral horn of the spinal cord. Their axons innervate skeletal muscle fibers. The cell bodies receive numerous synaptic connections from proprioceptors and from higher levels of the CNS including the brainstem, basal ganglia, cerebellum, and motor cortex.

2. Proprioceptors are mechanoreceptors situated within muscles and joints. They provide the CNS with information regarding muscle length, position and tension (force).

3. Muscle spindles lie in parallel with extrafusal muscle fibers. They are innervated by γ-motoneurons (efferents) and by group Ia and group II afferent fibers. The afferents respond to muscle stretch while γ-efferent activity regulates the sensitivity of the spindles.

4. Golgi tendon organs respond to the degree of tension within the muscle. Group Ib afferent fibers relay this information to the CNS (in particular the spinal cord and cerebellum).

9.4 Reflex action and reflex arcs

Reflexes represent the simplest form of irritability associated with the nervous system. Reflex arcs include at least two neurons, an *afferent* or *sensory* neuron and an *efferent* or *motor* neuron. The fiber of the afferent neuron carries information about the environment from a receptor towards the CNS while the efferent fiber transmits nerve impulses from the CNS to an effector. Reflexes may be (and often are) subject to modulation by activity in the CNS.

In the simplest reflex arc, there are two neurons and just one synapse. Therefore such reflexes are known as *monosynaptic reflexes*. Other reflex arcs have one or more neurons interposed between the afferent and efferent neurons. These neurons are called *interneurons* or internuncial neurons. If there is one interneuron, the reflex arc will have two synaptic relays and the reflex is called a *disynaptic reflex*. If there are two interneurons, there will be three synaptic relays so that the associated reflex would be trisynaptic. If many interneurons are involved, the reflex would be called a *polysynaptic reflex*. Examples are the stretch reflex (monosynaptic), the withdrawal reflex (disynaptic), and the scratch reflex (polysynaptic).

The knee jerk is an example of a dynamic stretch reflex

A classic example of a stretch reflex (also known as the myotactic reflex) is the *knee-jerk* or *tendon-tap reflex* which is used routinely in clinical neurophysiology as a tool for the diagnosis of certain neurological conditions. Hinge joints such as the knee and ankle are extended and flexed by extensor and flexor muscles, which act in an antagonistic manner. A sharp tap applied to the patellar tendon stretches the quadriceps muscle. The stretch stimulates the 'dynamic' nuclear bag receptors of the muscle spindles. As a result, there is an increase in the rate of firing of the group Ia afferents of the muscle spindles within the quadriceps. This informs the spinal cord that the quadriceps muscle has been stretched. The afferent fibers branch as they enter the spinal cord, and some enter the gray matter of the cord and make monosynaptic contact with the α-motoneurons supplying the quadriceps muscle, causing them to discharge in synchrony. The resulting contraction of this muscle abruptly extends the lower leg (hence the name knee jerk). Collaterals of the Ia fibers make synaptic contact with inhibitory interneurons, which in turn inhibit the antagonistic (flexor) muscles of the knee joint.

The stretch reflex arc is illustrated diagrammatically in Fig. 9.7(a). The reflex is lost if the lower lumbar dorsal roots of the spinal cord (through which the afferents from the quadriceps pass) are damaged. A similar reflex occurs when the Achilles tendon is struck (the ankle-jerk reflex). In this case, a plantar flexion of the foot is produced by contraction of the calf muscles.

The tonic stretch reflex

This reflex contributes to muscle tone and helps to maintain posture. Muscle spindles continue to relay information to the spinal cord even when the length of a muscle is kept more or less constant. When standing upright, for example, the slightest bending of the knee joint will stretch the quadriceps muscle and increase the activity in the primary muscle spindle endings. This

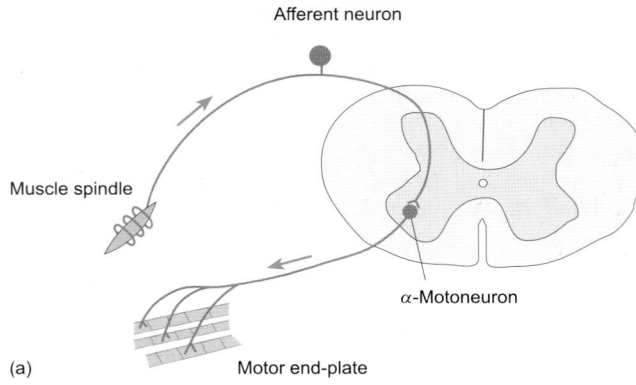

(a)

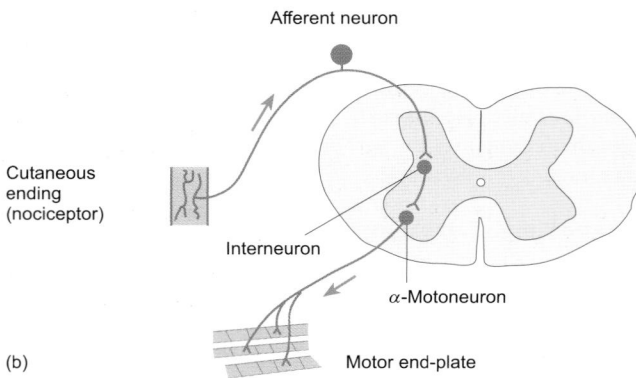

(b)

Fig. 9.7 (a) The stretch reflex arc. Note that this reflex arc comprises only two neurons and one synapse. Therefore it is a monosynaptic reflex. (b) The basic flexor (withdrawal) reflex arc. In this case, there are three neurons and two synapses in the basic arc. Therefore the reflex in its simplest form is disynaptic.

will result in stimulation of the α-motoneurons supplying the quadriceps. The tone of the muscle will increase and will counteract the bending so that the posture will be maintained. The converse will occur when there is excessive contraction of the muscle. The tonic stretch reflex therefore helps to stabilize the length of a muscle when it is under a constant load.

The flexion reflex

In this protective reflex, a limb is rapidly withdrawn from a threatening or damaging stimulus. It is more complex than the stretch reflex and usually involves large numbers of interneurons and proprio-spinal connections arising from many segments of the spinal cord. Withdrawal may be elicited by noxious stimuli applied to a large area of skin or deeper tissues (muscles, joints, and viscera) rather than from a single muscle as in the stretch reflex. The receptors responsible are called nociceptors (see Chapter 8, p. 109) and they give rise to the afferent impulses that are responsible for the flexion reflex.

To achieve withdrawal of a limb, the flexor muscles of one or more joints in the limb must contract while the extensor muscles relax. Afferent volleys cause excitatory interneurons to activate α-motoneurons that supply flexor muscles in the affected limb. At the same time, the afferent volleys activate interneurons, which inhibit the α-motoneurons supplying the extensor

muscles. The excitation of the flexor motoneurons coupled with inhibition of the extensor motoneurons is known as *reciprocal inhibition*.

The synaptic organization of the flexor reflex is illustrated in Fig. 9.7(b). As the figure shows, the basic reflex is disynaptic but in a powerful withdrawal reaction it is likely that several spinal segments will be involved and that all the major joints of the limb will show movement. The flexion reflex has a longer latency than the stretch reflex because it arises from a disynaptic arc. It is also non-linear in so far as a weak stimulus will elicit no response while a powerful withdrawal is seen when the stimulus reaches a certain level of intensity.

The crossed extensor reflex

Stimulation of the flexion reflex, as described above, frequently elicits extension of the contralateral limb about 250 ms later. This crossed extension reflex helps the subject to maintain posture and balance. The long latency between flexion and crossed extension represents the time taken to recruit interneurons. The reflex arc for crossed extension is illustrated in Fig. 9.8.

The Golgi tendon reflex

This reflex, which is the result of activation of Golgi tendon organs, complements the tonic stretch reflex and contributes to the maintenance of posture. The synaptic organization of this reflex is illustrated for the knee joint in Fig. 9.9. In this example, the tendon organ is located in the tendon of the rectus femoris muscle and its afferent fiber branches as it enters the spinal cord. One set of branches excites interneurons which inhibit the discharge of the α-motoneurons supplying the rectus femoris muscle, while another set activates interneurons that stimulate

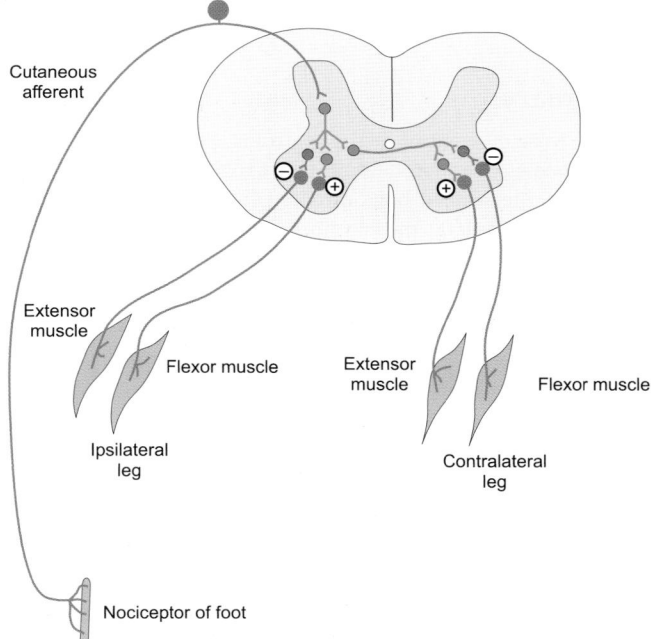

Fig. 9.8 The neuronal pathways participating in a crossed extensor reflex arc: ⊖ synaptic inhibition; ⊕ synaptic excitation. Many neurons and synapses are involved in this reflex, which is therefore polysynaptic.

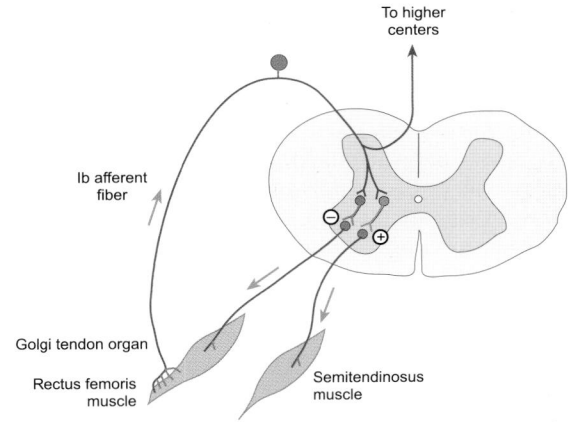

Ib afferent fiber

To higher centers

Golgi tendon organ

Rectus femoris muscle

Semitendinosus muscle

Fig. 9.9 A diagrammatic representation of the Golgi tendon organ reflex arc. This shows the synaptic basis of reciprocal inhibition. It is also seen in the crossed extensor reflex illustrated in Fig. 9.8.

activity in the α-motoneurons innervating the antagonistic semitendinosus (hamstring) muscles.

How does this reflex operate to maintain posture? During a maintained posture such as standing, the rectus femoris muscle will start to fatigue. As it does so, the force in the patellar tendon, monitored by Golgi tendon organs, will decline. As a result, activity in the afferent Ib fibers will decline and the

Summary

1. The simplest form of irritability associated with the nervous system is reflex activity. The neurons participating in a reflex form a reflex arc, which includes a receptor, an afferent neuron that synapses in the CNS, and an efferent neuron that sends a nerve fiber to an effector. Interneurons may be present between the afferent and efferent neurons. The number of synapses in the basic reflex arc is used to define the reflex as monosynaptic, disynaptic, or polysynaptic.

2. The simplest reflexes are the monosynaptic stretch reflexes such as the patellar tendon-tap reflex (the knee-jerk reflex). Here, stretching of a muscle stimulates the muscle spindle afferents. These excite the α-motoneurons supplying that muscle and cause it to contract. Stretch reflexes play an important role in the control of posture.

3. The protective flexion (withdrawal) reflexes are elicited by noxious stimuli. Their reflex arc possesses at least one interneuron, and so the most basic flexion reflex is disynaptic. More usually many muscles are involved through polysynaptic pathways, which may involve many spinal segments. Reciprocal inhibition ensures that the extensor muscles acting on a joint will relax while the flexor muscles contract.

4. Tendon organs monitor force in the tendons they supply. The inverse myotactic reflex is activated by discharge in the Golgi tendon organs. It also plays an important role in the maintenance of posture.

normal inhibition of the motoneurons supplying the rectus femoris will be removed. Consequently, the muscle will be stimulated to contract more strongly, thereby increasing the force in the patellar tendon once more.

9.5 The role of the muscle spindle in voluntary motor activity

The activity of the muscle spindles enables the nervous system to compare the lengths of the extrafusal and intrafusal muscle fibers. Whenever the length of the extrafusal fibers exceeds that of the intrafusal fibers, the afferent discharge of a muscle spindle will increase, and whenever the length of the extrafusal fibers is less than that of the intrafusal fibers, the discharge of the spindle afferents will decline. This decline is possible because the spindle afferents normally show a tonic level of discharge. In this way, the muscle spindles can provide feedback control of muscle length.

The role of the muscle spindles as comparators for the maintenance of muscle length is important during goal-directed voluntary movements. Studies of various movements (chewing and finger movements for example) have shown that when voluntary changes in muscle length are initiated by motor areas of the cerebral cortex, the motor command includes changes to the set-point of the muscle spindle system. To achieve this, both α- and γ-motoneurons are activated simultaneously by way of the neuronal pathways descending from higher motor centers. The simultaneous activation of extrafusal fibers (by way of α-motoneurons) and intrafusal fibers (by way of γ-motoneurons) is called α–γ *co-activation*. Its physiological importance appears to lie in the fact that it allows the muscle spindles to be functional at all times during a muscle contraction.

The benefits of incorporating length sensitivity into voluntary activities can be illustrated by the following example. Suppose that a heavy weight needs to be lifted. Before lifting, and from previous experience, the brain will estimate roughly how much force will be required to lift the weight and the motor centers will transmit the command to begin the lift. Both extrafusal and intrafusal fibers will be activated simultaneously. If the initial estimate is accurate, the extrafusal fibers will be able to shorten as rapidly as the intrafusal fibers and the activity of the spindle afferents will not change much during the lift. However, if the weight turns out to be heavier than expected, the estimate of required force will be insufficient and the rate of shortening of the extrafusal fibers will be slower than expected. Nevertheless, the intrafusal fibers will continue to shorten and their central region will become stretched. As a result, the activity in the spindle afferents will increase and will summate with the excitatory drive already arriving at the α-motoneurons via the descending motor pathways. The increased activity in the α-motoneurons will increase the force generated by the muscle until it matches that required to lift the load.

9.6 Transection of the spinal cord

Despite the protection afforded to the spinal cord by the vertebral column, spinal injuries are still relatively common. Motor accidents are the most frequent cause of spinal cord injury, fol-

Summary

1. The role of the muscle spindles as comparators for the maintenance of muscle length is important during goal-directed voluntary movements. When voluntary changes in muscle length are initiated by motor areas of the brain, the motor command includes changes to the set-point of the muscle spindle system. To achieve this, both α- and γ-motoneurons are activated simultaneously by way of the neuronal pathways descending from higher motor centers.

2. The simultaneous activation of extrafusal fibers (by way of α-motoneurons) and intrafusal fibers (by way of γ-motoneurons) is called α–γ *co-activation*. Co-activation of α- and γ-efferent motoneurons readjusts the sensitivity of muscle spindles continuously as the muscle shortens.

lowed by falls and sports injuries (particularly diving). The alterations in body function that result from such injuries depend on the level of injury and the extent of the damage to the spinal cord. Following damage, there is a loss of sensation due to interruption of ascending spinal pathways and loss of the voluntary control over muscle contraction as a result of damage to descending motor pathways below the level of the lesion. The loss of voluntary motor control is known as muscle paralysis. Both spinal reflex activity and functional activities such as breathing, micturition, and defecation may be affected. Paralysis may be spastic (in which the degree of muscle tone is increased above normal) or flaccid (in which the level of tone is reduced and the muscles are 'floppy'). The characteristics of motor dysfunction due to lesions of the corticospinal tract (*upper motoneuron lesions*) and those due to dysfunction of spinal motoneurons (*lower motoneuron lesions*) are discussed further in Box 9.1.

Immediately following injury to the spinal cord, there is a loss of spinal reflexes (areflexia). This is known as *spinal shock* or *neurogenic shock* and involves the descending motor pathways. The clinical manifestations are flaccid paralysis, a lack of tendon reflexes, and loss of autonomic function below the level of the lesion. Spinal shock is probably the result of the loss of the normal continuous excitatory input from higher centers such as the vestibulospinal tract, parts of the reticulospinal tract and the corticospinal tract (see below). Spinal shock may last for several weeks in humans. After this time, there is usually some return of reflex activity as the excitability of the undamaged spinal neurons increases. Occasionally this excitability becomes excessive and then spasticity of affected muscle groups is seen.

BOX 9.1 UPPER AND LOWER MOTONEURON LESIONS AND SENSORY DEFICITS FOLLOWING SPINAL LESIONS

Voluntary motor acts involve a large number of structures including the cerebral cortex, corticospinal tract, basal ganglia, cerebellum, and spinal cord. When any of these regions are damaged as a result of traumatic injuries or strokes, characteristic changes in the activity of the motor system are apparent.

Damage to the primary motor cortex or corticospinal tract is often referred to as an upper motoneuron lesion. Such lesions are characterized by spastic paralysis without muscle wasting. There is an increase in muscle tone and the tendon reflexes are exaggerated (hyper-reflexia). The Babinski sign is positive, i.e. in response to stimulation of the sole of the foot, the toes extend upwards instead of downwards as in a normal person. The muscles themselves are weak, although the flexors of the arms are stronger than the extensors. In the legs, the reverse is true, the extensors are stronger than the flexors. The muscles do not show fasciculation (muscle twitching in which groups of muscle fibers contract together).

Damage to the spinal cord that involves the motoneurons is called a lower motoneuron lesion. A lower motoneuron lesion results in denervation of the affected muscles, which show a loss of muscle tone. Tendon reflexes may be weak or absent. There is a loss of muscle bulk and fasciculation is usually present. Coordination is impaired. The Babinski sign is negative (normal), i.e. the toes flex in response to stimulation of the sole of the foot.

Lesions to the spinal cord affect sensation as well as motor activity. If the spinal cord is completely transected, there is a loss of both motor and sensory function below the level of the lesion. Partial loss of motor activity and sensation occurs when the spinal cord is compressed by either the protrusion of an intervertebral disk or a spinal tumor. The spinal cord may also be partially cut across as a result of certain traumatic injuries such as stabbing. In such partial transections, the pattern of impaired sensation provides important information regarding the site of the lesion.

In clinical examination, the primary senses tested are light touch, which is tested with a wisp of cotton wool, pinprick, temperature, vibration, and position sense. Reflex activity is tested with a patellar hammer.

On the side of the lesion, there is a loss of voluntary motor activity below the area of damage, which results from damage to the corticospinal tract. The immobility is associated with the signs of upper motoneuron damage described above. Motor activity is unimpaired on the unaffected side.

The spinothalamic tract is crossed at the segmental level and the dorsal column pathway crosses in the brainstem. Therefore the loss of sensation following hemisection of the spinal cord is characteristic. On the side of the lesion below the area of damage there is a loss of those sensations that reach the brain via the dorsal column pathway, namely proprioception, vibration sense, and fine touch discrimination. Pain and thermal sensations are transmitted to the brain via the crossed spinothalamic tract and so are preserved. On the side opposite the lesion, position and vibration sense below the area of damage are normal, as is tactile discrimination. Pain and thermal sensations are lost as the spinothalamic pathways are transected.

The term *tetraplegia* (sometimes also called quadriplegia) refers to impairment or loss of motor and sensory function in the arms, trunk, legs, and pelvic organs. *Paraplegia* refers to impairment of function of the legs and pelvic organs. The arms are spared and the degree of motor and sensory impairment will depend on the exact level of the spinal injury. Injuries above the level of around T12 will normally result in spastic paralysis of the affected skeletal muscle groups; control of bowel, bladder, and sexual functions are also affected. Injuries below the level of T12, especially those that involve damage to the peripheral nerves leaving the spinal cord, result most often in flaccid paralysis of the affected muscle groups and of the muscles controlling bowel, bladder, and sexual function.

Arteriolar tone is maintained by activity in axons that pass from the vasomotor regions in the brainstem to the intermediolateral column of the spinal cord via the lateral column of the white matter. Damage to the upper segments of the spinal cord will affect the sympathetic vasoconstrictor tone, causing a marked decrease in blood pressure. However, damage to the lumbar region of the spinal cord will have a relatively small effect.

Summary

1. Section of the spinal cord causes a loss of both sensation and voluntary control over muscle contraction (paralysis) below the level of the lesion. Immediately following such an injury, there is a loss of spinal reflexes (areflexia) and flaccidity of the muscles controlled by nerves leaving the spinal cord below the level of the injury. This is known as *spinal shock* or *neurogenic shock*. Both reflex somatic motor activity and involuntary activities involving bladder and bowel functions may be affected.

2. Spinal shock may last for several weeks. Over the ensuing weeks, spinal cord activity below the level of the lesion returns as the excitability of the undamaged neurons increases. This may give rise to spasticity of the paralyzed muscle groups.

9.7 Descending pathways involved in motor control

Although the spinal cord contains the neural networks required for reflex actions, more complex motor behaviors are initiated by pathways that originate at various sites within the brain. Furthermore, the activity of the neural circuitry within the spinal cord is modified and refined by the descending motor control pathways. There are five important brain areas which give rise to descending tracts, four of which lie within the brainstem and medulla. These are the reticular formation, the vestibular nuclei, the red nucleus, and the tectum. Their fibers constitute a set of descending pathways that are sometimes referred to as the *extrapyramidal tracts*. The fifth area lies within the cerebral cortex and gives rise to the *pyramidal* or *corticospinal tract*. Figure 9.10 shows the approximate positions of the major descending tracts and their location within the spinal cord.

Extrapyramidal pathways

The reticular system gives rise to two important descending tracts within the cord, the lateral reticulospinal tract and the medial reticulospinal tract. These tracts are largely uncrossed, terminating mainly on interneurons rather than on the α-motoneurons themselves. They influence mainly the muscles of the trunk and the proximal parts of the extremities. They are believed to be important in the control of certain postural mechanisms and in the startle reaction ('jumping' in response to a sudden and unexpected stimulus).

The vestibular nuclei are located just below the floor of the fourth ventricle close to the cerebellum. The medial and lateral vestibular nuclei give rise to a descending motor pathway called the vestibulospinal tract. Most of the neurons of this tract, like those of the reticulospinal tract, synapse with interneurons on the ipsilateral side of the cord. The vestibulospinal tracts are concerned mostly with the activity of extensor muscles and are important in the control of posture (acting to make adjustments in response to vestibular signals). Lesions of reticulospinal or

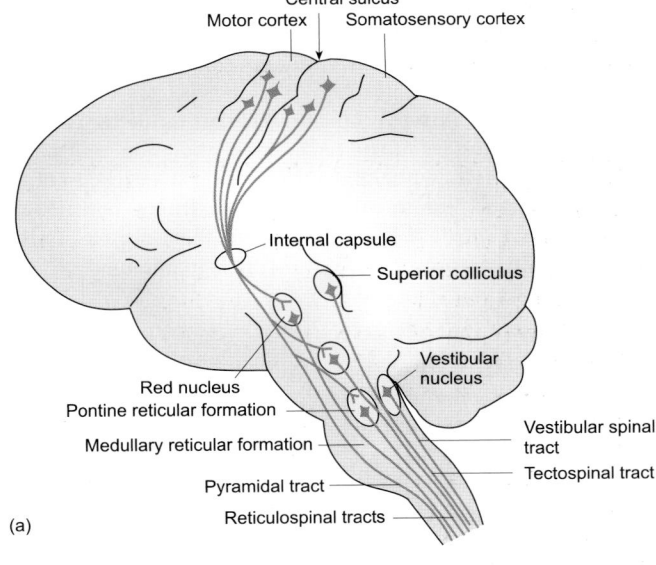

(a)

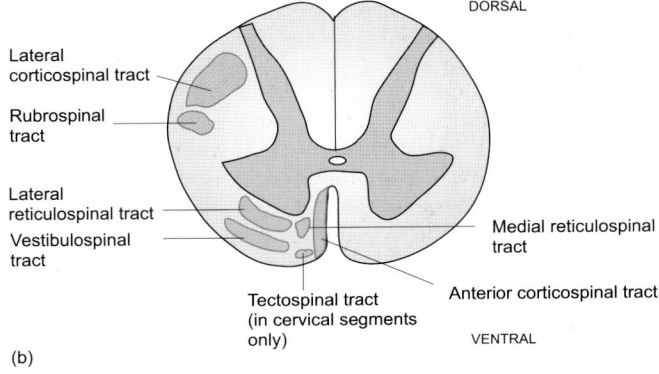

(b)

Fig. 9.10 The principal motor pathways arising in the brain.
(a) The arrangement of the major descending motor tracts and the approximate location of the motor nuclei. (b) The position of the major pyramidal and extrapyramidal descending pathways within the spinal cord.

vestibulospinal tracts affect the ability to maintain a normal erect posture.

The red nucleus is one of the most important 'extrapyramidal' structures. It has a topographic organization in which the upper limbs are represented dorsomedially and the lower limbs ventrolaterally. It receives afferent inputs from both the cortex and the cerebellum as well as from the globus pallidus, which is the major output nucleus of the basal ganglia. It gives rise to the rubrospinal tract. The fibers of this tract decussate (cross) and then travel to the spinal cord where they terminate in the lateral part of the gray matter (Fig. 9.10). Some make monosynaptic contact with α-motoneurons but most terminate on interneurons which excite both flexor and extensor motoneurons supplying the contralateral limb muscles. Lesions of the red nucleus or rubrospinal tract impair the ability to make voluntary limb movements while having relatively little effect on the control of posture.

The tectospinal tract has its origin in the tectum which forms the roof of the fourth ventricle and comprises the superior and inferior colliculi (see Chapter 6). These areas seem to be concerned with the integration of visual and auditory signals and may have a role in orientation. The tectospinal tract projects to cervical regions of the spinal cord. Its fibers are crossed and end on interneurons that influence the movements of the head and eyes.

In general, the extrapyramidal tracts that influence the axial muscles (the muscles of the neck, back, abdomen, and pelvis) are largely crossed, while those influencing the muscles of the limbs are mostly uncrossed. This arrangement permits the independent control of the limbs and the axial muscles so that manipulation can proceed while posture is maintained.

Cranial nerve motor nuclei

The neuronal circuitry of the spinal cord is responsible only for the motor activity of parts of the body below the upper part of the neck. Movements of the head and facial muscles are under the control of cranial nerve (CN) motor nuclei. These include the oculomotor nuclei (CN III), the trigeminal motor outflow (CN V), the facial nuclei (CN VII), and the nucleus ambiguus (CN X), all of which supply muscles that are part of a bilateral motor control system. Fibers from these nuclei make contact with interneurons in the brainstem which are thought to be organized in much the same way as those of the spinal cord which supply the axial and limb muscles.

The corticospinal (pyramidal) tract

The corticospinal tract has traditionally been described as the predominant pathway for the control of fine skilled manipulative movements of the extremities. While it is undoubtedly of considerable significance, it is now evident that many of its actions appear to be duplicated by the rubrospinal tract (see above) which can, if necessary, take over a large part of its motor function.

The corticospinal tract originates in the cerebral cortex and runs down into the spinal cord. Therefore many of its axons are very long. Around 80 per cent of the fibers cross as they run through the extreme ventral surface of the medulla (the medul-

lary pyramids) and pass down to the spinal cord as the lateral corticospinal tract. The uncrossed fibers descend as the anterior (or ventral) corticospinal tract and eventually cross in the spinal cord. As the fibers of the corticospinal tract pass the thalamus and basal ganglia, they fan out into a sheet called the internal capsule. Here they seem particularly susceptible to damage following cerebral vascular accidents (strokes), and the resulting ischemic damage produces a characteristic form of paralysis.

The axons that give rise to the pyramidal tract originate from a wide area of cerebral cortex. About 40 per cent of them come from the motor cortex proper, with the rest originating in either the frontal or the parietal lobes. The parietal lobe forms part of the somesthetic cortex and the function of fibers from this region may be to provide feedback control of sensory input. About 3 per cent of the fibers in the pyramidal tract are derived from large neurons called *Betz cells*, which were once thought to give rise to the entire pyramidal tract. In fact, the great majority of the fibers of the corticospinal tract are of small diameter and

Summary

1. The activity of the neural circuitry within the spinal cord is modified and refined by descending motor control pathways. These are the pyramidal (corticospinal) and extrapyramidal (reticulospinal, vestibulospinal, rubrospinal, and tectospinal) tracts.

2. The reticulospinal tracts are largely uncrossed and terminate on interneurons within the spinal cord which influence mainly the muscles of the trunk and proximal parts of the limbs. They seem to be important in the maintenance of certain postures and in startle reactions.

3. The fibers of the vestibulospinal tract mostly make synaptic contact with interneurons in the ipsilateral spinal cord. These interneurons control the activity of extensor muscles and are important in the maintenance of an erect posture, making adjustments in response to signals from the vestibular apparatus.

4. The red nucleus receives afferent information from the cortex, cerebellum, and basal ganglia. Fibers of the rubrospinal tract terminate in the contralateral gray matter of the spinal cord, and synapse with interneurons controlling both flexor and extensor muscles. Voluntary movements are impaired following lesions to this tract.

5. The tectospinal tract projects to contralateral cervical regions of the cord and makes synaptic contact with interneurons controlling head and eye movements.

6. The corticospinal tract originates over a wide area of cortex including both motor and somatosensory areas. The fibers of the corticospinal tract form the internal capsule as they pass the thalamus and basal ganglia. More than 80 per cent of the fibers cross to the contralateral side at the pyramids. Some fibers make monosynaptic contact with the α- and γ-motoneurons of the hand and finger muscles. Loss of precise hand movements is a feature of lesions to the corticospinal tract.

may be either myelinated or unmyelinated. These fibers conduct nerve impulses relatively slowly.

Most axons of the pyramidal tract synapse with interneurons in the contralateral spinal cord. Many also send collaterals to other areas of the brain, including the red nucleus, the basal ganglia, the thalamus, and the reticular formation of the brainstem. A few make monosynaptic contact with contralateral α- and γ-motoneurons. This is most noticeable in those supplying the hand and finger muscles and may reflect the large variety of movements of which these muscles are capable. In monkeys, for example, lesions to the pyramidal tract result in a loss of the precision grip, although the power grip and gross movements of the limbs and trunk remain unaffected. In humans with corticospinal tract damage (e.g. following a stroke) there are few overt motor deficits. However, there is weakness of the hand and finger muscles and a positive Babinski sign (a pathological reflex in which there is dorsiflexion of the big toe and fanning of the other toes in response to stroking the sole of the foot)

9.8 The control of posture

Maintenance of a stable upright posture is an active process. The skeletal muscles that maintain body posture (the axial muscles) function almost continuously and largely unconsciously, making one tiny adjustment after another, to enable the body to maintain a seated or erect posture despite the constant downward pull of gravity. When we stand, our center of gravity must lie within the area bounded by our feet; if it lies outside, we fall over. Furthermore, every voluntary movement we make must be accompanied by a postural adjustment to compensate for the shift in our center of gravity. Therefore the postural muscles maintain a certain level of tone so that they are in a constant state of partial contraction. Thus it is not surprising to find that these muscles contain a high proportion of slow-twitch (type 1) fibers (see Chapter 7 p. 90).

As with most forms of motor activity, assuming a posture and maintaining it depends upon an internally generated central program for action, which is then modified and regulated by peripheral feedback. The 'command' program for action is assembled within the CNS (particularly the basal ganglia, brainstem, and reticular formation), while information from four peripheral sources provides feedback about the action. These sources are:

(1) pressure receptors in the feet;

(2) the vestibular system;

(3) the eyes;

(4) proprioceptors in the neck and spinal column.

The last three provide information concerning the position of the head relative to the environment, while the pressure receptors in the feet relay information about the distribution of weight relative to the center of gravity.

Pressure receptors in the feet

While it is clearly very difficult to carry out a detailed mechanistic study of motor activity in human subjects, it is possible to do so in experimental animals. By using decerebrate preparations (animals in whom the cerebral cortex has been removed), it has been possible to demonstrate a variety of reflex mechanisms that act both to maintain postural stability and to track the center of gravity. One of these is the *positive supporting reaction*. Here, if the animal is held in mid-air and the sole of one of its feet is pressed, there is a reflex extension of the corresponding limb. This extension would act to support the weight of the body. Similarly, if the animal's body is pushed to the left, the changing pressures sensed by the feet result in an extension of the limbs on the left side and a retraction of the right limbs so that an upright posture can be maintained. If the animal were to be pushed to the right, the limbs on the right side would be extended. This is the *postural sway reaction*.

In the *righting reflex*, a decerebrate animal placed on its side will move its limbs and head in an attempt to right itself. In this case, cutaneous sensation of unequal pressures on the two sides of the body initiates the reflex activity. Other reflexes, which contribute to the maintenance of posture, include stepping reactions that can be demonstrated in humans very easily. It is impossible to fall over intentionally by leaning to one side, forwards, or backwards. Once the center of gravity has moved outside a critical point, reflex stepping or hopping will occur to prevent the fall.

Role of the vestibular system in the control of posture

The vestibular apparatus is the sensory organ that detects stimuli concerned with balance. Its anatomy and physiology are described in Chapter 8 (pp. 135–138). Patterns of stimulation of the different hair cells within the maculae of the utricles and saccules inform the nervous system of the position of the head in relation to the pull of gravity. Sudden changes in the orientation of the body in space elicit reflexes that help to maintain balance and posture. These are the vestibular reflexes which fall into three categories: the tonic labyrinthine reflexes, the labyrinthine righting reflexes, and dynamic vestibular reactions.

Tonic reflexes

These are elicited by changes in the spatial orientation of the head and result in specific responses (contraction or relaxation) of the extensor muscles of the limbs. In this way, extensor tone is altered in such a way as to maintain an erect posture.

Labyrinthine righting reflexes

These act to restore the body to the standing posture, for example from the reclining position. The first part of the body to move into position is always the head, in response to information from the vestibular system. As the head then assumes an abnormal position in relation to the rest of the body, further reflexes (notably the neck reflexes—see below) operate and the trunk follows the head into position.

Dynamic vestibular reactions

Because the semicircular canals of the vestibular apparatus are sensitive to angular velocity (see p. 136), they can give advance warning that one is about to fall over and avoiding action can be taken by the movement of the feet, as in the stepping reaction.

Neck reflexes

The neck is very flexible, and for this reason the position of the head is almost independent of that of the body. As described above, the head has its own mechanisms for preserving a constant position in space. Therefore it is important that there is a mechanism for informing the head and body about their relative positions. This is the role of the proprioceptors in the neck region. Movements of the head relative to the body result in a set of reflex movements of the limbs. For example, if the head is moved to the left, the left limbs extend and the right limbs flex to restore the axis of the head to the vertical. These reflexes modulate the basic stretch and tension reflexes described earlier to adjust the length and tone of the appropriate sets of muscles.

The role of visual information in control of posture

The eyes also provide information concerning head position. Visual information is used to make appropriate postural adjustments and responses. The visual system contains certain neurons that respond specifically to an image moving across the retina. When the head moves in relation to its surroundings, the relative position of objects in the visual field will change and this will excite those neurons sensitive to movement. This information is transmitted to the brain and used to supplement information originating in the semicircular canals.

9.9 Goal-directed movements

So far, the discussion has largely been concerned with the kinds of movements which rely heavily on reflexes. Throughout life, however, we perform a host of movements that are voluntary and goal-directed in nature. Regions of the brain which are of particular importance in the initiation and refinement of such movements are the motor regions of the cerebral cortex, the cerebellum, and the basal ganglia.

The motor cortex—its organization and functions

More than 120 years ago it was found that electrical stimulation of particular areas of the cerebral cortex of dogs could elicit movements of the contralateral limbs. Subsequent work by W. Penfield and his colleagues in the 1930s showed that stimulation of the precentral gyrus of the cerebral cortex elicited movements in the contralateral muscles of conscious human subjects. Penfield and his colleagues went on to establish in more detail the exact location of the areas of cortex in which stimulation led to coordinated movements.

Three important motor regions are recognized within the cortex.

1. The primary motor cortex (MI or Brodmann's area 4), which is located in the frontal lobe of the brain anterior to the primary somatosensory area of the parietal lobe (see Chapter 11). It extends into the depths of the central sulcus, over the medial edge of the hemisphere, and a little way anterior to the precentral gyrus.

2. The premotor cortex (Brodmann's area 6), which is situated in front of the primary motor cortex.

3. The secondary motor cortex (also called the supplementary motor cortex, MII, or Brodmann's areas 6 and 8), which lies anterior to the premotor area.

Both the primary and secondary motor areas also possess a sensory projection of the periphery of the body that is sometimes referred to as the motosensory cortex. Figure 9.11 shows the positions of the sensory and motor areas of the cerebral cortex.

Somatotopic organization of the motor cortex

Experimental observations have made it clear that both the primary and secondary motor cortices are somatotopically arranged just like the somatosensory areas. The motor represen-

Summary

1. Tension in the axial muscles is continuously adjusted to maintain and alter posture. The 'command' program for the posture required is assembled within the CNS (particularly the basal ganglia) and modified according to feedback from the vestibular system, the eyes, proprioceptors in the neck and vertebrae, and pressure receptors in the skin.

2. A variety of reflex reactions help to adjust posture and prevent falling. They occur in response to information regarding changes in the balance of pressures experienced by the feet. Such reactions include the positive supporting reaction, righting reflexes, and stepping reactions.

3. The vestibular apparatus provides information regarding the position of the head in relation to the pull of gravity. Vestibular reflexes, elicited by sudden changes in body position, help to maintain posture. They include tonic and dynamic reactions and labyrinthine righting reflexes.

4. Proprioceptors in the neck and vertebral column provide information about the positions of the head and body in relation to one another.

5. Information from the eyes concerning head position appears to supplement information from the semicircular canals of the vestibular system.

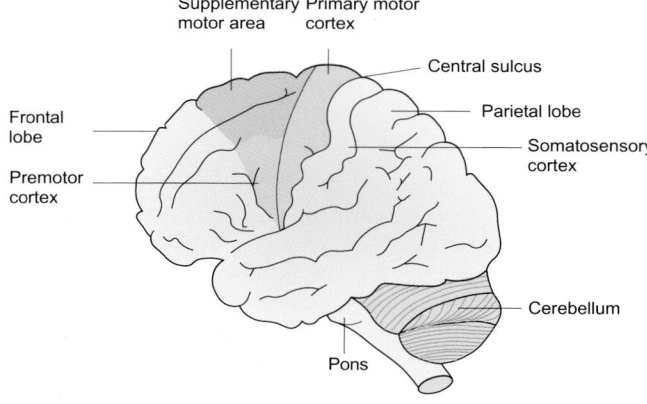

Fig. 9.11 The location of the motor areas of the cerebral cortex.

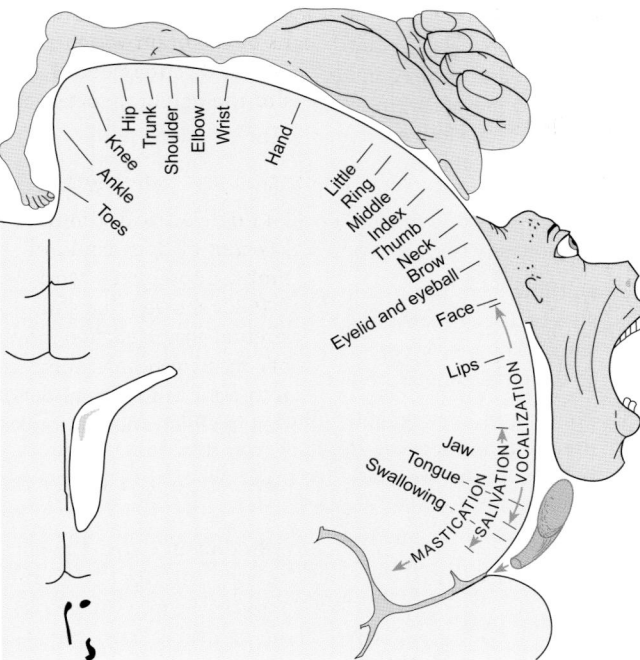

Fig. 9.12 Somatotopic organization of the motor cortex (motor homunculus) showing the relative size of the regions representing various parts of the body.

Connections of the motor cortex

Outflow from the motor cortex

The corticospinal (or pyramidal) tract has been discussed earlier (see p. 152). It is one of the major pathways by which motor signals are transmitted from the motor cortex to the anterior motoneurons of the spinal cord. Nevertheless, the extrapyramidal tracts also carry a significant proportion of the outflow from the motor cortex to the spinal cord. The pyramidal tract itself also contributes to these extrapyramidal pathways via numerous collaterals which leave the tract within the brain (see Fig. 9.10). Indeed, each time a signal is transmitted to the spinal cord to elicit a movement, these other brain areas receive strong signals from the pyramidal tract.

Inputs to the motor cortex

The motor areas of the cerebral cortex are interconnected and receive inputs from a number of other regions. The most prominent sensory input is from the somesthetic system. Information reaches the motor areas of the cerebral cortex either directly from the thalamus or indirectly by way of the somatosensory cortex of the postcentral gyrus. The posterior parietal cortex relays both visual and somatosensory information to the motor areas. The cerebellum and basal ganglia also send afferents to the motor cortex by way of the ventrolateral and ventroanterior thalamic nuclei. Figure 9.13 illustrates these pathways diagrammatically.

The pyramidal cells receive information from a wide range of sensory modalities including joint receptors, tendon organs, cutaneous receptors, and spindle afferents. It is believed that sensorimotor correlation of this information from the skin and muscles is carried out in the motor cortex and is used to enhance grasping, touching, or manipulative movements. The motor cortex is also believed to be involved in the generation of

tation of the body in the primary motor cortex is illustrated in Fig. 9.12 and it is apparent that those areas of the body with especially refined and complex motor abilities, such as the fingers, lips, and tongue, have a disproportionately larger representation in the primary motor cortex than those with poorer motor ability such as the trunk. While stimuli applied to the surface of the primary motor cortex will evoke discrete contralateral movements that involve several muscles, microstimulation within the cortex itself can elicit movement of single muscles. The motor cortex is made up of an overlapping array of motor points that control the activity of particular muscles or muscle groups. The motor homunculus shown in Fig. 9.12 should be compared with the sensory homunculus shown in Fig. 8.14.

The orderly representation of parts of the body in the motor cortex is dramatically illustrated in patients suffering from a form of epilepsy known as Jacksonian epilepsy (after the neurologist Hughlings Jackson). Jacksonian convulsions are characterized by twitching movements that begin at an extremity such as the tip of a finger and show a progressive and systematic 'march'. After the initial twitching is seen, there is clonic movement of the affected finger, followed by movement of the hand and then the arm. The process culminates in a generalized convulsion. This progression of abnormal movements reflects the spread of excitation over the cortex from the point at which the overactivity began (the epileptic focus).

Although the primary motor cortex controls the muscles of the opposite side of the body, the secondary motor cortex controls muscles on both sides and stimulation here may evoke vocalization or complex postural movements.

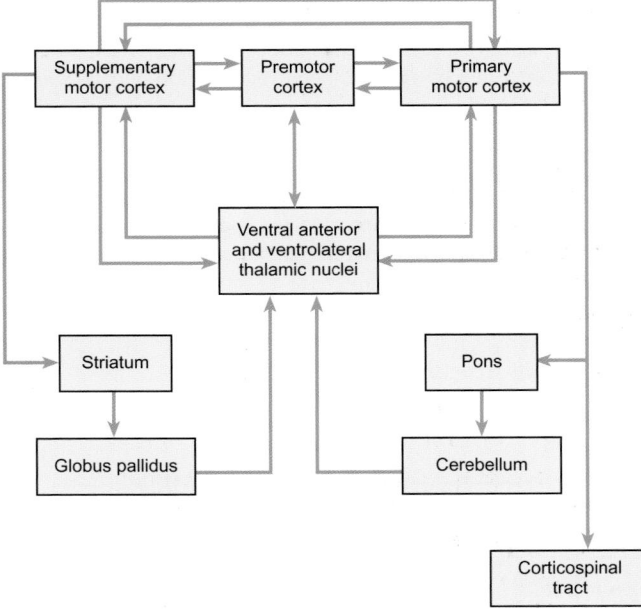

Fig. 9.13 The major connections of the motor areas of the cerebral cortex with subcortical structures.

'force commands' to muscles being used to counter particular loads. These commands adjust the force generated to match that of the load using information from pressure receptors in the skin and the load sensed by the tendon organs. Certainly, the rates of firing of pyramidal cells during voluntary movement correlate well with the force being produced to generate the movement.

Role of the motor areas in movement programming

For a voluntary movement to take place, an appropriate neuronal impulse pattern must first be generated. This pattern links the initial drive to perform the movement to the execution of the movement itself. Although the mechanisms underlying the generation of such patterns are not understood, some information regarding the kinds of neuronal activity that precede a movement has been obtained using conscious human subjects.

If a subject is told to make a voluntary movement such as bending a finger, an electrode placed over the cerebral cortex can record a slowly rising surface negative potential beginning some 800 ms before the start of the movement. This is known as the 'readiness potential'. This potential is particularly associated with the premotor area. As the movement gets under way, the electrode records a series of more rapid potentials which are particularly pronounced over the contralateral motor areas and may reflect their involvement in the execution of the movement.

Blood flow to the supplementary motor cortex increases both before and during complex voluntary movements. This further suggests a 'central command' role for this region in the planning or programming of actions as well as their execution. Motor activity is partly mediated by direct corticospinal connections. Furthermore, individual corticospinal neurons have also been shown to fire both before and during a movement (Fig. 9.14).

Effects of lesions of the motor cortex

The motor areas of the cortex are often damaged as a result of strokes (caused by a loss of blood supply to the cortex) and the muscles controlled by the damaged region show a corresponding loss of function (see Box 9.1). Rather than a total loss of movement, however, there is a loss of voluntary control of fine movements, with clumsiness and slowness of movement and an unwillingness to use the affected muscles.

Lesions of parts of the motor cortex (particularly the supplementary and premotor areas) impair the ability to prepare for voluntary movements, giving rise to a condition similar to apraxia (see Chapter 11) in which affected patients are unable to perform complex movements despite retaining both sensation and the ability to perform simple movements.

Fig. 9.14 The response pattern of a pyramidal tract neuron before and during the voluntary flexion of the wrist. The experiment was performed on a trained monkey and shows that the pyramidal tract neuron fibers discharge before the onset of the wrist flexion.

Summary

1. The cerebral cortex contains motor areas (primary, secondary, and premotor) in which the stimulation of cells will elicit contralateral movements. The primary and secondary areas are somatotopically arranged. Those areas of the body that are capable of especially refined and complex movements (i.e. fingers, lips, and tongue) have a disproportionately large area of representation.

2. Outflow from the motor cortex to the spinal cord is carried by the corticospinal (pyramidal) and extrapyramidal tracts. The motor cortex also sends numerous collaterals to the basal ganglia, cerebellum, and brainstem.

3. The motor areas receive inputs from many sources. The predominant sensory input is from the somatosensory system, which receives its input from the thalamus. Afferent information is also received from the visual system, cerebellum, and basal ganglia. This afferent information is used to refine movements, particularly to match the force generated in specific muscle groups to an imposed load.

4. In order that a voluntary movement can take place, an appropriate neural pattern must be generated. The secondary motor cortex is thought to play a role in this 'programming' of movements.

5. Motor areas of the cerebral cortex are often damaged by strokes, and muscles controlled by the damaged areas show a corresponding loss of function. However, recovery is often good, resulting in little more than clumsiness and a loss of fine muscle control.

6. Lesions to the secondary and premotor areas may give rise to apraxia, a loss of the ability to prepare for voluntary movement. The ability to execute simple movements is retained.

Many strokes cause widespread damage not only to the primary motor cortex but also to sensorimotor areas both anteriorly and posteriorly. These areas relay inhibitory signals to the spinal motoneurons via the extrapyramidal pathways (see Section 9.7). When the sensorimotor areas are damaged, there is a release of inhibition that can result in the affected contralateral muscles going into spasm. Spasm is particularly intense if the basal ganglia are also damaged since strong inhibitory signals are transmitted via this region (see below).

9.10 The role of the cerebellum in motor control

Electrical stimulation of the cerebellum causes neither sensation nor significant movement. However, loss of this area of the brain is associated with severe abnormalities of motor function. The cerebellum appears to play a particularly vital role in the coordination of postural mechanisms and in the control of rapid muscular activities such as running, playing a musical instrument, or typing.

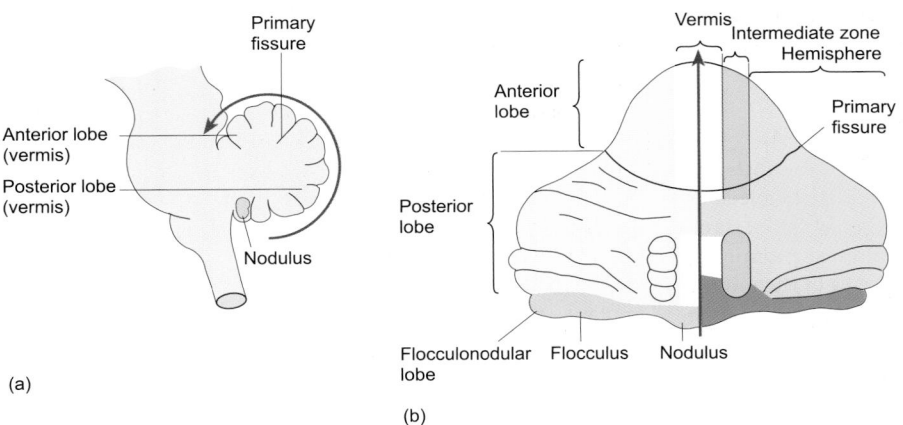

Fig. 9.15 The major lobes of the cerebellum. (a) A sectional view of the brainstem and cerebellum. (b) The surface of the cerebellum 'unfolded' along the axis of the arrow indicated in (a). The right-hand half shows the areas that receive projections from the motor cortex (cerebrocerebellum yellow), the vestibular system ('vestibulocerebellum' dark green), and the spinal cord ('spinocerebellum' white). In the intermediate zone (pale green) afferents from the motor cortex and spinal cord overlap.

The anatomical structures of the cerebellum

The cerebellum (literally little brain) is located dorsal to the pons and medulla and lies under the occipital lobes of the cerebral hemispheres (see Chapter 6, Fig. 6.1). The cerebellum is usually subdivided into two major lobes, the anterior and posterior lobes, separated by the primary fissure as shown in Fig. 9.15. The lobes are further subdivided into nine transversely orientated lobules, each of which is folded extensively. The folds are known as folia. The flocculonodular lobe is a small propeller-shaped structure, caudally situated with respect to the major lobes. As it is tucked under the main body of the cerebellum, it cannot be seen in a surface view. Along the midline is the vermis from which the cerebellar hemispheres extend laterally.

Like the cerebral cortex, the cerebellum is composed of a thin outer layer of gray matter, the cerebellar cortex, overlying internal white matter. Embedded within the white matter are the paired cerebellar nuclei. These are the dentate (which are the most lateral), globose, emboliform, and fastigial (the most medial) nuclei. They receive afferents from the cerebellar cortex

as well as sensory information via the spinocerebellar tracts. The cerebellar nuclei give rise to all the efferent tracts from the anterior and posterior lobes of the cerebellum. The flocculus and nodulus send efferent fibers to the lateral vestibular nucleus (Deiter's nucleus).

The cerebellum is attached to the brainstem by nerve fibers that run in three pairs of cerebellar peduncles: the inferior, middle, and superior peduncles. The *inferior peduncles* connect the cerebellum to the medulla, with afferents conveying sensory information to the cerebellum from muscle proprioceptors throughout the body and from the vestibular nuclei of the brainstem, which are concerned with equilibrium and balance. The cerebellum receives information from the pons via the *middle cerebellar peduncles* advising it of voluntary motor activities initiated by the motor cortex. The *superior cerebellar peduncles* connect the cerebellum and the midbrain. Fibers in these peduncles originate from neurons in the deep cerebellar nuclei and communicate with the motor cortex via the thalamus. (The cerebellum has no direct connections with the cerebral cortex.) In contrast

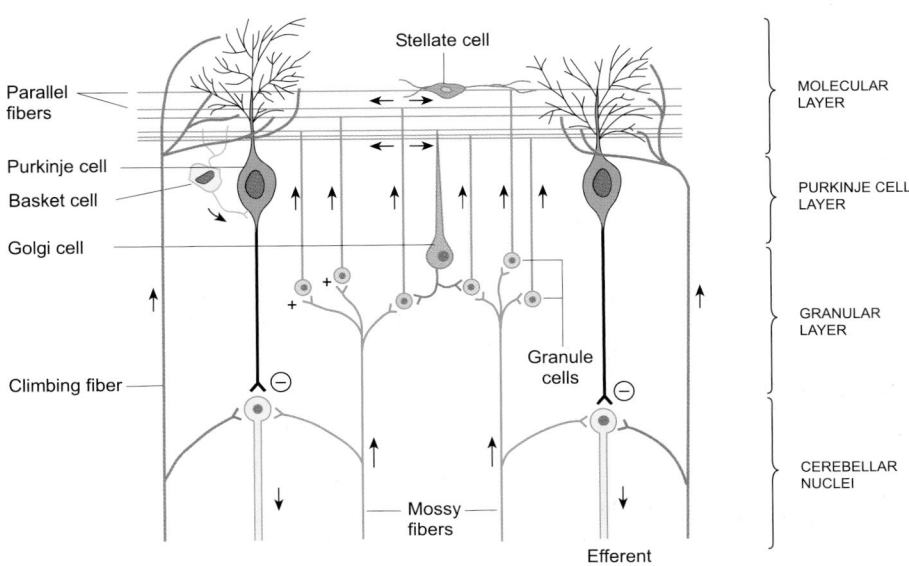

Fig. 9.16 The fundamental neural organization of the cerebellar cortex. The basket, Golgi, and stellate cells are inhibitory interneurons.

with the cerebral cortex, the two halves of the cerebellum each control and receive input from muscles on the ipsilateral side of the body.

The cellular organization of the cerebellar cortex

The cerebellar cortex is largely uniform in structure, consisting of three layers of cells: the granular, Purkinje cell, and molecular layers. The functional organization of the cerebellar cortex is relatively simple and is shown in Fig. 9.16. The most prominent cells of the cerebellar cortex are the Purkinje cells of the middle layer. A human brain contains about 15 million of these large neurons, each of which possesses an extensively branched dendritic tree. The axons of the Purkinje cells form the only output from the cerebellar cortex, passing to the deep cerebellar nuclei and, to a lesser extent, to the vestibular nuclei. The Purkinje cell output is entirely inhibitory, with the neurotransmitter being GABA.

Inputs to the cerebellar cortex are of two types, climbing fibers and mossy fibers. *Climbing fibers* originate in the inferior olive of the medulla and form part of the olivocerebellar tract, which crosses in the midline and enters the cerebellum via the contralateral inferior cerebellar peduncle. *Mossy fibers* form the greater proportion of the afferent input to the cerebellar cortex, originating in all the cerebellar afferent tracts apart from the inferior olive. Mossy fibers send collaterals to deep nuclear cells before ascending to the granular layer of the cortex where they branch to form large terminal structures making synaptic contact with the dendrites of many granule cells. These in turn give rise to the parallel fibers described earlier, which synapse with the dendrites of Purkinje cells.

The climbing fibers excite the Purkinje cells while the mossy fibers excite the granule cells which, in turn, make excitatory contact with the Purkinje cells. The actions of all the other cell types within the cerebellar cortex are inhibitory. Basket cells and stellate cells make inhibitory contact with the Purkinje cells while the Golgi cells inhibit the granule cells and so reduce the excitation of the Purkinje cells. The Purkinje cells exert a tonic inhibition on the activity of the neurons of the cerebellar nuclei. Since the axons of the Purkinje cells form the output of the cerebellar cortex, all excitatory inputs to the cerebellum will be converted to inhibition after two synapses at most. This has the effect of removing the excitatory influence of the cerebellar input and ensures that the neurons of the deep cerebellar nuclei are ready to process a new input. It is believed that this 'erasing' of inputs is important to the participation of the cerebellum in rapid movements.

Afferent connections of the cerebellum

With the exception of the olivocerebellar tract, all the cerebellar afferent tracts terminate in mossy fibers. Important tracts that give rise to mossy fibers include the corticocerebellar, vestibulocerebellar, reticulocerebellar, and spinocerebellar pathways. The *corticocerebellar tract* originates mainly in the motor cortex (but also in other cortical areas) and passes via the pontine nuclei to the cerebellum in the pontocerebellar tract. The *vestibulo-* and *reticulocerebellar tracts* originate in the brainstem, in the vestibular and reticular systems, respectively. Vestibulocerebellar fibers terminate mainly in the flocculonodular lobe, reticulocerebellar

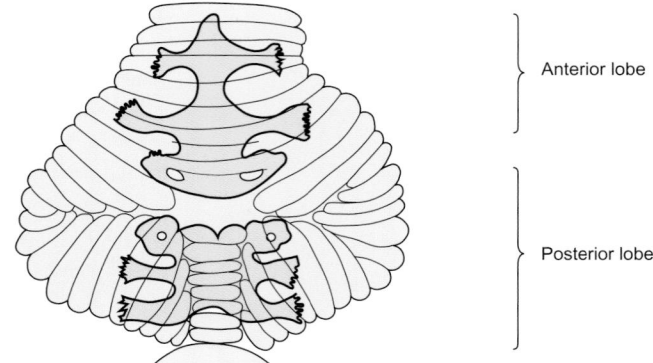

Fig. 9.17 Somatotopic maps in the anterior and posterior lobes of the cerebellum.

fibers terminate mainly within the vermis, and corticocerebellar fibers terminate mainly in the cerebellar hemispheres (see Fig. 9.15).

Sensory signals from the muscle spindles, tendon organs, and touch receptors of the skin and joints are received by the cerebellum directly by way of the ventral and dorsal spinocerebellar tracts. These signals inform the cerebellum of the moment-by-moment status of the muscles and joints. Input reaching the cerebellum via the spinocerebellar pathways is somatotopically organized. There are two maps of the body in the spinocerebellum, one in the anterior and one in the posterior lobe. These are shown in Fig. 9.17. In each map, the head is facing the primary fissure so that the two maps are inverted with respect to each other. Signals from the motor cortex, which is also arranged somatotopically, project to corresponding points in the sensory maps of the cerebellum.

The climbing fiber input to the cerebellar Purkinje cells originates in the inferior olive. Neurons in this region of the brain receive their input from a number of sources, including the motor cortex, the pretectal region, and the spino-olivary tracts (which relay information from cutaneous receptors as well as muscle and joint proprioceptors). Olivocerebellar fibers send collaterals to the deep cerebellar nuclei before ascending as climbing fibers to the cerebellar cortex. Like the mossy fiber input, the olivocerebellar projections are topographically organized.

Outputs from the cerebellum

The output neurons of the cerebellum lie in the deep cerebellar nuclei. The cells of the cerebellar nuclei receive inhibitory synapses from the Purkinje cells of the cerebellar cortex and excitatory synapses from collaterals of the mossy and climbing fibers. Therefore the final output from the cerebellum is the net result of the interaction between these excitatory and inhibitory inputs to the nuclei.

Crude topographical maps of the peripheral musculature can be demonstrated within the cerebellar nuclei. Neurons of the dentate nucleus project contralaterally through the superior cerebellar peduncle to neurons in the contralateral thalamus. From the thalamus, neurons project to the premotor and primary motor cortex where they influence the planning and initiation of voluntary movements. The neurons of the emboli-

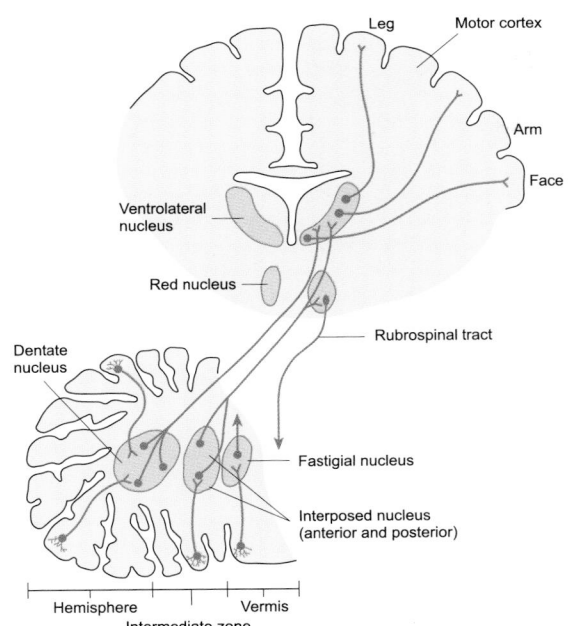

Fig. 9.18 Ascending connections from the cerebellum. The Purkinje neurons of the cerebellar cortex project to the deep cerebellar nuclei. The neurons of the dentate, emboliform, and globose nuclei project to the cerebral cortex via the ventrolateral nucleus of the thalamus and give off collaterals to the red nucleus. The neurons of the fastigial nucleus project to the lateral vestibular nuclei.

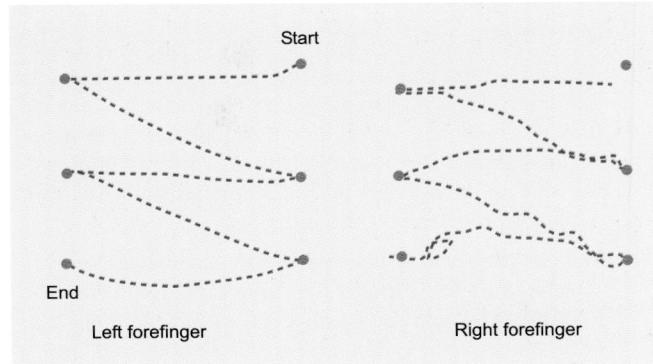

Fig. 9.19 Records illustrating cerebellar ataxia. In this task, a flashing light was fixed to each forefinger in turn and the subject asked to trace a path as accurately as possible between two sets of three lights spaced 75 cm apart (shown here as red circles). The movement was captured on photographic film. The patient, who had a right-sided cerebellar lesion, had great difficulty tracing the path with his right hand but no difficulty with his left.

form and globose nuclei project chiefly to the contralateral red nucleus, which is also topographically organized. There is also a small projection from these nuclei to the motor cortex. The red nucleus gives rise to the rubrospinal tract which is particularly important in the control of proximal limb muscles. The neurons of the fastigial nucleus project to the vestibular nuclei and to the pontine and medullary reticular formation which, in turn, give rise to the vestibulospinal and reticulospinal tracts respectively. A diagram of the major output pathways of the cerebellar nuclei is shown in Fig. 9.18.

How does the cerebellum contribute to the control of voluntary movement?

Although the neuronal architecture of the cerebellum is relatively well understood, the functions of the cerebellum are less clear cut. Much of our information concerning its role in the control of motor activity has been gathered from the effects of lesions and other damage to the cerebellum and from experimental stimulation of, and recording from, cerebellar neurons. On the basis of such work, it is now believed that the primary role of the cerebellum is to supplement and correlate the activities of the other motor areas. More specifically, it seems to play a role in the control of posture and the correction of rapid movements initiated by the cerebral cortex. The cerebellum may also contribute to certain forms of motor learning, as the frequency of nerve impulses in the climbing fibers almost doubles when a monkey is learning a new task.

Damage to the posterior cerebellum results in an impairment of postural coordination which is similar to that seen when the vestibular apparatus itself is damaged. For example, a patient

with a lesion in the area will feel dizzy, have difficulty standing upright, and may develop a staggering gait (*cerebellar ataxia*). Damage to other regions of the cerebellum produces more generalized impairment of muscle control. Ataxia is again a problem, coupled with a loss of muscle tone (*hypotonia*) and a lack of coordination (*asynergia*). Figure 9.19 shows an example of impaired motor control from a patient with a right-sided cerebellar ataxia. The patient was asked to trace a path between a set of lights with the forefinger of each hand. The left forefinger, which was unaffected, completes the task with ease, but the right forefinger traces the path in a very uncertain manner, with looping and obvious tremor.

The cerebellum exerts its control over motor activity on a moment-by-moment basis, using sensory information (especially from proprioceptors) to monitor body position, muscle tension, and muscle length. Inputs from the motor cortex to the cerebellum inform the cerebellum of an intended movement before it is initiated. Sensory information regarding the progress of the movement will then be received continually by the cerebellum via the spinocerebellar tract. This information is processed by the cerebellum to generate an error signal which is fed back to the cortex so that the movement can be adjusted to meet the precise circumstances of the situation. The cerebellum itself has no direct connection with the spinal motoneurons. It exerts its effect on motor performance indirectly.

Much of the clumsiness of movement seen in patients with cerebellar lesions seems to be the result of a slowness to respond to sensory information regarding the progress of the movement. There is often an intention tremor preceding goal-directed movements, in which the movement appears to oscillate around the desired position as in Fig. 9.19, or the movement may over-shoot when the patient reaches for an object. Speech may also become staccato ('scanning speech'), more labored, and less 'automatic'. In general, patients with cerebellar damage seem to require a great degree of conscious control over movements which normally require little thought. This may be because the cerebellum plays a role in the learning of complex motor skills.

Summary

1. The cerebellum is located dorsal to the pons and medulla and protrudes from under the occipital lobes. It is divided into an anterior, a posterior, and a smaller flocculonodular lobe. It consists of an outer cortical layer of gray matter and an inner layer of white matter in which are embedded the deep cerebellar nuclei.

2. The internal neuronal circuitry of the cerebellar cortex is largely uniform, comprising three layers of cells: the Purkinje cell layer, the granular layer, and the molecular layer. Inputs to the cerebellum are via climbing and mossy fibers. The climbing fibers originate in the inferior olive and make excitatory contact with the Purkinje cell dendrites. Mossy fibers form the bulk of the input, originating in the cortico-, vestibulo-, reticulo-, and spinocerebellar tracts. These excite the Purkinje cells indirectly by way of the parallel fibers of the granule cells.

3. The cerebellum is attached to the brainstem by nerves running in three pairs of cerebellar peduncles. Afferents from the vestibular nuclei, muscle proprioceptors, and pons run in these tracts and constantly inform the cerebellum about the progress of movements. It has no direct connections with the motor cortex but receives information about intended movements by way of the corticocerebellar tract.

4. Efferent tracts from the cerebellum originate in the deep nuclei and pass to the thalamic nuclei (and thence to the motor cortex), the red nucleus, the vestibular nuclei, pons, and medulla.

5. The cerebellum plays a vital role in the coordination of postural mechanisms and the control of rapid muscular activities, supplementing and correlating the activities of other motor areas. It may also contribute to motor learning. The two halves of the cerebellum control and receive inputs from ipsilateral muscles.

6. Lesions to the cerebellum may result in dizziness and postural difficulties or in a more generalized loss of muscle control involving intention tremor, clumsiness, and speech problems.

9.11 The basal ganglia

The basal ganglia are the deep nuclei of the cerebral hemispheres. They consist of the *caudate nucleus* and *putamen* (known collectively as the *corpus striatum*), the *globus pallidus*, and the *claustrum*. Functionally, the basal ganglia are associated with the *subthalamic nuclei* and the *substantia nigra* (so called because many of its cells are pigmented with melanin) of the midbrain and the *red nucleus*.

The location of the basal ganglia is illustrated in Fig. 9.20. The caudate nucleus and the putamen are separated by the anterior limb of the internal capsule. The basal ganglia form an important subcortical link between the frontal lobes and the motor cortex. Their importance in the control of motor function is

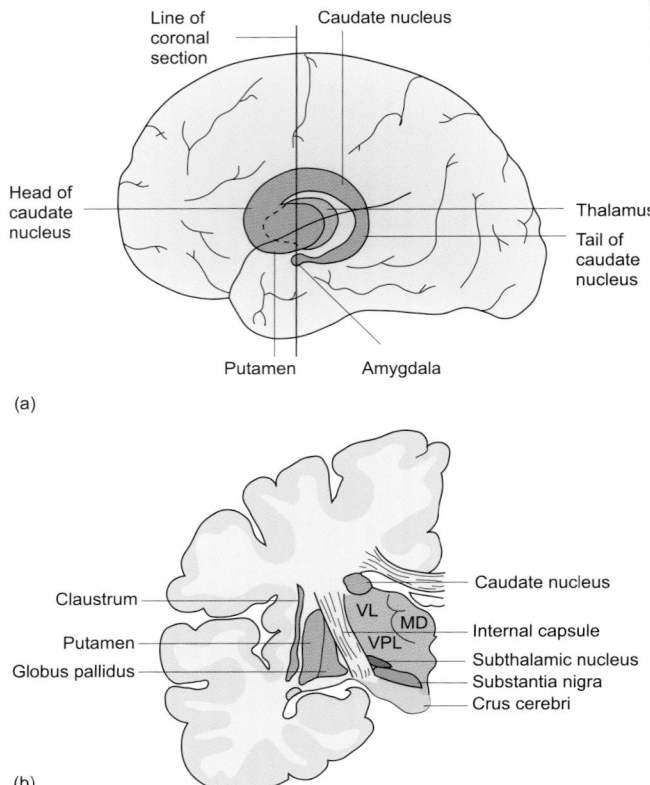

Fig. 9.20 The location of the basal ganglia and associated nuclei. (a) The position of the basal ganglia within the cerebral hemispheres. (b) A coronal section showing the detailed location of the various nuclei of the basal ganglia. MD, VL, and VPL are the mediodorsal, ventrolateral, and posterolateral thalamic nuclei.

clear from the severe disturbances of movement seen in patients with lesions of the basal ganglia. However, these deep structures are relatively inaccessible to experimental study and a detailed understanding of their actions is yet to emerge.

Afferent and efferent connections of the basal ganglia

The circuitry of the basal ganglia is very complex and not fully understood. There are numerous connections between the various structures but their significance is unknown. However, a large feedback loop seems to exist between the basal ganglia and the cerebral cortex. All parts of the cerebral cortex project to the corpus striatum of the basal ganglia (Fig. 9.21). This input is topographically organized and excitatory, and probably uses glutamate as its transmitter. From the putamen and caudate nucleus neurons project to the globus pallidus (a predominantly inhibitory projection), then to the thalamus, and from there back to the cerebral cortex (mainly to the motor area) to complete the loop. Such an arrangement suggests that the basal ganglia, like the cerebellum, are involved in the processing of sensory information, which is then used to regulate motor function.

The substantia nigra receives an input from the putamen and projects to both the putamen and the caudate nucleus (the nigrostriatal pathway). The pigmented neurons in the substantia nigra synthesize and store the neurotransmitter dopamine,

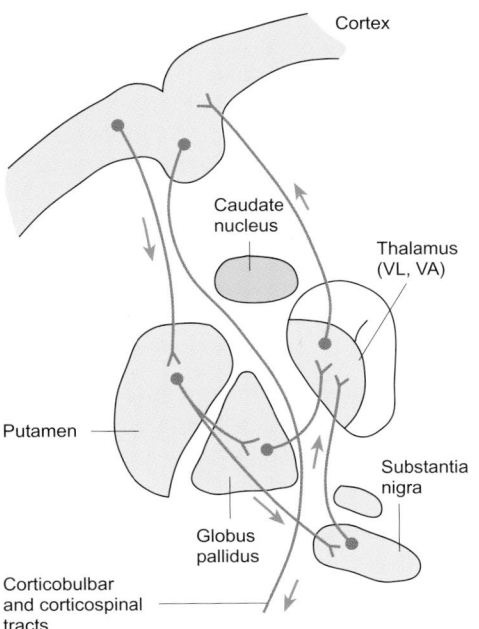

Fig. 9.21 The principal neural connections of the basal ganglia with other brain areas. VL and VA are the ventrolateral and ventro-anterior thalamic nuclei.

Summary

1. The basal ganglia are deep cerebral nuclei of which the corpus striatum and globus pallidus are involved in motor control in association with the substantia nigra of the midbrain and the red nucleus. All parts of the cerebral cortex project to the corpus striatum. The basal ganglia in turn project to the thalamus which, in turn, projects to the motor cortex, thus completing a loop. This loop is influenced by inhibitory connections between the substantia nigra and the corpus striatum.

2. A range of movement disorders (dyskinesias) result from damage to the basal ganglia or its connections. Parkinson's disease, in which movement is impoverished, is the most familiar. This disorder is due to degeneration of the dopaminergic neurons of the substantia nigra. Other disorders of the basal ganglia (e.g. the inherited disease Huntingdon's chorea) are characterized by the inappropriate or repetitive execution of movement patterns that are themselves often normal.

3. The basal ganglia are believed to generate basic patterns of movement, possibly representing motor 'programs' formed in response to cues from cortical association areas.

which is subsequently transported via the nigrostriatal fibers to receptors in the corpus striatum. It is believed that this pathway is inhibitory to the striatal neurons that in turn project to the motor cortex via the thalamus as shown in Fig. 9.21.

There are many other neuronal connections between the basal ganglia and other regions of the brain, such as the limbic system (which is concerned with motivation and emotion), but their role is unknown.

The role of the basal ganglia in the control of movement

Most of the information that has accumulated concerning the actions of the basal ganglia in humans has been derived from studies of the effects of damage to those brain areas. A range of movement disorders (*dyskinesias*) result from damage to the basal ganglia or their connections. The best known disease of the basal ganglia is Parkinson's disease (shaking palsy), a degenerative disorder that results in variable combinations of slowness of movement (*bradykinesia*), increased muscle tone (rigidity), resting (or 'pill-rolling') tremor, and impaired postural responses. Patients suffering from Parkinson's disease have difficulty both starting and finishing a movement. They also tend to have a mask-like facial expression. Parkinsonism appears to result primarily from a defect in the nigrostriatal pathway following degeneration of the dopaminergic neurons of the substantia nigra. Considerable success in relieving the symptoms of these patients has been achieved by the administration of carefully controlled doses of L-dopa, a precursor of the neurotransmitter dopamine.

In contrast with Parkinson's disease, in which movements are restricted, other disorders of the basal ganglia result in the spontaneous production of unwanted movements. These spontaneous movements include *hemiballismus* (a sudden flinging

out of limbs on one side of the body as if to prevent a fall or to grab something), *athetosis* (snake-like writhing movements of parts of the body, particularly the hands, which are often seen in cerebral palsy), and *chorea* (a series of rapid uncontrolled movements of muscles all over the body). In all cases, the disease process results in inappropriate or repetitive execution of normal patterns of movement. The hereditary disorder Huntingdon's chorea is perhaps the best known of this group of diseases.

While observations of the various movement disorders associated with damage to the basal ganglia cannot give direct information about their role in the control of normal movements, some tentative assumptions have been made. The basal ganglia seem to be involved in very basic innate patterns of movement, possibly elaborating relatively crude movement 'programs' in response to cues from the cortical association areas. These programs would then provide the basic structure of a movement which could subsequently be refined and updated in the light of sensory input from the periphery, with this latter function being carried out by the cerebellum. Furthermore, the basal ganglia may contain the blueprints for the particular sets of muscle contractions needed to effect or adjust certain postures—for example, the throwing out of an arm to prevent a fall.

9.12 Concluding remarks

While certain aspects of the mechanisms by which the nervous system controls movement are reasonably well understood, our understanding of the ways in which motor areas plan and initiate movements is far from complete. Nevertheless, a possible sequence of events might be as follows. Voluntary movements are planned in areas of the brain outside the primary motor

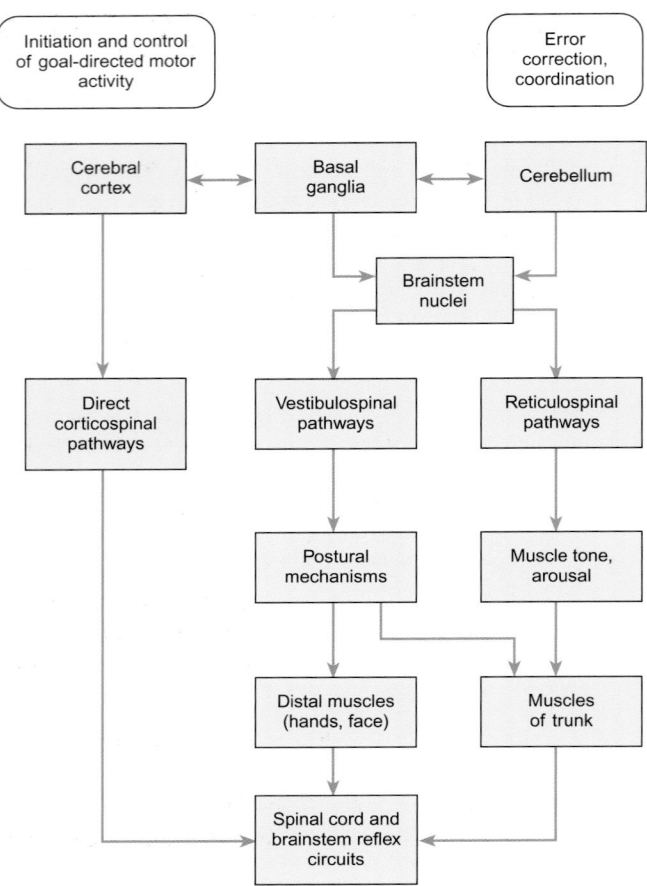

Fig. 9.22 A summary of possible interactions between the major motor areas of the brain.

movement, information from both cortical and peripheral regions constantly informs the cerebellum and other subcortical motor areas of the progress of the movement so that it can be modified appropriately to take account of changing circumstances. Figure 9.22 summarizes current ideas concerning the interactions between the various motor areas of the brain.

Recommended reading

Neuroanatomy of the motor system

Brodal, P. (2003). *The central nervous system. Structure and function* (3rd edn), Chapters 11–14. Oxford University Press, Oxford.

Pharmacology of neurodegenerative disorders

Rang, H.P., Dale, M.M., Ritter, J.M., and Moore, P. (2003). *Pharmacology* (5th edn), Chapter 31. Churchill-Livingstone, Edinburgh.

Physiology

Carpenter, R.H.S. (2002). *Neurophysiology* (4th edn), Chapters 9–12. Hodder Arnold, London.

Cody, F.W.J. (ed.) (1995). *Neural control of skilled human movement.* Portland Press, London.

Kandel, E.R., Schwartz, J.H., and Jessell, T.M. (1991). *Principles of neural science*, Chapters 35, 37–42. McGraw-Hill, New York.

Nicholls, J.G., Martin, A.R., Wallace, B.G., and Fuchs, P.A. (2001). *From neuron to brain* (4th edn), Chapter 22. Sinauer Associates, Sunderland, MA.

Shepherd, G.M. (1994) *Neurobiology* (3rd edn), Chapters 17, 20–22. Oxford University Press, Oxford.

Zigmond, M.J., Bloom, F.E., Landis, S.C., Roberts, J.L., and Squire, L.R. (1999). *Fundamental neuroscience*, Chapters 29–35. Academic Press, San Diego.

Clinical physiology

Campbell, E.J.M., Dickinson, C.J., Slater, J.D.H., Edwards, C.R.W., and Sikora, E.K. (eds.) (1984). *Clinical physiology* (5th edn), Chapter 11. Blackwell Scientific, Oxford.

Curtis, R.L., and McDonald, S.E. (1994). Alterations in motor function in *Pathophysiology—concepts of altered health states* (ed C.M. Porth), Chapter 50. J.B. Lippincott, Philadelphia, PA.

Medicine

Donaghy, M. (2005). *Neurology* (2nd edn), Chapters 5–9, 25–26. Oxford University Press, Oxford.

cortex. This motor planning is likely to involve the processing of sensory information and its integration with memory. An appropriate motor program will then be selected and preliminary commands relayed to the basal ganglia and cerebellum. These areas of the brain are thought to organize and refine the crude motor program and relay it back to the primary motor cortex which, in turn, activates the motor neurons (both α and γ) in order to bring about the desired sequence of muscle contractions. During the execution of the

Multiple choice questions

The following statements are either true or false. Answers are given below.

1. a. White matter is made up largely of nerve cell bodies.
    b. The anterior horns of the spinal cord gray matter contain somatic motor neurons.
    c. Ascending pathways in the spinal cord convey sensory information.
    d. The ventral and dorsal roots of the spinal cord unite to form the spinal nerves.
    e. Destruction of the anterior horn cells of the spinal cord results in a loss of sensory input from the area served.

2. a. Muscle spindles play an important role in the regulation of posture and movement.
    b. Muscle spindles measure primarily the tension of a muscle while tendon organs primarily measure the length.
    c. Muscle spindles receive both afferent and efferent nerve fibers.
    d. The dynamic sensitivity of secondary spindle endings is greater than that of primary spindle endings.
    e. The γ-efferent fibers of muscle spindles regulate the sensitivity of the spindles during a muscle contraction.

3. a. A monosynaptic reflex arc involves one or more interneurons.
    b. Withdrawal reflexes are lost following cervical section of the spinal cord.
    c. Spinal shock is characterized by flaccid paralysis of all muscles below the level of the spinal injury.
    d. The tendon tap reflex is an example of a stretch reflex.
    e. Motoneurons form the final common path for all reflexes.

4. a. The axons of the corticospinal tract mainly synapse directly with motoneurons in the spinal cord.
    b. Damage to the pyramidal tract, caused for example by a stroke, results in a loss of fine control of the hand and finger muscles.
    c. The rubrospinal tract exerts its major influence on the muscles of the extremities.
    d. The extensor muscles of the legs are under the control of the vestibulospinal tract.
    e. The reticulospinal tracts are largely decussated (crossed).

5. a. Muscle spindles and joint receptors in the neck signal misalignment of the head on the body.
    b. The muscles attached to the axial skeleton are of principal importance in the maintenance of an upright posture.
    c. Normal vision is required for the righting reflex.
    d. The cerebellum is the brain area most concerned with equilibrium and body posture.

6. a. The motor areas of the cortex are situated in the frontal lobe.
    b. Following a cerebral hemorrhage affecting the precentral gyrus of the right hemisphere, the patient feels no sensation on the left side of his body.
    c. The corticospinal tract provides the only connection between the motor cortical areas and the spinal cord.
    d. Motor areas of the cortex receive somatosensory input via the thalamus.

7. a. The cerebellum has a direct efferent projection to the motor cortex.
    b. Purkinje cells lie in the deep cerebellar nuclei.
    c. The synaptic connections of Purkinje cell axons are all inhibitory.
    d. Intention tremor is a characteristic sign of damage to the cerebellum.
    e. Hemiballismus is a sign of cerebellar damage.
    f. The cerebellar hemispheres control and receive inputs from ipsilateral muscles.

8. a. Disorders of the basal ganglia produce a marked loss of sensation.
    b. Acetylcholine is the predominant neurotransmitter of the substantia nigra.
    c. The globus pallidus projects to the cerebral cortex via the thalamus.
    d. Parkinsonism is caused by neuronal degeneration within the substantia nigra.

Answers

1. White matter is made up of nerve axons and oligodendrocytes. The anterior horn cells are motor neurons, the loss of which will lead to motor, not sensory, impairment.
   a. False
   b. True
   c. True
   d. True
   e. False

2. Muscle proprioceptors play an important role in both reflex and goal-directed movements. Muscle spindles monitor the length of muscle fibers while the tendon organs monitor tension. Spindle afferents relay information regarding stretch to higher centers. Primary spindle afferents have a greater dynamic sensitivity than secondary afferents. The efferent (fusimotor) fibers stimulate contraction of the intrafusal fibers, thereby altering their sensitivity to stretch throughout the duration of a muscle contraction.
   a. True
   b. False
   c. True
   d. False
   e. True

3. The sensory neuron of a monosynaptic reflex arc makes direct contact with the motor neuron. The withdrawal reflex is organized within the spinal cord. Flaccid paralysis is a characteristic of spinal shock, though recovery is often significant.
   a. False
   b. False
   c. True
   d. True
   e. True

4. Most corticospinal neurons make contact with spinal interneurons. The extensor muscles of the legs are under the control of the vestibulospinal tract, the output of which is determined by information arriving from the vestibular apparatus. These muscles are crucial to the maintenance of an upright posture. The neurons of the reticulospinal tracts are largely uncrossed.
   a. False
   b. True
   c. True
   d. True
   e. False

5. Normal vision is not required for the righting reflex. Tonic neck and labyrinthine reflexes are sufficient.
   a. True
   b. True
   c. False
   d. True

6. The precentral gyrus of the right cerebral hemisphere contains the motor areas that control the muscles of the left (contralateral) side of the body. Therefore the patient would experience a loss of movement on the left side. In addition to the corticospinal tract, descending extrapyramidal pathways connect the motor cortex with the spinal cord.
   a. True
   b. False
   c. False
   d. True

7. There is no direct efferent projection from the cerebellum to the motor cortex although a number of indirect pathways exist. Purkinje cells lie in the cerebellar cortex. Hemiballismus is a sign of damage to the basal ganglia.
   a. False
   b. False
   c. True
   d. True
   e. False
   f. True

8. The basal ganglia are involved in motor behavior, particularly via their influence on the motor cortex. The principal neurotransmitter of the substantia nigra is dopamine.
   a. False
   b. False
   c. True
   d. True

CHAPTER 10

The autonomic nervous system

Chapter contents

After reading this chapter you should understand:

♦ The anatomical organization of the autonomic nervous system and its separation into the sympathetic and parasympathetic divisions

♦ How the sympathetic division regulates the activity of the cardiovascular system, visceral organs and secretory glands

♦ How the parasympathetic nerves regulate the activity of the gut, heart and secretory glands

♦ The role of nicotinic and muscarinic receptors in both divisions of the autonomic nervous system

♦ The role of α- and β-adrenoceptors in the sympathetic nervous system

♦ The regulation of the autonomic nervous system by the CNS

10.1 Introduction

The autonomic nervous system regulates the operation of the internal organs to support the activity of the body as a whole. It is not a separate nervous system but is the efferent (motor) pathway that links those areas within the brain concerned with the regulation of the internal environment to specific effectors such as blood vessels, the heart, the gut, and so on.

Unlike the efferent fibers of the skeletal muscles, those of the autonomic nervous system do not pass directly to the effector organs; rather they pass to *autonomic ganglia* which are located outside the CNS. Furthermore, the autonomic nervous system is not under voluntary control. The fibers that project from the CNS to the autonomic ganglia are called *preganglionic fibers* and those that connect the ganglia to their target organs are called *postganglionic fibers*.

10.2 Organization of the autonomic nervous system

The autonomic nervous system is divided into two parts:

(1) the *sympathetic division*, which broadly acts to prepare the body for activity;

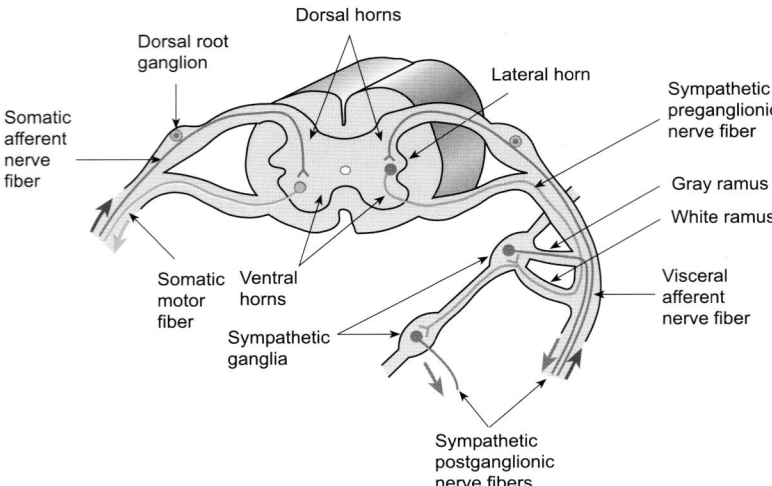

Fig. 10.1 Comparison of the arrangement of the sympathetic preganglionic and postganglionic neurons with the organization of the somatic motor nerves. Note that sympathetic preganglionic fibers may terminate in the sympathetic ganglion of the same segment or pass to another ganglion in the sympathetic chain. Some also terminate in prevertebral ganglia such as the celiac ganglion.

(2) the *parasympathetic division*, which is more discrete in its actions and tends to promote restorative functions such as digestion and a slowing of the heart rate.

The sympathetic nervous system

The sympathetic division originates in the cells of the intermediolateral column of the thoracic and lumbar regions of the spinal cord between segments T1 and L2 or L3. These neurons are called *sympathetic preganglionic neurons*. The axons of these neurons (sympathetic preganglionic fibers) pass from the spinal cord via the ventral root together with the somatic motor fibers. Shortly after the dorsal and ventral roots fuse, the sympathetic preganglionic fibers leave the spinal nerve trunk to travel to sympathetic ganglia via the *white rami communicantes* as shown in Fig. 10.1. The preganglionic fibers synapse on sympathetic neurons within the ganglia and the *postganglionic sympathetic fibers* project to their target organs mainly via the *gray rami communicantes* and the segmental spinal nerves (Fig. 10.1).

The majority of sympathetic ganglia are found on each side of the vertebral column and are linked together by longitudinal bundles of nerve fibers to form the two sympathetic trunks, as shown in Fig. 10.2. With the exception of the cervical region, the sympathetic ganglia are distributed segmentally down as far as the coccyx. The sympathetic preganglionic fibers to the abdominal organs join to form the splanchnic nerves, which pass to the celiac, superior, and inferior mesenteric ganglia where they synapse. Postganglionic sympathetic fibers then pass to the various abdominal organs as shown in Fig. 10.2.

The sympathetic innervation to the adrenal medulla is an exception to the general rule. The preganglionic fibers from the thoracic spinal cord pass to the adrenal glands via the splanchnic nerves where they synapse directly with the chromaffin cells of the adrenal medulla. The chromaffin cells of the adrenal medulla are homologous with sympathetic postganglionic neurons and share many of their physiological properties including the generation of action potentials and the secretion of catecholamines (see Section 10.3.).

Although the sympathetic preganglionic fibers are myelinated, the sympathetic postganglionic fibers are unmyelinated.

This explains the difference in appearance of the gray and white rami. As shown in Fig. 10.2, sympathetic postganglionic fibers innervate many organs including the eye, the salivary glands, the gut, the heart, and the lungs. They also innervate the smooth muscle of the blood vessels, the sweat glands, and the piloerector muscles of the skin hairs. As the sympathetic ganglia are located close to the spinal cord, most sympathetic preganglionic fibers are relatively short while the postganglionic fibers are much longer, as indicated in Fig. 10.3.

The parasympathetic nervous system

The preganglionic neurons of the parasympathetic division of the autonomic nervous system have their cell bodies in two regions: the brainstem and sacral segments S3–S4 of the spinal cord. Thus the parasympathetic preganglionic fibers emerge as part of the *cranial outflow* in CN III (oculomotor), CN VII (facial), CN IX (glossopharyngeal), and CN X (vagus) and from the *sacral outflow*.

The parasympathetic ganglia are usually located close to the target organ or even embedded within it. Thus the parasympathetic innervation is characterized by long preganglionic fibers and short postganglionic fibers, in contrast with the organization of the sympathetic nervous system (Fig. 10.3). Parasympathetic postganglionic fibers innervate the eye, the salivary glands, the genitalia, the gut, the heart, the lungs, and other visceral organs, as shown in Fig. 10.2. Parasympathetic innervation of blood vessels is confined to vasodilator fibers supplying the salivary glands, the exocrine pancreas, the gastrointestinal mucosa, the genital erectile tissue, and the cerebral and coronary arteries. Other blood vessels are innervated exclusively by sympathetic fibers.

Many organs receive a dual innervation from sympathetic and parasympathetic fibers

Most visceral organs (but not all) are innervated by both divisions of the autonomic nervous system. The specific actions of the autonomic nerves on the various organ systems of the body are discussed at length in the relevant chapters of this book. In

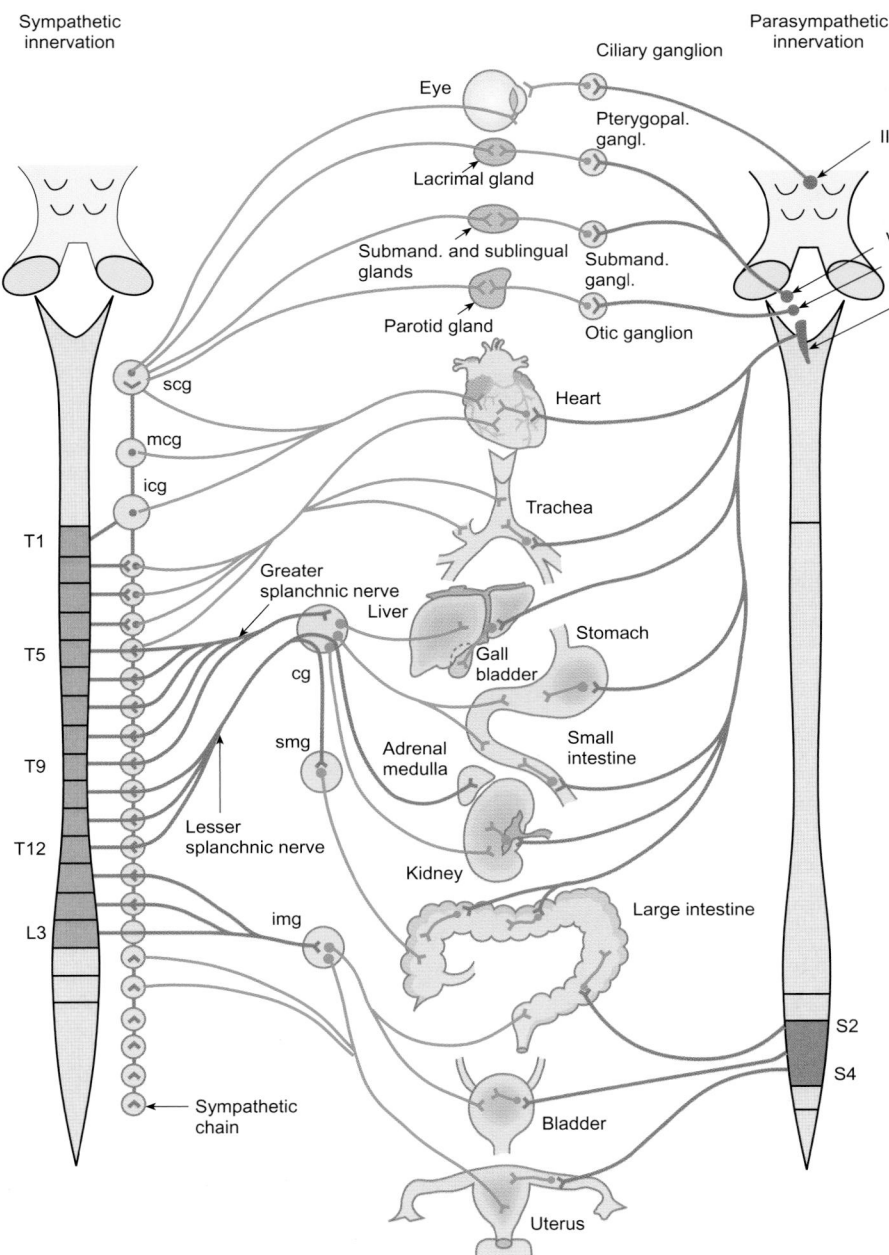

Fig. 10.2 The organization of the autonomic nervous system. In addition to the innervation of the principal organ systems, segmental sympathetic fibers also innervate blood vessels, piloerector muscles, and sweat glands. Parasympathetic preganglionic fibers are found in CN III (oculomotor), CN VII (facial), CN IX (glossopharyngeal), and CN X (vagus). scg, mcg and icg refer to superior, middle, and inferior cervical ganglion; cg celiac ganglion; smg and img refer to the superior and inferior mesenteric ganglia. Post-ganglionic fibres are shown in green.

many cases, the actions of the sympathetic and parasympathetic divisions are antagonistic, so that the actions of the two divisions provide a delicate control over the functions of the viscera. Thus, activation of the sympathetic nerves to the heart increases heart rate and the force of contraction of the heart muscle, while activation of the vagus nerve (parasympathetic) slows the heart. Activation of the parasympathetic supply to the gut enhances its motility and secretory functions, while activation of the sympathetic supply inhibits the digestive functions of the gut and constricts its sphincters.

Some organs only have a sympathetic supply. Examples are the adrenal medulla, the pilomotor muscles of the skin hairs,

the sweat glands, and the spleen. In humans, it is probable that most blood vessels are also exclusively innervated by the sympathetic nerves. The parasympathetic supply has exclusive control over the focusing of the eyes by the ciliary muscles and of pupillary constriction by the constrictor pupillae muscle of the iris. Sympathetic stimulation dilates the pupil of the eyes by its action on the dilator pupillae muscle of the iris. Thus, the iris provides an example of functional antagonism rather than dual antagonistic innervation of a specific smooth muscle. The effects of activation of the sympathetic and parasympathetic divisions of the autonomic nervous system on various organs are summarized in Table 10.1.

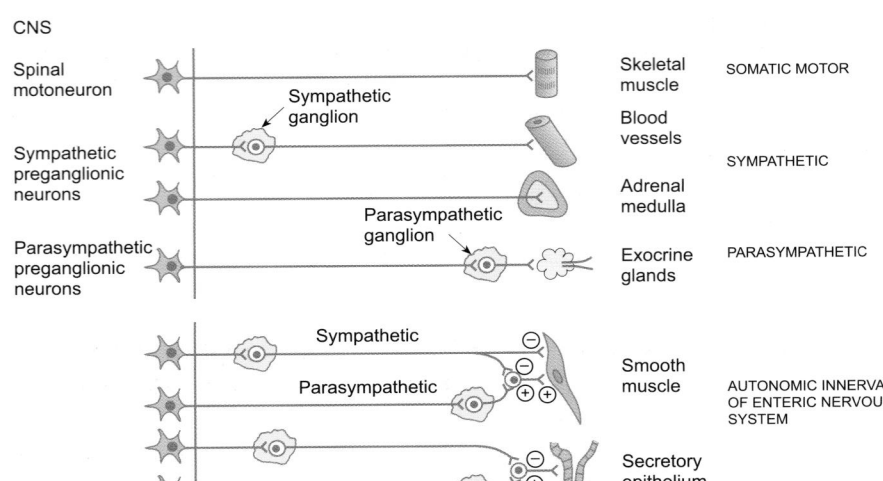

Fig. 10.3 Comparison of the organization of the sympathetic and parasympathetic divisions of the autonomic nervous system with that of somatic motor innervation. The lower part of the figure shows a simplified plan of the innervation of the gut by sympathetic and parasympathetic nerve fibers. (⊖ indicates an inhibitory action; ⊕ indicates an excitatory action.)

TABLE 10.1	The main actions of sympathetic and parasympathetic stimulation on various organ systems	
Organ	**Effect of sympathetic activation**	**Effect of parasympathetic activation**
Eye	Pupillary dilatation	Pupillary constriction and accommodation
Lacrimal gland	No effect (vasoconstriction?)	Secretion of tears
Salivary glands	Vasoconstriction, secretion of viscous fluid	Vasodilatation and copious secretion of saliva
Heart	Increased heart rate and force of contraction	Decreased heart rate, no effect on force of contraction
Blood vessels	Mainly vasoconstriction, vasodilatation in skeletal muscle	No effect except for vasodilatation of certain exocrine glands and the external genitalia
Lungs	Bronchial dilatation via circulating epinephrine	Bronchial constriction, secretion of mucus
Liver	Glycogenolysis, gluconeogenesis, and release of glucose into blood	No effect on liver but secretion of bile by gall bladder
Adrenal medullae	Secretion of epinephrine and norepinephrine	No innervation
Gastrointestinal tract	Decreased motility and secretion, constriction of sphincters, vasoconstriction	Increased motility and secretion, relaxation of sphincters
Kidneys	Vasoconstriction and decreased urine output	No effect
Urinary bladder	Inhibition of micturition	Initiation of micturition
Genitalia	Ejaculation	Erection
Sweat glands	Secretion of sweat by eccrine glands	No innervation
Hair follicles	Piloerection	No innervation
Metabolism	Increase	No effect

This table summarizes the main effects of activation of the sympathetic or parasympathetic nerves to various organs. The details of specific autonomic reflexes can be found in the relevant chapters of this book.

Autonomic nerves maintain a basal level of tonic activity

The autonomic innervation generally provides a basal level of activity in the tissues it innervates called *tone* (see also Chapter 7). The autonomic tone can be either increased or decreased to modulate the activity of specific tissues. For example, the blood vessels are generally in a partially constricted state as a result of *sympathetic tone*. This partial constriction restricts the flow of blood. If sympathetic tone is increased, the affected vessels become more constricted and this results in a decrease in blood flow. Conversely, if sympathetic activity is inhibited, tone decreases and the affected vessels dilate, thus increasing their blood flow (see Chapter 15).

The heart in a resting person is normally under the predominant influence of *vagal tone*. If the vagus nerves are cut, the heart rate rises. During the onset of exercise, the tonic parasympathetic inhibition of the heart declines and sympathetic activation increases, with a resulting elevation in heart rate.

The enteric nervous system

Sympathetic and parasympathetic nerve fibers act on neurons that are present in the walls of the gastrointestinal tract. These neurons have been considered to form a separate division of the nervous system—the *enteric nervous system*. The enteric neurons are organized as two interconnected plexuses:

(1) the *submucosal plexus* (also known as Meissner's plexus) which lies in the submucosal layer beneath the muscularis mucosae (see Chapter 18, Fig. 18.2);

(2) the *myenteric plexus* (or Auerbach's plexus) which lies between the outer longitudinal and the inner circular smooth muscle layers of the muscularis.

The enteric nervous system can function independently of its autonomic supply and its neurons play an important part in the regulation of the motility and secretory activity of the digestive system (see Chapter 18).

10.3 Chemical transmission in the autonomic nervous system

Within the autonomic ganglia, the synaptic contacts are highly organized and similar in structure to the other neuronal synapses described in Chapter 6. However, the synaptic contacts in the target tissues are more diffuse than those of the CNS or the neuromuscular junction of skeletal muscle. The postganglionic fibers have varicosities along their length which secrete neurotransmitters into the space adjacent to the target cells rather than onto a clearly defined synaptic region. However, in multi-unit smooth muscle, each varicosity is closely associated with an individual smooth muscle cell.

The main neurotransmitters secreted by the neurons of the autonomic nervous system are *acetylcholine* and *norepinephrine* (also known as *noradrenaline*). Within the ganglia of both the sympathetic and parasympathetic divisions, the principal transmitter secreted by the preganglionic fibers is acetylcholine. The postganglionic fibers of the parasympathetic nervous system also secrete acetylcholine onto their target tissues. The postganglionic sympathetic fibers secrete norepinephrine, except for those fibers that innervate the sweat glands and pilomotor muscles. These fibers secrete acetylcholine. The transmitters utilized by the different neurons of the autonomic nervous system are summarized in Fig. 10.4.

Acetylcholine activates nicotinic receptors in the autonomic ganglia but muscarinic receptors in the target tissues

The acetylcholine receptors of the postganglionic neurons in autonomic ganglia are called *nicotinic receptors* because they can also be activated by the alkaloid *nicotine*. They are similar in structure to the nicotinic receptors of the neuromuscular junction, but have a different response to various drugs and toxins. For example, they can be blocked by mecamylamine, which has no action at the neuromuscular junction, but not by α-bungarotoxin, which is a potent blocker of the nicotinic receptors of the neuromuscular junction. Activation of nicotinic receptors leads to the opening of an ion channel and rapid excitation as described in Chapter 6.

The acetylcholine receptors of the target tissues of both parasympathetic and sympathetic postganglionic fibers are *muscarinic receptors* as they can be activated by *muscarine*. Muscarinic receptors are also present at sympathetic nerve endings in sweat glands and pilomotor muscles, and, in some animal species, at

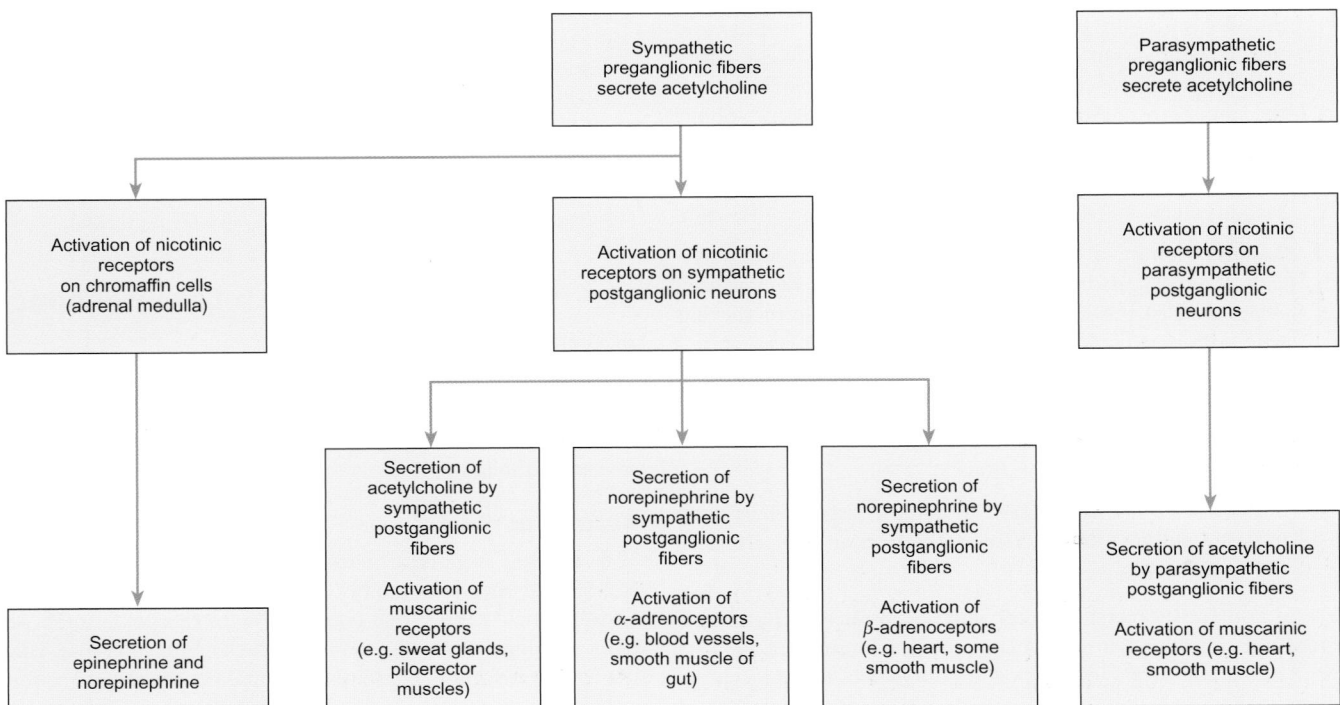

Fig. 10.4 The role of the cholinergic and adrenergic innervation in the autonomic nervous system.

the nerve endings of vasodilator fibers in skeletal muscle. They can be inhibited by low concentrations of atropine. Muscarinic receptors are linked to G proteins so that activation of these receptors leads to the modulation of the intracellular levels of IP_3 or cyclic AMP (see Chapter 5).

Although five different types have been identified by the techniques of molecular biology, muscarinic receptors can be grouped into three main classes based on their physiological actions.

1. M_1 receptors, which are mainly located on neurons in the CNS and peripheral ganglia. They are also found on gastric parietal cells. Activation of M_1 receptors generally has an excitatory effect. For example, they mediate the secretion of gastric acid that follows stimulation of the vagus nerve.

2. M_2 receptors, which occur in the heart and on the nerve terminals of both CNS and peripheral neurons. The slowing of the heart following stimulation of the vagus nerves is mediated by this class of muscarinic receptor.

3. M_3 receptors, which are located in secretory glands and on smooth muscle. Activation of these receptors generally has an excitatory effect. In visceral smooth muscle, for example, activation of M_3 receptors by acetylcholine leads to contraction.

Adrenergic receptors belong to two main classes

The receptors for norepinephrine are called *adrenoceptors*. Two main classes are known, *α-adrenoceptors* and *β-adrenoceptors*. As for muscarinic receptors, the adrenoceptors are coupled to second-messenger systems via a G protein. Each group is further subdivided so that five subtypes are presently recognized: α_1, α_2, β_1, β_2, and β_3.

1. α_1-Adrenoceptors are found in the smooth muscle of the blood vessels, bronchi, gastrointestinal tract, uterus, and bladder. Activation of these receptors is mainly excitatory and results in the contraction of smooth muscle. However, the smooth muscle of the gut wall (but not that of the sphincters) becomes relaxed after activation of these receptors.

2. α_2-Adrenoceptors are found in the smooth muscle of the blood vessels where their activation causes vasoconstriction.

3. β_1-Adrenoceptors are found in the heart where their activation results in an increased rate and force of contraction. They are also present in the sphincter muscle of the gut where their activation leads to relaxation.

4. β_2-Adrenoceptors are found in the smooth muscle of certain blood vessels where their activation leads to vasodilatation. They are also present in the bronchial smooth muscle where they mediate bronchodilatation.

5. β_3-Adrenoceptors are present in adipose tissue where they initiate lipolysis to release free fatty acids and glycerol into the circulation.

Although the diversity in both adrenoceptors and cholinergic receptors is somewhat bewildering, the development of agonists and antagonists that act on specific subtypes of these receptors has been of considerable clinical benefit in the treatment of diseases such as asthma and hypertension.

The adrenal medulla secretes epinephrine and norepinephrine into the circulation

Although the activation of the autonomic nerves provides a mechanism for the discrete regulation of specific organs, activation of the splanchnic nerve results in the secretion of epinephrine and norepinephrine from the adrenal medulla into the circulation. About 80% of the secretion is epinephrine and 20% is norepinephrine. These catecholamines exert a hormonal action on a variety of tissues (see Chapter 12), which forms part of the overall sympathetic response. Their release is always associated with an increase in the secretion of norepinephrine from sympathetic nerve terminals.

The response of a tissue to circulating epinephrine or norepinephrine will depend on the relative proportions of the different types of adrenoceptor it possesses. Epinephrine activates β-adrenoceptors more strongly than α-adrenoceptors, while norepinephrine is more effective at activating α-adrenoceptors than β-adrenoceptors. During exercise, increased activity in the sympathetic nerves will result in an increased heart rate and vasoconstriction in the splanchnic circulation. The circulating norepinephrine will have a similar effect, but the actions of circulating epinephrine will lead to relaxation of the smooth muscle that possesses a high proportion of β-adrenoceptors, such as that of the blood vessels of skeletal muscle. The increased levels of circulating epinephrine cause bronchodilatation and a vasodilatation in the skeletal muscle, thus favoring increased gas flow to the alveoli and blood flow to the exercising muscles.

Circulating catecholamines affect virtually every tissue, with the result that the metabolic rate of the body is increased. Indeed, maximal sympathetic stimulation may double the metabolic rate. The major metabolic effect of epinephrine and norepinephrine is to increase the rate of glycogenolysis within cells by activating adenylyl cyclase via β-adrenoceptors as described in Chapter 5 p. 55. The result is a rapid mobilization of glucose from glycogen and an increased availability of fatty acids for oxidation as a result of lipolysis occurring in adipose tissue. The increased availability of substrates for oxidative metabolism is important both in exercise and during cold stress, where an increase in metabolic rate is important for generating the heat required to maintain body temperature (see Chapter 26).

10.4 Central nervous control of autonomic activity

The activity of the autonomic nervous system varies according to the information it receives from both visceral and somatic afferent fibers. It is also subject to regulation by the higher centers of the brain, notably the hypothalamus.

Autonomic reflexes

The internal organs are innervated by afferent fibers that respond to mechanical and chemical stimuli. Some visceral afferents reach the spinal cord by way of the dorsal roots and enter the dorsal horn together with the somatic afferents. These fibers synapse at the segmental level and the second-order fibers ascend the spinal cord in the spinothalamic tract. They project to the nucleus of the solitary tract, various motor nuclei in the brainstem, and the thalamus and hypothalamus. Other visceral afferents such as those from the arterial baroreceptors, reach the brainstem by way of the vagus nerves.

Information from the visceral afferents elicits specific visceral reflexes which, like the reflexes of the somatic motor system, may be either segmental or involve the participation of neurons in the brain. Examples of autonomic reflexes are the baroreceptor reflex, the lung inflation reflex, and the micturition reflex. These are discussed in detail in the relevant chapters of this book.

In response to a perceived danger, there is a behavioral alerting that may result in aggressive or defensive behavior. This is known as the *defense reaction*, and has its origin in the hypothalamus. During the defense reaction there are marked changes in the activity of the autonomic nerves in which normal reflex control is overridden. Further details can be found in Chapter 15, p. 302.

Summary

1. The autonomic nervous system is a system of motor nerves that act to control the activity of the internal organs. It is divided into the sympathetic and parasympathetic divisions, which have different functions and different anatomical origins.

2. The main neurotransmitters secreted by the neurons of the autonomic nervous system are acetylcholine and norepinephrine. Acetylcholine is the principal transmitter secreted by the preganglionic fibers within the autonomic ganglia of both divisions. The parasympathetic postganglionic fibers also secrete acetylcholine onto their target tissues while norepinephrine is the principal neurotransmitter secreted by the postganglionic sympathetic fibers.

3. The sympathetic division broadly acts to prepare the body for activity ('fight or flight'). Increased sympathetic activity is associated with an increased heart rate, vasoconstriction in the visceral organs, and vasodilatation in skeletal muscle. It is also accompanied by bronchodilatation and by glycogenolysis and gluconeogenesis in the liver.

4. The parasympathetic division tends to promote restorative functions ('rest and digest'). Increased parasympathetic activity is associated with increased motility and secretion by the gastrointestinal tract and slowing of the heart rate.

5. Many organs receive innervation from both sympathetic and parasympathetic nerve fibers, which act in opposing ways. Nevertheless, the two divisions act together to regulate the activity of the internal organs according to the needs of the body at the time. The activity of the autonomic nerves is under the control of neurons in the brainstem and hypothalamus.

The hypothalamus regulates the homeostatic activity of the autonomic nervous system

Both the activity of the autonomic nervous system and the function of the endocrine system are under the control of the hypothalamus, which is the part of the brain mainly concerned with maintaining the homeostasis of the body. If the hypothalamus is destroyed, the homeostatic mechanisms fail. The hypothalamus receives afferents from the retina, the chemical sense organs, the somatic senses, and visceral afferents. It also receives many inputs from other parts of the brain including the limbic system and cerebral cortex. Hypothalamic neurons play important roles in thermoregulation, in the regulation of tissue osmolality, in the control of feeding and drinking, and in reproductive activity.

10.5 Autonomic failure

Autonomic failure is a relatively rare condition that generally affects older people. It is more common in men than in women and its clinical signs are dizziness, fatigue, and blackouts during exercise or after meals. The key feature is a persistent postural hypotension (low blood pressure) which results from the inability of the sympathetic nerves to constrict the peripheral blood vessels. The disease can occur on its own or in combination with neurodegenerative diseases, specifically Parkinson's disease and multiple system atrophy.

Recommended reading

Anatomy
Brodal, P. (2003). *The central nervous system. Structure and function* (3rd edn), Chapters 18 and 19. Oxford University Press, Oxford.

Kiernan, J.A. (2004). *Barr's The human nervous system: an anatomical viewpoint* (8th edn). Lippincott–Williams & Wilkins, Baltimore, MD.

Pharmacology
Rang, H.P., Dale, M.M., Ritter, J.M., and Moore, P. (2003). *Pharmacology* (5th edn), Chapters 9–11. Churchill-Livingstone, Edinburgh.

Physiology
Brading, A.S. (1999). *Autonomic nervous system and its effectors.* Blackwell Science, Oxford.

Shepherd, G.M. (1994). *Neurobiology* (3rd edn), Chapter 18. Oxford University Press. Oxford.

Zigmond, M.J., Bloom, F.E., Landis, S.C., Roberts, J.L., and Squire, L.R. (1999). *Fundamental neuroscience*, Chapters 37–39. Academic Press, San Diego, CA.

Medicine
Bannister, R. (2000). The autonomic nervous system. In *Concise Oxford textbook of medicine* (eds. J.G.G. Ledingham and D.A. Warrell), Chapter 13.7. Oxford University Press, Oxford.

Multiple choice questions

Each statement is either true or false. Answers are given below.

1. a. Sympathetic preganglionic neurons are found in spinal segments from T1 to L2.
   b. The sympathetic chain extends from the cervical to the sacral regions of the spinal cord.
   c. Sympathetic preganglionic fibers secrete norepinephrine.
   d. Acetylcholine is secreted by some sympathetic postganglionic fibers.
   e. The diameter of the blood vessels is entirely regulated by the sympathetic nervous system.

2. a. Parasympathetic preganglionic fibers are found in CN III (oculomotor).
   b. The vagus nerves provide the parasympathetic innervation to the heart.
   c. Parasympathetic vasoconstrictor fibers are present in the salivary glands.
   d. Parasympathetic postganglionic fibers secrete norepinephrine onto their target organs.
   e. Parasympathetic preganglionic fibers secrete acetylcholine.

3. a. Stimulation of the sympathetic nerves to the eyes causes pupillary constriction.
   b. Epinephrine secreted by the adrenal medulla causes glycogen breakdown in the liver.
   c. Stimulation of the vagus nerves increases the motility of the gastrointestinal tract.
   d. Activation of the sympathetic system causes vasoconstriction in the viscera and skin but vasodilatation in skeletal muscle.
   e. Stimulation of the vagus nerves slows the heart.

4. a. The acetylcholine receptors in both parasympathetic and sympathetic ganglia are nicotinic.
   b. Acetylcholine secreted by parasympathetic postganglionic fibers acts on muscarinic receptors.
   c. Norepinephrine secreted by sympathetic postganglionic fibers acts preferentially on β-adrenoceptors.
   d. The adrenal medulla secretes epinephrine in response to stimulation of sympathetic postganglionic fibers in the splanchnic nerves.

Answers

1. Although the sympathetic preganglionic neurons are found in the thoracic and upper lumbar spinal segments, paired sympathetic ganglia are found on both sides of the spinal cord from the cervical region to the sacral region. In the cervical region, only three pairs of ganglia are found (superior, middle, and inferior cervical ganglia) while from T1 to L3 there is a pair of ganglia for each spinal segment. Sympathetic preganglionic fibers secrete acetylcholine. Although the majority of blood vessels are innervated by sympathetic vasoconstrictor fibers, some, such as those in the salivary glands and in the exocrine pancreas, also have a vasodilator parasympathetic innervation.
   a. True
   b. True
   c. False
   d. True
   e. False

2. Parasympathetic preganglionic fibers are found in cranial nerves III, VII, IX, and X. The salivary glands are innervated by parasympathetic vasodilator fibers (parasympathetic stimulation promotes salivary secretion). Both parasympathetic preganglionic and parasympathetic postganglionic fibers secrete acetylcholine.
   a. True
   b. True
   c. False
   d. False
   e. True

3. Stimulation of the sympathetic nerves to the eyes results in dilation of the pupils. Although activation of the sympathetic nerves results in generalized vasoconstriction, epinephrine secreted by the adrenal medulla acts on the β-adrenoceptors of the blood vessels of skeletal muscle to cause vasodilatation.
   a. False
   b. True
   c. True
   d. True
   e. True

4. Norepinephrine acts preferentially on α-adrenoceptors; epinephrine acts preferentially on β-adrenoceptors. However, it is important to realize that norepinephrine and epinephrine will activate both types of adrenoceptor. Epinephrine and norepinephrine are secreted by the adrenal medulla in response to stimulation of sympathetic preganglionic fibers traveling in the splanchnic nerves. The chromaffin cells of the adrenal medulla are homologous with sympathetic postganglionic neurons.
   a. True
   b. True
   c. False
   d. False

Some aspects of higher nervous function

After reading this chapter you should understand:

- The functions of the association cortex in the cerebral hemispheres
- The separate roles of the two hemispheres
- The organization of the pathways that control speech
- The use of the EEG to monitor cerebral function in humans
- The physiology of sleep and other circadian rhythms
- The physiological basis of learning and memory

11.1 Introduction

The brain is not simply concerned with controlling movements and regulating the internal environment. It is also concerned with assessing aspects of the world around us so that we can adjust our behavior to assist our survival. Experiences are assimilated into our memory, which can be used to determine some future course of action. We communicate with our fellows via language. To achieve all this we require a certain level of awareness of ourselves and our environment that we call consciousness. This chapter will explore some aspects of these very complex phenomena.

11.2 The specific functions of the left and right hemispheres

The cerebral hemispheres are the largest and most obvious structures of the human brain. As discussed in earlier chapters, they receive information from the primary senses (somatic sensations, taste, smell, hearing, and vision) and control coordinated motor activity. The main somatosensory and motor pathways are crossed so that the left hemisphere receives information from, and controls the motor activity of, the right side of the body while the right hemisphere is concerned with the left side.

The two hemispheres are not symmetrical morphologically or in their functions. It is commonly recognized that people prefer to use one hand rather than another. Indeed, most people (about

90 per cent of the population) prefer to use their right hand for writing, holding a knife or tennis racquet, and so on. What is not always appreciated is that most people also prefer using their right foot (e.g. when kicking a ball) and they generally pay more attention to information derived from one eye and one ear. Therefore, the two halves of the brain are not equal with respect to the information to which they pay attention or with respect to the activities they control.

In the nineteenth century Paul Broca and Marc Dax described an association between the loss of speech and paralysis of the right side of the body in patients who had suffered a stroke. This showed that the left hemisphere controlled speech as well as the motor activity of the right side of the body. Later studies showed that specific deficits in the understanding of language, reading, and writing were also associated with damage to the left cerebral hemisphere. Other patients with damaged left hemispheres were unable to carry out specific directed actions, such as brushing their hair, although the muscles involved were not paralyzed. Thus, as the clinical evidence accumulated, it became apparent that the two hemispheres must have different capacities. The idea developed that one hemisphere, the left in right-handed people, was the leading or *dominant hemisphere* while the other was subordinate and lacked a specific role in higher nervous functions. This idea has had to be modified in the light of current evidence and it is now clear that both hemispheres have very specific functions in addition to the role that they play in sensation and motor activity.

The association areas of the frontal cortex

Early exploration of the human cerebral cortex showed that while stimulation of some areas elicited specific sensations or motor responses, stimulation of others had no detectable effect. These areas were called 'silent' areas and were loosely considered as *association cortex* (Fig. 11.1). Over the last century, the function of these areas has gradually become clearer.

Compared with the brains of other animals, including those of the higher apes, the frontal lobes of the human brain are very large relative to the size of the brain as a whole. Indeed, as Fig. 11.1 shows, the frontal lobes are the largest division of the human cerebral cortex. What is their role? One aspect of their

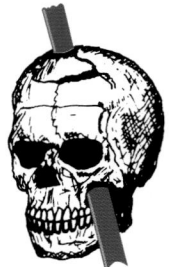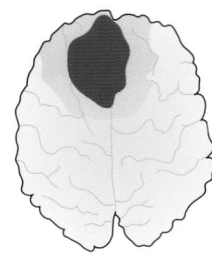

Fig. 11.2 The skull of Phineas Gage showing the course of the iron bar (left) and the likely extent of the damage to his frontal cortex.

function became clear in the last century when an American mining engineer called Phineas Gage received a devastating injury to his frontal lobes (Fig. 11.2). He was packing down an explosive charge when it detonated, driving the tamping iron through his upper jaw, into his skull, and through his frontal lobes. Remarkably, he survived the accident. Although he was able to live an essentially normal life, his personality had irrevocably changed. He became irascible, unpredictable, and much less inhibited in his social behavior. So changed was his behavior that his friends remarked that he was 'no longer Gage'.

Later work on monkeys showed that lesions to the frontal lobes appeared to reduce anxiety. This discovery was subsequently exploited clinically in an attempt to help patients suffering severe and debilitating depression. By cutting the fibers connecting the frontal lobes to the thalamus and other areas of the cortex (a *frontal leucotomy*) it was hoped to reduce the feeling of desperate anxiety and permit patients to resume a normal life. Early results were encouraging, in that the affected patients seemed less anxious than before, but it later became clear that the procedure could result in severe personality changes that perhaps might have been predicted from the case of Phineas Gage. Epilepsy (abnormal electrical activity of the brain—see Section 11.4) was another undesirable complication that often developed. In recent years this procedure has become replaced by less drastic and more reversible drug therapies.

Detailed psychological testing of patients who have had a frontal leucotomy showed that, although their general intelligence was little affected, they were less good at problem solving than people with intact frontal lobes. They tended to persevere with failed strategies. They also appeared to be less spontaneous in their behavior. Thus the association areas of the frontal lobes are implicated in the determination of the personality of an individual in all its aspects.

Damage to the parietal lobes leads to loss of higher-level motor and sensory performance

The parietal lobes extend from the central sulcus to the parieto-occipital fissure and from an imaginary line drawn from the angular gyrus to the parieto-occipital fissure (Fig. 11.1). As with the frontal lobes, most of our knowledge of the functions of the association areas of the parietal lobes in humans is derived from careful clinical observation. Damage to these parts of the brain is associated with specific deficits known as agnosias and apraxias.

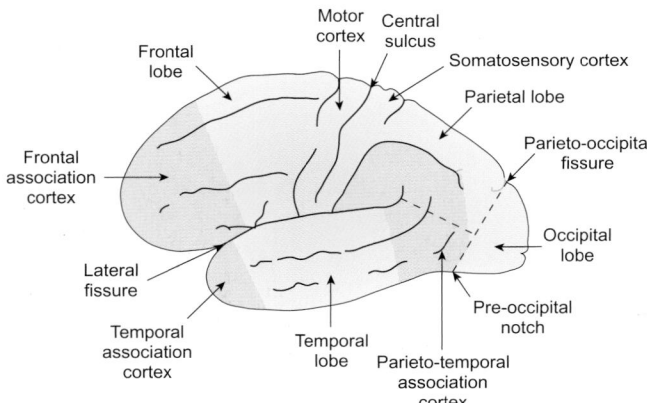

Fig. 11.1 A side view of the human brain showing the association areas of the frontal, parietal, and temporal lobes.

Agnosia is a failure to recognize an object even though there is no specific sensory deficit. It reflects an inability of the brain to integrate the information in a normal way. For this reason the agnosias are regarded as a failure of 'higher level' sensory performance. An example is visual object agnosia in which the visual pathways appear to be essentially normal but recognition of objects does not occur. If an affected person is allowed to explore the object with another sense, by touch for example, they can often name it. Nevertheless, they may not be able to appreciate its qualities as a physical object. Thus, they may see a chair but not avoid it as they cross a room. Another example is astereognosis in which there is a failure to recognize an object through touch. This disorder is associated with damage to regions in the parietal lobe adjacent to the primary cortical receiving areas of the postcentral gyrus.

Apraxia is the loss of the ability to perform specific purposeful movements even though there is no paralysis or loss of sensation. An affected person may be unable to perform a complex motor task on command (e.g. waving someone goodbye) but may be perfectly capable of carrying out the same act spontaneously. Other apraxias may result in failure to use an everyday object appropriately or to construct or draw a simple object (*constructional apraxia*). A deficit in the control of fine movements of one hand can be caused by damage to the premotor area of the frontal lobe on the opposite side. This is known as *kinetic apraxia*.

Fig. 11.3 Drawing by a patient suffering from a stroke in the posterior region of the right hemisphere leading to the neglect syndrome. The model is on the left of the figure and the patient's drawing is shown on the right. Note the failure to complete the left-hand parts of the original drawings.

The most bizarre effect of lesions to the parietal lobes is seen when the lesion affects the posterior part of the right hemisphere around the border of the parietal and occipital lobes. These lesions lead to neglect of the left side of the body. Affected individuals ignore the left side of their own bodies, leaving them unwashed and uncared for. They ignore the food on the left side of their plates and will only copy the right side of a simple drawing (Fig. 11.3). Many are blind in the left visual field although they are themselves unaware of the fact.

The corpus callosum plays an essential role in integrating the activity of the two cerebral hemispheres

Sensory information from the right half of the body is represented in the somatosensory cortex of the left hemisphere and vice versa. Equally, the left motor cortex controls the motor activity of the right side of the body. Despite this apparent segregation, the brain acts as a whole, integrating all aspects of neural function. This is possible because, although the primary motor and sensory pathways are crossed, there are many cross-connections between the two halves of the brain, known as *commissures*. As a result, each side of the brain is constantly informed of the activities of the other.

The largest of the commissures is the vast number of fibers that connect the two cerebral hemispheres, known as the *corpus callosum* (see Chapter 6, Figs 6.3 and 6.4). Damage to the corpus callosum was first reported for shrapnel injuries during the First World War. Amazingly, soldiers with these injuries showed remarkably little by way of a neurological deficit that could be attributed to the severance of so large a nerve tract.

Experimental work has shown that most of the nerve fibers that traverse the corpus callosum project to comparable functional areas on the contralateral side. It was subsequently found that epileptic discharges can spread from one hemisphere to the other via the corpus callosum and that major epileptic attacks can involve both sides of the brain. In a search for a cure for the severe bilateral epilepsy experienced by some patients, their corpus callosum was cut—the *split-brain operation*. This had the desired end result—a reduction in the frequency and severity of the epileptic attacks. It also offered the opportunity of careful and detailed study of the functions of the two hemispheres of the human brain.

The human split-brain operation showed that each hemisphere is specialized to perform a specific set of higher nervous functions

The key to the first experiments on split-brain patients was to exploit the fact that the visual pathway is only partially crossed at the optic chiasm while speech is located in the left hemisphere. Fibers from the temporal retina remain uncrossed while those from the nasal region of the retina project to the contralateral visual cortex as shown in Fig. 11.4. The effect of this arrangement is that the right visual field is represented in the left visual cortex while the left visual field is represented in the right visual cortex. By projecting words and images onto a screen in such a way that they would appear in either the right or the left visual field, R.W. Sperry and his colleagues were able to investigate the specific capabilities of each hemisphere. Subjects could then be asked what they had seen.

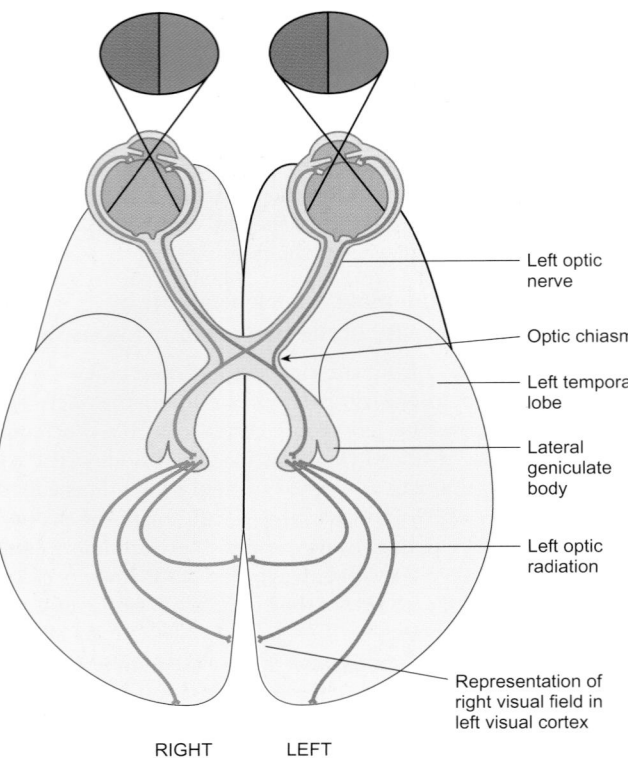

Fig. 11.4 A ventral view of the human brain showing the organization of the visual pathways. Note that the left cortex 'sees' the right visual field and vice versa.

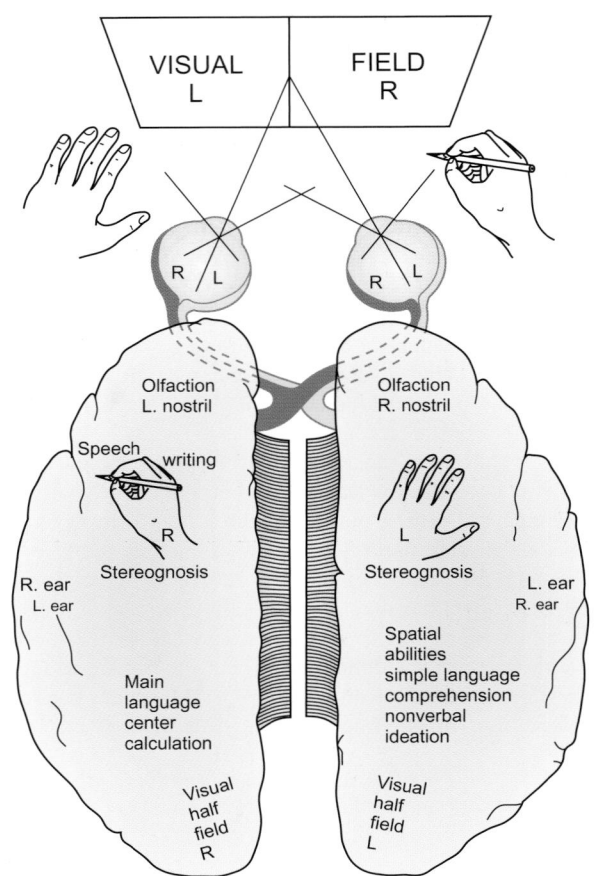

Fig. 11.6 The functional specialization of the two cerebral hemispheres. (Note that the areas of the cortex associated with the primary senses and those involved in motor control are not represented in this diagram.)

If, say, the word 'BAND' was briefly projected to the right visual field, the patient was able to report that to the investigator (see Fig. 11.5) as this word was represented in the visual cortex of the left hemisphere, which also controls speech. If the left visual field had a qualifying word, such as 'HAT', the subject was unaware of the fact. If asked what type of band had been mentioned, they were reduced to guessing. If a word such as 'BOOK'

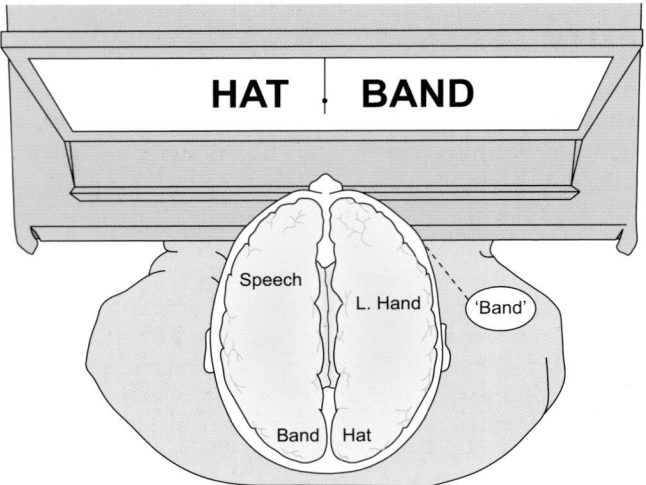

Fig. 11.5 The response of a split-brain patient to words projected onto the left and right visual fields.

was projected to the left field, they could only guess at the answer. If, however, the patient was then allowed to choose something that matched the word that had been projected onto the left visual field, they then made a suitable choice with their left hand (i.e. the hand that is controlled from the right hemisphere). This series of experiments showed that, while speech was totally lateralized to the left hemisphere, the right hemisphere was aware of the environment, was capable of logical choice, and possessed simple language comprehension.

In one patient there was a strong emotional response when a picture of a nude woman was projected onto the left visual field. The patient blushed and giggled, and when asked what she had seen she replied 'Nothing—just a flash of light'. When asked why she was laughing, she could not explain but replied 'Oh Doctor, you have some machine!'. From this observation, it would appear that emotional reactions begin at a lower level of neural integration than the cortex and involve both hemispheres.

Further studies showed that the right hemisphere could identify objects held in the left hand by their shape and texture (stereognosis) as efficiently as the right hand. Indeed, the left hand (and therefore the right hemisphere) proved to be much better at solving complex spatial problems. Overall, Sperry concluded that, far from being subordinate to the left hemisphere, the right hemisphere was conscious and was better at solving spatial problems and non-verbal reasoning. His conclusions are summarized in Fig. 11.6.

Summary

In addition to their role in motor control, the frontal lobes appear to play a significant part in shaping the personality of an individual. Lesions in the parietal lobe result in defects of sensory integration known as agnosias and in an inability to perform certain purposeful acts (apraxia). Severance of the corpus callosum, which interconnects the two hemispheres, has shown that the two hemispheres have very specific capabilities in addition to their role in sensation and motor activity. Speech and language abilities are mainly located in the left hemisphere, together with logical reasoning. The right hemisphere is better at solving spatial problems and non-verbal tasks.

11.3 Speech

Many animals have some means of communicating with their fellows. What distinguishes human communication is its range and subtlety of expression. Humans use *language*. The production of sound that has no specific meaning is called vocalization. A language consists of a specific vocabulary and a set of rules of expression (syntax). While most communication is by means of speech, expression in a given language is independent of the mode of communication. This text is written in English and obeys its grammatical rules, but the written word is only the same as the spoken word in its meaning. The representation of a word on the page is arbitrary.

Speech is a very complex skill. It involves knowledge of the vocabulary and grammatical rules of at least one language. It requires very precise motor acts to permit the production of specific sounds in their correct order. It also requires precise regulation of the flow of air through the larynx and mouth. For these reasons it is perhaps not surprising to find that large areas of the brain are devoted to speech and its comprehension. Because it is specifically a human characteristic, our knowledge of the systems that govern the production and comprehension of speech has largely come from careful neurological observations.

The principal areas of the brain controlling speech are located in the frontal and temporal lobes

Earlier in this chapter, the work of Broca and Dax was cited as evidence for the lateralization of speech to the left hemisphere. This was based on the study of patients who had difficulty in producing speech following a stroke that had damaged the left hemisphere. Such disabilities are known as *aphasias*. Broca studied a number of patients and noted that they had difficulty in producing speech but appeared to be quite capable of understanding spoken or written language. Characteristically their speech is slow, halting, and telegraphic in quality. Such a patient is able to name objects and describe their attributes but has difficulty with the small parts of speech that play such an important role in grammar (e.g. 'if', 'is', 'the', and so on). They also

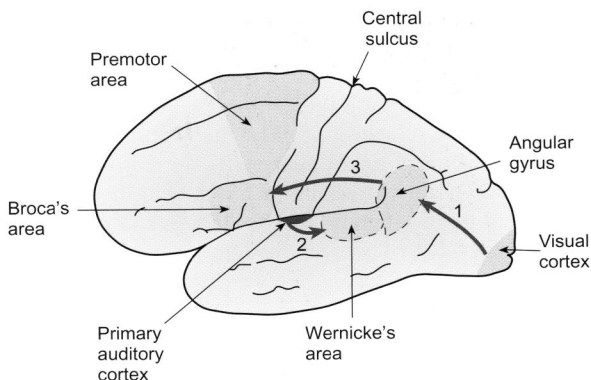

Fig. 11.7 A diagram of the left hemisphere to show the location of the principal regions involved in the control of speech.

have difficulty in writing. Since these patients apparently have a good understanding of language, this type of aphasia is sometimes called *expressive aphasia*. It is also known as *Broca's aphasia*.

Broca was able to examine the brains of several aphasic patients at post-mortem. He discovered that there was extensive damage to the frontal lobe of the left hemisphere, particularly in the region that lies just anterior to the motor area responsible for the control of the lips and tongue (Fig. 11.7). This is now known as Broca's area. Patients with Broca's aphasia do not have a paralysis of the lips and tongue. They can sing wordlessly. What they have lost is the ability to use the apparatus of speech to form words, phrases, and sentences.

Another type of aphasia, discovered by Carl Wernicke, was characterized by free-flowing speech that had little or no informational content (technically known as *jargon*). Patients suffering from this kind of aphasia tend to make up words (*neologisms*), e.g. 'lork', 'flieber', and often have difficulty in choosing the appropriate words to describe what they mean. They have difficulty in comprehending speech, and this type of aphasia is called *receptive aphasia* or *Wernicke's aphasia* after its discoverer. Wernicke's aphasia is associated with lesions to the posterior region of the temporal lobe adjacent to the primary auditory cortex and the angular gyrus (Fig. 11.7). Patients suffering from Wernicke's aphasia have difficulty with reading and writing.

The two brain regions responsible for speech are interconnected by a set of nerve fibers called the *arcuate fasciculus*. If these fibers are damaged, another type of aphasia occurs, called conduction aphasia. This is typified by the speech characteristics of a Wernicke's aphasic but comprehension of written and spoken language remains largely intact.

Electrical stimulation of the speech areas of the cerebral cortex of conscious human subjects does not lead to vocalization. However, if a subject is already speaking when one of the areas of cortex concerned with language is stimulated, their speech may be interrupted or there may be an inappropriate choice of words. Stimulation of the supplementary motor area on the medial surface of the left hemisphere may lead to speech, which is limited to a few words, or syllables that may be repeated, e.g. 'ba-ba-ba-'. Unlike lesions to Broca's area and Wernicke's area, lesions to the supplementary motor area do not result in permanent aphasia.

Although the areas that control speech production are located in the left hemisphere, the right hemisphere has a very basic language capability. More significantly, the posterior part of the right cerebral hemisphere seems to play an important role in the interpretation of speech. Unlike the written word, spoken language has an emotional content that reveals itself in its intonation. Consequently, many patients who have damage to their left temporal lobe are able to understand the intention of something said to them even though their comprehension of individual words and phrases is poor. Conversely, patients with damage to their right hemispheres will often speak in a flat monotone.

These observations suggest an organization of the pathways that control speech similar to that shown in Fig. 11.7. According to this model, speech is initiated in Wernicke's area and is passed to Broca's area via the arcuate fasciculus for execution. The neural pathway involved in naming an object that has been seen is also shown in Fig. 11.7, and disconnection of the visual association areas from the angular gyrus will lead to *word blindness* or *alexia*. *Word deafness* (auditory agnosia) occurs when lesions disconnect Wernicke's area from the auditory cortex. Comprehension of spoken, but not written, language is impaired.

Recent studies with imaging techniques have revealed some other areas of the cortex that are also involved in language (Fig. 11.8). As described above, normal speech involves Wernicke's area, Broca's area, and the premotor area. Reading aloud also involves the visual cortex and a region close to the end of the lateral fissure known as the angular gyrus. The angular gyrus interprets visual information that is then converted to speech patterns by Wernicke's area before being transmitted to Broca's area. Silent reading involves the visual cortex, the premotor area, and Broca's area, but not the auditory cortex or Wernicke's area, while silent counting involves the frontal lobe.

Perhaps not surprisingly, careful anatomical examination has found that the upper aspect of the left temporal lobe (the *planum temporale*) of most brains is larger that the right. Despite these important insights, it is important to remember that knowing which parts of the brain are involved in the production of speech and the comprehension of language is not the same as knowing how the brain encodes and executes the patterns of neural activity responsible for speech.

In the majority of people, speech is controlled from the left hemisphere

It is of some importance that a neurosurgeon is aware of which hemisphere controls speech before embarking on an operation. This is established by the *Wada test* which temporarily anesthetizes one hemisphere. The patient lies on his or her back while a cannula is inserted into the carotid artery on one side. The patient is then asked to count backwards from 100 and to keep both arms raised. A small quantity of a short-lasting anesthetic is then injected. As the anesthetic reaches the brain, the arm on the side opposite that of the injection falls and the patient may stop counting for a few seconds or for several minutes depending on whether speech is localized to the side receiving the injection.

These tests reveal which hemisphere controls speech in a person who has no aphasia. In 95 per cent of right handers speech $(85.5 \div 90)$ is controlled from the left hemisphere. This is also true of about 70 per cent of left handers. Speech is localized to

TABLE 11.1 The distribution of handedness and the location of speech in the cerebral hemispheres

Location of speech	Right-handed individuals (%)	Left-handed individuals (%)	Total (%)
Left hemisphere	85.5	7	92.5
Right hemisphere	4.5	1.5	6.0
Both hemispheres	0	1.5	1.5
Total	90	10	100

The table shows the location of speech as a percentage of the total adult population. Thus, 85.5 per cent of the population are right handed and have their speech localized to their left hemispheres, and so on.

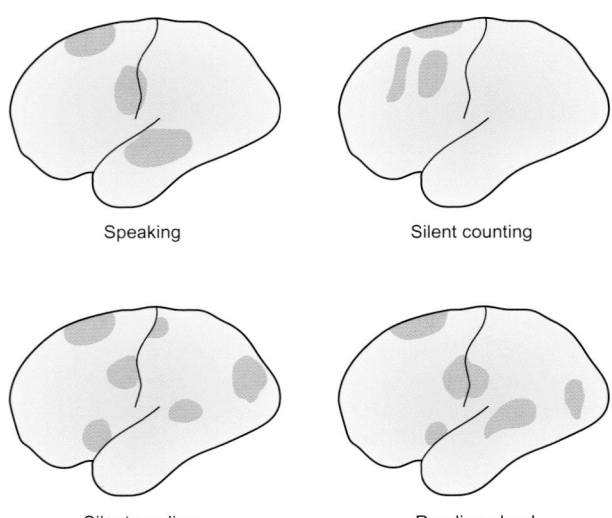

Fig. 11.8 Variations in local blood flow in a conscious human subject engaged in various language-related tasks.

Speaking

Silent counting

Silent reading

Reading aloud

Summary

Speech is localized to the left hemisphere in the majority of the population irrespective of whether they are right or left handed. The neural patterns of speech originate in the temporal lobe in Wernicke's area, which is adjacent to the auditory cortex. The neural codes for speech pass via the arcuate fasciculus to Broca's area for execution of the appropriate sequence of motor acts. Damage to the speech areas results in aphasia. Damage to Broca's area in the left frontal lobe results in expressive aphasia. Although patients with damage to Wernicke's area are able to speak fluently, they have poor comprehension of speech and their speech lacks clear meaning. This is known as receptive aphasia.

the right hemisphere in about 15 per cent of left handers and the remainder show evidence of speech being controlled from both hemispheres (Table 11.1). This distribution of the speech areas has recently been confirmed using a non-invasive imaging technique called Doppler ultrasonography.

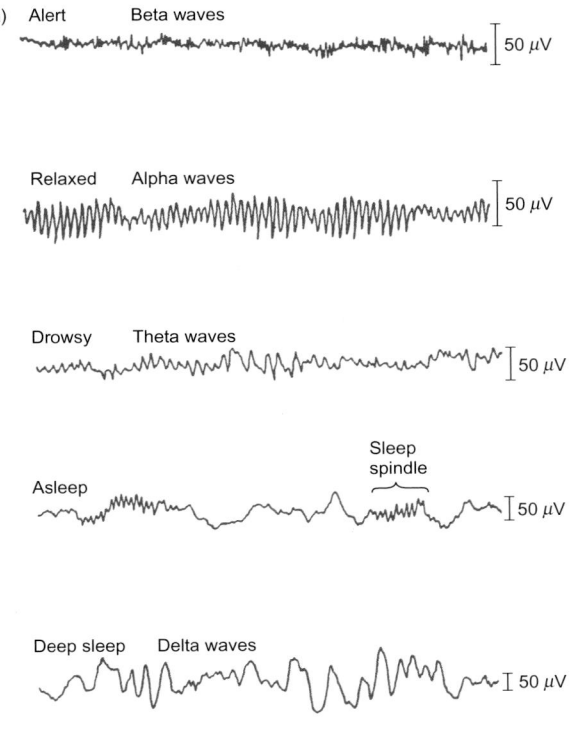

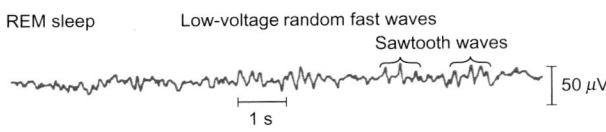

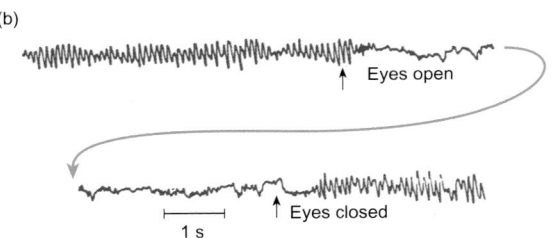

Fig. 11.9 Typical stretches of the EEG for various stages of awareness. (a) Changes in the EEG as a subject falls asleep. (b) Alpha block following opening of the eyes. Note the sudden loss of the alpha waves when the eyes open and their resumption when the eyes close again. The two lines in part (b) are of a continuous stretch of record from the occipital region of the left hemisphere. Note that during deep sleep the EEG changes to a pattern similar to that seen in the normal awake state but with characteristic 'sawtooth' waves. This is the REM phase of sleep.

11.4 The EEG can be used to monitor the activity of the brain

The brain is constantly involved in the control of a huge range of activities both when awake and during sleep. Its activity can be monitored indirectly by placing electrodes on the scalp. If this is done in such a way as to minimize electrical interference from the muscles of the head and neck, small oscillations are seen that can reflect the overall activity of the brain (Fig. 11.9). These electrical oscillations are known as the *electroencephalogram* or *EEG*. For normal subjects the amplitude of the EEG waves ranges from 10–150 μV.

The activity of the EEG is continuous throughout life but is not obviously related to any specific sensory stimulus. It reflects the spontaneous electrical activity of the brain itself. The appearance of the EEG varies according to the position of the electrodes, the behavioral state of the subject (i.e. whether awake or asleep), the subject's age, and whether there is any organic disease.

The specific state of the EEG is classified by the frequency of the electrical waves that are present. When a subject is awake and alert, the EEG consists of high-frequency waves (20–50 Hz) which have a low amplitude (about 10–20 μV). These are known as *beta waves* and they appear to originate in the cerebral cortex. As a subject closes his or her eyes, this low-amplitude high-frequency pattern gives way to a higher-amplitude lower-frequency pattern known as the alpha rhythm. The *alpha waves* have an amplitude of 20–40 μV and contain one predominant frequency in the range 8–12 Hz. As the subject becomes more drowsy and falls asleep, the alpha rhythm disappears and is replaced by slower waves of greater amplitude known as *theta*

TABLE 11.2	The characteristics of the principal waves of the EEG		
Wave type	**Frequency (Hz)**	**Amplitude (μV)**	**Notes**
Alpha	8–12	20–40	Best seen over the occipital pole when the eyes are closed
Beta	20–50	10–20	Normal awake pattern
Theta	4–7	40–80	Normal in children and in early sleep. Evidence of organic disease when seen in awake adults
Delta	<3	100–150	Seen during deep sleep. Evidence of organic disease when seen in awake adults

Note that the higher the frequency the lower the amplitude, i.e. the higher frequencies are less synchronized than the low frequencies.

waves (40–80 μV and 4–7 Hz). These are interspersed with brief periods of high-frequency activity known as 'sleep spindles'. In very deep sleep the EEG waves are slower still (*delta waves* which have a frequency of less than 3 Hz) and they have a relatively high amplitude (100–120 μV). The characteristics of the principal EEG waves are summarized in Table 11.2.

Alpha waves are best seen over the occipital region and they have a characteristic appearance, slowly growing in amplitude to a maximum and then slowly declining ('waxing and waning'). The alpha rhythm is disrupted when the subject concentrates their attention on a problem or when they open their eyes (Fig. 11.9). This is known as 'alpha block'. The alpha waves appear to be driven by feedback between the cerebral cortex and the thalamus. Theta waves are believed to originate in the hippocampus, while the delta waves probably originate from activity in the brainstem.

Epilepsy

Although detailed interpretation of the EEG is fraught with difficulties, it has proved to be of great practical value in the diagnosis and localization of organic brain disease. For this purpose, many pairs of electrodes are placed over the scalp and the pattern of activity between specific electrode pairs is scrutinized for the presence of abnormal activity. Its greatest use is in the diagnosis of *epilepsy*, which can be defined as a disorder of the electrical activity of the brain. Unlike many other disorders, epilepsy is intermittent; attacks occur infrequently although they may be triggered by specific stimuli. During an epileptic attack, consciousness may be lost and there may be overt seizures, as in a type of epilepsy called *grand mal*.

Epilepsy is caused by abnormal functioning of brain tissue that may arise from traumatic injury, from infections, or as a result of ischemic damage to the brain during birth. There may be no obvious precipitating cause (idiopathic epilepsy) and some cases of epilepsy appear to be of familial origin. Whether they are accompanied by gross seizures or by other mental changes, such as the lapses of concentration typical of *petit mal*, all epileptic attacks are associated with changes to the electrical activity

Summary

The EEG can be used to monitor the spontaneous electrical activity of the brain. The waves of the EEG are of low amplitude (10–150 μV) and their frequency and amplitude depend on the state of arousal of a person. The EEG is helpful in the diagnosis and location of abnormalities of the electrical activity of the brain such as epilepsy.

of the brain. The EEG changes that can be recorded are characteristic of particular disease processes. The EEG is not only useful in diagnosing the presence of epilepsy, but it can also be used to locate the site that triggers an epileptic attack (the *epileptic focus*). Some examples of the appearance of the EEG during an epileptic attack are shown in Fig. 11.10.

11.5 Sleep

It is common experience that, under normal circumstances, people are active during the day and sleep at night. In some other animals, the pattern is reversed and the main periods of activity occur during the night. This cyclical variation in activity is called the sleep–wakefulness cycle. Why do we sleep? What controls the sleep–wakefulness cycle? At present, these questions cannot be fully answered, but the work they have engendered has given some important information about the processes that occur during sleep.

The appearance of the EEG can be used to follow the main stages of sleep. As previously discussed, the EEG waves become slower and more synchronized as a subject passes from the awake state to deep sleep (Fig. 11.9). As sleep sets in, the EEG is dominated by theta waves and there may be periods of fast EEG activity with short periods of fast rhythmic waves called sleep spindles. As sleep becomes deeper, the sleep spindles are associated with occasional large waves to form K complexes and in deep sleep the EEG is dominated by large-amplitude low-frequency delta waves.

The depth of sleep, judged by how easy it is to arouse someone with a standard stimulus (e.g. a sound), correlates well with the state of the EEG. The deep phase of sleep associated with

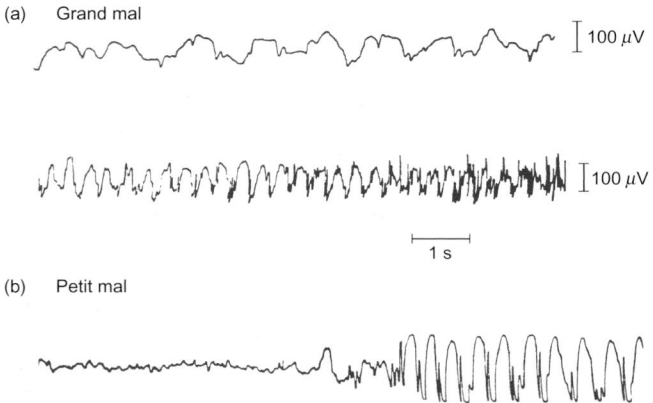

Fig. 11.10 The change in the EEG that can be seen during an epileptic attack. (a) The EEG has a preponderance of delta-wave activity, which is replaced by large spikes during a convulsive seizure. (b) The change in the EEG of another patient during a petit mal seizure. Note the characteristic spike and wave complex.

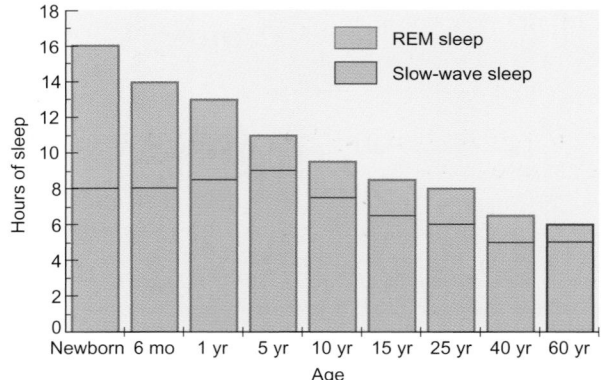

Fig. 11.11 Changes in the pattern of sleep with age. Note the decline in the total hours of sleep and in the proportion of REM sleep with increasing age.

delta-wave activity is called *slow-wave sleep* or *ortho-sleep*. It is associated with a slowing of the heart rate, a fall in blood pressure, slowing of respiration, low muscle tone, a fall in body temperature, and an absence of rapid eye movements. After 1 or 2 hours of slow-wave sleep, the EEG assumes a pattern similar to the awake state, with fast beta-wave activity (Fig. 11.9). This change in EEG pattern is associated with jerky movements of the eyes, the heart rate and respiratory rate increase, and penile erection may occur. There may also be clonic movements of the limbs, although most muscles are fully relaxed in this phase of sleep. During these changes, the subject is as difficult to arouse as during slow-wave sleep. Since the EEG has the appearance of an awake subject, this phase has been called *paradoxical sleep* although it is more often called rapid eye movement or *REM sleep*. During a single night slow-wave sleep is interrupted by four to six episodes of REM sleep, each of which lasts about 20 minutes.

The amount of time spent in sleep varies with age, as does the proportion of sleep spent as REM sleep. As Fig. 11.11 shows, very young children spend a large part of the day asleep and REM sleep accounts for nearly half of all their sleep. By the age of 20 years, only about 8 hours are spent in sleep of which less than a quarter is spent as REM sleep. The requirement for both kinds of sleep declines further with age and, by the age of 60, many people require as little as 6 hours. If a subject is woken during a REM episode, they are much more likely to report having dreams than if they are woken during periods of slow-wave sleep. Thus REM sleep is associated with dreaming. The content of dreams has received much attention over the millennia, from forecasting the future to revealing intimate details of a personality, but little can be said about the role of dreams with certainty. The vast majority are forgotten soon after they occur.

Why do we sleep? The answer to this question still eludes us. The notion that sleep is required to restore the body is an old one, but exactly what needs to be restored is not clear. It is known that the pattern of secretion of many hormones varies with the sleep–wakefulness cycle (see Chapter 12) and that sleep reflects a change in the pattern of brain activity. Sleep deprivation leads to irritability and a decline in intellectual performance. If it is prolonged, severe mental changes can occur bordering on psychosis. If subjects are deprived of REM sleep by the simple expedient of waking them every time a REM episode begins, they also become anxious and irritable. Following such a period, their sleep tends to have a higher frequency of REM episodes than normal. This suggests that the episodes of REM sleep serve an important function.

The sleep–wakefulness cycle is actively controlled from the diencephalon

In the 1930s, R. Hess found that electrical stimulation of the thalamus initiates sleep behavior in cats. During stimulation, they will circle their chosen resting place, yawn, and stretch their limbs, before curling up and going to sleep. In 1949 G. Morruzzi and H. Magoun showed that sleeping animals can be aroused by electrical stimulation of a region of the brainstem known as the ascending reticular formation. This observation suggested that sleep is controlled, at least in part, by the activity of the brainstem. Later work has suggested that some specific

groups of nerve cells containing serotonin and others that contain norepinephrine play a role in the regulation of sleep.

Serotonergic neurons are found in a series of nuclei called the raphe nuclei that are scattered along the ventral region of the brainstem. If these neurons are depleted of their serotonin, the affected animals are unable to sleep at all. However, sleep can be restored if 5-hydroxytryptophan (a precursor of serotonin) is administered. Noradrenergic neurons are located in the locus ceruleus, which lies beneath the cerebellum. If these neurons are depleted of their norepinephrine by administration of a metabolic inhibitor, the affected animals are able to enter slow-wave sleep but do not have normal REM sleep episodes.

An attractive, if simplified, explanation of these results is that activity in the serotonergic neurons initiates sleep by lowering the activity of the brainstem reticular formation. This leads to a lessening of cortical activity and results in sleep. Subsequent activity in the noradrenergic neurons gives rise to the REM episodes. However, as continued administration of inhibitors of serotonin synthesis does not cause permanent insomnia, even though serotonin levels remain low, this suggests that the system controlling the sleep–wakefulness cycle is much more complex than this simple model. Recent work has suggested that other neurotransmitters are involved.

Sleep promoting peptides are found in the CSF of sleep-deprived animals

Before the roles of the diencephalon and brainstem in the control of the sleep–wakefulness cycle were recognized, it had been suggested that, during wakefulness, the body might release some substance that accumulated during activity and would initiate sleep when its concentration had reached a certain threshold. In recent years, a number of peptides have been isolated from the CSF of sleep-deprived animals. When these peptides are injected into the cerebral ventricles of normal animals, they induce sleep. In addition, they also modulate body temperature and influence the immune response. For this reason, it has been suggested that part of the function of sleep may be to help counter infections.

11.6 Circadian rhythms

Many aspects of normal physiology are related to time. Indeed, more than 200 measurable physiological parameters have been shown to exhibit rhythmicity. Cycles of periodicity shorter than 24 hours are called *ultradian* rhythms (e.g. the heart beat, the respiratory rhythm) while those longer than 24 hours are called *infradian* rhythms (e.g. the menstrual cycle, gestation). However, most biological variables show a periodicity that roughly approximates to the Earth's rotational period—the 24-hour day. These are called *circadian rhythms* (from the Latin *circa*, meaning around, and *dies*, meaning 'a day'. Familiar examples include the core temperature, pulse rate, systemic arterial blood pressure, renal activity (as measured by the excretion of potassium), and the secretion of a number of hormones (e.g. the secretion of cortisol by the adrenal cortex).

Originally it was believed that the regular cycle of light and darkness provided the fundamental stimulus for synchronization, but tests carried out on animals and humans kept in wholly artificial environments have shown that the situation is rather more complex. If a normal subject is kept isolated from the outside world without any clues regarding the time of day, their sleep–wakefulness cycle tends to adopt a periodicity slightly longer than the 24-hour day. For most subjects the natural cycle appears to be nearer to 25 hours, so that a person isolated for 4 weeks would effectively lose a whole active day compared with someone living in a normal environment. Curiously, the daily variations in body temperature and cortisol secretion that normally closely follow the sleep–wakefulness cycle deviate from the 24-hour cycle far less, indicating that the mechanisms controlling these physiological processes may be different to those that control the sleep–wake cycle.

It is now suggested that internal biological clocks exist in the brain and other tissues, regulating many physiological processes to a roughly 24-hour periodicity, but that external environmental clues, both physical and social, may be able to entrain these biological clocks to a strict 24-hour cycle. The external clues are sometimes called 'zeitgebers' (the German for 'time-givers').

The exact nature of the intrinsic biological clocks is not clear, but recent experiments using laboratory rats suggest that the body possesses cell groups that may act as rhythm generators, in many tissues including the liver, white blood cells, salivary glands, and certain endocrine organs. These peripheral 'clocks' appear to be controlled, in turn, by cells in a region of the hypothalamus called the *suprachiasmatic nucleus* (SCN) sometimes referred to as the 'master clock'. Evidence for this comes from a variety of experiments. Brain slices or cultured neurons from the SCN retain synchronized, rhythmical firing patterns even though they are isolated from the rest of the brain. Furthermore, destruction of the SCN in laboratory rats causes a loss of circadian function which may be restored by the implantation in the SCN of cells from fetal rat brain. Circadian periodicity of cell division rates has also been demonstrated in cells isolated from other tissues, including skin and salivary glands, and most recently, genes have been identified that appear to be involved in the timing of cellular activity. Mutation or loss of these genes can lead to disruption of circadian rhythms.

The pineal gland has also been implicated in the regulation of circadian rhythmicity. This gland, which is attached to the dorsal aspect of the diencephalon by the pineal stalk, secretes a hormone called melatonin, which is synthesized from serotonin. The synthesis and secretion of melatonin is increased during darkness and inhibited by daylight. This pattern is controlled by sympathetic nerve activity that is itself regulated by light signals from the retina. Melatonin appears to induce drowsiness and loss of alertness, perhaps by modifying activity of the SCN neurons. Alterations in the secretion of melatonin, or other related neurochemicals, may be responsible for the condition known as SAD (*seasonal affective disorder*) in which the sufferer feels depressed and lethargic during the dark winter months and for which light therapy is often a successful treatment.

Although the physiological mechanisms that underlie biological rhythmicity are not established, there is no doubt that biological rhythms are of considerable significance both for normal daily activity and for clinical medicine. Shift work (particularly that which involves rotating day and night work) and travel through time zones, with its consequent jet lag, disrupt the normal circadian rhythms. This makes it difficult for an individual to maintain an active state at times when their natural biological rhythms are preparing the body for inactivity, rest, or sleep. Unsurprisingly, more errors of judgment are made at work in the early hours of the morning than at any other time. Conversely, when a worker is going home to rest or sleep following a night shift, the natural rhythms are stimulating a number of physiological processes in preparation for the activities of the day. Cortisol levels rise, along with core temperature, heart rate, and blood pressure. These changes interfere with normal sleep patterns.

Many of the physiological parameters that are routinely monitored by clinicians are influenced by the time of day. Heart rate, blood pressure, and temperature are obvious examples but there

Summary

1. Internal biological clocks exist in the brain and other tissues and regulate the activity of many physiological processes including heart rate, blood pressure, body core temperature, and the blood levels of many hormones. This intrinsic regulation is subject to environmental influences that entrain these biological clocks to a strict 24-hour cycle.

2. The pineal gland has been implicated in the regulation of circadian rhythms. This gland secretes a hormone called melatonin that is synthesized from serotonin. The synthesis and secretion of melatonin is increased during darkness and inhibited by daylight. This pattern is controlled by sympathetic nerve activity, which is regulated by light signals from the retina.

3. The sleep–wakefulness cycle is an example of a 24-hour or circadian rhythm. The stages of sleep can be followed by monitoring the EEG. During deep sleep the EEG is dominated by slow wave activity (slow-wave sleep) but this pattern is interrupted several times a night by bouts of REM sleep in which the EEG is desynchronized.

4. During slow-wave sleep, there is a slowing of the heart rate and respiratory rate, blood pressure falls, and there is extensive relaxation of the somatic muscles. During REM sleep, the heart rate and respiratory rate increase and penile erection may occur. There are rapid eye movements and there may be clonic movements of the limbs although most muscles are fully relaxed in this phase of sleep. REM episodes are associated with dreaming.

are many others, including blood cell numbers, respiratory values, enzyme activities, and blood gas levels. In addition, it is now clearly established that the effectiveness of many drugs differs according to the time of administration, as a result of differences in gastric emptying, gut motility, renal clearance rates, and the metabolic activity of cells. Indeed, the subject of chronopharmacology is becoming an increasingly important aspect of many drug therapies.

BOX 11.1 ALZHEIMER'S DISEASE

Spontaneous and progressive degeneration of neurons in specific areas of the brain or spinal cord is responsible for a number of disorders of the CNS. These include Parkinson's disease, Huntington's chorea, Creutzfeldt–Jakob disease (CJD), and Alzheimer's disease. Alzheimer's disease is the most common cause of dementia in the elderly, with more than 30 per cent of those over 85 years of age showing signs of the disorder, but it may also affect people in their thirties. Interestingly, degenerative changes characteristic of Alzheimer's disease are seen in almost all patients with Down's syndrome (see p.25) over the age of 40 years. The disease is characterized by progressive and inexorable disorientation and impairment of memory (generally over a period of 5–15 years) as well as by other defects, such as disturbances of language, visuo-spatial, and locomotor function.

The precise causes of Alzheimer's disease are not yet established, but post-mortem examination of the brains of patients has revealed some characteristic anatomical and microscopic histological changes. Gross inspection of an affected brain shows atrophy, with generalized loss of neural tissue, narrowing of the cerebral cortical gyri, and widening of the sulci. There is also dilation of the cerebral ventricles. These changes are most obvious in the frontal, temporal, and parietal lobes of the brain. The neocortex, basal forebrain, and hippocampal areas are particularly affected. The latter area is known to be associated with the formation of memories (see main text).

Microscopic examination of affected tissue reveals the presence of senile plaques which consist of a core of abnormal protein surrounded by neurofibrillary tangles within the neuronal cytoplasm. The abnormal protein of the senile plaque is β-amyloid protein (Aβ), which is formed from the breakdown of an amyloid precursor protein (APP). The neurofibrillary tangles are coarse filamentous aggregates of insoluble filaments made of a hyperphosphorylated form of a protein known as *tau protein*. Tau is a soluble, microtubule-associated protein, normally found in neurons and involved in neurite extension and maintenance. The hyperphosphorylated form found in Alzheimer's disease is insoluble and accumulates as aggregates

within the neurons, disrupting normal microtubule assembly and maintenance. Aβ is believed to increase the formation of neurofibrillary tangles derived from tau. Plaques and tangles are seen in the brains of normal elderly people but their increased prevalence and specific location within areas such as the hippocampus is diagnostic of Alzheimer's disease. Similar amyloid deposits are often seen post-mortem in the blood vessels of the brain (amyloid angiopathy) of patients who have had Alzheimer's disease.

Although the histological changes characteristic of Alzheimer's disease are well documented, the underlying causes are not established. However, a number of factors have been identified which may have a role in the development of the disease. Around 10 per cent of Alzheimer's disease cases are familial in origin. Furthermore, there is a clear association between Down's syndrome (trisomy 21) and neuronal degeneration similar to that seen in Alzheimer's disease. Thus it is likely that there is a genetic component to this disease. Indeed, it has been shown that mutations at four chromosomal loci are associated with Alzheimer's disease. Unsurprisingly, one of these is located on chromosome 21. The affected gene encodes the amyloid precursor protein, abnormal breakdown of which generates the β-amyloid plaques.

Further gene mutations also occur at loci on chromosomes 14 and 1, the so-called presenilin genes that are apparently associated with increased production of amyloid in the CNS and with an enhanced rate of apoptosis (programmed cell death). A fourth mutation on chromosome 19 occurs at the locus that encodes apolipoprotein E. The mutant form of apoE, expressed in Alzheimer's disease, seems to be involved in the transport and processing of APP. It also seems to bind Aβ more effectively than the normal form, and may therefore enhance the rate at which amyloid fibrils form.

There is no effective treatment for Alzheimer's disease at present, although considerable research effort is being directed towards the development of drugs that may be able to inhibit the activity of enzymes involved in the formation of β-amyloid.

11.7 Learning and memory

To survive in the world, all complex animals, including humans, need to learn about their environment. What is food? Where to find water? How to avoid danger, and so on. The process of learning is concerned with establishing a store of information that can be used to guide future behavior. The store of information gained through learning is known as memory. The importance of memory to normal human activity is evident from disabilities suffered by patients with various neurodegenerative diseases, notably Alzheimer's disease—see Box 11.1.

Although learning can, and often does, occur without any obvious immediate change in behavior, we can only be sure that a task has been learnt (i.e. has been stored in memory and can be retrieved) if there is some modification of behavior. For this reason, work on learning and memory requires an experimenter

to devise some external measure of performance. Therefore the study of the process of learning is bound up with the study of memory as a matter of practical necessity.

The various kinds of learning can be divided into three groups: simple learning, associative learning, and complex learning. *Simple learning* is concerned with the modification of a behavioral response to a repeated stimulus. The response may become weaker as the stimulus is perceived to have no particular importance. This process is called *habituation*. If an unpleasant or otherwise strong stimulus is given, the original reflex response is enhanced. This is called *sensitization*.

In *associative learning* an animal makes a connection between a neutral stimulus and a second stimulus that is either rewarding or noxious in some way. *Complex learning* is diverse in its nature. It includes *imprinting*, in which young birds learn to recognize their parents by some specific characteristic, *latent learning*, in which experience of a particular environment can hasten the learning of a specific task, such as finding food in a maze, and *observational learning* (*copying*).

In all types of learning, there needs to be a change in the strength of specific neural connections. In this way, a behavioral response to a given stimulus can be modified by experience to suit changing circumstances. Not all of our experiences are remembered. Many are forgotten or only partially remembered and so the central questions are:

1. What are the conditions that lead to learning?

2. Where is memory located? Are specific memories located to particular structures or are they stored in parallel pathways with overlap and significant redundancy?

3. How is memory laid down, i.e. what mechanisms are responsible for the changes in neural connectivity?

4. How is information recalled?

Associative learning

It is a matter of everyday experience that people will associate one thing with another. The chiming of a clock can be a reminder that lunch is due or that it is time to go home. In this type of association, a neutral stimulus is associated with some more important matter. This kind of behavior is not confined to humans. It can be found in many animals and was first systematically studied by I.P. Pavlov while he was working on salivary secretion in dogs.

In his experiments Pavlov paired a powerful stimulus, such as food, with a neutral stimulus—classically the ringing of a bell. He discovered that after a number of trials in which a dog was fed immediately after hearing the sound of a bell, the animal would begin to salivate on hearing the bell in anticipation of being fed. This is known as a *conditioned reflex* and the normal salivation in response to food is called the *unconditioned response*. After establishment of the conditioned response, the ringing of the bell is sufficient for the salivary response to occur. The process of establishing a conditioned reflex is known as *passive conditioning*.

In *operant conditioning* an animal learns to perform a specific task to gain a reward or avoid punishment. In this respect, the protocol is different from the establishment of a classical conditioned response where an animal responds passively to pairs of stimuli provided by the experimenter. During operant conditioning, an animal may learn to press a lever to gain a reward of a sweet drink or to avoid an electric shock. Initially, a naive animal is placed in the experimental chamber (often an apparatus known as a Skinner box). The animal then explores its new surroundings and by chance may find that pressing a lever will lead to the delivery of a food pellet or a drop of a sweet liquid. After a short period, the animal learns to associate the pressing of the lever with the delivery of food or some other powerful stimulus. In this way animals can be taught to perform quite complex tasks.

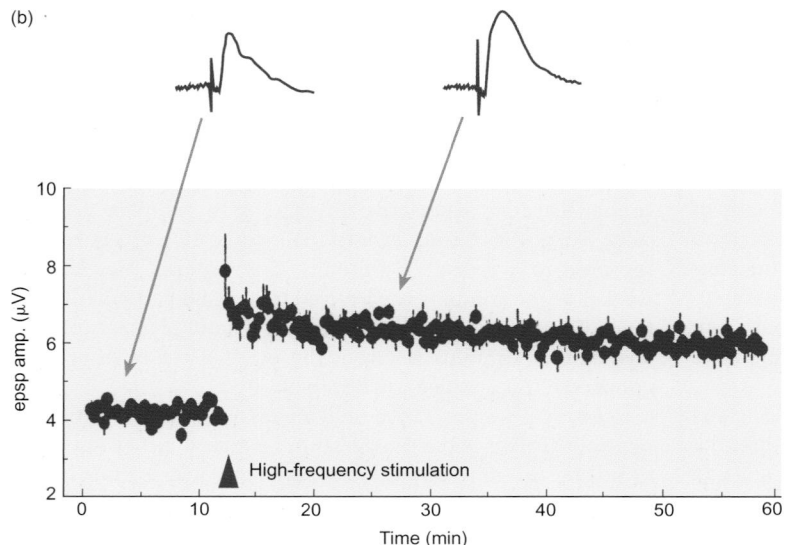

Fig. 11.12 Long-term potentiation of synaptic transmission in the hippocampus. (a) The organization of the hippocampus and the position of the electrodes for the records shown in (b). (b) Before the high-frequency stimulation, the epsps evoked by stimulation were about 4 μV in amplitude. After the period of high-frequency stimulation, they were about 6 μV (see inset examples). The increase in amplitude lasted for the duration of the experiment. (Courtesy of T.V.P. Bliss and D.A. Richards.)

In both passive conditioning and operant conditioning, the animal learns to *associate* one stimulus with another. This is also true of *aversive learning* where an animal learns to avoid unpleasant experiences, such as the eating of poisonous foods.

Cellular mechanisms of learning

All animal behavior is determined by the synaptic activity of the CNS and particularly that of the brain. In view of this, it would appear to be a logical step to look for changes in synaptic connectivity following some learnt task. This approach has proved feasible in certain lower animals that have simple nervous systems and stereotyped behavioral patterns. Changes in the strength of certain synaptic connections have been shown, for example during the habituation and sensitization of the gill withdrawal reflex of the sea-slug *Aplysia*.

In mammals, the CNS is so complex that this direct association is less easy to achieve. Nevertheless, in some brain regions, long-lasting changes in the strength of certain synaptic connections have been reliably found following specific patterns of stimulation. In some circumstances, synaptic connections are strengthened while in others they are weakened. These are precisely the kind of synaptic modulations that would be expected for the remodeling of a neural pathway during learning.

One region that has been the subject of intensive study is the hippocampus, which has been implicated in human memory (see below). The efficacy of synaptic connections between neurons of the hippocampus is readily modified by previous synaptic activity. This is precisely what is required of synaptic connections that are involved in learning. The basic neural circuit of the hippocampus consists of a trisynaptic pathway as shown in Fig. 11.12. Brief patterned stimulation leads to a very long-lasting increase in the efficacy of synaptic transmission in several of these pathways that lasts for many minutes or even hours (Fig. 11.12). This phenomenon has become known as long-term potentiation (LTP).

Since its discovery by T. Bliss and T. Lømo in 1971, LTP has become the focus of a great deal of experimental work to determine exactly how the changes in synaptic efficacy are brought about. One important factor is the increase in intracellular calcium in the postsynaptic neuron following repeated synaptic activation. Sadly, further discussion of the mechanisms involved is outside the scope of this book. Nevertheless, hippocampal LTP provides an example of synaptic transmission behaving as predicted for establishing associative memory.

Memory

Human memory does not act like a tape recorder or a computer hard disk in which our experiences are recorded in an orderly sequence which is then available for total recall. Only certain aspects of our experience are remembered; other items may be remembered for a short time and then forgotten. Yet other trivial incidents are soon beyond recall—the precise location of your pen or your spectacle case for example. Even our long-term memories are not exact—no two people will give identical versions of an event that they both witnessed. More usually, the salient points are recalled. This shows that memory is a representation of our past experience, not a record of it.

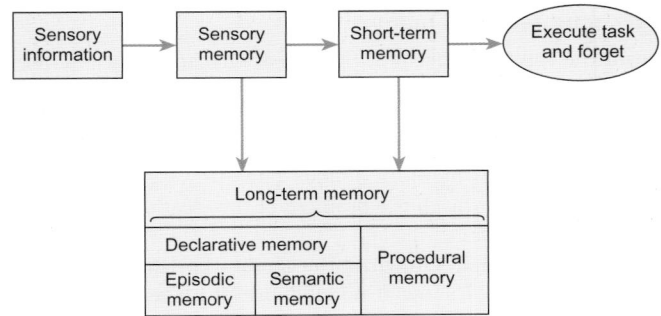

Fig. 11.13 One view of the organization of memory.

Memory must first begin with our sensory impressions—and at this early stage it will consist of the information passing through the sensory pathways. This will include that part of sensory experience that reaches the association cortex where information from the various senses is integrated to form our image of the world. This type of memory, sometimes called immediate memory, is very short lived as it is constantly being updated. For this reason, it is assumed that it is encoded in the electrical activity of networks of neurons.

More enduring information storage is classified as short-term memory and long-term memory. Information stored in short-term memory may either become incorporated into our permanent long-term memory store or discarded. Our long-term memory can be disrupted without loss of short-term memory capabilities, but the necessity of a long-term memory store becomes evident when our capacity to store information in short-term memory is exceeded, or when we are distracted in some way.

Examples of short-term memory include remembering an appointment, the rehearsal of a telephone number before dialing, or the need to buy our groceries—once the task has been executed, the incident is rapidly forgotten unless it is given some special significance. Our long-term memories include our name, the names and appearance of our family and friends, important events in our lives, and so on. In the absence of brain damage or disease, it lasts for our lifetime. This stage of memory requires the remodeling of specific neural connections. That this is the case is shown by the inability of experimental animals to learn specific tasks if inhibitors of protein synthesis are injected into their cerebral ventricles prior to training. The various stages of establishing a memory are summarized in Fig. 11.13.

Long-term memory is conveniently divided into procedural memory (also called reflexive memory) and declarative memory. *Procedural memory* has an automatic quality. It is acquired over time by the repetition of certain tasks and is evidenced by an improvement in the performance of those tasks. It is knowing how to do something. Examples are speech, in which the vocabulary and grammar of a language are acquired through experience, the playing of a game, and the mastery of a musical instrument. Once learnt, the performance of these tasks does not require a conscious effort of recall. *Declarative memory* is further subdivided into *semantic memory* (factual knowledge) and memory for events or *episodic memory* (our personal autobiography). Procedural memories are very resistant to disruption, while episodic memories (which relate to specific events) are relatively easily lost.

The neural basis of memory

From the foregoing account, it is clear that memory is a very complex phenomenon, but work carried out over the last half century has begun to reveal a little of its mysteries. The protocols for studying associative learning and careful clinical investigation of various memory deficits have provided some tools with which to approach the problem of the location of memory in the human brain.

Early animal studies in which the cerebral cortex was partially or totally removed showed no evidence of a specific memory being located at a specific site. In these experiments, animals were taught to find their way around a maze and their performance after part of their cortex was removed was compared with their performance before the operation. The results indicated that the loss of performance was related to the total area of cortex removed rather than to removal of a specific area. This suggested that memory is stored in many parallel pathways rather than at a specific site. This is, perhaps, not surprising as many parts of the brain will be involved in any task that requires extensive sensorimotor coordination.

The role of the temporal lobe

When the great Canadian neurosurgeon Wilder Penfield stimulated the temporal lobes of patients with an electrode prior to undertaking surgery for the relief of epilepsy, he found that stimulation of points in the temporal lobes, hippocampus, and amygdala apparently evoked specific memories of a vivid nature. These studies have been interpreted as indicating that specific episodic memories are located in the temporal lobes. Careful re-evaluation of this evidence suggests that this is not the case. The reported experiences tended to have a dream-like quality and may represent confabulations. Moreover, stimulation of the same site did not always elicit the same memory, and excision of the area under the electrode did not result in the loss of the memory item that had been elicited by stimulation.

How is memory retrieved?

At present, this question cannot be answered as we do not know where specific memories are laid down in the brain. For procedural memory it seems plausible that recall begins with the triggering of specific sequences of nerve activity by certain cues. The more a specific act has been rehearsed and refined, the more accurate is the performance. This is obviously true of motor acts, such as walking, running etc., but is also true of the use of language, reading music, and other similar tasks. Therefore the execution of the learnt pattern is the act of recall in this case.

The recall of declarative memory has been studied by observing how subjects recount stories they have been told. Rather than simply retelling the original with all its detail, they tend to shorten the story to its essentials—in a word, they reconstruct the story. This suggests that the recall of such memories is a specific decoding process.

A deficit in recall is called *amnesia* and is often observed in patients who have received a head injury. If the amnesia is of events before the injury, it is known as *retrograde amnesia*; if it is of events after the injury, it is *anterograde amnesia*. Total amnesia

Summary

1. Learning is the laying down of a store of knowledge that can be used as a guide to future activity. Memory is the name given to that store. Learning is a complex process that can be classified as associative or non-associative. All learning must involve changes to the connections between the neurons involved in specific tasks.

2. Memory is not, in general, localized to very specific sites but is usually distributed through a number of neural networks, each of which is involved in the task. Memory can be divided into short-term memory and long-term memory, which involves the remodeling of specific neural connections.

3. Loss of memory is called amnesia. Inability to recall events that occurred before an incident is retrograde amnesia, while inability to recall events that occurred after an incident is anterograde amnesia.

is rare and generally very short lived so that the amnesia that follows a head injury gradually passes and the period of amnesia is localized to the period immediately surrounding the incident.

The importance of the hippocampus and amygdala in memory was dramatically revealed by the case of a patient known as H.M. who underwent surgery for temporal lobe epilepsy that required bilateral removal of his hippocampus and amygdala. Following the operation he has had a profound anterograde amnesia (an inability to remember events that occurred after the operation). He is only able to remember items by rehearsal. If his repetition is interrupted, he is no longer able to recall any item. He does not know where he lives or what he last ate. In short, he has no short-term memory and effectively lives always in the present. In his own words, 'At this moment everything looks clear to me, but what happened just before? That's what worries me. It's like waking from a dream. I just don't remember'. Patients suffering degeneration of other related structures such as the mammillary bodies (which, with the hippocampus and amygdala, form part of the limbic system) also develop a profound anterograde amnesia.

A remarkable feature of H.M.'s condition is that he can learn new tasks such as how to read mirror writing. Like normal subjects, his performance improved with each training session even though he could not recall ever being trained in the task. The ability of other amnesic patients to learn new skills has been repeatedly demonstrated. This highlights the essential difference between episodic and procedural memory.

Recommended reading

Carpenter, R.H.S. (2002). *Neurophysiology* (4th edn), Chapters 13 and 14. Hodder Arnold, London.

Kandel, E.R., Schwartz, J.H., and Jessell, T.M. (2000). *Principles of neural science*, Chapters 50–4, 64, and 65. McGraw Hill, New York.

Levitan, I.B., and Kaczmarek, L.K. (2001). *The neuron* (3rd edn), Chapter 18. Oxford University Press, Oxford.

Mountcastle, V.B. (ed.) (1980). *Medical physiology* (14th edn), Chapters 10, 21, and 22. Mosby, St Louis, MO.

Nathan, P. (1988). *The nervous system* (3rd edn), Chapters 18–22. Oxford University Press, Oxford.

Sacks, O. (1986). *The man who mistook his wife for a hat.* Pan Books, London.

Shepherd, G.M. (1994). *Neurobiology* (3rd edn), Chapters 29 and 30. Oxford University Press, Oxford.

Springer, S.P., and Deutsch, G. (1989). *Left brain, right brain* (3rd edn). W.H. Freeman, New York.

Squire, L.R., and Kandel, E.R. (2002). *Memory: From mind to molecules. Scientific American Library.* W.H. Freeman, New York.

Zigmond, M.J., Bloom, F.E., Landis, S.C., Roberts, J.L., and Squire, L.R. (1999). *Fundamental neuroscience*, Chapters 50–9. Academic Press, San Diego, CA.

Medicine

Donaghy, M. (2005). *Neurology* (2nd edn), Chapters 21–4. Oxford University Press, Oxford.

Multiple choice questions

The following statements are either true or false. Answers are given below.

1. a. The frontal lobes are involved in motor control.
   b. Damage to the parietal lobes can lead to a failure of object recognition.
   c. In left-handed people the faculty of speech is located in the right hemisphere.
   d. Wernicke's aphasia results from damage to the frontal speech area.
   e. A patient with Broca's aphasia will have paralysis of the lips and tongue.

2. a. The normal EEG of an awake person is dominated by alpha waves.
   b. During deep sleep the EEG is always dominated by delta waves.
   c. The EEG can be used to monitor the health of the brain.
   d. The presence of delta waves in the EEG of an awake person is indicative of cerebral pathology.
   e. During normal sleep, the secretion of cortisol and some other hormones is increased.

Answers

1. The neural regions concerned with the control of speech are located in the left hemisphere in most people, including those who are left handed. In only a small minority of left-handed people are the regions concerned with speech located in the right hemisphere. Broca's aphasia results from damage to the frontal speech area (Broca's area).
   a. True
   b. True
   c. False
   d. False
   e. False

2. The EEG of a normal awake adult is dominated by beta waves. During deep sleep the EEG is dominated by delta-wave activity but this is interrupted by beta-like activity during episodes of REM sleep.
   a. False
   b. False
   c. True
   d. True
   e. True

CHAPTER 12

The hormonal regulation of the body

Chapter contents

After reading this chapter you should understand:

◆ The concept of glands, hormones and target tissues

◆ The principles of hormone action

◆ The role of the CNS in the regulation of the endocrine system via the hypothalamic–pituitary axis

◆ The actions of the hormones of the anterior pituitary gland, and the regulation of their secretion

◆ The actions of the hormones of the posterior pituitary gland—oxytocin and vasopressin (antidiuretic hormone)

◆ The importance of the thyroid gland—the chemistry, synthesis, storage, and secretion of the thyroid hormones

◆ The role of the adrenal cortical steroid hormones in metabolism and fluid balance

◆ The role of the adrenal medullary hormones in metabolism and their effects on the cardiovascular system

◆ The hormonal regulation of mineral metabolism with specific reference to whole body handling of calcium and phosphate

◆ The role of vitamin D, parathyroid hormone, and calcitonin

12.1 Introduction

Complex multicellular organisms require coordinating systems that can regulate and integrate the functions of different cell types. The two coordinating systems that have evolved are the nervous system and the endocrine system (see Chapter 5). The former uses electrical signals to transmit information very rapidly to discrete target cells, while the endocrine system uses chemical signaling to regulate the activity of particular cell populations. Chemical agents, *hormones*, are produced by a particular type of cell and travel in the bloodstream to other cells upon which they exert a regulatory effect. Therefore a hormone is a blood-borne chemical messenger. Endocrinology is the study of the endocrine glands and their hormones. The anatomical location of the principal endocrine glands is shown in Fig. 12.1. The figure also indicates the principal hormones secreted by each gland.

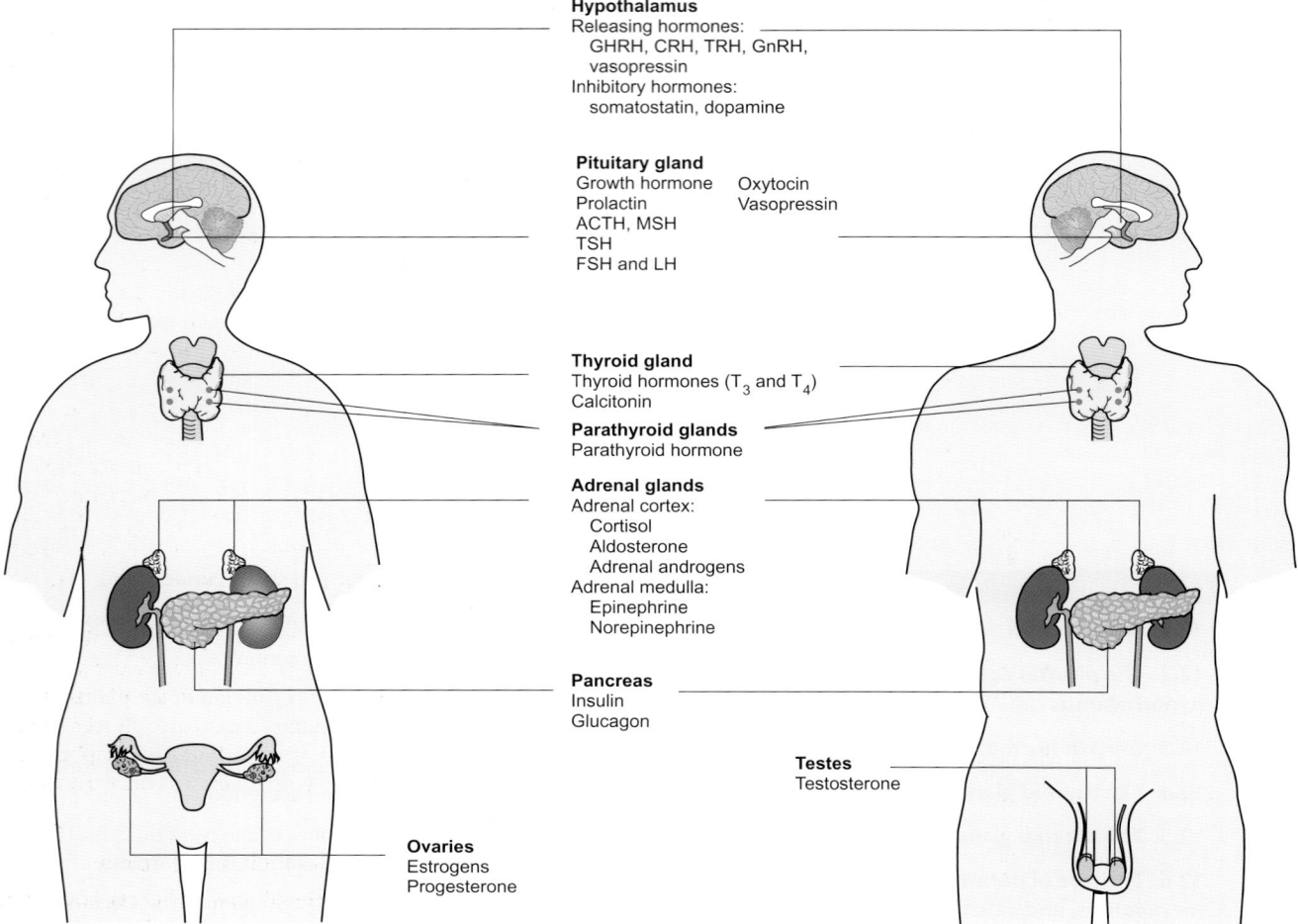

Hypothalamus
Releasing hormones:
  GHRH, CRH, TRH, GnRH,
  vasopressin
Inhibitory hormones:
  somatostatin, dopamine

Pituitary gland
Growth hormone   Oxytocin
Prolactin   Vasopressin
ACTH, MSH
TSH
FSH and LH

Thyroid gland
Thyroid hormones (T$_3$ and T$_4$)
Calcitonin

Parathyroid glands
Parathyroid hormone

Adrenal glands
Adrenal cortex:
  Cortisol
  Aldosterone
  Adrenal androgens
Adrenal medulla:
  Epinephrine
  Norepinephrine

Pancreas
Insulin
Glucagon

Testes
Testosterone

Ovaries
Estrogens
Progesterone

Fig. 12.1 The anatomical localization of the principal endocrine glands and the hormones they secrete.

Hormones play a crucial role in the activity of all the major physiological systems of the body. They are of particular importance in the regulation of growth, development, metabolism, the maintenance of a stable internal environment, and the reproductive processes of both men and women. The endocrine regulation of reproduction is discussed in Chapters 20–22.

Originally, hormones were considered to be chemical signals secreted by specific endocrine glands into the bloodstream for transport to their target cells. Although many hormones and glands fit this classical definition, it is now clear that such a description needs to be widened to include a variety of tissues which, while having other crucial roles within the body, also synthesize and secrete substances that exert effects on other cells. Examples of such organs include the heart, which secretes atrial natriuretic peptide (ANP), the liver, which secretes a number of growth factors, and the brain, which secretes specific hormones from the hypothalamus. Recently it has been shown that adipose tissue secretes a hormone (*leptin*) that travels to the brain where it may act to regulate food intake (see Chapter 19). A number of tumor cells are also known to be capable of secreting polypeptide hormones, which may result in specific disease processes. An example of a hormone-secreting tumor is the small-cell carcinoma of the lung that can secrete a variety of

hormones such as adrenocorticotrophic hormone (ACTH), vasopressin (antidiuretic hormone or ADH) and a parathyroid-hormone-like agent. A list of the major 'non-classical' endocrine organs is shown in Table 12.1.

While clearly differing in several respects, the nervous and endocrine systems are closely linked. Many neurons are capable of secreting hormones. This is known as *neurosecretion* and is seen, for example, in the hypothalamus where releasing hormones are secreted into specialized portal blood vessels. These vessels carry the releasing hormones to the anterior pituitary gland where they alter the rate of secretion of other hormones. The hormones of the posterior pituitary gland are also synthesized and secreted by neurons. Certain agents that were originally believed to act only as blood-borne hormones have since been shown to act also as neurotransmitters within the central nervous system. Examples are the gastrointestinal hormones gastrin and cholecystokinin.

Not all hormones enter the general circulation in appreciable concentrations. The hypothalamic releasing hormones, for example, are secreted into a local portal system of blood vessels (the hypophyseal portal vessels) that run between the hypothalamus and the anterior pituitary. Other hormones exert their effects at a still more local level, in some cases acting only on

TABLE 12.1 The secretion of hormones by non-classical endocrine tissues

Organ	Hormones secreted
Brain (particularly the hypothalamus)	Corticotropin releasing hormone (CRH)
	Thyrotropin releasing hormone (TRH)
	Luteinizing hormone releasing hormone (LHRH)
	Growth hormone releasing hormone (GHRH)
	Somatostatin
	Fibroblast growth factor
Heart	Atrial natriuretic peptide (ANP)
Kidney	Erythropoietin
	1,25-dihydroxycholecalciferol
Liver	Insulin-like growth factors (IGF-1, IGF-2)
	Hepcidin
Gastrointestinal tract	Cholecystokinin (CCK)
	Gastrin
	Secretin
	Pancreatic polypeptide
	Gastric inhibitory peptide
	Motilin
	Enteroglucagon
Platelets	Platelet-derived growth factor (PDGF)
	Transforming growth factor-β (TGF-β)
Lymphocytes	Interleukins
Adipose tissue	Leptin
Various sites	Epidermal growth factor
	Transforming growth factor-α (TGF-α)

contiguous cells (a so-called paracrine action) or even performing an autocrine function in modifying the secretory action of the cells that produce them (see Chapter 5). One such example is estradiol-17β which, in addition to its normal endocrine role, also modifies the activity of the follicular granulosa cells that secrete it (Chapter 20).

The chemical nature of hormones, their carriage in the blood and their modes of action

Hormones fall into three broad categories according to their chemical properties.

1. The steroids are derivatives of cholesterol and include the hormones of the gonads, the adrenal cortex, and the active metabolites of vitamin D.

2. The peptides form the largest group of hormones in the body. They vary considerably in size, from the hypothalamic thyrotropin releasing hormone (TRH), which consists of only three amino acid residues, to growth hormone (GH) and follicle-stimulating hormone (FSH), which have almost 200.

3. Derivatives of specific amino acids form the third hormonal category. This includes the catecholamines, which are derived from tyrosine, and the thyroid hormones, which are formed by the combination of two iodinated tyrosine residues.

The chemical nature of hormones influences the manner in which they are transported in the bloodstream. While the catecholamines and peptide hormones generally travel in free solution in the plasma, the steroids and thyroid hormones are very hydrophobic and are carried in the blood bound to a variety of plasma proteins including albumin. The binding proteins have a high affinity for specific hormones. Examples of these include testosterone-binding globulin, cortisol-binding globulin, and thyroid hormone-binding globulin.

Those hormones that are bound to carrier proteins in the plasma are cleared from the circulation much more slowly than those traveling in free solution. This means that their half-lives in the blood are much longer and helps to explain why the effects of steroids and thyroid hormones are more long lasting than those of peptide hormones. Furthermore, it ensures a relatively constant rate of tissue delivery and a regulated endocrine response.

All hormones act by binding to specific receptors on their target cells. These may be situated on the plasma membrane itself, within the cytoplasm, or in the nucleus. The details of the mechanisms of hormone action are discussed in Chapter 5. Briefly, steroid and thyroid hormones are thought to diffuse into cells before binding to intracellular receptors within the cytoplasm and nucleus. The receptor–hormone complex then alters gene expression. Water-soluble peptide hormones, catecholamines, and neurotransmitters bind to receptors on the plasma membrane where they exert their effects either through second messengers such as calcium, cyclic AMP, and inositol trisphosphate (IP$_3$) or by activating membrane-bound kinases which act to alter gene expression.

Measurement of hormone levels in body fluids

Most hormones are present in the plasma and other body fluids in very low concentrations. Nevertheless, modern assay techniques have made it possible to detect and quantify these hormone levels to give information concerning their rates of secretion, half-lives, and rates of clearance from the blood. Such information is valuable in the assessment of endocrine function.

The earliest hormone assays were mostly based on the biological response of a tissue or whole animal to a hormone, the *bioassay*. Most of these lacked precision, specificity, and sensitivity. Although chemical analysis of body fluids has been used to measure the plasma levels of the catecholamines, it has been of limited value in the measurement of plasma levels of other hormones. The whole field of endocrinology was revolutionized by the development of *competitive binding assays* and *radio-immunoassays*. While these assays measure immunological rather than biological activity, they are relatively quick, sensitive, and specific. Radio-immunoassays are now available for all the polypeptide, thyroid, and steroid hormones, and have been widely used for hormone assays. More recently, sensitive chemiluminescence assays have been developed for assaying peptides, and steroids have been measured using mass spectrometry.

Patterns of hormone secretion—circadian rhythms and feedback control

Many bodily activities show periodic or rhythmic changes that are controlled by the brain. Some of these patterns follow environmental cues such as the light–dark cycle or the sleep–

wakefulness cycle, while others appear to be driven by an internal biological clock that is independent of the environment. Such 24-hour rhythms are called *circadian rhythms* (see Chapter 11). The secretion of a number of hormones follows circadian rhythms. These include pituitary ACTH (and hence cortisol), growth hormone and prolactin. Knowledge of such patterns of hormone secretion is important when interpreting the results of assays performed on blood samples obtained at different times of the day.

Many biological systems, including the secretion of hormones, are regulated by negative feedback so that the response to a particular signal feeds back on the signal generator to inhibit the signal (see Chapter 1). In the case of endocrine systems, such feedback regulation may be seen when the target tissue itself secretes a hormone. Many of the anterior pituitary hormones show negative feedback control of this kind. Thus, for example, thyroid-stimulating hormone (TSH) stimulates the output of thyroid hormones from the thyroid gland. Once released into the bloodstream, thyroid hormones (T_3 and T_4) exert negative feedback on the anterior pituitary gland to inhibit the release of TSH. In this way, the secretion of TSH is kept within narrow limits.

Although negative feedback is the most widespread form of endocrine regulation, positive feedback is also seen under some conditions. During the ovarian cycle, for example, the pre-ovulatory gonadotrophin surge is brought about by the positive feedback actions of estrogens on the anterior pituitary (see Chapter 20). Positive feedback is also seen during parturition when oxytocin is secreted in response to stretching of the walls of the uterus and vagina (see Chapter 21).

12.2 The pituitary gland and the hypothalamus

The pituitary gland is situated in a depression of the sphenoid bone at the base of the skull called the sella turcica. It consists of two anatomically and functionally distinct regions, the *posterior lobe* and the *anterior lobe*. Between these lobes lies a small sliver of tissue called the *intermediate lobe*. The anterior and intermediate lobes are derived from embryonic ectoderm as an upgrowth from the pharynx. The posterior lobe is neural in origin.

Figure 12.2 illustrates the way in which the pituitary gland is formed during the first trimester of gestation. In the early embryo, the roof of the mouth lies adjacent to the third ventricle of the brain and both sheets of tissue bulge towards each other; the buccal cavity bulges upwards to form Rathke's pouch, and the neural ectoderm bulges downwards to form the infundibulum. Eventually, Rathke's pouch pinches off from the rest of the pharyngeal ectoderm and folds around the infundibulum.

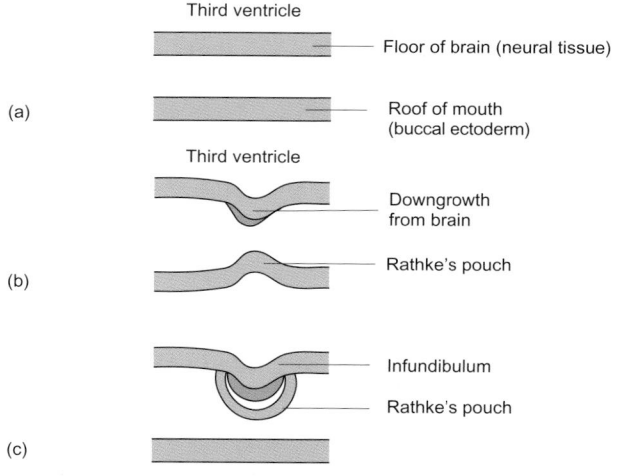

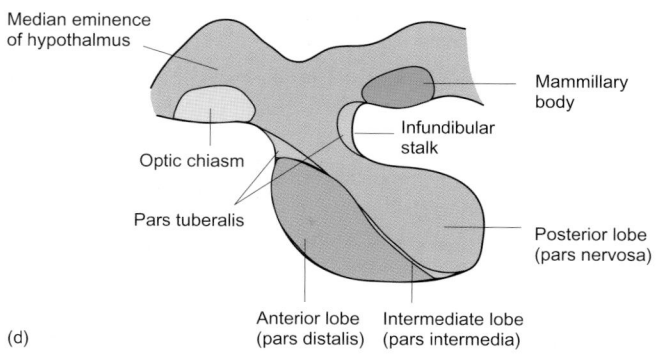

Fig. 12.2 The embryonic development of the pituitary gland. (a)–(c) Outgrowths of neural and ectodermal tissue, with separation of Rathke's pouch. (d) The structure of an adult gland in which the coloring corresponds to that shown in (a)–(c).

Summary

1. The endocrine and nervous systems coordinate the complex functions of the body. Most hormones are released by glands and travel in the blood to act on target cells at a distance. However, certain hormones exert a more localized (paracrine or autocrine) action. Some hormones are secreted by cells within organs such as the kidney, so-called 'non-classical' endocrine organs, and by tumor cells.

2. Hormones fall into three broad categories: steroids (derivatives of cholesterol), peptides (the largest group), and those derived from single amino acids (the thyroid hormones and catecholamines). The peptides and catecholamines travel in free solution in the plasma, while the steroids and thyroid hormones are largely bound to plasma proteins.

3. Hormones act by binding to specific receptor proteins in their target cells. Peptides and catecholamines bind to receptors on the plasma membrane while steroids and thyroid hormones enter cells and bind to intracellular receptors and act to modulate gene expression.

4. The secretion of most hormones is under negative feedback control but the secretion of some hormones is subject to positive feedback regulation e.g. the secretion of the gonadotrophins during the pre-ovulatory phase of the female reproductive cycle.

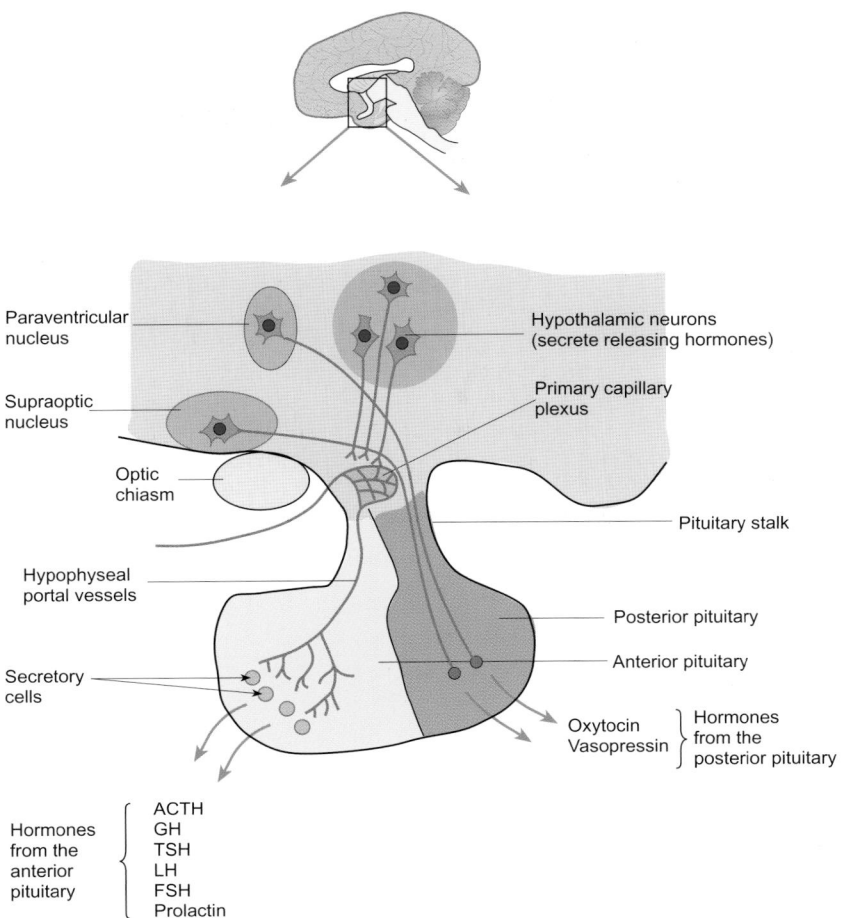

Paraventricular
nucleus

Supraoptic
nucleus

Optic
chiasm

Hypophyseal
portal vessels

Secretory
cells

Hypothalamic neurons
(secrete releasing hormones)

Primary capillary
plexus

Pituitary stalk

Posterior pituitary

Anterior pituitary

Oxytocin } Hormones
Vasopressin } from the
posterior pituitary

Hormones
from the
anterior
pituitary {
ACTH
GH
TSH
LH
FSH
Prolactin

Fig. 12.3 The relationship between the hypothalamus and the pituitary gland. Note the prominent portal system that links the hypothalamus to the anterior pituitary gland. The anterior pituitary has no direct neural connection with the hypothalamus. In contrast, nerve fibers from the paraventricular and supraoptic nuclei pass directly to the posterior pituitary where they secrete the hormones they contain into the blood stream.

Embryogenesis is complete at around 11 or 12 weeks of gestation in humans. The neural tissue, which remains as part of the brain, forms the posterior pituitary and the non-neural tissue forms the anterior pituitary.

The *neurohypophysis* strictly consists of three parts, the median eminence, which is the neural tissue of the hypothalamus from which the pituitary protrudes, the posterior pituitary itself, and the infundibular stem, which connects the two.

The *adenohypophysis* consists of two portions, the anterior pituitary itself (also called the pars distalis) and the much smaller pars tuberalis, which is wrapped around the infundibular stem to form the pituitary stalk. The principal features of the hypothalamo-pituitary system are shown in Fig. 12.3.

The anterior pituitary gland lies at the heart of the endocrine system. It secretes at least seven different hormones, many of which regulate the secretions of other endocrine organs. Although the anterior pituitary receives no direct neural input from the median eminence, it is now known that the hypothalamus plays a key role in regulating pituitary function. Between the hypothalamus and the anterior pituitary, there is a system of blood vessels known as the *hypothalamic–hypophyseal portal system*. This system originates in capillary loops in the median eminence—the primary plexus. Blood then flows in parallel veins, the long portal vessels, down the pituitary stalk to the anterior lobe. Here the portal veins break up into sinusoids which form the main blood supply of the anterior pituitary. This arrangement of blood vessels is shown in Fig. 12.3. It is now clear that the role of this system is to transport specific hormones secreted by the neurons of the median eminence to the anterior pituitary where they regulate the output of the pituitary hormones (see below). The hypothalamus, the pituitary, and the products of its target tissues form a complex functional unit.

The hormones of the anterior pituitary

The hormones of the anterior pituitary are listed in Table 12.2 together with their major target tissues and the cells that synthesize and secrete them. All the anterior pituitary hormones are proteins or polypeptides. The major hormones are:

(1) growth hormone (GH or somatotropin);

(2) prolactin;

(3) adrenocorticotrophic hormone (ACTH or corticotrophin);

(4) melanocyte-stimulating hormone (MSH), which is also secreted by the intermediate lobe;

(5) thyroid-stimulating hormone (TSH or thyrotrophin);

(6) the gonadotrophins—follicle-stimulating hormone (FSH) and luteinizing hormone (LH).

TABLE 12.2 The anterior pituitary hormones

Class of hormone	Specific hormones	Synthesized and secreted by	Target tissues	Number of amino acid residues
Somatotropic hormones These hormones have a single peptide chain	Growth hormone (GH), also known as somatotropin	Somatotrophs	Most tissues except CNS	191
	Prolactin (PRL)	Lactotrophs	Mammary glands	198
Corticotropin-related peptide hormones These are all derived from a single common precursor	Corticotropin (ACTH)	Corticotrophs	Adrenal cortex	39
	β-lipotropin (β-LPH)	Corticotrophs	? Adipose tissue	91
	β-endorphin (β-LPH 61–91)	Corticotrophs	Adrenal medulla, gut	31
	α-melanocyte-stimulating hormone (α-MSH)	Corticotrophs	Melanocytes	13
Glycoprotein hormones These are composed of a common α-peptide chain associated with a variable β-peptide chain	Thyrotropin (TSH)	Thyrotrophs	Thyroid gland	α-Chain 89 β-Chain 112
	Follicle-stimulating hormone (FSH)	Gonadotrophs	Ovaries (Granulosa cells) Testes (Sertoli cells)	α-Chain 89 β-Chain 115
	Luteinizing hormone (LH)	Gonadotrophs	Ovaries (Thecal and granulosa cells) Testes (Leydig cells)	α-Chain 89 β-Chain 115

The principal actions of these hormones will be discussed fully in the relevant sections dealing with their target glands. The various cell types that secrete the anterior pituitary hormones line the blood sinusoids. On the basis of microscopic examination of granule size and number, and of immunological staining reactions, the cells secreting certain pituitary hormones can be identified. At least five different endocrine cell types can be distinguished. Although many of the cells secrete only one type of hormone (e.g. lactotrophs secrete prolactin and somatotrophs secrete growth hormone), it is now known that some of the pituitary cells are able to produce more than one hormone. The best example of this is provided by the gonadotrophs, many of which secrete both FSH and LH. Furthermore, the corticotrophs, whilst chiefly secreting ACTH, also secrete β-lipotropin and α- and β-melanocyte-stimulating hormones.

Growth hormone (GH) and prolactin

These hormones have considerable structural similarities. They are both single peptide chains, with prolactin having 198 amino acid residues and GH 191. GH is synthesized and stored in somatotrophs, which are the most abundant pituitary cell type. It acts upon every tissue of the body, exerting powerful effects on growth and metabolism. The effects of growth hormone are discussed in detail in Section 12.3 and in Chapter 23. Prolactin is also weakly somatotropic (reflecting its structural closeness to GH), but its predominant action is to promote growth and maturation of the mammary gland during pregnancy to prepare it for milk secretion (for more details see Chapter 21).

Adrenocorticotrophic hormone (ACTH)

ACTH is a small polypeptide hormone consisting of a 39 amino acid residue chain which is part of a family of peptides derived originally from a much larger precursor molecule, pre-pro-opiomelanocortin. The relationships between these peptides derived from the common precursor are illustrated in Fig. 12.4. Pre-pro-opiomelanocortin splits to form β-lipotropin and a 146 amino acid peptide. The latter gives rise to ACTH and N-terminal peptide, while the β-lipotropin forms γ-lipotropin and β-endorphin (an endogenous opioid, part of which may further split to form metenkephalin). A variety of other peptides including α-MSH, corticotropin-like peptide (CLIP), and some with unknown physiological properties are also derived from the large precursor molecule.

ACTH regulates the function of the adrenal cortex, playing a crucial role in the stimulation of glucocorticoid secretion in response to stress. Its pattern of secretion varies during the day, showing a typical circadian rhythm. ACTH secretion is also markedly inhibited by glucocorticoids (steroid hormones secreted from the adrenal cortex—see Section 12.5). This inhibition provides a classic example of the negative feedback regulation of hormone secretion.

Melanocyte-stimulating hormone (MSH)

The chemical structure of MSH is very similar to that of ACTH (Fig. 12.4) but, although ACTH has some MSH-like activity, MSH does not appear to share any of the actions of ACTH. In certain species, it plays a role in skin pigmentation, through the stimulation of melanocytes in the epidermis, and in the control of sodium excretion, but the physiological significance of these effects in humans is unclear. Recently, however, it has been shown that α-MSH binds to a receptor (MC-1) on the human melanocyte membrane and that this binding activates tyrosinase, an enzyme required for the genesis of the pigment melanin. Furthermore, melanocytes taken from individuals who tan poorly (usually those with red hair) appear to have mutations in the MC-1 receptor.

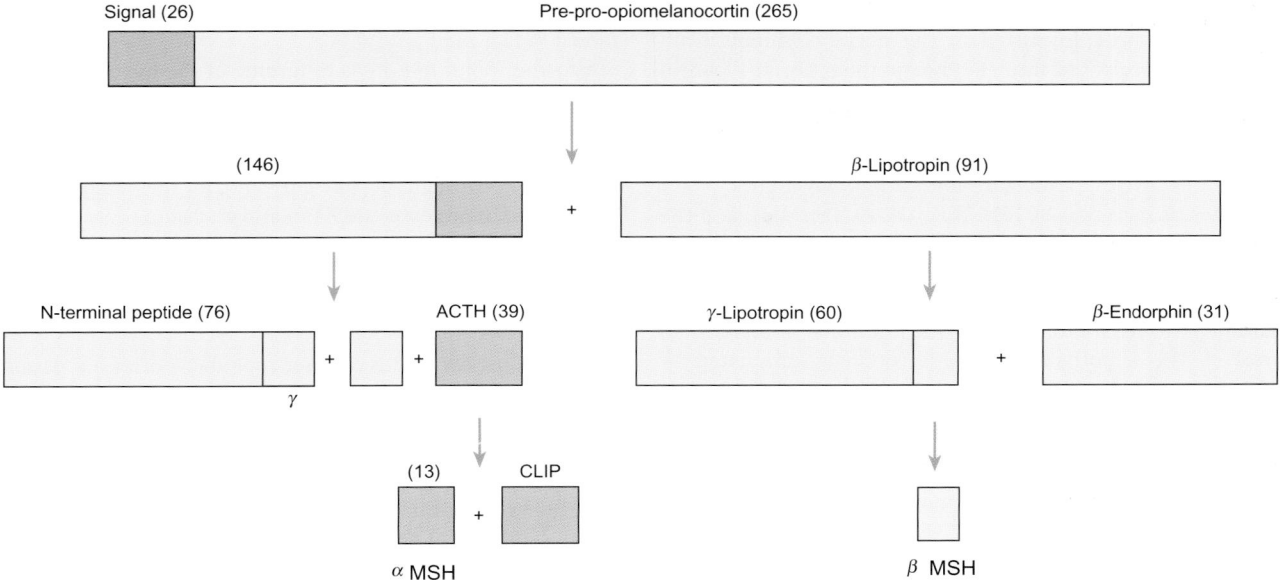

Fig. 12.4 The relationship between the amino acid sequences of the various hormones derived from pre-pro-opiomelanocortin. ACTH, β-lipoprotein, β-endorphin, and a 76 amino acid peptide are the end-products that are secreted by the anterior lobe of the pituitary gland. The numbers in parentheses after the peptide name give the number of amino acid residues in each peptide.

Pituitary glycoprotein hormones: thyroid-stimulating hormone (TSH), follicle-stimulating hormone (FSH), and luteinizing hormone (LH)

The pituitary glycoprotein hormones consist of two inter-connected amino acid chains (α- and β-subunits) containing the carbohydrates sialic acid, hexose, and hexosamine. The α-sub-units of all three hormones are identical, while the β-subunits confer biological specificity (another example of a 'family' of related hormones). They are all tropic hormones, which means that they not only regulate the secretions of their target glands but are also responsible for the maintenance and integrity of the target tissue itself. TSH controls the function of the thyroid gland, and the output of the thyroid hormones thyroxine (T_4)

and tri-iodothyronine (T_3) (see Section 12.4 for further details). The secretion of TSH is under strong negative feedback control.

The gonadotrophins FSH and LH control the cyclical activity of the ovaries. They play an important role in spermatogenesis by the testes and in the production of the sex steroids in both sexes. Both negative and positive feedback control mechanisms may operate to control their release (see Chapter 20).

The secretion of the anterior pituitary hormones is controlled by hormones released by the hypothalamus into the hypophyseal portal blood

The hypothalamus controls the secretory activity of the anterior pituitary gland. Specific hormones are synthesized in the cell

TABLE 12.3 The hypothalamic releasing and inhibitory hormones

Hypothalamic hormone	Alternative name	Structure	Stimulates	Inhibits
Vasopressin	Antidiuretic hormone (ADH)	9 Amino acid peptide	ACTH	–
Corticotropin-releasing hormone	CRH	41 Amino acid peptide	ACTH	–
Luteinizing hormone-releasing hormone	LHRH FSH-releasing factor Gonadotropin releasing hormone (GnRH)	10 Amino acid peptide	LH, FSH	–
Thyrotropin-releasing hormone	TRH	3 Amino acid peptide	TSH, prolactin	–
Growth hormone releasing-hormone	GHRH	44 Amino acid peptide	GH	–
Growth hormone-inhibiting factor	Somatostatin, GHIH	14 Amino acid peptide	–	GH, prolactin, TSH
Dopamine	Prolactin-inhibitory hormone (PIH)	Catecholamine	–	Prolactin

bodies of neurons lying within discrete areas of the median eminence and are transported to the nerve terminals from where they are released into the hypophyseal portal blood in response to neural activity. The hypothalamic hormones are then carried down the pituitary stalk to the anterior lobe where they act on specific pituitary cells to modify the secretion of one, or sometimes several, of the anterior pituitary hormones.

The rates of secretion of TSH, FSH, LH, ACTH, MSH, and the other peptides related to ACTH are all stimulated by hypothalamic hormones (known as *releasing hormones*), while the secretion of prolactin is mainly regulated by the inhibitory effect of dopamine. The release of GH (growth hormone) is under dual control by the hypothalamus. Its secretion is stimulated by growth hormone-releasing hormone (GHRH) and suppressed by another peptide, somatostatin also known as growth hormone-inhibiting hormone (GHIH). All the major hypothalamic releasing and inhibiting hormones, together with their target hormones and alternative names, are listed in Table 12.3.

The hypothalamic releasing hormones are localized to specific groups of neurons in the median eminence

The neurons that synthesize and store the various hypothalamic releasing hormones have been identified by immunocytochemistry (a specific staining technique that identifies substances by their immunological reactivity). Most of the releasing hormones seem to be produced by relatively discrete groups of neurons in the median eminence. For example, CRH is located mainly in neurons of the paraventricular nucleus, along with vasopressin, which also stimulates ACTH secretion. GnRH is found mainly in neurons of the medial preoptic area and in the arcuate nucleus. Dopamine is also present in neurons of the arcuate region, while TRH is located in both the preoptic and paraventricular nuclei.

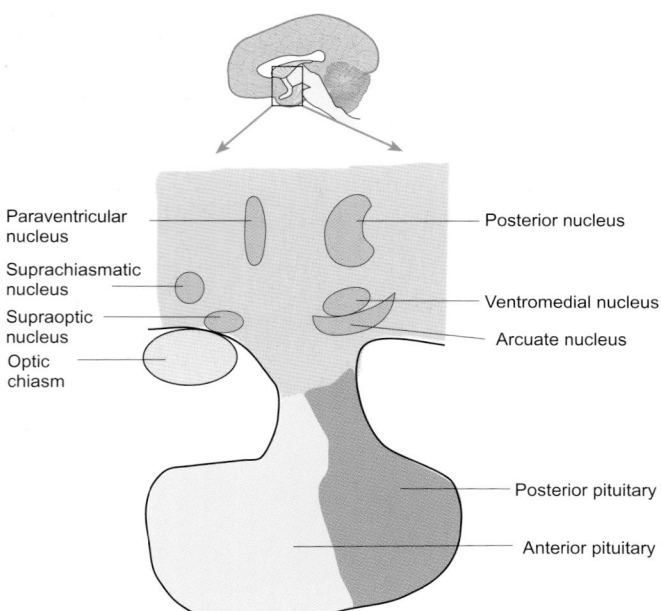

Fig. 12.5 A diagrammatic representation of the location of the principal nuclei of the hypothalamus that are associated with the production and secretion of releasing hormones.

A simple diagram showing the positions of these nuclei is shown in Fig. 12.5.

A further point to note is that some of the hypothalamic hormones can be found in parts of the body other than the hypothalamus, acting in different ways. For example, somatostatin is found acting as a neurotransmitter in other parts of the brain, as a hormone throughout the gut, and in the pancreas as an inhibitor of the release of insulin and glucagon.

Feedback mechanisms operate within the hypothalamo-pituitary–target tissue axes to ensure fine control of endocrine function

The introduction to this chapter includes a brief discussion of the regulatory role played by both negative and positive feedback mechanisms throughout the endocrine systems of the body. Such processes are of the utmost importance in determining the responsiveness of the anterior pituitary to the hypothalamic releasing hormones. They contribute to the overall control of the secretion of the anterior pituitary hormones and of its target glands.

In many cases, the output of a pituitary hormone is increased by the removal of its target gland. For example, removal of the thyroid gland, with the subsequent loss of the thyroid hormones, stimulates an increase in the output of TSH from the anterior pituitary. It is believed that the responsiveness of the TSH-secreting pituitary cells to hypothalamic TRH is enhanced in the absence of thyroid hormones. Similarly, administration of exogenous thyroxine depresses the output of TSH by reducing pituitary sensitivity to TRH. There may also be a direct effect of thyroid hormones on the output of TRH itself. Other feedback loops are thought to operate in a similar fashion to modulate pituitary function. For example, ACTH secretion is depressed by the adrenal steroids as a result of both a direct inhibition of CRH release and a reduction in the responsiveness of the ACTH-secreting cells of the anterior pituitary. Although the secretion of prolactin is largely controlled by the inhibitory action of dopamine, its release is stimulated by TRH. Like the other hormones discussed above, prolactin secretion is subject to negative feedback control. In this case, however, prolactin inhibits its own release by stimulating further output of dopamine.

The role of the posterior pituitary gland (neurohypophysis)

As described above, the posterior lobe of the pituitary gland develops as a downgrowth from the hypothalamus and, unlike the anterior pituitary, it is connected to the hypothalamus via a nerve tract (the hypothalamo-hypophyseal nerve tract). For this reason the posterior pituitary is also known as the neurohypophysis, pars nervosa, or neural lobe.

The posterior pituitary secretes two hormones. These are *oxytocin* and *vasopressin* (or ADH). The hormones are synthesized within the cell bodies of large (magnocellular) neurons lying in the supraoptic and paraventricular nuclei of the hypothalamus. They are transported in association with specific proteins, the neurophysins, along the axons of these neurons to end in nerve terminals that lie within the posterior lobe. Prior to secretion, these hormones are stored in secretory granules either in the terminals themselves or in varicosities (Hering bodies) that are

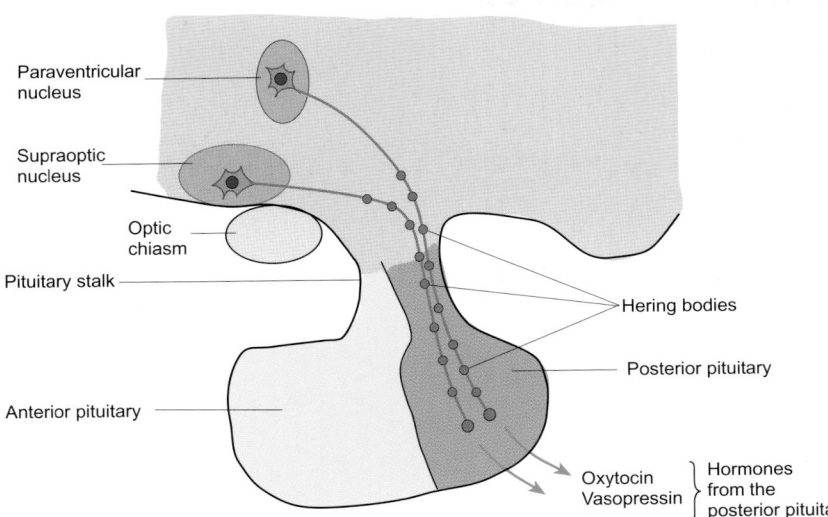

Fig. 12.6 A diagrammatic representation of the relationship between the supraoptic and paraventricular nuclei and the posterior pituitary gland (the neurohypophysis). The neurosecretory fibers originate in these nuclei and terminate in the posterior pituitary gland itself.

distributed along the length of the axons (Fig. 12.6). The hormones are secreted into the capillaries that perfuse the neural lobe in response to nerve impulses originating in the supraoptic and paraventricular nuclei. Both oxytocin and vasopressin are secreted by calcium-dependent exocytosis similar to the secretion of neurotransmitters at other nerve terminals (see Chapter 5).

Oxytocin and vasopressin (ADH) are closely related structurally but have different functions

Oxytocin and vasopressin are both nonapeptides. They differ in only two of their amino acid residues, as shown in Fig. 12.7. Although they are secreted along with the neurophysin molecules to which they are bound in the neurons, once released they circulate in the blood largely as free hormones. The kidneys, liver, and brain are the main sites of clearance of these peptides, which have a half-life in the bloodstream of around a minute.

Both oxytocin and vasopressin act on their target cells via G protein-linked cell surface receptors (see Chapter 5). Interaction of oxytocin with its receptors stimulates phosphoinositide turnover and thereby raises the level of intracellular calcium in the myoepithelial cells of the mammary gland. In turn, the increased intracellular calcium activates the contractile machinery to cause milk ejection (see Chapter 21). There are two main classes of vasopressin receptors, V_1 and V_2. Interaction

with V_1 receptors increases phosphoinositide turnover and elevates intracellular calcium. V_1 receptors mediate the effects of vasopressin on vascular smooth muscle. The renal actions of the hormone are mediated by V_2 receptors, with cyclic AMP as the second messenger (see Chapter 17).

Actions of vasopressin (ADH)

The principal physiological action of vasopressin is as an antidiuretic hormone. For this reason it is also known as ADH. This role is discussed in more detail in Chapter 17. Briefly, it facilitates the reabsorption of water from the final third of the distal tubule and the collecting ducts of the kidney by increasing the permeability of these cells to water. The net result of its actions is an increase in urine osmolality and a decrease in urine flow. Additional renal effects of vasopressin include stimulation of sodium reabsorption and urea transport from lumen to interstitial fluid in the medullary collecting duct. By this action vasopressin helps to maintain the osmotic gradient from cortex to papilla which is crucial for the elaboration of a concentrated urine (see Chapter 17).

Vasopressin, as its name suggests, is also a potent vasoconstrictor which acts particularly on the arteriolar smooth muscle of the skin and splanchnic circulation. Despite this, the increase in blood pressure brought about by vasopressin is small under normal circumstances because the hormone also causes bradycardia and a decrease in cardiac output, both of which tend to offset the increase in total peripheral resistance. However, the

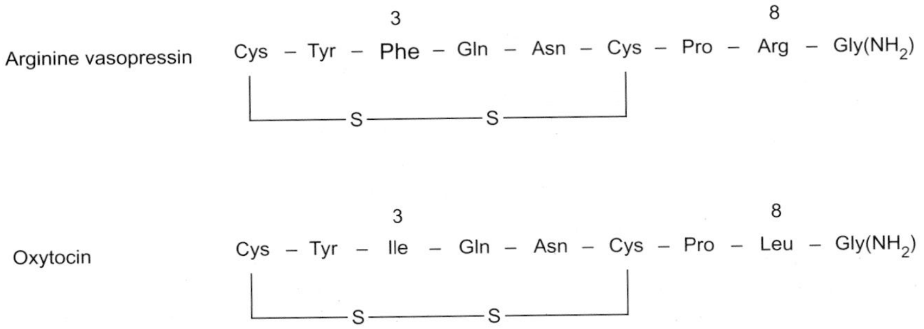

Arginine vasopressin

$$\text{Cys} - \text{Tyr} - \overset{3}{\text{Phe}} - \text{Gln} - \text{Asn} - \text{Cys} - \text{Pro} - \overset{8}{\text{Arg}} - \text{Gly(NH}_2)$$

Oxytocin

$$\text{Cys} - \text{Tyr} - \overset{3}{\text{Ile}} - \text{Gln} - \text{Asn} - \text{Cys} - \text{Pro} - \overset{8}{\text{Leu}} - \text{Gly(NH}_2)$$

Fig. 12.7 The amino acid sequences of arginine vasopressin and oxytocin. The small differences in structure result in molecules that have very different physiological effects.

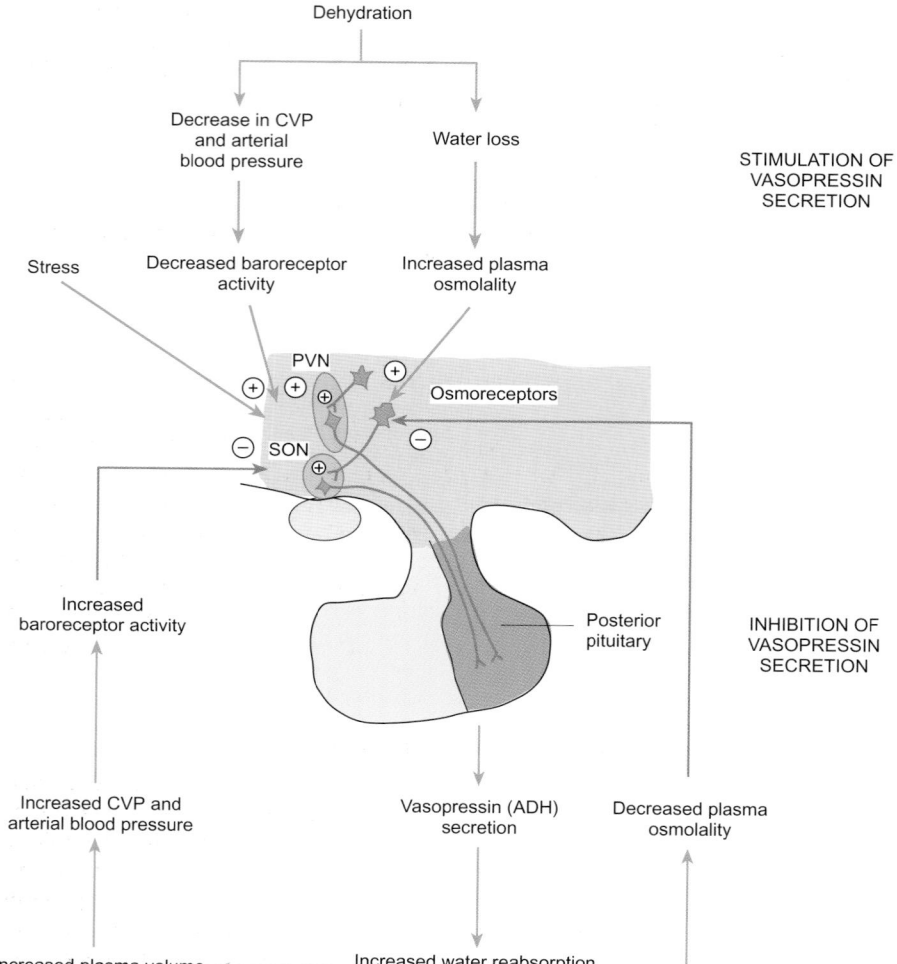

Fig. 12.8 A schematic diagram showing the factors that regulate vasopressin release in response to changes in plasma osmolality and blood volume. Green arrows directed towards the hypothalamus indicate stimulation; red arrows indicate inhibition. CVP, central venous pressure; PVN, paraventricular nucleus; SON, supraoptic nucleus.

vasoconstrictor effect of vasopressin is important during severe hemorrhage or dehydration (see Chapter 28). Vasopressin also exerts a CRF-like activity whereby it stimulates the release of ACTH from the anterior pituitary. It may also play a role in the control of thirst.

The circumstances under which vasopressin (ADH) is secreted are discussed in Chapters 17 and 28. A brief résumé will be given here. Figure 12.8 illustrates the changes in plasma osmolality and volume that control vasopressin release. The principal physiological stimulus for its release is an increase in the osmolality of the circulating blood. Osmoreceptors located in the hypothalamus detect this increase and activate neurons in the supraoptic and paraventricular nuclei. As a result of the increased rate of action potential discharge of these neurons, vasopressin secretion into the circulation is increased.

Vasopressin is secreted in response to a fall in the effective circulating volume (ECV), during hemorrhage for example (see Chapter 28), and in response to other factors including pain, stress, and other traumas. The amount of vasopressin secreted when there is a fall in the ECV increases proportionately as the central venous pressure and arterial pressure fall (see Chapter 28, Fig. 28.4). Central venous pressure is sensed by the low-pressure receptors (volume receptors) of the atria and great

veins, while the arterial blood pressure is sensed by the arterial baroreceptors which are located in the carotid sinuses and aortic arch (see Chapter 15 for further details).

Disorders of vasopressin secretion

The consequences of under- or overproduction of vasopressin may easily be predicted from the description of its actions given above. Abnormally high circulating levels of vasopressin may result from certain drug treatments, brain traumas, or vasopressin-secreting tumors. Such patients will have highly concentrated urine, with water retention, lowered plasma osmolality, and sodium depletion.

A lack of vasopressin leads to a condition known as *diabetes insipidus* in which the patient is unable to produce a concentrated urine or to limit the production of urine even when the plasma osmolality is raised. To counteract the loss of water via the kidneys, the sufferer has to drink a large amount of fluid. This condition may result from head injuries or tumors that damage the posterior pituitary. The deficiency is treated by the administration of synthetic analogs of vasopressin. Diabetes insipidus may also arise as a consequence of a loss of vasopressin receptors in the distal nephron (nephrogenic diabetes insipidus).

Actions of oxytocin

The main actions of oxytocin are described in some detail in Chapter 21. Briefly, this hormone stimulates the ejection of milk from the mammary glands in response to suckling—the milk-ejection or 'let down' reflex. It causes the myoepithelial cells surrounding the ducts and alveoli of the gland to contract, thus squeezing milk into the lactiferous sinuses and towards the nipple. Oxytocin may also play a role in expelling the fetus and placenta during labor (p. 463). It is known to promote contractions of the uterus and to increase the sensitivity of the myometrium to other spasmogenic agents. In males, oxytocin is believed to have a role in erection, ejaculation, and sperm progression.

Control of oxytocin secretion

Like vasopressin, oxytocin is released in response to afferent neural input to the hypothalamic neurons that synthesize the hormone. Although oxytocin is released in response to vaginal stimulation (particularly during labor), the most potent stimulus for release is the mechanical stimulation of the nipple by a suckling baby. Impulses from the breast travel to the hypothalamus via the spinothalamic tract and the brainstem (see Figure 21.18). Certain psychogenic stimuli can also influence the secretion of oxytocin. The milk-ejection reflex is known to be inhibited by certain forms of stress and to be stimulated by the cry of the hungry infant or during play prior to feeding.

Disorders of oxytocin secretion

Excessive oxytocin secretion has never been demonstrated, but oxytocin deficiency results in failure to breastfeed an infant because of inadequate milk ejection.

12.3 Growth hormone

The anterior pituitary secretes growth hormone in larger amounts than any other hormone. The major form of human growth hormone (hGH) is a large peptide containing 191 amino acid residues. The anterior pituitary contains around 10 mg of growth hormone which, in adults, is secreted at a rate of around 1.4 mg per day. However, it is important to note, that over a 24-hour period the rate of GH secretion fluctuates considerably. In the plasma, about 70 per cent of GH is bound to various proteins, including a specific GH-binding protein which is derived by cleavage of the extracellular region of the GH receptor present on target cells.

Actions of growth hormone

Growth hormone exerts a wide range of metabolic actions that involves virtually every type of cell except neurons. However, its major targets are the bones and skeletal muscles. It stimulates growth in children and adolescents but continues to have important metabolic effects throughout adult life. For the purposes of discussion, it is convenient to divide the actions of growth hormone into two categories. These are its direct effects on the metabolism of fats, proteins, and carbohydrates, and its indirect actions that result in skeletal growth. The principal actions of GH are illustrated in Fig. 12.9.

Direct metabolic effects of growth hormone

Essentially, growth hormone is anabolic, i.e. it promotes protein synthesis. It is also a glucose-sparing agent with an anti-insulin-like diabetogenic action. Growth hormone stimulates the uptake of amino acids by cells (particularly those of the liver, muscle, and adipose tissue) and the incorporation of amino acids into proteins in many organs of the body. There is an increase in the rates of synthesis of both RNA and DNA and ultimately of cell division. This effect is particularly important during the growing years when it contributes to the increase in bone length and soft-tissue mass. There is also an increase in the rate of chondrocyte differentiation from fibroblasts in cartilage. The net effects of GH on protein metabolism are an increase in the rate of protein synthesis, a decrease in plasma amino acid content, and a positive nitrogen balance (defined as the difference between daily nitrogen intake in food and excretion in urine and feces as nitrogenous wastes).

The actions of GH on lipid and carbohydrate metabolism are essentially diabetogenic. GH increases plasma glucose levels in two ways. It decreases the rate of glucose uptake by cells (largely

Summary

1. The anterior pituitary gland (adenohypophysis) is situated in a shallow depression at the base of the skull. It secretes growth hormone (GH), thyroid-stimulating hormone (TSH), adrenocorticotrophic hormone (ACTH), follicle-stimulating hormone (FSH), luteinizing hormone (LH), prolactin, and a number of related peptides.

2. A portal system of blood vessels (the hypophyseal portal vessels) carries hormones from the median eminence region of the hypothalamus to the anterior pituitary. These hormones control the release of the anterior pituitary hormones.

3. The hypothalamo–hypophyseal axis is regulated by feedback mechanisms involving many of its target organs.

4. The posterior pituitary gland (the neurohypophysis) is a downgrowth of the hypothalamus. It secretes two peptide hormones, vasopressin (antidiuretic hormone or ADH) and oxytocin. These are synthesized in the cell bodies of neurons within the paraventricular and supraoptic nuclei. Oxytocin and vasopressin are structurally similar but have very different actions.

5. Vasopressin stimulates the reabsorption of water from the medullary collecting ducts of the renal nephrons, thereby causing a decrease in urinary volume and an increase in urinary osmolality. It also exerts a pressor effect on vascular smooth muscle. It is released in response to an increase in the osmotic pressure of the plasma or a fall in blood volume.

6. Oxytocin stimulates the ejection of milk from the lactating breast. It also increases the contractile activity of the uterine myometrium and may play a role in expulsion of the fetus during parturition. In males oxytocin appears to play a role in erection, ejaculation, and sperm progression.

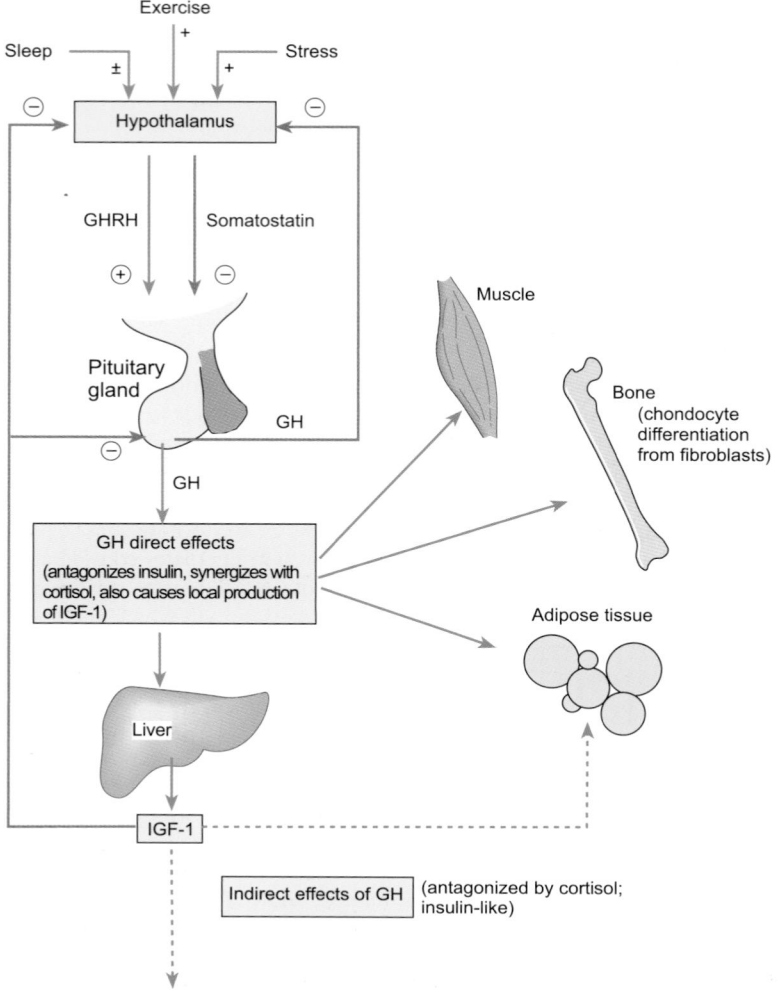

Fig. 12.9 A schematic diagram showing the principal actions of GH and the factors that regulate its secretion.

those of muscle and adipose tissue) and increases the rate of glycogenolysis by the liver. GH promotes the breakdown of stored fat in adipose tissue and the release of free fatty acids into the plasma. This action is important for providing a non-carbohydrate source of metabolic substrate for ATP generation by tissues such as muscle, and this action is enhanced during fasting. As a result of the increased oxidation of fat, there is a reduction in the respiratory quotient (see Chapter 24).

The direct actions of GH on the various metabolic substrates can be summarized as follows: essentially an anabolic hormone, GH promotes protein synthesis and encourages the use of fats for ATP synthesis, thus conserving glucose.

Indirect actions of GH on skeletal growth

The main indirect physiological actions of GH are the maintenance of tissues and the promotion of linear growth during childhood and adolescence. These actions, particularly the latter, are considered further in Chapter 23. It is important to remember that growth is a highly complex process that is under the

control of numerous agents, including a variety of growth factors and hormones in addition to GH itself.

The growth-promoting effects of GH are mediated partly by its actions on amino acid transport and partly by its stimulation of protein synthesis. Skeletal growth results from the stimulation of mitosis in the epiphyseal discs of cartilage present in the long bones of growing children. GH exerts direct actions on both cartilage and bone to stimulate growth and differentiation, which is aided by polypeptides called *insulin-like growth factors* (*IGFs*) that are synthesized chiefly by the liver. However, IGFs are also synthesized in the growing tissues themselves. These agents encourage the cartilage cells to divide and to secrete more cartilage matrix. As growing cartilage is eventually converted to bone, the growth of cartilage enables the bone to increase in length (see Chapter 23). Growth of the long bones in response to GH ceases once the epiphyseal disks themselves are converted to bone at the end of adolescence (a process known as fusion of the epiphyses). No further increase in stature can then occur despite the continued, though declining, secretion of GH throughout adulthood.

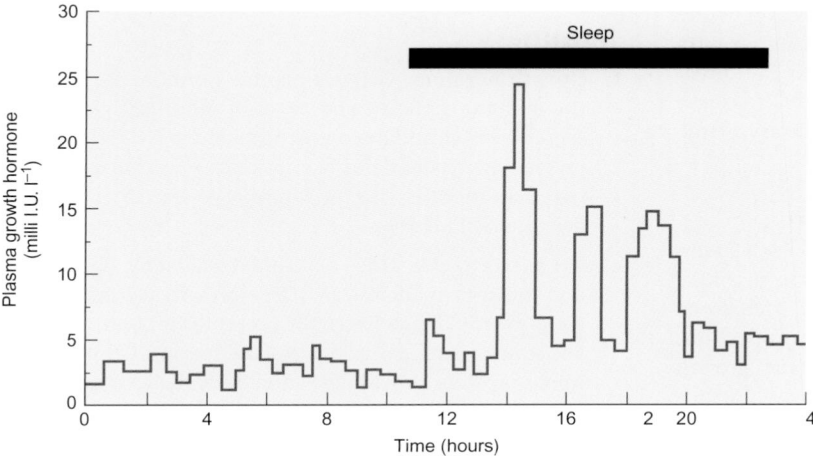

Fig. 12.10 Diurnal fluctuations in GH secretion in a normal 7-year old child. Note the marked pulsatile release of the hormone during sleep.

What are the IGFs and how do they promote growth?

The IGFs show significant amino acid sequence homology with the pancreatic hormone insulin and share some of its effects. Two such factors have been identified, IGF-1 and IGF-2, the former having the most potent growth-promoting effect and the latter the strongest insulin-like action. An important action of GH is to stimulate IGF-1 production from the liver. IGF-1 stimulates a variety of the cellular processes that are responsible for tissue growth. In many cells (e.g. fibroblasts, muscle, and liver), it stimulates the production of DNA and increases the rate of cell division. It also encourages the incorporation of sulfate into chondroitin in the chondrocytes (the cells which make cartilage for bones), and the synthesis of glycosaminoglycan for cartilage and collagen formation.

The growth-promoting effects of GH seem to be crucial for growth during childhood, particularly from the age of about 3 years to the end of adolescence. IGF-1 levels reflect the rate of growth during this time and there is a marked increase at puberty. GH and IGF-1 seem to be less important to growth during the fetal and neonatal periods when IGF-2 may be more significant.

Growth hormone secretion is governed by the hypothalamic secretion of GHRH and somatostatin

As described earlier, growth hormone is under dual control by hormones released from hypothalamic neurons. Its secretion is stimulated by GHRH and inhibited by somatostatin. Therefore the rate at which GH is released from the somatotrophic cells of the anterior pituitary is determined by the balance between these two hormones. Since GH secretion declines when all hypothalamic influences are removed (either experimentally or in disease), it can be assumed that the positive effects of GHRH on GH release are normally dominant.

GH, like all the anterior pituitary hormones, is released in discrete pulses, which are most frequent in adolescence. The detailed mechanism of pulsatile release is unclear, but peaks of output appear to coincide with peaks of GHRH output while troughs coincide with increased rates of somatostatin release. GH secretion also shows a definite circadian rhythm, with marked elevations in output associated with periods of deep sleep during which bursts of secretion occur every 1–2 hours. Serotonergic pathways in the brain are believed to mediate this response. The pulsatile and circadian nature of GH secretion is illustrated in Fig. 12.10. It is important to be aware of these characteristics of GH secretion when carrying out measurements of plasma GH levels in clinical practice. A single measurement will be insufficient for diagnostic purposes and it will be necessary to perform frequent serial assays.

A number of physical and psychological stresses promote the secretion of GH. Examples include anxiety, pain, surgery, cold, hemorrhage, fever, and strenuous exercise. Adrenergic and cholinergic pathways in the brain are believed to mediate these effects. The significance of the raised GH output is not fully understood, but it seems likely that the glucose-sparing effect of the hormone would be of value in circumstances of this kind.

Metabolic factors influencing GH secretion

The most potent metabolic stimulus for GH release is hypoglycemia. This is an appropriate homeostatic response since GH acts as a glucose-sparing hormone, promoting the breakdown of fats to make fatty acids available for oxidation. At the same time, GH inhibits the uptake of glucose by the peripheral tissues and conserves glucose for use by the brain. In contrast with the effect of hypoglycemia, an oral glucose load rapidly suppresses the release of GH.

GH is secreted during prolonged fasting. Again, its glucose-sparing effects and its effects on lipid metabolism are important here to ensure that those tissues, such as the CNS, which rely entirely on glucose as their metabolic substrate are adequately supplied. Other metabolic factors known to increase the rate of GH secretion include a rise in plasma amino acid levels and a reduction in the plasma concentration of free fatty acids. All these metabolic actions are mediated by changes in the output of GHRH and somatostatin.

Feedback actions of GH and the IGFs

The secretion of growth hormone appears to be influenced by the plasma concentration of GH itself. High plasma GH levels inhibit the release of further GH. This is an example of a short negative feedback loop in which GH depresses its own release by altering the rates of secretion of GHRH and somatostatin by the hypothalamus or by altering the sensitivity of the anterior pituitary to these hypothalamic factors. In addition to the direct feedback effects of GH itself, IGF-1 can also inhibit GH release via a feedback action on its synthesis by the pituitary gland. These interactions are shown in Fig. 12.9.

Disorders of growth hormone secretion

Both hypersecretion and hyposecretion of GH may result in disorders of growth. Hypersecretion in children results in *gigantism*, a condition in which growth is exceptionally rapid. Heights of 2.4 m (8 feet) may be reached, although the body proportions remain relatively normal as GH stimulates coordinated growth of the skeleton in children and adolescents.

GH deficiency in children results in *pituitary dwarfism* in which the growth of the long bones is slowed. Untreated individuals will attain a maximum height of no more than around 1.2 m (4 feet), but usually have fairly normal body proportions. Lack of GH is often accompanied by other pituitary hormone deficiencies, particularly of TSH and the gonadotrophins. If so, an affected child will be malproportioned and will fail to mature sexually. When GH deficiency is diagnosed prior to puberty, GH-replacement therapy (using genetically engineered human GH) can promote normal somatic growth (see also Chapter 23).

If excessive amounts of GH are secreted after adult height has been attained and the epiphyseal plates have closed, a condition known as *acromegaly* results. Acromegaly (literally translates as 'enlarged extremities') is characterized by enlargement and thickening of bony areas still responsive to GH, such as the hands, feet, face and jaw. Thickening of soft tissues leads to coarsening of the facial features (Fig. 12.11) and enlargement of

Summary

1. Growth hormone (GH) is a peptide hormone secreted by the somatotrophs of the anterior pituitary gland. GH secretion is stimulated by hypothalamic GHRH and suppressed by hypothalamic somatostatin. Secretion shows a circadian rhythm with the highest rates occurring during deep (non-REM) sleep.

2. GH exerts a wide range of metabolic actions, both direct and indirect (mediated by IGFs—growth factors synthesized mainly by the liver). GH promotes protein synthesis and its anti-insulin action is glucose sparing. GH promotes lipolysis, thus providing a non-carbohydrate source of substrate for ATP generation.

3. GH is crucial to normal skeletal growth between the ages of about 3 years and puberty. It helps to maintain tissues in adulthood. Skeletal growth occurs in response to IGFs stimulated by GH. These factors encourage cartilage cells to divide and enhance the deposition of cartilage at the epiphyses (growth plates).

4. The most important metabolic stimulus for GH secretion is hypoglycemia. Increased plasma levels of amino acids and reduced plasma levels of free fatty acids will also stimulate GH output.

5. GH deficiency in childhood results in pituitary dwarfism, while excess GH output results in gigantism. Excessive secretion of GH during adulthood causes acromegaly.

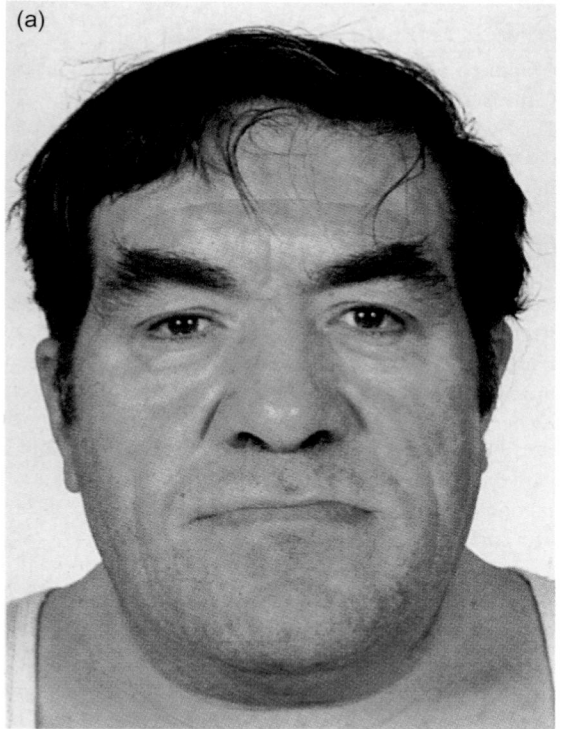

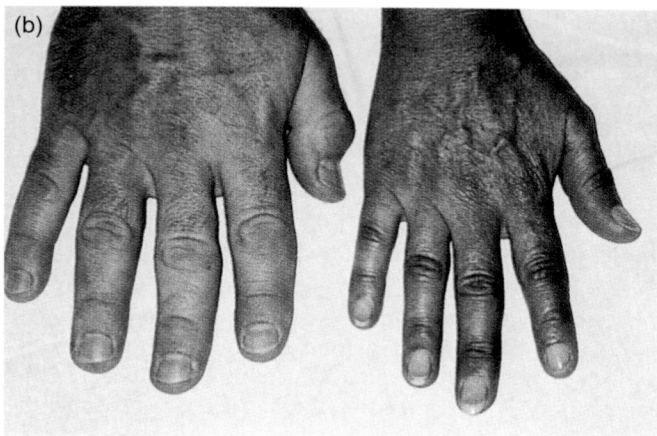

Fig. 12.11 A patient with acromegaly caused by the overproduction of GH during adulthood. Note the characteristically coarsened facial features (a) and the enlargement of the hand shown on the left in (b).

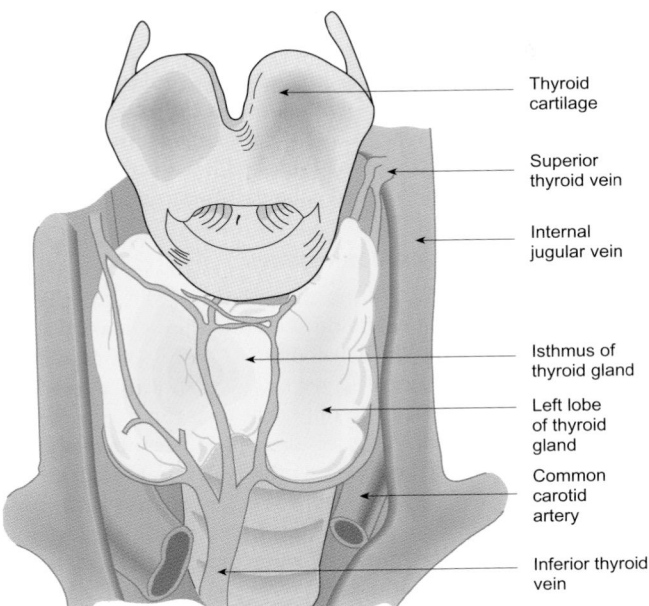

Fig. 12.12 A ventral view showing the location of the thyroid gland and its venous drainage.

the tongue. There is also overgrowth of certain visceral organs. Hypersecretion of GH usually results from a tumor of the anterior pituitary somatotropic cells. Such tumors can be removed surgically but the anatomical changes that have already occurred are irreversible.

Hyposecretion of GH in adults is not life threatening and has not been treated until recently. Further illustrations of patients suffering from pituitary disorders can be found in Chapter 23.

12.4 The thyroid gland

The thyroid gland is a discrete organ adhering to the trachea, just below the larynx. It consists of two flat lobes connected by an isthmus (Fig. 12.12). It secretes two iodine-containing hormones, *thyroxine* (T$_4$) and *tri-iodothyronine* (T$_3$), which exert a wide variety of actions throughout the body. They are chiefly concerned with the regulation of metabolism and the promotion of normal growth and development. A third hormone, *calcitonin*, is secreted by cells scattered throughout the gland (parafollicular cells or C-cells). Calcitonin is a peptide hormone that plays a role in the regulation of plasma calcium levels. It will be discussed further in Section 12.6.

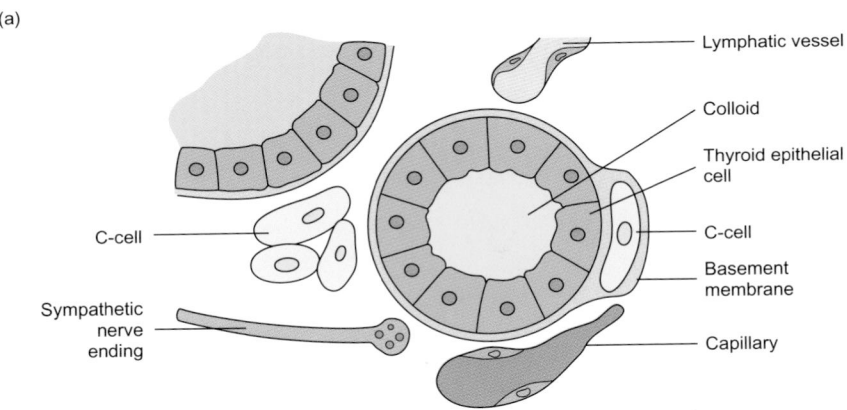

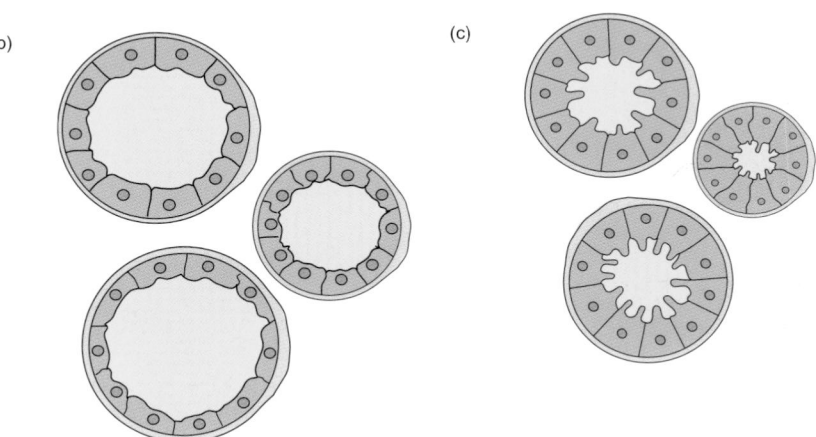

Fig. 12.13 The cellular organization of the thyroid gland. (a) The relationship between the follicles, C-cells, and other structures within the gland. (b) Relatively inactive follicles filled with colloid. Note the flattened appearance of the epithelial cells. (c) Highly active follicles containing little colloid. In this case, the epithelial cells are columnar in appearance.

Structure and development of the thyroid gland

The thyroid gland develops from endoderm associated with the pharyngeal gut before migrating to the front of the neck. It is capable of synthesizing and secreting thyroid hormones under the influence of fetal TSH by week 11 or 12 of gestation. Fetal thyroid hormones are thought to be crucial for the subsequent growth and development of the fetal skeleton and central nervous system (see p. 507).

The adult thyroid gland weighs between 10 and 20 g and has a rich blood supply (around 5 ml g^{-1} min^{-1}). It also has a rich autonomic innervation from the vagus nerve and the cervical sympathetic trunks. These autonomic nerves regulate the blood supply to the gland.

Histology of the thyroid gland

The chief histological components of the thyroid gland are illustrated in Fig. 12.13. The functional unit is the follicle, many thousands of which are present in the gland. Each follicle consists of a central colloid-filled cavity lined by an epithelial layer of follicular cells. Colloid is composed largely of an iodinated glycoprotein, *thyroglobulin*. A basement membrane surrounds each follicle and the parafollicular or C-cells lie within this membrane or scattered between the follicles.

The follicles range from 20 to 900 μm in diameter and vary in appearance according to the prevailing level of thyroid gland activity. Figure 12.13 shows that when the thyroid is highly active the follicles appear to contain rather little colloid and the follicular epithelial cells are tall columnar in form. However, during periods of inactivity the follicles fill with colloid and the epithelial cells have a flattened cuboidal appearance.

Synthesis of the thyroid hormones

T_3 and T_4 are the principal hormones secreted by the thyroid gland. They are iodinated amino acids and their chemical structures, as well as those of their iodinated precursors and of the biologically inactive form of T_3, *reverse* T_3, are shown in Fig. 12.14.

Thyroid hormones are the only substances in the body that contain iodine. Consequently, an adequate iodide intake is essential for normal thyroid hormone synthesis. The minimum dietary requirement is about 75 μg a day, but in most parts of the world intake considerably exceeds this level. The synthesis of thyroid hormones involves the uptake of iodine from the blood and the incorporation of iodine atoms into tyrosine residues of thyroglobulin, the glycoprotein that is synthesized by the follicular cells. The principal steps involved in the synthesis and secretion of the thyroid hormones are shown in Fig. 12.14 and discussed below.

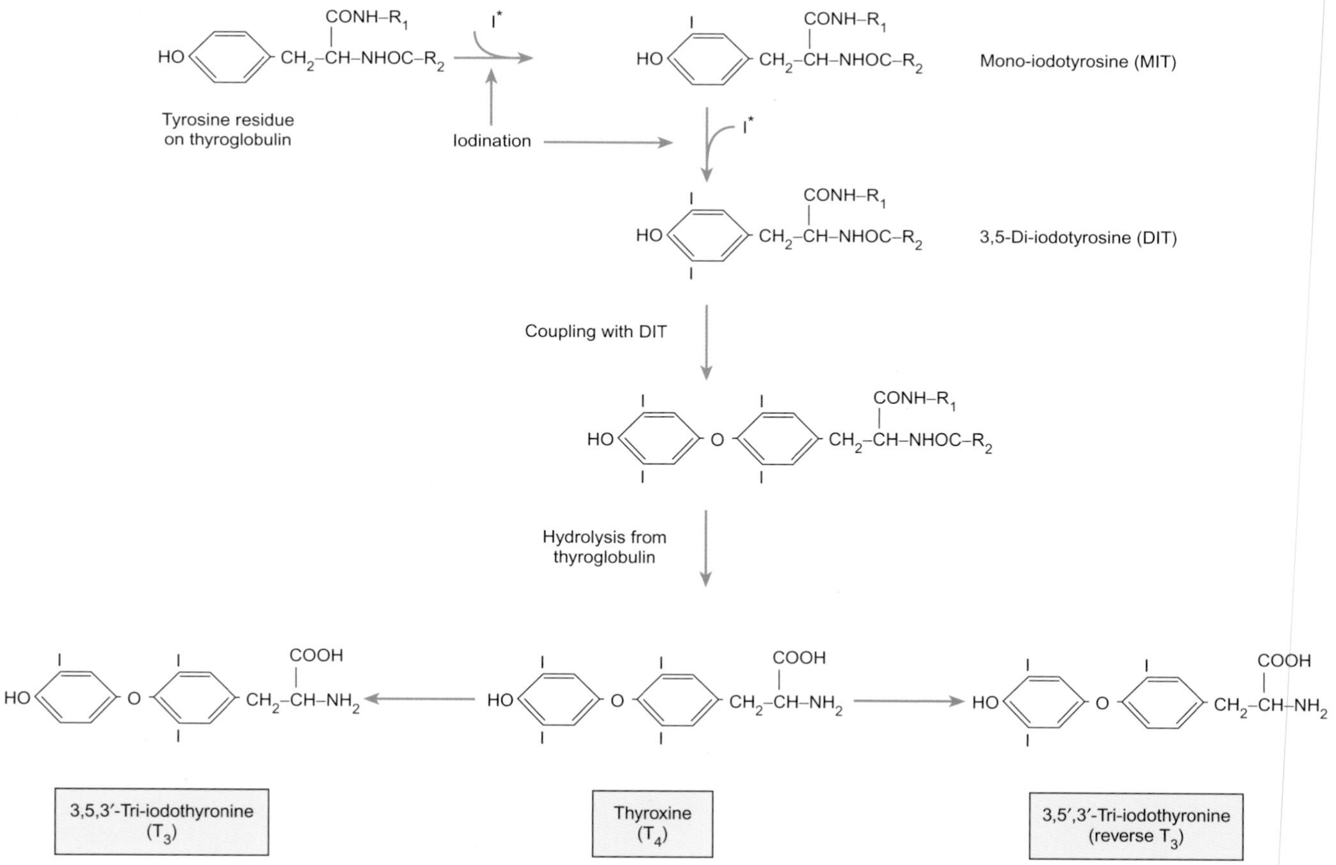

Fig. 12.14 A simplified scheme showing the principal steps in the synthesis of the thyroid hormones. Note that the thyroid gland secretes about 80 μg of T_4 a day compared with only about 4 μg of T_3 and about 2 μg of reverse T_3 (rT_3). Most T_3 and rT_3 is formed by de-iodination of T_4 in the tissues. I* indicates the participation of an iodide free radical.

Iodide trapping—the iodide pump

Iodine from food and drinking water is absorbed by the small intestine as inorganic iodide and is taken up by the thyroid gland. The basal plasma membrane of the follicular cells contains a sodium–iodide cotransport system (a symporter) which actively transports iodide from the blood into the cells against a steep electrochemical gradient. In this way, iodide is concentrated 20–100-fold by the follicular cells. The energy for iodide trapping is generated by oxidative phosphorylation and the pump is stimulated by TSH. The iodide pump may be inhibited competitively by anions such as perchlorate, thiocyanate, bromide, and nitrite.

Iodide oxidation

Before dietary iodide can enter into organic combination, it must be converted to free iodine (oxidized). This conversion to reactive iodine is carried out in the presence of the enzyme peroxidase, located mainly near to the apical membrane of the follicular cell. Hydrogen peroxide, generated by peroxisomes, acts as the electron acceptor for the oxidation reaction. Free iodine is released at the interface between the follicular cell and the colloid contained within the lumen of the follicles. Subsequent iodination of thyroglobulin takes place within the lumen.

Thyroglobulin synthesis

Thyroglobulin is a glycoprotein with a molecular mass of 670 kDa. It is synthesized on the rough endoplasmic reticulum of the follicular cells as peptide units of molecular mass 330 kDa. These combine and the carbohydrates are added. The completed protein is packaged into small vesicles which move to the apical plasma membrane where they are released into the follicular lumen by exocytosis (see Chapter 4) to be stored as colloid.

Iodination of thyroglobulin

Contained within the colloid are tyrosine residues held to the thyroglobulin molecules by peptide linkages. Free iodine becomes attached to the 3 position of a tyrosine residue to form mono-iodotyrosine (MIT). A second iodination at position 5 gives rise to di-iodotyrosine (DIT). Although each thyroglobulin molecule contains around 125 tyrosine residues, only about a third of these are available for iodination because they are located at or near the surface of the glycoprotein. Following the iodination, coupling reactions occur between the mono- and di-iodotyrosines to form the hormonally active forms, tri-iodothyronine (T_3) and thyroxine (T_4), as indicated in Fig. 12.14.

The iodinated protein is stored within the lumen of the thyroid follicle. In contrast with most endocrine glands, which do not store appreciable amounts of hormone, the thyroid gland contains several weeks supply of thyroid hormones. This can be important for maintaining normal circulating levels of thyroid hormones should there be a temporary fall in the supply of dietary iodine.

Secretion of thyroid hormones

Thyroglobulin must be hydrolyzed before the thyroid hormones T_3 and T_4 can be released into the circulation. When the thyroid is stimulated by TSH from the anterior pituitary, droplets of colloid are taken up into the follicular cells across the apical membranes by endocytosis. The endocytotic vesicles containing

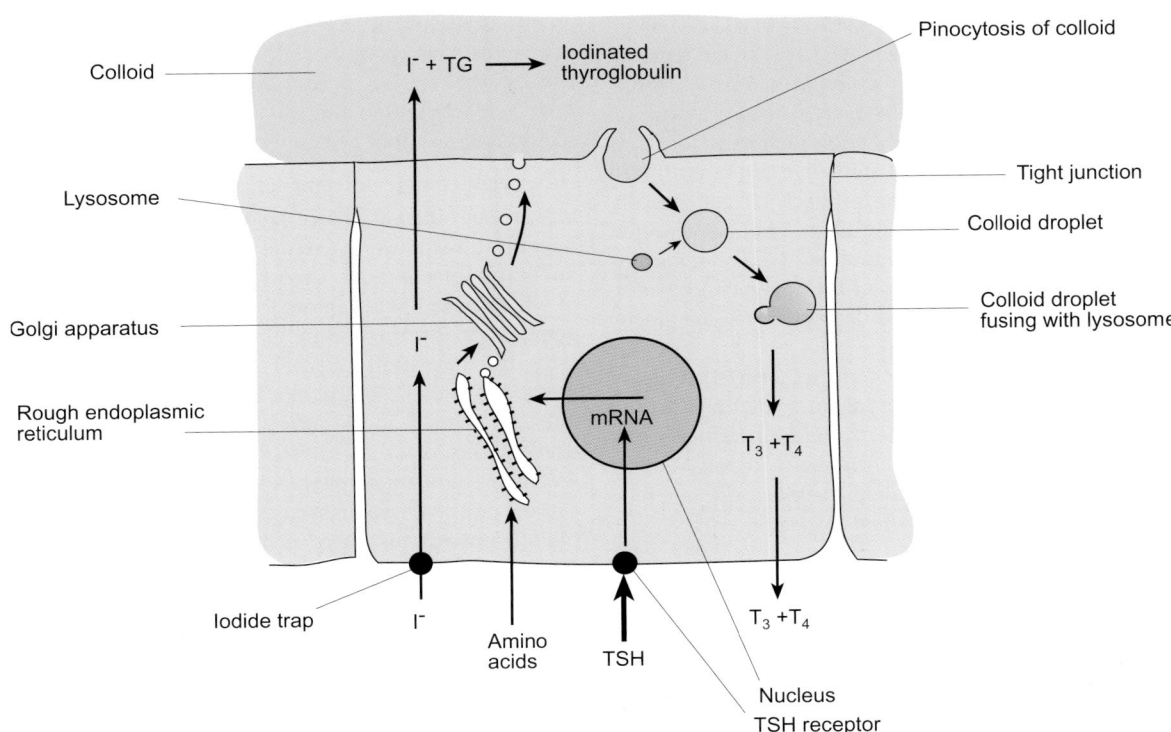

Fig. 12.15 The cellular processes involved in the synthesis and subsequent release of the thyroid hormones. Note that TSH stimulates both the synthesis of thyroglobulin (TG) and the secretion of T_3 and T_4.

colloid then move away from the apical membrane, assisted by the cytoskeletal structures (microtubules and microfilaments—see Chapter 3). At the same time, lysosomes move towards the endocytotic vesicles. Fusion then occurs between the membranes of the lysosomes and of the colloid-containing vesicles, and the thyroglobulin is degraded. As a result, iodotyrosines, amino acids, and sugars are liberated. The amino acids and sugars are recycled within the gland, while MIT and DIT are de-iodinated so that the iodine can be used again. T_3 and T_4 are released into the fenestrated capillaries that surround the follicle, probably by diffusion. Thyroglobulin itself is not normally released into the circulation unless inflammation or damage to the gland occurs. Figure 12.15 illustrates the processes involved in the secretion of thyroid hormone.

Regulation of thyroid hormone secretion

As explained in Section 12.2, the activity of the thyroid gland is controlled by TSH from the anterior pituitary gland. The secretion of TSH is, in turn, controlled by the hypothalamic TRH. TSH acts by binding to specific receptors on the thyroid follicular cell surface and stimulating membrane-bound adenylyl cyclase, thereby increasing the level of cyclic AMP. The elevated cyclic AMP activates protein kinase A, which leads to an increase in gene transcription.

TSH regulates most aspects of thyroid hormone synthesis and secretion—see Figure 12.15. These include stimulation of iodide trapping, peroxidase activity, thyroglobulin synthesis, and the iodination of the tyrosine residues on thyroglobulin. TSH also stimulates the coupling of mono- and di-iodotyrosine molecules and all the events leading to thyroid hormone secretion. The net result of the actions of TSH is to increase the synthesis of fresh thyroid hormone for storage within the follicles as well as to increase the secretion of thyroid hormones into the circulation. TSH is a tropic hormone, which means that it exerts a tonic maintenance effect on the thyroid gland and its blood supply. In the absence of TSH the thyroid gland rapidly atrophies.

The circulating thyroid hormones influence the rate of TSH secretion by means of their negative feedback effects on both the hypothalamus and the anterior pituitary. Other hormones also appear to be able to alter the output of TSH. For example, estrogens increase the responsiveness of the TSH-secreting cells of the anterior pituitary to TRH, while high levels of glucocorticoids inhibit TSH release.

In addition to chemical control mechanisms, a variety of nervous inputs to the hypothalamus play a role in regulating the output of TRH and thus of TSH and the thyroid hormones. Cold stress is known to stimulate thyroid hormone secretion. Within 24 hours of entering a cold environment there is a rise in circulating T_4 levels which reach a peak a few days later.

Transport, tissue delivery, and metabolism of thyroid hormones

T_3 and T_4 travel in the bloodstream largely bound to plasma proteins. Only about 0.5 per cent of circulating T_3 is unbound (free), while less than 0.05 per cent of T_4 is in the unbound form. About 75 per cent of the circulating T_4 is bound to thyronine-binding globulin, 15–20 per cent is bound to prealbumin, and the remainder is bound to albumin. Virtually all the T_3 is bound to thyronine-binding globulin. Because T_4 binds to plasma proteins with an affinity 10 times that of T_3, its metabolic clearance rate is slower. Consequently, the half-life of T_4 in the plasma is significantly longer than that of T_3 (7 days for T_4 and less than 1 day for T_3).

T_3 and T_4 appear to enter their target cells by a carrier-mediated process that requires energy and seems to be dependent upon extracellular sodium. Once inside cells, much of the T_4 undergoes de-iodination to T_3; most of this conversion takes place in the liver and kidneys. For this reason T_4 is sometimes called a prohormone for T_3. Nevertheless, it should be remembered that, although biologically less active than T_3, T_4 does exert hormonal effects in its own right and probably has its own cellular receptors.

In addition to being de-iodinated to the active form of T_3, T_4 may also undergo conversion to inactive reverse T_3 (rT_3) in the tissues. In this case, the iodide is removed from the inner tyrosine residue of the molecule (Fig. 12.14). The production of rT_3 from T_4 seems to increase when calorie intake is restricted. From the following discussion of the effects of the thyroid hormones, it will become clear that this acts as an energy-conserving mechanism.

Actions of the thyroid hormones

Thyroid hormones have effects on virtually every tissue. Information about these effects has largely been obtained from studies of human and animal subjects with over- or underactive thyroid glands (hyper- and hypothyroidism), and from studies of the *in vitro* actions of T_3 and T_4. In general terms, the thyroid hormones can be thought of as tissue growth factors because, even if growth hormone levels are normal, overall whole body growth is impaired in the absence of thyroid hormones. Thyroid hormones have specific tissue effects as well as increasing the basal rate of oxygen consumption and heat production by the body as a whole (see also Chapter 26). The principal effects of the thyroid hormones are summarized in Fig. 12.16

Heat production (calorigenesis)

Most tissues increase their oxygen consumption and heat production in response to thyroid hormone. The main exceptions are brain, spleen, testes, uterus, and anterior pituitary gland, all of which have few receptors for thyroid hormones. This calorigenic action is important in temperature regulation and particularly in adaptation to cold environments, although basal metabolic rate (BMR) does not begin to rise until a day or more after the increase in the secretion of thyroid hormone. BMR is a useful indicator of the thyroid status of an individual. A high BMR may indicate an overactive thyroid gland, while thyroidectomy or hypothyroidism causes BMR to fall. The measurement of BMR and a discussion of its hormonal control will be found in Chapter 24.

What causes the increase in oxygen consumption in response to thyroid hormone?

The precise cellular mechanisms involved in the effects of thyroid hormone on oxygen consumption are not clear. However, associated with the increased BMR is an increase in the size and number of mitochondria within the cells of responsive tissues and a rise in the concentrations of many of the enzymes of the

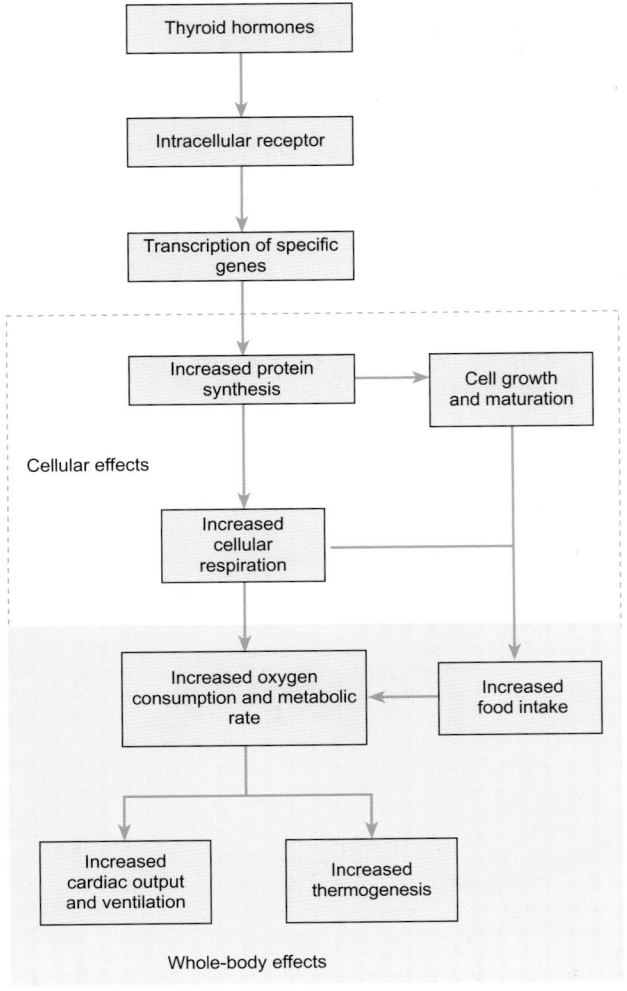

Fig. 12.16 The principal actions of the thyroid hormones.

respiratory chain. In addition, both T_3 and T_4 stimulate the activity of the Na$^+$, K$^+$-ATPase, thereby increasing the rate of Na$^+$ and K$^+$ transport across the cell membrane. As a result of this action, ATP hydrolysis is enhanced. This in turn stimulates oxygen consumption by the mitochondria.

The increase in oxygen consumption is, of necessity, accompanied by an increase in body temperature, though this is somewhat offset by a compensatory increase in heat loss through increased blood flow to the surface vessels, sweating, and ventilation (also mediated by thyroid hormone). Cardiac output is increased in response to thyroid hormones which act to enhance the effects of the catecholamines. Both the resting heart rate and the stroke volume are raised (although if thyroid hormone levels remain chronically elevated, there may be a loss of muscle mass and a reduction in the strength of the heart beat).

The thyroid hormones modify the metabolism of fats and carbohydrates to provide substrates for oxidation

Thyroid hormones exert a number of effects on the metabolism of carbohydrates. They act directly on the enzyme systems involved and work indirectly by potentiating the actions of other hormones such as insulin and catecholamines. Thyroid

hormones enhance the rate of intestinal absorption of glucose and increase the rate of glucose uptake by peripheral cells such as muscle and adipose tissue. Low doses of thyroid hormones appear to potentiate the effects of insulin, stimulating both glycogen synthesis and glucose utilization. However, higher thyroid hormone levels seem to enhance the calorigenic effects of epinephrine, increasing the rates of both glycogenolysis and gluconeogenesis.

Thyroid hormones affect most aspects of lipid metabolism. They exert a powerful lipolytic action on fat stores, thereby increasing plasma levels of free fatty acids. Part of this action is probably due to a potentiation of catecholamine activity that increases lipolysis via the adenylyl cyclase–cyclic AMP system. There is also increased oxidation of the free fatty acids released and this contributes to the calorigenic effect. The overall effect of these changes in fat metabolism is a depletion of the body's fat stores, a fall in weight, and a fall in plasma levels of cholesterol and other lipids.

Thyroid hormones stimulate both protein synthesis and degradation

The actions of the thyroid hormones on protein metabolism are somewhat complex. At low concentrations, T_3 and T_4 stimulate the uptake of amino acids into cells and the incorporation of these into specific structural and functional proteins, many of which are involved in calorigenesis (see above). Protein synthesis is depressed in hypothyroid individuals. Conversely, high levels of thyroid hormones are associated with protein catabolism. This effect is particularly marked in muscle and can lead to severe weight loss and muscle weakness (thyrotoxic myopathy). The breakdown of protein results in an increased level of amino acids in the plasma and increased excretion of creatine.

Whole-body actions of thyroid hormones

Thyroid hormones have major effects on growth and maturation. While linear growth itself appears to be independent of thyroid hormone levels in the fetus, T_3 and T_4 seem to be essential for the normal differentiation and maturation of fetal tissues, particularly those of the skeleton and the nervous system. After birth, the thyroid hormones stimulate the linear growth of bone until puberty. They promote ossification of the bones and maturation of the epiphyseal growth regions. More details of skeletal growth and the effects of hormones upon growth are given in Chapter 23.

Adequate thyroid hormone levels during the late fetal and early prenatal periods are vital to normal development of the CNS. Inadequate hormone concentrations at this time result in severe mental retardation (*cretinism*) which, if not recognized and treated quickly, is irreversible (see below). The exact role of thyroid hormones in maturation of the nervous system is unclear, but it is known that in their absence there is a reduction in both the size and number of cerebral cortical neurons, a reduction in the degree of branching of dendrites, deficient myelination of nerve fibers, and a reduction in blood supply to the brain.

A number of the effects of the thyroid hormones, including the increase in BMR, calorigenesis, increased heart rate, and CNS excitation, are shared by the sympathetic system. Indeed, synergism between the catecholamines and thyroid hormones may be essential for maximum thermogenesis, lipolysis, glycogenolysis, and gluconeogenesis to occur. This physiological link is emphasized by the fact that the administration of drugs which block β-adrenoceptors has the effect of lessening many of the cardiovascular and central nervous manifestations of hyperthyroidism.

Mode of action of the thyroid hormones

The thyroid hormones act at the level of the cell nucleus. They bind to specific receptors in the nucleus of their target cells. After binding, the hormone–receptor complexes modulate the transcription of specific genes as described in Chapter 5. The genes may be either activated or repressed (as in the negative feedback of TSH production by the anterior pituitary thyrotrophs). Because they act on the nucleus of their target cells, the effects of the thyroid hormones are slow in onset. Biological effects normally take between a few hours and several days to appear following the stimulation of hormone secretion. The first intracellular change to occur is an increase in nuclear RNA, followed by an increase in protein synthesis which, in turn, leads to raised intracellular levels of specific enzymes.

Principal disorders of thyroid hormone secretion

Disorders of the thyroid may represent a congenital defect of thyroid development or they may develop later in life, with a gradual or a sudden onset. The thyroid gland may be overactive, leading to *hyperthyroidism*, or underactive, leading to *hypothyroidism*. Defects of thyroid function are among the most common endocrine disorders, affecting 1–2 per cent of the adult population. The chief manifestations of both these extremes can largely be predicted from the preceding account of normal thyroid function.

The clinical features of thyroid hormone excess

Hyperthyroidism or *thyrotoxicosis* results from excessive delivery of thyroid hormone to the peripheral tissues. The most common cause is *Graves' disease*, which is believed to be an autoimmune disease because the patient's serum often contains abnormal antibodies that mimic TSH (so-called thyroid-stimulating antibodies (TSAb)). This disorder is characterized by a swelling of the thyroid gland (goiter) and protrusion of the eyeballs (exophthalmos) caused by inflammation and edema of the extra-ocular muscles coupled with lid retraction. Figure 12.17 shows a patient with the typical physical characteristics of hyperthyroidism.

Most of the manifestations of hyperthyroidism are related to the increase in oxygen consumption and increased utilization of metabolic fuels associated with a raised metabolic rate, and to a parallel increase in sympathetic nervous activity. The principal symptoms of hyperthyroidism are:

(1) loss of weight;

(2) excessive sweating;

(3) palpitations and an irregular heartbeat;

(4) nervousness;

(5) protrusion of the eyeballs (exophthalmos).

On clinical examination, there is a raised BMR and oxygen consumption, increased heart rate, hypertension, and possibly

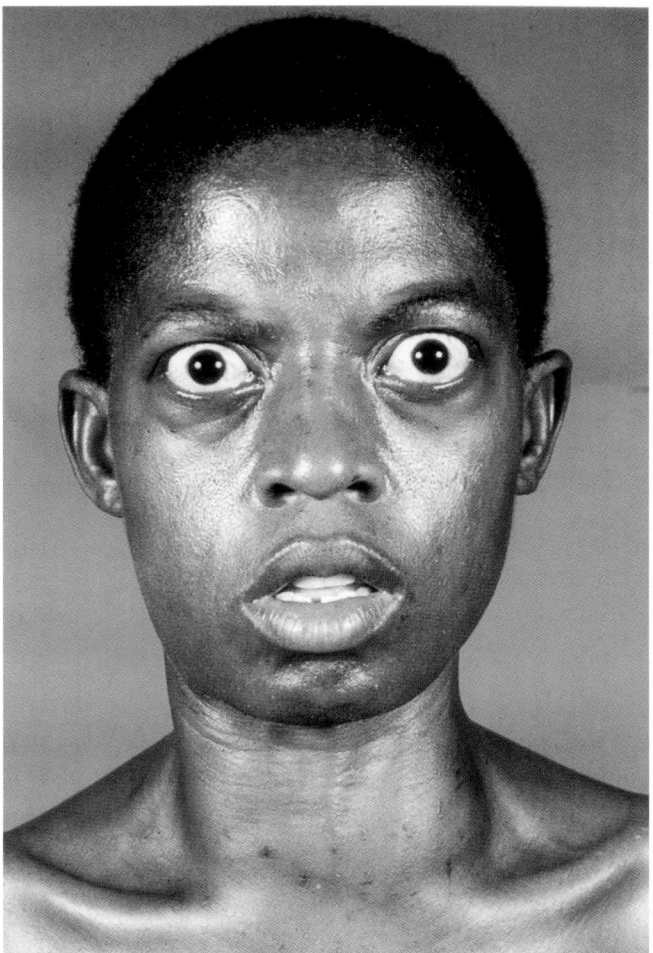

Fig. 12.17 Typical features of a patient with hyperthyroidism. Note the exophthalmos and retraction of the eyelids.

atrial fibrillation. The diagnosis is confirmed by raised plasma levels of T_3 and T_4. Treatments for hyperthyroidism include surgical removal of all or part of the thyroid gland or ingestion of radioactive iodine (^{131}I) which selectively destroys the most active thyroid cells. The output of thyroid hormones can also be decreased by a number of drugs that are structurally related to thiourea (the thioureylenes).

The clinical features of hypothyroidism

In the absence of sufficient dietary iodine, the thyroid cannot produce sufficient amounts of T_3 and T_4. As a result, there is a reduction in the negative feedback inhibition of TSH secretion by the anterior pituitary gland. Therefore TSH secretion is abnormally high and this results in abnormal growth of the thyroid gland due to the trophic effect of TSH. This is known as an *iodine-deficiency* or *endemic goiter* (see Chapter 19, Fig. 19.5). Hypothyroidism may also result from a primary defect of the thyroid gland or as a secondary consequence of reduced secretion of either TSH or TRH. The most common cause is an autoimmune disorder called *Hashimoto's thyroiditis*, which causes destruction of the thyroid by an autoimmune process.

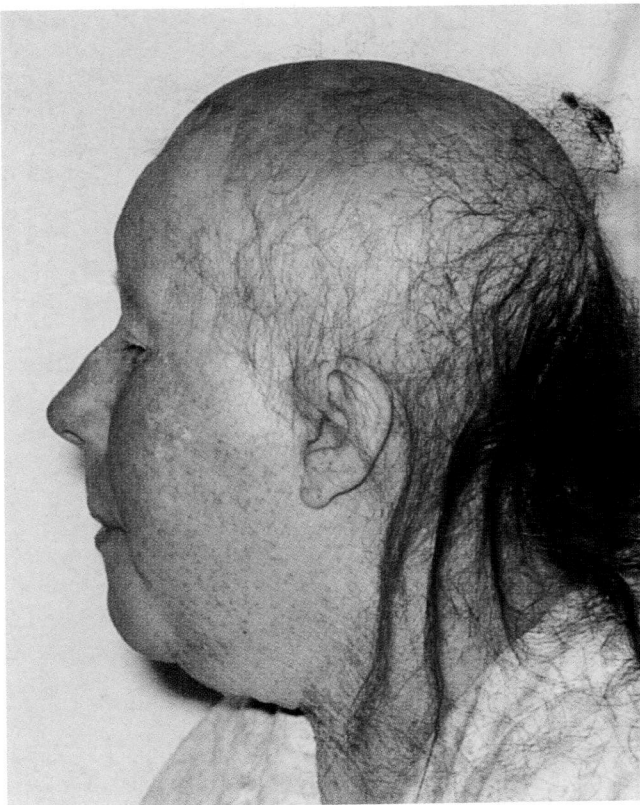

In adults, the full-blown hypothyroid syndrome is called *myxedema* (literally 'mucous swelling'). The symptoms of myxedema include:

(1) puffiness of the face and swelling around the eyes;

(2) dry cold skin;

(3) loss of hair (known as alopecia);

(4) sensitivity to cold;

(5) weight gain despite a loss of appetite;

(6) constipation;

(7) impaired memory;

(8) mental dullness;

(9) lethargy.

Summary

1. The follicular cells of the thyroid gland secrete two iodine-containing hormones, tri-iodothyronine (T_3), and thyroxine (T_4). A further hormone, calcitonin, is secreted by parafollicular cells. T_3 and T_4 play an important role in the control of metabolic rate, maturation of the skeleton and development of the central nervous system.

2. Iodide is concentrated in the follicular cells where it is oxidized to iodine for incorporation into thyroglobulin. Thyroid hormones are released from the gland following enzymatic hydrolysis of the iodinated thyroglobulin. T_3 and T_4 travel in the plasma bound to carrier proteins including a specific thyronine-binding globulin. In the tissues most of the T_4 is converted to T_3.

3. TSH secreted by the anterior pituitary controls all aspects of the activity of the thyroid gland. Its rate of secretion is inhibited by the negative feedback effects of thyroid hormone. TSH is itself regulated by hypothalamic TRH.

4. Most tissues increase their oxygen consumption in response to thyroid hormones. This increase in metabolic rate is important in the maintenance of body temperature. Thyroid hormones also increase ventilation, cardiac output, and the rate of red blood cell production.

5. The metabolic actions of thyroid hormones are somewhat complex and dose dependent. Low concentrations tend to be hypoglycemic, while higher doses stimulate glycogenolysis and gluconeogenesis. Thyroid hormones are powerfully lipolytic and stimulate the oxidation of free fatty acids. Low levels of thyroid hormone stimulate protein synthesis while higher concentrations are catabolic.

6. Excessive thyroid hormone secretion (hyperthyroidism) results in an increased BMR, sweating, increased heart rate, nervousness, and weight loss. Insufficient secretion of thyroid hormones (hypothyroidism) results in a reduced BMR, bradycardia, lethargy, cold sensitivity, and a range of other metabolic abnormalities.

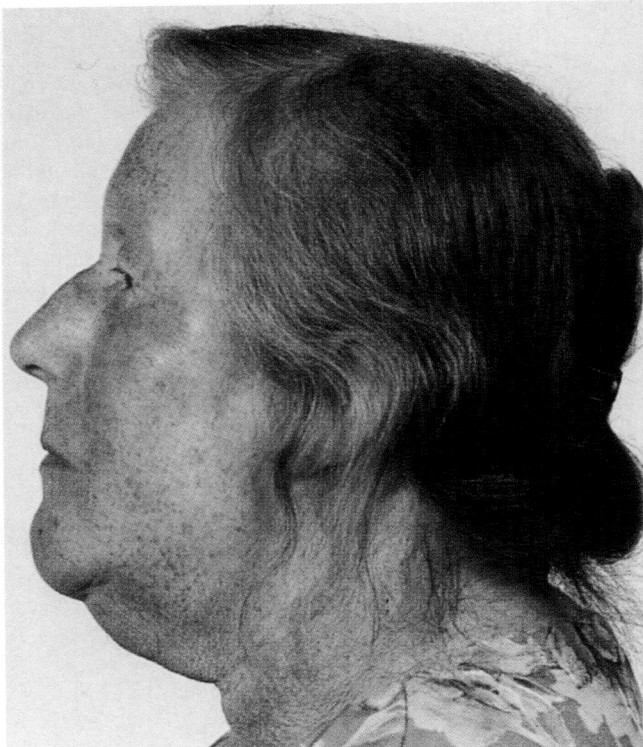

Fig. 12.18 A patient suffering from hypothyroidism (myxedema) manifesting typical facial puffiness, thinning hair, and dull appearance (top). The same patient after treatment with thyroid hormone (bottom), which has reversed the physical symptoms.

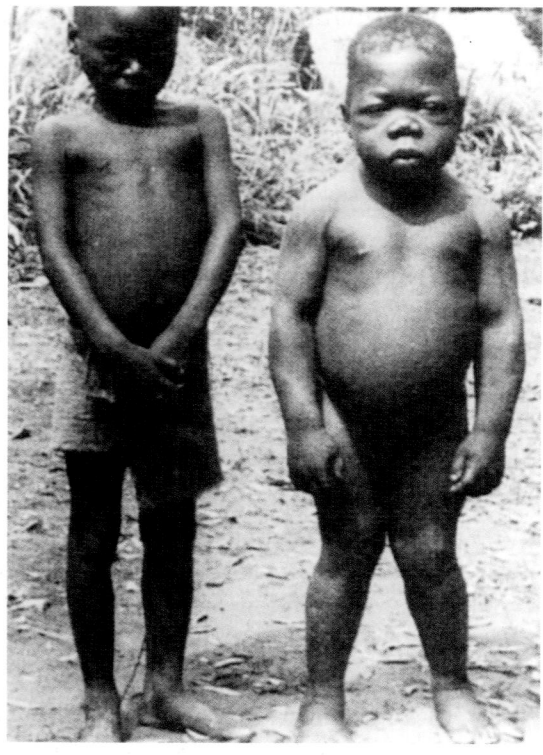
(a)

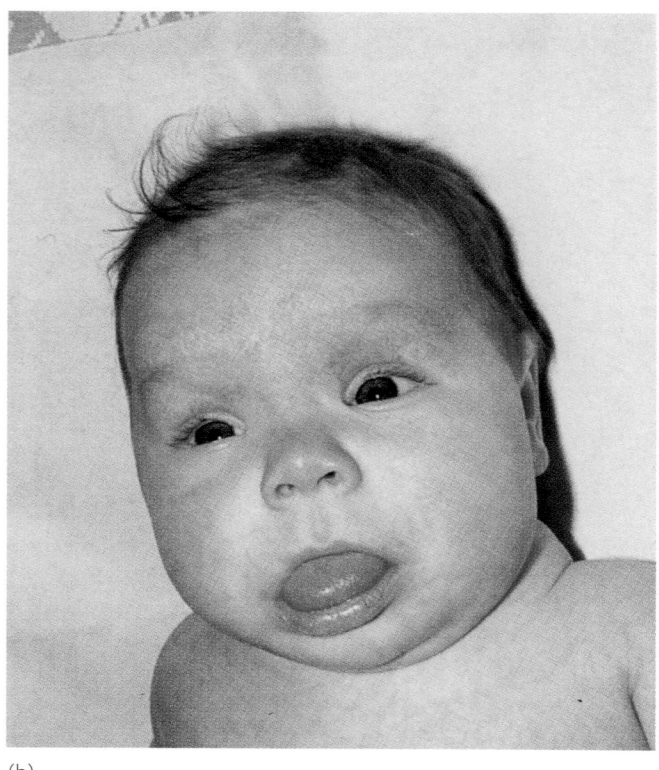
(b)

Fig. 12.19 Typical examples of children suffering from hypothyroidism. (a) A 17-year-old girl cretin (on the right) standing next to a normal 6-year-old child. (b) A baby suffering from congenital hypothyroidism which was successfully treated by hormone replacment.

On clinical examination, there is a lowered BMR and oxygen consumption, hypothermia, a slowed pulse rate, and reduced cardiac output. The diagnosis is confirmed by low plasma levels of T_3 and T_4. Depending upon the cause, hypothyroidism can be reversed by iodine supplements or by hormone replacement therapy. Figure 12.18 shows the typical physical appearance of a patient with hypothyroidism and the effects of hormone replacement.

Thyroid hormones are essential for normal brain development and growth. Severe hypothyroidism in infants is called *cretinism*. Congenital hypothyroidism is one of the most common causes of preventable mental retardation, affecting approximately 1 in every 4000 infants. In addition to being mentally retarded, affected children often have a short disproportionate body, a thick tongue and neck, and obesity. In Western countries, neonatal tests to screen for congenital hypothyroidism are now carried out routinely between the first and fifth days of life. Affected infants can be treated successfully by hormone replacement therapy. Figure 12.19 shows examples of a hypothyroid infant, a 17-year-old cretin (untreated hypothyroidism), and a normal 6-year-old child.

12.5 The adrenal glands

The paired adrenal (or suprarenal) glands are roughly pyramid-shaped organs lying above the kidneys, where they are enclosed in a fibrous capsule surrounded by fat (Fig. 12.20). Each adrenal gland is, in effect, structurally and functionally two endocrine

glands in one. The inner *adrenal medulla* derives from ectodermal cells of the embryonic neural crest and secretes the cate-cholamines *epinephrine* (adrenaline) and *norepinephrine* (noradren-aline) in response to activation of its sympathetic nerve supply. It thereby acts as part of the sympathetic nervous system. The outer *adrenal cortex*, which encapsulates the medulla and forms the bulk of the gland, is derived from embryonic mesoderm and

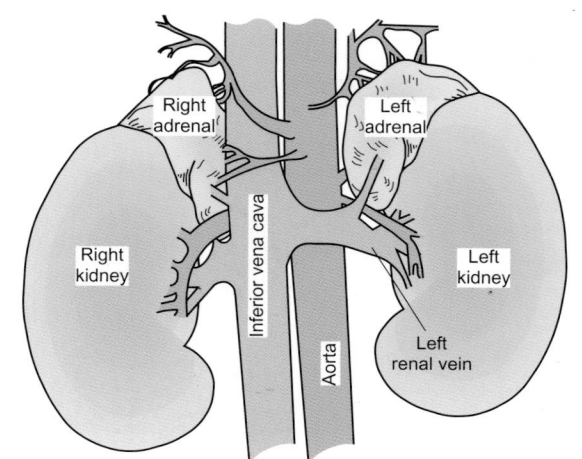

Fig. 12.20 The anatomical location of the adrenal glands and the organization of their blood supply. Note that the arterial supply is via many small arteries that originate from the aorta. The venous drainage is via a large central vein.

secretes a number of steroid hormones. Unlike those of the adrenal medulla, the secretions of the adrenal cortex are controlled hormonally. Loss of the adrenal cortical hormones will result in death within a few days.

The adult adrenal glands each weigh between 6 and 10 g. They receive arterial blood from branches of the aorta, the renal arteries, and the phrenic arteries. They have an extremely rich blood supply. The arterial blood enters from the outer cortex, flows through fenestrated capillaries between the cords of cells, and drains inwardly into venules in the medulla. This arrangement is important for interactions between the different regions of the gland. The right adrenal vein drains directly into the inferior vena cava while the left drains into the left renal vein. In the fetus, the adrenal glands are much larger, relative to body size, than in the adult. Their functions during fetal life are described in Chapter 22.

The adrenal cortex

There are three morphologically distinct zones of cells within the adrenal cortex (Fig. 12.21). These are the outer *zona glomerulosa* (occupying around 10 per cent of the adrenal cortex), the *zona fasciculata* (around 75 per cent), and the *zona reticularis*, which lies closest to the adrenal medulla. The zona reticularis does not differentiate fully until between 6 and 8 years of age. In the adult gland, the cells of the glomerulosa continually migrate down through the zona fasciculata to the zona reticularis, changing their secretory pattern as they go. The purpose of this migration is not clear.

Synthesis and secretion of adrenal cortical hormones

The cells of the three zones secrete different steroid hormones: the cells of the zona glomerulosa secrete the *mineralocorticoids*;

those of the zona fasciculata secrete *glucocorticoids*, while the cells of the zona reticularis secrete *sex steroids*.

Cholesterol is the starting point for all steroid biosynthesis (see also Chapter 20 which discusses the synthesis of gonadal steroids). Figure 12.22 summarizes the pathways of steroid biosynthesis in the adrenal cortex and shows that the cortical hormones are obtained by modification of the cholesterol by a series of hydroxylation reactions. Most of the reactions from cholesterol to active hormones involve *cytochrome P-450 enzymes*, which are situated in the mitochondrial cristae and the endoplasmic reticulum. Small amounts of cholesterol are synthesized within the adrenal gland from acetyl CoA, but most is obtained by the receptor-mediated endocytosis of low-density lipoproteins circulating in the bloodstream. The cholesterol is stored mainly within cytoplasmic lipid droplets in the adrenal cortical cells.

Aldosterone is the principal mineralocorticoid secreted by the adrenal cortex. It is synthesized exclusively by the cells of the zona glomerulosa. It is not stored within the adrenal cortical cells to any significant extent but diffuses rapidly out of the cells after synthesis. Consequently, the rate of aldosterone production must increase whenever there is a need for increased circulating levels of the hormone. Normally between 0.1 and 0.4 μmol of aldosterone are secreted each day.

In humans, *cortisol* is the dominant glucocorticoid, but corticosterone and cortisone are other glucocorticoids that are produced in small amounts. The synthesis of the glucocorticoids takes place largely in the zona fasciculata and, like aldosterone, these hormones are released rapidly from the cells following synthesis, by diffusion across the plasma membrane. The adrenal cortex secretes much more cortisol than aldosterone; between 30 and 80 μmol of cortisol are secreted each day.

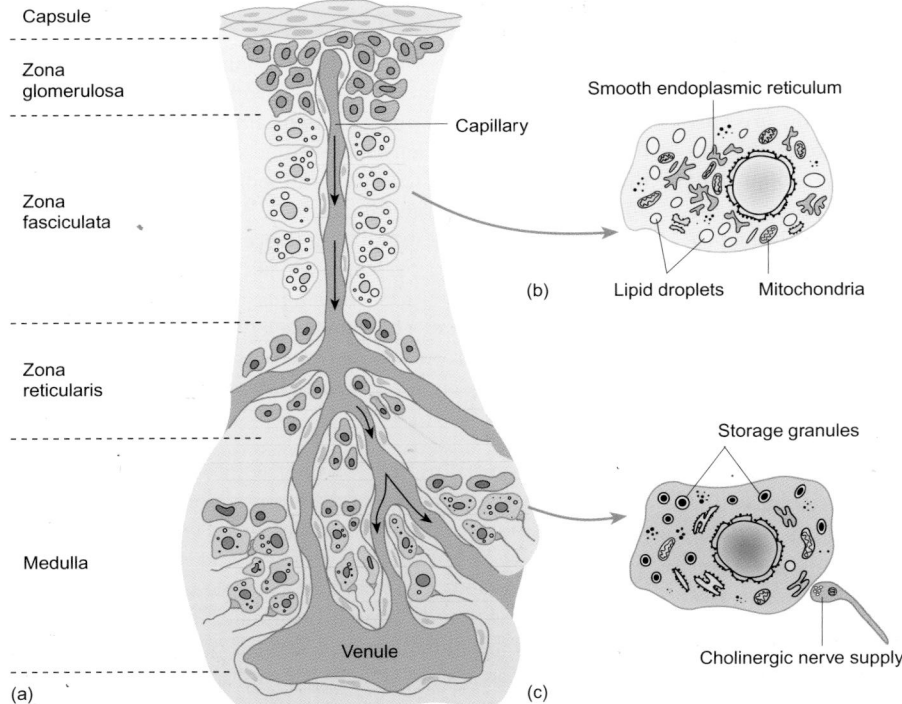

Fig. 12.21 (a) A diagrammatic representation of the cortex and medulla of the adrenal gland. Note the three zones of the adrenal cortex, the cells of which secrete steroid hormones. (b) The appearance of steroid-secreting cells. (c) A single catecholamine-secreting chromaffin cell.

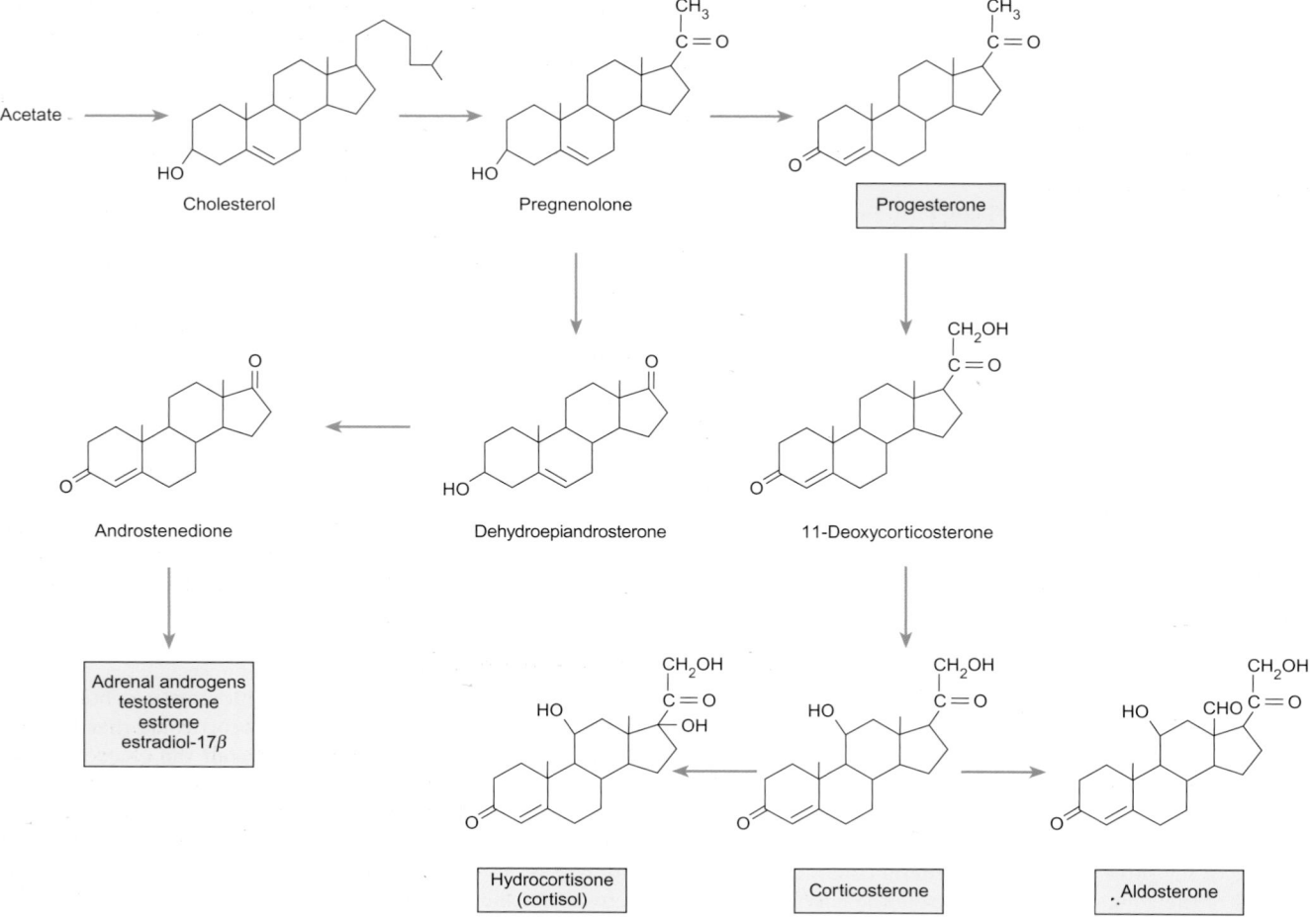

Fig. 12.22 The principal steps in the synthesis of the adrenal steroid hormones from cholesterol. Cortisol is the principal glucocorticoid and aldosterone is the principal mineralocorticoid.

The *principal sex steroids* produced by the cells of the zona reticularis are dehydroepiandrosterone and androstenedione. Although they are weak androgens themselves, they are converted to the more powerful androgen testosterone in peripheral tissues. In females the adrenals secrete about half the total androgenic hormone requirements, but in males the amounts produced are insignificant in comparison with the production of testosterone by the testes. Similarly, the estrogenic hormones secreted by the cortex are quantitatively insignificant in women (at least until after the menopause).

Plasma contains a specific corticosteroid-binding globulin called *transcortin* which is a glycoprotein synthesized in the liver. Between 70 and 80 per cent of the circulating cortisol is reversibly bound to transcortin while a further 15 per cent or so binds to serum albumin. Only 5–10 per cent of plasma cortisol is in a free or active form. Transcortin also binds progesterone with a fairly high affinity, but it has a much lower affinity for aldosterone which is carried mainly in combination with albumin.

Principal actions of the adrenocorticosteroids
Effects of aldosterone
The mineralocorticoid secretions of the adrenal cortex, of which aldosterone is the most important, are essential to life. In their

absence, death occurs within a few days unless prevented by the therapeutic administration of salt or by the replacement of hormones. The secretion and effects of aldosterone are discussed fully in Chapter 17. Briefly, aldosterone secretion is regulated by the plasma levels of sodium and potassium via the renin-angiotensin system (see Chapter 17). Aldosterone acts to conserve body sodium by stimulating the reabsorption of sodium in the distal nephron in exchange for potassium. Failure of aldosterone secretion results in a rise in plasma potassium levels and a fall in sodium and chloride levels. The extracellular fluid volume and the blood volume fall, with a subsequent drop in cardiac output which, if uncorrected, may prove fatal.

Effects of cortisol
Cortisol is essential to life. Together with the other glucocorticoids, it has a wide range of different actions throughout the body. The most important physiological effects are on the metabolism of carbohydrates, proteins, and, to a lesser extent, fats. Glucocorticoids also play a crucial role in the responses of the body to a variety of stressful stimuli. They are immunosuppressive, anti-inflammatory, and anti-allergic, and they possess weak mineralocorticoid activity. This mineralocorticoid activity is important, as plasma levels of cortisol are much higher than

those of aldosterone. Recent evidence suggests that enzymes in different target tissues are very important in both inactivating and activating different steroids. For example, the enzyme 11β-sterol dehydrogenase in the cells of the distal tubule inactivates circulating cortisol, thereby blocking its mineralocorticoid actions on the kidney.

Metabolic actions of cortisol

In general terms, the metabolic effects of cortisol can be said to oppose those of insulin. Its effects vary with the target tissue but result in a rise in plasma glucose concentration. The most important overall action of cortisol is to facilitate the conversion of protein to glycogen. It stimulates the mobilization of protein from muscle tissue in particular, thereby increasing the rate at which amino acids are presented to the liver for *gluconeogenesis* (the synthesis of glucose from non-carbohydrate precursors). As a result of this, glycogen stores are initially built up while any excess glucose is released into the plasma. At the same time, cortisol seems to inhibit the uptake and utilization of glucose by those tissues in which glucose uptake is insulin dependent.

Cortisol also stimulates the appetite and influences the metabolism of fats, particularly when it is released in greater than normal amounts. It seems to stimulate lipolysis in adipose tissue, both directly and because it enhances the lipolytic actions of other hormones such as growth hormone and the catecholamines. The major metabolic actions of cortisol (and other glucocorticoids) are summarized in Fig. 12.23.

Other actions of cortisol

Cortisol acts to counteract many of the effects of stress throughout the body. Although the concept of stress is difficult to define, it includes physical trauma, intense heat or cold, infection, and mental or emotional trauma. The exact nature of the glucocorticoid response is unclear, but may include cardiovascular, neurological, and anti-inflammatory effects as well as effects on the immune system. Cortisol increases vascular tone, possibly by promoting the actions of catecholamines, and blocks the processes which lead to inflammation in damaged tissues. Although this action is only apparent at high concentrations of the hormone, it has proved to be of value in the treatment of inflammatory conditions, such as rheumatoid arthritis, and in the treatment of severe asthma.

High levels of cortisol also suppress the normal immune response to infection. There is a gradual destruction of lymphoid tissue, leading to a fall in antibody production and in the number of circulating lymphocytes. It is thought that this action may be important physiologically in preventing the immune system from causing damage to the body.

Glucocorticoid hormones are known to influence the CNS. Although the underlying mechanisms are unclear, cortisol acts on the CNS to produce euphoria, an effect that may also have a role in helping to mitigate the effects of stress. This also explains why the sudden withdrawal of prescribed steroidal drug treatments can lead to severe depression. Fetal cortisol plays a vital role in the maturation of many organs and is particularly important in the stimulation of pulmonary surfactant production (see Chapter 22).

Effects of adrenocortical sex steroids

The production of sex steroids (androgens in men, estrogens and progesterone in women) is mainly from the gonads (see Chapter 20). Secretion of these hormones by the adrenal cortex is minor by comparison. Indeed, the role of the adrenal sex steroids is not fully understood. They may be important in the development of certain female secondary sexual characteristics, particularly in the growth of pubic and axillary hair. They may also play a role in the growth spurt seen in middle childhood. The actions of the adrenal sex steroids become more significant in disease, particularly when there is hypersecretion by adrenal tumors or

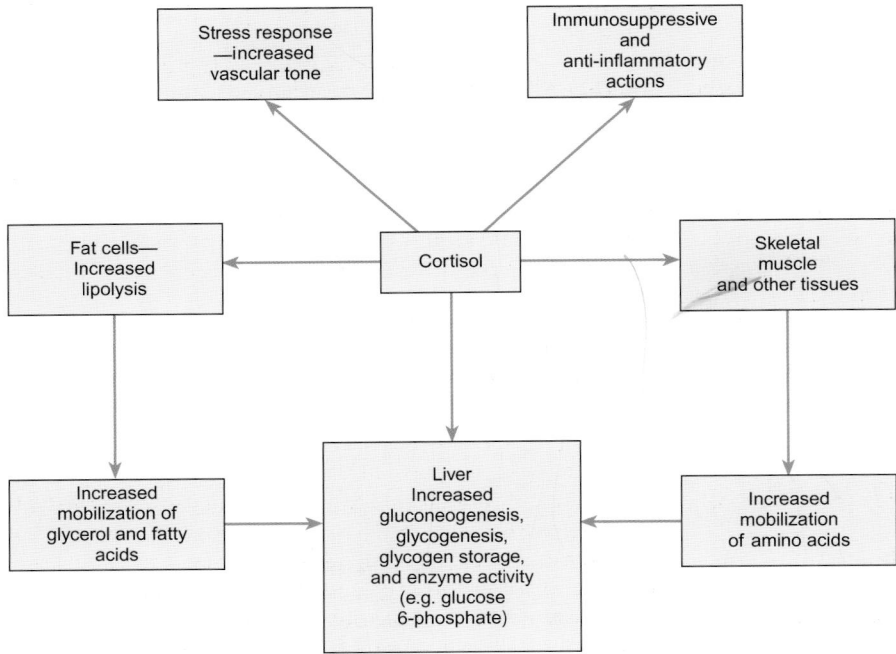

Fig. 12.23 The principal physiological actions of the glucocorticoid hormone cortisol.

enzyme defects in the pathways that normally synthesize cortisol.

The regulation of steroid hormone secretion from the adrenal cortex

The regulation of aldosterone secretion is discussed fully in Chapter 17. It is known that the synthesis and secretion of the glucocorticoids and, to a lesser extent, the adrenal sex steroids are under the control of ACTH secreted by the anterior pituitary in response to hypothalamic CRH (see Section 12.2). ACTH interacts with G protein-linked cell-surface receptors and stimulates the production of cortisol by increasing its rate of synthesis from cholesterol. Glucocorticoid secretion is regulated by a typical negative feedback system. Rising cortisol levels act on the anterior pituitary and probably also the hypothalamus to inhibit CRH release and release of ACTH, thereby reducing the rate of secretion of cortisol. Loss of ACTH or removal of the pituitary (hypophysectomy) results in gradual atrophy of the fasciculata and reticularis regions of the adrenal cortex while chronically high levels of ACTH cause the adrenal cortex to hypertrophy. These effects underline the trophic nature of the actions of ACTH on the structure and function of the fasciculata and reticularis zones.

ACTH secretion shows a distinct circadian rhythm related to the sleep–wakefulness cycle, and this is reflected in the pattern of cortisol secretion. The concentration of cortisol in the plasma is minimal at around 3 a.m. and rises to a maximum between 6 and 8 a.m. before falling slowly during the rest of the day (Fig. 12.24). About half of the total daily output of cortisol is released during the pre-dawn surge. Superimposed on this cycle is an episodic pattern of release characterized by short-lived fluctuations in output (pulsatile secretion)

The normal rhythm of cortisol release is interrupted by acute stress of any kind as a result of the direct stimulation of CRH secretion. The mechanisms involved in the regulation of cortisol secretion are shown diagrammatically in Fig. 12.25.

Principal disorders of adrenal cortical hormone secretion
Overproduction of cortisol

Excessive cortisol release, also known as *Cushing's syndrome*, may result from a tumor of the adrenal gland itself, from increased output of ACTH resulting from a pituitary tumor, or hypersecretion of CRH. It may also occur following the overadministration of steroid medications. In some cases, there is ectopic produc-

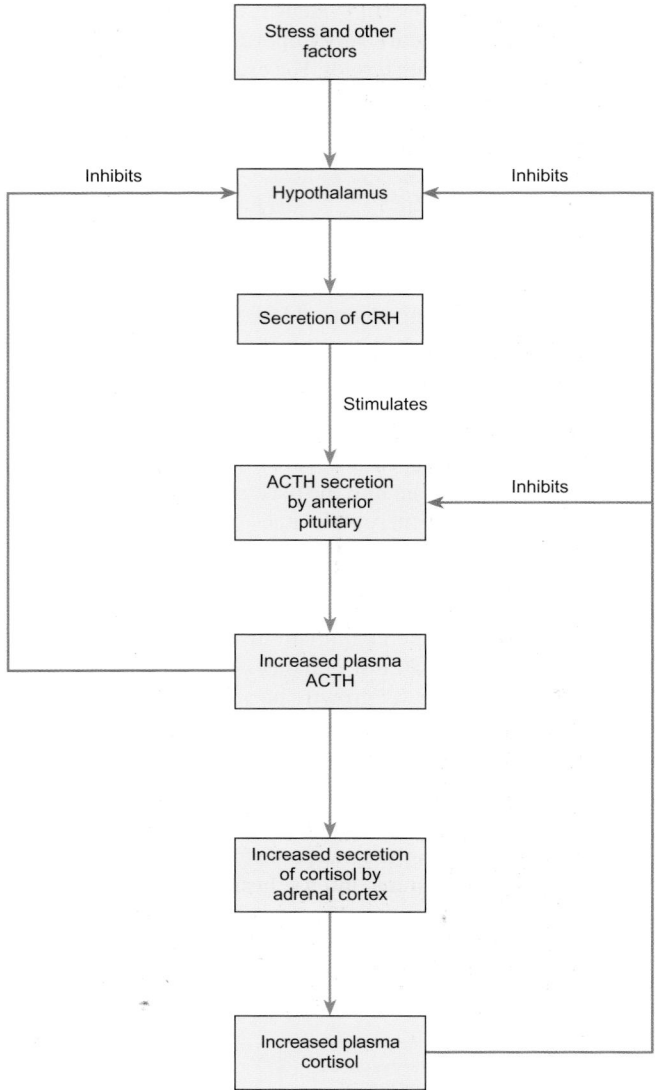

Fig. 12.25 A flow chart showing the factors that regulate the secretion of glucocorticoids. The red arrows represent negative feedback inhibition. Circulating glucocorticoids inhibit the secretion of CRH by the hypothalamus and that of ACTH by the anterior pituitary. Circulating ACTH probably also inhibits the secretion of CRH.

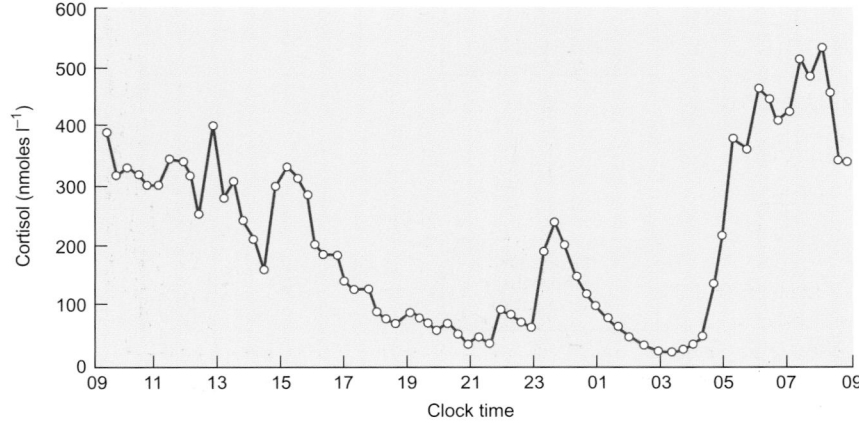

Fig. 12.24 The diurnal variation in plasma cortisol. The plasma levels rise during the early part of the day and are lowest around 3 a.m. This reflects the pattern of ACTH secretion by the anterior pituitary.

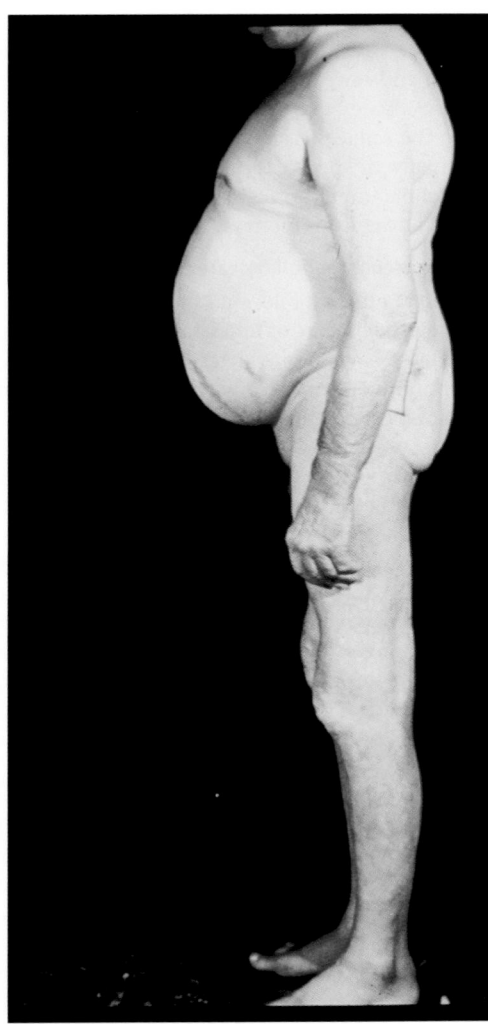

Fig. 12.26 A patient exhibiting the typical physical characteristics associated with Cushing's syndrome.

tion of ACTH, for example from certain types of lung tumors. Patients with Cushing's syndrome have a characteristic appearance. There is often a redistribution of body fat, which includes increased deposition of fat in the face and trunk. This, combined with the wasting of muscle tissue resulting from increased protein mobilization, gives the typical 'melon on toothpicks' appearance seen in Fig. 12.26. The skin of patients with Cushing's syndrome becomes thin and bruises easily, and there may be abnormal pigmentation. The disorder is also characterized by changes in carbohydrate and protein metabolism, hyperglycemia, and hypertension. Similar effects are sometimes seen in patients with chronic inflammatory disorders who are receiving prolonged treatment with corticosteroids.

Mineralocorticoid excess

Overproduction of aldosterone can arise either from hyperactivity of the cells of the zona glomerulosa (*Conn's syndrome*) or as a result of excessive renin secretion. The consequences of aldosterone excess include retention of sodium, loss of potassium (hypokalemia), and alkalosis. Hypertension results from

the expansion of plasma volume, which follows the increased reabsorption of sodium.

Excessive production of adrenal androgens

The effects of excess androgen secretion can be very distressing to both men and women. It occurs most often because of overproduction of ACTH by either the anterior pituitary itself or an ACTH-secreting tumor. Occasionally an adrenal tumor or a fault in steroid biosynthesis, as in congenital adrenal hyperplasia, will lead to androgen excess. The principal signs of androgen excess are acne, frontal baldness, and hirsutism (excessive growth of abdominal hair). Males may experience a reduction in testicular volume due to negative feedback inhibition of pituitary gonadotrophins by the androgens, while females may show enlargement of the clitoris and virilization of the secondary sex characteristics. Excessive adrenal sex steroid production in children may bring about precocious puberty.

Deficiency of adrenal cortical hormones

The major deficiency disorder of the adrenal cortex is called *Addison's disease*. It is a comparatively rare condition that may arise because of damage to the adrenal glands, autoimmune disease, or pituitary damage. It usually involves deficits in both glucocorticoids and mineralocorticoids, and the symptoms of the disease can be predicted from the preceding account of the actions of these steroids. People with Addison's disease tend to show progressive weakness, lassitude, and loss of weight. Their plasma glucose and sodium levels drop, while potassium levels rise. Severe dehydration and hypotension are common. A characteristic sign is pigmentation of the skin and mucosal membranes of the mouth. Corticosteroid replacement therapy at physiological doses is the usual treatment. Figure 12.27 shows the characteristic appearance of a patient suffering from Addison's disease.

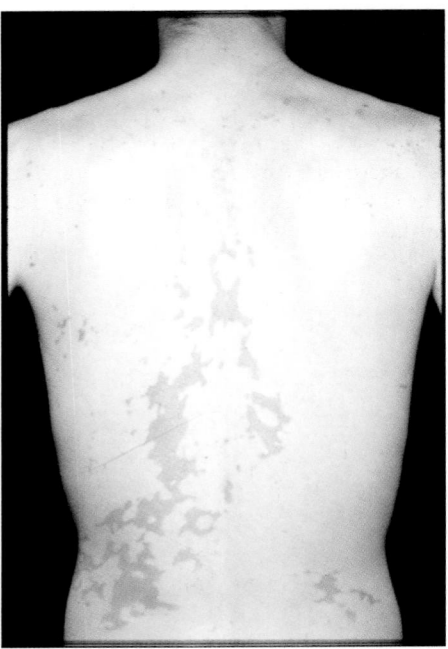

Fig. 12.27 The characteristic pigmentation of the skin of a patient suffering from Addison's disease.

The adrenal medulla

In essence, the adrenal medulla represents an enlarged and specialized sympathetic ganglion that secretes the catecholamine hormones *epinephrine* (adrenaline) and *norepinephrine* (noradrenaline) into the circulation. It is chiefly composed of *chromaffin cells*, which can be considered as specialized sympathetic postganglionic neurons. As for other preganglionic nerves the transmitter secreted by the splanchnic nerve fibers onto the chromaffin cells is acetylcholine (see Chapter 10).

The chromaffin cells are filled with storage granules that contain either epinephrine or norepinephrine, dopamine β-hydroxylase (one of the enzymes important in the synthesis of catecholamines (Fig. 12.28)), ATP, and a variety of opioid peptides such as met-enkephalin and leu-enkephalin. These are secreted by exocytosis into the bloodstream in response to acetylcholine released from the splanchnic nerve terminals. About 80 per cent of the secreted catecholamine is epinephrine while the remaining 20 per cent is norepinephrine. Catecholamine release occurs as part of a general sympathetic stimulation and is of particular importance in preparing the body for coping with acute stress (the fight–flight–fright response).

Epinephrine and norepinephrine secreted from the adrenal medulla are inactivated extremely rapidly and have half-lives in the plasma in the range of 1 to 3 minutes. They are either taken up into sympathetic nerve terminals or inactivated by the enzymes catechol-*O*-methyltransferase and monoamine oxidase in tissues such as the liver, kidneys, and brain.

The actions of adrenal medullary catecholamines

The physiological effects of adrenal medullary catecholamines must be considered as part of an overall sympathetic response since their release is always associated with an increase in the secretion of norepinephrine from sympathetic nerve terminals. While the adrenal medulla is not vital for survival, it does contribute to the response of the body to stress. Epinephrine and norepinephrine exert slightly different effects, which are summarized in Table 12.4.

Both epinephrine and norepinephrine raise the systolic blood pressure by stimulating heart rate and contractility, thereby increasing cardiac output. However, epinephrine reduces diastolic pressure as a result of causing vasodilatation of certain vessels, particularly those of skeletal muscle, while norepinephrine raises diastolic pressure by causing a more generalized vasoconstriction. Both catecholamines cause piloerection and dilatation of the pupils. Epinephrine also acts as a bronchodilator and reduces the motility of the gut.

Epinephrine exerts important metabolic effects. It promotes the breakdown of glycogen in the liver (glycogenolysis), lipolysis, oxygen consumption, and calorigenesis. In this respect, its actions are similar to those of the thyroid hormones with which it synergizes (see Section 12.4). Norepinephrine is also a potent stimulator of lipolysis but has little effect on glycogenolysis.

Epinephrine and norepinephrine act on different types of adrenergic receptors (or *adrenoceptors*). These are α- and β-receptors, which are further subdivided into α_1, α_2, β_1, β_2, and β_3 receptors (see Chapter 10). Epinephrine interacts primarily with β-receptors while norepinephrine binds preferentially to α- and β_1-receptors. The receptors that mediate the various actions of the adrenal medullary hormones are indicated where known in Table 12.4.

The various classes of adrenergic receptors provide a mechanism by which the same adrenergic hormone or neurotransmitter can exert differing effects on different effector cells. Both agonists and antagonists of the α- and β-adrenoceptors are widely used in clinical medicine. For example, β-antagonists ('β-blockers'), such as atenolol, are often prescribed to reduce cardiac output in the treatment of high blood pressure. β_2-agonists, such as salbutamol, are administered to asthmatic patients to bring about bronchodilatation.

Control of adrenal medullary secretion

While basal secretion of adrenal medullary catecholamines is probably very small, the gland may be stimulated to release its hormones in response to a number of stressful situations. These include exercise, hypoglycemia, cold, hemorrhage, and hypotension. Secretion may also accompany emotional reactions, such as fear, anger, pain, and sexual arousal, while the fetal adrenal medulla seems to respond directly to hypoxia. With the exception of this direct response, catecholamine secretion is

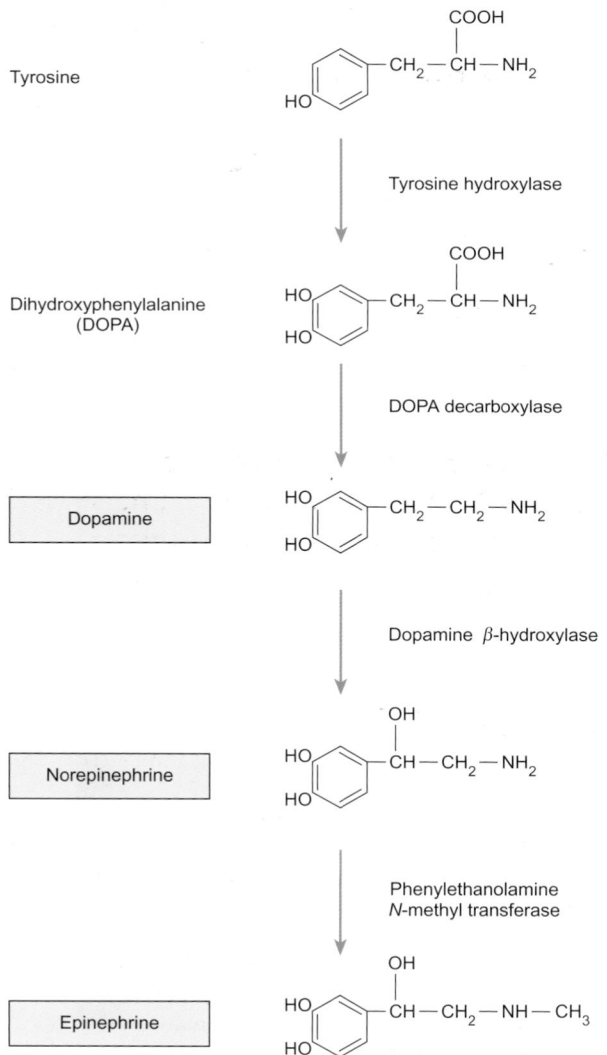

Fig. 12.28 The synthesis of epinephrine and norepinephrine from tyrosine in the adrenal gland.

TABLE 12.4 The efficacy of the catecholamine hormones in modulating various physiological processes

Epinephrine > norepinephrine	Norepinephrine > epinephrine
↑ Glycogenolysis (β_2)	↑ Gluconeogenesis (α_1)
↑ Lipolysis (β_3)	
↑ Calorigenesis (β_1)	
↑ Insulin secretion (β_2)	↓ Insulin secretion (α_2)
↑ Glucagon secretion (β_2)	
↑ K^+ uptake by muscle (β_2)	
↑ Heart rate (β_1)	
↑ Contractility of cardiac muscle (β_1)	
↓ Arteriolar tone in skeletal muscle (β_2)	↑ Arteriolar tone in non-muscle vascular beds (α_1) leading to vasoconstriction and ↑ BP
	↑ Tone in gastrointestinal sphincters (α_1)
↓ Tone in non sphincter GI smooth muscle (β_1)	↓ Tone in non-sphincter smooth muscle (α_1)
↓ Tone in bronchial smooth muscle (bronchodilatation) (β_2)	↑ Tone in bronchial smooth muscle (bronchoconstriction) (α_1)

The adrenoceptors mediating the various effects are shown in parentheses. ↑ Indicates an increase and ↓ indicates a decrease in the specified physiological process. For a discussion of adrenergic receptor subtypes, see Chapter 10.

Summary

1. The adrenal gland is a composite gland consisting of an outer cortex and inner medulla. The cortex secretes glucocorticoids, mineralocorticoids, and small quantities of sex steroids. The adrenal medulla secretes the catecholamines epinephrine (adrenaline) and norepinephrine (noradrenaline).

2. In humans, aldosterone is the chief mineralocorticoid and cortisol is the dominant glucocorticoid. Adrenal cortical hormones travel in the blood in combination with plasma proteins such as albumin and the more specific transcortin.

3. Aldosterone acts to conserve body sodium by stimulating its reabsorption in exchange for potassium in the distal nephron. A lack of aldosterone would be fatal within days without sodium supplementation and/or hormone replacement therapy.

4. Cortisol is essential to life. It has crucial metabolic actions and forms a vital part of the body's response to stress of all kinds. It stimulates gluconeogenesis and glycogen production as well as lipolysis. High levels of cortisol suppress the immune response to infection by reducing the mass of lymphoid tissue. Glucocorticoids are also used clinically as anti-inflammatory drugs.

5. Glucocorticoid secretion is regulated by pituitary ACTH, which shows a distinct circadian rhythm. Plasma cortisol levels are lowest around 3 a.m. and rise to a peak around the time of waking. Glucocorticoid secretion is regulated by a typical negative feedback system.

6. The adrenal medulla is composed of chromaffin cells that secrete epinephrine and norepinephrine as part of a general sympathetic response to stress. Epinephrine and norepinephrine cause increases in heart rate, contractility, and cardiac output. Epinephrine is also a bronchodilator and reduces gut motility. It promotes glycogenolysis, lipolysis, and oxygen consumption.

mediated by the activity of the splanchnic nerves. The adrenal medulla becomes non-functional if these nerves are sectioned.

Disorders of adrenal medullary secretion

While underproduction of the adrenal medullary hormones is not a clinical problem, excessive catecholamine output can have serious consequences. It can arise as the result of a tumor of the chromaffin tissue, a *pheochromocytoma*. The principal symptoms of such a tumor are severe hypertension, which may be episodic or sustained, hyperglycemia, and a raised metabolic rate. There is often anxiety, tremor, arrhythmias, and sweating as well. Surgical removal of the tumor, though a difficult procedure, is the usual treatment, although most of the symptoms can be alleviated by the administration of drugs that block the catecholamine receptors.

12.6 The role of parathyroid hormone, vitamin D metabolites, and calcitonin in the regulation of calcium and phosphate

About 99 per cent of the body's calcium is found in the bones in the form of calcium phosphate salts (hydroxyapatites) which give strength and rigidity to the skeleton. Therefore the skeleton acts as a large reservoir of calcium that can be mobilized by various hormones. Calcium is also a major component of the teeth and connective tissue. It plays a central role in blood clotting and in many cellular functions such as stimulus–secretion coupling, muscle contraction, cell–cell adhesion, and the control of neural excitability. Calcium also acts as a second messenger to regulate the activity of many enzymes. Because of its importance as a regulatory ion, it is essential that the level of free calcium in both the intra- and extracellular fluids is maintained within narrow limits.

The intracellular free calcium concentration is normally held at values of about 0.1 μmol l^{-1} by a variety of mechanisms. These include a plasma membrane calcium pump, which extrudes calcium from the intracellular fluid, and the sequestration of calcium by intracellular storage sites such as mitochondria and the endoplasmic (or sarcoplasmic) reticulum (see Chapter 4). Total intracellular calcium (i.e. bound plus free calcium) is much higher, around 1–2 mmol kg^{-1} of tissue.

Total plasma calcium is 2.3–2.4 mmol l^{-1} of which around half is ionized and half is bound, either to plasma proteins (albumin and globulins) or as inorganic complexes with anions, particularly phosphate. The concentration of free calcium ions in the extracellular fluid is around 1.4 mmol l^{-1}, more than 10 000 times higher than intracellular free calcium.

Hypocalcemia results in tetany

Low plasma calcium (*hypocalcemia*) will be reflected in a low calcium concentration in the extracellular fluid. This may result in *tetany*, an abnormal excitability of the nerves and skeletal muscles which is manifested as muscular spasms, particularly in the feet and hands (*carpopedal spasm*). The increased neural excitability may even result in convulsions. Tetany occurs when plasma calcium falls from its normal level of 2.3–2.4 mmol l^{-1} to around 1.5 mmol l^{-1}. The hyperexcitability caused by a fall in plasma calcium can be detected before the onset of tetany by tapping the facial nerve as it crosses the angle of the jaw. In normal people this has no effect, but in hypocalcemia the muscles on that side of the face will twitch or even go into spasm. (This is Chvostek's sign of *latent tetany*).

Tetany should be carefully distinguished from tetanus, the maintained contraction of skeletal muscle, and tetanus, the disease caused by the toxin (tetanus toxin) of the bacillus *Clostridium tetani*.

Calcium balance

The adult human body contains around 1 kg of calcium, the vast majority of which exists in the form of hydroxyapatite crystals within the bones and teeth. Only about 10 g of calcium is available for other cellular processes. Daily oral intake of calcium is very variable but falls between 200 and 1500 mg per day. In Western diets, the major sources of calcium are dairy products and flour, to which calcium is often added. For a daily intake of 1000 mg calcium, roughly 350 mg will be absorbed into the extracellular fluid from the small intestine. However, net absorption is reduced to around 150 mg a day because the intestinal secretions themselves contain calcium and around 850 mg of calcium are lost from the body each day in the feces.

Although the absolute amounts of calcium excreted vary according to the prevailing calcium balance of the body, almost 99 per cent of filtered calcium is normally reabsorbed along the length of the renal tubules. Nevertheless, close to 150 mg of calcium is excreted in the urine per day in the form of inorganic salts. Small amounts of calcium are also lost in the saliva and sweat. Thus the intestine and the kidneys are important organs in the regulation of the entry and exit of calcium from the plasma.

The bones of the skeleton provide an enormous reservoir of calcium within the body. About 99 per cent of the skeletal calcium forms so-called stable bone, which is not readily exchangeable with the calcium in the extracellular fluid. The remaining 1 per cent is in the form of simple calcium phosphate salts that form a readily releasable pool of calcium and can act as a buffer system in response to alterations in plasma calcium. Even during adult life, all the bones are in dynamic equilibrium where

deposition (accretion) and resorption of bone are balanced while allowing remodeling of the skeleton in response to changing mechanical requirements. The cells responsible for the accretion of new bone are called *osteoblasts*, while those responsible for the resorption of bone are large multinucleated cells called *osteoclasts* ('bone-eating' cells). Osteoblasts secrete the organic constituents of bone (osteoid) that later becomes mineralized provided that there is sufficient calcium and phosphate in the extracellular fluid. Bone resorption results in the release of calcium and phosphate into the plasma. The physiology of bone is discussed in more detail in Chapter 23.

When whole-body calcium balance is normal, resorption and accretion occur at a similar rate. Under these conditions:

[dietary Ca^{2+}] + [Ca^{2+} resorbed from bones]
= [Ca^{2+} loss in feces and urine] + [Ca^{2+} added to new bone].

Whole-body handling of phosphate

The plasma level of inorganic phosphate is around 2.3 mmol l^{-1} although its concentration is not as closely regulated as that of calcium. The total body phosphate content of a 70 kg man is about 770 g, of which between 75 and 90 per cent is contained within the skeleton in combination with calcium. Part is in the form of hydroxyapatite crystals and part is in the readily exchangeable pool as calcium phosphate. Daily intake of phosphate is roughly 1200 mg, of which about a third is excreted in the feces. The principal route of phosphate loss from the plasma is the urine, where it is excreted along with calcium.

Like calcium, phosphate is required for a wide variety of cellular functions. It is an important component of the phospholipids of the cell membranes and is needed for the synthesis of DNA and RNA. Phosphate metabolites play a central role in energy metabolism and the activity of numerous enzymes is regulated by phosphorylation.

Hormonal regulation of the plasma levels of calcium and phosphate

The minute-to-minute regulation of plasma mineral levels, particularly that of calcium, is achieved by the combined effects of three different hormones acting on the bone, kidney, and intestine. These are:

(1) metabolites of cholecalciferol (vitamin D);

(2) parathyroid hormone;

(3) calcitonin.

These hormones are released in response to various physiological stimuli related to changes in plasma calcium.

Vitamin D and its metabolites

Vitamin D or *cholecalciferol* acts as the precursor for a group of steroid compounds that behave as hormones and play a crucial role in the regulation of plasma calcium levels. Vitamin D is not itself biologically active but undergoes hydroxylation reactions to form active hormones. The first hydroxylation takes place in the liver, giving rise to 25-hydroxycholecalciferol which is the major form of vitamin D in the circulation. A further hydroxyl group is added in the kidney to give either 1,25-dihydroxycholecalciferol (also called *calcitriol*) or 24,25-dihydroxy-

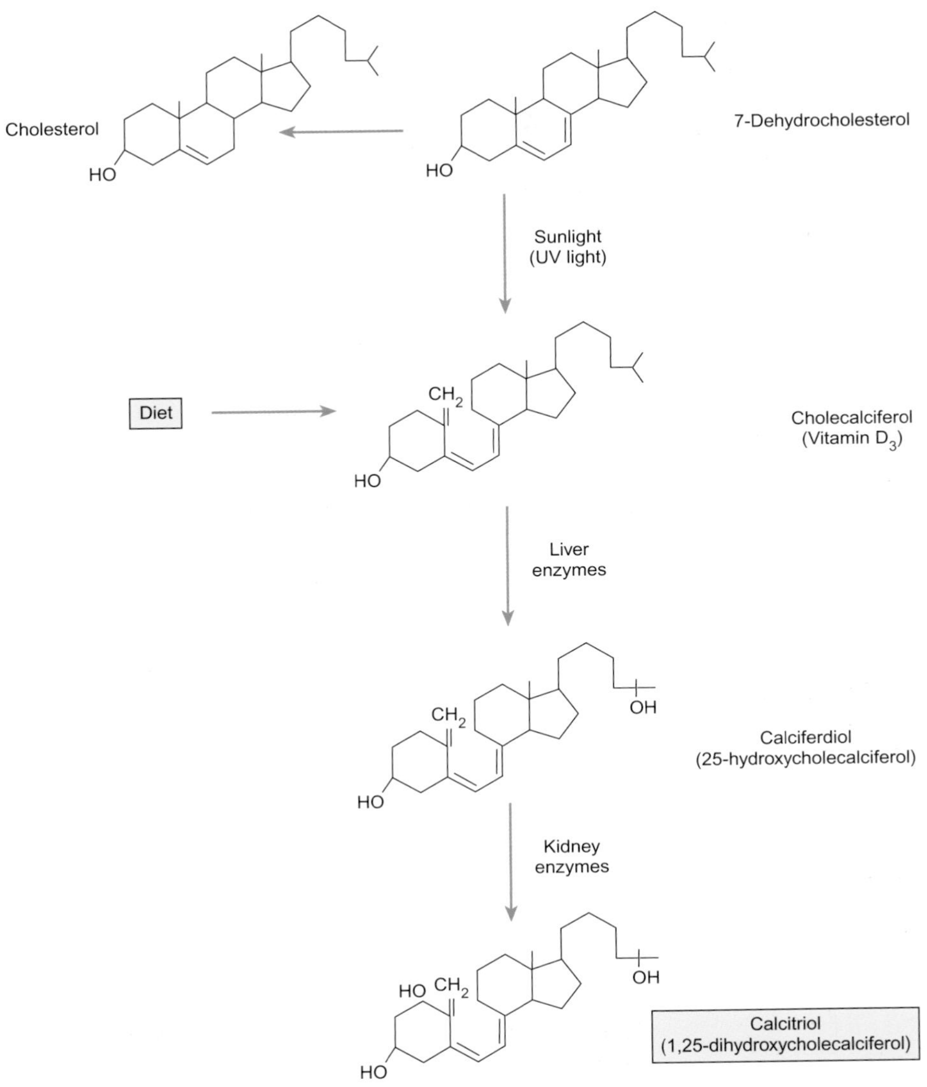

Fig. 12.29 The synthesis of 1,25-dihydroxycholecalciferol from vitamin D.

cholecalciferol, which is inactive. The synthesis of the active 1,25-derivative predominates at low levels of plasma calcium. Figure 12.29 shows the hydroxylation steps involved in the synthesis of the hormone.

Although much of the daily vitamin D requirement (400 IU for children and 100 IU for adults) is obtained from the diet, particularly from oily fish and eggs, cholecalciferol can be synthesized in the skin from 7-dehydrocholesterol in the presence of sunlight.

Actions of 1,25-dihydroxycholecalciferol

The main action of this hormone is to stimulate the absorption of ingested calcium. This occurs by a direct effect on the intestinal mucosa. The hormone binds to specific nuclear receptors and this interaction leads to an increase in the rate of synthesis of the calcium-binding proteins believed to transport calcium across the cell. Phosphate absorption is also enhanced by 1,25-dihydroxycholecalciferol.

The actions of vitamin D on bone are rather poorly understood and much of our current knowledge has been derived from observations of vitamin D deficiency. 1,25-dihydroxycholecalciferol stimulates the calcification of the bone matrix. While part of this effect is probably an indirect result of increased plasma levels of calcium and phosphate, there also seems to be a direct stimulation of both osteoblast and osteoclast activity. The combined effect of these actions will be to facilitate the remodeling of bone.

Effects of vitamin D deficiency

It has been known for many years that vitamin D deficiency gives rise to the conditions known as *rickets* in children and *osteomalacia* in adults. In both these disorders, bone remodeling, whereby old bone is resorbed and new bone is laid down, is impaired and there is a failure of calcification of the bone matrix. In children, this leads to skeletal deformities, particularly of the weight-bearing bones, giving rise to the characteristic bowing of the tibia. A picture of a child with rickets may be found in Chapter 23 (Fig. 23.15). In adults the chief feature is a reduction in bone density, with increased susceptibility to fracture and bone pain.

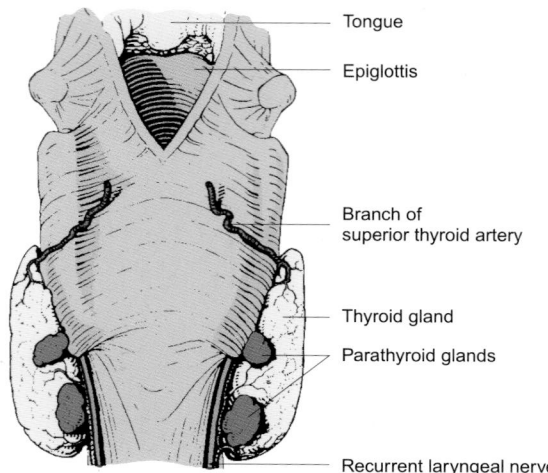

Fig. 12.30 The anatomical location of the parathyroid glands and their arterial blood supply. This view is from the *dorsal* (posterior) aspect of the pharynx.

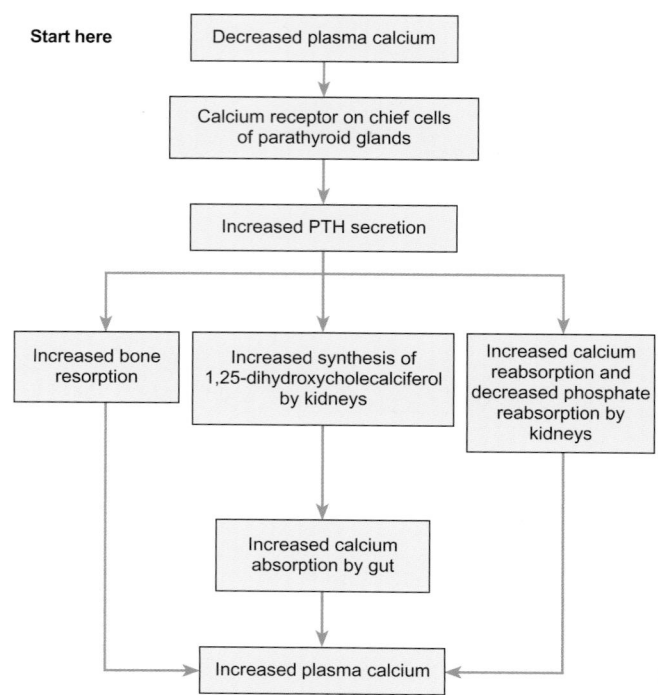

Fig. 12.31 A flow diagram summarizing the principal actions of PTH.

Patients with vitamin D deficiency may have low plasma calcium levels because of the reduction in intestinal calcium absorption. This may lead to increased excitability of nervous tissue, causing paresthesia ('pins and needles') or attacks of tetany. However, hypocalcemia in such patients is unlikely to become severe since there will be a compensatory rise in parathyroid hormone secretion (see below).

Effects of vitamin D excess

Large doses of vitamin D can give rise to a condition called vitamin D intoxication. This is characterized by nausea, vomiting, and dehydration due to elevated plasma calcium (*hypercalcemia*). If the plasma calcium level remains chronically elevated, renal function may be impaired as soft tissue within the kidneys becomes calcified.

Parathyroid hormone and plasma calcium homeostasis

Parathyroid hormone (PTH) is secreted by the parathyroid glands. There are usually four of these (although supernumerary glands are fairly common) lying embedded in the posterior surfaces of the lateral lobes of the thyroid gland. Each gland measures 3–8 mm in length, 2–5 mm in width, and about 1.5 mm in depth. Figure 12.30 shows the anatomical location of the parathyroids. The glands receive a rich blood supply derived mainly from the inferior thyroid arteries. The adult gland contains two major cell types, chief cells and oxyphilic cells. The chief cells secrete PTH but the function of the oxyphilic cells is uncertain.

PTH is a polypeptide hormone consisting of 84 amino acid residues, with a relative molecular mass of 9500 kDa. It is derived initially from a 115 amino acid polypeptide precursor called pre-pro-PTH. Two cleavages of the peptide chain give rise to active PTH which is packed into secretory granules for storage and eventual secretion.

Parathyroid hormone acts to mobilize calcium from bone and increase calcium reabsorption by the kidneys

PTH exerts effects on bone, gut and kidneys. It acts to raise plasma calcium levels and reduce plasma phosphate levels. These effects are summarized diagrammatically in Fig. 12.31.

Normal levels of PTH are necessary for the maintenance of the skeleton. It fosters the production of osteoblasts and the calcification of the bone matrix. However, when plasma calcium levels fall, PTH secretion rises. High levels of PTH have a biphasic action on bone metabolism. There is an initial rapid loss of calcium from the readily releasable pool of calcium on the bone surface. There is some evidence that PTH acts in combination with vitamin D to bring about this effect. In the longer term (hours to days), high circulating PTH levels stimulate resorption of stable bone by the osteoclasts, thereby adding large amounts of mineral to the extracellular fluid.

PTH stimulates the reabsorption of calcium in the distal tubules and decreases reabsorption of phosphate in the proximal tubules. The net effect of these actions will be an increase in plasma calcium and a fall in plasma phosphate. Although the latter effect may appear to be undesirable, the fall in plasma phosphate will further raise free plasma calcium levels by reducing the amount of phosphate ions available to bind with calcium. It is the rise in calcium which seems to be of primary importance to the body.

PTH does not appear to exert a direct effect on the intestine. However, it does stimulate the synthesis of 1,25-dihydroxycholecalciferol in the kidney. As discussed earlier, this metabolite of vitamin D enhances the intestinal absorption of ingested calcium.

Control of PTH secretion

PTH is cleared rapidly from the plasma, having a half-life in blood of only around 5 minutes. In order to maintain its basal concentration in the circulation, PTH is secreted continuously at a low rate. There appears to be a direct feedback relationship between plasma calcium ions and PTH secretion. This is mediated via a specific calcium receptor that is present on the cell

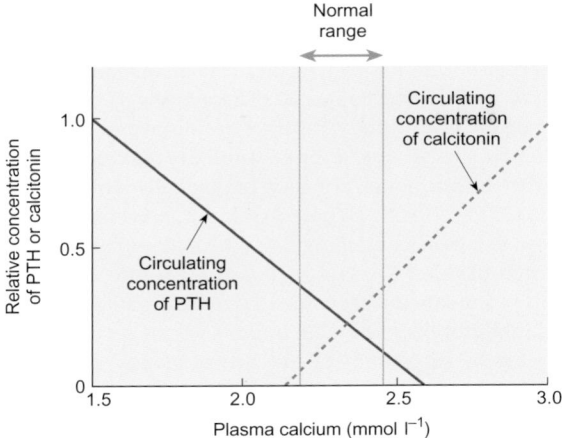

Fig. 12.32 The relationship between the plasma calcium concentration and the secretion of both PTH and calcitonin. As calcium rises, the secretion of PTH falls while that of calcitonin rises.

membrane of the PTH-secreting chief cells. As a result, the most potent stimulus for increased PTH secretion is hypocalcemia (lowered plasma calcium concentration), which also encourages the biosynthesis of the hormone. Conversely, a rise in plasma calcium inhibits the release of PTH. The inverse relationship between plasma calcium levels and PTH secretion is illustrated in Fig. 12.32. PTH secretion is also stimulated by catecholamines and dopamine but is suppressed by 1,25-dihydroxychole-calciferol. The secretion of PTH in hypocalcemia appears to require the presence of normal circulating levels of magnesium. Premature babies, who often have low plasma magnesium levels, may also become hypocalcemic because of low PTH output.

Disorders of parathyroid hormone secretion

Effects of excess PTH secretion (primary hyperparathyroidism)
Hypersecretion of PTH occurs when a tumor of the parathyroid glands develops. Certain malignant tumors originating in other cell types, such as the lungs, also secrete PTH and can give rise to the symptoms of PTH excess. The principal result of excess secretion of PTH is hypercalcemia associated with a reduction in plasma phosphate concentration. Despite the increased rate of reabsorption of calcium in the tubules, there is still a rise in the amount of calcium excreted in the urine because the filtered calcium load is very high. Kidney stones are a common finding in this disorder and may lead to severely impaired renal function, which is the most common cause of death in untreated hyperparathyroidism.

The skeletal effects of high PTH levels vary considerably between patients, but demineralization of the skeleton is often found. In such cases, there is bone pain, fractures of the long bones, and compression fractures of the spine. Cysts composed of osteoclasts ('brown tumors') may also be present. The hypercalcemia that results from overproduction of PTH has other consequences, including fatigue, weakness, mental aberrations, CNS depression, constipation, and anorexia.

Effects of insufficient secretion of PTH (hypoparathyroidism)
Hypoparathyroidism, due to damage, failure, or removal of the parathyroid glands (which may occasionally occur accidentally

during thyroid surgery), results in a gradual decline in plasma calcium which, if untreated, will eventually prove fatal. The major consequences of hypocalcemia are described above. Treatment of the disorder includes an initial infusion of calcium, to restore normal levels, followed by the administration of vitamin D metabolites which will stimulate the intestinal absorption of ingested calcium.

The role of calcitonin in the regulation of plasma calcium
Calcitonin (also known as thyrocalcitonin) is a peptide hormone secreted by the parafollicular or C-cells of the thyroid gland that lie between the follicles. Although it is known that calcitonin is able to lower the level of free calcium in the plasma, its significance in the overall regulation of mineral levels is unclear. Indeed, plasma calcium remains unaffected in patients with tumors of the C-cells in whom calcitonin levels are elevated.

Actions of calcitonin
The primary action of this hormone appears to be the inhibition of the activity of the bone osteoclasts. Therefore bone resorption is reduced and mineral is not released into the plasma. Calcitonin is sometimes administered to patients with hypercalcemia associated with malignancy and in Paget's disease in which there is excessive turnover of bone. There are receptors

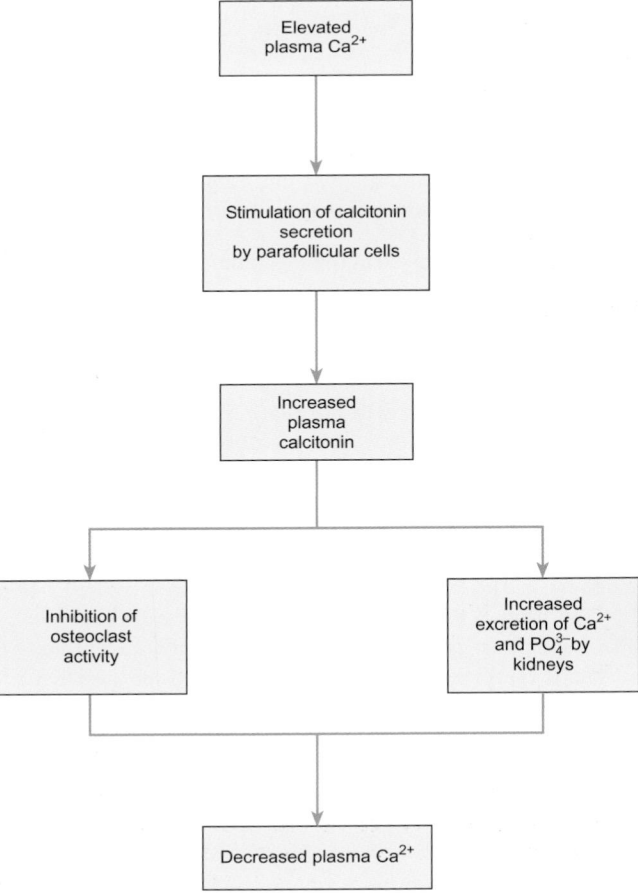

Fig. 12.33 The principal actions of calcitonin and the factors thought to regulate its secretion.

Summary

1. Calcium plays a vital role in many aspects of cellular function. It is also a major structural component of the bones of the skeleton. Plasma levels of calcium and phosphate are regulated by the actions of three hormones. These are the active metabolites of vitamin D, PTH, and calcitonin. They act on bone, gut, and kidney to regulate the entry and exit of calcium into and out of the extracellular pool.

2. Dihydroxycholecalciferol is a metabolite of vitamin D which is active in the regulation of plasma calcium. Its main effect is to enhance the absorption of dietary calcium by the intestine. It also seems to stimulate the turnover of bone. Vitamin D deficiency causes rickets in children and osteomalacia in adults.

3. Parathyroid hormone (PTH) is a peptide hormone secreted by the parathyroid glands. It is released in response to a fall in plasma calcium and acts to maintain normal plasma calcium. Normal PTH levels are needed for the maintenance of the skeleton. PTH also stimulates the reabsorption of calcium in the distal tubules of the kidney.

4. Calcitonin is secreted by the parafollicular cells of the thyroid gland. It is a hypocalcemic agent, secreted in response to elevated plasma calcium, but its physiological significance in the whole-body handling of calcium is unclear.

for calcitonin on kidney cells and calcitonin may produce a transient increase in the rates of excretion of calcium, phosphate, sodium, potassium, and magnesium. Figure 12.33 summarizes the actions of calcitonin and the regulation of its secretion.

Other hormones involved in the regulation of plasma calcium

Although PTH, metabolites of vitamin D, and possibly calcitonin are the major hormonal regulators of plasma mineral concentrations, a number of other hormones are known to exert effects on the way in which calcium and phosphate are handled by the body. These include growth hormone, adrenal glucocorticoids, and the thyroid hormones, but of considerably more significance are the estrogens and androgens, particularly the former.

In adult females, estrogens appear to inhibit the PTH-mediated resorption of bone and to stimulate the activity of the osteoblasts. Indeed, following removal of the ovaries, or after the menopause, the fall in estrogens results in an increased rate of bone resorption that can lead to a condition called osteoporosis. This is characterized by increased bone fragility, susceptibility to vertebral compression fractures, and fractures of the wrist and hip. Estrogen replacement therapy has been shown to reduce the rate of progress of postmenopausal osteoporosis.

12.7 The hormones of the gastrointestinal tract

A large number of peptide hormones, which regulate the activity of the stomach, intestine and accessory organs, are secreted by the gastrointestinal tract (see Table 12.1). Furthermore, two pancreatic hormones, insulin and glucagon, play a crucial role in the regulation of plasma glucose levels. Their actions and those of the other hormones involved in the control of blood sugar are discussed further in Chapter 27. The actions of the other hormones of the gastrointestinal tract are discussed more fully in Chapter 18 and so only a brief summary will be given here. A number of hormones are secreted by cells within the small intestine. These include cholecystokinin (secreted by I-cells in response to a meal containing fat), which stimulates contraction of the gall bladder. Secretin and vasoactive intestinal polypeptide (VIP) stimulate the secretion of pancreatic alkaline fluid, and gastric inhibitory peptide (GIP) inhibits gastric secretion. Gastrin is secreted by the G-cells of the gastric glands. This hormone stimulates the secretion of gastric acid.

Recommended reading

Biochemistry
Berg, J.M., Tymoczko, J.L., and Stryer, L. (2002). Biochemistry (5th edn), Chapters 15, 30, and 31. Freeman, New York.

Histology
Junquieira, L.C., and Carneiro, J. (2003). Basic histology (10th edn), Chapters 20 and 21. McGraw-Hill, New York.

Pharmacology
Rang H.P., Dale, M.M., Ritter, J.M., and Moore, P. (2003). Pharmacology (5th edn), Chapter 5, 27–8. Churchill-Livingstone, Edinburgh.

Endocrine physiology
Brook, C., and Marshall, N. (2001). Essential endocrinology (4th edn). Blackwell Science, Oxford.

Campbell, E.J.M., Dickinson, C.J., Slater, J.D.H., Edwards, C.R.W., and Sikora, E.K. (eds.) (1984). Clinical physiology (5th edn). Blackwell Scientific, Oxford.

Griffin, J.E., and Ojeda, S.R. (2000). Textbook of endocrine physiology (4th edn). Oxford University Press, Oxford.

Kettle, W.M., and Arky, R.A. (1998). Endocrine pathophysiology. Lippincott–Raven, Philadelphia, PA.

Laycock, J., and Wise, P. (1996). Essential endocrinology. Oxford Medical Publications, Oxford.

Medicine
Hall, R., and Evered, D.C. (1990). A colour atlas of endocrinology (2nd edn). Wolfe Medical, London.

Kumar, P., and Clark, M. (1998). Clinical medicine (4th edn), Chapter 16. W.B. Saunders, Edinburgh.

Ledingham, J.G.G., and Warrell, D.A. (eds.) (2000). Concise Oxford textbook of medicine, Chapters 7.1–7.16. Oxford University Press, Oxford.

Maitra, A., and Kumar, V. (2003). Chapter 20 in Robbins basic pathology (7th edn), Kumar, V., Cotran, R.S., and Robbins, S.L. (eds). Saunders, New York.

Porth, C.M. (1994). Pathophysiology: concepts of altered health states. J.B. Lippincott, Philadelphia, PA.

Multiple choice questions

Each statement is either true or false. Answers are given below.

1. a. Hypothalamic releasing hormones are synthesized and secreted by neurons.

   b. Blood flows from the anterior pituitary to the hypothalamus in portal vessels.

   c. The hypothalamic releasing hormones reach the general circulation in significant amounts.

   d. Loss of dopaminergic neurons in the hypothalamus is likely to lead to a rise in the secretion of prolactin.

   e. Growth hormone secretion is regulated by a single hypothalamic hormone.

2. A 10-year-old child in whom anterior pituitary function is deficient is likely to:

   a. Develop acromegaly.

   b. Be of short stature but have relatively normal body proportions.

   c. Be in constant danger of becoming dehydrated.

   d. Become sexually mature at a later age than normal.

   e. Have a low basal metabolic rate.

3. a. Chromaffin cells are found in the adrenal medulla.

   b. Excess secretion of catecholamines will lead to hypertension.

   c. The heart rate is reduced by circulating epirephrine and norepinephrine.

   d. Increased catecholamine secretion stimulates lipolysis.

   e. Epinephrine and norepinephrine are secreted by the adrenal cortex.

4. a. Normal plasma levels of PTH stimulate osteoblast activity.

   b. PTH decreases calcium excretion from the body.

   c. PTH directly increases calcium absorption by the gut.

   d. PTH is secreted in response to elevated plasma calcium levels.

   e. High levels of circulating PTH demineralize bone and elevate plasma calcium.

5. a. Oxytocin stimulates the synthesis of milk by the mammary glands.

   b. Lack of ADH will result in excessive production of urine.

   c. Both oxytocin and ADH are secreted in response to neuroendocrine reflexes.

   d. The secretion of oxytocin and ADH are regulated by releasing hormones secreted by the hypothalamus.

   e. ADH acts by binding to receptors on the plasma membrane of cells in the collecting ducts.

6. a. The adrenal cortex secretes both peptide and steroid hormones.

   b. The adrenal cortex will atrophy following removal of the anterior pituitary gland.

   c. Aldosterone plays a role in the regulation of plasma calcium.

   d. The secretion of cortisol peaks at around 6 a.m. each day.

   e. Cortisol is a hyperglycemic hormone.

7. a. Thyroid hormones are essential for the early development and maturation of the central nervous system.

   b. T_3 and T_4 stimulate the secretion of TSH by the anterior pituitary.

   c. People who have an underactive thyroid gland have a low BMR.

   d. A resting pulse rate of 65 b.p.m would suggest a diagnosis of thyrotoxicosis.

   e. Most of the iodide in the body is present in the thyroid gland.

Answers

1. Blood flows from the hypothalamus to the anterior pituitary gland via the hypothalamo-hypophyseal portal system. Although the releasing hormones are secreted into the portal system, negligible quantities reach the general circulation. Dopamine acts as an inhibitory hormone on the pituitary lactotrophs that secrete prolactin. GH secretion is under the control of both somatostatin and GHRH. The latter hormone increases GH secretion.

   a. True

   b. False

   c. False

   d. True

   e. False

2. Acromegaly is a disorder characteristic of hypersecretion of GH in adults. Deficient secretion of GH in childhood results in pituitary dwarfism in which body proportions are normal. Lack of ADH from the *posterior* pituitary increases the risk of dehydration. Lack of pituitary gonadotrophins will lead to delay or failure of sexual maturation due to lack of stimulation of the gonads. Deficient anterior pituitary function will lead to low TSH levels and so thyroid hormone secretion will be depressed, leading to a reduction in BMR.

   a. False

   b. True

   c. False

   d. True

   e. True

3. Chromaffin cells are found in the adrenal medulla as well as in some other locations. Epinephrine and norepinephrine raise arterial blood pressure through their actions on both the heart and vasculature. The catecholamines increase heart rate (and acetylcholine released by the parasympathetic post-ganglionic fibers decreases heart rate). Catecholamines activate lipases which mobilize fats to increase the production of free fatty acids and glycerol. Epinephrine and norepinephrine are secreted by the adrenal medulla.

   a. True

   b. True

   c. False

   d. True

   e. False

4. When plasma calcium levels are normal, PTH secretion stimulates bone formation through its action on osteoblasts (cells which secrete bone matrix). While PTH stimulates calcium absorption by the renal tubules, it has no direct action on the gut. However, it does stimulate the production of 1,25-dihydroxycholecalciferol, which stimulates calcium absorption from the gut. Low plasma calcium levels stimulate PTH secretion; high plasma levels stimulate the secretion of calcitonin. When PTH levels are elevated, osteoclast activity is stimulated which leads to the demineralization of bone and release of calcium into the plasma.

   a. True

   b. True

   c. False

   d. False

   e. True

5. Oxytocin stimulates the ejection of milk from the lactiferous ducts. Synthesis of milk is controlled by prolactin. In the absence of ADH, very little water is reabsorbed by the collecting ducts of the kidney, leading to the production of a large volume of dilute urine. Oxytocin is released in response to the mechanical stimulus of suckling while ADH is released in response to changes in the volume and osmolality of the plasma, which are detected by specific receptors. The releasing hormones regulate the secretion of hormones from the anterior pituitary. ADH exerts its actions by binding to G protein-coupled V_2 receptors on the P-cells of the collecting ducts.

   a. False

   b. True

   c. True

   d. False

   e. True

6. All the hormones secreted by the adrenal cortex are steroids. Pituitary ACTH acts to maintain the structural integrity and function of the adrenal cortex. Aldosterone plays an important role in the regulation of plasma sodium and thereby in the regulation of total body water. Cortisol secretion shows a circadian rhythm that is controlled by ACTH release from the anterior pituitary. Secretion is at its lowest around midnight and rises to a peak around dawn. It then declines during the course of the day. Cortisol stimulates gluconeogenesis and maintains reserves of glycogen for use in periods of fasting.

   a. False

   b. True

   c. False

   d. True

   e. True

7. Lack of TH in fetal and early neonatal life leads to a condition called cretinism in which there is mental retardation. Thyroid hormones *inhibit* the secretion of TSH by negative feedback. Thyroid hormones stimulate metabolism and when TH levels are low, BMR is reduced. Thyrotoxicosis is caused by an overactive thyroid gland. High levels of TH will increase BMR and elevate the heart rate. The value given is normal. The thyroid gland avidly takes up iodine and incorporates it into thyroglobulin.

   a. True

   b. False

   c. True

   d. False

   e. True

CHAPTER 13

The properties of blood

After reading this chapter you should understand:

- The principal roles of the blood, its chief constituents, and the hematocrit
- The physical and chemical characteristics of the plasma
- The role of the red cells, white cells, and platelets
- The origin of blood cells—hematopoiesis
- The main features of the metabolism of iron and its role in the formation of hemoglobin
- The carriage of oxygen and carbon dioxide by the red cells
- Some major disorders of the blood—anemia, leukemia and thrombocytopenia
- Blood clotting (hemostasis), clot retraction, and dissolution
- Blood groups and their importance in blood transfusions

13.1 Introduction

The blood is a vital vehicle of communication between the tissues of multicellular organisms. Its numerous functions include the following:

(1) delivery of nutrients from the gut to the tissues;

(2) gas exchange—the carriage of oxygen from the lungs to the tissues, and carbon dioxide from the tissues to the lungs;

(3) transport of the waste products of metabolism from the sites of production to the sites of disposal;

(4) carriage of hormones from endocrine glands to specific target tissues;

(5) protection against invading organisms—its immunological role.

Blood consists of a fluid called *plasma* in which are suspended the so-called formed elements—the *red cells* (erythrocytes), *white cells* (leukocytes), and *platelets* (thrombocytes). It is possible to demonstrate the nature of the suspension by centrifuging a sample of whole blood in a test tube for a short time at low speed. After centrifugation, the heavier red cells are packed at

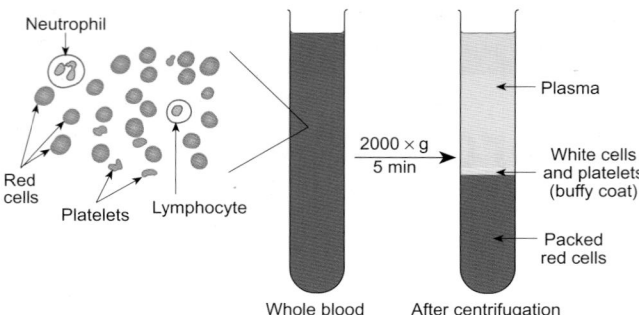

Fig. 13.1 Separation of blood into cells and plasma by centrifugation. On the left is a diagrammatic view of the appearance of whole blood showing red cells, leukocytes, and platelets.

the bottom of the tube while the plasma can be seen as a clear pale yellow fluid above them, as illustrated in Fig. 13.1. A thin layer of white cells and platelets (the 'buffy coat') separates the packed red cells from the plasma.

The circulating blood volume is about 7–8 per cent of body weight, so that blood volume for a 70 kg man will be around 5 liters, but for a newborn baby weighing 3.2 kg (7 lb), it will only be around 250 ml—an important point to remember when considering a blood transfusion on a small baby. At any one time, assuming a blood volume of 5 liters, about 0.6 liters will be in the lungs, about 3 liters will be in the systemic venous circulation, and the remaining 1.4 liters will be in the heart, systemic arteries, arterioles, and capillaries (see Chapter 15 p. 287).

13.2 The physical and chemical characteristics of plasma

The total blood volume and the plasma volume can be measured using the dilution techniques described in Chapter 28. Normal adults have 35–45 ml of plasma per kilogram body weight, so the plasma volume is 2.8–3.0 liters in men and around 2.4 liters in women.

The plasma accounts for about 4 per cent of body weight in both sexes. It consists of 95 per cent water, with the remaining 5 per cent being made up by a variety of substances in solution and suspension. These include mineral ions (e.g. sodium, potassium, calcium, and chloride), small organic molecules (e.g. amino acids, fatty acids, and glucose) and plasma proteins (e.g. albumin). Typical values for a number of important constituents of the plasma are given in Table 13.1. The major constituents of the plasma are normally present at roughly constant levels. These include the inorganic ions and the plasma proteins. However, the plasma also contains a number of substances that are in transit between different cells of the body and may be present in varying concentrations according to their rates of removal or supply from various organs. Such substances include enzymes, hormones, vitamins, products of digestion (e.g. glucose), and dissolved excretory products.

The composition of the plasma is normally kept within biologically safe limits by a variety of homeostatic mechanisms. However, this balance can be disturbed by a variety of disorders, particularly those involving the kidneys, liver, lungs, cardio-

TABLE 13.1	Principal constituents of the plasma		
Constituent	**Quantity**	**Units**	**Remarks**
Water	945	g l^{-1}	
Bicarbonate	25	mmol l^{-1}	Important for the carriage of CO_2 and for H$^+$ buffering
Chloride	105	mmol l^{-1}	The principal extracellular anion
Inorganic phosphate	1.0	mmol l^{-1}	
Calcium	2.5	mmol l^{-1}	This is total calcium; ionized calcium is about 1.25 mmol l^{-1}
Magnesium	0.8	mmol l^{-1}	
Potassium	4	mmol l^{-1}	
Sodium	144	mmol l^{-1}	The principal extracellular cation
Hydrogen ions	40	nmol l^{-1}	This corresponds to a pH value of c. 7.4
Glucose	4.5	mmol l^{-1}	Major source of metabolic energy, particularly for the CNS
Cholesterol	2.0	g l^{-1}	
Fatty acids (total)	3.0	g l^{-1}	
Total protein	70–85	g l^{-1}	
Albumin	45	g l^{-1}	Principal protein of the plasma; binds hormones and fatty acids
α-Globulins	7	g l^{-1}	
β-Globulins	8.5	g l^{-1}	
γ-Globulins	10.6	g l^{-1}	Immunoglobulins (antibodies)
Fibrinogen	3	g l^{-1}	Blood clotting (factor I)
Prothrombin	1	g l^{-1}	Blood clotting (factor II)
Transferrin	2.4	g l^{-1}	Iron transport

Note that these values are approximate mean values and that even in health there is considerable individual variation.

vascular system, or endocrine organs. For this reason, accurate analysis of the plasma levels of a host of variables forms an essential part of diagnosis and treatment.

The ionic constituents of plasma

The chief inorganic cation of plasma is sodium (see Table 13.1), which has a concentration of 140–145 mmol l⁻¹. There are much smaller amounts of potassium, calcium, magnesium, and hydrogen ions. Chloride is the principal anion of plasma (around 100 mmol l⁻¹); electroneutrality is achieved by the presence of other anions, including bicarbonate, phosphate, sulfate, protein, and organic anions.

The ionic components of the plasma maintain both its osmolality (280–300 mOsm per kg water) and its pH (7.35–7.45) within physiological limits. Further information concerning the homeostatic mechanisms responsible for the regulation of plasma pH, volume, and osmolality can be found in Chapters 17, 28, and 29.

Plasma proteins

In normal healthy individuals, 7–9 per cent of the plasma is made up of plasma proteins. There are a great many different proteins in plasma, but the principal proteins can be divided into three categories: the albumins, the globulins, and the clotting factors including fibrinogen and prothrombin.

The *albumins* are the smallest and the most abundant, accounting for about 60 per cent of the total plasma protein. They are transport proteins for lipids and steroid hormones that are synthesized by the liver. They are also important in body fluid balance since they provide most of the colloid osmotic pressure (the *oncotic pressure*) that regulates the passage of water and solutes through the capillaries (see Chapters 15, 17, and 28).

Globulins account for about 40 per cent of total plasma protein and can be further subdivided into *alpha* (α), *beta* (β), and *gamma*

(γ) *globulins*. The α- and β-globulins are made in the liver and they transport lipids and fat-soluble vitamins in the blood. The γ-globulins are antibodies produced by lymphocytes in response to exposure to antigens (agents usually foreign to the body that evoke the formation of specific antibodies). They are crucial in defending the body against infection.

Fibrinogen is an important clotting factor produced by the liver (see Section 13.8). It accounts for about 2–4 per cent of the total plasma protein and is generally grouped with the globulins.

Summary

1. Blood is a fluid consisting of plasma in which are suspended red cells, white cells, and platelets. It is the vehicle of communication between the tissues and serves to transport the respiratory gases, nutrients, hormones, and waste materials around the body.

2. Plasma is about 95 per cent water, with the rest consisting of a variety of proteins, including albumins, globulins, and fibrinogen, mineral ions (chiefly Na⁺ and Cl⁻), small organic molecules (e.g. glucose), and a number of substances in transit between tissues (hormones, products of digestion, and excretory products).

3. Plasma albumins carry lipids and steroid hormones in the plasma. The α- and β-globulins transport lipids and fat-soluble materials, while the γ-globulins are antibodies and play an essential role in defense against infection.

13.3 The formed elements of the blood

The formed elements of blood include red cells, five classes of white cells (recognized according to their morphology and staining reactions), and platelets (see Fig. 13.1 and 13.2). Of these, the

TABLE 13.2	The cellular elements of whole blood		
Cell type	**Site of production**	**Typical cell count (l⁻¹)**	**Comments and function**
Erythrocytes (red cells)	Bone marrow	5 × 10¹² (men) 4.5 × 10¹² (women)	Transport of O_2 and CO_2
Leukocytes (differential count) Granulocytes		7 × 10⁹	
Neutrophils	Bone marrow	5.0 × 10⁹ (40–75%)	Phagocytes—engulf bacteria and other foreign particles
Eosinophils	Bone marrow	100 × 10⁶ (1–6%)	Congregate around sites of inflammation—have antihistamine properties Very short lived in blood
Basophils	Bone marrow	40 × 10⁶ (<1%)	Circulating mast cells—produce histamine and heparin
Agranulocytes			
Monocytes	Bone marrow	0.4 × 10⁹ (2–10%)	Phagocytes—become macrophages when they migrate to the tissues
Lymphocytes	Bone marrow, lymphoid tissue, thymus, spleen	1.5 × 10⁹ (20–45%)	Production of antibodies
Platelets	Bone marrow	250 × 10⁹	Aggregate at sites of injury and initiate hemostatis

Note that, while mean values are given, these are subject to considerable individual variation. The approximate percentage of individual types of leukocyte are given after the number per liter—this is called the differential white cell count.

red cells are by far the most numerous. Table 13.2 lists the cellular components and their concentration in whole blood.

The Hematocrit

The *hematocrit ratio* or hematocrit describes the proportion of the total blood volume occupied by the erythrocytes. For any blood sample, the hematocrit can be obtained by centrifuging a small volume of blood in a capillary tube until the cellular components become packed at the bottom of the tube (see Fig. 13.1). For this reason the hematocrit is also known as the *packed cell volume*. By measuring the height of the column of red cells relative to the total height of the column of blood, and correcting for the plasma which remains trapped between the packed red cells, it is possible to determine the volume occupied by the packed red cells as a percentage of the total blood volume. In adult males, the average hematocrit determined in this way from a sample of venous blood is around 0.47 liter per liter of whole blood (it ranges from 0.4 to 0.54 L), while in females it is closer to 0.42 (normal range 0.37–0.47). However, the ratio of cells to plasma is not uniform throughout the body. In the capillaries, arterioles, and other small vessels, it is lower than in the larger arteries and veins as a result of *axial streaming* of blood cells in vessels (see Chapter 15). This is the tendency for red cells not to flow near to the walls of vessels but to remain near the center. In large vessels the wall surface area to volume/ratio is smaller than in the tiny vessels and so they contain relatively more cells.

Red blood cells—erythrocytes

The red cells (also called *erythrocytes*) are the most numerous cells in the blood—each liter of normal blood contains 4.5–6.5×10^{12} red cells. Their chief function is to transport the respiratory gases oxygen and carbon dioxide around the body. The red cells are small circular biconcave disks of diameter 7–8 μm and they do not possess a nucleus. They are very thin and flexible and can squeeze through the narrow bore of the capillaries, which have internal diameters of only 5–8 μm. Their shape gives red cells a large surface area to volume ratio which promotes efficient gas exchange (see Section 13.6). In a mature erythrocyte, the principal protein constituent of the cytoplasm is *hemoglobin*, an oxygen-binding protein, which is synthesized by the red cell precursors in the bone marrow.

White blood cells—leukocytes

Leukocytes are larger than the red blood cells, possess a nucleus, and are present in smaller numbers—normal blood contains around 7×10^9 white cells per liter (Table 13.2). These cells have a vital role in the protection of the body against disease—they are the mobile units of the body's protective system, being transported rapidly to specific areas of inflammation to give powerful defense against invading organisms. They possess several characteristics that enhance their efficacy as part of the body's defense system. They are able to pass through the walls of capillaries and to enter the tissue spaces in accordance with the local needs. This process is known as *diapedesis*. Once within the tissue spaces, leukocytes (particularly the polymorphonucleocytes) have the ability to move through the tissues by an amoeboid motion at speeds of up to 40 μm min^{-1}. Furthermore, they seem

to be attracted by certain chemical substances released by bacteria or inflamed tissues (*chemotaxis*). For more information concerning the immune system, see Chapter 14.

There are three major categories of white blood cells:

1. the *granulocytes* (or polymorphonuclear leukocytes, so-called because their nuclei are divided into lobes or segments);

2. the *monocytes* (macrophages);

3. the *lymphocytes*.

The monocytes and lymphocytes are sometimes also referred to collectively as agranulocytes or mononuclear leukocytes. The granulocytes are further subdivided into neutrophils, eosinophils, and basophils, according to their staining reactions. Although all the white blood cells are concerned with defending the tissues against disease-producing agents, each class of cell has a slightly different role to play. Consider first the *granulocytes*, which account for around 70 per cent of the total number of white cells in the blood.

Neutrophils are by far the most numerous of the granulocytes. They are *phagocytes*, which are able to enter the intercellular spaces by diapedesis to engulf and destroy disease-producing bacteria. Enzymes within the cytoplasmic granules then digest the phagocytosed particles. This action of the neutrophils forms the first line of defense against infection. They are so named because their cytoplasm does not stain with eosin or with basophilic dyes such as methylene blue.

Eosinophils are so-called because their granules stain red in the presence of the dye eosin. Normally they represent only about 1.5 per cent of the total number of white blood cells but in people with allergic conditions such as asthma or hay fever, their population greatly increases. These cells have antihistamine properties and they congregate around sites of inflammation. Their lifespan is very short (12–20 hours).

Basophils possess granules that stain blue in the presence of basic dyes such as methylene blue. They are considered to be circulating mast cells and represent only about 0.5 per cent of the white cell population. They produce heparin and histamine and are responsible for some of the phenomena associated with local immunological reactions such as local vasodilatation and increased permeability of blood vessels, resulting in local edema. They are stimulated by certain antigen complexes bound to immunoglobulin E (IgE).

Monocytes are larger than the other classes of white cells, having a diameter of 15–20 μm. Their nuclei are kidney shaped. They are formed in the bone marrow where they mature before being released into the circulation. Within 2 days, they have migrated to tissues such as the spleen, liver, lungs, and lymph nodes. These cells are *macrophages* and act in much the same way as the neutrophils, ingesting bacteria and other large particles. They also participate in immune responses both by presenting antigens so that they will be recognized by the *T-lymphocytes* and this stimulates the production of *B lymphocytes* (see Chapter 14 p. 254).

Lymphocytes represent around 25 per cent of the total white cell population in adults (although in children they are much more numerous) and vary from 6 to 20 μm in diameter. They are of two types, the *B lymphocytes*, which mature in the lymphoid

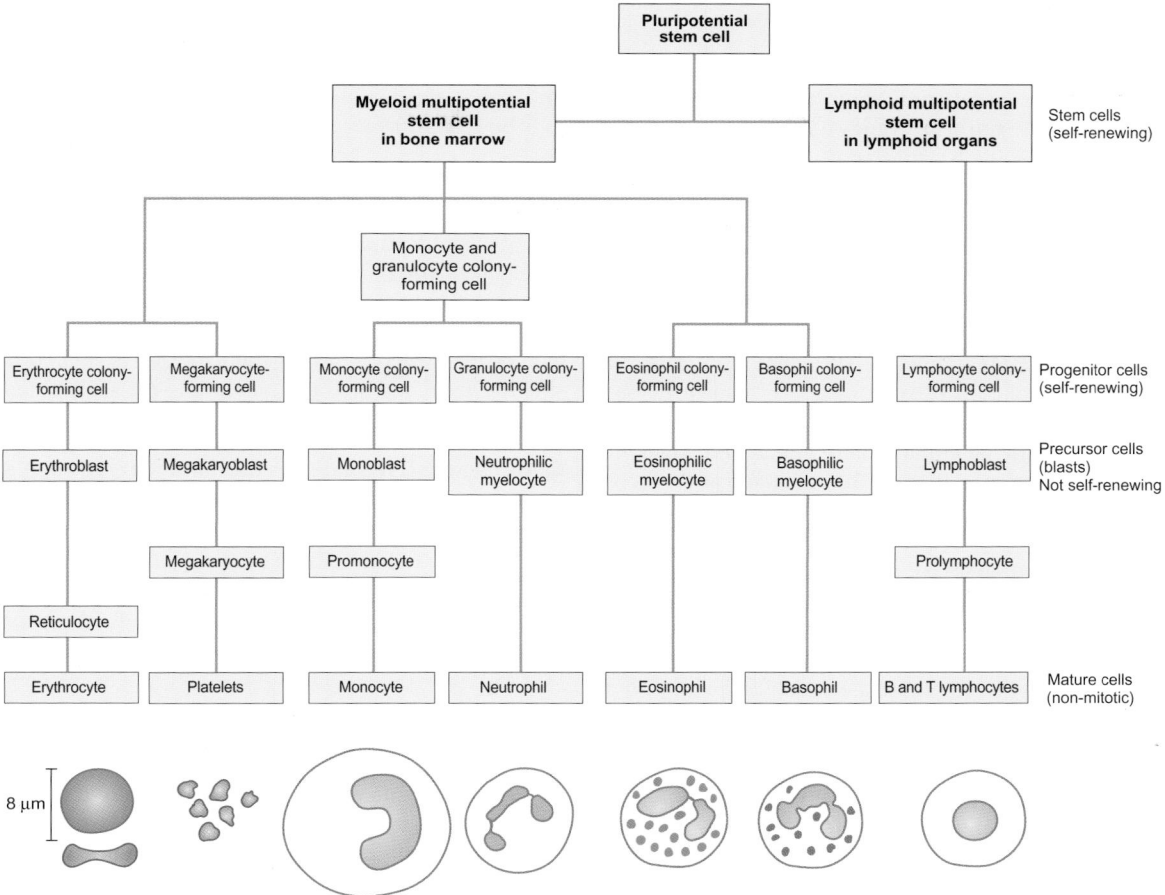

Fig. 13.2 An outline of the cellular differentiation that occurs during hematopoiesis to give rise to the cellular elements of the blood. The appearance of the different types of mature blood cell after staining is given at the foot of the figure.

Summary

1. The formed elements of blood include the erythrocytes, five types of leukocyte, and the platelets.

2. The formed elements of the blood can be separated from the plasma by centrifugation. The red cells become packed at the bottom of the tube, with the white cells and platelets forming a thin line above them. The packed cell volume (the hematocrit) can be measured in this way.

3. Red blood cells are small non-nucleated biconcave disks whose function is to transport oxygen and carbon dioxide between the lungs and tissues. They contain a protein, hemoglobin, which has a high affinity for oxygen.

4. Leukocytes are present in fewer numbers than red cells but play a crucial role in mediating the body's immune responses. They employ a variety of mechanisms to achieve this. These include phagocytosis, antibody production, and antihistamine reactions according to the cell type.

5. Platelets (thrombocytes) play an essential role in hemostasis. They are cell fragments derived from the megakaryocytes of the bone marrow.

tissue such as the lymph nodes, tonsils, and spleen, and to a lesser extent in the bone marrow, and the *T lymphocytes*, which are formed in the thymus. B cells have a very short life in the circulation (a few hours) but T cells can live for 200 days or more. Each has a very important part to play in the protection of the body against infection either by producing *antibodies* (B cells) or by participating in cell-mediated immune responses (T cells). The functions of the lymphocytes will be discussed in detail in Chapter 14.

Platelets (thrombocytes)

Strictly speaking, platelets are not cells at all. They are irregularly shaped membrane-bound cell fragments which are formed in the bone marrow by budding off from the cytoplasm of large polyploid cells called *megakaryocytes*. Megakaryocytes are derived from primitive hematopoietic stem cells (see Fig. 13.2 and Section 13.4). Platelets rarely possess a nucleus, are 2–4 μm in diameter, and have a lifespan in the blood of around 10 days. Normal blood contains $(150–400) \times 10^9$ platelets l^{-1}. Platelets have an important role in the control of bleeding (hemostasis—see Section 13.8) and in the maintenance of integrity of the vascular endothelium.

13.4 Hematopoiesis—the formation of blood cells

Mature blood cells have a relatively short lifespan in the blood-stream and therefore must be renewed continuously. The replacement of blood cells is achieved by a process known as hematopoiesis. The term *erythropoiesis* refers to the formation of erythrocytes (red blood cells), and *leukopoiesis* to the formation of leukocytes (white blood cells).

Pluripotent stem cells give rise to all the blood cell types

Despite the fact that the blood contains many different cells with a variety of functions, they are all generated ultimately from a common population of stem cells present within the hematopoietic tissue of the bone marrow. These cells are said to be *pluripotent* (having the potential to differentiate into any kind of blood cell), and give rise to all the differentiated types of blood cell through a series of cell divisions, which are shown diagrammatically in Fig. 13.2.

The pluripotent stem cells are particularly abundant in the bone marrow of the pelvis, ribs, sternum, vertebrae, clavicles, scapulae, and skull. They proliferate to form two distinct cell lines, the *lymphoid cells* and the *myeloid cells*. The lymphoid cells migrate to the lymph nodes, spleen, and thymus, where they differentiate to become lymphocytes. The myeloid cells remain within the bone marrow to develop as granulocytes, monocytes, erythrocytes, and megakaryocytes.

As can be seen from Fig. 13.2, in general terms the pluripotent stem cells divide (infrequently under normal conditions) to give rise to more stem cells as well as various types of 'committed' cells, each capable of giving rise to one or a few types of blood cell. These are called *progenitor cells*. These cells, in turn, generate *precursor cells* in which the morphological characteristics of the mature cell are evident for the first time. Full differentiation and maturation of the blood cells then occurs following a further series of cell divisions.

The maturation of erythrocytes

Those precursor cells that are committed to becoming red blood cells are called *erythroblasts*. These subsequently undergo a further series of cell divisions, each of which produces a smaller cell, as they mature into erythrocytes. During these divisions, the cells synthesize hemoglobin. Finally, they lose their nuclei to become *reticulocytes*. Development from erythroblast to reticulocyte normally takes around 7 days.

Most red cells are released into the circulation as reticulocytes. Within a day of being released into the circulation they mature and become erythrocytes. During this transition, they lose their mitochondria and ribosomes. Consequently, they also lose the ability to synthesize hemoglobin and carry out oxidative metabolism. The mature red blood cell relies on glucose and the glycolytic pathway for its metabolic needs. They also produce large amounts of 2,3-bisphosphoglycerate (2,3-BPG), which reduces the affinity of the hemoglobin for oxygen thereby facilitating the release of oxygen at the tissues (see Section 13.6).

Red blood cells have a lifespan of about 120 days

Once it has entered the general circulation, the average lifespan of a red blood cell is around 120 days, after which time it is destroyed in the spleen, the liver, or lymph nodes by large phagocytic cells known as *macrophages*. The protein portion of the erythrocyte is broken down into its constituent amino acids. The iron in the heme group is stored in the liver as ferritin and may be re-used later (see Section 13.5), while the remainder of the heme group is broken down into the two bile pigments *bilirubin* and *biliverdin*, both of which are eventually excreted into the gut by way of the bile. If red cell destruction, and therefore bilirubin production, is excessive, unconjugated bilirubin may build up in the blood and give a yellow colour to the skin—*hemolytic jaundice*. This condition can arise following a hemolytic blood transfusion reaction (see Section 13.9), in hemolytic disease of the newborn, or in genetic disorders such as hereditary spherocytosis in which the membrane of the red blood cells is defective.

Erythropoiesis is regulated by the hormone erythropoietin

Each liter of blood contains around 5 million million (5×10^{12}) erythrocytes, although this figure varies according to the age, sex, and state of health of the individual. Because most red cells enter the circulation as reticulocytes, the rate of red cell production is indicated by the relative numbers of reticulocytes in the circulation (normally 1–1.5 per cent). The rate of erythropoiesis is closely matched to the requirement for new red cells within the circulation and is controlled by a glycoprotein hormone *erythropoietin*, which is secreted mainly by the kidneys (probably by cells in the endothelium of the peritubular capillaries). This hormone acts by accelerating the differentiation of stem cells in the marrow to form erythroblasts. In addition to erythropoietin, iron, folic acid, and vitamin B_{12} are also essential for normal red blood cell production. Vitamin B_{12} is absorbed from the small intestine in combination with *intrinsic factor*, which is secreted by the parietal cells in the gastric mucosa (vitamin B_{12} was formerly known as extrinsic factor). If the diet is deficient in B_{12} or there is a lack of intrinsic factor, red cell development is impaired, resulting in pernicious anemia (see Section 13.7).

A variety of stimuli may cause an increase in the rate of production of new erythrocytes, including loss of red cells through hemorrhage or donation of blood, and chronic hypoxia such as that experienced while living at high altitude. In all cases, it appears that the secretion of erythropoietin is stimulated by a fall in tissue P_{O_2}.

Maturation of white cells
Granulocytes

There are three types of granulocytes: neutrophils, basophils, and eosinophils. Multipotential progenitor cells (*promyelocytes*) in the hematopoietic tissue give rise to three kinds of precursor cells: neutrophilic, basophilic, and eosinophilic myelocytes. These mature further, with condensation of the nucleus and an increase in specific granule content over the next 10 days or so, before appearing in the circulation. During an infection, the rate of granulocyte production (especially of neutrophils) increases considerably.

Maturation of monocytes

The committed precursor cell of the monocyte is the *monoblast* which differentiates further to generate the *promonocyte* which is a large cell (18 μm in diameter) containing a large nucleus and nucleoli. Promonocytes divide twice more to become monocytes containing a large amount of rough endoplasmic reticulum, Golgi complex, and lysosomes (see Chapter 3). After entering the blood, mature monocytes circulate for about 8 hours before entering the connective tissues where they mature into *macrophages*.

Maturation of lymphocytes

Circulating lymphocytes originate mainly in the thymus and peripheral lymphoid organs (spleen, lymph nodes, tonsils, etc.). The first identifiable progenitor is the *lymphoblast*. These cells divide several times to become smaller *prolymphocytes*. These subsequently synthesize the cell surface receptors that distinguish them as T or B lymphocytes (see Chapter 14).

Production of platelets

The precursor cells that ultimately give rise to the platelets are known as *megakaryoblasts*. These cells are 15–20 μm in diameter and possess a large ovoid or kidney-shaped nucleus. Differentiation of the megakaryoblasts gives rise to the *megakaryocytes*, which are giant cells (35–150 μm in diameter) whose cytoplasm contains numerous mitochondria, rough endoplasmic reticulum, and Golgi complex. As these cells mature within the bone marrow, invaginations of the plasma membrane become evident, eventually branching throughout the entire cytoplasm. These form the so-called *demarcation membranes* that define the areas which will be shed as platelets. The factors that control the rate of production of platelets are poorly understood, although it is known that the maturation of megakaryocytes and the release of platelets is stimulated by a hormone called *thrombopoietin*.

13.5 Iron metabolism

Iron is an essential component of hemoglobin and myoglobin, as well as of certain pigments and enzymes. The body of an adult man contains, in total, about 4.5 g of iron, of which about 65 per cent is within the hemoglobin of red blood cells. A further 5 per cent or so is contained within myoglobin and enzymes, while the remainder is stored in the form of *ferritin*, largely by the liver, but to a lesser extent also by the spleen and intestine.

When red blood cells become senescent, they are removed by phagocytes in the liver and spleen. Much of the iron derived from hemoglobin is recycled by the body, as illustrated in Fig. 13.3. The iron from the digested hemoglobin is either returned to the plasma, where it binds to *transferrin* (an iron-carrying protein), or is stored in the liver as ferritin. The iron–transferrin complex is carried to the erythropoietic tissue in the bone marrow, where it is either used immediately in the production of hemoglobin for developing red cells, or it is stored within the marrow itself. If blood loss occurs, iron contained within the stores of the marrow is utilized and there is an increase in the rate of uptake of iron from the circulation. A few days later, the balance is restored by an increased rate of iron absorption in the gut.

Because the recycling of iron is so efficient, the need for dietary iron in adults arises mainly from loss by bleeding and the death of intestinal cells. It therefore follows that the dietary requirement for iron is greater in menstruating women than in

Summary

1. Mature blood cells are renewed continuously by hematopoiesis. All the cell types are generated ultimately from a common population of pluripotent stem cells in the bone marrow. These form two distinct cell lines, myeloid and lymphoid cells. The myeloid cells remain in the marrow and form red cells and leukocytes other than lymphocytes. Lymphoid stem cells migrate to the lymph nodes, spleen, and thymus where they develop into lymphocytes.

2. The stem cells divide to form committed precursor cells which differentiate, via a series of cell divisions, into one of the mature cell types. The precursor cells for erythrocyte production, for example, are called erythroblasts. Through successive divisions, these start to synthesize hemoglobin before losing their nuclei to become reticulocytes, in which form they are released into the circulation.

3. Erythropoiesis is closely matched with the requirement for red cells in the circulation. This is controlled by erythropoietin, a hormone secreted by the kidneys. After about 120 days in the circulation, red cells are destroyed by macrophages in the spleen, liver, or lymph nodes.

4. Granulocytes, monocytes, and lymphocytes mature from precursor cells in a fashion similar to the red cells.

5. Platelets bud off from giant cells (megakaryocytes), which are themselves derived from the pluripotent stem cells in the bone marrow.

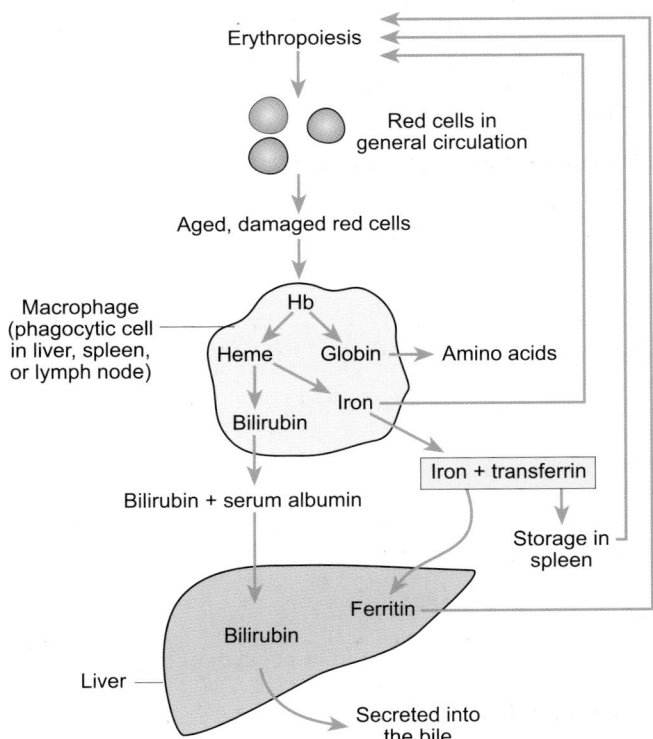

Fig. 13.3 The principal stages in the recycling of Fe^{2+} from red cell hemoglobin.

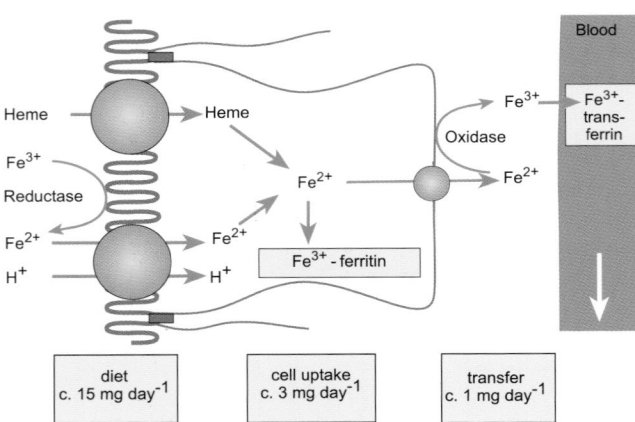

Fig. 13.4 An outline of the mechanism by which iron is absorbed by the intestine.

men, being about 1 mg day^{-1} in men and 2 mg day^{-1} in women of child-bearing age. Furthermore, children and pregnant women need relatively more iron because of their expanding circulatory volume. Dietary sources of iron include meat (particularly the myoglobin of the muscle), vegetables, and fruits. The normal Western diet contains adequate quantities of iron (around 15 mg day^{-1}) but strict vegetarians and vegans risk iron deficiency as much of their dietary iron is unavailable for absorption.

How is iron absorbed in the intestine?

Ionized iron can exist in two oxidation states, ferrous (Fe^{2+}) and ferric (Fe^{3+}). The low pH of the stomach lumen caused by the secretion of hydrochloric acid by the gastric mucosa solubilizes both ferrous and ferric iron salts. In addition, ascorbate (vitamin C) reduces Fe^{3+} to Fe^{2+}, which is less likely to form insoluble complexes with other constituents of the diet (particularly the fiber of cereal grains).

A simple scheme of the mechanism by which the epithelial cells of the upper intestine absorb iron is shown in Fig. 13.4. Any remaining ferric iron (Fe^{3+}) entering the duodenum is reduced to ferrous iron by an enzyme on the brush border (ferric oxido-reductase). The Fe^{2+} is taken up with hydrogen ions into the enterocytes by a transporter known as DMT1. Iron complexed with heme derived from dietary meat is absorbed directly by a separate pathway. Within the cell, heme oxygenase liberates Fe^{2+} from the heme molecule. Once inside the cell, iron can follow one of two pathways: either it leaves the enterocytes by the baso-lateral membrane transporter, ferroportin, or it becomes bound to specific cytoplasmic proteins, the best known of which is apo-ferritin. The path taken is determined by the body's demand for iron. Normally, around two-thirds of absorbed iron is bound to ferritin within the enterocytes. Ferrous iron leaving the entero-cytes across their basolateral membrane is converted to ferric iron before being absorbed into the blood and transported to the liver bound to its specific transport protein, transferrin.

Iron absorption is regulated in accordance with the body's needs

In iron deficiency, or following hemorrhage, the capacity of the small intestine to absorb iron is increased. After severe blood

Summary

1. About two-thirds of the total body iron is within the hemoglobin of red blood cells, 5 per cent is within myo-globin and enzymes, while the rest is stored, mainly in the liver, as ferritin.

2. When red blood cells are phagocytosed, most of their iron is recycled and either reused immediately, or stored as ferritin within the liver and bone marrow.

3. Most of the iron in the diet is absorbed as ferrous iron (Fe^{2+}). Duodenal and jejunal epithelial cells take up iron from the intestinal lumen by a carrier-mediated process. Iron is stored within the enterocytes bound to iron-binding proteins, including ferritin. Absorbed iron is released into the blood across the basolateral membrane, where it combines with transferrin in the plasma to be transported to the tissues.

4. Iron absorption is regulated in accordance with the body's requirement by a hormone, hepcidin. Following a hemor-rhage, for example, the capacity of the small intestine to absorb iron is enhanced.

loss, there is a time lag of 3 or 4 days before absorption is enhanced. This is the time needed for the enterocytes to migrate from their sites of origin in the mucosal glands to the tips of the villi, where they are best able to participate in iron absorption. The enterocytes of iron-deficient animals are able to absorb iron from the intestinal lumen more rapidly than normal, a process controlled by a hormone synthesized by the liver called *hepcidin*, which acts to inhibit iron uptake. When demand for iron is high, for example following hemorrhage, circulating levels of hepcidin fall and iron uptake from the small intestine is increased. When demand is low, hepcidin levels are high and iron absorption deceases. Hepcidin appears to regulate the num-ber of iron transporters (both DMT1 and ferroportin) in the membrane of the enterocytes.

Excess iron absorption is as undesirable as iron deficiency since high levels of iron can be toxic. This can become a problem if the diet is excessively rich in iron or in the genetic disease *idio-pathic hemochromatosis*, in which excessive amounts of iron are absorbed even from a healthy diet. This situation is normally prevented by the binding of iron to ferritin within the cyto-plasm of the enterocyte. This binding is almost irreversible, so any iron bound in this way is unavailable for absorption into the plasma. Instead, it is lost in the feces when the intestinal cell desquamates. The amount of iron held in the so-called storage pool increases when dietary intake rises, to maintain homeo-stasis. It is thought that the level of iron in the plasma in some way regulates the synthesis of ferritin.

13.6 The carriage of oxygen and carbon dioxide by the blood

The blood transports the respiratory gases around the body. Oxygen is carried from the lungs to all the tissues of the body while the carbon dioxide produced by metabolizing cells is

TABLE 13.3 Standard values for the partial pressures of blood gases

	Arterial blood	Mixed venous blood
Oxygen	13.3 kPa (100 mmHg)	5.33 kPa (40 mmHg)
Carbon dioxide	5.33 kPa (40 mmHg)	6.12 kPa (46 mmHg)

Note that the capacity of the blood to carry oxygen will depend on its hemoglobin content. In males the hemoglobin content is about 15 g dl⁻¹ while in females the value is usually lower at about 13.5 g dl⁻¹. See text for further details.

transported back to the lungs for removal from the body. The principles governing the exchange of gases in the lungs and the tissues are discussed fully in Chapter 16 (see p. 331). Briefly, oxygen passes from the alveoli to the pulmonary capillary blood by diffusion because the partial pressure of oxygen (P_{O_2}) in the alveolar air is greater than that of the pulmonary blood. In the peripheral tissues, P_{O_2} is lower in the cells than in the arterial blood entering the capillaries and so oxygen diffuses out of the blood, through the interstitial spaces, and into the cells. Conversely, the partial pressure of carbon dioxide (P_{CO_2}) in metabolizing cells is higher than that of the capillary blood. Carbon dioxide diffuses down its concentration gradient into the blood, which transports it to the lungs. Here, the P_{CO_2} of the pulmonary capillary blood is greater than that of the alveoli and carbon dioxide diffuses across the capillary and alveolar membranes. It is removed from the body during expiration. Standard values for the partial pressures of the blood gases are given in Table 13.3.

Hemoglobin increases the capacity of the blood to transport oxygen: each gram of hemoglobin can bind 1.34 ml of oxygen

At rest, oxygen is consumed by the body at a rate of around 250 ml min⁻¹ and this must be supplied by the blood. The solubility of oxygen in the plasma water is very low—only 0.225 ml of oxygen are dissolved in every liter of plasma for each kPa P_{O_2} (equivalent to 0.03 ml mm Hg⁻¹). Therefore at the normal arterial P_{O_2} of 13.3 kPa (100 mmHg), each liter of plasma will contain just 3 ml of dissolved oxygen. If this were the only means of transporting oxygen to the tissues, the heart would need to pump more than 80 liters of blood each minute to supply the required 250 ml min⁻¹. In fact, the blood is able to carry far more oxygen than this. At a P_{O_2} of 13.3 kPa (100 mmHg), the oxygen content of whole blood is about 20 ml dl⁻¹ (i.e. 200 ml l⁻¹). As a result, the normal resting cardiac output (about 5 l min⁻¹) is more than sufficient to meet the oxygen requirements of the body at rest.

The vast majority of the oxygen in the blood is carried in chemical combination with *hemoglobin*, an oxygen-binding protein contained within the red cells. Each hemoglobin molecule consists of a protein part (globin) consisting of four polypeptide chains, and four nitrogen-containing pigment molecules called heme groups. Each of the four polypeptide groups is combined with one heme group (see Fig. 13.5). In the center of each heme group is one atom of ferrous (Fe^{2+}) iron that can combine loosely with one molecule of oxygen. Therefore each molecule of hemoglobin (Hb) can combine with four molecules of oxygen to form *oxyhemoglobin* (often written as HbO_2). The reaction for the binding of oxygen can be expressed as

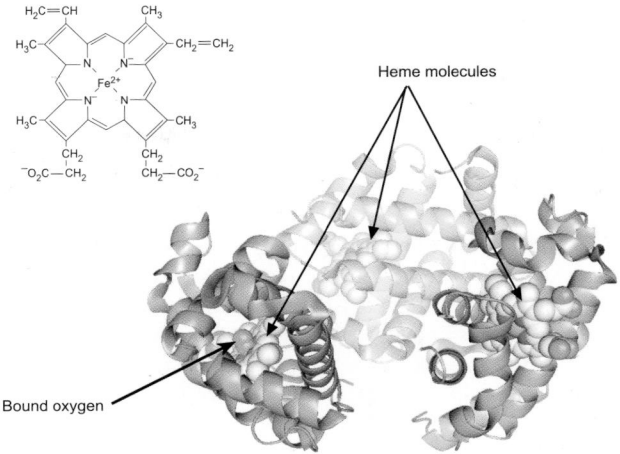

Fig. 13.5 The structure of hemoglobin. The hemoglobin molecule consists of four peptide chains—two α and two β chains. Each peptide chain has a single heme group (shown left) which binds a single molecule of oxygen. Thus one molecule of hemoglobin can carry four molecules of oxygen. (One α chain is omitted for clarity.)

$$Hb + O_2 \rightleftharpoons HbO_2.$$

When oxyhemoglobin dissociates to release oxygen to the tissues, the hemoglobin is converted to deoxyhemoglobin—also called reduced hemoglobin. Combination of oxygen with hemoglobin to form oxyhemoglobin occurs in the alveolar capillaries of the lungs where the P_{O_2} is high (13.3 kPa or 100 mmHg). Where the P_{O_2} is low (as in the capillaries supplying metabolically active cells), oxygen is released from oxyhemoglobin and is then able to diffuse down its concentration gradient to the cells via the interstitial space.

Hemoglobin that is fully saturated with oxygen is bright red, while hemoglobin that has lost one or more oxygen molecules (deoxyhemoglobin) is darker in appearance. When it has lost most of its oxygen, hemoglobin becomes deep purple in color. As the blood passes through the tissues it gives up its oxygen and the proportion of deoxyhemoglobin increases. For this rea-

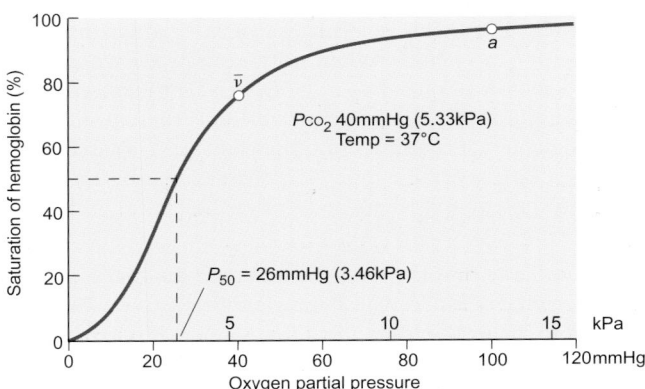

Fig. 13.6 The oxyhemoglobin dissociation curve for a P_{CO_2} of 5.3 kPa (40 mmHg) at 37°C. Under these conditions, the P_{50} value is 3.46 kPa (26 mmHg). *a* is the P_{O_2} in arterial blood (97 per cent saturated) and *v̄* is the P_{O_2} for mixed venous blood (5.33 kPa or 40 mmHg) at which value the hemoglobin is still 75 per cent saturated. Note that as the P_{O_2} falls below 8 kPa (60 mmHg) the curve becomes progressively steeper.

son venous blood is much darker in color than arterial blood. When the quantity of deoxyhemoglobin exceeds 5 g dl⁻¹, the skin and mucous membranes appear blue—a condition known as *cyanosis*.

The ease with which hemoglobin accepts an additional molecule of oxygen depends on how many of the binding sites are already occupied by oxygen molecules. There is cooperation between the binding sites such that occupancy of one of the four sites makes it easier for a second oxygen molecule to bind and so on. As a result, the amount of oxygen bound to hemoglobin increases in an S-shaped (sigmoid) fashion as P_{O_2} increases (Fig. 13.6). This is known as the *oxyhemoglobin dissociation curve* (or the *oxygen dissociation curve*). The sigmoid nature of the dissociation curve is physiologically significant because, as P_{O_2} falls from 13.3 kPa (100 mmHg)—the value in arterial blood—to about 8 kPa (60 mmHg), the saturation of the hemoglobin with oxygen decreases by only about 10 per cent. However, as the P_{O_2} falls below 8 kPa, the curve becomes relatively steep so that small changes in P_{O_2} cause large changes in the degree of hemoglobin saturation.

The quantity of oxygen in a given volume of blood must be carefully distinguished from the percentage saturation, which only indicates what proportion of the available hemoglobin is saturated. This distinction should be clear from the following definitions.

- *Oxygen content* is the quantity of oxygen in a given sample of blood, whether obtained from an artery or a vein. It represents the quantity of oxygen combined with hemoglobin plus that physically dissolved in the plasma.

- *Oxygen capacity* is the maximum quantity of oxygen that can combine with the hemoglobin of a given sample of blood. It can be determined in two ways. The first approach is by equilibrating a sample of blood at 20 kPa (150 mmHg) at 37°C and determining the quantity of oxygen in the sample. This will provide a value for the amount of oxygen combined with hemoglobin *plus* that physically dissolved in the plasma. At a P_{O_2} of 20 kPa about 0.5 ml O_2 is dissolved per deciliter of blood. This must be subtracted from the total to obtain the value for the oxygen capacity. Alternatively, and more conveniently, the hemoglobin concentration of the blood sample is determined first. This is normally about 15 g dl⁻¹ in males and 13.5 g dl⁻¹ in females. When fully saturated, each gram of hemoglobin will bind 1.34 ml of O_2 at STP; the oxygen capacity in milliliters O_2 per deciliter blood is then given by the hemoglobin concentration multiplied by 1.34. Therefore the oxygen capacity of a sample of blood depends on the hemoglobin content and is independent of the partial pressure of oxygen.

- *Oxygen saturation* is the term given to the ratio of the quantity of oxygen combined with hemoglobin in a given sample of blood to the oxygen capacity of that sample. It is expressed as a percentage thus:

$$\% \text{ saturation} = \frac{O_2 \text{ content} - \text{dissolved } O_2}{O_2 \text{ capacity}} \times 100.$$

For a normal adult male, when the arterial P_{O_2} is close to 13.3 kPa (100 mmHg) and the hemoglobin is 97 per cent saturated, the oxygen content of the blood will be 15 × 1.34 × 0.97 = 19.5 ml O_2 dl⁻¹ bound to hemoglobin *plus* 13.3 × 0.0225 ml = 0.3 mL O_2 in physical solution, giving a total oxygen content of 19.8 ml dl⁻¹. In the case of an anemic patient (see Section 13.7) with, say, a hemoglobin concentration only half of normal (7.5 g dl⁻¹ blood) at a P_{O_2} of 13.3 kPa (100 mmHg), the amount of oxygen bound to hemoglobin will be 7.5 × 1.34 × 0.97 = 9.7 ml plus 0.3 ml oxygen in solution, giving a total of only 10 ml dl⁻¹ blood—about half the total content of normal arterial blood.

The affinity of hemoglobin for oxygen is influenced by pH, P_{CO_2}, 2,3-BPG, and temperature

So far, the oxyhemoglobin dissociation curve has been considered as though the percentage saturation of hemoglobin remained constant for a given P_{O_2}. In reality, the position of the curve varies with temperature, pH, P_{CO_2}, and the concentration of certain metabolites such as 2,3-bisphosphoglycerate (2,3 BPG, formerly known as 2,3 diphosphoglycerate). In view of this, the dissociation curve is usually given for a pH of 7.4, a P_{CO_2} of 5.3 kPa (40 mmHg), and a temperature of 37°C. It is worth noting that, under these conditions, the hemoglobin in normal red cells is 50 per cent saturated with oxygen at a P_{O_2} of 3.4 kPa (26 mmHg). This is also expressed as a P_{50} value of 3.4 kPa (or 26 mmHg).

Both an increase in P_{CO_2} (above 5.3 kPa or 40 mmHg) and a reduction in pH (i.e. an increase in H⁺ ion concentration) shift the hemoglobin dissociation curve to the right (Fig. 13.7). This effect is known as the *Bohr shift* or the *Bohr effect*. Physiologically this effect is very important as the affinity of hemoglobin for oxygen becomes less as P_{CO_2} rises. Thus, in the tissues where the P_{CO_2} is relatively high, the affinity of hemoglobin for oxygen is lower than in the lungs (i.e. less oxygen is bound for a given P_{O_2}).

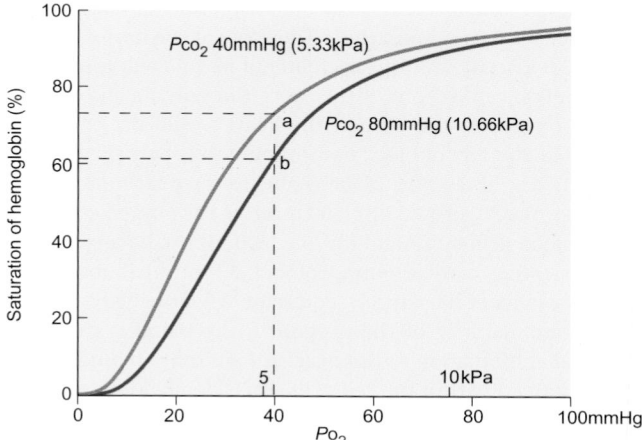

Fig. 13.7 The effect of increasing P_{CO_2} on the oxyhemoglobin dissociation curve. As P_{CO_2} increases the P_{50} value for the dissociation curve is shifted to the right. This is known as the Bohr shift. The dissociation curve is affected in a similar manner by a fall in pH or an increase in 2,3-DPG or temperature. The effect of the rightward shift is to decrease the affinity of hemoglobin for oxygen. This is shown by the difference in hemoglobin saturation when P_{O_2} is 5.33 kPa (40 mmHg) as P_{CO_2} increases from 5.33 kPa (40 mmHg) (point a) to 10.66 kPa (80 mmHg) (point b).

Consequently, oxygen delivery to actively metabolizing tissues is facilitated. In the lungs, as the P_{CO_2} of the pulmonary capillary blood falls, the Bohr shift acts to increase the affinity of hemoglobin for oxygen. In this way, the uptake of oxygen is facilitated during the passage of blood through the lungs.

As the temperature increases, the affinity of the hemoglobin for oxygen is also reduced and the dissociation curve for hemoglobin shifts to the right. Consequently, for a given level of P_{O_2}, the percentage saturation of hemoglobin will be less than at 37°C. This may be of benefit during heavy muscular exercise, for example, since oxygen will be delivered more readily from the blood to the active tissues as body temperature rises.

The affinity of purified hemoglobin for oxygen is much greater than that seen in whole blood—indeed, purified hemoglobin has an affinity for oxygen similar to that of myoglobin which has a P_{50} of 0.13 kPa (1 mmHg) (see below). However, in normal red cells hemoglobin has a P_{50} of 3.4 kPa at a P_{CO_2} of 5.3 kPa (i.e. P_{50} is 26 mmHg at a P_{CO_2} of 40 mmHg). This difference in the affinity of hemoglobin for oxygen is attributable to 2,3-BPG which is synthesized by the red cells during glycolysis. The 2,3-BPG binds strongly to hemoglobin and decreases its affinity for oxygen (i.e. it causes the oxyhemoglobin dissociation curve to be shifted to the right). The concentration of 2,3-BPG is about 4 mM in normal red cells, but may be increased in anemia or when living at high altitude where the P_{O_2} of the inspired air is significantly reduced.

Myoglobin

Myoglobin is an oxygen-binding protein present in cardiac and skeletal muscle that has a much higher affinity for oxygen than the hemoglobin of the red cells. It is half saturated at a P_{O_2} of only 0.13 kPa (1 mmHg). Nevertheless, myoglobin acts as a store of oxygen for situations where the oxygen supply from the capillaries is insufficient to meet the demands of aerobic metabolism in an exercising muscle. The muscles of diving mammals such as seals contain very large quantities of myoglobin. The shape of the oxygen dissociation curve for myoglobin is shown in Fig. 13.8. Oxygen is not liberated in significant quantities until the P_{O_2} falls below 0.65 kPa (5 mmHg). This situation may arise both in skeletal muscles during heavy exercise and during the contraction of the heart when the capillary circulation is temporarily interrupted. During periods of severe tissue hypoxia, the oxygen bound by myoglobin can be used to maintain the production of ATP by the mitochondria until the local circulation is restored.

Carbon monoxide binds strongly to hemoglobin

Carbon monoxide is able to bind to hemoglobin. Indeed, the affinity of carbon monoxide for hemoglobin is more than 200 times that of oxygen. This would mean that breathing air containing a P_{CO} of only 0.13 kPa (1 mmHg) would quickly result in virtually all the hemoglobin in the blood being bound to carbon monoxide (as carboxyhemoglobin). Moreover, carbon monoxide tends to shift the oxygen–hemoglobin dissociation curve to the left and this impairs the unloading of oxygen from the blood. For these reasons, carbon monoxide is a highly toxic gas. Treating patients suffering from CO poisoning requires a means of overcoming the high affinity of hemoglobin for CO. This is achieved by ventilating the patients with a gas mixture containing 95 per cent O_2, to drive the CO from its binding sites on the hemoglobin, and 5 per cent CO_2, to stimulate breathing.

Carbon dioxide is carried in the blood in three different forms: as dissolved gas, as bicarbonate, and as carbamino compounds

Chemical determination shows that arterial blood contains much more CO_2 than O_2 (49 ml dl^{-1} compared with 19.8 ml dl^{-1} for O_2). The CO_2 is carried in the blood in several forms. These are:

(1) in physical solution as dissolved CO_2;

(2) as bicarbonate ions;

(3) as carbamino compounds—a combination between CO_2 and free amino groups on proteins.

At first sight, this gives the impression that the carriage of carbon dioxide by the blood is far more complex than that of oxygen. In reality, the principles involved are quite straightforward. In what follows each form of transport will be considered briefly.

As with all gases, the concentration of dissolved CO_2 in the blood is determined by its solubility and its partial pressure. For plasma at normal body temperature, the solubility of CO_2 is 0.526 ml dl^{-1} kPa^{-1} (0.07 ml dl^{-1} $mmHg^{-1}$). Therefore, at a normal arterial P_{CO_2} of 5.3 kPa (40 mmHg), the amount of CO_2 transported in solution is $5.3 \times 0.526 = 2.8$ ml dl^{-1} (this is equivalent to 1.2 mmol CO_2 per liter of blood). Mixed venous blood has a P_{CO_2} of around 6.12 kPa (46 mmHg) and therefore will contain $6.12 \times 0.526 = 3.2$ ml CO_2 dl^{-1}. Because of its high solubility, between 5 and 7 per cent of total blood carbon dioxide is in physical solution (in normal arterial blood only 1.5 per cent of oxygen is in solution).

The carbon dioxide which is produced as a result of tissue metabolism combines with water in the reaction

$$CO_2 + H_2O \rightleftharpoons H_2CO_3 \qquad (13.1)$$

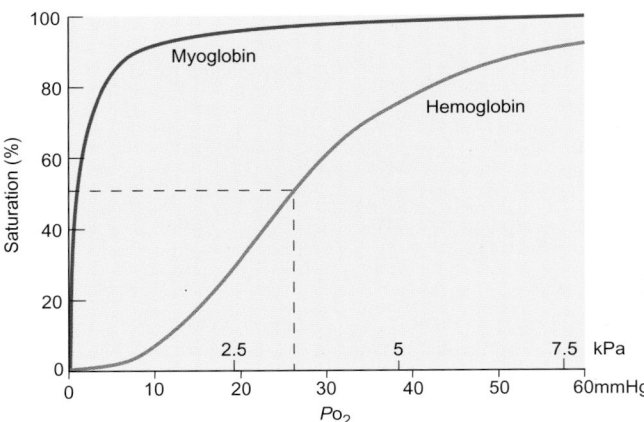

Fig. 13.8 A comparison between the oxygen dissociation curves for myoglobin and hemoglobin. Myoglobin has a P_{50} value of about 0.13 kPa (1 mmHg) while hemoglobin has a P_{50} of 3.46 kPa (26 mmHg).

to form *carbonic acid*. This readily dissociates to form hydrogen ions (H⁺) and bicarbonate ions (HCO_3^-) as follows:

$$H_2CO_3 \rightleftharpoons H^+ + HCO_3^-. \qquad (13.2)$$

Reaction (13.1) takes place only very slowly in the plasma but in the red cells it is catalyzed by an enzyme called *carbonic anhydrase*. Consequently, as carbon dioxide diffuses into the red blood cells, carbonic acid is formed but it immediately dissociates to yield bicarbonate and hydrogen ions. The latter are buffered mainly by hemoglobin while much of the bicarbonate moves back out of the cell in exchange for chloride ions. This is known as the *chloride shift* or *Hamburger effect* and accounts for the fact that plasma Cl^- is lower in venous blood than in arterial blood. About 90 per cent of the total blood CO_2 is transported in the form of bicarbonate ions.

The buffering of the hydrogen ions formed by dissociation of carbonic acid by hemoglobin is extremely important since it allows large amounts of carbon dioxide to be carried in the blood (as HCO_3^-) without the pH of the blood altering by more than about 0.05 pH units (see Chapter 29, Section 29.3).

Although the majority of the carbon dioxide that enters the red blood cells from the tissues is hydrated to form carbonic acid which dissociates into H⁺ and HCO_3^- as described above, about a third combines with amino groups on the hemoglobin molecules in the reaction

$$HbNH_2 + CO_2 \rightleftharpoons HbNHCOO^- + H^+.$$
$$\text{(carbaminohemoglobin)}$$

In addition, a very small amount of carbon dioxide is carried in the blood combined with α-amino groups on plasma proteins in the form of carbamino compounds formed by the general reaction

$$R\text{-}NH_2 + CO_2 \rightleftharpoons R\text{-}NHCOO^- + H^+.$$

The reactions involved in the carriage of carbon dioxide in the form of both bicarbonate ions and carbamino compounds are illustrated diagrammatically in Fig. 13.9.

To summarize, each deciliter of arterial blood has a P_{CO_2} of 5.3 kPa (40 mmHg) and contains about 2.8 ml of CO_2 in solution, 43.9 ml as HCO_3^-, and 2.3 ml as carbamino compounds making a total of 49 ml dl^{-1}. Mixed venous blood has a P_{CO_2} of 6.1 kPa (46 mmHg) and each deciliter contains approximately 3.2 ml of CO_2 in solution, 47 ml as HCO_3^-, and 3.8 ml as carbamino compounds (mainly carbamino hemoglobin) equivalent to a total of 54 ml CO_2 per deciliter.

The carbon dioxide dissociation curve

The amount of carbon dioxide present in solution depends on the P_{CO_2} and this in turn will determine the amount of HCO_3^- and carbamino compounds that will be formed in the blood. The relationship between P_{CO_2} (in kPa or mmHg) and the total CO_2 (ml CO_2 dl^{-1} blood) is called the CO_2 dissociation curve. It differs from the oxyhemoglobin dissociation curve in that it does not become saturated even at high P_{CO_2} (see Fig. 13.10). Across the physiological range of P_{CO_2} for whole blood (5.3 kPa (40 mm Hg) in arterial blood to 6.13 kPa (46 mm Hg), in mixed venous blood) the CO_2 dissociation curve is roughly linear. However, the quantity of CO_2 carried in the blood is dependent on the degree of oxy-

(a) CO_2 uptake by red cells as the blood perfuses active tissues

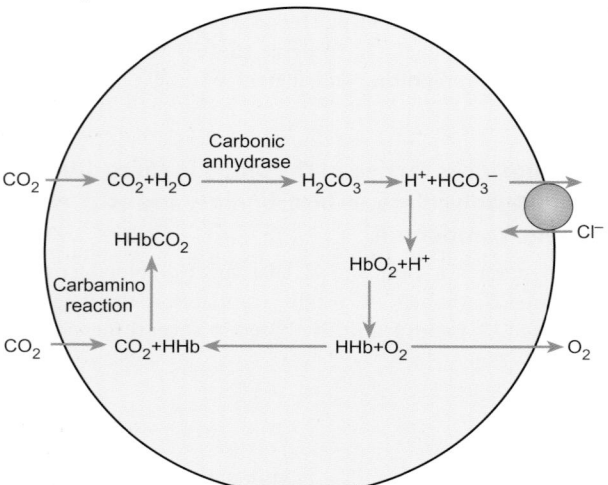

(b) O_2 uptake by red cells as the blood passes through the lungs

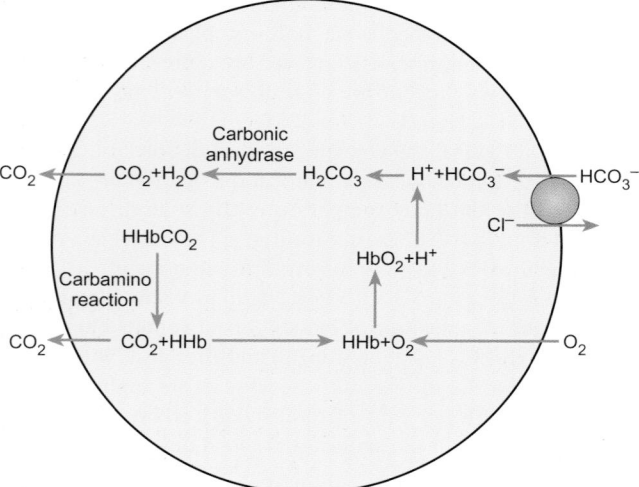

Fig. 13.9 A schematic representation of CO_2 and O_2 transport in the blood: (a) the exchange of CO_2 and O_2 that occurs between the blood and the tissues; (b) the exchange that occurs in the lungs between the blood and the alveolar air.

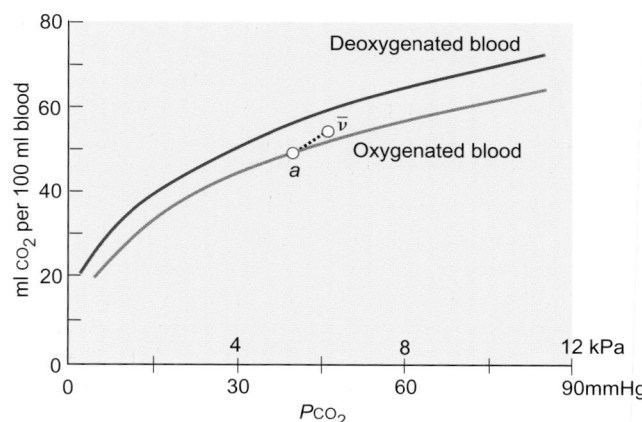

Fig. 13.10 The carbon dioxide dissociation curve for whole blood and the Haldane effect. *a* arterial blood; $\bar{v}$ mixed venous blood.

genation of hemoglobin. This is called the *Haldane effect* and is illustrated in Fig. 13.10.

Two main factors are responsible for the changes in carbon dioxide affinity of the blood seen when HbO_2 levels vary.

1. Oxyhemoglobin is less able to form carbamino compounds than reduced hemoglobin.

2. Oxyhemoglobin is a less efficient buffer of hydrogen ions than reduced hemoglobin. Consequently, hydrogen ions are more readily buffered by hemoglobin in the tissues (where less of the hemoglobin is in the form of HbO_2). This favors the formation of bicarbonate ions in the venous blood by driving the reaction

$$H_2O + CO_2 \rightleftharpoons HCO_3^- + H^+$$

to the right. This permits more carbon dioxide to be carried as bicarbonate ion.

In the lungs, where about 97 per cent of the hemoglobin is in the form of oxyhemoglobin, the carbon dioxide content of the red cells is relatively lower than it is in the tissues where oxy-

hemoglobin makes up around 75 per cent of the total Hb. In other words, more carbon dioxide may be carried when HbO_2 is low. This makes good sense physiologically, as a major purpose of blood gas transport is to load the blood with carbon dioxide in the tissues and unload it for expiration in the lungs.

13.7 Major disorders of the red and white blood cells

This section focuses on the consequences arising from changes in the rate of production or destruction of the cellular elements of the blood. Broadly, blood cell disorders fall into two categories: proliferative disorders (where there is an excess of cells, often with abnormal function) and deficiency disorders (where there are too few). Red and white cells will be considered here, while abnormalities of platelet function will be considered in Section 13.8.

Red cell abnormalities
Anemia

This term covers a variety of blood disorders characterized by a reduced number of red cells, a reduced hemoglobin concentration, or both. All types of anemia result in a reduction in the oxygen-carrying capacity of the blood. Anemia may arise for a number of reasons.

1. A reduction in red cell number—this can arise as a result of acute hemorrhage after which plasma volume is restored in a short time but red cell production takes much longer (see also Chapter 28, Section 28.5).

2. A reduction in the hemoglobin content of the red cells, for example as a result of iron deficiency due to chronic blood loss or to pregnancy.

3. A reduction in red cell size. Normally, mean corpuscular (red cell) volume is generally $(80–95) \times 10^{-15}$ l (80–95 fL). This condition is also seen in cases of iron deficiency and is known as *microcytic anemia.*

4. *Pernicious* or *macrocytic anemia* is seen in patients lacking vitamin B_{12} (cyanocobalamin), which is essential for the normal maturation of erythrocytes in the bone marrow (see Section 13.4). This situation arises when there is inadequate absorption of vitamin B_{12} due to a lack of intrinsic factor in the gastric mucosa (see Chapter 18). In this disorder, the red cells that are produced are much larger and contain more hemoglobin than normal (megaloblasts), but are present in greatly reduced numbers.

5. Occasionally the bone marrow fails to function normally. This results in so-called *aplastic anemia* and can arise spontaneously or as a consequence of damage to the marrow, for example by irradiation with X-rays.

6. Abnormalities in hemoglobin structure can lead to acceleration of red cell destruction. One such abnormality is *sickle cell anemia* in which there is a defect in one of the chains of the hemoglobin molecule. Sickle hemoglobin (HbS) is transmitted by recessive autosomal inheritance and the disease is prevalent in black African populations. In homozygous individuals, the HbS becomes sickled when deoxygenated, caus-

Summary

1. The blood must supply oxygen to all the tissues of the body and transport the carbon dioxide produced by metabolism to the lungs for removal from the body.

2. Only a small amount of oxygen is carried by the plasma in physical solution; most is carried loosely bound to hemoglobin within the red blood cells. The amount of oxygen carried in the blood depends on the partial pressure of oxygen and is described by the oxyhemoglobin dissociation curve which has a sigmoid shape

3. The position of the dissociation curve with respect to P_{O_2} (i.e. the affinity of hemoglobin for oxygen) varies with temperature, pH, P_{CO_2}, and the concentration of 2,3-BPG in the red cells. The curve is shifted to the right by an increase in P_{CO_2}, an increase in the level of 2,3-BPG, an increase in temperature, and a fall in pH. This is known as the Bohr shift.

4. Carbon dioxide carried in the blood in three forms: in physical solution, as bicarbonate ion, and as carbamino compounds. Carbon dioxide combines with water to form carbonic acid. This reaction is catalyzed by carbonic anhydrase in the red blood cells. The carbonic acid dissociates to H^+ and HCO_3^-. The H^+ is buffered by hemoglobin and other blood buffers while the HCO_3^- diffuses out of the red cells in exchange for Cl^- (the chloride shift).

5. The carbon dioxide dissociation curve is virtually linear in the physiological range of blood P_{CO_2}. The exact position of the dissociation curve (i.e. the affinity of the blood for carbon dioxide) is determined by the degree of oxygenation of the hemoglobin. More carbon dioxide may be carried by the blood as the level of oxyhemoglobin falls. This is the situation for blood perfusing the tissues. This is known as the Haldane effect.

ing deformation of the red cells. The deformed cells obstruct the blood flow in the capillaries, causing tissue hypoxia with subsequent damage and intense pain. Virtually every organ is affected, but the liver, spleen, heart, and kidneys are especially vulnerable to damage because of the increased risk of blood clot formation caused by the sluggish blood flow. A small but significant advantage of this disease is that people who carry the HbS gene have a high resistance to malaria. This is because the parasite that causes malaria cannot live in blood cells containing HbS.

7. *Thalassemia* is the name given to a group of anemias caused by the hereditary inability to produce either the α or the β chain of hemoglobin. It is found predominantly amongst Mediterranean, African, and black American populations. It is characterized by a reduction in Hb synthesis, damage to red cell membranes, and abnormal oxygen-binding characteristics.

Polycythemia

This condition is the result of overstimulation of red blood cell production. It brings about an increase in the hematocrit value (to as much as 60–80 per cent) and a rise in blood viscosity. It is often seen in people living at high altitude who experience chronic hypoxia because of the low prevailing atmospheric oxygen tension (see Chapter 30 for further details), though it can also arise under other circumstances. The increase in red blood cell numbers increases the oxygen-carrying capacity of the blood, but, at the same time, it increases the viscosity of the blood and this places an extra load upon the heart. Over time, the heart hypertrophies (enlarges) to adapt to the increased work load.

White cell abnormalities

As with the red cells, disorders of the leukocytes fall into two broad categories: deficiency disorders and proliferative disorders.

Leukopenia

This term describes an absolute reduction in the numbers of white blood cells. It may affect any of the different types of leukocyte but most often involves the neutrophils, which are the predominant type of granulocyte. In this case, the disorder is known as *neutropenia*. It can result from defective neutrophil production or from an increase in the rate of removal of neutrophils from the circulation. The former may arise as part of a genetic impairment of the regulation of neutrophil production, aplastic anemia, in which all the myeloid stem cells are affected, or following certain types of chemotherapy. It may also be a consequence of the overgrowth of neoplastic cells characteristic of some forms of leukemia, which suppresses the function of the neutrophil precursor cells.

Occasionally, leukopenia arises because of an accelerated rate of neutrophil removal from the circulation rather than a reduction in the rate of production. This is most usually a consequence of chemotherapy but may also be seen in certain infections or autoimmune disorders in which neutrophils are destroyed. The neutrophils are essential in the inflammatory response. Therefore infections are common in people with neutropenia and these may be severe or even life threatening.

Proliferative disorders of the white blood cells

Malignant proliferative diseases of the blood include the leukemias, lymphomas, and myelomas. Self-limiting proliferative disorders such as infectious mononucleosis (glandular fever) can also occur.

Leukemia is characterized by greatly increased numbers of abnormal white blood cells circulating in the blood. There are several different types of leukemia, classified according to their cells of origin (lymphocytic or myelocytic) and whether the disease is acute or chronic. Lymphocytic leukemias, which are most commonly seen in children, involve the lymphoid precursors that originate in the bone marrow. Cancerous production of lymphoid cells then spreads to other tissues such as the spleen, lymph nodes, and CNS. Myelocytic disease, which is more common in adults, involves the pluripotent myeloid stem cells in the bone marrow. The maturation of all the blood cell types, including granulocytes, erythrocytes, and thrombocytes, is affected.

Leukemic cells are usually non-functional and therefore cannot provide the normal protection associated with white blood cells. Common consequences of the disease include the development of infections, severe anemia, and an increased tendency to bleed because of a lack of platelets (thrombocytopenia). Furthermore, the leukemic cells of the bone marrow may grow so rapidly that they invade the surrounding bone itself. This causes pain and an increased risk of fractures.

Almost all forms of leukemia spread to other tissues, particularly those which are highly vascular such as the spleen, liver, and lymph nodes. As they invade these regions the grow-

Summary

1. Disorders of both the red and white cells fall into two broad categories; deficiency and excess production.

2. Anemia is a general term to describe disorders of the red blood cells characterized by a reduced hematocrit. It may arise from reductions in red blood cell number or size, reduction of the hemoglobin content of red cells, or abnormalities of hemoglobin structure.

3. An important consequence of all types of anemia is a reduction in the oxygen-carrying capacity of the blood.

4. Polycythemia results from overstimulation of red blood cell production and leads to a rise in both hematocrit and blood viscosity.

5. Leukopenia is defined as an absolute reduction in white blood cell numbers and may be due to either defective production or accelerated removal of white cells from the circulation. Infections are common in patients suffering from leukopenia.

6. Proliferative disorders of the white blood cells include leukemias, lymphomas, and myelomas. In leukemias there are high numbers of abnormal white blood cells which are usually non-functional. Patients suffer from severe anemia, infections, weight loss, and excessive fatigue.

ing cancerous cells cause extensive tissue damage and place heavy demands on the metabolic substrates of the body, especially amino acids and vitamins. Thus the energy reserves of the patient are depleted and the body protein is broken down. Hence, weight loss and excessive fatigue are characteristic symptoms of leukemia.

13.8 Mechanisms of hemostasis

When a blood vessel is damaged by mechanical injury of some kind, excessive blood loss from the wound is prevented by a process called *hemostasis*. This involves a series of events—vasoconstriction, platelet aggregation, and blood coagulation (clot formation). Later, blood vessel repair, clot retraction, and dissolution complete the healing process.

Vasoconstriction

When the vascular endothelium (see Chapter 15, Section 15.9) is damaged, there is a localized contractile response by the vascular smooth muscle, causing the vessel to narrow. This may be mediated by humoral factors or directly by mechanical stimulation. In arterioles and small arteries closure may be virtually complete. However, this response lasts for only a short time and, to prevent serious loss of blood, further hemostatic mechanisms are initiated.

The role of platelets

Within seconds of a vascular injury, platelets start to build up and adhere to the site of damage. This process is self-perpetuating as the adhering platelets secrete ADP and 5-hydroxytryptamine. They also synthesize arachidonic acid and thromboxane A_2. These factors trigger a change in the surface characteristics of the platelets (see Chapter 5), which causes them to adhere to the walls of damaged vessels and to each other. This process results in the formation of a *platelet plug*, which may be sufficient to stem the flow of blood from minor wounds.

In addition to sealing damaged vessels, the platelets play a continuous role in maintaining normal vascular integrity. This is illustrated by the increased capillary permeability seen in people suffering from platelet deficiency (*thrombocytopenia*). Such individuals often develop spontaneous tiny hemorrhages in the skin and mucous membranes (petechiae), giving the patient a curious blotchy appearance, with further bleeding into subcutaneous tissue.

Blood coagulation

This is the process by which fibrin strands create a mesh that binds blood components together to form a blood clot. It is a complex process that involves the sequential activation of a number of factors that are normally present in the blood in an inactive form. A cascade of reactions occurs by which one activated factor activates another according to the following scheme:

Activated clotting factor
↓
Inactive clotting factor → Active clotting factor.

Many *clotting factors* are synthesized in the liver and their manufacture is dependent upon vitamin K. The major reactions in

TABLE 13.4 The nomenclature of the clotting factors of blood

Factor	Names and synonyms
I	Fibrinogen
II	Prothrombin
IIa	Thrombin
III	Tissue factor, tissue thromboplastin
IV	Calcium (Ca^{2+})
V	Proaccelerin, labile factor, accelerator globulin
VI	Not assigned
VII	Proconvertin, SPCA, stable factor, autoprothrombin I
VIII	Antihemophilic factor, antihemophilic globulin, antihemophilic factor A, platelet cofactor I
IX	Plasma thromboplastic component, Christmas factor, antihemophilic factor B, platelet cofactor II
X	Stuart–Prower factor, autoprothrombin III
XI	Plasma thromboplasitin antecedent (PTA), antihemophilic factor C
XII	Hageman factor
XIII	Fibrin stabilizing factor, Laki–Lorand factor

Note that Factors I–IV are generally known by their names while Factors V–XIII are generally referred to by their roman numeral. Activated factors are designated by the letter a after the numeral, e.g. activated Factor X is called Xa.

the clotting process are shown in Fig. 13.11, from which it is evident that there are two pathways which may lead to the formation of a fibrin clot. These are the *intrinsic* and *extrinsic* pathways; both are needed for normal hemostasis and both involve a number of enzyme factors which are mostly tissue proteases related to trypsin. Throughout the medical and scientific literature, these enzymes are known by a variety of names and/or roman numerals (Table 13.4). In this account the factors are assigned the nomenclature by which they are most commonly known.

Both systems are activated when blood passes out of the vascular system. The intrinsic system (which is the slower of the two) is activated as blood comes into contact with the injured vessel wall, while the extrinsic system is activated when blood is exposed to the products of damaged tissue—specifically *tissue factor* or *thromboplastin*. The intrinsic pathway is so called because all the elements required to activate it are present in normal blood, while the extrinsic pathway is activated by a factor from outside the blood, i.e. tissue factor. Both pathways lead to the formation of activated Factor X (Factor Xa) at the end of the first stage of coagulation. Further steps in the clotting reaction are common to both pathways and involve the enzymatic conversion of inactive prothrombin to *thrombin*. This then initiates the polymerization of *fibrinogen* to *fibrin* strands within which plasma and blood cells are trapped to form a clot.

The intrinsic pathway

The initial step in this series of reactions is dependent upon the activation of Factor XII (Hageman factor). When there is vascular damage and blood is exposed to collagen, Factor XII is converted to 'activated Factor XII'. At the same time, platelets release phos-

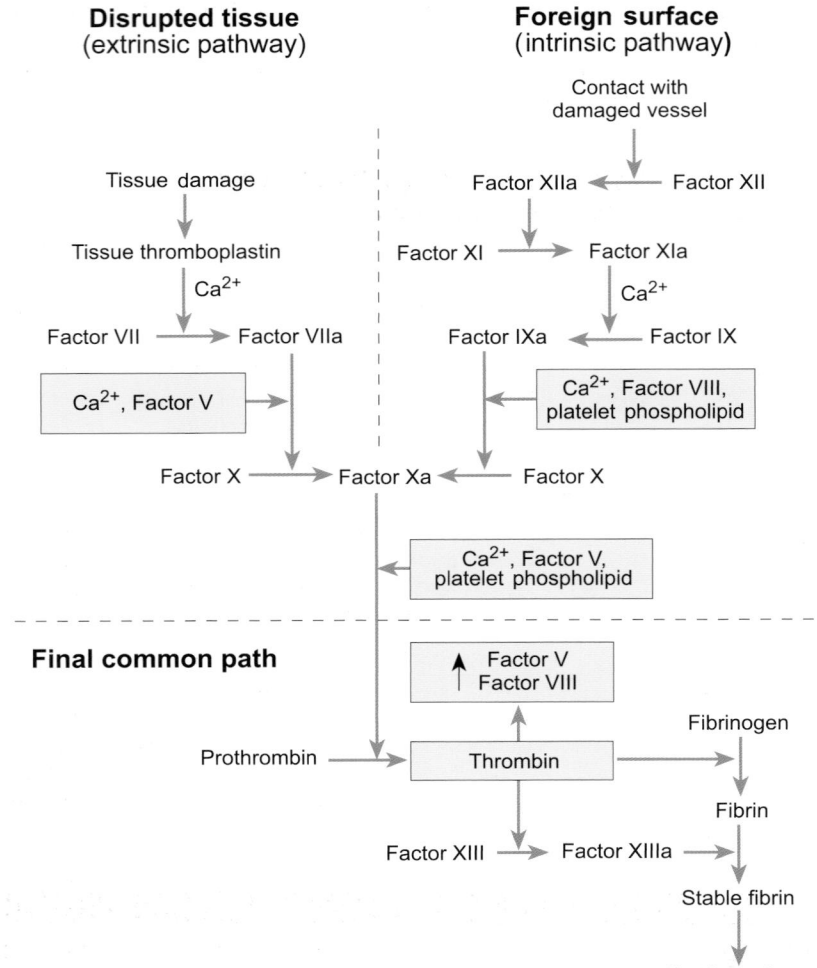

Fig. 13.11 The extrinsic and intrinsic pathways leading to the formation of a blood clot. Note the central roles played by Factor Xa and thrombin in the process of blood coagulation.

pholipid, which plays a role in subsequent steps of the process.

Activated Factor XII converts Factor XI to activated Factor XI which subsequently converts Factor IX (Christmas factor) to activated Factor IX by a calcium-dependent process. Activated Factor IX then acts together with the phospholipids from the traumatized platelets and with Factor VIII (antihemophilic factor) to activate Factor X. Factor VIII is the factor missing in people (chiefly males) who suffer from hemophilia. Activated Factor X combines with Factor V and platelet phospholipids to form a complex which, in turn, quickly initiates the cleavage of prothrombin (an inactive enzyme) to thrombin.

The extrinsic pathway

The extrinsic pathway, initiated when blood comes into contact with damaged tissue, occurs in three basic steps (Fig. 13.11, top left).

1. The damaged tissue releases a protein called tissue factor (also known as tissue thromboplastin), and phospholipids. These set the clotting process in motion.

2. The tissue thromboplastin combines with Factor VII and, in the presence of Ca^{2+}, activates Factor X.

3. This step is identical with the final step in the intrinsic pathway. Activated Factor X combines with tissue phospholipids and Factor V to form thrombin from prothrombin as described earlier for the intrinsic pathway.

The common end-point of both the intrinsic and extrinsic pathways for blood clotting is the conversion of prothrombin to thrombin. In the next step, the thrombin brings about the polymerization of the soluble plasma protein fibrinogen to form long strands of the insoluble protein fibrin. These strands of fibrin form a mesh-like structure that traps the blood constituents (plasma and formed elements) to form the clot and bind the edges of the damaged vessel together.

Clot retraction

Following the coagulation of blood, the clot gradually shrinks as *serum* is extruded from it. The exact mechanism of this process is not understood, but it appears to be initiated by the action of thrombin on platelets. One idea is that thrombin causes the release of intracellularly stored calcium into the platelet cytoplasm. This calcium then triggers the contraction of contractile proteins within the platelets by a process resembling the contraction of muscle. The contractile process may then cause the extrusion of pseudopodia from the platelets. These stick to the fibrin strands within the clot and, as they contract, the fibrin strands are pulled together, at the same time squeezing out the entrapped fluid as serum.

Dissolution of the clot

Once the wall of the damaged blood vessel is repaired, the blood clot is removed by lysis. Activated Factor XII stimulates the production of a substance in the plasma known as *kallikrein*. In turn, kallikrein promotes the conversion of inactive *plasminogen* into active *plasmin*, an enzyme that digests fibrin and thus brings about dissolution of the clot.

A variety of other plasminogen activators are used clinically to promote the dissolution of clots. These include *streptokinase*, a substance produced naturally by certain bacteria, and an endogenous substance called *tissue plasminogen activator* (TPA), which can now be produced commercially by genetic engineering. These substances can be injected either into the general circulation or into a specific blood vessel to promote lysis of a clot.

The role of calcium in hemostasis

As Fig. 13.11 shows, calcium ions are required for each step in the clotting process except for the first two reactions of the intrinsic pathway. Therefore, adequate levels of calcium ions are necessary for normal clotting. In reality, plasma calcium levels never fall low enough to impair the clotting processes since death would have resulted from other causes (most notably tetany of the respiratory muscles) long before. However,

it is possible to prevent the coagulation of blood removed from the body and stored *in vitro* by reducing the calcium ion concentration of the plasma. This can be achieved by the addition of substances such as EDTA or citrate, which bind calcium.

Inappropriate clotting of blood is prevented by endogenous anticoagulants

Normally, blood is prevented from clotting inappropriately by a variety of mechanisms. The preceding account showed that clotting is initiated when blood encounters an abnormal or damaged tissue surface. In contrast, undamaged vascular endothelial cells generally prevent clotting by releasing substances that inhibit coagulation—*anticoagulants*.

1. *Prostacyclin* is a potent inhibitor of platelet aggregation which acts as an antagonist of thromboxane A_2 (which causes platelet aggregation—see above).

2. *Heparin*, a negatively charged proteoglycan, is present in the plasma and on the surface of the endothelial cells of the blood vessels. It inhibits platelet aggregation and interacts with antithrombin III to inhibit the action of thrombin.

3. Normal endothelial cells express a protein called *thrombomodulin* which binds thrombin. The thrombomodulin–thrombin complex activates *protein C,* which inhibits the actions of Factors Va and VIIIa. In addition, protein C stimulates the production of the proteolytic enzyme *plasmin* from plasminogen. Plasmin disperses any clots that have begun to form by dissolving fibrin.

BOX 13.1 DEEP VEIN THROMBOSIS

The blood normally flows freely around the circulation. The integrity of the walls of the blood vessels, the anticoagulation proteins of the plasma, and normal flow dynamics all ensure that unwanted thromboses do not occur. However, if the vascular wall is damaged or defective, or if blood flow is low, or even static, a local accumulation of platelets and thrombin may occur and lead to the formation of a blood clot.

In arteries, local stasis of blood flow only occurs in the vicinity of an atheromatous plaque which facilitates the formation of a clot by reducing the linear flow rate. If pieces of the clot break off, they will be carried along by the blood and block its passage through smaller vessels. If a coronary artery is affected, the myocardium supplied by that artery becomes ischemic and the heart's ability to pump blood is impaired. This is commonly known as a heart attack, which is usually serious and may be fatal. The blocking of a cerebral artery results in local brain ischemia, neuronal death, and impairment of brain function.

Prolonged immobility results in failure of the leg muscles to pump blood back to the heart by the normal rhythmic contraction that accompanies walking (this is known as the muscle pump, see Chapter 15). The result is sluggish blood flow and pooling of the blood in the deep veins of the legs resulting in an increased likelihood of clot formation. Long periods of inactivity, including car journeys and intercontinental

flights, also favor clot formation. The most common sites of thrombosis are the deep veins of the calf. Imaging studies indicate that the clots first form in the valve pockets where flow is static. If pieces of the clot break off, they will lodge in the lungs, causing a pulmonary embolism which may be life threatening.

Other circumstances that favor the development of a deep vein thrombosis include polycythemia, malignant cancer, and surgery. The best treatment for deep vein thrombosis is prevention, which is easily achieved by gentle exercise involving the legs. The use of elastic stockings is also beneficial, particularly for the period just before and after surgery, and during long flights. These reduce swelling of the legs and so prevent venous pooling.

Dispersion of venous thromboses relies on anticoagulants such as heparin, and on fibrinolytic agents such as streptokinase or, most recently, recombinant tissue plasminogen activator. Oral anticoagulants such as warfarin are used for the long-term management of patients at risk of developing deep vein thromboses. They act as vitamin K inhibitors and reduce the synthesis of clotting factors by the liver (particularly prothrombin, and factors VII, IX, and X). The effectiveness of the therapy is monitored by measuring the prothrombin clotting time, which is expressed as the international normalized ratio (or INR).

Clot formation at inappropriate sites within the circulation is potentially lethal

It is clearly highly undesirable for blood clots to form in the circulation at inappropriate sites. If such a clot does occur, it is known as a *thrombus* and may block the vessel in which it forms. If a thrombus forms or lodges within a coronary artery the result is a heart attack, where a part of the heart muscle becomes ischemic (i.e. receives insufficient blood to meet its oxygen requirements) and dies. A clot forming in one of the blood vessels supplying the brain deprives the affected area of oxygen. The ischemia that results leads to the death of the neurons in the affected area, giving rise to a stroke. Sometimes small clots (emboli) break off from larger thrombi and travel to other parts of the circulation where they may block small vessels. Such a block is called an *embolism*. Clots that form in the systemic vessels often travel to the lungs where they lodge in pulmonary vessels to produce pulmonary emboli, as discussed in Box 13.1.

Summary

1. Following damage to the vascular endothelium, a cascade of events is initiated leading ultimately to the formation of a blood clot (hemostasis). Platelet aggregation at the site of damage occurs within seconds of an injury, leading to the formation of a platelet plug. This is followed by the formation of a blood clot. Clot retraction and dissolution complete the healing process.

2. In the formation of a blood clot, a soluble plasma protein called fibrinogen is converted into insoluble threads of fibrin that trap blood cells and plasma. This reaction is catalyzed by the enzyme thrombin which is derived from an inactive precursor (prothrombin) by either an intrinsic or an extrinsic pathway. A number of clotting factors participate in the events leading to the formation of thrombin. The clotting mechanism requires calcium ions and the phospholipids present in the membranes of the platelets.

3. Following coagulation, the clot retracts by shrinkage. The blood clot is then dissolved by an enzyme called plasmin. Undamaged vascular endothelial cells prevent inappropriate clotting by synthesizing anticoagulants, such as heparin and prostacyclin, and by expressing thrombomodulin, a protein which binds thrombin and thereby activates protein C, an activator of plasmin.

4. If a clot forms in an undamaged blood vessel, that vessel will become obstructed and tissue supplied by it will become ischemic. This is potentially lethal if it occurs in vessels such as the coronary arteries or those supplying the brain.

5. Failure of the normal clotting reactions may occur for a variety of reasons. These include thrombocytopenia (a reduction in platelets), structural disorders of the vasculature, and hereditary deficiency of clotting factors, such as lack of Factor VIII in hemophilia.

Certain conditions favor the formation of clots within blood vessels. These include damage to the vascular wall, sluggish blood flow (stasis), and alterations in one or more of the components of blood that renders it more easily coagulated. One of the most common causes of thrombosis is a condition known as *atherosclerosis*. It is characterized by the formation of fibro-fatty lesions (plaques) in the intimal lining of large or medium-sized arteries such as the aorta, the coronary arteries, and the large vessels that supply the brain. As the lesions increase in size, they gradually occlude the vessel, causing a reduction in blood flow and a predisposition to thrombus formation. Major risk factors in atherosclerosis (apart from genetic factors, being male, and getting older) appear to include cigarette smoking, high blood pressure, and a high blood cholesterol level.

Failure of the normal clotting mechanisms may result from a reduction in the number of platelets or from a deficiency of one of the clotting factors

Bleeding disorders or failure of coagulation may result from defects in any of the factors that are involved in the normal process of hemostasis, i.e. platelets, clotting factors, or vascular integrity. These will be considered briefly in turn.

Thrombocytopenia is a decrease in the number of circulating platelets. The depletion of platelets must be relatively severe before problems with clotting are seen. Hemorrhagic tendencies become evident when the platelet count falls to around 30×10^9 l^{-1} compared with the normal $(150–300) \times 10^9$ l^{-1}. Characteristics of the condition include the appearance of bruised areas and tiny reddish spots on the arms and legs (petechiae), and bleeding from the mucous membranes of the nose, mouth, and gastrointestinal tract.

Certain drugs as well as pathological states can result in thrombocytopenia. Aplastic anemia (where bone marrow function is impaired) or invasion of the bone marrow by malignant cells, as in leukemia, results in decreased production of platelets.

In addition to a reduction in platelet numbers, impairment of the clotting process may result from a deficiency of platelet function—*thrombocytopathia*. Such a defect may be inherited, as with the disorder of platelet adhesion known as von Willebrand's disease, or acquired following disease or drug treatment.

Hereditary disorders of blood clotting—the hemophilias

As can be deduced from the cascade mechanism of clotting illustrated in Fig. 13.11, impairment of blood coagulation can result from deficiencies in one or more of the clotting factors involved. Such disorders may be inherited, or they may arise from a reduction in synthesis of one or more of the clotting factors.

Although there are known to be hereditary defects associated with each of the ptotein clotting factors, most are extremely rare. By far the most common forms of hemophilia, each affecting 1 in 10 000 males, are Factor VIII deficiencies and von Willebrand's disease (in which there is a loss of Factor VIII associated with a loss of platelet adhesion).

Hemophilia is a sex-linked recessive trait primarily affecting males, although many cases are thought to arise as new mutations. The disease may be mild or severe in form. In severe cases, spontaneous bleeding into soft tissues, joints, and the gastrointestinal tract occurs that can lead to serious disability. In such cases, Factor VIII replacement therapy is essential. Recent advances in recombinant DNA technology have enabled pure Factor VIII to be produced, thereby eliminating the risk of disease transmission from Factor VIII extracted from donated blood.

Impaired synthesis of clotting factors

Prothrombin, fibrinogen, and Factors V, VII, IX, X, XI, and XII are all synthesized in the liver. Furthermore, the activity of Factors VII, IX, X, and prothrombin, requires the presence of vitamin K. Therefore vitamin K deficiency or liver disease may result in a loss of factors or a loss of activity. Either of these will produce impairment of the clotting mechanism, with abnormal bleeding.

Vascular disorders

Abnormal bleeding may occur from vessels that are structurally weak or have been damaged by inflammation or immune responses. Examples include vitamin C deficiency (scurvy), in which the vessels become fragile due to a lack of adhesion between the endothelial cells, and Cushing's disease, in which the excess corticosteroid hormones cause a loss of protein and reduction in support for the vascular tissue.

13.9 Blood transfusions and the ABO system of blood groups

Early attempts to restore heavy loss of blood by transfusion of blood from another person were frequently disastrous. The transfused cells aggregated together in large clumps that were sufficiently large to block minor blood vessels. This clumping is known as *agglutination*. Following the agglutination reaction, the red cell membranes broke down and hemoglobin was liberated into the plasma (this is known as *hemolysis*). The liberated hemoglobin was converted to bilirubin by the liver and this resulted in jaundice (yellowish skin coloration). In addition, the high plasma levels of bilirubin adversely affected urine production by the kidney. When such clinical signs follow the transfusion of blood, the transfused blood, is said to be *incompatible* with that of the recipient. Death frequently occurred following the transfusion of incompatible blood.

What is the basis of this incompatibility and why is some blood compatible while other blood is not? It is now known that agglutination results from an antibody–antigen interaction. Normal human plasma (and the corresponding serum) may contain antibodies that cause red cells to stick together in large clumps (i.e. to agglutinate) (Fig. 13.12). The antibodies that cause the reaction are known as *agglutinins*. Unlike most other antibodies, the agglutinins have not arisen as the result of a specific antibody reaction. They occur naturally and are inherited according to Mendelian

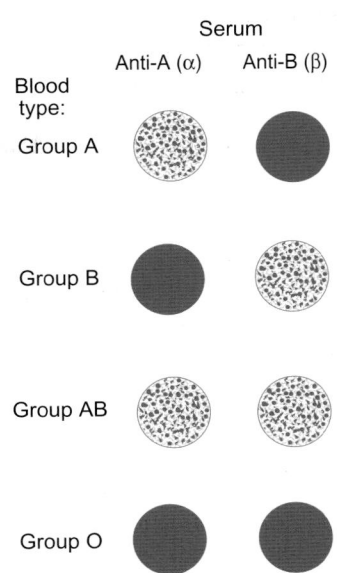

Fig. 13.12 The agglutination reaction of incompatible blood types. Drops of anti-A and anti-B serum are placed in shallow wells on a porcelain plate as shown in the figure. A drop of the test sample of blood is added to each well and mixed. If the blood is compatible, the mixed blood sample appears uniform, but if the blood is incompatible with the serum, it aggregates and precipitates as shown.

laws. Clearly, if red cells agglutinate in response to a particular kind of plasma or serum, they must possess the corresponding antigen, which is known as an *agglutinogen*.

To account for the known cross-reactivity of blood from different people, K. Landsteiner proposed that two kinds of agglutinogen are present on human red cells. These agglutinogens are called A and B and they may be present separately or together, or be completely absent, so giving rise to four groups: A, B, AB, and O (Table 13.5). In addition, human plasma may contain antibodies to one or both agglutinogens. The plasma antibodies are known as anti-A and anti-B or as agglutinins α and β. Where the blood contains red cells with a particular agglutinogen, the corresponding agglutinin is absent from the plasma. Thus, people with agglutinogen A on their red cells do not have anti-A in their plasma, as they do not agglutinate their own blood. Nevertheless, this group of people do have anti-B in their plasma. Conversely, group B have agglutinogen B on their red cells but anti-A in their plasma. Group AB have both agglutinogens A and B on their red cells but no agglutinins in their plasma, and group O have neither agglutinogen but both anti-A and anti-B agglutinins. Table 13.5 gives the relationships between the different groups and their approximate frequency of occurrence in the general population of the United Kingdom and the United States.

TABLE 13.5 ABO blood group characteristics

Blood group	Percentage of population	Agglutinogen on red cells	Agglutinin in plasma
A	41	A	Anti-B (β)
B	10	B	Anti-A (α)
AB	3	A and B	None
O	46	None	Anti-A and anti-B (α and β)

The rhesus blood group system

In 1940 Landsteiner and Wiener found that the serum of rabbits that had been immunized against the blood of rhesus monkeys could agglutinate human blood. Using this antibody, they identified two groups in the general population. Those whose blood could be agglutinated by this serum—rhesus (or Rh) positive (about 85 per cent of the population), and those whose blood could not be agglutinated—Rh negative. Rh-positive persons have a specific antigen on their red cells known as the D-antigen (also known as the rhesus factor).

Since the D-antigen is inherited like the AB agglutinogens, anti-Rh antibody can occur in the serum of Rh-negative mothers who have had Rh-positive children. During pregnancy, a Rh-negative mother may form anti-Rh antibodies in response to the leakage of fetal red cells into her circulation. This immunization of the mother by the baby's red cells may occur at any time during pregnancy, but is most likely when the placenta is separating from the wall of the uterus while the mother is giving birth. For this reason anti-Rh antibodies generally arise after a second or third pregnancy. The anti-Rh antibodies are IgG antibodies of about 150 kDa and are sufficiently small to pass across the placenta into the fetal circulation. If this happens, they may cause a severe agglutination reaction. The resulting disorder is known as hemolytic disease of the newborn and, in the absence of suitable preventative measures, it occurs in about 1 in every 160 births. As indicated above, this problem usually arises after a woman's second or third pregnancy. About half of the affected babies will require a partial replacement of their blood by transfusion. In many countries, this problem is now largely avoided by removing anti-Rh antibodies from the plasma of Rh-negative mothers by injecting them with anti-D immunoglobulin (IgG) after delivery.

Although hemolytic disease can occur as a result of an anti-A antibody in the blood of group O mothers, ABO blood group incompatibility generally causes no problems during pregnancy. This reflects the fact that the plasma agglutinins are IgM antibodies of high relative molecular mass (about 900 kDa) and proteins of this size do not readily cross the placenta.

Other blood group types

The blood group antigens (agglutinogens) are found on the surface of the red cell membrane and many kinds of antigen have been discovered in addition to the fundamental ABO system. For example, soon after the original description of the ABO system of blood groups it was discovered that group A could be further subdivided into two groups, A_1 and A_2. Other blood groups such as the M, N, P, and Lewis groups are also known. Nevertheless, the A_1 and A_2 subdivisions and other blood groups are not generally of significance in blood transfusion.

Blood must be cross-matched for safe transfusions

To prevent the problems of blood group incompatibility, blood for transfusion is cross-matched to that of the recipient. In this process, serum from the recipient is tested against the donor's cells. If there is no reaction to the cross-match test, the transfusion will be safe. Note that this test screens for all serum agglutinins and not just those of the ABO system. Although correct matching of blood groups of both donor and recipient is preferable, in emer-

Summary

1. For successful blood transfusion, the blood of the donor must be compatible with that of the recipient. If it is not compatible, the red cells will agglutinate following transfusion. This situation arises because normal human plasma contains antibodies (agglutinins) to certain red cell membrane proteins known as agglutinogens.

2. In the ABO system, two kinds of agglutinogen may be present on human red cells. These agglutinogens are called A and B, and they may be present separately or together, or be completely absent, so giving rise to four groups: A, B, AB, and O. In addition, human plasma may contain agglutinins (anti-A and anti-B) to one or both agglutinogens. When plasma containing an agglutinin (e.g. anti-A) is mixed with red cells possessing an agglutinogen with which it can react (in this case A), the red cells agglutinate.

gencies group O blood can be transfused into people of other groups because group O red cells have neither A nor B antigens. For this reason a group O person is sometimes called a universal donor. As the plasma of group AB has neither anti-A nor anti-B antibodies, other blood groups can be transfused into a group AB patient. Such a patient is known as a universal recipient. The plasma agglutinins present in the blood of a donor do not generally cause adverse reactions because they become diluted in the recipient's circulation.

Recommended reading
Biochemistry of hemoglobin and myoglobin
Berg. J.M., Tymoczko, J.L., and Stryer, L. (2002). Biochemistry (5th edn), pp. 269–74. Freeman, New York.

Hematopoiesis and histology
Alberts, B., Johnson, A., Lewis, J., Raff, M., Roberts, K., and Walter, P. (2002). Molecular biology of the cell (4th edn), Chapter 22, pp. 1283–96. Garland, New York.
Junquieira, L.C., and Carneiro, J. (2003). Basic histology (10th edn), Chapters 12 and 13. McGraw-Hill, New York.

Physiology of gas transport
Levitzky, M.G. (1999). Pulmonary physiology (5th edn), Chapter 7. McGraw-Hill, New York.

Pharmacology of the blood
Rang H.P., Dale, M.M., Ritter, J.M., and Moore, P. (2003). Pharmacology (5th edn), Chapters 20 and 21. Churchill-Livingstone, Edinburgh.

Hematology and immunology
Hoffbrand, A.V., Pettit, J.E., and Moss, P.A.H. (2001). Essential hematology (4th edn). Blackwell Science, Oxford.
Ledingham, J.G.G., and Warrell, D.A. (2000). Concise Oxford textbook of medicine, Chapters 3.1–3.41. Oxford University Press, Oxford.
Linch, D.C. (2003). Hematological disorders. In Textbook of medicine (4th edn) (ed. R.L. Souhami and J. Moxham), Chapter 22. Churchill-Livingstone, Edinburgh.

Multiple choice questions

Each statement is either true or false. Answers are given below.

1. The plasma of a normal adult:
   a. Accounts for 10% of body weight.
   b. Has an osmolality of about 290 mOsm kg^{-1}.
   c. Contains about 140 mmol l^{-1} of sodium.
   d. Contains about 4 mmol l^{-1} of potassium.
   e. Contains about 5 per cent by weight of albumin.
   f. Is about 60 per cent water.

2. For a normal healthy adult of 60–70 kg body weight:
   a. The blood volume would be about 5 liters.
   b. The blood pH is 7.0.
   c. There are about 4.5×10^{12} red cells per liter.
   d. There are about 1×10^9 leukocytes per liter.
   e. The blood contains about 10 per cent of total body iron

3. Normal red cells:
   a. Are about 7 μm in diameter.
   b. Do not have a nucleus.
   c. Are produced in the bone marrow from erythroblasts.
   d. Are released into the circulation as immature cells called reticulocytes.
   e. Live for about 10 days.

4. The following statements refer to gas carriage by the blood:
   a. After leaving the lungs, each deciliter of blood contains about 20 ml of oxygen.
   b. Each deciliter of mixed venous blood contains about 50 ml of carbon dioxide.

c. When the P_{CO_2} is 5.3 kPa (40 mmHg), hemoglobin is half saturated with oxygen when the P_{O_2} is about 3.3 kPa (25 mmHg).
   d. The affinity of hemoglobin for carbon monoxide is lower than its affinity for oxygen.
   e. Most of the carbon dioxide in blood is carried as bicarbonate.
   f. As the P_{CO_2} rises, the affinity of hemoglobin for oxygen is increased.

5. Platelets:
   a. Are produced in the bone marrow from megakaryocytes.
   b. Are present in greater number than red cells.
   c. Will adhere to the walls of damaged blood vessels.
   d. Secrete the main clotting factors.
   e. Aggregate in the presence of thromboxane A$_2$.

6. The following statements refer to the mechanisms of hemostasis:
   a. Both extrinsic and intrinsic pathways lead to the activation of Factor X.
   b. Coagulation can be prevented by adding EDTA or citrate to a sample of blood.
   c. All the clotting factors are produced by the liver.
   d. Hemostasis is impaired when the platelet count falls below 20×10^9 l^{-1}.
   e. Hemostasis is initiated when tissue factor comes into contact with blood.
   f. Failure of hemostasis is always caused by a deficiency of Factor VIII.

Quantitative problems

Answers are given below.

1. A young male gives a sample of blood that contains 4.8×10^{12} red cells l^{-1}, 14.8 g.dl^{-1} hemoglobin and has a hematocrit of 0.45. Do these values fall within the normal range?

2. Using the values given for question 1 calculate the following (to two significant figures):
   a. The average volume of the red cells.
   b. The total amount of hemoglobin in each red cell.
   c. The amount of oxygen carried by 1 dl blood (100 ml) when P_{O_2} is 13.3 kPa (100 mmHg).
   d. The amount of oxygen carried by 1 dl blood when P_{O_2} is 5.3 kPa (40 mmHg).

      For parts (c) and (d) refer to Fig. 13.6 for the change in Hb saturation with P_{O_2}. Assume that P_{CO_2} is 5.3 kPa (40 mmHg); 1 g Hb binds 1.34 ml O$_2$ when it is fully saturated and O$_2$ solubility is 0.225 ml l^{-1} kPa^{-1} (0.03 ml l^{-1} mmHg^{-1}).)

Answers to the multiple choice questions

1. While blood volume accounts for 7–8 per cent of body weight, the plasma accounts for just over half of this, i.e. about 4 per cent. Albumins account for 60 per cent of total plasma protein or about 5 per cent of plasma by weight. Plasma is 95 per cent water.
   a. False
   b. True
   c. True
   d. True
   e. True
   f. False

2. Blood pH varies between 7.35 (mixed venous blood) and 7.4 (arterial blood) under normal circumstances. There are about $(5–7) \times 10^9$ leukocytes l^{-1} and the red cells account for about 65 per cent of total body iron.

a. True

b. False

c. True

d. False

e. False

3. Healthy red cells survive in the circulation for about 120 days before being destroyed by macrophages in the spleen.

   a. True

   b. True

   c. True

   d. True

   e. False

4. The affinity of hemoglobin for carbon monoxide is greater than its affinity for oxygen. As P_{CO_2} increases, the affinity for O_2 decreases (this is the Bohr effect).

   a. True

   b. True

   c. True

   d. False

   e. True

   f. False

5. The platelet count is normally about 250×10^9 (i.e. about one platelet for every 20 red cells). While platelets play an essential role in clotting, the main clotting factors are plasma proteins secreted by the liver.

a. True

b. False

c. True

d. False

e. True

6. Deficiency of Factor VIII will impair the clotting process. Deficiencies of other factors or a reduction in platelet count will also result in a failure of clotting. While most clotting factors are synthesized in the liver, calcium is one of the major ionic species present in plasma.

   a. True

   b. True

   c. False

   d. True

   e. True

   f. False

Answers to the quantitative problems

1. True for all variables given

2. a. 94×10^{-15} liters

   b. 31×10^{-12} g

   c. 19.5 ml O_2

   d. 15 ml O_2

CHAPTER 14

Defense against infection: inflammation and immunity

After reading this chapter you should understand:

♦ The passive mechanisms by which the body resists infection

♦ How the body recognizes invading organisms: self and non-self

♦ The natural immune system

♦ The inflammatory response

♦ The adaptive immune system and the role of the lymphocytes

♦ Disorders of the immune system

♦ The need for tissue matching in transplantation

14.1 Introduction

As animals move through their environment to feed and reproduce, they are inevitably brought into contact with other organisms. Some of these will be food, while others may attempt to invade the body. Of those that invade the body, some will live in harmony with the host. When this is of benefit to the host, it is known as *mutualism*. If it is neither beneficial nor harmful, it is called *commensalism*. When the presence of the invading organism compromises the health of the host, the relationship is known as *parasitism*. All infectious diseases are due to parasites of one kind or another. In the developed countries, most infections are caused by bacteria, fungi, and viruses, but infections by protozoa and worms of various kinds are also very common in poorer regions of the world.

To defend themselves against infections, animals have two basic strategies: they use passive barriers to prevent parasites entering the body, and they actively attack those organisms that have become lodged in the tissues. To eliminate an invading organism, the host must first be able to distinguish it from its own cells. Secondly, it must neutralize or kill it. Finally, it must dispose of the remains in such a way that does no further harm. These functions are performed by the *immune system*, which can be conveniently divided into the *natural immune system* and the *adaptive immune system*.

The immune system is a complex network of organs, cells, and circulating proteins. The principal organs of the immune

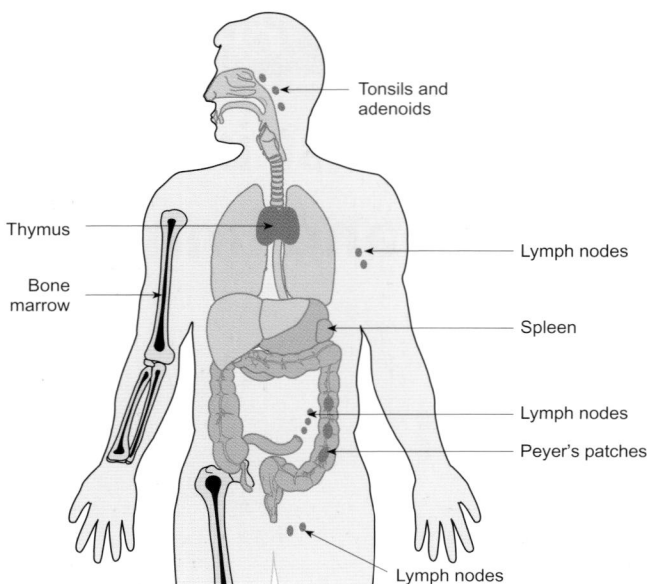

Fig. 14.1 The location of the major lymphoid organs. The thymus and bone marrow are the primary lymphoid tissues; the remainder are secondary lymphoid tissues.

system are the bone marrow, the thymus, the spleen, the lymph nodes, and the lymphoid tissues associated with the epithelia that line the gut and airways (known as mucosal associated lymphoid tissue (MALT)). Collectively they are known as the lymphoid organs (Fig. 14.1). The cells of the immune system include the leukocytes of the blood (see Chapter 13), mast cells, and various accessory cells that are scattered throughout the body. The accessory cells include phagocytic cells that are found in many organs, including the lungs, liver, spleen and kidneys, together with cells known as antigen-presenting cells, which are particularly associated with the lymphoid organs. The proteins of the immune system are *antibodies* and *complement*.

The cells of the immune system recognize foreign materials by their surface molecules. Those molecules that generate an immune response are called *antigens*. The recognition may be relatively non-specific (e.g. the binding of complement to bacterial cell walls) or highly specific in which a small part of a particular molecule is precisely recognized. This type of interaction is characteristic of the antibodies secreted by the cells of the adaptive immune system in response to a particular antigen.

14.2 Passive barriers to infection

The first barrier encountered by an invading organism is the skin. Its pseudostratified and keratinized epidermis forms an effective physical barrier to infiltration by micro-organisms. In addition, the sweat glands and sebaceous glands secrete fatty acids that inhibit the growth of bacteria on the skin surface. When the skin is broken by either abrasion or burns, infection may become a significant problem.

The skin forms a continuous sheet with the membranes that line the airways, gut and urogenital tract. The epithelia of these membranes are less rugged than that of the skin but they still provide an effective barrier to invasion by micro-organisms. For

example, the epithelia that line the airways are protected by a thick layer of mucus which traps many bacteria and viruses and prevents them adhering to the underlying cells. The mucus is then eliminated via the mucociliary escalator and coughed up (see Chapter 16, Section 16.8).

Other regions that are vulnerable to infection are regularly flushed by sterile fluid (e.g. the urinary tract) or by fluids that contain bactericidal agents. For example, the external surface of the eyes is washed by fluid from the tear glands, which both flushes the surface to remove foreign materials and contains the bactericidal enzyme lysozyme. Other body secretions, such as semen and milk, also contain antibodies and bactericidal substances.

The food we eat is inevitably contaminated by bacteria and other micro-organisms. Indeed, some organisms are deliberately introduced into certain foods, such as cheese, to flavor them. The gut has several stratagems to combat infection arising from this source. The mucous membranes of the mouth and upper gastrointestinal tract are protected by lysozyme and antibodies of the IgA class (see below) secreted by the salivary glands. Many bacteria are killed by the low pH of the gastric juice. The mucosal surface of the gut also possesses mucous glands that secrete a layer of mucus that both lubricates the passage of food and protects the surface epithelium from infection. Despite these barriers, the lumen of the intestine contains a healthy bacterial population. These are commensal organisms, which provide the body with a further line of defense. The normal bacterial flora both compete with potential pathogens for essential nutrients and secrete inhibitory factors (*bactericidins*) that kill invading pathogens.

Clearly, these passive mechanisms do not always prevent the ingress of pathogens. The skin can be penetrated by ectoparasites such as ticks and mosquitoes, which may themselves be infected with micro-organisms such as *Plasmodium* (the organism that causes malaria). Small pathogens such as bacteria and viruses may penetrate the body's defenses via the internal epithelia, such as those of the airways. Those that enter the gut may overwhelm the defenses afforded by the natural commensal bacteria, for example in typhoid fever. When infection occurs the active processes of immunity come into play.

14.3 Self and non-self

Before the body can mount a defense against infection, it first needs to know the difference between the normal cells of the body and those of invading parasites. How does the immune system recognize 'self' from 'non-self'? It is now known that mammalian cells possess markers on their surface that identify them as host cells. So, just as red cells possess surface proteins that determine particular blood groups, other cells possess integral membrane proteins that identify them as being host cells. By using these markers, the immune system can distinguish host cells from those of invading organisms. The molecules of the immune system that detect a general 'non-self' characteristic are called *non-specific*, while those that can detect a particular invading organism amongst the thousands of possible candidates are called *specific* recognition molecules. As we shall see, non-specific recognition is characteristic of the natural immune system, while the adaptive immune system can identify and destroy a specific type of invading organism.

The proteins that identify host cells are known as the major histocompatability complex or MHC. Their rather unfortunate name arises from the history of their discovery. They were first detected as the proteins responsible for the rejection of tissue grafts between a donor and a recipient animal. In human immunology the MHC complex is known as the HLA complex (for human leukocyte antigen). It is now known that MHC (or HLA) consists of a large number of genes that encode two classes of proteins which are:

♦ MHC class I proteins. These are integral membrane proteins found on the plasma membrane of all nucleated cells and on platelets. However, they are not found on red cells.

♦ MHC class II proteins. These are found on dendritic cells, macrophages, and a class of lymphocytes known as B cells (see below).

There are three kinds of MHC class I proteins (HLA A, B, and C) and three kinds of MHC class II proteins (known as HLA DP, DQ, and DR). The MHC proteins function to expose parts of foreign proteins type this to a class of lymphocytes known as T cells, to stimulate an immune response to infection.

There are many genetic variants of the MHC proteins and as each MHC protein is a combination of two different polypetide chains, there are around 10^{13} possible variations in the MHC. With such a large degree of difference, except for identical twins, every individual has a unique signature in their MHC. As mentioned above, the role of the MHC is to expose parts of a foreign protein to T cells. This is known as antigen presentation and it permits the T-cell population to identify infected cells prior to initiating an appropriate immune response.

14.4   The natural immune system

The natural immune system consists of innate defense mechanisms that do not change very much either with age or following infections. It consists of four kinds of cells and three different classes of proteins. The cells of the natural immune system are:

♦ phagocytes

♦ natural killer cells

♦ mast cells

♦ eosinophils.

The classes of protein are:

♦ complement

♦ interferons

♦ acute-phase proteins.

The principal phagocytes of the natural immune system are the neutrophils and the macrophages

The *neutrophils* are the most common white cell in the blood (see Chapter 13, Section 13.3). They contain two main types of granule called the primary azurophil granules and the secondary specific granules. The primary azurophil granules contain an enzyme called myeloperoxidase, a range of bactericidal proteins, and a protease called cathepsin G. The secondary specific gran-

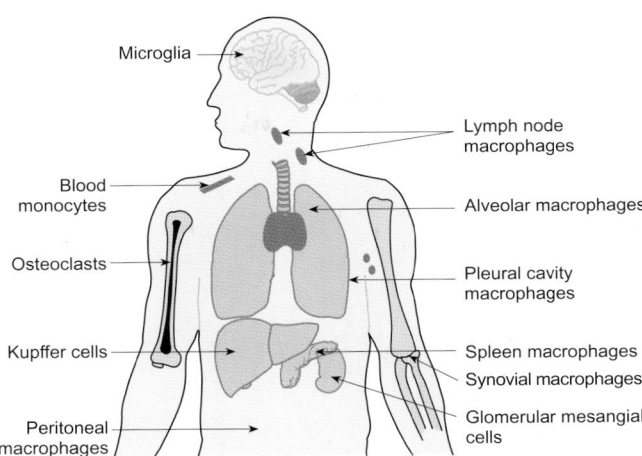

Fig. 14.2 The mononuclear phagocyte system. Monocytes are formed in the bone marrow where they mature before being released into the circulation. They migrate to tissues such as the spleen, liver, lungs, and lymph nodes, where they take up residence as mature macrophages. The neutrophils (the other main class of phagocyte) remain in the circulation until they participate in an inflammatory reaction.

ules contain lysozyme, alkaline phosphatase, and a peculiar form of cytochrome (cytochrome b_{558}) that can be inserted in the plasma membrane.

Neutrophils are able to pass from the blood into the intercellular spaces by diapedesis (see below) and actively phagocytose and engulf disease-producing bacteria. The enzymes within the cytoplasmic granules then kill the invading organisms and digest them. As a result of this action, the neutrophils form the first line of defense against infection.

The *macrophages* are formed in the bone marrow and released into the blood as monocytes. Within 2 days, they migrate to tissues such as the spleen, lungs, and lymph nodes, where they mature. Macrophages contain a large number of lysosomes and phagocytic vesicles containing the remains of ingested materials. They are found in all tissues, even in the brain where they are known as microglia (Fig. 14.2). Macrophages are situated around the basement membrane of small blood vessels. They also line both the spleen sinusoids and the medullary sinuses of the lymph nodes, where they are able to remove particulate matter from circulation. In the liver, they are known as *Kupffer cells*.

The phagocytosis and killing of microbes

Before a phagocyte can kill an invading bacterium, it must first recognize it as a foreign body and engulf it. This is the process of phagocytosis, which is described in Chapter 4, p. 45. Phagocytes are non-specific immune cells that will attack a wide variety of invading organisms and cell debris.

After a bacterium has been engulfed, the phagocyte will kill it. This is achieved by a variety of methods. Following phagocytosis, the macrophage or neutrophil produces a number of reactive oxygen intermediates via cytochrome b_{558}. Molecular oxygen is first converted to the superoxide anion which, under the influence of an enzyme called superoxide dismutase, produces hydrogen peroxide. The hydrogen peroxide then gives rise to a

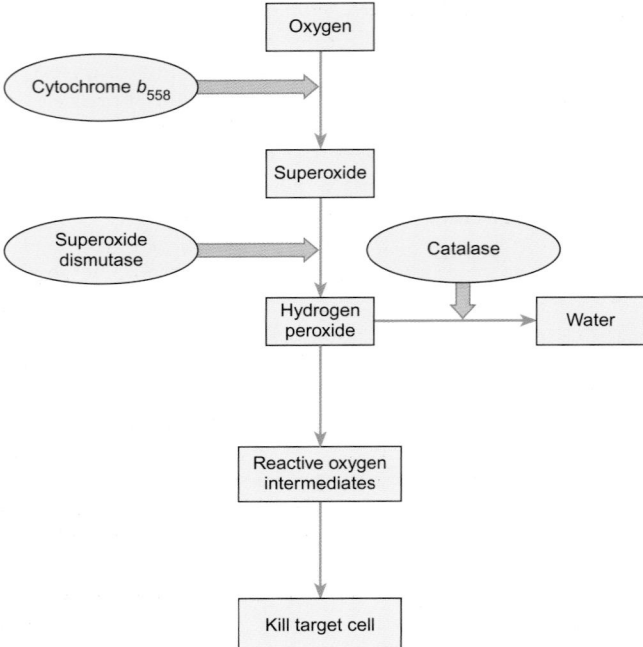

Fig. 14.3 The processes leading to the generation of reactive oxygen intermediates by the phagocytes. The host cell is protected by various mechanisms from damage (e.g. vitamins C and E, glutathione, and the enzyme catalase).

number of highly reactive intermediates that kill the bacteria trapped within the phagosome (Fig. 14.3). These events cause a marked increase in oxygen uptake by the activated cell. This increase is called a *respiratory burst*. Both macrophages and neutrophils also produce nitric oxide and other reactive nitrogen intermediates to kill bacteria. In addition, bactericidal proteins (called *defensins*) are inserted into the bacterial cell membrane and cause it to rupture. Various enzymes then digest the remains.

Natural killer cells and the interferons

Viruses lack the ability to replicate by themselves. Instead, they subvert the genetic machinery of host cells to make copies of themselves. For this reason, it is important that those cells that become infected are destroyed before the virus has time to replicate and infect neighboring cells. The cells that perform this vital function are known as *natural killer cells*. They are large granular lymphocytes, which are believed to recognize virus-infected cells from modified cell-surface markers (compare this with T cell recognition of infected cells—see below). When a natural killer cell has recognized a target, it is activated and positions specific granules between its nucleus and the target cell. The granule contents are then released by exocytosis onto the target cell, which responds by undergoing a pre-programmed cell death (*apoptosis*) that prevents viral replication.

When cells are infected by a virus, they make *interferons* (of which there are many different kinds) and secrete them into the extracellular fluid. The interferons then bind to receptors on neighboring cells, which respond by reducing their rate of mRNA translation. This results in the infected cell being surrounded by a layer of cells that cannot replicate the virus so

forming a barrier that prevents the spread of the infection. Finally, the natural killer cells seek out and destroy any infected cells.

Eosinophils

Eosinophils are the least numerous of the white cells of the blood. The granules that take up the dye eosin (eosinophil-specific granules) contain a protein rich in arginine residues called *major basic protein*. The cells are also able to secrete membrane-penetrating proteins called *perforins* and a battery of enzymes including peroxidase and phospholipase D.

Eosinophils appear to play an important role in combating helminth infections. These organisms are too large to be phagocytosed by a single cell so that they must be attacked extracellularly. Eosinophils are particularly attracted to parasites whose outer membranes have been coated with antibody of the IgE class (see below). Major basic protein, perforins, peroxidase, and phospholipase D attack the outer membrane of the parasites to inactivate or kill them. Eosinophils are attracted to sites of infection or inflammation by chemical signals (*interleukins*) released from mast cells. They are also found in the connective tissues underlying the epithelia of the skin, bronchi, gut, and other hollow organs.

Complement

Complement is the name given to a group of about 20 plasma proteins that play an important part in the control of infections, particularly those caused by bacteria and fungi. Like the clotting factors, the complement proteins are a series of enzymes that can be sequentially activated.

Activation of the complement system occurs by one of three pathways, known as the classical, alternative, and lectin pathways (see below). In each case, a component of the complement system known as C3 is cleaved to form two fragments known as C3a and C3b. The larger fragment, C3b, binds to the surface of microbes and facilitates their uptake by phagocytes. It also initiates a series of reactions that can lead to lysis of the invading bacterium. The smaller fragment, C3a, activates phagocytes which kill phagocytosed bacteria and interact with other complement proteins. Both fragments, together with another complement fraction known as C5, play an important role in the initiation of the inflammatory response (see p. 251).

The classical pathway was the first to be discovered and acts to complement the actions of the antibody molecules secreted by the B lymphocytes, which is why the proteins are called complement. It is activated by IgG or IgM antibodies bound to the surface of the invading organism. They bind to a complement component known as C1q which combines with two other components C1r and C1s to form a complex that activates two further components, C2 and C4. In turn, these activate C3 convertase which cleaves C3 to form C3a and C3b, which have the actions described below and summarized in Fig. 14.4.

The alternative pathway depends on the spontaneous formation of low levels of C3b and C3a from C3. This process is markedly accelerated at the surfaces of invading organisms such as bacteria and fungi. Normal host cells produce a series of proteins that inhibit the activation of the complement system on their surfaces.

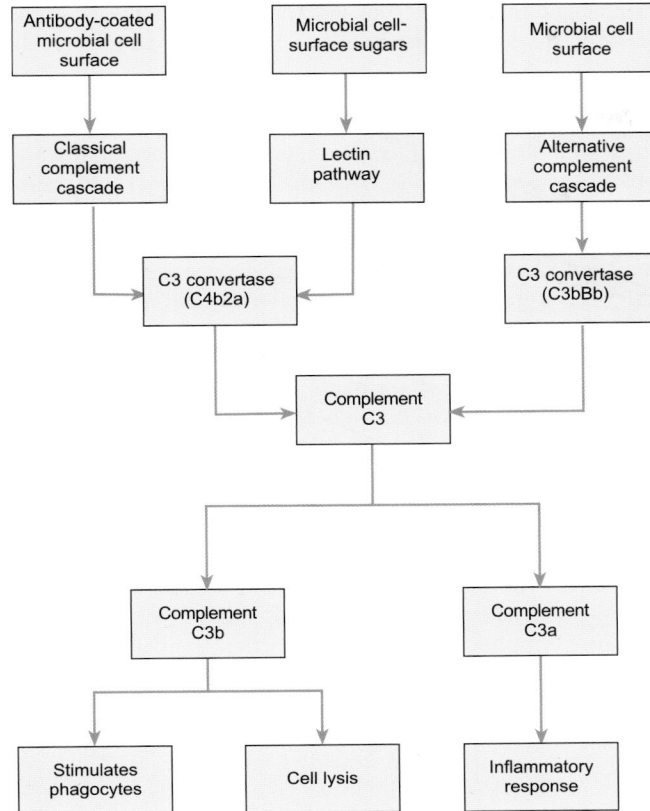

Fig. 14.4 An outline of the complement system showing the pathways of activation and involvement in the immune response.

The lectin pathway is activated when a protein known as mannose-binding lectin (MBL) becomes attached to mannose or fucose residues present on bacterial cell walls (mannose and fucose are carbohydrates not normally associated with mammalian cell surfaces). The MBL binds to repeated sequences of these sugars and activates two proteases (MASP-1 and MASP-2) that cleave C2 and C4, which then form a complex known as C4b2a that acts as a C3 convertase.

Acute-phase proteins

The acute-phase proteins are a group of plasma proteins synthesized by the liver that show a huge increase in concentration during an infection. They include C-reactive protein and complement components—including C3. As described above, C3b binds to the surface of invading organisms. This surface coating is known as *opsonization*. Both C-reactive protein and antibodies also opsonize foreign organisms. As the phagocytes have receptors for the coating proteins, they are able to recognize opsonized particles and engulf them.

The acute inflammatory response

When the body becomes injured or infected, a number of physiological changes occur in the affected area. There is local vasodilatation, increased permeability of the capillaries, and infiltration of the damaged tissues by white cells. These changes constitute the *inflammatory response*, which appears to be geared to bringing plasma proteins and cells to the point of injury.

Inflammation can be caused by a variety of stimuli, including traumatic injury, infection, and cellular necrosis.

For an injury to the skin the stages of the inflammatory response are as follows: There is a reddening of the skin at the site of injury, which results from vasodilatation. This is known as the *acute vascular response*. This is rapidly followed by local tissue swelling due to the accumulation of fluid by the affected tissues. The skin of the surrounding area becomes flushed (the flare). These three components of the inflammatory response constitute the *triple response* discussed in Chapter 8, Section 8.3.

If the infection or trauma is sufficiently extensive, the acute vascular phase is followed by the *acute cellular phase* in which the injured tissues become infiltrated by polymorphonuclear leukocytes, particularly neutrophils. The vascular endothelium in the injured area becomes modified and the neutrophils attach themselves to the capillary wall in a process called *margination*. They then squeeze between the endothelial cells and pass into the tissues (diapedesis). This brings them into direct contact with invading organisms or cell debris where they can undertake their normal phagocytic role.

A *chronic cellular response* then follows in which macrophages and lymphocytes invade the damaged area. Like the neutrophils, the macrophages dispose of the cellular debris. They also seem to play an important role in the healing process.

Finally, the inflammatory response declines as the damaged tissue becomes healed. This phase is known as *resolution*. If it has not been possible to eliminate the invading organism or any particles that triggered the inflammatory response, the offending material is sealed off by a layer of macrophages, lymphocytes, and other cells to form a *granuloma*. Injury to internal organs is accompanied by a similar sequence of events.

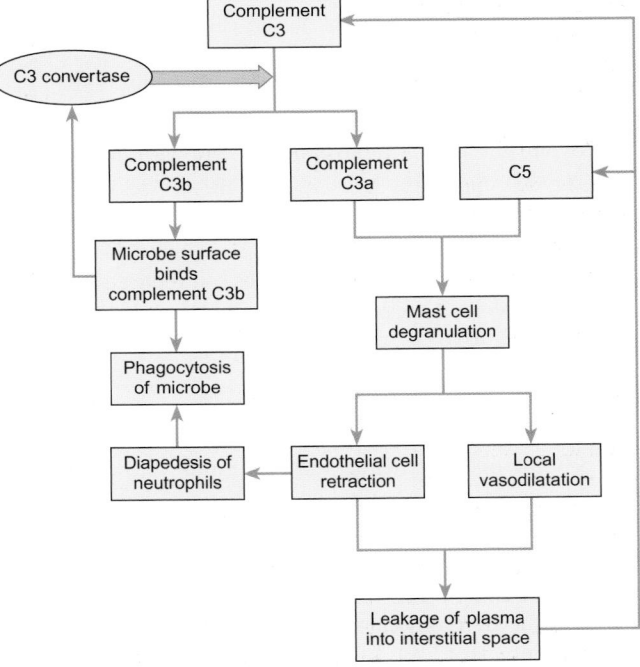

Fig. 14.5 The mechanism by which an invading organism triggers an inflammatory response. The initiation of this series of events begins with the activation of an enzyme known as C3 convertase.

What triggers the inflammatory response?

The processes involved in the inflammatory response are summarized in Fig. 14.5. In the first stages of a response to infection, the invading organism becomes coated with small amounts of complement C3b which is found in normal plasma. The immobilized C3b then generates a form of complement called C3 convertase. This complex splits C3 into C3a and C3b. The C3b molecules bind to the organism that triggered the initial response (a positive feedback loop). It also activates another complement component called C5. Together with C3a, C5a stimulates the tissue mast cells to degranulate and release their contents into the interstitial space.

The mast cell granules contain a wide variety of substances including histamine, chemotactic agents that attract polymorphonuclear cells from the blood, and signaling molecules called *interleukins*. In addition, activated mast cells synthesize prostaglandins and leukotrienes. The histamine and prostaglandins elicit a local vasodilatation and this, coupled with the retraction of the capillary endothelial cells, leads to leakage of plasma into the interstitial space. The plasma contains complement and antibodies, both of which aid the countermeasures against infection (the interstitial fluid normally has very little plasma protein).

The histamine and interleukins modify the surface of the capillary endothelial cells so that the neutrophils adhere to them. The neutrophils then squeeze between the endothelial cells and migrate to the site of infection. Once in position, they begin engulfing the invaders and killing them by means of the mechanisms described earlier.

14.5 The adaptive immune system

It is well known that, while exposure to certain disease-producing organisms (e.g. the chickenpox virus) will cause disease on the first exposure, a subsequent exposure will not generally result in infection. Nevertheless, the resistance to that infection does not extend to other diseases. Experience of chickenpox does not prevent infection by the measles virus. These facts highlight two important features of our immune system. First, resistance is acquired by one exposure to an invading organism and then lasts for many years—even for a whole lifetime. Secondly, the resistance is specific for that organism. In immunological terminology, the response is *specific* and has *memory*. These characteristics distinguish the response of the adaptive immune system from that of the natural immune system.

Lymphocytes

The cells of the adaptive immune system are the lymphocytes. The *lymphoid system* consists of the total mass of tissue associated with the lymphocytes and their function. It is disseminated throughout the body (see Fig. 14.1). The tissues in which the lymphocytes mature (the bone marrow and thymus) are known as *primary lymphoid tissue*, while the lymph nodes, spleen, and other lymphoid tissues are *secondary lymphoid tissue*. There are two principal classes of lymphocyte, which are known as B cells and T cells. The B cells mature in the bone marrow and secrete antibody, while the T cells mature in the thymus gland and secrete signaling molecules known as *cytokines*, cytotoxic substances, or both.

Summary

1. The natural immune system consists of the innate defense mechanisms. It consists of four kinds of cells—the phagocytes, natural killer cells, mast cells, and eosinophils—plus three different classes of proteins—complement, interferons, and acute-phase proteins.

2. The macrophages and neutrophils engulf small invading organisms (e.g. bacteria) and kill them with highly reactive oxygen and nitrogen intermediates. They then digest the remains and release the contents for use by the host.

3. Cells that become infected with a virus are destroyed by natural killer cells before the virus can replicate. The natural killer cells are large granular lymphocytes which are believed to recognize cells infected with a virus via modified cell-surface markers.

4. When the body becomes injured or infected, there is local vasodilatation, increased permeability of the capillaries leading to local edema, and infiltration of the damaged tissues by white cells. These changes constitute the *inflammatory response*.

5. The trigger for the inflammatory response is mast cell degranulation. The histamine that is released, together with newly synthesized prostaglandins, elicits local vasodilatation that leads to leakage of plasma into the interstitial space. The granules also release chemotactic agents that attract neutrophils to the site of injury.

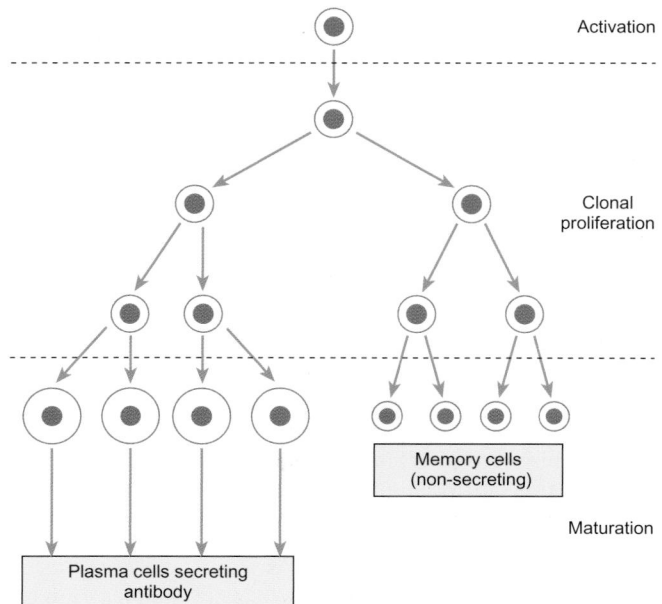

Activation

Clonal proliferation

Memory cells (non-secreting)

Maturation

Plasma cells secreting antibody

Fig. 14.6 The generation of a B-lymphocyte clone. An antigen activates a naive lymphocyte, which then proliferates. Some of the clonal cells mature and secrete antibody of the same specificity as that of the receptor (i.e. it is capable of binding to the initiating antigen), while other cells remain in the lymphoid tissues as memory cells.

Lymphocytes are stimulated by antigens that bind to their surface receptors. Individual lymphocytes respond only to one antigen and, when they are stimulated, they proliferate by mitosis to form a population of cells with an identical specificity called a *clone*. As shown in Fig. 14.6, some of the cells continue to proliferate and carry out their specific immunological function (see below), while others remain in the lymphoid tissue as memory cells, able to respond quickly to the same antigen in the future.

Lymphocytes continuously circulate through the tissues

To monitor the tissues of the body for invading organisms, the lymphocytes continuously circulate throughout the tissues. They migrate from the blood, through the tissues, to the lymph nodes, which they enter by way of the afferent lymphatic vessels (see Chapter 15, pp. 297–298, for an explanation of the circulation of the lymphatic fluid). After they have entered the lymph nodes, they pass into the efferent lymphatics and return to the blood via the thoracic duct. In addition, some lymphocytes enter the lymph nodes directly from the post-capillary venules.

The first stage of a lymphocyte's migration from the blood is its adhesion to the wall of the blood vessel. Normally, like the red cells, the lymphocytes remain in the center of a blood vessel, but when they reach a target tissue some of them become attached to the vessel wall. This process is guided by homing receptors that are specific for particular tissues. The cells then flatten and squeeze between the endothelial cells and move into the surrounding tissues (diapedesis).

Lymph nodes

The lymph nodes consist of an outer capsule beneath which lies the subcapsular sinus which is fed by the afferent lymphatics (Fig. 14.7). Below the subcapsular sinus lies the cortex, which is organized into primary follicles (which contain B cells) and secondary follicles, also known as germinal centers, which contain a class of B cells known as *memory cells*. The space between the follicles is called the paracortex and is populated by T cells. At the center of the gland lies the medulla, which contains antibody-secreting B cells and macrophages. The arrangement of the lymph nodes allows the afferent lymph to percolate through the

tissue to the efferent lymphatic. In this way, any antigens that may be present are exposed to cells that are capable of mounting an appropriate immune response.

B lymphocytes and antibody

A resting B lymphocyte has little cytoplasm, a darkly staining nucleus, and few mitochondria or ribosomes. After it has been stimulated by antigen, it becomes transformed into a *plasma cell* in which the cytoplasm is greatly expanded and full of ribosomes. It also has a well-developed Golgi apparatus. The ribosomes are the site of synthesis of the antibodies, which are then secreted into the plasma. As the antibody has the same specificity as the B cell antigen receptor, a particular antigen will stimulate only those B cells that will respond by secreting an appropriate antibody. In this way, the secretion of antibody is tailored to the nature of the infection.

There are five classes of antibody

The antibodies are globulins (immunoglobulins) which have the same basic structure, consisting of two identical light chains and two identical heavy chains linked to give a Y-shaped molecule as shown in Fig. 14.8. The antigen-binding domains are formed by one heavy and one light chain and are located at the two tips of the Y so that each antibody molecule has two identical antigen-binding sites. This part of the molecule is highly variable in structure. This variable region permits an antibody to distinguish one antigen from another. It is important to realize that an antibody will recognize and bind to a particular part of an antigen. Antibodies do not bind to the whole molecule. Rather, they recognize specific structural motifs.

The stem of an immunoglobulin molecule is known as the Fc region and this is used by the cells of the immune system to recognize particles that have been coated with antibody, e.g. viruses or bacteria.

There are five classes of antibody: IgA, IgD, IgE, IgG, and IgM.

1. IgA is the most common immunoglobulin in secretions such as saliva and bile. It is also found in colostrum (the milk secreted during the first week of lactation—see Chapter 21, p. 473).

2. IgD acts as a surface receptor on B cells together with IgM.

3. IgE binds to Fc receptors on mast cells to facilitate an inflammatory response to antigen.

4. IgG is the most abundant immunoglobulin in plasma. In addition, this protein can cross the placental membrane and so provide the fetus with ready-made antibodies to protect it *in utero* and for some months after birth.

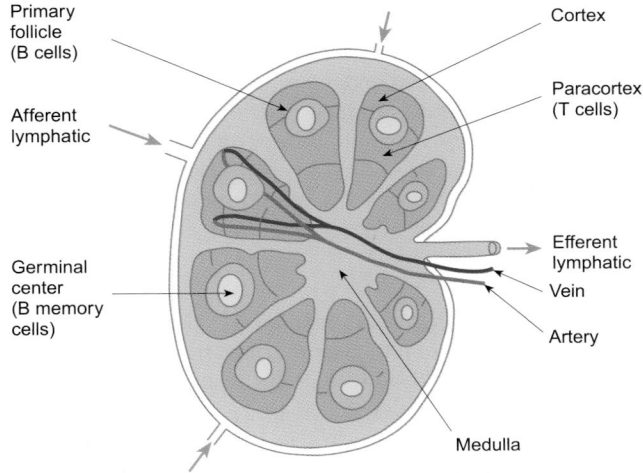

Fig. 14.7 The structure of a typical lymph node. Note the compartmentation of the B and T lymphocytes.

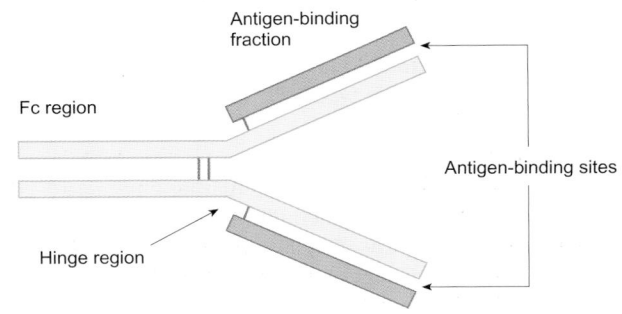

Fig. 14.8 The structure of antibody.

5. IgM is the first antibody to be produced during development and during the primary immune response. It activates complement.

Antibodies have two main functions: to bind an antigen, and to elicit a response that results in the removal of the antigen from the body. The antibody acts together with complement to stimulate the phagocytes, with the result that the organism carrying the antigen is killed and digested. The combination of antibody, complement, and phagocyte is very effective, but a lack of any one component seriously compromises the ability of the body to mount an adequate immune response.

In some cases, the binding of the antibody to its antigen is sufficient to inactivate the invading organism. Thus, when viruses are coated with the antibody, they cannot infect the host cells and so are prevented from proliferating.

Antibody diversity

Although antibodies can be grouped into five main classes, the number of different antibody molecules is immense. Estimates suggest that there may be more than a thousand million million (10^{15}) different kinds of antibody molecule in a single individual, each with its own specificity. These antibodies constitute the pre-immune antibody repertoire. It would seem that this number would be sufficient to ensure that the body could mount an antibody response to any potential antigen.

How is such a large number of different proteins generated when human DNA consists of only about 30 000 (3×10^4) genes? The answer to this problem lies in the way in which antibody molecules are encoded by DNA. Each antibody consists of four polypeptide chains, two light chains and two heavy chains. The genomic DNA encodes for each of these polypeptides by combining different gene segments. Therefore there are many different DNA sequences from which the polypeptide chains of the antibodies can ultimately be synthesized. Moreover, during normal clonal proliferation, point mutations occur in the DNA, which generate still further antibody diversity. This fact explains why the affinity of an antibody for a particular antigen increases with time after the initial exposure.

T lymphocytes and cell-mediated immunity

Unlike B cells, T cells do not enlarge greatly or secrete antibody when they encounter their specific antigen. Instead, they secrete cytokines or cytotoxic molecules (or both) onto neighboring cells. Since these substances act in a paracrine fashion, the effects of T cell activation are local and usually only one cell responds. Those T cells that secrete cytokines to stimulate B cells to proliferate and secrete antibody and are called *helper* T cells. Those that secrete cytotoxic substances to kill a target cell are called *cytotoxic* T cells. Each type has a different role to play.

Antigen processing and presentation

Unlike B cells, which respond to circulating antigens, T cells respond to cells with altered chemistry—such as a tumor cell or a cell infected with a virus. T cells respond to MHC molecules that have bound a foreign peptide. An individual T cell therefore recognizes only a particular MHC molecule incorporating a par-ticular foreign peptide. In effect, the MHC molecules are exposing the foreign peptide to the T cell, i.e. they are *presenting* it to the T cell.

The sequence of events leading to antigen presentation is as follows. When cells become infected by viruses or intracellular bacteria, foreign proteins are expressed within their cytoplasm. These foreign proteins are broken down into peptides by the cell's endosomes. The resulting peptides are transported to the endoplasmic reticulum where they become complexed with MHC class I proteins before being inserted in the plasma membrane where they can be recognized by cytotoxic T cells.

Extracellular antigens are taken up by endocytosis and are degraded into small peptides. The endosomes then fuse with vesicles containing MHC class II proteins. The MHC proteins then complex with foreign peptides and become inserted in the plasma membrane where they can be recognized by helper T cells.

Dendritic cells play a central role in presenting antigens to the lymphocyte population. They are derived from the monocyte colony-forming cell line (see Chapter 13) and take their name from their characteristic appearance. They are covered with numerous thin processes similar to the dendrites of neurons. This gives them a large surface area with which to interact with foreign antigens or with other cells of the immune system. Dendritic cells that have captured a foreign antigen migrate to the various lymphoid organs where they present the antigen to helper T cells.

Although all dendritic cells perform an essentially similar role, they are known by different names according to their location. They are termed *Langerhans cells* in the skin, *veiled cells* in the lymph, *interdigitating dendritic cells* in the medulla of the thymus and secondary lymphoid tissue, and *interstitial dendritic cells* in other tissues.

All dendritic cells express high levels of class II MHC molecules, which complex with foreign antigen as follows. Extracellular antigens are taken up by endocytosis and degraded to form small peptides. The endosomes then fuse with vesicles containing MHC class II proteins, which complex with the foreign peptides before becoming inserted in the plasma membrane where they can be scrutinized by helper T cells.

A further class of dendritic cells, the *follicular dendritic cells*, behave in a rather different fashion. They do not express class II MHC and do not present antigen to helper T cells. Instead they remain located within lymphoid follicles, the areas of a lymph node populated by B lymphocytes. Follicular dendritic cells express membrane receptors for antibody and complement (FcRs and CRs). Binding of immune complexes to these cells appears to promote the activation of B lymphocytes.

T cell receptors

The T cell receptor has two components, a part that recognizes specific cell markers and a part that investigates the MHC for the presence of foreign antigen. Helper cells identify cells bearing MHC class II molecules, which are B cells and macrophages. When they detect one of these cells bearing a foreign peptide, they secrete cytokines to stimulate them. The B cells proliferate, so increasing the size of the particular clone and the amount of

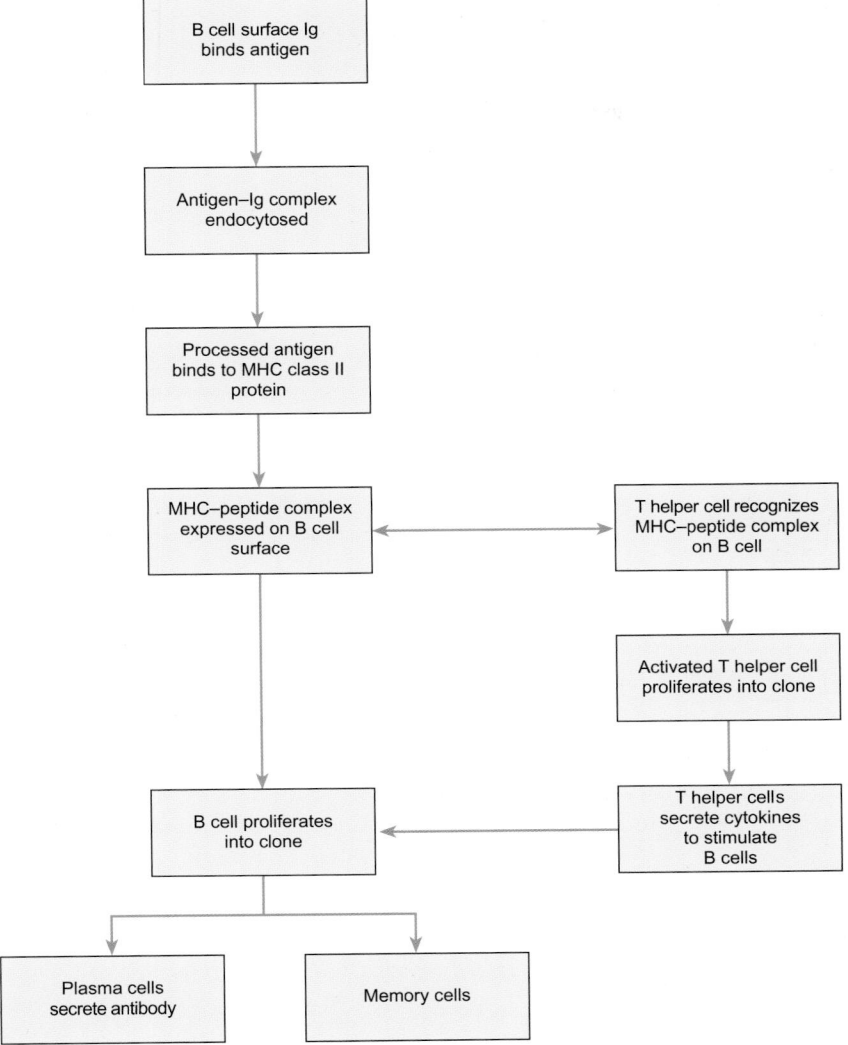

Fig. 14.9 The cellular basis of the antibody response. Both pathways of T cell activation are shown, although the antigen-presenting cells would be of greatest importance in the primary response and the B cells in secondary responses.

antibody secreted (Fig. 14.9), with the result that the antibody response is enhanced. Those T cells that respond to the MHC complex on macrophages do so by proliferating and stimulating the affected macrophages to activate their normal killing mechanisms as shown in Figure 14.10(a). This process is important for those cells that have become infected with intracellular bacteria (e.g. the bacillus that causes tuberculosis). Cytotoxic T cells respond to MHC class I molecules bearing an antigenic peptide by killing the target cell (Fig. 14.10(b)).

The antibody response to infection

When people are first exposed to an infectious agent, they rely on the natural immune system for their defense. If the infection is severe, B cells begin to make an antibody but this only occurs after a lag phase during which B and T cells are activated. Initially, the B cells secrete mostly IgM, which reaches a peak after about a week (Fig. 14.11). This antibody is able to activate complement and so enhances the ability of the natural immune system to combat the infection. As the disease proceeds, some B cells switch to the production of IgG and the level of this antibody in plasma continues to rise after the IgM level begins to decline. IgG levels peak after about 2 weeks and then slowly decline as the infection wanes. This pattern of antibody secretion is known as the *primary response*.

A subsequent infection by the same organism is met by a much more prompt response—the *secondary response*. IgG levels rapidly rise above those seen during the primary response and remain elevated for a longer period (Fig. 14.11). In contrast, the IgM levels follow approximately the same time course as the primary response. The rapid and augmented secondary response is due to the recruitment of specific memory cells—cells that arose during the development of the initial clone but which did not mature into antibody-secreting plasma cells. The B-lymphocyte memory cells, in particular, respond to very small levels of antigen once they have been primed. This memory function of the adaptive immune system makes possible artificial immunization against specific diseases such as poliomyelitis and smallpox ('vaccination')—see Box 14.1.

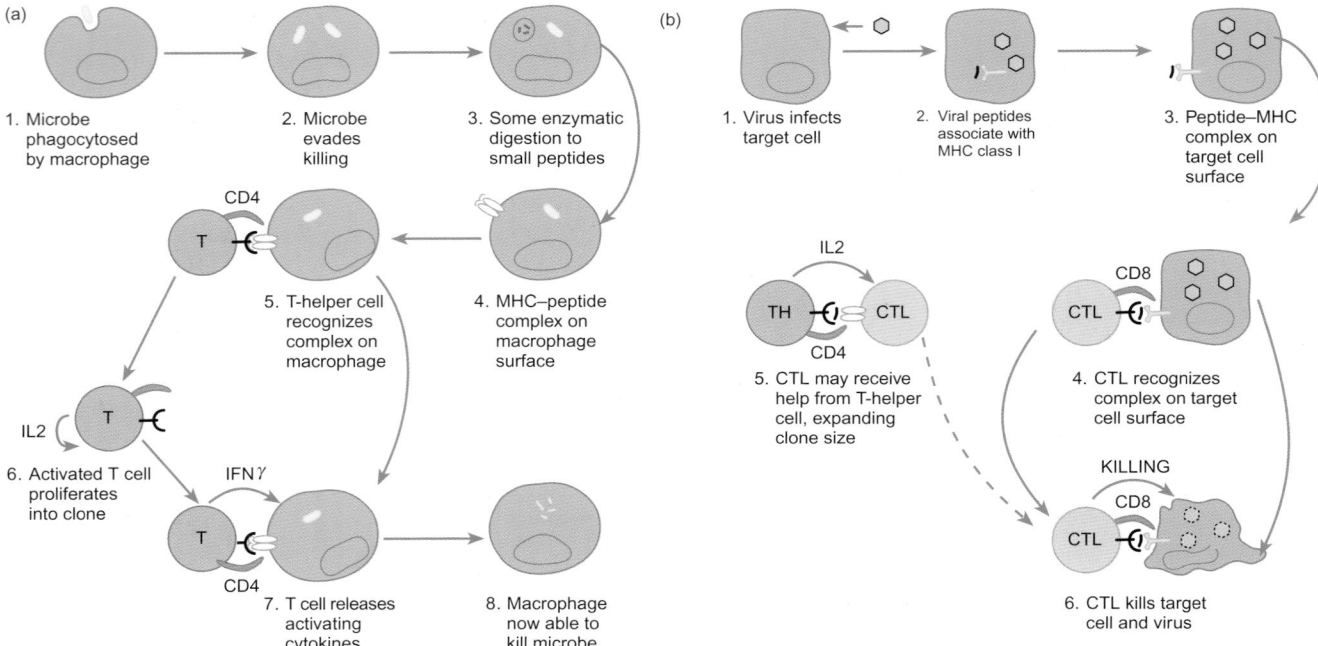

Fig. 14.10 Cell-mediated immune response: (a) activation of macrophages by T cells; (b) the mechanism by which cytotoxic T cells (CTL) kill cells infected with a virus. IL2, interleukin 2; CD4, CD8 are co-receptors for the invariant part of the MHC protein. IFN γ: interferon γ

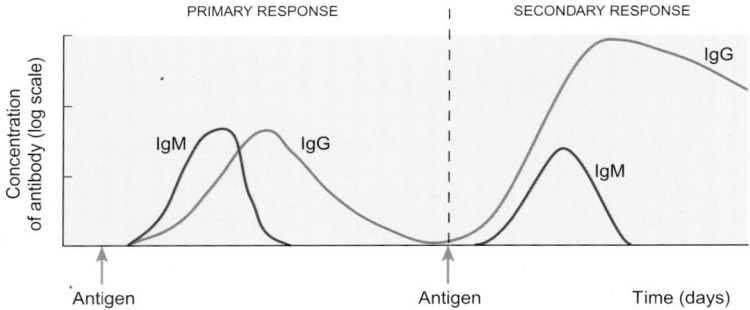

Fig. 14.11 Primary and secondary antibody responses to an infection. Note the relatively slow antibody response to the first exposure and the early secretion of IgM. Compare this with the very rapid IgG response to the second challenge.

BOX 14.1 PEYER'S PATCHES AND THE ORAL POLIO VACCINE

One component of the mucosal-associated lymphoid tissue (MALT) is the organized gut-associated lymphoid tissue (GALT) that includes the Peyer's patches. These are lymphoid aggregates in the gut mucosa, found predominantly in the ileum. They house large numbers of naïve T and B lymphocytes and are covered by an epithelial layer containing cuboidal cells known as M-cells. These are specialized antigen-transporting cells that do not express MHC II. They lie over the phagocytic dendritic cells that take up the antigens before migrating to the local lymphoid tissue, where they activate the lymphocytes. Once exposed to the antigen, lymphocytes within the Peyer's patch become sensitized and begin to secrete IgA on to the epithelial surface. IgA binds to, and inactivates, the anti-gen. In this way the Peyer's patches play an important part in the immunological homeostasis of the gastrointestinal tract.

A further example of the importance of the gut-associated lymphoid tissue is provided by the effectiveness of the oral polio vaccine. The polio virus is normally transmitted via the fecal–oral route, and may cause paralysis as a result of damage to neurons of the anterior horn of the spinal cord. The polio vaccine is administered routinely to children in three doses during the first year of life (with boosters in childhood and adolescence). An attenuated form of the virus is given to the child orally. Once administered, the attenuated virus colonizes the intestine where it triggers an immune response mediated by cells of the Peyer's patches and other GALT cells.

Summary

1. The adaptive immune system provides a mechanism for defending the body against an extraordinarily wide range of organisms. Unlike the response of the natural immune system, that of the adaptive immune system is specific and has memory. Indeed, once resistance has been acquired it usually lasts for many years.

2. The cells of the adaptive immune system are the lymphocytes. There are two principal classes of lymphocyte, B cells and T cells, which mature in the bone marrow and thymus, respectively. To scan the body for invading organisms, the lymphocytes pass from the blood vessels, pass through the tissues, and re-enter the venous blood by way of the lymph nodes and the thoracic duct.

3. When lymphocytes are stimulated by antigens, they proliferate by mitosis to form a population of cells with an identical specificity called a *clone*. Some of the clonal cells continue to proliferate and carry out their specific immunological function, such as antibody production, while others remain in the lymphoid tissue as memory cells, able to respond to a similar challenge in the future.

4. After it has been stimulated by antigen, a B cell becomes transformed into a *plasma cell*, which secretes antibody into the circulation. Antibodies have two main functions—to bind an antigen and to elicit a response that results in the removal of the antigen from the body. The antibody acts together with complement to stimulate the phagocytes, with the result that the organism carrying the antigen is killed and digested.

5. Activated T cells secrete cytokines or cytotoxic molecules (or both) onto neighboring cells. T cells may be *cytotoxic* or *helper* cells. The effects of T cell activation are local and usually only one target cell responds. Unlike B cells, which respond to circulating antigens, T cells respond to those cells that possess MHC molecules that have bound a foreign peptide. Thus, T cells respond to tumor cells or to cells infected with a virus.

6. On initial exposure to an antigen, the B cells secrete IgM and IgG. Plasma levels of these antibodies peak after 1–2 weeks and then slowly decline. A subsequent infection by the same organism is met by a much more prompt and long-lasting increase in the plasma levels of the appropriate IgG antibody.

14.6 Disorders of the immune system

Like other organ systems, the function of the immune system may become disturbed. Broadly speaking the disorders are due to inappropriate immune activity (*hypersensitivity*) or to a failure of the immune system (*immunodeficiency*). Each will be discussed in turn.

Hypersensitivity

The adaptive immune system is very powerful and capable of responding to a wide variety of antigens. In normal people, the response is appropriate and correctly targeted against the invading organism. Sometimes, however, the activity of the immune system leads to pathological changes in the host tissues. Such reactions are called *hypersensitivity* or immunopathology.

Hypersensitive reactions can be grouped under one of four headings:

- Type I are *allergic reactions*, which are mediated by IgE antibodies and mast cells. Examples are hay fever and asthma. The steps that lead to an allergic response are:

  (1) The generation of an IgE that is specific for a particular allergen (e.g. pollen);

  (2) the binding of the IgE to mast cells and basophils;

  (3) the binding of further molecules of allergen to the *bound* IgE which results in

  (4) the degranulation of the mast cells and the secretion of vasoactive materials.

This sequence leads to the generation of the inflammatory response by the mechanisms discussed in pp. 251–252. The subsequent clinical events depend very much on the site at which the allergen initiates the reaction. In the airways, the release of histamine and the formation of leukotrienes causes the bronchoconstriction characteristic of asthma. If the allergen reaches the circulation, it may provoke a generalized inflammatory reaction that results in a circulatory collapse known as *anaphylactic shock*. This is a potentially fatal condition (see Chapter 28).

- Type II is *antibody-dependent cytotoxic hypersensitivity*. In this case, IgG reacts with host cells, so initiating the processes that lead to their destruction. This unwanted reaction may follow a transfusion of incompatible blood (see Chapter 13, Section 13.9). It may also follow skin or organ transplants, leading to the rejection of the graft or transplant (see below). Occasionally the lymphocytes attack the host cells themselves (i.e. they respond to the normal host antigens). This is known as *autoimmunity*. If antibodies react with normal hormone receptors, they may either stimulate the particular receptor as in hyperthyroidism (Grave's disease) or they may block the action of the normal hormone as in hypothyroidism (myxedema).

- Type III is *immune-complex-mediated hypersensitivity*. When an antibody reacts with an antigen, it forms an immune complex that may precipitate. If these complexes are not phagocytosed rapidly (their usual fate), they may accumulate in the small blood vessels. When this happens, they are attacked by complement and neutrophils. The ensuing reactions may cause damage to the endothelium at a critical point in the circulation and compromise the function of the affected organ. One example is *glomerulonephritis* (inflammation of the renal glomeruli), which results from the deposition of immune complexes in the glomerular capillaries and is a common cause of renal failure.

◆ Type IV or *cell-mediated delayed hypersensitivity* can manifest itself as an allergic reaction to certain parasites (e.g. bacteria and fungi), as a contact dermatitis resulting from sensitization to a chemical agent, or by the rejection of a tissue graft or transplant. This type of hypersensitivity is mediated principally by the T cells.

Defects of the complement system

These appear to be due mainly to deficiency of particular components of the complement system. Low levels of C1, C2, or C4 lead to difficulty in clearing antigen–antibody complexes from the plasma. They are a common feature of systemic lupus erythematosus, in which deposition of immune complexes in the renal glomeruli leads to acute renal failure, as discussed above. If complement components C5–C9 are low or even absent, there is a predisposition to serious infections by bacteria of the *Neisseria* class. These organisms cause disseminated gonorrhea and menigococcal meningitis. Reduced levels of C1s esterase inhibitor can result in overactivity of the classical complement cascade. In this situation, there are recurrent inflammatory attacks of mucosal tissues that may cause obstruction of the intestine or larynx.

Self-tolerance

In autoimmunity, the lymphocytes of the immune system mount an attack on normal host cells. Although these reactions are frequently triggered by an infection, their very existence raises an important question. Why are the cells of the immune system normally tolerant of the host cells? The explanation is as follows. Only those T cells that react to self-MHC molecules that incorporate a foreign peptide are released into the circulation. If a T cell reacts with unaltered MHC molecules, it will not be released from the thymus but will be destroyed. It is thought that many B cells do express antibody receptors that react with host cells. Those that have a high affinity for the host cell-surface markers are deleted in the bone marrow and never reach the circulation (this is comparable with the fate of T cells that react with self-MHC molecules). Any remaining B cells that react with host antigens will do so only weakly, and even these are ineffective because they do not receive help from the T cells. Such B cells are said to be *anergic*.

Immunodeficiency and human immunodeficiency virus (HIV)

Immunodeficiency of genetic origin is called *primary immunodeficiency*. If it is due to some other cause, it is *secondary immunodeficiency*. Primary immunodeficiencies are rare and are usually due to a defect in a single gene. They include defects in phagocyte function, defects in complement, and defects of lymphocyte function. In all cases, there is an increased susceptibility to infection.

Secondary immunodeficiency is far more common, particularly in adults. The most common cause is malnutrition, but damage to the immune system by certain infections (notably HIV—see below), tumors, traumas, and some medical interventions also impair the function of the immune system (e.g. treatment with immunosuppressive drugs or exposure of the bone marrow to high levels of X-irradiation).

Summary

1. The immune system may react powerfully to an antigen (hypersensitivity) or it may fail to mount an adequate immune response (immunodeficiency). Host cells are not normally attacked by the cells of the immune system, which are able to differentiate self from non-self.

2. Hypersensitive reactions may be grouped under one of four headings:

   (a) Allergic reactions, e.g. hay fever and asthma.

   (b) Cytotoxic hypersensitivity, which may occur after a tissue transplant or when the lymphocytes attack the host cells themselves (autoimmunity).

   (c) The deposition of immune complexes in small blood vessels, leading to an inappropriate inflammatory reaction (e.g. glomerulonephritis).

   (d) Cell-mediated hypersensitivity—a delayed allergic reaction that follows exposure to certain antigens.

3. Immunodeficiency may be of genetic origin (primary immunodeficiency) or due to some other cause (secondary immunodeficiency). Primary immunodeficiencies are usually due to a defect in a single gene. Secondary immunodeficiency is more common and may result from malnutrition or from damage to the immune system caused by certain infections, tumors, traumas, or some medical treatments.

The human immunodeficiency virus (HIV) is perhaps the best known infectious agent that can compromise the function of the immune system. Unlike other agents, HIV attacks the cells of the immune system, particularly the T cells. The explanation for this alarming state of affairs is that the T cell receptor is also the receptor to which HIV binds. HIV is a retrovirus that can insert its genetic material into the host cell DNA. Therefore the stimulation of an infected T cell results in replication of the virus. This leads to a slow depletion of the T cell population and an increased susceptibility to infection. The resulting disease is now known as acquired immunodeficiency syndrome (AIDS).

14.7 Transplantation and the immune system

Tissue transplantation from one person to another to treat organ failure has long been a major goal of medicine. The first such procedure to be wholly successful was the transfusion of blood. This depended on the correct identification of the agglutinins (antibodies) and agglutinogens (antigens) present in both donor and recipient blood as described in Chapter 13, p. 243. In the same way, successful skin grafts or organ transplants require a close match between the specific cell markers of both donor and recipient.

The problems associated with transplantation are very well illustrated by the difficulties experienced in successfully grafting skin. If the skin becomes severely damaged, for example as a result of extensive burns, the healing process cannot make good the lost germinal tissue and the wound contracts as it becomes infiltrated by connective tissue. This results in disfigurement and distortion of the neighboring tissue. If the affected area is large, there may also be a continued loss of fluid from the damaged area, which will also be susceptible to infection. For these reasons it is sometimes desirable to graft some healthy skin from another part of the body onto the site of injury. Such grafts (which are known as *autografts*) are usually successful. The transplanted tissue is quickly infiltrated by blood vessels and heals into place. The donor area also heals rapidly. However, if the damaged area is very extensive, it may be impossible to find sufficient undamaged skin to act as a source for the grafts. In this case, it is necessary to consider grafting skin from someone else—to use an *allograft*.

A skin graft from another individual will initially take quite well, but after about a week it will be rejected. Moreover, a second graft from the same donor will be rejected immediately. However, if the donor is an identical twin, the first graft will not be rejected as both twins have the same genetic make-up and their tissue antigens (MHC or, in humans, HLA) will be identical. These discoveries were crucial evidence that tissue rejection is an immunological phenomenon and paved the way for successful skin grafts and organ transplants between individuals of different genetic backgrounds.

The solution to graft rejection is to match the tissue of the donor as closely as possible with that of the recipient (tissue typing) and to inhibit the activity of the immune system with immunosuppressant drugs. This approach has proved very successful with skin grafts and with kidney and heart transplants. More than 80 per cent of kidney transplants will survive for more than 5 years provided that the HLA is well matched. For heart transplants the figure is more than 70 per cent. The success rate for the transplantation of other organs is also showing considerable improvement. Nearly half of all liver transplant patients will survive for more than 5 years, although lung transplants are less successful at present. Close tissue matching is not required for corneal grafts because the cornea does not become vascularized and so is not subject to attack by the lymphocytes.

Recommended reading

Biochemistry and cell biology

Alberts, B., Johnson, A., Lewis, J., Raff, M., Roberts, K., and Walter, P. (2002). *Molecular biology of the cell* (4th edn), Chapter 24. Garland, New York.

Berg. J.M., Tymoczko, J.L., and Stryer, L. (2002). *Biochemistry* (5th edn), Chapter 33. Freeman, New York.

Immunology

Austyn, J.M., and Wood, K.J. (1993). *Principles of cellular and molecular immunology*. Oxford University Press, Oxford.

Playfair, J.H.L., and Bancroft, G. (2004). *Infection and immunity* (2nd edn). Oxford University Press, Oxford.

Roitt, I.M., and Delves, P.J. (2001). *Essential immunology* (10th edn). Blackwell Science, Oxford.

Medicine

Souhami, R.L., and Moxham, J. (eds) (2003). *Textbook of medicine* (4th edn), Chapter 4. Churchill-Livingstone, Edinburgh.

Pharmacology

Rang H.P., Dale, M.M., Ritter, J.M., and Moore, P. (2003). *Pharmacology* (5th edn), Chapters 15 and 16. Churchill-Livingstone, Edinburgh.

Multiple choice questions

Each statement is either true or false. Answers are given below.

The natural immune system

1. a. The monocytes of the blood are the precursors of tissue macrophages.
   b. Neutrophils are only able to destroy bacteria circulating in the blood.
   c. Macrophages use nitrous oxide to kill bacteria they have engulfed.
   d. Complement facilitates the uptake of bacteria by phagocytes.
   e. Natural killer cells act by destroying cells infected with a virus.

2. a. Tissue swelling is the first stage of the inflammatory response.
   b. The capillaries at a site of inflammation exude plasma.
   c. Mast cells secrete chemotactic agents that attract lymphocytes.
   d. Macrophages secrete vasoactive materials during the inflammatory response.
   e. The inflammatory response can be triggered by complement binding to the surface of micro-organisms.

The adaptive immune system

3. a. The lymphocytes are the most abundant white cells of the blood.
   b. Lymphocytes only leave the blood at sites of inflammation.
   c. B lymphocytes mature in the bone marrow.
   d. B lymphocytes respond to antigens in the extracellular fluid.
   e. All B lymphocytes become plasma cells after they have been stimulated.

4. a. B lymphocytes secrete IgG when they are activated.
   b. T cells respond to cells infected with a virus.
   c. T lymphocytes mature in the thyroid gland.

d. All T cells secrete cytotoxic materials.

e. T cells respond to proteins on the host cells.

5. a. The antigen receptor on a lymphocyte has the same specificity as the antibody it secretes.

   b. Antibody molecules have two identical antigen-binding sites.

c. B lymphocytes secrete IgM and IgG in response to an infection.

d. IgM crosses the placenta and provides a fetus with antibody protection.

e. Mast cells play an essential role in allergy.

Answers

1. Neutrophils pass from the circulation into the tissues in response to chemotactic stimuli. Once in the tissues, they phagocytose and kill invading micro-organisms. Macrophages use nitric oxide (and reactive oxygen intermediates) to kill organisms that they have ingested.
   a. True
   b. False
   c. False
   d. True
   e. True

2. The first stage of the inflammatory response is vasodilatation (the red reaction of the triple response). Mast cells (not macrophages) secrete vasoactive substances (prostaglandins and leukotrienes) during the inflammatory response.
   a. False
   b. True
   c. True
   d. False
   e. True

3. The neutrophils are the most abundant of the white cells (about 70 per cent). Lymphocytes account for about 25 per cent of the white cells of the blood. Lymphocytes leave the circulation to migrate through the tissues, finally re-entering the blood via the thoracic duct. Although some B cells become transformed into plasma cells and secrete antibody, others remain in the lymphoid tissue as memory cells.
   a. False
   b. False
   c. True
   d. True
   e. False

4. T lymphocytes are derived from stem cells in the bone marrow and mature in the thymus gland. T cells are of two main kinds: helper cells and cytotoxic cells. Only the cytotoxic cells secrete cytotoxic materials to kill target cells. Helper cells secrete cytokines. T cells respond to MHC molecules that incorporate a foreign peptide.
   a. True
   b. True
   c. False
   d. False
   e. True

5. IgM is the largest of the plasma proteins and cannot cross the placenta. However, IgG can cross the placenta and it is this immunoglobulin that helps to protect the fetus and neonate from infection.
   a. True
   b. True
   c. True
   d. False
   e. True

The heart and circulation

After reading this chapter you should understand:

- The basic organization of the circulation
- How the heartbeat is initiated and regulated
- How the ECG can be used to assess the electrical activity of the heart
- The cardiac cycle and the heart sounds
- The measurement of cardiac output
- The factors that regulate the cardiac output (cardiodynamics)
- Pressure, resistance, and flow in the circulation (hemodynamics)
- Systemic arterial blood pressure—its regulation and clinical measurement
- Intrinsic and extrinsic factors controlling the caliber of blood vessels
- The principle of autoregulation
- Blood–tissue exchange: the microcirculation and lymphatic drainage
- The role of the CNS in the control of the heart and circulation
- The specialized features of the circulation in various vascular beds

15.1 Introduction

In unicellular organisms and simple animals such as sponges, the exchange of nutrients and waste products between the cells and the environment can be accomplished by simple diffusion across the cell membranes. However, since diffusion is a random process, the time required for equilibration increases disproportionately with increasing distance. Consequently, in more complex animals, where most cells are separated from the external environment by a considerable distance, diffusion by itself would not suffice to permit adequate exchange. To overcome this problem, animals have evolved a circulatory system which serves three primary functions:

1. It promotes the carriage of nutrients such as oxygen and glucose to the cells and removes the products of metabolism.

2. It regulates the volume and composition of the extracellular fluid via the renal circulation. This is essential for the proper function of the cells.

3. Because the blood is distributed to all parts of the body, the circulation plays an important role in the regulation of a wide variety of physiological functions.

   It acts as the vehicle for distributing hormones and thereby contributes to hormonal control.

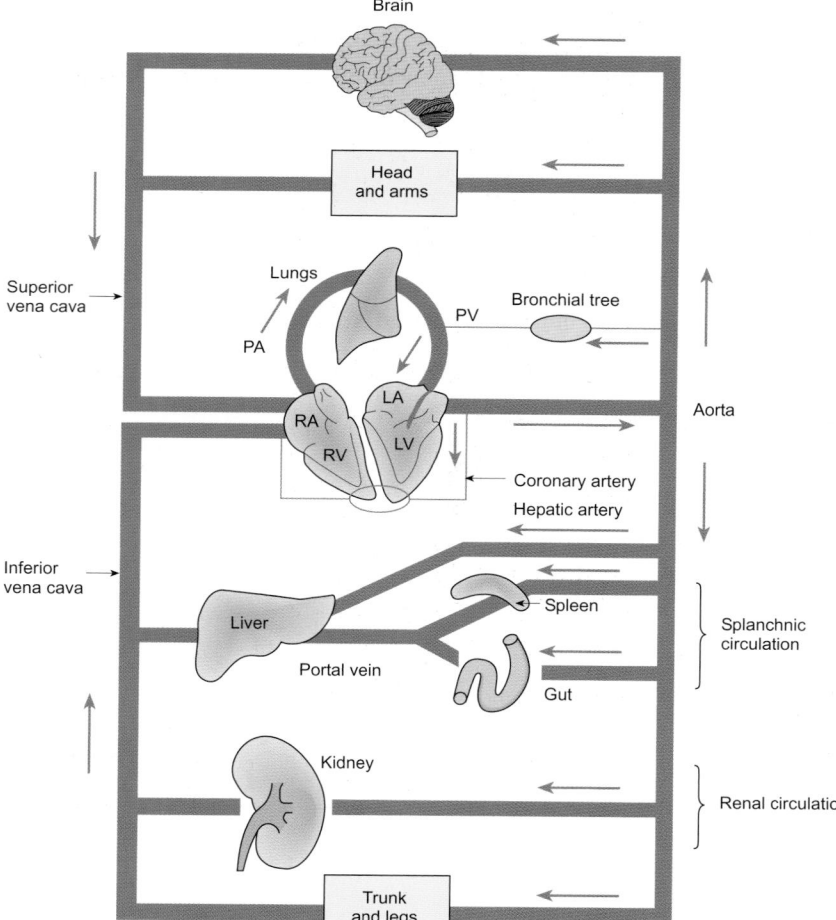

Fig. 15.1 A schematic drawing of the circulation. The arrows indicate the direction of blood flow. Note that the blood returning to the heart enters the right atrium. It then enters the right ventricle, which pumps the blood through the lungs. After leaving the lungs, the blood enters the left atrium and then passes to the left ventricle, which pumps it through the rest of the body via the systemic circulation. Thus the pulmonary circulation is in series with the systemic circulation. PA, pulmonary artery; PV, pulmonary vein; RA, LA, right and left atria; RV, LV, right and left ventricles.

The circulating white cells and immunoglobulins of the blood provide the principal means of defense against infection (Chapter 14).

By adjusting blood flow to the skin, the circulation also plays a significant role in the regulation of body temperature (see Chapter 24).

In what follows it will be useful to keep a number of key questions in mind.

◆ How is the circulation organized?

◆ How does the heart pump the blood and how is it able to adjust its output according to the needs of the body?

◆ How is the blood distributed to the tissues and how is that distribution regulated to meet the changing demands of the cells?

◆ How is exchange between the blood and the tissues accomplished?

The consequences of circulatory failure and other disorders of the cardiovascular system will be discussed in Chapters 28 and 31.

15.2 The organization of the circulation

The circulation consists of a pump (the heart) and a series of interconnected pipes (the blood vessels). As the blood is pumped from the right side of the heart through the lungs (the *pulmonary circulation*) and then from the left side of the heart to the body (the *systemic circulation*), the overall arrangement is of two circulations in series (Fig. 15.1). The pumping activity of the heart raises the pressure in the aorta above that of the large veins, which are at atmospheric pressure. It is this pressure (the *systemic arterial blood pressure* or, more commonly, the *blood pressure*) that causes the flow of blood around the systemic circulation. Equally, blood flows through the lungs because the pressure in the pulmonary arteries is greater than that in the pulmonary veins.

The gross anatomy of the heart

The arrangement of the heart and major vessels of the vascular tree can be seen in Fig. 15.2, which is derived from functional magnetic imaging (fMRI) of the circulation of a healthy young adult male. The aorta is clearly visible, as are all the major arteries and the great veins. Figure 15.3 shows the heart and major vessels in slightly greater detail.

The heart lies in the lower part of the left side of the chest and consists of four muscular chambers: two atria and two ventricles (see Fig. 15.4). The muscle of the heart is called the *myocardium*. In an adult, the heart is about the size of a clenched fist. The atria are thin-walled chambers that receive blood from the large veins and deliver it to the ventricles. The walls of the ventricles are much thicker than those of the atria—the wall of the left ventricle being thickest. The two ventricles are separated by a muscular sheet known as the interventricular septum. The contraction of the ventricles provides the force required to pump the blood round the pulmonary and systemic circulations.

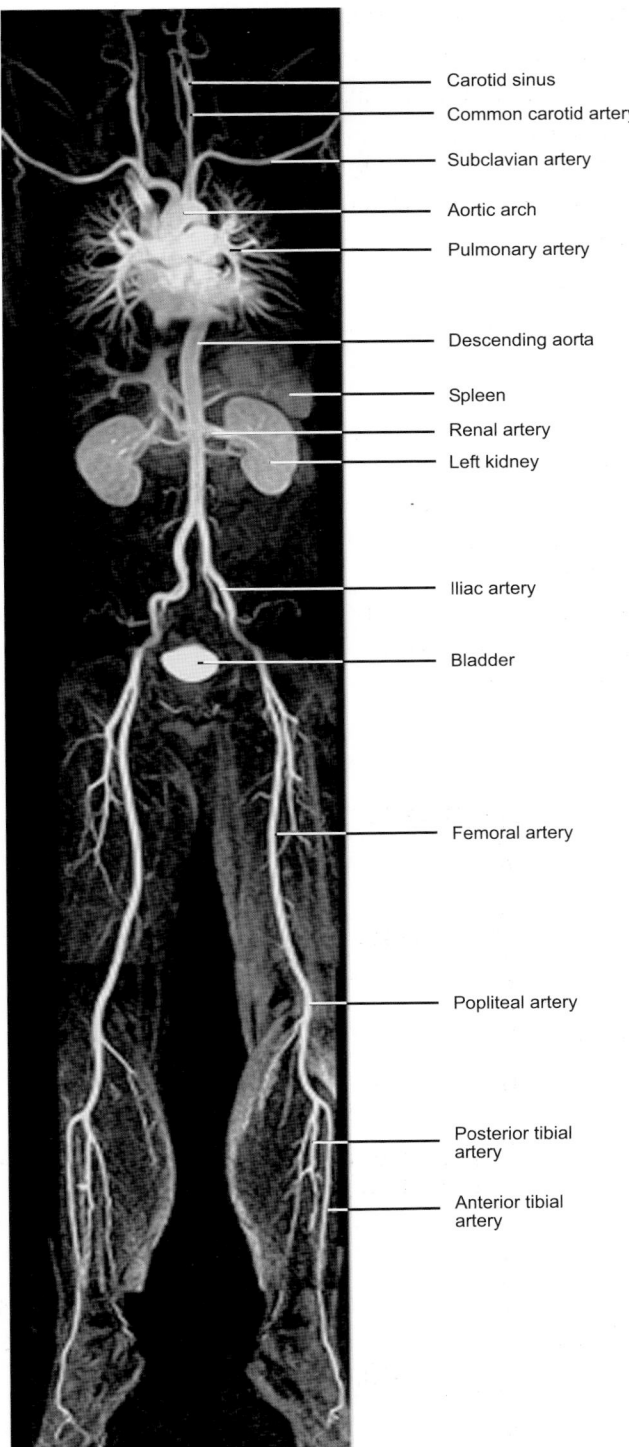

Fig. 15.2 An fMRI image of the circulation of a healthy young adult male. A magnetic contrast medium was injected prior to acquiring the magnetic resonance image in which the heart and the main arterial vessels are clearly seen. The contrast medium has accumulated in the bladder during the imaging process.

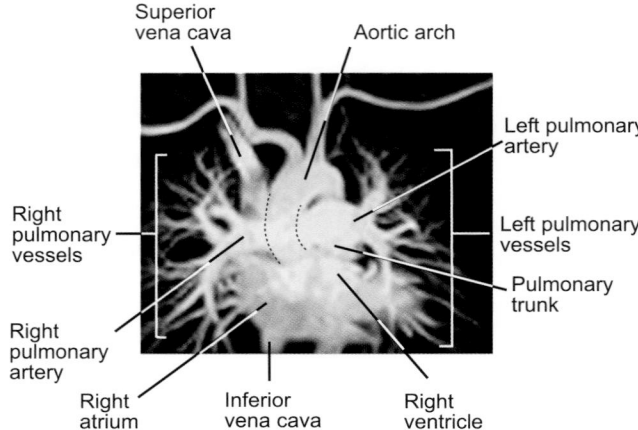

Fig. 15.3 An enlarged image of the chest region from Fig. 15.2 to show the position of the heart and the origin of the aorta and of the pulmonary arteries more clearly.

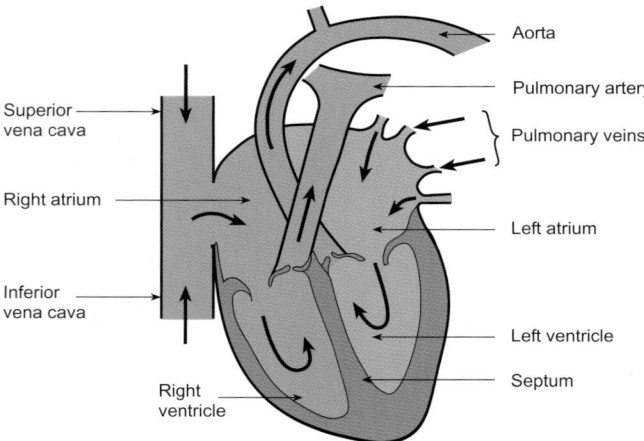

Fig. 15.4 The arrangement of the chambers and the direction of blood flow through the heart. The blood passing through the right side of the heart contains deoxygenated blood while that passing through the left side of the heart contains oxygenated blood.

The atria are separated from the ventricles by a fibrous skeleton on which the four heart valves are located. These serve to ensure the unidirectional flow of blood from the atria to the ventricles and from the ventricles into the pulmonary artery and aorta. Back flow of blood from the ventricles into the atria is prevented by the atrio-ventricular valves. On the right side of the heart, the atrio-ventricular valve (the *tricuspid valve*) consists of three roughly triangular flaps of fibrous connective tissue. On the left side, the atrio-ventricular valve consists of two such flaps and is known as the *mitral (or bicuspid) valve*. Back flow of blood from the pulmonary artery into the right ventricle is prevented by the pulmonary valve, while back flow of blood from the aorta into the left ventricle is prevented by the aortic valve. The pulmonary and aortic valves are also known as the *semilunar valves*. The valves and chambers of the heart are lined by a cellular layer known as the endocardium.

The heart lies in a tough fibrous sac known as the *pericardium* that prevents the heart from expanding excessively due to over-filling with blood. The pericardium is attached to the diaphragm so that the apex of the heart is relatively fixed. When the ventricles contract, the atria move towards the apex. This has the effect of expanding the atria as the ventricles contract.

The right ventricle occupies most of the anterior surface of the heart. It gives rise to the pulmonary trunk which divides after a short distance to give rise to the left and right pulmonary arteries which can be seen clearly in Fig. 15.2. After leaving the left ventricle, the aorta emerges from beneath the right side of the pulmonary trunk and ascends, giving rise to the brachiocephalic artery which subsequently divides to form the common carotid and subclavian arteries. The ascending aorta curves backwards to the left, over the right pulmonary artery, as indicated in Fig. 15.3, to form the aortic arch. When the aorta reaches the left side of the 4th vertebra, it descends to supply the lower regions of the body. This is known as the descending aorta (see Fig. 15.2), which supplies all the major organs of the abdomen, as well as the skeletal muscles of the trunk and lower limbs. The renal arteries and those of the legs are clearly visible in Fig. 15.2

The structure of the blood vessels

The blood vessels are divided into four broad categories: arteries, arterioles, capillaries and veins (Table 15.1). The walls of the larger blood vessels consist of three layers—the tunica intima, the tunica media, and the tunica adventitia—whose thickness varies according to the type of vessel (Fig. 15.5).

♦ The *tunica intima* consists of a layer of flat endothelial cells overlying a thin layer of connective tissue. It is separated from the tunica media by the internal elastic lamina. The endothelial cells of the tunica intima are in direct contact with the blood.

♦ The *tunica media* consists of a circular layer of smooth muscle containing elastin and collagen. The smooth muscle of the tunica media is innervated by sympathetic nerve fibers. The tunica media provides the mechanical strength of the blood vessel.

♦ The *tunica adventitia* consists of a loosely formed layer of elastic and collagenous fibers oriented along the length of the vessel.

TABLE 15.1 The total cross-sectional area of the different types of blood vessel in the systemic and pulmonary circulations relative to that of the aorta

Vessels	Relative area
Aorta	1
Systemic arteries	14
Arterioles	20
Capillaries	485
Venules	280
Veins	225
Venae cavae	1.1
Pulmonary arteries	50
Pulmonary capillaries	485
Pulmonary venules and veins	75

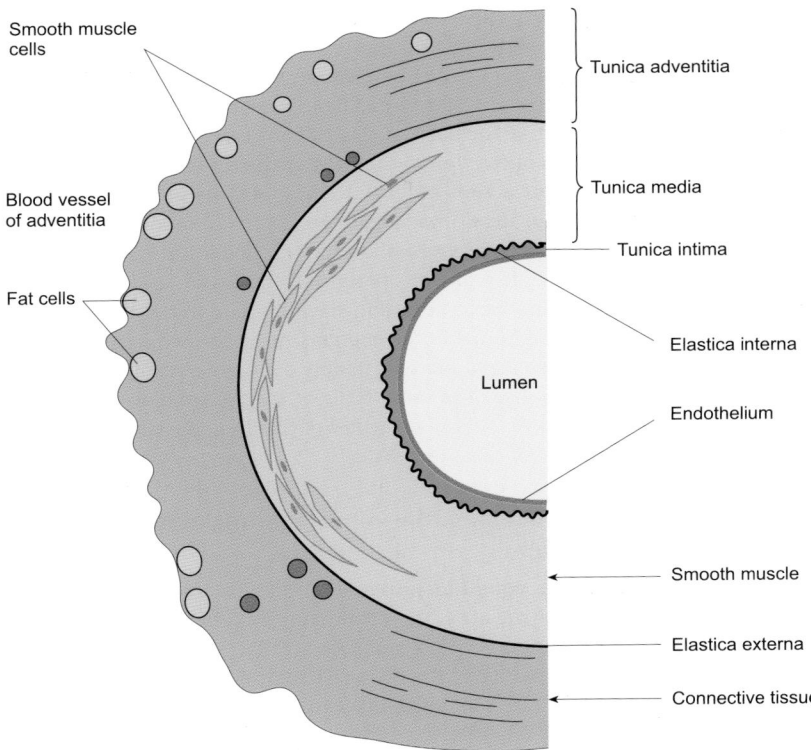

Fig. 15.5 The arrangement of the principal layers of a muscular artery (the tunica intima, tunica media, and tunica adventitia). The smooth muscle of the tunica media is arranged in a circular manner.

It serves to anchor the blood vessel in place. The tunica adventitia is separated from the tunica media by the external elastic lamina.

The arteries are the primary distribution vessels and can be subdivided into two groups.

1. The *elastic arteries* which, in humans, are large vessels 1–2 cm in diameter. They include the aorta and pulmonary arteries together with their major branches. The walls of the elastic arteries are very distensible because their tunica media contains a high proportion of elastin (up to 40 per cent compared with about 10 per cent for a muscular artery).

2. The *muscular arteries* which range in size from about 1 mm to 1 cm in diameter. The tunica media of the muscular arteries contains a higher proportion of smooth muscle than that of elastic arteries and the thickness of the tunica media relative to the diameter of the lumen is greater. This makes them very resistant to collapse at sharp bends such as occur at the joints. The principal function of these vessels is the efficient distribution of blood to the various vascular beds. Examples of muscular arteries are the cerebral, popliteal (in the legs), and brachial (in the upper arm) arteries.

The muscular arteries give rise to the *arterioles*, which are the resistance vessels responsible for regulating blood flow through particular vascular beds. The arterioles branch repeatedly and the final branches (the terminal arterioles) give rise to *capillaries*, which are thin-walled vessels about 5–8 μm in diameter. The capillaries are the principal *exchange vessels*. They have no smooth muscle in their walls, and the walls themselves consist of a single layer of endothelial cells lying on a basement membrane across which solutes move (see Section 15.10). The capillaries coalesce to form *postcapillary venules* about 20 μm in diameter, which also lack smooth muscle. In turn, these give rise to the *true venules* and the *veins* that merge to form the venae cavae that return the blood to the heart.

Summary

1. The circulation is organized so that the right side of the heart pumps blood through the lungs (the pulmonary circulation) and the left side of the heart pumps blood round the rest of the body (the systemic circulation). Thus the two circulations are arranged in series.

2. The heart itself consists of four chambers: two atria and two ventricles. The atria are separated from the ventricles by the mitral and tricuspid valves which prevent the reflux of blood into the atria when the ventricles contract. Reflux of blood from the pulmonary artery and aorta into the ventricles is prevented by the pulmonary and aortic valves.

3. The principal types of blood vessel are the arteries, the arterioles, the capillaries, the venules, and the veins. Except for the capillaries and the smallest venules, the walls of the blood vessels have three layers: the tunica intima, the tunica media, and the tunica adventitia. The smooth muscle of the blood vessels is innervated by sympathetic nerve fibers, which regulate its tone.

The walls of the veins and venules are similar in structure to those of the arteries but they are much thinner in relation to the overall diameter of the vessel. Consequently, the veins are much more distensible than the arteries. Unlike other blood vessels, the larger veins of the limbs possess valves at intervals along their length. These are arranged so that blood can pass freely towards the heart while back-flow is prevented.

In a few tissues, notably the skin, there are some direct connections between the arterioles and venules. These specialized vessels are known as arteriovenous shunt vessels (or *anastomoses*) and they have relatively thick muscular walls that are richly supplied by sympathetic nerve fibers. When these vessels are open, some blood can pass directly from the arterioles to the venules without passing through the capillaries (see Section 15.12).

15.3 The initiation of the heartbeat

The rhythmic pulsation of the heart is maintained by excitatory signals generated within the heart itself. Indeed, under appropriate conditions, the heart will continue to beat rhythmically following removal from the body. This property is called *autorhythmicity*. For the heart to be an effective pump, the contractions of the myocardial cells of the atria and ventricles must be coordinated. This is achieved by means of specialized conducting tissue.

Pacemaker potentials and myocardial excitation

The structure of cardiac muscle cells and the mechanism of their contraction has been discussed in Chapter 7. To summarize, all the cells of the myocardium can show spontaneous electrical activity given suitable conditions, i.e. they are all potential pacemaker cells. Normally, however, only the cells of the *sinoatrial (SA) node* (or pacemaker region) show such activity. These cells are located within the wall of the right atrium near the junction with the superior vena cava. The activity of these cells initiates the electrical impulse that is subsequently conducted throughout the whole myocardium (Fig. 15.6).

In the absence of any extrinsic nervous input, the SA node cells will drive the heart at a rate of around 100 beats per minute (b.p.m.). In other words, the cells of the SA node spontaneously generate an action potential every 600 ms or so. The action potentials of the SA node cells are relatively small, slowly rising depolarizations brought about by the inward movement of calcium ions. The action potentials of atrial and ventricular myocytes are quite different, having a fast initial rise followed by a prolonged period of depolarization known as the plateau phase. The plateau phase is mainly due to the inward movement of calcium ions. The ionic basis of the pacemaker activity and action potential of the myocytes of the SA node has already been discussed (Chapter 7, p. 93).

The conduction of the impulse throughout the myocardium

The action potential initiated in the SA node is propagated throughout the atria, the conducting system, and the ventricles via gap junctions in a manner similar to that employed by unmyelinated nerve fibers (see Chapter 6, Fig. 6.13). This process continues until the entire myocardium has been excited.

From the SA node, excitation spreads first across the whole of both atria. It then passes to the ventricles via the atrioventricular (AV) node, which forms the only bridge of conducting tissue between the atria and ventricles. The AV node consists of a narrow bundle of small-diameter fibers which have relatively few gap junctions linking them to their neighbors. As a result, conduction through the AV node is relatively slow and the spread of excitation to the ventricles is delayed for around 0.1 s at this point. This ensures that the atria have time to contract before the ventricular muscle is excited.

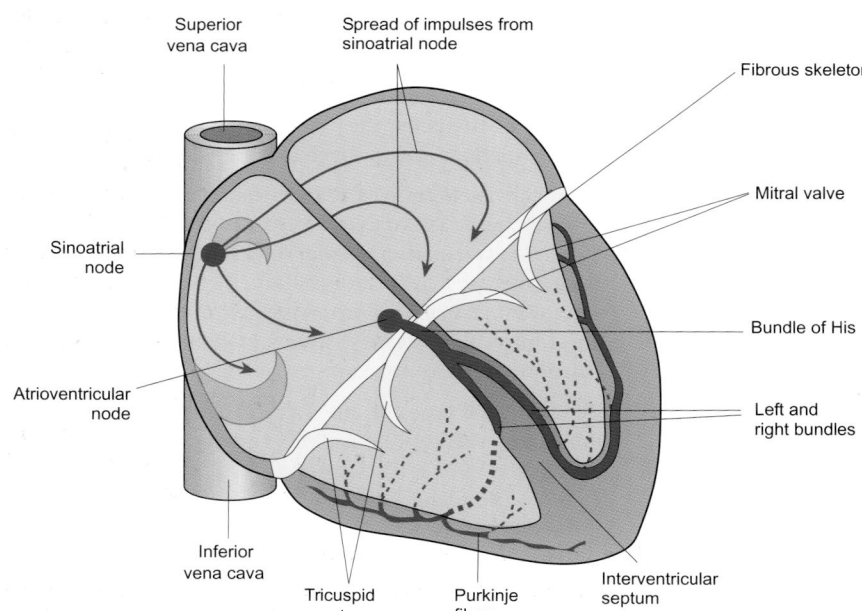

Fig. 15.6 The arrangement of the specialized conducting fibers of the heart and its fibrous skeleton to which the heart valves are attached.

Conduction through the remainder of the system is rapid (around 1 m s^{-1}) and is mediated by the *bundle of His* which divides into the left and right bundle branches supplying the left and right ventricles as shown in Fig. 15.6. The bundle fibers are specialized large-diameter cardiac myocytes arranged end to end to permit the rapid conduction of the wave of excitation from the AV node to an extensive network of large fibers (*Purkinje fibers*) that lies just beneath the endocardium. These fibers then spread the excitation to the ventricular myocytes via gap junctions. Conduction through the Purkinje fiber network is much faster (3–5 m s^{-1}) than that through the myocardium itself, with the result that all parts of the ventricles are excited at much the same time.

The myocyte membrane, like that of a nerve axon, is refractory during and immediately following an action potential (see also Chapter 7). This means that it cannot be re-excited during the relaxation phase of the heart and ensures that the conduction of the cardiac impulse is unidirectional. Finally, the membrane repolarizes to regain its original resting value. Figure 15.7 shows the action potentials of myocytes in various parts of the heart and illustrates some important differences between them.

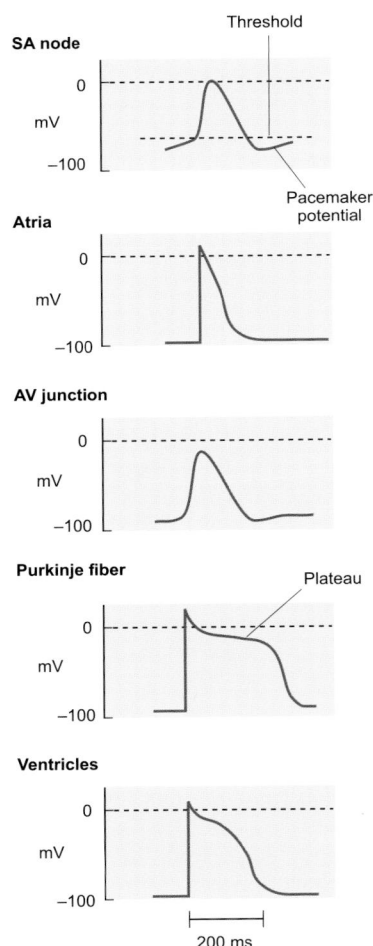

Fig. 15.7 The characteristic appearance of action potentials recorded from various types of myocardial cell.

Summary

1. The heart beats spontaneously. It shows an inherent rhythmicity that is independent of any extrinsic nerve supply. Excitation is initiated by a group of specialized cells in the sinoatrial node that lies close to the point of entry of the great veins into the right atrium. A wave of depolarization is then conducted throughout the myocardium.

2. The cells of the sinoatrial node have an unstable resting potential. The membrane potential between successive action potentials shows a progressive depolarization. This is the pacemaker potential. When threshold is reached, an action potential is triggered to initiate a heartbeat.

3. The myocytes of the atria, ventricles, and conducting system have action potentials with different characteristics. Although these action potentials vary in duration, they all show a fast initial upstroke followed by a plateau phase of depolarization prior to repolarization. The plateau phase is due to the inward movement of calcium ions.

4. The plateau phase ensures that the action potential lasts almost as long as the contraction of the cell. Because the muscle is refractory both during and shortly after the passage of an action potential, the long plateau phase ensures the unidirectional excitation of the myocardium.

5. Repolarization of the myocardial cells occurs when the voltage-dependent calcium channels inactivate.

15.4 The electrocardiogram can be used to monitor the electrical activity of the heart

As explained above, the spread of excitation through the myocardium occurs in much the same way as conduction in unmyelinated nerve fibers. In both cases, local electrical circuits are established and small currents flow through the extracellular fluid. The passage of these currents through the extracellular fluid creates small potential differences that can be detected by appropriately positioned electrodes and recorded as an *electrocardiogram* or ECG (see Box 15.1). In acknowledgement of the pioneering work of the Dutch physiologist Einthoven, the ECG is sometimes called the EKG (from the German *elektrokardiogramm*).

The ECG provides a means of recording the electrical activity of the heart *in situ* and is of clinical value in the investigation of cardiac arrhythmias and myocardial ischemia. Amongst the most common arrhythmias are atrial fibrillation, extra contractions of the ventricles (ventricular extrasystoles or ectopic beats), and progressive stages of heart block in which excitation from the atria to the ventricles is impaired. The ECG can also be used to determine the electrical axis of the heart as a whole, which can itself give information about certain pathological cardiac conditions such as ventricular hypertrophy.

How is the ECG trace recorded?

Although the most direct way of recording the ECG is by placing electrodes directly on the cardiac muscle itself, this is

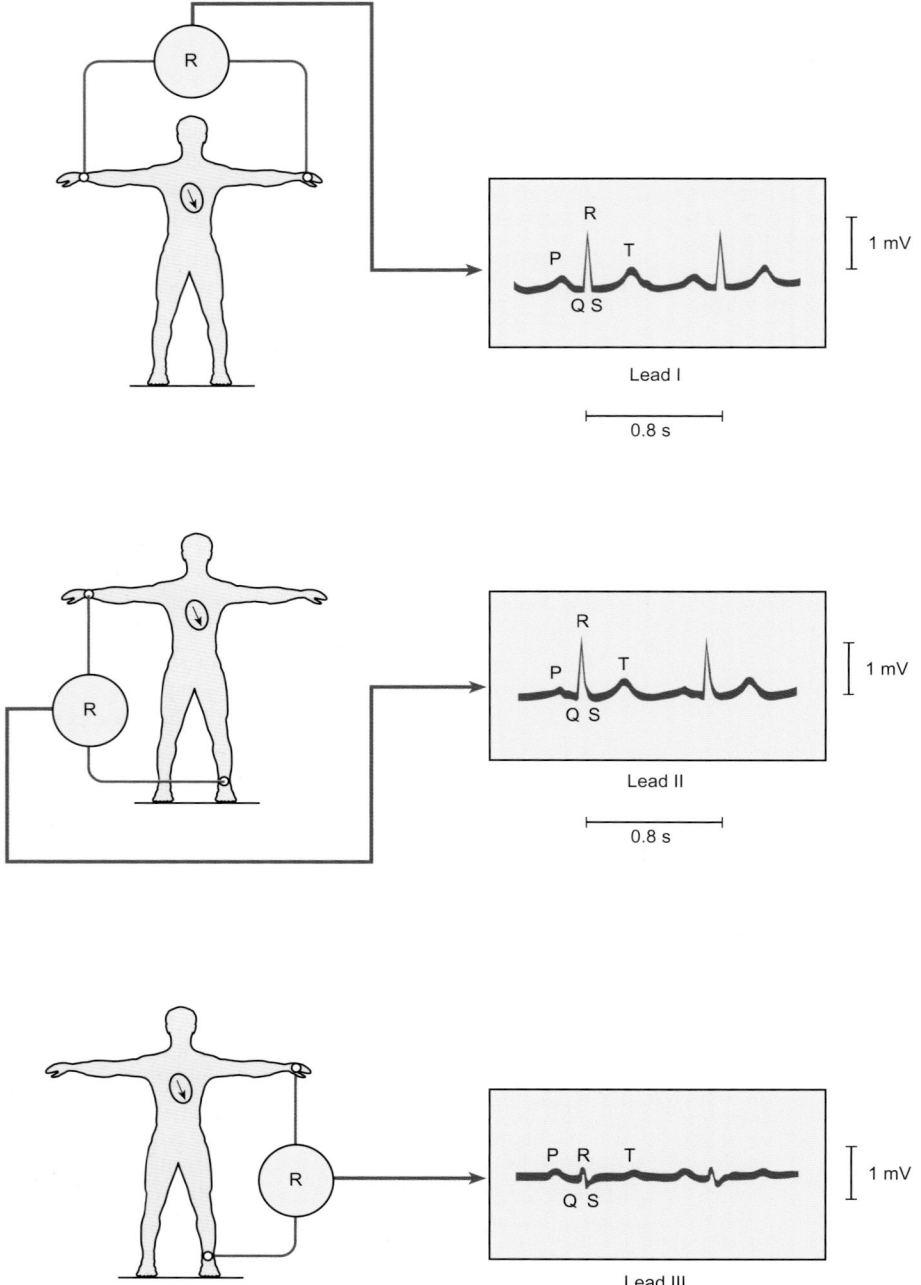

Fig. 15.8 The arrangement of the limb leads used to record the ECG. The appearance of the principal waves for the various leads is also shown. Lead II is orientated along the main atrioventricular axis of the heart and gives rise to the most prominent R wave.

rarely done except during certain surgical procedures in which the thorax is open. Instead, the ECG is recorded by placing electrodes at different points on the body surface and measuring voltage differences between these points with the aid of an electronic amplifier. Any particular position of a pair of electrodes on the body surface will detect a particular portion of the current flow during depolarization and repolarization, so that a complete picture of the spread of excitation requires information from a number of electrode placements or *leads*.

There are two types of ECG leads, *bipolar* and *unipolar*. Bipolar leads record the voltage between electrodes placed on the wrists and left ankle (with the right ankle acting as the earth). Unipolar leads record the voltage between a single electrode placed on the body surface and an electrode that is maintained at zero potential (earth).

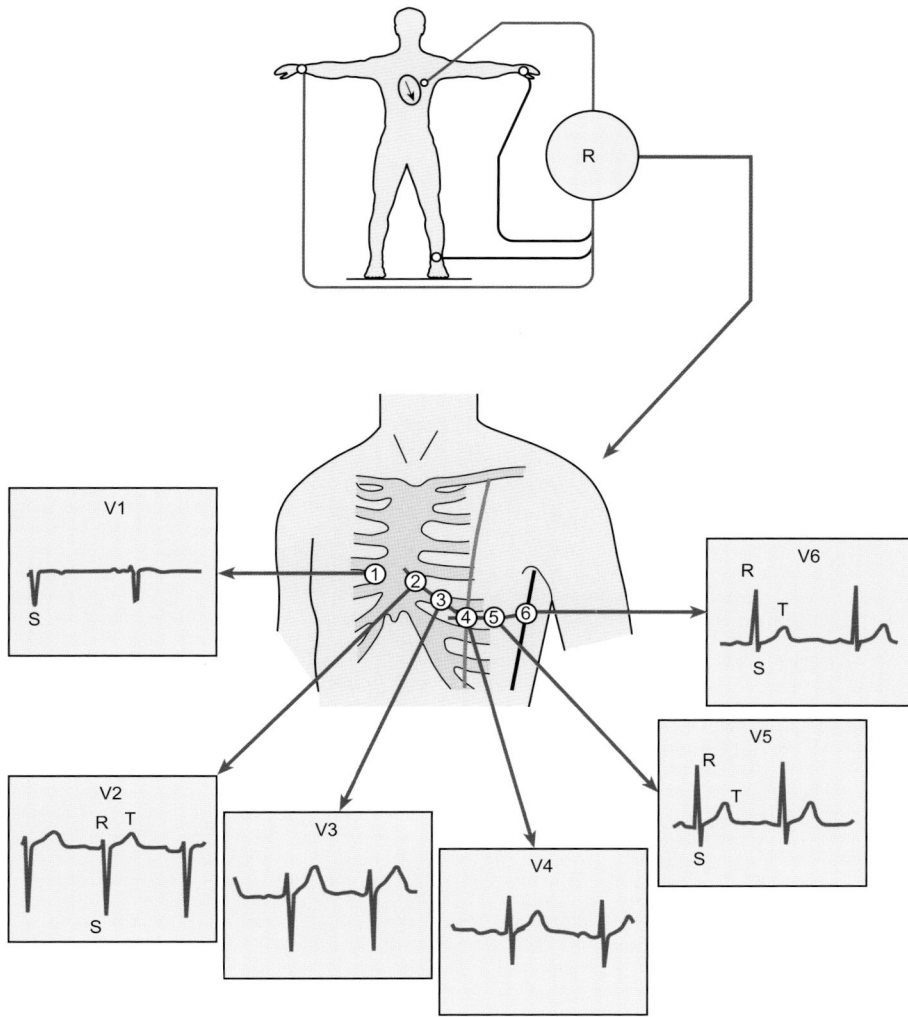

Fig. 15.9 Unipolar recording of the ECG with the standard chest leads. The limb leads are connected together to provide a virtual earth as shown in the inset figure at the top. The exploring electrode is then placed in one of six positions on the chest (leads V1 to V6) as shown. Recordings from leads V1 and V2 usually show a pronounced S wave, while leads V5 and V6 show a large R wave. Lead V1 is placed at the right margin of the sternum in the fourth intercostal space. Lead V2 is placed at the left margin of the sternum in the fourth intercostal space. Lead V3 is placed midway between leads V2 and V4. Lead V4 is placed on the mid-clavicular line (shown here in red) in the fifth intercostal space. Lead V5 is placed at the same level as lead 4 in the anterior axillary line. Lead V6 is also placed at the same level as lead 4 but in the mid-axillary line (shown here in blue).

The three standard bipolar limb leads are, by convention, known as limb leads I, II, and III. These are illustrated in Fig. 15.8.

- In lead I, the positive terminal of the amplifier is connected to the left arm and the negative terminal to the right arm. With this placement of electrodes, the amplifier records the component of excitation that is moving along an axis between the right and left sides of the heart.

- In lead II, the right arm is the negative terminal and the left leg the positive, so that the component of excitation moving from the right upper portion of the heart to the tip of the ventricles is recorded.

- In lead III, the left leg is the positive terminal and the left arm the negative. This lead records the component of excitation spreading along an axis between the left atrium and the tip of the ventricles.

Two types of unipolar leads are used in electrocardiography: the *augmented limb leads* and the *chest (precordial) leads*. When the ECG is recorded using the unipolar chest leads, a reference elec-

trode is produced by joining the three limb leads while an exploring electrode records from specific points on the chest at the level of the heart as shown in Fig. 15.9. The exploring electrode is then placed in one of six positions on the chest. These positions are known as leads V1 to V6. The exact locations of the electrode positions (or leads) are described in the legend to Fig. 15.9.

The voltage that can be recorded from the two arms and the left ankle with a unipolar exploring electrode can be increased using the augmented limb leads. In this configuration, two of the limb leads are used to produce a reference electrode. The signal is then recorded as the difference between these electrodes and the remaining limb electrode as shown in Fig. 15.10. Lead aV_R is recorded from the right arm with the reference electrode formed by adding the signals from left arm and left ankle (Figure 15.10(a)). Lead aV_L is formed by recording from the left arm with the reference electrode formed by adding the signals from right arm and left ankle (Fig. 15.10(b)). Lead aV_F is formed by recording the signal from the left ankle with the reference electrode formed by adding the signals from the two arms (Figure 15.10(c)).

(a)

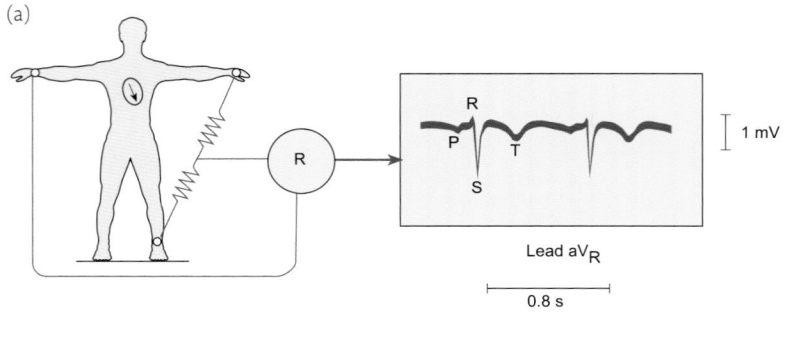

Lead aV$_R$

0.8 s

(b)

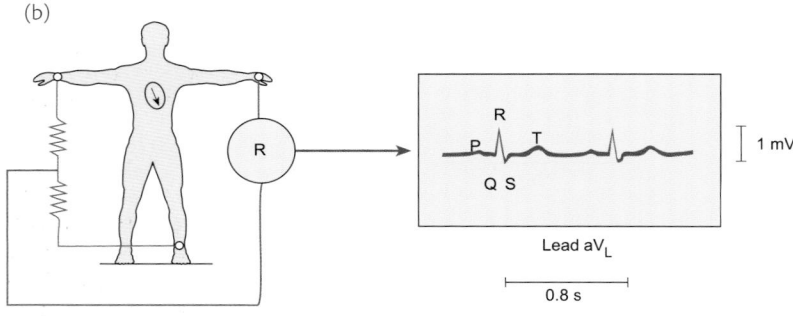

Lead aV$_L$

0.8 s

(c)

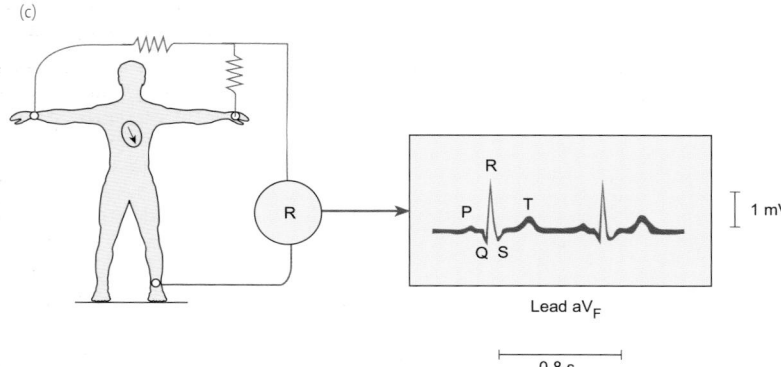

Lead aV$_F$

0.8 s

Fig. 15.10 The arrangement for recording the augmented limb leads and the typical appearance of the ECG recorded from these leads.

The characteristics of the normal ECG recorded with the limb leads

The normal ECG shows three main deflections in each cardiac cycle (Fig. 15.11). These are the *P wave* which corresponds to electrical currents generated as the atria depolarize prior to contraction, the *QRS complex* which corresponds to ventricular depolarization, and the *T wave* which corresponds to ventricular repolarization. Atrial repolarization occurs during ventricular depolarization and is masked by the QRS complex. Figure 15.12 shows the timing of the ECG waves in comparison with the underlying cardiac action potentials, and in Fig. 15.14 (see below) the ECG trace is shown alongside the mechanical events of the heart.

Although each cardiac cycle is initiated by depolarization of the SA node, this electrical event is not seen in the ECG trace because the mass of tissue involved is very small. The first dis-

cernible electrical event, the P wave, lasts about 0.08 s and coincides with the depolarization of the atria. The isoelectric line (i.e. the part of the record in which there are no measurable deflections) between the P wave and the start of the QRS complex coincides with the depolarization of the AV node, the bundle branches, and the Purkinje system. Atrial contraction occurs during the P–R interval, which lasts 0.12–0.2 s.

The next electrical event of the cardiac cycle is reflected in the QRS complex of the ECG trace which lasts about 0.08–0.1 s. This is normally seen as a large deflection from the isoelectric line because it is produced by the excitation (depolarization) of the ventricles, which represent a large mass of muscle tissue. The Q and S waves are downward deflections and the R wave is an upward deflection, although the exact pattern and size of the components of the complex depend upon the position of the electrodes being used to record the ECG (see Figs 15.8–15.10 and

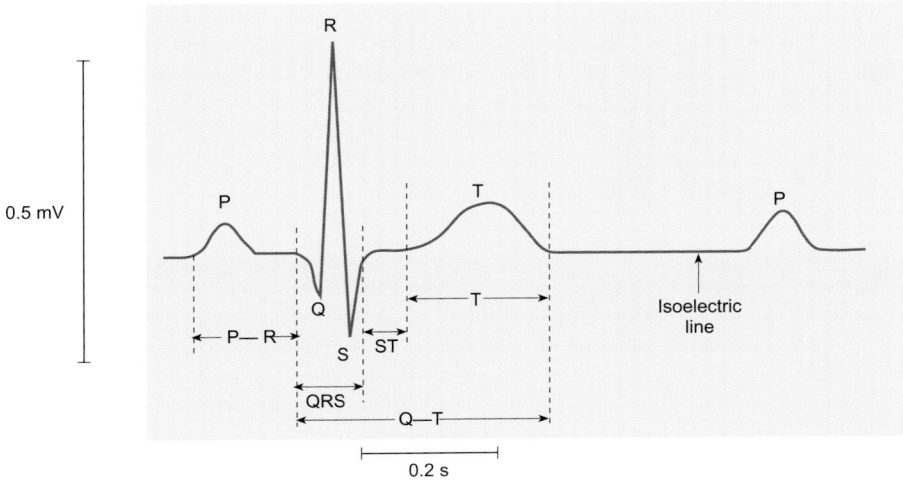

Fig. 15.11 A typical ECG trace recorded with a limb lead labeled to show the principal waves and the intervals that are normally measured.

Box 15.1). The atria repolarize during the QRS complex but, as the muscle mass of the ventricles is so much larger than that of the atria, this event is not seen as a separate wave in the ECG.

During the interval between the S and the T waves the entire ventricular myocardium is depolarized and the ventricles contract. Because all the myocardial cells are at about the same potential, the S–T segment lies on the isoelectric line. This corresponds to the long plateau phase of the cardiac action potential.

The final major event of the ECG trace is the T wave, which normally appears as a broad upward deflection and represents the repolarization of the ventricular myocardium which precedes ventricular relaxation. The action potential of the ventri-

cular muscle usually lasts for 0.2–0.3 s so that the interval between the start of the QRS complex and the end of the T wave is around 0.3 s. The T wave is relatively broad because some ventricular fibers begin to repolarize earlier than others, and thus the whole process of repolarization is rather prolonged. Occasionally the T wave is followed by a low-amplitude wave known as the U wave. Its origin remains unclear.

Why does the T wave have the same polarity as the R wave? As Box 15.1 explains, the polarity of an ECG wave reflects the predominant direction of current flow sensed by a particular ECG electrode. During ventricular depolarization, the cardiac cells closest to the Purkinje cells depolarize first and the wave of depolarization passes from the inside of the ventricles to the

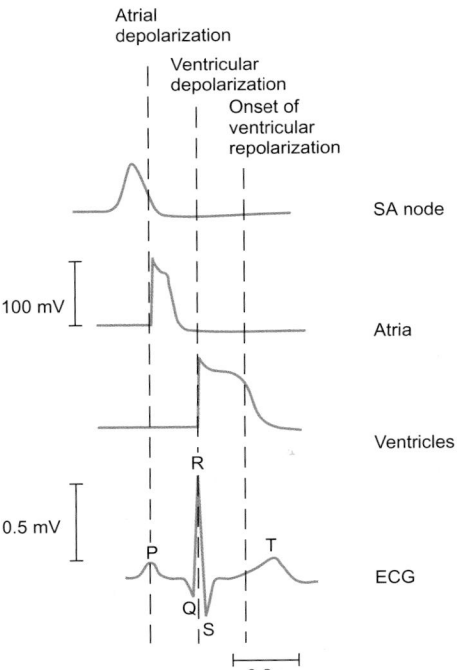

Fig. 15.12 The relationship between the onset and duration of the action potentials of cardiac cells during a single cardiac cycle and the ECG trace.

Summary

1. The electrocardiogram (ECG) is a recording of the electrical activity of the heart, giving insight into both normal and abnormal cardiac function. The ECG records small potential differences (around 1 mV) arising from the sequential electrical depolarization and repolarization of the heart muscle.

2. The ECG is recorded by placing electrodes at different points on the body surface and measuring voltage differences between them. Standard limb leads I, II, and III, augmented leads, or chest leads are used for this purpose. The precise appearance of the ECG traces depends on the position of the individual lead with respect to the electrical activity of the heart.

3. The P wave of the ECG is due to atrial depolarization, the QRS complex is due to ventricular depolarization, and the T wave is due to ventricular repolarization. Atrial repolarization is hidden within the QRS complex. The PR interval reflects the delay in transmission through the AV node.

4. The ECG gives information about a number of cardiac abnormalities including arrhythmias such as ectopic beats, heart block, myocardial ischemia, and anatomical abnormalities.

outside. The cells on the outer surface of the ventricles have shorter action potentials than those on the inside and begin to repolarize first. Thus the wave of repolarization passes from the outside to the inside of the ventricles. Consequently, the T wave has the same polarity as the R wave.

The electrical signals that give rise to the ECG are greatly attenuated as they are conducted through the body to the skin surface and are very small compared with the amplitude of the cardiac action potential (c.120 mV). The largest QRS complexes are only of the order of 3 or 4 mV even when one of the recording leads is placed directly over the heart. When the electrodes are placed on the body surface the voltages are smaller still, and the amplitude of a typical R wave recorded with a standard limb lead is 0.5–1 mV.

BOX 15.1 HOW THE ELECTRICAL ACTIVITY OF THE HEART GIVES RISE TO THE ECG

The ECG arises because, during the cardiac cycle, the myocytes of one part of the heart may be at their normal resting potential while others are in a depolarized state. Since the heart muscle is a syncytium, the myocytes are electrically connected and current will flow between these regions. The extracellular currents that arise because of this state of affairs can be detected by appropriately placed electrodes. The situation is analogous to a battery placed in a conducting medium where current will flow from one pole to the other (see Fig. 1). Current flowing towards an electrode will give rise to a positive potential, while that flowing away from an electrode will give rise to a negative potential. Current flowing at right angles to an electrode will not be detected.

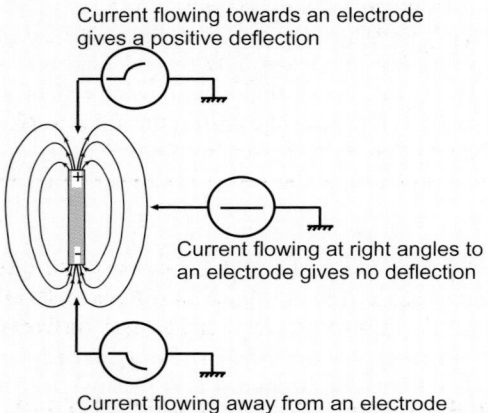

Current flowing towards an electrode gives a positive deflection

Current flowing at right angles to an electrode gives no deflection

Current flowing away from an electrode gives a negative deflection

Fig. 1 A simple diagram to show the pattern of current flow between the positive and negative poles of an electrical dipole in a conducting medium. The voltage deflections recorded when current flows towards an electrode are positive and those recorded when current flows away from an electrode are negative. The changing voltage deflections of the ECG reflect progressive changes in the direction of current flow during the cardiac cycle.

During the cardiac cycle the situation is more complex, as the pattern of current flow resulting from depolarization of the heart muscle continually changes. Each electrode placement of the ECG detects the average current flowing towards or away from it at any instant of time. Thus a particular wave may be positive in one electrode and negative in another.

The component waves of the ECG were arbitrarily named P, Q, R, S, and T in the early days of electrocardiography. The P and T waves are distinct entities, but the naming of the QRS complex follows the following rules: If the first deflection of the QRS complex is downward, it is called a Q wave. An upward deflection is always called an R wave irrespective of whether it is preceded by a Q wave. A negative deflection immediately following the R wave is called an S wave.

It is an unfortunate fact that there are many minor variations in the appearance of the ECG recorded by the standard leads from normal individuals. In what follows, only the changes in potential that can be detected by the chest leads (V1–V6) will be considered, but the same principles apply to the ECG recorded with the limb leads and the augmented limb leads.

The heart lies mainly in the lower part of the left side of the chest and is normally oriented so that the ventricles point diagonally downwards away from the atria which lie close to the midline.

- As the atria depolarize, extracellular current flows between the depolarized regions of the atria and the rest of the heart. This current flow can be detected by an electrode (or lead) placed on the chest wall. The chest leads show a small positive deflection (the P wave) which reaches a peak and then declines as the atrial muscle becomes fully depolarized (Fig. 2(a)).
- After a brief delay at the AV node, the septum depolarizes and current flows from left to right. This is at right angles to V4 and is not sensed by this lead. However, the current flows away from V5 and V6 so that the depolarization of the septum is detected as a small negative deflection (the Q wave) in these leads (Fig. 2(b)).
- Depolarization of the ventricular wall follows. It starts from the endocardium and moves outwards. The current flows towards V3–V6 and a large positive wave (the R wave) is recorded in these leads (Fig. 2(c)).
- As the depolarization passes up the ventricular wall towards the atria, the current flows away from V2 and V3 and a prominent negative wave (the S wave) is recorded in these leads (Fig. 2(d)). The S wave is less prominent in V5 and may be absent in V6.
- By the end of the S wave, the ventricles are fully depolarized and there is no current flow (the S–T segment) (Fig. 2(e)).
- The duration of the action potential of the myocytes near the endocardium is longer than that of the outer myocytes. As a result, the outermost cells repolarize first and the extracellular current flow is towards V3–V6, which thus records a positive deflection (the T wave) (Fig. 2(f)). Note that, depending on the orientation of the heart, the current may flow away from V1, so that the T wave may be inverted in this lead.

When the cardiac cycle is complete, no myocytes are depolarized and the ECG returns to baseline (the isoelectric line).

BOX 15.1 (CONTINUED)

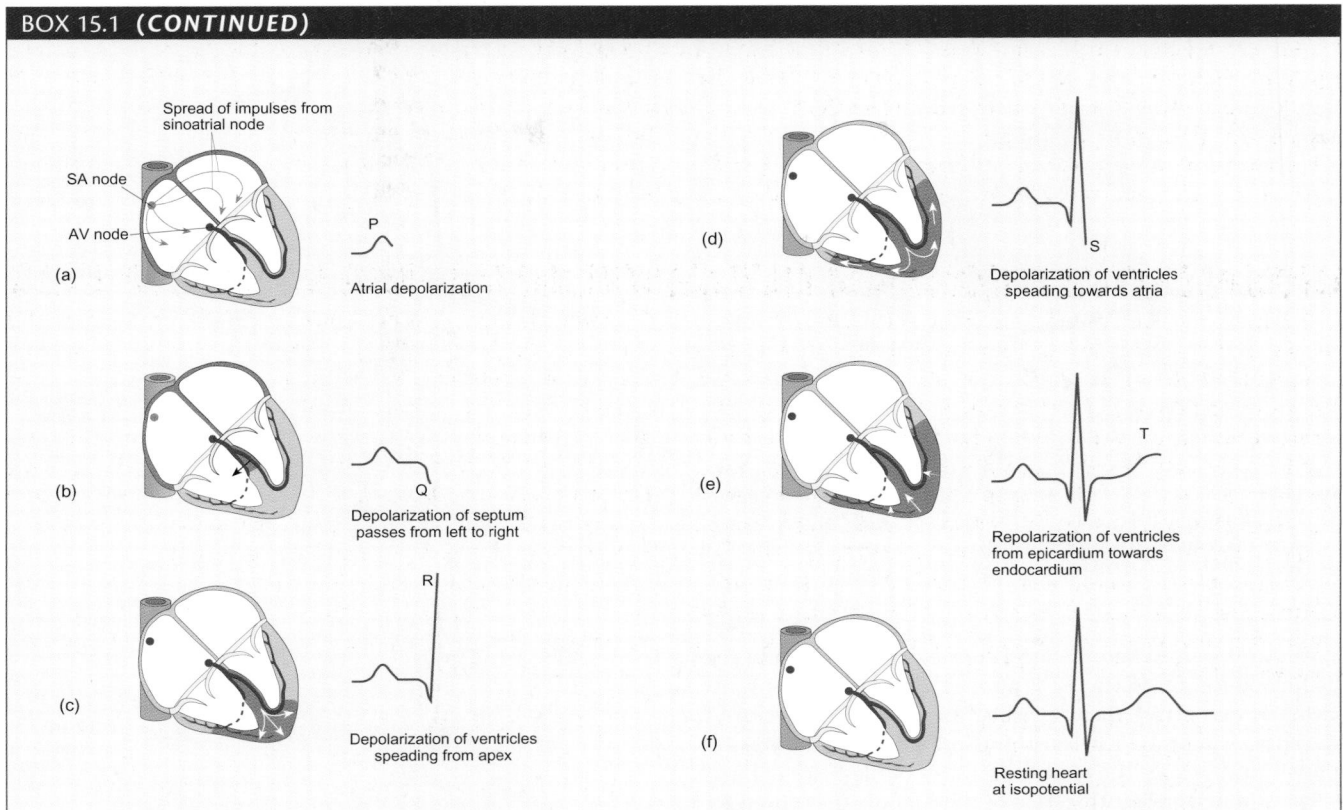

Fig. 2 The relationship between the activity of the cardiac myocytes and the component waves of the ECG. The arrows indicate the direction of current flow at different points of the cardiac cycle. Note that the ECG shown is generic in form and does not represent the trace recorded by a particular lead.

15.5 The heart as a pump—the cardiac cycle and the heart sounds

The electrical events described in the preceding section govern the mechanical activity of the heart. Excitation causes the myocardial cells to contract, while repolarization of the membrane brings about their relaxation. The alternating contraction and relaxation of the myocardium allows the heart to act as a pump that propels blood from the venous to the arterial systems. The *cardiac cycle* refers to this repeating pattern of contraction and relaxation of the heart. It consists of two major phases—*diastole* during which the chambers relax and fill with blood, and *systole* during which the heart contracts, ejecting blood into the pulmonary and systemic circulations. The heart has a two-step pumping action. The right and left atria contract virtually simultaneously (*atrial systole*), followed 0.1–0.2 s later by contraction of the right and left ventricles (*ventricular systole*).

During diastole, the atria and ventricles are relaxed. The right atrium and ventricle fill with the blood of the venous return and the left atrium and ventricle fill with oxygenated blood from the lungs. The volume filling a ventricle just before it contracts is known as the *end-diastolic volume*. When a ventricle contracts, it eject about two-thirds of the blood that it contains. This is called the *stroke volume*. The volume of blood remaining in a ventricle after systole is the *residual* or *end-systolic volume*. The proportion of blood ejected during systole is called the *ejection fraction*. When an adult human is at rest, the stroke volume is about 75 ml and the residual volume is about 50 ml, giving an ejection fraction of about 0.6.

The action of the heart valves is driven by pressure differences

The ability of the ventricles to fill under low pressure and to eject blood against high arterial pressures is critically dependent on the precise operation of the two sets of valves which cover the inlets and outlets of both ventricles—the atrioventricular (AV) and semilunar valves (Fig. 15.13). The AV valves (the bicuspid or mitral valve on the left and the tricuspid valve on the right) are composed of membranous leaflets or cusps that are flexible flaps of fibrous connective tissue which protrude into the ventricle. Their free edges are attached to a set of tendons (the *chordae tendinae*) that connect them to a set of conical muscles on the ventricular wall known as the papillary muscles. This arrangement prevents the valves from being pushed into the atria during systole.

Opening and closing of the AV valves occurs as a result of the pressure differences between the atria and ventricles that occur during the cardiac cycle. When the ventricles are relaxed, the pressure in the atria exceeds that in the ventricles (because blood is entering the atria from the venous circulation) and the

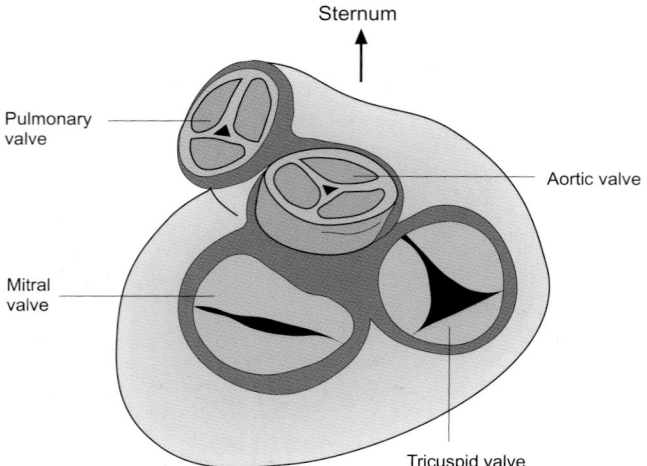

Fig. 15.13 The valves of the heart viewed from above after removal of the atria, aorta, and pulmonary artery. Note that the fibrous tissue that separates the atria from the ventricles (shown here in red) is continuous with that which supports the AV valves and the aortic and pulmonary valves.

valves are open. When the atria contract, atrial pressure rises and blood is forced into the ventricles. However, as the ventricles start to contract, the ventricular pressure rises above that in the atria. As soon as ventricular pressure exceeds atrial pressure, the AV valves close. The total surface area of the cusps of the AV valves is much greater than that of the opening they cover. As a result the upper (atrial) surfaces are firmly pressed together. This arrangement ensures reliable closure of the valves while the ventricles change size in systole.

The aortic and pulmonary (semilunar) valves are located at the origin of the aorta and the pulmonary artery where they form three crescent-shaped pockets. When the valves are closed, the three cusps are pressed against each other as shown in Fig. 15.13. During systole, the pressure in the ventricles rises until it exceeds that in the arteries and this difference in pressure causes the valves to open. At the commencement of diastole, the pressure in the ventricles falls below that in the aorta and pulmonary artery and the valves close. This prevents reflux of blood from the arterial system into the ventricles during diastole.

Note that during diastole, the AV valves are open and the aortic and pulmonary valves are closed. In the ejection phase of systole the situation is reversed: the AV valves close and the aortic and pulmonary valves open. Thus the valves ensure the one-way flow of blood essential to the efficient operation of the circulatory system.

The pressure changes of the cardiac cycle

Figure 15.14 illustrates the major events occurring during a single cardiac cycle (the period from the end of one contraction of the heart to the end of the next). In a resting individual, the heart rate is around 70 b.p.m. and each cardiac cycle lasts for about 0.8 s, where systole lasts for 0.3 s and diastole 0.5 s. In addition to the pressure changes in the aorta, left ventricle, and left atrium, the figure shows the changes in ventricular volume, the ECG, and the *phonocardiogram* (a recording of the heart sounds). The first two heart sounds can be heard easily using a stethoscope.

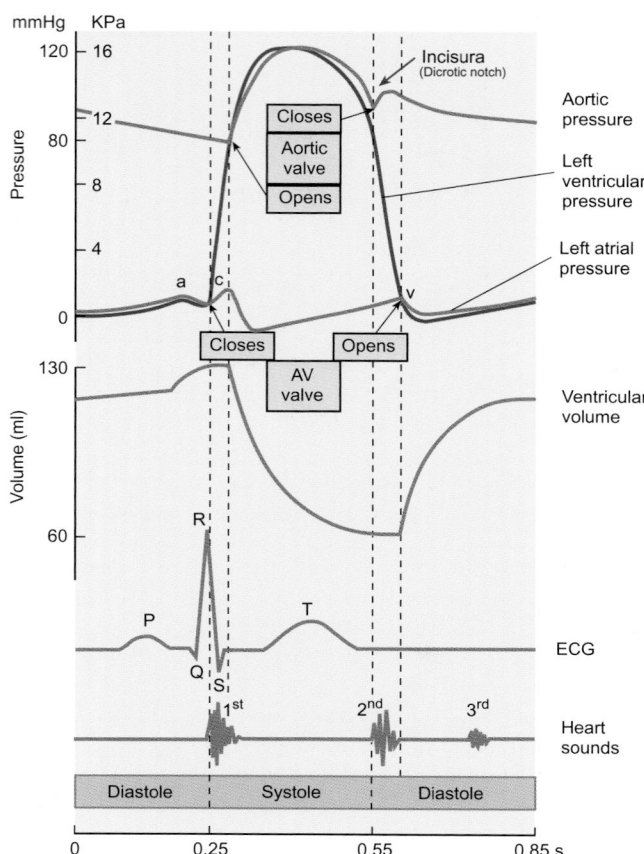

Fig. 15.14 The major mechanical and electrical events of the cardiac cycle. The pressure changes shown are for the left side of the heart and reflect the underlying mechanical events. The heart sounds are also shown. It is particularly important to note the relative timing of the various events. For example, the QRS complex (which reflects ventricular depolarization) largely precedes ventricular contraction, while the first heart sound is heard as the AV valves close following the start of the rise in intraventricular pressure. See text for further explanation.

Blood normally flows continually from the great veins (the superior and inferior venae cavae) into the right atrium, and most of it flows directly into the right ventricle. Only around 20 per cent is pumped actively into the right ventricle during contraction of the right atrium. A similar pattern is seen in the left side of the heart.

Figure 15.14 shows the pressure changes during the cardiac cycle for the left side of the heart. Three major increases in pressure can be seen in the left atrium: the a, c, and v waves. The a wave occurs during atrial contraction which itself is preceded by atrial depolarization (the P wave of the ECG). When the AV valves close just after the start of ventricular contraction (i.e. when ventricular pressure exceeds atrial pressure), there is a second 'bump' in the atrial pressure curve, the c wave. This is due mainly to the bulging of the mitral valve into the left atrium. While the ventricles are contracting, blood continues to enter the left atrium from the pulmonary veins causing atrial pressure to rise slowly over this period. This rise is reflected in the v wave of the atrial pressure curve. Pressure falls once more

as soon as the AV valves open at the end of ventricular contraction. A similar pattern of pressure changes is seen in the
right side of the heart and can be recorded as the jugular pulse
(Fig. 15.15).

Ventricular contraction starts at the peak of the R wave of the
ECG (remember that the QRS complex represents ventricular
depolarization). The AV valves close at the start of contraction to
produce the first heart sound. For a short time (0.02–0.03 s) the
semilunar valves at the entrance to the pulmonary artery and
the aorta are also shut, so that this is a period of *isovolumetric
contraction*—the intraventricular pressure rises but the volume
does not change. This can be seen in the ventricular volume
curve of Fig. 15.14.

The *ejection phase* of ventricular systole begins when the semilunar valves open (the point at which left ventricular pressure
exceeds aortic pressure and right ventricular pressure rises
above that in the pulmonary artery). There is a period of rapid
ejection followed by a period of rather slower emptying (see ventricular volume curve in Fig. 15.14). During the rapid ejection
phase, both ventricular and aortic pressures rise steeply. For
about the last quarter of ventricular systole very little blood
flows from the ventricles to the aorta even though the ventricular myocardium remains contracted.

At the end of ventricular systole, the ventricles repolarize (the
T wave of the ECG) and begin to relax so that intraventricular
pressures fall rapidly. The higher pressures in the pulmonary
artery and aorta cause the semilunar pulmonary and aortic
valves to close, preventing back-flow of blood into the ventricles.
There then follows a brief period of *isovolumetric relaxation* during
which the ventricular myocardium continues to relax (with a
concomitant fall in ventricular pressure) but the ventricular volume remains constant because the AV valves are not yet open.
This phase lasts for 0.03–0.06 s before the ventricular pressures
fall below those of the atria and the mitral and tricuspid valves
open. Ventricular relaxation continues but now blood enters the
ventricles from the atria and ventricular volume increases
quickly. About two-thirds of the way through ventricular diastole the atria depolarize (the P wave of the ECG) and subsequently contract to thrust additional blood into the ventricles.
This can be seen clearly in Fig. 15.14, which shows a second
component to the rising phase of the ventricular volume curve
corresponding to atrial systole.

The aortic pressure curve

Figure 15.14 also shows that when the left ventricle contracts,
the pressure in the left ventricle rises quickly. When this pressure exceeds that in the aorta, the aortic valve opens and blood
begins to flow rapidly into the aorta. As it does so, it causes the
walls of the aorta to stretch, thereby increasing the pressure
within it to around 16 kPa (120 mmHg) (systolic pressure).
When the left ventricle starts to relax and blood is no longer
flowing into the aorta, so that the pressure in this vessel falls
a little. When the aortic valve closes there is a brief surge in
aortic pressure. This gives rise to the *incisura* or dichrotic notch.
After this, aortic pressure falls slowly during diastole as the
stretched elastic tissue of the artery forces the blood into the
systemic circulation. By the time the ventricles contract again,
the aortic pressure has fallen to around 10.6 kPa (80 mmHg)

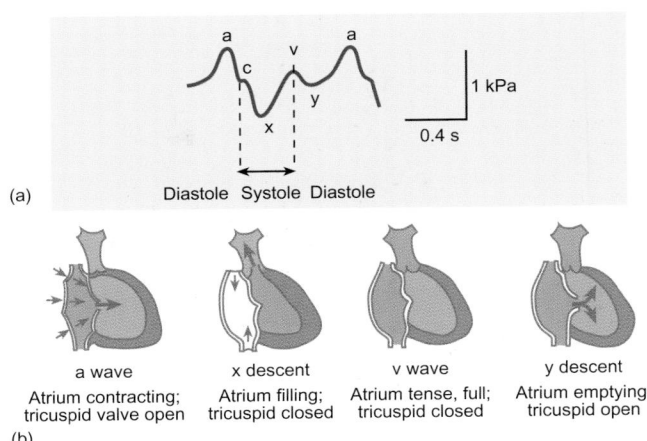

(a)

(b)

Fig. 15.15 The origin of pressure waves recorded from the internal jugular vein:
(a) a normal jugular pulse; (b) the events that give rise to the component waves
of the jugular pulse.

(diastolic pressure).

The pressure curve for the pulmonary artery has similar characteristics to that of the aorta except that the pressures are lower.
The systolic pressure in the pulmonary artery is around 3.3 kPa
(25 mmHg) and diastolic pressure is around 1 kPa (8 mmHg).

Jugular pulse

As there are no valves between the right atrium and the internal
jugular vein, the pressure changes of the right atrium are
reflected in the jugular pulse (Fig. 15.15). Four components can
be detected in the waveform: the a, c, x, and y waves. The origin
of these waves is explained in Figure 15.15(b). As the pressures
are very low (<0.4 kPa or 4 cmH$_2$O), normally no pulse can be
felt. However, during heart failure (see Chapter 31) and certain
other conditions in which the venous pressure is elevated, pulsation of the jugular vein can be clearly observed if the patient's
upper body is supported at 45° to the horizontal with the head
slightly inclined to the left. The shape of the pulse wave is
changed in certain disease states. For example, if the tricuspid
valve does not close properly, regurgitation of blood into the
right atrium reduces the amplitude of the x wave, and the c and
v waves merge to create one large wave.

Stroke work and the pressure–volume loop

The pumping action of the heart is achieved by the mechanical
work of the myocardium. In Chapter 7 we saw that to perform
mechanical work a load must be moved through a distance. For
a three-dimensional system such as the heart, the work done is
equal to the change in pressure multiplied by the change in volume, so that the work performed by the heart each time it beats
is given by the area of the pressure–volume curve for ventricular
contraction. This is known as the *stroke work*. Ventricular pressure varies during the ejection phase of the cardiac cycle.
Therefore, in order to determine stroke work accurately, it is
necessary to construct a graph of ventricular pressure against
volume. The total external work carried out by the left ventricle
during a single cardiac cycle is given by the area of the pressure–volume loop ABCDA in Fig. 15.16.

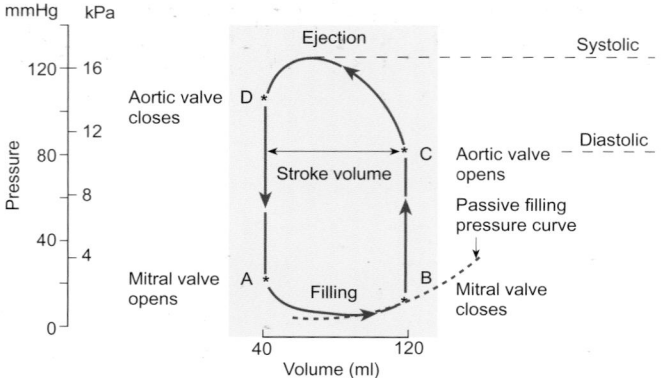

Fig. 15.16 The pressure–volume cycle of the human left ventricle. The area bounded by the pressure–volume curve gives the stroke work of the heart. The lower confine of the pressure–volume loop shown by the dotted line is defined by the passive stretch of the relaxed ventricular muscle by the returning blood. See text for further details.

Between points A and B the ventricle is filling. Pressure falls at first due to the suction effect of the relaxing muscle. This is analogous to the way a rubber bulb fills with liquid after it has been emptied of air. Subsequently, the pressure starts to rise passively as the volume of blood in the ventricle increases and the tension in the ventricular wall rises. Between points B and C the ventricle contracts but, since the aortic valve is closed, there is no change in volume and the pressure rises steeply. This is the phase of isovolumetric contraction. During this phase of the cardiac cycle the heart is performing no external work (there is no change in volume). The aortic valve opens at point C, and between points C and D, blood is being ejected, the volume of the ventricle decreases, and the heart performs external work. The valve closes again at point D and the phase between points D and A represents isovolumetric relaxation.

The shape of the pressure–volume curve is similar for the right and left ventricles, though the left has slightly higher filling pressures because it has thicker and less distensible walls. As it has to eject blood against a higher arterial pressure, the left ventricle performs significantly more external work than the right even though both ventricles pump the same volume of blood. If either the stroke volume or the systolic pressure is increased, the area of the pressure–volume loop will increase, indicating an increase in stroke work. This is true for both ventricles.

The heart sounds

The bottom part of Fig. 15.14 shows the timing of the heart sounds in relation to the electrical and mechanical events of the cardiac cycle. These sounds are caused by the closure of the heart valves, which sets up oscillations in the blood both within the chambers of the heart and, the wall of the heart itself.

During routine clinical examination with a stethoscope only two heart sounds are generally audible in normal adults. The first heart sound ('1st' in Fig. 15.14) begins to be heard at the onset of ventricular systole and is associated with the closure of the AV valves. This sound begins immediately after the R wave of the ECG. The second heart sound ('2nd' in Fig. 15.14) occurs on closure of the aortic and pulmonary valves at the start of ventricular relaxation. It is quieter and has a more snapping quality

Summary

1. The alternating contraction and relaxation of the heart muscle is known as the cardiac cycle. The chambers of the heart relax and fill with blood during diastole, and contract to eject blood during systole.

2. During diastole, both the atria and ventricles fill with blood. Towards the end of diastole, the atria contract, increasing the volume of blood contained in the ventricles by about 20 per cent. The total amount of blood filling the ventricles at the end of diastole is called the end-diastolic volume (EDV).

3. Nearly two-thirds of the blood in the ventricles is expelled during systole. The volume ejected by each ventricle is known as the stroke volume and is roughly 70 ml in an adult at rest.

4. The work performed by the heart during each beat (the stroke work) is the product of the rise in ventricular pressure occurring during systole and the stroke volume.

5. During the cardiac cycle various sounds can be heard when a stethoscope is applied to the chest. The first heart sound arises as the AV valves close and the second heart sound corresponds to closure of the pulmonary and aortic valves. Two other heart sounds, the third and fourth heart sounds, can occasionally be heard in normal subjects.

than the first sound (hence the commonly used terms 'lub' and 'dup' for the first and second heart sounds). It coincides with the end of the T wave of the ECG. As the second heart sound is due to the closure of the aortic and pulmonary valves, it consists of two components designated A_2 (aortic) and P_2 (pulmonary). During inspiration, these two components may be heard at slightly different times (splitting of the second heart sound). This is quite normal and is caused by the changes in venous return during the respiratory cycle.

Two further heart sounds may be detected in some normal individuals, particularly with the aid of electronic amplification to provide a graphical record known as a phonocardiogram. The third heart sound ('3rd' in Fig. 15.14) occurs as a result of opening of the AV valves at the end of the period of isovolumetric relaxation. It is thought to be due to rapid turbulent entry of blood into the ventricles at the onset of filling and occurs 140–160 ms after the second heart sound. It is heard most often in children in whom the chest wall is thin. A fourth heart sound is occasionally audible in normal people. It is a very-low-frequency sound thought to be caused by oscillations in the blood flow following atrial contraction and occurs just before the first heart sound. Abnormal heart sounds are usually (but not invariably) of pathological origin and are called *murmurs* (see p. 591).

15.6 The measurement of cardiac output

The volume of blood pumped from one ventricle each minute is known as the *cardiac output*. It is the product of the heart rate (beats per minute) and the stroke volume. Thus

cardiac output = heart rate × stroke volume.

In an adult human at rest, the cardiac output is between 4 and 7 l min^{-1}, but throughout normal life the output varies continually according to the oxygen requirements of the body tissues. The output is reduced in sleep, for example, and raised following a heavy meal, in fear, or during periods of excitement. A much larger rise in cardiac output (by as much as a factor of 6) is seen during periods of strenuous exercise.

The *venous return* is the volume of blood returning to the heart from the vasculature every minute, and it is inextricably linked with cardiac output. For the closed circulatory system to function efficiently it is essential that, except for very small transient alterations, the heart must be able to pump a volume equivalent to that which it receives, i.e. cardiac output must be equal to venous return and vice versa over any significant period of time.

Since cardiac output is determined by heart rate and stroke volume, changes in either (or more often both) of these variables will bring about alterations in cardiac output. Regulation of cardiac output is achieved by both autonomic nerves and mechanisms that are intrinsic to the cardiovascular system. These will be discussed in Section 15.7.

Measurement of the cardiac output

A variety of methods have been devised to measure the cardiac output. They either measure the cardiac output as a whole (Fick principle and dilution methods), or they measure stroke volume and heart rate separately (pulsed ultrasound and radionuclide methods).

The *Fick principle* states that the total uptake or release of a substance by an organ is equal to the blood flow through that organ multiplied by the difference between the arterial and venous concentrations of the substance. To measure cardiac output, this principle is applied to the uptake of oxygen by blood flowing through the lungs (remember that the outputs of the right and left sides of the heart are the same and that the entire cardiac output flows through the lungs). The cardiac output (CO) is then given by the following equation:

$$CO = \frac{\text{oxygen uptake (ml min}^{-1})}{Cvo_2 - Cao_2}$$

where Cvo_2 is the oxygen content of the blood in the pulmonary veins, which can be measured by taking a sample of arterial blood, and Cao_2 is the oxygen content of the blood in the pulmonary arteries, which is the same as that of the mixed venous blood of the right ventricle. An example of this way of calculating the cardiac output is given in Box 15.2.

It is worth noting that the Fick principle is quite general and applies to any organ through which blood flows and in which a substance is exchanged at a steady rate. It is widely used to calculate the rate at which an organ consumes substances such as fats or glucose. This is done by measuring both the rate of blood flow through the organ in question and the arteriovenous difference in concentration of the substance under investigation.

Dilution methods for estimating cardiac output

The dye dilution method is the most accurate method of measuring cardiac output in humans. In general terms, a known quantity (I) of an indicator is rapidly injected intravenously. The

BOX 15.2 APPLICATION OF THE FICK PRINCIPLE TO THE DETERMINATION OF CARDIAC OUTPUT

The Fick equation for the uptake of oxygen by the blood as it passes through the lungs is as follows

$$CO = \frac{\text{oxygen uptake (ml min}^{-1})}{Cvo_2 - Cao_2}$$

where Cvo_2 is the oxygen content of the blood in the pulmonary veins (which is the same as that of normal arterial blood) and Cao_2 is the oxygen content of the blood in the pulmonary arteries (which is the same as that of fully mixed venous blood).

In practice, oxygen consumption is determined using spirometry (Chapter 24) and the oxygen content of pulmonary venous blood can be obtained from a sample of blood obtained from the radial, brachial, or femoral artery. The oxygen content of pulmonary arterial blood is more difficult to determine since it requires a sample of fully mixed venous blood which can only be obtained from the right ventricle or pulmonary artery itself. Normally a catheter is introduced through the antecubital vein and passed into the outflow of the right ventricle or the pulmonary artery. The blood samples are then analyzed and the cardiac output calculated using the equation given above.

Worked example

Assume that a person consumes 250 ml of oxygen each minute from the inspired air, that the oxygen content of the pulmonary arterial blood is 15 ml per 100 ml of blood and that the oxygen content of the pulmonary venous blood is 20 ml per 100 ml blood. Therefore each 100 ml of blood passing through the lungs must have taken up 5 ml of oxygen. To supply an oxygen demand of 250 ml min^{-1}, 50×100 ml of blood must traverse the lungs each minute, so the cardiac output is 5000 ml min^{-1}.

concentration of dye in the arterial blood is then continuously monitored until it effectively disappears from the circulation as it becomes diluted in the circulating blood volume. This follows an exponential time course, from which the average concentration C of the dye can be determined. If the time between the first appearance of the indicator in the arterial blood and its disappearance (passage time t) is known, then the cardiac output (CO) is given by

$$CO = \frac{I \times 60}{C \times t}.$$

If I is measured in milligrams, the mean concentration C in milligrams per liter, and the passage time t in seconds, the cardiac output will be given as liters per minute. Further details of the method are given in Box 15.3.

Thermodilution is a widely used variant of the dilution method. In this case, a known volume of isotonic saline at room temperature is rapidly injected into the right heart and the temperature of the blood is continuously measured in the distal pul-

BOX 15.3 MEASUREMENT OF CARDIAC OUTPUT BY THE DYE DILUTION METHOD

One way of measuring the volume of an irregularly shaped vessel is to fill it with liquid and add a known quantity of dye. The concentration of the dye can then be used to calculate the volume in which it has been diluted from the simple formula

$$\text{volume} = \frac{\text{mass of indicator}}{\text{concentration}}$$

This principle can be extended to measure the flow of liquid through a tube, as follows. A quantity of dye is injected into the stream of liquid. At a remote point samples are taken at regular intervals and the concentration of dye is measured. At first there is no dye but, after a delay, the dye concentration rises to a maximum and then decays exponentially with time as shown in the following Fig. 1.

The flow can be calculated from the time taken for the dye to be cleared by the flowing liquid using the following formula (remember that flow = volume per unit time):

$$\text{Flow} = \frac{I}{C \times t}$$

where I is the mass of dye injected, C is its average concentration as it passes the point of measurement, and t is the time taken for the dye to pass the point of measurement.

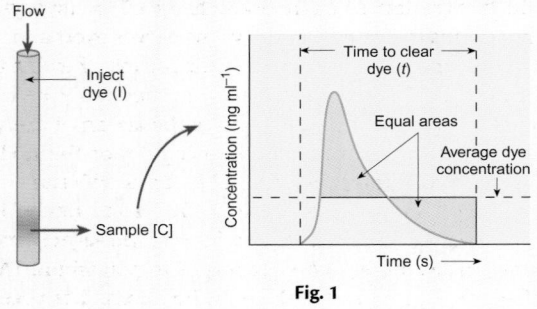

Fig. 1

This principle can be used to measure cardiac output. A known amount of dye such as indocyanine green is injected rapidly into a vein. Its appearance in the arterial blood is then measured either by continuously sampling the arterial blood or, more usually, by a microcolorimeter clipped to the ear lobe. A plot of dye concentration against time has a characteristic appearance, as shown in the top panel of Fig. 2. Once the injected dye reaches the point of sampling, its concentration rises rapidly to a peak. The concentration then declines exponentially but rises again before reaching zero. This secondary rise is due to the recirculation of dye that took the shortest path through the circulation.

The time taken to clear the dye requires knowledge of the time it would have taken for the concentration to fall to zero had recirculation of the dye not taken place. This requires extrapolation of the falling phase of the dye concentration curve to zero, a process that is greatly aided by plotting the dye concentration on a logarithmic scale, as shown in the bottom panel of Fig. 2. This type of plot exploits the fact that an exponential curve plotted on logarithmic coordinates is a straight line that can readily be extrapolated to the time axis, as shown, to give the clearance time, t.

The average concentration of dye is given by the area under the curve as shown in the figure (the area under the curve has units of milligrams per liter). From this information the cardiac output (CO) can be calculated from the formula

$$CO = \frac{I \times 60}{C \times t}$$

where the factor 60 converts the cardiac output to liters per minute.

The ideal indicator for measuring cardiac output should possess a variety of characteristics: it should be non-toxic, it should mix completely and virtually instantaneously with the blood, it should remain within the circulation for the period of the determination, and it should not itself alter cardiac output. Several types of indicators meet these requirements, including dyes such as indocyanine green.

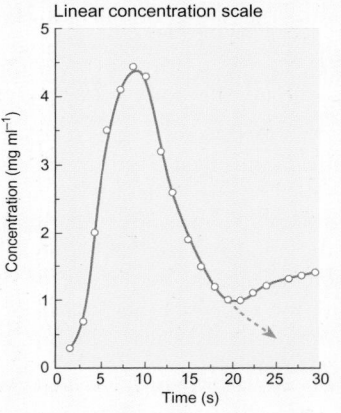

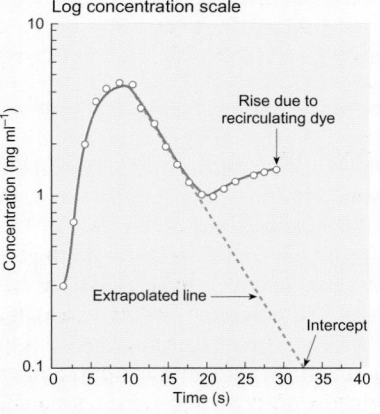

Fig. 2

Summary

1. The cardiac output is the volume of blood pumped each minute by either ventricle. It is the product of heart rate and stroke volume. At rest, it is around $5 \, l \, min^{-1}$ for an adult man, but it varies with the size and sex of the individual. It also varies according to the metabolic demands of the body.

2. Changes in either heart rate or stroke volume will alter cardiac output. Within very narrow limits, the amount of blood pumped by the ventricles is equal to the amount received (the venous return).

3. Cardiac output is often measured clinically and for the purposes of research. A variety of methods are used including indicator dilution techniques, ultrasonography, and methods based on the Fick principle.

monary artery via a suitable temperature probe. A curve of temperature against time is plotted from which the cardiac output can be calculated. As the cold saline is innocuous and rapidly warms to body temperature, repeated measurements can be made.

Other methods used for measuring cardiac output

While the Fick principle and the dilution methods described above measure cardiac output, there are a variety of other methods available which estimate cardiac output by measuring heart rate and stroke volume separately. These include *echocardiography*, which compares the end-diastolic and end-systolic diameters of the ventricle, and *pulsed Doppler ultrasonography*, which measures the velocity of the blood flowing in the aorta. While these methods are non-invasive and are sufficiently fast to permit the cardiac output to be measured beat by beat, they are not as accurate as the dye-dilution method. Finally, it is possible to implant an electromagnetic flow probe around the aorta of an experimental animal. This is commonly done under anesthesia and the animal is allowed to recover. The flow probe can then be used to provide an accurate measure of aortic blood flow. Since the heart rate can be measured simultaneously, the cardiac output can be readily calculated. This is the preferred method for experimental studies.

15.7 Cardiodynamics

At first sight, it seems logical to suppose that an increase in heart rate will necessarily bring about an increase in cardiac output. Up to a point, this is true, but as the heart rate increases, the time for filling of the ventricles falls. As a result, the stroke volume does not increase in proportion to the increase in heart rate. In fact, it tends to level off when cardiac output approaches 50 per cent of its maximum value (see Chapter 25). The following section is concerned with cardiodynamics, the regulation of the activity of the heart by intrinsic and extrinsic factors.

Nervous and hormonal control of heart rate

Although cardiac muscle has an inherent rhythmicity, it is supplied with parasympathetic and sympathetic autonomic nerves, both of which are able to influence the heart rate. Physiological changes in heart rate are known as *chronotropic effects*. The parasympathetic supply to the heart is via the vagus nerves which, when activated, slow the heart (*negative chronotropy*), while stimulation of the sympathetic nerves increases heart rate (*positive chronotropy*). Clinically, a heart rate at rest that is below 60 b.p.m. is called a bradycardia, while a high heart rate (above 100 b.p.m.) is called a tachycardia. These terms are also used more generally to describe any significant slowing down (bradycardia) or speeding up (tachycardia) of the heart rate.

The resting heart is dominated by the parasympathetic innervation

Vagal nerve fibers synapse with postganglionic parasympathetic neurons in the heart itself (see Chapter 10) and the short postganglionic fibers synapse mainly on the cells of the SA and AV nodes. They release acetylcholine from their terminals, which acts to slow the heart rate. Vagal stimulation also reduces the rate of conduction of the cardiac impulse from the atria to the ventricles by decreasing the excitability of the AV bundle fibers.

The heart rate of a normal healthy adult at rest is about 70 b.p.m. A denervated heart beats around 100 times per minute, which is the intrinsic rate of discharge of the myocytes of the SA node. This indicates that the vagus nerves exert a tonic inhibitory action on the SA node to slow the intrinsic heart rate. This effect may also be demonstrated experimentally by the application of atropine, which antagonizes the action of acetylcholine on the muscarinic cholinergic receptors of the nodal cells. Under these conditions the resting heart rate will rise.

How does vagal stimulation decrease the heart rate? The acetylcholine released by the nerve terminals of the vagal nerves increases the permeability of the nodal cells to potassium. This has two effects: it decreases the slope of the pacemaker potential, and it hyperpolarizes the membrane potential. These changes increase the time taken for the pacemaker potential to reach the threshold so that the interval between successive action potentials is longer and the heart rate falls. The effect of vagal stimulation on the heart rate is illustrated in Fig. 15.17(a).

Stimulation of the sympathetic nerves increases the heart rate

The sympathetic preganglionic nerves that supply the heart originate in spinal segments T1–T6 (mainly T1–T3). They synapse in the ganglia of the thoracic sympathetic chain from where they project to the heart via long postganglionic fibers. The varicosities of the postganglionic fibers secrete norepinephrine, and the effect of stimulating these nerves is to increase the rate at which the heart beats. When physiological circumstances require the heart to beat more rapidly, as in exercise, the activity of the parasympathetic nerves is inhibited while that of the sympathetic nerves is enhanced. Maximal sympathetic stimulation can almost triple the resting heart rate.

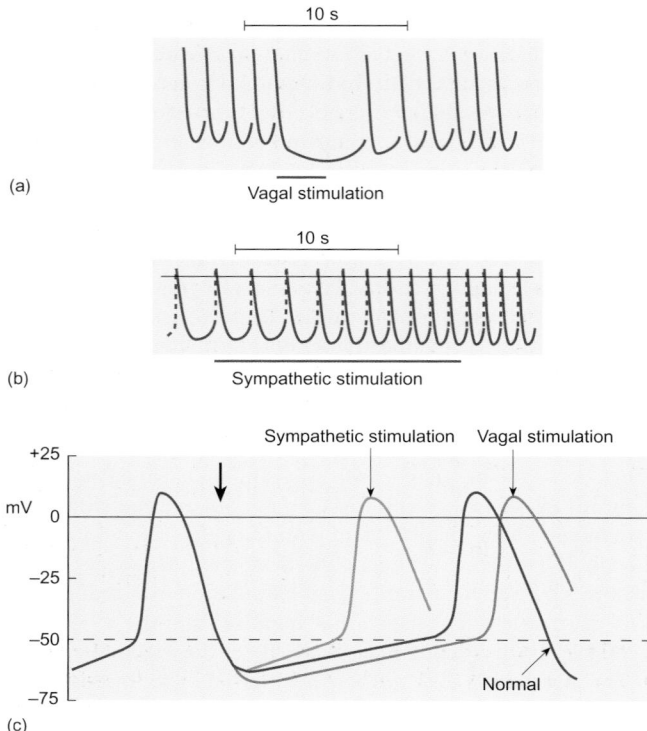

(a)

(b)

(c)

Fig. 15.17 The effect of stimulation of the sympathetic and the parasympathetic (vagal) nerves on the pacemaker activity of the frog heart. (a) The effect of vagal stimulation. Note that the hyperpolarization of the pacemaker cell stops the heart. After the period of stimulation, the slope of the pacemaker potential is reduced and the heart rate is slowed compared with the period before stimulation. (b) The effect of stimulating the sympathetic nerves. Note the increased slope of the pacemaker potential and increased heart rate following stimulation. (c) A diagrammatic representation of the effects of sympathetic and parasympathetic stimulation on the pacemaker potential.

The precise mechanism by which norepinephrine increases the rate of discharge of the cells of the SA node is still not entirely certain, but it is believed that it increases the permeability of the pacemaker cell membrane to sodium and calcium ions. This increases the slope of the pacemaker potential so that the cells of the SA node reach threshold more quickly and the interval between successive action potentials is reduced, as illustrated in Fig. 15.17(b). Conduction time through the AV node is also reduced by sympathetic stimulation.

The regulation of stroke volume

The stroke volume is regulated by two different mechanisms:

(1) *intrinsic regulation* of the force of contraction, which is determined by the degree of stretch of the myocardial fibers at the end of diastole;

(2) *extrinsic regulation*, which is determined by the activity of the autonomic nerves and the circulating levels of various hormones.

Intrinsic regulation of the stroke volume: the Frank–Starling relationship

The fundamental relationship between the initial length of skeletal muscle and the force of contraction discussed in Chapter 7 was shown by Otto Frank to be valid for cardiac muscle. As the blood returns to the heart in diastole, it begins to fill the ventricle. As it does so, the pressure rises and this stretches the myocardial fibers, placing them under a degree of tension known as *preload*. It can be shown experimentally that an increase in filling pressure leads to both an increase in end-diastolic volume and an increase in the subsequent stroke volume (Fig. 15.18). This response can be explained by the ability of cardiac muscle to respond to increased stretch with a more forceful contraction (see Chapter 7). On the basis of this kind of evidence, Starling formulated his law of the heart which states that 'the energy of contraction of the ventricle is a function of the initial length of the muscle fibers comprising its walls'. This is now often referred to as the Frank–Starling relationship. Stated simply, this means that during systole the ventricle will eject the volume of blood that entered it during diastole. Consequently, *the heart automatically adjusts its cardiac output to match its venous return.*

The most important function of the Frank–Starling mechanism is to balance the outputs of the right and left ventricles

The left and right sides of the heart are arranged as two pumps in series. It is crucial that the output of each side of the heart is the same. While transient imbalances (a few beats) may occur, more prolonged imbalances, even very small ones, will quickly prove disastrous. To demonstrate this, consider a situation in which the output of the right ventricle exceeds that of the left ventricle by 1 per cent. If the output of the left ventricle is 5 l min^{-1}, that of the right will be 5.05 l min^{-1}. After about 30 min, pulmonary blood volume will have increased from its normal level of about 600 ml to about 2.1 liters, while the volume in the systemic circulation will have fallen from 3.9 to

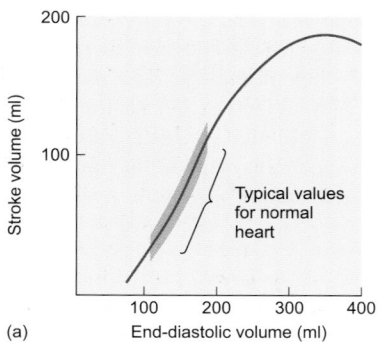

(a)

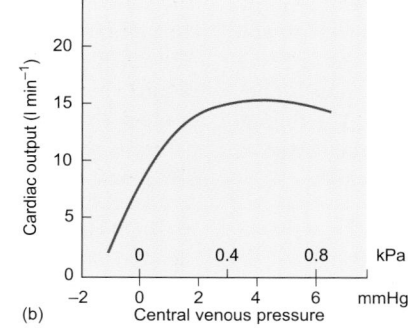

(b)

Fig. 15.18 (a) The relationship between end-diastolic volume and stroke volume determined using an isolated heart–lung preparation. (b) The relationship between cardiac output and central venous pressure in the intact heart.

2.4 liters. It is clear from this that, unless the output of the two ventricles is matched, the distribution of the blood between the pulmonary and systemic circulations will be altered, with potentially serious consequences.

The Frank–Starling mechanism normally ensures that the output from the two ventricles is closely matched. If the output of the right ventricle exceeds that of the left, after a few beats the volume of blood in the pulmonary circulation will be increased slightly and pressure in the pulmonary veins will rise. As a result, the venous return to the left side of the heart will be increased, resulting in a greater end-diastolic volume in the left ventricle. This, in turn, will lead to an increase in its stroke volume via the Frank–Starling mechanism, so restoring the balance in the output of the two ventricles.

The pressure in the aorta opposes the ejection of blood from the ventricles and represents the load against which the heart must pump the blood. For this reason, it is known as the *afterload*. An increase in afterload causes only a transient fall in stroke volume. Why does the stroke volume recover? As the venous return remains unchanged, the left ventricle becomes more distended (i.e. the preload is increased). This increases the force with which its muscle contracts. The result is a restoration of the stroke volume.

What factors determine the end-diastolic volume of the ventricles?

From the preceding account it should be clear that the end-diastolic volume (EDV) and therefore the degree of stretch of the ventricles is an important determinant of the cardiac output and the work performed by the heart. Therefore it is important to understand the factors that affect this volume. Essentially these fall into two categories: factors affecting pressure outside the heart, and factors affecting pressure inside the heart.

The pressure outside the heart is the intrathoracic pressure. This is altered by breathing (see Chapter 16). During inspiration, contraction of the diaphragm increases the volume of the chest and decreases the volume of the abdominal cavity. This results in a fall in the pressure within the thorax and a rise in the pressure within the abdominal cavity. This pressure difference favors the flow of blood from the abdominal to the thoracic veins and enhances the filling of the right ventricle.

TABLE 15.2   The factors controlling the venous return

The pressure at the end of the capillaries
The right atrial pressure
The total blood volume
The venous tone
The inotropic state of the heart
The muscle pump
The respiratory pump
The abdominal pump
The drawing of blood into the atria during ventricular systole due to descent of the AV fibrous ring
The suction of the ventricles as they relax during diastole, drawing blood into the heart

---

The pressure inside the right ventricle at the end of diastole is equal to the pressure in the superior and inferior vena cavae as they enter the heart. This is called the *central venous pressure*. Since ventricular filling is largely determined by the central venous pressure, it follows from the Frank–Starling relationship that the central venous pressure is an important determinant of cardiac output. This is shown in Fig. 15.18(b).

The central venous pressure is determined by a variety of factors, a number of which will be discussed more fully in later sections. These include gravity, respiration (as described above), the muscle pump (compression of the deep veins during exercise displaces blood into the central veins), peripheral venous tone, and the blood volume (Table 15.2). Since about two-thirds of the total blood volume is located within the highly distensible veins (the capacitance vessels—see below), any change in venous tone or blood volume will also bring about alterations in the central venous pressure. For example, hemorrhage will lower the blood volume and the central venous pressure. Consequently, cardiac output will fall. In contrast, strong sympathetic activation causes venoconstriction, an increased venous return, and increased cardiac output.

Extrinsic regulation of the stroke volume—the modulation of myocardial contractility by the sympathetic nervous system

The contractile energy of the heart muscle may be altered by factors other than the initial resting length of the fibers. A change in contractile energy (i.e. the force of contraction) that is mediated by such extrinsic factors is referred to as a change in the *inotropic state* of the myocardium. This change in contractile energy is independent of the initial length of the cardiac muscle fibers (see Chapter 7). It is often, perhaps somewhat misleadingly, called a change of contractility.

Both sympathetic and parasympathetic nerve fibers innervate the SA and AV nodes. In contrast, the ventricles receive a rich innervation of sympathetic fibers but have little or no parasympathetic innervation. Increased sympathetic activity increases the heart rate and enhances the inotropic state of the myocardium both in the atria and in the ventricles. This is achieved in two ways.

1. Via sympathetic nerves which are widely distributed throughout the heart.

2. By the action of circulating epinephrine and norepinephrine secreted from the adrenal medulla. These hormones reach all parts of the heart by way of the coronary vessels.

The adrenergic receptors of cardiac muscle are mainly of the β_1 type (see Chapter 10). Activation of these receptors by circulating epinephrine or, less powerfully, by norepinephrine results in a *positive inotropic effect*—an increase in the force of contraction of the ventricles during systole. The contractile energy produced at a given end-diastolic volume is enhanced. This effect is illustrated in Fig. 15.19 where the abscissa shows the left ventricular end-diastolic pressure, which is directly related to the end-diastolic volume. (The relationship between the end-diastolic pressure and the end-diastolic volume is shown by the line AB in Fig. 15.16). At the same time as increasing contractility, sympathetic activity shortens the duration of systole. While this does not affect the stroke volume (because the rate of contraction is

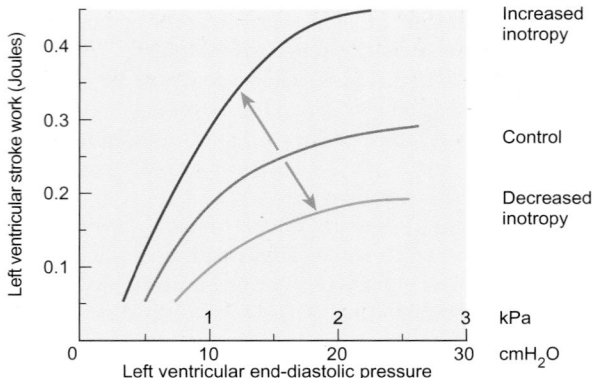

Fig. 15.19 Positive and negative inotropic effects on the relationship between stroke work and end-diastolic pressure.

also enhanced), it does help to offset the reduction in the time available for filling during diastole. Without this speeding up of systole, the filling of the heart might otherwise be compromised by the increase in heart rate that results from sympathetic stimulation (see below).

The positive inotropic effect of sympathetic stimulation results in a more complete emptying of the ventricles and an increase in systolic pressure, with the result that stroke work is increased (Fig. 15.19). The greater the degree of sympathetic stimulation, the greater the increase in stroke work.

Other circulating positive inotropic agents

Although the catecholamines, particularly epinephrine, are by far the most potent positive inotropic agents acting upon the heart, other substances circulating in the bloodstream can also exert effects on the contractility of cardiac muscle. These include calcium ions, thyroxine, and certain drugs such as caffeine, theophylline, and the cardiac glycoside digoxin. All these are thought to act by raising levels of free intracellular calcium either directly (in the case of calcium itself) or by stimulating the release of calcium from intracellular stores.

Negative inotropic agents

Substances that reduce the force with which the heart contracts are known as negative inotropic agents. As there is virtually no parasympathetic innervation to the ventricular myocardium in humans, parasympathetic activation has no direct effect on the inotropic state of the ventricles. However, β-receptor antagonists such as propranolol, calcium-channel blockers, and many general anesthetics, including the barbiturates and halothane, are negative inotropic agents. A low blood pH (acidemia) also has a negative inotropic effect.

Variation of cardiac output in the normal and denervated heart

The cardiac output is the product of the heart rate and the stroke volume. The heart rate can increase from 70 b.p.m. at rest to as much as 200 b.p.m. during heavy exercise. The stroke vol-

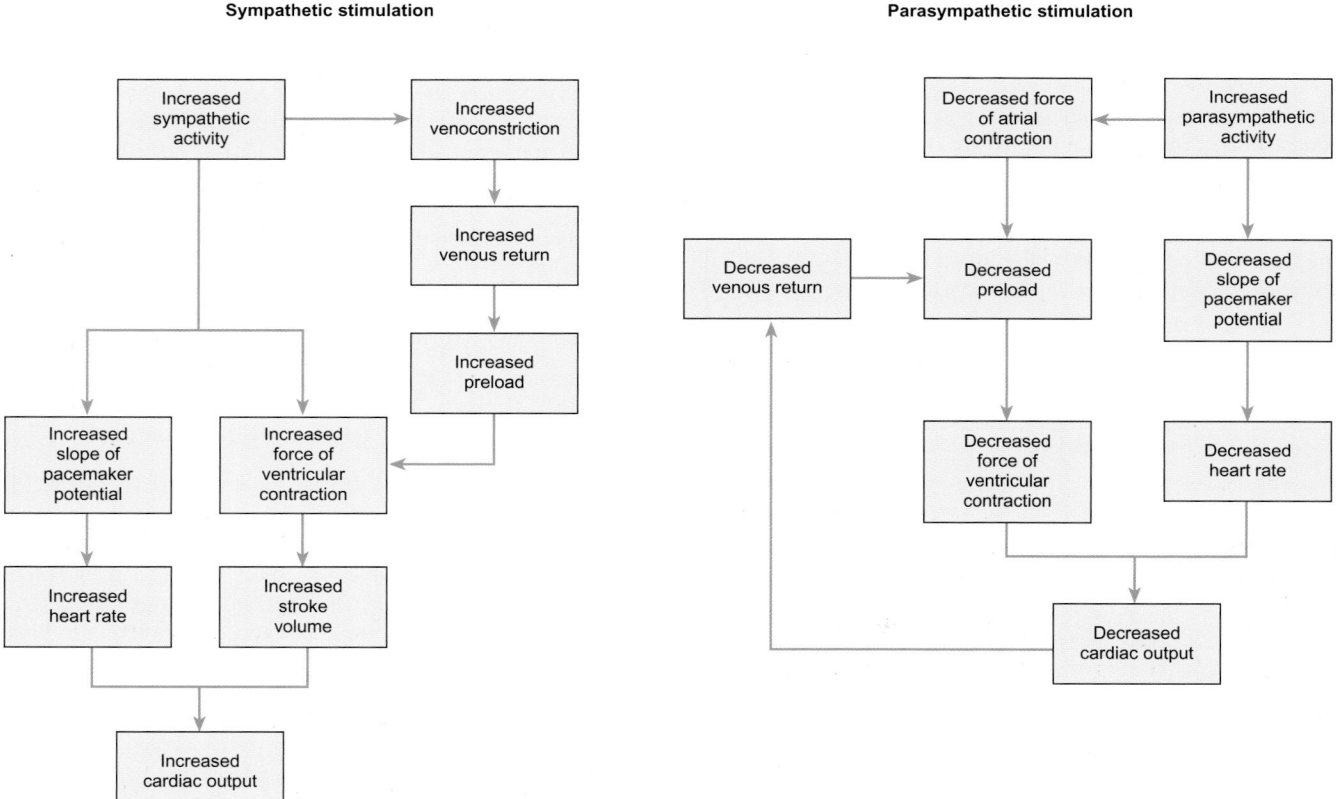

Fig. 15.20 Flow diagram summarizing the effects of sympathetic and parasympathetic activation on the cardiac output.

ume is normally about 70 ml, but it too can increase during exercise. As a consequence of these changes, cardiac output can vary from 5 l min^{-1} at rest up to about 25 l min^{-1} in severe exercise. In trained athletes, the range is much wider. The resting heart rate may be as low as 40–50 b.p.m. while stroke volume is about 120 ml. During maximal effort, the cardiac output of such individuals can exceed 35 l min^{-1}. The effects of the sympathetic and parasympathetic nervous divisions of the autonomic nervous system on cardiac output are summarized in Fig. 15.20.

Heart transplant patients are able to increase their cardiac output to meet the demands of heavy exercise almost as efficiently as normal subjects, despite the lack of sympathetic innervation to the heart itself. Part of this increase is due to the sympathetic effects on the heart rate of catecholamines released from the adrenal medulla, but to a large extent the increased cardiac output reflects the ability of the heart muscle to contract more forcefully as it is stretched (i.e. to the intrinsic regulation of stroke volume).

Summary

1. Cardiac output is the product of heart rate and stroke volume. It varies from 5 l min^{-1} at rest to over 25 l min^{-1} in severe exercise. The stroke volume is regulated by intrinsic and extrinsic mechanisms.

2. The intrinsic regulation is expressed by Starling's law of the heart which states that 'the energy of contraction of the ventricle is a function of the initial length of the muscle fibers comprising its walls'. The degree of stretch is determined by the venous return. Over any significant period, the output of the heart matches the venous return.

3. The activity of the sympathetic nerves supplying the heart and circulating catecholamines secreted from the adrenal medulla are the main factors that are involved in the extrinsic regulation of the stroke volume. Increased sympathetic activity and increased secretion of epinephrine and norepinephrine both cause an increase in the force of contraction during systole for any given end-diastolic volume. This is known as a positive inotropic effect.

4. Heart rate is governed by the influence of the autonomic nerves on the rate of discharge of the pacemaker cells of the SA node. Parasympathetic stimulation slows the heart rate (negative chronotropy), while sympathetic stimulation increases it (positive chronotropy). Circulating catecholamines from the adrenal medulla also increase the heart rate.

15.8 Hemodynamics: the relationship between blood flow and pressure in the circulation

The flow of blood through any part of the circulation is driven by the difference in pressure between the arteries that supply the region in question and the veins that drain it. This pressure difference is known as the *perfusion pressure*. The resistance offered by the blood vessels to the flow of blood is known as the *vascular resistance*. Mathematically, the relationship between perfusion pressure, blood flow, and vascular resistance is described by the following equations:

perfusion pressure = arterial pressure – venous pressure

and

$$blood\ flow = \frac{perfusion\ pressure}{vascular\ resistance}.$$

Thus blood flow will increase if the perfusion pressure is increased or the vascular resistance is decreased. Conversely, if perfusion pressure falls or vascular resistance rises, the blood flow will fall. This relationship can be applied to any vascular bed and can be stated in terms which apply to the systemic circulation or pulmonary circulation acting as a single unit.

For the systemic circulation as a whole, the driving force for blood flow is the difference between the arterial pressure and the pressure in the right atrium—the *central venous pressure*. Since the blood is pumped into the arteries intermittently, the pressure in the arterial system varies with the cardiac cycle. The pressure at the peak of ejection is called the *systolic pressure*, while at its lowest point, during ventricular relaxation, it is known as the *diastolic pressure*.

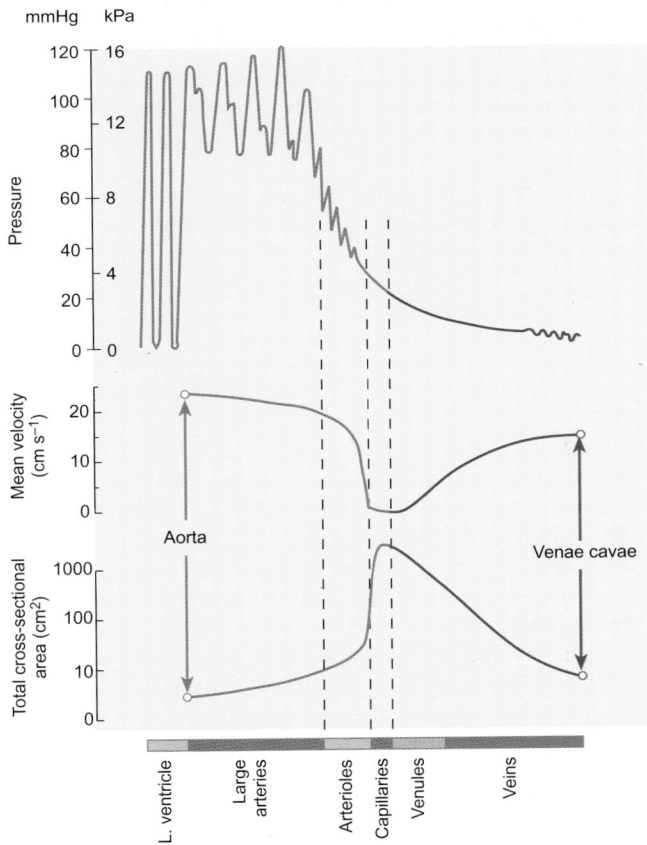

Fig. 15.21 The changes in blood pressure and the velocity of blood flow in the various parts of the systemic circulation. Note that the greatest fall in pressure occurs as the blood traverses the arterioles, which are the main site of vascular resistance.

The cardiac output represents a volume of blood flowing round the circulation each minute (milliliters per minute), while the velocity of blood flow is expressed in centimeters per second and represents the time taken to traverse the distance between two points. For a given flow rate, the velocity varies inversely with the cross-sectional area so that the velocity of blood flow in the aorta and major arteries is much greater than that in either the capillaries or veins, both of which have a larger total cross-sectional area (Fig. 15.21).

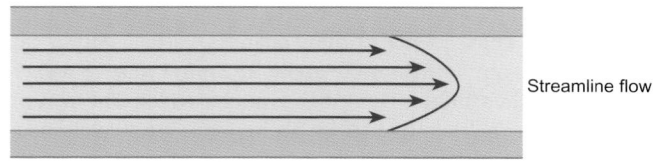

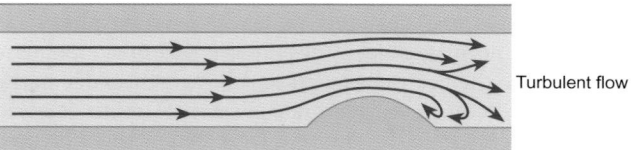

Fig. 15.22 Streamline (or laminar) flow and turbulent flow. In this case the turbulence occurs as the blood passes a distortion of the vessel wall such as an atheromatous plaque.

BOX 15.4 POISEUILLE'S LAW AND BLOOD FLOW

Poiseuille's law quantitatively relates the flow of a liquid through a rigid tube to the driving pressure. The equation describing this relationship is

$$\dot{Q} = (P_o - P_1)\frac{\pi r^4}{8\eta l}$$

where $\dot{Q}$ is the flow rate, P_0 and P_1 are the pressures at the beginning and end of the tube, respectively (i.e. the perfusion pressure), r is the tube radius, l is the tube length, and η us the viscosity of the liquid. Thus, for a given fluid (e.g. blood), the larger the diameter of a tube the higher will be the flow rate for a given pressure difference. As the flow depends on the fourth power of the radius, *doubling the diameter will increase the flow rate 16-fold.* Conversely, halving the diameter will lead to a 16-fold reduction in flow rate. The longer the tube, or the higher the viscosity, the smaller will be the flow rate.

The equation for Poiseuille's law can be rearranged in a manner analogous to Ohm's law to provide a description of hydraulic resistance R which is the conventional way of describing the relationship between blood pressure and blood flow in the circulation:

$$R = \frac{(P_0 - P_1)}{\dot{Q}} = \frac{8\eta l}{\pi r^4}$$

This rearrangement makes it clear that the resistance to blood flow is directly related to the length of a blood vessel and to the viscosity of the blood, but is inversely related to the fourth power of the radius, so that the smaller the blood-vessel diameter, the higher is the resistance.

Poiseuille's law relates strictly to laminar flow (sometimes called streamline flow), but if the pressure gradient along the tube is increased, the flow will eventually become irregular and *turbulence* occurs. Once turbulence has occurred, proportionately more pressure is required to achieve a given increase in flow. The critical pressure at which flow ceases to be laminar is determined by a coefficient known as the *Reynolds number.* Studies have shown that turbulence is more likely to occur at high flow rates in wide tubes that have an irregular cross-section (as in a large branching blood vessel such as the aorta). The Reynolds number is not normally reached in healthy blood vessels, except in the aorta during peak flow.

The vascular resistance depends on the diameter of the vessels and the viscosity of the blood

The resistance offered by a blood vessel to the flow of blood is described by Poiseuille's law (see Box 15.4), which states that the flow of blood is proportional to the fourth power of the radius of the vessel and is inversely proportional to the viscosity. Thus, if the radius of a vessel is reduced by half, it will carry one-sixteenth the quantity of blood for the same pressure difference. In other words, after the blood vessel radius has been reduced by half, it offers 16 times the resistance to blood flow.

The flow of blood is not uniform across the diameter of a blood vessel. The layer of fluid next to the vessel wall tends to adhere to it and the neighboring layer tends to adhere to this static layer and so on. Thus the flow velocity is fastest in the middle of the vessel and slowest next to the wall. When the different layers slip smoothly past each other flow is said to be laminar (or streamline) (Fig. 15.22). This is the situation that normally exists in the blood vessels and Poiseuille's law holds. If this smooth pattern of flow is broken up, for example by an irregularity in the vessel wall (such as an atheromatous plaque), eddies form and the flow is said to be turbulent.

Turbulence is, in general, undesirable as blood clots are more likely to form, but it occurs naturally in the ventricles and the aorta during peak flow. In both these situations, turbulence promotes the mixing of blood before its distribution to the systemic vessels. When the flow is turbulent, vibrations are set up which can be heard as sounds with a stethoscope. These are known as 'bruits' or murmurs. These sounds can be useful in the diagnosis of cardiovascular disease (see Chapter 31 p. 591) and p. 276. Laminar flow is silent.

The apparent viscosity of blood falls in vessels of small diameter

The resistance to flow is affected not only by the caliber of the smaller blood vessels, but also by the viscosity of the blood. When measured in a conventional viscometer, the apparent viscosity of blood is about 2.5 times that of water. In living tissues, however, the apparent viscosity of the blood is about half this

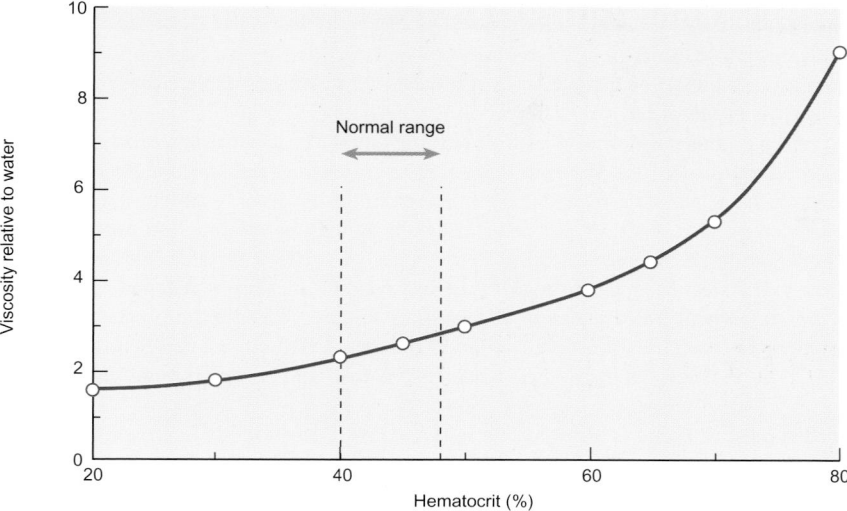

Fig. 15.23 The effect of hematocrit on the viscosity of blood relative to that of water. Note that the viscosity rises steeply when the hematocrit rises above about 60 per cent.

value. This anomalous behavior is due to the tendency of red blood cells to flow along the central axis of the smaller blood vessels—a phenomenon known as axial streaming. While the mechanisms responsible are not fully understood, it appears that flexibility of the erythrocytes is an important factor. At the low velocities found in the capillaries, rigid particles tend to remain uniformly distributed across the vessel while flexible particles migrate towards the central axis.

Since blood is made up of plasma and formed elements, it is perhaps not surprising to find that the viscosity varies with the hematocrit. The higher the hematocrit, the greater is the viscosity (Fig. 15.23). The hematocrit can increase both in disease (e.g. polycythemia) and as a result of physiological adaptation to life at high altitude. The associated increase in viscosity has significant effects on the work needed to pump a given quantity of blood round the circulation and can lead to a persistent elevation of the blood pressure (hypertension). Conversely, a fall in hematocrit (as a result of anemia or hemorrhage) will lower the viscosity of the blood.

Blood flow and pressure in the systemic arteries

The systemic arterial pressure fluctuates during the cardiac cycle. It is at its maximum during systole and at its minimum during diastole. Its maximum value depends on the rate of ejection from the ventricles, the distension of the arterial walls, and the rate at which it becomes distributed throughout the circulation. During systole, the pressure rises rapidly as the rate at which blood is being pumped into the arterial tree is greater than the rate at which it can be distributed. As a result, the pressure rises and the arterial walls become distended. As the ventricle begins to relax, the flow of blood into the aorta declines and the pressure falls. When the pressure in the aorta exceeds that in the ventricles, the aortic valve closes and this generates a small pressure wave giving rise to the dichrotic notch or incisura (Figs 15.14 and 15.24). Following this, the pressure declines to its diastolic value before the next systole causes another pulse wave.

The pressure wave in the arteries travels as a pulse that can be felt at various parts of the body. This pressure wave distends the

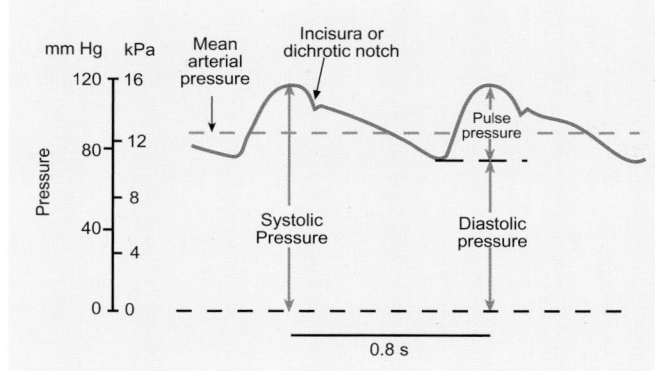

Fig. 15.24 The arterial pressure wave showing the systolic, diastolic, and mean arterial pressures.

arterial walls and, as the pressure begins to decline at the end of systole, the relaxation of the arterial walls provides a further source of energy for the propulsion of the blood. The shape and magnitude of the pulse wave changes as it passes from the aorta to the peripheral arteries. As Figure 15.21 shows, the peak pressure increases in the large arteries. This increase in pressure is believed to be due to the reflection of pressure waves from the distal arterial tree. The peak pressure in the small arteries and arterioles progressively declines as the vessels become narrow.

How is blood pressure measured?

Although it is possible to measure the blood pressure in various parts of the arterial system by direct insertion of a cannula connected to a pressure transducer, this is normally performed in humans only during cardiac catheterization or heart surgery. Pressure is much more often measured indirectly by *auscultation*. This method relies on the fact that turbulent blood flow creates sounds within the blood vessels that can be heard by means of a stethoscope, whereas streamlined flow is silent. Further details of the auscultatory method for measuring blood pressure are

BOX 15.5 MEASUREMENT OF BLOOD PRESSURE BY AUSCULTATION

Auscultation means 'listening to'. Therefore measuring blood pressure by auscultation means making use of the sounds that are heard when the blood flow through an artery is gradually restored after it has been occluded by an inflatable rubber cuff. The device used to record the pressures is known as a *sphygmomanometer*.

Normally, the blood pressure of the brachial artery is measured. Initially, an inflatable rubber cuff within a cotton sleeve is placed around the upper arm of the person whose blood pressure is to be measured. *The cuff is inflated until the radial pulse can no longer be felt so that the pressure within the cuff is in excess of the systolic pressure.* A stethoscope is then positioned over the brachial artery at the antecubital fossa (the inside of the elbow). The precise position can be established by feeling the brachial pulse before inflating the cuff. The pressure in the cuff is then gradually lowered. Initially, as the pressure of the cuff occludes blood flow through the artery, no sounds will be heard through the stethoscope. At the point where systolic pressure is reached (i.e. at the peak of the ejection phase of the cardiac cycle), pressure within the artery will be just sufficient to overcome the pressure within the cuff. There will be a brief spurt of blood into the artery which will cause vibration of the vessel wall that can be heard as a tapping sound through the stethoscope (**phase 1**). This sound is known as the first **Korotkoff sound** and is conventionally accepted to represent systolic pressure.

Cuff pressure is then lowered further. As more and more blood is allowed through the artery the sounds heard through the stethoscope become louder. However, as the diastolic pressure is approached, the artery remains open for almost all of the cardiac cycle and blood flow starts to become less turbulent and more streamlined. Streamlined flow creates less vibration and therefore less noise in the artery, and so the Korotkoff sounds diminish in volume fairly abruptly as diastolic pressure is reached (**phase 4**). The pressure is allowed to fall still further until the sounds disappear (**phase 5**). By convention, the point at which complete silence occurs (phase 5) is taken as diastolic pressure. Between the systolic and diastolic pressures, the Korotkoff sounds may disappear (**phase 2**) and reappear (**phase 3**)—this is known as the auscultatory gap. Therefore it is important not to mistake phase 2 for the diastolic pressure or phase 3 for the systolic pressure.

Normal systolic pressure measured in this way is usually less than 150 mmHg in a healthy adult and diastolic pressure should be less than 90 mmHg. In young adults and children, the pressures tend to be lower. In elderly people, there tends to be an increase in systolic pressure without a proportionate increase in diastolic pressure.

given in Box 15.5. Another method for the indirect measurement of blood pressure is the Finapres. This device permits the continuous non-invasive measurement of systolic and diastolic pressure by monitoring the pressure that needs to be applied to a small artery to keep its diameter constant. It is a research instrument employed where fast accurate measurements of blood pressure are required, for example while studying the effects of *G*-forces on military pilots.

What is normal arterial blood pressure?

In a healthy young adult at rest, systolic pressure is around 16 kPa (120 mmHg) while diastolic pressure is around 10.7 kPa (80 mmHg). This is normally written as 16/10.7 kPa or 120/80 mmHg. The difference between the systolic and diastolic pressures (normally about 5.3 kPa or 40 mmHg) is called the *pulse pressure* (Fig. 15.24).

Although the figure of 16/10.7 kPa (120/80 mmHg) is a useful one to remember, it is also important to realize that a number of factors will influence the blood pressure, even at rest. Probably the most obvious effect is that of age. Mean blood pressure tends to increase with age, so that by the age of 70, blood pressure averages 24/12 kPa (180/90 mmHg). This increase in blood pressure is due to a reduction in the elasticity of the arteries (*arteriosclerosis* or hardening of the arteries). Consistently high blood pressure (diastolic pressure above 13 kPa or 100 mmHg) is known as *hypertension* and is very common. The vascular complications associated with hypertension include stroke, heart disease, and chronic renal failure. For this reason regular screening is essential to avoid serious organ damage. For a more detailed discussion of hypertension and its causes, see Chapter 31 p. 588.

The *mean arterial pressure* (MAP) is a time-weighted average of the arterial pressure over the whole cardiac cycle. It is not a simple arithmetic average of the diastolic and systolic pressures because the arterial blood spends relatively longer near the diastolic pressure than near the systolic pressure (Fig. 15.24). However, for most working purposes an approximation to MAP can be obtained by applying the equation

$$MAP = diastolic\ pressure + (pulse\ pressure)/3.$$

Using the pressure in the brachial artery as an example, since this is normally the pressure measured in clinical practice, if the systolic pressure is 14.7 kPa (110 mmHg) and the diastolic pressure is 10.7 kPa (80 mmHg),

$$MAP = 10.7 + (14.7 - 10.7)/3\ kPa$$
$$= 12\ kPa\ (90\ mmHg)$$

The central venous pressure is close to zero and does not alter significantly during the cardiac cycle, so it does not need to be averaged. Thus the perfusion pressure of the systemic circulation is equal to the mean arterial pressure.

The flow through the circulation is the cardiac output, and the relationship between the mean blood pressure, the cardiac output, and the total peripheral resistance (TPR) is given by:

$$mean\ blood\ pressure = cardiac\ output \times TPR.$$

The total peripheral resistance is the sum of all the vascular resistances within the systemic circulation. It is determined by the viscosity of the blood and the total cross-sectional area of those vessels that are being perfused.

Short-term rises in arterial blood pressure can be brought about by so-called pressor stimuli such as pain, fear, anger, and sexual arousal. Conversely, pressure falls significantly during sleep, sometimes to as little as 9.3/5.3 kPa (70/40 mmHg), and to a much lesser and more gradual extent during normal pregnancy. Gravity also affects blood pressure. On rising from a lying to a standing position, there is a transient fall in blood pressure followed by a small reflex rise (see below).

The arterioles are the main source of vascular resistance

Measurement of the pressures in the different kinds of blood vessel show that the largest fall in pressure in the systemic circulation occurs as the blood passes through the arterioles (Fig. 15.21). Since the flow is the same throughout a given vascular bed, the greatest fall in pressure must occur in the region of greatest resistance. This occurs as the blood passes through the arterioles, showing that they are the principal sites of vascular resistance.

The majority of arterioles are in a state of tonic constriction due to the activity of the sympathetic nerves that supply them. As a result, their effective cross-sectional area is much less than the total cross-sectional area they would offer if they were fully dilated. Since the resistance of a vessel is dependent on the fourth power of its radius (Poiseuille's law), major changes in blood flow to a particular region can be achieved by modest adjustment of the caliber of the arterioles. This adaptation is important in regulating the distribution of the cardiac output between the various vascular beds. The mechanisms by which this regulation is achieved are discussed on p. 289.

The resistance of the systemic circulation (i.e. the total peripheral resistance) in humans is around 2.6 Pa ml^{-1} min^{-1} (0.02 mmHg ml^{-1} min^{-1}). The total resistance of the pulmonary circulation is much lower, around 0.4 Pa ml^{-1} min^{-1} (0.003 mmHg ml^{-1} min^{-1}), which is why a lower pressure is needed to drive the cardiac output through the lungs.

The capillary pressure

On the basis of Poiseuille's law, one might expect that, since the capillaries have the smallest diameter, they would be the principal site of vascular resistance. However, the overall resistance to blood flow depends both on the diameter of the vessels and on the total cross-sectional area available for the passage of the blood. The cross-sectional area offered by the capillaries is about 25 times that of the arterioles. Since the capillaries have no smooth muscle in their walls, they cannot be constricted. As a result they offer relatively little resistance to blood flow. The blood flow in the capillaries is steady, not pulsatile (the fluctuations in capillary blood flow result from the vasomotion of the arterioles—see Section 15.10). The pressure at the arteriolar end of the capillaries is about 4.3 kPa (32 mmHg). This falls to 1.5–2.7 kPa (12–20 mmHg) by the time that the blood has reached the venous end of the capillary bed. The small pressure at the venous end of the capillaries is sufficient to drive the blood back to the heart because the veins offer little resistance to blood flow unless they are collapsed.

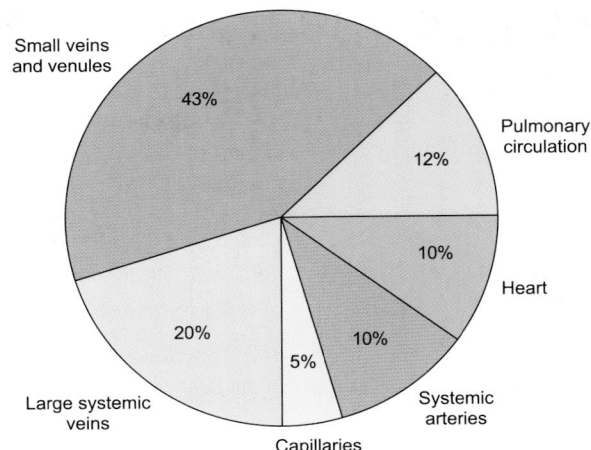

Fig. 15.25 The approximate distribution of the blood between the different parts of the circulation of a man at rest. Note the very high proportion of the blood in the systemic veins (c. 60 per cent).

Venous pressure

The blood volume of a normal adult is about 5 liters but the distribution of this blood is not even throughout the circulation (Fig. 15.25). The heart and lungs each contain about 600 ml of blood and the systemic arteries account for a further 500 ml while the capillaries have still less (about 250 ml). The bulk of the blood (about 3–3.5 liters) is found in the veins. Thus the veins, particularly the large veins, act as a reservoir for blood and are called *capacitance vessels*.

The walls of the veins are relatively thin and possess little elastic tissue, so that blood returning to the heart can pool in the veins simply by distending them. The degree of venous pooling is regulated by the tone of the smooth muscle (known as *venomotor tone*) which is in turn governed by the activity of the sympathetic nerves supplying the veins. During periods of activity when the cardiac output is high, venomotor tone is increased and the diameter of the veins is correspondingly reduced. Consequently, blood stored in the large veins is mobilized for distribution to exercising tissues and the velocity of the blood returning to the heart is increased.

Although veins hold so much blood, the average venous pressure measured at the level of the heart is only around 0.27 kPa (2 mmHg) compared with the average arterial pressure of about 10.3 kPa (100 mmHg). Venous pressure is highest in the venules; blood enters them at a pressure of about 1.6–2.7 kPa (12–20 mmHg) and falls to about 1 kPa (8 mmHg) by the time the blood reaches larger veins such as the femoral vein. This pressure head is sufficient to drive the blood into the central veins and thence into the right side of the heart where pressure is essentially zero (i.e. equal to that of the atmosphere).

The effect of gravity, the skeletal muscle pump, and breathing on venous pressure

When a subject stands up, pressure is increased in all the veins below the heart and reduced in all those above the heart as a result of the effects of gravity (Fig. 15.26). In an adult, the pressure in the veins of the foot increases by about 12 kPa

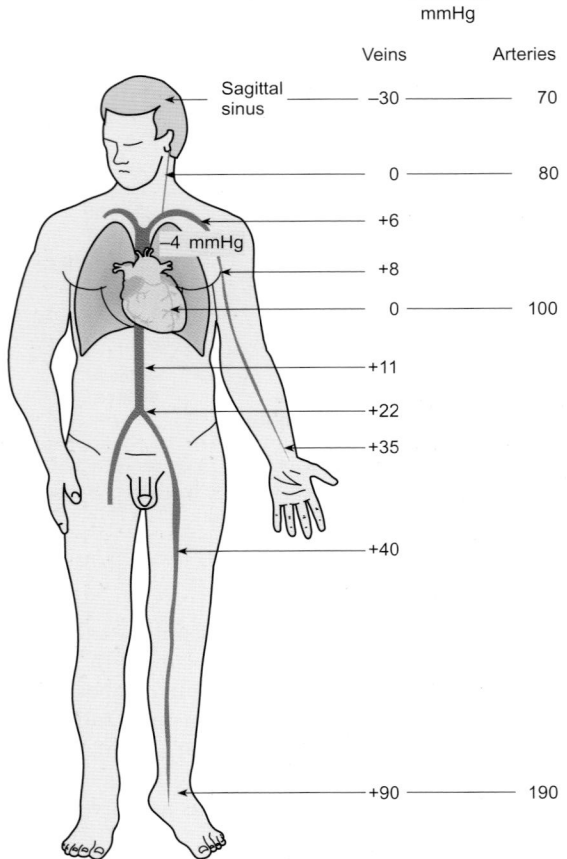

Pressure
mmHg

Veins   Arteries

Sagittal sinus ———— −30 ———— 70

0 ———— 80

+6

−4 mmHg

+8

0 ———— 100

+11

+22

+35

+40

+90 ———— 190

Fig. 15.26 The effect of hydrostatic pressure on the venous and arterial pressures in an adult male who is standing quietly. The figures are approximate and will depend on the height of the individual.

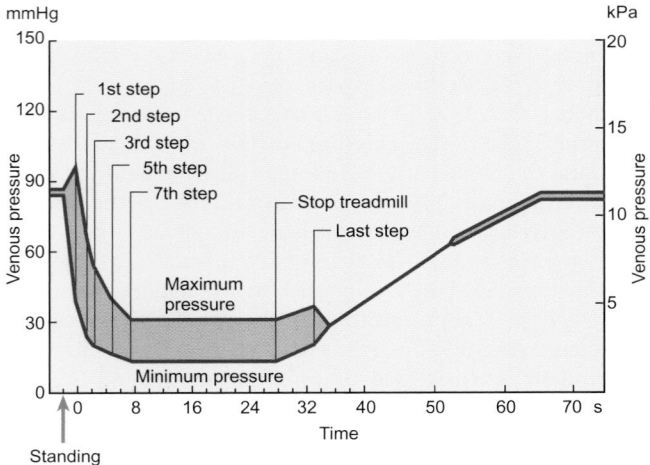

Fig. 15.27 The pressure changes in a dorsal vein of the foot when a subject is first standing and then begins to walk on a treadmill. During walking, the active muscles help to 'pump' the blood towards the heart. As a result, venous pressure falls and stabilizes at a lower level where it remains until exercise ceases. The pressure then progressively rises towards its original level.

(90 mmHg) on standing. Consequently, the veins in the lower limbs become distended and accumulate blood (an effect sometimes referred to as venous pooling). The additional blood comes mostly from the intrathoracic compartment and so the central venous pressure falls. By the Frank–Starling mechanism discussed above, stroke volume falls and there is a transient arterial hypotension known as postural hypotension. This is rapidly corrected by the baroreceptor reflex (see below). Since gravity also affects the arterial pressure in an exactly similar manner, the difference in pressure between the arteries and veins does not change significantly.

When a skeletal muscle contracts, it compresses the veins within it. Since the limb veins contain valves that prevent the backward flow of blood, the compression of the veins forces the blood towards the heart. This is known as the skeletal muscle pump. During exercise, the central venous pressure may rise slightly due to this effect. As Fig. 15.27 shows, the squeezing action of the muscles on the veins leads to a progressive decline in venous pressure measured at the level of the foot. Once exercise ends, venous pressure begins to rise once more. When the

Summary

1. Blood flows through the systemic circulation from the aorta to the veins because the pressure in the aorta and other arteries is higher than that in the veins. This pressure is known as the arterial blood pressure and it is derived from the pumping activity of the heart.

2. The arterial blood pressure is determined by both the cardiac output and the resistance offered by the arterioles (the total peripheral resistance), which in turn is determined by the total cross-sectional area offered by the arterioles to blood flow.

3. Blood flow in the arteries is pulsatile. Pressure at the peak of ejection is called the systolic pressure, while the lowest pressure occurs at the end of diastole (the diastolic pressure). The difference between the systolic and diastolic pressures is called the pulse pressure. In a healthy young adult, arterial blood pressure at rest will be around 16/11 kPa (120/80 mmHg). The mean arterial pressure is a time-weighted average, which is calculated as the sum of the diastolic pressure plus one third of the pulse pressure.

4. The capillaries offer little resistance to blood flow. The chief determinant of capillary blood flow is the caliber of the arterioles supplying a particular capillary bed. At the level of the heart, the capillary pressure is about 4.3 kPa (32 mmHg) at the arteriolar end and declines to about 1.6 kPa (12 mmHg) at the venous end.

5. The veins are the capacitance vessels, which contain around two-thirds of the total blood volume. Average venous pressure at the level of the heart is only around 0.3 kPa (2 mmHg). It is about 1.3 kPa (10 mmHg) in the venules and falls to around zero in the right atrium (central venous pressure). Respiration, gravity, and the pumping action of the skeletal muscles can all influence venous return and the central venous pressure.

muscle pumps are less active, as in a bedridden subject, blood tends to accumulate in the veins and there is a risk of deep vein thrombosis (see Box 13.1). A similar situation occurs during long periods of standing or sitting. In all three situations, there is an increase in the peripheral venous pressure and a reduced venous return to the heart. Cardiac output falls as a result.

Venous return is also influenced by respiration. During inspiration, the fall in intrathoracic pressure expands the intrathoracic veins and lowers central venous pressure. In addition, there is an increase in the pressure in the abdominal veins caused by the compression of the abdominal contents. These two factors tend to favor the movement of blood from the abdomen to the thorax. The situation is reversed during expiration. As a result, right ventricular stroke volume increases during inspiration and falls during expiration. However, left ventricular stroke volume falls during inspiration and rises during expiration. Over the course of one complete respiratory cycle, the outputs of the two sides of the heart are equalized. The various factors influencing the venous return are summarized in Table 15.2 (p. 281).

15.9 The mechanisms that control the caliber of blood vessels

The smooth muscle of all blood vessels exhibits a degree of resting tension known as 'tone'. Changes in vascular tone alter the caliber of the blood vessels and so alter vascular resistance. If the tone is increased (i.e. if the smooth muscle contracts further), *vasoconstriction* occurs and vascular resistance increases. If tone decreases, there is *vasodilatation* and a fall in vascular resistance. The level of resting or basal tone varies between vascular beds. In areas where it is important to be able to increase blood flow substantially, such as skeletal muscle, basal tone is high while in the large veins basal tone is much lower.

The tone of a blood vessel is controlled by a variety of factors. These fall into two broad categories: *intrinsic* and *extrinsic* mechanisms.

◆ Intrinsic (or local) control of blood vessels is brought about by the response of the smooth muscle to stretch, temperature, and locally released chemical factors.

◆ Extrinsic control is exerted by the autonomic nervous system and by circulating hormones.

The major arteries (except the aorta) and veins are mainly under extrinsic control, while the arterioles and small veins are subject to both mechanisms. As the capillaries and postcapillary venules have no smooth muscle, their diameter is not regulated.

Local control of blood vessels

Autoregulation maintains a relatively constant blood flow in the face of changes in perfusion pressure

Figure 15.28 shows that, over a certain range, maintained changes in blood pressure have little effect on the flow of blood through a particular vascular bed. When the pressure is first changed, the blood flow changes in the same direction. Thus, if the pressure is raised, blood flow increases at first but then returns close to its original level. Equally, if pressure falls, blood flow initially declines before returning to its previous level. This relative stability of blood flow is known as *autoregulation*. It occurs independently of the nervous system and is the result of direct changes in vascular tone in response to changes in perfusion pressure.

The mechanism of this effect is believed to be as follows. A rise in pressure within a vessel causes it to distend slightly. The smooth muscle of the vessel wall is initially stretched and responds by contracting (the myogenic response). This narrows the vessels, increases their resistance, and restores blood flow to its previous level. If the pressure falls, the smooth muscle relaxes and the vessels dilate, so restoring blood flow.

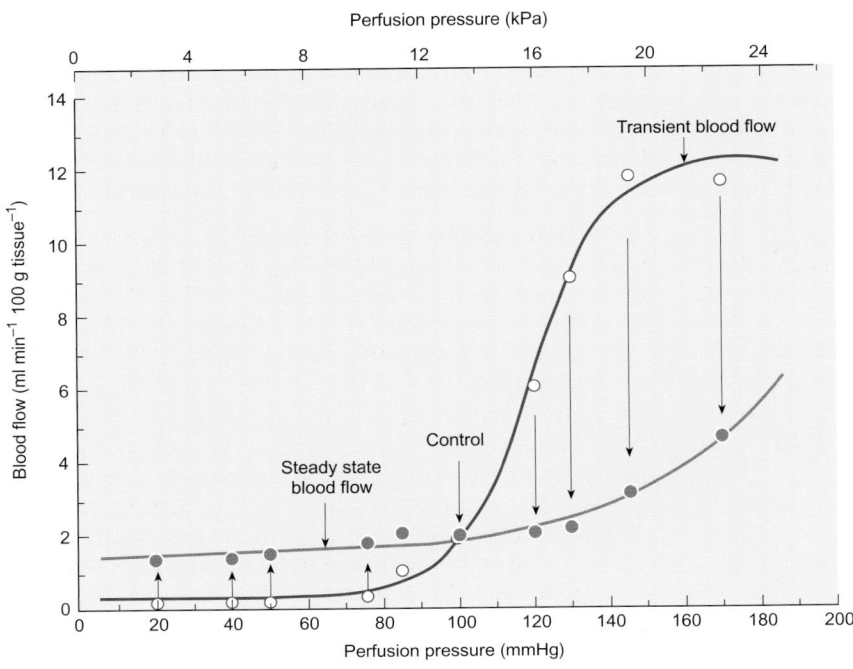

Fig. 15.28 Graph showing the autoregulation of blood flow in an isolated perfused skeletal muscle of the dog. The white circles represent the blood flow measured immediately after the perfusion pressure had been raised or lowered from the control level. As perfusion pressure is altered, there is a transient rise or fall in blood flow but autoregulatory mechanisms quickly restore blood flow to levels close to control (shown by the red symbols).

The exact mechanism of the myogenic response to stretching is unclear. In some vessels, the action potential frequency of the smooth muscle fibers is thought to increase in direct response to the stimulus, while in others there are believed to be calcium channels that open specifically in response to mechanical stress.

Although autoregulation is seen in most vascular beds (though not the lungs—see Chapter 16), blood flow in specific organs will vary with the physiological requirements. Indeed, changes in sympathetic drive and metabolic rate frequently act to reset the autoregulatory mechanism so that it will operate at a new level.

Vasodilatation occurs in response to a variety of metabolic by-products

Metabolism within cells gives rise to a number of chemical by-products. Many of these cause relaxation of the vascular smooth muscle and, therefore, vasodilatation. This has the effect of increasing the flow of blood through the vascular bed. The phenomenon is known as *functional hyperemia* (or as *metabolic* or *active hyperemia*). It has the important effect of facilitating the removal of potentially toxic waste products from the vicinity of the actively metabolizing cells. It is particularly significant in tissues such as exercising muscle and the myocardium (see below), and in the brain. Chemicals that are known to induce vasodilatation include carbon dioxide, lactic acid, potassium ions, and the breakdown products of ATP (adenosine and inorganic phosphate). Local tissue hypoxia can also bring about relaxation of the vascular smooth muscle, although this is not the case in the pulmonary circulation (see Chapter 16).

If the artery supplying blood to a tissue is compressed, blood flow is interrupted and the tissue becomes *ischemic*. When the compression is relieved, the blood flow becomes greater than normal for a short time. This response is called *reactive hyperemia*. It arises partly because of the myogenic response described earlier, and partly because of the vasodilatatory effects of the tissue metabolites that accumulated during the period of ischemia. The result of this enhanced perfusion is the rapid supply of nutrients and oxygen to the deprived tissue and the rapid removal of metabolic waste products.

The vasodilatation that occurs in response to locally released chemicals may also play a role in autoregulation. When pressure increases, blood flow also increases and this will tend to flush away the vasodilator chemicals. Consequently, vascular tone will increase and blood flow will decline to its original level. When perfusion pressure and blood flow both fall, there will be an accumulation of local metabolites which will act to cause vasodilatation.

Local hormones influence blood flow

A number of so-called local hormones or *autocoids*, released and acting locally, are believed to alter blood flow through their role in processes such as inflammation and blood clotting. Such agents include prostaglandins, leukotrienes, and platelet-activating factor. In addition to these, histamine plays an important role in inflammation, producing both vasoconstriction and vasodilatation, depending upon the type of receptors present on the target vessel (arterioles dilate while veins constrict). Like histamine, bradykinin is released during the inflammatory response. It is a strong vasodilator, which causes the release of nitric oxide (see below). The inflammatory response is discussed in more detail in Chapter 14.

The vascular endothelium produces both vasodilator and vasoconstrictor chemicals

The endothelial cells of arteries and veins synthesize a substance which is able to cause the vessel to dilate. When first discovered, this substance was known as *endothelium-derived relaxing factor*. It is now known that the factor is nitric oxide (NO) and that it is produced in response to a wide variety of stimuli. These include bradykinin (see above), acetylcholine, and the shear stress exerted on the endothelium by flowing blood. Nitric oxide is produced by the cleavage of arginine by an enzyme present in the endothelial cells called nitric oxide synthase. In its turn, the activity of this enzyme is regulated by the level of intracellular calcium (see Chapter 5). Nitric oxide synthesis can be inhibited by certain analogs of arginine, and administration of such inhibitors to human subjects brings about vasoconstriction. This suggests that nitric oxide exerts a continuous or tonic vasodilator influence on the vasculature.

Several vasoconstrictor agents derived from the vascular endothelium have recently been discovered. One of these, endothelin, is a peptide that brings about a relatively long-lasting constriction. Its physiological significance remains unclear, though much research is in progress to clarify the role of this and other endothelial agents in the control of blood vessels.

Extrinsic mechanisms of blood vessel control

The mechanisms described above all exert local control over particular vascular beds. Superimposed upon these mechanisms is the overall control of the heart and circulation exerted by the nervous and endocrine systems. The purpose of this extrinsic regulation is to provide for the needs of the body as a whole by diverting blood to where it is needed. It regulates arterial blood pressure and maintains an adequate blood supply to the brain. Receptors of many types, located throughout the cardiovascular system, provide information regarding the blood pressure and blood volume via afferent nerves that travel to the medulla in the brainstem. The efferent arm of the regulatory loop is formed by the autonomic nervous system and by a variety of hormones that act on the heart and blood vessels to initiate appropriate responses. The extrinsic regulation of the heart (its rate and strength of contraction) has already been discussed in Section 15.7, and so the following discussion is mainly concerned with the nervous and endocrine control of the vasculature.

The nervous control of the blood vessels

Figure 15.29 is a highly simplified diagram of the sympathetic innervation of the cardiovascular system. The pathway begins in the medulla where afferent fibers from the glossopharyngeal and vagus nerves terminate in an elongated nucleus called the *nucleus of the tractus solitarius* (NTS). The neurons of this tract send axons to cells in the rostral ventrolateral medulla. In addition, these cells also receive projections from the cerebral cortex and hypothalamus, which may override the reflex regulation.

The neurons of the rostral ventrolateral medulla project to the sympathetic preganglionic neurons in the gray matter of segments T1 to L3 in the intermediolateral cell column of the spinal cord. These so-called bulbospinal fibers may have either an inhibitory or an excitatory action on the sympathetic preganglionic neurons. The level of activity in the sympathetic

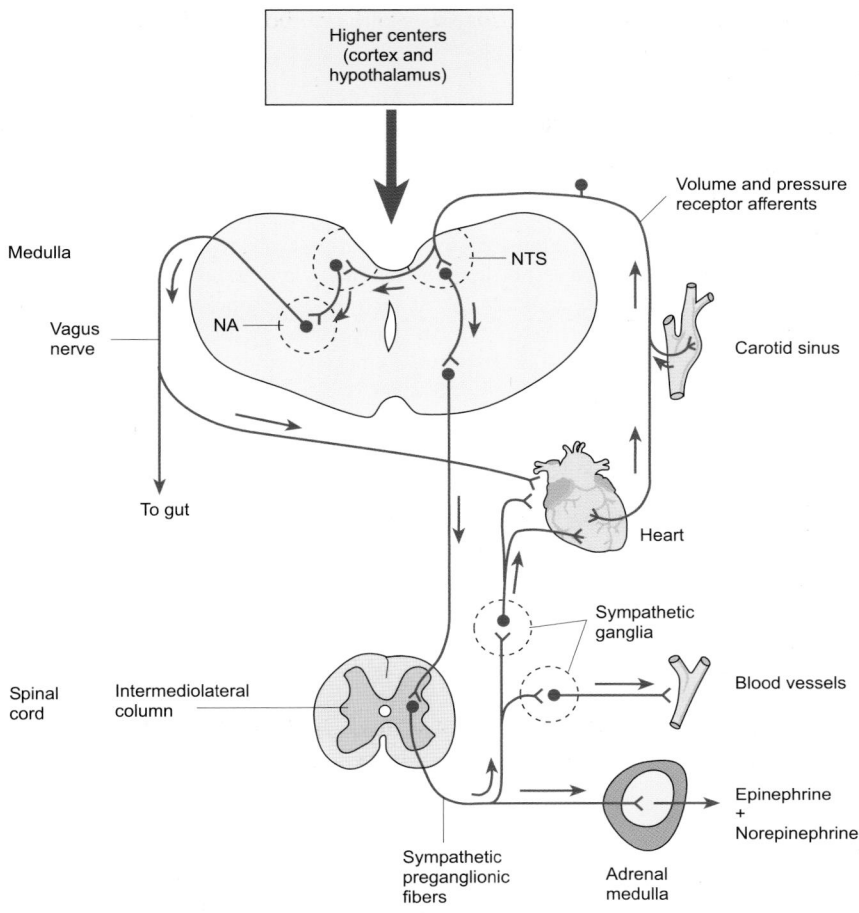

Fig. 15.29 A highly simplified schematic diagram to show the sympathetic and parasympathetic innervation of the cardiovascular system. Afferent nerve fibers relay pressure information from the baroreceptors and volume receptors of the heart to the nucleus of the tractus solitarius (NTS). Fibers from this nucleus project to the rostral ventrolateral medulla from where other neurons send axons to the sympathetic preganglionic neurons in the spinal cord. The neurons of the NTS also send axons to cardiovagal motoneurons in the nucleus ambiguus of the medulla (NA). The vagal fibers innervate the heart while the postganglionic sympathetic fibers also innervate the blood vessels and adrenal medulla. Superimposed on this organization are projections from the hypothalamus and cerebral cortex to the medulla.

preganglionic fibers is determined by the balance of nervous traffic in the descending fibers from the medulla and the local spinal inputs.

The sympathetic preganglionic neurons travel to the sympathetic chains via the ventral roots of the spinal cord where most synapse with the cell bodies of postganglionic neurons located in the sympathetic ganglia. A few sympathetic preganglionic fibers synapse in the more distant celiac and hypogastric ganglia or in the adrenal medulla (see Chapter 10).

Axons from the postganglionic cell bodies are non-myelinated and travel in the mixed peripheral nerves to the blood vessels where they terminate mainly in the outer parts of the tunica media. Terminal arterioles tend to be controlled largely by local mechanisms and are poorly innervated, but arteries and the larger arterioles receive a rich supply of sympathetic vasoconstrictor fibers. Veins are generally more sparsely innervated. Indeed, the veins draining the skeletal muscles receive no innervation.

Autonomic nerves that alter the caliber of blood vessels fall into three groups:

(1) sympathetic vasoconstrictor fibers;

(2) sympathetic vasodilator fibers;

(3) parasympathetic vasodilator fibers.

Sympathetic vasoconstrictor fibers predominate in most vascular beds

The sympathetic vasoconstrictor fibers show tonic activity that contributes to the resting tone of blood vessels. The nonepinephrine released by the sympathetic postganglionic nerve fibers acts on the α-adrenoceptors of the vascular smooth muscle to cause it to contract, and this contraction results in vasoconstriction. Interruption of the tonic activity of the sympathetic nerves (for example by the administration of α-blockers or by cutting the sympathetic nerves) leads to a significant rise in blood flow in the vessels of many tissues.

Vasodilatation induced by a fall in sympathetic vasoconstrictor fiber activity is also important physiologically. It contributes to the regulation of arterial blood pressure via the baroreceptor reflex (see p. 300) and is partly responsible for producing vasodilatation in the vessels of the skin during temperature regulation (Chapter 24).

Sympathetic vasodilator fibers are cholinergic

Human sweat glands are innervated by sympathetic cholinergic fibers whose stimulation brings about both an increase in sweat production and vasodilatation of the skin vessels via M_3 muscarinic receptors. However, the response is not totally blocked by muscarinic antagonists such as atropine, and it is believed that another transmitter, the neuropeptide VIP (vasoactive intestinal peptide), may also play a role in regulating cutaneous vasodilatation and sweating.

Cholinergic vasodilator fibers are thought by some authors to supply the arterioles of human skeletal muscle. However, the evidence for the importance of this mechanism in humans is not compelling and the vasodilatation elicited by circulating epinephrine acting on β-adrenoceptors is probably of much greater significance. Many non-human mammalian species certainly do possess such fibers and here the vasodilatation is thought to be involved in the alerting response of the animal (the preparation for 'fight or flight'—see Section 15.11).

Parasympathetic vasodilator nerves

As described in Chapter 10, the preganglionic fibers of the parasympathetic nerves leave the CNS via the cranial and sacral spinal outflows. They synapse with postganglionic fibers in ganglia situated within the end-organs themselves (compare this with the sympathetic organization described earlier). Parasympathetic vasodilator fibers innervate the salivary glands, the exocrine pancreas, the gastrointestinal mucosa, the genital erectile tissue, and the cerebral and coronary arteries. They are not tonically active but release acetylcholine when they are stimulated. The acetylcholine then acts on the muscarinic receptors of the vascular smooth muscle to cause membrane hyperpolarization and relaxation. Blood flow to the tissue will then increase following the vasodilatation. The role of the parasympathetic vasodilator innervation of the erectile tissue is described in Chapter 20.

Hormonal control of the blood vessels

Under normal conditions the circulation is under both nervous and endocrine control. Long-term regulation of blood pressure requires the cooperation of a number of hormonal mechanisms that work alongside the neural controls to regulate both plasma volume and vascular tone. The major regulatory hormones are epinephrine, vasopressin (ADH), atrial natriuretic peptide (ANP), and the renin–angiotensin–aldosterone system.

The adrenal medullary hormones—epinephrine and norepinephrine

Epinephrine and norepinephrine are secreted by the adrenal medulla in response to acetylcholine released from the splanchnic nerve terminals during stress of all kinds, including exercise. In humans about 80 per cent of the secretion is epinephrine. Both epinephrine and norepinephrine act on *adrenoceptors*. These are of two main types, known as α- and β-adrenoceptors (see also Chapters 5 and 10). Interaction of the catecholamines with the α-adrenoceptors leads to vasoconstriction, while their interaction with β-adrenoceptors brings about vasodilatation.

Norepinephrine has a much greater affinity for α- than for β-adrenoceptors and therefore will normally cause vasoconstriction. Epinephrine interacts with the β-adrenoceptors present in the vascular smooth muscle of skeletal muscle, the heart, and the liver to produce vasodilatation. For this reason, sympathetic stimulation of the kind seen during the alerting reaction allows the blood vessels of the heart and skeletal muscle to dilate, thereby increasing the blood flow to these tissues, while vessels elsewhere constrict. Thus blood is diverted preferentially to those tissues with an important role to play in the alerting response.

Vasopressin (antidiuretic hormone or ADH)

More details of the secretion and actions of this hormone can be found in Chapters 12, 17, and 28. Briefly, although vasopressin (ADH) is chiefly concerned with the regulation of fluid excretion by the kidneys, it also exerts powerful effects on the vasculature. It is secreted into the circulation from the posterior pituitary gland in response to the fall in blood pressure that follows a substantial hemorrhage. At such times, vasopressin causes a powerful vasoconstriction in many tissues, which helps to maintain arterial blood pressure (hence its name). In the cerebral and coronary vessels, however, vasopressin seems to elicit vasodilatation. The net effect is a redistribution of the blood to the essential organs—the heart and brain.

The renin–angiotensin–aldosterone system

Renin is a proteolytic enzyme that is secreted by the kidneys in response to a fall in the sodium concentration in the distal tubule. It acts on an inactive peptide in the blood called angiotensinogen to form angiotensin I, which is then converted in the

Summary

1. The diameter of a blood vessel is determined by the degree of constriction of the smooth muscle in its wall. Vasoconstriction or vasodilatation may be superimposed on the resting tone of a vessel by either intrinsic or extrinsic mechanisms.

2. Intrinsic factors include the myogenic contraction seen in response to stretch of a vessel, the vasodilator actions of tissue metabolites, and the effects of local vasoactive substances. The intrinsic regulation of the arterioles is thought to contribute to resting tone and to account in large part for the autoregulation of blood flow.

3. Nerves and hormones exert extrinsic control over the heart and circulation in response to information arising from cardiovascular receptors of many kinds.

4. Sympathetic vasoconstrictor fibers are the most widespread and important of the nerves that alter the caliber of blood vessels, but parasympathetic vasodilator fibers also play a part.

5. Postganglionic sympathetic vasoconstrictor fibers secrete norepinephrine, which interacts with α-adrenoceptors of vascular smooth muscle to cause vasoconstriction. Widespread activation of α-adrenoceptors leads to an increase in arterial blood pressure. Some arterioles, such as those of muscle, liver, and heart, also possess β-adrenoceptors. Interaction of norepinephrine with these receptors causes vasodilatation.

6. A number of hormonal mechanisms work alongside the neural mechanisms to provide extrinsic regulation of plasma volume and vascular tone. These include epineprine, ADH, atrial natriuretic peptide and the renin–angiotensin–aldosterone system.

lungs to its active form angiotensin II. This hormone has two important actions. It stimulates the secretion of aldosterone from the adrenal cortex and it causes vasoconstriction. The former will result in the reabsorption of an increased amount of salt and water by the distal tubule, which may be particularly important when the blood volume is low (e.g. following a hemorrhage). The latter brings about a rise in arterial blood pressure. Further details of this important hormonal system can be found in Chapter 17.

Atrial natriuretic peptide (ANP)

This hormone is secreted by atrial myocytes in response to high cardiac filling pressures. Its action is the opposite to that of aldosterone in that it stimulates the excretion of salt and water by the renal tubules. It also has a weak vasodilator action on the resistance vessels.

15.10 The microcirculation and tissue fluid exchange

In this section, the processes responsible for the exchange of fluid, nutrients, and metabolites between the blood and the tissues are examined in detail. In general terms, blood flows from the arterioles to the venules via the capillaries, which are the main exchange vessels. As part of the process of tissue exchange, a small volume of fluid passes from the plasma to the interstitial space. While a small fraction of this fluid is returned directly to the circulation, the remainder is returned via the afferent lymphatic vessels. For this reason, the basic organization of the lymphatic system and its role in the regulation of the volume of the interstitial fluid will also be discussed here.

The microcirculation is organized in functional units

The arterioles that branch directly from the arteries are known as primary arterioles and they are extensively innervated by sympathetic nerve fibers. The primary arterioles progressively give rise to secondary and tertiary arterioles, which have less smooth muscle and are more sparsely innervated. The final degree of branching gives rise to the terminal arterioles, which have a very sparse innervation, their caliber being regulated largely by the local concentration of tissue metabolites (see discussion of the local control of blood vessels above).

The terminal arterioles are 10–40 μm in diameter and each arteriole gives rise directly to a group of capillaries known as a cluster or module (Fig. 15.30). The flow through the capillary cluster is regulated by the caliber of the terminal arterioles. The capillaries themselves are 5–8 μm in diameter and about 0.5–1 mm long. They drain into postcapillary venules whose walls do not contain smooth muscle. As the walls of these vessels contain cells known as pericytes, these small venules are also known as *pericytic venules*. The walls of these vessels, like those of the capillaries, are sufficiently thin to permit the passage of fluid between the plasma and interstitial space. Thus, the pericytic venules are able to act as exchange vessels. The postcapillary venules coalesce into larger vessels. Venules with a diameter greater than about 30 μm have smooth muscle in their walls.

The structure of the capillaries

The walls of the capillaries consist of a single layer of flattened endothelial cells (Fig. 15.31). This thin layer is partly covered by pericytes and the whole structure is surrounded by a basement membrane. The capillary wall is so thin (about 0.5 μm) that the diffusion path between plasma and tissue fluid is extremely short. There are three types of capillary.

1. *Continuous capillaries* in which the wall consists of a continuous endothelial layer perforated only by narrow clefts between the cells. This is the most common type of capillary.

2. *Fenestrated capillaries* in which the endothelial cells are perforated by small circular pores, the *fenestrae*, which permit a relatively free passage of salts and water from the plasma to the tissues. This type of capillary is found in tissues that are specialized for bulk fluid exchange such as exocrine glands and the capillaries of the renal glomerulus.

3. *Discontinuous capillaries*, which have gaps in the capillary wall that are sufficiently large to permit the passage of plasma proteins. Discontinuous capillaries are found in the liver, spleen, and bone marrow.

Vasomotor activity in the arterioles causes blood flow in the capillaries to vary

Direct observation of the pattern of blood flow in a capillary bed shows that it is constantly varying. The blood flow alternates between near stasis and rapid flow according to the changes in the caliber of the arterioles. This variation in capillary blood flow is known as *vasomotion*. When there is increased metabolic activity, the demand for oxygen and nutrients is increased and the arterioles become dilated by the action of local metabolites.

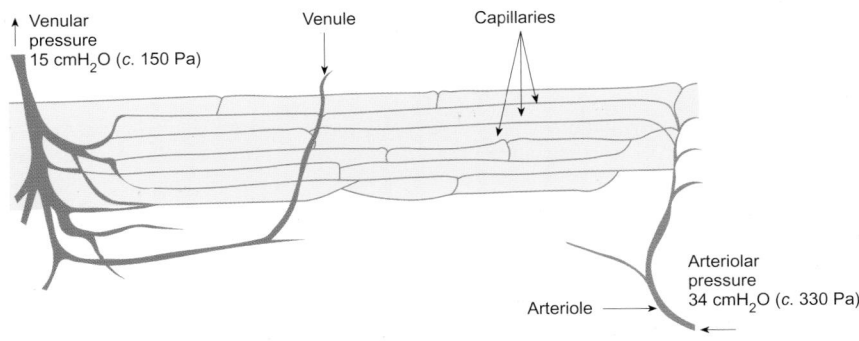

Venular pressure 15 cmH₂O (*c.* 150 Pa)

Venule   Capillaries

Arteriole

Arteriolar pressure 34 cmH₂O (*c.* 330 Pa)

Fig. 15.30 A diagrammatic representation of the capillary bed in a relaxed cremaster muscle of a rat. A terminal arteriole feeds a module of capillaries which eventually coalesce to form postcapillary venules. These in turn drain into the venules.

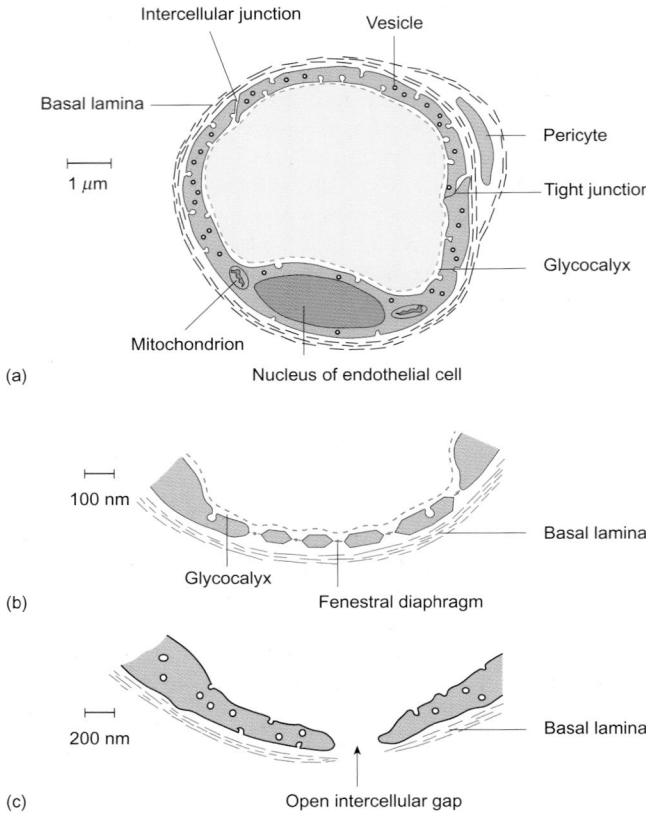

Intercellular junction
Vesicle
Basal lamina
1 μm
Pericyte
Tight junction
Glycocalyx
Mitochondrion
Nucleus of endothelial cell
(a)

100 nm
Basal lamina
Glycocalyx
Fenestral diaphragm
(b)

200 nm
Basal lamina
(c)
Open intercellular gap

Fig. 15.31 (a) A cross-sectional view of the capillary wall for a continuous capillary, (b) a fenestrated capillary, and (c) a discontinuous capillary. Note the different scale bars on the left.

BOX 15.6 FICK'S LAW OF DIFFUSION

Substances move from the plasma to the tissues or from the tissues to the plasma by diffusion down their concentration gradients. Diffusion is a passive process and results from the random movement of molecules. The amount of a substance moving from one region to another can be expressed by Fick's law of diffusion, which states that the amount of substance moved depends on the area available for diffusion, the concentration gradient, and a constant known as the diffusion coefficient. Thus

amount moved = area × concentration gradient
              × diffusion coefficient.

or

$$J = DA \frac{dC}{dx}$$

where J is the quantity moved, D the diffusion coefficient, A the area over which diffusion can occur, and dC/dx the concentration gradient. The negative sign indicates that diffusion occurs from a region of high concentration to one of low concentration. The diffusion coefficient becomes smaller as the molecular size increases so that large molecules diffuse more slowly than small ones.

This results in the recruitment of more capillaries and an increase in the surface area for exchange.

Most solute exchange between the plasma in the capillaries and the interstitial fluid occurs by diffusion

As capillaries are the principal exchange vessels, their density determines the total surface area for exchange between tissue and blood. If the density of capillaries is high, there will be a large surface area for exchange and a relatively short distance between capillaries. Consequently, the delivery of oxygen, glucose, and other nutrients will be very efficient. The exchange of solutes between the blood and the surrounding tissues is entirely passive and occurs chiefly by simple physical diffusion. This is described by Fick's law of diffusion, which states that the rate of diffusion of any substance depends on the area available for diffusion, the concentration gradient, and the *diffusion coefficient* (see Box 15.6). Substances with small molecular weights diffuse more readily (i.e. have higher diffusion coefficients) than substances with large molecular weights such as proteins.

In free solution these three factors (area, concentration gradient, and diffusion coefficient) are sufficient to describe the rate of diffusion of a particular substance. However, diffusion

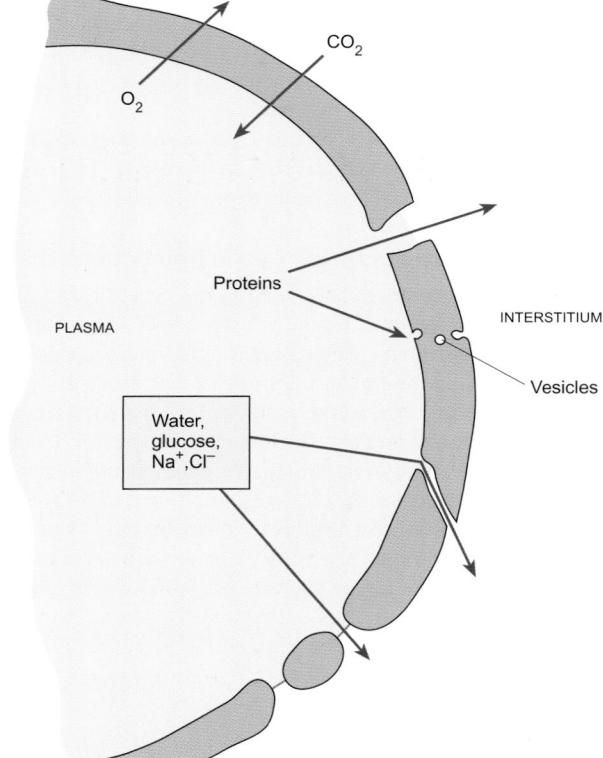

CO₂
O₂
Proteins
PLASMA
INTERSTITIUM
Vesicles
Water,
glucose,
Na⁺, Cl⁻

Fig. 15.32 A schematic diagram illustrating how different types of molecule can cross the capillary wall. Gases and small lipophilic molecules can penetrate the capillary wall directly and exchange with the extracellular fluid via the walls of the endothelial cells (transcellular exchange). Water and small water-soluble molecules such as glucose and inorganic ions exchange only with the extracellular fluid via small pores in the capillary wall (the intercellular junctions and the fenestrae). Large molecules such as proteins have a low capillary permeability except in discontinuous capillaries where they pass through large intercellular spaces.

TABLE 15.3 Relative permeability of continuous capillaries to various molecules

Substance	Molecular weight	Permeability relative to water
Oxygen	32	~3000
Water	18	1.00
NaCl	58	0.96
Urea	60	0.8
Glucose	180	0.6
Inulin	5000	0.2
Albumin	69 000	0.0001

Note that water-soluble substances pass through spaces between the endothelial cells and that, as a group, their permeability is related to their molecular weight. Lipophilic substances such as oxygen and other gases diffuse freely through the walls of the endothelial cells.

between the plasma in the capillaries and the fluid of the surrounding tissues is also limited by the permeability of the capillary wall. Very lipophilic substances such as carbon dioxide and oxygen pass through the endothelial cells relatively freely (*transcellular exchange*) and therefore the capillaries are very permeable to them. Water-soluble substances such as electrolytes and glucose are not soluble in the lipids of the cell membranes. Consequently, they do not pass across the endothelial cells themselves but diffuse through the small spaces between the cells (*paracellular exchange*) and through the fenestrations when these are present (Fig. 15.32). Water passes into the interstitial space by bulk flow. Measurements with radioactive tracers indicate that the capillaries allow relatively free exchange of both water and small water-soluble molecules between the plasma and the interstitial fluid (Table 15.3).

The rate of diffusion through the endothelial clefts and fenestrae falls sharply as molecular weight increases (Table 15.3). Consequently, large molecules like the plasma proteins pass from the plasma to the interstitial space very slowly. Despite this, the passage of proteins from the plasma to the interstitium is functionally important as it permits the delivery of protein-bound substances with low water solubility to their target tissues while maintaining the colloid osmotic pressure in the interstitial fluid below that of the plasma. Examples of substances that are delivered in this way are the steroid hormones (e.g. estradiol-17β, testosterone), thyroxine, and essential fatty acids. In addition, the movement of immunoglobulins between the plasma and the fluid surrounding the cells forms part of the body's defense against infection (see Chapter 14).

What are the factors that govern the movement of fluid between the capillaries and the surrounding tissues?

Net fluid movement across the capillary wall is driven by flow down pressure gradients. The direction of fluid movement

---

BOX 15.7 THE PLASMA ONCOTIC PRESSURE AND DONNAN EQUILIBRIA

The oncotic pressure of the plasma is around 3.5 kPa (25 mmHg). Simple calculation shows that the osmotic pressure of the plasma proteins can account for only about 60 per cent of the total. As the smallest and most abundant of the plasma proteins, albumin contributes around 1.6 kPa, while all the other plasma proteins together only contribute around 0.5 kPa at most, partly because of their relatively low abundance and partly because of their greater molecular mass (for example, the relative molecular mass of the γ-globulins (IgGs) is around 150 000 while that of fibrinogen is 340 000).

How can the missing 40 per cent be accounted for? The plasma proteins, especially albumin, are negatively charged at normal blood pH. To maintain electroneutrality, they must be associated with a positive counter-ion which, in the case of the plasma, is sodium. This is the key to understanding why the oncotic pressure of the plasma is higher than can be accounted for by the plasma proteins alone.

A particular type of equilibrium, known as a Donnan equilibrium, exists when two compartments are separated by a membrane that is permeable to ions (e.g. Na^+ and Cl^-) and to water, but not to large molecules such as proteins. This is the situation for the equilibrium that exists across the walls of most capillaries. The ions can diffuse freely between the two compartments, so in the absence of proteins, the composition of the two compartments will eventually become the same. If a negatively charged protein is present in one compartment (compartment I) but not the other, the situation changes.

As before, the ions can move freely between the compartments but their distribution is affected by the presence of the protein in compartment I. At equilibrium, the concentration of the diffusible cation (Na^+ in this example) is greater in compartment I which contains the protein, while that of chloride is less. In compartment II the situation is reversed, so the concentration of sodium is less while that of chloride is greater. It can be shown from thermodynamic arguments that the relationship between the concentrations of the ions in compartments I and II is given by:

$$[Na^+]_I \times [Cl^-]_I = [Na^+]_{II} \times [Cl^-]_{II}$$

In words, the product of the concentrations of any pair of diffusible ions is the same on each side of the membrane. However, the *sum* of the concentrations of the diffusible ions is greater in compartment I, the side containing the protein. Thus the osmotic pressure of compartment I must be greater than that of compartment II, not only because of the presence of the protein but also because of the greater concentration of diffusible ions. The concentration of diffusible cations is about 6 mmol l^{-1} higher in the plasma than it is in the interstitial fluid, and the concentration of the diffusible anions is slightly less, giving a net excess of diffusible ions of about 0.4 mmol l^{-1}. This corresponds to an osmotic pressure of around 1.3 kPa and will bring the total oncotic pressure to around 3.5 kPa. The main text and Box 15.8 explain the importance of the oncotic pressure in the exchange of fluid between plasma and the tissues.

between a capillary and the surrounding interstitial fluid depends on four pressures:

(1) the pressure within the capillary (the capillary pressure);

(2) the pressure within the tissue surrounding the capillary (the interstitial pressure);

(3) the osmotic pressure exerted by the plasma proteins (the oncotic pressure);

(4) the osmotic pressure exerted by the proteins present within the interstitial fluid.

The difference between the capillary pressure and that of the interstitial fluid is the hydraulic pressure acting on the fluids. The greater this value, the greater is the tendency for fluid to pass from the capillary to the interstitium. For most tissues, this physical movement of fluid amounts to about 0.1 ml min^{-1} for every kilogram of tissue and must be carefully distinguished from tissue exchange by diffusion, which is a two-way process in which no net movement occurs.

Since the capillary wall has only a low permeability to proteins, it behaves as a semipermeable membrane and the osmotic pressure exerted by the plasma proteins (the oncotic pressure)—see Box 15.7—acts to oppose the hydrostatic pressure, tending to force the fluid from the capillaries to the surrounding tissues. This is offset to some extent by the osmotic pressure due to the escaped plasma proteins in the interstitial fluid, which acts to draw fluid from the capillaries. The algebraic sum of the various pressures is called the *net filtration pressure*. These relationships are summarized by the following equation:

net filtration pressure = net pressure forcing fluid from capillary – net pressure in tissues.

This relationship summarizes the forces governing the circulation of fluid between the capillaries and the interstitium and is called the Starling or Starling–Landis principle. The pressures involved are often called *Starling forces* (see Box 15.8 for further details). For most tissues, the net filtration pressure is positive and the hydrostatic forces favor fluid movement from the capil-

---

BOX 15.8 CALCULATION OF STARLING FORCES FOR CAPILLARY FILTRATION

The pressures exerted on the capillaries of human skin can be readily measured and provide a clear example of the forces that determine the net filtration pressure (P_f) in a capillary bed. At the level of the heart, the hydrostatic pressure at the arterial end of the capillaries (P_c) is about 5 kPa (c. 37 mmHg). At the venular end, the pressure is about 2 kPa (c. 15 mmHg). The oncotic pressure of human plasma (π_p) is about 3.5 kPa (c. 26 mmHg) and that of the interstitial fluid (π_i) is about 2 kPa (15 mmHg). The hydrostatic pressure of interstitial fluid (P_i) is 0.2 kPa (1.5 mmHg) below that of the atmosphere (i.e. –0.2 kPa or –1.5 mmHg).

The filtration pressure is equal to the difference between the pressure forcing fluid from a capillary and that tending to oppose filtration, and is given by the following equation:

$$P_f = (P_c - P_i) - (\pi_p - \pi_i).$$

At the arteriolar end

$$P_f = [5 - (-0.2)] - (3.5 - 2) = +3.7 \text{ kPa } (c. +28 \text{ mmHg}).$$

At the venular end

$$P_f = [2.0 - (-0.2)] - (3.5 - 2) = + 0.7 \text{ kPa } (c. +5 \text{ mmHg}).$$

Thus, at the level of the heart, the pressure in the capillaries favors filtration along their length (as the net filtration pressure is positive) and so movement of fluid from the capillaries to the interstitial fluid occurs. If the capillary pressure falls to zero (as it does when the arterioles constrict fully during vasomotion or following hemorrhage), the filtration pressure becomes negative (c. –1.5 kPa) and absorption of fluid is favored. However, fluid reabsorption is quickly limited by the rise in the oncotic pressure of the interstitial fluid. Thus, in this case, fluid absorption is transient.

In those tissues where the continuous absorption of fluid by the capillaries is an important part of their function (e.g. the peritubular capillaries of the nephron or the capillaries of the small intestine), the net filtration pressures are negative, at least at the venular end. For example, when isotonic fluid is being reabsorbed from the gut lumen, the filtration pressure at the venular end of the capillaries is between – 0.5 and – 1 kPa (c. –7 mmHg). The interstitial oncotic pressure is kept low because the absorbed fluid is replaced by fluid transported across the epithelial lining of the intestine.

Calculations such as these reflect the forces tending to move fluid in one direction or another across the capillary wall, assuming a steady state and that the capillary wall is impermeable to the plasma proteins. The quantities of fluid that move into or out of a capillary will depend on the hydraulic permeability of the capillary wall, the area available for fluid movement, and its permeability to plasma proteins. The full relationship can be expressed mathematically as

$$J_v = L_p A[(P_c - P_i) - \sigma (\pi_p - \pi_i)]$$

where J_v is the volume of fluid filtered or absorbed per unit time, L_p is a constant that reflects the hydraulic permeability of the capillary wall, A is the area of the capillary wall, and σ is the reflection coefficient, which is a measure of how freely proteins can cross the capillary wall. If the capillary wall is completely impermeable to proteins, $\sigma = 1$, and if it is fully permeable to proteins $\sigma = 0$. In most capillaries $\sigma \approx 0.9$. P_c and P_i are the capillary and interstitial pressures, respectively, and π_p and π_I are the oncotic pressures of the plasma and interstitial fluid, respectively.

lary to the interstitium, a process known as *filtration*. If the net filtration pressure is negative, the osmotic forces favor the movement of fluid from the interstitium to the capillary. Movement of fluid in this direction is *absorption* and usually results from a fall in capillary pressure (e.g. after hemorrhage—see Chapter 28).

The *rate of fluid filtration* depends on both the net filtration pressure and the filtration coefficient which takes into account the permeability of the capillary wall (see Box 15.8). Continuous capillaries have low filtration coefficients and the rate of filtration is correspondingly modest. In those capillary beds where there is significant transepithelial movement of fluid (e.g. the exocrine secretory glands and the renal glomerulus), the capillaries are fenestrated and have large filtration coefficients. This specialization allows the relatively rapid movement of fluid from the plasma to the interstitial space.

Capillary pressure depends both on the resistance of the arterioles and on the pressure in the veins

Clearly, while a capillary module is being perfused, the pressure at the arteriolar end of the capillaries will be higher than that at the end terminating in the venules (the venular end). In many capillary beds, such as those of the skin and skeletal muscle, the capillary pressure is greater than the oncotic pressure of the plasma, the net filtration pressure is positive, and Starling forces favor filtration (Box 15.8 and Fig. 15.33). This difference is accentuated by gravity. The pressure in the capillaries of the foot increases dramatically on standing up. When a subject is lying down, the mean capillary pressure in the toes is around 4 kPa (30 mmHg). On standing, this increases to around 13 kPa (97.5 mmHg). Note that the capillary pressures above the heart do not fall by a corresponding amount, as venous pressure remains close to zero (Fig. 15.26).

Since the microcirculation of an individual organ has a particular role, it is not surprising to find that the average pressure in the capillaries varies from one organ to another. In some organs, the capillaries absorb large volumes of fluid. Examples are the capillaries of the intestinal mucosa and the peritubular capillaries of the kidneys. In these tissues, the capillary pressure is low. Moreover, the bulk transport of fluid across the adjacent epithelium keeps the interstitial oncotic pressure low and increases interstitial pressure. As a result, the Starling forces favor fluid uptake by these capillaries (Box 15.8).

How do the capillaries reabsorb interstitial fluid? For most capillary beds, the net filtration pressure is positive except when the arterioles constrict during vasomotion. During these periods, the pressure within the capillaries declines towards zero and the net filtration pressure becomes negative. Under these conditions, fluid is reabsorbed into the capillary mainly because the oncotic pressure of the plasma is higher that that of interstitial fluid for the reasons explained in Box 15.7. However, as fluid is reabsorbed, the oncotic pressure of the interstitial fluid rises and the driving force for absorption declines to zero. Thus fluid reabsorption is transient, not sustained. Therefore it appears that the role of vasomotion is to provide periods of fluid absorption, so reducing the accumulation of fluid in the interstitial space.

Taking the body as a whole, a very small fraction of the plasma passing through the capillaries is filtered (around 0.1–0.2 per cent), but since about 4000 liters of plasma pass through the systemic capillaries every day this results in the movement of approximately 4–8 liters of fluid from the capillaries to the interstitial space each day. Between a half and two-thirds of this fluid is reabsorbed in the lymph nodes and the remainder is returned to the circulation via the efferent lymph (see below).

The role of the lymphatic circulation

In the majority of capillary beds, filtration of fluid exceeds any absorption by the capillaries and pericytic venules. If the excess fluid was not removed, it would accumulate in the tissues causing them to swell. Such swelling does occur in some pathological states and is called *edema* (see Chapter 28, p. 557).

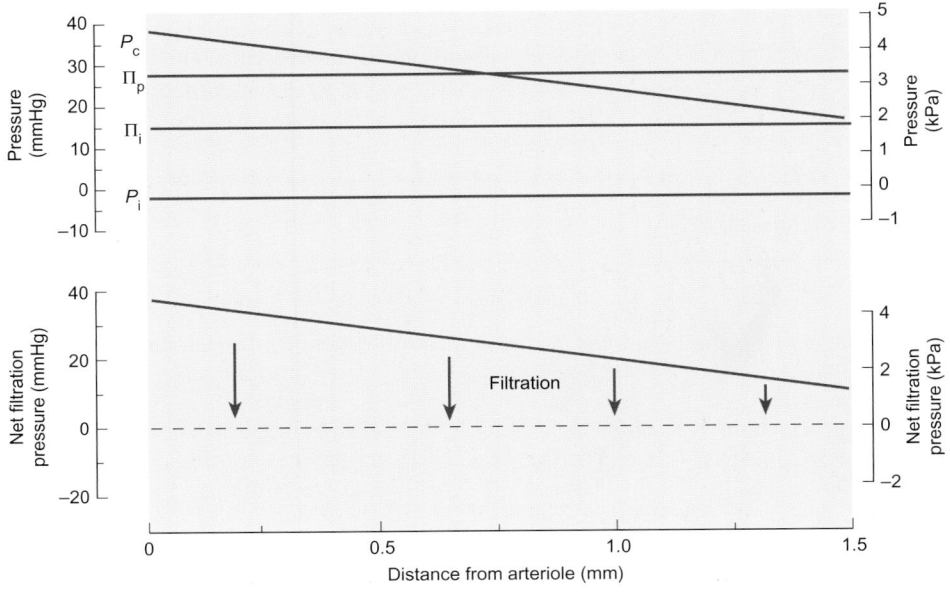

Fig. 15.33 The factors that determine the direction of fluid movement across the wall of a capillary in the skin of the hand. At the arteriolar end, the hydrostatic pressure in the capillaries (P$_c$) is high, but as the blood passes along the capillary the pressure progressively falls. Nevertheless, filtration is favored along the whole length of the capillary. See Box 15.8 for further details.

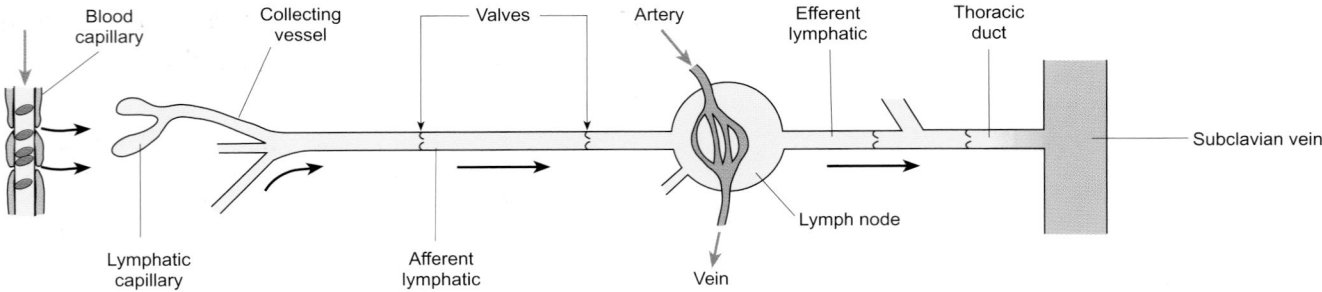

Fig. 15.34 The principal features of the organization of the lymphatic system.

Normally, the excess fluid leaves the interstitial spaces by draining into specialized vessels known as lymphatics. The fluid in the lymphatics is known as *lymph*. The lymph directly derived from the tissues is known as *afferent lymph* as it flows towards the *lymph nodes* (also known as lymph glands). The afferent lymph has the same composition as the interstitial fluid, i.e. it has the same ionic composition as the plasma but a lower protein content. It also has few cells. After the lymph has passed through the lymph nodes, it has a different composition (see below) and contains many lymphocytes that have migrated through the tissues to the lymph nodes (see Chapter 14). It is then known as *efferent lymph*.

The lymphatic capillaries either begin as small tubes closed at their distal end (as in the villi of the small intestine) or as a network of small tubes with diameters in the range 10–50 μm. Their walls are very thin and consist of only a single layer of endothelial cells on a basement membrane. Unlike most blood capillaries, the intercellular junctions of the walls of the lymphatic capillaries have clefts eaQ)ly large enough to permit the passage of plasma proteins. As with blood capillaries, the lymphatic capillaries join together to form collecting vessels, which in turn empty into afferent lymph trunks that are analogous to, and run alongside, major veins. The afferent lymph trunks possess semilunar valves at intervals along their length that prevent the back flow of lymph. The afferent lymph trunks empty into lymph nodes, which receive a blood supply of their own. The lymph leaving the nodes in the efferent lymphatics drains into the thoracic duct, which empties into the left subclavian vein. This organization is shown diagrammatically in Fig. 15.34.

How is the lymph formed?

The exact processes responsible for drawing fluid into the lymphatics are not known with certainty. Contraction of the walls of the large lymphatic vessels is thought to drive the lymph forward, and this is assisted by the compression and subsequent recoil of the lymphatics by the surrounding tissues during muscular activity. These forces drive the lymph forward towards the lymph nodes. As the lymphatic capillaries are closed at one end and can only empty into the larger lymphatic vessels, the recoil of the larger lymphatics sucks fluid from the lymphatic capillaries. Therefore the process of filling the lymphatics is somewhat analogous to the filling of a rubber bulb with liquid after expelling air.

Summary

1. The capillaries have thin walls that consist of a single layer of endothelial cells. The capillaries provide a large surface area for the exchange of solutes between the blood and the tissues. Their blood flow is governed by vasomotor activity in the arterioles and varies continuously as a result of changes in arteriolar tone (vasomotion).

2. Most solute exchange between blood and tissues occurs by diffusion. Oxygen and carbon dioxide diffuse through the endothelial cells and equilibrate rapidly by transcellular exchange. Water-soluble substances cannot pass across the endothelial cells and diffuse through the intercellular clefts of the capillary wall (paracellular exchange). Smaller molecules diffuse more freely than large molecules. Consequently, the capillary wall is almost impermeable to proteins.

3. The bulk flow of fluid between the plasma and the interstitial fluid is determined by the net filtration pressure, which varies somewhat from one capillary bed to another according to the physiological requirements. The rate of fluid movement depends on the permeability of the capillary wall.

4. Filtration occurs when the net filtration pressure is positive, and absorption is favored when the net filtration pressure is negative. In general, filtration exceeds absorption and the excess fluid is returned to the circulation via the lymph.

Lymph nodes are known to modify the volume and protein composition of the lymph. The afferent lymph generally has a low oncotic pressure, but after passing through the lymph nodes the protein content is increased substantially—probably as a result of absorption of water and electrolytes by the vascular capillaries of the lymph nodes. It has been shown experimentally that as much as 50 per cent of the volume of the afferent lymph can be reabsorbed during its passage through the lymph nodes.

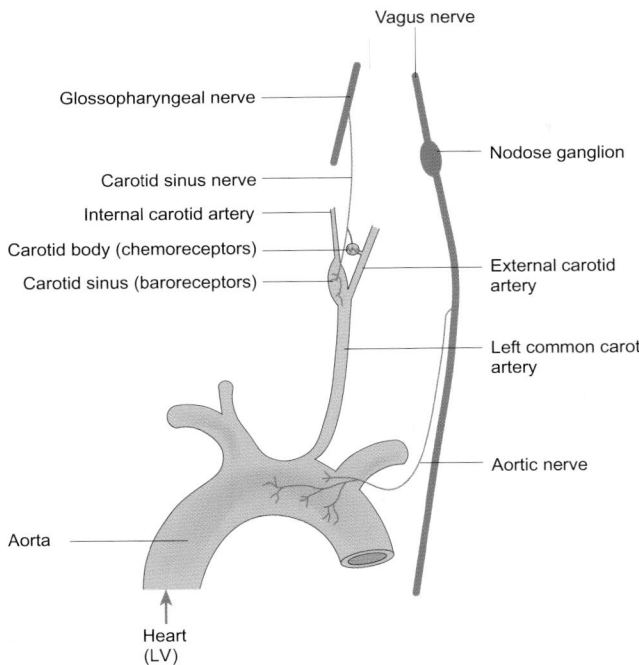

Fig. 15.35 A highly simplified diagram illustrating the position and nerve supply of the aortic and carotid baroreceptors.

15.11 The role of the central nervous system in the control of the heart and the circulation

The activity of the heart and the tone of the blood vessels are both regulated by the autonomic nervous system, which is in turn subject to control by the CNS. The principal regions of the CNS that are concerned with the control of the cardiovascular system are the hypothalamus and the medulla oblongata. The areas of the medulla concerned with control of the cardiovascular system are sometimes referred to as cardiac control 'centers', although this term is somewhat misleading since the medullary neurons are themselves regulated by higher brain areas and by afferent input from pressure receptors. Figure 15.29 shows the basic organization of those areas that are concerned with the control of the cardiovascular system.

Afferent information concerning the pressures within the circulation is supplied by two groups of receptors: high-pressure receptors (*baroreceptors*), which are located in the aorta and carotid arteries, and low-pressure receptors (or *volume receptors*), which are located in the walls of the atria and ventricles. The afferent fibers of these receptors relay information to the brainstem regarding arterial pressure and cardiac filling pressure. This information, coupled with other inputs from higher brain

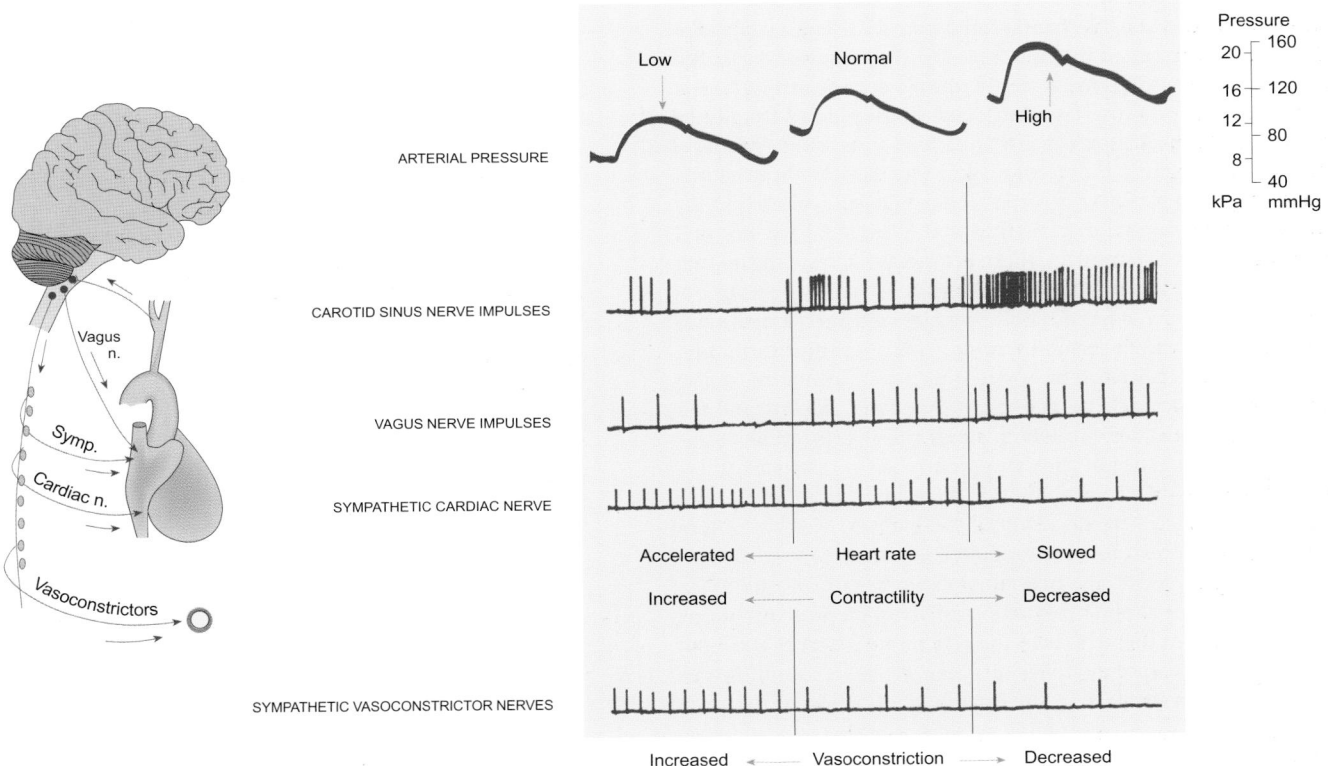

Fig. 15.36 The discharge pattern of single nerve fibers in various nerves as arterial pressure changes. The basic organization of the nervous pathways is shown on the left. The top panel on the right shows the aortic pressure wave for low, normal, and elevated arterial pressure. Below is shown the discharge pattern for a baroreceptor afferent in the carotid sinus nerve, the discharge pattern for a cardiovagal nerve fiber, the discharge pattern for a sympathetic cardiac nerve fiber, and the pattern for a sympathetic vasoconstrictor nerve fiber. At low arterial pressure, vagal activity is inhibited while sympathetic activity is increased, resulting in an increased heart rate and vasoconstriction. At high pressures, vagal activity is increased while sympathetic activity is decreased resulting in a slowing of the heart rate and vasodilatation.

areas, chemoreceptors, and other sensors, is used to initiate the appropriate cardiovascular responses. These responses are mediated by the parasympathetic and sympathetic nervous pathways to the heart and blood vessels.

Stimulation of the arterial baroreceptors elicits a reflex vasodilatation and a fall in heart rate

The arterial blood pressure is very closely regulated by the body. In young adults, systolic pressure is maintained close to 16 kPa (120 mmHg) during the day, although it may fall somewhat during sleep. There are two types of regulatory mechanisms. These are the rapid regulation by nerves and hormones, and the longer-term control of blood volume, which is largely mediated by the kidneys (and is described in Chapter 28, pp. 549–557).

The pressure in the arterial circulation is monitored by baroreceptors which are most abundant in the walls of the carotid sinuses and the aortic arch. The baroreceptors are mechanoreceptors that sense the degree of stretch of the walls of the vessels in which they are located and thus are able to monitor blood pressure. Impulses from the baroreceptors are transmitted to the brainstem where they terminate in the nucleus of the tractus solitarius (NTS). Afferents from the aortic arch baroreceptors travel in the vagus nerves, while those from the carotid sinus receptors are carried by the carotid sinus nerves which merge with the glossopharyngeal nerve. The afferent nerve supply of the baroreceptor system is illustrated diagrammatically in Fig. 15.35.

The response of the baroreceptors to a changing arterial pressure is illustrated in Fig. 15.36. The frequency of discharge in baroreceptor afferents varies in phase with the arterial pulse wave. As the pressure rises, so does the discharge frequency and vice versa. At normal levels of blood pressure the baroreceptor afferents show tonic activity. Below about 8 kPa (60 mmHg) there is very little activity in the baroreceptor afferents but, as pressure rises over the range between 8 and 24 kPa (60 and 180 mmHg), there is a steep increase in the rate of discharge.

A sharp rise in arterial blood pressure will increase the discharge of the baroreceptor afferents. This afferent information is relayed to the brainstem where it elicits an inhibition of the sympathetic activity to the heart and vasculature and an increase in vagal activity which slows the heart (bradycardia). The reduced activity of the sympathetic vasoconstrictor fibers results in vasodilatation, which in turn leads to a fall in peripheral resistance. The net result of these effects will be a fall in arterial blood pressure which rapidly 'buffers' the initial rise.

Conversely, a fall in arterial blood pressure (acute hypotension) brings about a reduction in baroreceptor discharge. There is a rise in the activity of the sympathetic nerves and a fall in that of the parasympathetic nerves. The heart rate increases and there is an increase in the force of contraction of the myocardium. The vessels supplying skeletal muscle, skin, kidneys, and splanchnic circulation constrict. As a result, blood pressure will rise to counteract the initial fall.

These changes in peripheral vascular tone and in the force and rate of cardiac contraction that occur in response to changes in baroreceptor activation form the *baroreceptor reflex* or *baroreflex* which is illustrated in Fig. 15.36. The reflex vasomotor responses occurring in response to a change in blood pressure are due entirely to alterations in the frequency of discharge in sympathetic vasoconstrictor fibers. The sympathetic vasodilator fibers are not involved.

Resetting of the baroreceptor reflex

If arterial blood pressure is raised for a period of about 15 minutes, the threshold for baroreceptor activity rises to a higher value. This property of the baroreceptors makes them ineffective monitors of the absolute pressure of the blood passing to the brain. The baroreceptors are short-term regulators of blood pressure rather than long-term controllers. They can be reset so that they respond to higher pulse pressures. This occurs during exercise, for example. Consequently, the heart rate does not fall in response to the increase in systolic blood pressure that accompanies exercise, so that cardiac output is maintained. Similarly, during the alerting response in which the body is prepared for intense activity, the baroreceptor mechanism appears to be inhibited and there is a sharp rise in blood pressure.

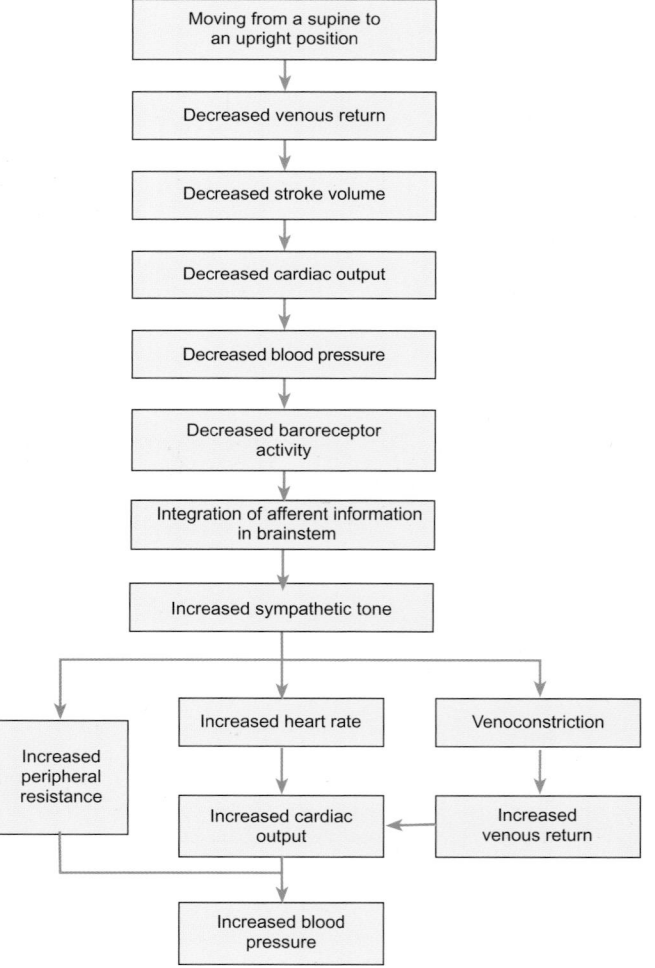

Fig. 15.37 The sequence of cardiovascular changes initiated by the baroreceptors following a fall in arterial blood pressure and resulting in restoration of normal pressure.

The baroreceptor reflex stabilizes blood pressure following a change in posture

When a person moves quickly from a lying to a standing position, there is a significant fall in venous return to the heart (as a result of a shift of blood from the cardiopulmonary region to veins in the legs). In turn, this leads to a fall in stroke volume (by Starling's Law), cardiac output, and therefore blood pressure. This is known as *postural hypotension* and the baroreceptors play an important role in the rapid restoration of normal pressure. When pressure falls, baroreceptor discharge is proportionately decreased and this results in an increase in sympathetic discharge to the heart and vasculature. Heart rate is increased and there is a rise in total peripheral resistance. Both these effects restore the blood pressure to its normal level. Because the baroreceptor reflex normally takes a few seconds to become fully effective, the blood supply to the brain is briefly reduced when standing up quickly and, as a result, it is quite common to experience momentary dizziness. This sequence of events is summarized in Fig. 15.37.

The Valsalva maneuver

The Valsalva maneuver is essentially an attempt to expire against a closed glottis. It can be carried out voluntarily and is associated with forceful defecation, the lifting of heavy weights and childbirth. The maneuver raises intrathoracic pressure which elicits a complex cardiovascular response that is illustrated in Fig. 15.38.

The initial response is a rise in blood pressure which is due to the normal contraction of the left ventricle to which is added the additional force of the raised intrathoracic pressure acting on the myocardium. This is followed by a transient fall in heart rate. At this stage, the increased intrathoracic pressure impedes

the venous return, with the result that cardiac output and mean arterial pressure fall. As the arterial pressure falls, the heart rate increases and this, together with an increase in total peripheral resistance, stabilizes the blood pressure. When the intrathoracic pressure falls following opening of the glottis, the blood pressure initially falls but as the venous return is rapidly restored, end-diastolic volume and cardiac output increase, and blood pressure rises. This rise is sensed by the baroreceptors, which cause a reflex bradycardia. This, together with a fall in peripheral resistance, restores the blood pressure to normal.

Low-pressure receptors in the atria sense central venous pressure and provide information about cardiac distension

Stretch receptors are present in the walls of the right and left atria of the heart. The atrial receptors respond to the central venous pressure and cardiac distension. They are stimulated when venous return is increased. This elicits a reflex rise in heart rate and contractility that is mediated via the sympathetic nervous system. Activation of this reflex rapidly reduces the initial cardiac distension. In contrast, activation of the mechanoreceptors in the left ventricle induces a reflex bradycardia and vasodilatation in response to a rise in end-diastolic pressure. The exact function of this response is unclear, but it may serve to assist the baroreceptor reflex in the regulation of arterial blood pressure.

Hormonal control of blood volume provides long-term regulation of blood pressure

Although the baroreceptors are concerned with minute-to-minute regulation of the arterial blood pressure, the dynamic nature of their responses makes them unsuitable for long-term regulation of absolute blood pressure. This depends on the maintenance of a normal extracellular volume and composition by the kidneys. The chief mechanisms involved are hormonal and include the reflex control of pituitary ADH secretion by osmoreceptors in the hypothalamus, the operation of the renin–angiotensin–aldosterone system, and the role of the atrial natriuretic peptide. All these processes act to maintain a constant circulating volume. The underlying mechanisms are discussed more fully in Chapter 28.

Activation of the peripheral arterial chemoreceptors causes an increase in blood pressure

While the most familiar role of the peripheral arterial chemoreceptors is in the control of respiration (see Chapter 16 for a detailed discussion), they also play a part in the reflex elevation of blood pressure seen during hypoxia. At normal blood gas tensions, they have little influence on the circulation, but during both hypoxia and hypercapnia they elicit a reflex vasoconstriction in the resistance vessels and in the large veins of the splanchnic circulation. The constriction of the arterioles and the mobilization of blood from the splanchnic circulation elevate the arterial blood pressure and increase the amount of oxygen carried to the tissues, particularly the brain. This chemoreceptor reflex is particularly powerful at low arterial blood pressures, when the baroreceptor reflex is relatively inactive.

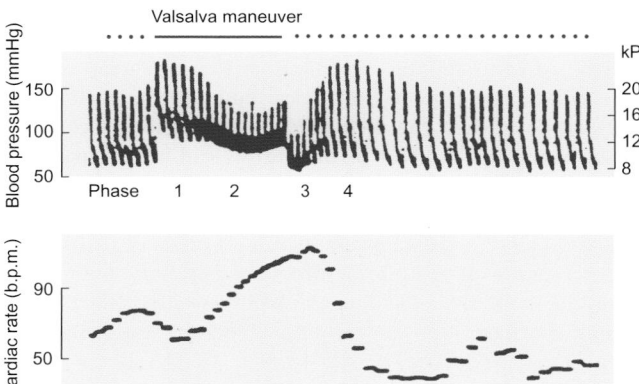

Fig. 15.38 Changes in blood pressure and heart rate occurring during and after the Valsalva maneuver. When the intrathoracic pressure first rises there is an increase in mean arterial pressure as venous return is initially favored by compression of the great veins (phase 1). Cardiac output and pulse pressure rise. This is followed by a decline in mean arterial pressure and pulse pressure as venous return is impeded and cardiac output falls (phase 2). At the end of the maneuver the intrathoracic pressure falls, which briefly results in a drop in blood pressure (phase 3). This is followed by a sharp increase in venous return leading to an increase in stroke volume and pulse pressure (phase 4). The changes in heart rate are a reflex response to the changes in pressure. When pressure rises, the heart rate falls and vice versa.

Respiratory sinus arrhythmia

During inspiration there is a small increase in heart rate which is followed by a decrease in rate during expiration. This variation in heart rate with the respiratory cycle is called *respiratory sinus arrhythmia* and is almost entirely due to changes in the activity of cardiovagal motoneurons whose normal tonic activity is partly dependent on the arterial baroreceptor input (Fig. 15.29). During inspiration, the cardiovagal motoneurons are inhibited by central inspiratory neurons and by an increased activity of the slowly adapting stretch receptors that are present within the larger airways of the lungs. Consequently, they become less sensitive to their baroreceptor inputs. This results in a decrease in vagal tone and an increase in heart rate. During expiration, this effect is reversed; the excitability of the cardiovagal motoneurons returns to normal, vagal tone increases and the heart slows.

Activation of 'work receptors' elicits an increase in cardiac output

The final group of receptors which can reflexly influence the cardiovascular system are the 'work receptors' found in skeletal muscle. They are of two types: chemoreceptors that respond to potassium and hydrogen ions produced during exercise, and mechanoreceptors sensitive to the active tension generated within a muscle. Activation of these receptors, during exercise, causes an increase in heart rate, vasoconstriction of vessels other than those supplying the working muscles (notably the renal and splanchnic circulations), and increased myocardial contractility. All of these serve to raise blood pressure and increase perfusion of the active muscles.

The defense reaction

In response to a perceived danger, all mammals, including humans, show a behavioral alerting reaction that may lead to aggressive or defensive behavior if the stimulus is strong enough. This is accompanied by marked changes in sympathetic activity in which normal reflex control is overridden. The pupils dilate, the skin hair becomes erect (piloerection), the teeth are bared, and a defensive posture is adopted. In addition, and of more concern here, there is an increase in heart rate, cardiac output, and blood pressure. The cardiovascular responses involve resetting of the baroreceptor reflex to a higher level of pressure and there is a redistribution of blood flow. There is vasoconstriction in the viscera and skin while the blood flow to the muscles is increased. In humans, this increase is due to an inhibition of sympathetic vasoconstrictor tone while in other species there may also be an activation of sympathetic cholinergic vasodilator activity. This complex sequence of events is termed the *defense reaction*. The changes that occur during normal behavioral altering are more modest and are called the alerting response.

15.12 Special circulations

This section deals briefly with the circulation of blood to specific organs. It will illustrate the ways in which some of the regulatory mechanisms described in earlier sections are used in localized regions of the cardiovascular system to match blood flow to tissue requirements. The specific features of the pulmonary circulation are discussed in Chapter 16, pp. 328–332, those of the renal circulation are considered in Chapter 17, pp. 350–352 and those of the gut are described in Chapter 18, p. 382.

The coronary circulation

Figure 15.39 illustrates the coronary blood supply to the heart. The myocardial blood supply is provided by the right and left coronary arteries, which arise at the root of the aorta. The right coronary artery supplies principally the right atrium and right ventricle, while the left coronary artery supplies principally the left atrium and left ventricle. The venous drainage is principally to the right atrium via the coronary sinus.

Blood flow through the coronary arteries varies greatly during the cardiac cycle. Flow to the myocardium is at its peak during early diastole when the mechanical compression of coronary vessels is minimal and aortic pressure is still high. During the isovolumetric contraction phase of the cycle, the coronary blood vessels become compressed as the ventricular pressure rises and coronary blood flow declines to its minimum value. During the ejection phase of systolic contraction, coronary flow varies according to the rise and fall in aortic pressure. The effect of ventricular contraction on the coronary blood flow is illustrated in Fig. 15.40.

It will be evident from the earlier discussions that the work carried out by the heart varies considerably with the demands of the circulation. Clearly, a greater workload requires a greater supply of oxygen and nutrients to be delivered to the myocardium. Indeed, blood flow through the coronary circulation can increase from its resting level of around 75 ml min⁻¹ per 100 g tissue to as much as 400 ml min⁻¹ per 100 g tissue during maximal cardiac work.

As Figure 15.41 shows, the coronary blood flow increases as the metabolic activity of the heart is increased. How is the blood flow matched to the metabolic demands? The coronary sinus

Summary

1. Arterial blood pressure is very closely regulated by the autonomic nerves, hormones, and, on a more long-term basis, changes in blood volume.

2. Blood pressure is monitored by baroreceptors present in the walls of the aortic arch and the carotid sinuses. A rise in blood pressure results in increased firing of the baroreceptor afferents that leads to a reflex slowing of the heart, peripheral vasodilatation, and a fall in blood pressure.

3. Long-term regulation of the blood pressure is achieved through the maintenance of normal extracellular volume and composition, largely by means of the renin–angiotensin–aldosterone system and atrial natriuretic peptide.

4. Arterial chemoreceptors and muscle work receptors also play a role in the control of blood pressure.

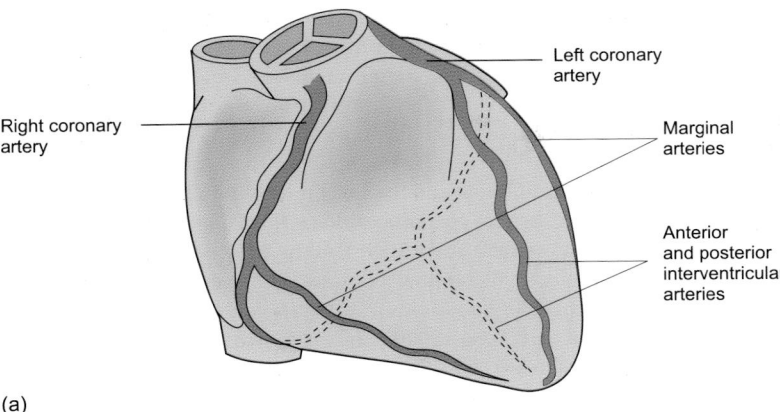

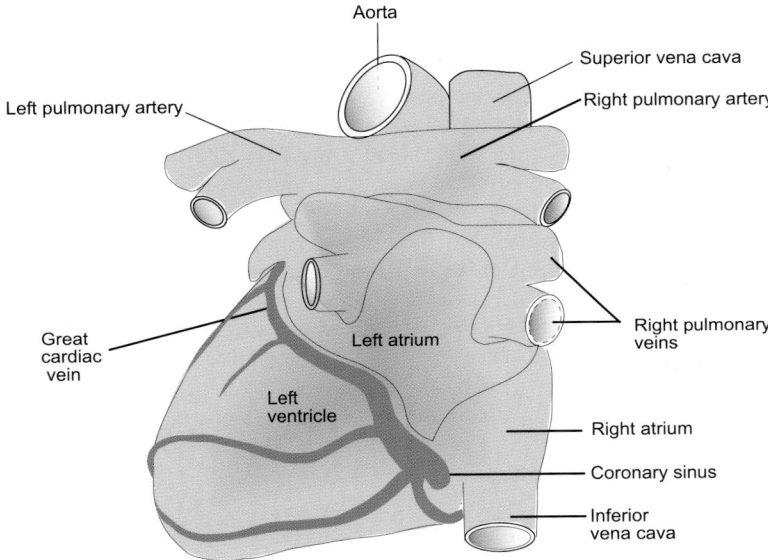

Fig. 15.39 The organization of the coronary blood supply. (a) The arterial supply; note the origin of the coronary arteries from the aortic sinuses at the base of the aorta. (b) The venous drainage of the coronary circulation viewed from the posterior aspect.

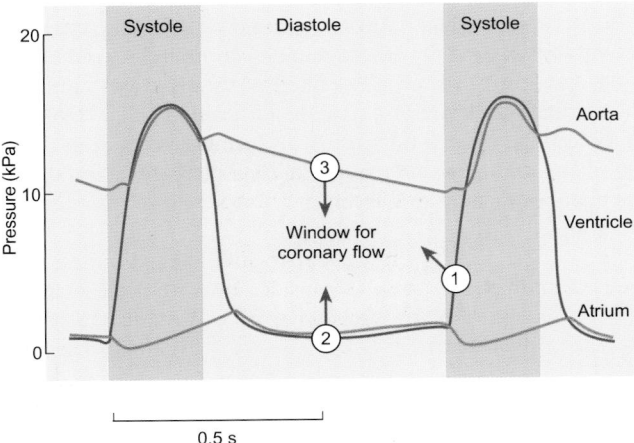

Fig. 15.40 The pressure changes that affect coronary blood flow during the cardiac cycle. The period during which coronary blood flow occurs will be shortened by (1) a reduction in the diastolic interval, (2) a rise in ventricular end-diastolic pressure, or (3) a fall in arterial pressure.

blood has a P_{O_2} of about 2.7 kPa (20 mmHg). Therefore oxygen extraction is very high. An increase in oxygen demand elicits an increase in coronary blood flow as a result of vasodilatation of the coronary vessels. This vasodilatation is thought to be mediated principally by adenosine, whose production increases during increased workload. Increased concentrations of other metabolites such as carbon dioxide, hydrogen ions, and potassium ions may also play a role, as may interstitial hypoxia. The sympathetic nerve supply to the coronary arteries is of lesser inportance in the control of coronary blood flow. Nevertheless, stimulation of these fibers causes some vasodilatation, as does the interaction of circulating epinephrine on the β_2-adrenoceptors of the vascular smooth muscle. However, the most important determinant of coronary blood flow is metabolic hyperemia.

During periods of high cardiac work, the time available for blood to flow through the coronary circulation will be much reduced because the heart is contracting more powerfully and more often. Under these conditions, it is crucial that blood flow

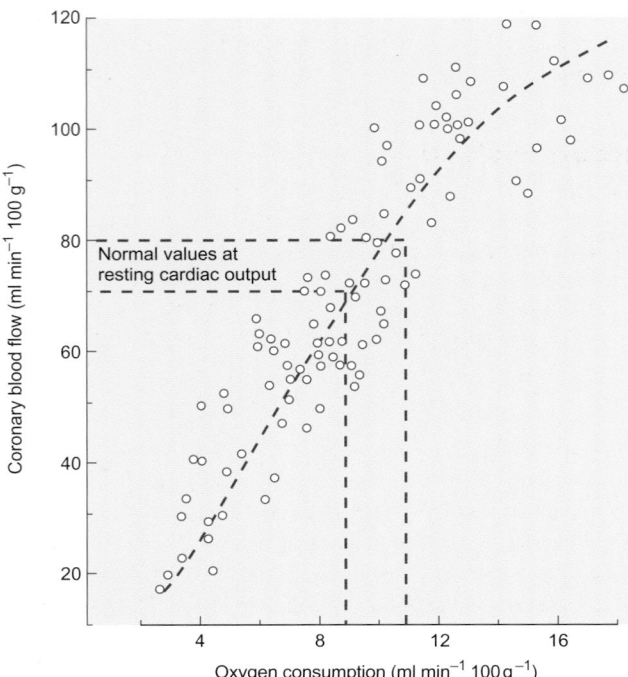

Fig. 15.41 The relationship between the coronary metabolism (as measured by oxygen consumption) and coronary blood flow in the dog. In these experiments, the cardiac work was increased by the administration of epinephrine and decreased by producing a controlled hemorrhage.

to the myocardium during diastole is adequate to provide sufficient oxygen for the whole of the cardiac cycle. This need is met in part by the release of oxygen from the myoglobin present in the cardiac muscle fibers.

The circulation of the skin

The skin possesses two types of resistance vessels, the arterioles which are similar to those found elsewhere, and *arteriovenous (AV) anastomoses*. These vessels provide a route for blood to pass from the arterioles directly to venules, bypassing the capillaries. They are found predominantly in the skin of the hands, feet, ears, nose, and lips. A simplified diagram of the cutaneous circulation in an extremity is shown in Fig. 15.42. The AV anastomoses are richly innervated with sympathetic fibers, which can cause substantial vasoconstriction, but they do not appear to be under local metabolic control. In contrast, the arterioles of the skin show autoregulation of blood flow and reactive hyperemia.

The most important regulator of blood flow to the skin is not the supply of oxygen and nutrients as it is in most other tissues. Indeed, its nutritional requirements are small. The principal role of the cutaneous circulation is in the regulation of body temperature (see also Chapter 26). Consequently, there are wide fluctuations in the blood flow to the skin. When ambient temperature is low and the body must retain heat, the cutaneous vessels constrict to reduce blood flow. In contrast, when the ambient temperature is high and the body needs to lose heat, the vessels dilate, blood flow increases, and heat is lost from the skin surface. Under conditions of extreme cold, cutaneous blood flow may fall to as little as 0.01 l min^{-1} kg^{-1} while in hot environments it may increase to 2.0 l min^{-1} kg^{-1}. Flow to the skin under normal (thermoneutral) conditions is around 0.15 l min^{-1} kg^{-1}.

In response to information from the temperature receptors in the skin, the level of sympathetic activity is adjusted according to the need to retain heat. The cutaneous vessels do not receive any parasympathetic vasodilator fibers. If the ambient temperature falls and heat production is insufficient to maintain normal body temperature, sympathetic activity is increased, the AV anastomoses close, and the arterioles constrict. This vasoconstriction shifts blood away from the body surface. When body temperature rises, the reduction in sympathetic activity opens up the AV anastomoses and the arterioles dilate. Blood flow to the skin is increased and there is greater heat loss. In circumstances that require a greater heat loss, sweating is stimulated. The sweat glands are innervated by cholinergic sympathetic vasodilator fibers. Sweat contains an enzyme whose action produces the potent vasodilator bradykinin. This acts locally to bring about vasodilatation of the arterioles. The arterioles also dilate in response to increased local levels of metabolites, a situation likely to occur when body temperature is increased.

The circulation of skeletal muscle

The muscle mass is very large and the skeletal muscle represents the largest vascular bed in the body. Consequently, the caliber of the resistance vessels of skeletal muscle plays an important role in determining arterial blood pressure.

Blood flow to skeletal muscle varies directly with contractile activity. Total blood flow through quiescent muscle is rather low (0.01–0.04 l min^{-1} kg^{-1}) because many of the arterioles supplying the resting muscle are in a state of constriction and a large part of the capillary bed is not perfused. During exercise, the arterioles dilate and blood flow to the muscle increases significantly (up to 20 times its resting level).

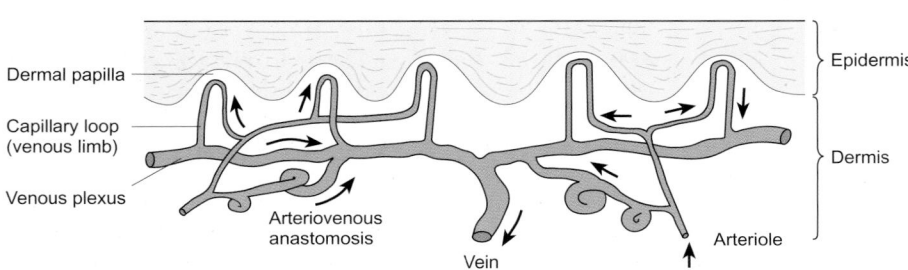

Fig. 15.42 A diagrammatic representation of the cutaneous circulation from a region possessing AV anastomoses.

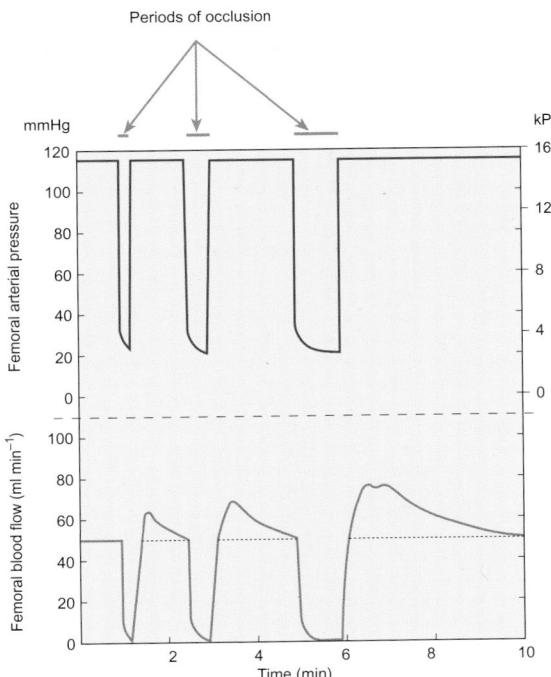

Fig. 15.43 Reactive hyperemia in the hind limb of the dog following periods of occlusion of the femoral artery. Note that the magnitude and duration of the hyperemia is related to the duration of the occlusion.

Local and neural factors both contribute to the regulation of skeletal muscle blood flow. At rest, neural and myogenic control predominates but, during exercise, control by local metabolites is much more important. The arterioles of skeletal muscle have a high resting level of tone. This is partly myogenic in origin and partly the result of tonic sympathetic vasoconstrictor activity, which is controlled reflexly by the baroreceptors. As a result, blood flow is low. During exercise, there is a marked vasodilatation of the arterioles of the active muscle with greatly increased perfusion of the capillary beds. This dilatation is due mainly to the increased production of metabolites and represents an example of functional hyperemia. In addition to this, however, there is also some vasodilatation in response to circulating epinephrine secreted by the adrenal medulla (interacting with β_2-adrenoceptors).

Exercising muscle has a problem similar to that of the myocardium with respect to its perfusion (see p. 302). A strongly contracting leg muscle, for example, squeezes its blood vessels with each contraction, thereby reducing the amount of blood that can flow to it. In continuous rhythmical exercise, such as walking, there will be time between contractions for the dilated vessels to provide an adequate supply of nutrients to the muscle. Myoglobin is able to supply oxygen to the muscle when the blood vessels are being compressed. However, during a sustained strong contraction, blood flow will practically cease and metabolites, particularly lactic acid, will quickly build up. When the contraction ceases, reactive hyperemia occurs because of the accumulation of these metabolites. The change in blood flow in the hind limb of a dog following a period of vascular

compression is illustrated in Fig. 15.43. After each period of compression, the blood flow is significantly increased above the normal resting level and the extent of this increase is proportional to the duration of the occlusion.

Cerebral circulation

Blood flows to the brain via the internal carotid and vertebral arteries. These join together (*anastomose*) around the optic chiasm to form the circle of Willis from which the anterior, middle, and posterior cerebral arteries arise. These branch to form the pial arteries, which run over the surface of the brain. From here, smaller arteries penetrate into the brain tissue itself to give rise to short arterioles and an extensive capillary network. Brain tissue, particularly the gray matter, has a very high capillary density.

In contrast with most other organs, the blood flow to the brain is kept within very closely defined limits (around 0.55 l min^{-1} kg^{-1} or about 15 per cent of the resting cardiac output). The brain is very sensitive to ischemia, and loss of consciousness occurs if blood flow is interrupted for only a few seconds. Irreversible tissue damage may occur if blood flow is stopped for several minutes. If cardiac output falls (e.g. in hemorrhage), the cerebral blood flow will be maintained as far as possible even though perfusion of many peripheral organs may be compromised.

Cerebral blood flow is closely controlled by a combination of autoregulation, local metabolic regulatory mechanisms, and reflexes initiated by the brain itself which either act locally or by altering systemic blood pressure (e.g. Cushing's reflex—see below). The cerebral arteries receive some sympathetic vasoconstrictor innervation, though its effect on cerebral perfusion is fairly weak. There are also some vasodilator fibers (which are probably parasympathetic in origin) but their role is unclear. As the brain is enclosed within the rigid structure of the skull, changes in the volume of blood or extracellular fluid are minimal. Any change in the volume of blood flowing to the brain must be matched by a change in the venous outflow.

Regional functional hyperemia is well developed in the brain—an increase in neuronal activity evokes a local increase in blood flow. Indeed, modern techniques of computerized mapping of regional blood flow have exploited this property to localize sites of mental function or epileptic foci in the brains of conscious subjects. The cause of the metabolite-induced increase in blood flow is thought to be a rise in extracellular potassium ions resulting from the outward potassium currents of active neurons. Local increases in hydrogen ion concentration may also play a role.

Cerebral blood flow is highly sensitive to arterial carbon dioxide. An increase in PCO_2 (*hypercapnia*) brings about vasodilatation and a rise in blood flow. Vasoconstriction occurs during hypocapnia (which is a consequence of hyperventilation). The fall in blood flow leads to a feeling of dizziness and may cause disturbed vision. There is also a degree of vasodilatation in response to hypoxia. It is now believed that the arterial responses to hypoxia and hypercapnia in the brain are mediated by nitric oxide released from the vascular endothelium.

As the brain is encased in the skull, which provides a rigid enclosure except in very young children, anything that increases the volume of the brain tissue, such as a cerebral tumor or a

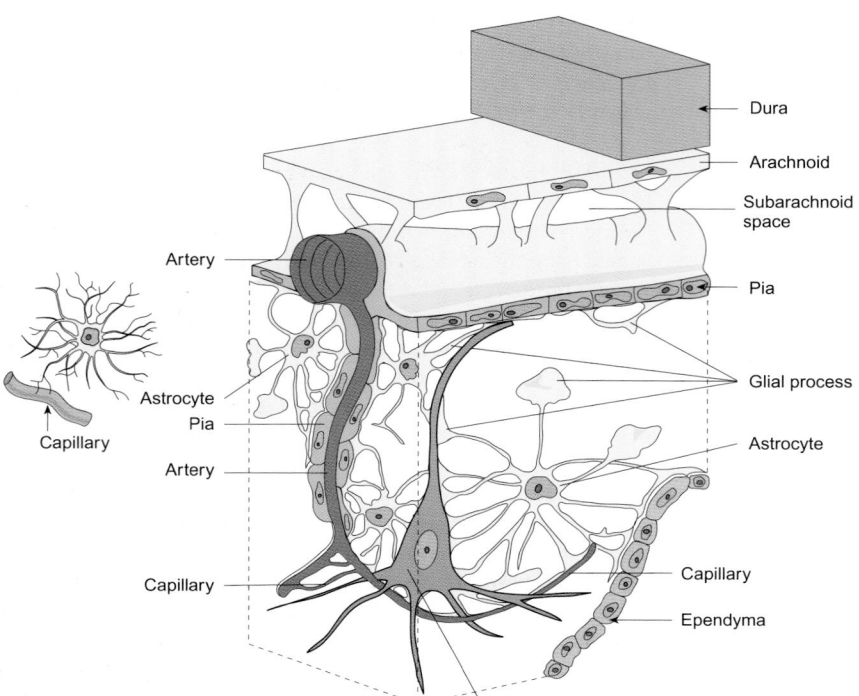

Fig. 15.44 The arrangement of the capillary endothelium and astrocytic end-feet in the cerebral circulation.

cerebral hemorrhage, will increase the pressure within the skull (the *intracranial pressure*). This will tend to reduce the blood flow to the brain and force the brainstem into the opening of the skull through which the spinal cord passes (the foramen magnum). The resulting compression of the brainstem elicits a marked rise in arterial blood pressure, which acts to offset the fall in cerebral blood flow. This is known as Cushing's reflex (sometimes also called Cushing's response). In addition to the rise in blood pressure, there is a fall in heart rate mediated by the baroreceptor reflex. This combination of a marked rise in blood pressure and a slow heart rate is an important clinical indicator of a space-occupying lesion within the skull. If the increase in intracranial pressure is very high, cerebral blood

flow will decline and the sufferer will experience mental confusion, leading to coma and death, unless prompt action is taken.

The blood–brain barrier

Although lipid-soluble molecules are able to pass freely from the blood to the interstitium of the brain, ionic solutes are unable to do so. Because of this, the neuronal environment can be tightly controlled and the cells protected from fluctuating levels of ions, hormones and tissue metabolites circulating in the plasma. This restricted movement has been attributed to a specific 'blood–brain' barrier. The restriction appears to be at least partly due to the nature of the tight junctions between the capillary endothelial cells of the cerebral circulation. In addition, astrocytes

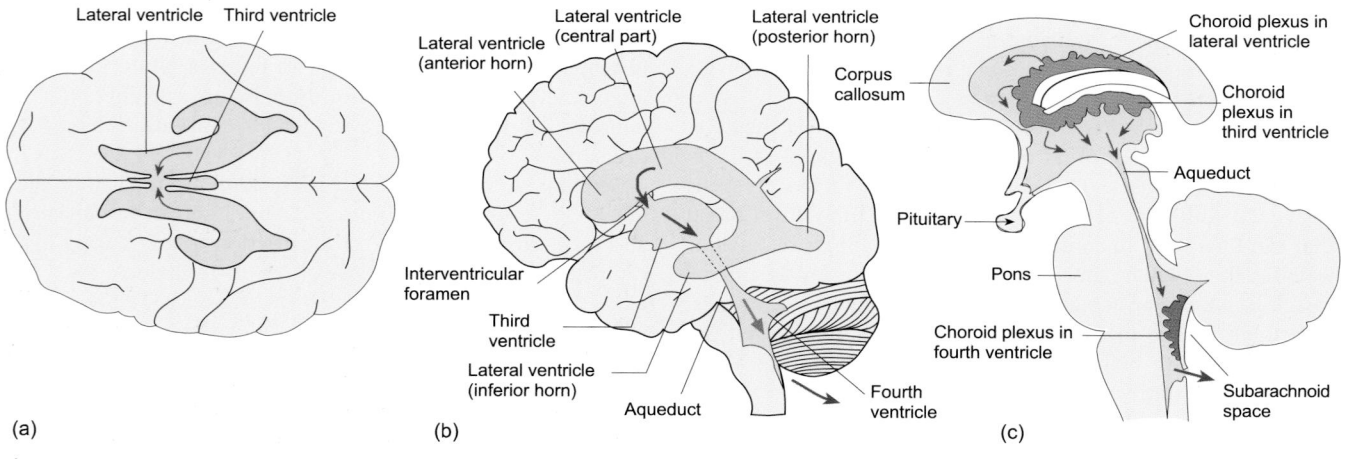

(a)   (b)   (c)

Fig. 15

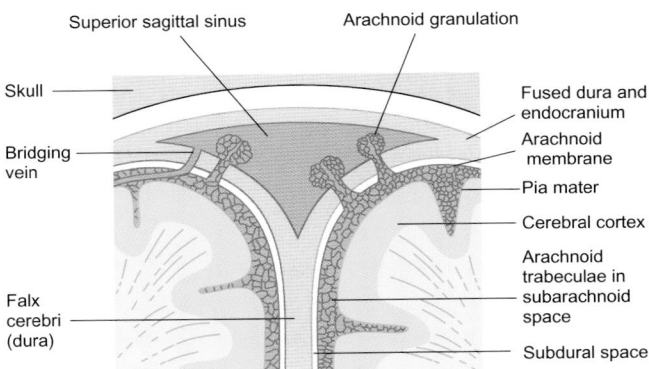

Superior sagittal sinus   Arachnoid granulation

Skull

Bridging vein

Falx cerebri (dura)

Fused dura and endocranium
Arachnoid membrane
Pia mater
Cerebral cortex
Arachnoid trabeculae in subarachnoid space
Subdural space

Fig. 15.46 The relationship between the meninges, the arachnoid villi, and the superior sagittal sinus. The CSF drains into the venous blood via the arachnoid granulations that penetrate the large venous sinuses formed by folds in the dura.

extend processes that contact the endothelial cells (astrocytic end-feet), helping to seal the interstitial space of the brain from the circulating plasma (Fig. 15.44).

The blood–brain barrier is breached at a few sites along the mid-line of the brain. These sites are located along the third and fourth cerebral ventricles and are known as the circumventricular organs. Here the capillaries are fenestrated, allowing relatively free exchange between the plasma and the interstitial fluid of the neural space. The function of these areas is still being elucidated, but they appear to play a significant role in the regulation of fluid balance.

The cerebral ventricles and the cerebrospinal fluid

The fluid-filled spaces in the brain (the cerebral ventricles) and the central canal of the spinal cord contain a clear fluid known as cerebrospinal fluid (CSF) The CSF also fills the space between the dura mater (the tough membranous outer covering of the brain and spinal cord) and the surface of the brain and spinal cord (see Chapter 6, Fig. 6.1). In all, the total volume of the CSF is about 130 ml.

About 500 ml of CSF is formed each day by the choroid plexus in the cerebral ventricles. The choroid plexus consists of networks of capillaries surrounded by an epithelial layer. The CSF

TABLE 15.4 The composition of the plasma and cerebrospinal fluid

	CSF	Plasma	Ratio CSF to plasma
Na$^+$ (mmol l^{-1})	141	141	1.0
K$^+$ (mmol l^{-1})	3.0	4.5	0.67
Ca^{2+} (mmol l^{-1})	1.15	1.5	0.77
Mg^{2+} (mmol l^{-1})	1.12	0.8	1.4
Cl$^-$ (mmol l^{-1})	120	105	1.14
HCO$_3^-$ (mmol l^{-1})	23.6	24.9	0.95
Glucose (mmol l^{-1})	3.7	4.5	0.82
Protein (g l^{-1})	0.18	75	0.002
pH	7.35	7.42	—

flows from the lateral ventricles into the third ventricle and thence through the cerebral aqueduct to the fourth ventricle as shown in Fig. 15.45. It leaves the cerebral ventricles via apertures in the roof of the fourth ventricle (the foramens of Magendie and Luschka) and flows into the subarachnoid space which lies between the arachnoid membrane and the pia mater. From the subarachnoid space, the CSF passes into the venous blood via the *arachnoid granulations*, which project into the venous sinuses as shown in Fig. 15.46. The veins connecting the subarachnoid space and the venous sinuses are known as *bridging veins*. They form part of the attachments between the brain and the skull and may become ruptured during head injuries. This results in venous bleeding and the formation of a chronic subdural hematoma. Over time this raises intracranial pressure, with consequent alterations to behavior such as confusion, impaired motor function, and headache.

Samples of CSF are normally obtained by inserting a needle between the third and fourth lumbar vertebrae of a subject who is lying on his or her side. Although the tip of the needle lies in the subarachnoid space, there is no risk of damage to the spinal cord, which only extends as far as the first lumbar vertebra. This procedure is known as lumbar puncture. The CSF is not an ultrafiltrate of plasma as it differs in its ionic composition, having relatively less potassium and calcium, and more chloride than plasma (Table 15.4). It also has a very low protein content, no red cells, and very few leukocytes. The CSF serves two main functions: it acts as a hydraulic buffer to cushion the brain against damage resulting from movements of the head, and, as it is in direct contact with the extracellular fluid of the brain, it helps to provide a stable ionic environment for neuronal function.

Consequences of raised intracranial pressure

As with the circulation of the blood, the circulation of the CSF is driven by the difference in pressure between the cerebral ventricles (the site of production) and the venous sinuses. The hydrostatic pressure of the CSF can be measured by lumbar puncture and is normally about 0.5–1.5 kPa (approximately 4–12 mmHg). Obstruction of the CSF drainage leads to an increase in intracranial pressure. This can occur following hemorrhage, head injury, inflammation of the aqueduct, or the growth of a tumor. In an adult, the raised intracranial pressure elicits Cushing's reflex discussed above. In babies and young children (where the skull sutures are not fused), the increased pressure leads to a disproportionate increase in the size of the head known as *hydrocephalus*. If this is severe and remains untreated, brain tissue may be damaged, resulting in mental impairment.

A further cause of raised intracranial pressure is *meningitis* in which the membranes covering the brain (the meninges) are invaded by viruses or bacteria. Under these circumstances, the meninges become inflamed, leading to a potentially life-threatening condition. In meningitis, the intracranial pressure is increased due to restricted drainage of the CSF, giving rise to characteristic clinical signs including headache, vomiting, photophobia, and rigidity of the neck. In severe cases convulsions and coma may occur. Because the CSF flows freely around the brain and spinal cord, a sample of CSF obtained by lumbar puncture can be used to ascertain the nature of the invading organism.

Summary

1. Specific regulatory mechanisms are used in localized regions of the cardiovascular system to match blood flow to tissue requirements.

2. The coronary circulation has to provide the myocardium with sufficient blood to meet its metabolic requirements despite the compression of coronary vessels that occurs during each contraction. Coronary blood flow is regulated largely by local metabolites including adenosine, carbon dioxide, hydrogen ions, and potassium ions. There is a close parallel between metabolic activity and coronary blood flow.

3. The cutaneous circulation plays an important role in thermoregulation which is achieved by varying the blood flow to the skin. Specialized vessels, the arteriovenous anastomoses, constrict in response to sympathetic nerve activity to shunt blood from the arterioles directly to the venules. When this occurs, blood bypasses the capillaries and is diverted away from the skin surface. This acts to conserve body heat.

4. Blood flow to skeletal muscle varies directly with contractile activity. During exercise, vasodilatation occurs in response to the local buildup of metabolites—functional hyperemia. A sustained contraction may compress the vessels so much that blood flow to the muscle practically ceases. Metabolites then accumulate quickly and cause reactive hyperemia when the circulation is restored.

5. Brain tissue has a very high capillary density. Cerebral blood flow is kept within narrow limits by autoregulation, local metabolic control, and reflexes initiated by the brain itself. Regional functional hyperemia is evident. Hypercapnia causes vasodilatation and an increase in blood flow.

6. The cerebral ventricles and the central canal of the spinal cord are filled with cerebrospinal fluid, formed continuously by the choroid plexus. It provides a hydraulic cushion for the brain and a stable chemical environment for its neurons. Obstruction of the circulation of the cerebrospinal fluid has serious consequences for brain function.

Recommended reading

Anatomy

MacKinnon, P.C.B., and Morris J.F. (2005). *Oxford textbook of functional anatomy*, Vol. 2, *Thorax and abdomen* (2nd edn), pp. 65–79. Oxford University Press, Oxford.

Histology of the heart and blood vessels

Junquieira, L.C., and Carneiro, J. (2003). *Basic histology* (10th edn), Chapter 11. McGraw-Hill, New York.

Physiology of the circulatory system

Berne, R.M., and Levy, M.N. (2001). *Cardiovascular physiology* (8th edn). Mosby, St Louis, MO.

Jordan, D., and Marshall, J. (eds.) (1995). *Cardiovascular regulation*. Portland Press, London.

Levick, J.R. (2003). *An introduction to cardiovascular physiology* (4th ed). Hodder Arnold, London.

Pharmacology of the heart and circulation

Grahame-Smith, D.G., and Aronson, J.K. (2002). *Oxford textbook of clinical pharmacology* (3rd edn), Chapter 23. Oxford University Press, Oxford.

Rang H.P., Dale, M.M., Ritter, J.M., and Moore, P. (2003) *Pharmacology* (5th edn), Chapters 17 and 18. Churchill-Livingstone, Edinburgh.

Multiple choice questions

Each statement is either true or false. Answers are given below.

1. a. The action potentials in the heart are about 100 times longer than those of skeletal muscle.

   b. The cells of the sinoatrial node have a steady resting potential of –90 mV.

   c. The cardiac action potential is conducted through the myocardium entirely via specialized conducting fibers.

   d. The spread of cardiac excitation is delayed by about 0.1 s at the atrioventricular node.

   e. The conducting tissue of the heart is composed of specialized cardiac myocytes linked by gap junctions.

2. a. The P wave of the ECG reflects atrial contraction.

   b. The QRST complex of the ECG reflects the time during which ventricular fibers are depolarized.

   c. The peak amplitude of the ECG is about 1 mV.

   d. The T wave reflects the repolarization of the ventricular fibers.

   e. The P–Q interval is normally about 0.1 s.

3. a. During ventricular diastole the pressure in the left ventricle is close to zero.

   b. During ventricular systole, the pressure in the left ventricle reaches a maximum of about 16 kPa (120 mmHg)

   c. During ventricular systole, all the blood in the ventricles is ejected.

   d. During the initial stage of ventricular contraction the volume of the ventricle does not change.

   e. The mitral valve closes because the pressure in the left ventricle exceeds that in the left atrium.

4. a. The first heart sound corresponds to the closure of the mitral and tricuspid valves.

   b. The first heart sound occurs just before the R wave of the ECG.

   c. The second heart sound is due to closure of the aortic and pulmonary valves.

   d. The second heart sound occurs during the T wave.

5. a. The cardiac output is, on average, the same for both left and right sides of the heart.

   b. The cardiac output can be measured by dividing the oxygen consumption by the difference in oxygen content between the mixed venous and arterial blood.

   c. The end-diastolic volume is an important factor in determining the stroke volume.

   d. The stroke volume is increased during strong sympathetic stimulation.

   e. Stimulation of the vagus nerve increases heart rate.

   f. Cardiac output can be increased by circulating catecholamines.

6. a. The arterial blood pressure is normally 16/10.6 kPa (120/80 mmHg).

   b. The systemic arterial blood pressure depends solely on the cardiac output.

   c. The mean blood pressure is the arithmetic average of the systolic and diastolic pressures.

   d. The main source of vascular resistance is the capillaries as they have the smallest diameter.

   e. The flow of blood in the large veins is pulsatile

7. a. The diameter of the arterioles is entirely regulated by the sympathetic nervous system.

   b. Autoregulation refers to the nervous control of the blood vessels.

   c. Reactive hyperemia is due to vasodilatation caused by accumulation of metabolites during a period of occluded blood flow.

   d. Activation of the sympathetic system causes vasoconstriction in the viscera and skin but vasodilatation in skeletal muscle.

   e. Parasympathetic vasodilator fibers innervate the blood vessels of the gastrointestinal tract.

8. a. Solute exchange between the capillaries and tisues occurs mainly by diffusion.

   b. The plasma proteins play an important role in tissue fluid exchange.

   c. The capillaries provide a significant source of vascular resistance.

   d. All of the fluid that passes from the capillaries to the tissues is returned to the blood via the lymphatic circulation.

   e. The lymph has the same ionic composition as the plasma but has a lower oncotic pressure.

9. a. If the arterial pressure suddenly falls, the baroreceptor reflex increases the heart rate.

   b. An increase in arterial pressure elicits a peripheral vasoconstriction that is mediated via the sympathetic nervous system.

   c. The cardiac volume receptors are mainly responsible for the long-term regulation of systemic blood pressure.

   d. The coronary blood flow is mainly regulated by metabolic hyperemia.

Quantitative problems

Answers are given below.

1. If the systolic pressure is 16 kPa (120 mmHg) and the diastolic pressure is 9.33 kPa (70 mmHg). what is the mean arterial pressure?

2. A subject has an oxygen consumption of 250 ml min^{-1}. A sample of arterial blood shows it to contain 20.1 ml O_2 dl^{-1} and a sample of mixed venous blood contains 13.5 ml O_2 dl^{-1}. What is the subject's cardiac output?

3. A subject has a left–right shunt (i.e. blood from the right side of the heart mixes with that from the left). A sample of blood from the pulmonary vein shows it to contain 20 ml O_2 dl^{-1} while the oxygen content of blood in the radial artery is only 18 mL O_2 dl^{-1}. If the mixed venous blood has an oxygen content of 13 ml dl^{-1}, what proportion of the cardiac output passes through the shunt?

4. A series of measurements were made of the blood flow through an isolated skeletal muscle at different perfusion pressures. The following values were obtained:

Pressure (kPa)	4	8	10.6	13.3	16	18.6	21
Blood flow (ml min^{-1})	1.8	1.9	1.95	2.0	2.05	2.2	3.0

Plot the data and calculate the resistance in each case. What phenomenon could account for the results? What would happen to blood flow if the pressure were suddenly raised from 10.6 to 21 kPa?

Answers to multiple choice questions

1. The action potential of skeletal muscle is about 2 ms in duration while the action potentials of the heart cells range from about 150 ms in the cells of the SA node to about 300 ms in a Purkinje fiber. The membrane potential of the SA node cells is low, about –60 mV, while that of ventricular cells is about –90 mV. Although the conduction of the cardiac impulse through the atria occurs preferentially via certain fiber bundles, these cells are normal atrial myocytes. In the ventricles, the specialized conducting cells are the bundle cells and the

Purkinje fibers, which transmit their action potentials to ventricular myocytes via gap junctions.
a. True
b. False
c. False
d. True
e. True

2. The P wave reflects atrial depolarization and precedes atrial contraction.
a. False
b. True
c. True
d. True
e. True

3. The volume of blood in the ventricles at the end of diastole is about 120 ml. Of this about 70 ml is ejected (the stroke volume). The ratio of the stroke volume to the end-diastolic volume is called the ejection fraction and is usually about 60 per cent at rest.
a. True
b. True
c. False
d. True
e. True

4. Since the first heart sound is caused by closure of the atrio-ventricular valves, it must occur after the onset of ventricular contraction, which follows the R wave. The second heart sound is caused by the closure of the aortic and pulmonary valves. This occurs at the end of systole when the ventricles have repolarized. Therefore the second heart sound occurs just after the T wave.
a. True
b. False
c. True
d. False

5. Stimulation of the vagus nerve slows the heart.
a. True
b. True
c. True
d. True
e. False
f. True

6. Arterial blood pressure depends on cardiac output and the total peripheral resistance. Mean blood pressure is given by the sum of the diastolic pressure and one-third of the pulse pressure (the difference between the diastolic and systolic pressures. The main site of vascular resistance is the arterioles. The pressure changes in the right atrium give rise to distinct pressure waves in the large veins—see Fig. 15.75.

a. True
b. False
c. False
d. False
e. True

7. The arterioles are under nervous, hormonal, and local control. Autoregulation occurs in denervated blood vessels and is an intrinsic property. Two mechanisms appear to be involved—local metabolites and the myogenic response of the smooth muscle to stretch.
a. False
b. False
c. True
d. True
e. True

8. The main resistance vessels are the arterioles. A fraction of the fluid that leaves the capillaries at the arteriolar end is reabsorbed by the capillaries during constriction of the arterioles. The fraction that is not absorbed is returned to the circulation via the lymph.
a. True
b. True
c. False
d. False
e. True

9. When arterial blood pressure falls, the baroreceptors are less active and this increases sympathetic activity in the cardiac and vasoconstrictor nerves. Heart rate and peripheral vascular resistance increase, restoring the blood pressure to normal.
a. True
b. False
c. True
d. True

Answers to quantitative problems

1. 11.5 kPa (86.7 mmHg).
2. 3.78 l min^{-1}.
3. 28.6 per cent.
4. 2.2, 4.2, 5.4, 6.6, 7.8, 8.4 and 7 kPa ml^{-1} min.

The resistance rises as perfusion pressure rises due to the autoregulation of blood flow. At the highest perfusion pressure, the hydrostatic forces begin to overcome the vasoconstriction. A sudden rise in pressure from 10.6 kPa to 21 kPa would result in a transient rise in blood flow that would decline to the value given in the table as the autoregulatory mechanisms come into play.

CHAPTER 16

The respiratory system

After reading this chapter you should understand:

- The gas laws and their application to respiratory physiology
- The structure of the respiratory system
- The lung volumes and the mechanics of ventilation
- Dead space and its measurement
- The principles of gas exchange in the alveoli and the role of surfactant
- The pulmonary circulation
- The factors that determine the ratio of ventilation to blood flow in different parts of the lung
- The origin and control of the respiratory rhythm
- The chemical regulation of respiration—the role of the central and peripheral arterial chemoreceptors
- The lung defense systems
- Some common disorders of respiration
- Hypoxia, its origins and consequences. Oxygen therapy

16.1 Introduction

The energy needed by animals for their normal activities is mainly derived from the oxidative breakdown of foodstuffs, particularly carbohydrates and fats. During this process, which is called *internal* or *cellular respiration*, oxygen is utilized by the mitochondria and carbon dioxide is produced. The oxygen needed for this process is ultimately derived from the atmosphere by the process of *external respiration*, which also serves to eliminate the carbon dioxide produced by the cells. The processes that govern external respiration and determine its efficiency are the subjects of this chapter.

The key process of external respiration is gas exchange between the air deep in the lungs and the blood that perfuses them. In addition to their role in gas exchange, the lungs have a variety of non-respiratory functions such as their role in trapping blood-borne particles (e.g. small fragments of blood clots) and the metabolism of a variety of vasoactive substances.

To be able to understand fully the process of breathing and gas exchange the following questions need to be addressed:

1. What mechanisms are employed to cause air to move in and out of the lungs?

2. How is oxygen taken up in the lungs and carried in the blood?

3. How is carbon dioxide carried in the blood and eliminated from the body via the lungs?

4. How efficient are the lungs in matching their ventilation to their blood flow?

5. How is the respiratory rhythm generated?

6. What factors determine the rate and depth of respiration?

7. What mechanisms prevent the lungs becoming clogged with particles from the air?

Most of these issues are discussed in this chapter, but gas transport via the blood is discussed in Chapter 13 and the role of the lungs in acid–base balance is discussed in Chapter 29.

Respiratory physiology employs a large number of standard abbreviations which are often used to calculate respiratory data. The most common are given in Box 16.1 together with the conventions for their use.

16.2 The application of the gas laws to respiratory physiology

External respiration involves the exchange of oxygen and carbon dioxide between the blood and the air in the lungs. To understand the factors that determine their uptake and loss from the lungs, it is useful to have a knowledge of the physical properties of gases.

The air we breathe and that in our lungs is a mixture of gases consisting mainly of nitrogen, oxygen, carbon dioxide, and water vapor. *Dalton's law of partial pressures* states that the total pressure is the sum of the pressures that each of the gases would exert if it were present on its own in the same volume. Thus

$$P_T = P_{N_2} + P_{O_2} + P_{CO_2} + P_{H_2O}$$

where P_T is the total pressure of the gas mixture and P_{N_2}, P_{O_2}, P_{CO_2}, and P_{H_2O} are the partial pressures of nitrogen, oxygen, carbon dioxide, and water vapor respectively.

Boyle's law states that the pressure exerted by a gas is inversely proportional to its volume so that

$$P \propto 1/V.$$

Charles' law states that the volume occupied by a gas is directly related to the absolute temperature T:

$$V \propto T.$$

These two laws are combined with Avogadro's law in the *ideal gas law* which states that

$$PV = nRT$$

where n is the number of moles of gas and R is the gas constant (8.31 J K^{-1} mol^{-1}). Each mole of gas occupies 22.4 liters at standard temperature and pressure (STP).

BOX 16.1 THE USE OF SYMBOLS IN RESPIRATORY PHYSIOLOGY

Respiratory physiology makes extensive use of standard symbols to express concepts concisely and allow simple algebraic manipulations. The primary variables are given as capital letters in italic (e.g. pressure P or volume V), while the location to which they apply are given by a suffix (e.g. Pa_{O_2} is the partial pressure of oxygen in the arterial blood while PA_{O_2} is the partial pressure of oxygen in the alveolar air). In addition, the prefix, s, is used for specific, e.g. sR_{aw} is the specific airway resistance.

Variable	Symbol used
Pressure, partial pressure, or gas tension	P
Volume of gas	V
Flow of gas (l min^{-1})	$\dot{V}$
Volume of blood	Q
Flow of blood (l min^{-1})	$\dot{Q}$
Fractional concentration of dry gas	F
Resistance	R
Conductance	G

Variable	Suffix
Arterial	a
Capillary	c
Venous	v
Mixed venous	$\bar{v}$
Alveolar	A
Inspired	I
Expired	E
End-tidal air (≡ alveolar gas)	E′
Tidal	T
Lung	L
Barometric	B
Dead space	D
Pleural space	pl
Airway	aw
Chest wall	w
Elastic	el
Resistive	res
Total	tot

From the ideal gas law the pressure and volume of a given mass of gas are related to the absolute temperature by the following relationship:

$$\frac{P_1 V_1}{T_1} = \frac{P_2 V_2}{T_2}$$

BOX 16.2 CONVERSION OF GAS VOLUMES TO BTPS AND STPD

Conversion from ambient temperature and pressure to BTPS

As air is breathed in, it is heated and humidified. This leads to an increase in its volume, which can be calculated from the universal gas law:

$$\frac{P_{ATP}V_{ATP}}{T_{ATP}} = \frac{P_{BTPS}V_{BTPS}}{T_{BTPS}}.$$

The following formula can be derived from this relationship:

$$V_{BTPS} = V_{ATP} \times \frac{273 + 37}{273 + T_A} \times \frac{P_B - P_{H_2O}}{P_B - 6}.$$

where V_{BTPS} is the volume at body temperature and pressure saturated, V_{ATP} is the volume of air inhaled at ambient temperature and pressure, T_A is the ambient temperature, P_B is the barometric pressure and P_{H_2O} is the water vapor pressure of the ambient air. The numerical constants 37 and 6 are the body temperature in degrees celsius and the saturated water vapor pressure in kPa (equivalent to 47 mmHg).

Thus, for a liter of room air inhaled when the temperature is 20°C, the barometric pressure is 100 kPa (c. 750 mmHg), and the water vapor pressure is 2 kPa (c. 17 mmHg), the change in the volume of the thorax will be

$$V_{BTPS} = 1 \times \frac{310}{293} \times \frac{100 - 2}{100 - 6} = 1.103 \text{ litres.}$$

Therefore the expansion of the chest will be about 10 per cent more than the volume of gas inhaled.

Conversion from ambient temperature and pressure to STPD

The volume of oxygen absorbed or carbon dioxide exhaled is expressed as STPD because, in this case, the need is to express the number of moles of gas exchanged. The volume of 1 mole of gas is 22.4 liters at STP (0 °C and 101 kPa (760 mmHg)). In this case the conversion uses the formula

$$V_{STPD} = V_{ATP} \times \frac{273}{273 + T_A} \times \frac{P_B - P_{H_2O}}{101}.$$

Thus 1 liter of humidified oxygen at the same ambient temperature and pressure as the previous example would occupy a volume at STPD of

$$V_{STPD} = 1 \times \frac{273}{298} \times \frac{100 - 2}{101} = 1.89 \text{ litres.}$$

The above calculations are for air saturated with water vapor. This is normally the case for gas samples taken from a spirometer. For normal ambient air, the percentage saturation of the air (the relative humidity) must be taken into account. To do this, simply multiply the saturated water vapor pressure for the ambient temperature by the relative humidity expressed as a fraction (70 per cent humidity is 0.7, etc.).

---

which makes clear that the volume of gas will depend on both the temperature and the pressure. Thus a volume of gas collected from a subject in a gas sampling bag (known as a Douglas bag) will depend on both the atmospheric pressure and the room temperature. To be able to compare samples of gas collected at different times, the volume of a gas can be expressed in one of two ways.

- As STPD—standard temperature and pressure dry. This gives the volume of gas after removal of water vapor at standard temperature (273 K or 0°C) and pressure (101 kPa or 760 mmHg).

- As BTPS—body temperature and pressure saturated with water vapour, i.e. at 37°C (310 K) and a water vapour pressure of 6.2 kPa (47 mmHg). This would be the volume of gas expired from the lungs.

Because a given mass of gas will occupy significantly different volumes at 0°C and at 37°C, it is essential to state which standard is being employed. Box 16.2 gives the formulae for converting a volume of gas at ambient temperature and pressure (ATP) to BTPS or STPD.

Like liquids, gases flow from regions of high pressure to regions of lower pressure. Moreover, since the partial pressure of a gas is a direct measure of its molar concentration, a gas will diffuse from a region of high partial pressure to one of a lower partial pressure even though the total pressure in the gas phase is uniform. The rate of diffusion under these circumstances is inversely proportional to the molecular weight of the gas, i.e. the greater the molecular mass, the slower the rate of diffusion. This is known as *Graham's law*. However, since the molecular weights of oxygen, carbon dioxide and nitrogen are 32 Da, 44 Da and 28 Da, respectively, all three gases diffuse in the gas phase at very similar rates.

Solubility of gases

To reach the cells where it is required, oxygen must first dissolve in the aqueous lining of the lung. The amount of oxygen that is dissolved is proportional to its partial pressure in the gas phase. This is *Henry's law*, which can be written as

$$V = sP$$

where s is the solubility coefficient (ml l^{-1} kPa^{-1} or ml l^{-1} mmHg^{-1}), V is the volume of dissolved gas in a liter of the liquid phase and P is the partial pressure of the gas under consideration. For oxygen at body temperature (37°C), $s = 0.225$ ml l^{-1} kPa^{-1} (0.03 ml l^{-1} mmHg^{-1}). Thus the volume V of oxygen dissolved in 1 liter of water or plasma when P_{O_2} is 13.33 kPa (100 mmHg) is

$$V = 0.225 \times 13.33$$
$$= 3 \text{ ml.}$$

Similar calculations can be performed for carbon dioxide, where $s = 5.1$ ml l^{-1} kPa^{-1} (0.68 ml l^{-1} mmHg^{-1}), and for nitrogen, where $s = 0.112$ ml l^{-1} kPa^{-1} (0.015 ml l^{-1} mmHg^{-1}).

Note that this relationship only applies to dissolved gas. *Where the gas enters into chemical combination, the total amount in the liquid phase is the sum of that chemically bound plus that in physical solution.*

In respiratory physiology the concentration of dissolved gases is usually given as their partial pressures even when they are present in a solution with no gas phase (e.g. in the arterial blood). The partial pressure of a gas can readily be converted to the equivalent molar concentration using Avogadro's law. For example, when carbon dioxide has a partial pressure of 5.33 kPa (40 mmHg) each liter of plasma will dissolve

$$5.33 \times 5.1 = 27.2 \text{ ml.}$$

Since at STP 1 mole of carbon dioxide occupies 22.4 l, this corresponds to

$$27.2 \times 10^{-3}/22.4 = 1.2 \times 10^{-3} \text{ mol } l^{-1} \text{ (1.2 mmol } l^{-1}).$$

Diffusion of dissolved gases in the lungs

When oxygen, for example, is taken up by the blood it must first dissolve in the aqueous phase that lines the lungs and then diffuse across the alveolar membrane into the blood. The rate at which oxygen and carbon dioxide diffuse from the aqueous lining of the alveoli to the blood is governed by *Fick's law of diffusion* (see Chapter 15, Box 15.6).

The extreme thinness of the alveolar membranes and their large area helps to optimize the diffusion of the respiratory gases. In addition to these factors, the rate at which the respiratory gases diffuse will depend on their solubility and their concentration gradient. The importance of solubility in determining the rate of diffusion is evident with carbon dioxide. This gas is about 20 times more soluble in the alveolar membranes than oxygen. Thus, although it has a concentration gradient one-tenth that of oxygen, it diffuses from the blood to the alveolar air about twice as fast as oxygen diffuses from the alveoli to the blood. The ability of a given gas to diffuse between the alveolar air and the blood is measured by its *diffusing capacity*, also known as its *transfer factor*, which is discussed below (p. 331).

TABLE 16.1 Standard values for respiratory gases

	N_2	O_2	CO_2	H_2O
Inspired air (kPa)	79.6	21.2	0.04	0.5
mm Hg	597	159	0.3	3.7
% total	78.5	20.9	0.04	0.5
Expired air (kPa)	75.5	16	3.6	6.3
mm Hg	566	120	27	47
% total	74.5	15.8	3.5	6.2
Alveolar air (kPa)	75.9	13.9	5.3	6.3
mm Hg	569	104	40	47
% total	74.9	13.7	5.2	6.2

The composition of expired air

Expired air contains less oxygen and more carbon dioxide than inspired air. Standard values for the partial pressures of the gases present in expired and alveolar air are given in Table 16.1. Note that, although nitrogen is not exchanged with the blood, its partial pressure changes as it becomes diluted by the water vapor and carbon dioxide from the lungs.

The ratio of the carbon dioxide produced divided by the oxygen uptake is called the *respiratory exchange ratio* (R) which, under steady state conditions, is equal to the metabolic respiratory quotient (RQ) (see Chapter 24 for further details):

$$\text{respiratory exchange ratio} = \frac{\text{volume of } CO_2 \text{ produced}}{\text{volume of } O_2 \text{ taken up}}$$

Under normal resting conditions, the respiratory exchange ratio varies according to the type of food being metabolized to produce ATP. It ranges from 0.7 when fats are the principal substrate to 1.0 for carbohydrates. Usually R is around 0.75–0.8 as both carbohydrates and fats are metabolized. During starvation, protein becomes an important source of energy and R has a value of about 0.8.

Summary

1. The volume of a given quantity of gas depends on both its temperature and its pressure. It can be expressed by the ideal gas law. In respiratory physiology, the volume of a gas is expressed as standard temperature and pressure dry (STPD) or as body temperature and pressure saturated with water vapor (BTPS).

2. The amount of a gas in solution is proportional to its partial pressure in the gas phase (Henry's law). Its rate of diffusion in body fluids is governed by Fick's law.

3. The expired air has less oxygen and more carbon dioxide than room air. The ratio of the amount of carbon dioxide expired to the amount of oxygen taken up is known as the respiratory exchange ratio. It depends on the nature of the foodstuffs being metabolized.

16.3 The structure of the respiratory tree

The lungs are the principal organs of the respiratory system. They form the surface over which oxygen is absorbed and carbon dioxide is excreted. As the lungs are situated in the chest, air from the atmosphere must pass through the nose or mouth and enter the airways before it can be directed to the respiratory surface where gas exchange occurs. The airways are supplied with blood via the pulmonary and bronchial circulations, which are discussed in detail in Section 16.6.

During quiet breathing air is normally taken in via the nose, but during heavy exercise it is taken in via the mouth which offers much less resistance to air flow. Although the nasal passages offer a high resistance to airflow, they moisten and warm the air during its passage to the lungs. After entering the nose or mouth, the air passes through the pharynx to the larynx. Like the nose, the larynx is a significant source of resistance to the flow of air, and this property is exploited in vocalization.

The *trachea* links the larynx to the lungs. In an adult, it is about 1.8 cm in diameter and 12 cm in length. It is the first component of the respiratory tree—the branching set of tubes that link the respiratory surface to the atmosphere.

In the upper chest, the trachea branches to form the two main *bronchi*, one for each lung. The right bronchus has a larger diameter than the left. In turn, the bronchi branch to give rise to two smaller branches on the left and three on the right, corresponding to the lobes of the lung. (The right lung has three lobes, while the left has two.) Within each lobe, the bronchi divide into two smaller branches and these smaller branches also divide into two, and so on until the final branches reach the respiratory surface. A diagram of the arrangement of the trachea and lungs within the chest is shown in Fig. 16.1.

In all, there are 23 generations of airways between the atmosphere and the alveoli. The trachea is generation 0. It bifurcates asymmetrically to give rise to the two main bronchi, which form generation 1. The bronchi serving the lobes of the lungs (the lobar bronchi) form generations 2 and 3. Generation 4 serves the segments within the lobes. Small bronchi form generations 5–11. From generation 12 to generation 19 the airways are known as *bronchioles*. The airways in generation 16, which link the bronchioles to the respiratory surface, are known as the *terminal bronchioles*. The airways as far as the terminal bronchioles are concerned with warming and moistening the air on its way to the respiratory surface. They are known as *conducting airways* and play no significant part in gas exchange. From generation 17 to generation 19, the airways begin to participate in gas exchange and these are known as the *respiratory bronchioles*. These give rise to the *alveolar ducts* from which the principal gas exchange structures arise. These are the *alveolar sacs*, which consist of two or more *alveoli*. The respiratory bronchioles, alveolar ducts, and alveoli comprise the *transitional* and *respiratory airways*, which provide a total area for gas exchange of about 60–80 m² in an adult.

The structure of the airways

The trachea and the primary bronchi are held open by C-shaped rings of cartilage. In the smaller bronchi, this role is taken by overlapping plates of cartilage. The bronchioles, which are less than 1 mm diameter, have no cartilage. The absence of cartilage from the bronchioles allows them to be easily collapsed when the pressure outside the lung exceeds the pressure in the airways; this happens during a forced expiration (see below). Smooth muscle is found in the walls of all the airways, including the alveolar ducts, but not in the walls of the alveoli themselves. In the terminal bronchioles, the smooth muscle accounts for

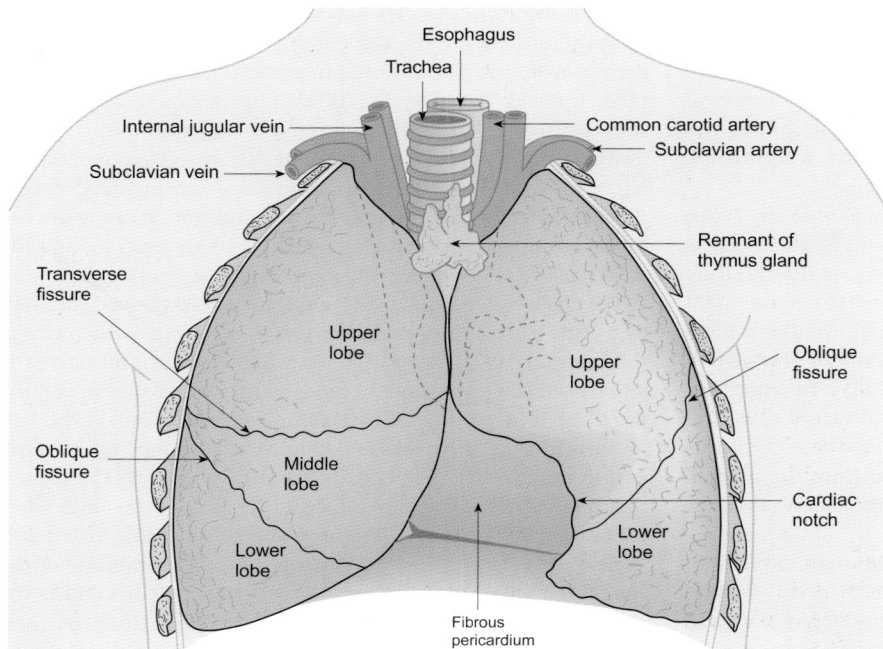

Fig. 16.1 The disposition of the lungs within the thorax.

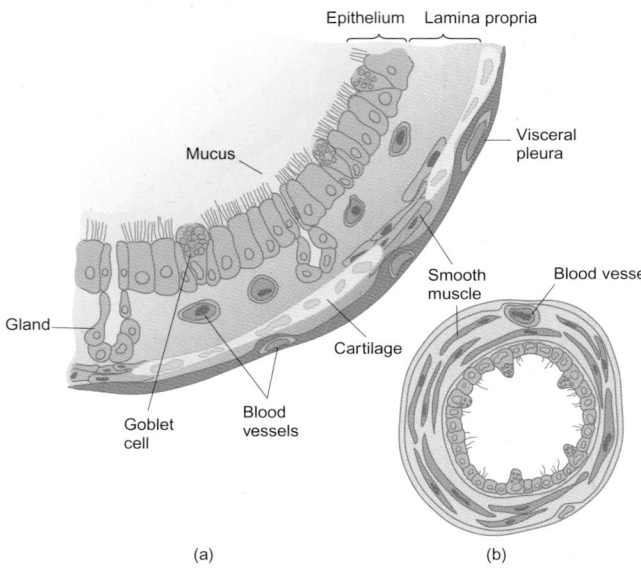

Fig. 16.2 The structure of (a) the bronchi and (b) the bronchioles. Note that the bronchus has a thicker epithelium and lamina propria compared with the bronchiole. It also has plates of cartilage while the bronchioles have a higher proportion of smooth muscle.

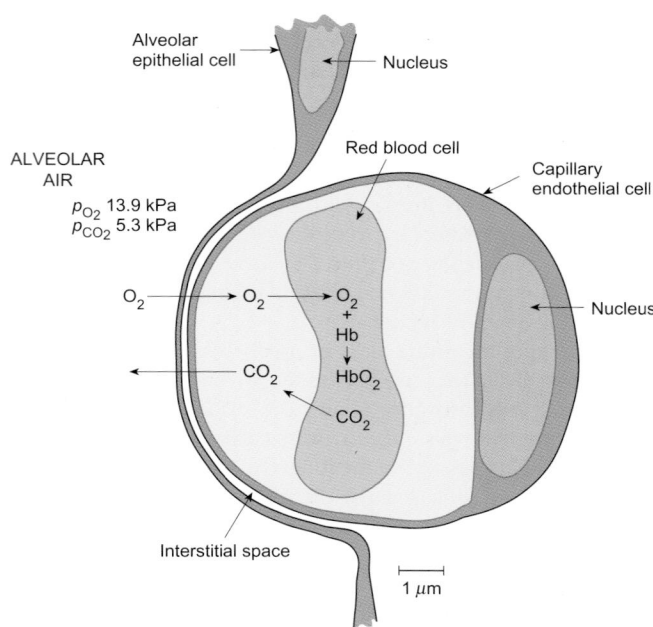

Fig. 16.3 Diagrammatic representation of the layers separating the alveolar air space from the blood in the pulmonary capillaries.

much of the thickness of the wall. The outermost part of the bronchiolar wall—the adventitial layer—is composed of dense connective tissue including elastic fibers. The structure of the bronchi and bronchioles is illustrated in Fig. 16.2.

From the nasal passages to the small bronchi, the airways are lined with a pseudostratified columnar ciliated epithelium that contains many mucus-secreting goblet cells. Beneath the epithelial layer, there are numerous submucosal glands that discharge their secretions into the bronchial lumen. In the bronchioles, the epithelium progressively changes to become a simple ciliated cuboidal epithelium. The cilia beat continuously and slowly move the mucus secreted by the goblet cells and submucosal glands towards the mouth. This arrangement is known as the *mucociliary escalator*, and plays an important role in the removal of inhaled particles (see Section 16.8). The epithelium of the bronchioles also contains non-ciliated cells that are probably secretory in function.

The site of gas exchange is the *alveolar–capillary unit*. There are about 300 million alveoli in the adult lung and each is almost completely enveloped by pulmonary capillaries (see Fig. 16.20, p. 328). Estimates suggest that there are about 1000 pulmonary capillaries for each alveolus. This provides a huge area for gas exchange by diffusion. The walls of the alveoli (the alveolar septa) consist of a thin epithelial layer comprising two types of cell called the alveolar type I and type II cells. The type I cells are squamous epithelial cells while the type II cells are thicker and produce the fluid layer that lines the alveoli. The type II cells also synthesize and secrete *pulmonary surfactant* (see below).

Beneath the alveolar epithelium lie the pulmonary capillaries. The cell membranes of the alveolar epithelial cells and pulmonary capillary endothelial cells are in close apposition, and the pulmonary blood is separated from the alveolar air by as lit-

tle as 0.5 μm (Fig. 16.3). Interspersed between the capillaries in the walls of the alveoli are the elastic and collagen fibers that form the connective tissue of the lung. This connective tissue links the alveoli together to form the lung *parenchyma*, which is sponge-like in appearance. Neighboring alveoli are interconnected by small air passages called the pores of Kohn.

The structure of the chest wall

The lungs are not capable of inflating themselves; inflation is achieved by changing the dimensions of the chest wall by means of the respiratory muscles (see Section 16.4). The principal respiratory muscles are the *diaphragm* and the *internal* and *external intercostal muscles*. The external intercostal muscles are arranged in such a way that they lift the ribs upwards and outwards as they contract. The internal intercostal muscles pull the ribs downwards, in opposition to the external intercostal muscles. In addition, some other muscles, which are not involved during normal quiet breathing, may be called upon during exercise. These are the *accessory muscles*, which assist inspiration, and the *abdominal muscles*, which assist expiration (Fig. 16.4).

The chest wall is lined by a membrane called the *parietal pleura* (Fig. 16.5). This is separated from the *visceral pleura* (which covers the lungs) by a thin layer of liquid which serves to lubricate the surfaces of the pleural membranes as they move during respiration. The total volume of intrapleural fluid is about 10 ml. It is an ultrafiltrate of plasma and is normally drained by the lymphatic system that lies beneath the visceral pleura. The pleural membranes themselves are joined at the roots of the lungs. They consist of two layers of collagenous and elastic connective tissue. Beneath the visceral pleura lies the limiting membrane of the lung itself which, together with the visceral pleura, limits the expansion of the lungs. The lungs are separated from the chest wall only by the pleural membranes and, in health, they occupy

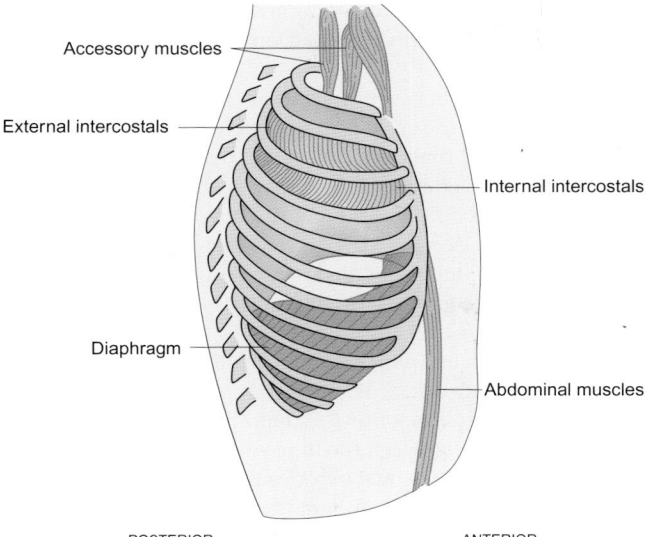

Fig. 16.4 The arrangement of the respiratory muscles of the human chest. Between each pair of ribs there are two layers of muscle: the external intercostal muscles and the internal intercostal muscles. The figure illustrates the orientation of the muscle fibers in these muscle groups. Note that the angle of the external intercostal muscles allows them to lift the rib cage when they shorten, so expanding the chest. The internal intercostal muscles act to lower the rib cage. The contraction of the accessory muscles also acts to lift the rib cage, while contraction of the abdominal muscles tends to force the diaphragm upwards into the chest, assisting expiration.

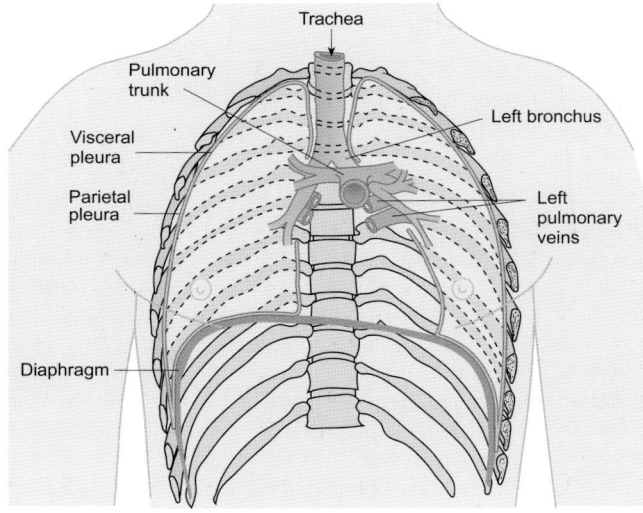

Fig. 16.5 The arrangement of the pleural membranes.

the entire cavity of the chest except the mediastinum (which contains the heart and great vessels).

Innervation of the respiratory system

The respiratory muscles do not contract spontaneously. Rhythmical breathing depends on nerve impulses in the phrenic and intercostal nerves, which are the motor nerves serving the diaphragm and intercostal muscles. The rhythmical discharge of

Summary

1. The airways consist of the nasopharynx, larynx, trachea, bronchi, and bronchioles. From the trachea, the airways branch dichotomously for 23 generations to reach the alveoli. The first 16 generations play no significant part in gas exchange. They are known as the conducting airways. The remaining seven generations (the respiratory bronchioles, alveolar ducts, and alveoli) comprise the transitional and respiratory airways.

2. The trachea and bronchi are kept open by rings or plates of cartilage. The bronchioles have no cartilage and their walls consist mainly of smooth muscle. The airways are lined by a ciliated epithelium that contains many mucus-secreting cells.

3. The alveoli are the principal site of gas exchange. Their walls consist of a very thin epithelium beneath which lies a dense network of pulmonary capillaries. The alveolar walls also contain some connective tissue.

4. The chest wall is formed by the rib cage, the intercostal muscles, and the diaphragm. It is lined by the pleura and forms a large gas-tight compartment that contains the lungs. Thus, the chest wall is an integral part of the respiratory system.

5. The muscles of respiration receive their motor innervation via the phrenic and intercostal nerves. The smooth muscle of the bronchi and bronchioles is innervated by cholinergic parasympathetic fibers. The lung parenchyma and bronchial tree have stretch receptors and receptors that respond to irritants.

these nerves is governed by the activity of specific groups of nerve cells that are located in the brainstem. The neural control of respiration will be discussed in Section 16.7 p. 332.

The smooth muscle of the bronchi and bronchioles is innervated by cholinergic parasympathetic fibers which reach the lungs via the vagus nerves. Activation of these nerve fibres causes bronchoconstriction. Sympathetic nerve fibers innervate the blood vessels of the bronchial circulation but there is no direct sympathetic innervation of the bronchiolar smooth muscle. Bronchodilatation occurs in response to circulating epinephrine and norepinephrine, which act on β-adrenergic receptors to cause relaxation of the smooth muscle. Inhalation of β-adrenergic drugs such as salbutamol is used to overcome the bronchospasm that occurs during asthmatic attacks. In addition to their cholinergic and adrenergic innervation, the tone of the smooth muscle of the airways is also regulated by autonomic fibers that secrete nitric oxide. Agents that constrict the airways include substance P and neurokinin A.

The lungs themselves contain slowly adapting stretch receptors, irritant receptors and pulmonary C-fiber endings which send information to the CNS via visceral afferent fibers in the vagus nerves. These receptors play an essential role in the respiratory reflexes, which are discussed below.

16.4 The mechanics of breathing

It is common knowledge that breathing is associated with changes in the volume of the chest. During inspiration, the chest is expanded and air enters the lungs. During expiration, the volume of the chest decreases and air is expelled from the lungs. This movement of air into and out of the lungs is known as *ventilation*. In this section, the mechanisms responsible for the changes in the dimensions of the chest will be examined, followed by a discussion of the factors governing the flow of air in the airways.

The lung volumes

The volume of air that moves in and out of the chest during breathing can be measured with the aid of a *spirometer*, which consists of an inverted bell that has a water seal to form an air-

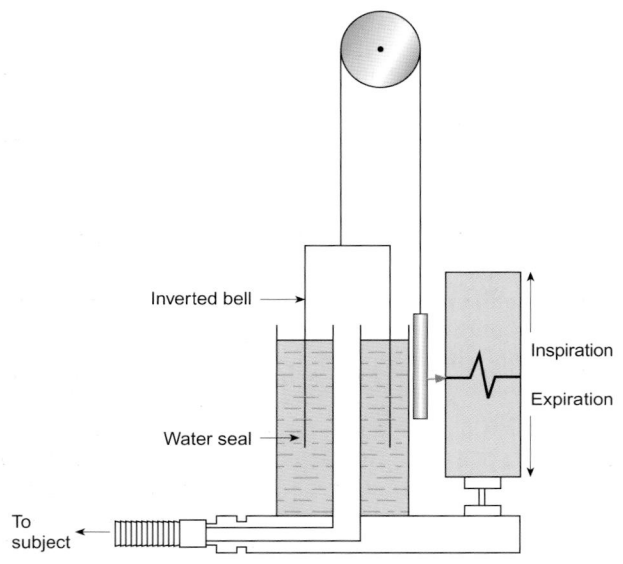

Inverted bell

Inspiration

Expiration

Water seal

To subject

Fig. 16.6 A recording spirometer. The subject breathes in and out through the mouth via the flexible tube shown at the bottom left. Inspiration draws air from the bell and the volume of air trapped within the bell decreases. It increases during expiration. These changes in volume are recorded on a calibrated chart as shown or by a computer. In use the spirometer would normally be connected to a more elaborate gas circuit with a soda lime canister to absorb exhaled carbon dioxide.

tight chamber. The bell is free to move in the vertical direction and the movements are recorded on a chart or logged by a computer (Fig. 16.6).

The relationship between the various lung volumes is shown in Fig. 16.7. When the chest is expanded to its fullest extent and the lungs allowed time to inflate fully, the amount of air they contain is at its maximum. This is the *total lung capacity*. If this is followed by a maximal expiration, the lungs will still contain a volume of air that cannot be expelled. This is called the *residual volume* (RV). The amount of air breathed out during a maximal expiration following a maximal inspiration is called the *vital capacity* (VC).

The air taken in and exhaled with each breath is known as the *tidal volume* (V_T). At rest, with normal quiet breathing, the tidal volume is much less (about 0.5 l) than the vital capacity (about 5 l). The difference between the lung volume at the end of a normal inspiration and the vital capacity is known as the *inspiratory reserve volume* (IRV) and the amount of air that can be forced from the lung after a normal expiration is called the *expiratory reserve volume* (ERV). The *functional residual capacity* (FRC) is the volume of air left in the lungs at the end of a normal expiration. The various lung volumes depend on height (they are larger in tall people), age, sex (the volumes tend to be smaller in women than in men of similar body size), and training. Typical values for healthy young adult males are given in Table 16.2.

The tidal volume varies according to the requirements of the body for oxygen. Consequently, the inspiratory and expiratory

TABLE 16.2 Typical values for respiratory variables in healthy young adult males at rest

Total lung volume (liters)	6.0
Vital capacity (liters)	4.8
Residual volume (liters)	1.2
Tidal volume (liters)	0.6
Respiratory frequency (breaths min^{-1})	12
Minute ventilation (l min^{-1})	7.2
Functional residual capacity (liters)	2.2
Inspiratory capacity (liters)	3.8
Inspiratory reserve volume	2.5
Expiratory reserve volume (liters)	1.0

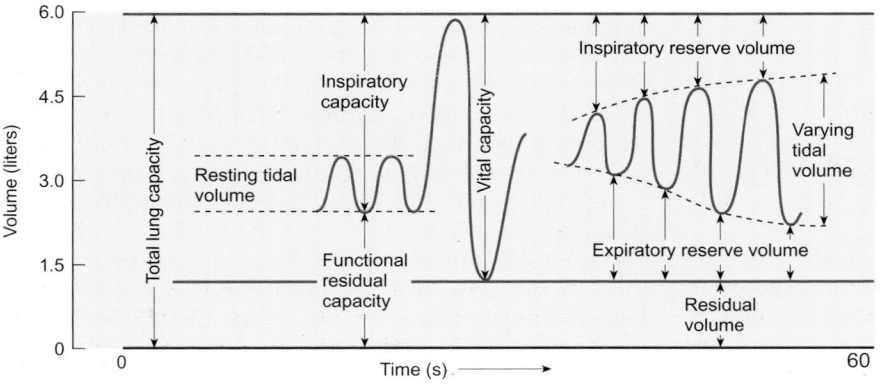

Fig. 16.7 The subdivisions of the lung volumes. The record shows an idealized spirometry record of the changes in lung volume during normal breathing at rest (resting tidal volume) followed by a large inspiration to total lung capacity and a full expiration to the residual volume. This measures the vital capacity. Note that the residual volume and functional residual capacity cannot be measured by a spirometer. As shown on the right-hand side of the figure, the tidal volume can be varied (e.g. during exercise). The inspiratory and expiratory reserve volumes become smaller as tidal volume increases.

BOX 16.3 DETERMINATION OF FUNCTIONAL RESIDUAL CAPACITY AND RESIDUAL VOLUME BY THE HELIUM DILUTION METHOD

The various subdivisions of the lung volume are of interest in a number of respiratory diseases. Therefore it is desirable that they can be measured with some accuracy. While the *change* in the volume of the lungs during various breathing maneuvers can be measured directly by a spirometer, the volume of air left in the lungs at the end of a normal expiration (the functional residual capacity (FRC)) and the residual volume (RV) cannot be measured in this way. Instead they are measured by the helium dilution method.

To determine the FRC, a subject is asked to breathe out normally and then is asked to inspire from a spirometer filled with a known volume of air containing a known concentration of the inert gas helium. As the helium-containing gas is breathed in and out the helium is diluted by the volume of air left in the lungs. A subsequent measurement of the helium concentration permits this additional volume to be calculated from the following formula:

$$\text{FRC} = \left(\frac{\text{initial He concentration}}{\text{final He concentration}} - 1 \right) \times \text{volume of spirometer.} \tag{1}$$

A similar procedure is used to measure the RV, but the subject is asked to perform a maximal expiration before breathing from the spirometer containing the helium.

The difference in volume between the FRC and RV is the expiratory reserve volume (ERV), while the total lung capacity (TLC) is attained after a full inspiration. The volume taken in from the FRC to reach TLC is called the inspiratory capacity (IC). The inspiratory reserve volume (IRV) is the maximum volume of air that can be inspired after a normal breath. These quantities are related by the following equations:

$$\text{FRC} = \text{ERV} + \text{RV} \tag{2}$$
$$\text{TLC} = \text{FRC} + \text{IC} \tag{3}$$
$$\text{IC} = V_T + \text{IRV} \tag{4}$$

where V_T is the tidal volume. The vital capacity (VC) is the volume of air that can be expired from the total lung volume:

$$\text{VC} = \text{IC} + \text{ERV} \tag{5}$$

If either the FRC or the RV is known, all the other subdivisions of the lung volume (including the total lung volume) can be determined with the aid of a spirometer.

reserve volumes are also variable—the larger the tidal volume, the smaller the inspiratory and expiratory reserve volumes. In contrast, for a given individual, the vital capacity and residual volume are relatively fixed.

With the exception of the residual volume and functional residual capacity, all the lung volumes can be directly measured by spirometry. The functional residual capacity and residual volume can be measured by the helium dilution method (Box 16.3).

Inspiration is caused by contraction of the diaphragm and external intercostal muscles; expiration is largely passive and due to elastic recoil of the lungs and chest wall

The various muscles of respiration are called on at different times. The diaphragm is the principal muscle of respiration and, during quiet breathing, it is normally the main active muscle of respiration. The diaphragm forms a continuous sheet that separates the thorax from the abdomen. At rest, it assumes a dome-like shape. When it contracts during inspiration, the crown of the diaphragm descends, thereby increasing the volume of the chest. During expiration, the diaphragm smoothly relaxes and the elastic recoil of the chest wall and lungs results in passive expiration.

When the demand for oxygen increases, the other muscles of inspiration are called into play. The chest wall is lifted upward and outward by the activity of the external intercostal muscles and the diaphragm contracts more strongly so that the volume of the chest is increased further. In severe exercise, the accessory muscles (scalenes and sternocleidomastoids) are called on, and they act to lift the chest wall still further as shown in Fig. 16.4. During exercise, the internal intercostal muscles contract to assist in decreasing the volume of the chest so that expiration under these conditions becomes partly an active process. Powerful expiration may also be assisted by contraction of the abdominal muscles that force the abdominal contents against the diaphragm, pushing it upwards and reducing the volume of the chest.

The intrapleural pressure

In health, the lungs are expanded to fill the thoracic cavity because the pressure outside the lungs (the *intrathoracic pressure*) is less than that of the air in the alveoli. The intrathoracic pressure is also known as the *intrapleural pressure*, which is measured clinically as described in Box 16.4. At the end of a quiet expiration, it is found to be about 0.5 kPa (5 cmH$_2$O) *below* that of the atmosphere. By convention, pressures less than that of the atmosphere are called negative pressures and those above atmospheric pressure are called positive pressures. Thus, the intrathoracic pressure at the end of a quiet expiration is –0.5 kPa (–5 cmH$_2$O). During inspiration the chest wall expands and the intrathoracic pressure becomes more negative, reaching a maximum of about –1 kPa (–10 cmH$_2$O).

If the chest wall is punctured by a hollow needle so that the tip of the needle lies in the intrapleural space, there is an inrush of air from the atmosphere into the cavity of the chest (*pneumothorax*). Under these conditions, the intrapleural pressure

BOX 16.4 MEASUREMENT OF INTRAPLEURAL PRESSURE AND THE ABSORPTION OF GAS FROM THE INTRAPLEURAL SPACE

In principle, the intrapleural pressure can be measured directly by inserting a hollow needle into the intrapleural space and recording the pressure. This is both technically difficult and invasive. A simpler and much less invasive method depends on the fact that the pressure in the lumen of the intrathoracic region of the esophagus is the same as intrapleural pressure. This situation arises because the esophagus is a collapsible tube and, as it passes though the thorax, it will experience the same pressure as the outside of the lungs—the intrapleural pressure.

To measure the intrapleural pressure, a small air-filled balloon is usually placed in the lower third of the esophagus and connected to a manometer. With this technique, the intrapleural pressure at the end of quiet expiration is found to be about –0.5 kPa (–5 cmH$_2$O) relative to that of the atmosphere (which by convention is taken as zero). During inspiration, the chest wall expands and the intrapleural pressure becomes more negative, reaching a maximum of about –1 kPa (–10 cmH$_2$O). During forced expiration, the intrapleural pressure may become positive and can reach values of 10 kPa (100 cmH$_2$O).

It may seem surprising that the intrapleural pressure is always less than that of the alveolar pressure. Why is there no accumulation of gas in the intrapleural space? Why does the gas of a pneumothorax eventually disappear from the intrapleural space? The total pressure of the gases in the venous blood (~93 kPa or 700 mmHg) is lower than that of the alveoli (101 kPa or 760 mmHg), as there is a considerable drop in P_{O_2} but only a small rise in P_{CO_2}. The P_{N_2} is unchanged. In trapped air, P_{O_2} and P_{CO_2} equilibrate with the surrounding tissue and P_{N_2} rises as the oxygen is absorbed. This favors the absorption of the nitrogen down its concentration gradient. The consequence of absorption of the nitrogen is a rise in the partial pressures of oxygen and carbon dioxide, which again equilibrate with the surrounding tissue, so raising the P_{N_2} once more. The cycle repeats itself until all the gas has been reabsorbed. Strictly speaking, once the gas has been absorbed, there is no intrapleural pressure, as the pleural membranes are held together by intermolecular forces. However, the concept of intrapleural pressure provides a convenient way of describing the magnitude of the forces acting on the lungs during normal respiration.

The balance of the Starling forces explains why there is normally almost no liquid in the intrapleural space. The osmotic pressure of the plasma proteins is greater than the transmural capillary pressure (the capillary pressure minus the intrapleural pressure). Therefore the net filtration pressure favors fluid absorption. However, in certain disease states, the Starling forces favor fluid movement into the intrapleural space, resulting in a pleural effusion.

---

becomes equal to that of the atmosphere and the lungs will collapse. This shows that the negative value of the intrapleural pressure is due to the elastic recoil of the lungs.

Pressure changes during the respiratory cycle

As with all liquids and gases, air will flow from a region of high pressure to one of low pressure. Therefore it follows, that air enters the lungs during inspiration because the pressure within the lungs (the *alveolar pressure* or *intrapulmonary pressure*) is less than that of the atmosphere. Conversely, during expiration, the alveolar pressure exceeds atmospheric pressure and air is expelled.

Figure 16.8 shows the variation in the pressure in the lungs during a single respiratory cycle. As the chest expands, the intrapleural pressure falls and this leads to expansion of the lungs. As the lungs expand, the alveolar pressure falls below that of the atmosphere and air flows into the lungs until the alveolar pressure becomes equal to that of the atmosphere. At this time, the intrapleural pressure remains at its most negative. During expiration, intrapleural pressure rises (i.e. becomes less negative) and the tension in the elastic fibers of the parenchyma causes the volume of the lungs to decrease. This results in an increase in alveolar pressure which forces air from the lungs. As the air is expelled, the alveolar pressure falls until it reaches atmospheric pressure once more and the cycle begins again with the next breath.

How much does the intrapleural pressure have to change for a given amount of air to enter the lungs?

The change in the volume of the chest that results from a given change in intrapleural pressure is called the *compliance*. It is a measure of the ease with which the chest volume can be changed and is determined when there is no movement of air into or out of the lungs (*static compliance*). If compliance is high, there is little resistance to expansion of the chest; conversely, if it is low, the chest is expanded only with difficulty. For healthy young subjects, the static compliance has a typical value of 1.0 liter kPa^{-1} (0.1 liter cmH$_2$O^{-1}) as shown by the straight black line in Fig. 16.9.

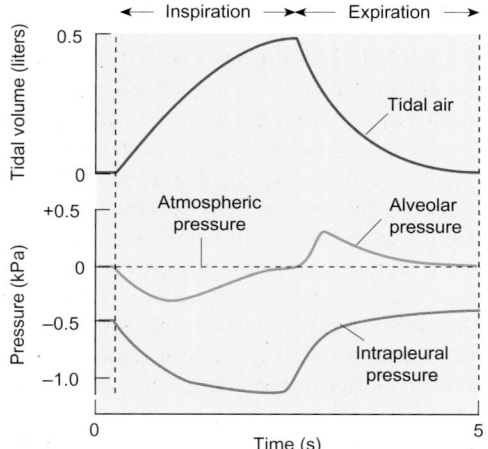

Fig. 16.8 The changes in intrapleural and alveolar pressure during a single respiratory cycle. Note that the changes in intrapleural pressure occur before the change in alveolar pressure.

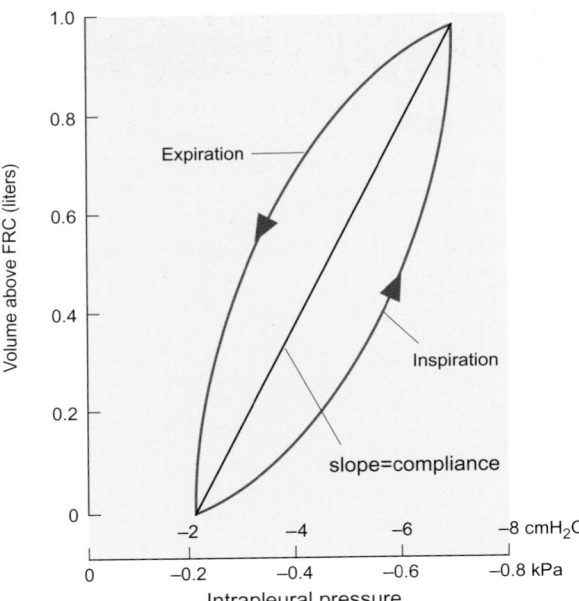

Fig. 16.9 The pressure–volume relation for a single respiratory cycle. The compliance of the respiratory system is given by the slope of the line. This represents the change in volume if the work of respiration is against purely elastic resistances. During inspiration, additional pressure is needed to overcome the airflow resistance and other resistive forces. This is shown by the curve to the right of the compliance line. The curve to the left of the compliance line shows the resistive work done during passive expiration.

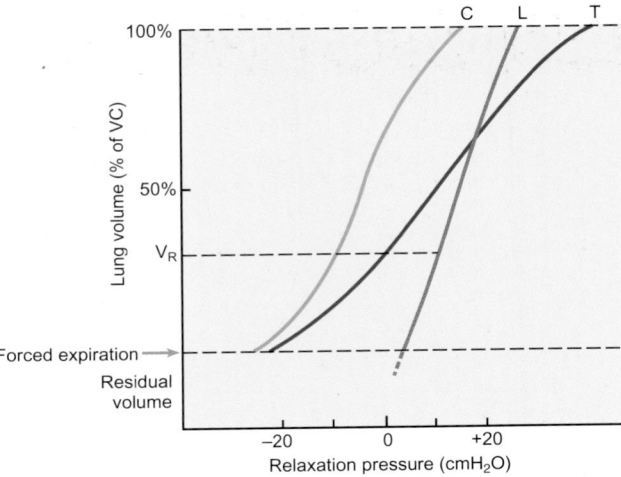

Fig. 16.10 The pressure–volume relationships for the intact chest (curve T), the chest wall (curve C) and the lungs (curve L). The ordinate shows their volumes as a percentage of the vital capacity, while the abscissa is the pressure relative to barometric pressure. Note that the inflation of the lungs alone requires positive pressures, but the chest wall naturally assumes about two-thirds of its maximum volume. Smaller volumes require negative pressure. The pressure required for a given volume of the intact chest is the sum of the pressures of the chest wall and the lungs. At the relaxation volume V_R the pressure in the lungs exactly balances that of the chest wall.

During normal breathing, a larger change in pressure is required to move a given volume of air into the chest than would be expected from the static compliance of the chest. This is shown by the curve labeled 'inspiration' in Fig. 16.9. The additional pressure is required to overcome additional *non-elastic resistances*, which are:

(1) the resistance of the airways to the movement of air (the *airways resistance*);

(2) the frictional forces arising from the viscosity of the lungs and chest wall (*tissue resistance*);

(3) the inertia of the air and tissues.

The airways resistance is by far the most significant of these. The tissue resistance is about a fifth of the total while the inertia of the respiratory tract (air plus tissues) is only of significance when there are sudden large changes in airflow such as in coughing or sneezing. During expiration, particular lung volumes are reached at lower pressures. Thus the pressure–volume relationship during a single respiratory cycle is a closed loop as shown in Fig. 16.9. The physiological reason for this property (known as *hysteresis*) will be discussed below.

The compliance of the intact chest is determined by the elasticity of the chest wall and the lungs. (An elastic body is one which resumes its original dimensions after the removal of an external force by which it has become deformed.) If the respiratory muscles are relaxed and the glottis is open, the volume of the intact chest is about 30 per cent of the total vital capacity. This is known as the relaxation volume for the intact chest. Chest volumes greater or smaller than this value can only be

attained by muscular effort. If the chest wall is cut open when it is at its relaxation volume, the ribs spring outwards and the lungs collapse inwards. The volume assumed by the chest wall is greater, and that of the lungs is smaller, than that of the intact chest. This shows that the dimensions of the chest at rest reflect the balance of the forces acting on the chest wall and the lungs. The elasticity of the chest wall (and therefore its compliance) is determined mainly by that of its muscles, ligaments, and tendons. The elasticity of the lungs is determined by two major factors: the elastic fibers of the lung parenchyma and the surface tension of the liquid film that lines the alveoli. For this reason, the pressure–volume relationships of the chest wall and the lungs differ significantly as shown in Fig. 16. 10.

What are the sites of airways resistance?

In a highly branched set of tubes such as the human bronchial tree, it is difficult to obtain precise knowledge of the patterns of airflow throughout the whole structure. Like the flow of blood in the circulatory system, the flow of air through the airways may be either laminar or turbulent. Laminar flow occurs at low linear flow rates (flow rate in volume per second divided by the cross-sectional area), but when the linear flow rate increases beyond a critical velocity the orderly pattern of flow breaks down, eddies form, and the flow becomes turbulent.

Turbulent flow is more likely to occur in large-diameter irregularly branched tubes when the flow rate is high. Unlike laminar flow, in which resistance to flow is constant for a tube of given dimensions (see Chapter 15, Box 15.4), when airflow is turbulent, resistance increases with the flow rate. This is the situation in the upper airways (the nose, pharynx, and larynx), which account for about a third of the total airways resistance. Upper

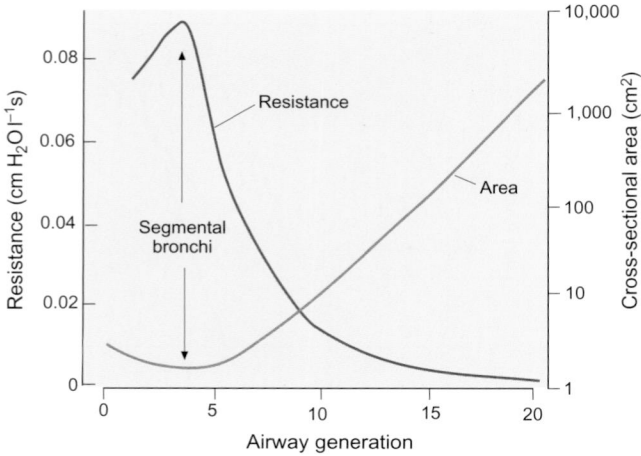

Fig. 16.11 The resistance of the airways plotted as a function of airway generation. Note that the resistance is highest in the segmental bronchi, which also have the smallest cross-sectional area. The resistance falls sharply as the cross-sectional area increases.

airway resistance can be significantly reduced by breathing through the mouth—a fact that is widely exploited during heavy exercise. The remaining two-thirds of the airways resistance is located within the tracheobronchial tree. The greatest resistance to airflow is found in the segmental bronchi (generation 4) where the cross-sectional area is relatively low and the airflow is high (and turbulent). As the airways branch further, their cross-sectional area increases and the linear flow rate falls. By the time air reaches the smallest airways, the flow is laminar and the resistance becomes very small (Fig. 16.11).

Airways resistance falls as the volume of the lungs increases. This change is chiefly due to the forces acting on the bronchioles which have no cartilage in their walls. These airways are attached to the lung parenchyma, which contains connective tissue. As the lungs expand, the connective tissue of the parenchyma pulls on the bronchioles so that their diameter increases and their resistance to the flow of air decreases.

How much work is done by the respiratory muscles?

The inflation and deflation of the lungs and chest wall requires that work is performed by the respiratory muscles. Mechanical work is performed when a load is moved through a distance (see Chapter 7) and, for a three-dimensional system such as the respiratory system, the work done is equal to the change in pressure multiplied by the change in volume (Fig. 16.12). In quiet breathing, the volume changes are modest and the work done is small. If the depth of breathing is increased, as in exercise, the pressure–volume loop has a greater area and the energy cost of each breath increases.

During inspiration, the work of breathing consists of two components: the work needed to overcome the elastic forces of the chest wall and lungs, and the work needed to overcome non-elastic resistances. During inspiration, the elastic elements of the lungs and chest wall are stretched and the work done during this phase of respiration is shown by the area marked A in Fig. 16.12(a). The additional work required to overcome the airways resistance is given by area B. In quiet expiration, the chest relaxes passively and the energy required is derived from the elastic elements stretched during inspiration. Thus the work of quiet expiration is done by the muscles of inspiration. In forced

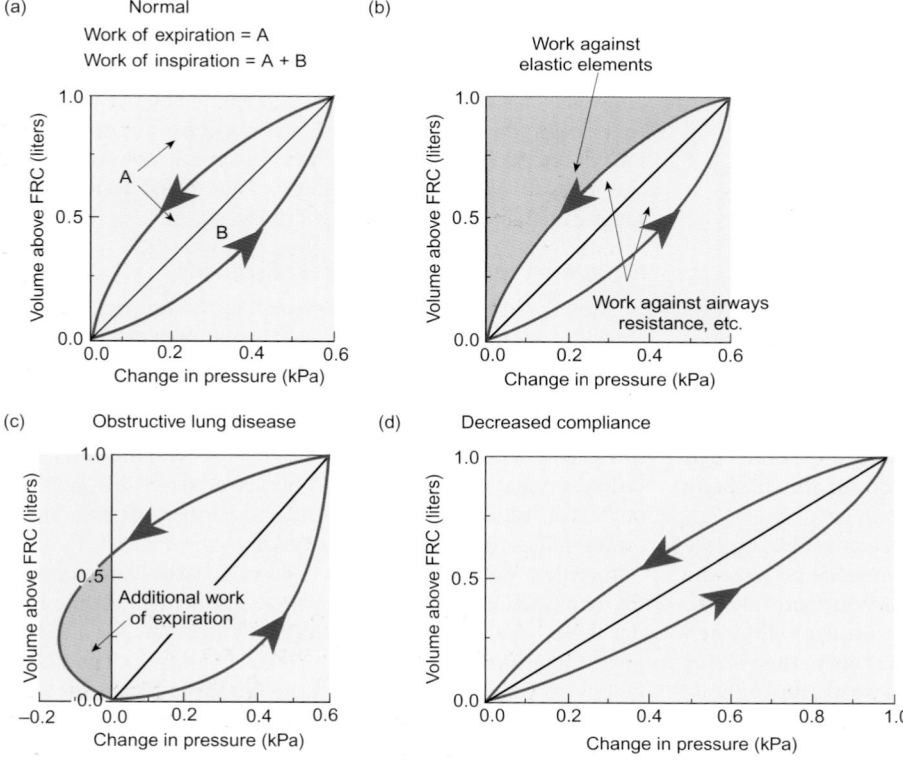

Fig. 16.12 The work of respiration. The figure shows the increase in volume above FRC plotted as a function of the change in intrapleural pressure. The work involved in changing the volume of the chest is given by the area of the pressure–volume curve.
(a) The work of inspiration is greater than the work of expiration. The area of the loop labeled B represents the energy required to overcome the airways resistance during inspiration. The energy for expiration is largely derived from the stretching of the elastic elements of the lungs and chest during inspiration. (b) The two principal components of the work of inspiration. (c) People with obstructive lung disease must breathe against an increased airways resistance. Thus, a greater pressure change is required to move a given volume of air into and out of the lungs. This results in the performance of extra work.
(d) If the compliance of the chest is decreased, a larger pressure change is required to move a given volume of air and this requires extra work to be done. (Compare the area of this pressure–volume curve with that shown in (a).)

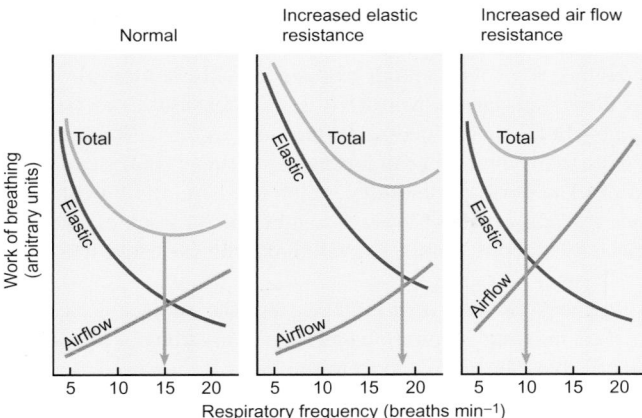

Fig. 16.13 The work of breathing at different frequencies in normal subjects, those with increased elastic resistance, and those with increased airflow resistance. In normal subjects the work of breathing is a minimum at a respiratory frequency of about 15 breaths min⁻¹. When the elastic resistance is increased, work is minimized by frequent shallow breaths. In contrast, slow deep breathing minimizes the work required when airflow resistance is increased.

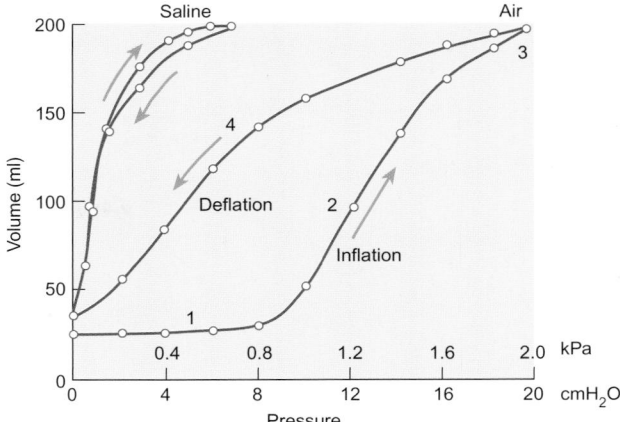

Fig. 16.14 Pressure–volume relationships for isolated cat lungs when inflated by air or with isotonic saline. Note the low pressures required to expand the saline-filled lungs and that the curve for inflation is virtually the same for deflation. For air-filled lungs, much greater pressures are required for a given volume change. The curve shows hysteresis—a greater pressure is required to inflate the lungs to a given volume compared with the pressure required to hold that volume during deflation. See text for further explanation.

expiration, however, an additional muscular effort will be required, such as that shown for obstructive lung disease (Fig. 16.12(c)). If the compliance of the chest is decreased (Fig. 16.12(d)), a greater intrapleural pressure will be required to cause a given change in lung volume so that the work of breathing is increased.

To move a given volume of air into and out of the lungs, breathing can be either deep and slow or fast and shallow. Deep, slow breathing results in an increased work against the elastic forces of the lungs and chest wall (the elastic resistance), while rapid shallow breathing results in an increased airflow resistance. Therefore the work of breathing for normal subjects is at a minimum when the respiratory rate is about 15 breaths min⁻¹. If the elastic resistance increases (as in pulmonary fibrosis), the work of breathing is minimized by increasing the rate and decreasing the depth of respiration. If airflow resistance is increased, the work of breathing is minimized by increasing the depth of breathing and decreasing the rate (Fig. 16.13).

Surface tension in the alveoli contributes to the elasticity of the lungs

During the initial stages of inflation with air, collapsed lungs require a considerable pressure before they begin to increase in volume (about 0.8–1.0 kPa—phase 1 in Fig. 16.14). The lungs then expand roughly in proportion to the increase in pressure (phase 2) until they approach their maximum capacity (phase 3). Once they have been fully expanded, the volume of the lungs changes slowly during deflation until the pressure holding them open decreases to about 0.8 kPa (phase 4) whereupon their volume declines more steeply as the pressure falls. The unequal pressure required to maintain a given lung volume during inflation with air as opposed to deflation accounts for the hysteresis in the pressure–volume relationship seen during the respiratory cycle (Fig. 16.9). However, if the lungs are inflated with isotonic saline (0.9% NaCl), the pressures required to expand them to a given volume are much reduced and there is little or no hysteresis.

Why is it more difficult to inflate the lungs with air than with isotonic saline? When the lungs are inflated with saline, the only force opposing expansion is the tension in the elastic elements of the parenchyma that become stretched as the lungs expand. However, when the lungs are inflated by air, the surface tension at the air–liquid interface in the alveoli also opposes their expansion. As with bubbles of gas in a liquid, the magnitude of the surface tension in the alveoli is given by *Laplace's law* which states that the pressure P inside a hollow sphere is twice the surface tension T divided by the radius r:

$$P = \frac{2T}{r}$$

The alveoli are about 100 μm in diameter after a normal quiet expiration. If they were lined with normal interstitial fluid which has a surface tension of about 70 mN m⁻¹, Laplace's law would require that the pressure gradient across the alveolar wall would need to be about 3 kPa just to prevent collapse—expansion would require still higher pressures. However, as Fig. 16.12 shows, the lungs can be inflated with much lower pressures than this. Maximum inflation is achieved at less than 2 kPa and the lungs can be held open during deflation by pressures well below 1 kPa. Thus the alveoli cannot be lined with interstitial fluid. This anomaly was resolved when it was discovered that the alveoli are lined with a fluid layer containing *lung surfactant* (also called *pulmonary surfactant*) which is secreted by the type II cells of the alveoli. The surfactant lowers the surface tension of the liquid lining the air spaces of the lung. Consequently, the pressures needed to hold the alveoli open are much reduced.

Lung surfactant stabilizes the alveoli by reducing the surface tension of the air–liquid interface

By lowering surface tension, lung surfactant minimizes the tendency of the small alveoli to collapse and tends to stabilize the alveolar structure. In addition, the lowering of surface tension

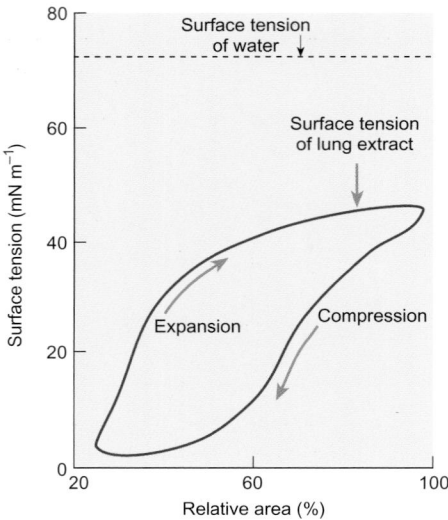

Fig. 16.15 The relationship between surface tension and surface area for water and lung surfactant. Note that lung surfactant greatly reduces the surface tension and that the reduction in surface tension is greatest during compression (equivalent to lung deflation). The surface tension of water does not vary with area.

increases the compliance of the lungs and this reduces the work of breathing. This effect of lung surfactant on surface tension is particularly important at birth when the lungs first expand (see p. 483, Fig. 22.5). Finally, as surfactant lowers surface tension it helps to prevent fluid accumulating in the alveoli and so plays an important role in keeping the alveolar air space dry.

Lung surfactant consists mainly of phospholipid molecules, which form a separate phase in the air–liquid interface that exists over the alveolar epithelium. The phospholipids are aligned so that their polar headgroups remain in the aqueous phase while their long hydrocarbon chains are oriented to the air space of the alveoli. When alveoli close during expiration, the phospholipid molecules become compressed against each other to form a monolayer, water molecules are excluded, and the surface tension falls dramatically (Fig. 16.15). Once the monolayer is formed, a subsequent increase in area causes the surface tension to rise quite rapidly at first. During this stage, the phospholipid molecules tend to remain packed together. As the area increases, they begin to separate and surface tension increases more slowly until it reaches its maximum value. This property of pulmonary surfactant explains the hysteresis of air-inflated lungs (Fig. 16.14).

Tests of ventilatory function

In the diagnosis and treatment of respiratory diseases, assessment of pulmonary function is of considerable importance. Key tests of ventilatory function are:

(1) the vital capacity;

(2) the forced vital capacity;

(3) the maximal expiratory flow rate;

(4) maximal ventilatory volume.

To measure the *vital capacity* (VC), a subject is asked to make a maximal inspiration and then to breathe out as much air as possible. The volume of air exhaled is measured by spirometry as described earlier. Note that in this test the time taken to expel the air is not taken into account and that it is usual to estimate the vital capacity during expiration, rather than inspiration. The normal values depend on age, sex, and height. For a healthy adult male of average height and 30 years of age, vital capacity is about 5 liters. For women of the same age, it is about 3.5 liters.

In the *forced vital capacity* (FVC) test the subject is asked to make a maximal inspiration and then to breathe out fully as fast as possible. The air forced from the lungs is measured as a function of time (with an instrument called a *pneumotachograph*). After the final quantity of air is forced from the lungs, only the residual volume (RV) is left. In healthy young subjects, around 85 per cent of the vital capacity is forced from the lungs within the first second. This is known as the forced expiratory volume at 1 s or FEV_1. The remaining volume is expired over the next few seconds (Fig. 16.16). The FEV_1 declines with increasing age. Even so, a healthy 60-year-old man should have a value of around 70 per cent. In contrast, a patient with an obstruction of the airways (e.g. during an asthmatic attack) would have a much lower FEV_1. In severe cases, FEV_1 can be less than 40 per cent.

The FVC test is also useful in the diagnosis of restrictive lung diseases such as fibrosis of the lung. In restrictive disorders, the ability of the lung to expand normally is compromised. As a result, FEV_1 may be normal but the vital capacity will be much reduced.

The *maximal expiratory flow rate* (also known as the *peak expiratory flow*) is also used to distinguish between obstructive and restrictive diseases. The maximal airflow is measured with a pneumotachograph during a forced expiration following an inspiration to the total lung volume. The maximum flow rate is normally reached in the first tenth of a second of the forced expiration and is measured in liters per second. Healthy young

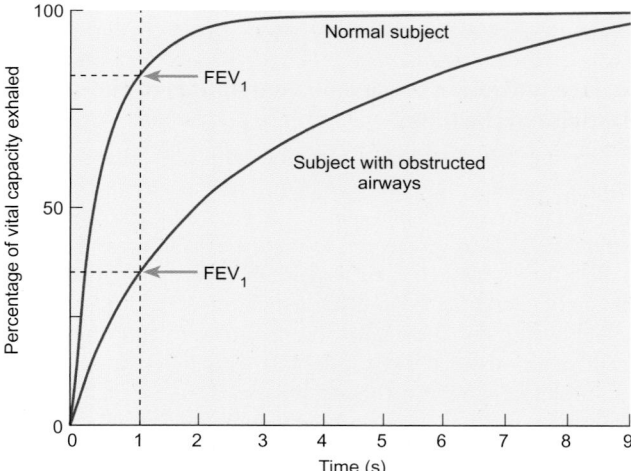

Fig. 16.16 Forced vital capacity test for a normal subject and for a subject with obstructed airways. Note the marked difference in the FEV_1 values.

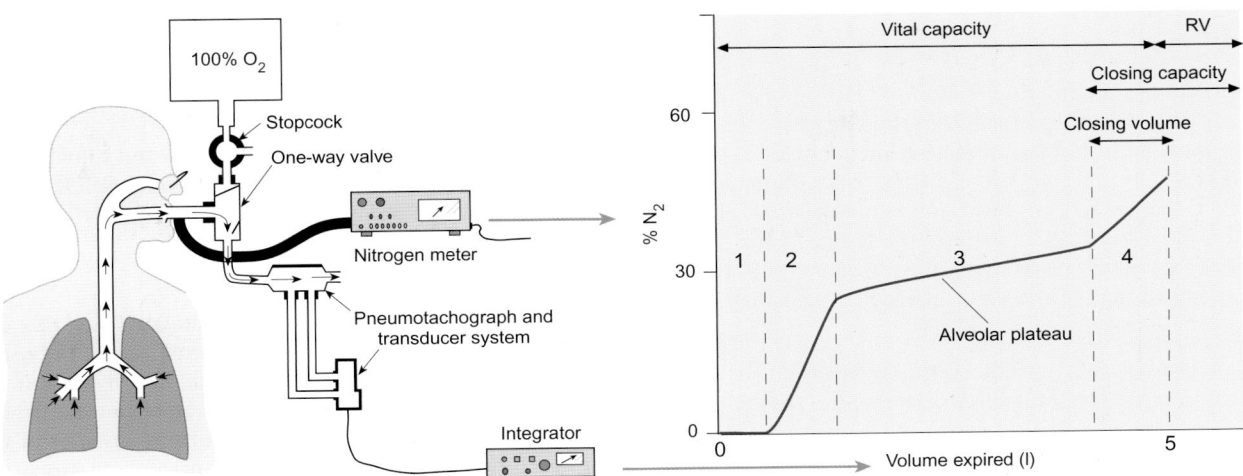

Fig. 16.17 The determination of closing volume by the single-breath nitrogen washout curve. The subject breathes out to the residual volume and then inspires pure oxygen until the vital capacity is reached. Finally, the subject breathes out to the residual volume while the fractional content of nitrogen of the expired air is measured continuously. The closing volume is indicated by the upturn in the nitrogen washout curve (see text for further details).

adults are able to achieve flow rates of 8–10 l s^{-1}. Obstructive airway disease results in a reduced peak flow, as for FEV$_1$, but this is not the case in restrictive lung disease.

The maximum minute volume attainable by voluntary hyperventilation is known as the *maximal ventilatory volume* (MVV) or *maximum breathing capacity* (MBC). The subject is asked to breathe in and out as fast and as deeply as possible through a low-resistance circuit for 15–30 s. This test involves the whole respiratory system during inspiration and expiration. As for other respiratory variables, the MVV varies with the age and sex of the subject. Healthy young men of 20 years of age can attain an MVV equal to 150 l min^{-1}. By the age of 60 years, however, the MVV for normal men has fallen to about 100 l min^{-1}. For women the equivalent values are 100 l min^{-1} at age 20 falling to about 75 l min^{-1} by age 60. The MVV is dependent on airways resistance, the compliance of the lungs and chest wall, and the activity of the muscles of respiration. As a result, it is a sensitive measure of ventilatory function. It is profoundly reduced in patients with obstructed airways (e.g. asthma) and in those with decreased compliance (e.g. pulmonary fibrosis).

The closing volume

The airways resistance increases as lung volume decreases. This situation arises because the decrease in lung volume is accompanied by a reduction in the volume of both the alveoli and the airways. However, to reach the residual capacity the internal intercostal muscles and the abdominal muscles generate a positive intrapleural pressure, which can be over 10 kPa (100 cmH$_2$O). This pressure adds to the alveolar pressure to drive air from the alveoli. However, when the intrathoracic pressure exceeds the pressure in the airways, the small airways become compressed because they have no supporting cartilage. Consequently, air becomes trapped within the lungs. This is known as

dynamic airways compression. The lung volume at which the airways begin to collapse during a forced expiration is known as the *closing capacity*. The *closing volume* is equal to the closing capacity minus the residual volume.

The closing volume can be measured as follows: the subject is asked to breathe out to the residual volume and to take a breath of pure oxygen to vital capacity. The subject is then asked to breathe out to residual volume once more while the fractional content of nitrogen in his or her expired air is measured as shown in Fig. 16.17. At first, the oxygen is forced from the dead space and the fractional nitrogen content is zero (phase 1). As the oxygen is cleared from the conducting airways, the nitrogen content rises rapidly (phase 2) until it reaches a gently rising plateau (phase 3). The slow rise in the nitrogen content reflects differences in ventilation in different parts of the lungs. As the lung volume approaches the residual volume, the lower airways become compressed and more of the exhaled gas comes from the upper region of the lungs, which have a higher fractional content of nitrogen because they were less well ventilated during the breath of oxygen. (The reasons for the difference in ventilation between the upper and lower parts of the lung are discussed below.) The nitrogen content therefore begins to rise more sharply (phase 4). The point at which this begins to occur marks the closing volume.

In normal healthy young people, the closing volume measured by this method is about 10 per cent of the vital capacity (i.e. about 500 ml). It is greater when lying down and increases with age. By the age of 65 years, the closing volume may be as much as 40 per cent of vital capacity. In emphysema, the loss of the parenchymal tissue results in decreased traction on the airways so that the small airways begin to collapse at a higher lung volume than normal. Thus an increased closing volume is an early sign of small airways disease.

Summary

1. Ventilation is the volume of air moved into and out of the lungs. It is driven by changes in the dimensions of the chest arising from the contraction and relaxation of the muscles of respiration. As the chest expands during inspiration, the pressure in the alveoli falls below that of the atmosphere causing air to enter the lungs. As the chest volume falls during expiration, the pressure in the alveoli rises above that of the atmosphere and air is expelled from the lungs.

2. Inspiration is an active process that depends on the contraction of the diaphragm and external intercostal muscles. Expiration is largely passive and is due to the elastic recoil of the lungs and chest wall.

3. During breathing, the total pressure required to inflate the chest is the pressure required to expand the elastic elements of the chest (measured by the compliance) plus the pressure required to overcome the airways resistance. Most of the resistance to the flow of air is located in the upper airways.

4. The compliance of the lungs is determined both by the elastic elements in the lung parenchyma and by the surface tension of the air–liquid interface of the alveoli. The surface tension is reduced below that of water by pulmonary surfactant secreted by the type II alveolar cells.

5. The work of breathing is equal to the change in pressure multiplied by the change in volume. Thus diseases that reduce lung compliance or increase airways resistance increase the work of breathing. The presence of pulmonary surfactant in the alveoli significantly reduces the work of breathing.

6. Clinical assessment of ventilatory function can be made using a variety of tests. These include measurement of vital capacity, FEV_1, peak expiratory flow rate, and the maximal ventilatory volume. The extent of small airways disease can be assessed by measuring the closing volume.

16.5 Alveolar ventilation and dead space

Broadly, the respiratory system can be considered to consist of two parts: the conducting airways and the area of gas exchange. In dividing the respiratory system in this way it becomes obvious that not all the air taken in during a breath reaches the alveolar surface. Some of it must occupy the airways that connect the respiratory surface to the atmosphere. This air does not take part in gas exchange and is known as the *dead space*.

The remaining fraction of the tidal volume enters the alveoli. Thus

tidal volume = dead space + volume of air entering the alveoli

or, in symbols,

$$V_T = V_D + V_A$$

where V_D is the volume of the dead space and V_A is the volume of air entering the alveoli.

Since not all of the air that enters the alveoli takes part in gas exchange, two different types of dead space are distinguished:

(1) the *anatomical dead space*, which is strictly the volume of air taken in during a breath that does not mix with the air in the alveoli;

(2) The *physiological dead space*, which is the volume of air taken in during a breath that does not take part in gas exchange.

As with other lung volumes, such as vital capacity, the anatomical and physiological dead space depends on the body size, age, and sex of the individual. In a normal healthy person the anatomical and physiological dead spaces are about the same—150 ml for a tidal volume of 500 ml. In some diseases of the lung, such as emphysema, the physiological dead space can greatly exceed the anatomical dead space (see below).

The anatomical dead space can be measured in a similar manner to the closing volume. The subject is asked to take a breath of pure oxygen and hold it for a second before exhaling. By this simple maneuver, the air in the airways is made to have a different composition to that of the alveoli. Air in the conducting system will be pure oxygen while that in the alveoli will also contain nitrogen. All that is then required to determine the dead space is to ask the subject to breathe out while the nitrogen content and volume of the expired air are continuously monitored (Fig. 16.18). The airways contain pure oxygen and this volume must first be displaced before the alveolar air is exhaled. Thus the level of nitrogen will rise from zero to reach a plateau value, which is the same as that in the alveoli. The anatomical dead space is taken as the volume exhaled between the beginning of expiration and the mid-point between the zero level and the plateau value.

The physiological dead space is equal to the volume of the non-respiratory airways plus the volume of air that enters those alveoli that are not perfused with blood, since these alveoli cannot participate in gas exchange. The physiological dead space can be estimated by measuring the carbon dioxide content of the expired and alveolar air using the Bohr equation (see Box 16.5).

The amount of air that is taken in during each breath (the tidal volume V_T) multiplied by the frequency f of breathing is known as the *minute volume* $\dot{V}_E$. The fraction of the minute volume that ventilates the alveoli is known as the *alveolar ventilation* $\dot{V}_A$. Therefore

$$fV_T = fV_D + fV_A$$

or

$$\dot{V}_E = \dot{V}_D + \dot{V}_A$$

Note that the dots over the volumes indicate flow in liters per minute. In the case of a subject breathing a tidal volume of 0.5 liters 12 times a minute,

$$\dot{V}_E = 12 \times 0.5$$
$$= 6 \, l \, min^{-1}.$$

If the dead space is 150 ml, the alveolar ventilation is

$$12 \times (0.50 - 0.15) = 12 \times 0.35$$
$$= 4.2 \, l \, min^{-1}.$$

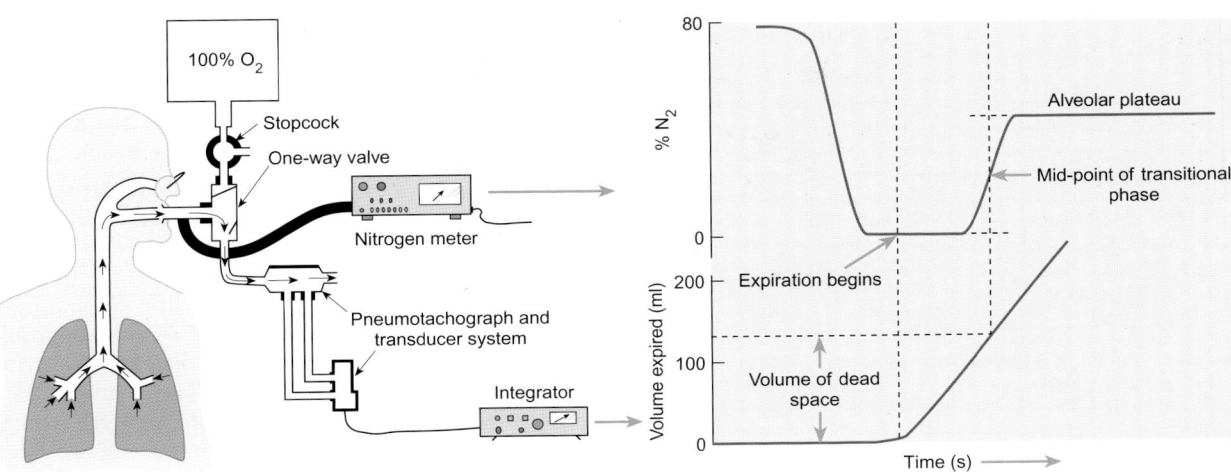

Fig. 16.18 Fowler's method for the determination of the anatomical dead space. The subject is asked to take a breath of pure oxygen and then breathe out. The concentration of nitrogen in the expired air is continuously measured. The flat portion of the nitrogen trace represents the alveolar gas. The expired volume at the mid-point of the transition from pure oxygen to alveolar gas is taken as the volume of the dead space.

BOX 16.5 DERIVATION OF THE BOHR EQUATION FOR CALCULATING THE PHYSIOLOGICAL DEAD SPACE

All the CO_2 of the expired air comes from the alveoli. The amount of CO_2 in any volume of gas is simply the fractional content of CO_2 in the sample (F) multiplied by the volume. Since the volume of air expired (V_E) is the sum of the air displaced from the dead space (V_D) plus that expelled from the alveoli (V_A):

$$V_E = V_D + V_A \tag{1}$$

and

$$F_E V_E = F_D V_D + F_A V_A. \tag{2}$$

Since there is no CO_2 in the dead space (as it is filled with atmospheric air which has a negligible CO_2 content) $F_D = 0$ and eq (2) becomes

$$F_E V_E = F_A V_A. \tag{3}$$

which can be rewritten as

$$V_A = \frac{V_E F_E}{F_A}. \tag{3a}$$

but since $V_A = V_E - V_D$ (from eq (1)),

$$V_E - V_D = \frac{V_E F_E}{F_A} \tag{4}$$

which can be rewritten as the *Bohr equation*

$$V_D = V_E \left(1 - \frac{F_E}{F_A} \right) \tag{4a}$$

Therefore the physiological dead space can be calculated from the volume and fractional CO_2 content of the expired gas and the fractional concentration of CO_2 in the alveolar air.

The volume and CO_2 content of the expired air can readily be measured and a sample of alveolar air can be obtained by asking a subject to expire fully through a long, thin tube. The last part of the expired volume will have the same composition as the alveolar air. This gas can be sampled and its F_{CO_2} determined. This will give an average for the composition of the alveolar air. A more accurate estimate of the CO_2 content of alveolar air can be obtained by measuring the P_{CO_2} of arterial blood.

Worked example

If the percentages of CO_2 in the expired and alveolar air were 3.6 per cent and 5.2 per cent and the tidal volume was 500 ml, what is the physiological dead space?

Using the Bohr equation,

$$V_D = 500 \times [1 - (3.6/5.2)]$$
$$= 500 \times (1 - 0.69)$$
$$= 155 \text{ ml.}$$

The alveolar ventilation is not uniform throughout the lung

The ventilation of the lung is its *change in volume relative to its resting volume during a single respiratory cycle*; the greater the relative change in volume, the greater is the ventilation. Measurements with tracer gases show that the inspired air is not distributed evenly to all parts of the lung. The pattern of ventilation depends on posture (i.e. whether the subject is upright or lying down) and on the amount of air inspired. In an upright subject, during a slow inspiration following a normal expiration, the base of each lung is ventilated about 50 per cent more than the apex (Fig. 16.19). If the subject inspires after a forced expiration

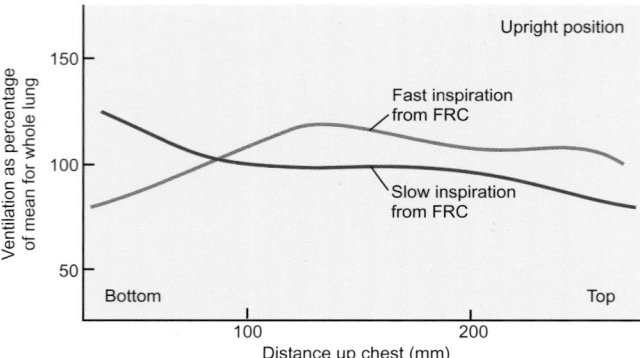

Fig. 16.19 The distribution of ventilation in the normal upright human lung. The data shown are for a slow inspiration following a normal expiration.

to the residual volume, the base of the lung is ventilated nearly three times as much as the apex. This difference is reduced if the subject lies down. Until recently, the variation in ventilation was attributed to the influence of gravity on the lung parenchyma. However, recent measurements on astronauts during space flight have shown that uneven ventilation persists in zero gravity.

What are the causes of uneven pulmonary ventilation? First, during inspiration, the volume of the lower part of the chest increases significantly more than the upper part. This situation arises because the lower ribs are more curved and mobile than the upper ribs. Secondly, the descent of the diaphragm expands the lower lobes of the lungs more than the upper ones, which are attached to the main bronchi. (The bronchi are much less easily stretched than the lung parenchyma.) Thirdly, the compliance of the lung is not uniform. The peripheral lung tissue is more compliant than the deeper tissue, which is attached to the stiffer airways. The combination of these factors results in differing regions of the lung exhibiting differing amounts of ventilation.

Summary

1. The anatomical dead space is the volume of air taken in during a breath that does not mix with the air in the alveoli. It is a measure of the volume of the conducting airways. The physiological dead space is the volume of air taken in during a breath that does not take part in gas exchange.

2. The minute ventilation is the tidal volume multiplied by the frequency of breathing. The alveolar ventilation is the volume of air entering the alveoli per minute.

3. The ventilation of the lung is not uniform, being somewhat greater at the base than at the apex. This situation arises in part because the base of the lungs expands proportionately more than the apex.

16.6 The bronchial and pulmonary circulations

The lungs receive blood from two sources: the bronchial circulation and the pulmonary circulation (Fig. 16.20).

1. The *bronchial circulation*. Bronchial arteries arise from the aortic arch, the thoracic aorta, or their branches (mainly the intercostal arteries). These arteries supply oxygenated blood to the smooth muscle of the principal airways (the trachea, bronchi, and bronchioles as far as the respiratory bronchioles), the intrapulmonary nerves, nerve ganglia, and the interstitial lung tissue. The blood draining the airways is deoxygenated. The blood from the upper airways (as far as the second-order bronchi) drains into the right atrium. The venous return from the later generations of airways flows into the pulmonary veins where it mixes with the oxygenated blood from the alveoli. The bronchial circulation normally accounts for only a very small part of the output of the left ventricle (about 2 per cent), but in certain rare clinical conditions it can be as much as 20 per cent.

2. The *pulmonary circulation*. The output of the right ventricle passes into the pulmonary artery, which subsequently branches to supply the individual lobes of the lung. The pulmonary arteries branch along with the bronchial tree until they reach the respiratory bronchioles. Here they form a dense capillary network to provide a vast area for gas exchange that is similar in extent to that of the alveolar surface (Fig. 16.20). The capillaries drain into pulmonary venules, which arise in the septa of the alveoli. Small veins merge with larger veins that are arranged segmentally. Finally, two large pulmonary veins emerge from each lung to empty into the left atrium.

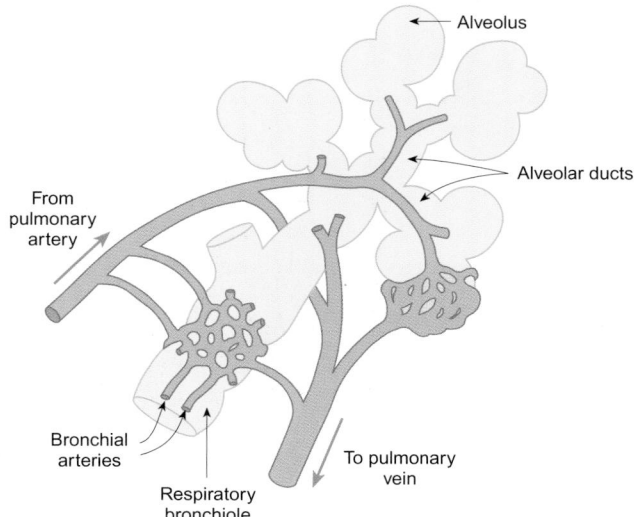

Fig. 16.20 The arrangement of the pulmonary and bronchial circulations in relation to the alveoli. Note that the bronchial circulation does not supply the alveoli. The blood returning from lower airways drains into the pulmonary veins having bypassed the alveoli.

Pulmonary blood flow and its regulation

The output of the right ventricle is equal to that of the left ventricle so that, at rest, about 5 liters of blood pass through the pulmonary vessels each minute. Since the pulmonary capillaries contain only about 100 ml of blood and the stroke volume is about 70 ml (see Chapter 15), a large part of the blood in the pulmonary capillary bed is replaced with each beat of the heart.

The pulmonary arterioles do not appear to be subject to autonomic regulation to any large degree, but the caliber of the small vessels is regulated by the alveolar P_{O_2} and P_{CO_2}. In areas of the lung that have a low P_{O_2} (hypoxia) or a high P_{CO_2} (hypercapnia), the arterioles constrict and blood is diverted to better oxygenated areas. This response is the opposite of that seen in other vascular beds (where low P_{O_2} causes a dilatation of the arterioles) and is not abolished by section of the autonomic nerves. Therefore it is a local response.

Blood flow through the upright lung is greatest at the base and least at the apex

As in other vascular beds, the flow of blood through the lungs is determined by the perfusion pressure and the vascular resistance. Compared with the systemic circulation, however, the pressures in the pulmonary arteries are rather low; the systolic and diastolic pressures are about 3.3 kPa and 1.0 kPa, respectively (c. 25/8 mmHg). Since a column of blood 1 cm high will exert a pressure of about 0.1 kPa, the pressure in the pulmonary artery during systole is sufficient to support a column of blood 33 cm high. During diastole, however, the pressure is sufficient only to support a column about 10 cm high. As a result, differences in the hydrostatic pressures of the blood in various parts of the pulmonary circulation will exert a considerable influence on the distribution of pulmonary blood flow when the body is erect. Moreover, because the pressures in the pulmonary circulation are low, the pressure of the air in the alveoli has a marked effect on vascular resistance and hence on blood flow. Thus the flow of blood to different parts of the lung is determined by three pressures:

(1) the hydrostatic pressure in the pulmonary arteries in different parts of the lungs;

(2) the pressure in pulmonary veins;

(3) the pressure of the air in the alveoli.

When the body is erect, the base of the lung lies below the origin of the pulmonary artery and the hydrostatic pressure of the blood sums with the pressure in the main pulmonary artery. Above the origin of the pulmonary artery the pressure is diminished by the hydrostatic pressure so that the blood flow falls with distance up the lung until, at the apex of the lung which is about 15 cm above the origin of the pulmonary artery, blood flow occurs only during systole and not during diastole. The influence of these pressures on the distribution of blood flow in the upright lung is shown in Fig. 16.21.

For simplicity we shall assume that the pressure in the pulmonary veins (P_V) is equal to that of the atmosphere (taken as zero by convention) so that the perfusion pressure will be equal to the pressure in the pulmonary arteries (P_a).

◆ At the apex of the lung, the alveolar pressure (P_A) is similar to the pressure in the pulmonary arteries (P_a). The pulmonary capillaries will be relatively compressed and vascular resistance will be high. As a result, blood flow is relatively low (zone 1 of Fig. 16.21). Indeed, if the arterial pressure falls below the alveolar pressure, the capillaries will be collapsed (as $P_A > P_a$) and there will be no blood flow. This may happen during diastole.

◆ In the middle zone (zone 2 of Fig. 16.21), the pressure in the pulmonary arteries is higher and exceeds the alveolar pressure so that blood flow progressively increases down the zone.

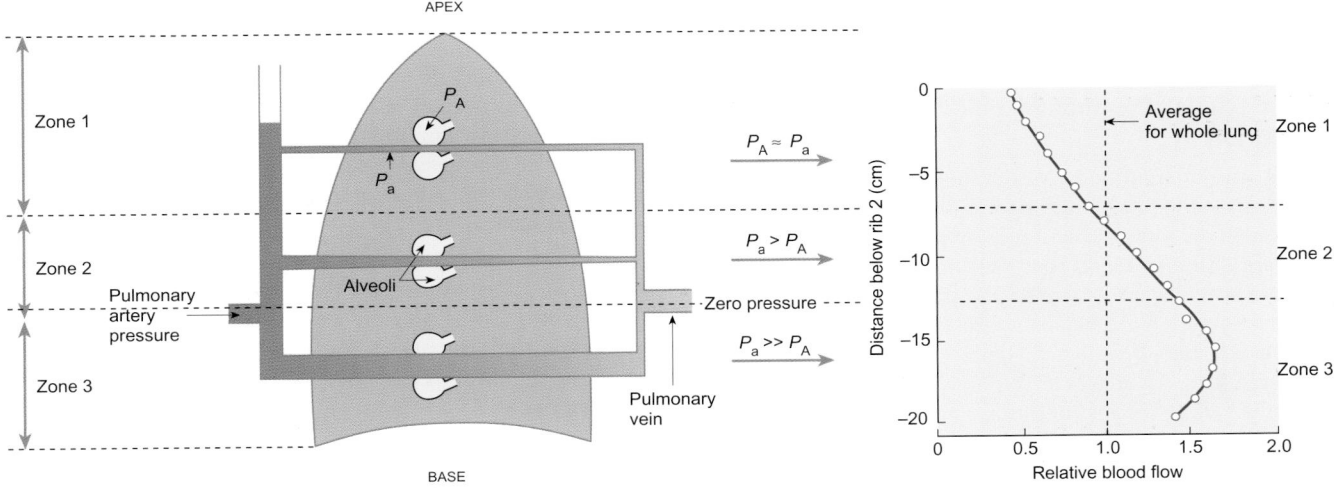

Fig. 16.21 The influence of the hydrostatic and alveolar pressures on the distribution of blood flow in the upright lung. Blood flow depends on the perfusion pressure (here assumed to be proportional to the pressure P_a in the pulmonary arteries for simplicity) and on the vascular resistance. When the perfusion pressure exceeds the alveolar pressure P_A, vascular resistance is low and blood flow is high (zone 3). When perfusion pressure is low and is similar in value to the alveolar pressure, the pulmonary vessels will be compressed and vascular resistance will be increased (zone 1).

◆ At the base of the lung (zone 3 in Fig. 16.21), the arterial pressure exceeds the alveolar pressure by a considerable margin and the pulmonary vessels are fully open. Blood flow is relatively high.

It is important to understand that the regional variation in blood flow is mainly due to the effect of gravity. It disappears when the subject lies down, as the anteroposterior difference is much smaller. When gravity is reduced during simulated space flights in aircraft, apical blood flow is increased.

When cardiac output increases, the pulmonary vascular resistance falls

As in other vascular beds, the blood flow through the lungs depends on the perfusion pressure and the vascular resistance. During exercise, cardiac output rises considerably but the pressure in the pulmonary arteries shows only a small increase. For example, if cardiac output increases threefold, the mean pressure in the pulmonary artery rises by less than 1 kPa (8 mmHg). Since

$$\text{pressure} = \text{flow} \times \text{resistance},$$

the small increase in pressure coupled with the large increase in flow indicates that the resistance of the pulmonary vessels must fall with increased cardiac output.

This fall in the resistance of the pulmonary vasculature is not due to an autonomic reflex or to circulating hormones, but is thought to be a passive response to the increased perfusion pressure. Two mechanisms are believed to be responsible: the recruitment of additional vessels and the distension of those vessels that are already open.

Passive recruitment of the pulmonary vessels is possible because many are closed at the resting values of pressure in the pulmonary arteries. This occurs because the pressure of the air in the alveoli acts directly on the walls of the pulmonary capillaries, tending to collapse them (see above). However, as cardiac output rises during exercise the pressure in the pulmonary vessels exceeds that in the alveoli with the result that more capillaries are recruited. Moreover, the distribution of blood flow to the various parts of the lungs becomes more even.

The matching of pulmonary blood flow to alveolar ventilation

In the upright lung both ventilation and perfusion fall with height above the base. Since the local blood flow falls more rapidly than ventilation with distance up the lung, the ratio of alveolar ventilation to blood flow ($\dot{V}_A/\dot{Q}$ ratio) will vary. Since the physiological purpose of ventilating the lung is to promote gas exchange between the blood and the alveolar air, this variation has considerable physiological significance.

For the lungs as a whole the alveolar ventilation is about 4.2 l min^{-1} while the resting cardiac output is about 5.0 l min^{-1} so that the average value for the $\dot{V}_A/\dot{Q}$ ratio is 4.2/5.0 = 0.84. The base of the lung is relatively well perfused and ventilated, and estimates suggest a $\dot{V}_A/\dot{Q}$ ratio of about 0.6. The ratio rises slowly with distance from the base (measured by rib number in Fig. 16.22). About two-thirds of the way up, the ratio is close to one—theoretically perfect matching. Above this, the ratio rises

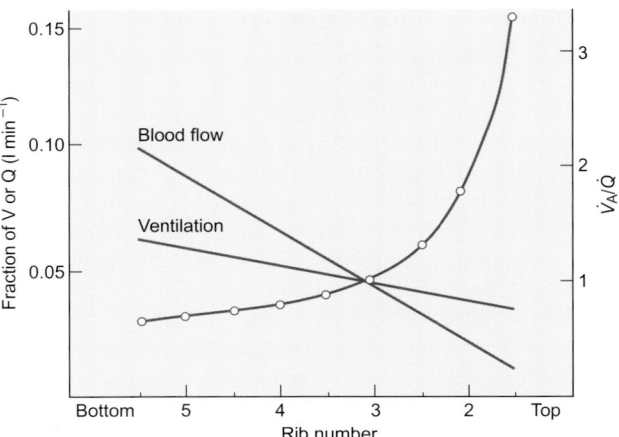

Fig. 16.22 The distribution of ventilation, blood flow, and the ventilation–perfusion ratio in the normal upright lung. Straight lines have been drawn through the data for ventilation and blood flow (data shown in Figures 16.19 and 16.21). Note that the ventilation–perfusion ratio rises slowly at first, and then rapidly towards the top of the lung.

steeply, reaching a value of about 3 in the apex. These are average figures for the various segments of the lung but they are not constant. In exercise, for example, the ventilation increases disproportionately more than the pulmonary blood flow (see Chapter 25).

The $\dot{V}_A/\dot{Q}$ ratio can vary considerably from infinity (ventilated alveoli that are not perfused) to zero (for blood that passes through the lung without coming into contact with the alveolar air). For present purposes it is convenient to distinguish three situations.

1. Well-ventilated alveoli that are well perfused with blood ($\dot{V}_A/\dot{Q} \approx 1$). In this situation, the blood will equilibrate with the alveolar air and will become arterialized (i.e. it will have the $P\text{O}_2$ and $P\text{CO}_2$ of arterial blood). This is the optimum matching of ventilation and perfusion.

2. Poorly ventilated alveoli that are well perfused with blood ($\dot{V}_A/\dot{Q} \ll 1$). In this case, $P\text{O}_2$ and $P\text{CO}_2$ in the alveolar air will tend to equilibrate with the blood. As a result, $P\text{O}_2$ will be lower than normal but $P\text{CO}_2$ will be close to normal. The extent of this equilibration will obviously depend on the extent of the alveolar ventilation, but when this is zero the alveolar air will have the same $P\text{O}_2$ and $P\text{CO}_2$ as mixed venous blood.

3. Well-ventilated alveoli that are poorly perfused with blood ($\dot{V}_A/\dot{Q} \gg 1$). In this case, the blood leaving the alveoli will have a low $P\text{CO}_2$ as the pressure gradient favors the loss of carbon dioxide from the blood. The *oxygen content* of the blood will not be significantly increased as the hemoglobin will be fully saturated.

The lowered $P\text{O}_2$ of blood leaving poorly ventilated parts of the lungs will not be compensated by well-oxygenated blood leaving relatively overventilated parts. This situation arises because the oxygen content of blood from the overventilated alveoli is not significantly higher than that from well-matched alveoli, while that from poorly ventilated areas will be substantially below normal. Thus, when blood from well-ventilated and poorly ventilated regions becomes mixed in the left side of the heart, the oxygen contents of the two streams of blood are averaged but the P_{O_2} of the mixed arterial blood will be below average because of the shape of the oxygen dissociation curve. Thus it is the $\dot{V}_A/\dot{Q}$ ratio in the different parts of the lungs that is important in determining the P_{O_2} of the arterial blood and not the average $\dot{V}_A/\dot{Q}$ ratio for the whole of the lungs.

The mixing of venous blood with oxygenated blood is known as *venous admixture*. It occurs naturally when blood from the bronchial circulation drains into the pulmonary veins. When venous blood completely bypasses the lungs it is called a right–left shunt and is commonly seen in congenital heart disease where deoxygenated blood from the right side of the heart mixes with oxygenated blood from the pulmonary veins. As for the case when blood from poorly ventilated alveoli mixes in significant quantities with arterialized blood from well-ventilated alveoli, a right–left shunt will reduce the P_{O_2} and the oxygen content of the blood reaching the systemic circulation.

Gas exchange across the alveolar membrane occurs by physical diffusion

As the inspired air passes through the passages of the lung its velocity falls steeply with each airway generation until it arrives at the alveoli where it equilibrates with the gas in the alveoli by simple diffusion. To be able to oxygenate the blood, an oxygen molecule must first dissolve in the aqueous layer covering the alveolar epithelium. It then diffuses across the thin membranes that separate the alveolar air spaces from the blood (see above). The rate of diffusion across the alveolar membranes is governed by Fick's law.

When a subject is resting, the blood takes about 1 s to pass through the pulmonary capillaries, but during severe exercise it takes as little as 0.3 s. Despite the short time available, in healthy subjects the blood becomes fully equilibrated with the alveolar air during its transit through the pulmonary capillaries. This ability of the lungs to ensure equilibration between the blood of the pulmonary capillaries and the alveolar air is measured by its *diffusing capacity* (sometimes called its *transfer factor*).

The diffusing capacity is defined as the volume of gas that diffuses through the alveolar membranes per second for a pressure difference of 1 kPa. The formal definition is given by

$$D_{LX} = \frac{V_X}{(P_{AX} - P_{CX})}$$

where D_{LX} (ml min^{-1} kPa^{-1}), is the diffusing capacity of gas X, V_X is the volume of gas diffusing between the alveoli and the blood, P_{AX} is the partial pressure of the gas in the alveolar air, and P_{CX} is the average partial pressure of the gas in the pulmonary capillaries. The diffusing capacity depends on the method of measurement. It is normally estimated by asking the subject to take a breath of a gas mixture containing a small amount of carbon monoxide. As this gas binds strongly to hemoglobin, the partial pressure in the alveolar capillaries is effectively zero. The equation for calculating the diffusing capacity becomes

$$D_{LCO} = \frac{V_{CO}}{P_{ACO}}.$$

In normal healthy young people the diffusing capacity for carbon monoxide measured by the single-breath method averages 225 ml min^{-1} kPa^{-1} (30 ml min^{-1} mmHg^{-1}) at rest. The oxygen diffusing capacity is equal to that for carbon monoxide multiplied by 1.23.

The diffusing capacity for carbon dioxide is about 20 times that for oxygen, as it is much more soluble in the alveolar membranes. Nevertheless, the overall rate of equilibration of carbon dioxide between the blood and the alveolar air is similar to that for oxygen. This is partly because the pressure gradient for carbon dioxide is much smaller (0.8 kPa or 6 mmHg) and partly because most of the carbon dioxide in the blood is in chemical combination (as carbamino compounds and bicarbonate), from which it is released relatively slowly.

The diffusing capacity increases with body size and lung volume. (Diffusing capacity depends directly on the area available for gas exchange while the partial pressures of the respiratory gases do not.) It increases significantly during exercise when previously closed pulmonary capillaries open. It also increases when the subject lies down. This change probably reflects the more uniform distribution of pulmonary blood flow. Diffusing capacity declines with age. If the alveolar membranes become thickened by disease (as in emphysema or fibrosis) or if they become filled with fluid (pulmonary edema), the diffusing capacity will be significantly reduced.

Fluid exchange in the lungs

In health, the alveoli are dry. What prevents fluid leaking from the pulmonary capillaries into the alveoli? Like other vascular beds, the exchange of fluid between the capillaries and the interstitial fluid is governed by Starling forces (see Box 15.8). The pressure in the pulmonary artery is low (the mean pressure is only about 1.5 kPa (12 mmHg)) and the pressure in the pulmonary capillaries is still lower—between 0 and 2 kPa (0–15 mmHg) depending on the distance above the heart. The interstitial pressure is about 0.5 kPa (3.8 mmHg) below that of the atmosphere. Although the oncotic pressure of the plasma is the same as that in the systemic circulation (c. 3.6 kPa or 27 mmHg), that of the pulmonary interstitial fluid is relatively high (c. 2.2 kPa or 16 mmHg). Consequently, the net filtration pressure in the lower part of the lungs is about 1 kPa (~7.5 mmHg). Normally the filtered fluid is returned to the circulation via the pulmonary lymphatics. Under normal conditions, the pulmonary lymph flow is about 10 ml h^{-1} and the alveoli are free of fluid.

As the capillary pressure in the upper part of the lungs is low, there is little fluid formation in this part of the lung. However, as the Starling forces favor filtration in the dependent regions, a small increase in the pressure within the pulmonary capillaries will lead to a greater filtration of fluid. If this exceeds the drainage capacity of the lymphatics, fluid will accumulate in the interstitium, resulting in pulmonary edema. This occurs during left-sided heart failure (see also Chapter 31, p. 602) and following mechanical or chemical damage to the lining of the alveoli. The fluid first accumulates within the pulmonary interstitium and the lymphatic vessels. Above a critical pressure, fluid will enter the alveoli themselves, flooding them and seriously compromising their ability to participate in gas exchange.

Summary

1. The lungs receive their blood supply via the bronchial and the pulmonary circulations. The bronchial circulation is part of the systemic circulation and supplies the trachea, bronchi, and bronchioles as far as the respiratory bronchioles. The pulmonary circulation is supplied by the output of the right ventricle and the blood in this vascular bed participates in gas exchange.

2. The pressures in the pulmonary artery are much lower than those in the aorta. Systolic pressure is about 3.3 kPa (25 mmHg) and diastolic pressure is only 1 kPa (8 mmHg).

3. Because the systolic and diastolic pressures in the pulmonary arteries are low, the effects of gravity on regional blood flow are very significant. As a result, there is considerable variation in blood flow in the upright lung. The base of the lung is relatively well perfused compared with the apex.

4. In the upright lung, both ventilation and perfusion fall with height above the base, but the blood flow falls significantly faster than ventilation. This results in an increase in the $\dot{V}_A/\dot{Q}$ ratio with height above the base of the lung. In regions that are underventilated or overperfused the blood will tend to have a lower than normal P_{O_2}. In healthy lungs, a low P_{O_2} tends to cause a local vasoconstriction, thus diverting the blood to better-ventilated areas.

5. The ability of the lungs to ensure equilibration between the blood of the pulmonary capillaries and the alveolar air is measured by its diffusing capacity. While the diffusing capacity for carbon dioxide is about 20 times that for oxygen, the overall rate of equilibration is similar for both gases.

6. The pulmonary circulation plays a significant role in the metabolism of many vasoactive substances.

Metabolic functions of the pulmonary circulation

All of the venous return from the systemic circulation passes through the lungs on its way to the left side of the heart. As a result, the pulmonary circulation is ideally situated to metabolize vasoactive materials released by specific vascular beds. Many substances such as bradykinin, norepinephrine, and prostaglandin E_1 are almost completely removed in a single pass through the lungs. In contrast, the lungs convert angiotensin I to its active form angiotensin II which stimulates aldosterone secretion by the adrenal cortex (see Chapter 17). The site of this metabolic activity is the endothelium of the pulmonary circulation.

16.7 The control of respiration

No normal person has to think about when or how deeply to breathe. Breathing is an automatic rhythmical process that is constantly adjusted to meet the everyday requirements of life such as exercise and speech. To account for this remarkable fact it is necessary to consider three important questions.

1. Where does this rhythmical activity originate?
2. How is it generated?
3. How is the rate and depth of respiration controlled?

The respiratory rhythm is established by specific groups of neurons that lie within the lower part of the brainstem

If the brainstem of an anesthetized animal is completely cut through above the pons, the basic rhythm of respiration continues. The basic respiratory rhythm is maintained even if all the afferent nerves are sectioned. Section of the spinal cord below the outflow of the phrenic nerve (C3–C5) leads to paralysis of the intercostal muscles but not of the diaphragm (which is innervated by the phrenic nerve). However, section of the lower region of the medulla will block all respiratory movements. From these observations two things are clear.

1. The respiratory muscles themselves have no intrinsic rhythmic activity.
2. The caudal part of the brainstem has all the neuronal mechanisms required to generate and maintain a basic respiratory rhythm.

After the vagus nerves have been cut, respiration becomes slower and deeper. If the brainstem is subsequently cut across between the medulla and pons, there is little change in the respiratory pattern. However, sections through the pons alter the pattern of respiration so that inspiration becomes relatively prolonged with brief episodes of expiration. Stimulation of specific groups of nerve cells in the pons synchronizes the discharge of the phrenic nerves with the stimulus. From these and other experimental observations, the pons has been shown to have an important role in regulating the respiratory rhythm. It is here that afferent information concerning the state of the lungs is believed to act to modulate the rate and depth of respiration.

How is the respiratory rhythm generated?

There are two groups of neurons in the medulla that discharge action potentials with an intrinsic rhythm that corresponds to that of the respiratory cycle. These are known as the *dorsal respiratory group* and the *ventral respiratory group*. The dorsal respiratory group mainly discharge action potentials just prior to and during inspiration and are therefore mainly inspiratory neurons. They are upper motoneurons which project to the lower respiratory motoneurons of the contralateral phrenic nerve. The ventral respiratory group consists of both inspiratory and expiratory neurons and receives inputs from the dorsal respiratory group. They are the upper respiratory motoneurons for both the contralateral phrenic and intercostal nerves.

The dorsal and ventral respiratory groups receive a variety of inputs from higher centers in the brain including the cerebral cortex and pons. They also receive inputs from the carotid and aortic bodies (which are the peripheral chemoreceptors that sense the P_{O_2}, P_{CO_2}, and pH of the arterial blood—see below) and the vagus nerve (which carries afferent nerve fibers from the lungs). It appears that inspiration is initiated by neurons of the dorsal respiratory group. This intrinsic activity subsequently sums with afferent activity coming from lung stretch receptors

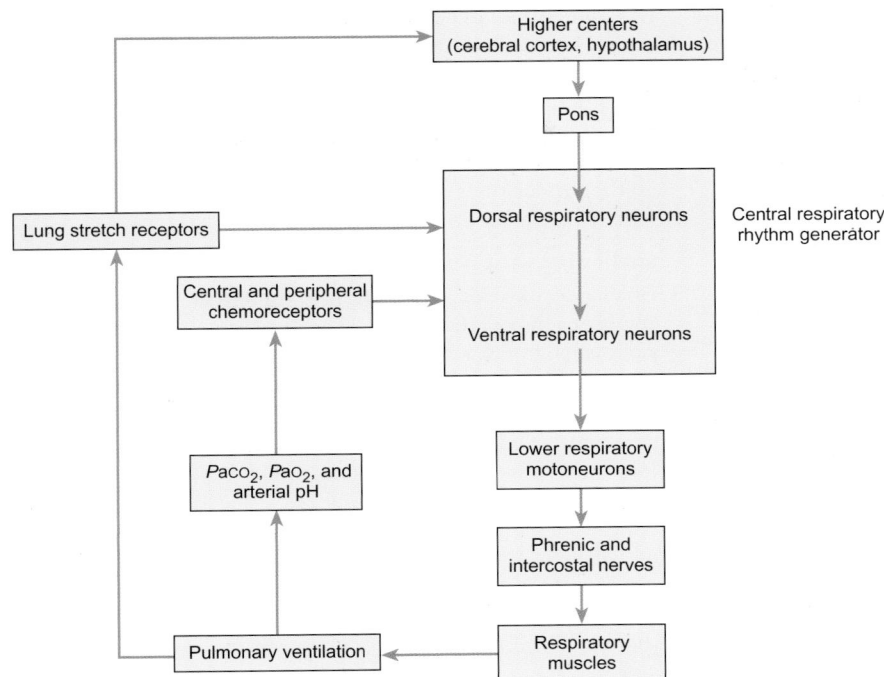

Fig. 16.23 The interrelations between the main neural elements that regulate the rate and depth of respiration.

to switch off inspiration and commence expiration. A simple diagram of the arrangement of the respiratory control pathway is shown in Fig. 16.23.

The activity of the expiratory respiratory motoneurons in the spinal cord (the lower respiratory motoneurons) is inhibited during inspiration, while that of the inspiratory motoneurons is inhibited during expiration. This pattern of reciprocal inhibition has its origin in the dorsal and ventral respiratory groups and is not a local spinal reflex.

The efferent activity of the thoracic respiratory neurons can be monitored by recording the electrical activity of the intercostal muscles and diaphragm (Fig. 16.24). Throughout the respiratory cycle, the motor units of the respiratory muscles are active. During inspiration, the activity of the inspiratory muscles (the diaphragm and external intercostal muscles) progressively increases, additional motor units are recruited and the muscles shorten progressively thereby expanding the volume of the

chest (see Section 16.4). During expiration, the activity of the inspiratory muscles gradually declines allowing the chest to return to its resting volume (FRC). The expiratory muscles show a reciprocal pattern, with increasing activity during expiration and falling activity during inspiration. The progressive modulation of the tone of the respiratory muscles provides a smooth transition from expiration to inspiration and forms part of the work of breathing.

The smooth muscle of the upper airways (trachea, bronchi, and bronchioles) possesses slowly adapting stretch receptors. When the lung is inflated, these receptors send impulses to the dorsal respiratory group via the vagus nerves. This afferent information tends to inhibit respiratory activity and so acts to limit inspiration. This is known as the *Hering–Breuer lung inflation reflex*. If the lungs are inflated by positive pressure, the frequency of respiratory movements falls and may cease altogether (*apnea*).

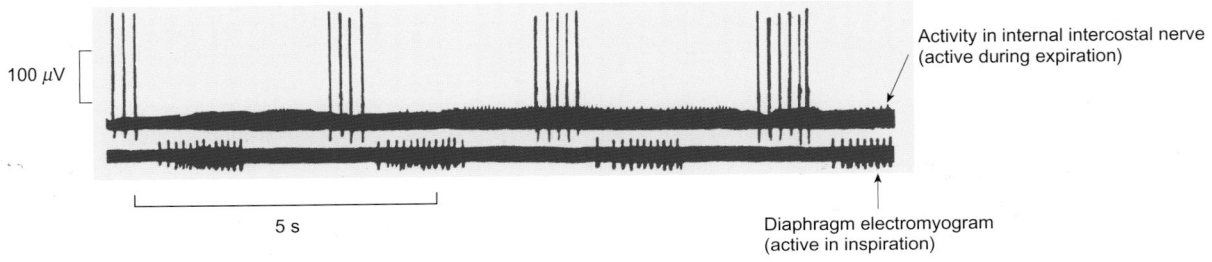

Fig. 16.24 An example of the reciprocal motor activity in the diaphragm and the electrical activity of a nerve serving an internal intercostal muscle. The upper record shows the action potential activity of a motor nerve fiber that supplies an internal intercostal muscle. Note the increased activity during expiration. The lower record shows the electrical activity of the diaphragm, which is active during inspiration.

In animals such as the cat and rabbit, the Hering–Breuer reflex appears to play a significant part in the control of the respiratory rhythm. In humans, this reflex is not activated at normal tidal volumes. However, it is activated when tidal volumes exceed about 0.8–1 liter. For this reason it is thought that the Hering–Breuer reflex may play a role in regulating inspiration during exercise.

Voluntary control of respiration

Normal regular breathing (or *eupnea*) is an automatic process, although the rate and depth of breathing can be readily adjusted by voluntary means. For example, it is possible to suspend breathing for a short period. This breath-holding is known as *voluntary apnea* and its duration is normally limited by the rise in arterial P_{CO_2}. Equally, it is possible to increase the rate and depth of breathing deliberately during *voluntary hyperventilation* (also known as *voluntary hyperpnea*). This voluntary control affects both lungs—it is not possible to rest the left lung voluntarily while ventilating the right. The pathways involved in voluntary regulation are not known with any certainty but presumably have their origin in the motor cortex. Quite fine degrees of control over the muscles of respiration are possible. This is important during speech, singing, or the playing of wind instruments.

The reflex control of respiration

Cough and sneeze

In addition to their stretch receptors, the airways possess receptors that respond to irritants. Activation of these receptors elicits a range of reflexes. When the irritant receptors of the upper air-ways are stimulated, they elicit a cough (Fig. 16.25) or, in the case of irritants on the nasal mucosa, a sneeze. The initial phase of either response is a deep inspiration followed by a forced expiration against a closed glottis. As the pressure in the airways rises, the glottis suddenly opens and the trapped air is expelled at high speed. This dislodges some of the mucus covering the epithelium of the airways and helps to carry the irritant away with it via the mouth or nose.

Swallowing

During swallowing, respiration is inhibited. This is part of a complex reflex pattern. As food or drink passes into the oropharynx, the nasopharynx is closed by the upward movement of the soft palate and the contraction of the upper pharyngeal muscles. Respiration is inhibited at the same time and the laryngeal muscles contract, closing the glottis. The result is that aspiration of food into the airways is avoided. The act of swallowing is followed by an expiration that serves to dislodge any food particles lying near the glottis. These actions are coordinated by neural networks in the medulla. If particles of food are accidentally inhaled, they stimulate irritant receptors in the upper airways and elicit a cough reflex. If water is aspirated into the larynx, there is a prolonged apnea which prevents water entering the airways.

Pulmonary chemoreflex

Inhalation of smoke and noxious gases, such as sulfur dioxide and ammonia, stimulates irritant receptors (also known as rapidly adapting receptors) within the tracheobronchial tree and elicits a powerful pulmonary chemoreflex in which there is a

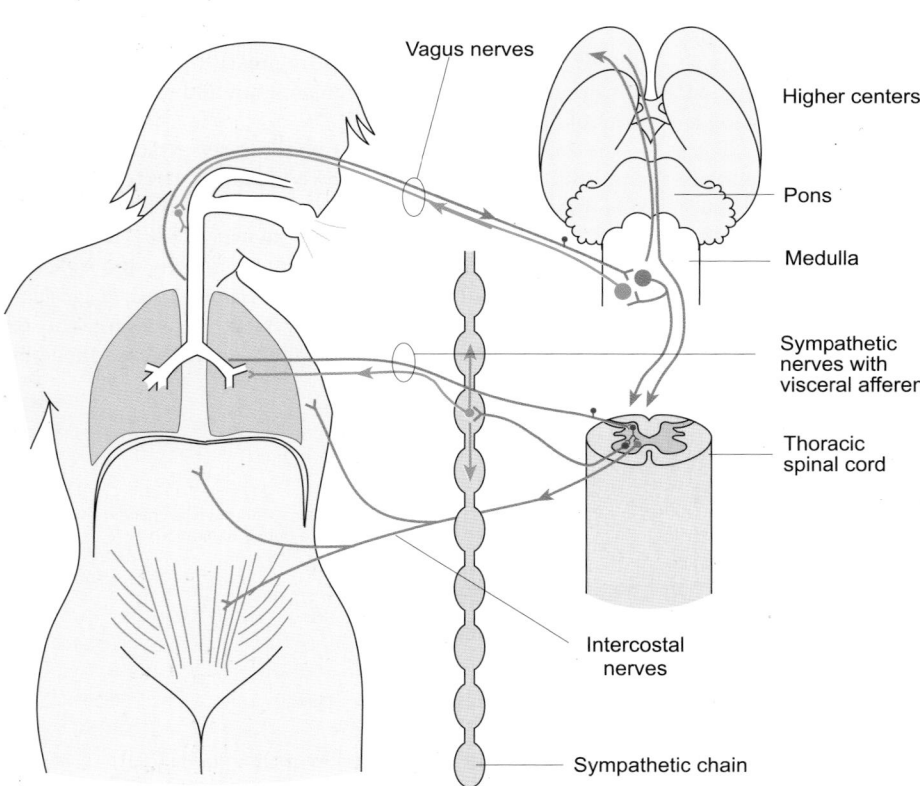

Fig. 16.25 The pathways subserving the cough reflex. Irritation of the trachea or bronchi stimulates the nerve endings of the vagus nerve and visceral afferent fibers (which run alongside the sympathetic motor fibers). Impulses pass to the medulla and spinal cord, from which the excitation spreads to higher regions of the brain. The glottis is closed by activity in vagal afferents and the internal intercostal and abdominal muscles contract. The glottis is suddenly opened and the airflow resulting from the rapid change in pressure in the airways dislodges the offending material.

constriction of the larynx and bronchi and an increase in mucus secretion. If the lungs become congested, breathing becomes shallow and rapid (*pulmonary tachypnea*). The receptors that mediate this response are C-fiber endings located in the interstitial space of the alveolar walls, previously known as J receptors (for *juxtapulmonary capillary receptors*). The role that these receptors play in normal breathing is not known.

Other reflex modulations of respiration

The normal pattern of breathing is modified by many other factors. For example, passive movement of the limbs results in an increase in ventilation that is believed to occur as a result of stimulation of proprioceptors in the muscles and joints (Fig. 16.26). This reflex may play an important role in the increase in ventilation during exercise (see Chapter 25). Pain results in alterations to the normal pattern of respiration. Abdominal pain (e.g. postoperative pain) can be so sharp that it causes a reflex inhibition

of inspiration and apnea. Prolonged severe pain is associated with fast shallow breathing. Immersion of the face in cold water elicits the diving response in which there is apnea, bradycardia, and peripheral vasoconstriction (see Chapter 30).

The blood gases are of major importance in the control of ventilation and are sensed by peripheral and central chemoreceptors

The purpose of respiration is to provide the tissues with oxygen and to remove the carbon dioxide derived from oxidative metabolism. This is achieved by close regulation of the P_{CO_2} and P_{O_2} of the arterial blood (i.e. the Pa_{CO_2} and Pa_{O_2}), which are maintained within very close limits throughout life. Indeed, Pa_{CO_2} and Pa_{O_2} vary little between deep sleep and severe exercise, where the oxygen consumption and carbon dioxide output of the body may increase more than tenfold. Clearly, to achieve such remarkable stability the body needs some means of sensing Pa_{CO_2} and Pa_{O_2} and relaying that information to the neurons that determine the rate and depth of ventilation. This role is performed by the peripheral and central chemoreceptors.

The *peripheral arterial chemoreceptors*—the carotid bodies—are small organs about 7 mm × 5 mm in size that are located just above the carotid bifurcation on each side of the body. The carotid bodies are anatomically and functionally separate from the arterial baroreceptors, which are located in the wall of the carotid sinus (Fig. 16.27). Nevertheless, afferent fibers from the carotid body and the ipsilateral carotid sinus run in the same nerve—the carotid sinus nerve, which is a branch of the glossopharyngeal nerve (cranial nerve IX). The carotid bodies have a very high blood flow relative to their mass (about 20 l kg⁻¹ min⁻¹), which is derived from the external carotid arteries.

The carotid bodies respond to changes in the Pa_{O_2}, Pa_{CO_2}, and pH of the arterial blood. Afferents from the carotid body increase their rate of discharge very significantly as Pa_{O_2} falls below about 8 kPa (60 mmHg), as shown in Fig. 16.30 below. The *aortic bodies* are diffuse islets of tissue scattered around the arch of the aorta that have a similar microscopic structure to the carotid bodies. However, there is no evidence to suggest that they act as chemoreceptors in humans, although they may do so in other species (see below).

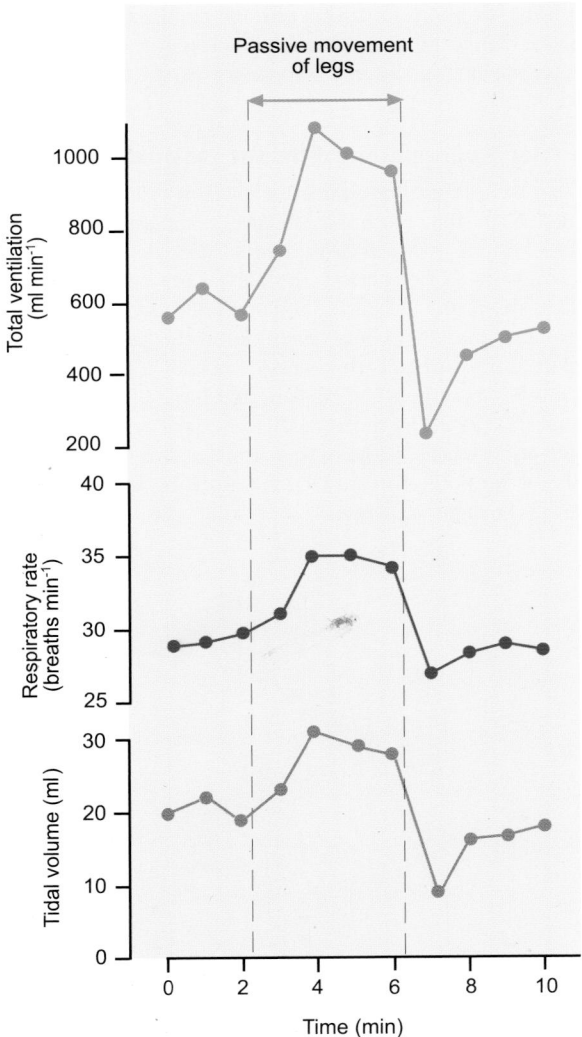

Fig. 16.26 Proprioceptive reflexes increase ventilation. In this experiment, on an anesthetized cat, the motor nerves to the legs were cut, leaving the afferent fibers intact. Vigorous movement of the limbs resulted in a marked increase in ventilation, which declined to resting levels after movement ceased.

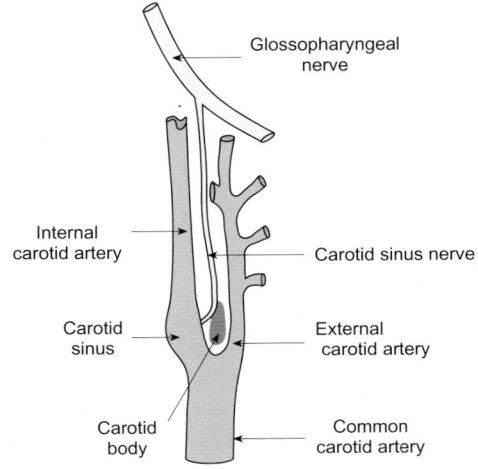

Fig. 16.27 The relative positions of the carotid body and the carotid sinus.

The carotid bodies are the only receptors that are able to elicit a ventilatory response to hypoxia. Thus, after they have been surgically removed for therapeutic reasons, the ventilatory response to hypoxia is lost, even though the aortic bodies remain intact. When breathing normal room air, the influence of the carotid bodies on the rate of ventilation is small. For example, if a subject suddenly switches from breathing room air to breathing 100 per cent oxygen, the minute volume falls by about 10 per cent for a brief period before returning to its previous level. The transient nature of this response can be explained as follows: Breathing pure oxygen reduces the respiratory drive from the peripheral chemoreceptors and this has the effect of reducing the minute volume. During this period, Pa_{CO_2} rises slightly and this acts on the central chemoreceptors to cause the minute volume to return to its original value.

During hypoxia, however, the carotid bodies play an important role in stimulating ventilation. This can be shown in anesthetized, spontaneously breathing animals. If Pa_{O_2} is lowered by giving the animal a gas mixture of 8 per cent oxygen and 92 per cent nitrogen, the minute volume increases by about 50 per cent. This is a reflex response that can be blocked by cutting the aortic and carotid sinus nerves. The increase in minute volume seen following administration of a gas mixture containing 5 per cent carbon dioxide (in 21 per cent oxygen and 74 per cent nitrogen) is unaffected by section of these nerves, showing that the response to hypercapnia is mediated by the central chemoreceptors.

The *central chemoreceptors* respond to changes in the pH of the CSF resulting from alterations in Pa_{CO_2}. They are located on or close to the ventral surface of the medulla near to the origin of the glossopharyngeal and vagus nerves, and provide most of the chemical stimulus to respiration under normal resting conditions. The mechanism by which they sense the Pa_{CO_2} is illustrated in Fig. 16.28. Increased Pa_{CO_2} results in an increase in the P_{CO_2} of the CSF and the hydration reaction for carbon dioxide is driven to the right, leading to the increased liberation of hydrogen ions:

$$CO_2 + H_2O \rightleftharpoons H_2CO_3 \rightleftharpoons H^+ + HCO_3^-.$$

Unlike blood, the CSF has little protein and so the hydrogen ions produced by this reaction are not buffered to any great extent. As a result, the pH falls in proportion to the rise in P_{CO_2} and stimulates the chemoreceptors. Conversely, during hyperventilation, CO_2 is lost from the blood and this causes a reduction in the P_{CO_2} of the CSF. The hydration reaction is driven to the left, the pH of the CSF rises, and ventilation decreases.

If the Pa_{CO_2} were to be persistently above or below its normal value of 5.3 kPa (40 mmHg), the central chemoreceptors would be less sensitive to changes in Pa_{CO_2} than normal. In these situations, the bicarbonate concentration of the CSF is regulated by exchange with chloride ions derived from the plasma. This compensation is important during chronic changes in Pa_{CO_2} arising from residence at high altitude (where the Pa_{CO_2} falls—see Chapter 30) or from chronic respiratory disease (where the Pa_{CO_2} rises).

The effects of breathing different gas mixtures

When air containing a significant amount of carbon dioxide is inhaled, its partial pressure in the alveoli and arterial blood rises. This is known as *hypercapnia*. If a subject deliberately hyperventilates for a brief period, the partial pressure of carbon dioxide in the alveoli and arterial blood falls, as it is lost from the lungs faster than it is being generated in the tissues. This fall in the partial pressure of carbon dioxide is known as *hypocapnia*.

If subjects breathe a gas mixture that has a P_{O_2} lower than the normal 21.2 kPa (159 mmHg), the arterial P_{O_2} will fall. This is known as *hypoxemia*. If the oxygen content is insufficient for the needs of the body, the subject is said to be *hypoxic*. The total absence of oxygen is *anoxia*.

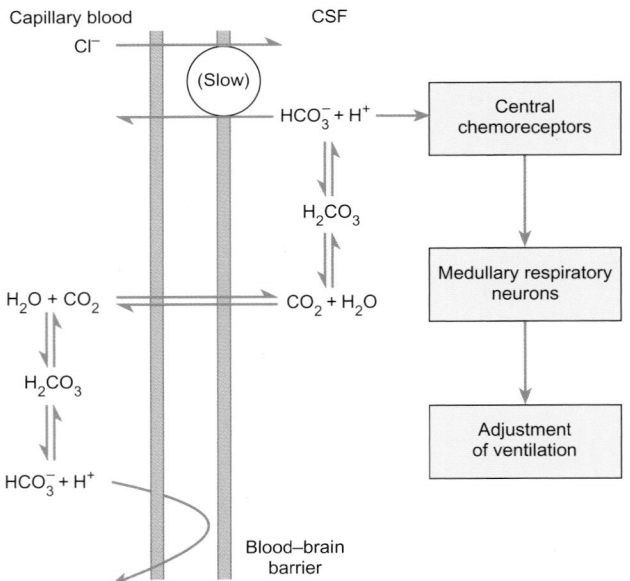

Fig. 16.28 How the P_{CO_2} of the capillary blood in the brain stimulates the central chemoreceptors. A rise in plasma CO_2 leads to increased CO_2 uptake into the brain where it is converted to bicarbonate and hydrogen ions via carbonic acid. The hydrogen ions stimulate the central chemoreceptors and this increases the rate and depth of respiration. A fall in plasma CO_2 has the opposite effect.

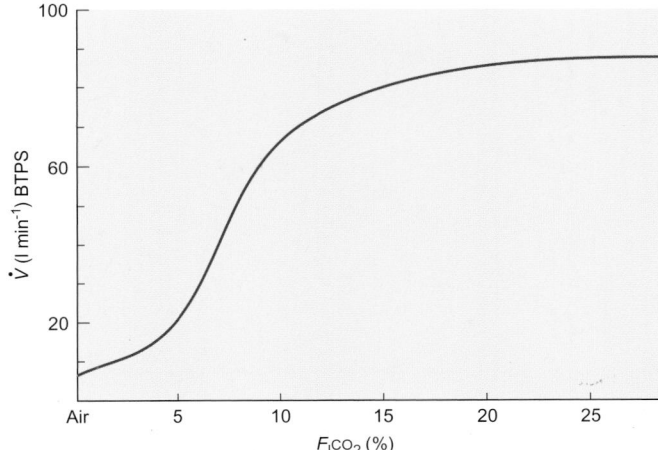

Fig. 16.29 The effect of breathing CO_2 on ventilation. The figure shows the relationship between the concentration of CO_2 in the inspired air ($F_{I_{CO_2}}$) and the total ventilation for a normal subject. Note the steep rise as $F_{I_{CO_2}}$ increases from 5 to 10 per cent.

The relative importance of carbon dioxide and oxygen in determining the ventilatory volume is readily investigated by asking subjects to breathe different gas mixtures. If a normal healthy subject breathes a gas mixture containing 21% oxygen, 5% carbon dioxide, and 74% nitrogen for a few minutes, ventilation increases about threefold. A higher fraction of carbon dioxide in the inhaled gas mixture will stimulate breathing even more. Even a single breath of air containing an elevated carbon dioxide is sufficient to increase ventilation for a short time. Conversely, if a subject hyperventilates for a brief period, the subsequent ventilation is temporarily decreased. Thus, any maneuver that alters the partial pressure of carbon dioxide in the alveolar air (P_{ACO_2}) results in a change in ventilation that tends to restore the P_{ACO_2} to its normal value (5.3 kPa or 40 mmHg). The relationship between the partial pressure of carbon dioxide in the inspired air and total ventilation is shown in Fig. 16.29.

In contrast, if the same subject breathes a mixture of 15% oxygen and 85% nitrogen there is little change in the rate of ventilation at normal barometric pressure. Indeed, hypoxia only tends to stimulate ventilation strongly when the alveolar P_{O_2} falls below about 8 kPa (60 mmHg) (Fig. 16.30). As the P_{AO_2} falls further, ventilation increases steeply.

From these observations, it appears that the principal chemical stimulus to respiration is the P_{CO_2} of the alveolar air rather than the P_{O_2}. At first, this may appear strange, as the main purpose of gas exchange is to maintain the oxygenation of the tissues. The reason for the relatively small ventilatory effect of mild hypoxia can be understood by looking at the oxyhemoglobin dissociation curve (see Fig. 16.30) which shows that at a P_{O_2} of 8 kPa (60 mmHg) the hemoglobin is still about 90 per cent saturated. Below this value, the percentage saturation rapidly falls. Consequently, at normal atmospheric pressure (101 kPa or 760 mmHg), hypoxia would be a relatively weak stimulus to ventilation.

As Fig. 16.31 shows, the increase in ventilation with increasing alveolar P_{CO_2} becomes steeper as the P_{O_2} falls. This shows that the sensitivity of the respiratory drive to carbon dioxide is greater during hypoxia than it is when the P_{O_2} is normal. The effects of hypoxia and hypercapnia are not simply additive; they have a strong synergistic interaction. This is of some importance during breath-holding and asphyxia, where hypoxia and hypercapnia occur together. In this context is should be noted that breathing air containing more than 5 per cent carbon dioxide is unpleasant and causes mental confusion. Prolonged breathing of air containing a high concentration of carbon dioxide or breathing air with a very low P_{O_2} may lead to loss of consciousness.

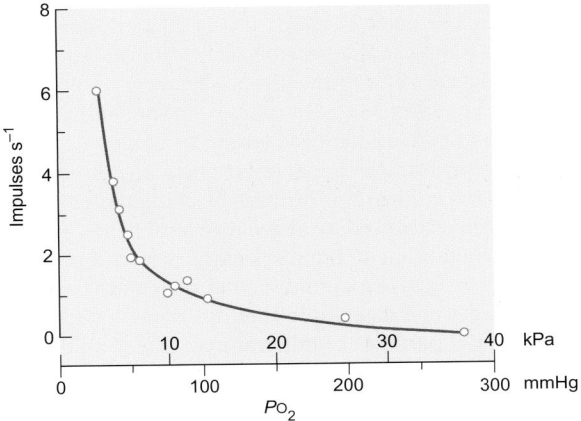

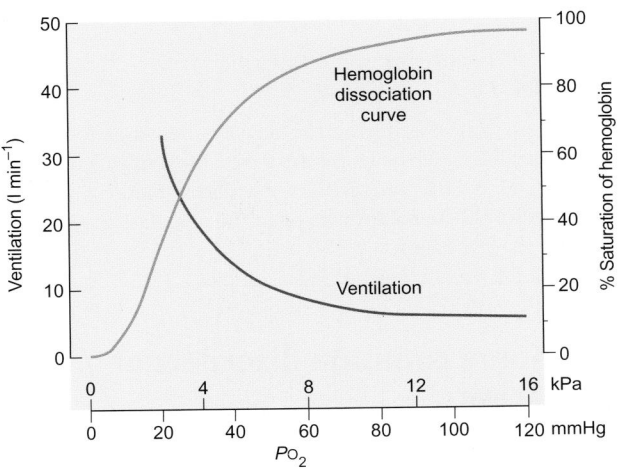

Fig. 16.30 The effect of acute hypoxia on pulmonary ventilation compared with the oxyhemoglobin dissociation curve. The sensitivity of ventilation to the inspired P_{O_2} becomes much steeper below about 8 kPa (60 mmHg) which is the point at which the oxyhemoglobin dissociation curve also becomes very steep. The upper panel shows the rate of carotid chemoreceptor discharge (impulses s^{-1}) plotted against the inspired P_{O_2}. Note that the rate of discharge progressively increases below 13 kPa (100 mmHg). The discharge shows little adaptation.

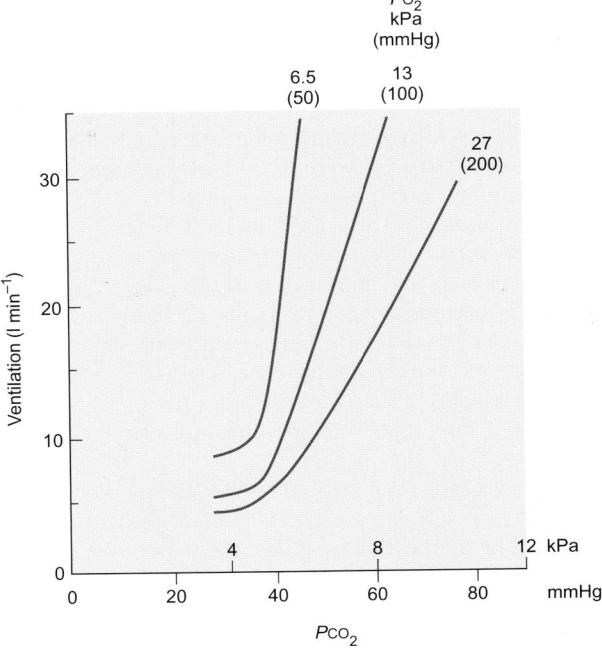

Fig. 16.31 The ventilatory response to hypercapnia at different values of P_{O_2}. Note the elevated basal level of ventilation as P_{O_2} falls and that the increase in ventilation becomes more sensitive to P_{CO_2} as P_{O_2} falls from 27 kPa (200 mmHg) to 6.5 kPa (50 mmHg).

Summary

1. The diaphragm and intercostal muscles have no inherent rhythmic activity themselves but contract in response to efferent activity in the phrenic and intercostal nerves. The basic respiratory rhythm originates in the medulla.

2. A number of reflexes directly influence the pattern of breathing. These include the cough reflex, the Hering–Breuer lung inflation reflex and swallowing.

3. In broad terms, respiration is stimulated by a lack of oxygen (hypoxia) and by an increase in carbon dioxide (hypercapnia).

4. The partial pressures of the blood gases are sensed by the peripheral and central chemoreceptors. The peripheral chemoreceptors are the carotid and aortic bodies. They respond to changes in the Pa_{O_2}, Pa_{CO_2}, and pH of the arterial blood and are the only receptors that respond to hypoxia. The central chemoreceptors are located in the brainstem and are responsible for most of the chemical stimulus to breathing. They respond to the changes in the pH of the CSF brought about by alterations in arterial P_{CO_2}.

While the emphasis in this section has been on the role of Pa_{CO_2} as a ventilatory stimulus, it is important to realize that changes in Pa_{CO_2} cause very significant changes to blood pH. Indeed, one important consequence of the changes in ventilation in response to altered Pa_{CO_2} is that the pH change in the blood is limited. The pulmonary control of blood pH and its role in acid–base balance is discussed in detail in Chapter 29.

16.8 Pulmonary defense mechanisms

As all city dwellers know, the air we breathe is full of particulate matter, some of which is inhaled with each breath. Even if the concentration of particles were to be only 0.001 per cent (10 parts per million), a respiratory minute volume of 6 l min^{-1} results in the intake of over 8500 liters of air each day. This would include 85 ml of particulate matter. Clearly, unless some mechanism existed for the removal of this matter, our lungs would rapidly become clogged with dust and debris. Moreover, not all the inhaled material is biologically inert. Some will be infectious agents (bacteria and viral particles) and some will be allergenic (e.g. pollen). Therefore the lungs need to remove the inert material and inactivate the infectious and allergenic agents.

The airflow through the nose and upper airways is rapid and turbulent. As a result, large particles (>10–15 μm) are brought into contact with the mucus lining these passages and are entrapped by it. This results in filtration of the air and removal of most of the large particles before they reach the trachea. In addition to filtering the incoming air, the upper airways also warm and moisten it. As in the upper airways, the airflow in the trachea and bronchi is turbulent, and this brings the incoming air stream into contact with the wall of the airways. As a result, most of the remaining large particles (5–10 μm) become lodged in the mucus lining of the upper respiratory tree. Further down the airways the airflow becomes slow and laminar. In these regions of the lung, smaller particles (0.2–5 μm) settle on the walls of the airways under the influence of gravity. Only the smallest particles reach the alveoli, where most remain suspended as aerosols and are subsequently exhaled. Nevertheless, about one-fifth of these small particles become deposited in the alveolar ducts or in the alveoli themselves where they are phagocytosed by the alveolar macrophages.

As described earlier (Section 16.3), the respiratory tract from the upper airways to the terminal bronchioles is lined with a ciliated epithelium that is covered by a layer of mucus. The ciliated cells make up about half of the epithelium of the upper airways (trachea and bronchi) but this proportion declines as the airways branch. By generation 5, the ciliated cells only account for about 15 per cent of the total. The cilia beat steadily towards the pharynx and this slowly moves the mucus upwards. As it reaches the pharynx, the mucus is swallowed or coughed up and expectorated. The mucociliary escalator is very efficient in removing those inhaled particles that became trapped in the mucus layer during the passage of air to the alveoli. By using small particles labeled with a radioactive tracer, it has been shown that most of the trapped particles are removed within 24 hours.

The respiratory bronchioles, alveolar ducts, and alveoli do not possess a ciliated epithelium. Particles reaching these regions of the lung are phagocytosed by alveolar macrophages which are found in the fluid lining of the respiratory airways. Bacteria and other biologically active material such as pollen are ingested by the macrophages and digested in their lysosomes. Macrophages also ingest small particles of mineral matter (e.g. silica) which they cannot digest. In these cases, the material is stored within the cell and is removed via the mucociliary escalator when the cell dies. If bacteria or viruses penetrate these defenses and enter the interstitial space, they are dealt with by the immune system (see Chapter 14).

Summary

Particulate matter entering the airways becomes lodged in the mucus lining of the respiratory tree. Most of this material is removed by the mucociliary escalator. Material that is deposited in the alveolar ducts or in the alveoli is ingested by alveolar macrophages.

16.9 Some common disorders of respiration

Normally, respiration continues unnoticed and uneventfully. It is only when things go wrong that we become aware of our breathing. Difficulty in breathing causes distress known as *dyspnea*. It is a subjective phenomenon during which a patient may report being breathless. It may be quite normal, as in a subject who has rapidly climbed to a high altitude (where the atmospheric P_{O_2} is low), or it may reflect some organic disease. In both cases, it is the sense of breathlessness that limits the ability to undertake exercise.

The ability of the lungs to provide enough air for the body's needs is known as the *ventilatory capacity*. If this is less than normal, some form of respiratory disorder is present. These can be divided into those in which the airways are obstructed, those in which the expansion of the lungs is restricted, and those in which the respiratory muscles are weakened and unable to expand the chest fully. Some of the commonly occurring respiratory disorders are discussed here.

Asthma is a condition in which a person has difficulty in breathing, particularly expiration. Asthmatic attacks are characterized by the sudden onset of dyspnea. This is the result of bronchospasm, which usually occurs in response to an allergen that is present in the environment, although it may also occur in response to exercise. The initial phase is probably triggered by an interaction between an allergen and antibodies present on the mast cells of the pulmonary interstitium. These cells then secrete inflammatory mediators including histamine and leukotrienes. These cause spasm of the smooth muscle of the bronchi and an increased secretion of mucus into the airway, both of which effectively reduce the diameter of the airways.

During an asthmatic attack, the FRC and RV are increased although vital capacity is normal. The FEV_1 and peak flow are markedly reduced, in severe cases by more than half (see Fig. 16.16). This limits the ventilatory capacity and results in the dyspnea. In chronic asthma, there is destruction of the bronchial epithelium, and FRC and RV become permanently elevated.

The bronchospasm that occurs in the initial phase of an asthmatic attack is of relatively short duration. For this reason, the airways obstruction in asthma is considered reversible even though the inflammatory condition of the bronchi persists long after the acute attack has subsided.

Emphysema is a condition in which the alveoli are increased in size due to destruction of the lung parenchyma. It is frequently (but not invariably) caused by the smoking of tobacco. The pathophysiology of the disease is not entirely clear. Bronchial lavage shows that air spaces of the lungs of smokers become invaded by neutrophils. It is now thought that these cells secrete proteolytic enzymes that damage the parenchyma of the lungs. In addition, the smoke inhibits the movement of the bronchial cilia, slowing the removal of particulate matter from the airways. The irritant effect of the smoke is the probable cause of the increased secretion of mucus in the larger airways. These effects combine to increase the chances of infection, which results in chronic inflammation of the bronchiolar epithelium. As a result, the diameter of the airways is reduced and, as in asthma, it becomes difficult to exhale, leading to the entrapment of air (from which the disease takes its name).

As a result of the loss of parenchymal tissue, the traction on the airways is reduced and their resistance is increased. By itself, this will limit ventilation, but the problem is compounded as the destruction of the alveoli also restricts the diffusing capacity of the lung. Moreover, the destruction is not uniform, the phys-

iological dead space is increased, and there are abnormal $\dot{V}_A/\dot{Q}$ ratios in different parts of the lung. The result is inadequate gas exchange, hypoxemia (poor oxygenation of the blood), and chronic dyspnea.

As for asthma, the progress of the disease is reflected in respiratory function tests. The entrapment of air leads to high lung volumes but vital capacity, peak flow, and FEV_1 are all reduced. Because the underlying cause is the destruction of the parenchyma, the increased resistance is always present, unlike that seen in the acute stages of asthma.

In *pulmonary fibrosis*, the alveolar wall becomes thickened and this decreases the diffusing capacity of the lung. The diffusing capacity is also reduced in *pneumonia*. In this case bacterial infection leads to fluid accumulation in the alveoli. In severe cases a pulmonary lobe may become filled with fluid ('consolidated') containing bacterial toxins with the result that there is a local increase in blood flow and $\dot{V}_A/\dot{Q}$ abnormalities.

Cystic fibrosis is a recessive inherited disorder in which the mucus of the airways is abnormally thick and difficult to dislodge. It results from the failure of an epithelial chloride channel to open normally in response to cyclic AMP. This results in a decreased secretion of both sodium and chloride into the lumen of the airways. Consequently, less water passes across the epithelial membrane and the viscosity of the mucus is greatly increased. The normal clearance of mucus via the mucociliary escalator is reduced, resulting in obstruction of the small airways and frequent bronchial infections.

Sleep apnea describes a number of conditions in which breathing temporarily stops. Two principal types can be recognized: obstructive sleep apnea and central sleep apnea.

Obstructive sleep apnea results from physical obstruction of the upper airway. During deep sleep the muscles of the mouth, larynx and pharynx relax. As a result, there is a tendency for the pharynx to collapse, so obstructing the airway. Periods of increasing obstruction are accompanied by loud snoring followed by a period of silence when obstruction is complete. Pa_{O_2} falls and Pa_{CO_2} rises—both stimulating respiratory effort. Breathing movements then increase in intensity until the intrathoracic pressure overcomes the obstruction and ventilation resumes, often with a loud snore. When the period of apnea is prolonged (and it may last for up to 2 minutes) the patient wakes and breathing resumes, but the sleep pattern has been interrupted. If this occurs frequently, the characteristic signs of sleep deprivation appear which include irritability and loss of concentration.

Central sleep apnea occurs when the respiratory drive is not able to initiate breathing movements. During the apneustic phase, the Pa_{CO_2} rises and the patient wakes. Once awake the patient resumes breathing, the Pa_{CO_2} returns to normal, and they fall asleep once again. As this cycle may be repeated many times during the night, sleep may become severely disturbed.

Periodic breathing. The importance of the chemical control of breathing is revealed by a pattern known as *Cheyne–Stokes breath-*

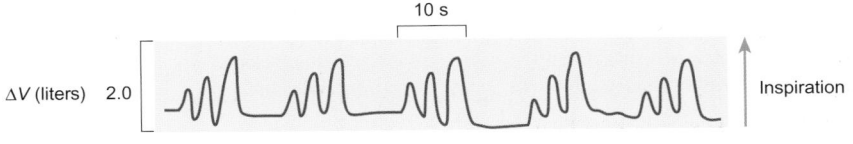

Fig. 16.32 A record of respiratory movements to illustrate the pattern of Cheyne–Stokes breathing. In this example, the record was obtained from a healthy subject who had recently moved to a high altitude.

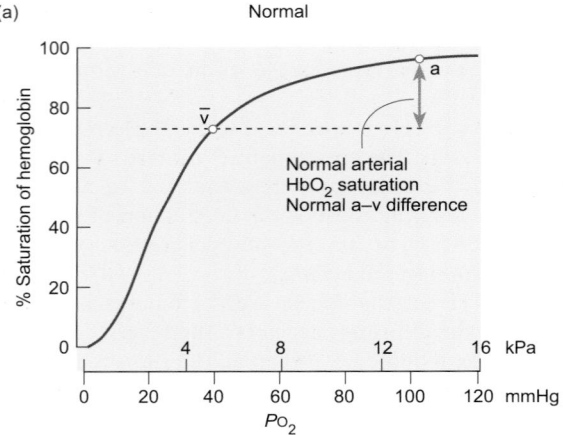

(a) Normal

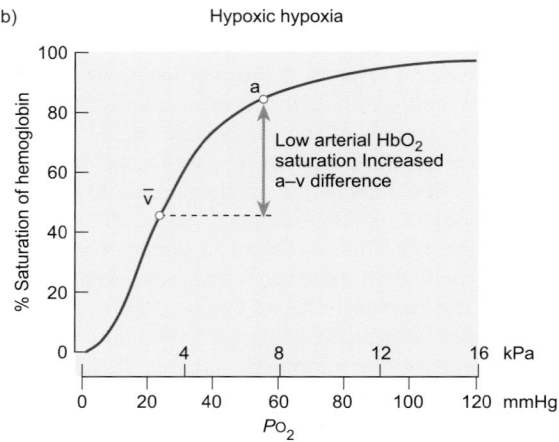

(b) Hypoxic hypoxia

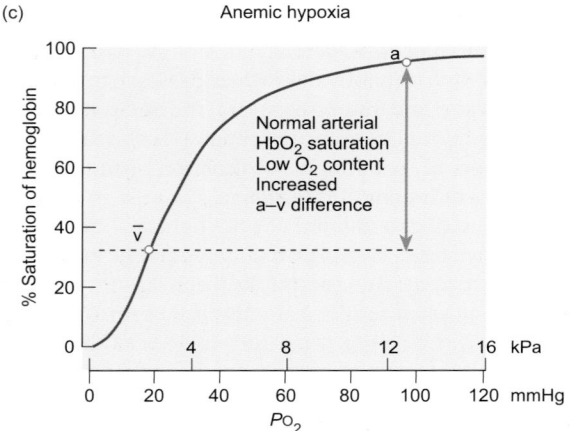

(c) Anemic hypoxia

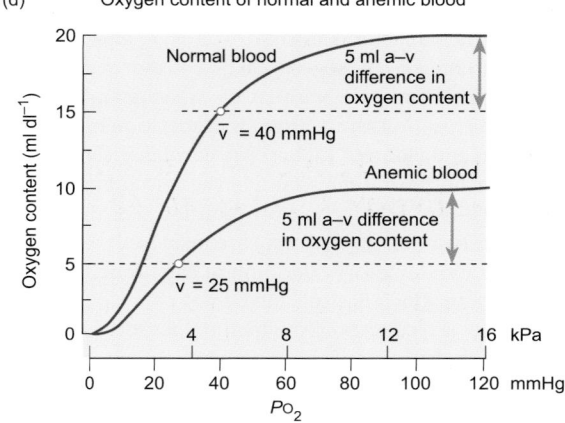

(d) Oxygen content of normal and anemic blood

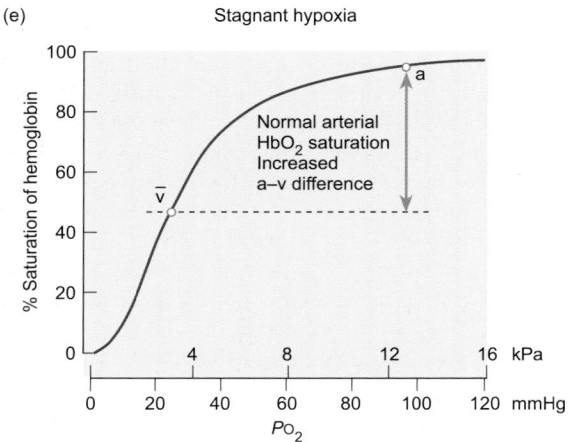

(e) Stagnant hypoxia

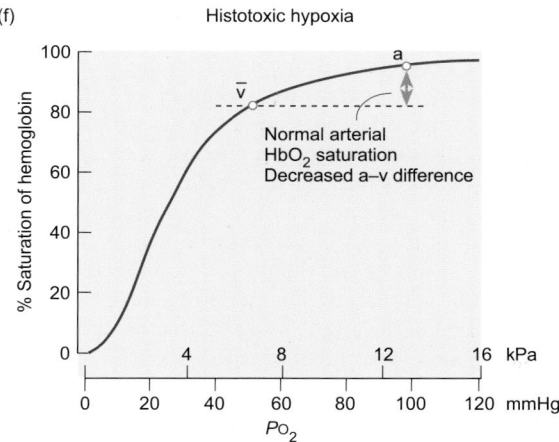

(f) Histotoxic hypoxia

Fig. 16.33 The effect of different types of hypoxia on the percentage saturation of hemoglobin with oxygen in arterial and venous blood. Note that in anemic hypoxia the total oxygen content of arterial blood is less than normal despite being fully saturated with oxygen as shown in (d).

ing. This consists of a periodic change in the frequency and tidal volume. The pattern of breathing alternates between apnea and mild hyperventilation (Fig. 16.32). Breathing begins with slow shallow breaths; the frequency and tidal volume gradually increase to a maximum before slowly subsiding into apnea. This cyclical pattern is seen in patients who are terminally ill or who have suffered brain damage. It is sometimes seen in normal people during sleep, especially at high altitude.

The pattern in terminally ill patients is explained as follows. During the period of low ventilation Pa_{CO_2} rises and this stimulates the central chemoreceptors which progressively increase the respiratory drive so that both frequency and depth increase. The increased ventilation results in a fall in Pa_{CO_2} and a reduction of respiratory drive so both rate and depth of respiration fall. The Pa_{CO_2} increases and the cycle recommences. In order for this pattern to occur, there must be an abnormal delay in the ability of the central chemoreceptors to respond to the change in Pa_{CO_2}. This may reflect either a reduced sensitivity to carbon dioxide or a sluggish blood flow to the brain, for example during heart failure.

Periodic breathing at high altitude does not indicate a pathological state. It can be induced by a brief period of hyperventilation and is abolished by the inhalation of pure oxygen.

16.10 Insufficient oxygen supply to the tissues—hypoxia and its causes

The respiratory system is principally concerned with gas exchange—the uptake of oxygen from the air and the elimination of carbon dioxide from the pulmonary blood. If the oxygen content of the blood is reduced, there may be insufficient oxygen to support the aerobic metabolism of the tissues. This condition is known as *hypoxia*. There are four principal types of hypoxia: hypoxic hypoxia, anemic hypoxia, stagnant hypoxia, and histotoxic hypoxia. Each will be discussed in turn.

Hypoxic hypoxia results from a low arterial P_{O_2}. There are a number of causes but each results in lowering of the oxygen content of the systemic arterial blood. If the alveolar P_{O_2} is low, the arterial P_{O_2} will inevitably follow and so will the oxygen content. As a result, more oxygen is extracted from the blood to support the oxidative metabolism of the tissues (Fig. 16.33(b)). This is quite normal following ascent to high altitude, as the barometric pressure and P_{O_2} both fall with increasing altitude.

As discussed below (Section 16.11), reduced ventilation (hypoventilation) will lead to a reduced alveolar P_{O_2} and an increased P_{CO_2} (hypercapnia). Hypoventilation may result from respiratory depression caused by a drug overdose (e.g. barbiturate poisoning). It may also arise because of severe weakness of the muscles that support respiration, for example in poliomyelitis and myasthenia gravis. Airway obstruction will also lead to hypoventilation.

A right–left shunt will allow some venous blood to bypass the lungs completely. (A shunt is the passage of blood through a channel that diverts it away from its normal route.) Although the hemoglobin of the blood that has passed through the alveoli is virtually fully saturated, the shunted blood will have the same P_{O_2} as mixed venous blood. As a result, the P_{O_2} and oxygen content of the blood in the systemic arteries is reduced. Equally,

ventilation–perfusion inequality will lead to hypoxic hypoxia if the $\dot{V}_A/\dot{Q}$ ratio is low in a significant portion of the lung. This occurs in many respiratory diseases and is the most common cause of central cyanosis (i.e. cyanosis caused by inadequate oxygenation of the blood). Another significant cause of hypoxic hypoxia is a reduced diffusing capacity due to fibrosis of the lung parenchyma or to pulmonary edema.

Anemic hypoxia is caused by a decrease in the amount of hemoglobin available for binding oxygen so that the oxygen content of the arterial blood is abnormally low (Fig. 16.33(d)). It may be due to blood loss, reduced red cell production, or the synthesis of abnormal hemoglobin because of a genetic defect. It may also be caused by carbon monoxide poisoning since the affinity of hemoglobin for this gas is greater than its affinity for oxygen (see Chapter 13 for further discussion of the anemias).

In anemic hypoxia, the arterial P_{O_2} is normal but, since the oxygen content of the blood is low, a higher proportion of the available oxygen will need to be extracted from the hemoglobin to support the metabolism of the tissues. Consequently, the venous P_{O_2} is much lower than normal (Figure 16.33(c), (d)).

Stagnant hypoxia is the result of a low blood flow. It may occur peripherally due to local vasoconstriction (e.g. exposure of the extremities to cold) or it may result from a reduced cardiac output. In this case, the alveolar and arterial P_{O_2} may be normal but, since the blood flow through the metabolizing tissues is very slow, excessive extraction of the available oxygen occurs and venous P_{O_2} is very low (Fig. 16.33(e)). This gives rise to peripheral cyanosis.

Histotoxic hypoxia refers to poisoning of the oxidative enzymes of the cells, by cyanide for example. In this situation, the supply of oxygen to the tissues is normal but they are unable to make full use of it. As a result, venous P_{O_2} is abnormally high (Fig. 16.33(f)).

Oxygen therapy in hypoxia

The administration of a high partial pressure of oxygen can be beneficial in the treatment of hypoxic hypoxia. By increasing

Summary

1. Hypoxia is a condition in which the metabolic demand for oxygen cannot be met by the circulating blood. There are many causes of hypoxia but four principal types of hypoxia are recognized: hypoxic hypoxia, anemic hypoxia, stagnant hypoxia, and histotoxic hypoxia. Of these, hypoxic hypoxia and stagnant hypoxia are the most commonly seen in clinical medicine. Hypoxic hypoxia can be treated by administration of pure oxygen.

2. Respiratory failure occurs when the respiratory system fails to maintain normal values of arterial P_{O_2} and P_{CO_2}. It may result from a major right–left shunt of deoxygenated blood or because of an abnormal $\dot{V}_A/\dot{Q}$ ratio (type I respiratory failure). Respiratory failure also occurs when alveolar ventilation is not sufficient to excrete the metabolically derived carbon dioxide. This is known as type II respiratory failure (or ventilatory failure) in which Pa_{O_2} is low and Pa_{CO_2} is elevated.

the partial pressure of oxygen in the alveoli, the oxygen content of the blood leaving the lungs is raised. This will lessen any central cyanosis and alleviate any dyspnea. Oxygen therapy will be of less value in other forms of hypoxia.

To increase the amount physically dissolved in the plasma, oxygen is occasionally administered for short periods at pressures higher than that of the atmosphere. This is known as hyperbaric oxygen therapy, which can be helpful in the treatment of carbon monoxide poisoning. The high P_{O_2} acts to displace the carbon monoxide bound to the hemoglobin and provides much needed oxygen for the tissues. Newborn infants should not be exposed to partial pressures of oxygen greater than about 40 kPa (c. 300 mmHg) as they are particularly sensitive to its toxic effects (see Chapter 30).

16.11 Respiratory failure

Respiratory failure occurs when the respiratory system fails to maintain normal values of arterial P_{O_2} and P_{CO_2}. Normal Pa_{CO_2} is 5.1±1.0 kPa (38.5±7.5 mmHg). This encompasses 95 per cent of the normal population. Normal Pa_{O_2} depends on the inspired P_{O_2} and decreases with age. For healthy young subjects (<30 years of age) inspiring air at sea level, the mean value is 12.5±1.3 kPa (94±10 mmHg). However, the Pa_{O_2} declines with age, and the mean value in healthy subjects over 60 years of age is 10.8±1.3 kPa (81±10 mmHg). A diagnosis of respiratory failure is made if Pa_{O_2} is less than 8 kPa (60 mmHg) or the Pa_{CO_2} is greater than 7 kPa (55 mmHg).

*   In *type I respiratory failure*, Pa_{O_2} is low while Pa_{CO_2} is normal or low. This occurs when there is a major right to left shunt of deoxygenated blood or when the $\dot{V}_A/\dot{Q}$ ratio is abnormal (see cases 2 and 3 on p. 330). Such a situation may arise during pneumonia, pulmonary edema, or adult respiratory distress syndrome (ARDS) (see below).

*   In *type II respiratory failure* Pa_{O_2} is low while Pa_{CO_2} is elevated. This situation occurs when alveolar ventilation is not sufficient to excrete the carbon dioxide produced by the normal metabolism of the body. This is known as *ventilatory failure* and may be caused by a number of different factors (see below). The most common cause is chronic obstructive pulmonary disease.

Ventilatory failure and its causes

Ventilatory failure may occur as a result of one or more of the following:

*   failure of neural control of the respiratory muscles
*   neuromuscular blockade
*   pneumothorax
*   decreased compliance of the chest or lungs
*   increased airways resistance.

As discussed above, the activity of the respiratory muscles is neurogenic in origin. The drive from the respiratory neurons of the medulla may be depressed during hypoxia or exposure to respiratory depressants such as anesthetics. In this case, ventilation will inevitably be diminished. Traumatic damage to the spinal column below cervical segment C4 may interrupt the flow of information from the medulla to the lower respiratory motoneurons, resulting in paralysis of the intercostal muscles. Lesions above C4 may damage the phrenic outflow. Loss of lower motoneurons following poliomyelitis or other neurological disorders (e.g. motoneuron disease) may also result in ventilatory failure. A decrease in neuromuscular transmission caused by myasthenia gravis may have the same effect.

Ventilation is reduced following a pneumothorax, as the lungs are unable to expand properly. As discussed earlier, the compliance of the chest depends on the elasticity of the lungs and of the chest wall itself. The compliance of the chest will be reduced if there is a reduction in the elasticity of the lungs, if the pleural cavity becomes infiltrated with fibrous tissue, or if the elasticity of the chest wall itself is reduced. Decreased elasticity of the chest wall can occur in various postural disorders such as scoliosis (sideways curvature of the spine) and kyphosis (abnormal backward curvature of the spine). These are examples of restrictive disorders.

Increased airways resistance may arise from the presence of foreign material in the airways (e.g. refluxed gastric contents entering the larynx) or from narrowing of the airways themselves, as in asthma and emphysema.

Adult respiratory distress syndrome (ARDS)

This is a condition in which the lung parenchyma is severely damaged—so much so that more than half of all cases are fatal. There is no entirely satisfactory definition but ARDS is characterized by a severe hypoxemia (hence its alternative name of *acute respiratory failure*), the presence of diffuse shadows in chest radiographs (probably due to patches of fluid accumulation), low pulmonary compliance, and pulmonary edema that is not due to left-sided heart failure. Precipitating causes include septic shock (see Chapter 28), aspiration of the gastric contents, near drowning, and inhalation of toxic gases or smoke. The condition seems to arise from damage to alveolar–capillary membranes, which results in fluid accumulating in the air spaces. This leads to a redistribution of pulmonary blood flow, partly as a result of the normal response to local hypoxia and partly as a result of compression of the pulmonary vessels by the local edema. Subsequently, the release of chemical mediators may result in further constriction of the pulmonary vasculature and the development of pulmonary hypertension. Within a week of the onset of the condition, the lungs become infiltrated by fibroblasts which lay down fibrous tissue in the pulmonary interstitium. There is a loss of elastic tissue and emphysema develops. This is reflected in an increase in the physiological dead space.

Recommended reading

Anatomy

MacKinnon, P.C.B., and Morris J.F. (2005). *Oxford textbook of functional anatomy*, Vol. 2, *Thorax and abdomen* (2nd edn), pp. 45–63. Oxford University Press, Oxford.

Histology

Junquieira, L.C., and Carneiro, J. (2003). *Basic histology* (10th edn), Chapters 12 and 13. McGraw-Hill, New York.

Pharmacology of the respiratory system

Rang H.P., Dale, M.M., Ritter, J.M., and Moore, P. (2003). *Pharmacology* (5th edn), Chapter 22. Churchill-Livingstone, Edinburgh.

Physiology of the respiratory system

Andrews, P., and Widdicombe, J. (1993). *Pathophysiology of the gut and airways. An introduction*, Chapters 2, 4, 6, 8 and 10. Portland Press, London.

Hlastala, M.P., and Berger, A.J. (2001). *Physiology of respiration* (2nd edn). Oxford University Press, New York.

Levitzky, M.G. (2003). *Pulmonary physiology* (6th edn). McGraw-Hill, New York.

Lumb, A.B. (2000). *Nunn's applied respiratory physiology* (5th edn). Butterworth-Heinemann, Oxford.

Slonim, N.B., and Hamilton, L.H. (1987). *Respiratory physiology*. Mosby, St Louis, MO.

West, J.B. (2004). *Respiratory physiology: the essentials* (7th edn). Lippincott–Williams & Wilkins, Philadelphia, PA.

Widdicombe, J., and Davies, A. (1991). *Respiratory physiology* (2nd edn). Edward Arnold, London.

Multiple choice questions

Each statement is either true or false. Answers are given below.

1. The following relate to the pulmonary circulation:
   a. The whole of the cardiac output passes through the lungs.
   b. The pressures in the pulmonary arteries are similar to those in the systemic arteries.
   c. The mean pressure in the pulmonary arteries rises as cardiac output increases.
   d. The resistance of the pulmonary circulation falls as the pulmonary blood flow increases.
   e. In an upright man, pulmonary blood flow is greatest at the base of lung.
   f. The pattern of pulmonary blood flow is dependent on posture.
   g. The lungs inactivate all circulating vasoactive materials.

2. The following relate to the airways and alveoli:
   a. The trachea, bronchi, and bronchioles are all prevented from collapsing by the cartilage in their walls.
   b. The upper airways play an important role in protecting the lungs from airborne particles.
   c. The small-diameter airways are the main sites of airways resistance.
   d. The alveoli are the only site of gas exchange.
   e. In a healthy lung, the distance between the alveolar air and the blood in the pulmonary capillaries is less than 1 μm.

3. In a normal healthy individual with a total lung capacity of 6 liters:
   a. The tidal volume at rest is about 1 liter.
   b. The vital capacity is equal to the total lung capacity.
   c. The functional residual capacity is about 2 liters.
   d. The expiratory reserve volume at rest would be about 1 liter.
   e. The FEV_1 would be about 3.5 liters.

4. The following relate to the mechanics of ventilation:
   a. The change in the volume of the lungs with pressure is the compliance.
   b. The total compliance of the chest is determined solely by the compliance of the lungs.
   c. The recoil of the lungs assists inspiration.
   d. At the functional residual volume, the elastic recoil of the lungs is balanced by the elastic forces tending to expand the chest.
   e. The compliance of the lungs is determined by surface tension forces in the alveoli and by the elastic tissues of the lung parenchyma.
   f. Pulmonary surfactant maintains a constant surface tension in the alveoli.

5. The following relate to ventilation and gas exchange:
   a. The anatomical dead space is the volume of air taken in during a breath, that does not enter the alveoli.
   b. The physiological dead space is always greater than the anatomical dead space.
   c. The anatomical dead space is independent of the tidal volume.
   d. In an upright lung, the alveolar ventilation is always highest at the base.
   e. The $\dot{V}_A/\dot{Q}$ ratio for a subject in a sitting position increases with distance above the base of the lung.
   f. The regional variation in $\dot{V}_A/\dot{Q}$ ratios is much smaller in subjects who are lying down.
   g. The diffusing capacity for carbon dioxide in the lungs is similar to that of oxygen.

6. The following relate to the respiratory rhythm:
   a. The respiratory muscles have an intrinsic rhythmical activity.
   b. The basic neural machinery for the generation of the respiratory rhythm is located in the lower medulla.
   c. Respiration will continue even if all afferent nerves are cut.
   d. The principal muscle of respiration is the diaphragm.
   e. During expiration, the activity of the external intercostal muscles is inhibited.

7. The chemical regulation of respiration:
   a. The depth and rate of ventilation is increased when a subject breathes air containing 5 per cent carbon dioxide.
   b. The central chemoreceptors sense the carbon dioxide tension of the arterial blood.
   c. The peripheral chemoreceptors are located in the carotid sinus and aortic arch.
   d. The peripheral chemoreceptors only respond to changes in the partial pressure of oxygen in the arterial blood.
   e. Ventilation is markedly increased when the partial pressure of oxygen in the arterial blood falls below 8 kPa (60 mmHg).

Quantitative problems

Answers are given below.

1. A subject climbs a high mountain. What is the partial pressure of oxygen in the upper airways during inspiration (a) at an altitude of 3000 m (P_B70 kPa) and (b) at the summit, which is at 4350 m (P_B56 kPa). The fractional oxygen content of atmospheric air is 21 per cent and saturated water vapor pressure is 6.2 kPa.

2. In an estimate of functional residual capacity, a subject breathes an air–helium mixture from a spirometer. The spirometer has a volume of 5 liters and the concentration of helium before the subject breathes the gas mixture is 5.0 per cent. After the subject has breathed from the spirometer, the fractional concentration of helium has fallen to 3.56 per cent. What is the FRC?

3. A subject is trained to breathe at a series of tidal volumes without hyperventilating. From the data in the table shown below, calculate the dead space for each value of tidal volume (to two significant figures). Assume that alveolar carbon dioxide remains constant at 5.20 per cent. What conclusions can be drawn from these data?

Tidal volume (liters)	Expired F_{CO_2}
0.50	3.64
1.10	4.49
1.56	4.50
2.45	4.60

4. The following relationship was found between blood flow and perfusion pressure for an isolated dog lung.

Perfusion pressure (kPa)	Blood flow (l min⁻¹)
2.0	0.27
2.7	1.1
3.3	2.2
4.0	3.0

Calculate the pulmonary vascular resistance for each pair of values. What mechanisms can account for the changes seen?

5. Equal volumes of blood from hypoventilated and normally ventilated regions of a lung become mixed in a pulmonary vein. If the P_{O_2} of the blood draining the hypoventilated region is 8 kPa (60 mmHg) and that of the normally ventilated region is 13.3 kPa (100 mmHg), will the P_{O_2} of the mixed blood be (a) 12 kPa (90 mmHg), (b) 10.7 kPa (80 mmHg), or (c) 9.3 kPa (70 mmHg)?

6. A subject has been breathing steadily and then takes a deep inspiration at time zero, exhales, and continues to breathe spontaneously. The following values for minute ventilation are recorded. What factors are responsible for the changes in ventilation?

Time (s)	Ventilation (l min⁻¹)
–10	5.8
–5	5.8
0	Deep inspiration
5	5.8
10	5.6
15	5.1
20	4.7

25	4.65
30	4.9
35	5.3

7. A subject has been breathing a hypoxic mixture containing 14 per cent oxygen and 86 per cent nitrogen. At time zero, he inhales pure oxygen and the following values for minute ventilation are recorded as he continues to breathe pure oxygen normally. Plot the data and explain the changes observed.

Time (s)	Ventilation (l min⁻¹)
–20	8.5
–10	8.5
0	8.0
10	6.8
20	5.2
30	6.2
40	6.8
50	6.3
60	6.4

Answers to multiple choice questions

1. All the output of the right ventricle passes through the pulmonary circulation and the output of both right and left heart will be the same over any significant period. The systolic and diastolic pressures in the pulmonary arteries are about 25 and 8 mmHg respectively compared with 120/80 for the systemic arteries. Although the pressure in the pulmonary arteries rises with increasing cardiac output, the increase is small as the resistance of the pulmonary circulation falls. While the lungs inactivate many vasoactive materials such as bradykinin, they convert angiotensin I to its active form angiotensin II.

   a. True
   b. False
   c. True
   d. True
   e. True
   f. True
   g. False

2. The trachea and bronchi have cartilage rings or plates in their walls that allow them to stay open despite changes in intrapleural pressure. The bronchioles have no cartilage and may collapse when the intrapulmonary pressure exceeds the pressure in the airway. The principal sites of airways resistance are the upper airways (mainly the nose and airway generations 1–6). The fall in resistance with the branching of the airways is the result of a very large increase in the total cross-sectional area. While the alveoli are the principal site of gas exchange, the respiratory bronchioles and alveolar ducts also contribute to gas exchange.

   a. False
   b. True
   c. False
   d. False
   e. True

3. The tidal volume at rest would be about 500 ml. The vital capacity is equal to the total lung volume *minus* the residual volume and the expiratory reserve volume is equal to the functional residual capacity *minus* the residual volume.

    a. False
    b. False
    c. True
    d. True
    e. True

4. The compliance of the chest is determined by that of the chest wall and the lungs. Lung recoil assists expiration not inspiration. Pulmonary surfactant reduces the work of breathing by lowering the surface tension of the air–liquid interface in the alveoli. The surface tension is related to the degree of inflation—the more the lungs are expanded, the higher is the surface tension.

    a. True
    b. False
    c. False
    d. True
    e. True
    f. False

5. The physiological dead space is always greater than the anatomical dead space as it includes the volume of air taken up by alveoli that are not perfused by blood. These alveoli cannot take part in gas exchange. In healthy subjects, the difference is not great. When the lungs are expanded, the traction on the airways is increased and this dilates them. As a result, the dead space increases with tidal volume. Alveolar ventilation is greatest at the base of the lungs for lung volumes at or above FRC. For lung volumes less than FRC, the ventilation at the apex may be greater than that at the base. While the volumes of carbon dioxide and oxygen exchanged during respiration are similar, the pressure gradient for carbon dioxide is much smaller. This is offset by the much greater diffusing capacity of carbon dioxide (about 20 times that for oxygen).

    a. True
    b. True
    c. False
    d. False
    e. True
    f. True
    g. False

6. The respiratory muscles are skeletal muscle and have no intrinsic rhythm. Their rhythmical activity is derived from the activity of their motoneurons.

    a. False
    b. True
    c. True
    d. True
    e. True

7. The peripheral chemoreceptors are the carotid and aortic bodies, which sense plasma pH as well as the partial pressures of oxygen and carbon dioxide. They are the only chemoreceptors that respond to the partial pressure of oxygen in the arterial blood. The receptors in the aortic arch and carotid sinus are baroreceptors that monitor the blood pressure.

    a. True
    b. True
    c. False
    d. False
    e. True

Answers to quantitative problems

1. a. 8.5 kPa
   b. 5.6 kPa

2. 2.0 liters

3. The values for the dead space are 0.15, 0.15, 0.21, and 0.28 liters. The dead space increases as the tidal volume increases. This is caused by the connective tissue of the parenchyma pulling on the airways, so dilating them.

4. The values for pulmonary vascular resistance are 7.4, 2.45, 1.5 and 1.33 kPa l^{-1} min. The vascular resistance falls as pressure increases. There is no autoregulation. The fall in resistance is the result of additional recruitment of vessels and of further dilatation of those vessels that were already open.

5. Approximately 9.3 kPa (70 mmHg) as it is the *oxygen content* of each stream of blood that is averaged, not their partial pressures.

6. This demonstrates the importance of carbon dioxide in determining ventilation. The deep breath results in mild hyperventilation and a temporary reduction in alveolar carbon dioxide. The chemical stimulus to breathe is lessened and ventilation falls. The reduction in ventilation allows a build-up of carbon dioxide, which stimulates breathing and so increases ventilation.

7. This experiment demonstrates the hypoxic response to breathing. Note that the initial ventilation is high due to the prevailing hypoxemia. Breathing pure oxygen reduces the hypoxic stimulus and ventilation falls. However, since the initial ventilation was high, there will be a degree of hypocapnia and the stimulus from carbon dioxide will be less then normal. The initial reduction in minute ventilation from 8.4 to 5.2 liters reflects this. The subsequent rise in minute ventilation to 6.8 liters occurs as the carbon dioxide tension returns to its normal level.

CHAPTER 17

The kidneys and the regulation of the internal environment

After reading this chapter you should understand:

- The basic structure of the kidneys and the renal circulation
- The structure of the nephron and the organization of its blood supply
- The concept of autoregulation and the regulation of renal blood flow
- The formation of the glomerular filtrate
- The concept of renal clearance
- The transport processes in the kidneys including tubular re-absorption and secretion
- The role of the distal tubule in the regulation of the ionic balance of the body
- The establishment of the osmotic gradient in the renal medulla and its role in the regulation of plasma osmolality
- Bladder function

17.1 Introduction

Human beings, like all animals, feed on other organisms both to provide material for tissue growth and maintenance and to provide themselves with the resources for other activities such as reproduction. This lifestyle inevitably leads to the intake of variable quantities of essential body constituents such as sodium, potassium, and water, and to the production of metabolic waste products. Nevertheless, the body needs to maintain a close control over the composition of the body fluids and this is the principal role of the kidneys. They achieve this by regulating the composition of the blood and, since the plasma equilibrates with the interstitial fluid of the tissues, the kidneys effectively regulate the composition of the extracellular fluid. The process of regulation requires the kidneys to excrete the excess water, salts, and metabolic waste products. Therefore the production of urine of variable composition is a necessary part of the homeostatic role of the kidney.

About 1–1.5 liters of urine containing 50–70 g of solids, chiefly urea and sodium chloride, are normally produced each day. Urine volume and osmolality vary both with fluid intake and with fluid loss through sweating and in the feces. Moreover, the chemical composition of the urine is very variable (Table 17.1) and changes with the diet. Although the urine contains traces of most of the plasma constituents, some substances such as protein, glucose, and amino acids are not normally detected. Other substances are much more concentrated in the urine than they are in plasma (e.g. creatinine, phosphate, and urea).

TABLE 17.1 Comparison between the composition of the plasma and that of urine

	Plasma	Urine	Units
Na^+	140–150	50–130	mmol l^{-1}
K^+	3.5–5	20–70	mmol l^{-1}
Ca^{2+}	1.35–1.50	10–24	mmol l^{-1}
HCO_3^-	22–28	0	mmol l^{-1}
Phosphate	0.8–1.25	25–60	mmol l^{-1}
Cl^-	100–110	50–130	mmol l^{-1}
Creatinine	0.06–0.12	6–20	mmol l^{-1}
Urea	4–7	200–400	mmol l^{-1}
NH_4^+	0.005–0.02	30–50	mmol l^{-1}
Protein	65–80	0	g l^{-1}
Uric acid	0.1–0.4	0.7–8.7	mmol l^{-1}
glucose	3.9–5.2	0	mmol l^{-1}
pH	7.35–7.4	4.8–7.5	($-\log_{10}[H^+]$)
Osmolality	281–297	50–1300	mOsm kg^{-1}

The data given in the table are the values that would encompass 95 per cent of a population of normal healthy adults. Note that while the concentration of the principal constituents of the plasma is maintained relatively constant, the composition of the urine is subject to considerable variation. Moreover, some important constituents of plasma such as protein, glucose, and bicarbonate are absent from normal urine. Others, such as creatinine, NH_4^+, phosphate, and urea, are present in far higher concentrations in the urine.

The urine is normally somewhat acid compared with the plasma or extracellular fluid. Its pH may range between 4.8 and 7.5 but, for people eating a normal mixed diet, it usually lies between 5 and 6 (cf. the range for plasma which is 7.35–7.4). Normal fresh urine has a slight odor that can readily be masked by aromatic compounds from certain foodstuffs (e.g. coffee or garlic), but decomposition of urine by bacterial action generates an unpleasant fetid odor due to the production of ammonia. The characteristic yellow color is principally due to the presence of pigments known as urochromes. In various disease states the color of the urine changes in characteristic ways giving important diagnostic clues. It is almost colorless in diabetes insipidus (see Section 17.8) but is strongly colored during infections and has a brownish red color when hemoglobin is present.

In addition to their primary regulatory and excretory roles, the kidneys also produce the hormone *erythropoietin*, which regulates the production of red blood cells (see Chapter 13), and an enzyme *renin*, which is important in the regulation of sodium balance via aldosterone secretion (see Chapter 28). They synthesize *1,25-dihydroxycholecalciferol* (also called calcitriol) from vitamin D, which stimulates the absorption of calcium from the gut and the calcification of bone (see Chapter 12, Section 12.6). Together with the liver, they also synthesize glucose from amino acids during fasting (gluconeogenesis—see Chapter 2).

17.2 The anatomical organization of the kidneys and urinary tract

The kidneys lie high in the abdomen on its posterior wall, either side of the vertebral column (T11–L3). In adults, each kidney is about 11 cm long and 6 cm wide and weighs about 140 g. A simple diagram of the gross structure of the kidney is shown in Fig. 17.1. Facing the midline of the body is an indentation called the *hilus* through which the renal artery enters and the

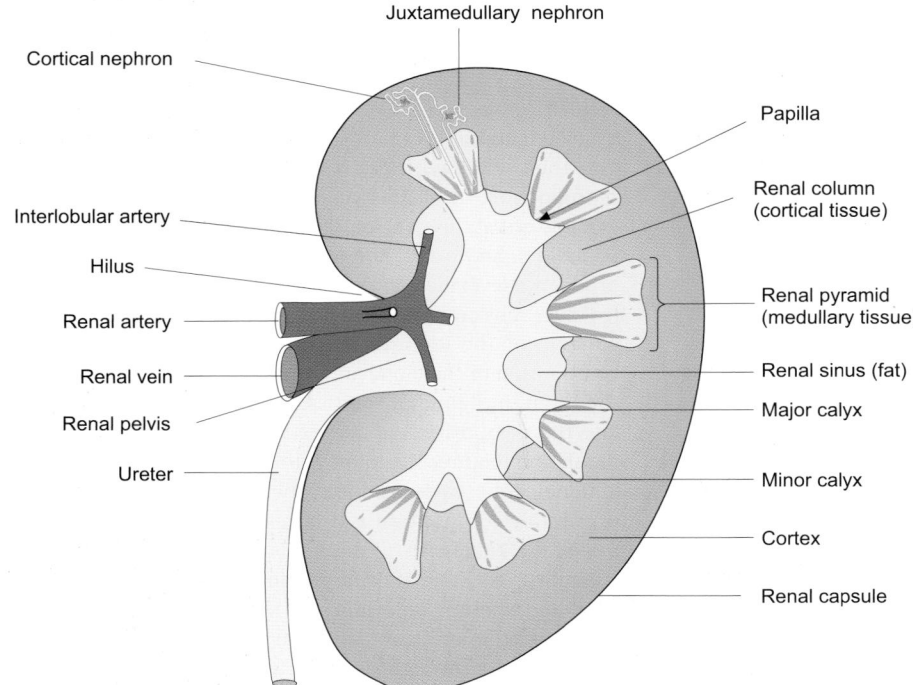

Fig. 17.1 Sectional view of the left kidney showing the principal anatomical features.

renal vein and ureter leave. Each kidney is covered by a tough fibrous inelastic capsule which serves to restrict changes in volume in response to any increase in blood pressure.

If a kidney is cut open, two regions are easily recognized—a dark brown *cortex* and a pale inner region, which is divided into the *medulla* and the *renal pelvis*. The renal pelvis contains the major renal blood vessels and is the region where the ureter originates. The medulla of each human kidney is divided into a series of large conical masses known as the *renal pyramids* which have their origin at the border between the cortex and the medulla. The apex or *papilla* of each pyramid lies in the central space of the renal pelvis, which collects the urine prior to its passage to the bladder. The central space can be divided into two or three large areas known as the *major calyxes* (or *calyces*) that divide, in turn, into the minor calyxes that collect the urine from the renal papillae.

As befits their role as regulators of the internal environment, the kidneys receive a copious blood supply from the abdominal aorta via the renal arteries. Venous drainage is via the renal veins into the inferior vena cava. The renal circulation is regulated by nerves from both the parasympathetic and sympathetic divisions of the autonomic nervous system (see below). The kidneys form the upper part of the urinary tract and the urine they produce is delivered to the bladder by a pair of ureters. The bladder continuously accumulates urine and periodically empties its contents via the urethra under the control of an external sphincter—a process known as *micturition*. Like the kidneys, the lower urinary tract receives innervation from both divisions of the autonomic nervous system.

The nephron is the functional unit of the kidney

Each human kidney contains about 1.25 million *nephrons*, which are the functional units of the kidney. Each nephron consists of a *renal* or *Malpighian corpuscle* attached to a long thin convoluted tube and its associated blood supply. The renal corpuscle is about 200 μm in diameter while the tubules are about 55 μm in diameter and 50–65 mm long. This arrangement provides a very large surface area for the exchange of solutes between the tubular fluid and the cells of the tubular epithelium (Fig. 17.2). The renal corpuscle consists of an invaginated sphere (*Bowman's capsule*) that envelops a tuft of capillaries known as a *glomerulus*. The glomerular capillaries originate from an afferent arteriole and recombine to form an efferent arteriole (Fig. 17.3). Between the glomerular capillaries are clusters of phagocytes called mesangeal cells.

The *proximal tubule* arises directly from Bowman's capsule. It is about 30–60 μm in diameter and 15 mm long. The epithelial cells of the proximal tubule are cuboidal in appearance and rich in mitochondria. They are closely fused with one another via tight junctions near their apical surfaces, which are densely covered by microvilli giving rise to a prominent brush border (Fig. 17.4). As each square micron of the apical surface has approximately 100 microvilli, each of which is about 3 μm in height, the brush border increases the area available for transport by a factor of about 200. The lateral surfaces of the basolateral membranes also have an adaptation that greatly increases their surface area. In this case, it is an extensive series of deep infoldings of the plasma membrane. The space between the basolateral regions of the cells is called the *lateral intercellular space*.

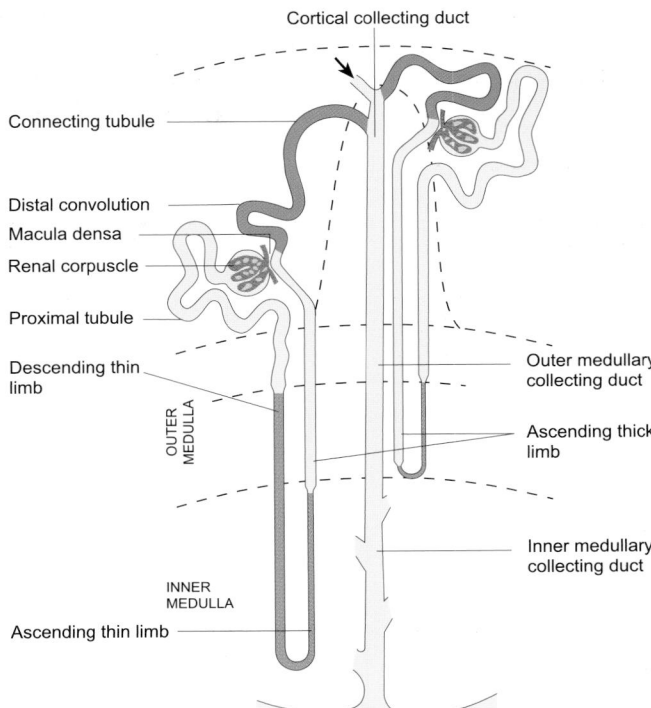

Fig. 17.2 The basic organization of short-looped (cortical) and long-looped (juxtamedullary) nephrons. Note that the early distal tubule of each type of nephron is in contact with the afferent arteriole of its own glomerulus.

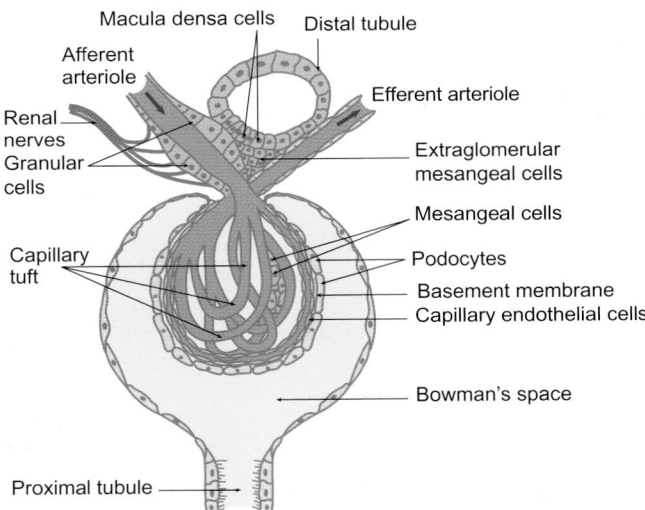

Fig. 17.3 The principal features of a renal glomerulus and the juxtaglomerular apparatus. The wall of the afferent arteriole is thickened close to the point of contact with the distal tubule where the juxtaglomerular (or granular) cells are located. These cells secrete the enzyme renin in response to low sodium in the distal tubule.

The proximal tubule connects with the *intermediate tubule* which is also known as the descending *loop of Henle*. Here the epithelial cells are thin and flattened (the wall thickness is only 1–2 μm). Compared with the cells of the proximal tubule, these cells have few mitochondria. The thin descending limb turns and ascends towards the cortex, finally merging with the thick

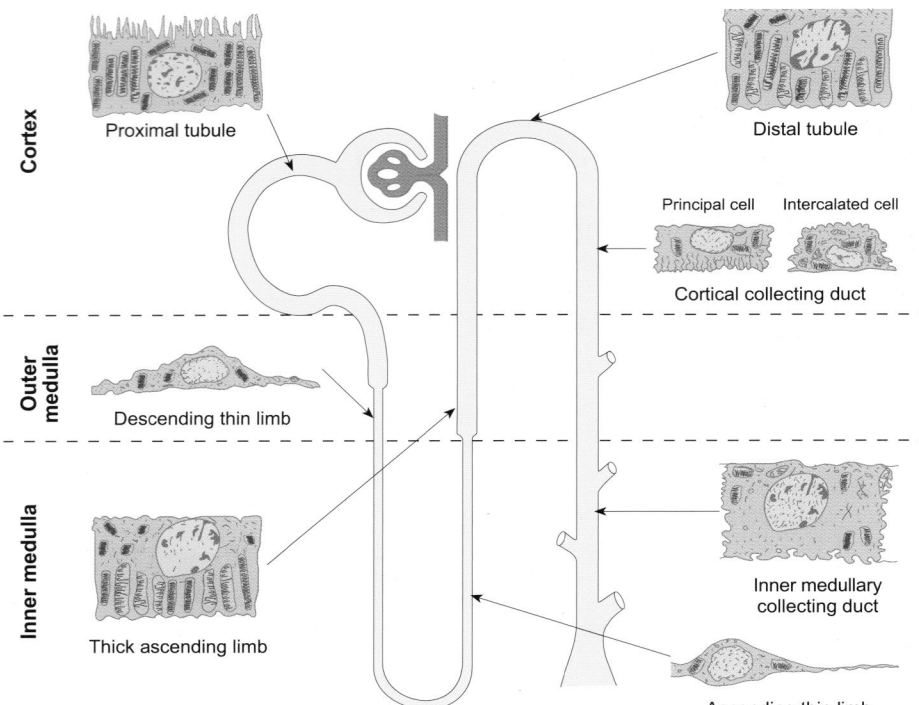

Fig. 17.4 The ultrastructure of the cells that constitute a nephron.

segment, which is about 12 mm long. The cells of the thick segment are cuboidal with extensive invaginations of the basolateral membrane. Like the cells of the proximal tubule, they are rich in mitochondria, suggestive of a major role in active transport. The nephrons of the outer renal cortex (the *cortical* or *superficial nephrons*) have short loops of Henle (the thin segment is as little as 2 mm long), some of which lie entirely within the cortex. In contrast, the nephrons nearest the medulla (the *juxtamedullary nephrons*) have long loops that penetrate deep into the medulla (Fig. 17.2). About 15 per cent of the nephrons in man have long loops and, in these nephrons, the thin segments (which may be as much as 14 mm long) pass deep into the renal papillae.

The terminal part of the thick ascending limb of the loop of Henle and the initial segment of the distal tubule contact the afferent arteriole close to the glomerulus from which the tubule originated. The tubular epithelium is modified to form the *macula densa* and the wall of the afferent arteriole is thicker due to the presence of *juxtaglomerular* or *granular cells*. These are modified smooth muscle cells containing secretory granules. The juxtaglomerular cells, macula densa, and associated mesangeal cells form the *juxtaglomerular apparatus* (Fig. 17.3). The juxtaglomerular cells of the arteriole secrete an enzyme called *renin* that has an important role in the regulation of aldosterone secretion from the adrenal cortex. In this way the juxtaglomerular apparatus plays an important role in the maintenance of sodium balance (see Section 17.7 and Chapters 12 and 28).

The *distal tubule* arises from the ascending loop of Henle and is about 5 mm long. Here the tubular wall is composed of cuboidal cells similar in appearance to those of the ascending thick limb. The distal tubules of a number of nephrons merge via *connecting*

tubules to form *collecting ducts* which are up to 20 mm long and pass through the cortex and medulla to the renal pelvis. The epithelium of the collecting ducts consists of two cell types, principal cells (P cells) which play an important role in the regulation of sodium balance and intercalated cells (I cells) which are important in regulating acid–base balance. A diagrammatic representation of the cellular organization of the nephron is shown in Fig. 17.4.

The renal circulation is arranged in a highly ordered manner

The renal artery enters the hilus and branches to form the interlobular arteries. These subsequently give rise to the arcuate arteries which course around the outer medulla. The arcuate arteries lead to cortical radial arteries (sometimes called the interlobular arteries) that ascend towards the renal capsule, branching *en route* to form the afferent arterioles of Bowman's capsule (Fig. 17.5). The afferent arterioles give rise to tufts of capillaries within Bowman's capsule that recombine to form the efferent arterioles.

The efferent arterioles of the outer cortex give rise to a rich supply of capillaries that cover the renal tubules—the *peritubular capillaries* (Fig. 17.6). Blood from the peritubular capillaries first drains into stellate veins and thence into the cortical radial veins and arcuate veins. In contrast, the efferent arterioles close to the medulla (a juxtamedullary efferent arteriole) give rise to a series of straight vessels known as the descending *vasa recta* (from the Latin for straight vessel) that provide the blood supply of the outer and inner medullary regions. Blood from the ascending vasa recta drains into the arcuate veins.

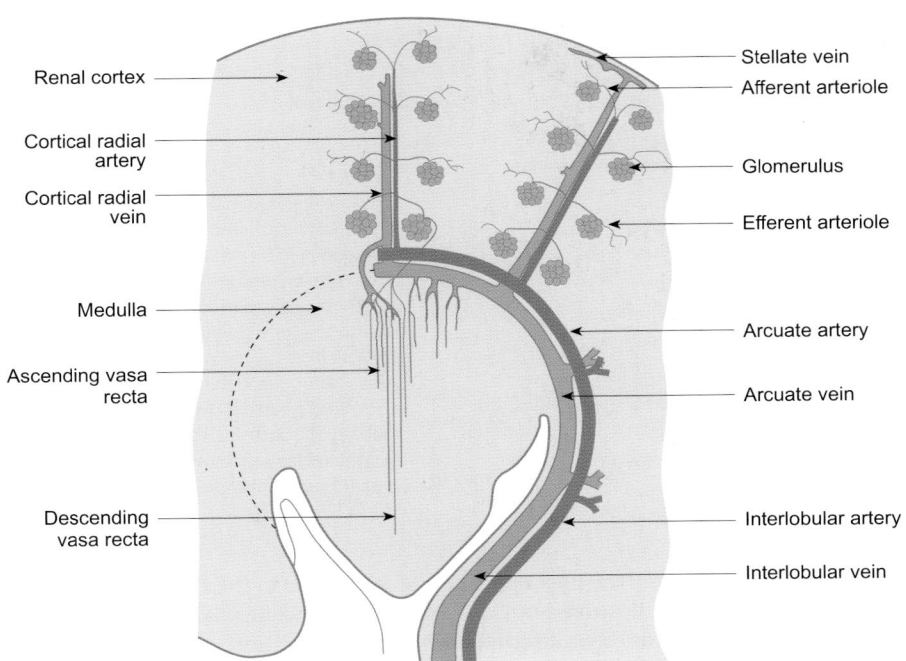

Renal cortex

Cortical radial artery

Cortical radial vein

Medulla

Ascending vasa recta

Descending vasa recta

Stellate vein
Afferent arteriole

Glomerulus

Efferent arteriole

Arcuate artery

Arcuate vein

Interlobular artery

Interlobular vein

Fig. 17.5 The principal features of the renal circulation. The diagram shows the circulation after the renal artery has branched to form the interlobular artery. Arterial blood is shown in red and venous blood in blue. Note that the efferent arterioles of the superficial nephrons give rise to the peritubular capillaries, while those of the juxtaglomerular nephrons give rise to straight vessels that pass deep into the renal medulla (the vasa recta).

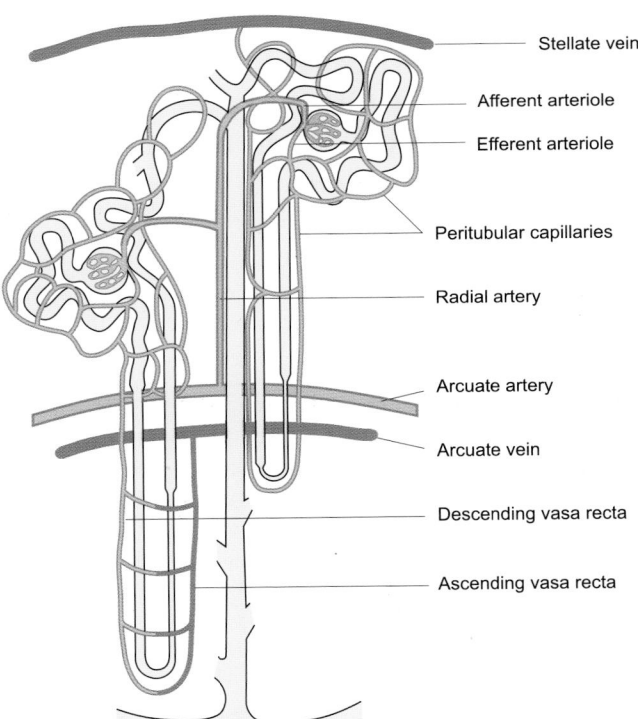

Stellate vein

Afferent arteriole
Efferent arteriole

Peritubular capillaries

Radial artery

Arcuate artery

Arcuate vein

Descending vasa recta

Ascending vasa recta

Fig. 17.6 The blood supply of the cortical and juxtamedullary nephrons.

The kidneys are innervated by sympathetic and parasympathetic nerve fibers

The kidneys have a rich nerve supply and are innervated by both sympathetic postganglionic fibers from the sympathetic paravertebral chain (T12–L2) and efferent fibers from the vagus nerve. The postganglionic sympathetic nerve fibers travel along-side the major arteries supplying the renal cortex as far as the afferent arterioles. The vagal parasympathetic fibers synapse in a ganglion in the hilus and appear to innervate the efferent arterioles. The sympathetic supply is adrenergic, while the parasympathetic fibers are cholinergic. This innervation provides extrinsic control for the renal circulation that can override the intrinsic autoregulation of blood flow (see below).

17.3 Renal blood flow is kept constant by autoregulation

In normal adults, the total renal blood flow (i.e. the blood flow for both kidneys) has been measured by a number of methods and is about 1.25 l min⁻¹ or about 25 per cent of the resting cardiac output. The renal cortex has the highest blood flow—about five times that of the outer medulla and 20 times that of the inner medulla. If systemic arterial pressure is altered over the range 10–26 kPa (80–200 mmHg), the renal blood flow remains remarkably constant (Fig. 17.7). This stability of renal blood flow persists even after the renal nerves have been cut and can be observed in isolated perfused kidneys. Therefore it is due to mechanisms intrinsic to the kidneys and is called *autoregulation*.

What are the mechanisms that underlie autoregulation? Two hypotheses have been advanced: the *myogenic hypothesis* and the *metabolic hypothesis*. The myogenic hypothesis proposes that autoregulation is due to the response of the renal arterioles to stretch. An increase in pressure will distend the arteriolar wall and stretch the smooth muscle fibers, which then contract after a short delay. The resulting vasoconstriction will increase vascular resistance and decrease blood flow (see also Chapter 15, Section 15.9). The metabolic hypothesis proposes that metabolites from the renal tissue maintain a degree of vasodilatation. An increase in perfusion pressure will lead to an increased blood flow which, in turn, will leach out more metabolites and so

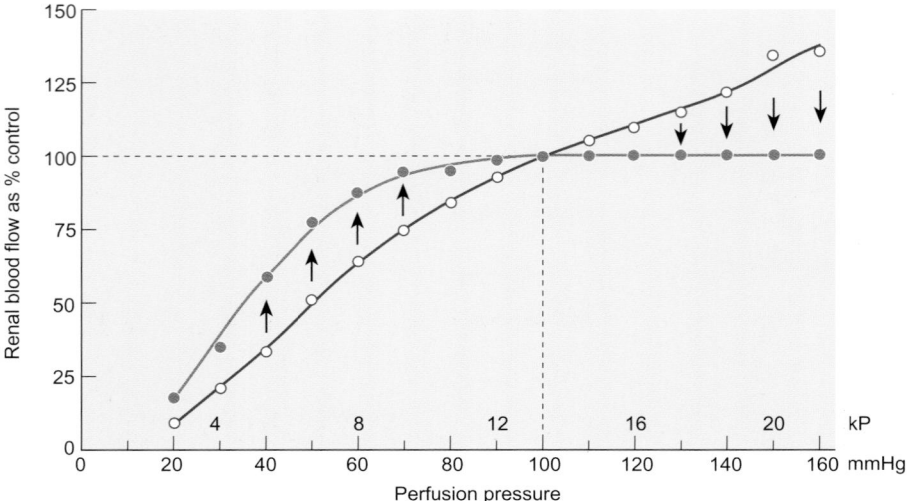

Fig. 17.7 The autoregulation of renal blood flow. The renal blood flow for an isolated dog kidney was first allowed to stabilize at a perfusion pressure of 13.3 kPa (100 mmHg). The perfusion pressure was then abruptly altered to a new value and the blood flow measured immediately after the change in pressure (open circles and the blue line). After a short period the blood flow stabilized at a new level (shown by the filled circles and the red line). The data show that the steady state renal blood flow remains essentially constant once the renal artery pressure rises above about 10 kPa (75 mmHg).

decrease the vasodilatation. In addition to tissue metabolites, humoral factors such as prostaglandins and nitric oxide may also act as vasodilators. Additionally, the macula densa of the juxtaglomerular apparatus has been postulated to play a role in maintaining the vasomotor tone of the afferent and efferent arterioles of the glomerulus (see Section 17.5). In summary, it is probable that both myogenic and humoral factors are responsible for the maintenance of blood flow in the kidney.

Despite powerful autoregulation, renal blood flow is subject to modulation by extrinsic factors. This additional regulation is achieved by both the activity in the renal nerves and humoral factors circulating in the blood. Sympathetic stimulation causes vasoconstriction of the afferent arterioles and thus reduces renal blood flow. Circulating epinephrine and nerepinephrine also cause vasoconstriction in the renal circulation, but norepinephrine acts mainly on the renal cortex. These are the factors responsible for the fall in renal blood flow observed during exercise. Angiotensin II and antidiuretic hormone (vasopressin) are other powerful vasoconstrictors which play a significant role in regulating blood flow through the kidneys, particularly following severe hemorrhage. The role of the cholinergic parasympathetic nerve fibers is less clear but they may act as vasodilators.

The mean pressures at key points in the renal circulation are illustrated in Fig. 17.8. Note that, while vasoconstriction in the afferent arterioles will reduce the pressure in the glomerular capillaries, vasoconstriction in the efferent arterioles will increase it.

17.4 The nephrons regulate the internal environment by ultrafiltration followed by selective modification of the filtrate

Regulation of the plasma composition occurs via three key processes:

(1) filtration

(2) reabsorption

(3) secretion.

First, some of the plasma flowing through the glomerular capillaries is forced through the capillary wall, by the hydrostatic pressure of the blood, into Bowman's space (filtration). Then, as this fluid passes along the renal tubules, its composition is modified by both the reabsorption of some substances and the secretion of

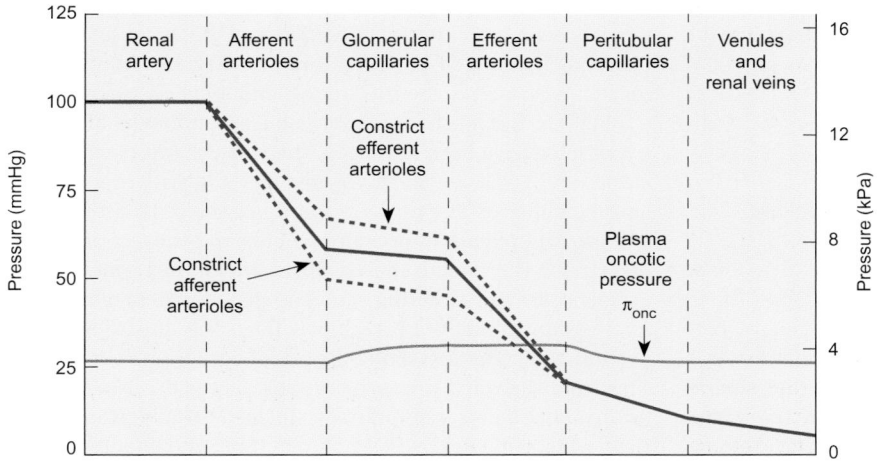

Fig. 17.8 The fall in mean blood pressure across the renal circulation. Note the relatively high pressure in the glomerular capillaries and that the pressure in the peritubular capillaries (2 kPa or 15 mmHg) is lower than the plasma oncotic pressure. Consequently, fluid reabsorption occurs along their length, unlike the capillaries of other vascular beds. Note that when the afferent arterioles constrict, the pressure in the glomerular capillaries falls, and when the efferent arterioles constrict the pressure in the glomerular capillaries rises. The oncotic pressure of the plasma rises as fluid is filtered.

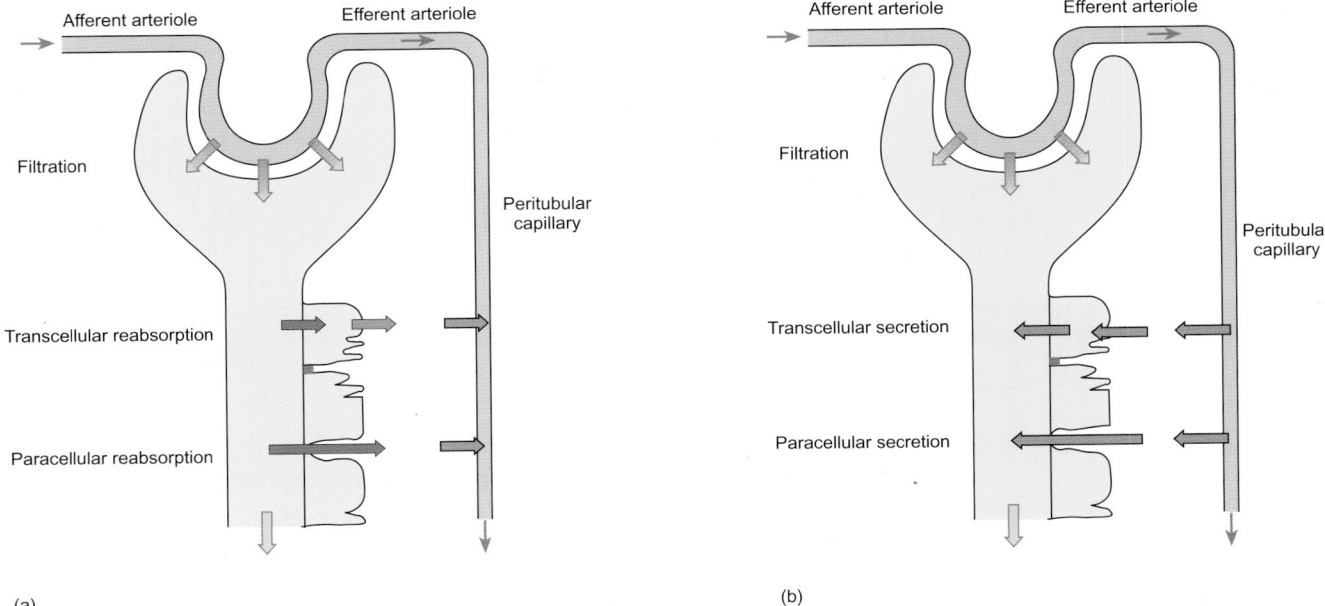

(a)   (b)

Fig. 17.9 The key processes that occur as fluid flow through a nephron. The process begins with filtration of part of the plasma flowing through the glomerular capillaries. Subsequently, some substances are reabsorbed (a) while others are secreted (b). Note that both reabsorption and secretion can take place via the cells (the transcellular route) or via the tight junctions between the cells (the paracellular route). Both secretion and reabsorption may occur by either passive or active transport.

others. *Reabsorption* is defined as movement of a substance from the tubular fluid to the blood, and this process occurs either via the tubular cells—the *transcellular route*—or between the cells—the *paracellular route* (Fig. 17.9) Tubular *secretion* is defined as movement of a substance from the blood into the tubular fluid. The reabsorption and secretion that occur via the transcellular route are largely the result of secondary active transport of solutes by the tubular cells. Paracellular reabsorption occurs as a result of concentration or electrical gradients that favor movement of solutes out of the tubular fluid. Paracellular secretion results when such forces favor movement into the tubular fluid.

The main stages of urine formation are illustrated in Fig. 17.10 and can be summarized as follows. First, about a fifth of the plasma is filtered into Bowman's space from where it passes along the proximal tubule. Here, many substances are reabsorbed while others are secreted. The proximal tubule reabsorbs all the filtered glucose and amino acids as well as most of the sodium, chloride, and bicarbonate. The reabsorption of these substances is accompanied by an osmotic equivalent of water so that, by the end of the proximal tubule, about two-thirds of the filtered fluid has been reabsorbed. As this phase of reabsorption is not closely linked to the ionic balance of the body, it is often called the *obligatory phase* of reabsorption. The next limb of the nephron, the loop of Henle, is concerned with establishing and maintaining an osmotic gradient in the renal medulla. It does this by transporting sodium chloride from the tubular fluid to the tissue surrounding the tubules (the *interstitium*) without permitting the osmotic uptake of water. As a result, the osmolality of the fluid leaving the loop of Henle is lower than the plasma while that of the interstitium is higher. The distal tubule regulates the ionic balance of the body by adjusting the

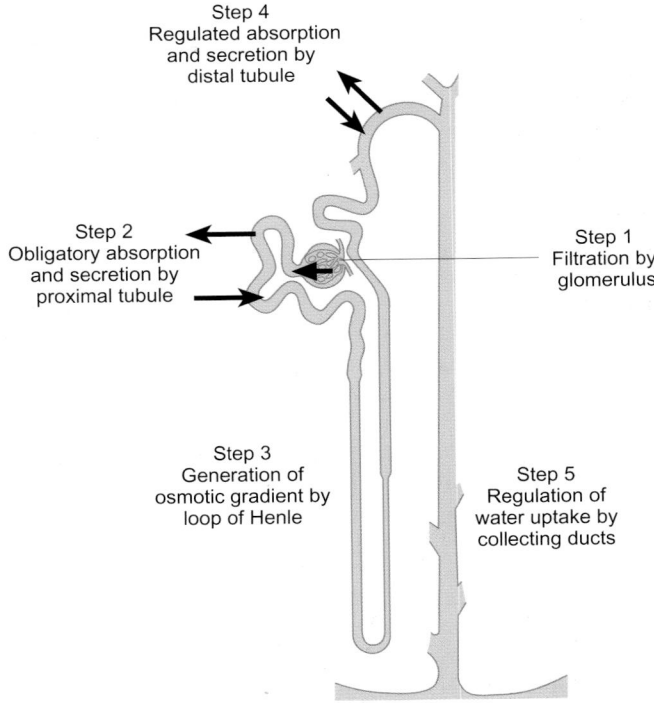

Fig. 17.10 The main stages of urine formation by the kidney. Fluid is first filtered by the glomerulus. As it passes along the proximal tubule, it is modified by selective reabsorption and secretion. The reabsorption of ions by the loop of Henle without an osmotic equivalent of water leads to the generation of an osmotic gradient in the medulla, which is exploited to regulate the uptake of water by the collecting ducts.

amount of sodium and other ions it reabsorbs according to the requirements of the body. It also secretes hydrogen ions, which leads to the acidification of the urine. The fluid leaving the distal tubule is relatively dilute. As it passes through the collecting ducts, if the osmolality of the body fluids is relatively high (>290 mOsm kg⁻¹) water is absorbed under the influence of antidiuretic hormone (ADH) and concentrated urine is excreted. If the osmolality of the body fluids is relatively low (<285 mOsm kg⁻¹), little ADH is secreted and dilute urine is excreted.

BOX 17.1 DIRECT MEASUREMENTS OF THE COMPOSITION OF TUBULAR FLUID CAN BE ACHIEVED BY MICROPUNCTURE METHODS

Important insights into the function of the renal tubules can be gained by knowing the composition of the tubular fluid at different parts of the nephron. This can be achieved by penetrating the tubule wall with a very small micropipette and sucking up a sample of the fluid for chemical analysis. This is known as micropuncture. If the micropipettes contain pressure transducers, it is possible to measure the pressures in the afferent and efferent arterioles as well as that in Bowman's space. These micropuncture methods have been used extensively to study tubular function *in situ* in anesthetized animals.

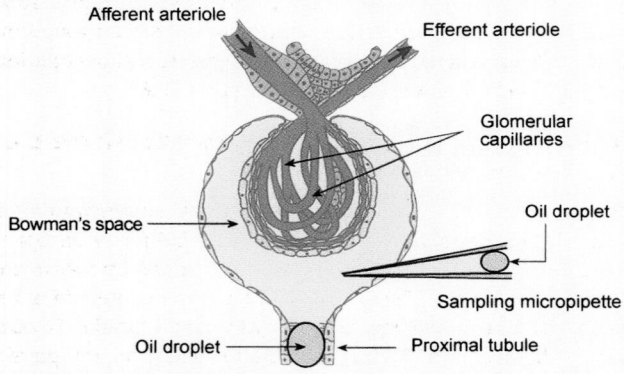

Fig. 1. A simple schematic diagram to illustrate the principle of micropuncture

In the first application of this method a droplet of oil was injected at the top of the proximal tubule and a micropipette inserted into Bowman's capsule (Fig. 1). The fluid in Bowman's space was aspirated and analyzed for protein and electrolytes. Later variants of the technique have employed oil droplets to isolate sections of tubule. A fluid of known composition is then injected into the isolated segment and aspirated after a set period of time. Finally, this fluid is analyzed for changes in its composition.

The methods outlined above have provided a great deal of information regarding the function of superficial nephrons but, by their nature, they cannot provide information about the deeper segments of the nephron or about the juxtamedullary nephrons. To overcome this difficulty, methods have been developed that permit the isolation of particular segments of individual nephrons. The transport processes occurring in these isolated segments can then be studied by microperfusion techniques similar to those employed in intact nephrons.

In 1921 J. T. Wearn and A. N. Richards proved that the nephron works in this way by taking samples of the fluid in Bowman's capsule by micropuncture (see Box 17.1). Analysis of this fluid showed that it had the same composition as plasma, except that there was very little protein.

The simplest explanation of this key observation is that the capsular fluid is formed by filtering out the plasma proteins while allowing the free passage of ions and small molecules such as sodium and glucose. The role of the subsequent segments has been studied by obtaining small samples of tubular fluid by micropuncture, which can then be subjected to chemical analysis. In a variant of this approach, isolated segments of the nephron are perfused with a small volume of liquid, which can then be analyzed to determine which substances have been reabsorbed or secreted.

Formation of the glomerular filtrate

As discussed above, the pressure in the glomerular capillaries forces a small proportion of the plasma into Bowman's space. During this process, small molecules and ions pass across the capillary wall, leaving the plasma proteins behind. This process is known as *ultrafiltration* and the fluid formed in this way is called the *glomerular filtrate*. The rate at which the two kidneys form the ultrafiltrate is known as the *glomerular filtration rate (GFR)* and is expressed in units of milliliters per minute.

The barrier that restricts the passage of fluid from the glomerular capillary into Bowman's capsule consists of three components (Fig. 17.11). First, there is the capillary wall itself that consists of endothelial cells pierced by small gaps known as *fenestrae*. This arrangement makes the glomerular capillaries very much more permeable to water and to solutes than capillaries in other vascular beds. Secondly, the endothelial cells abut a basement membrane, which consists of fibrils of negatively charged glycoproteins. Finally, the epithelial cells or *podocytes* of the capsular membrane do not form a continuous layer but extend thin processes known as *pedicels* over the basement membrane leaving gaps that provide the filtration slits.

Measurements of glomerular filtration have shown that substances with a low molecular weight are freely filtered while the passage of large molecules is severely restricted (Table 17.2). Myoglobin (M_r 17 kDa) passes through the filter relatively easily —it has about three-quarters the permeability of a small molecule like glucose (M_r 180 Da). Hemoglobin (M_r 68 kDa) passes through the filter only with great difficulty, and albumin, the smallest of the plasma proteins (M_r 69 kDa), has about a tenth of the permeability of hemoglobin, i.e. less than a hundredth of the permeability of a small molecule like glucose.

Even the largest of the plasma proteins are much smaller than the filtration slits that can be seen with the aid of the electron microscope. The barrier to the passage of proteins appears to be the meshwork of protein fibrils that forms the basement membrane. Experiments with charged and neutral dextrans (high-molecular-weight carbohydrates) show that the barrier depends on both the size of a molecule and its charge. The barrier is more permeable to neutral or positively charged molecules than it is to negatively charged ones. Thus albumin, which has a strong negative charge at physiological pH, is retained in the plasma both because of its size and by mutual repulsion between its negative charge and that of the glycoproteins of the

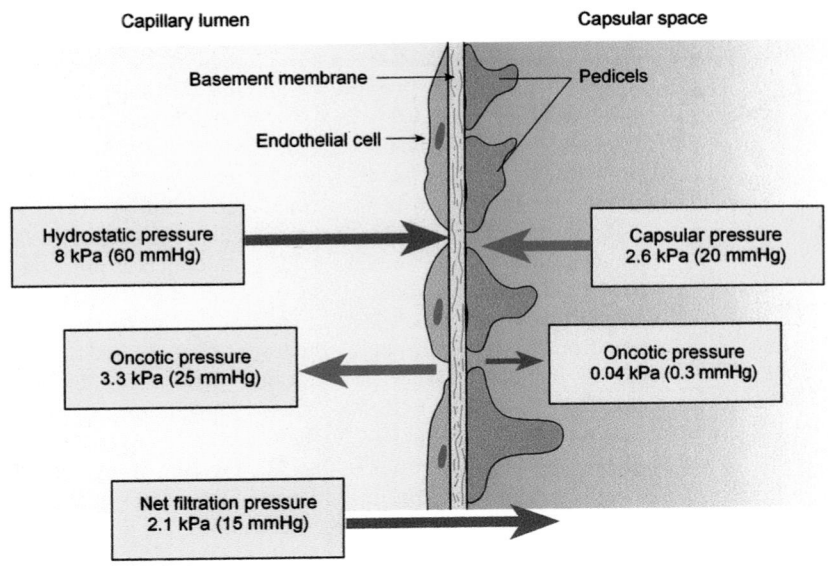

Fig. 17.11 The filtration barrier in the glomerulus and the hydrodynamic forces that determine the rate of ultrafiltration. (For calculation of the Starling forces see Box 17.2.) Fluid from the plasma is forced out by the pressure in the glomerular capillaries, but this is opposed by the pressure in Bowman's capsule and the osmotic pressure exerted by the plasma proteins (the oncotic pressure).

basement membrane. In contrast, hemoglobin, which has a similar molecular size but is not strongly charged at physiological pH, passes through the glomerular filter over five times more easily than albumin. Fortunately, unlike albumin, hemoglobin is contained within the red cells and is not normally present in the plasma.

The amount of fluid passing into Bowman's capsule of a single nephron is known as the single nephron glomerular filtration rate (snGFR). It is governed by the balance of the forces acting on the glomerular capillaries (Fig. 17.11 and Box 17.2). Hydrostatic pressure acts to force the plasma out of the capillaries but, as discussed above, the plasma proteins cannot pass into the glomerulus and are retained within the blood. The osmotic pressure they exert (the *oncotic pressure*) opposes the hydrostatic force due to the pressure within the capillaries. In addition, there is a small pressure within the capsule itself that also opposes the hydrostatic pressure with the capillaries. The sum of these opposing pressures is called the *net filtration pressure*.

The pressure within Bowman's capsule (the *capsular pressure*) is about 2.6 kPa (20 mmHg) and provides the force required to propel the filtrate through the nephron. It arises from the hydrostatic pressure forcing the ultrafiltrate from the glomerular capillaries and the restricted movement of the fluid through the tubules. (This is directly analogous to the pressure–flow relationship for blood flow—see Chapter 15, Section 15.8.)

Inulin clearance can be used to measure the glomerular filtration rate

Measurement of the GFR is important to an understanding of renal physiology, but it cannot be measured directly except in isolated nephrons. However, it can be estimated by measuring the rate of excretion of substances that are freely filtered but are then neither absorbed nor secreted by the renal tubules. In addition, such substances should have no influence on any physiological parameter that may alter renal function such as blood pressure or blood flow. These criteria are met by the plant polysaccharide *inulin* (5.2 kDa) which is excreted by the kidneys in direct proportion to its plasma concentration over a very wide range (Fig. 17.12(a)).

The following explanation shows why the excretion of such substances can be used to measure the GFR: the rate at which a substance is excreted is simply its concentration in the urine (U_c) multiplied by the amount of urine produced per minute ($\dot{V}$)

$$\text{rate of excretion} = U_c \times \dot{V} \text{ mg min}^{-1}.$$

For a substance that is neither reabsorbed nor secreted by the renal tubules, the rate of excretion must be the same as the plasma concentration (P_c) multiplied by the rate at which it was filtered:

$$\text{rate of filtration} = P_c \times \text{GFR mg ml}^{-1}.$$

This is known as the *filtered load*. Combining the two equations:

$$P_c \times \text{GFR} = U_c \times \dot{V}.$$

Rearranging thus gives

$$\text{GFR} = \frac{U_c \times \dot{V}}{P_c} \text{ ml min}^{-1}.$$

TABLE 17.2 The relationship between molecular radius and glomerular permeability

Substance	Molecular mass (Da)	Effective radius (nm)	Filtrate/ plasma
Water	18	0.1	1.0
Urea	60	0.16	1.0
Glucose	180	0.36	1.0
Sucrose	342	0.44	1.0
Inulin	5 200	1.48	0.98
Myoglobin	17 000	1.95	0.75
Hemoglobin	68 000	3.25	0.03
Serum albumin	69 000	3.55	~0.005

The ratio of the concentration of a substance in the filtrate to that found in the plasma is a direct measure of the ease with which it passes through the glomerular filter. A ratio of 1 indicates free passage, and a ratio of 0 would indicate a complete inability to pass the filter.

BOX 17.2 CALCULATION OF THE STARLING FORCES GOVERNING THE FORMATION OF THE GLOMERULAR FILTRATE

The relationship between the glomerular filtration rate (GFR) and the hydrodynamic forces responsible for the formation of the filtrate is given by the relationship

$$GFR = K_f[(P - \Pi_{BC}) - (P_{BC} - \Pi_{onc})]$$

where K_f is the *filtration coefficient* and represents the amount of fluid filtered for each unit of pressure in a minute, P is the hydrostatic pressure in the capillaries, Π_{BC} is the osmotic pressure exerted by the protein in the capsular fluid, P_{BC} is the pressure within Bowman's capsule, and Π_{onc} is the oncotic pressure of the plasma. The net filtration pressure P_f is the difference between the forces tending to force fluid from the glomerulus and those opposing filtration:

$$P_f = (P - \Pi_{BC}) - (P_{BC} - \Pi_{onc})$$

and

$$GFR = K_f P_f.$$

The pressure in the glomerular capillaries is about 8 kPa (60 mmHg), significantly higher than that of the capillaries of other internal organs where average capillary pressures range from 1.3 to 4 kPa (approximately 10–30 mmHg). The pressure within Bowman's capsule is about 2.6 kPa (about 20 mmHg) and the oncotic pressure in normal plasma is about 3.3 kPa (25 mmHg) as the blood enters the afferent arteriole and about 4.7 kPa (35 mmHg) as it leaves the glomerulus via the efferent arteriole. The osmotic pressure attributable to the proteins in Bowman's capsule is *c.* 0.04 kPa (0.3 mmHg) in normal subjects. Thus the net filtration pressure at the afferent end of the glomerular capillaries (i.e. the pressure forcing fluid from the glomerular capillaries) is (8 + 0.04) – (3.3 + 2.6) = 2.14 kPa (*c.* 15 mmHg). As the blood traverses the glomerular capillaries, the net filtration pressure declines as the oncotic pressure rises. By the time the blood leaves the glomerulus the net filtration pressure is much less, as the forces tending to drive fluid from the blood into the capsular space are opposed by the rise in oncotic pressure. In addition, the pressure in the glomerular capillaries falls progressively along their length and is about 7.75 kPa (58 mmHg) by the time they merge with the efferent arteriole. Thus the net filtration pressure at the end of the glomerular capillaries is (7.75 + 0) – (4.7 + 2.6) = 0.45 kPa (*c.* 3 mmHg). It follows from these calculations that if the capillary pressure falls below about 6 kPa (45 mmHg), the forces tending to force fluid from the capillaries will be balanced by those opposing filtration. This can occur during constriction of the afferent arterioles (see Fig. 17.8). Where the fall in renal capillary pressure is large, during hemorrhage for example, little or no plasma will be filtered and urine production will decline.

As the GFR in humans is about 125 ml min⁻¹, K_f has a value of about 60 ml min⁻¹ kPa⁻¹ (or 8 ml min⁻¹ mmHg⁻¹). This is more than a hundred times greater than the permeability of those vascular beds that do not have fenestrated capillaries.

The GFR measured by the rate of inulin excretion is called the *inulin clearance*. It is generally about 120–130 ml min⁻¹ for adult men and about 10 per cent less than this for women of similar size. Despite its advantages, the use of inulin is not very convenient for clinical purposes as a steady concentration needs to be maintained in the plasma for accurate measurement. For this reason it is desirable to use a substance that is normally present in the plasma, is freely filtered, but is neither secreted nor reabsorbed by the renal tubules. In addition, the plasma concentration of such a substance should not fluctuate rapidly. These criteria are largely met by creatinine (a metabolite of creatine) and the *creatinine clearance* is generally used to measure GFR in clinical practice. Nevertheless, as a small component of the excreted creatinine is secreted by the tubules, creatinine clearance overestimates the GFR when renal blood flow is low. (An increased fraction of the excreted creatinine is the result of tubular secretion rather than filtration.)

The concept of clearance can be extended to other substances that are secreted or reabsorbed by the renal tubules. As discussed above, the rate at which a substance is excreted is simply its concentration in the urine multiplied by the amount of urine produced per minute, and the ratio of the rate of excretion to the plasma concentration (P_c) represents the *minimum* volume of plasma from which the kidneys could have obtained the excreted amount. This volume is called the clearance (C) and is expressed in milliliters per minute. Thus

Summary

1. The nephron regulates the internal environment by first filtering the plasma and then reabsorbing substances from, or secreting substances into, the tubular fluid.

2. The barrier that restricts the passage of fluid from the glomerular capillaries into Bowman's capsule consists of the capillary endothelial cells, a basement membrane, and the podocytes of the epithelial cells of the capsular membrane. These components prevent the passage of large molecular weight substances while allowing the free filtration of substances with a low molecular weight.

3. The amount of fluid passing into Bowman's capsule is governed by the net filtration pressure, which is determined by the balance of hydrodynamic forces acting on the glomerular capillaries. The sum of the opposing pressures is called the net filtration pressure. The rate at which the kidneys form the ultrafiltrate is known as the glomerular filtration rate or GFR and has units of milliliters per minute.

4. Renal clearance is defined as the volume of plasma completely cleared of a given substance in 1 min. The clearance of inulin or creatinine is commonly used to estimate the GFR.

$$C = \frac{U_c \times \dot{V}}{P_c} \text{ ml min}^{-1}.$$

Therefore renal clearance can be defined as the volume of plasma completely cleared of a given substance in 1 min. If a substance has a clearance smaller than the GFR, either the kidneys must reabsorb it (e.g. glucose) or it is not freely filtered (e.g. plasma proteins). If a substance has a clearance larger than the GFR, it must be secreted by the renal tubules.

17.5 Tubular absorption and secretion

When a substance is simply filtered and excreted unchanged, the amount excreted is directly proportional to the plasma concentration (see Fig. 17.12(a)). The amount excreted represents the filtered load. If it is reabsorbed, the amount excreted will be less than the filtered load (see Fig. 17.12(b)) and, if filtration is followed by tubular secretion, the amount appearing in the urine will be greater than the filtered load (see Fig. 17.12(c)). As discussed in Box 17.3, the difference between the filtered load and the amount excreted by the kidney is the amount that has either been reabsorbed or secreted. For example, glucose and amino acids are freely filtered but healthy people have virtually no glucose or free amino acids in their urine so there must be tubular mechanisms for the reabsorption of these substances. (Remember that the filtered load is equal to the plasma concentration multiplied by the GFR.) Conversely, as much as 70 per cent of the dye phenolsulfonphthalein is removed from the blood in a single pass through the kidneys. Since 75 per cent is bound to plasma proteins, only 5 per cent of the dye could have appeared in the urine by filtration. The remainder must have been secreted into the tubule. The detailed cellular mechanisms by which individual substances are absorbed or secreted are discussed below.

When polar substances such as glucose are reabsorbed via the transcellular route, carrier molecules are required to permit their movement across the apical and basal membranes of the tubular cells. Since each cell has a limited number of carriers, it is possible to saturate the transport capacity of the tubule if the plasma concentration rises above its physiological level. The amount of solute delivered to the tubule per minute that just saturates its transport process is called the *transfer* or *transport maximum* (Tm).

The concentration dependence of glucose clearance provides a clear example of the Tm concept as applied to tubular transport. If glucose is infused into the blood, the plasma concentration rises above its normal level of about 4.5 mmol l^{-1}. Initially no glucose appears in the urine. However, when the plasma concentration exceeds 10–12 mmol l^{-1} it begins to do so. This value is known as the *renal threshold*. Above the renal threshold the amount of glucose appearing in the urine increases slowly at first but then, as the transport process responsible for glucose reabsorption becomes fully saturated (above about 17 mmol l^{-1}), the increase in urinary glucose becomes directly proportional to the increase in the plasma concentration as shown in Fig. 17.12b.

The transport maximum for glucose (Tm$_g$) can be determined by a simple calculation. Suppose that the GFR is 120 ml min^{-1}

and that plasma glucose is 20 mmol l^{-1}. Further suppose that the urine flow rate is 1.6 ml min^{-1} and that the concentration in the urine is 200 mmol l^{-1}. Under these conditions, the glucose carriers are fully saturated and the amount reabsorbed is equal to the transport maximum. This is equal to the filtered load minus the amount excreted in the urine:

$$\text{filtered load} = 20 \times 120 \times 10^{-3} = 2.4 \text{ mmol min}^{-1}$$

and

$$\text{amount excreted} = 200 \times 1.6 \ 10^{-3} = 0.32 \text{ mmol min}^{-1}.$$

Therefore

$$\begin{aligned} Tm_g &= \text{amount reabsorbed} \\ &= 2.4 - 0.32 \\ &= 2.08 \text{ mmol min}^{-1} \text{ (or 374 mg min}^{-1}). \end{aligned}$$

Plasma glucose levels capable of saturating the transport capacity of the kidneys often occur in patients with inadequately controlled *diabetes mellitus* (see Chapter 27). This leads to the appearance of glucose in the urine (*glycosuria*). As the excreted

---

BOX 17.3 CALCULATING THE AMOUNT OF A SUBSTANCE TRANSPORTED BY THE RENAL TUBULES

The difference between the filtered load and the amount excreted by the kidneys is the amount of a substance that has been reabsorbed or secreted by the tubules, and it can be simply calculated using the following relationship:

$$T_S = U_S \dot{V} - \text{GFR} \times P_S \text{ mg min}^{-1}$$

where T_S is the amount of substance (e.g. glucose) transported, GFR is the glomerular filtration rate, P_S and U_S are the plasma and urine concentrations (mg ml^{-1}), and $\dot{V}$ the urine flow rate. When T_S is negative, the substance has been reabsorbed; when T_S is positive, the substance has been secreted into the tubular fluid.

Consider the data relating to the excretion of phosphate and PAH summarized in the following table:

	Phosphate	PAH	Units
Plasma concentration (P)	0.9	0.05	mg ml^{-1}
Urine concentration (U)	30	25	mg ml^{-1}
Urine flow rate (V)	1.0	1.0	ml min^{-1}
Inulin clearance (GFR)	100	100	ml min^{-1}

The amount of phosphate transported is

$$(30 \times 1) - (0.9 \times 100) = -60 \text{ mg min}^{-1}.$$

Therefore the amount of phosphate excreted is 60 mg min^{-1} less than the filtered load. This difference represents the phosphate that has been reabsorbed by the tubules.

For PAH, the amount transported is

$$(25 \times 1) - (0.05 \times 100) = 20 \text{ mg min}^{-1}.$$

In this case 20 mg min^{-1} more PAH appears in the urine than can be accounted for by simple filtration. Thus it must have been secreted by the tubules.

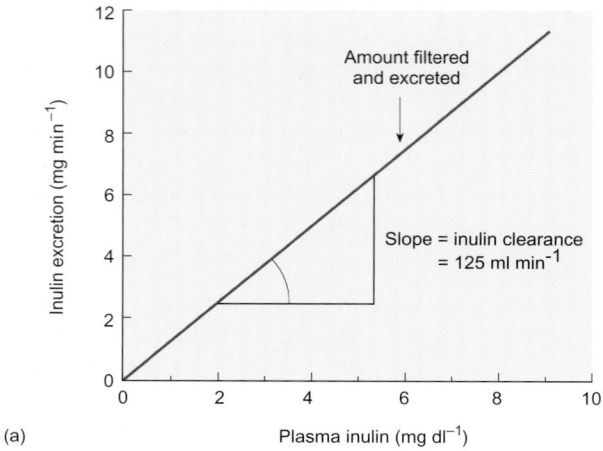

(a)

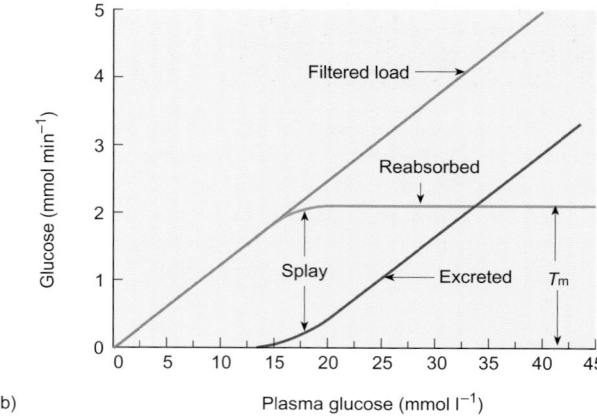

(b)

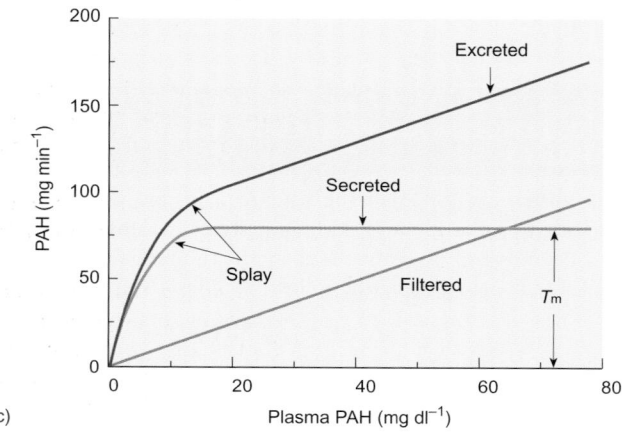

(c)

Fig. 17.12 The relationship between the plasma concentration and the amount excreted in the urine for substances that are subject (a) to filtration only, (b) to filtration followed by reabsorption, and (c) to filtration plus secretion. (a) The amount of inulin excreted plotted as a function of the plasma inulin concentration. The amount excreted is the product of urine concentration and urine flow rate ($U \times \dot{V}$) and the slope of the line is the clearance (C), which is 125 ml min^{-1} in this case. Note that the amount of inulin excreted is directly proportional to the plasma concentration. (b) The quantities of glucose reabsorbed and excreted plotted as a function of the plasma concentration. The blue line shows that glucose appears in the urine when the plasma concentration exceeds about 12 mmol l^{-1}. This is known as the renal threshold for glucose. Above about 18 mmol l^{-1} the extra amount of glucose excreted is directly proportional to the plasma concentration. The red line shows the filtered load in mmoles per minute and this is calculated from the GFR (here taken as 120 ml min^{-1}) and the plasma concentration. The green line indicates the amount of glucose reabsorbed by the tubules, which is the difference between the filtered load and the amount excreted. Reabsorption reaches a maximum when the glucose load exceeds 2 mmol min^{-1} (360 mg min^{-1}). This is the transport maximum for glucose or Tm_{g}.

(c) The amount of PAH excreted plotted as a function of the plasma concentration. The blue line indicates the amount excreted (in mg min^{-1}) while the red line indicates the filtered load. The amount secreted by the tubules (indicated by the green line) is given by the difference between the filtered load and the amount excreted. Tubular secretion of PAH reaches a maximum when the plasma concentration is about 18 mg dl^{-1} and Tm_{PAH} is about 80 mg min^{-1}.

glucose is accompanied with an osmotic equivalent of water, gly-cosuria is associated with an increase in urine production known as an osmotic diuresis. The line relating the amount of glucose absorbed to the plasma concentration is curved at its upper end (Fig. 17.12(b)). This curvature is known as *splay*. While differences in the transport capacity of different nephrons may partly account for the splay, it is an inherent feature of carrier-mediated transport.

The clearance of *p*-aminohippurate can be used to estimate renal plasma flow

As discussed earlier, tubular secretion (movement of a substance from the blood to the tubular lumen) may be either active or passive. Active tubular secretion occurs by carrier-mediated processes analogous to those discussed above for glucose re-absorption, which operate in the opposite direction. It is known that many organic substances are secreted by the tubules, including *p*-aminohippurate (PAH), penicillin, and other organic anions and cations. The relationships between filtration, secretion, and excretion of PAH are illustrated in Fig. 17.12(c).

Moreover, analysis of arterial blood and of blood taken from the renal vein shows that PAH is almost totally removed from the plasma as it passes through the renal circulation. This property is exploited to estimate the renal plasma flow (RPF) in a relatively non-invasive way using the Fick principle (see Chapter 15, Section 15.6):

$$RPF = \frac{\text{amount appearing in urine per minute}}{\text{arteriovenous difference for PAH}} \text{ ml min}^{-1}.$$

Since we can readily measure both the amount of urine produced per minute and the concentration of PAH in the urine, the amount of PAH excreted per unit time can be calculated. The PAH content in a sample of venous blood taken from a peripheral vein can be measured and will be the same as that of arterial plasma. If we neglect the small amount of PAH in the venous blood leaving the kidney, the total RPF (i.e. the total for both kidneys) can be calculated. However, it should be noted, that this way of estimating renal plasma flow is reliable only while the tubular transport mechanism is not saturated—a situation that applies only when the plasma concentration is low (Tm_{PAH} is

BOX 17.4 THE USE OF INULIN AND PAH CLEARANCE TO ESTIMATE GFR AND RENAL PLASMA FLOW

Two experiments were performed on the same subject to determine GFR and renal plasma flow.

Experiment 1:

	Inulin	PAH
Plasma (mg ml^{-1})	0.25	0.08
Urine (mg l^{-1})	12.5	20.0
Urine flow rate (ml min^{-1})	2.5	2.5

Experiment 2:

	Inulin	PAH
Plasma (mg ml^{-1})	0.24	0.36
Urine (mg ml^{-1})	6	25.0
Urine flow rate (ml min^{-1})	5.0	5.0

Calculating the clearances for inulin and PAH:

Experiment 1:
inulin clearance = $(12.5 \times 2.5)/0.25 = 125$ ml min^{-1}
PAH clearance   = $(20.0 \times 2.5)/0.08 = 625$ ml min^{-1}

Experiment 2:
inulin clearance = $(6.0 \times 5.0)/0.24 = 125$ ml min^{-1}
PAH clearance   = $(25.0 \times 5.0)/0.36 = 347$ ml min^{-1}

In experiment 1 normal values are obtained for GFR (inulin clearance) and renal plasma flow (PAH clearance; see text), but in experiment 2 the GFR is normal but the apparent renal plasma flow is 60 per cent of normal. Obviously, the assumption that PAH clearance is a measure of renal plasma flow must be questioned. Inspection of the data shows plasma PAH to be nearly five times higher in experiment 2 than in experiment 1. Here is the clue to the problem. The amount of PAH delivered to the kidneys is 50 mg min^{-1} in experiment 1 and 225 mg min^{-1} in experiment 2. In experiment 1 all of this is excreted, but in experiment 2 only 125 mg min^{-1} is excreted. Of this, 45 mg min^{-1} is filtered. The difference is a measure of the total transport capacity—in this case 80 mg min^{-1}.

about 80 mg min^{-1}; see Box 17.4). From estimates of GFR and renal plasma flow, we can determine the proportion of plasma filtered. This is called the *filtration fraction*. If the GFR is 125 ml min^{-1} and renal plasma flow is 625 ml min^{-1}, the filtration fraction is $125/625 = 0.20$.

The GFR is regulated by glomerulotubular feedback; this regulation may be overridden by sympathetic nerve activity

For efficient operation of the kidney the GFR must be well matched to the transport capacity of the tubules. If the GFR is too small, the kidneys may be unable to regulate the internal environment; if it is too large, the transport capacity for amino

acids, glucose, and ions will be exceeded, resulting in the loss of vital nutrients from the body.

Changes in the GFR will markedly alter the filtered load of sodium in particular. Under normal conditions, the daily filtered load of sodium is about 26 mol (180 liters of plasma are filtered daily and plasma sodium concentration is approximately 145 mmol l^{-1}). Less than 1 per cent of this is usually excreted and sodium balance is maintained (i.e. sodium intake via the diet equals sodium loss in the sweat, feces, and urine). If GFR increases, the filtered load of sodium increases. For example, suppose that the GFR rose from 125 to 130 ml min^{-1}; the total filtered load of sodium would increase to 27 mol a day. Unless the tubules can transport the extra mole of sodium that is filtered, it will be lost in the urine. Evidently, unless the filtered load is regulated or the transport capacity is adjusted, sodium balance will be greatly disturbed.

The matching of the transport capacity to the filtered load is called *glomerulotubular balance*. Amongst the factors that contribute to glomerulotubular balance are the Starling forces in the peritubular capillaries. As the filtrate is formed there is an increase in the oncotic pressure of the plasma (Fig. 17.8). The extent of this increase will depend on the filtration fraction. A larger filtration fraction will result in a greater oncotic pressure in the peritubular capillaries. As a result, the Starling forces in the peritubular capillaries augment movement of fluid from the tubular lumen into the capillaries via the lateral intercellular spaces.

Experiments with perfused single nephrons have shown that when an increase in the GFR for an individual nephron (the snGFR) is mimicked by an increased rate of perfusion of the loop of Henle, the snGFR for that nephron decreases. If the rate of perfusion is reduced, the snGFR increases. This relationship between the rate of fluid flow and glomerular filtration is known as *tubuloglomerular feedback*. This phenomenon, which probably contributes to glomerulotubular balance, is regulated by the cells of the *macula densa*, which form part of the juxtaglomerular apparatus. Whether it is the osmolality, the concentration of sodium, or some other variable that is sensed by these cells remains unclear—as does the mechanism that brings about the change in snGFR.

In response to an elevated systemic blood pressure the renal blood flow and GFR remain remarkably stable. As explained above, an increase in the diameter of the afferent arterioles following a rise in blood pressure would quickly lead to arteriolar constriction. This would offset the rise in pressure and maintain the net filtration pressure and GFR nearly constant. Conversely, following a modest fall in blood pressure, the renal blood flow would fall and this would be countered by a vasodilatation of the afferent arterioles, resulting in a compensatory increase in capillary pressure. The net effect would be to maintain the filtration pressure and GFR at near normal values as before. Therefore the autoregulation of renal blood flow is essential to the maintenance of the GFR.

However, there are circumstances in which renal blood flow and GFR are reduced. For example, during exercise, increased activity in the sympathetic nerves together with an elevation in the concentration of circulating catecholamines (epinephrine and norepinephrine) leads to vasoconstriction in the afferent arterioles. The net filtration pressure falls and with it the GFR. Hemorrhage also leads to pronounced vasoconstriction in the afferent arterioles and a decreased GFR. In severe hemorrhage, the release of antidiuretic hormone (ADH) from the posterior pituitary gland greatly increases and adds to the vasoconstrictor effect of sympathetic activation (for this reason ADH is also known as vasopressin). The resulting intense vasoconstriction can lead to a complete failure of urine production, which is called *anuria*. The actions of ADH during hemorrhage (increased water reabsorption and vasoconstriction) contribute to the maintenance of an adequate circulating plasma volume and so help to offset the fall in blood pressure that would otherwise ensue.

The proximal tubule absorbs water and solutes by active transport and by facilitated diffusion

The proximal tubule reabsorbs about two-thirds of the filtered water, sodium, potassium, chloride, bicarbonate, and other solutes. Under normal circumstances, it removes virtually all the filtered glucose, lactate and amino acids. The sodium pump (Na+, K+ ATPase) of the basolateral surface of the epithelial cells of the proximal tubule provides the driving force for the re-absorption of all of these substances. Specific carrier proteins that are located in the brush-border membrane take up glucose and other organic substances. Some of these carriers have now been cloned and their primary structure determined. They behave in many respects like enzymes. They can be saturated by large amounts of substrate and can be inhibited by appropriate agents. For example, glucose transport can be inhibited by other sugars such as galactose. In addition to the glucose carrier, there are five different carriers for the amino acids, a lactate trans-porter, and at least one carrier for inorganic anions such as phos-phate and sulfate. The existence of these carriers and their similarity to enzymes has provided a simple logical explanation for the existence of the transport maximum for the reabsorp-tion and secretion of many substances.

How do these carriers work? To take glucose as an example, it is clear that its uptake is not favored by a transepithelial gra-dient as the filtrate leaving the glomerulus has the same glucose concentration as the plasma. Moreover, as glucose is a neutral molecule, it cannot be absorbed along an electrical gradient by itself. Experiments have demonstrated that the uptake of glu-cose is sodium dependent. Since the concentration of sodium in the cells of the proximal tubule is about 10–20 mmol l^{-1}, the movement of sodium from the tubular lumen (where the sodium concentration is 145 mmol l^{-1}) into these cells can occur along a favorable electrochemical gradient. It is this gradient that is exploited by the tubular cells to permit glucose uptake, even though there is no concentration gradient in its favor. Both glucose and sodium bind to the carrier. Thus the inward move-ment of glucose is coupled to the movement of sodium down its electrochemical gradient and both are transported into the cell. The glucose leaves the cell via another carrier protein that is not sodium dependent. The Na+, K+ ATPase of the basolateral membrane pumps the absorbed sodium into the lateral and basal extracellular spaces. The arrangement of these carriers on the cell therefore permits the *secondary active transport* of glucose. The reabsorption of amino acids also occurs via sodium-linked

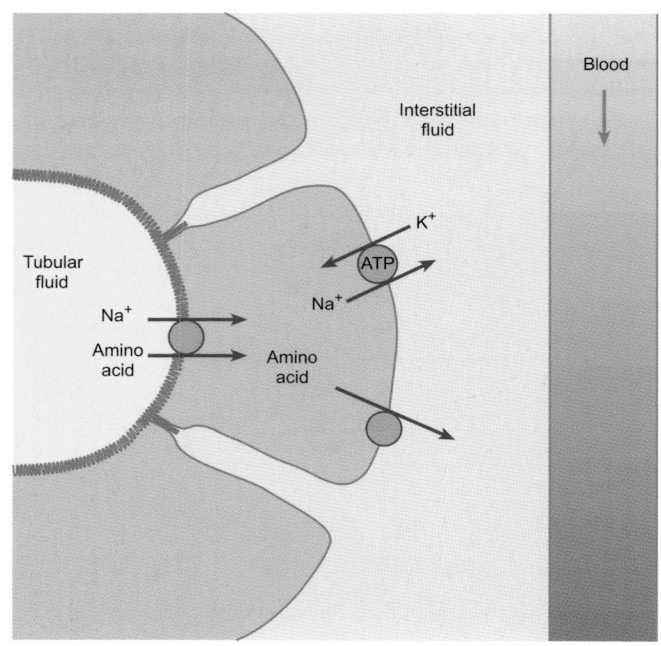

Fig. 17.13 The processes responsible for the reabsorption of amino acids in the proximal tubule. Amino acids and sodium cross the apical membrane via a carrier protein. The sodium gradient provides the energy for the uptake process. The amino acids leave the cell via another carrier protein, which is not linked to sodium. The sodium gradient in the cells of the proximal tubule is maintained by the activity of the sodium pump (Na+, K+ ATPase). Similar processes allow the reabsorption of glucose and organic acids such as lactate.

carrier molecules (Fig. 17.13). This active transport is so effective that all the glucose and amino acids are normally removed from the tubular fluid during its passage along the first half of the proximal tubule. Essentially the same processes permit the efficient absorption of glucose and the amino acids by the small intestine—see Chapter 4, p. 39).

The reabsorption of bicarbonate ions is also linked to that of sodium. A Na+–H+ antiporter is present in the brush-border membrane. This exchanges sodium ions in the tubular fluid for intracellular hydrogen ions. The resulting secretion of hydrogen ions into the lumen favors a shift of the carbonic acid–bicarbon-ate equilibrium towards carbonic acid. This is then rapidly con-verted into carbon dioxide and water by the enzyme carbonic anhydrase, which is located on the brush border of the proximal tubule cells. This locally raises the partial pressure of carbon dioxide in the tubular fluid and favors its diffusion into the tubu-lar cells where it is re-formed into carbonic acid by intracellular carbonic anhydrase. The bicarbonate formed by this reaction leaves the cell via the basolateral membrane in exchange for chloride and passes into the circulation. The processes involved in bicarbonate absorption are summarized in Fig. 17.14.

The uptake of sodium in the first half of the proximal tubule is coupled chiefly with the uptake of organic solutes and anions other than chloride (e.g. bicarbonate). As a result, the concentra-tion of chloride in the tubular fluid passing through the second half of the proximal tubule rises to about 140 mmol l^{-1} com-pared with about 105 mmol l^{-1} for the plasma. Therefore some

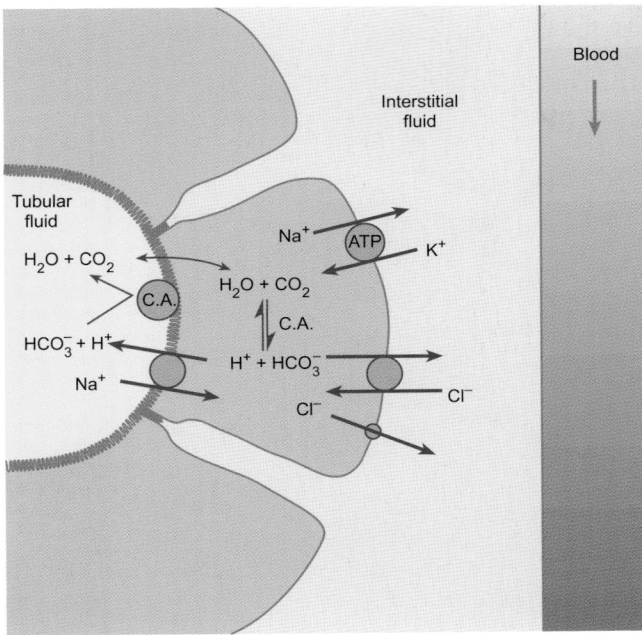

Fig. 17.14 Bicarbonate reabsorption in the proximal tubule. H^+ secreted into the lumen lowers the pH of the tubular fluid and this favors the conversion of HCO_3^- to carbonic acid which is converted to CO_2 and water by the carbonic anhydrase (CA) of the brush-border membrane. The CO_2 diffuses down its concentration gradient into the tubular cell where some is reconverted to carbonic acid by intracellular carbonic anhydrase and ionizes to form HCO_3^-, which leaves the basolateral surface of the cell in exchange for chloride. The H^+ that is generated is secreted into the lumen via Na^+–H^+ exchange to promote further HCO_3^- reabsorption.

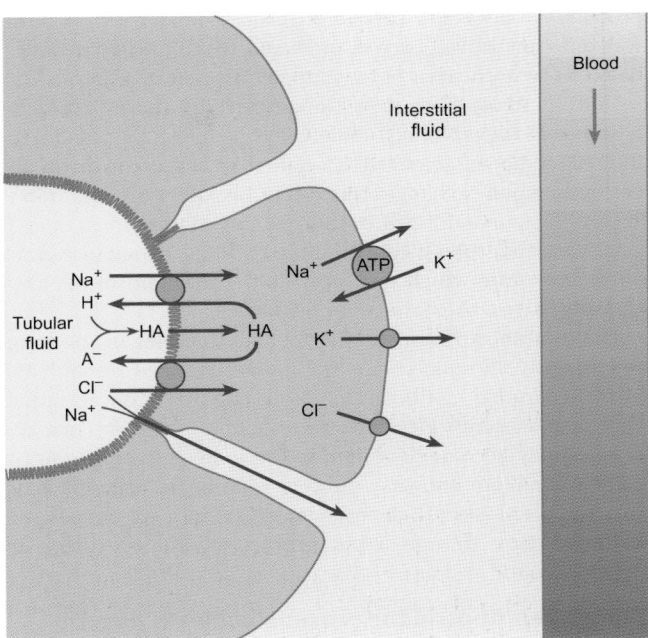

Fig. 17.15 The processes involved in chloride reabsorption in the proximal tubule. Sodium and chloride enter the cell across the apical membrane via the parallel activity of an Na^+–H^+ exchanger that couples chloride movement to the efflux of an organic anion (here represented as A^-) which is recycled. Chloride leaves the cell via chloride channels in the basolateral surface. Some sodium and chloride are reabsorbed via the paracellular pathway as a consequence of water movement (solvent drag).

chloride is able to diffuse, together with sodium, through the tight junctions, down its concentration gradient and into the lateral intercellular spaces. In addition, sodium and chloride are transported into the tubular cells via the parallel action of Na^+–H^+ and Cl^-–anion exchangers as shown in Fig. 17.15.

Phosphate absorption

The phosphate concentration in the plasma plays an important role in determining the rate of bone formation and reabsorption. Excess phosphate can bind to plasma calcium. This binding has a number of undesirable consequences: it decreases the ionized calcium in the plasma, and it also leads to a decreased production of calcitriol that results in a decreased calcium uptake by the gut. The fall in the plasma calcium concentration leads to an increased secretion of parathyroid hormone and demineralization of bone (see Chapter 12). In addition, phosphate ions are involved in many aspects of cellular function. For these reasons, plasma phosphate must be closely regulated. A person eating a typical Western diet absorbs between 800 and 1500 mg (c. 25–50 mmol) of phosphate from the gut each day. To maintain phosphate balance, the kidneys excrete a similar amount. The excreted phosphate also serves to buffer hydrogen ions secreted by the distal tubule.

For a normal person with a plasma phosphate concentration of 1 mmol l^{-1}, about 180 mmol of phosphate are filtered each day. Of this about 80 per cent is reabsorbed by the proximal

tubule by a carrier-mediated process. Like the uptake of glucose and amino acids, phosphate reabsorption by the tubular cells is linked to sodium. The absorbed phosphate leaves the cells via the basolateral membrane by a poorly characterized anion exchange process. No phosphate is absorbed by the loop of Henle or collecting ducts and only half of the remaining phosphate is reabsorbed by the distal tubule, so that about 10–15 per cent of the filtered load is excreted under normal circumstances.

The mechanism by which the kidney regulates plasma phosphate is unusual. The renal threshold for phosphate is just below the normal plasma levels at about 1.1 mmol l^{-1}; the filtered load is $1.25 \times 1.1 = 1.127$ mmol of phosphate per minute. As a result, 0.027 mmol of phosphate will be excreted per minute because the transport capacity of the tubules has been exceeded. If plasma phosphate rises, all the excess phosphate is excreted. If it falls, less phosphate will be excreted.

The Tm_{phos} varies according to the physiological circumstances. Parathyroid hormone regulates phosphate transport by the proximal tubule by a cyclic AMP-dependent mechanism. When the plasma level of parathyroid hormone is high, phosphate reabsorption is inhibited and phosphate secretion is increased. In this way the kidney is able to regulate phosphate balance.

Urate transport by the proximal tubule

Uric acid is the end product of purine metabolism. It circulates in blood as the anion urate, very little of which is bound to plasma proteins (<5 per cent). There are considerable differences in the way different species excrete urate and the following account is based on recent studies of renal urate transport in man.

Urate is freely filtered by the glomerulus and is largely reabsorbed in the proximal tubule, only about 5–10 per cent of the filtered load being excreted. It is taken up across the brush border by an anion exchanger that can utilize other organic anions, such as lactate and β-hydroxybutyrate as counter-ions. In addition, proximal tubule cells possess a voltage-dependent transporter for urate which may provide the mechanism for urate transport across the basolateral membrane.

The clearance of urate in normal subjects lies in the range 6–9 ml min^{-1}. Although urate excretion can exceed 1 g day^{-1}, it is normally around half this value. If plasma urate concentrations become persistently elevated above about 0.4 mmol l^{-1}, uric acid crystals may become deposited in the joints, giving rise to the painful condition known as gout. In addition, as uric acid is not very soluble at low pH, deposits may form in the renal pelvis as kidney stones. Around 5 per cent of all kidney stones are deposits of uric acid or its salts.

Water absorption in the proximal tubule is directly linked to solute uptake

The uptake of sodium, chloride, glucose, and other solutes by the tubular cells results in a transfer of osmotically active particles from the tubular lumen to the extracellular space. This transport is not accompanied by a full osmotic equivalent of water. As a result, there is a slight increase in the osmolality of the fluid surrounding the renal tubules. To maintain osmotic equilibrium, water moves from the tubular lumen to the extracellular space. While some of this water moves via the cells to maintain osmolality (the transcellular pathway), some apparently passes to the lateral extracellular space via the tight junctions (the paracellular pathway). It is important to recognize that the water absorbed in the proximal tubule is not regulated independently of the reabsorption of solutes (unlike water reabsorption in the distal nephron—see Section 17.8); consequently this is sometimes known as the obligatory phase of water reabsorption. Some solutes such as potassium, magnesium, and calcium are partially reabsorbed via the paracellular pathway along with the osmotically driven uptake of water (a process known as solvent drag).

Protein lost in the glomerular filtrate is reabsorbed by pinocytosis

The glomerular filtrate contains a small amount of protein (about 40 mg l^{-1} compared with 65–80 g l^{-1} in plasma) but, as the kidneys form 180 l of filtrate daily, about 7 g of protein is nevertheless filtered. Such a loss would adversely affect the body's nitrogen balance. To avoid this, the cells of the proximal tubule engulf the proteins in the filtrate by pinching off a small volume of filtrate containing protein and absorbing it into the cell by endocytosis. This process is also called *pinocytosis*. The vesicles formed within the cells (the endocytotic vesicles) fuse with the lysosomes of the cell and the proteolytic enzymes of the lysosomes degrade the proteins they contain. The amino acids released by this process are then transported across the basolateral membrane and absorbed back into the blood stream (Fig. 17.16).

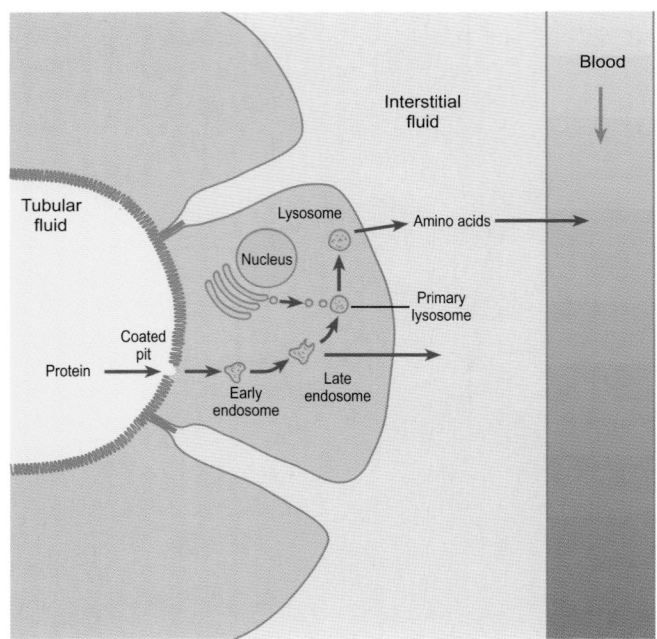

Fig. 17.16 Endocytosis and protein breakdown in a proximal tubule cell. Plasma proteins escaping into the filtrate become bound to the surface membrane of the brush border. The membrane-bound proteins are absorbed into coated pits from where they become absorbed into endosomes prior to their breakdown by lysosomes.

Under normal circumstances, virtually no protein is found in the urine. However, if the glomeruli become diseased, significant amounts of protein may pass into the filtrate. Since the proximal tubule has a very limited capacity for reabsorbing protein, some will be lost to the urine which will have a frothy appearance as the protein lowers the surface tension. This is known as *proteinuria* and is a sign of some abnormality of kidney function. As hemoglobin is small enough to be filtered by the glomeruli, proteinuria also occurs following hemolysis of the red cells (see Section 17.4). Furthermore, the excreted hemoglobin will give the urine a reddish brown appearance.

Secretion by the cells of the proximal tubule

In addition to their reabsorptive activities, the cells of the proximal tubule actively secrete a variety of substances into the tubule lumen. Many metabolites are eliminated from the blood in this way including bile salts, creatinine, hippurates, prostaglandins, and urate. In addition, the kidney also eliminates many foreign substances by secretion, including drugs such as penicillin, quinine, and salicylates (aspirin). These substances are ionized at physiological pH and two transport systems are involved, one for anions such as PAH and one for cations such as creatinine. In common with the transport of amino acids and glucose from the lumen into the tubular cells, these carriers are proteins and their transport capacity can be saturated. Since infusion of penicillin can depress the secretion of PAH and other organic anions (and vice versa), these molecules appear to be secreted by the same transport system. As there is little structural similarity between the various organic anions, the transport is said to be of

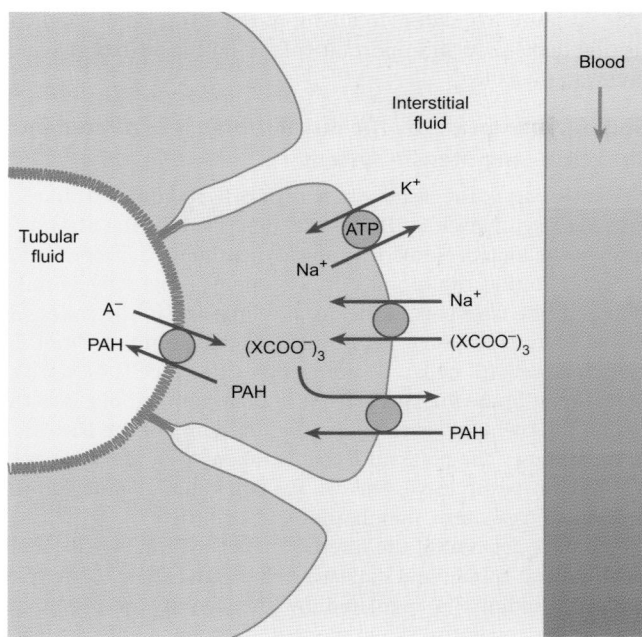

Fig. 17.17 The secretion of organic anions (e.g. PAH) into the lumen of the proximal tubule. PAH crosses the basolateral membrane in exchange for di- and tricarboxylate anions (here represented as (XCOO⁻)₃) and is secreted into the lumen down its chemical gradient in exchange for another organic anion such as urate. The di- and tricarboxylates are reabsorbed into the cell via a sodium-dependent symporter.

low specificity. Organic cations are secreted by a separate low-specificity carrier system. Active secretion provides the kidney with a very efficient means of elimination of protein-bound substances that could otherwise only be eliminated very slowly by filtration.

The secretion of organic anions such as PAH occurs by a two-stage process. The anion is taken up into the tubular cell across the"asolateral membrane in exchange for a-ketoglutarate and other di- or tricarboxylate anions that diffuse down their concentration gradients. As the concentration of PAH (or other organic anion) in the cell rises, it passes into the tubule lumen via an anion exchange protein located in the apical membrane. The di- and tricarboxylates re-enter the cells of the proximal tubule via a sodium-dependent symporter that is located in the basolateral membrane (Fig. 17.17).

17.6 Tubular transport in the loop of Henle

On entering the descending thin limb of the loop of Henle, the tubular fluid is still isotonic with the plasma. As it passes down the descending limb, the tubular fluid becomes increasingly hypertonic. As the cells of the descending thin loop are thin and flattened, they do not actively transport significant amounts of salts. From this it follows that the change in the osmolality of the tubular fluid is the result of passive movement of water out of the tubule into the medullary interstitium, and of sodium,

Summary

1. As the filtrate passes along the proximal tubule, all the protein, amino acids, and glucose contained in the fluid are reabsorbed. The absorption of amino acids and glucose is linked to the sodium gradient across the apical membrane and the driving force for their uptake is the sodium pump of the basolateral membrane.

2. Almost all the essential organic constituents in the tubular fluid are absorbed in the first half of the tubule. Additionally, about 80 per cent of the filtered bicarbonate is also reabsorbed in the first half of the proximal tubule. As a result, sodium absorption in the second half of the proximal tubule is mainly coupled to that of chloride.

3. In addition to reabsorbing solutes, the proximal tubule actively secretes some organic anions and cations into the tubular fluid. Like the uptake of glucose and amino acids, these are carrier-mediated processes, each with its own transport maximum.

4. The movement of an osmotic equivalent of water accompanies the absorption of solutes so that the fluid leaving the proximal tubule is isotonic with the plasma. About two-thirds by volume of the filtrate is absorbed in this part of nephron. Water movement occurs both through the epithelial cells (the transcellular pathway) and via the tight junctions (the paracellular pathway).

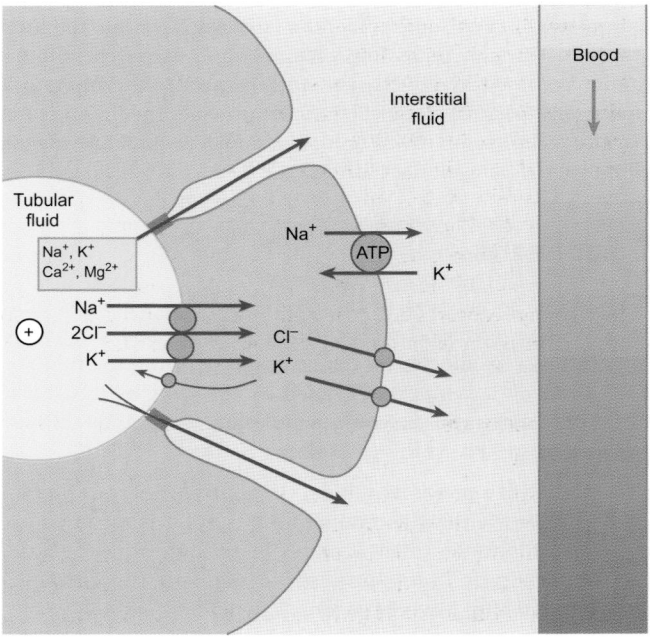

Fig. 17.18 The transport processes responsible for the uptake of sodium and chloride in the thick ascending limb of the loop of Henle. Sodium, potassium and chloride are transported into the tubular cells by an electroneutral cotransporter. The diffusion of potassium from the tubular cells into the lumen via potassium channels leads to the development of a lumen positive potential which provides the driving force for the paracellular absorption of cations. NB. In this section of the nephron the ion movements occur without the osmotically driven uptake of water.

chloride, and urea from the interstitium into the tubule. The movement of solute and water occurs through the tight junctions of the epithelium. The osmolality of the tubular fluid reaches its peak at the hairpin bend. The osmolality for the longest loops, which reach the tips of the renal papillae, may reach 1200 mOsm kg^{-1}. For loops that do not penetrate so deeply into the medulla, the peak osmolality will be less.

The thin ascending limb does not actively pump sodium chloride across its wall, but the wall is impermeable to water so that the fluid it contains does not equilibrate with the interstitium. Therefore the fluid entering the thick portion of the loop of Henle is strongly hypertonic with respect to the plasma. The epithelial cells of the thick part of the ascending limb possess an electroneutral symporter that transports sodium, potassium, and chloride ions into the cell across the apical membrane (Fig. 17.18). Like the thin ascending limb, the wall of the thick ascending limb is impermeable to water. Consequently, the osmolality of the tubular fluid falls as the cells of the thick segment transport sodium, potassium, and chloride into the interstitium. By the time that the tubular fluid has reached the beginning of the distal tubule, it has an osmolality of about 150 mOsm kg^{-1} (i.e. about half that of the plasma). The way in which this transport is exploited to generate the osmotic gradient in the renal medulla will be discussed later (Section 17.8) as part of the process of osmoregulation.

As with the other epithelial cells of the kidney, the driving force for the uptake of sodium, potassium, and chloride by the cells of the thick limb of the loop of Henle is provided by the sodium gradient established by the sodium pump of the basolateral membrane. Some potassium ions leak back into the tubular fluid through potassium channels and cause the tubular lumen to be positively charged with respect to the interstitial space. This positive electrical gradient provides the driving force for the reabsorption of sodium, potassium, calcium, and magnesium via the paracellular pathway.

Summary

1. About 20 per cent of the filtered sodium, chloride, and water are reabsorbed by the loop of Henle. Sodium, potassium, and chloride ions are transported from the tubular fluid by a symporter located in the cells of the thick ascending limb. Calcium, magnesium, and other cations are absorbed via the paracellular pathway.

2. Unlike the proximal tubule, the transport of ions by the cells of the thick ascending limb is not accompanied by an osmotic equivalent of water. As a result, while the fluid entering the descending loop of Henle is isotonic with plasma, that leaving the loop is hypotonic.

17.7 The distal tubules regulate the ionic balance of the body

The reabsorption of sodium, potassium, and water by the proximal tubule and ascending loop of Henle largely takes place regardless of the ionic balance of the body. However, in the distal tubule and collecting ducts, the uptake and secretion of these ions is closely regulated. In addition, the distal tubule and collecting ducts play an important role in both acid–base balance and water balance.

Sodium ion uptake by the distal tubule is regulated by the renin–angiotensin system

The early part of the distal tubule reabsorbs sodium and chloride ions via a symporter similar to that described for the ascending limb of the loop of Henle. The apical membrane is impermeable to water so that the tubular fluid becomes progressively more dilute. In the later part of the distal tubule and in the collecting ducts, sodium reabsorption is linked to potassium secretion by principal cells (P cells). Sodium enters the P cells across the apical membrane via channels. It is then pumped out into the lateral intercellular space by the sodium pump of the basolateral membrane. The potassium that is taken up by the activity of the sodium pump leaves the cell via potassium channels in the apical and basolateral membranes.

About 12 per cent of the filtered load of sodium is reabsorbed in the distal tubule and collecting ducts and their capacity to reabsorb sodium is regulated by the activity of the juxtaglomerular apparatus. When the sodium of the fluid in the distal tubule is low, the cells of the macula densa cause the granular cells of the afferent arteriole to secrete the proteolytic enzyme renin into the blood. The exact process by which the macula densa cells stimulate the granular cells to secrete renin is not yet known. Renin converts a plasma peptide called angiotensinogen into angiotensin I. This, in turn, is converted to angiotensin II by *converting enzyme*, which is found on the capillary endothelium of the lungs and some other vascular beds. Angiotensin II acts on the zona glomerulosa cells of the adrenal cortex to stimulate the secretion of the hormone aldosterone (Fig. 17.19).

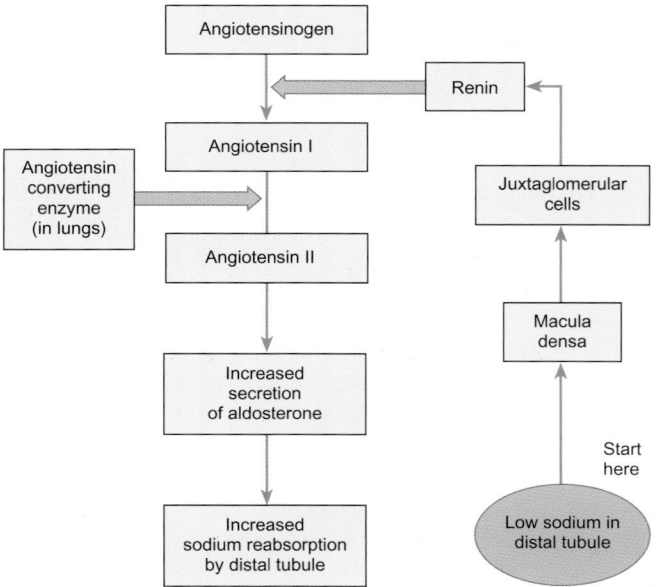

Fig. 17.19 The regulation of aldosterone secretion by the sodium load in the distal tubule. The cells of the macula densa sense the sodium load delivered to the distal tubule and regulate the level of circulating renin according to the body's sodium requirements. If the sodium concentration in the tubular fluid is low, the juxtaglomerular cells increase their secretion of renin. This ultimately results in an increased sodium reabsorption.

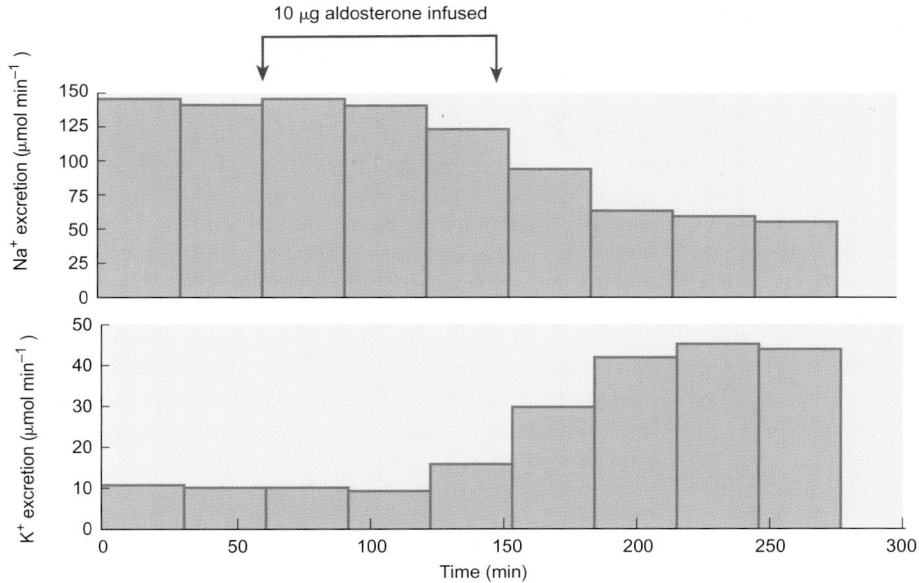

Fig. 17.20 The effect of infusing 10 μg aldosterone on sodium and potassium excretion in the adrenalectomized dog. Note the time delay before significant changes in sodium and potassium excretion are seen and that the effect of aldosterone considerably outlasts the period of infusion.

Aldosterone stimulates the production of sodium channels that become inserted in the apical membranes of the P cells. It also stimulates the synthesis of Na^+, K^+ ATPase molecules, which are inserted in the basolateral membrane. By increasing the number of available channels for sodium uptake, aldosterone promotes sodium reabsorption. Sodium moves into the cells down its concentration gradient and is transported to the interstitial fluid by the sodium pump of the basolateral membrane. Increasing the activity of the sodium pump also increases intracellular potassium which can pass into the tubular fluid down its concentration gradient. These adjustments act to increase the ability of the nephron to reabsorb sodium and to secrete potassium. As the action of aldosterone requires the synthesis of new proteins, its effect is not immediate but is delayed by an hour or so and it reaches its maximum after about a day (Fig. 17.20).

When the plasma volume is expanded due to an increase in total body sodium, the secretion of renin is inhibited by atrial natriuretic peptide (ANP). This results in an increase in sodium excretion. The interplay between the renin–angiotensin system and ANP is important for the regulation of body fluid volume (see Chapter 28).

Body potassium balance is regulated by the distal tubules and collecting ducts

The diet is rich in potassium, with about 4 g (100 mmol) being ingested each day in a normal diet. As the distribution of potassium ions across the plasma membrane of cells is the principal determinant of their membrane potential, there is a need for the extracellular potassium concentration to be closely regulated. This is achieved by the distal tubule and collecting duct, which actively regulate potassium excretion. Potassium reabsorption is increased in deficiency and secreted in normal and potassium-retaining (*hyperkalemic*) states.

By the time the filtrate reaches the distal tubule, nearly 90 per cent of the filtered potassium has been reabsorbed. About two-thirds is normally absorbed by the proximal tubule via the para-

cellular route and some 20 per cent is absorbed by the ascending thick limb of the loop of Henle by cotransport with sodium and chloride ions. In the remainder of the nephron, both potassium absorption and potassium secretion can occur and the balance between them determines how much potassium is lost in the urine.

Potassium secretion into the tubular fluid occurs via a transcellular pathway (Fig. 17.21). It is taken up into the P cells by the activity of the Na^+, K^+ ATPase located in the basolateral membrane and is secreted into the tubular fluid via potassium channels located in the apical membrane.

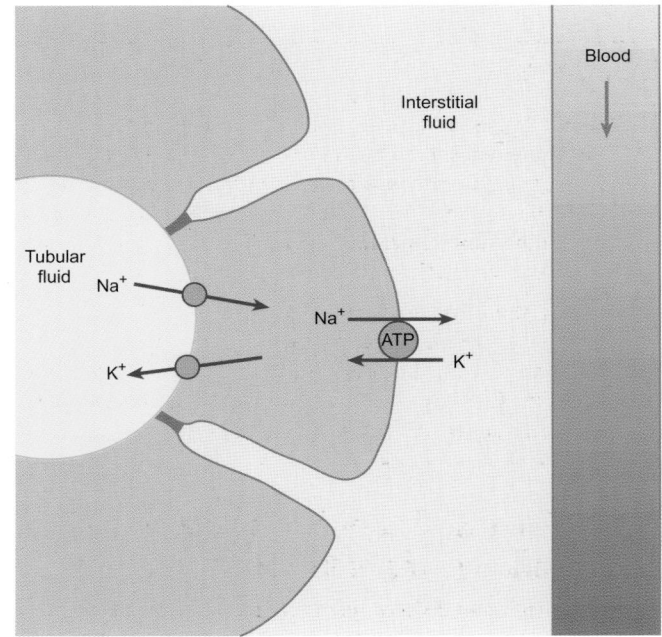

Fig. 17.21 The tubular secretion of potassium ions in the distal tubule and collecting duct.

Under normal circumstances the amount of potassium secreted is determined by its concentration in the plasma. If plasma potassium is elevated, this will directly increase potassium uptake into the cells via the Na$^+$, K$^+$ ATPase. Additionally, if plasma potassium is elevated by as little as 0.2 mmol l^{-1}, the secretion of aldosterone by the zona glomerulosa cells of the adrenal cortex is increased. The aldosterone, in turn, stimulates the P cells to synthesize more sodium channels and Na$^+$, K$^+$ ATPase molecules, which become inserted in the apical membranes and basolateral membranes respectively. These changes augment the uptake of sodium from the tubular fluid and increase the secretion of potassium by the P cells. The effect is to restore plasma potassium to its normal level.

The intercalated cells regulate acid–base balance by secreting hydrogen ions

Although most of the filtered bicarbonate is reabsorbed in the proximal tubule and the loop of Henle, there is 1–2 mmol l^{-1} in the fluid entering the distal tubule. Under normal circumstances, all this bicarbonate is reabsorbed and acid urine is excreted. The reabsorption of bicarbonate by the intercalated cells differs from that of the proximal tubule. The intercalated cells actively secrete hydrogen ions into the lumen via an ATP-dependent pump (Fig. 17.22). As in the proximal tubule, the decrease in the pH of the tubular fluid favors the conversion of bicarbonate ions to carbon dioxide and water and the liberated carbon dioxide diffuses down its concentration gradient into the tubular cells where carbonic anhydrase catalyzes the reformation of carbonic acid within the intercalated cells. The carbonic acid dissociates into hydrogen ions and bicarbonate ions. The

hydrogen ions are secreted into the lumen and the bicarbonate ions exit the tubular cells via a bicarbonate–chloride exchanger.

Active secretion of hydrogen ions into the lumen of the distal tubule and collecting ducts results in a fall in the pH of the tubular fluid which can reach values as low as 4–4.5, much lower than elsewhere in the nephron. The apical membranes of the cells of the distal tubule and collecting ducts have a very low passive permeability to protons, which are thereby prevented from diffusing back into the tubular cells. As pH 4 corresponds to a free hydrogen ion concentration of only 0.1 mmol l^{-1}, only 0.15 mmol of free hydrogen ion can be excreted each day for a normal daily urine output of 1.5 l. Nevertheless, the body produces about 50 mmol of non-volatile acid daily that it needs to excrete. This is achieved by the physicochemical buffering of hydrogen ions with phosphate and by the secretion of ammonium ions (see pp. 569–570).

Calcium ions are absorbed from the distal tubule by an active process that can be stimulated by parathyroid hormone

About 70 per cent of the filtered load of calcium is reabsorbed in the proximal tubule, 20 per cent is absorbed by the ascending loop of Henle, and most of the remainder is absorbed by the distal tubule and cortical collecting duct. Only about 1 per cent is normally excreted.

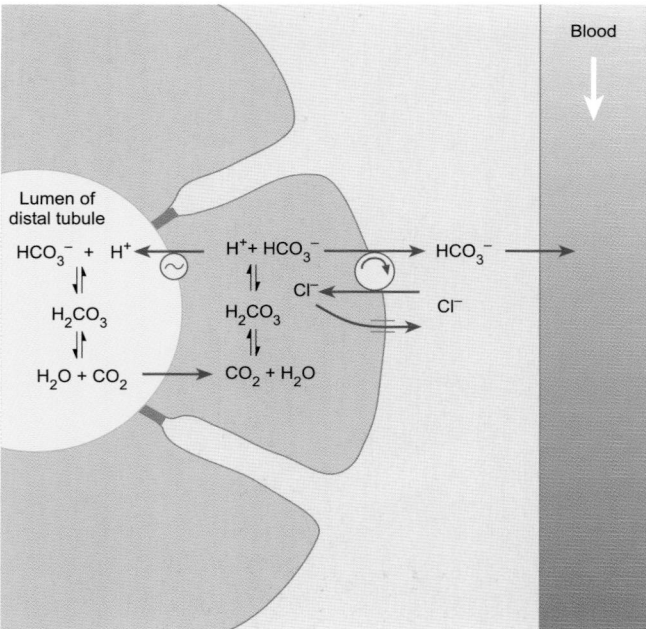

Fig. 17.22 The mechanism by which the intercalated cells of the late distal tubule and cortical collecting ducts secrete hydrogen ions.

Summary

1. The first part of the distal tubule continues the dilution of the tubular fluid that began in the ascending thick limb of the loop of Henle.

2. In the second part of the distal tubule and in the collecting duct, the epithelium consists of two cell types: the P cells and the I cells. The P cells absorb sodium and water. They also secrete potassium into the tubular fluid. The intercalated cells secrete hydrogen ions and reabsorb bicarbonate. As elsewhere in the nephron, these ion movements ultimately depend on the sodium gradient established by the sodium pump of the basolateral membrane.

3. The efficacy of sodium uptake and potassium secretion is regulated largely by the hormone aldosterone which is, in turn, regulated by the secretion of the enzyme renin from the juxtaglomerular cells of the afferent arterioles.

4. Although the transcellular movement of calcium accounts for about a third of the uptake of calcium in the proximal tubule, in the distal tubule the entire calcium uptake is via the transcellular route. This uptake is driven by passive influx of calcium down its electrochemical gradient into the tubular cells coupled to active extrusion of calcium across the basolateral membrane. This calcium uptake is stimulated by parathyroid hormone (PTH). In contrast, phosphate reabsorption is decreased by an increase in PTH secretion.

Calcium reabsorption occurs by both transcellular and para-cellular routes in the proximal tubule and the ascending limb of the loop of Henle. In the proximal tubule, calcium transport occurs mainly via the paracellular pathway by solvent drag. The transcellular movement of calcium accounts for only about a third of its uptake. In contrast, in the distal tubule all the calcium uptake is via the transcellular route. This uptake is driven by passive influx of calcium down its steep electrochemical gradient into the tubular cells coupled to active extrusion of calcium across the basolateral membrane.

Calcium uptake by the distal tubule and cortical collecting ducts is stimulated by parathyroid hormone (PTH), which plays a major role in calcium homeostasis. Conversely, an increase in PTH secretion decreases the reabsorption of phosphate by the proximal tubule. (The regulation of calcium and phosphate balance by PTH and other hormones is considered in greater detail in Chapter 12.)

17.8 The kidneys regulate the osmolality of the plasma by adjusting the amount of water reabsorbed by the collecting ducts

In a normal individual, water intake varies widely according to circumstances. As a result the osmolality of urine can range from as little as 50 mOsm kg^{-1} following a large water load, to around 1200 mOsm kg^{-1} in severe dehydration. How do the kidneys produce urine with such a wide range of osmolality?

Three key facts are fundamental to understanding how the kidney regulates water balance.

1. Between the outer border of the renal medulla and the papilla of the renal pyramids, the osmolality of the interstitium progressively increases from about 300 mOsm kg^{-1} to 1200 mOsm kg^{-1}.

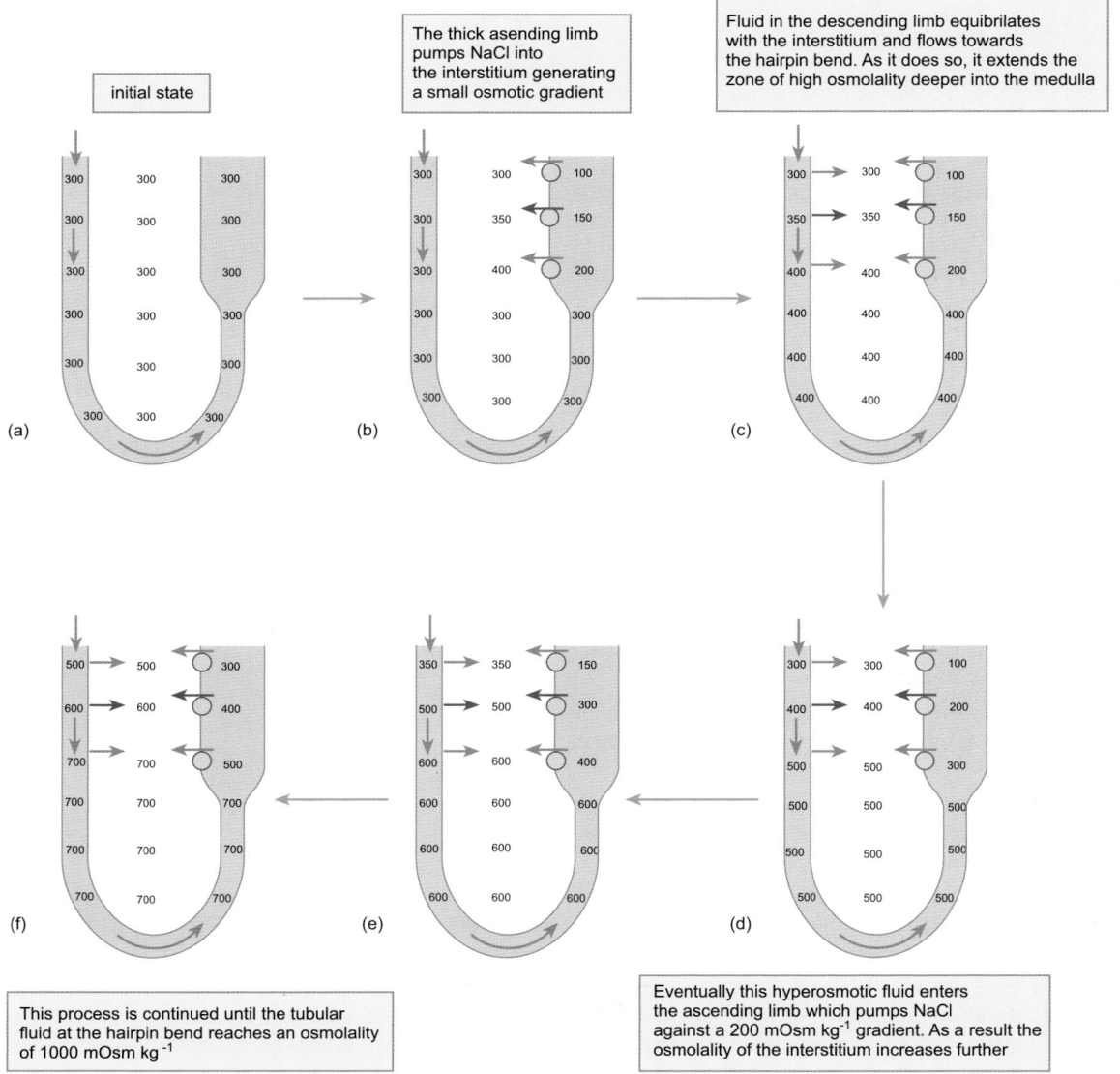

Fig. 17.23 The operation of the countercurrent multiplier in the establishment of the osmotic gradient of the renal medulla (see text for further details).

2. The flow of fluid in different parts of the nephron runs in opposite directions (there is a countercurrent arrangement).

3. The collecting ducts are impermeable to water unless antidiuretic hormone (ADH) is present.

In essence, the kidney generates an osmotic gradient within the medulla by active transport of sodium, potassium and chloride from the lumen of the ascending loop of Henle into the interstitial space. This gradient is then used to reabsorb water from the urine as it passes through the medulla. The amount of water reabsorbed is regulated by the level of ADH circulating in the blood. When the plasma osmolality is low, little ADH is produced and copious dilute urine is produced. Conversely, when plasma osmolality is high ADH secretion is increased. Under such circumstances water reabsorption in the distal nephron is increased with the result that a small volume of concentrated urine is produced.

In what follows, the mechanisms responsible for establishing the medullary osmotic gradient will first be discussed and this will be followed by a simple account of the way ADH regulates water reabsorption in the distal nephron.

Salt transport by the ascending limb of the loop of Henle leads to the generation of a large osmotic gradient between the renal cortex and the inner medulla

While the fluid in the space between the tubules (the interstitium) in the renal cortex is approximately iso-osmotic with the plasma, the osmolality of the interstitial fluid of the medulla increases progressively from the border with the cortex to the renal papilla. In the outer medulla the osmolality is about 290 mOsm kg^{-1}, chiefly attributable to sodium and chloride ions, but in the inner medulla the osmolality of the interstitium can reach 1200 mOsm kg^{-1}, about half of which is attributable to these ions and half to urea. This remarkable osmotic gradient is formed primarily by the active transport of sodium and chloride by the thick ascending limb of the loop of Henle without the reabsorption of an osmotic equivalent of water.

The crucial factors for the generation of the osmotic gradient are listed below.

1. As shown in Fig. 17.23, the fluid in the loop of Henle follows a countercurrent arrangement so that fluid returning from the deeper reaches of the medulla flows in the opposite direction to that entering the medulla.

2. Sodium transport by the thick ascending limb of the loop of Henle can occur against an osmotic gradient of about 200 mOsm kg^{-1}.

3. The walls of the proximal tubule and descending thin limb are freely permeable to water and have a high passive permeability to sodium, chloride and urea.

4. In contrast with the proximal tubule and the descending thin limb, the ascending thick limb of the loop of Henle, the distal tubule, and the collecting ducts have little passive permeability to ions or to urea.

5. Although the ascending thin and thick limbs of the loop of Henle and the first third of the distal tubule are impermeable to water, they actively transport sodium and chloride. As a result, water is separated from solute in this part of the nephron.

The mechanisms by which the osmotic gradient of the medulla is established are outlined in Fig. 17.23. Before the gradient is established, the osmolality is the same throughout the nephron (Fig 17.23(a)). The active transport of sodium and chloride across the tubular epithelium of the ascending thick limb of the loop of Henle and the first part of the distal tubule occurs without concomitant movement of water. The result of this transport is a *decrease* in the osmolality of the tubular fluid in the thick ascending limb of the loop of Henle and an *increase* in the osmolality of the fluid surrounding the tubule (i.e. the fluid of the medullary interstitium) (Fig. 17.23(b)).

As the tubular fluid passes down the descending limb of the loop of Henle, it loses water to the medullary interstitium and gains sodium and chloride ions so that its osmolality progressively rises as it flows towards the hairpin bend (Fig. 17.23(c)). Moreover, as it equilibrates with the interstitium, the flow of fluid carries sodium and chloride deeper into the medulla, raising its osmolality. Since the ascending limb is impervious to water, the hypertonic fluid (400 mOsm kg^{-1}) in the thin ascending limb now enters the thick limb, which transports sodium and chloride into the interstitium until there is a transepithelial osmotic gradient of 200 mOsm kg^{-1}. This causes the osmolality of the interstitium to increase from 400 to 500 mOsm kg^{-1} (Fig. 17.23(d)). As more fluid enters the descending loop, it equilibrates with the interstitium which is now more hypertonic than before. This hypertonic solution enters the ascending thick limb which transports sodium and chloride until there is a transepithelial osmotic gradient of 200 mOsm kg^{-1} once more and the osmolality of the interstitium rises from 500 to 600 mOsm kg^{-1} (Fig. 17.23(e)). This process continues until the osmolality of the tubular fluid at the hairpin bend reaches about 1200 mOsm kg^{-1}. This high osmolality is chiefly attributable to sodium and chloride ions (1000 mOsm kg^{-1}) but urea contributes about 200 mOsm kg^{-1}.

Although the fluid entering the thick limb of the loop of Henle is hypertonic, sodium, potassium, and chloride are progressively removed from it as it flows towards the cortex. Consequently, the fluid entering the distal tubule is hypotonic with respect to the plasma. Indeed, by the time the fluid has reached the middle of the distal tubule so much sodium and chloride has been lost that its osmolality is less than 100 mOsm kg^{-1} (i.e. less than a third of that of plasma). Overall, the progressive transport of sodium and chloride from the tubular fluid into the interstitium results in the establishment of a longitudinal osmotic gradient in the medulla. *Thus the countercurrent arrangement of the loop of Henle multiplies a relatively small transepithelial osmotic gradient into a large longitudinal gradient.* The transport processes of the loop of Henle and the resulting changes in osmolality are summarized in Fig. 17.23.

Urea is concentrated in the medullary interstitium by a passive process

Chemical analysis shows that the osmotic pressure of the interstitial fluid of the inner medulla (which may be 1200–1400 mOsm kg^{-1}) is almost equally attributable to sodium chloride and urea. (However, the osmolality of the tubular fluid is mainly due to sodium and chloride.) Like other small solutes, urea is freely filtered at the glomerulus and a significant fraction of the filtered load is reabsorbed passively together with its osmotic equivalent of water in the proximal tubule. As the tubular fluid reaches the loop of Henle, the urea concentration is still essentially the same as that of plasma but, by the time the fluid reaches the hairpin bend, urea contributes about 200 mOsm kg^{-1}. This increase in urea concentration is due to passive secretion of urea from the medullary interstitium.

In the thick ascending limb, sodium and chloride transport occurs without the movement of water so that the fluid in the distal tubule is hypotonic with respect to plasma. When the urine is concentrated as a result of the action of ADH on the cortical collecting ducts, the osmolality in this part of the nephron can reach that of the plasma (290 mOsm kg^{-1}). However unlike the fluid entering the nephron, sodium and chloride ions account for much less of the osmolality and urea contributes much more because the powerful transport mechanisms of the thick ascending limb and distal tubule have removed most of the sodium and chloride ions.

As the fluid flows into the inner medulla, more of the water in the collecting ducts is reabsorbed under the influence of ADH and the urea concentration in the urine rises until, in the inner medullary collecting ducts, it exceeds that of the interstitium. Thus urea is concentrated by the abstraction of water. When the urea concentration in the urine is high, its movement from the urine into the medullary interstitium is favored (Fig. 17.24). Moreover, in the inner medulla, ADH not only increases the permeability of the collecting ducts to water, but it also increases their permeability to urea. This adaptation optimizes the conservation of osmotically active urea during dehydration and minimizes its loss from the interstitium during diuresis. There is a further advantage of this adaptation: although urea is the primary end-product of nitrogen metabolism, it can be excreted in large amounts without large volumes of water. This situation arises because urea is in osmotic equilibrium across the wall of the collecting ducts when the urine is concentrated.

The vasa recta provide blood flow to the medulla without depleting the osmotic gradient

As described above, the renal medulla receives its blood supply by way of the vasa recta. The blood flow is much less than that of the renal cortex but is sufficient to provide the medullary tissue with nutrients and oxygen. The vasa recta play a further important role: they act to maintain the osmotic gradient of the medulla whilst removing the ions and water that have been reabsorbed. As the vasa recta are derived from the efferent arterioles of the juxtamedullary glomeruli, the blood entering them is isotonic with normal plasma (280–290 mOsm kg^{-1}). Like the descending limb of the loop of Henle, the walls of the vasa recta are permeable to salts and water so that the blood they contain progressively increases in osmolality as it passes into the inner medulla, by gaining salt and by losing water (Fig. 17.24). By the time it reaches the deepest parts of the medulla, the blood has an osmolality equal to that of the surrounding interstitium. As the blood returns towards the cortex, the reverse sequence occurs and the blood leaving the vasa recta is only slightly hyperosmotic to normal plasma. The countercurrent arrangement of the vasa recta together with their relatively low blood flow thus helps to maintain the osmolality of the renal medulla. During the course of its passage through the medulla, the blood has removed the excess salt and water that have been added by the transport processes occurring in the deeper regions of the medulla.

Antidiuretic hormone (ADH) regulates the absorption of water from the collecting ducts

The absorption of ions by the ascending limb of the loop of Henle results in the tubular fluid becoming hypo-osmotic as it approaches the distal tubule. During its passage along the first third of the distal tubule, sodium, potassium and chloride continue to be transported from the lumen to the interstitium. Moreover, the tubular epithelium is still relatively impermeable to water and so the tubular fluid becomes progressively more

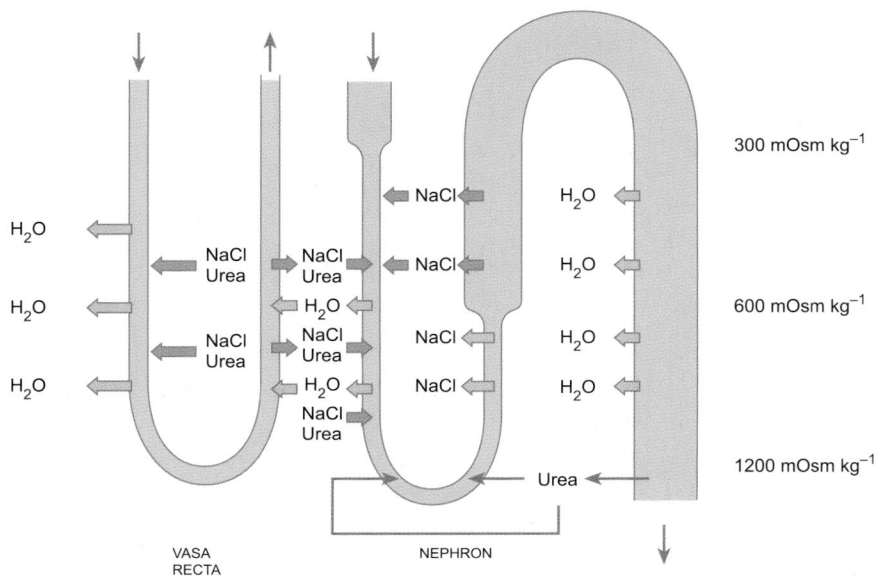

Fig. 17.24 The mechanism of urea concentration in the renal medulla and the role of the vasa recta in preserving the osmotic gradient. As urine flows towards the renal pelvis, water is reabsorbed by the collecting ducts under the influence of ADH. As a result, the concentration of urea in the urine rises until it exceeds that of the interstitium. In the inner regions of the medulla, ADH increases the permeability of the collecting ducts to urea, which can then be passively reabsorbed down its concentration gradient.

dilute. By the time the fluid reaches the last third of the distal tubule there is a substantial osmotic gradient in favor of water reabsorption. Both this part of the distal tubule and the collecting ducts are impermeable to water unless ADH (vasopressin) is present. This hormone is secreted by the posterior pituitary gland in response to increased plasma osmolality. Its secretion is regulated by sensors known as *osmoreceptors* which are located in the hypothalamus, close to the supra-optic and paraventricular nuclei which produce the hormone and transport it to the posterior pituitary (see Chapter 12, pp. 197–198). When plasma osmolality is below 285 mOsm kg⁻¹, ADH secretion by the posterior pituitary is very low and plasma levels are below 1 pg ml⁻¹. An increase in plasma osmolality of as little as 3 mOsm kg⁻¹ is sufficient to stimulate ADH secretion and the degree of stimulation depends on the increase in osmolality above the threshold of 285 mOsm kg⁻¹ (Fig. 17.25). This osmotic regulation of ADH secretion is central to the control of plasma osmolality. The secretion of ADH is inhibited by atrial natriuretic hormone and some drugs such as ethanol.

ADH increases the permeability of the last third of the distal tubule and of the whole of the collecting duct to water. The result is a movement of water down its osmotic gradient into the tubular cells and thence into the interstitial fluid and the plasma. *This water movement is independent of solute uptake* and therefore results in an increase in urine osmolality and a fall in plasma osmolality that is directly related to the amount of water reabsorbed. Consequently, when the body has excess water and the plasma osmolality is less than 285 mOsm kg⁻¹, ADH secretion will be suppressed. Under these circumstances, water will not be reabsorbed during its passage through the collecting ducts so that a large volume of dilute urine will be produced (giving rise to a *diuresis*). In contrast, ADH secretion is stimulated during dehydration as the plasma osmolality is greater than 285 mOsm kg⁻¹. The secreted ADH acts on the collecting ducts to increase their permeability to water and a smaller volume of more concentrated urine is produced.

Only some 10–15 per cent of the total filtered load of water is subject to regulation by ADH as the remainder has been reabsorbed along the nephron together with salts and other solutes. If the posterior pituitary is unable to secrete ADH or if the collecting ducts are unable to respond to it, a large volume of dilute urine is produced (*polyuria*)—in excess of 15 liters of urine can be excreted per day and must be made up by increased water intake if life-threatening dehydration is to be avoided. This condition is known as *diabetes insipidus*.

The osmolality of the extracellular fluid is controlled independently of its volume, which is determined by the total body sodium. The interrelationship between the maintenance of osmolality and sodium balance will be discussed in Chapter 28.

ADH controls the insertion of water channels into the apical membrane of the P cells of the collecting ducts

How does ADH regulate the permeability of the collecting ducts to water? It now appears that ADH circulating in the blood binds to receptors on the basolateral surface of the P cells of the collecting ducts. These receptors are coupled to adenylyl cyclase and the resulting increase in cyclic AMP activates a protein kinase. The protein kinase then initiates the fusion of vesicles containing water channels with the apical membrane. When ADH levels fall, part of the apical membrane is endocytosed and the water channels are taken back into the cells to await recycling. Therefore the permeability of the apical membrane of the P cells to water is regulated by the insertion and removal of specific water channels as shown in Fig. 17.26. The water that enters the cell across the apical membrane passes freely into the lateral intercellular space and thence to the plasma. Similar mechanisms are thought to regulate the permeability of the membrane to urea.

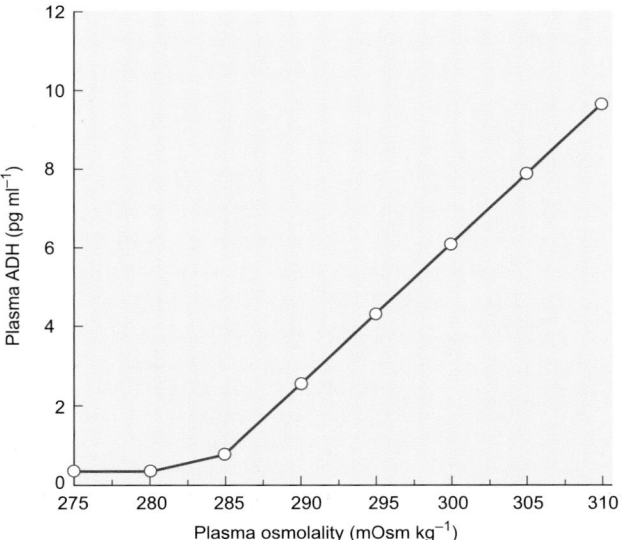

Fig. 17.25 The effect of changes in plasma osmolality on circulating levels of ADH.

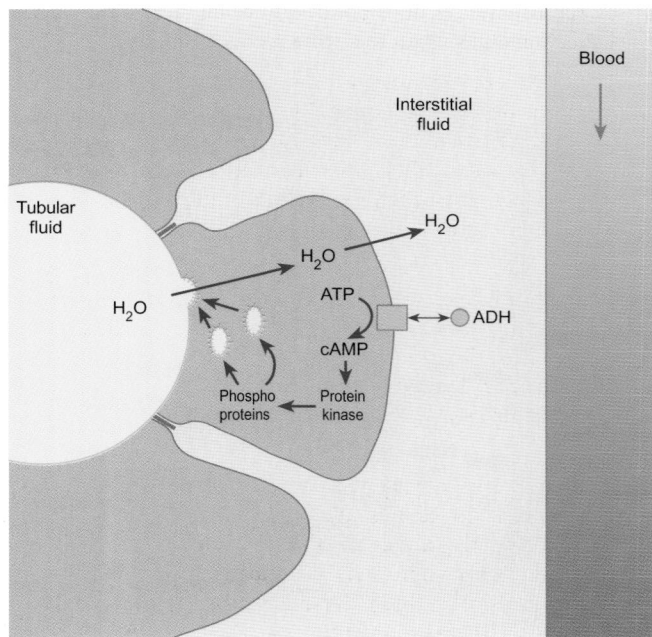

Fig. 17.26 The control of water uptake by ADH in the distal tubule and collecting duct.

The clearance of 'free water'

According to the prevailing water balance of the body, the kidneys excrete urine that may be hypertonic or hypotonic with respect to the plasma. The ability of the kidneys to excrete a concentrated or a dilute urine is sometimes expressed by the unsatisfactory concept of the 'free-water' clearance. This is the amount of pure water that must be added to, or subtracted from, the urine to make it isotonic with the plasma. Therefore, it is not calculated in the same way as the clearance of other substances but as follows:

$$C_{H_2O} = \dot{V} - C_{osm}$$

where C_{H_2O} is the free-water clearance, $\dot{V}$ is the urine flow rate and C_{osm} is the osmolar clearance which, from the standard definition of clearance, is given by

$$C_{osm} = U_{osm} \times \dot{V}/P.$$

Suppose that the kidneys are excreting urine with an osmolality of 120 mOsm kg^{-1} at a rate of 5 ml min^{-1} and plasma osmolality is 285 mOsm kg^{-1}:

$$C_{H_2O} = 5 - (120 \times 5/285) = 5 - 2.1 = +2.9$$

Therefore the free-water clearance is 2.9 ml min^{-1} (i.e. the kidneys are excreting 2.9 ml min^{-1} more water than is required to keep the urine iso-osmotic with the plasma). If the urine osmolality is 450 mOsm kg^{-1} and the urine flow rate is 1 ml min^{-1},

$$C_{H_2O} = 1 - (450 \times 1/285) = -0.6 \text{ ml min}^{-1}.$$

In this case 0.6 ml min^{-1} less water is excreted than is required to maintain the urine iso-osmotic with the plasma. From these calculations, it is evident that if hypotonic urine is excreted, the kidneys have a positive free-water clearance. If the urine is hypertonic, they have a negative free-water clearance. When the free-water clearance is zero, the urine is iso-osmotic with the plasma.

In the kidneys 'free water' is generated by the absorption of salts in the ascending thick limb of the loop of Henle and distal tubule. Therefore it represents the extent to which the fluid in the distal tubules is diluted with respect to the plasma. How much of the 'free water' is excreted will, of course, depend on the circulating level of ADH as discussed above.

17.9 The collection and voiding of urine

The renal calyxes, ureters, urinary bladder, and urethra comprise the urinary tract which is concerned with collecting the urine formed by the kidneys and storing it until a convenient time occurs for the bladder to be emptied. The epithelia that line the urinary tract are impermeable to water and solutes and so do not modify the composition of the urine.

Urine passes from the kidney to the bladder by the peristaltic action of the muscle in the wall of the urethra

The ureters are tubes about 30 cm long, which consist of an epithelial layer surrounded by circular and longitudinal bundles of smooth muscle. In addition, some muscle fibers are disposed in a spiral arrangement around the ureter. When the renal calyxes and upper regions of the ureters become distended due to the accumulation of urine, peristaltic contractions occur in the ureters that propel the urine towards the bladder. These contractions are almost certainly myogenic in origin and are sufficiently powerful to propel urine towards the bladder against pressures of 6–13 kPa (50–100 mmHg). They normally occur at intervals of 10 s to 1 min, but their frequency can be modified by the activity of the pelvic nerves. The ureters pass obliquely through the bladder wall for 2–3 cm in a region known as the *trigone* before they finally empty into the bladder just above the neck. This arrangement closes the ends of the ureters when the pressure within the bladder rises above that in the ureter, thus preventing reflux of urine.

The bladder itself consists of two principal parts: the *body* or *fundus*, which serves to collect the urine, and the bladder neck or posterior urethra, which is 2–3 cm long. The bladder is lined by a mucosal layer which becomes greatly folded when the bladder is empty. The bladder wall consists of smooth muscle and elastic tissue. The muscle is of the single-unit type and is known as the *detrusor muscle*. The wall of the bladder neck consists of a higher proportion of elastic tissue interlaced with detrusor muscle. The tension in the wall of the bladder neck keeps this part of the bladder empty of urine during normal filling and so the posterior urethra behaves as an internal sphincter. The urethra passes through the urogenital diaphragm, which contains a layer of striated muscle called the external sphincter which is under voluntary control via the pudendal nerves.

The micturition reflex is responsible for the voiding of urine

As the bladder fills it becomes distended, and the detrusor muscle becomes stretched and contracts. This basal tone results in a

Summary

1. The osmolality of the renal medulla is largely due to the high concentrations of sodium ions, chloride ions, and urea found in the fluid of the medullary interstitium.

2. The osmotic gradient arises from the countercurrent arrangement of the loop of Henle and the active transport of sodium and chloride from the lumen of the ascending limb into the medullary interstitium. This is augmented by passive urea movement from the tubular fluid to the medullary interstitium under the control of ADH when the urine is concentrated.

3. Active transport of sodium and chloride ions by the ascending thin and thick limbs of the loop of Henle and by the initial part of the distal tubule leads to dilution of the tubular fluid. The circulating level of ADH regulates the subsequent reabsorption of water by the collecting ducts.

4. When there is a significant water load, little ADH is secreted and a copious dilute urine is produced. Under these conditions, the 'free-water' clearance is positive. During dehydration ADH secretion is increased and the collecting ducts reabsorb water so that a small volume of concentrated urine is produced. The 'free-water' clearance will be negative.

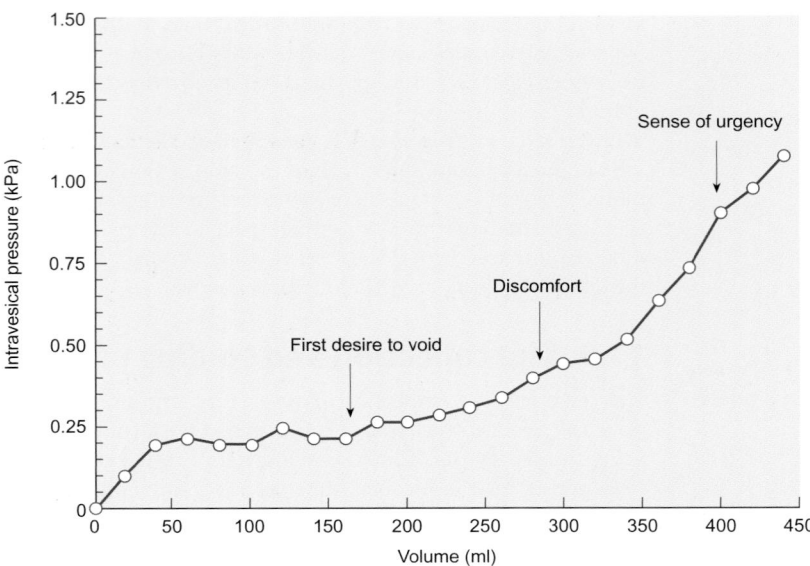

Fig. 17.27 Pressure–volume relationship for the normal human bladder. Note that after the initial phase of filling, the volume increases three- to fourfold with little increase in intravesical pressure. As the volume increases further the pressure rises progressively more and more steeply.

pressure within the bladder (the *intravesical pressure*) of about 250 Pa (2.5 cmH$_2$O) (Fig. 17.27). Further filling results in little change in pressure until the bladder volume reaches 150–250 ml when the first desire to void is experienced. Further increases in volume lead to increased intravesical pressure until at about 300–350 ml the pressure begins to rise steeply as more urine

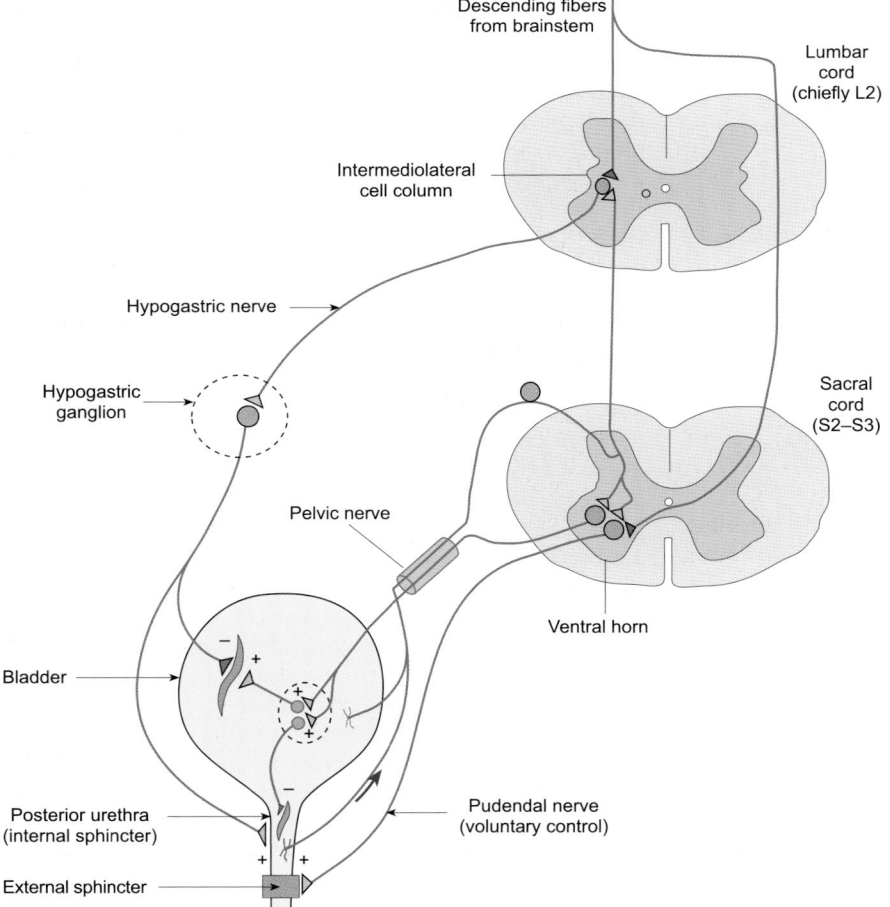

Fig. 17.28 The principal nervous pathways that control micturition. Micturition is inhibited by activity in the hypogastric (sympathetic) nerves and pudendal nerves. It is facilitated by activity in the pelvic (parasympathetic) nerves.

accumulates in the bladder. At these volumes the bladder undergoes periodic reflex contractions (see below) and there is an urgent need to urinate. The process by which the bladder normally empties is known as *micturition*. It is a reflex that is controlled by the sacral segments of the spinal cord.

The innervation of the lower urinary tract is shown schematically in Fig. 17.28. The ureters, bladder, and internal sphincters receive information from stretch receptors in their walls. This information is carried in the visceral afferent fibers of the pelvic nerves. The motor fibers that control bladder function are derived from both divisions of the autonomic nervous system. Parasympathetic fibers from the sacral outflow (S2 and S3) innervate the bladder and internal sphincter via the pelvic nerves. Parasympathetic fibers also control the external sphincter (via the pudendal nerves) while sympathetic postganglionic fibers derived chiefly from spinal segment L2 run in the hypogastric nerve to innervate the bladder and posterior urethra. The sympathetic fibers act to inhibit micturition by decreasing the excitability of the detrusor muscle and exciting the muscle of the internal sphincter. The parasympathetic fibers act to initiate micturition by exciting the detrusor muscle and inhibiting activity of the smooth muscle of the internal sphincter.

During the storage of urine, the progressive distension of the bladder causes stimulation of afferent nerve fibers in the pelvic nerves. The activity in these afferents stimulates activity in the sympathetic fibers of the hypogastric nerve and this leads to inhibition of the detrusor muscle and constriction of the neck of the urethra. In addition, the external sphincter is held closed by activity in the pudendal nerves. These responses are 'guarding reflexes' which act to promote continence. At the initiation of micturition intense activity in the afferent fibers from the bladder wall activates neurons in the brainstem that send impulses to the spinal cord to inhibit the guarding reflexes and permit voiding.

When the bladder wall becomes distended due to the accumulation of urine, stretch receptors initiate periodic reflex contractions of the detrusor muscle known as *micturition contractions*. These contractions involve the whole of the muscle and greatly increase the intravesical pressure, thus triggering a strong desire to urinate. Nevertheless, voiding will not occur unless the muscles surrounding the urethra relax. If voiding does not occur, the micturition reflex becomes suppressed and intravesical pressure falls. The cycle will repeat itself after an interval of some minutes. If the micturition reflex succeeds in overcoming the tension in the wall of the posterior urethra, a further stretch reflex is activated that inhibits the external sphincter and urination occurs. Voiding occurs as a result of activity in the parasympathetic fibers, leading to stimulation of the detrusor muscle and inhibition of the muscle of the internal sphincter.

Normally, micturition is a voluntary act that is controlled by corticospinal impulses sent to the lumbrosacral region of the spinal cord, and the basic spinal reflexes outlined above are normally inhibited by impulses from the brainstem. During urination, they become facilitated. Urination is further aided by contraction of the abdominal muscles, which raises intra-abdominal pressure, and by relaxation of the muscles of the urogenital diaphragm, which permits the dilation of the urethra. Once urination occurs the bladder is normally almost fully emptied, with only a few milliliters of urine remaining.

Section of the spinal cord causes loss of bladder control

In the period that immediately follows section of the spinal cord above the sacral segments (e.g. through a crush injury to the spine in the thoracic region), there is a complete suppression of the micturition reflex. This inhibition is due to spinal shock and the loss of descending control. Clinically, the first priority is to prevent damage to the bladder wall through overdistension. This is achieved by draining the bladder via a catheter. Over time, the bladder reflexes gradually return. Despite this recovery, such patients have no control over when they urinate and micturition is triggered when the pressure within the bladder reaches a threshold level. This is known as *automatic bladder*. In this situation, the urine does not dribble constantly from the urethra but is periodically voided as in a normal subject. A similar situation exists in infants before they have learnt voluntary control.

If the spinal cord is partially crushed, the fibers that normally inhibit the micturition reflex may be damaged while those that facilitate it may remain intact. In this situation, the micturition reflex is facilitated and small volumes of urine are frequently voided in an uncontrolled manner. This is known as *neurogenic uninhibited bladder*.

Some crush injuries and compression of the dorsal roots of the sacral region of the spinal cord can lead to loss of afferent nerve fibers from the bladder. If this happens, information from the stretch receptors of the bladder wall will not reach the spinal cord and the normal tone of the detrusor muscle is lost. In addition, the micturition reflex will be abolished even though the efferent fibers may be intact. This gives rise to a condition known as *atonic bladder* in which the bladder cannot contract by either voluntary or reflex mechanisms. Instead, it fills to capacity and urine is lost in a continuous dribble through the urethra.

Summary

1. Urine is transported from the kidneys to the bladder by the peristaltic action of the muscle in the wall of the ureters. The intravesical pressure (the pressure within the bladder) increases very slowly at first until 200–300 ml of urine has been accumulated. Thereafter the intravesical pressure rises progressively more steeply.

2. This rise in pressure is associated with a sensation of fullness, and when intravesical pressure reaches about 250 Pa the stretch receptors in the wall of the bladder trigger contractions of the detrusor muscle of the bladder wall known as micturition contractions. If these contractions succeed in overcoming the resistance of the posterior urethra, the micturition reflex is facilitated and voiding occurs. During normal voiding there is a voluntary relaxation of the external sphincter.

Recommended reading

Anatomy of the urinary tract

MacKinnon, P.C.B., and Morris J.F. (2005). *Oxford textbook of functional anatomy*. Vol. 2, *Thorax and abdomen* (2nd edn), pp. 169–178. Oxford University Press, Oxford.

Histology of the urinary tract

Junquieira, L.C., and Carneiro, J. (2003). *Basic histology* (10th edn) Chapter 19. McGraw Hill, New York.

Pharmacology and the kidneys

Grahame-Smith, D.G., and Aronson, J.K. (2002). *Oxford textbook of clinical pharmacology and drug therapy* (3rd edn), Chapter 26. Oxford University Press, Oxford.

Rang H.P., Dale, M.M. Ritter, J.M., and Moore, P. (2003). *Pharmacology* (5th edn), Chapter 23. Churchill-Livingstone, Edinburgh.

Renal physiology

Koeppen, B.M., and Stanton, B.A. (2001). *Renal physiology* (3rd edn). Mosby, St Louis, MO.

Lote, C.J. (2000). *Principles of renal physiology* (4th edn). Kluwer Academic Publications, London.

Seldin, D.W., and Giebisch, G. (2000). *The kidney—physiology and pathophysiology* (3rd edn). Lippincott-Williams & Wilkins, Baltimore, MD.

Valtin, H., and Schafer, J.A. (1995). *Renal function* (3rd edn). Little, Brown, Boston, MA.

Multiple choice questions

Each statement is either true or false. Answers are given below.

1. a. The urine concentration of potassium is greater than that of plasma.
   b. The plasma concentration of calcium is greater than that of urine.
   c. Urine normally contains about 2 mmol l^{-1} of bicarbonate.
   d. Urine pH is normally greater than that of plasma.
   e. Urine normally contains no measurable quantity of protein.

2. a. Renal plasma flow can be measured by the clearance of PAH.
   b. The kidneys receive about a fifth of the resting cardiac output.
   c. As cardiac output increases during exercise renal blood flow rises.
   d. Renal blood flow is maintained within narrow limits by autoregulation.
   e. Increased activity in the renal sympathetic nerves results in increased blood flow to the kidneys.

3. a. The glomerular filtrate is identical in composition to plasma.
   b. The glomerular filtration rate can be determined by measuring the clearance of inulin.
   c. A substance that has a clearance less than that of inulin must have been reabsorbed by the renal tubules.
   d. The glomerular filtration rate is about 125 ml min^{-1} in an adult male.
   e. The glomerular filtration rate depends on the pressure in the afferent arterioles.

4. a. The absorption of small molecules by the renal tubules always occurs via specific transport proteins.
   b. The appearance of glucose in the urine reflects a saturation of the glucose carriers of the proximal tubule.
   c. Glucose transport and amino acid transport across the epithelia of the renal tubules is linked to sodium transport.
   d. The transport maximum for glucose is about 360 mg min^{-1}.
   e. If the renal clearance of a substance exceeds that of inulin then it must be secreted into the tubules.

5. a. The GFR is closely matched to the transport capacity of the tubules.
   b. In a healthy person the proximal tubules reabsorb all of the filtered glucose.
   c. The filtered bicarbonate is absorbed by anion transport.
   d. All the filtered bicarbonate is normally reabsorbed in the first half of the proximal tubule.
   e. The proximal tubules secrete bile salts into the tubular fluid.

6. a. All nephrons have loops of Henle that penetrate deep into the medulla.
   b. The descending thin limb of the loop of Henle is permeable to water.
   c. The osmolality of the fluid at the hairpin bend of the loop of Henle may exceed 1200 mOsm kg^{-1}.
   d. The thick ascending limb of the loop of Henle transports sodium ions in the same way as the proximal tubule.
   e. The thin ascending limb of the loop of Henle is impermeable to water.

7. a. The uptake of sodium ions is regulated by the proximal tubule.
   b. Macula densa cells secrete renin when plasma sodium is low.
   c. Angiotensin II is formed from renin by the action of an enzyme found on the endothelium of the pulmonary blood vessels.
   d. Angiotensin II acts on the adrenal medulla.
   e. The adrenal glands secrete aldosterone in response to stimulation by angiotensin II.

8. a. Plasma potassium is regulated by the sodium pump activity of the proximal tubules.

   b. Calcium absorption by the distal tubules is decreased by parathyroid hormone.

   c. Phosphate absorption by the proximal tubule is regulated by parathyroid hormone.

   d. The intercalated cells of the distal tubule secrete hydrogen ions to reabsorb bicarbonate from the tubular fluid.

   e. Without the buffering action of phosphate, the kidneys would be able to excrete less than 1 mmol of hydrogen ions each day.

9. a. Following the intake of a large volume of water, a normal person can produce urine with an osmolality of less than 100 mOsm kg^{-1}.

   b. The renal medulla has an osmotic gradient that increases from the border with the cortex to the renal papilla.

   c. The osmotic gradient in the renal medulla results from the transport of sodium and chloride ions from the thick ascending limb of the loop of Henle to the interstitial fluid.

   d. ADH is secreted by the anterior pituitary in response to a decrease in the osmolality of the blood.

   e. ADH acts on the collecting ducts to increase their permeability to water.

Quantitative problems:

1. The following data relating blood flow to perfusion pressure were obtained from an intact perfused dog kidney.

Time (s)	0	15	30	45	60	90	120	150
Perfusion pressure (mmHg)	140	135	140	180	190	190	190	190
Renal blood flow (ml/min)	135	133	136	155	145	130	134	135

The pressure in the renal vein was unchanged throughout at 20 mmHg. Plot and interpret the data. Would you expect to see a closely similar pattern for blood flow through a denervated kidney?

2. Given that the pressure in the glomerular capillaries is 60 mmHg, the plasma oncotic pressure is 32 mmHg, and the pressure in Bowman's capsule is 18 mmHg:

   a. Calculate the net filtration pressure (ignore the oncotic pressure of the filtrate).

   b. If the concentration of inulin is 0.25 mg/ml and 12.5 mg/ml in the plasma and urine, respectively, and the urine flow rate is 2.5 ml/min, calculate the glomerular filtration rate (GFR) and filtration coefficient.

   c. If the pressure in the glomerular capillaries is increased to 65 mmHg what would the GFR be?

3. A subject was given glucose infusions and the concentration of glucose in the plasma and urine was measured together with urine flow rate. Calculate the clearance of glucose in each case and interpret the results of your calculations. Creatinine clearance was unchanged throughout at 120 ml/min.

Plasma glucose (mg/ml)	0.8	1.5	3	6	9
Urine glucose (mg/ml)	0	0	8	60	67
Urine flow rate (ml/min)	2.5	2.0	3.5	6.0	10.7

Answers to multiple choice questions

1. Urine pH is normally much lower than that of plasma and contains no bicarbonate. If plasma pH is elevated (alkalemia), the kidneys will excrete bicarbonate so that urine pH rises.

   a. True

   b. True

   c. False

   d. False

   e. True

2. As cardiac output rises in exercise, renal blood flow does not increase. It usually falls due to vasoconstrictor activity in the sympathetic nerves supplying the smooth muscle of the afferent arterioles.

   a. True

   b. True

   c. False

   d. True

   e. False

3. The concentration of small molecules in the glomerular filtrate is the same as in plasma but the filtrate contains very little protein (unlike plasma). A clearance less than that for inulin can be due to either a barrier to filtration (e.g. plasma proteins) or reabsorption by the renal tubules (e.g. glucose).

   a. False

   b. True

   c. False

   d. True

   e. True

4. Although most small molecules and ions are absorbed via the transcellular route (and therefore utilize carrier proteins or ion channels), some are absorbed via the paracellular route

by solvent drag or as a result of charge differences between the lumen and the interstitial space.

a. False

b. True

c. True

d. True

e. True

5. The GFR is closely matched to the transport capacity of the tubules by glomerulotubular feedback. Tubular bicarbonate is converted to carbon dioxide and water by the secretion of hydrogen ions by an Na^+–H^+ antiporter in the proximal tubule and an H^+ ion ATPase in the distal tubule. It then enters the tubular cells as CO_2 and water where it is reconverted to bicarbonate. It passes from the cells to the blood in exchange for chloride ions. While most of the filtered bicarbonate is reabsorbed in the proximal tubule, the fluid reaching the distal tubule contains about 2 mmol l^{-1}.

a. True

b. True

c. False

d. False

e. True

6. Superficial cortical nephrons do not have long loops of Henle. Only about 15 per cent of all nephrons extend deep into the medulla. These are the juxtamedullary nephrons. The thick ascending limb of the loop of Henle transports sodium, potassium, and chloride from the tubular fluid by a specific electroneutral carrier. In the proximal tubule, most sodium is reabsorbed with small organic molecules or bicarbonate.

a. False

b. True

c. True

d. False

e. True

7. Sodium balance is regulated by the activity of the cells of the distal tubules. The sodium uptake by the proximal tubules is obligatory and linked to the absorption of small organic molecules, chloride and bicarbonate. Renin is an enzyme that is secreted by the granulosa cells of the juxtaglomerular apparatus in response to low sodium in the fluid reaching the distal tubule. It cleaves angiotensin I from a precursor molecule called angiotensinogen which is always present in the plasma. Angiotensin II acts on the zona glomerulosa of the adrenal cortex, which responds by secreting aldosterone.

a. False

b. False

c. False

d. False

e. True

8. Plasma potassium is regulated by the activity of the cells of the distal tubule. Calcium absorption by the cells of the distal tubule is increased by parathyroid hormone. This hormone acts on the cells of the proximal tubule to regulate phosphate uptake.

a. False

b. False

c. True

d. True

e. True

9. ADH is secreted by the posterior pituitary in response to an increase in plasma osmolality.

a. True

b. True

c. True

d. False

e. True

Answers to quantitative problems

1. The curve should show a sharp increase in blood flow at the point where the perfusion pressure increases followed by a region in which blood flow returns to its initial value. This demonstrates the active nature of autoregulation. A similar response would be seen in a denervated kidney.

2. a. 10 mmHg

   b. 125 ml min^{-1} (GFR); 12.5 ml min^{-1} mm Hg^{-1} (filtration coefficient)

   c. 187.5 ml min^{-1}

3. The initial set of values reveal that the urine flow rate increases when glucose is lost in the urine (osmotic diuresis). Glucose appears in the urine when plasma glucose reaches a concentration of 3 mg ml^{-1} (16.6 mmol l^{-1}). This is the renal threshold for glucose. As plasma glucose rises to 6.0 and then 9.0 mg ml^{-1}, the amount of glucose reabsorbed remains close to 360 mg min^{-1} which is therefore the transport maximum for the renal transport of glucose.

Plasma glucose (mg ml^{-1})	Glucose clearance (ml min^{-1})	Filtered load (mg min^{-1})	Glucose lost in urine (mg min^{-1})	Glucose reabsorbed (mg min^{-1})
0.8	0	96	0	96
1.5	0	180	0	180
3.0	9.33	360	28	332
6.0	60	720	360	360
9.0	79.6	1080	717	363

CHAPTER 18

The digestive system

After reading this chapter you should understand:

- The basic anatomical organization of the gastrointestinal tract
- The nervous and hormonal regulatory mechanisms operating within the gut
- Salivary secretion and the functions of saliva
- The functions of the stomach—gastric secretion, motility, and its control by nerves and hormones
- Why the gut does not digest itself
- The control of secretion and motility in the small intestine
- The role of the exocrine part of the pancreas
- The digestive functions of the liver and gall bladder
- The portal circulation
- The absorption of nutrients and disorders of absorption
- The role of the large intestine in the absorption of water and electrolytes and the importance of the intestinal flora
- The control of defecation

18.1 Introduction

Food is required by the body for the production of energy and for the growth and repair of tissues. Each day an average adult consumes around 1 kg of solid food and 1–2 liters of fluid. As the majority of this material is in a form that cannot be used immediately by the body for cellular metabolism, it must be broken down into simple molecules which can be absorbed into the bloodstream for distribution to the tissues. The digestive or gastrointestinal (GI) system performs this task. Specifically, the major functions of the digestive system are:

(1) the ingestion of food;
(2) transport of food through the GI tract at a rate that allows optimal digestion and absorption;
(3) the secretion of fluid, salts, and digestive enzymes;
(4) digestion;
(5) absorption of the products of digestion;
(6) removal of indigestible remains from the body (defecation).

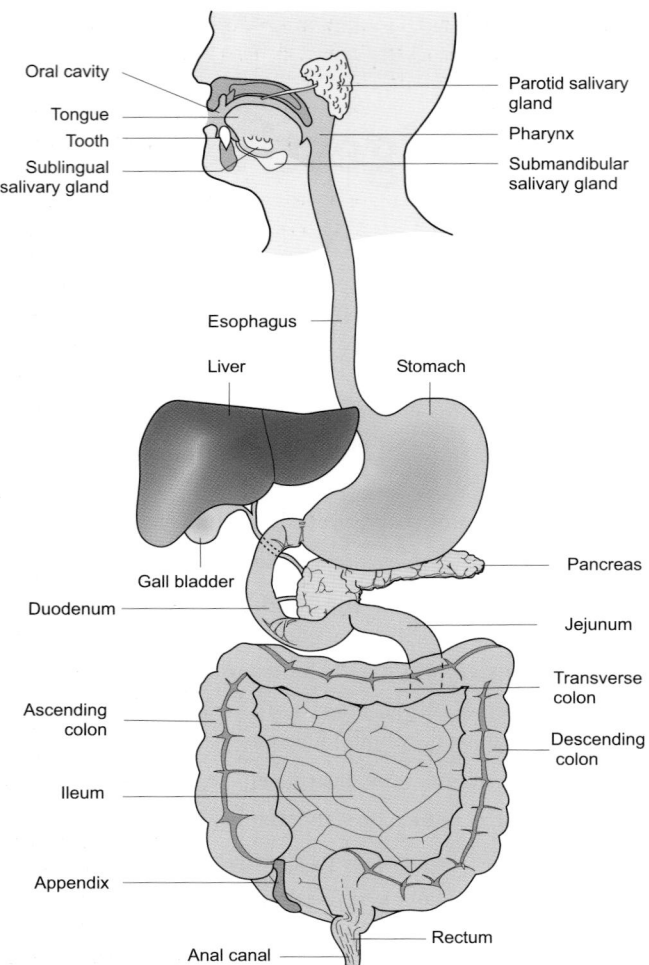

Fig. 18.1 The gastrointestinal tract and its principal accessory organs.

18.2 The general organization of the gastrointestinal system

The major anatomical components of the GI tract and its accessory organs are illustrated in Fig. 18.1. Although the tract is located within the body, it is, in reality, a hollow convoluted tube that is open at both ends (mouth and anus). Therefore the lumen of the tube is an extension of the external environment. The GI tract includes the oral cavity, pharynx, esophagus, stomach, small intestine (duodenum, jejunum, and ileum), large intestine (ascending, transverse, and descending colon), rectum, and anal canal. In life, the GI tract is about 5.5 m (18 ft) long, although its length increases after death as smooth muscle tone is lost. The accessory organs are the teeth, tongue, salivary glands, pancreas, liver, and gall bladder. Each organ or part of the tract is adapted to carry out specific functions.

Histological features of the wall of the gastrointestinal tract

Although the detailed structure of the GI tract varies throughout its length according to the particular functions of each region, there are common features in the overall organization of the tissues of the gut wall. Figure 18.2 illustrates the basic plan of the layers of the gut wall. From the outside inwards these layers are the serosa, a layer of longitudinal smooth muscle, a layer of circular smooth muscle, the submucosa, and the mucosa. A layer of smooth muscle fibers, the muscularis mucosa, lies deep within the mucosa.

The *serosa* (also called the adventitia) is the covering of the GI tract. It forms the visceral peritoneum and is continuous with the parietal peritoneum, which lines the body cavity. It is a binding and protective layer that consists mainly of loose connective tissue covered by a layer of simple squamous epithelium.

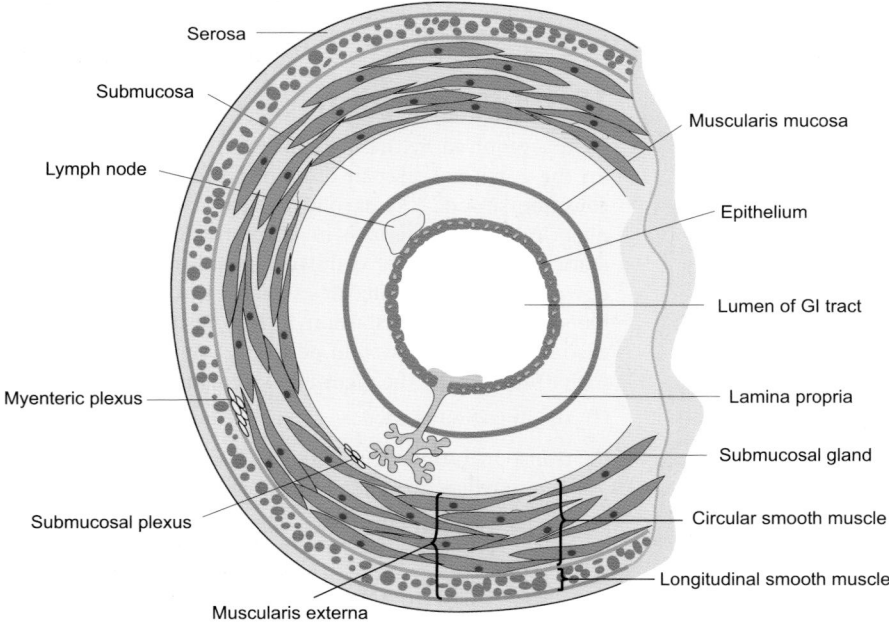

Fig. 18.2 Transverse section through a portion of small intestine to illustrate the general organization of the gut wall. Although this figure shows the basic organization, there are individual variations in structure in different regions of the gut.

The smooth muscle is arranged in two distinct layers, one of which runs longitudinally while the other is circularly arranged. The smooth muscle layers are together known as the *muscularis externa*. Contraction of these muscle layers mixes the food with digestive enzymes and propels it along the GI tract. The circular layer is three to five times thicker than the longitudinal layer. At intervals throughout the GI tract, the circular smooth muscle layer is thickened and modified to form a ring of tissue called a sphincter. Sphincters control the rate of movement of the GI contents from one part of the gut to another. They are present, for example, at the junctions between the esophagus and the stomach, the stomach and the small intestine, and the ileum and the cecum. The internal and external anal sphincters control the elimination of feces.

The *submucosa* is made up of loose connective tissue with collagen and elastin fibrils, blood vessels, lymphatics, and, in some regions, submucosal glands. The innermost layer of the gut wall, the *mucosa*, is further subdivided into three regions: a layer of epithelial cells, their basement membrane, the *lamina propria*, and the *muscularis mucosa*, which consists of two thin layers of smooth muscle—an inner circular layer and an outer longitudinal layer. The characteristics of the epithelium vary greatly from one region of the GI tract to another. For example, it is smooth in the esophagus but is thrown into finger-like projections called villi in the small intestine.

The nature of the smooth muscle of the gastrointestinal tract

With the exception of the mouth and tongue, the motor functions of the GI tract are performed almost entirely by the action of smooth muscle. Motor activity is responsible for the mixing of food with digestive juices and the propulsion of food along the GI tract at a rate that allows the optimal digestion of food and absorption of the digestion products. The general anatomical and electrophysiological characteristics of smooth muscle are discussed fully in Chapter 7. Only the specific characteristics of the smooth muscle of the GI tract will be considered here.

The smooth muscle in the gut is of the *single-unit* or *visceral* type. It operates as a functional syncytium whereby electrical signals originating in one fiber are propagated to neighboring fibers so that sections of smooth muscle contract synchronously. The smooth muscle fibers maintain a level of tone that determines the length and diameter of the GI tract. Various types of contractile responses are superimposed upon this basal tone. The most important of these are segmentation and peristaltic contraction.

Segmentation, which occurs principally in the small intestine, facilitates the mixing of food with digestive enzymes and exposes the digestion products to the absorptive surfaces of the GI tract. It is characterized by closely spaced contractions of the circular smooth muscle layer separated by short regions of relaxation as shown in Fig. 18.3(a). Segmentation occurs at a rate of around 12 contractions per minute in the duodenum and increases in both frequency and strength when chyme enters the small intestine. Segmental contractions are less frequent in the jejunum and ileum where they seem to occur in bursts, in which there may be around eight contractions per minute, interspersed with periods of inactivity.

Peristaltic contractions are concerned mainly with the propulsion of food along the tract and consist of successive waves of contraction and relaxation of the smooth muscle as shown in Fig. 18.3(b). The longitudinal smooth muscle contracts first, and halfway through its contraction the circular muscle begins to contract. The longitudinal muscle relaxes during the latter half of the circular muscle contraction. This pattern of contraction is repeated, resulting in the slow but progressive movement of material along the GI tract. The normal trigger for a wave of peristalsis is local distension by food bulk.

The contractile properties of smooth muscle are determined by the underlying electrical activity of its cells. The resting membrane potential of the cells shows spontaneous rhythmical fluctuations (the *basic electrical rhythm* or *slow-wave rhythm*). The frequency of these fluctuations varies along the length of the GI tract, generally decreasing with distance from the mouth. For example, in the duodenum there are 11 or 12 slow waves per minute but in the colon there are only three or four. These differences in rate result in different degrees of contraction along the GI tract. In consequence, a gradient of pressure is created which contributes to the steady movement of the gut contents towards the ileocecal sphincter.

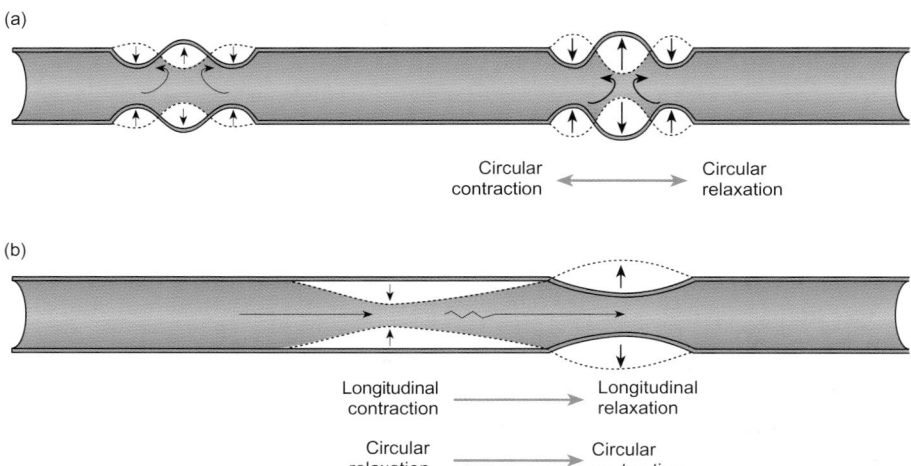

(a)

Circular contraction ⟷ Circular relaxation

(b)

Longitudinal contraction → Longitudinal relaxation

Circular relaxation → Circular contraction

Fig. 18.3 Principal patterns of contractile activity in the smooth muscle of the GI tract: (a) segmentation movements; (b) peristaltic contraction.

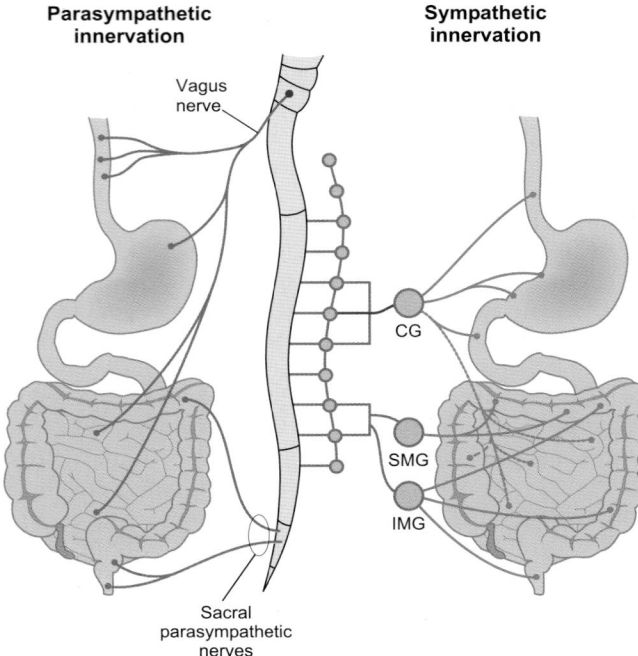

Parasympathetic innervation

Vagus nerve

Sympathetic innervation

CG

SMG

IMG

Sacral parasympathetic nerves

Fig. 18.4 The innervation of the gut. CG celiac ganglion, SMG, IMG superior and inferior mesenteric ganglia

Although the smooth muscle layers of the gut can contract in the absence of action potentials, the depolarizing phases of the slow waves are sometimes accompanied by bursts of action potentials. Such bursts are associated with vigorous propulsive movements such as those seen in the antral region of the stomach.

Innervation of the gastrointestinal tract

The complex afferent and efferent innervation of the GI tract allows for fine control of secretory and motor activity via intrinsic (enteric) and extrinsic (sympathetic and parasympathetic) pathways which show a considerable degree of interaction. Figure 18.4 illustrates the extrinsic innervation of the gastrointestinal tract

The enteric nervous system

Two well-defined networks of nerve fibers and ganglion cell bodies are found within the wall of the GI tract from the esophagus to the anus. The networks are called intramural plexuses and constitute the *enteric nervous system* (ENS). The *myenteric plexus* (also known as Auerbach's plexus) lies between the circular and longitudinal smooth muscle layers of the muscularis externa, while the less extensive *submucosal plexus* (or Meissner's plexus) lies within the submucosa (Fig. 18.2). The myenteric plexus is largely motor in function and regulates motility of the GI tract. In Hirschsprung's disease, an inherited developmental disorder of the colon, ganglion cells are missing from part of the myenteric plexus. The condition is characterized by severe constipation caused by the resulting reduction in gut motility. The submucosal plexus is chiefly concerned with controlling secretory activity and blood flow to the gut. It also receives signals from the intestinal epithelium and from stretch receptors in the gut wall.

The organization of the ENS is rather complex, and for this reason it is sometimes referred to as the 'gut brain'. The neuronal cell bodies within each plexus are organized into ganglia and the nerves that form the networks (intrinsic nerves) synapse with the ganglion cells to connect the two plexuses. They also innervate the blood vessels, glands, and smooth muscle of the GI tract. The ENS is responsible for coordinating much of the secretory activity and motility of the GI tract through intrinsic pathways often called 'short-loop' reflexes. These loops utilize many neurotransmitters and neuromodulators including cholecystokinin (CCK), substance P, vasoactive intestinal polypeptide (VIP), somatostatin, and the enkephalins. Furthermore, some intrinsic neurons also synapse with autonomic and sensory neurons of the extrinsic nerve supply (see below), thereby allowing interaction between the intrinsic and extrinsic innervations.

Extrinsic innervation of the gastrointestinal tract

Although much of the activity of the GI tract is controlled through the intrinsic nerves of the ENS, the nerve plexuses are themselves linked to the central nervous system (CNS) via afferent fibers and receive efferent input from the autonomic nervous system.

Afferent innervation of the GI tract

Chemoreceptor and mechanoreceptor endings are present in the mucosa and the muscularis externa. Chemoreceptors may be stimulated by substances present within foods or by products of digestion, while mechanoreceptors respond to distension or to irritation of the gut wall. Some of these receptors send afferent axons back to the CNS to mediate reflexes via the CNS ('long-loop' reflexes), while the axons of others synapse with cells within the plexuses to mediate local reflexes. Many of the extrinsic sensory afferents travel in the vagus nerve (vagal afferents) and have their cell bodies in the nodose ganglion of the brainstem. Some, particularly those forming part of the reflex arcs that control motility, travel to the spinal cord via the sympathetic nerves ('visceral afferents'). The cell bodies of these fibers lie within the dorsal root ganglia. It is important to realize that at least 80 per cent of fibers in the vagus and up to 70 per cent of splanchnic nerve fibers are afferent. Indeed, in vagovagal reflexes, both the afferent and efferent fibers travel in the vagus nerve. Such reflexes play a significant role in the control of motility and secretion in the GI tract.

Efferent innervation of the gastrointestinal tract

Sympathetic innervation   The preganglionic fibers of the gastrointestinal sympathetic innervation arise from segments T8–L2 in the thoracic region of the spinal cord. The cell bodies of the postganglionic fibers lie within the celiac, superior, and inferior mesenteric and hypogastric plexuses (see Chapter 10). Some of the sympathetic fibers innervate the smooth muscle of arterioles within the GI tract, causing vasoconstriction, constriction of precapillary sphincters (see Chapter 15), and redirection of blood away from the splanchnic bed, while others enter glandular tissue and innervate secretory cells. These responses are mediated by α-adrenoceptors. In addition, a large number of sympathetic fibers terminate within the submucosal and myenteric plexuses where they appear to inhibit synaptic transmission, possibly by presynaptic inhibition. There is also some innervation of the circular smooth muscle layers of the small and large intestine by sympathetic fibers. In general, increased sympathetic discharge reduces GI activity. Indeed, powerful stimulation of the sympathetic innervation of the gut can produce almost complete inhibition of motility. In contrast, the sphincters of the GI tract are supplied with adrenergic fibers whose actions are usually excitatory and serve to constrict the circular smooth muscle of these regions.

Parasympathetic innervation   Parasympathetic input to the gut stimulates both its motility and secretory activity and thus opposes the actions of sympathetic stimulation. The vagus nerve relays the parasympathetic innervation to the esophagus, stomach, small intestine, liver, pancreas, cecum, appendix, ascending colon, and transverse colon. The remainder of the colon receives parasympathetic innervation from pelvic nerves via the hypogastric plexus. All the parasympathetic fibers terminate within the myenteric plexus and are predominantly cholinergic. Further details of the autonomic innervation of the gut are given in Chapter 10.

Hormonal regulation of the gastrointestinal tract

In addition to its extensive innervation, the gastrointestinal tract is regulated by a number of peptide hormones that act through endocrine and/or paracrine pathways (see Chapter 5). The hormone-secreting cells are scattered throughout the mucosa and are known by the acronym APUD cells, which stands for **a**mine **p**recursor **u**ptake and **d**ecarboxylation and relates to the role of these cells in hormone synthesis. The GI tract utilizes at least 20 different regulatory peptides. Eight polypeptides that are known to act as circulating (endocrine) hormones are listed below together with the region of the gut from which they are secreted:

- gastrin (antrum of the stomach)
- secretin (duodenum)
- cholecystokinin (CCK) (duodenum)
- pancreatic polypeptide (pancreas)
- gastric inhibitory polypeptide (GIP) (jejunum and duodenum)
- motilin (jejunum and duodenum)
- glucagon-like peptides GLP-1 and GLP-2 (ileum and colon)
- neurotensin (the lower small intestine).

GLP-1 and GLP-2 were formerly known as enteroglucagons. CCK and neurotensin also exert a neurocrine action, producing their effects close to their site of secretion by nerve fibers. Paracrine agents acting on the gut include somatostatin and histamine. In general, the effects of these agents on motility and secretory activity supplement those of the GI innervation, although the relative importance of nervous and hormonal influences differs throughout the tract. In the salivary glands, for example, nervous control is the dominant influence on secretion, while in other areas, such as the stomach, nervous and endocrine influences are of equal importance. Hormones are the principal regulators of secretion from the exocrine pancreas.

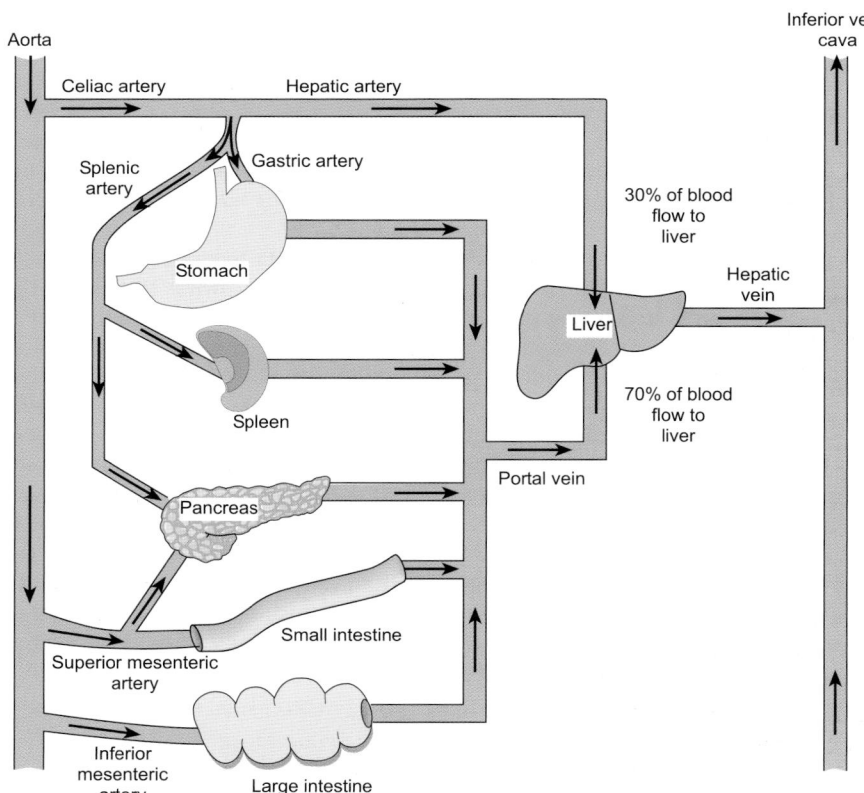

Fig. 18.5 A diagrammatic representation of the blood supply to the liver and GI tract.

General characteristics of the blood flow to the gastrointestinal tract

The various digestive functions of the gut require a rich and highly organized blood supply. The combined circulation to the stomach, liver, pancreas, intestine, and spleen (which has no digestive function) is called the *splanchnic circulation*. Throughout the postabsorptive phase the splanchnic vessels receive 20–25 per cent of the cardiac output but, during the digestion and absorption of food, blood flow to the gut increases considerably. This increase is partly mediated by the secretion of gastrin and CCK although certain products of digestion, principally fatty acids and glucose, also act as powerful vasodilators.

The splanchnic circulation is derived from the celiac artery and the superior and inferior mesenteric arteries. The celiac artery branches from the aorta immediately after it enters the abdomen and the gastric and splenic arteries arise from this vessel. The gastric arteries supply the stomach while branches of the splenic artery supply the capillary beds of the spleen and pancreas.

The intestine is supplied by the superior and inferior mesenteric arteries, small arterial branches of which form a vascular network in the submucosal layer that penetrates the longitudinal and circular smooth muscle. Venous blood leaving the stomach, pancreas, intestine and spleen flows ultimately into the portal vein, which provides about 70 per cent of the blood supply to the liver. The remainder of the hepatic blood supply is provided by the hepatic artery which supplies most of the oxygen required by the liver. The chief purpose of the portal circulation is to allow rapid delivery of the products of digestion from the intestine to the liver where they will undergo further processing. The liver is drained by the hepatic vein, which enters the inferior vena cava. The general organization of the splanchnic and portal circulations is illustrated in Fig. 18.5

18.3 Intake of food, chewing, and salivary secretion

Food is ingested via the mouth (also called the oral or buccal cavity), the only part of the GI tract that has a bony skeleton. In the mouth, the food is broken into smaller pieces by the process of chewing (mastication) and mixed with saliva, which softens and lubricates the food mass. As food moves around the mouth, the taste buds are stimulated and odors are released from the food. The presence of food in the mouth and the sensory stimuli of taste and smell play a part in the stimulation of gastric secretion (see p. 391).

The mouth is divided into two regions, the *vestibule* and the *oral cavity*. The vestibule is the region between the teeth, lips, and cheeks. The oral cavity is the inner area bounded by the teeth. The epithelium here is typically 15–20 layers of cells thick and is structurally adapted to withstand the frictional forces generated during mastication.

The teeth

A child has 20 deciduous (milk) teeth which normally erupt during the first 2–3 years of life. This set has no premolars. Between the ages of about 6 and 15 these teeth are gradually shed and replaced by 32 permanent (adult) teeth. The teeth are embedded in the alveoli or sockets of the alveolar ridges of the mandible (lower jaw) and maxilla (upper jaw). In adults, the upper and lower jaws each possess four incisors, two canines, four premolars and six molars. The incisors and canines are the cutting teeth, while the premolars and molars have broad flat surfaces for grinding or chewing food.

Although the shapes of the different teeth vary, they share a similar basic structure (Fig. 18.6). The crown is the part of the tooth that protrudes from the gum (gingiva) while the root is embedded in the bone of the mandible or maxilla. In the center of the tooth is the pulp cavity, which contains blood vessels, lymphatics, and nerves, and surrounding this is the hard dentin. Outside the dentine of the crown is a layer of even harder translucent enamel. The root of the tooth is surrounded by softer cement which fixes the tooth into its socket. The periodontal ligament holds the tooth in place, while still allowing a little movement during chewing. Blood vessels and nerves pass to the tooth through a small opening at the tip of each root. The nerve supply to the upper teeth is by branches of the maxillary nerves and to the lower teeth by branches of the mandibular nerves, both of which arise from the trigeminal nerve (cranial nerve V).

The tongue

The tongue, which is formed from skeletal (voluntary) muscle, is inserted into the hyoid bone and attached to the anterior floor of

Summary

1. The GI tract consists of the oral cavity, pharynx, esophagus, stomach, small intestine, large intestine, and anal canal. Accessory organs include the teeth, tongue, salivary glands, exocrine pancreas, liver, and gall bladder.

2. The major functions of the GI tract include the ingestion of food and its transport along the tract, the secretion of fluids, salts and digestive enzymes, the absorption of digestion products, and the elimination of indigestible remains.

3. Certain structural features are common to all regions of the gut. The basic layers are (from the outside) the serosa, a layer of longitudinal smooth muscle, a layer of circular smooth muscle, the submucosa, and the mucosa. Contractile activity of the smooth muscle promotes mixing and propulsion of food.

4. The GI tract is regulated by extrinsic nerves and by the enteric nervous system, a system of intramural plexuses which mediate a number of intrinsic reflexes that control secretory and contractile activity. Afferent and efferent extrinsic nerves, endocrine hormones, and paracrine hormones also play an important role in regulating the activities of the GI tract.

5. The gut receives a rich blood supply. The combined circulation to the stomach, liver, pancreas, intestine, and spleen is called the splanchnic circulation, and accounts for 20–25 per cent of the cardiac output at rest.

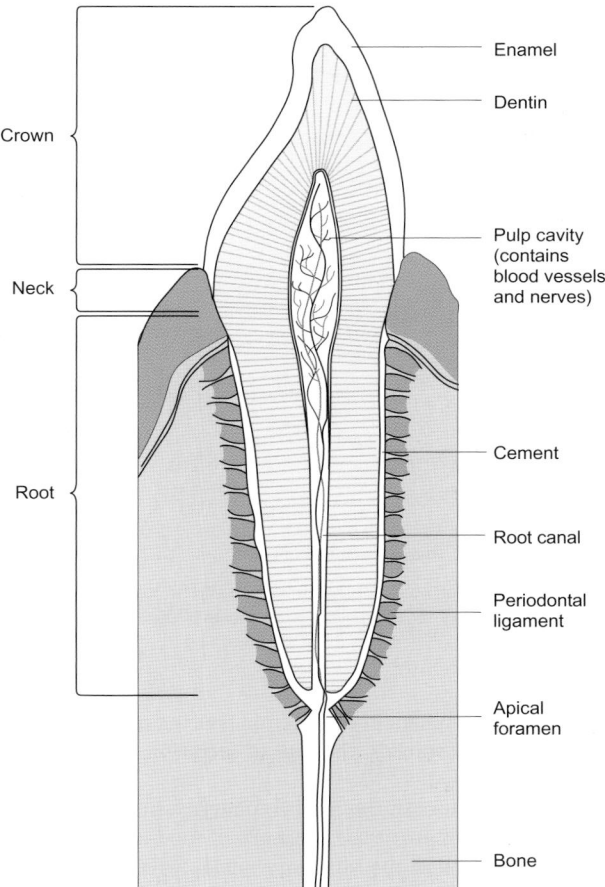

Crown

Neck

Root

Enamel

Dentin

Pulp cavity
(contains
blood vessels
and nerves)

Cement

Root canal

Periodontal
ligament

Apical
foramen

Bone

Fig. 18.6 A longitudinal section through a canine tooth to illustrate the major structural features.

the mouth, behind the lower incisor teeth, by a fold of its mucous membrane covering called the frenulum. The role of the tongue in the perception of taste has been described in Chapter 8. It also plays an important role in speech and is necessary for swallowing.

Mastication

A crushing force of 50–80 kg can be generated on the molars during chewing, a value which far exceeds the forces generally required for a normal diet. Both the tongue and the cheeks also have important roles to play in the process of chewing. Their movements help to keep the food in the correct position for effective chewing while the sensory receptors on the tongue provide information regarding the readiness of the food for swallowing.

The secretion of saliva

Approximately 1500 ml of saliva is produced each day by the salivary glands. It performs several important functions. It lubricates the food to facilitate swallowing, it aids in speech, and it contains an enzyme, salivary α-amylase (ptyalin), that begins the process of starch digestion. The saliva dissolves certain substances in foods making them available to the taste cells. Saliva also contains IgA and an enzyme called lysozyme which act on the cell

walls of certain bacteria to cause lysis and death. The bacteriostatic actions of saliva contribute significantly to oral comfort and reduce the risk of infection developing in the mouth after oral or dental surgery. In individuals who lack functional salivary glands or in whom salivary secretion has been inhibited by medication or irradiation, there is a predisposition to dental caries and infections of the buccal mucosa. This condition is known as *xerostomia*.

There are three main pairs of salivary glands; the parotid, submandibular, and sublingual glands. Other smaller glands exist over the surface of the palate and tongue and inside the lips, though these do not appear to be under nervous control. Each gland consists of a number of lobules surrounded by a fibrous capsule. Each lobule or acinus is made up of balls of cells (acinar cells). The acini are drained by ductules that join to form larger ducts leading into the mouth.

All the major salivary glands receive both sympathetic and parasympathetic innervation. Noradrenergic sympathetic fibers from the superior cervical ganglion are distributed to both blood vessels and acinar cells. Preganglionic parasympathetic fibers arrive by way of the facial and glossopharyngeal nerves and synapse with postganglionic neurons close to the salivary glands themselves. Both the secretory cells and the duct cells receive parasympathetic postganglionic fibers.

The *parotid glands* are situated at the angle of the jaw, lying posterior to the mandible and inferior to the ear (see Fig. 18.1). The main ducts from the parotid glands open into the mouth opposite the second molars on each side. They are the largest of the salivary glands and produce an entirely serous secretion, a watery fluid lacking mucus. Saliva from the parotid glands accounts for around 25 per cent of the total output of the salivary glands. It contains α-amylase and a small amount of immunoglobulin A.

The *submandibular glands* lie below the lower jaw (mandible). These glands contain acinar cells that secrete mucoproteins, as well as cells producing serous fluid. Therefore their secretion is more viscid than that of the parotid glands. Overall, the submandibular glands secrete about 70 per cent of the daily output of saliva.

The *sublingual glands* lie in the floor of the mouth below the tongue. They contribute the remaining 5 per cent or so of the total salivary output, producing a secretion that is rich in mucoprotein and gives the saliva its somewhat sticky character.

The mechanism of salivary secretion

The formation of saliva is a two-stage process. An isotonic fluid (primary secretion) is formed by the acinar cells as a result of the active transport of electrolytes followed by the passive transfer of water. Secondary modification of this fluid then occurs by means of ion transport processes occurring in the epithelial cells lining the ducts. The concentrations of sodium, chloride and bicarbonate ions in the primary secretion resemble those of plasma. Modification of the primary secretion involves the active reabsorption of sodium and bicarbonate by the epithelial cells lining the ducts as the saliva flows past them. Figure 18.7 illustrates diagrammatically the ion movements believed to occur during the formation of saliva.

The final electrolyte content of saliva depends upon the rate at which it is secreted and flows along the salivary ducts. At low

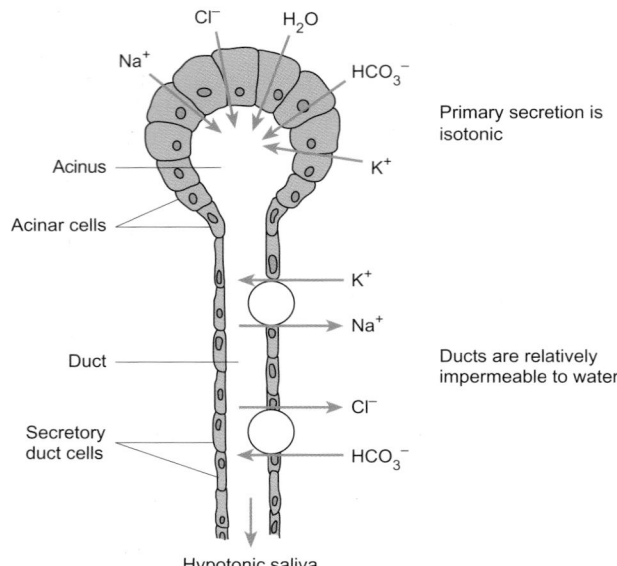

Fig. 18.7 Water and electrolyte transport leading to the formation of saliva by acinar and ductal cells of a salivary gland. This primary secretion is subsequently modified as fluid passes along the ducts.

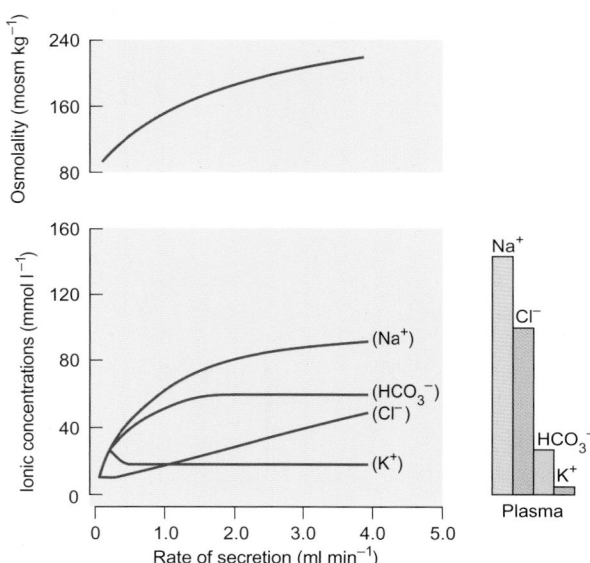

Fig. 18.8 The variation of the ionic composition and osmolality of the saliva with flow rate. The composition of plasma is shown in the bar diagram on the right for comparison.

rates of secretion, there is ample time for ductal modification of the primary secretion. Electrolytes are reabsorbed without an osmotic equivalent of water so that the osmolality of the saliva is relatively low. As the flow rate increases, there is less time for the reabsorption of electrolytes, and the salivary content of sodium, bicarbonate and chloride increases while there is a small drop in the potassium concentration (Fig. 18.8). The pH of saliva rises from 6.2 to 7.4 with increasing rates of secretion because of the increase in bicarbonate concentration.

The regulation of salivary secretion

The rate of salivary secretion is controlled primarily by reflexes mediated by the autonomic nervous system. The resting rate of

salivary secretion is about 0.5 ml/ min. During maximal stimulation by sapid substances, the rate of secretion may increase to 7 ml min⁻¹.

Sensory receptors in the mouth, pharynx, and olfactory area relay information about the presence of food in the mouth, its taste and its smell to the salivatory nuclei, which are located in the medulla. The medulla also receives both facilitatory and inhibitory impulses from the hypothalamic appetite area and regions of the cerebral cortex concerned with the perception of taste and smell. Most salivary responses are mediated by parasympathetic efferent fibers originating in the salivatory nuclei. The reflex pathway responsible for the stimulation of salivary secretion in response to food is illustrated in Fig. 18.9.

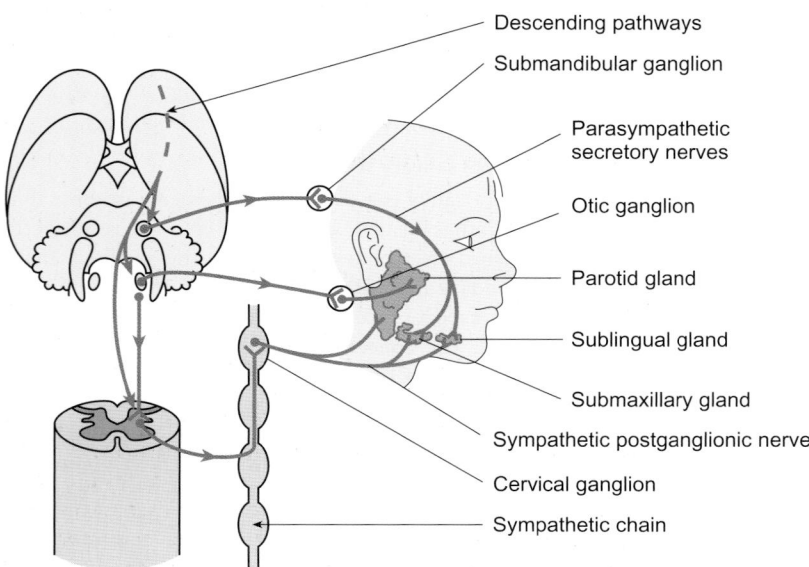

Fig. 18.9 The pathways subserving the salivary reflex.

Parasympathetic stimulation promotes an abundant secretion of watery saliva that is rich in amylase and mucins. The transporting characteristics of the ductal epithelium are also altered so that bicarbonate secretion is stimulated while the reabsorption of sodium and the secretion of potassium are inhibited. These changes are mediated by muscarinic receptors and are inhibited by atropine (see Chapter 10).

The response to parasympathetic stimulation also includes a significant increase in blood flow to the salivary glands. This effect is not atropine sensitive. Several mechanisms are thought to contribute to the increase in blood flow. These include the release of vasoactive intestinal polypeptide (VIP) from parasympathetic nerve terminals and the release of the proteolytic enzyme kallikrein into the interstitial fluid from the acinar cells themselves. Kallikrein, in turn, promotes the production of the powerful vasodilator, bradykinin, from plasma α-2 globulin.

The response of the salivary glands to sympathetic stimulation is variable. Sympathetic fibers stimulate the secretory cells and enhance the output of amylase. At the same time, however, blood flow to the glands is usually reduced through vasoconstriction and the net result is a fall in the rate of salivary secretion. Indeed, a dry mouth is a characteristic feature of the sympathetic response to fear and stress. A summary of the mechanisms involved in the control of salivary secretion is shown in Fig. 18.10.

Digestive actions of saliva

The digestive enzyme salivary amylase is stored within zymogen granules in the serous acinar cells. It is able to degrade complex polysaccharides, such as starch and glycogen, to maltose, maltotriose, and dextrins, working optimally at pH 6.9. Although food remains in the mouth for only a short time and the contents of the stomach are highly acidic, salivary amylase is believed to

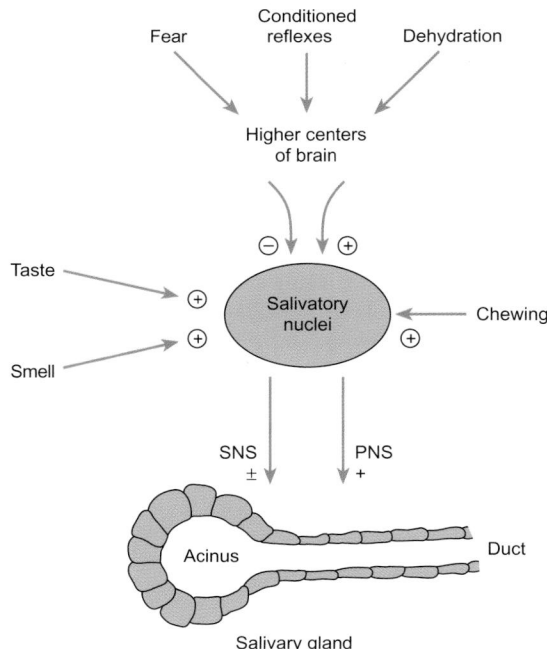

Fig. 18.10 A summary of the major factors which influence the secretion of saliva: PNS, parasympathetic nervous system; SNS, sympathetic nervous system.

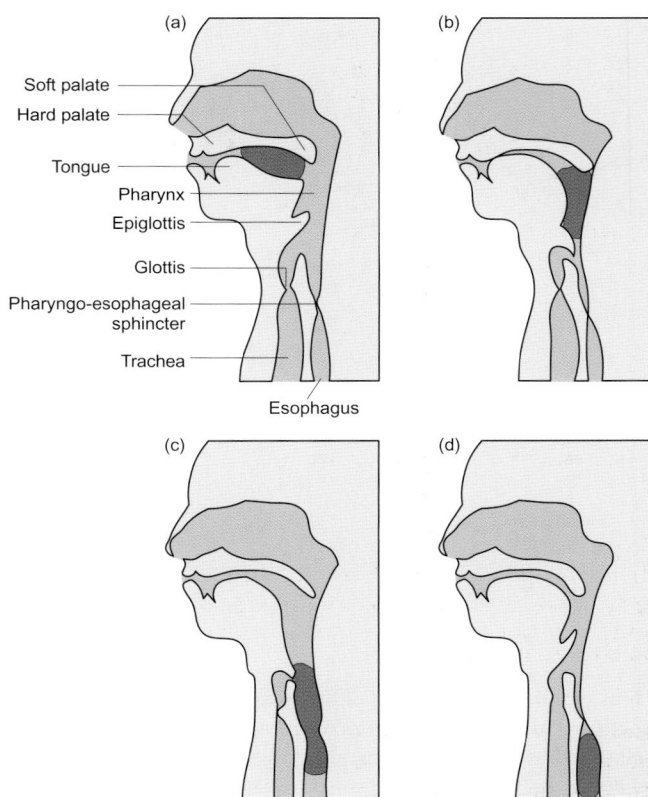

Fig. 18.11 The events occurring during swallowing as a food bolus (shown in red) moves from the mouth to the esophagus.

continue working within the food bolus for some time after the entry of food into the stomach. It is probably inactivated only after complete mixing of the bolus contents with the gastric juice.

Swallowing (deglutition)

Once mastication is complete and the lubricated food bolus has been formed, it is swallowed. From the mouth, food passes posteriorly into the oropharynx and then into the laryngopharynx, both of which are common passageways for food, fluids and air.

The first (oral) phase of swallowing is voluntary, but the subsequent pharyngeal and esophageal stages of the process are under reflex autonomic (involuntary) control. Neurons in the medulla and lower pons mediate the involuntary phase of swallowing. During the voluntary oral phase, the tip of the tongue is placed against the hard palate and the tongue is then contracted to force the food bolus into the oropharynx (the part of the pharynx lying immediately behind the mouth). The pharynx is richly endowed with mechanoreceptors. When food stimulates these receptors, a complex sequence of events is initiated which results in completion of the swallowing process. Information from the mechanoreceptors passes via afferents in the glossopharyngeal nerve (cranial nerve IX) to sensory nuclei within the tractus solitarius. It also spreads via interneurons to the motor nucleus ambiguus in the medulla. Motor impulses travel from the brain to the muscles of the pharynx, palate, and upper esophagus via cranial nerves including the glossopharyngeal and

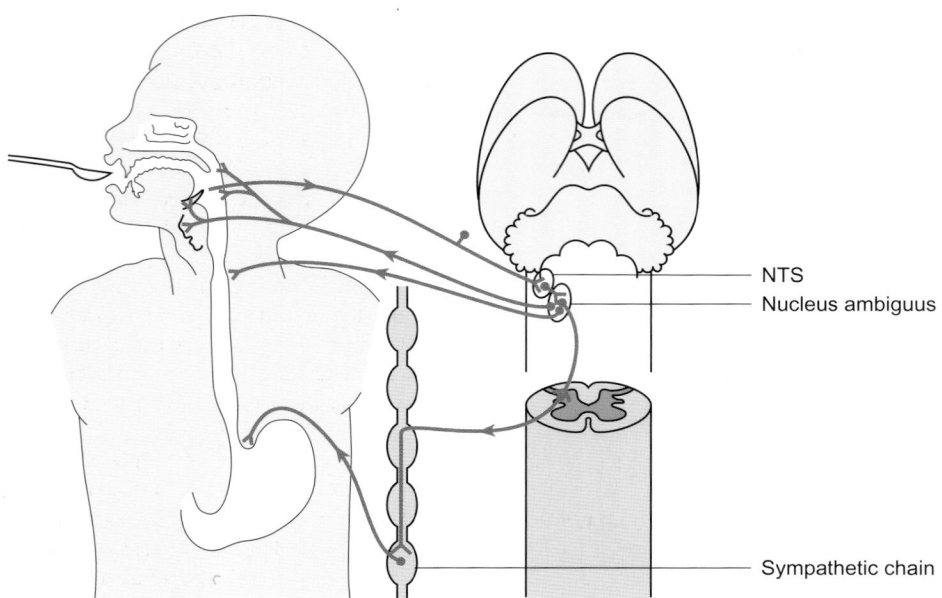

NTS
Nucleus ambiguus

Sympathetic chain

Fig. 18.12 The pathways subserving the swallowing reflex. NTS, nucleus of the tractus solitarius.

vagus (cranial nerve X). The efferent response begins with contraction of the superior constrictor muscle, which raises the soft palate towards the posterior pharyngeal wall to prevent food entering the nasopharynx. It also initiates a wave of peristaltic contraction that propels the bolus through the relaxed upper esophageal sphincter into the esophagus. The larynx rises so that the epiglottis covers the opening of the nasopharynx into the trachea while the opening of the esophagus is stretched. At the same time, respiration is inhibited (*deglutition apnea*). In this way, food is prevented from entering the trachea.

Lesions of the area of the brain that control swallowing or of the glossopharyngeal and vagal nerves carrying the efferent impulses result in difficulty in swallowing (*dysphagia*). The major components of the swallowing reflex are illustrated in Fig. 18.11.

In the final (or esophageal) phase of swallowing, the wave of peristaltic contraction that was initiated in the pharynx continues along the length of the esophagus. The wave lasts for 7–10 seconds and is usually sufficient to propel the bolus to the stomach. If it fails to do so, the resulting distension of the esophagus initiates a vago-vagal reflex which triggers a secondary peristaltic wave. The esophageal submucosa contains glands that secrete mucus in response to pressure from a food bolus. This mucus helps to lubricate the esophagus and facilitates the transport of food. The reflex arc that brings about swallowing is illustrated in Fig. 18.12.

The time taken for food to travel down the esophagus depends on the consistency of the food and the position of the body. Liquids take only 1–2 seconds to reach the stomach, but solid foods can take much longer. The process is normally aided by gravity. At the junction between the esophagus and the stomach, there is a slight thickening of the circular smooth muscle of the GI tract called the lower esophageal sphincter (also called the cardiac or gastro-esophageal sphincter). This acts as a valve and remains closed when food is not being swallowed, preventing regurgitation of food and gastric juices. Just before the peri-

staltic wave (and the food) reaches the end of the esophagus, the lower esophageal sphincter relaxes to permit the entry of the bolus into the stomach.

Summary

1. In the mouth, food is mixed with saliva as it is chewed. Three pairs of salivary glands (parotid, submandibular, and sublingual) secrete about 1500 ml of saliva each day. This contains mucus, which helps to lubricate the food and an α-amylase that initiates the breakdown of carbohydrate.

2. Isotonic fluid is formed by the acinar cells of the salivary glands by the secretion of electrolytes and water. This undergoes modification as it flows along the salivary ducts so that the final composition of saliva depends upon its flow rate.

3. Salivary secretion is controlled by reflexes mediated by the autonomic nervous system. Parasympathetic stimulation promotes an abundant secretion of watery saliva rich in amylase and mucus. Blood flow to the salivary glands is also enhanced. Sympathetic stimulation promotes the output of amylase but reduces blood flow to the glands. The overall effect of sympathetic stimulation is generally a reduction in the rate of salivary secretion.

4. Swallowing occurs in three phases. The first oral phase is voluntary but subsequent phases are under reflex autonomic control. As swallowing occurs, a wave of peristalsis is initiated which propels the food bolus through the upper esophageal sphincter into the esophagus. This wave of contraction continues for several seconds and moves the bolus down to the lower esophageal sphincter, which relaxes to permit the entry of food into the stomach.

Gastro-esophageal reflux of acidic stomach contents can occur under certain conditions. If the lower esophageal sphincter remains relaxed for long periods or if the abdominal pressure exceeds that in the thorax, acidic stomach contents may enter the esophagus (reflux). Prolonged or excessive reflux may cause inflammation of the esophagus, resulting in burning retrosternal pain in the region between the epigastrium and the throat (heartburn). Symptoms are often exacerbated by pregnancy, stooping, lying down, or following a heavy meal, all situations in which intra-abdominal pressure is raised. Many older people with gastro-esophageal reflux are found to have a **hiatus hernia**, a condition in which the proximal region of the stomach herniates through the opening in the diaphragm through which the esophagus passes (the esophageal hiatus) into the thoracic cavity.

18.4 The stomach

Below the esophagus, the GI tract expands to form the stomach which lies in the left side of the upper abdominal cavity. The functions of the stomach include:

- the temporary storage of food
- mechanical breakdown of food into small particles
- chemical digestion of proteins to polypeptides by pepsins
- regulated passage of processed food (chyme) into the small intestine
- the secretion of intrinsic factor, which is essential for the absorption of vitamin B_{12}.

The stomach is continuous with the esophagus at the lower esophageal (cardiac) sphincter and with the duodenum at the pyloric sphincter. Its position in relation to other abdominal organs is shown in Fig. 18.1. It is around 25 cm long, and roughly J-shaped, although its size and shape varies between individuals and with its degree of fullness. When the stomach is empty, it has a volume of about 50 ml and its mucosa and submucosa are thrown into large longitudinal folds called *rugae* which flatten out as the stomach fills with food. When fully distended, the stomach has a volume of around 4 liters.

The major regions of the stomach are illustrated in Fig. 18.13. The cardiac region (cardia) surrounds the cardiac orifice through which food enters the stomach. The part of the stomach that extends above the cardiac orifice is called the *fundus* while the mid-portion is called the *body*. This is continuous with the funnel-shaped pyloric region. The upper widest portion of the pyloric region is the *antrum*, which narrows to form the pyloric canal terminating in the *pylorus* itself. The convex portion of the stomach is called the greater curvature, while the concave region is the lesser curvature.

The fundus and body of the stomach act as a temporary reservoir for food. As indicated above, they can accommodate large increases in volume without appreciable changes in intragastric pressure because the smooth muscle in these areas relaxes in response to the presence of food. This reflex is mediated, at least in part, by afferent and efferent fibers running in the vagus nerve causing an inhibition of muscle tone. Contractile activity in the fundus and body is relatively weak so that food remains here largely undisturbed for fairly long periods. In contrast, vigorous contractions take place in the antral region where food is broken down into smaller particles and mixed with gastric juice to form chyme, the semiliquid form in which it is passed to the duodenum.

Blood supply and innervation of the stomach

Arterial blood is supplied to the stomach by the gastric arteries. These are branches of the celiac artery which form a plexus of vessels within the submucosa. From here, the vessels branch extensively to provide the mucosal layer with a rich vascular network. Venous drainage is via the gastric veins. These empty into the portal vein, which conveys blood to the liver. Blood flow to the gastric mucosa increases significantly when the stomach secretes gastric juice in response to a meal. The lymphatics of the stomach lie along the arteries and drain into lymph nodes around the celiac artery.

The stomach is richly innervated with both intrinsic and extrinsic nerves (see p. 380). The intrinsic neurons of the enteric nervous system supply the gastric smooth muscle and secretory cells. They also make numerous synaptic connections with extrinsic neurons. The extrinsic innervation includes sympathetic fibers from the celiac plexus and parasympathetic vagal neurons. A number of afferent fibers also innervate the stomach. These are sensitive to a variety of stimuli including distension and pain. Many afferent fibers leave the GI tract via vagal and sympathetic nerves. Others form the afferent arms of intrinsic reflex arcs mediated by the enteric nervous system.

Structure of the gastric mucosa

The basic organization of the stomach wall resembles that shown in Fig. 18.2. As well as the longitudinal and circular muscle layers, there is, on the anterior and posterior sides of the stomach, an additional layer of muscle between the mucous

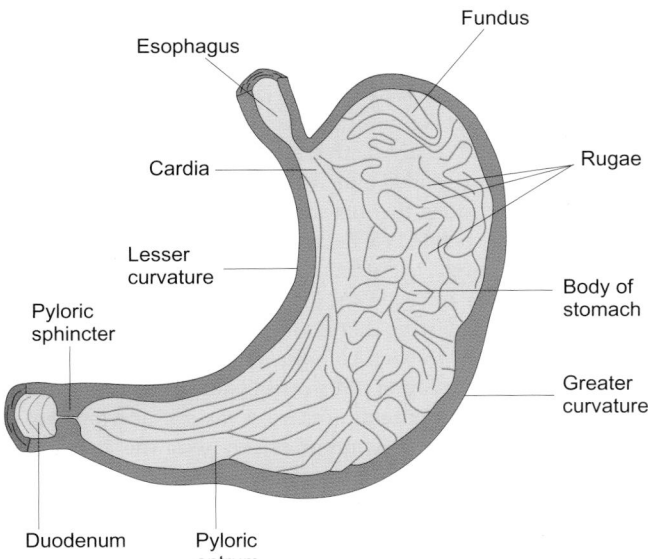

Fig. 18.13 A longitudinal view showing the major anatomical regions and rugae of the stomach.

membrane and the circular layer. The smooth muscle cells in this layer are obliquely orientated and may play a role in the grinding and churning movements displayed by the stomach. The circular smooth muscle layer varies in thickness in different parts of the stomach. It is relatively thin in the fundus and body but thicker in the antral region where the most vigorous contractile activity occurs. Further thickening is seen in the pylorus where the circular muscle forms a sphincter that regulates gastric emptying.

The gastric mucosa contains a variety of secretory cells. The surface epithelium of the gastric mucosa is of the simple columnar type, composed almost entirely of mucus cells which produce a protective alkaline fluid containing mucus. This epithelial layer is dotted with millions of deep *gastric pits*. These are depressions into which the secretions of gastric glands empty. There are around 100 gastric pits per square millimeter of the mucosa occupying about 50 per cent of its total surface. The *gastric glands* contain several types of cells, the exact nature of which differs according to the region of the stomach. A diagrammatic representation of a gastric gland, showing the locations of the various cell types, is given in Fig. 18.14

The gastric glands contain four types of cells.

1. *Mucous neck cells* which are situated at the opening of the gastric glands. These cells secrete a mucus distinct from that secreted by the surface epithelial cells. Its special significance is unclear.

2. *Chief cells* which are located in the basal regions of the gastric glands. These cells secrete *pepsinogen*, the inactive form of the proteolytic enzyme pepsin.

Summary

1. The functions of the stomach include storage of food, mixing, churning, and kneading the food to produce chyme, and the secretion of acid, enzymes, mucus, and intrinsic factor.

2. In addition to the circular and longitudinal smooth muscle layers, the stomach wall possesses a third obliquely arranged muscle layer that promotes churning movements.

3. The surface epithelium of the gastric mucosa is composed almost entirely of cells that secrete an alkaline fluid containing mucus. Gastric glands empty into gastric pits in the epithelium. The glands contain mucous cells, chief cells, which secrete pepsinogens, and parietal cells, which produce gastric acid and intrinsic factor. A variety of endocrine cells are also present, for example the G cells which secrete gastrin.

3. *Parietal* or *oxyntic cells* are scattered amongst the chief cells. They secrete hydrochloric acid and *intrinsic factor*.

4. *Entero-endocrine cells* (G cells) secrete gastrin, which enters the bloodstream and exerts effects on motility and secretory processes within the GI tract.

Both chief and oxyntic cells are numerous in the fundus and body of the stomach. In the antral and pyloric regions, oxyntic cells are much less numerous, with mucus, pepsinogens, and gastrin the predominant secretions. In the cardiac region of the stomach, the gastric glands consist almost entirely of mucus-secreting cells. Because of these regional variations in structure, the exact effects of partial gastrectomy will depend upon the area that is removed.

18.5 The composition of gastric juice

The fluid secreted by the gastric glands is called *gastric juice*. It contains salts, water, hydrochloric acid, pepsinogens, and intrinsic factor. The exact composition and flow rate of gastric juice is determined by the relative activities of the different types of cell within the gastric glands and will vary according to the time that has elapsed since the ingestion of food. Some 2–3 liters of gastric juice are secreted each day in adults.

Clinical tests of gastric secretory function, involving the collection of gastric juice by means of a swallowed tube, have shown that during fasting there is little or no secretion of gastric juice and that the stomach contains only about 30 ml of fluid. Following the ingestion of a test meal (such as thin gruel), the pH of the stomach contents first rises, as the acid present is neutralized, and then falls progressively over the ensuing 90 minutes or so as hydrochloric acid is secreted by the parietal (oxyntic) cells of the gastric glands. The output of all the other constituents of gastric juice also increases following a meal.

The electrolyte composition of gastric juice depends upon its rate of secretion. As the rate of secretion rises, the sodium concentration falls while that of hydrogen ions is increased. The

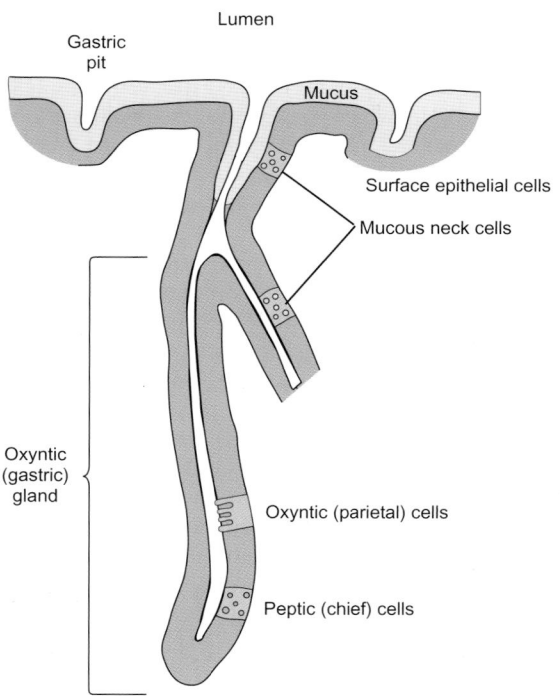

Fig. 18.14 A diagrammatic representation of a gastric gland showing the various secretory cell types present.

level of potassium ions in gastric juice is always higher than that of the plasma, which is why prolonged vomiting may lead to *hypokalemia* (low plasma potassium).

The formation of stomach acid

The pH of gastric juice is very low (pH 1–3). Although HCl is not essential for the overall digestive process, the highly acidic environment it creates is important for several reasons.

1. It helps in the breakdown of connective tissue and muscle fibers of ingested meat.

2. It activates inactive pepsinogen.

3. It provides optimal conditions for the activity of pepsins.

4. By combining with calcium and iron to form soluble salts, HCl aids in the absorption of these minerals.

5. It acts as an important defense mechanism for the stomach, killing many of the micro-organisms that may cause infection (e.g. typhoid, salmonella, cholera, and dysentery).

Gastric acid is secreted by the parietal cells of the gastric glands, predominantly in the fundus and body of the stomach. The majority of these cells only secrete HCl after they have been stimulated following the arrival of food. The cytoplasm of un-stimulated cells is filled with an elaborate branching system of tubular structures derived from the endoplasmic reticulum. These are lined by microvilli, which possess apparatus for hydro-gen ion secretion. When the parietal cells are stimulated to se-crete by food entering the stomach, the tubular structures fuse to form deep invaginations of the apical membrane which are known as secretory canaliculi. The formation of canaliculi results in a large (more than 10-fold) increase in the surface area of the parietal cell membrane and brings large numbers of hydrogen ion pumping sites into close proximity with the luminal fluid.

The metabolic steps involved in the production of acid by a parietal cell are shown in Fig. 18.15. The secretion of both hydro-gen and chloride ions occurs by active transport. Chloride ions are moved from plasma to lumen against an electrochemical gradient. Hydrogen ions move down an electrical gradient but against a massive concentration gradient (as much as a million to one). Therefore the process of acid production is highly energy dependent and parietal cells contain numerous mito-chondria.

A unique membrane transport system is now known to be located on the canalicular membrane of parietal cells. The sys-tem is driven by a H^+, K^+-ATPase that uses energy derived from the hydrolysis of ATP to pump hydrogen ions out of the cell in exchange for potassium ions. Chloride ions can leave the cell by two routes. There is a chloride channel in the secretory canali-culi through which chloride ions diffuse. In addition, the canali-culi have a potassium–chloride symporter that transports chloride across the membrane. Therefore the potassium ions shuttle in and out of the cells, leaving with chloride ions and re-entering in exchange for hydrogen ions.

The hydrogen ions themselves are derived from the dissoci-ation of water within the cell (Fig. 18.15). Hydroxyl ions are also produced as a result of this reaction. These combine with carbonic acid to generate bicarbonate ions, which leave the cell in exchange for chloride ions at the basolateral membrane. This provides the chloride ions which will leave the cell at the canalicular membrane via the chloride channel and the potas-

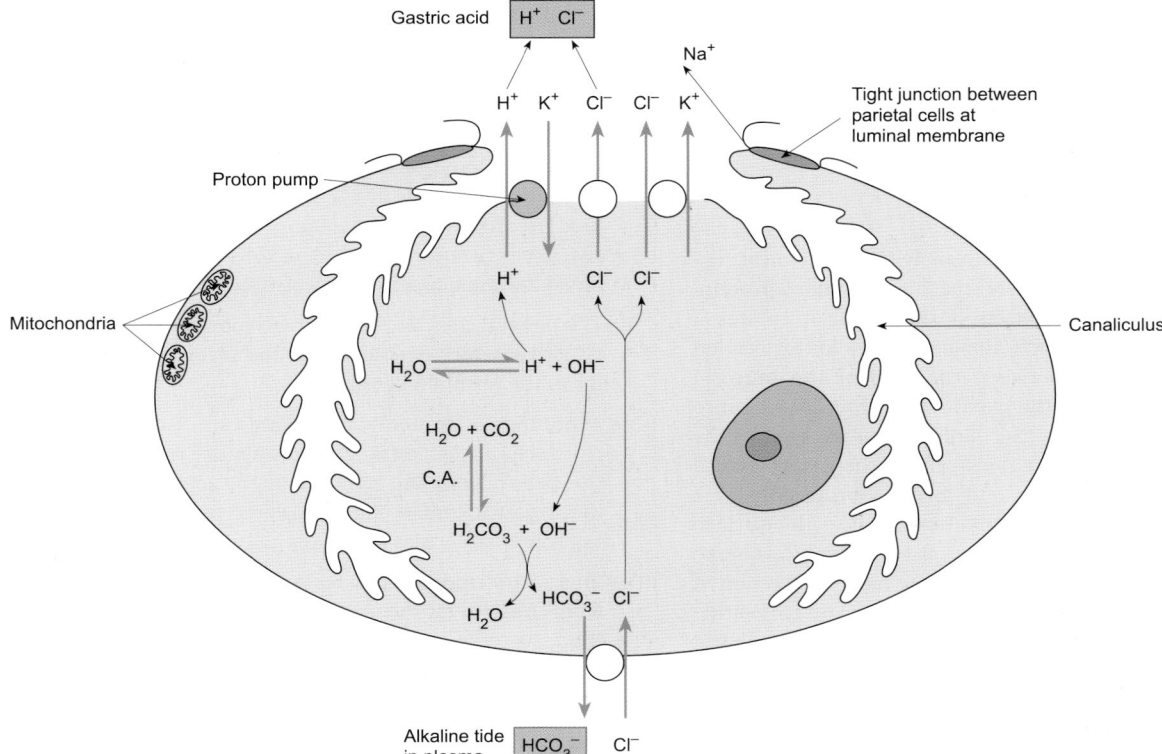

Fig. 18.15 The steps involved in the secretion of gastric acid by a parietal cell. C.A. carbonic anhydrase

sium chloride symporter. As bicarbonate is added to the plasma during the secretion of acid by the stomach, the venous blood draining the stomach is more alkaline than the arterial blood. This is known as the 'alkaline tide'.

The secretion of enzymes by the gastric glands

Gastric acid secretion is accompanied by the release of a number of proteolytic enzymes from the chief cells of the gastric glands. These are collectively known as pepsin. They are secreted in the form of inactive precursors called pepsinogens, which are contained in membrane-bound zymogen granules that are secreted when the gastric glands are stimulated. In the acid environment of the stomach, pepsinogens are converted to active pepsins which show their greatest proteolytic activity at pH values below 3. The gastric pepsins are endopeptidases, i.e. they hydrolyze peptide bonds within the protein molecule to liberate polypeptides and a few free amino acids.

Although fat digestion by the stomach is probably negligible, the gastric glands do secrete a lipase that is stable at the very low pH levels found in the gastric juice. This lipase works over the broad pH range 4–7. It is most active against the short-chain triglycerides found in milk and therefore is probably more important in children than in adults.

The secretion of intrinsic factor by the gastric glands

In addition to secreting acid, the parietal cells of the stomach secrete a glycoprotein known as intrinsic factor which is essential for life. Intrinsic factor is secreted in response to the same stimuli that promote acid secretion. It binds to vitamin B_{12} (cobalamin) in the upper small intestine and protects it from the enzymatic actions of the gut. The complex of B_{12} and intrinsic factor is absorbed by the mucosal epithelial cells of the lower ileum. Vitamin B_{12} is needed for the production of mature red blood cells and its absence gives rise to *pernicious anemia* (see Chapter 13, p. 237). Lifelong treatment by intramuscular injections of vitamin B_{12} reverses this anemia and can enable patients to survive even following total gastrectomy.

Why doesn't the stomach digest itself?

The gastric mucosa is exposed to extremely harsh chemical conditions. Gastric juice is corrosively acidic and contains protein-digesting enzymes. To protect itself the stomach creates a so-called *mucosal barrier*. Three factors contribute to this barrier. First, the tight junctions between the cells of the mucosal epithelium help to prevent the gastric juice from leaking into the underlying layers of tissue. Secondly, mucus secreted by the surface epithelial cells and the neck cells of the gastric glands adheres to the gastric mucosa and forms a protective layer 5–200 µm thick. This mucus is alkaline because the surface epithelial cells secrete a watery fluid that is rich in bicarbonate and potassium ions. When food is eaten, the rates of secretion of both the mucus and the alkaline fluid from the surface epithelial cells are enhanced. Consequently, the surfaces of the gastric epithelial cells themselves remain bathed in their own protective fluid. Therefore they are shielded from direct contact with the potentially damaging gastric contents. Finally, prostaglandins, particularly those of the E series, appear to play an important role in the protection of the gastric mucosa. They increase the thickness of the mucus gel layer, stimulate the production of bicarbonate and cause vasodilatation of the microvasculature of the mucosa. This improves the supply of nutrients to any damaged areas of mucosa while the increased bicarbonate content of the fluid neutralizes the gastric acid, thus optimizing conditions for tissue repair.

The epithelial cells of the gastric mucosa are in a dynamic state of growth, migration, and desquamation (shedding). Indeed, the gastric epithelium is continuously renewed, providing further protection against damage from the harsh environment. Damaged epithelial cells are shed and replaced by new cells derived from relatively undifferentiated stem cells, which migrate up from the necks of the gastric glands.

Anything that breaches the mucosal barrier produces inflammation of the underlying tissue, a condition known as *gastritis*. Persistent erosion of the stomach wall can lead to the formation of gastric ulcers. Common predisposing factors for the formation of gastric ulcers include hypersecretion of acid and/or reduced mucus secretion. Many drugs promote ulcer formation by altering the rates of acid and mucus production. These include caffeine, nicotine, and non-steroidal anti-inflammatory drugs such as ibuprofen and aspirin. The latter act by interfering with the production of prostaglandins. Occasionally, bile acids are regurgitated from the small intestine via the pyloric sphincter. Their detergent action may break down the mucosal barrier, rendering it susceptible to erosion by the gastric acid. Stress may also contribute to the development of gastric ulcers in some individuals. However, it is now thought that many ulcers are caused by an acid-resistant bacterium, *Helicobacter pylori*, which adheres to the gastric epithelium and destroys the mucosal barrier to expose large unprotected areas of mucosa.

Summary

1. Gastric juice contains salts, water, hydrochloric acid, pepsinogen, and intrinsic factor. Gastric acid is secreted by the parietal cells of the gastric glands in response to food. Hydrogen ions move out of the cells against a massive concentration gradient. Chloride ions move from blood to lumen against an electrochemical gradient.

2. The parietal cells also secrete a glycoprotein called intrinsic factor, which is essential for the absorption of vitamin B_{12} in the terminal ileum.

3. A number of proteolytic enzymes are secreted by the chief cells of the gastric glands. They are released as inactive pepsinogens and activated by the acidic environment in the gastric lumen. They hydrolyze peptide bonds within protein molecules to liberate polypeptides.

4. The gastric mucosa creates a 'barrier' to protect itself from erosion by the gastric juice. Alkaline mucus-rich fluid, secreted by the gastric epithelial cells, provides a protective coating for the mucosa. Prostaglandins of the E series increase the thickness of this layer and stimulate the secretion of bicarbonate ions.

18.6 The regulation of gastric secretion

The secretion of hydrochloric acid and the secretion of pepsinogens by the glands of the gastric mucosa are regulated largely in parallel, by the same factors. Both nervous and endocrine mechanisms are involved and these interact at many levels. Gastric secretion is normally considered to occur in three phases, the timing of which overlaps considerably. These are the cephalic, gastric, and intestinal phases.

The cephalic phase of gastric secretion

This takes place even before food enters the stomach and results from the anticipation of food and its sight, smell, and taste. The relative contribution of the cephalic phase to overall gastric secretion in response to a meal is variable and dependent upon mood and appetite, but may amount to as much as 30 per cent. Neurogenic signals originating in the cerebral cortex, or in the appetite centers of the amygdala and hypothalamus, are relayed to the stomach via efferent fibers whose cell bodies lie within the dorsal motor nuclei of the vagus nerve.

Parasympathetic vagal activity influences gastric secretion both directly and indirectly, as indicated in Fig. 18.16. Postganglionic parasympathetic fibers in the myenteric plexus release acetylcholine and stimulate the output of the gastric glands. Vagal stimulation also causes the release of gastrin from the G cells of the antral glands. Gastrin reaches the gastric glands by way of the bloodstream and stimulates them to secrete acid and pepsinogens. Furthermore, both vagal activity and gastrin stimulate the release of histamine from mast cells. Histamine acts on parietal cells via H_2 receptors to stimulate hydrogen ion secretion. Thus, acetylcholine, gastrin, and histamine all enhance the secretion of gastric juice.

The gastric phase of gastric secretion

The arrival of food in the stomach stimulates the gastric phase of acid, pepsinogen, and mucus production, which accounts for more than 60 per cent of total gastric secretion. The two principal triggers are distension of the stomach wall and the chemical content of the food.

Distension of the stomach activates mechanoreceptors and initiates both local (short-loop) myenteric reflexes and long-loop vago-vagal reflexes. Both reflexes lead to the secretion of acetylcholine, which stimulates the output of gastric juice by the secretory cells of the stomach. The importance of the vagally mediated reflexes is revealed by the 80 per cent reduction in acid production in response to distension which is seen following vagotomy. Emotional stress, fear, anxiety, or any other state that triggers a sympathetic response will inhibit gastric secretion because the parasympathetic controls over the GI tract are temporarily overridden.

In addition to its direct cholinergic action, the vagus stimulates the output of gastrin from G cells in response to distension of the body of the stomach. Gastrin is a powerful stimulus for acid secretion from the parietal cells. It also enhances the release of enzymes and mucus from the gastric glands. Although intact proteins are without effect on the rate of gastric secretion, peptides and free amino acids stimulate the output of gastric juice through a direct action on the G cells. The amino acids tryptophan and phenylalanine are particularly potent secretagogues, as are bile acids and short-chain fatty acids.

Gastrin secretion is inhibited when the pH of the gastric contents falls to between 2 and 3. Thus gastrin secretion is maximal soon after entry of food into the stomach, when the pH is relatively high, but declines as acid secretion and protein digestion get under way and the pH of the gastric contents falls. The inhibition of gastrin secretion is mediated by an increase in the secretion of somatostatin from cells (D cells) of the gastric mucosa. Thus gastric acid secretion is self-limiting and the gastric phase of gastric secretion normally lasts for only about an hour.

The intestinal phase of gastric secretion

A small proportion (5 per cent or so) of the total gastric secretion in response to food takes place as partially digested food starts to enter the duodenum. This is believed to be due to the secretion of gastrin from G cells in the duodenal mucosa, which encourages the gastric glands to continue secreting. However, this effect is short-lived and, as acid chyme distends the duodenum, an enterogastric reflex is initiated whereby gastric secretory activity is suppressed. Several hormones contribute to this reflex.

Secretin is secreted by the duodenal mucosa in response to acid. It reaches the stomach via the bloodstream and inhibits the release of gastrin. It also exerts a direct inhibitory action on the parietal cells to reduce their sensitivity to gastrin.

Two hormones are released in response to the presence of products of fat digestion in the duodenum and proximal jejunum. These are *cholecystokinin* (*CCK*) and *gastric inhibitory peptide* (*GIP*) (also called glucose-dependent insulinotropic peptide). Both inhibit the release of gastrin and gastric acid, though their relative importance is not clear. Fig. 18.16 summarizes the factors regulating gastric secretion.

Disorders of gastric acid secretion

Reduced gastric secretion is a relatively rare condition, generally restricted to elderly patients with atrophy of the gastric mucosa. *Achlorhydria* (a decrease in hydrochloric acid secretion) may occur because of a loss of parietal cells. Although digestive processes are normally unaffected, achlorhydria can cause impaired absorption of substances requiring an acid environment. Patients with achlorhydria normally have a high circulating level of gastrin because the stomach contents are never acidic enough to inhibit secretion (see above).

A number of disorders, including stress in some individuals, are associated with abnormally high rates of gastric acid secretion, and a variety of drugs and constituents of foods are known to stimulate the production of acid (e.g. caffeine and alcohol). A rare condition, the Zollinger–Ellison syndrome, is caused by a gastrin-secreting tumor of the non-beta cells of the pancreatic islets. Here, gastric acid secretion reaches such high values that erosion of the gastric mucosal barrier occurs, leading to ulceration of the stomach wall.

Several strategies have been developed to treat excessive acid production and promote healing of the gastric mucosa. Specific H_2 receptor antagonists such as cimetidine and ranitidine,

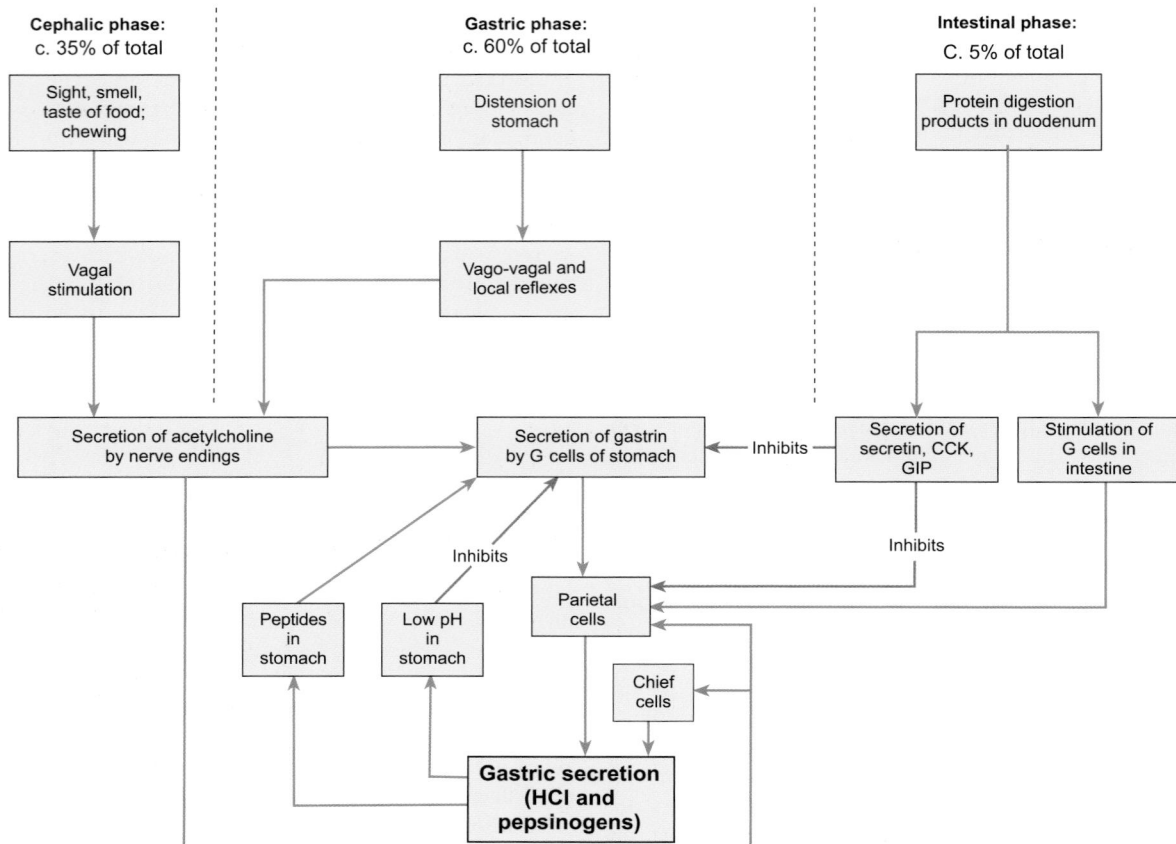

Fig. 18.16 The major factors involved in the regulation of gastric secretion. Secretin, CCK, and GIP are secreted by entero-endocrine cells in the epithelium of the upper small intestine and have an inhibitory action on gastrin secretion, as does a low pH in the lumen of the stomach. The stimulatory action of gastrin on mucus and enzyme secretion is omitted for clarity.

Summary

1. Nervous and endocrine mechanisms combine to regulate gastric secretion. Secretion occurs in three phases: cephalic, gastric, and intestinal.

2. The cephalic phase of secretion occurs in response to the sight, smell and anticipation of food, and as a result of chewing. Parasympathetic vagal fibers stimulate secretion via the release of acetylcholine both directly and by stimulating gastrin secretion.

3. The arrival of food in the stomach initiates the gastric phase of secretion in which distension and the presence of amino acids and peptides stimulates the output of HCl and pepsinogen. Gastrin is an important mediator of this phase. Secretion is inhibited when the pH of the gastric contents falls to around 2 or 3.

4. As partially digested food enters the duodenum some gastrin is secreted from G cells in the duodenal mucosa. This stimulates some further gastric secretion. However, secretin, CCK, and GIP all contribute to the inhibition of gastric secretion.

which block parietal cell histamine receptors, may be used to inhibit acid secretion. Other agents such as benzimidazoles, which are weak bases, are known to inhibit the activity of the proton pumps on the apical surface of the parietal cells. Drugs based on agents of this kind, such as omeprazole, are increasingly used to treat patients with ulcers caused by the hypersecretion of gastric acid. Individuals whose gastric ulcers are caused by the bacterium *H. pylori* are now treated with a course of antibiotics combined with antacid therapy.

18.7 The storage, mixing, and propulsion of gastric contents

In addition to the important secretory activity of the stomach, its motor functions play an important role in the overall process of digestion. The stomach stores food until it can be accommodated by the lower regions of the GI tract. It also mixes the food with gastric secretions and, through its grinding action, breaks down food particles into smaller pieces to form semiliquid chyme. The stomach contents are then delivered to the duodenum at a rate compatible with their complete digestion and absorption. Gastric motility and emptying are controlled by complex interactions between the enteric nervous system, the autonomic nervous system, and a number of hormones.

For the purposes of its motor function, the stomach can be divided into two parts. These are the proximal motor unit, consisting of the fundus and body of the stomach, and the distal motor unit, consisting of the antral and pyloric regions. The proximal motor unit carries out the reservoir functions of the stomach, while the distal motor unit is responsible for the mixing of food and its propulsion into the duodenum.

Storage function of the stomach

The empty stomach has a volume of around 50 ml and an intragastric pressure of 0.6 kPa (5 mmHg) or less. Although it can stretch to accommodate large volumes of food, little increase in intragastric pressure is seen until the stomach volume exceeds 1 liter. There are several reasons for this. The smooth muscle of the stomach wall is able to increase its length significantly without altering its tone, a property known as *plasticity* (see Chapter 7). The stomach exhibits receptive relaxation. As it is stretched, a vagal reflex is triggered which inhibits muscle activity in the body of the stomach. This reflex is coordinated by the regions of the brainstem responsible for swallowing. Finally, the shape of the stomach itself contributes to its effectiveness as a reservoir. As the diameter of the stomach increases during filling, the radius of curvature of its walls also increases. At a given pressure, the stretching force on the walls increases in proportion to this radius of curvature (Laplace's law; see also Chapter 16, p. 323). As a result, intragastric pressure rises only slightly despite significant distension.

Contractile activity of the full stomach—mixing and propulsion

Gastric motility results from the coordinated contraction of the three smooth muscle layers which lie within the stomach wall. The different orientations of the longitudinal, circular and oblique layers allow the stomach to perform a wide variety of different movements, including grinding, churning and kneading, as well as propulsion.

During fasting, the stomach shows only weak contractile activity (though in extreme hunger there may be short periods of intense contractile activity experienced as hunger 'pangs'). After a meal, peristaltic contractions begin in the body of the stomach. These are very weak rippling movements but as the contractions approach the pylorus, where the musculature is thicker, they become much more powerful, reaching a maximum close to the gastroduodenal junction. Thus the contents of the fundus remain relatively undisturbed while those of the pyloric regions receive a vigorous pummeling and mixing. Although the intensity of these peristaltic contractions may be modified by many factors, their rate remains constant at around three contractions a minute. The rate of propagation of the peristaltic wave accelerates as it nears the distal regions so that the smooth muscle of the antrum and pylorus contracts virtually simultaneously, pushing the gastric contents ahead of the peristaltic wave. As pressure in the antrum rises, the pyloric sphincter opens and a few milliliters of chyme are squirted through it into the first part of the duodenum, the duodenal bulb. The sphincter closes almost immediately thus preventing further emptying, and as a result of the continued high pressure in the antrum, some of the gastric contents are forced back into more proximal regions. This is called retropulsion. It increases the effectiveness of mixing and breakdown of food within the stomach as food particles rub against each other.

Gastric motility is enhanced by many of the same nervous and hormonal factors that stimulate gastric secretion. Although intragastric pressure rises only a little, distension of the stomach wall by food activates stretch receptors and, as a result, both the force of peristaltic contractions and the overall level of smooth muscle tone are enhanced. Gastrin-secreting cells are stimulated by the presence of food in the stomach. This hormone also increases gastric motility. Consequently, distension of the stomach by food increases the efficiency of mixing and emptying movements. The nervous control of gastric motility is not completely understood. Both parasympathetic (vagal) and sympathetic fibers supply the smooth muscle and, in general, it appears that parasympathetic activity increases motility while sympathetic activity decreases it.

The rate of gastric emptying is carefully controlled

If digestion and absorption are to proceed with optimum efficiency, it is essential that chyme is delivered to the duodenum at a rate which enables the small intestine to process it fully. Furthermore, it is important that duodenal contents are prevented from being regurgitated into the stomach. The gastric and duodenal environments are very different. The gastric mucosa is resistant to acid but may be eroded by bile, while the duodenum is resistant to the effects of bile but is unable to tolerate low pH. Consequently, gastric emptying which is too rapid may result in the formation of duodenal ulcers, while regurgitation of duodenal contents may result in gastric ulceration.

Many factors contribute to the regulation of gastric emptying

Emptying of the stomach depends upon the factors that influence motor activity throughout the GI tract—the inherent excitability of smooth muscle, intrinsic and extrinsic nervous pathways, and hormones. In general, the stomach empties at a rate that is proportional to gastric volume, i.e. the fuller the stomach, the more rapidly it empties. In addition, the physical and chemical nature of the gastric contents affects the rate of emptying. Fats and proteins in the ingested food, a very acidic juice, and a hypertonic mixture of juice and food will all delay emptying. In general, the closer the contents are to isotonic saline, the more rapidly they will leave the stomach. The half-time for liquids remaining in the stomach is about 20 minutes compared with about 2 hours for solids. Receptors of various kinds are present within the duodenum and contribute to the regulation of gastric emptying via the so-called *enterogastric reflex*, a collective term used to describe all the hormonal and neural mechanisms that mediate intestinal control of gastric emptying.

The presence of fatty acids or monoglycerides in the duodenum causes an increase in the contractility of the pyloric sphincter. This reduces the rate at which the gastric contents are propelled through the sphincter into the small intestine and ensures that fats are not delivered to the duodenum more quickly than the bile salts can emulsify them (see p. 404). How fats exert this effect is unclear, but both CCK and GIP are released from the small intestine in response to fats and their digestion products, and both hormones have been shown to delay gastric emptying.

The products of protein digestion are also believed to exert their inhibitory effect on gastric emptying through endocrine pathways. Gastrin is secreted from G cells in both the antrum and duodenum in response to peptides and amino acids. The action of gastrin on motility is twofold. Although it stimulates contraction of the antrum, it also increases the degree of constriction of the pyloric sphincter, so that the net effect of gastrin secretion is normally to delay emptying of the stomach.

The delay in gastric emptying seen when acid enters the duodenum is probably mediated by both nervous and hormonal factors. Vagotomy reduces the response, suggesting a role for the vagus nerve in the regulation of gastric emptying by acid. Secretin also appears to be involved. This hormone, which also stimulates the secretion of alkaline pancreatic juice (p. 402), is released in response to acid in the duodenum and delays gastric emptying by inhibiting contraction of the antrum.

The electrical activity underlying gastric contractions

The regular peristaltic contractions shown by the stomach are the mechanical consequence of the basic electrical rhythm (BER) of the smooth muscle cells. This basic rhythm is set by the spontaneous activity of pacemaker cells in the longitudinal smooth muscle layer of the stomach wall in the region of the greater curvature near to the middle of the body of the stomach (the pacemaker zone). The cells here show spontaneous depolarization and repolarization every 20 seconds or so to establish the BER or 'slow wave' rhythm of the stomach. The pacemaker cells are electrically coupled to the rest of the stomach muscle sheet by means of gap junctions and therefore their rhythm is transmitted to the entire muscularis.

The electrical properties of the gastric slow wave and its relationship with the contractile force generated by the smooth muscle are illustrated in Fig. 18.17. The potential change is triphasic and similar to a cardiac muscle action potential, although it is about 10 times as long. The inward current responsible for initial depolarization is probably carried by calcium ions moving through voltage-gated channels, while the plateau is maintained by the inward movement of both sodium and calcium ions. Repolarization is associated with a delayed outward potassium current.

In the body and fundus of the stomach, action potentials are not normally associated with the gastric pacemaker potentials.

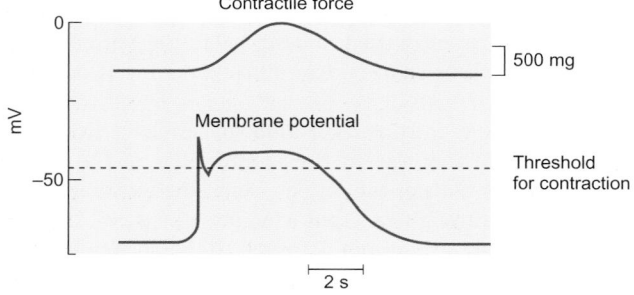

Fig. 18.17 The relationship between contraction and the electrical activity of the smooth muscle of the stomach.

Nevertheless, contraction of the smooth muscle occurs when the depolarization phase of the potential reaches the so-called mechanical threshold. The force of this contraction is related to the degree of depolarization and the time for which the potential exceeds threshold. In the distal antrum and pyloric regions of the stomach, trains of action potentials are superimposed on the BER. These are associated with vigorous propulsive movements, which may result in gastric emptying.

Vomiting

Vomiting (or *emesis*) is the sudden and forceful oral expulsion of the contents of the stomach and sometimes the duodenum. It is frequently preceded by *anorexia* (loss of appetite) and *nausea* (a feeling of sickness). Immediately before vomiting, it is common to experience characteristic autonomic responses such as copious watery salivation (waterbrash), vasoconstriction with pallor, sweating, dizziness, and tachycardia. Retching often precedes vomiting. During the process of vomiting, respiration is inhibited. The larynx is closed and the soft palate rises to close off the nasopharynx and prevent the inhalation of vomited material (*vomitus*). The stomach and pyloric sphincter relax, and contraction of the duodenum reverses the normal pressure gradient so that intestinal contents are allowed to enter the stomach (a process sometimes referred to as reverse peristalsis). The diaphragm and abdominal wall then contract powerfully, the gastro-esophageal sphincter relaxes, and the pylorus closes. The resulting rise in intragastric pressure causes the expulsion of the gastric contents.

The vomiting reflex is coordinated by the dorsal portion of the reticular formation of the medulla which lies close to the respiratory and cardiovascular control areas of the brainstem. Afferent impulses arrive at this region from many parts of the body including the pharynx and other areas of the GI tract, viscera such as the liver, gall bladder, urinary bladder, uterus and kidneys, the cerebral cortex, and the semicircular canals of the ears (giving rise to motion sickness). Furthermore, a variety of chemical agents, including general anesthetics, opiates, and digitalis, appear to trigger vomiting by stimulating receptors in the floor of the fourth ventricle. The motor impulses responsible for the action of vomiting are transmitted from the vomiting 'center' via the trigeminal, facial, glossopharyngeal, vagus, and hypoglossal nerves (cranial nerves V, VII, IX, X, and XII). Figure 18.18 illustrates the nervous pathways involved in the vomiting reflex.

Although vomiting is generally a protective mechanism whereby potentially toxic substances are removed from the body, prolonged vomiting can lead to a state of metabolic alkalosis due to the continued loss of acid from the stomach (see Chapter 29). It also may result in hypokalemia.

Absorption by the stomach

Very little absorption occurs in the stomach. Ethyl alcohol is the only water-soluble substance normally absorbed in significant amounts and even this can only be absorbed because its lipid solubility enables it to diffuse readily through the plasma membranes of the gastric mucosal cells. Certain organic substances, which are un-ionized at the acidic pH of the stomach, may be absorbed here. An example is aspirin, which has a pK_a of 3.5 so that it remains largely un-ionized in the stomach. Molecules of

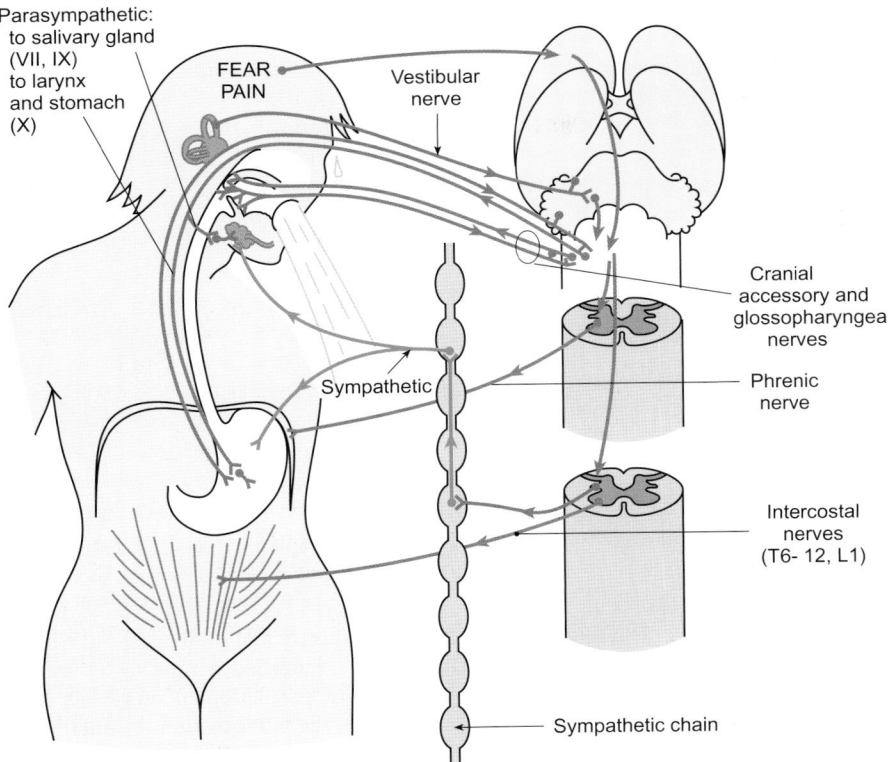

Parasympathetic:
to salivary gland
(VII, IX)
to larynx
and stomach
(X)

FEAR
PAIN

Vestibular
nerve

Cranial
accessory and
glossopharyngeal
nerves

Phrenic
nerve

Sympathetic

Intercostal
nerves
(T6- 12, L1)

Sympathetic chain

Fig. 18.18 The neural pathways of the vomiting reflex.

Summary

1. The stomach stores food, mixes it with gastric juice, and breaks it into smaller pieces. Eventually semiliquid chyme is formed. The stomach then delivers chyme to the duodenum in a controlled fashion.

2. The stomach is able to store large amounts of food since intragastric pressure rises very little despite significant distension of the stomach wall

3. The fasting stomach shows only weak contractile activity. After a meal peristaltic contractions begin, increasing in power as they approach the antrum where mixing is most vigorous. Contractions are the mechanical consequence of the basic electrical rhythm (BER) or slow-wave rhythm of the gastric smooth muscle. Gastric motility is enhanced by mechanical distension and by gastrin.

4. The stomach normally empties at a rate compatible with full digestion and absorption by the small intestine. Many factors contribute to this regulation. Distension of the stomach increases the rate of emptying. The presence in the chyme of fats, proteins, high acidity, and hypertonicity all delay the rate of emptying.

5. Vomiting is a protective mechanism whereby noxious or potentially toxic substances are expelled from the GI tract. The vomiting reflex is coordinated in the medulla of the brain. Prolonged vomiting can cause hypokalemia as well as metabolic alkalosis through the loss of gastric acid.

aspirin diffuse through the mucosal barrier into the intracellular compartment in which the pH is closer to neutral. The aspirin molecules become ionized and are therefore unable to diffuse back into the gastric lumen. They pass out of the cells and are absorbed into the circulation.

18.8 The small intestine

The small intestine is the major site of both digestion and absorption in the GI tract. In life, the small intestine is a tube about 4 meters long with a diameter of about 2.5 cm and is divided into three segments: the duodenum, the jejunum, and the ileum.

The *duodenum* is about 25 cm long (literally 'duodenum' means 12 finger-widths) and has no mesentery. It takes the form of an arc within which lies the head of the pancreas. Chyme, produced by the chemical and mechanical actions of the stomach, is emptied into the duodenum where it is mixed with secretions from the liver and exocrine pancreas as well as from the duodenum itself. The ducts, which deliver bile and pancreatic secretions, unite close to the duodenum at the hepatopancreatic ampulla, which opens into the duodenum at the major duodenal papilla. The *sphincter of Oddi* controls the entry of bile and pancreatic juice into the small intestine. The junction between the duodenum and the jejunum is formed by a sharp bend, the duodeno-jejunal flexure (Fig. 18.19). The *jejunum* is about 1.5 meters long and extends from the duodenum to the *ileum*, a coiled tube about 2.5 meters long. The jejunum and ileum are supported by a mesentery that contains branches of the superior mesenteric artery and the venous and lymphatic drainage vessels of the

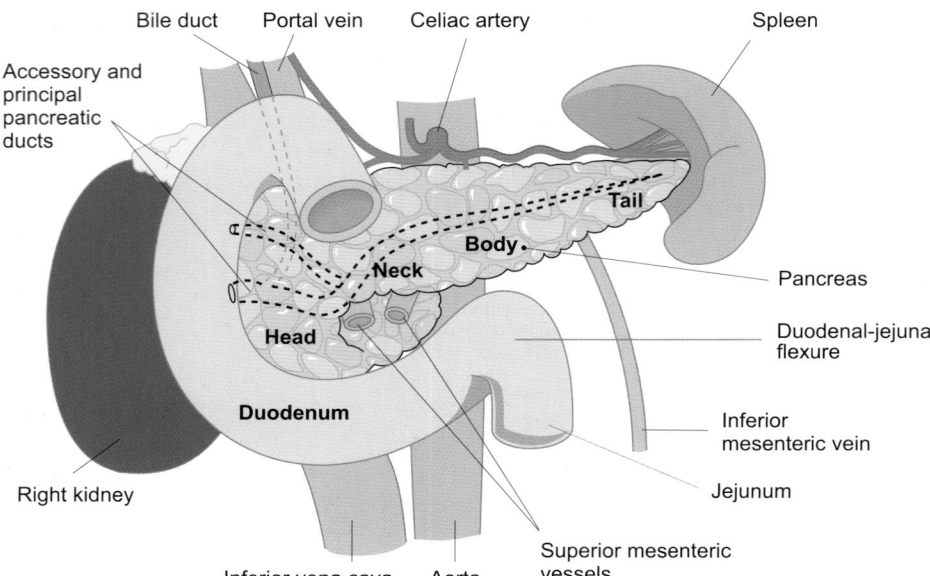

Fig. 18.19 The anatomical arrangement of the duodenum, jejunum, and pancreas.

small intestine. There is no clear anatomical distinction between the jejunum and the ileum, but throughout the length of the small intestine there is a gradual reduction in the thickness of the mucosal wall and subtle differences in its histological characteristics.

Special histological characteristics of the small intestine

The small intestine is exquisitely adapted for nutrient absorption, particularly in the proximal portion. It presents a huge surface area (estimated at around 200 m²) both because of its length and the structural modifications of its wall. The mucosa and submucosa, particularly of the jejunum, are thrown into deep folds called *circular folds* (Fig. 18.20(a)) which, because of their shape, force chyme into a spiral motion as it passes through the lumen. This spiraling slows down the rate of passage of chyme and facilitates its mixing with intestinal juices, thereby optimizing conditions for digestion and absorption.

The folded surface of the small intestine is covered with finger-like projections or *villi*, which are between 0.5 and 1.5 mm high depending upon their location (Fig. 18.20(b)). The surface of each villus is formed mainly by columnar absorptive epithelial cells (enterocytes) bound by tight junctions at their apical surfaces. The mucosal surface of these cells consists of many tiny processes or *microvilli* (roughly 1.0 μm long and 0.1 μm in diameter) which constitute the *brush border*. This adaptation increases the surface area of the small intestine still further. Other epithelial cells are endocrine or paracrine in character. Although the functions of some are unknown, others have been shown to produce somatostatin (D cells), secretin (S cells), neurotensin (N cells), CCK (I cells), and 5-HT (enterochromaffin cells). CCK and secretin are secreted by cells in the wall of the upper small intestine in response to the presence of fat digestion products and acid, respectively.

The villi themselves differ in appearance throughout the small intestine. They are broad in the duodenum, slender and leaf-like in the jejunum, and shorter and more finger-like in the ileum. Within each villus is a modified lymph vessel (lacteal) opening into the local lymphatic circulation, blood vessels, some smooth muscle (which enables the villus to alter its length), and connective tissue. The arterioles supplying the villi branch extensively to form a capillary network that collects into veins at the base.

Between the villi are simple tubular glands, 0.3–0.5 mm deep, called the *crypts of Lieberkuhn*. They are found throughout the small intestine but are most numerous in the mucosa of the duodenum and jejunum. A number of different cell types have been identified within the crypts, including Paneth cells, which secrete lysozyme, and undifferentiated cells which proliferate to replace lost enterocytes. In addition to the crypts of Lieberkuhn, the duodenum (but not the jejunum or ileum) also contains submucosal *Brunner's glands*, which secrete a mucus-rich alkaline fluid.

Peyer's patches, large isolated clusters of lymph nodules similar to the tonsils, are found in the wall of the ileum. They form part of the collection of small lymphoid tissues of the GI and respiratory tracts referred to as the mucosa-associated lymphatic tissue (MALT) (see Chapter 14).

The epithelium of the small intestine is self-renewing

The small intestine has a very rapid rate of cell turnover. In humans, the entire epithelium is renewed every 6 days. This rapid turnover is important because the epithelial cells are sensitive to hypoxia and other irritants. Epithelial cells are formed by the mitotic proliferation of a population of undifferentiated stem cells within the crypts. These new cells then migrate upwards towards the tip of the villus from where they are shed into the lumen of the intestine. As the cells migrate and leave the crypts, they become fully mature and the brush-border enzymes and transporter proteins develop. The rate at which cells proliferate can be altered by a number of factors. Irradi-

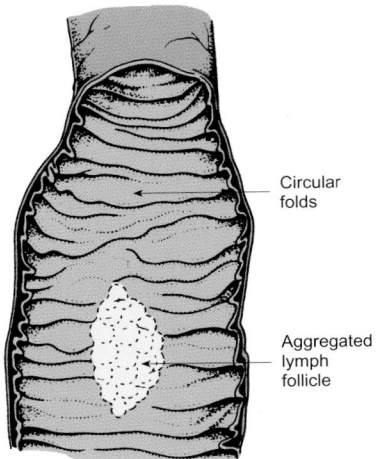

(a)

Circular folds

Aggregated lymph follicle

(b)

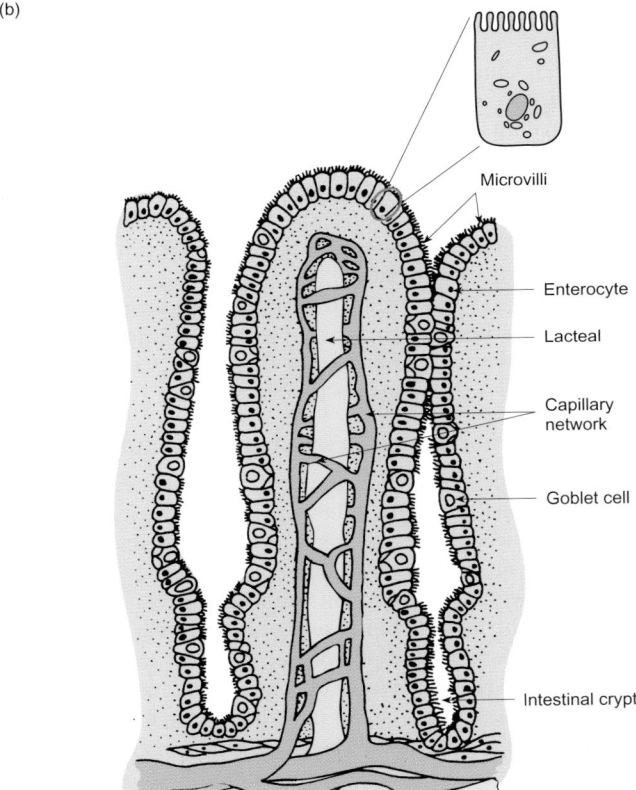

Microvilli

Enterocyte

Lacteal

Capillary network

Goblet cell

Intestinal crypt

Lymphatic

Fig. 18.20 Structural characteristics of the small intestine. (a) A section of ileum showing circular folds and the position of a lymph nodule (Peyer's patch). (b) An intestinal villus in longitudinal section.

ation, starvation, and prolonged intravenous feeding, for example, cause atrophy of the cells and a reduction in proliferation. Certain medications such as methotrexate and other drugs used in the treatment of cancer may also slow the rate of production of enterocytes.

Secretion of fluid and enzymes by the small intestine

The cells of the crypts are responsible for the secretion of about 1.5 liters of isotonic fluid each day. Secretion occurs because of transcellular chloride movement from the interstitial fluid to the lumen followed by paracellular movement of sodium and water. The principal stimulant of fluid secretion is distension of the intestine by acidic or hypertonic chyme. In the duodenum, Brunner's glands secrete a bicarbonate-rich alkaline fluid containing mucus which, together with the secretions of the crypts, protects the duodenal mucosa from mechanical damage and erosion by the acid and pepsin contained within chyme arriving from the stomach. Although the glands secrete spontaneously when acid chyme enters the duodenum, their secretion may be further stimulated by vagal activity, endogenous prostaglandins, and the hormones gastrin, secretin and CCK. However, sympathetic stimulation causes a marked decrease in the rate of mucus production, leaving the duodenum more susceptible to erosion. Indeed, three-quarters of peptic ulcers occur in this region of the gut, and many are related to stress, which is characterized by a generalized increase in sympathetic activity.

The fluid secreted by the small intestine is known as intestinal juice or *succus entericus* and it was once believed to contain most of the enzymes required for the complete digestion of food. However, it has now been established that the only enzymes derived from the small intestine itself (rather than the pancreas) are the *brush-border enzymes*. The principal brush-border enzymes are disaccharidases (maltase, sucrase, etc.), peptidases and phosphatases. One of the duodenal brush-border peptidases, enteropeptidase (commonly, though wrongly, called *enterokinase*), breaks down pancreatic trypsinogen to activate it.

Children under 4 years of age also express the enzyme lactase, which promotes the digestion of lactose (milk sugar). The enzyme is less active in older individuals. Lactose intolerance is caused by a lack of this enzyme (see Section 18.12). The major hormones, electrolytes, and enzymes produced by the small intestine are listed in Table 18.1.

TABLE 18.1   The secretions of the small intestine

Secretory product	Source
Hormones	
Cholecystokinin (CCK)	I cells of villus
Neurotensin	N cells of villus
Secretin	S cells of villus
Serotonin (5-HT)	Enterochromaffin cells
Somatostatin	D cells of villus
Other secretory products	
Lysozyme	Paneth cells of crypts
Mucus	Goblet cells of villus
Isotonic fluid (1.5 l day^{-1})	Crypts
Alkaline mucus	Brunner's glands (duodenum only)

Summary

1. The small intestine is the major site of both digestion and absorption in the GI tract. Here chyme is mixed with bile, pancreatic juice and the intestinal secretions.

2. The small intestine provides a huge surface area for nutrient absorption. The folded mucosal surface is covered with projections called villi. The brush-border membranes of the mucosal epithelial cells house enzymes. Between the villi lie simple tubular glands—the crypts of Lieberkuhn. These contain many different cell types including mucus-secreting goblet cells and phagocytic cells.

3. The small intestinal epithelium is self-renewing, replacing itself completely every 6 days or so. Loss of cells at the tips of the villi releases enzymes from the brush border of the enterocytes into the intestinal lumen. One of these, enteropeptidase, activates pancreatic trypsin which then activates other proteolytic enzymes.

4. Crypt cells secrete 1.5 liters of isotonic fluid each day. Chloride is transported out of the cells, and sodium and water follow passively by the paracellular route. In the duodenum, Brunner's glands secrete an alkaline mucus which helps to protect the epithelium from the corrosive effects of acidic chyme arriving from the stomach. Secretion is stimulated by vagal neurons and by CCK, secretin, gastrin, and endogenous prostaglandins.

18.9 Motility of the small intestine

Typically, chyme traverses the length of the small intestine in 3–5 hours (although under certain conditions it may take as long as 10 hours). The rate of movement is normally such that the last part of one meal is leaving the ileum as the next meal enters the stomach. The most important types of movement in the small intestine are segmentation and peristalsis, described on pp. 379–380. Segmentation is of great importance in mixing the chyme with the digestive enzymes present in the small intestine and in facilitating the absorption of the products of digestion. The villi and microvilli of the intestinal mucosa also exhibit mixing movements.

The peristaltic waves rarely travel more than about 10 cm and are known as 'short-range' peristaltic contractions. Exceptions to this are the so-called 'housekeeper' contractions described later. Waves of peristalsis are initiated by distension of the small intestine.

Segmentation and peristalsis are inherent properties of the intestinal smooth muscle

The BER of the small intestine is independent of extrinsic innervation, and both segmentation and peristaltic contractions of the intestinal smooth muscle are inherent properties of the intramural plexuses of the enteric nervous system. However, the excitability of the smooth muscle and the strength of its contraction can be modified by extrinsic nerves as well as by the variety of hormones utilized as neurotransmitters by the intramural nerve plexuses. Parasympathetic stimulation increases

the excitability of the smooth muscle, while sympathetic stimulation depresses it. These autonomic effects are exerted principally via the enteric nerve plexuses.

Extrinsic nerves play a role in certain long-range intestinal reflexes. These include the so-called ileogastric and gastro-ileal reflexes, which describe the reflex interactions that operate between the stomach and the terminal ileum. The *ileogastric reflex* refers to the reduction in gastric motility that occurs in response to distension of the ileum. The *gastro-ileal reflex* describes the increase in motility of the terminal ileum (particularly of segmentation) that occurs whenever there is an increase in secretory and/or motor activity of the stomach. Both of these will have the overall effect of matching emptying of the small intestine with the arrival of chyme in the duodenum.

Movements of the villi of the mucosa contribute to absorption and mixing

The intestinal villi show piston-like contraction and relaxation movements, which are thought to facilitate the removal of the digestion products of fats from the lacteals (the lymphatic vessels which course through the villi). One possible sequence of events is that, when the villus is relaxed, absorption takes place via intercellular channels. As the villus contracts, these intercellular channels are cut off and the absorbed material is forced into more distal parts of the lymphatic system. This is sometimes referred to as 'milking' the lacteals. Strands of smooth muscle within the lamina propria are thought to give rise to these pumping movements.

The villi also show pendular (swaying) movements that may contribute to the mixing of chyme within the intestinal lumen. These movements are enhanced by the presence of amino acids and fatty acids within the intestinal lumen.

Patterns of motility in the small intestine during fasting

The patterns of contractility described above relate to the behavior of the small intestine following a meal. During periods of fasting, or once a meal has been processed, the smooth muscle of the small intestine shows a different characteristic pattern in which segmentation movements wane and waves of peristalsis, initiated at the duodenal end, sweep slowly along the length of the small intestine. Individual waves travel up to 70 cm before dying out, and the entire wave of contraction takes 1–2 hours to travel the length of the small intestine. The electrical activity that underlies this contractile behavior is known as the *migrating motility complex* (MMC) and is repeated every 70–90 minutes. The purpose of these waves of peristalsis appears to be to sweep out the last remains of the digested meal together with bacteria and other debris into the large intestine. For this reason, the contractions are sometimes called 'housekeeper' contractions.

The mechanisms that initiate and control the MMC are not understood. Both vagal and hormonal mechanisms (particularly *motilin*, another gut hormone) have been implicated.

Emptying of the small intestinal contents into the large intestine is controlled by the ileocecal sphincter

The first part of the large intestine is called the *cecum*, and the junction between the terminal ileum and the cecum is the ileocecal sphincter. This normally regulates the rate of entry of

Summary

1. The rate at which chyme moves through the small intestine is carefully controlled to ensure adequate time for the completion of digestion and absorption. Two types of movement are inherent properties of the intestinal smooth muscle. Segmentation is responsible for the mixing of chyme with enzymes and for exposing it to the absorptive mucosal surface. Peristaltic contractions propel chyme along towards the ileocecal valve.

2. Segmentation is characterized by closely spaced contractions of the circular smooth muscle whose frequency coincides with the rate of slow wave activity in each part of the gut. Peristaltic contractions are less frequent and usually propel chyme only short distances.

3. The motility of the intestinal smooth muscle is influenced by both intrinsic and extrinsic neurons and the neurotransmitters of the intramural plexuses. Parasympathetic activity enhances intestinal motility.

4. The intestinal villi exhibit both piston-like contractions and swaying pendular movements. The latter may contribute to the mixing of chyme, while the former serve to facilitate the removal of fatty digestion products from the lacteals of the villi.

5. In the fasting intestine, segmentation wanes and periodic bursts of peristaltic activity are seen in which the contents of the gut are swept long distances along the tract. These are called 'housekeeper' contractions.

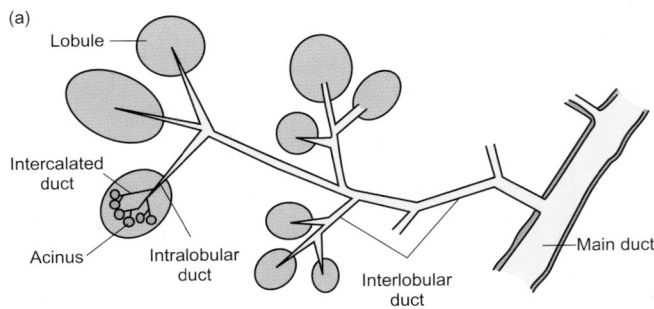

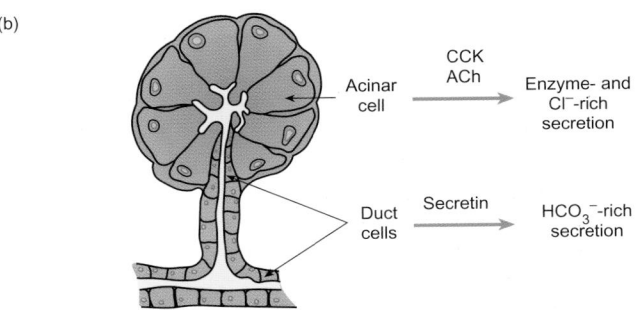

Fig. 18.21 (a) A schematic representation of the structure of the exocrine pancreas. (b) A summary of the sites of action and effects of secretin, cholecystokinin (CCK), and acetylcholine (ACh) on secretion by the acinar and duct cells of the exocrine pancreas.

chyme into the large intestine to ensure that water and electrolytes are fully absorbed from it in the colon. Its activity is governed by the neurons of the intramural plexuses. The sphincter is normally closed, but short-range peristaltic contractions of the terminal ileum cause the sphincter to relax and allow a small amount of chyme to pass through. The long-range reflexes ensure that the rate of emptying is matched to the ability of the colon to deal with the volume of chyme delivered. After a meal, for example, ileal emptying is enhanced through the operation of the gastro-ileal reflex.

18.10 The exocrine functions of the pancreas

The pancreas performs two distinct functions in the body. It acts both as an endocrine gland, secreting the hormones insulin and glucagon into the bloodstream, and as an accessory digestive (exocrine) organ secreting enzyme-rich fluid into the duodenum. The endocrine role of the pancreas is discussed fully in Chapter 27. Only its exocrine function will be described here.

Gross and fine structure of the pancreas

The pancreas lies deep to the stomach and extends across the abdomen for about 20 cm. The tail of the pancreas lies close to the spleen, while its head is encircled by the duodenum. Its anatomical situation is illustrated in Figs 18.1 and 18.19.

A schematic representation of the structure of the exocrine pancreas is shown in Fig. 18.21. It is similar to the salivary glands, being made up of lobules consisting of acinar cells that secrete enzymes and fluid into a system of microscopic (intercalated) ducts lined with epithelial cells which also secrete fluid. These drain into larger intralobular ducts which in turn empty into interlobular ducts. Finally they empty into the main pancreatic duct which extends along the axis of the pancreas. In most people, the main pancreatic duct fuses with the bile duct before emptying into the duodenum. There is also a smaller pancreatic duct (duct of Santorini) that drains directly into the duodenum. Acinar cells occupy more than 80 per cent of the total pancreatic volume and duct cells about 4 per cent. The endocrine islet cells occupy about 2–3 per cent of the gland, and the remainder consists of connective tissue, blood vessels, and so on.

The pancreas is supplied by branches of the celiac and superior mesenteric arteries and its venous drainage is via the portal vein. It is innervated by preganglionic parasympathetic vagal fibers, which synapse with cholinergic postganglionic fibers within the pancreas. Pancreatic blood vessels receive sympathetic innervation from the celiac and superior mesenteric plexuses.

The composition of pancreatic juice

The exocrine pancreas secretes about 1500 mL of fluid each day, the aqueous component of which is rich in bicarbonate and has a pH of about 8. Together with the intestinal secretions, it helps to neutralize the acidic chyme as it arrives in the duodenum. All the major enzymes needed to complete the digestion of fats, proteins and carbohydrates are also contained within the pancreatic juice (the enzyme component).

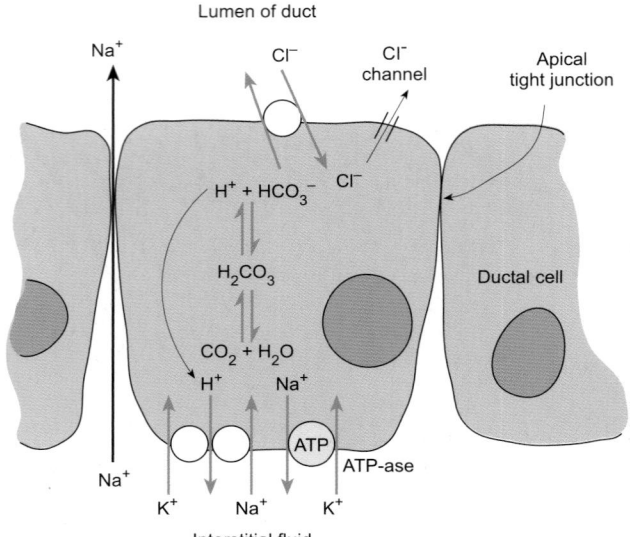

Fig. 18.22 The ionic movements underlying the secretion of an alkaline fluid by duct cells of the exocrine pancreas.

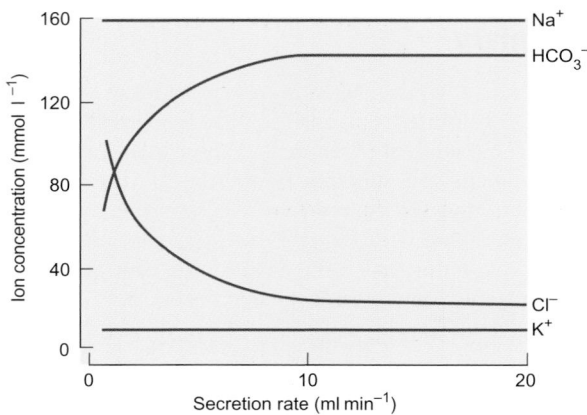

Fig. 18.23 The electrolyte composition of rabbit pancreatic juice as a function of the rate of secretion. The greater the rate of secretion, the higher is the bicarbonate concentration.

The aqueous component of pancreatic juice

This is formed almost entirely by the columnar epithelial cells which line the ducts. Resting secretion is chiefly from the intercalated and intralobular ducts but, during stimulation, the interlobular ducts also secrete pancreatic fluid. The fluid secreted by the duct cells is slightly hypertonic. It is rich in bicarbonate ions and has sodium and potassium concentrations similar to those of plasma.

Precise details of the ionic mechanisms underlying the secretion of the alkaline fluid have not been clarified. A possible sequence of events is illustrated in Fig. 18.22. Hydrogen ions are transported out of the cell into the interstitial fluid and thence to the plasma in exchange for either sodium or potassium ions. Bicarbonate ions are then transported out of the duct cell across its luminal membrane in exchange for chloride. Sodium diffuses from the interstitial fluid to the duct lumen via a paracellular

pathway to maintain electroneutrality. Water follows osmotically, moving transcellularly or paracellularly into the duct lumen.

Secretion of the aqueous component of pancreatic juice appears to be regulated by cyclic AMP, which increases the time for which the apical membrane Cl$^-$ channels are open. It may also stimulate the activity of the proton pumps in the basolateral membrane.

The ionic composition of the pancreatic fluid depends upon its rate of secretion as shown in Fig. 18.23. As it flows along the ducts, the primary secretion of the ductal epithelial cells undergoes modification. Bicarbonate ions are reabsorbed from the fluid in exchange for chloride ions. The result of this is that, at low flow rates, the bicarbonate content of the pancreatic fluid is lower (down to 20–30 mmol l^{-1}) than it is at high flow rates, when the fluid spends little time in the ducts and is therefore scarcely modified. At maximal flow rates, the bicarbonate concentration of human pancreatic juice is around 140 mmol l^{-1}. It can be seen from Fig. 18.23 that as the bicarbonate concentration falls, chloride levels rise correspondingly.

TABLE 18.2 The pancreatic enzymes			
Enzyme	**Zymogen**	**Activator**	**Action**
Trypsin	Trypsinogen	Enteropeptidase	Cleaves internal peptide bonds
Chymotrypsin	Chymotrypsinogen	Trypsin	Cleaves internal peptide bonds
Elastase	Pro-elastase	Trypsin	Cleaves internal peptide bonds
Carboxypeptidase	Procarboxypeptidase	Trypsin	Attacks peptides at C-terminal end
Amylase			Digests starch to maltose and oligosaccharides
Lipase			Cleaves glycerides, liberating fatty acids and glycerol
Colipase	Procolipase	Trypsin	Binds to micelles to anchor lipase to lipid
Phospholipase A$_2$	Prophospholipase	Trypsin	Cleaves fatty acids from phospholipids
Cholesterol esterase		Trypsin	Releases esterified cholesterol
RNAase			Cleaves RNA into short fragments
DNAase			Cleaves DNA into short fragments

The enzyme components of pancreatic juice

Pancreatic juice contains a wide array of digestive enzymes including proteolytic, amylolitic, and lipolytic agents as well as many others such as ribonuclease, deoxyribonuclease, and elastases. A list of the principal enzymes and their actions is given in Table 18.2.

Proteolytic enzymes of the pancreas

Proteolytic enzymes including *trypsin*, a number of *chymotrypsins*, and *carboxypeptidases* are stored within the acinar cells as zymogen granules. They are secreted in this inactive form (trypsinogen, chymotrypsinogens, and procarboxypeptidases) and activated inside the lumen of the small intestine. In this way, the pancreas, like the stomach, avoids self-digestion. Activation of trypsinogen may occur spontaneously, in response to the alkaline environment of the small intestine, or in response to enteropeptidase, one of the brush-border enzymes (see p. 397). Chymotrypsinogens are then activated by trypsin itself. Trypsin and chymotrypsins are endopolypeptidases which hydrolyze peptide bonds within the protein molecule to release some free amino acids and polypeptides of varying size. Carboxypeptidases (activated by trypsin), elastase, and aminopeptidases are then able to digest these further to release small peptides and amino acids.

It is important that trypsinogen is not activated within the acinar cells themselves or as it passes along the ducts. Activation is normally prevented by the maintenance of an acid environment within the zymogen granules (probably through the action of a proton pump) and by the presence of *trypsin inhibitor* in the pancreatic juice. The latter binds to any active trypsin that may be present to form an inactive complex. Acute necrotizing *pancreatitis* is a life-threatening disorder often caused by the reflux of bile into the pancreas or because of alcoholism. It is characterized by autodigestion of the pancreatic tissue, with inflammation and tissue damage caused by the escape of activated enzymes from the pancreas.

Pancreatic amylase

Although salivary amylase may initiate the digestion of starch in the mouth and possibly the stomach, pancreatic α-amylase is responsible for the majority of starch digestion in the duodenum. This enzyme is secreted in its active form and is stable between pH 4 and 11, although its optimal pH is 6.9. Like salivary amylase, it splits the α-1,4-glycosidic bond (see Chapter 2) but, unlike the salivary enzyme, it is able to attack uncooked as well as cooked starch. Within 10 minutes or so of entering the small intestine, starch is entirely converted to various oligosaccharides, chiefly maltose and maltriose. The intestinal brush-border disaccharidases then hydrolyze them into glucose. In certain pathological conditions including acute pancreatitis, blood concentrations of pancreatic amylase start to increase. Measurement of plasma enzyme levels may provide useful diagnostic information about the extent and progression of injury to pancreatic tissue.

Lipolytic enzymes of the pancreas

Pancreatic juice contains several lipases, secreted in the form of inactive zymogens, among the most important of which are colipase, cholesterol esterase, and phopholipase A2. These are activated by trypsin in the duodenal lumen. Pancreatic lipase (triacylglycerol hydrolase) is probably secreted in its active form. It hydrolyzes water-insoluble triglycerides to release free fatty acids and monoglycerides. Colipase anchors the lipase close to the oil:water interface so that it is able to act more effectively. Phospholipase A2 digests phospholipids to release free fatty acids and lysolecithin. The role of bile in the digestion and absorption of fats is discussed below.

The regulation of pancreatic secretion

Like gastric secretion, pancreatic secretion is regulated both by the activity of the vagus nerves and by hormones. However, the endocrine control of pancreatic secretion is the more important. A list of the chief modulators of exocrine pancreatic secretion is given in Table 18.3. As for the stomach, the process of secretion can be considered in three phases: cephalic, gastric, and intestinal. The cephalic phase is under nervous control, while the gastric and intestinal phases are controlled chiefly by hormones. Figure 18.24 illustrates the principal factors that regulate pancreatic secretion.

TABLE 18.3 Chemical regulators of exocrine pancreatic secretion

Agent	Action on pancreas
Cholecystokinin (CCK) / Gastrin / Acetylcholine / Substance P	Increased secretion of pancreatic enzymes and chloride-rich fluid by acinar cells
Secretin / Vasoactive intestinal polypeptide (VIP) / Histidine / Isoleucine	Increased secretion of bicarbonate-rich fluid from duct cells
Insulin / Insulin-like growth factors (IGFs)	Increased enzyme synthesis and secretion, trophic effects
Somatostatin	Inhibits secretion from acinar and duct cells

The stimuli are given in order of importance.

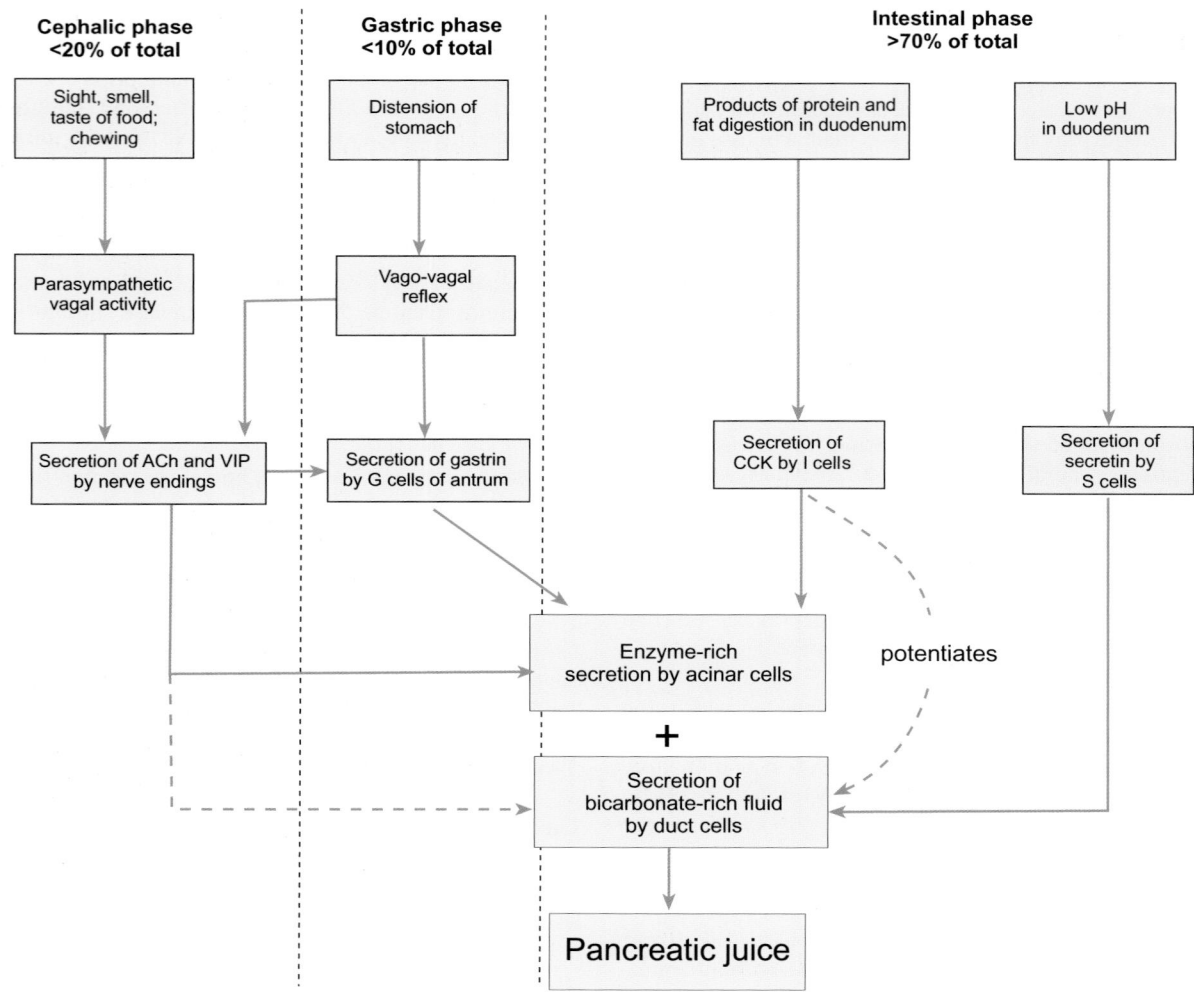

Fig. 18.24 The major factors involved in the regulation of pancreatic secretion during the different phases of digestion.

The cephalic phase of pancreatic secretion

The acinar cells and the smooth muscle cells of the ducts and blood vessels are innervated by parasympathetic vagal efferent fibers. Stimulation of these fibers causes the release of zymogen granules from the acinar cells into the ducts and an increase in blood flow. The blood vessels also receive some sympathetic vasoconstrictor fibers whose activity causes a reduction in blood flow.

Parasympathetic vagal activity is enhanced by the sight, smell, and taste of food. The neurotransmitters acetylcholine and VIP are released and act synergistically to increase blood flow and promote the secretion of pancreatic juice. In addition to the direct action of vagal efferents, a small part of the cephalic phase of pancreatic secretion is mediated by gastrin released from the antral cells in response to vagal stimulation.

The gastric phase of pancreatic secretion

Gastrin is chiefly responsible for this relatively small component of pancreatic secretion. Gastrin is secreted in response to distension of the stomach and the presence of amino acids and peptides in the antrum. Distension also elicits secretion via a vagovagal reflex.

The intestinal phase of pancreatic secretion

This phase accounts for more than 70 per cent of total secretion by the exocrine pancreas. It occurs in response to hormones secreted by the upper intestinal mucosa. CCK is secreted when the mucosal surface is bathed in monoglycerides, fatty acids, peptides, and amino acids (especially tryptophan and phenylalanine). Secretin is secreted in response to low pH. CCK stimulates the production of an enzyme-rich fluid from the acinar cells, while secretin increases the rate of flow of bicarbonate-rich fluid from the ductal cells. Furthermore, CCK appears to potentiate the secretory effects of secretin.

18.11 The role of the liver and gall bladder in the function of the gastrointestinal tract

The liver and gall bladder are accessory organs associated with the small intestine. The liver is the largest internal organ, weighing 1.3 kg on average. It receives and processes the nutrient-rich venous blood reaching it from the GI tract and performs many

Summary

1. The exocrine portion of the pancreas consists of acinar cells which secrete enzymes and fluid into a system of tiny ducts lined with epithelial cells. The epithelial cells secrete alkaline fluid and modify the primary acinar secretion.

2. All the major enzymes required to complete the digestion of fats, carbohydrates, and proteins are contained within the pancreatic juice. The ionic composition of the pancreatic juice depends upon its rate of secretion. At high rates of secretion, the bicarbonate content of the juice is higher than at lower rates.

3. Most of the proteolytic enzymes (trypsins) are stored in the acinar cells as inactive precursors (zymogen granules) to avoid self-digestion. Activation of these enzymes takes place in the duodenum.

4. Pancreatic α-amylase is responsible for starch digestion to oligosaccharides in the duodenum. It is secreted in its active form.

5. Several lipases are present in pancreatic juice. They hydrolyze water-insoluble triglycerides to release free fatty acids and monoglycerides.

6. Control of exocrine pancreatic secretion is chiefly hormonal, although the initial cephalic phase of secretion is under the control of parasympathetic nerves. Gastrin contributes to the gastric phase of secretion but about 70 per cent of secretion occurs during the intestinal phase in response to secretin and CCK. These hormones are released by the upper intestinal mucosa in response to H$^+$ ions and the products of fat and protein digestion.

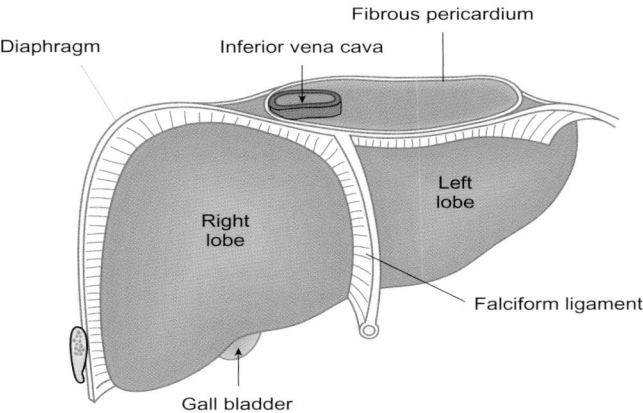

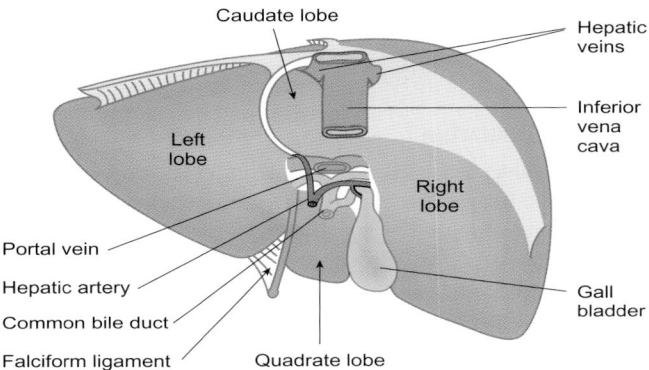

Fig. 18.25 The main anatomical features of the liver as seen from the front (top) and from the back (bottom).

vital metabolic and homeostatic functions, which are summarized briefly here (see also Chapter 31, p. 604). The liver plays an extremely important role in energy metabolism. It stores glucose as glycogen, converts amino acids to glucose, and degrades lipids. It is also important in biosynthesis. It synthesizes all the plasma proteins except for the immunoglobulins, including complement and clotting factors. It also manufactures protein carriers for cholesterol and the triacylglycerols. The liver secretes bile, which contains the bile salts that are crucial for both the emulsification of fats and their dispersion as micelles prior to their absorption. Finally, the liver converts ammonia to the much less toxic urea and adds polar groups to many drugs, some hormones, and certain metabolites so that they can be excreted in the urine and the bile.

The structure of the liver

Figure 18.25 illustrates the main structural features of the liver. Situated in the upper right quadrant of the abdominal cavity (see Fig. 18.1), the liver consists of four lobes surrounded by a tough fibroelastic capsule called Glisson's capsule. The falciform ligament, which attaches the liver to the diaphragm and anter-

ior abdominal wall, separates the major right and left lobes. The smaller visceral and quadrate lobes are on the posterior surface of the liver. A dorsal mesentery, the lesser omentum, attaches the liver to the lesser curvature of the stomach. The gall bladder rests in a recess on the inferior surface of the right lobe of the liver. Bile leaves the liver through *terminal bile ducts* which fuse to form the large common hepatic duct. As this duct passes towards the duodenum it fuses with the *cystic duct* draining the gall bladder to form the *bile duct*. Separating the bile duct from the duodenum is the sphincter of Oddi, a ring of muscle which prevents the reflux of bile.

Microscopically, the liver consists of between 50 000 and 100 000 *lobules* separated by septa. These are roughly hexagonal structures, 1–2 mm in diameter, which form the functional units of the organ. Each lobule consists of a central vein that empties into the hepatic vein, from which single columns of *hepatocytes* (liver cells) radiate towards the surrounding layer of thin connective tissue. Between the hepatocytes lie small *bile canaliculi* that empty into the bile ducts and then the terminal bile ducts. At each of the six corners of a lobule lies a *portal triad*, so called because three structures are always present there. These are a branch of the hepatic artery, a branch of the portal vein, and a bile duct. A simplified view of the structure of a liver lobule is shown in Fig. 18.26

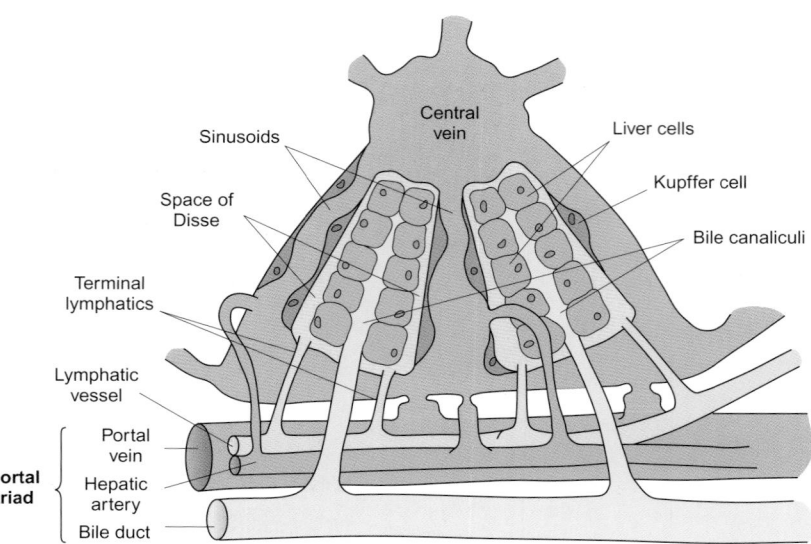

Fig. 18.26 The basic histological structure of a liver lobule.

Hepatic circulation

At rest the liver normally receives about 25 per cent of the cardiac output. It is unique among the abdominal organs in having a dual blood supply—the hepatic artery carrying about 400 ml min⁻¹ and the portal vein supplying about 1000 ml min⁻¹ of nutrient-rich blood. Small portal venules lying in the septa between lobules receive blood from the portal veins. From the venules, blood flows into branching *sinusoids* between the columns of hepatocytes. The sinusoids essentially form a leaky capillary network from which blood flows into the central vein of the lobule. Deoxygenated blood from the central veins empties into the hepatic veins, which join the inferior vena cava just below the level of the diaphragm. The pressure in the portal vein is about 1.3 kPa (10 mmHg), while that in the hepatic vein is only slightly lower (about 0.6 kPa or 5 mmHg). Consequently, 200–400 ml of blood are stored within the capacitance vessels of the liver. This blood can be shunted back into the systemic circulation during periods of hypovolemia or shock.

The interlobular septa also contain hepatic arterioles derived from branches of the hepatic artery, many of which drain directly into the sinusoids supplying them with blood fully saturated with oxygen.

The sinusoids are lined with two types of cells: typical endothelial cells and phagocytic Kupffer cells. Since the epithelium is fenestrated (see Chapter 15), there is almost no barrier to the exchange of materials of molecular weights up to 250 000 between the sinusoids and the hepatocytes. Microvilli on the area of hepatocyte membrane facing the sinusoid increase the surface area for exchange. The space between the hepatocytes and the sinusoidal wall is called the space of Disse (see Figs 18.26 and 18.28). This contains a system of supporting collagen fibers and is drained by terminal lymphatic vessels.

The production of bile

The hepatocytes secrete a fluid known as *hepatic bile* into the bile canaliculi. This is an isotonic fluid with a pH between 7 and 8 that resembles plasma in its ionic composition. It also contains bile salts, bile pigments, cholesterol, lecithin, and mucus. As it passes along the bile ducts, the ductal epithelial cells modify this primary secretion by secreting a watery bicarbonate-rich fluid. This adds considerably to the volume of the bile, so that overall the liver produces 600–1000 ml of bile each day. The bile may be continuously discharged into the duodenum or stored in the gall bladder, during which time its composition changes (see below).

The chemical nature of bile acids and bile salts

The bile acids are derived from the metabolism of cholesterol. Cholic acid and chenodeoxycholic acid are formed in the hepatocytes themselves and are known as the *primary bile acids*. In the intestine, the secondary bile acids, deoxycholic acid and lithocholic acid, are formed in small amounts from the primary acids by the dehydroxylating action of bacteria. The primary bile acids are conjugated (by means of a peptide linkage) to amino acids such as glycine and taurine in a complex with sodium to form water-soluble *bile salts* prior to secretion into the bile. Figure 18.27 illustrates the structures of the bile acids and the conjugation of cholic acid with glycine.

Bile salts are amphipathic, that is they have both hydrophobic and hydrophilic regions. The bile salts form aggregates called *micelles* when they reach a certain concentration in the bile. (This is known as the *critical micellar concentration*.) The micelles are organized so that the hydrophilic groups of the bile salts face the aqueous medium while the hydrophobic groups face each other to form a core. This chemical characteristic of the bile salts is of key importance to their role in the emulsification of fats.

Bile-acid-dependent and bile-acid-independent components of bile secretion

Two distinct secretory mechanisms are involved in the elaboration of bile by the liver, giving rise to the so-called bile-acid-dependent and bile-acid-independent components of bile.

♦ The rate at which bile salts are actively secreted into the canaliculi depends upon the rate at which bile acids are returned from the small intestine to the hepatocytes via the *enterohepatic circulation*. Therefore this component of bile secretion is referred to as the bile-acid-dependent fraction.

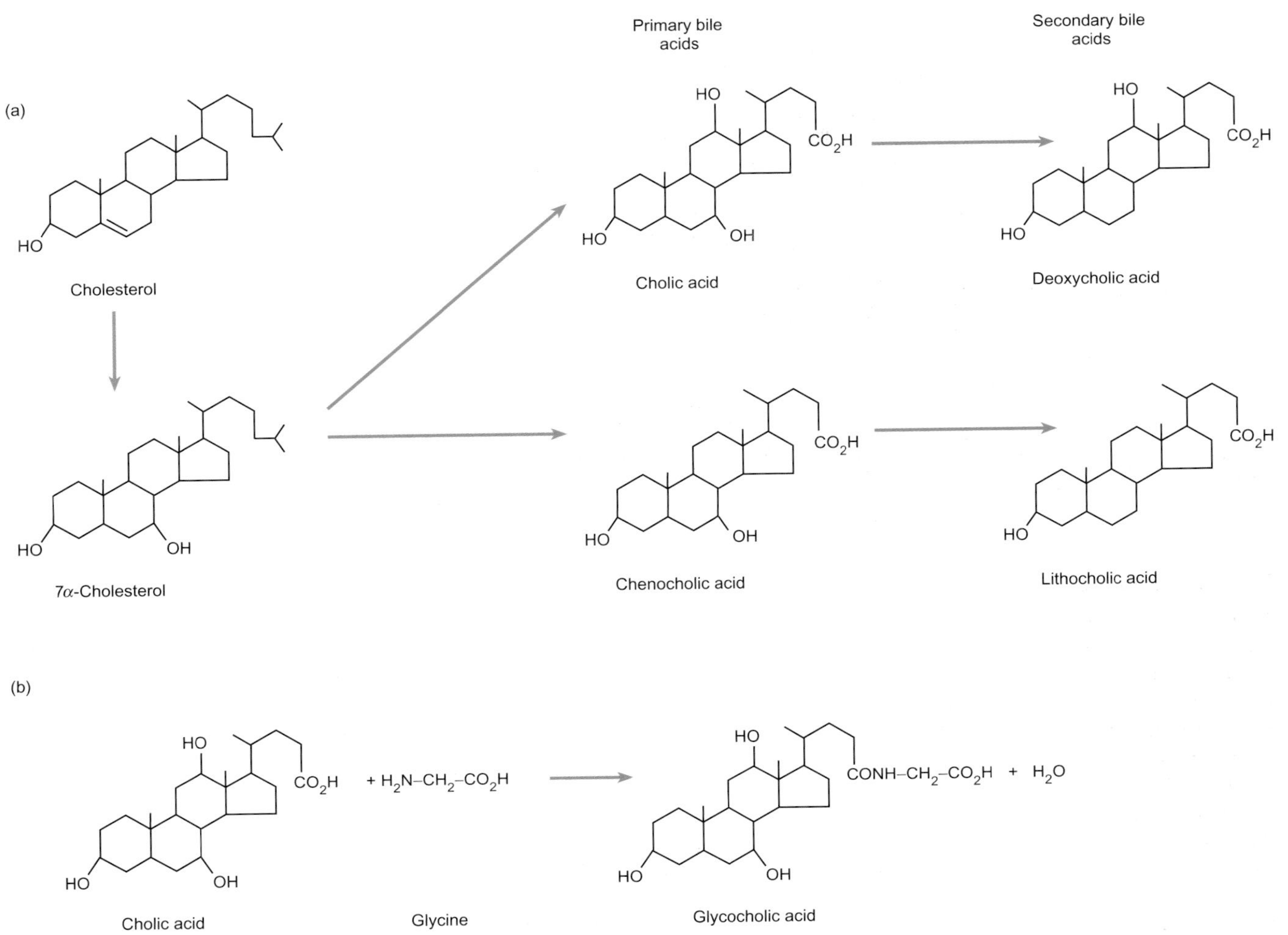

Fig. 18.27 (a) The structure of the bile acids and their formation from cholesterol and (b) the conjugation of cholic acid with glycine to form the bile salt glycocholic acid.

◆ The bile-acid-independent fraction of bile secretion refers to the secretion of water and electrolytes by the hepatocytes and the ductal epithelial cells. Sodium is transported actively into the bile canaliculi and is followed by the passive movement of chloride ions and water. Bicarbonate ions are actively secreted into the bile by the ductal cells and are followed by the passive movement of sodium and water. The processes involved in the formation of hepatic bile are summarized in Fig. 18.28.

The enterohepatic circulation

About 94 per cent of the bile salts that enter the intestine in the bile are reabsorbed into the portal circulation by active transport from the distal ileum. Many of the bile salts return to the liver unaltered and are recycled. Some are deconjugated in the gut lumen and returned to the liver for reconjugation and recycling. A small fraction of the deconjugated bile acids may undergo modification by intestinal bacteria to secondary bile acids. Some of these, particularly lithocholic acid, are relatively insoluble and are excreted in the feces. It is estimated that bile acids

may be recycled up to 20 times before finally being excreted. A schematic illustration of the enterohepatic circulation of the primary bile acids is shown in Fig. 18.29.

The regulation of bile secretion

Substances which increase the rate of bile secretion are called *choleretics*. *Cholagogues* are substances such as CCK that increase the flow of bile by stimulating the emptying of the gall bladder. The major regulator of the rate of production of hepatic bile is the return of bile salts to the hepatocytes via the enterohepatic circulation. This provides the driving force for fluid transport into the biliary system. Although the production of hepatic bile is not under hormonal control, the secretion of bicarbonate-rich watery fluid by the ductal epithelial cells is enhanced by secretin and, to a lesser extent, by glucagon and gastrin. A further stimulus to hepatic production of bile is thought to be the increase in liver blood flow that follows a meal. A meal will also result in an increase in the rate of reabsorption of bile salts via the enterohepatic circulation. This in turn will stimulate the bile-acid-dependent fraction of bile secretion.

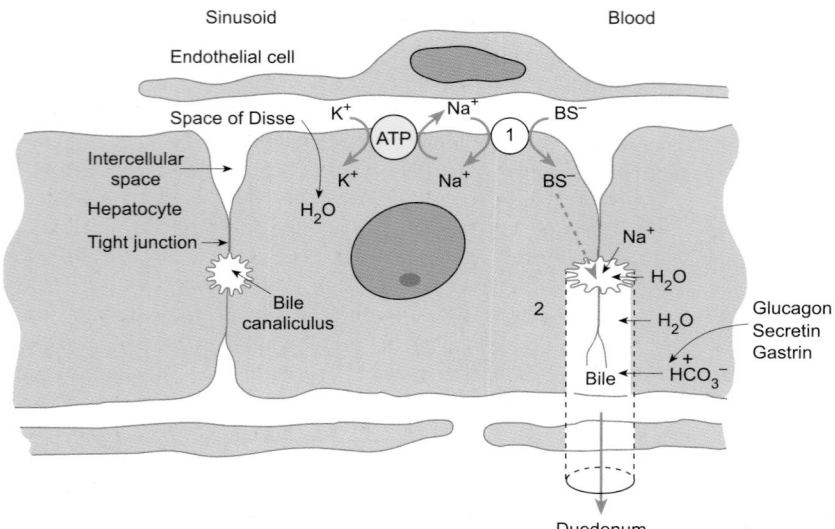

Fig. 18.28 The processes involved in the formation of bile by the hepatocytes (BS, bile salt).

Other important constituents of bile include phospholipids, cholesterol, and bile pigments

Bile is the major route for the excretion of cholesterol from the body. Phospholipids, especially lecithin, are also secreted into the bile. Both are secreted as lipid vesicles, which then form micelles following emulsification with bile salts. The cholesterol partitions into the hydrophobic core while the lecithin, which is amphipathic, lies partly in the core and partly near the outer surface of the micelle. Any excess cholesterol that cannot be dispersed into micelles may form crystals in the bile. These may contribute to the formation of gallstones in the hepatic ducts or the gall bladder by acting as nuclei for the deposition of calcium and phosphate salts. If the common bile duct becomes blocked by a gallstone, bile cannot enter the duodenum. There is disten-

sion and a build up of pressure within the gall bladder that can result in severe pain (*biliary colic*) and jaundice (see below).

Bile is a vehicle for the elimination of bile pigments and other waste products, particularly less polar molecules of high molecular weight that are not excreted by the kidneys. *Bile pigments* are the excretory products of heme and are responsible for the characteristic colors of both bile and feces. They form about 0.2 per cent of the total bile composition and are formed from the breakdown of old red blood cells in the spleen. The major bile pigment is *bilirubin*, which is relatively insoluble and is carried to the liver mainly in combination with plasma albumin. In the hepatocytes, about 80 per cent of the bilirubin is conjugated with glucuronic acid by an enzyme, glucuronyl transferase, to form bilirubin diglucuronide. This is water soluble and enters the bile, giving it its characteristic greenish-yellow color. The remaining bilirubin is conjugated with sulfate to form bilirubin sulfate, or with a variety of other hydrophilic agents.

In the intestine, particularly the colon, bilirubin diglucuronide is hydrolyzed by bacteria to form urobilinogen, which is extremely water soluble and colorless, as well as stercobilin and urobilin which give the feces their characteristic brown color. Some of the urobilinogen is reabsorbed from the intestine into the blood. From there, it is either re-secreted back into the bile by the liver, or excreted by the kidneys into the urine. The processes of bilirubin formation, circulation, and elimination are shown diagrammatically in Fig. 18.30.

Accumulation of bilirubin in the blood causes jaundice

Jaundice (icterus) is due to an abnormal level of bilirubin in the blood (hyperbilirubinemia). It is characterized by yellow discoloration of the skin, the sclera of the eyes, and the deep tissues. There are many causes of jaundice, the most important of which are excessive hemolysis of red cells, impaired uptake of bilirubin by hepatocytes, and obstruction of bile flow either through the bile canaliculi or the bile ducts. Excessive hemolysis may occur following a poorly matched blood transfusion or in certain hereditary disorders. Jaundice is also seen in newborns

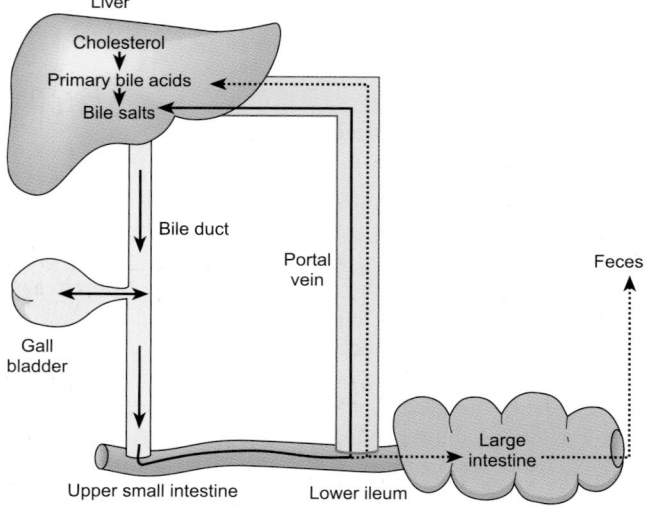

Fig. 18.29 The enterohepatic circulation. Only the recirculation of the primary bile acids is illustrated. The transport of conjugated bile acids is shown by the solid line and that of the deconjugated bile acids by the broken line.

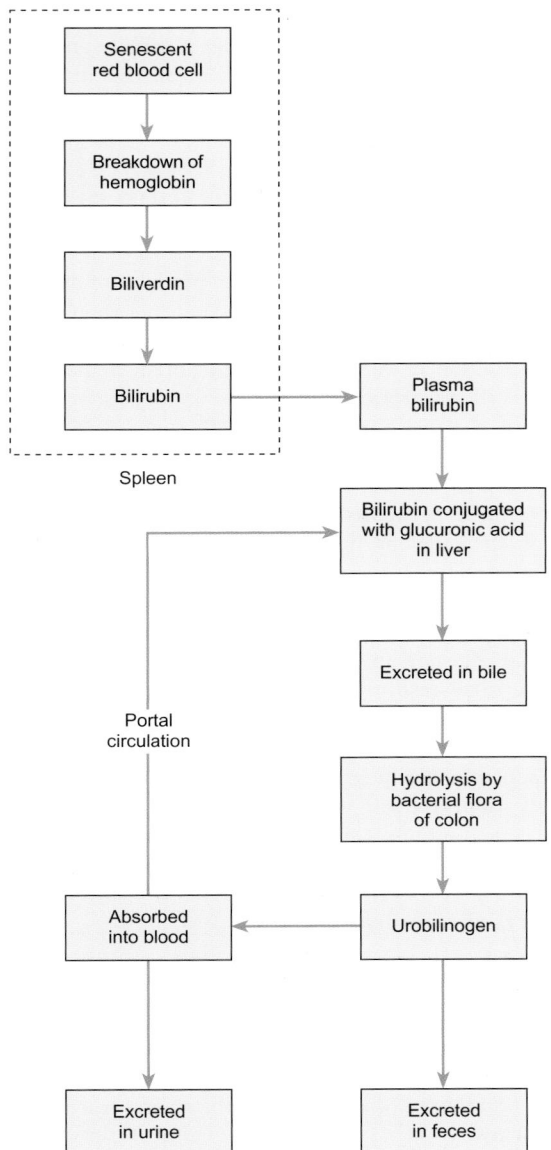

Fig. 18.30 The processes of bilirubin formation, circulation, and elimination.

The role of the gall bladder

The gall bladder is a thin-walled muscular sac 10 cm long protruding from the inferior margin of the liver (see Fig. 18.25). It stores bile that is not required immediately for digestion. The mucosa of the gall bladder, like that of the stomach, is thrown into folds when the organ is empty. These can expand to accommodate up to 60 ml of bile during the period between meals.

Between meals, most of the bile produced by the liver is diverted into the gall bladder because of the relatively high level of tone in the sphincter of Oddi. The gall bladder concentrates the bile by absorbing sodium, chloride, bicarbonate, and water from it. As a result, the bile salts present in the bile of the gall bladder may be concentrated as much as 20-fold. Active transport of sodium by the mucosa from the lumen to the blood is the primary mechanism involved in the concentration of the bile. The anions chloride and bicarbonate are absorbed to maintain electroneutrality, and water follows passively. Potassium concentrations rise as water is absorbed, but subsequently fall as potassium diffuses passively down the concentration gradient

TABLE 18.4 The relative concentrations of some constituents of hepatic bile and gall bladder bile

Solute	Solute concentration ratio (gall bladder bile/hepatic bile)
Na^+	1.7
Ca^{2+}	5.0
HCO_3^-	0.2
Cl^-	0.2
Bile acids	8.9
Bile pigments	4.0
Cholesterol	8.3
Lecithin	8.0
Bile osmolality	290–300 mOsm kg^{-1}
Secreted volume of gall bladder bile	500 ml day^{-1}

whose fetal red cells are hemolyzing more quickly than the immature liver can process the bilirubin.

Jaundice resulting from a failure of the liver to take up or conjugate bilirubin is known as *hepatic jaundice*. Hepatitis and cirrhosis are the most common causes of this disorder.

Obstructive jaundice occurs if bile is prevented from flowing from the liver to the intestine. Gallstones, strictures or tumors of the bile duct, and pancreatic tumors are common causes. Pruritus (itching) often accompanies this type of jaundice, which is caused by the accumulation of bile acids in the blood. The feces are pale in color due to the absence of bilirubin in the bile and often contain fatty streaks due to the lowered absorption of dietary fat. However, the urine is darker than normal due to the increased excretion of bilirubin via the kidneys.

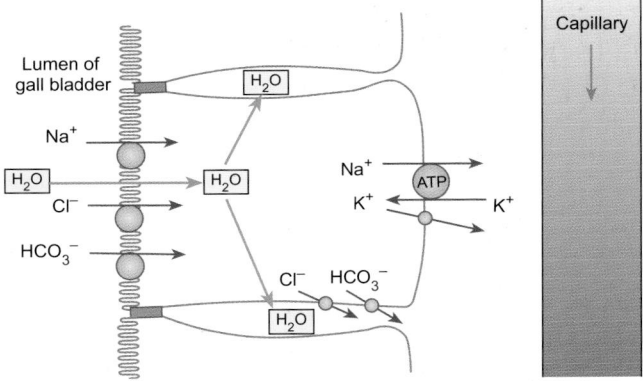

Fig. 18.31 The absorption of salt and water in the gall bladder, illustrating the standing-gradient hypothesis. Sodium, chloride, and bicarbonate are transported across the basolateral membrane into the intercellular space. Water follows passively. NB. The Na$^+$, K$^+$ ATPase is distributed all over the basolateral surface of the epithelial cells.

established. Table 18.4 shows the solute concentration ratios for gall bladder and hepatic bile resulting from absorptive processes occurring in the gall bladder.

Water moves out of the gall bladder by osmosis, despite the fact that the mucosa is not highly permeable to water. The mucosa consists of a single layer of tall columnar epithelial cells bound at their apical regions by tight junctions so that long lateral channels form between them. As salts are transported into these channels, local regions of high osmotic pressure are created, with tonicity highest at the apical regions of the channel. This sets up a standing osmotic gradient that permits the continuous absorption of water from the gall bladder to the interstitial fluid. Figure 18.31 illustrates the mechanism by which solutes and water are absorbed by the gall bladder.

Contraction of the gall bladder forces bile into the duodenum

Within a few minutes of starting a meal, particularly one that is rich in fats, the muscle of the gall bladder contracts, providing a pressure which forces bile towards the duodenum. This initial response is mediated via the vagal nerves but the major stimulus for contraction is CCK. This hormone is secreted in response to the presence of fatty and acidic chyme in the intestine. CCK also stimulates pancreatic enzyme secretion and relaxes the sphincter of Oddi so that bile and pancreatic juice can enter the duodenum. Parasympathetic vagal activity makes a relatively minor contribution to the stimulation of gall bladder contrac-

tion. Conversely, emptying of the gall bladder is suppressed by sympathetic activity. The gall bladder normally empties completely about an hour after a fat-rich meal. This maintains the level of bile acids in the duodenum above the critical micellar concentration.

18.12 Absorption of digestion products in the small intestine

Absorption is the process by which the products of digestion are transported into the epithelial cells that line the GI tract and from there into the blood or lymph draining the tract. Each day about 8–10 liters of water and up to 1 kg of nutrients pass across the gut wall. Absorption through the GI mucosa occurs by the processes of active transport and facilitated diffusion. Because the epithelial cells of the intestinal mucosa are joined at their apical (luminal) surfaces by tight junctions, products of digestion cannot move between the cells. Instead, they must move through the cells and into the interstitial fluid abutting their basal membranes if they are to enter the capillary blood. This process is called transcellular transport. The physical principles of active and passive transport are described together with epithelial transport in Chapter 4. The mechanisms involved in the intestinal absorption of iron are described in detail in Chapter 13. The intestinal absorption of calcium and its regulation by parathyroid hormone and the metabolites of vitamin D are described in Chapter 12. The specific mechanisms by which the digestion products of the principal nutrients (fats, carbohydrates, and proteins) are absorbed will be discussed here.

The absorption of monosaccharides, the digestion products of carbohydrate

As described earlier, starch and other large polysaccharides are broken down by the pancreatic and brush-border enzymes to form glucose. Galactose and fructose are produced by the breakdown of lactose and sucrose respectively. These are absorbed largely in the upper small intestine, entering the blood of the hepatic portal vein. None remain in the chyme reaching the terminal ileum.

The absorption of glucose depends on its concentration in the gut lumen. When its concentration is low, glucose is absorbed into the enterocytes against its concentration gradient by the sodium-dependent cotransporter SGLT1, described in detail in Chapter 4 (p. 39). The glucose is transported across the basolateral surface via the glucose carrier GLUT2, as illustrated in Fig. 18.32(a). The sodium gradient that drives this form of glucose transport is maintained by the Na$^+$, K$^+$ATPase and enables the gut to scavenge all the available glucose.

After a meal, the glucose concentration in the gut lumen rises substantially, and may exceed 100 mmol l^{-1}, and this saturates the SGLT1 carriers. However, the increased activity of SGLT1 activates an intracellular signaling cascade that results in the insertion of preformed GLUT2 carriers from intracellular vesicles into the apical membrane (cf. the insertion of water channels into the apical membrane of the P cells of the collecting ducts). The GLUT2 carriers of the apical membrane permit glucose to diffuse into the cells down its concentration gradient, as shown in Fig. 18.32(b). As before, glucose leaves the enterocytes

Summary

1. The liver secretes 600–1000 ml of bile each day. Bile is vital for the processing of fats by the small intestine. It is stored and concentrated in the gall bladder, which contracts to deliver bile to the duodenum following a meal.

2. Bile acids are important constituents of bile. They become conjugated to amino acids to form bile salts, which are amphipathic (having both hydrophobic and hydrophilic regions). At high concentration, bile salts aggregate together to form micelles. The enterohepatic circulation returns about 94 per cent of the bile salts that enter the small intestine to the liver.

3. The formation of bile is stimulated by bile salts, secretin, glucagon, and gastrin circulating in the blood. The release of bile stored in the gall bladder is stimulated by CCK secreted in response to the presence of chyme in the duodenum. It is also enhanced in response to an increase in the rate of bile salt recirculation. This occurs following a meal because hepatic blood flow increases.

4. Bile pigments (the excretory products of heme) and other waste products are excreted in the bile. Bilirubin is the principal pigment. In the hepatocytes it is conjugated with glucuronic acid to form the water-soluble bilirubin diglucuronide which enters the bile. Failure to excrete bile pigments leads to their accumulation in the blood and the development of jaundice.

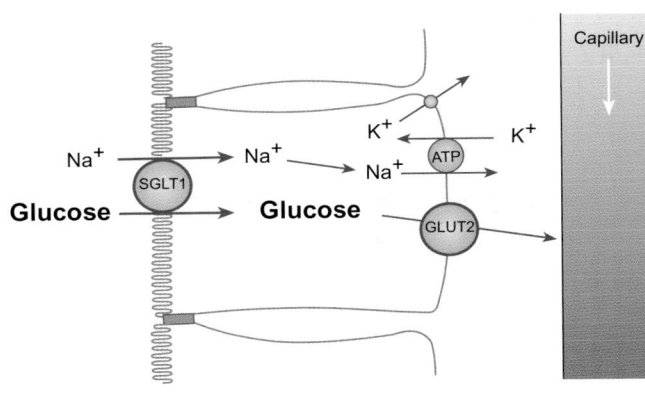

(a) Low glucose

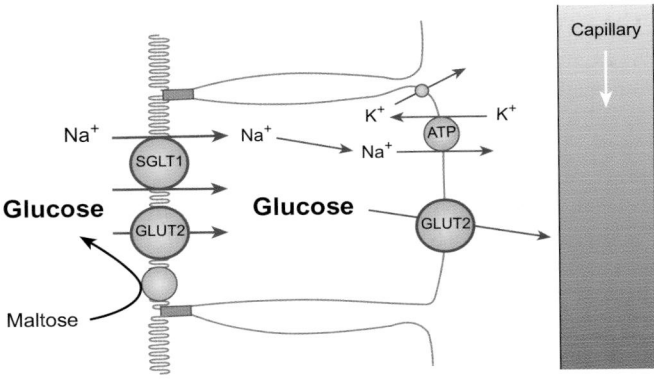

(b) After a meal

Fig. 18.32 Glucose absorption by the small intestine. (a) When the glucose concentration in the lumen of the small intestine is low, glucose is absobed by secondary active transport using the SGLT1 cotransporter. This permits effective absorption of all the available glucose. (b) After a meal, the glucose concentration in the lumen is high (in excess of 100 mmol l⁻¹). The glucose transporter GLUT2 becomes inserted in the apical membrane, allowing glucose to diffuse into the enterocytes down its concentration gradient (see text for further detail). In both situations, glucose is transported to the extracellular space across the basolateral membrane by the GLUT2 transporter. From there it diffuses into the blood and is transported to the liver via the portal vein.

via the GLUT2 carriers of the basolateral surface. It then diffuses from the extracellular space into the blood and is transported to the liver via the portal vein. Galactose is absorbed by the same processes as glucose.

Fructose is absorbed from the intestinal lumen by sodium-independent facilitated diffusion. It is taken up into the enterocytes mainly via a specific carrier called GLUT5 and leaves the cells via the GLUT2 carriers of the basolateral membrane (Fig. 18.33). As the plasma concentration of fructose is very low, the gradient for passive fructose absorption is always favorable.

The absorption of peptides and amino acids, the digestion products of proteins

Dietary protein is broken down by gastric and pancreatic protease enzymes to small peptides and amino acids. Each day approximately 200 g of amino acids and small peptides are

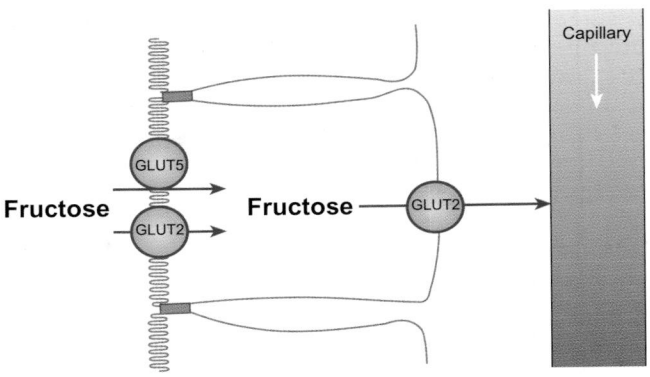

Fig. 18.33 Fructose absorption by the small intestine occurs by facilitated diffusion and uses the carrier GLUT5, which is selective for fructose. Fructose can also enter the enterocytes via GLUT2 when this carrier has become inserted in the apical membrane.

absorbed from the small intestine of an adult eating a normal mixed diet. In order to maintain a positive nitrogen balance and to meet the needs of an adult body for tissue growth and repair, it is essential that at least 50 g of amino acids are absorbed each day. Large peptides and whole proteins are not normally absorbed, although some may enter the bloodstream. For example, the immunoglobulins present in colostrum (see Chapter 21) appear to be absorbed intact across the intestinal epithelium of the gut of neonates. Amino acids are absorbed at the brush border of the intestinal epithelial cells by a sodium-dependent cotransport mechanism similar to that utilized for the absorption of monosaccharides (Fig. 18.34). At least 10 separate transporters have now been characterized for the transport of amino acids. Seven of these are located in the brush-border membrane and three in the basolateral membrane. Once the amino acids have entered the enterocyte, they cross the basolateral surface by carrier-mediated transport. They then enter the capillaries of the villus from where they travel to the liver via the portal vein. Most of the amino acids are absorbed in the first part of the small intestine. A few may enter the colon where they are metabolized by the colonic bacteria.

Small peptides (mainly dipeptides) are transported into the enterocytes by another carrier that is not linked to sodium but is believed to be linked to the influx of hydrogen ions. This carrier

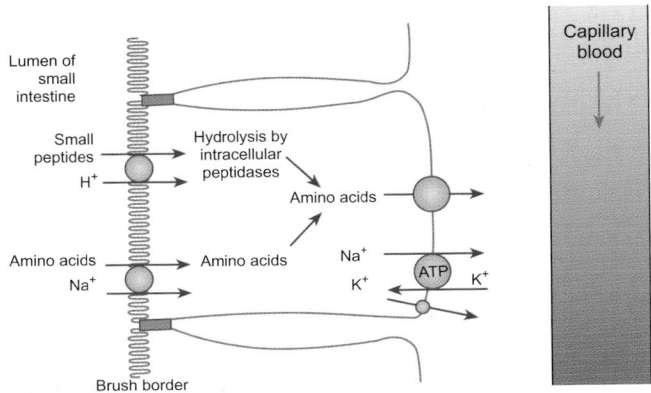

Fig. 18.34 The mechanisms by which amino acids and small peptides are absorbed in the small intestine.

is also responsible for the rapid uptake by the gut of certain drugs such as the antihypertensive drug captopril. Once the peptides enter the intracellular compartment, they are broken down to their constituent amino acids. These leave the enterocytes via the amino acid carrier system of the basolateral surface (Fig. 18.34). Up to half of the ingested protein is now believed to be absorbed in this way.

The absorption of monoglycerides and free fatty acids, the digestion products of fats

Because of their insolubility in water, fats pose a special problem for the GI tract in terms of both their digestion and their absorption. Bile salts play an essential part in each of these processes. In the stomach, ingested fats form large fat globules. As these globules enter the duodenum, they are coated with bile salts. The non-polar regions of the bile salts cling to the fat molecules while their hydrophilic polar regions allow them to repel each other and interact with water. As a result, fatty droplets are pulled off the large fat globules and a stable emulsion is created. (An emulsion is an aqueous suspension of fatty droplets each about 1 μm in diameter.) This dispersal of fat molecules greatly increases the number of triglycerides exposed to the pancreatic lipases and facilitates their breakdown to monoglycerides and free fatty acids.

Each day about 80 g of fat are absorbed from the small intestine, largely in the jejunum. The monoglycerides and free fatty acids liberated by the activity of the pancreatic lipases become associated with bile salts and lecithin to form micelles. The non-polar core of the micelle also contains cholesterol and fat-soluble vitamins. The hydrophilic outer region of the micelle enables it to enter the aqueous layer surrounding the microvilli that form the brush border of the enterocytes. Monoglycerides, free fatty acids, cholesterol, fat-soluble vitamins, and lecithin then diffuse passively into the duodenal cells while the bile salt portion of the micelle remains within the lumen of the gut until the terminal ileum where it is reabsorbed. The majority of the bile salts entering the small intestine are recycled by the entero-hepatic circulation (see p. 406).

Small amounts of short-chain fatty acids are absorbed directly from the intestinal epithelial cells into the capillary blood by passive diffusion. However, the majority of the products of fat digestion undergo further chemical processing inside the enterocytes. In the smooth endoplasmic reticulum, triglycerides are reformed by the re-esterification of monoglycerides, phospholipids are resynthesized, and much of the cholesterol undergoes re-esterification. The lipids accumulate in the vesicles of the smooth endoplasmic reticulum to form *chylomicrons*, which are released from the cells by exocytosis at the basolateral membrane. From here, they enter the lacteals of the villi and leave the intestine in the lymph from where they are released into the venous circulation via the thoracic duct. Thus lipids avoid the hepatic portal vein and bypass the liver in the short term. Figure 18.35 illustrates the processes involved in the absorption of fat digestion products.

The feces contain about 5 per cent fat, most of which is derived from bacteria. Increased amounts of fat are found in the feces if bile production is diminished or if bile is prevented from entering the duodenum due to biliary obstruction.

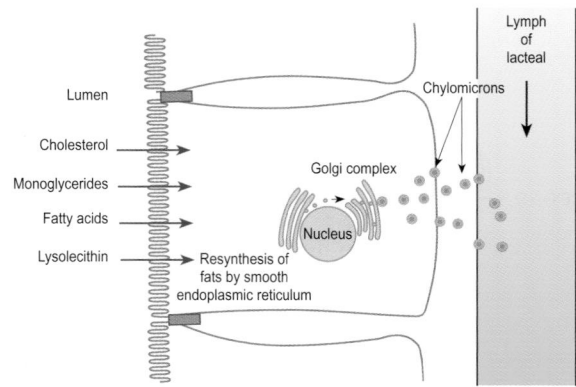

Fig. 18.35 The key steps involved in the absorption of lipids by the small intestine.

The absorption of fluid and electrolytes

Approximately 2 liters of fluid are ingested each day (although this may vary considerably depending upon thirst and social factors). The secretion of digestive juices and intestinal fluids adds a further 8.5 liters of fluid to the GI lumen. Almost all of this fluid is absorbed from the small and large intestines; only 50–200 ml leave the body with the feces. About 5–6 liters a day are absorbed in the jejunum, 2–3 liters a day in the ileum, and between 400 ml and 1 liter a day in the colon. A summary of overall fluid balance in the GI tract is shown in Fig. 18.36.

Absorbed electrolytes originate from both ingested foods and GI secretions. Most are actively absorbed along the length of the small intestine, though the absorption of calcium and iron is restricted mainly to the duodenum. As described earlier, the absorption of sodium ions is coupled with the transport of both sugars and amino acids. There are numerous active sodium–potassium pumps in the basal membranes of the intestinal epithelial cells that pump sodium out of the cells, thereby creating a gradient that draws sodium passively into the cell from the intestinal lumen. Some potassium is actively secreted into the gut, particularly in mucus. Potassium is absorbed passively along a concentration gradient set up by the absorption of water. For the most part, anions passively follow the electrical potential generated by the active transport of sodium. Chloride ions are also actively transported and, in the lower ileum, bicarbonate ions are actively secreted into the intestinal lumen in exchange for chloride.

The absorption of water by the intestine occurs by osmosis into blood vessels in response to the gradients established by the absorption of nutrients and electrolytes. Water can also be transported by osmosis from the blood to the intestinal lumen when the chyme entering the duodenum becomes hypertonic as a result of the digestion of nutrients. In this way, isotonicity of the chyme is rapidly established and is then maintained along the length of the intestine. As nutrients and electrolytes are progressively absorbed, water follows almost instantaneously.

The absorption of vitamins

The fat-soluble vitamins are absorbed in the same way as the products of fat digestion, partitioning into micelles and passing into the lymph as described above.

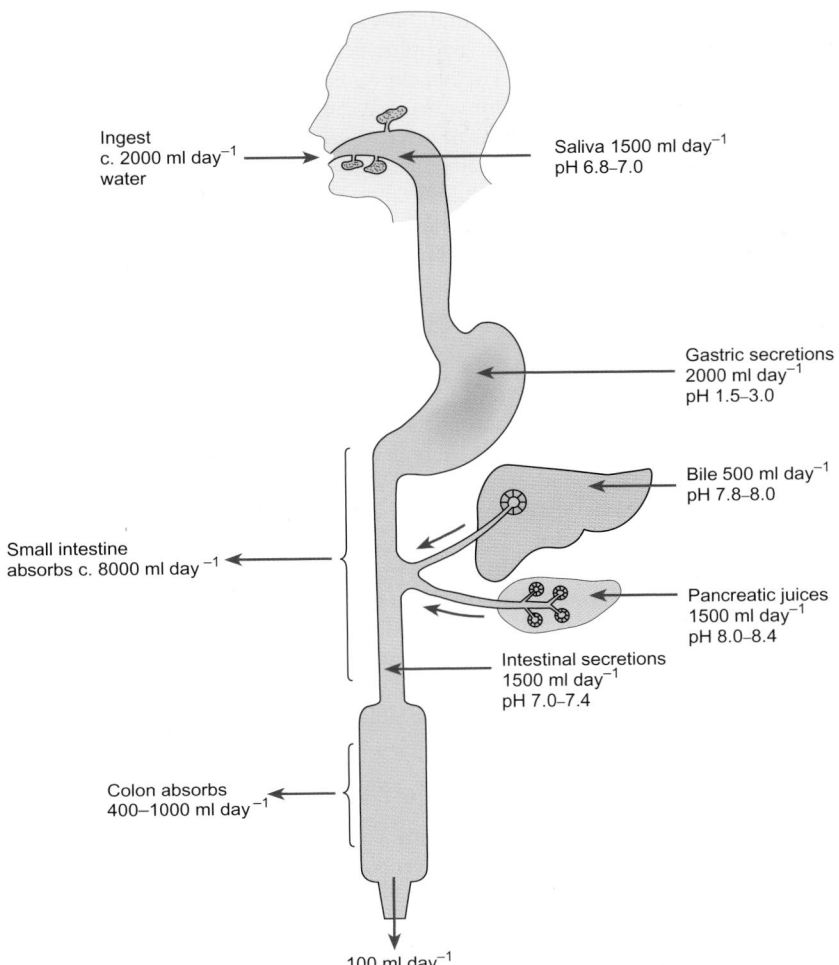

Ingest
c. 2000 ml day^{-1}
water

Saliva 1500 ml day^{-1}
pH 6.8–7.0

Gastric secretions
2000 ml day^{-1}
pH 1.5–3.0

Bile 500 ml day^{-1}
pH 7.8–8.0

Small intestine
absorbs c. 8000 ml day^{-1}

Pancreatic juices
1500 ml day^{-1}
pH 8.0–8.4

Intestinal secretions
1500 ml day^{-1}
pH 7.0–7.4

Colon absorbs
400–1000 ml day^{-1}

100 ml day^{-1}
water excreted

Fig. 18.36 The overall balance between secretion and absorption of fluid in the GI tract.

Specific transport molecules have been identified for most of the water-soluble vitamins and they can enter the intestinal epithelial cells by passive diffusion, facilitated diffusion, or active transport. Vitamin C, for example, is absorbed in the jejunum by sodium-dependent active transport via a mechanism similar to that described earlier for amino acids and monosaccharides.

Vitamin B_{12} (cyanocobalamin) is absorbed in the terminal ileum by a specific mechanism involving intrinsic factor, a glycoprotein secreted by parietal cells in the gastric mucosa. In the lumen of the jejunum, intrinsic factor binds to vitamin B_{12}. The resulting complex is recognized by a receptor protein in the brush-border membrane of ileal cells to which the complex binds before slowly entering the cell and eventually the blood. The mechanisms by which vitamin B_{12} crosses both the luminal membrane and the basal membrane of the ileal cell remain unclear. Vitamin B_{12} appears in the blood mostly bound to transcobalamin II.

Disorders of absorption (malabsorptive states)

A number of disease states may lead to a failure to absorb nutrients from the small intestine. In some cases there is direct impairment of the absorption of nutrients. In others, impaired absorption arises as a consequence of problems with digestion. A range of symptoms are common to many kinds of malabsorption, including weight loss, flatulence with abdominal distension and discomfort, and glossitis (painful loss of the normal epithelium covering the tongue). Symptoms that are more specific are indicative of particular malabsorption states. Poor digestion and absorption of fats, for example, will produce greasy, soft, and malodorous stools. Anemia is often seen if the absorption of iron or folic acid is impaired, while calcium deficiency (manifest as demineralization of the skeleton and possibly tetany) may be the result of vitamin D deficiency. Failure to absorb vitamin K may lead to an increased tendency to bleed. A detailed account of the many specific malabsorptive states is beyond the scope of this book, but a few important examples are described below

Carbohydrate intolerance

An inability to digest carbohydrate may arise because of an inherited or acquired deficiency in one or more of the necessary intestinal enzymes. Several types of deficiency may occur. One of the most common is lactase deficiency in which the enzyme responsible for digesting lactose (the disaccharide milk sugar) is lacking. Undigested disaccharide remains in the lumen of the bowel, causing fluid to be retained there by its osmotic effect. This causes distension and diarrhea. The colonic bacteria ferment the lactose and gaseous acidic stools are produced. Children with lactase deficiency will fail to thrive and may suffer diarrhea each time they ingest milk. Adults will suffer nausea, flatulence, and abdominal discomfort. The condition is readily controlled by the avoidance of foods containing lactose. A very rare congenital form of carbohydrate intolerance arises from an inability to absorb glucose and galactose. This is due to the absence of the normal intestinal sodium-dependent carrier protein for these sugars.

Celiac disease (non-tropical sprue)

This is a hereditary chronic intestinal disorder of absorption caused by intolerance to gluten, a cereal protein found in wheat, rye, oats and barley. Part of the gluten molecule forms an immune complex within the intestinal mucosa promoting aggregation of killer T lymphocytes (see Chapter 14) which release toxins promoting lysis of the enterocytes. There is progressive atrophy of the villi (in particular of the duodenum and proximal jejunum) because enterocytes cannot be replaced quickly enough by stem cell replication in the crypts (see Section 18.8). Furthermore, many of the cells that are present are relatively immature and cannot absorb nutrients effectively.

The symptoms of celiac disease vary in severity and in presentation, but steatorrhea (fatty stools), bloating, and discomfort are common. Children fail to thrive and pass soft pale malodorous stools after eating foods containing gluten. Nutritional supplementation of a gluten-free diet may be necessary in extreme cases.

Crohn's disease

This condition affects the terminal ileum and ascending colon. It is characterized by chronic inflammatory lesions with accumulation of macrophages (granulomatous deposits) and thickening of the bowel. There is enlargement of the lymph nodes and inflammation across the entire thickness of the gut wall. The causes of the disease are unclear but there may be a genetic component. The disease shows periods of remission and renewed activity, and symptoms are extremely variable both in type and in severity. There may be pain in the right iliac fossa, diarrhea, steatorrhea, and rectal passage of blood and mucus. Complications such as gallstones may arise because of malabsorption of bile acids in the terminal ileum. There may also be deficiencies of the fat-soluble vitamins A,D,E, and K.

18.13 The large intestine

Around 500 ml of chyme pass via the ileocecal valve from the ileum into the cecum every day. Material then passes in sequence through the ascending colon, transverse colon, descending colon, sigmoid colon, rectum, and anal canal (Fig. 18.37). Semisolid waste material (feces) is eliminated from the body through the anus. In adults, the large intestine is approximately 1.3 m long and its diameter is greater than that of the small intestine. It has a variety of functions, which include the storage of food residues prior to their elimination, the secretion of mucus, which lubricates the feces, and the absorption of most of the water and electrolytes remaining in the residue. In addition, bacteria which live in the colon synthesize vitamin K and some B vitamins.

Special histological features and innervation of the large intestine

Structurally, the wall of the large intestine follows the basic plan of the GI tract. However, the longitudinal smooth muscle layer of the muscularis externa is thickened to form longitudinal bands called *taeniae coli*. Three of these are present in the cecum and the colon. The longitudinal muscle layer between the taeniae is relatively thin. The tone of the smooth muscle in the taeniae causes the wall of the large intestine to pucker into pocket-like sacs called *haustra*. In the rectum, there are two broad bands of longitudinal muscle but haustra are absent.

The mucosal surface of the cecum, colon, and upper rectum is smooth and has no villi. However, large numbers of crypts are present. The mucous membrane consists of columnar absorptive cells, with many mucus-secreting goblet cells.

The anal canal, which is about 3 cm long and lies entirely outside the abdominal cavity, has internal and external sphincters

Summary

1. Absorption is the process by which the products of digestion are transported into the epithelial cells of the GI tract and thence into the blood or lymph draining the gut. Almost all the absorption of water, electrolytes, and nutrients occurs in the small intestine.

2. Monosaccharides are absorbed in the duodenum and upper jejunum by sodium-dependent cotransport driven by the sodium–potassium pump. Amino acids utilize similar mechanisms, though at least 10 separate transporters exist

3. The products of fat digestion are incorporated into micelles along with bile salts, lecithin, cholesterol, and fat-soluble vitamins. In this way, they are brought close to the enterocyte membrane and the fatty components of the micelle diffuse into the cells. Bile salts are recycled and the fats are reprocessed by the smooth endoplasmic reticulum to form chylomicrons. These are exocytosed across the basolateral cell membrane and enter the lacteals of the villi.

4. The GI tract absorbs 8–10 liters of fluid and electrolytes each day. The active transport of sodium and nutrients is followed by anion movement and the absorption of water by osmosis. Failure to absorb fluid results in potentially life-threatening diarrhea.

5. Fat-soluble vitamins are absorbed along with the products of fat digestion. Most water-soluble vitamins are absorbed by facilitated transport. A specific uptake process involving gastric intrinsic factor is responsible for the absorption of vitamin B_{12}.

6. A number of conditions result from malabsorption of nutrients. Specific syndromes include carbohydrate intolerance, celiac disease, and Crohn's disease.

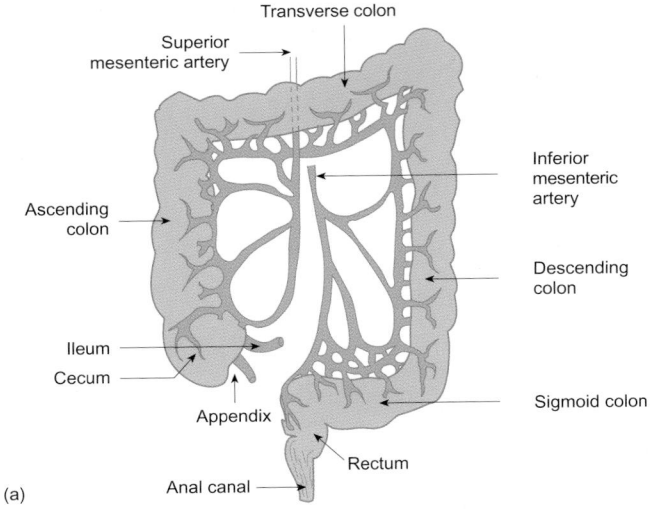

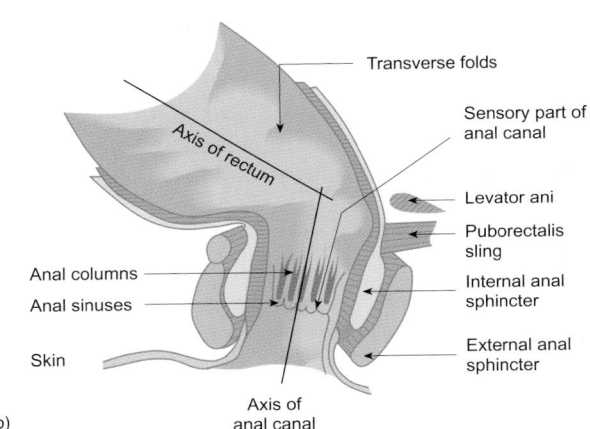

Fig. 18.37 (a) The major structures of the large intestine showing the principal blood supply. (b) Longitudinal section through the rectum and anal canal.

(Fig. 18.37) which remain closed (acting rather like purse strings) except during defecation. The mucosa of the anal canal reflects the greater degree of abrasion that this area receives. It hangs in long folds called anal columns and contains stratified squamous epithelium. Between the columns are the anal sinuses. Two superficial venous plexuses are associated with the anal canal. If these become inflamed, itchy varicosities called *hemorrhoids* form.

The large intestine receives both parasympathetic and sympathetic innervation. Vagal fibers supply the cecum and the colon as far as the distal third of the transverse region. Parasympathetic fibers supplying the rest of the colon, rectum and anal canal originate from the pelvic nerves of the sacral spinal cord (*nervi erigentes*). The parasympathetic fibers end chiefly on neurons of the intramural plexuses. The sympathetic input is from the celiac and superior mesenteric ganglia (cecum, ascending and transverse colon), and from the inferior mesenteric ganglion (descending and sigmoid colon, rectum, and anal canal), see Fig. 18.4. The

external anal sphincter receives branches of somatic nerves arising from the sacral region of the spinal cord.

The cecum and appendix

The cecum is a blind-ended tube about 7 cm long leading from the ileocecal valve to the colon. Although it is important for cellulose digestion in herbivores, it has no significant digestive role in humans. Attached to the posteromedial surface of the cecum is the vermiform appendix, a small blind pouch about the size of a finger, containing lymphoid tissue. Although it is part of the mucosa-associated lymphoid tissue (MALT), it has no essential function in humans. Inflammation of the appendix is known as *appendicitis*. To prevent its rupture it is necessary to remove the appendix surgically. If the appendix does rupture, the more serious condition of peritonitis develops because of the presence of fecal material containing bacteria in the abdominal cavity.

The colon

The colon is around 1.20 m long and has a diameter of around 6 cm. It acts as a reservoir, storing unabsorbed and unusable food residues. Although most of the residue is excreted within 72 hours of ingestion, up to 30 per cent of it may remain in the colon for a week or more.

Electrolyte and water absorption in the colon

Although large amounts of fluid are absorbed from chyme as it passes along the small intestine, the chyme entering the colon still contains appreciable quantities of water and electrolytes. Indeed, the colon absorbs 400–1000 ml of fluid each day. Failure to do so results in severe diarrhea. Sodium ions are transported actively from the lumen to the blood. This absorption is sensitive to aldosterone. As in the lower ileum, the absorption of chloride ions is linked to the secretion of bicarbonate. This bicarbonate may help to neutralize the acidic end-products of bacterial action (see below). Water is absorbed by osmosis.

The role of intestinal bacterial flora

A variety of bacteria colonize the large intestine, living symbiotically with their human host. Some of these, such as *Clostridium perfringens* and *Bacteroides fragilis* are anaerobic species, while others, like *Enterobacter aerogenes*, are aerobic. The intestinal flora perform a number of functions within the large intestine. One of these is the fermentation of indigestible carbohydrates (notably cellulose) and lipids that enter the colon. As a result of these fermentation reactions, short-chain fatty acids are produced, along with a number of gases (e.g. hydrogen, nitrogen, carbon dioxide, methane, hydrogen sulphide), which form about 500 ml of flatus each day (more if the diet is rich in indigestible carbohydrates such as cellulose). Short-chain fatty acids, including acetate, propionate, and butyrate, are absorbed readily by the colon, stimulating water and sodium uptake at the same time. The colonocytes appear to utilize the short-chain fatty acids for energy.

Another action of the intestinal flora is the conversion of bilirubin to non-pigmented metabolites—the urobilinogens (see above). They are also able to degrade cholesterol and some drugs.

Finally, intestinal bacteria are able to synthesize certain vitamins (e.g. vitamin K, vitamin B_{12}, thiamine, and riboflavin). Vitamin B_{12} can only be absorbed in the terminal ileum, so that which is synthesized in the colon is usually excreted and is of no value to the body.

Movements of the colon

The movements of the colon, like those in the small intestine, can be defined as either mixing or propulsive movements. Since the role of the colon is to store food residues and to absorb water and electrolytes, the propulsive movements of the colon are relatively sluggish. Characteristically, material travels along the colon at 5–10 cm h^{-1} and typically remains within the colon for 16–20 hours.

Mixing movements (haustrations)

Contraction of the circular smooth muscle layer of the colon serves to constrict the lumen in much the same way as described for the small intestine above. This kind of segmental movement is called haustration in the colon because the segments correspond to the smooth muscle thickenings called haustra (see above). It is the predominant type of movement seen in the cecum and the proximal colon. The purpose of haustration seems to be to squeeze and roll the fecal material around so that every portion of it is exposed to the absorptive surfaces of the colonic mucosa, thus aiding the absorption of water and electrolytes.

Propulsive movements—peristalsis and mass movements

In addition to haustral contractions, short-range peristaltic waves are seen in the more distal parts of the colon (transverse and descending regions). These serve to propel the intestinal contents, now in the form of semisolid fecal material, towards the anus.

Several times a day, usually after meals, a more vigorous propulsive movement of the colon occurs in which a portion of the colon remains contracted for rather longer than during a peristaltic wave. This is called a mass movement and results in the emptying of a large portion of the proximal colon. Mass movements are also seen in the transverse and descending colon. When they force a mass of fecal material into the rectum, the desire for defecation is experienced.

Mass movements are initiated, at least in part, by intrinsic reflex pathways resulting from distension of the stomach and duodenum. These are termed the gastrocolic and duodenocolic reflexes. These intrinsic motor patterns are modified by autonomic nerves and by hormones. For example, vagal stimulation enhances colonic motility while both gastrin and CCK increase the excitability of the colon and facilitate ileal emptying by causing the ileocecal sphincter to relax.

Analgesics such as morphine, codeine and pethidine decrease the frequency of colonic mass movements. Other drugs, including aluminum-based antacids, have the same effect. Therefore people taking these drugs may become constipated.

The role of dietary fiber in the large intestine

The time taken for food residues to be expelled from the body after eating varies considerably, but appears to be directly related to the amount of dietary fiber ingested. Dietary fiber (or 'roughage') consists largely of cellulose. Humans are unable to digest this and so it remains in the intestine, adding bulk to the food residues. It tends to exert a hygroscopic effect, absorbing water so that stools with a high fiber content tend to be bulkier and softer making them easier to expel. A shorter mouth-to-anus transit time is also believed to reduce the risk of developing carcinoma of the large intestine and rectum. This may be due partly to a reduction in the time for which bacterial toxins and potentially harmful metabolites are in contact with the gut wall.

The rectum and defecation

The rectum is a muscular tube about 12–15 cm long. It is normally empty, but when a mass movement forces feces into the rectum the urge to defecate is initiated. The rectum opens to the exterior via the anal canal, which has both internal and external sphincters. The internal sphincter is not under voluntary control. It is supplied by both sympathetic and parasympathetic nerves. Contraction of the smooth muscle of the internal sphincter is initiated by sympathetic stimulation and relaxation by parasympathetic stimulation. The external anal sphincter is composed of skeletal muscle. It is supplied by the pudendal nerve and is under learned voluntary control from the age of about 24 months. Both the anal sphincters are maintained in a tonic state of contraction.

About 100–150 g of feces, consisting of 30–50 g of solids and 70–100 g of water, are normally eliminated each day. The solid portion consists largely of cellulose, epithelial cells shed from

Summary

1. The large intestine consists of the cecum (which plays no significant role in humans), colon, rectum, and anal canal. Its main functions are to store food residues, secrete mucus, and absorb remaining water and electrolytes from the food residue. Feces are eliminated via the anus.

2. The colon absorbs 400–1000 ml of fluid each day. Sodium is actively transported from the lumen to the blood. Chloride moves in exchange for bicarbonate, and water follows by osmosis.

3. Intestinal flora perform fermentation reactions that produce short-chain fatty acids and flatus. The short-chain fatty acids are absorbed by the colonocytes, stimulating salt and water uptake. Intestinal bacteria also synthesize certain vitamins such as vitamin K.

4. The colon exhibits mixing movements (haustrations) and sluggish propulsive movements. 'Housekeeper' contractions occur several times daily, serving to move intestinal contents over longer distances. These contractions move fecal material into the rectum and the resulting stretch of its wall elicits the urge to defecate.

5. Between 100 and 150 g of feces are eliminated each day. Defecation involves both voluntary and involuntary contractions of the anal sphincters and muscles of the abdominal wall and diaphragm.

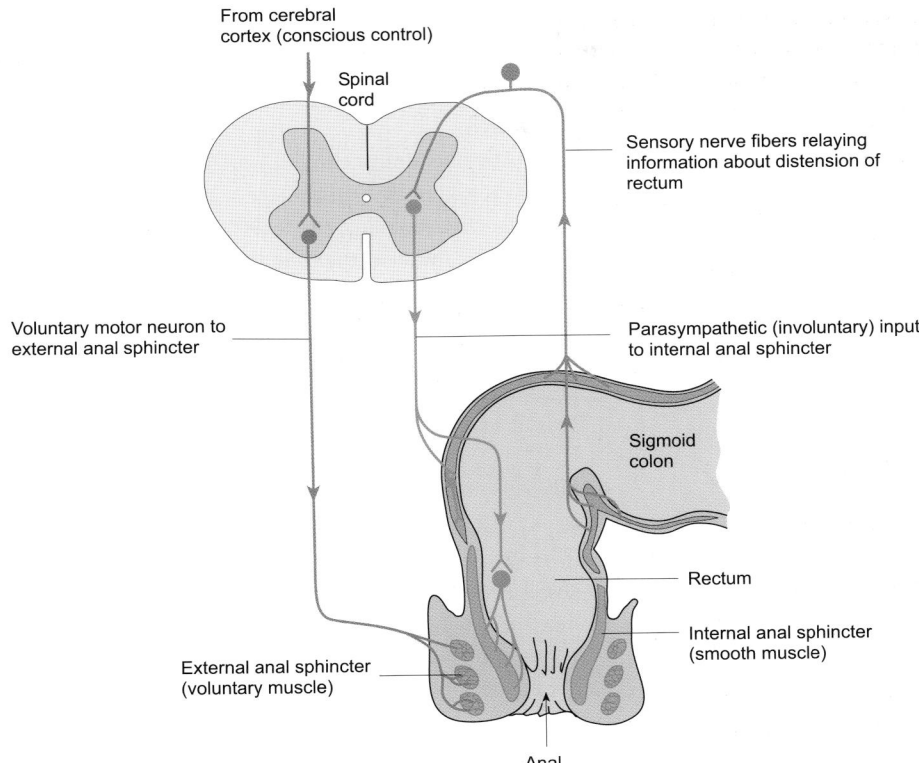

Fig. 18.38 Neural pathways involved in the defecation reflex. Sensory axons are shown in red and motor axons in blue.

the lining of the GI tract, bacteria, some salts, and the brown pigment stercobilin. The characteristic odor of feces is due to the presence of hydrogen sulfide and organic sulfides.

Defecation itself is a complex process involving both reflex and voluntary actions. It is possible to inhibit the reflex consciously if the circumstances are not convenient and, under such conditions, the urge to defecate will often subside until reinitiated by the arrival of further fecal material in the rectum. Eventually, however, the urge to defecate will become overwhelming and the reflex will proceed. Under the influence of the parasympathetic nervous system, the walls of the sigmoid colon and the rectum contract to move feces towards the anus. The anal sphincters relax to allow feces to move through the anal canal. Expulsion of the fecal material is aided by voluntary contractions of the diaphragm and the muscles of the abdominal wall as well as closure of the glottis. As a result, intra-abdominal pressure rises and helps to force feces through the relaxed sphincters. The muscles of the pelvic floor relax to allow the rectum to straighten, thus helping to prevent rectal and anal prolapse. The components of the defecation reflex are depicted diagrammatically in Fig. 18.38.

Recommended reading

Anatomy
MacKinnon, P.M., and Morris, J. (2005). *Oxford textbook of functional anatomy*. Vol. 2, *Thorax and abdomen* (2nd edn), pp. 129–166. Oxford University Press, Oxford.

Biochemistry
Elliott, W.H., and Elliott, D.C. (2005). *Biochemistry and molecular biology* (3rd edn), Chapters 9 and 10. Oxford University Press, Oxford.

Histology
Junquieira, L.C., and Carneiro, J. (2003). *Basic histology* (10th edn), Chapters 15 and 16. McGraw-Hill, New York.

Physiology
Johnson, L.R. (2001). *Gastrointestinal physiology* (6th edn). Mosby, St Louis, MO.

Sandford, P.A. (1992). *Digestive system physiology* (2nd edn). Edward Arnold, London.

Smith, M.E., and Morton, D.G. (2001). *The digestive system: basic science and clinical conditions*. Churchill-Livingstone, London.

Medicine
Ledingham, J.G.G., and Warrell, D.A. (eds.), (2000). *Concise Oxford textbook of medicine*. Chapters 5.1–5.1.2

Pharmacology
Grahame-Smith, D.G., and Aronson, J.K. (2002). *Oxford textbook of clinical pharmacology and drug therapy* (3rd edn), Chapter 25. Oxford University Press, Oxford.

Rang, H.P., Dale, M.M., Ritter, J.M., and Moore, P. (2003). *Pharmacology* (5th edn), Chapter 24. Churchill-Livingstone, Edinburgh.

Multiple choice questions

The following statements are either true or false. Answers are given below.

1. a. The ileum contains Brunner's glands.
   b. Villi are present in the large intestine.
   c. The myenteric plexus lies between the longitudinal and circular smooth muscle layers in the gut wall.
   d. The GI tract absorbs about 8–10 liters of fluid each day.
   e. The serosa is the innermost layer of the gastrointestinal wall.

2. a. Swallowing is a purely voluntary activity.
   b. The food bolus is propelled down the esophagus by segmentation movements.
   c. The saliva contains an enzyme that digests starch.
   d. The pH of saliva rises as its rate of secretion increases.
   e. Nerves are more important than hormones in the regulation of salivary secretion.

3. a. Intrinsic factor is secreted by G cells in the gastric glands.
   b. Cholecystokinin inhibits gastric secretion.
   c. Gastric secretion does not begin until food enters the stomach.
   d. Gastric acid is secreted by parietal cells of the gastric glands.
   e. Secretion of most of the acid and pepsinogen by the stomach occurs during the intestinal phase of gastric secretion.

4. a. The most vigorous mixing movements in the stomach take place in the antral region.
   b. Gastric emptying is inhibited by the enterogastric reflex.
   c. Venous blood draining the stomach after a meal has a higher pH than blood in the right atrium.
   d. The plasma bicarbonate concentration will be lower than normal following prolonged vomiting.
   e. Persistent vomiting often leads to metabolic alkalosis.

5. a. Pancreatic acinar cells contain trypsin.
   b. Cholecystokinin inhibits secretion from the exocrine pancreas.
   c. Loss of pancreatic enzymes will result in weight loss due to poor protein digestion.
   d. The introduction of acid into the duodenum stimulates pancreatic secretion.
   e. The chloride content of pancreatic juice falls as the rate of secretion rises.

6. a. Seventy per cent of the blood flow to the liver is via the portal vein.
   b. The cystic duct drains the gall bladder.
   c. Bile is diluted in the gall bladder.
   d. In obstructive jaundice, the feces are pale and fatty.
   e. Micelles are found in the lacteals after a fat-rich meal.

7. a. Bile salts are the breakdown products of hemoglobin.
   b. The reabsorption of bile salts in the intestine stimulates bile secretion.
   c. Most bile salts are absorbed in the terminal ileum.
   d. Loss of bile salts will lead to poor absorption of vitamin E.
   e. Bile salts are hydrophobic molecules.

8. a. The Na^+, K^+ ATPase of the basolateral membrane of intestinal epithelial cells plays an important role in the absorption of salts and water.
   b. Intestinal digestive enzymes are secreted by cells of the crypts of Lieberkuhn.
   c. About half of the digested carbohydrate is absorbed in the small intestine
   d. Amino acids are absorbed in the small intestine by cotransport with sodium.
   e. Cellulose cannot be digested or absorbed by the human small intestine.
   f. The first organ to receive the blood-borne products of digestion is the liver.

9. a. Total gastrectomy leads to malabsorption of vitamin B_{12}.
   b. Total gastrectomy will lead to a reduction in plasma osmolality after meals.
   c. Parasympathetic activity inhibits intestinal motility.
   d. Distension of the ileum inhibits gastric motility.
   e. The presence of large amounts of fat in the chyme will accelerate gastric emptying.

10. a. The mucosa of the anal canal is covered by stratified squamous epithelial cells.
    b. Aldosterone stimulates the absorption of sodium and water by the large intestine.
    c. Dietary fiber reduces the rate at which food residues move through the colon.
    d. Vitamin K is synthesized by the intestinal flora.
    e. Gastrin facilitates ileal emptying.

Quantitative problems

1. An experiment was carried out in which the concentration of gastrin in the plasma of a human volunteer was measured every 15 minutes for a period of 2 hours. At time zero (0 min) the subject ate a protein-rich meal. The data are tabulated below.

Time (min)	Plasma gastrin concentration (pg ml−1)
−60	75
−45	60
−30	60
−15	70
0	80
15	100
30	130
45	160
60	180
75	110
90	80
105	60
120	40

  a. Plot the data on linear graph paper.
  b. Discuss the shape of the graph in relation to the regulation of gastric secretion.
  c. How might the shape of the graph differ if the subject had eaten a meal rich in fats rather than protein?

2. The pH of the gastric contents of a subject was measured following a meal (time zero).
  a. Plot a graph to illustrate the changes in pH with time.
  b. What causes the initial rise in pH?
  c. Why does the pH fall between 0.5 and 2.5 hours?
  d. What mechanism is responsible for the stable value of intragastric pH after 2.5 hours?
  e. What function does the low intragastric pH serve?

Answers to multiple choice questions

1. Brunner's glands are only found in the duodenum. Villi are only present in the small intestine. The serosa is the outermost layer of the GI tract wall.
  a. False
  b. False
  c. True
  d. True
  e. False

2. Swallowing is a complex reflex, the first part of which is under voluntary control. The bolus is moved along the esophagus by a wave of peristaltic contraction. The saliva contains an α-amylase. The pH of saliva rises as its secretion accelerates due to an increased bicarbonate concentration.

Time (hours)	pH
0	1.9
0.5	4.0
1.0	3.1
1.5	2.2
2.0	1.9
2.5	1.6
3.0	1.5
3.5	1.5
4.0	1.5

  a. False
  b. False
  c. True
  d. True
  e. True

3. Intrinsic factor is secreted by the parietal cells. CCK inhibits gastric secretion as part of the enterogastric reflex. Gastric secretion begins before food reaches the stomach (the cephalic phase of secretion). The intestinal phase contributes little to overall gastric secretion.
  a. False
  b. True
  c. False
  d. True
  e. False

4. During vomiting acid is lost from the stomach. As a result metabolic alkalosis is seen, with a rise in plasma pH and bicarbonate.
  a. True
  b. True
  c. True
  d. False
  e. True

5. Pancreatic acinar cells store proteolytic enzymes as inactive precursors (trypsinogens) to avoid self-digestion. CCK stimulates pancreatic secretion. Pancreatic juice contains the enzymes required for protein digestion. In their absence, protein utilization by the body will be impaired.
  a. False
  b. False
  c. True
  d. True
  e. True

6. Bile is concentrated many times in the gall bladder. Chylomicrons are found in the lacteals after a fatty meal. Micelles are the aggregations of bile salts, fats, and fat-soluble vitamins formed in the small intestine. They form a stable complex that permits the absorption of fat-soluble substances into the enterocytes.

a. True
b. True
c. False
d. True
e. False

7. Bile salts are formed in the liver. Bile pigments are the break-down products of hemoglobin. Reabsorption of bile salts stimulates further bile secretion via the enterohepatic circulation. Fat-soluble vitamins including vitamin E are absorbed along with fats and therefore depend upon bile salts in the same way. Bile salts are amphipathic, i.e. they possess both hydrophobic and hydrophilic regions.
   a. False
   b. True
   c. True
   d. True
   e. False

8. The enzymes of the small intestine are associated with the brush border of the epithelial cells. Virtually all the digestion products of ingested carbohydrate are absorbed in the small intestine.
   a. True
   b. False
   c. False
   d. True
   e. True
   f. True

9. Intrinsic factor secreted by the parietal cells is needed for the absorption of vitamin B_{12}. Following gastrectomy, large amounts of hypertonic chyme enter the duodenum. Water enters the gut lumen down the osmotic gradient, leaving the plasma more concentrated. This is known as 'dumping'. Parasympathetic activity stimulates intestinal motility. Fatty chyme leaves the stomach more slowly than chyme that contains little fat.

a. True
b. False
c. False
d. True
e. False

10. Dietary fiber increases the transit time through the colon. Vitamin K synthesis by the intestinal bacteria is important for the maintenance of normal hemostasis.
   a. True
   b. True
   c. False
   d. True
   e. True

Answers to quantitative problems

1. b. The initial rise in plasma gastrin levels is the result of local reflexes in response to distension of the stomach wall. Subsequently, protein digestion products (mainly peptides) act directly on the G cells to stimulate their secretion. The subsequent fall in gastrin levels is the result of inhibition of gastrin secretion by the low pH of the gastric contents.
   c. As fats inhibit gastric emptying, the secretion of gastrin will be more prolonged but the plasma levels will probably not rise so high.

2. b. The initial rise in intragastric pH is due to buffering of the gastric contents by the ingested food.
   c. The subsequent fall in pH reflects the secretion of gastric acid by the parietal (oxyntic) cells in response to local and vagal reflexes. It is also the result of increased acid secretion in response to gastrin secretion by the G cells.
   d. The fall in intragastric pH inhibits the secretion of gastrin and limits the secretion of acid by the parietal cells.
   e. Low intragastric pH denatures dietary protein and renders it susceptible to attack by pepsin. It kills bacteria and other micro-organisms present in the stomach and solubilizes iron and calcium salts.

The nutritional needs of the body

After reading this chapter you should understand:

- The concept of nutrition and the importance of a mixed diet
- The role of the principal nutrients, carbohydrates, fats, and proteins
- The importance of vitamins, minerals, and trace elements
- Some important aspects of malnutrition
- The factors that regulate hunger, appetite, and satiety
- Eating disorders, including obesity, anorexia nervosa, and bulimia nervosa
- The measurements used to determine nutritional status
- Enteral and parenteral nutrition

19.1 Introduction

The selection of foods eaten by an individual is called the diet. A nutrient is any substance that is absorbed and utilized to promote the activities of cells and, in turn, the various functions of the body. Nutrients include carbohydrates, proteins, fats, vitamins, mineral salts, and water. All are essential to health. Therefore, a balanced diet should contain appropriate amounts of each nutrient. In certain diseases nutritional requirements, nutrient supplies and metabolism may be altered. The concept of the energy content of foodstuffs is explored in Chapter 24, which also discusses the energy requirements of individuals in different circumstances. The chemical characteristics of carbohydrates, fats, and proteins, are considered in Chapter 2, while the basic biochemical principles of energy metabolism are treated in Chapter 3. These aspects of nutrition will not be discussed further here.

19.2 The principal requirements for a balanced diet

In addition to the carbohydrates, fats, and proteins that form the bulk of the diet, certain micronutrients are classed as *essential* because they are necessary for cellular metabolism but cannot be made by the body. They must therefore be included in the diet. Such nutrients include vitamins, essential amino acids, and essential fatty acids. By convention, essential nutrient requirements are assessed by determining the amount per day needed to prevent clinical deficiency. An additional 30–100 per cent is added to this value to give a figure for the 'recommended daily amount' (RDA). Since for many nutrients, deficiencies develop progressively, it is often difficult to define RDAs with precision. For this reason, published figures often show a wide range and may differ significantly between different countries. Furthermore, a number of factors influence an individual's nutritional requirements. These include age, gender, size, activity level, pregnancy, lactation, and state of health.

Carbohydrates

These are found in a wide variety of foods, such as cereals, bread, pasta, fruit, and in vegetables such as potatoes. Animal carbohydrate is present in the glycogen of meat and liver. Polysaccharides (which are complex molecules consisting of large numbers of monosaccharide residues) include starches, cellulose, glycogen, and dextrins (see also Chapter 2). Some of these, celluloses and lignins for example, which are components of plant cell walls, are not digested by the human gut and pass through virtually unchanged. Such carbohydrates are collectively known as dietary fiber and are thought to be important in facilitating the movement of material through the gut. Dietary fiber normally accounts for around 10 per cent of the total intake of carbohydrate. The remaining 90 per cent is digested in the gastrointestinal tract. After digestion, carbohydrates are absorbed in the form of monosaccharides which are utilized by cells to provide energy. Most people consume around 50 per cent of their total energy requirement in the form of carbohydrate (see Chapter 24). The body also stores carbohydrate as glycogen, most of which is found in the liver. Once glycogen stores are full, excess dietary carbohydrate is laid down in the form of fat.

Proteins

Most mixed diets provide about 15 per cent of the body's total energy requirements in the form of protein. Proteins are broken down into small peptides and amino acids before being absorbed by the small intestine. The body's requirement for protein is determined by its need for amino acids as the building blocks for structural protein (used for growth, maintenance, and repair of tissues) and functional proteins such as enzymes and hormones.

Of the 20 α-amino acids which make up the proteins of the body, 12 can be synthesized by the body itself and therefore need not be included in the diet These are alanine, arginine, asparagine, aspartic acid, cysteine, glutamic acid, glutamine, glycine, histidine, proline, serine, and tyrosine. They are called *non-essential amino acids*. The other eight amino acids cannot be synthesized by the body and must be included in the diet. These *essential amino acids* are isoleucine, leucine, lysine, methionine, phenylalanine, threonine, tryptophan, and valine.

The nutritional value of a protein depends upon the amino acids it contains. Protein foods that contain all the essential amino acids in the proportions required to maintain health, are termed *first-class (complete) proteins*. These include meat, fish, soya beans, milk, and eggs. *Second-class (incomplete) proteins* do not contain all the essential amino acids in the correct proportions. They are mainly of vegetable origin and include cereals and pulses such as peas, beans, and lentils. By eating a wide variety of incomplete proteins, it is possible to avoid amino acid deficiencies. This is an especially important consideration for those following a strictly vegetarian or vegan diet.

During illness, both protein synthesis and degradation are increased so that there is an increase in total protein turnover. In most cases, there is an overall loss of body protein with muscle wasting, which is restored during recovery from the illness.

A diet deficient in both energy and protein may result in a range of clinical syndromes collectively known as protein-energy malnutrition (PEM). In Western societies it usually only occurs in patients who have or have had severe illness such as GI disease or surgery. In developing countries, it is more often caused by a poor diet and may be exacerbated by infections that cause diarrhea. Chronic deprivation may result in stunting of height as well as inadequate weight gain. *Marasmus* and *kwashiorkor* are clinical syndromes resulting from a diet lacking in protein. In the case of marasmus, the energy content of the diet is also inadequate. This condition typically presents in infants under a year old. The child is less than 60 per cent of median body weight for age. Both lean tissue and subcutaneous fat are lost and the skin becomes characteristically wrinkled (Fig. 19.1). Because of the typical appearance of the child, marasmus may be confused with dehydration. Head hair is often thin, but there may be long fine silky hair (laguno) on the arms. The child is generally alert and keen to eat. Plasma albumin is usually nor-

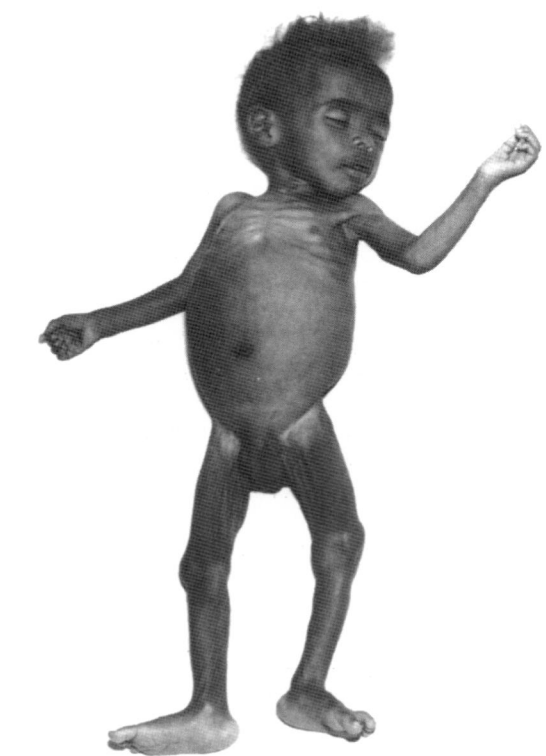

Fig. 19.1 The typical appearance of a child with marasmus. Note the characteristic wrinkling of the skin and thinning of head hair.

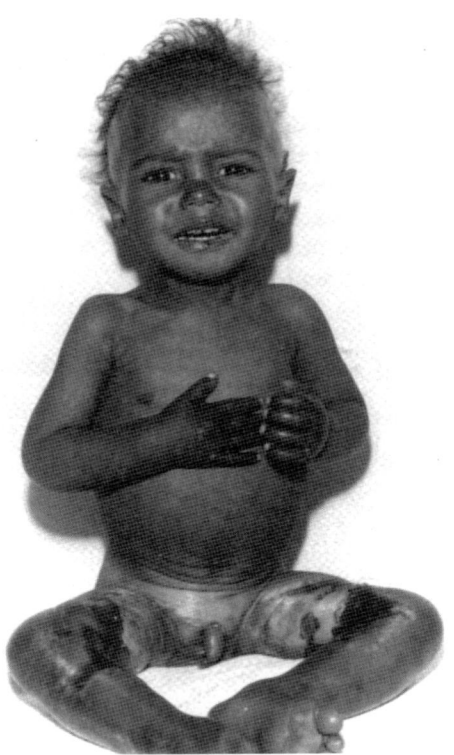

Fig. 19.2 The typical appearance of a child with kwashiorkor in which edema masks the loss of lean tissue. Note also the cracked, peeling, and discolored skin.

TABLE 19.1 Recommended daily intake of dietary protein (in grams) for different age groups

Age (years)	Males	Females
1–3	15	15
4–6	20	20
7–10	28	28
11–14	42	41
15–18	55	45
19–50	56	45
Over 50	53	47

weight loss may be due to cancers of the GI tract, malabsorption diarrhea, pyrexia (which increases the demand for nutrients), or chronic conditions such as AIDS and chronic obstructive airways disease (COAD). However, in many cases weight loss occurs because of inadequate food intake. This may occur for a variety of reasons, including depression, pain, social deprivation, or anorexia nervosa.

Table 19.1 lists the daily protein requirements of males and females at different ages. If protein is eaten in excess of the body's requirements, the nitrogenous part is detached (deamination) in the liver and excreted by the kidneys as urea. The rest of the molecule may be converted to fat for storage in adipose tissue or used for the synthesis of glucose (gluconeogenesis).

Fats

Fats are divided into two groups, saturated and unsaturated (see also Chapter 2). Saturated fat is the principal type found in milk, cheese, butter, eggs, meat, and oily fish such as herring and cod. Vegetable oils contain predominantly unsaturated fat. Linoleic, linolenic, and arachidonic acid are polyunsaturated fats that cannot be synthesized by the body. Therefore they are known as *essential fatty acids* and must be included in the diet. Plant oils such as sunflower, corn, walnut, and linseed are good sources of linoleic and linolenic acid, while animal tissues contain small amounts of arachidonic acid. Cholesterol is synthesized in the body. It is also present in fatty meat, egg yolk, and full-fat dairy products.

Fats serve a number of important functions in the body. As well as providing the fatty tissue which supports and protects organs such as the kidneys and eyes, fat is an important constituent of nerve sheaths and cell membranes and is an essential component of some hormones (see also Chapter 2). It plays an important role in cell signaling through arachidonic acid, it is an important source of heat and energy in metabolism, and it stores the fat-soluble vitamins A, D, E, and K.

In a typical Western adult diet, fats contribute around 30 per cent to total energy intake, of which around one-third is in the form of saturated fats and the rest is present as mono- and polyunsaturated fats. However, in populations whose diets are relatively low in fat (contributing less than 15 per cent to total energy) adequate subcutaneous stores are deposited, indicating that many people's diets contain amounts of essential fatty acids which are considerably in excess of actual requirements. Consequently, fatty acid deficiencies are relatively rare.

mal, but plasma sodium and potassium levels are low. Although internal organs are relatively unaffected in the early stages of the disease, atrophy of the heart and reduced brain weight may occur in advanced cases of marasmus.

Kwashiorkor most often affects children at the time of weaning (i.e. in the second year of life) as milk is replaced by solid foods that contain very little protein such as cassava. This condition is characterized by edema that may mask the loss of lean tissue and create the false impression that the child is well nourished (Fig. 19.2). As protein intake is inadequate, there is a loss of plasma proteins that results in a fall in plasma oncotic pressure. Thus fluid leaks out of capillaries and into the interstitium (see also Chapter 28, Section 28.6). Edema is first seen in the legs but later becomes generalized. The hair may become thin and discolored and there are characteristic skin changes with areas of pigmentation which later crack and peel. Also typical of kwashiorkor is enlargement of the liver as a result of fatty deposits. This occurs because the hepatocytes fail to synthesize sufficient very-low-density lipoproteins normally required for the transport of fats out of the liver. Many children with kwashiorkor also have an infection such as measles that may exacerbate the edema. Some pathogens produce toxins that generate free radicals. These in turn may damage capillaries and increase the leakage of fluid into surrounding tissue.

Although protein-energy malnutrition in children is relatively common in third world societies, it is also seen in adults in Western communities. For example, protein-energy malnutrition is present in many patients on admission to hospital. Severe

Vitamins

Although required only in very small quantities by the body, vitamins are essential for normal metabolism and health. They are found in a wide range of foods and are subdivided into two categories, fat-soluble vitamins (A, D, E, and K) and water-soluble vitamins (C and B complex). Table 19.2 lists the different vitamins, their recommended daily requirement, their major sources, and their functions. Deficiencies of particular vitamins result in the progressive development of specific deficiency disorders. Furthermore, some vitamins, such as vitamin C and folic acid, are involved in cellular processes in a more generalized way and deficiencies will have more wide-reaching consequences. Some of the more important deficiency syndromes are discussed below and, in cases where excess intake is known to be damaging to health, toxic effects are also considered.

Fat-soluble vitamins (A, D, E, and K)

Vitamin A (retinol) is required for the formation of visual pigments, and *xerophthalmia* caused by vitamin A deficiency is a major cause of blindness in the tropics. The first visual symptoms are a loss of sensitivity to green light, followed by a loss of acuity in dim light and then total night-blindness. The conjunctival membranes of the eyes become dry and develop oval or triangular spots (Bitot's spots). The corneas become cloudy and soft, and in severe cases keratinization of the cornea with erosion and ulceration may lead to total blindness. Other epithelia such as those of the respiratory tract, GI tract, and genito-urinary tract are also affected by vitamin A deficiency, which may cause respiratory disease and diarrhea. Vitamin A is also thought to play a part in the efficient operation of the immune system and a mild deficiency seems to increase susceptibility to infectious disease. Vitamin A is needed for normal limb development, so that deficiency during pregnancy may result in developmental abnormalities of the fetus.

Vitamin A toxicity is potentially serious. It was first seen in explorers who had eaten polar bear liver, which is very rich in vitamin A. Its acute effects include vomiting, vertigo, headache, blurred vision, loss of muscle coordination, and raised CSF pres-

TABLE 19.2 Actions and daily requirements of vitamins

Vitamin	Daily requirement (adults)	Major dietary sources	Functions	Deficiency disorders
A (retinol)	700 μg	Dairy products, oily fish, eggs, liver	Formation of visual pigments, development of bone cells	Night-blindness, epithelial atrophy, susceptibility to infection
D (cholecalciferol)	10 μg	Fish oils, dairy products. Synthesized in skin	Normal bone development,	Rickets (children), osteomalacia (adults) absorption of calcium from gut
E (α-tocopherol)	8 μg	Nuts, egg yolk, wheatgerm, milk, cabbage	Antioxidant	Hemolytic anemias
K (coagulation vitamin)	100 μg	Green vegetables, pig liver. Synthesized by intestinal flora	Formation of clotting factors and some liver proteins	Bruising, bleeding
B_1 (thiamine)	1.6 mg	Lean meat, fish, eggs, legumes, green vegetables	Carbohydrate metabolism	Beriberi with many neurological symptoms, disturbances of metabolism
B_2 (riboflavin)	1.8 mg	Milk, liver, kidneys, heart, meat, green vegetables	Constituent of flavine coenzymes	Dermatitis, hypersensitivity to light
B_3 (niacin or nicotinamide)	15 mg	Most foods. Can be synthesized from tryptophan	Constituent of nicotinamide coenzymes	Pellagra, listlessness, nausea, dermatitis, neurological disorders
B_6 (pyridoxine)	2 mg	Meat, fish	Amino acid metabolism, synthesis of hemoglobin and antibodies	Irritability, convulsions, anemia, vomiting, skin lesions
Pantothenic acid	5–10 mg	Most foods	Component of coenzyme A	Neuropathy, abdominal pain
Biotin (vitamin H)	100 μg	Liver, egg yolk, nuts, legumes	Fatty-acid synthesis	Muscle pain, scaly skin, elevated blood cholesterol
B_{12} (cyanocobalamin)	1.2 mg	Liver, meat, fish (*not* plants)	Erythrocyte production and amino acid metabolism	Pernicious anemia
Folic acid	250 μg	Liver, dark green vegetables. Synthesized by intestinal bacteria	Hematopoiesis, nucleic acid synthesis, development of neural tube	Anemias, GI disturbances, diarrhea
C (ascorbic acid)	50 mg	Fresh fruits (especially citrus fruits), vegetables	Protein metabolism, collagen synthesis	Scurvy, liability to infection, poor wound healing, anemia

sure. The chronic effects are variable and include hyperlipidemia, bone and muscle pain, and skin disorders. Excess vitamin A is now known to be teratogenic (i.e. a cause of fetal malformations) in the first trimester of pregnancy. It may lead to spontaneous abortion, abnormalities of the cranium, face (harelip), heart, kidneys, thymus, and CNS (including deafness and learning difficulties).

The metabolism of dietary cholecalciferol to *calcitriol* (*vitamin D*) and its importance in the maintenance and health of the skeleton are described in Chapter 12 (pp. 217–222). Vitamin D deficiency has different consequences at different stages of life. In toddlers, it causes a condition known as *rickets* in which bones are under-mineralized because of poor calcium absorption. When the child starts to walk, the long bones of the legs become deformed and bowed (see Chapter 23, Fig. 23.15). There may also be collapse of the rib cage and pelvic deformities.

Osteomalacia is the adult equivalent of rickets and results from the demineralization of bone rather than failure to mineralize in the first place. The elderly, and those women who have had several pregnancies but little exposure to sunlight are at most risk of experiencing skeletal problems caused by vitamin D deficiency.

Vitamin D is toxic if ingested in amounts far in excess of the RDA. Symptoms of toxicity are related to the elevated plasma calcium concentration resulting from enhanced intestinal calcium absorption (see Chapter 12). Hypercalcemia may lead to contraction of blood vessels and dangerously high blood pressure, and to calcinosis—the calcification of soft tissues including the kidney, heart, lungs, and blood vessel walls.

Vitamin E (*tocopherol*) is a powerful antioxidant that protects cell membranes and plasma lipoproteins from free-radical damage. Deficiency can lead to damage to nerve and muscle membranes. Children may develop ataxia, loss of reflexes, and changes in gait, with loss of proprioception. As vitamin E is fat soluble, it relies on normal fat digestion and absorption. Disorders such as cystic fibrosis, celiac disease, and biliary obstruction all reduce the intestinal absorption of vitamin E. Premature infants are also at risk of deficiency as they are often born with inadequate reserves. Their red blood cell membranes are very fragile due to attacks by free radicals, which could lead to hemolytic anemia unless vitamin E supplements are given. Deficiency in experimental animals causes infertility but there is no evidence of a similar role for vitamin E in humans.

Vitamin K is a fat-soluble vitamin that is found in green leaves. It is also made by bacteria found in the large bowel and absorbed in small amounts in the cecum. It promotes the synthesis in the liver of a special amino acid (γ-carboxyglutamic acid) which is an essential component of four of the coagulation factors of the clotting cascade (prothrombin and factors VII, IX, and X). Thus it is needed for the normal coagulation of blood (see Chapter 13) and vitamin K deficiency causes bleeding disorders characterized by hypoprothrombinemia. The widely used anticoagulant drug warfarin is a vitamin K antagonist.

Vitamin K is not transported readily across the placenta from mother to fetus and the neonatal gut is sterile. For this reason vitamin K levels in the newborn may be very low. Hemorrhagic disease of the newborn is a risk, particularly for premature infants, and such babies are often given an intramuscular injection of vitamin K immediately after delivery as a precaution. Deficiency may also occur in patients with malabsorption of fats and following prolonged treatment with antibiotics, which destroy the colonic bacteria. Elderly patients admitted to hospital with fractured neck of the femur are often found to be low in vitamin K, which suggests that deficiency may predispose to osteoporosis. This vitamin is known to be required for the synthesis of osteocalcin, a protein that enhances osteoblastic activity and is an important calcium-binding protein of the bone matrix (see Chapter 23).

Water-soluble vitamins (B group and vitamin C)

Vitamin B$_1$ (*thiamin*) is essential for the normal metabolism of carbohydrates, and deficiency gives rise to widespread metabolic disturbances including acidosis as a result of lactic acid accumulation. These changes trigger a variety of clinical responses collectively known as *beriberi*. Beriberi occurs in different forms classified as wet or dry. In so-called wet beriberi vasodilatation occurs which leads to right-sided heart failure and general edema. In dry beriberi (usually a chronic deficiency state) a variety of neuropathies are seen including nystagmus, ophthalmoplegia, ataxia, abnormal pupillary reactions, and altered consciousness. There may be headaches, vomiting, and confusion. If untreated, this may progress to an irreversible condition known as Korsakoff's psychosis in which there is an inability to form new memories. This condition occurs most often in alcoholics because ethanol metabolism requires thiamine. It may also be seen, though rarely, in patients suffering from hyperemesis gravidarum (extreme sickness of pregnancy)

Vitamin B$_2$ (*riboflavin*) is a constituent of flavine coenzymes which participate in a variety of oxidation and reduction reactions. Its main dietary sources are milk products, meat, and eggs. Deficiency is comparatively rare and non-fatal. Symptoms include cracking at the edges of the lips (cheilosis) and corners of the mouth (angular stomatitis), glossitis (loss of the epithelium covering the tongue), and skin lesions.

Vitamin B$_3$ (*niacin*) is a constituent of nicotinamide coenzymes, and deficiency gives rise to a clinical syndrome known as *pellagra* (literally sour skin) which is fatal if left untreated. Niacin is present in many foods and deficiency is normally restricted to populations whose chief food is maize, a cereal that contains very little tryptophan from which niacin can be synthesized. In developed countries with a rich protein diet and a variety of cereal crops, pellagra is very rare.

Diarrhea, dementia, and dermatitis are typical symptoms of pellagra. The skin becomes inflamed when exposed to sunlight giving the appearance of severe sunburn (Fig. 19.3). Lesions become pigmented and later crack and peel. These changes most often involve the skin of the neck (Casal's collar). The tongue may also be inflamed.

Neurological presentations of severe niacin deficiency include neuropathies, tremor, fits, depression, ataxia, and rigidity. Depressive psychoses may be due to a lack of serotonin as a result of tryptophan deficiency.

Vitamin B$_6$ (*pyridoxine*) exists in three forms that are interconvertible in the body. These are pyridoxine, pyridoxal, and pyridoxamine. The main form in the body is pyridoxal 5′-phosphate (PLP), which is involved in amino acid metabolism and

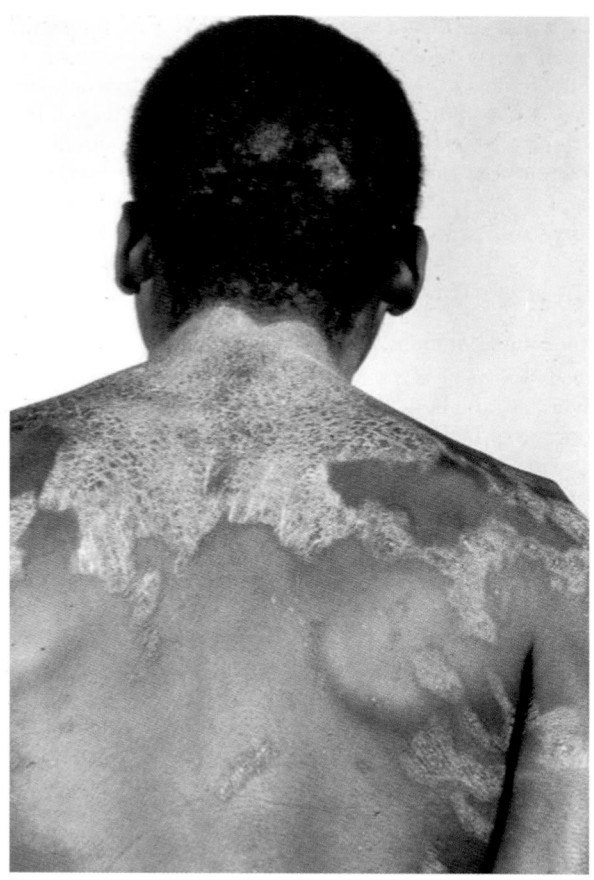

Fig. 19.3 A patient with pellagra showing Casal's collar.

the release of glucose from glycogen stores, particularly those of muscle. It may also be involved in the modulation of steroid hormone receptors. In deficiency, there is increased sensitivity of target tissues to actions of low concentrations of hormones such as estrogens, androgens, cortisol, and vitamin D. It is also needed for the synthesis of GABA, an important inhibitory neurotransmitter of the CNS (see Chapter 6). Deficiency is often associated with convulsions, sleeplessness, and other neurological changes.

Extreme use of vitamin supplements may lead to a toxic overdose of vitamin B_6. Moderate overdose may cause tingling in the fingers and toes, reversible on cessation of supplements, while very high intake may result in damage to peripheral nerves and partial paralysis.

Vitamin B_{12} (cobalamin) is the collective name for a group of cobalt-containing substances known as cobalamins. They are present in meat, fish, and dairy products, and deficiency is normally only seen in those following a vegan diet or in individuals deficient in intrinsic factor (due to gastrectomy or malabsorption disorders). Pernicious anemia (see also Chapter 13) and neuropathies are characteristic symptoms of vitamin B_{12} deficiency. There may be loss of sensation in the lower limbs, some spinal nerve demyelination, and a loss of motor power. Infants who are breast-fed by vegan mothers may be at risk of impaired neurological development and anemia.

Vitamin C is needed for the synthesis of collagen, the principal connective tissue of tendons, arteries, bone, skin and muscle. It is also required for the synthesis of norepinephrine from dopamine. The disorder caused by vitamin C deficiency is called scurvy. In this disease, new connective tissue cannot be made in sufficient amounts to replace aging or injured tissue. Therefore even minor injuries may cause bleeding. There may be bleeding from gums, nose, or hair follicles, or into joints, the bladder, or the gut. Other symptoms include listlessness, anorexia, weight loss, halitosis, gingivitis (gum disease), and loose teeth (Fig. 19.4).

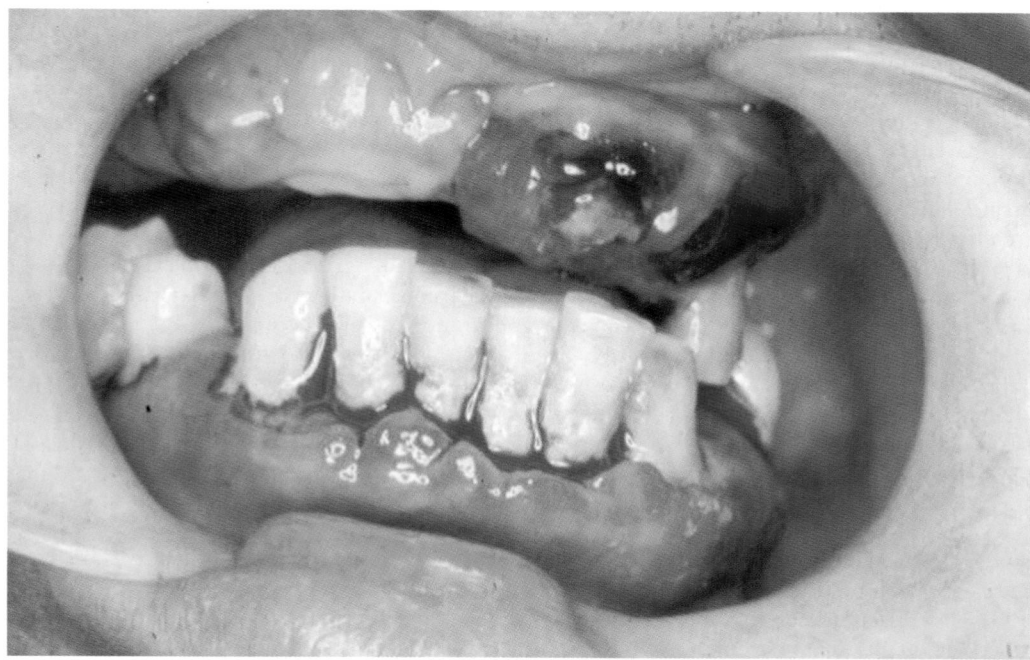

Fig. 19.4 The typical symptoms of scurvy. Note the bleeding and swollen gums and the loss of teeth from the upper jaw.

Minerals

A wide variety of inorganic ions are required for normal cellular activities. These include calcium, phosphorus, sodium, potassium, iron, and iodine. Furthermore, a number of other elements such as zinc, copper, manganese, selenium, and vanadium are required in tiny amounts. These are the so-called trace elements, many of which act as cofactors for enzyme-catalyzed cellular processes. Calcium is found in milk, eggs, green vegetables, and some fish. Sources of phosphorus include cheese, oatmeal, liver, and kidney. Calcium and phosphorus are needed for the normal mineralization of bone. Calcium is involved in secretion, muscle contraction, and blood clotting, while phosphorus is an important component of cell membranes and ATP.

Sodium is found in most foods, especially meat, fish, eggs, milk, bread, and as table and cooking salt. The adult daily requirement for sodium is about 1.6 g, though most people ingest much more. It is the major extracellular cation and plays a crucial part in volume regulation, muscle contraction, and nervous conduction. Potassium is widely distributed in all foods, especially fruit and vegetables. It is the major intracellular cation and is involved in many cellular processes.

Iron as a soluble compound is found in liver, kidney, beef, egg yolk, green vegetables, and wholemeal bread. About 1–2 mg of iron is required each day to replace that lost from the body. A higher intake is needed by women, particularly during pregnancy. Iron is essential for the formation of hemoglobin (see Chapter 13) and is necessary for cellular respiration.

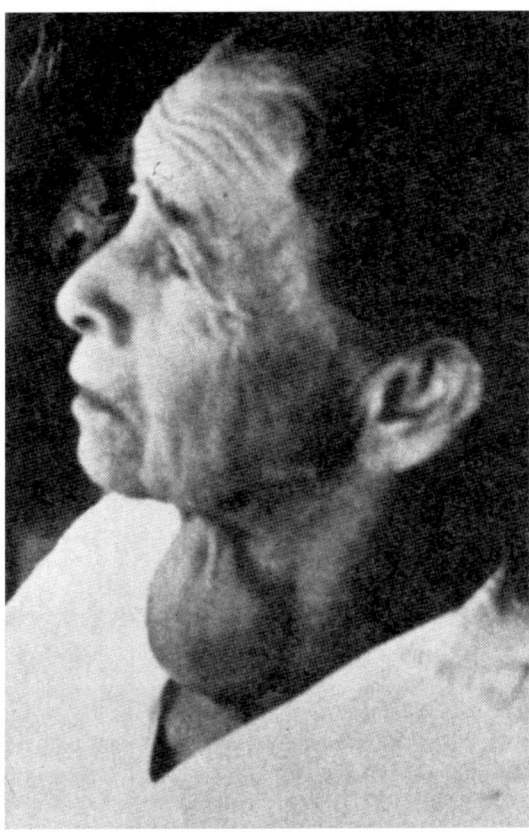

Fig. 19.5 A man with endemic goiter due to iodine deficiency.

Iodine is found in saltwater fish and in vegetables grown in iodine-rich soil. The daily requirement is 140 μg. In areas of the world in which naturally occurring iodine is deficient, small quantities may be added to table salt. Iodine is required for the synthesis of thyroid hormones. Individuals whose diets lack sufficient iodine develop an enlarged thyroid gland (goiter) in an attempt to trap any available iodine from the plasma (see Fig. 19.5).

19.3 Regulation of dietary intake

Although, to a large extent, the content and size of our meals is dictated by social factors and the daily pattern of activity, both hunger and appetite are important regulators of dietary intake. *Hunger* refers to a physiological sensation of emptiness, usually accompanied by contraction of the stomach. *Appetite* refers to the feelings associated with the anticipation of forthcoming food. It may be affected by an individual's emotional state. Nervousness or fear often suppress the appetite. Therefore hunger and appetite, although related, are different sensations.

The hypothalamus plays an important role in the regulation of food intake. The lateral hypothalamus contains a region known as the feeding center. Lesions here in rats produce *aphagia* (lack of feeding), leading to starvation and death. Increased food intake (*hyperphagia*) can be induced by lesions of the ventromedial hypothalamus, and this region is called the satiety center. The exact role of these areas in the control of food intake requires further clarification.

Overweight and obesity

These are common conditions which are increasing in prevalence, especially in affluent societies. Overweight or obese people have excessive fat stores in relation to their height, gender, and race (see below for more information on body mass index (BMI)). Obesity has many consequences, both physical and physiological. These include metabolic changes such as diabetes, fatty liver, gallstones, infertility in women, changes in plasma lipid levels, cardiovascular problems such as hypertension, coronary heart disease, varicose veins, peripheral edema, osteoarthritis, spinal problems, and obstructive sleep apnea. Furthermore, obese people often have low self-esteem. The distribution of the excess fat is thought to be significant in relation to certain health risks. Abdominal fat deposition, with waist circumference measurements of over 100 cm in men and 95 cm in women, is particularly associated with an increased risk of cardiovascular and metabolic disorders such as heart attack and diabetes.

There is considerable interest in the likely causes of obesity and the possibility of a 'cure' for overweight and obese individuals. Obesity reflects an imbalance between energy expenditure and intake, so clearly the amount and type of food eaten and the amount of work done (both to maintain body functions and in exercise) are key issues. High-fat diets predispose to obesity, and some studies have shown that weight and body mass index can go on increasing even when total energy consumed is reduced if the proportion of fat in the diet is increased.

It is a commonly held belief that excessive weight can arise because of glandular disturbances. However, endocrine disorders are not generally associated with obesity, although the following conditions may contribute to weight gain (see also Chapter 12):

◆ hypothyroidism

◆ acromegaly

◆ Cushing's syndrome

◆ insulin resistance.

There is some evidence to suggest that the tendency to be overweight is at least partly determined by genetic factors, probably in a multigene, multifactorial fashion. There is currently much interest in a gene (the ob gene) responsible for the production of a satiety factor in obese (ob/ob) mice. Its protein product, *leptin*, has been shown to reduce the body weight of obese mice and there is some suggestion that the administration of leptin to obese humans could be used to modulate homeostatic mechanisms of energy balance and lead to weight loss. However, leptin is produced in adipocytes and its levels are related to the amount of fat tissue in the body. Women have higher levels of leptin than men, reflecting their higher proportion of body fat, and leptin is also elevated in obese individuals. These findings would indicate that control of appetite and therefore weight is not just a simple matter of altering leptin concentrations. Clearly, more work is required in this important area of nutrition.

Anorexia nervosa and bulimia nervosa

Together, anorexia nervosa and bulimia nervosa represent a significant source of morbidity in young Western women. *Anorexia nervosa* is mainly confined to women aged between 10 and 30 years, with males accounting for less than 10 per cent of all cases. The incidence of the disorder in girls aged 11–18 years is thought to be between 0.2 and 1.1 per cent.

Anorexic patients show a number of characteristic symptoms. There is commonly a very small food intake, which is usually between 2.5 and 3.37 MJ day^{-1} (600 and 900 kcal day^{-1}). BMI is generally below 17.5 and the patient will actively strive to maintain an unduly low weight, often exacerbating her condition by pursuing a vigorous exercise regime. There seems to be a fear of fatness and a distorted body image, often with a feeling of worthlessness and depression. Amenorrhea (cessation of menstrual periods) is also typical of postmenarchical anorexic patients who are not taking oral contraceptives, presumably because body fat stores are no longer sufficient to sustain normal sex steroid production (see Chapter 20). Anorexia is associated with many other physiological abnormalities related to reduced energy intake and low body weight. These include an increased susceptibility to cold, gastrointestinal problems including constipation, and abdominal pains. Blood pressure and heart rate are often low in anorexic patients. Furthermore, there may also be skeletal changes such as osteoporosis. A severely anorexic patient is illustrated in Fig. 19.6.

Bulimia nervosa has many similarities with anorexia in that the patient has extreme concerns about weight, a distorted body image, and often depression and anxiety. However, weight and BMI in bulimics is usually in the normal range. Typically, bulimic patients starve themselves and then binge on enormous amounts of 'forbidden' foods, often consuming the energy equivalent of 8–12 MJ (*c.* 2000–3000 kcal) at a time. After this,

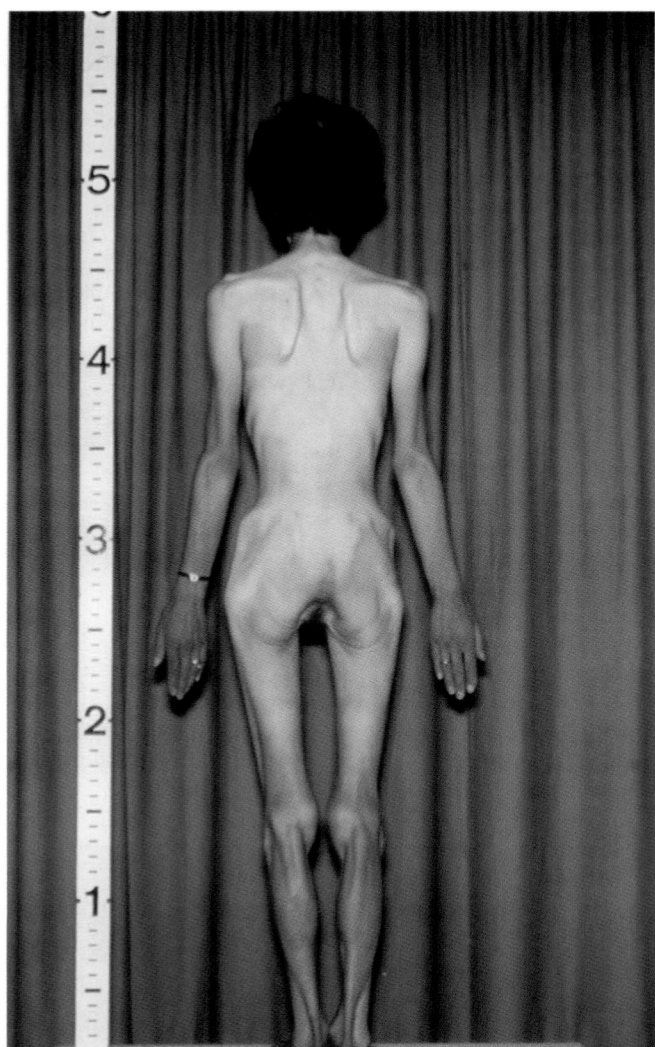

Fig. 19.6 A severely anorexic patient. The height scale is in feet.

they feel very guilty and may vomit or take large doses of laxatives in an attempt to purge the calories. Although there are few physical complaints associated directly with bulimia, persistent vomiting can cause erosion of the enamel on the lingual surface of the upper front teeth. Frequent vomiting or laxative abuse may also lead to electrolyte imbalances.

19.4 Measurements used to monitor nutritional status

In individuals who are being treated for malnourishment of any kind, it is important to have ways of monitoring progress in terms of either weight loss or gain. In addition to physical examination and detailed medical history, a variety of measurements of nutritional status may be employed. Although a single determination of weight is of relatively little value, regular sequential weighing over a period of time is more helpful. It is also important to view weight in the context of a patient's height. This is the BMI or body mass index that is calculated as follows:

TABLE 19.3	Body fat estimated from skinfold thickness		
Skinfold thickness	**Percentage body fat**		
(mm)	**Age 17–19 years**	**Age 30–39 years**	**Age 50+ years**
Male subjects			
20	8.3	12	12.5
30	13	16	18.5
40	16.5	19	23
50	19	21.5	26
60	21	23.5	29
70	23	25	32
Female subjects			
20	14	17	21
30	19	22	26.5
40	23	25.5	30
50	26	28	33
60	28	31	35.5
70	30.5	32.5	37.5

Skinfold thickness is measured at four locations: the biceps, the triceps, below the shoulder blade, and at the waist (supra-iliac skinfold). The values are added together to determine the final skinfold thickness. For males the desirable percentage of body fat lies between 15 and 21 per cent depending on age. For females, the desirable range is 17–25 per cent.

$$\text{BMI} = \frac{\text{weight (kg)}}{(\text{height in m})^2}.$$

Values between 20 and 25 kg m^{-2} are considered normal, while values below 20 are classified as underweight for height and may be associated with health problems in some individuals. Values in excess of 25 are considered increasingly overweight. When BMI exceeds 30, the individual is classified as obese. As BMI increases beyond this value, obesity is further classified as moderate to severe. However, it is important to realize that BMI is of much less value in children. Many healthy children have a calculated BMI well below 20 because of the changes in build which occur at various times during development (e.g. the growth spurt at puberty). Longer-term monitoring of growth (both height and weight) as well as cognitive function is of greater value.

BMI may also be misleading in people who have a large muscle mass, since muscle weighs more than the same volume of fat. For example, highly trained athletes may well have a calculated BMI in excess of 25 because of their lean body mass but could not be considered overweight or obese.

Additional information regarding nutritional status can be obtained from measurements of skinfold thickness at various sites such as the triceps, biceps, subscapular, and supra-iliac regions. About one-third of body fat is subcutaneous, so skinfold thickness reflects the size of the subcutaneous fat depot. This is in turn related to total body fat. Such measurements are made using calipers and tables of normal ranges are published. Table 19.3 shows some typical values for men and women.

Other measurements include mid upper-arm circumference and mid-arm circumference which can be helpful in monitoring the progress of severely undernourished individuals. These measurements are particularly useful indicators of the body's protein reserves. Coupled with such measurements, analysis of body composition may be helpful in determining nutritional status. Devices are available which determine the relative proportions of fat, lean tissue, and water in the body. However, it should be remembered that all measurements are subject to considerable errors in patients who have edema.

Certain biochemical measurements may also contribute to an overall assessment of nutritional status. For example, the appearance of ketones in the urine indicates that fats are being broken down for use in metabolism. This will reflect weight loss and may be desirable in obese people who are trying to lose weight, though not of course in a malnourished individual. Nitrogen balance is a further indicator of the extent to which protein synthesis and breakdown are taking place.

19.5 Enteral and parenteral nutritional support

There are a number of circumstances in which people are unable to eat normally. These may include an inability to swallow (dysphagia) or to absorb food across the intestine, a reduced level of consciousness, pain, or the effects of therapy. Furthermore, the emotional and psychological factors associated with being in hospital may modify the appetite. Nutritional support may be required in these cases. Enteral nutrition (via the GI tract) is possible if the gut is healthy, while parenteral nutrition (bypassing the GI tract) may be necessary if it is not.

Enteral support can take the form of high-energy liquid feeds taken orally or formula feeds administered via a nasogastric tube. Such feeds will contain appropriate amounts of protein, carbohydrates (including glucose), fats, vitamins, and minerals. Different formulas are available to meet the requirements of different patients. If the nasogastric tract is obstructed or the gut

Summary

1. A balanced diet is essential for health. A mixed diet contains adequate amounts of the essential nutrients and strictly vegetarian diets must be carefully controlled.

2. The main foodstuffs are the carbohydrates, fats, and proteins. The carbohydrates provide an important energy source. Fat is an important source of energy. It is also an important constituent of cell membranes. Certain fatty acids are classed as essential since they cannot be manufactured by the body. The proteins are broken down into their constituent amino acids before being absorbed and used to make structural protein, enzymes, hormones etc. Eight essential amino acids must be included in the diet.

5. In addition to the main foodstuffs, a wide variety of accessory food factors are required for health. These are organic substances known as vitamins and minerals. Vitamin deficiencies are associated with characteristic symptoms (e.g. rickets, pellagra, and beriberi).

6. Hunger and appetite are important regulators of food intake. Feeding and satiety 'centers' are located in the hypothalamus. Obesity, anorexia nervosa, and bulimia nervosa are examples of inappropriate food intake and carry a range of associated health problems.

8. A variety of anthropometrical and physiological tests can be carried out to assess nutritional status. These include measurements of skinfold thickness, arm circumference, body composition, and nitrogen balance.

9. Under some circumstances, it is necessary to provide nutritional support either by a nasogastric tube (enteral nutrition) or by infusing nutrients directly into the bloodstream via a cannula inserted into a large vein (parenteral nutrition).

is unable to process food adequately, parenteral feeding may be needed. Here the gut is bypassed and nutrients are infused directly into a large central vein, usually the subclavian or jugular. Nutrients must be in a form in which they can be used without digestion, for example pure amino acids, glucose, and an emulsion of triglyceride. If parenteral feeding is prolonged, vitamins, minerals, and trace elements will also need to be included in the feed. Feeds of this kind are hypertonic and must be given along with appropriate amounts of fluid. Furthermore, regular monitoring of the patient's weight and hydration status is essential.

Recommended reading

Bender, D.A. (1997). *Introduction to Nutrition and Metabolism* (2nd edn). Taylor and Francis, London.

Mann, J.A., and Truswell, S. (eds.) (2002). *Essentials of Human Nutrition* (2nd edn). Oxford Medical Publications, Oxford.

Multiple choice questions

The following statements are true or false. Answers are given below.

1. a. Within the body, carbohydrates are stored in the form of glycogen.
    b. All essential amino acids are found in meat.
    c. Edema is a characteristic feature of marasmus.
    d. Plant oils mainly consist of unsaturated fats.
    e. Ketones are likely to be present in the urine of an individual who is eating a very low carbohydrate diet.

2. a. Excess intake of vitamin A can cause severe toxic reactions.
    b. Vitamin K deficiency may increase the risk of thrombosis.
    c. Vitamin B_3 deficiency may cause severe neurological disorders.
    d. Vitamin B_{12} deficiency results in aplastic anemia.
    e. Rickets is caused by vitamin C deficiency.

3. a. A diet that is deficient in iodine may cause goiter.
    b. The hypothalamus plays an important role in the regulation of food intake.
    c. A person suffering from bulimia nervosa is likely to be severely underweight.
    d. An individual with a BMI of 25 kg m^{-2} is overweight.
    e. A growing child may have a BMI significantly below 20 kg m^{-2} and yet be perfectly healthy.

Answers

1. Animal protein is first class protein and contains all 20 of the amino acids used by the body to synthesize new proteins. Edema is a feature of kwashiorkor. Children with marasmus generally have wrinkled skin with no edema. Plant oils are rich in unsaturated fat. Ketones are by-products of fat metabolism and appear in the urine of individuals who are eating little or no carbohydrate.
    a. True
    b. True
    c. False
    d. True
    e. True

2. Vitamin A deficiency causes night-blindness and xerophthalmia. Excess is highly toxic and can cause a variety of neurological problems. It also causes abnormalities of development. Vitamin K is required for normal blood clotting. Bleeding disorders are associated with deficiency. Vitamin B_3 deficiency results in pellagra. This disease is associated with a variety of neurological symptoms and with skin disorders. Vitamin B_{12} deficiency leads to pernicious anemia. Rickets is caused by a deficiency of vitamin D. Scurvy is caused by vitamin C deficiency.

a. True

b. False

c. True

d. False

e. False

3. Iodine deficiency may lead to the development of a goiter as the thyroid gland enlarges to trap any available iodine from the plasma. Individuals with bulimia are usually of normal weight. BMI is a measurement that relates weight to height. As a measure of bodily health it is of most value in adults.

Normal values range from 20–25 kg m^{-2}. Values in excess of 30 represent increasing levels of obesity. During periods of accelerated growth in childhood, the body is in positive nitrogen balance, which means that the synthesis of new protein exceeds protein breakdown or fat deposition.

a. True

b. True

c. False

d. False

e. True

CHAPTER 20

The physiology of the male and female reproductive systems

Chapter contents

After reading this chapter you should understand:

◆ The significance of sexual reproduction

◆ The principal structures of the male reproductive system and their function

◆ The formation of mature sperm by spermatogenesis and spermiogenesis

◆ The regulation of testicular function by the anterior pituitary hormones and testosterone

◆ The principal structures of the female reproductive system and their function

◆ The ovarian (or menstrual) cycle and its hormonal regulation

◆ The role of the pituitary and ovarian hormones in the regulation of the female reproductive system

◆ Puberty and the menopause

◆ The peripheral actions of the testicular and ovarian steroids in the adult

20.1 Introduction

Reproduction—the ability to produce a new generation of individuals of the same species—is one of the fundamental characteristics of living organisms. Genetic material is transmitted from parents to their offspring to ensure that the characteristics of the species are perpetuated. The essential feature of sexual reproduction is the mixing of chromosomes from two separate individuals to produce offspring that differ genetically from their parents. At the core of the process lies the creation and fusion of the male and female gametes: the spermatozoa (sperm) and ova (eggs).

Gametes are specialized sex cells produced by the gonads, which provide a link between one generation and the next. Spermatozoa, the male gametes, are produced by the testes, while ova, the female gametes, are produced by the ovaries. The nuclei of these cells are *haploid*, i.e. they contain a set of 23 unpaired chromosomes. Haploid cells are created when a diploid cell divides by *meiosis*, a process in which the genes are parceled out afresh in single chromosome sets (for more details see pp. 24–25). During meiotic division, old combinations of genes are broken and new combinations formed by chromosomal exchange so that the genetic composition of each chromosome is modified. At fertilization, the gametes fuse to form a new cell possessing a full set of chromosomes, half of which originate from the sperm and half from the ovum. This is called a *zygote*. The reshuffling of genes during sexual reproduction helps to create a genetically diverse population which is able to show greater resilience in the face of environmental challenge.

This chapter will discuss the processes that lead to the production of the male and female gametes. It will also discuss the hormonal mechanisms that regulate their activity. In addition, the neural and endocrine control of reproductive activity, including puberty and the menopause, will be discussed.

Reproductive physiology of the male

20.2 The anatomy of the male reproductive system

Figure 20.1 is a simple diagram of the adult male reproductive tract showing the major organs within it. Figure 20.2 illustrates the internal structure of the testis, the male gonad responsible for the production of spermatozoa and the male sex hormones (the *androgens*). The testes lie outside the abdominal cavity, within the scrotal sac. Each testis is about 4.5 cm in diameter and weighs around 40 g. The organ is made up of a large number of coiled *seminiferous tubules* containing *Sertoli cells* where the sperm are made. Between these tubules lies supportive connective tissue which contains the *interstitial* or *Leydig cells* which are responsible for the synthesis and secretion of the testicular

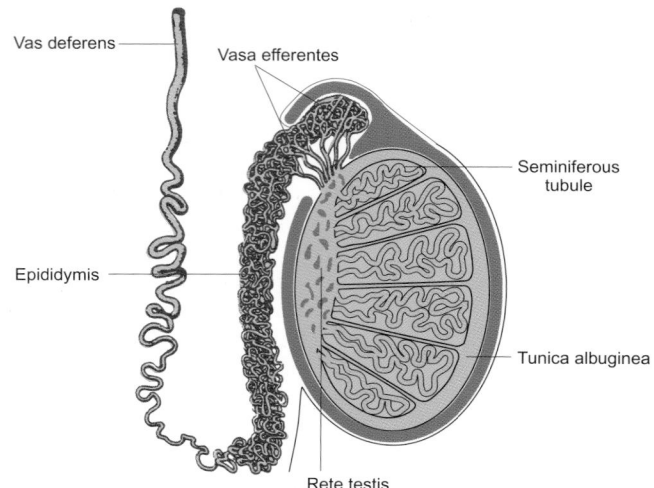

Fig. 20.2 The adult testis, epididymis, and vas deferens.

androgens, particularly testosterone. This anatomical arrangement gives the testes a lobular structure in which each lobule contains two or three tubules. At the apex of each lobule, its seminiferous tubules join and pass into the first section of the excretory ducts, the *tubuli recti*. These are short straight tubes which enter the dense connective tissue of the mediastinum testis and form within it a system of irregular epithelium-lined spaces, the *rete testis*. From here the tubules drain into another coiled tube, the *epididymis*, which in turn leads into the *vas deferens*, a tubular structure 30–35 cm long which terminates in the ejaculatory duct close to the prostate. The seminal vesicles are located on either side of the prostate. They empty their

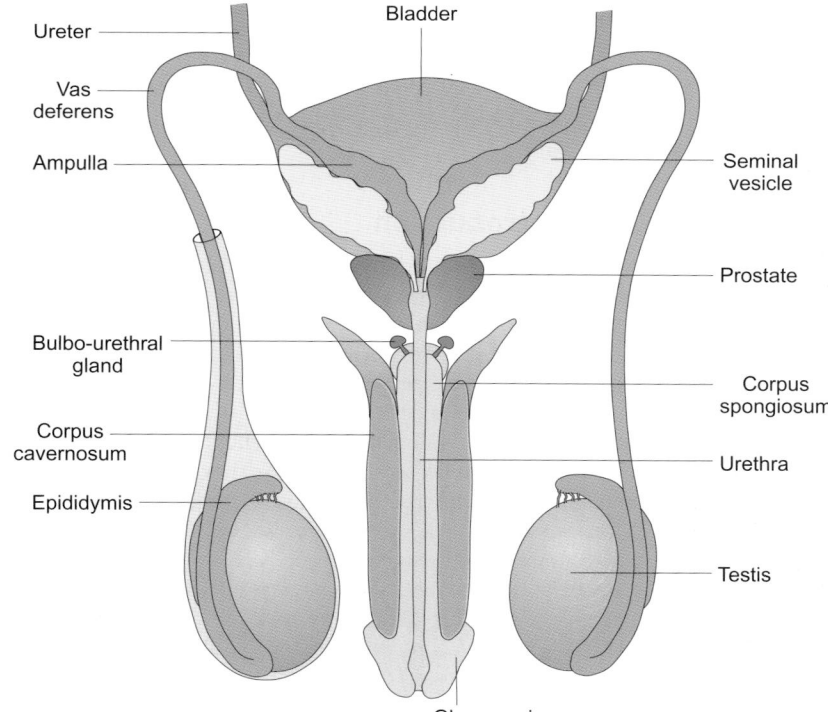

Fig. 20.1 A posterior view of the adult human male reproductive system to show the principal structures.

secretions into the *ejaculatory duct* together with the sperm and prostatic secretions to form semen. From the ejaculatory duct, the semen enters the penis, from which it is released during sexual intercourse.

The development of the testis takes place within the abdominal cavity of the fetus (see Chapter 22). However, by the time of birth or soon afterwards, the testes are lying within the scrotal sac outside the body cavity and experience an ambient temperature which is some 2–3°C lower than the core temperature. To arrive here, the testes descend, migrating posteriorly through the abdominal cavity and over the pelvic brim. Failure to migrate in this way results in a condition known as *cryptorchidism* which, if it persists until puberty, leads to an arrest of spermatogenesis and thus infertility since the testes cannot function normally at body temperature.

20.3 The adult testis makes gametes and androgens

The testis performs two fundamental roles in the mature male, both of which are vital to his fertility and sexual competency:

- the production of sperm which will carry his genes and fertilize an ovum;
- the secretion of testicular androgens, particularly testosterone, which bring about full masculine development.

The two major products of the male gonads, the spermatozoa and the androgenic steroid hormones, are synthesized in separate compartments within the testis. The sperm are made in the seminiferous tubules themselves, while the androgens are synthesized and secreted by the Leydig cells that lie between the tubules. Indeed, these two compartments of the testis appear to be separated functionally as well as anatomically since there is a barrier between them that prevents the free exchange of water-soluble materials. This is known as the blood–testis barrier and arises as a result of the extremely tight junctional complexes which exist between the basal regions of adjacent Sertoli cells (Fig. 20.3). This barrier protects the developing sperm from any noxious blood-borne agents and so maintains a suitable environment for their maturation. It also prevents antigenic materials (e.g. proteins) that arise in the course of spermatogenesis from passing into the circulation and triggering an autoimmune response to the sperm. Where this does happen—for example,

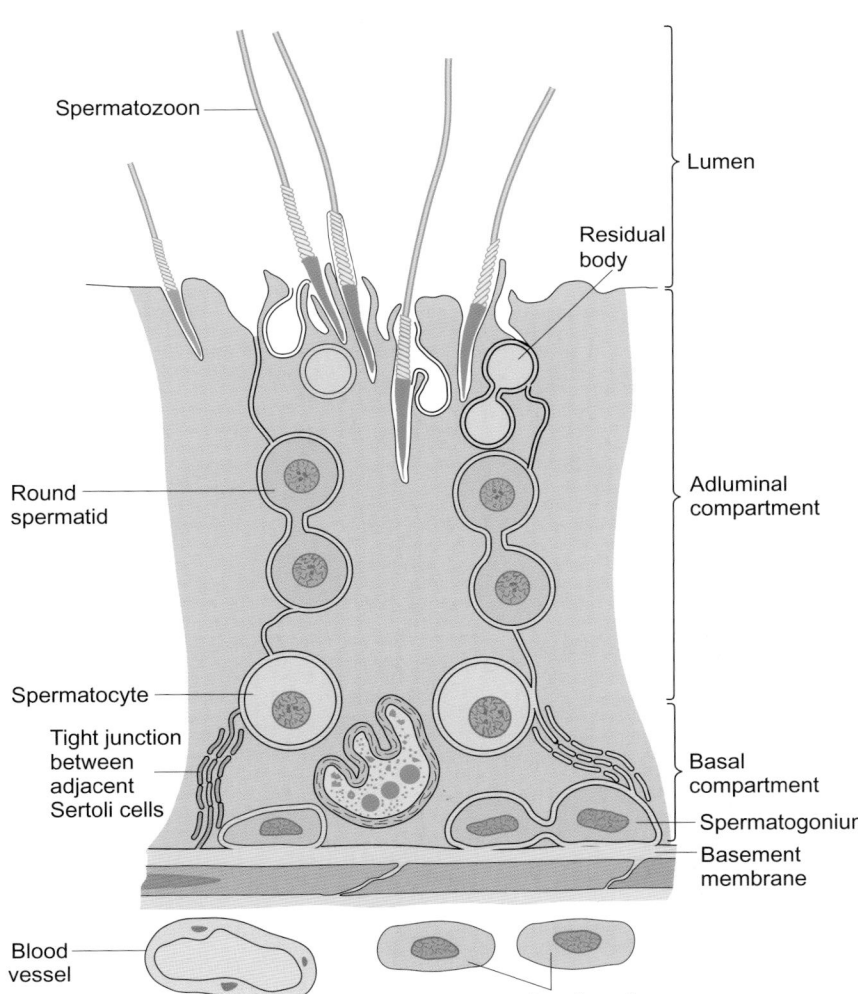

Fig. 20.3 Sectional view of the wall of a seminiferous tubule to show the relationship between the Sertoli cells and the developing spermatozoa. Note the tight junctions between the basal regions of the Sertoli cells separating the basal compartment from the adluminal compartment.

as a consequence of traumatic injury to the testis—it can lead to male infertility. Although the manufacture of sperm and androgens occurs in separate compartments, their production is closely related functionally because the production of mature sperm is only possible if androgen secretion is normal.

Testosterone is the major testicular androgen

The Leydig cells of the testis synthesize and secrete testosterone, the principal testicular androgen, from acetate and cholesterol as described in Box 20.1. Adult males secrete about 4–10 mg of testosterone each day, most of which passes into the blood. However, a small amount enters the seminiferous tubules, where it binds to an androgen-binding protein secreted by the Sertoli cells and subsequently plays a crucial role in the development of the spermatozoa (see below). Since it is a steroid and thus fat soluble, testosterone is able to cross the blood–testis barrier by passive diffusion.

The peripheral actions of testosterone

The mode of action of steroid hormones has been described previously (see Chapters 5 and 12) and the general rules are applicable to testosterone. The hormone circulates in the plasma bound either to a sex-steroid-binding globulin or to other plasma proteins. It enters cells freely where it may be converted to *dihydrotestosterone* or *5α-androstenedione*. All three androgens bind to specific intracellular receptor proteins to form a steroid–receptor complex that interacts with chromosomal DNA to regulate gene expression as described in Chapter 5. In addition, they can bind to receptors on the plasma membrane or interact directly with nuclear DNA to modify gene expression. Androgen receptors are most numerous in the tissues which are specific targets for the hormone, i.e. those which depend upon androgens for their growth, maturation, and/or function. Such tissues include the accessory organs of the male reproductive tract—the prostate, seminal vesicles, and epididymis—as well as nonreproductive tissues such as the liver, heart, and skeletal muscle.

Dihydrotestosterone is important in the fetus for the differentiation of the external genitalia and, at puberty, for the growth of the scrotum, prostate, and sexual hair. In addition to its role in the production of sperm, testosterone stimulates the fetal development of the epididymis, vas deferens, and seminal vesicles, and, at puberty, is responsible for enlargement of the penis,

---

BOX 20.1   BIOSYNTHESIS OF THE MAJOR SEX STEROID HORMONES FROM CHOLESTEROL

seminal vesicles, and larynx together with the changes in the skeleton and musculature characteristic of the male.

Spermatogenesis—the production of sperm by the testes

A sexually mature man produces around 200 million spermatozoa each day. *Spermatogenesis* is a complex process that involves the generation of huge numbers of cells by *mitosis* and the halving of their chromosomal complement by *meiosis*. Furthermore, it involves the formation of a highly specialized cell that is designed to carry the genetic material a considerable distance within the female reproductive tract to maximize the chances of fertilization. The major steps in the process of spermatogenesis are shown in Fig. 20.4.

At puberty, the male germ cells commence mitotic division, an event which marks the beginning of spermatogenesis and which results in the formation of a population of spermatogonia lying within the basal compartment of the seminiferous tubules.

The first two mitotic divisions of each germ cell give rise to four cells which in fact remain connected to one another by a thin bridge of cytoplasm (Fig. 20.4). Three of these undergo further division to form spermatogonia, while the fourth arrests at this stage and will later serve as a stem cell for a subsequent generation of sperm. The three active cells divide twice more to give

rise to a population of so-called primary spermatocytes. Up until now, the cell divisions have taken place within the basal compartment of the tubule (Fig. 20.3) but at this point the primary spermatocytes enter the adluminal tubular compartment. They apparently achieve this by transiently disrupting the tight junctions between neighboring Sertoli cells. Following a period of growth, each of the primary spermatocytes then undergoes two meiotic divisions. The first of these gives rise to haploid secondary spermatocytes, which immediately divide again to form spermatids each of which possesses 22 *autosomes* (i.e. chromosomes not associated with the determination of sex) and either an X or a Y sex chromosome. The progeny of a single spermatogonium still remain connected by cytoplasmic bridges. The genetic events of spermatogenesis are now complete. The final stages of the process involve the conversion of the round spermatids to mature motile spermatozoa, a process known as *spermiogenesis*.

Spermiogenesis involves major cytoplasmic remodeling of the spermatid

Figure 20.5 shows the essential structures of a mature motile human spermatozoon, and it is clear that there are considerable differences between this and the round spermatid. The process of spermiogenesis involves reorganization of both the nucleus and the cytoplasm of the cell as well as the acquisition of a flagellum. The whole process takes place in close association with the Sertoli cells.

The immature sperm consists of two regions that are morphologically and functionally distinct: the head region and the tail.

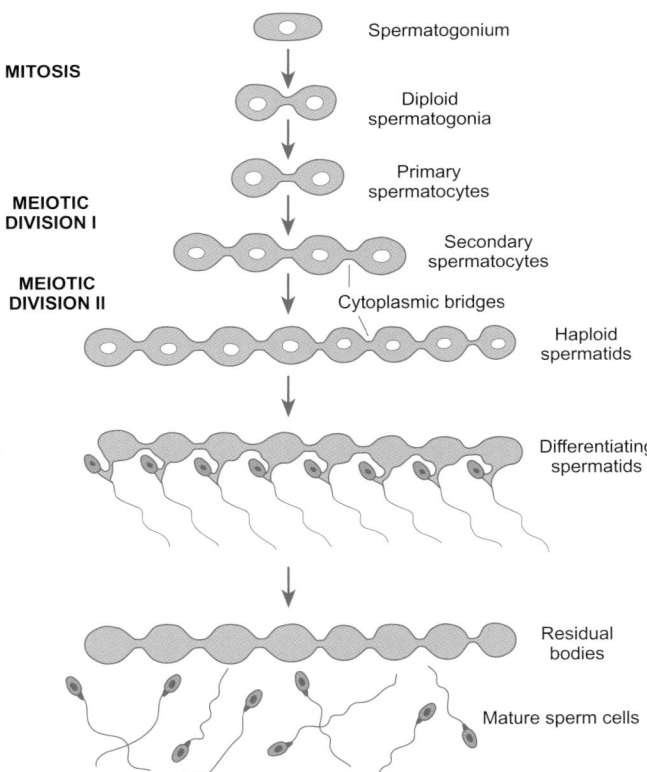

Fig. 20.4 The principal stages of spermatogenesis. The primordial cells divide to form spermatogonia which undergo two more divisions to form primary spermatocytes. The primary spermatocytes undergo meiotic divisions to form secondary spermatocytes and spermatids. During meiosis the chromosome number is halved. Note the cytoplasmic bridges between the differentiating spermatids.

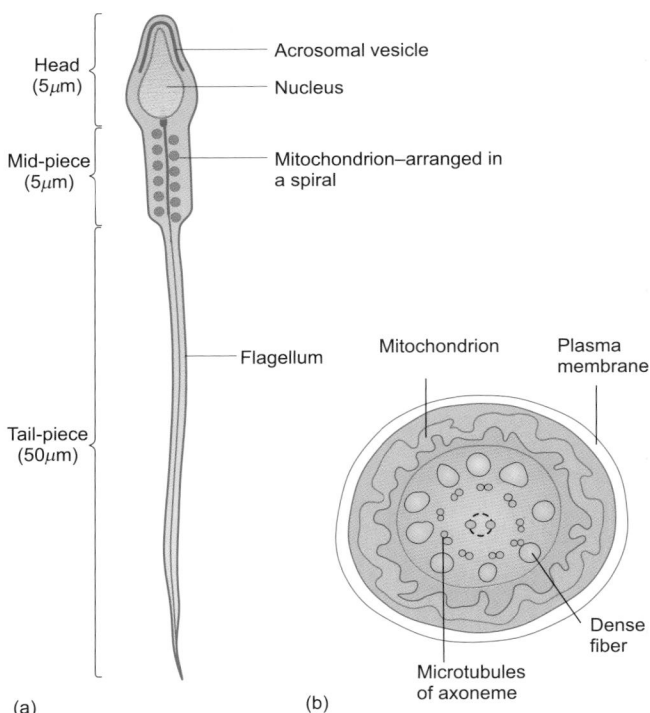

Fig. 20.5 Diagram of a mature spermatozoon in (a) longitudinal section and (b) transverse section through the midpiece. Note the spiral mitochondria and the arrangement of the microtubules.

The head contains a haploid nucleus (i.e. possessing half the full chromosomal complement) and a specialized secretory vesicle called the *acrosomal vesicle*, which contains hydrolytic enzymes that will help the sperm to penetrate the oocyte prior to fertilization.

The tail region of the sperm is motile. It is a long flagellum that has essentially the same internal structure as that seen in all cilia or flagellae from green algae to humans. It consists of a central *axoneme* originating from a *basal body* situated just posterior to the nucleus. The axoneme consists of two central microtubules surrounded by nine evenly spaced pairs of microtubules. Active bending of the tail is caused by the sliding of adjacent pairs of microtubules past one another, and movement is powered by the hydrolysis of ATP generated by mitochondria present in the first part of the tail, the *mid-piece*.

The process of differentiation of a spermatocyte to a motile sperm takes approximately 70 days in the human. After this time the newly formed sperm are released from the adluminal compartment of the Sertoli cells into the lumen of the seminiferous tubule and from there into the epididymus where they undergo further maturation, in particular acquiring the capacity for sustained motility. As can be seen from Fig. 20.4, connected residual bodies of cytoplasm are left behind by the released motile sperm.

The formation of seminal fluid

The epididymis can serve as a reservoir for sperm, with their passage through this coiled tube taking anything from 1 to 21 days. The sperm and other testicular secretions are then transported along the vas deferens and into the ejaculatory ducts. The seminal fluid is markedly increased in volume by contributions from the seminal vesicles (about 60 per cent of total volume) and the prostate (about 20 per cent). The fluids secreted by these glands provide nutrients for the sperm. The seminal fluid is alkaline in nature, helping to counteract the normally acidic fluid of the vagina and thus to increase the motility and fertility of the sperm, both of which are optimal at a pH of around 6.5.

20.4 The hormonal control of spermatogenesis—the pituitary–testis axis

Some of the general mechanisms involved in the regulation of hormone secretion within the body, including the importance of the hypothalamic releasing hormones and the concept of negative feedback control, were discussed in Chapter 12. These key regulatory processes have been shown to operate in the endocrine control of male reproductive function and are summarized in Fig. 20.6.

Gonadotrophin-releasing hormone (GnRH) is synthesized by neurons in the hypothalamus. It is secreted into the vessels of the hypophyseal portal tract and transported to the anterior pituitary. Here it stimulates the pituitary gonadotrophs to secrete the gonadotrophins follicle-stimulating hormone (FSH) and luteinizing hormone (LH) into the systemic circulation. LH acts particularly on the Leydig cells to bring about secretion of testosterone, while FSH acts mainly on the Sertoli cells to cause the release of androgen-binding protein and a further hormone known as inhibin. It also promotes the synthesis of the aromatase complex

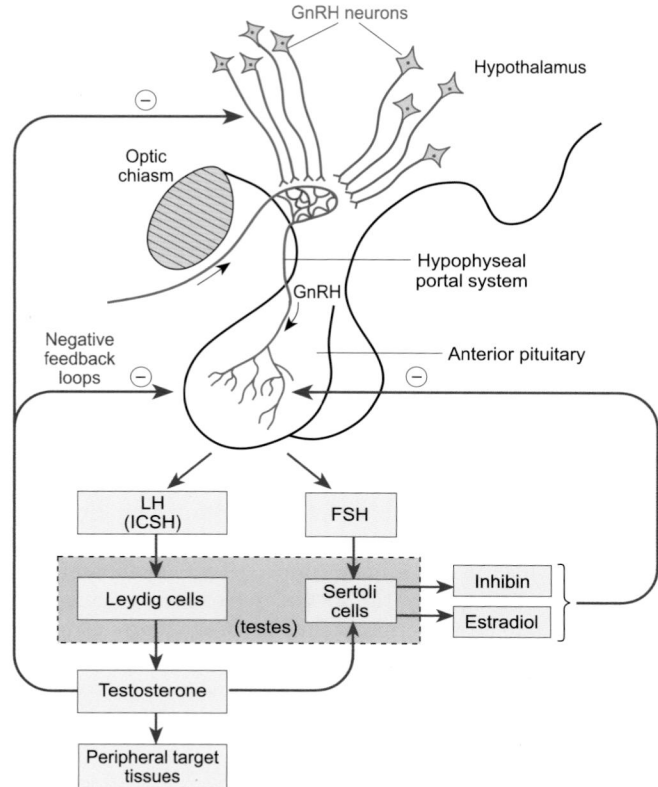

Fig. 20.6 The relationship between the hormonal secretions of the hypothalamus, pituitary gland, and testes.

Summary

1. The adult testis produces sperm, which carry the male genes. It also secretes steroid hormones known as androgens, which bring about full masculine development. The principal androgen is testosterone.

2. Spermatogenesis takes place in the Sertoli cells of the seminiferous tubules, while the androgens are secreted by the Leydig cells which lie between the seminiferous tubules.

3. Spermatogenesis is a complex process involving the generation of huge numbers of cells by mitosis and the halving of the chromosomal complement by meiosis. It culminates in the formation of a highly specialized cell—the mature motile sperm.

4. The sperm are then mixed with secretions from the seminal vesicles and prostate to form seminal fluid. This is released as semen from the penis at ejaculation during sexual intercourse.

5. Spermatogenesis is regulated by a variety of hormones including FSH, LH, and testicular testosterone. In turn, hormone levels are regulated by negative feedback loops operating within the hypothalamic–pituitary–testicular axis.

that is responsible for the conversion of testosterone to estradiol. In turn, testosterone inhibits the secretion of LH by exerting a negative feedback action at the level of both the anterior pituitary itself and the hypothalamus. It is thought that, at the same time, inhibin and estradiol depress the further secretion of FSH by a similar feedback mechanism. These negative feedback loops provide an important internal control system for maintaining fairly constant levels of both gonadotrophic and androgenic hormones in the systemic circulation.

As discussed earlier, testosterone exerts a variety of important effects throughout the body of the male, including the development of secondary sex characteristics at puberty. It is also essential for normal sperm production. As it is lipid soluble, some testosterone secreted by the Leydig cells enters the intratubular compartment where it is bound by androgen-binding protein secreted by the Sertoli cells. In some way, which is as yet poorly understood, this bound testosterone helps to maintain the production of spermatozoa. While testosterone is needed for the maintenance of spermatogenesis, pituitary FSH is required both to initiate the process and to mediate the differentiation of spermatids into spermatozoa.

Reproductive physiology of the female

20.5 Introduction

Like the testis in the male, the ovary produces haploid gametes and a variety of hormones. The production of gametes by the ovary is coordinated with its endocrine activity. However, unlike the testes, which release an enormous number of gametes in a continuous stream, the ovaries produce relatively few oocytes which are normally released only once every 4 weeks or so at ovulation. This regular release of ova from the ovary is controlled by physical, neural, and, above all, endocrine mechanisms involving a complex interplay between the hypothalamic, pituitary, and ovarian hormones. These will be discussed in some detail later, but briefly the ovarian steroids (estrogens and progesterone) are secreted in a cyclical fashion. A period of estrogen dominance characterizes the first half of each cycle, during which one ovarian follicle (see later) reaches full maturity and the body is prepared for gamete transport and fertilization. This period culminates in ovulation roughly halfway through the cycle and is followed by a period of progesterone dominance during which the genital tract is maintained in a state favorable for the implantation and early development of a zygote.

20.6 The anatomy of the female reproductive tract

Figure 20.7 shows a simplified diagram of the adult female reproductive organs. A more detailed diagram of the ovary, showing the important stages in follicular development, is given in Fig. 20.8.

The ovaries are about 3–4 cm long, each weighing around 15 g, and lie in the ovarian fossa of the pelvis. They are attached to the posterior wall of the abdomen by the mesovarium (or ovarian mesentery). The adult organ is composed of stromal tissue, which contains the primary oocytes housed within primordial follicles and glandular interstitial cells. The physiology of the ovaries will be considered in more detail in subsequent sections.

The remainder of the reproductive tract is concerned not with gamete production itself, but with the processes of fertilization and nurture of an embryo. To achieve this, both the male and female gametes must be transported to the site of fertilization (in the Fallopian tubes) and a favorable environment must be

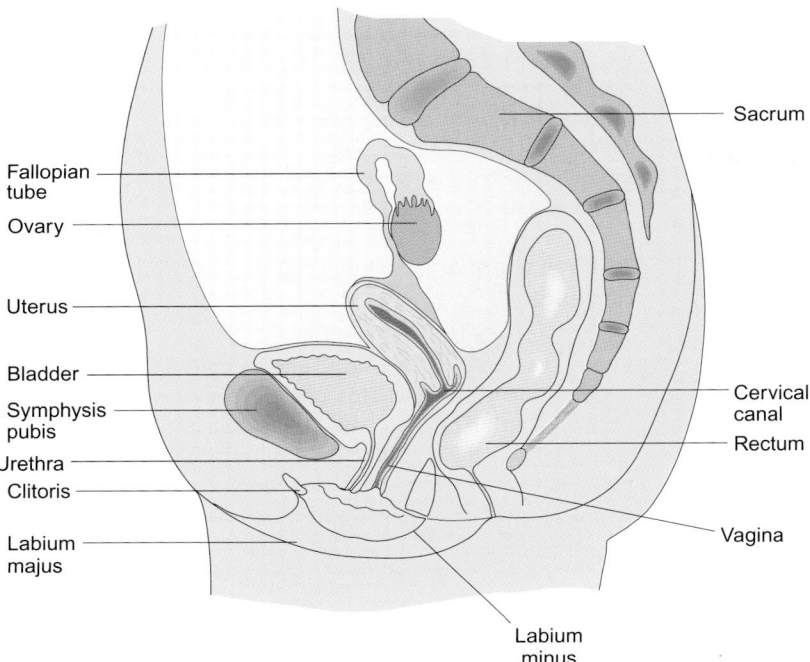

Fig. 20.7 The gross anatomy of the female reproductive organs and their spatial relationship with other structures.

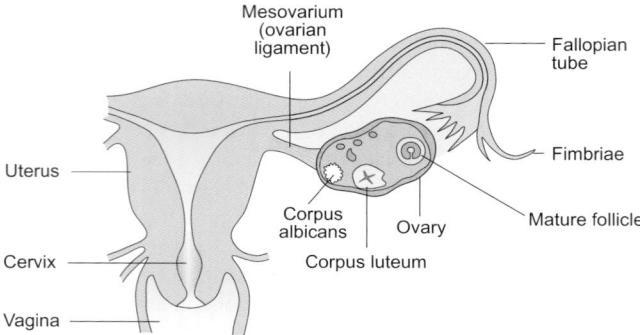

Fig. 20.8 The relationship between the ovary, the Fallopian tube, and the uterus. The cross-section through the ovary shows a follicle, a corpus luteum, and a corpus albicans.

created both for implantation and for the subsequent development of the embryo. Each ovarian cycle reflects these two roles.

The *Fallopian tubes* are thin tubes about 12 cm long that serve to transport the ovum released at ovulation from the ovary to the uterus. The opening of the tube is expanded and split into fringes or fimbriae, which move nearer to the ovary at ovulation. There are numerous cilia on the fimbriae and these create currents in the peritoneal cavity so that after ovulation the egg is directed towards the mouth or *ostium* of the Fallopian tube. The tube itself is muscular and covered with peritoneum. Internally it has a layer of stromal tissue that is overlaid by ciliated high-columnar secretory epithelial cells.

The non-pregnant *uterus* is about 7.5 cm long and 5 cm wide and is the organ that houses the fetus during its gestation period of 38 weeks. It must be adapted to receive the early embryo, and permit implantation and the formation of a placenta. While it has to contract powerfully to expel the fetus at birth, it must remain quiescent throughout gestation to allow full fetal development. The uterus consists of an outer covering or serous coat, a middle layer of smooth muscle, the *myometrium*, which forms the bulk of the thick wall, and an inner endometrial layer or *endometrium*. This is composed essentially of epithelial cells, simple tubular glands, and the spiral arterioles that supply the cells. The characteristics of the endometrium are altered considerably during each ovarian cycle.

The neck of the uterus is formed by the *cervix*, a ring of smooth muscle containing many mucus-secreting cells. It forms the start of the birth canal, which is traversed by both sperm and fetus. The mucus-secreting cells undergo important changes in activity during each monthly cycle that optimize conditions for fertilization.

The final internal structure of the female reproductive tract is the *vagina*. The cells which line the vagina and the vaginal fluids also show cyclical variations. The cyclical changes in the vaginal fluid (particularly its pH) can be useful in the treatment of infertility as they can be used to determine which stage of the cycle has been reached.

The vaginal orifice, urethral orifice, and clitoris are protected by folds of tissue called the *vulva* composed of the labia majora and labia minora. Within the walls of the vulva lie the vestibular glands, which secrete mucus during sexual arousal and help to lubricate the movement of the penis within the vagina during

sexual intercourse. The clitoris is a small erectile structure that is homologous with the male penis.

20.7 The ovarian cycle

Like the testis in the male, the ovary plays a fundamental role in the reproductive physiology of the female. It releases mature fertilizable ova at regular intervals throughout the reproductive years of a woman, and it secretes a number of hormones which are involved in the regulation not only of the ovaries themselves but also of the rest of the reproductive tract. These are steroid hormones and include progesterone and a number of estrogenic hormones. In the non-pregnant woman, the predominant estrogenic hormone is estradiol-17β, while during pregnancy estrone and oestriol are also produced (particularly by the placenta). In the following account of ovarian function the term 'estrogen' will be used to refer to the estrogenic agents secreted by the follicular cells. In the discussion of placental function (see pp. 456–461), it will be used as a collective term for the variety of estrogenic hormones of physiological importance during pregnancy.

In the following discussion of ovarian function it will be important to bear in mind two questions.

1. What are the mechanisms that ensure the regular release of ova?

2. How does the endocrine activity of the ovary prepare the rest of the reproductive tract for successful fertilization and the ensuing pregnancy?

It is well known that during the fertile years of a woman's life the activity of her ovaries occurs in a cyclic fashion. The orderly sequence of events which underlies this cyclical behavior is called the *ovarian cycle* or, more commonly, the *menstrual cycle*. During this time, there is remarkable coordination between the physical changes in various organs and hormone secretion. The interplay between morphological and endocrine events is rather complicated. To simplify matters, this account of the ovarian cycle will be divided broadly into two sections: first, a description of the physical changes leading up to and following the release of an ovum at mid-cycle, and, secondly, the changes that follow ovulation. The mechanisms that regulate each half of the cycle will be considered following the descriptions of the physical changes.

At birth the ovary already contains its full complement of gametes

The fundamental functional unit of the ovary is the follicle—indeed, the bulk of the ovary is made up of follicles at various stages of development. This can be seen diagrammatically in Fig. 20.9. During fetal life, the primordial germ cells of the ovary are laid down and continue their mitotic proliferation throughout gestation. However, mitosis is complete by the time of birth, so at this time a female will possess all the gametes she will ever have. During fetal life, the primordial germ cells are known as oogonia and, once mitosis is complete, they enter their first meiotic division and become known as oocytes. They also become surrounded by mesenchymal cells on a basement membrane (the basal lamina) to form the primordial follicles. The oocytes

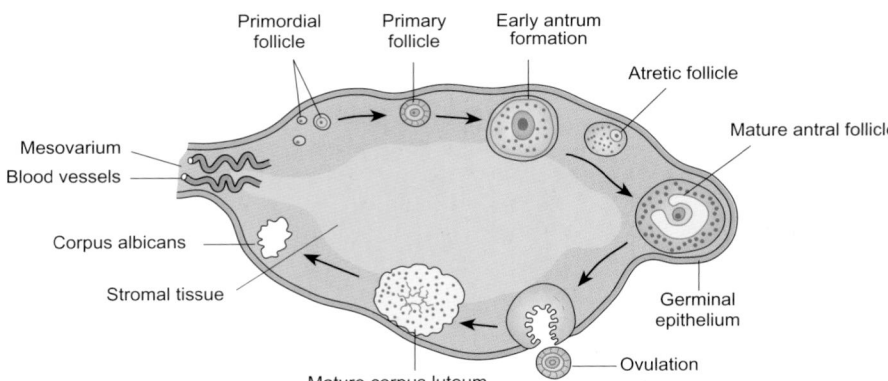

Fig. 20.9 The internal structure of the ovary showing the stages of follicular development, ovulation, the formation of the corpus luteum, and its subsequent regression. In reality, not all stages would be seen at the same time.

arrest in the diplotene of the first meiotic prophase (see Chapter 3, pp. 24–25) and then remain in this arrested state until signaled to resume further development. This might occur at any time during the reproductive life of a woman.

The pool of primordial follicles established during fetal life is gradually depleted throughout the years between puberty and the climacteric (menopause) as, each day, follicles are recruited in a steady trickle to undergo further development. It is believed that one to four primordial follicles begin this process each day. During each ovarian cycle, a follicle progresses through a series of developmental stages that include growth and maturation, ovulation, corpus luteum formation, and, in the absence of fertilization, degeneration. In most women, the menstrual cycle is between 25 and 35 days in length, although wider variations occasionally occur. This represents the time that it takes for an ovary to complete one cycle of follicular activity. If no pregnancy occurs, the cyclical activity is obvious, ending with the occurrence of menstruation so that in the time between successive menstrual periods one ovarian cycle is completed.

The development of a follicle may conveniently be split up, for the purposes of explanation, into several distinct stages. These stages have names that reflect either the changing structure or function of the follicle. The first half of the cycle consists of the preantral, antral, and preovulatory stages, which are concerned with follicular growth and development. At mid-cycle, ovulation occurs, after which the collapsed follicle is converted to a corpus luteum—a process known as *luteinization* (this is the

luteal phase). During the final stage of the cycle, the corpus luteum involutes and regresses—a process called *luteolysis*. These events are summarized in Fig. 20.10.

The preantral follicle

Once a primordial follicle has been triggered to recommence development, it undergoes conversion to a preantral follicle (Fig. 20.11(a)). This involves a considerable increase in diameter from about 20 μm to 200–400 μm. The primary oocyte within the follicle also increases in size to around 120 μm. During this phase of growth, an enormous amount of synthetic activity is going on within the oocyte in order to load its cytoplasm with the nutrient materials that it will require during its subsequent maturation. The stromal cells surrounding the oocyte divide to form several layers of *granulosa cells* and secrete a glycoprotein which forms a cell-free region around the oocyte known as the *zona pellucida*. In addition, the cells adjacent to the basal lamina multiply and differentiate to form concentric layers around the primary follicle called the *theca*. The outermost layers of thecal cells are flattened and fibromuscular in nature (the theca externa) while the inner layers are more cuboidal (the theca interna). Figure 20.11(b) shows the appearance of the primary follicle by the end of the preantral stage of development.

Very little is known about the factors that control the entry of primordial follicles into the preantral stage. Even the duration of the phase is uncertain, although it is probably around 2 days. The regular recruitment of follicles appears to take place inde-

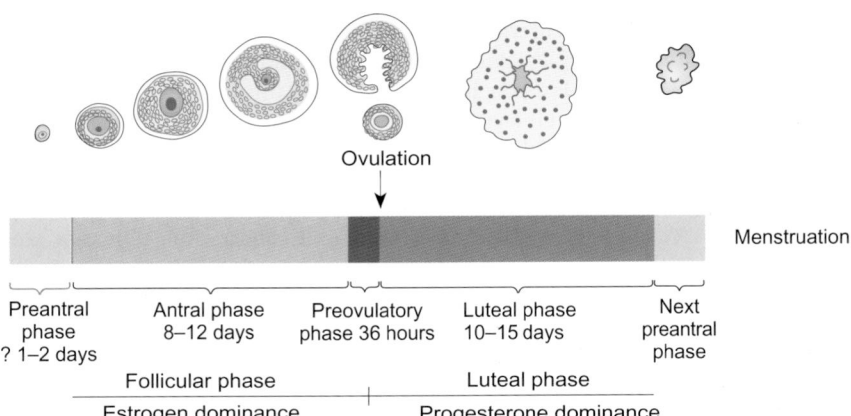

Fig. 20.10 A summary of the principal phases of the ovarian cycle. The top panel shows the sequence of follicular development, ovulation, and formation of the corpus luteum. The lower part of the figure shows the relationship between the follicular and luteal phases of the cycle in relation to follicular development.

Early preantral follicle   **Late preantral follicle**

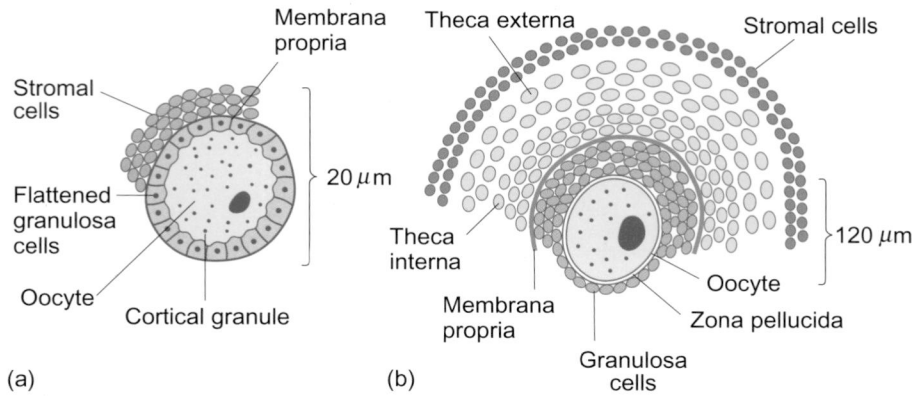

(a)   (b)

Antral follicle

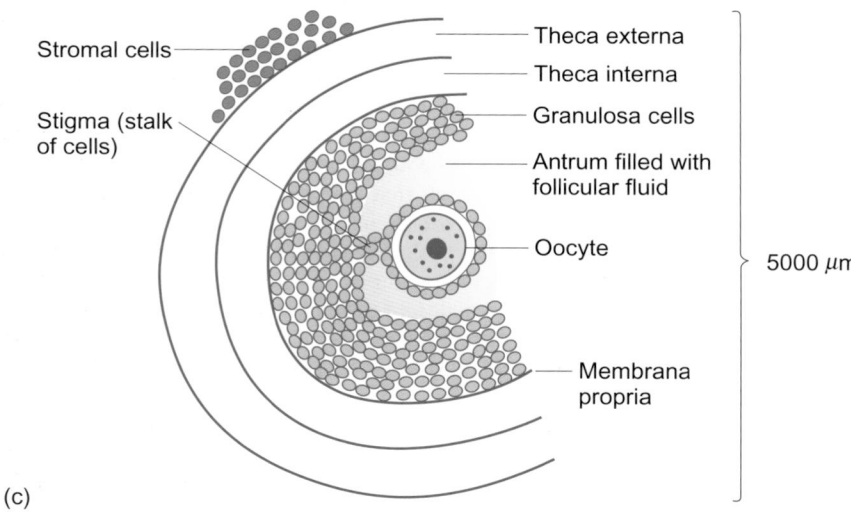

(c)

Fig. 20.11 Stages in the development of the ovum: (a) an early pre-antral follicle; (b) a late preantral follicle; (c) a late antral (or Graafian) follicle. Note the proliferation of stromal and granulosa cells and the development of the fluid-filled antrum.

pendently of hormonal control as removal of the anterior pituitary gland has no effect on the process. However, towards the end of the preantral stage an event occurs which is crucial for further follicular development. The follicular cells acquire receptors for certain hormones; the granulosa cells develop receptors for estrogens and for pituitary FSH, while the thecal cells develop receptors for pituitary LH. This acquisition of hormone sensitivity is a prerequisite for continued follicular development since each subsequent stage is absolutely dependent on hormonal control.

The antral follicle

The continuous trickle of follicles through the hormone-independent preantral stage ensures that, at any one time, there are always several follicles that have completed their preantral growth and possess the appropriate receptors for gonadotrophins and estrogens. Further development depends upon the endocrine status of the body at the time. Provided that there are adequate

levels of FSH and LH in the circulation, any follicles with the appropriate receptors enter the next, *antral*, stage of development. Preantral follicles that do not possess hormone receptors undergo a process of *atresia*, i.e. they degenerate and die.

The anterior pituitary gonadotrophins FSH and LH convert preantral to antral follicles. The antral stage of development normally lasts for 8–10 days. During this time, the granulosa and thecal cell layers continue to increase in thickness. The granulosa cells also start to secrete follicular fluid all round the oocyte. This fluid forms the antrum, which gives the stage its name. Figure 20.11(c) shows the general appearance of an antral follicle towards the end of this stage. The entire follicle is now much larger (around 5 mm in diameter), although the oocyte itself remains much the same size (120 μm). The oocyte, surrounded by a few granulosa cells, is virtually suspended in follicular fluid and remains attached to the main rim of granulosa cells by a thin stalk. A fully developed antral follicle is also known as a Graafian follicle.

Under the influence of gonadotrophins, the cells of the antral follicle start to secrete large quantities of hormones. Both the granulosa and thecal cells take on the characteristics of steroid-secreting tissue, with many lipid droplets, microtubules, and smooth endoplasmic reticulum. Pituitary LH stimulates the cells of the theca interna to synthesize and secrete the androgens testosterone and androstenedione. They also produce small amounts of estrogens. The granulosa cells, which possess receptors for FSH, appear to respond to this hormone by converting androgens to estrogens (particularly estradiol-17β). The overall result of this secretory activity is a substantial increase in the circulating levels of both androgens and estrogens, especially the latter during the antral phase of the menstrual cycle. A simplified diagram showing the synthesis of the main sex hormones is given in Box 20.1.

The estrogens secreted at this time seem to exert a significant effect within the follicle itself. As well as converting androgens to estrogens, the granulosa cells of the antral follicle possess receptors for estrogens. The estrogens produced by the follicular cells bind to these receptors and stimulate proliferation of further estrogen-sensitive granulosa cells. Thus there are more granulosa cells available for converting androgens to estrogens and this internal potentiation mechanism results in a substantial increase in circulating levels of estrogens throughout the antral phase. Indeed, during the final 2–3 days of this stage (around days 10–12 of the cycle), estrogen levels rise rapidly (the

estrogen surge). A profile of estrogen secretion during the menstrual cycle illustrating this peak is shown in Fig. 20.12. The estrogens secreted during this time have many important actions throughout the reproductive tract which will be discussed later.

The pre-ovulatory follicle

As the follicle approaches the end of its antral phase of development, around the time of the estrogen surge, two important events must coincide if it is to progress further and enter the brief but dramatic preovulatory stage:

(1) acquisition of receptors for pituitary LH by the granulosa cells;

(2) a sharp rise in circulating levels of LH.

LH receptors are synthesized in response to pituitary FSH (for which the granulosa cells already possess receptors) and estrogen. An estrogen surge also seems to be required for the rise in LH secretion.

If an antral follicle is to proceed to the preovulatory stage, with subsequent ovulation at mid-cycle, its acquisition of appropriate receptors must coincide with high circulating gonadotrophin levels. Any follicles that do not have LH receptors at this time will become atretic. Therefore, although several primordial follicles begin to develop every day during fertile life, usually only one (the so-called dominant follicle) proceeds to ovulation. As a result, there is considerable wastage of follicles during each cycle as some undergo atresia at each stage of development.

The preovulatory stage lasts for only about 36 hours, but during that time the follicle shows marked changes that culminate in its rupture and release of the oocyte. This is the process of *ovulation*, which occurs approximately halfway through the menstrual cycle. All the changes that characterize the preovulatory stage are critically dependent upon pituitary gonadotrophins, particularly LH.

Soon after the rise in LH output at the start of the preovulatory stage, the oocyte completes its first meiotic division. This culminates in a rather peculiar division in which half the chromosomes and virtually all the cytoplasm are contained within one cell, the secondary oocyte, while the remaining chromosomes are discarded in the form of the first polar body. Meiosis then arrests again and the secondary oocyte is ovulated in this stage of development. The mechanism by which LH initiates the recommencement of meiosis is not understood—perhaps it antagonizes the activity of a meiotic inhibitory factor.

During the antral stage, the granulosa cells of the follicle are mainly concerned with converting androgens to estrogens under the influence of pituitary FSH. In the preovulatory stage, LH stimulates these cells to start synthesizing progesterone instead. As a result, estrogen levels begin to fall slightly while progesterone output rises. At the same time, the granulosa cells lose their receptors for FSH and estrogen.

At ovulation the follicle ruptures and the secondary oocyte enters the Fallopian tube

By the end of the preovulatory stage of development, the volume of follicular fluid has increased substantially and the oocyte remains attached to the outer rim of granulosa cells by a thin

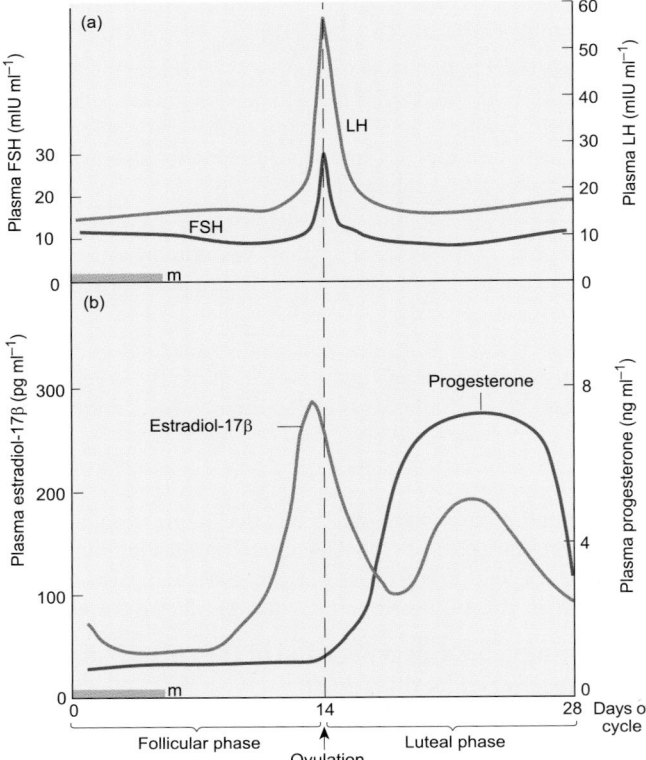

Fig. 20.12 The changes in hormone levels during the menstrual cycle. (a) The pattern of secretion shown by the gonadotrophins (FSH and LH); (b) the changes in the plasma levels of estradiol-17β and progesterone. The solid bar marked 'm' represents the period of menstruation.

stalk (Fig. 20.11(c)). At the time of ovulation, under the influence of LH, the cells of the stalk dissociate and the follicle ruptures. The detailed biochemistry of this process is not understood, but it is widely suspected that follicular rupture is in some way dependent upon the switch away from estrogen production towards progesterone production that occurs in the granulosa cells just prior to ovulation.

At ovulation, the follicular fluid flows out onto the surface of the ovary carrying with it the secondary oocyte with a few surrounding cells. The egg mass is swept into the Fallopian tube by currents set up by the movements of cilia on the fimbriae of the ostium (see Section 20.8). The first half of the ovarian cycle is now complete.

After ovulation the follicle forms a corpus luteum whose activity is regulated by pituitary LH

After the departure of the oocyte and follicular fluid, the remainder of the follicle collapses into the space left behind, and a blood clot forms within the cavity. Therefore the postovulatory follicle consists of a fibrin core surrounded by collapsed layers of granulosa cells enclosed within the fibrous thecal capsule. This collapsed follicle then undergoes transformation to become a *corpus luteum* (from the Latin meaning 'yellow body') which, in the event of fertilization, will secrete the appropriate balance of steroid hormones to ensure implantation and maintenance of the embryo during the early weeks of pregnancy. The second half of the ovarian cycle is often referred to as the *luteal phase*.

Formation of the corpus luteum is entirely dependent upon the surge of pituitary LH that occurs during the preovulatory stage to bring about ovulation itself. The factors that maintain the corpus luteum following the steep decline in gonadotrophin levels after ovulation are not clear. In some animals, a luteotrophic complex of LH, prolactin, and possibly other hormones seems to be important, but the situation in the human female is unclear. Normal basal levels of LH may be sufficient for luteal function.

In the first hours following expulsion of the egg from the ovary, the remaining follicular cells undergo the process of *luteinization*. They enlarge and develop lipid inclusions that give the corpus luteum the yellowish color from which it takes its name. The corpus luteum can grow to between 15 and 30 mm in size by about 8 days after ovulation. At this time, it shows peak secretory capacity. The cells of the corpus luteum contain increased amounts of Golgi apparatus, endoplasmic reticulum, and mitochondrial protein, and they secrete large amounts of progesterone (Fig. 20.12). Progesterone levels now increase dramatically from about 1 ng ml^{-1} to around 6–8 ng ml^{-1}. In addition, a considerable amount of estrogen is secreted by the corpus luteum and a second estrogen peak is seen around the middle of the luteal phase. Nevertheless the dominant steroid at this time is progesterone.

In the absence of fertilization, the corpus luteum has a finite lifespan

If the oocyte that was released at ovulation remains unfertilized, the corpus luteum degenerates after 10–14 days. This process is known as *luteolysis*. It involves collapse of the luteinized cells, ischemia, and cell death, with a resulting fall in the output of

Summary

1. The first half of the ovarian or menstrual cycle is known as the *follicular phase* and is the period during which a follicle undergoes the growth and development that culminates in ovulation—rupture of the follicle and release of the oocyte from the ovary.

2. A number of follicles start to develop each day but normally only one, the dominant follicle, matures to ovulation in each cycle. The remainder become atretic and die.

3. The physical changes occurring as the follicle develops are closely regulated by hormones, particularly the anterior pituitary gonadotrophins FSH and LH, and by estrogens produced by the follicle itself.

4. The follicular phase may be subdivided into preantral, antral, and preovulatory stages. The preantral stage of growth lasts for about 2 days and appears to be hormone independent. The antral stage, during which considerable further growth, occurs, is dependent upon FSH and LH.

5. Under the influence of FSH and LH, the follicle secretes large amounts of estrogens. Large amounts of fluid are also secreted by the follicular cells so that by the end of the antral stage the oocyte is suspended in fluid and attached to the outer rim of follicular cells by a thin stalk.

6. During the preovulatory stage, under the influence of high circulating LH levels, the first meiotic division of the oocyte is completed, progesterone secretion begins, and the follicle ruptures to release the egg mass—ovulation.

7. The second half of the ovarian cycle, following ovulation, is known as the *luteal phase*. The postovulatory follicle is transformed into a corpus luteum under the influence of anterior pituitary LH and the luteal cells change both their structure and function.

8. The luteal phase is characterized by the secretion of large amounts of progesterone that has important effects throughout the reproductive tract. Estrogens are also produced.

9. In the absence of fertilization, the corpus luteum degenerates after 10–14 days and steroid output falls to very low levels. This is the process of luteolysis, which marks the end of one ovarian cycle.

estrogen and progesterone. This rapid decline in steroid output can be seen in Fig. 20.12. The degenerated corpus luteum leaves a whitish scar within the ovarian stroma, which persists for several months. This is known as the *corpus albicans* (white body).

What brings about degeneration of the corpus luteum in the absence of fertilization?

The mechanisms that underlie the regression of the luteal cells after 12 days or so remain unclear. Estrogen has been implicated in the control of luteal regression in humans for two reasons. First, the start of degeneration roughly coincides with the estrogen peak seen 6–8 days after ovulation (Fig. 20.12) and, secondly, injections of estrogens given prior to the naturally occurring peak hasten luteal decline. An alternative explanation is that luteolysis simply occurs gradually as gonadotrophin support slowly declines during the luteal phase (Fig. 20.12).

20.8 Hormonal regulation of the female reproductive tract

During the follicular phase, estrogens prepare the reproductive tract for fertilization

The follicular phase of the ovarian cycle is characterized by the secretion of increasing amounts of estrogens. This pattern of estrogen output can be seen in Fig. 20.12 and 20.13. Levels of estradiol-17β rise gradually during the antral stage of follicular development reaching 'surge' levels of up to 300 pg ml⁻¹ just prior to ovulation (Fig. 20.12). The estrogens secreted during the first half of the cycle perform the crucial task of preparing the reproductive tract to receive and transport gametes, and to create a favorable environment for fertilization and implantation.

Estrogens increase ciliary activity in the Fallopian tubes

The influence of ovarian hormones on the Fallopian tubes appears to be quite significant. Under the influence of the high levels of estrogens seen in the follicular phase, tubal ciliary and contractile activity is enhanced in preparation for recovering the oocyte from the peritoneal cavity after ovulation, and transporting it towards the uterus. Similarly, contractile and ciliary activity may help to transport sperm towards the egg. Removal of the

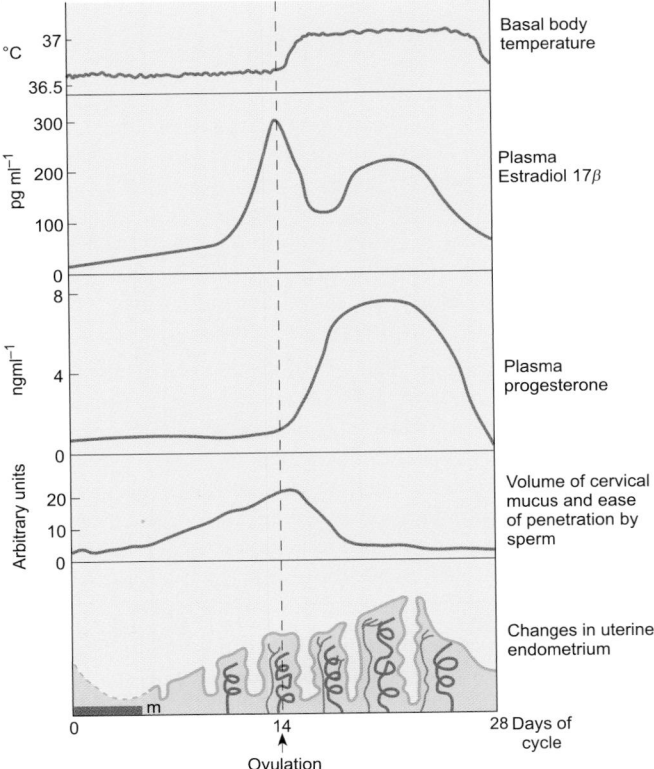

Fig. 20.13 The cyclical changes shown by body temperature, cervical secretions, and the uterine endometrium in relation to the circulating levels of estradiol-17β and progesterone.

ovaries results in a loss of tubal cilia and a reduction in both secretory and contractile activity of the tubal cells. These effects can be reversed by the subsequent administration of estradiol-17β, which suggests that estrogens are important for ciliary and muscular activity in the Fallopian tubes. This makes sense when one considers the reproductive role that the tubes perform.

Estrogens stimulate endometrial proliferation and increase myometrial excitability

Both the myometrium and the endometrium of the uterus are extremely sensitive to the ovarian steroids, and the changes in structure and function occurring in response to these hormones reflect the different roles that the uterus must fulfill during each cycle. The uterus prepares first to receive and transport sperm from the cervix to the Fallopian tubes and, later on, to receive and nourish the embryo.

Steroids secreted from the follicular cells act on the uterus to enable it to fulfill these tasks. The estrogens secreted during the follicular phase of the cycle exert a uterotrophic (stimulatory) effect on the endometrium (Fig. 20.13). As a result, the endometrial stroma proliferates and the surface epithelium increases in surface area; the estrogen-primed epithelial cells secrete an alkaline watery fluid. At the same time, the spiral arteries that permeate the stroma start to enlarge. By the time of ovulation, the endometrial thickness has increased to around 10 mm (from about 2 or 3 mm just after menstruation). This phase of the endometrial cycle, corresponding to the estrogen-dominated follicular phase, is known as the *proliferative phase*. During this phase, the uterus is being prepared to receive a fertilized egg. Estrogens also stimulate the development of progesterone receptors on the endometrial cells so that, by the end of the follicular phase, the endometrium is primed to respond to progesterone.

The uterine myometrium is also under the influence of the ovarian hormones. Estrogens appear to increase the excitability of the myometrial smooth muscle and therefore its spontaneous contractility.

Estrogens also affect non-reproductive tissues

The estrogens have widespread and generalized effects throughout the body in addition to those specific actions within the reproductive tract discussed above. In particular, they exert effects on metabolism and the cardiovascular system.

- Estrogens are mildly anabolic and tend to depress the appetite.
- They reduce plasma levels of cholesterol, which may explain why premenopausal women have a lower risk of heart attacks than both postmenopausal women and men of comparable age.
- They reduce capillary fragility.
- Estrogens appear to have profound effects on mood and behavior but the underlying mechanisms are not clear.
- Estrogens cause proliferation of the ductal system of the mammary tissue (see Chapter 21).
- They have important effects on the maintenance of the skeleton (see Chapter 23).

Progesterone secreted by the corpus luteum optimizes conditions for implantation within the uterus

The uterus houses the embryo for the entire period of its development (gestation). There are two elements to this task.

1. The endometrial layer must permit implantation of the newly fertilized egg and subsequently must participate in the formation of the placenta (placentation).

2. The myometrium must remain quiescent during gestation to guard against premature expulsion of the fetus.

Progesterone plays a key role in each of these elements. Indeed, *adequate levels of progesterone are essential throughout the entire period of gestation to ensure a successful outcome.*

During the follicular phase of the cycle, estrogen secreted by the antral follicle brings about proliferation of the uterine endometrium along with an increase in the number of glandular structures (see above). Estrogen also stimulates the acquisition of progesterone receptors by the cells of the endometrium. As progesterone levels rise during the luteal phase, the stromal proliferation continues and the spiral arteries develop fully. In the event of pregnancy, the spiral arteries will form the blood supply to the maternal side of the placenta. The endometrial glands start to secrete a thick fluid rich in sugars, amino acids, and glycoprotein. For this reason, the second half of the uterine cycle is often referred to as the *secretory phase*. It coincides with the luteal phase of the ovarian cycle. All these progesterone-mediated changes help to create a favorable environment for a newly fertilized egg and to optimize conditions for implantation and placental formation.

In the absence of a fertilized egg, the corpus luteum regresses after 10–14 days and steroid output falls precipitously (Figs 20.12 and 20.13). Once the endometrium is deprived of its steroidal support, its elaborate secretory epithelium collapses. The endometrial layers are shed together with blood from the ruptured spiral arteries, which contract to reduce bleeding. This process is known as *menstruation*. The onset of menstrual bleeding is taken to mark the start of a new ovarian cycle. Contraction of the spiral arteries can lead to the pain experienced by some women at the start of menstruation (dysmenorrhea). Bleeding continues for 3–7 days during which the total blood loss is between 30 and 200 ml. After this time the endometrial epithelium has been repaired completely.

Progesterone also has a very important effect on the uterine myometrium. As described earlier, the estrogen-dominated myometrium shows a fair degree of excitability and spontaneous contractility. Clearly, although this may be helpful in assisting gamete transport, it is highly undesirable once an embryo has entered the uterus. Too much excitability could result in spontaneous abortion ('miscarriage') of the fetus. Progesterone tends to relax the smooth muscle of the myometrium, probably by reducing its excitability. This reduces the likelihood of spontaneous contractions.

Some non-reproductive tissues are influenced by progesterone

Like estrogens, progesterone exerts widespread effects throughout the whole body, most of which are poorly understood. For example, it is a mildly catabolic steroid that stimulates the appetite. Increased levels of progesterone during the luteal phase cause a rise in basal body temperature of 0.2–0.5°C (Fig. 20.13). This rise is a useful indicator that ovulation has occurred for women who are trying to conceive—and for those who are trying not to!

Progesterone promotes development of the lobules and alveoli of the breast (see Chapter 21) and causes the breasts to swell because of fluid retention by the mammary tissue. This is probably the reason for the breast discomfort experienced by many women during the premenstrual period.

The cervical secretions and vaginal epithelium show hormone-dependent cyclical changes

The endocervical glands secrete mucus whose characteristics vary considerably during the ovarian cycle. These changes are regulated by the ovarian hormones and have important consequences for fertility. Under the influence of the high circulating levels of estrogens seen during the follicular phase, the cervical epithelium increases its secretory activity and mucus is produced in large amounts, up to 30 times the quantity secreted in the absence of estrogen (Fig. 20.13). The mucus is thin, watery, and clear and exhibits a characteristic 'ferning' pattern if dried on a glass slide. It also shows increasing elasticity, so that a drop of mucus may be stretched to a length of 10–12 cm. The peak volume and elasticity coincide with the estrogen surge just prior to ovulation. Mucus with these characteristics is most readily penetrated by sperm, and this action of estrogen on the cervical

Summary

1. The follicular phase of the ovarian or menstrual cycle is dominated by estrogen secreted by the developing follicle. This estrogen acts within the tissues of the reproductive tract to prepare it for gamete transport, fertilization, early embryonic development, and implantation. Ciliary and contractile activity in the Fallopian tubes is enhanced, the uterine endometrium proliferates, and the glands of the cervix secrete large volumes of thin stretchy mucus that is easily penetrated by sperm.

2. During the second half of the ovarian cycle, the luteal phase, the major steroid hormone secreted by the corpus luteum is progesterone. Maximum progesterone secretion occurs about 8 days after ovulation. Progesterone prepares the uterus to receive and nourish an early embryo in the event of fertilization, and maintains the endometrium in a favorable condition for implantation and placentation. Progesterone also renders the myometrium less excitable to guard against premature expulsion of the embryo.

3. In the absence of an embryo, the corpus luteum degenerates after 10–14 days and steroid output falls steeply. As progesterone levels fall, the elaborate endometrium which was built up during the cycle is sloughed off and shed together with blood from the spiral arteries. This process is called menstruation and its onset marks the beginning of a new ovarian cycle.

glands is a good example of the way in which the endocrine activity of the ovary optimizes the conditions for successful reproduction; when an oocyte is likely to be present, the passage of sperm through the female tract is facilitated.

During the luteal phase, with its high progesterone levels, cervical mucus is produced in far smaller volumes and becomes much thicker, stickier, and relatively hostile to sperm. Thus it is less likely that sperm will reach the uterus and Fallopian tubes during the luteal phase. This action of progesterone forms part of the mechanism of action of the progesterone-only contraceptive pill (see also Box 21.1).

The stratified squamous epithelium lining the vagina also changes in appearance in response to the ovarian hormones. Indeed, the histological appearance of the vaginal epithelial cells can be used as an indicator of the phase of the menstrual cycle reached. In the follicular phase, increased secretion of estrogens stimulates proliferation of the epithelial layers. As the superficial layers move farther away from the blood supply they keratinize and many slough off. At mid-cycle, a vaginal smear will show a preponderance of such cells.

20.9 Why do the plasma concentrations of gonadotrophins and ovarian steroids vary during the ovarian cycle?

The previous sections have focused particularly on the structural and functional changes which occur throughout the 28 days or so which span one complete ovarian cycle. The cyclical alterations in the plasma levels of FSH and LH (shown in Fig. 20.12) are crucial in controlling the cellular and endocrine activity of the ovary, i.e. the growth of follicles, the formation of the corpus luteum and its endocrine activity. How do these fluctuations come about and how do they regulate the follicular cells? To answer these questions it is important to realize not only that the gonadotrophins regulate ovarian function but also that the ovarian steroid hormones themselves, the estrogens and progesterone, in turn influence gonadotrophin secretion. This feedback interaction between the anterior pituitary gland, the hypothalamus, and the ovary is illustrated in Fig. 20.14.

The ovarian steroids can exert both negative and positive feedback control on the output of FSH and LH, depending upon the concentration of hormone present and the time for which it has been present. Low or moderate levels of estrogens, particularly estradiol-17β, exert negative feedback on gonadotrophin output, i.e. they tend to inhibit the secretion of FSH and LH. However, if estrogens are present in high concentrations for several days, the effect switches to one of positive feedback in which the output of FSH and LH is stimulated. The feedback actions of progesterone are roughly opposite to those of the estrogens. High concentrations of progesterone inhibit gonadotrophin release, while low levels appear to enhance the positive feedback effects of estrogens.

The feedback effects of the ovarian steroids are mediated primarily at the level of the anterior pituitary itself, probably by alterations in the sensitivity of the gonadotrophin-secreting cells to hypothalamic GnRH. There may also be a direct effect of the steroids on GnRH output by the hypothalamic neurons, though

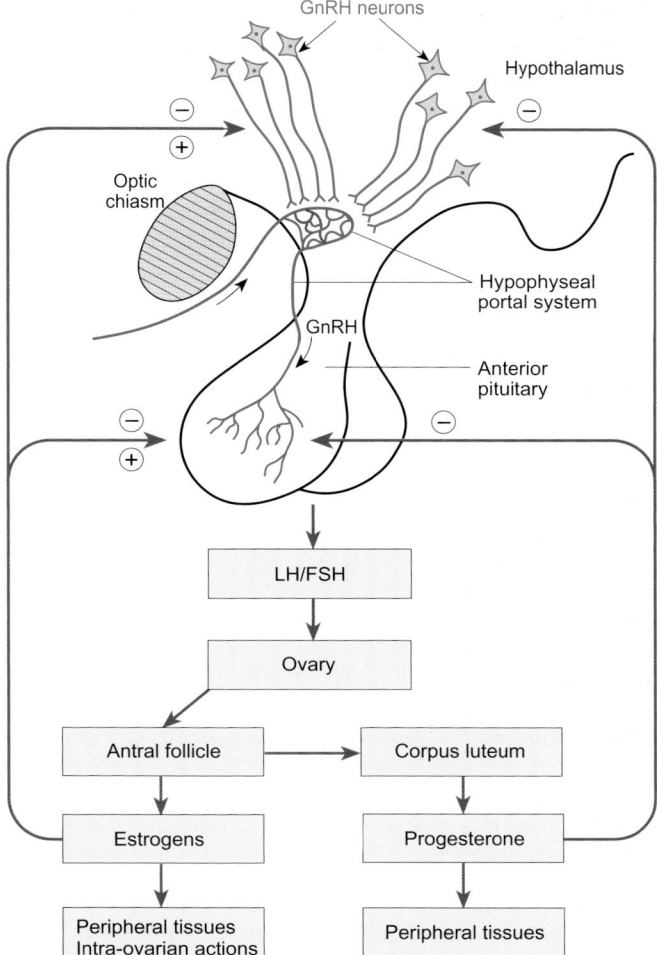

Fig. 20.14 Schematic diagram illustrating the positive (+) and negative (−) feedback control of the hormonal secretions of the hypothalamus, pituitary gland, and ovaries.

this is hard to establish conclusively because of the difficulty in measuring the tiny quantities of GnRH in the portal blood.

The ovarian cycle begins on day 1 of menstruation. Just prior to this, levels of both estrogens and progesterone have fallen as the previous corpus luteum declined. Released from the negative feedback inhibition of the ovarian steroids, FSH levels start to rise slowly, followed shortly afterwards by LH. These events coincide with the initiation of the antral phase of follicular development. Towards the end of the hormone-independent preantral phase, the thecal cells gain receptors for LH while the granulosa cells become responsive to both FSH and estrogens. The coincidence of receptor acquisition with steadily rising levels of FSH and LH allows the follicle to enter the hormone-dependent antral phase.

During the preantral and early antral phases of the cycle, ovarian steroid output does not change very much (Fig. 20.12). However, over the next 6–8 days or so, the maturing follicle starts to produce large quantities of estrogens under the influence of FSH and LH. During this period, gonadotrophin levels themselves remain low because of the negative feedback effect of low and moderate levels of estrogens. This steady rise in estrogen secretion culminates in an 'estrogen surge' during the latter days of the antral phase, when plasma concentrations of estradiol-17β reach values between 200 and 400 pg ml^{-1}. This estrogen surge initiates important changes in the output of gonadotrophins. After about 36 hours, the negative feedback effects of estrogens are replaced by positive feedback that results in a sharp increase in the output of both the gonadotrophins but especially LH. This constitutes the so-called LH surge, which is responsible for the events occurring during the preovulatory phase and for ovulation itself (see above).

Once ovulation has taken place, the level of estrogens falls sharply as the luteal cells switch to progesterone production. Consequently, the gonadotrophins are released from the positive feedback effects of high estrogen levels and output of FSH and LH drops as negative feedback reasserts control (Fig. 20.12). Although estrogen levels may rise to values similar to those seen during the preovulatory surge, this second, luteal peak fails to elicit a further LH surge because the high circulating levels of progesterone seem to block the positive feedback effects of the estrogens. Instead, negative feedback continues to predominate and gonadotrophin secretion remains low throughout the luteal phase.

In the absence of fertilization, the corpus luteum regresses after 10–14 days and steroid output declines quickly. Deprived of steroid support, the specialized uterine endometrial layers are shed during menstruation. Soon afterwards, FSH and LH levels begin to rise slowly as the anterior pituitary is released from the negative feedback inhibition of estrogen, and a new cycle gets under way as preantral follicles enter the gonadotrophin-sensitive antral phase.

The menstrual cycle may be influenced by neural factors

Although the interactions between the ovarian steroids and the pituitary gonadotrophins are well documented, and appear to offer a reasonably complete explanation for the cyclical activity of the ovaries, it is known that a variety of other factors, both neural and hormonal, also affects the control of fertility.

The reproductive behavior of many farm, domestic, and laboratory animals illustrates how environmental stimuli mediated by the CNS, such as olfactory, tactile, coital, and visual cues, play a role in regulating gonadal function. For example, the fertile periods of certain animals such as sheep appear to be controlled by the relative lengths of day and night. Others, such as cats and rabbits, are reflex ovulators, i.e. they ovulate in response to coitus. Such regulatory mechanisms must be mediated via the CNS, with the afferent information being integrated in the hypothalamus in order to control the output of GnRH and thus of the gonadotrophins.

While human gonadal function is clearly not subject to external control mechanisms in such a rigid fashion, the possibility of neurally mediated influences on gonadotrophin release cannot be ruled out. Indeed, the wealth of neural inputs to the GnRH-secreting neurons of the hypothalamus would argue strongly in favor of such mechanisms. It is certainly well established that factors such as anxiety and emotional stress can upset cyclical ovarian activity and fertility.

Summary

1. Cyclical variations in the levels of steroid and gonadotrophic hormones act together to ensure the regular release of mature ova and to prepare the body for fertilization and pregnancy.

2. While the gonadotrophins control ovarian function, the estrogens and progesterone secreted by the ovaries regulate the secretion of pituitary FSH and LH by both negative and positive feedback.

3. Very high levels of estrogens stimulate the anterior pituitary to initiate an LH surge which is crucial to the events of the preovulatory stage and to ovulation itself. For the remainder of the cycle, negative feedback prevails and gonadotrophin output is relatively low.

4. The menstrual cycle appears to be sensitive to both neural and hormonal influences in addition to the fundamental regulation exerted by the interactions between the pituitary gonadotrophins and the ovarian steroids.

5. A variety of emotional and physical factors, mediated via neural inputs to the hypothalamus, can influence cyclicity.

6. Prolactin, another anterior pituitary hormone, appears to inhibit ovulation.

Physiologically, the role of prolactin in the normal menstrual cycle is unclear, but clinically it is well known that oversecretion of prolactin (hyperprolactinemia) is a common cause of female infertility. The condition may be physiological, as in lactating women (see Chapter 21), or pathological, arising from a pituitary tumor. It is often associated with anovular cycles (cycles in which ovulation fails to occur) or a complete loss of cyclical ovarian activity. High levels of prolactin seem to impair the response of the anterior pituitary to GnRH so that LH surges are lost and ovulation fails to take place.

20.10 Activation and regression of the gonads—puberty and the menopause

Menarche and the menopause

In the female, the fertile years are defined by two events. These are the onset of menstruation at puberty (*menarche*), and the cessation of cyclical ovarian activity which occurs at around the age of 50 (the *menopause* or *climacteric*).

Puberty is a collective term which includes a variety of changes taking place within the body of an adolescent girl as her ovaries mature. Menarche, the onset of menstruation, is the outward sign that these changes have taken place and that cyclical secretion of ovarian steroids has begun. The other changes that take place during the 2 or 3 years preceding menarche include the adolescent growth spurt, the development of secondary sexual characteristics (pubic hair and mammary development), and changes in body composition—adult females have about twice as much body fat as males and a smaller mass of skeletal muscle.

Circulating levels of the pituitary gonadotrophins FSH and LH rise gradually up to the age of around 10 years. After this, with the approach of menarche, pulsatile release is established, with spurts of secretion during sleep. With this increase in gonadotrophin secretion there is also a rise in the output of ovarian estrogens and, under the influence of these steroids, budding of the breasts occurs. This is usually the first outward physical sign of puberty.

From 2 to 3 years prior to the onset of menstruation, androgen secretion from the adrenal cortex is increased (adrenarche) and these hormones are important in the stimulation of pubic hair growth. It has also been suggested that androgen secretion plays a part in the control of menarche, though no clear link has been established. What is clear, however, is that the increasing synthesis and secretion of FSH and LH eventually result in the onset of menstruation, although the first few cycles are usually anovular and progesterone is not produced in large amounts. Therefore bleeding tends to be light and to occur irregularly at first.

The exact trigger for menstruation is still not fully understood, but various suggestions have been made. The ovaries might become more sensitive to gonadotrophins, or the anterior pituitary may become more sensitive to the positive feedback effects of estrogens.

The average age for menarche in the United Kingdom is around 12 years, with the first ovulatory cycles occurring 6–9 months later. However, the range of normality extends from 10 to 16 years of age (Table 20.1). There has been a trend towards earlier menarche in the last 150 years, possibly because of improved health care and nutrition. The latter may be particularly important since it is thought that menarche requires the attainment of either a critical body mass (around 47 kg) or possibly a critical ratio of fat to lean mass. It is certainly true that regular menstrual cycles are disrupted in girls who lose large amounts of weight through anorexia, excessive exercise, or starvation.

The cessation of menstrual cycles (the climacteric)

The menopause or climacteric marks the end of a woman's fertile years. It is the progressive failure of the reproductive system and usually occurs between the ages of 45 and 55. The number of oocytes in the ovaries has been depleted by atresia and the ovarian responsiveness to gonadotrophins declines. Cycles often become anovular and irregular before ceasing altogether. Pituitary FSH and LH levels are high in postmenopausal women because of the loss of negative feedback inhibition by estrogen, though LH surges are no longer seen.

Many somatic and emotional changes accompany the loss of ovarian steroids. The uterine muscle becomes fibrous, the vagina may become dry, and there is a loss of breast tissue. Depression, night sweats, hot flushes, and an increased susceptibility to myocardial infarction also accompany the menopause. There is often a reduction in bone strength due to increased bone resorption. Most of these changes are attributable to the loss of ovarian estrogen and can be treated successfully by hormone replacement therapy (HRT) should they be sufficiently serious to warrant medical intervention.

Puberty in the male

Testosterone is the key to reproductive function in the male. It plays a crucial role in sexual differentiation during embryonic life (see Chapter 21) and its concentration in the plasma rises at the onset of puberty to reach adult levels by about 17 years of age. Between early infancy and the start of puberty, testosterone secretion by the testes is low. Pituitary gonadotrophin levels are also low. During the years between 10 and 16 (on average), boys develop their full reproductive capability. At the same time, they acquire secondary sexual characteristics and adult musculature and undergo a linear growth spurt, which is halted by closure of the epiphyses when adult height is reached (Table 20.1).

The first endocrinological event of puberty is an increase in the secretion of pituitary LH. As a result, there is maturation of the Leydig cells and the initiation of spermatogenesis. Testosterone production is also enhanced and this hormone is responsible for the anatomical changes that are characteristic of puberty. These include enlargement of the testes, growth of pubic hair, starting at the base of the penis, reddening and wrinkling of the scrotal sac, and, later on, an increase in size of the penis. Facial hair begins to appear, the scalp hair takes on the masculine pattern, and there is deepening of the voice due to thickening of the vocal cords and enlargement of the larynx. These maturational changes take place over a period of several years and Table 20.1 shows the average timing of the major events that occur during male puberty.

Is there a male menopause?

While there is no obvious event marking the end of reproductive capacity in the male comparable to the female menopause, sperm production does decline between the ages of 50 and 80. There is also a reduction in plasma testosterone levels in men over 70 and a parallel increase in plasma levels of FSH and LH, though this is much less marked than in women. Nevertheless,

TABLE 20.1	Summary of the principal changes of puberty	
Characteristic	**Age range of first appearance (years)**	**Principal hormones responsible for development**
Girls		
Breast bud	8–13	Estrogens, progesterone, GH
Pubic hair	8–14	Adrenal androgens
Menarche	10–16	Estrogens, progesterone
Growth spurt	10–14	Estrogens, GH
Boys		
Growth of testis	10–14	Testosterone, FSH, GH
Growth of penis	11–15	Testosterone
Pubic hair	10–15	Testosterone
Facial and axillary hair	12–17	Testosterone
Enlargement of larynx	11–16	Testosterone
Growth spurt and male pattern of skeletomuscular development	12–16	Testosterone, GH

these changes are relatively small, and many elderly men maintain active sex lives and retain their reproductive capacity.

Recommended reading

Anatomy

MacKinnon, P., and Morris, J. (2005). *Oxford textbook of functional anatomy* (2nd edn), Vol. 2, pp. 181–196. Oxford University Press, Oxford.

Cell biology of germ cells

Alberts, B., Johnson, A., Lewis, J., Raff, M., Roberts, K., and Walter, P. (2002). *Molecular biology of the cell* (4th edn), Chapter 20. Garland, New York.

Histology

Junquieira, L.C., and Carneiro, J. (2003). *Basic histology* (10th edn), Chapters 22 and 23. McGraw-Hill, New York.

Physiology

Case, R.M., and Waterhouse, J.M. (eds.) (1994). *Human physiology: age, stress and the environment.* Oxford Science Publications, Oxford.

Ferin, M., Jewckwicz, R., and Warres, M. (1993). *The menstrual cycle.* Oxford University Press, Oxford.

Griffin, N.E., and Ojeda, S.R. (2000). *Textbook of endocrine physiology* (4th edn). Oxford University Press, Oxford.

Johnson, M.H., and Everitt, B.J. (1999). *Essential reproduction* (5th edn), Chapters 2–5 and 7. Blackwell Scientific, Oxford.

Summary

1. In the female, the fertile years are defined by menarche and the menopause, the commencement and cessation of cyclical ovarian activity. They occur at around 12 years and 50 years respectively.

2. Many changes take place within the body of an adolescent girl in addition to the onset of menstrual cycles. These include a growth spurt and the development of secondary sexual characteristics.

3. FSH and LH are secreted in increasing amounts prior to menarche but the exact trigger for the onset of cyclical ovarian activity is unclear.

4. The menopause marks the progressive failure of the reproductive system and is due to depletion of the oocyte pool by atresia and a reduced ovarian responsiveness to pituitary gonadotrophins. Many somatic and emotional changes accompany the loss of ovarian steroids.

5. Between 10 and 16 years of age, boys show a growth spurt and develop their full reproductive capacity.

6. Leydig cells mature under the influence of pituitary LH and start to produce sperm. Testosterone output rises and this hormone is responsible for the development of secondary sexual characteristics.

Multiple choice questions

Each statement is either true or false. Answers are given below.

1. The following statements apply to sexual reproduction in humans.
   a. The male gametes are known as sperm.
   b. A mature sperm contains a full complement of chromosomes.
   c. At birth, an ovary contains all the oocytes it will ever have.
   d. Primary and secondary spermatocytes divide by mitosis to give rise to spermatids.

2. In the testis:
   a. Leydig cells secrete testosterone.
   b. Testosterone synthesis requires both FSH and LH.
   c. Sertoli cells prevent free diffusion of water-soluble substances between the seminiferous tubules and the blood.
   d. Sertoli cells respond to testosterone by synthesizing androgen-binding protein.
   e. Sertoli cells line the seminiferous tubules and directly give rise to the developing sperm.

3. During the ovarian cycle:
   a. The initiation of the preantral phase of follicular development is under the control of LH.
   b. The development of the antral follicle depends on the expression of receptors for FSH and LH
   c. Ovulation occurs after about 14 days.
   d. Ovulation occurs in response to a sudden increase in plasma LH.
   e. The period before ovulation is known as the luteal phase.
   f. The luteal phase is associated with a large increase in plasma progesterone.
   g. The myometrium proliferates under the influence of estrogens.

4. The following questions relate to the hormonal control of the menstrual cycle.
   a. Estrogens are synthesized mainly by the cells of the theca interna.
   b. In the absence of receptors for FSH and LH, preantral follicles undergo atresia.
   c. After ovulation, the ruptured follicle is converted into a corpus albicans which secretes progesterone.
   d. Progesterone promotes full development of the endometrium.
   e. If no egg is fertilized, progesterone levels fall and this is the trigger for menstruation.

Answers to multiple choice questions

1. During spermatogenesis, the number of chromosomes is halved by meiosis. The primary spermatocytes undergo two meiotic divisions to give rise to spermatids. As a result, mature sperm are haploid. This is also true of a mature ovum which undergoes meiosis prior to ovulation. Following fertilization, the full complement of chromosomes is restored.

   a. True
   b. False
   c. True
   d. False

2. Sertoli cells synthesize androgen-binding protein in response to FSH. Testosterone binds to the androgen-binding protein and the hormone–protein complex enables the Sertoli cells to maintain the production of spermatozoa. While the spermatozoa are nurtured by the Sertoli cells until they develop into mature sperm, they are derived from germ cells known as spermatogonia which lie within the basal compartment of the seminiferous tubules.

   a. True
   b. True
   c. True
   d. False
   e. False

3. The initial recruitment of primordial follicles into the preantral phase is hormone independent. During the preantral phase, the follicular cells acquire receptors for various hormones. The granulosa cells develop receptors for FSH and estrogen, while the thecal cells acquire receptors for LH. The period before ovulation is known as the follicular phase, while the period after ovulation is called the luteal phase.

   a. False
   b. True
   c. True
   d. True
   e. False
   f. True
   g. True

4. The cells of the theca interna secrete mainly androgens which are converted to estrogens by the granulosa cells in response to FSH. After ovulation, the follicle collapses around a blood clot to form a corpus luteum. If no pregnancy occurs, the corpus luteum regresses to form a corpus albicans which has no secretory role.

   a. False
   b. True
   c. False
   d. True
   e. True

Fertilization, pregnancy, and lactation

After reading this chapter you should understand:

◆ The sexual reflexes of males and females

◆ The process of fertilization, implantation, and the maternal recognition of pregnancy

◆ Placental formation and function

◆ The role of the placental hormones in the maintenance of pregnancy and preparation for delivery and lactation

◆ Parturition

◆ Important changes in maternal physiology during gestation

◆ The nutritional demands of pregnancy

◆ The structure of the non-pregnant mammary gland

◆ Hormonal control of the development of the mammary gland during puberty and pregnancy

◆ The hormonal control of milk synthesis and milk ejection

◆ The composition of milk and how it changes in the first weeks post-partum

◆ Involution of the mammary gland after weaning

21.1 Introduction

The discussion of female reproductive physiology in the previous chapter assumed a situation in which cyclical ovarian activity continues uninterrupted by pregnancy, i.e. with regression of the corpus luteum. However, if an oocyte is fertilized, an entirely different set of events must be initiated. Loss of the uterine endometrium must be prevented, and the uterus must be maintained in a quiescent state to allow implantation of the embryo, formation of the placenta, and gestation. This chapter will be concerned with the events surrounding fertilization, pregnancy and parturition, and with the important physiological changes that take place in the body of a pregnant woman during gestation. Finally, the physiology of lactation will be considered.

21.2 The sexual reflexes

Fertilization of an ovum requires that sperm are deposited high in the vagina of a woman close to the time of ovulation. Except for the case of fertilization by artificial insemination, this is achieved through the act of sexual intercourse. For successful intercourse to take place, the penis of the male must become erect and ejaculation of seminal fluid must occur within the vagina.

The sexual response in both males and females can be divided into four main phases: excitement, plateau, orgasm, and resolution. Each phase is under the control of autonomic and somatic nerves originating in the lumbar and sacral regions of the spinal cord. The innervation of the sexual organs is derived from both the parasympathetic (S2–S4) and the sympathetic (T11–L2) divisions of the autonomic nervous system together with somatic fibers running in the pudendal nerves (see Figs. 21.1 and 21.2).

The main stages of the male sexual act are:

(1) the erection of the penis;

(2) the secretion of mucus by the bulbourethral gland;

(3) the emission of fluid from the seminal vesicles, vas deferens and prostate gland;

(4) the expulsion of seminal fluid from the penis—ejaculation.

Penile erection is the result of a parasympathetic reflex

Penile erection results from either descending nerve activity originating in the higher centers of the brain (*psychogenic erection*) or from stimulation of the skin in the genital region (*reflexogenic erection*). The afferent nerve fibers from the genital region run in the pudendal nerves to the sacral region of the spinal cord. The efferent fibers are parasympathetic in origin, derived from spinal segments S2–S4. When they leave the spinal cord, the efferent fibers run in the pelvic nerve and synapse in the pelvic plexus. The neurons of the pelvic plexus also receive sympathetic fibers from the hypogastric nerves. The cavernous nerve provides the final limb of the efferent pathway (Fig. 21.1).

The erectile tissue of the penis consists of the two *corpora cavernosa* and the *corpus spongiosum* that surround the urethra (see Fig. 20.1 for more detail). The erectile tissues are essentially large venous sinusoids surrounded by a coat of strong fibrous tissue. Erection is a simple hydraulic process mainly controlled by the parasympathetic nerves. Impulses in the cavernous nerve cause the internal pudendal artery and its main branches to dilate. Consequently, the blood flow to the penis is increased but the venous outflow remains unchanged so that blood becomes pooled in the erectile tissue, causing the penis to enlarge and extend. As the volume of blood within the erectile tissue increases the pressure rises, partially occluding the emissory veins, so that the penis becomes rigid and erect. Erection is also known as *tumescence*.

Dilation of the internal pudendal artery and the associated arterioles is mediated chiefly by nitric oxide derived from fibers running in the parasympathetic nerves. Somatic afferent pudendal nerves provide sensory feedback to maintain erection during intercourse. Most sensory input is derived from touch receptors in the skin of the most distal part of the penis (the glans penis). *Detumescence* (reversal of erection) is probably mediated by sympathetic nerves arising from the sacral ganglia.

Emission and ejaculation are the final stages of the male sexual act. When sexual stimulation becomes very intense, rhythmic contractions in the vas deferens and ampulla begin to drive sperm towards the ejaculatory duct. This is followed by secretion of prostatic fluid and contraction of the seminal vesicles. This stage is called *emission*; the expulsion of this fluid from the penis is called *ejaculation*. During ejaculation, the internal urethra becomes tightly closed to prevent the seminal fluid entering the bladder, prostate, or seminal glands. Contractions of the bulbocavernous

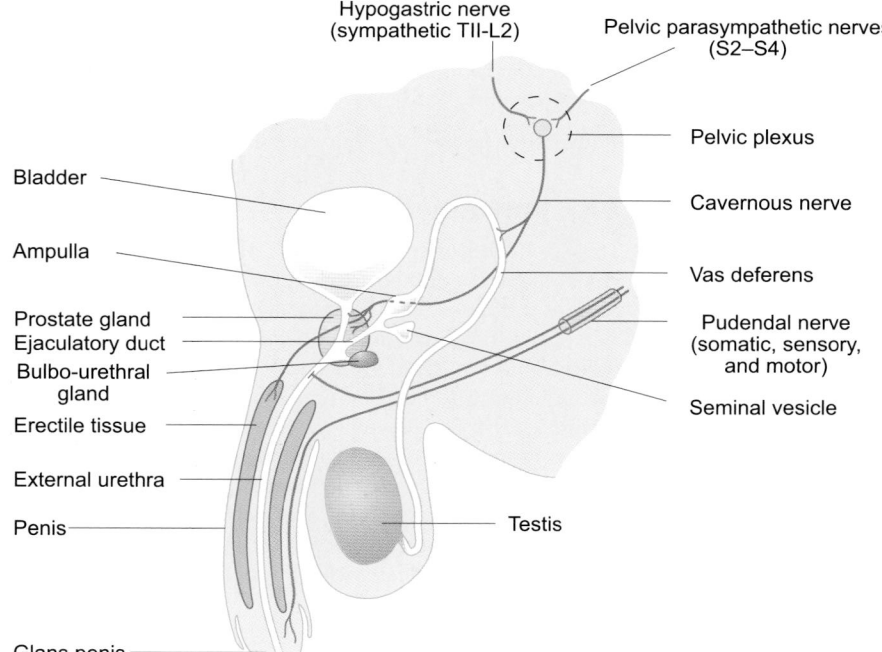

Fig. 21.1 The nerve supply to the penis and accessory sex organs.

and ischiocavernous muscles drive the seminal fluid along the urethra. Together, these adaptations result in forward movement of the semen and its expulsion from the penis. Ejaculation is accompanied by an intense sensation known as *orgasm*.

In the female, the neural pathways, both afferent and efferent, involved in controlling the sexual responses are the same as for males. The innervation of the female genitalia is illustrated in Fig. 21.2. Afferent impulses are relayed from the genitalia to the sacral region of the spinal cord, where sexual reflexes are integrated. The spinothalamic tract carries sensory information to the brain from which descending pathways transmit impulses back to the sacrum. In this way, sexual reflexes are modified by cerebral influences. During sexual excitement, erectile tissue within the clitoris and around the vaginal opening becomes engorged with blood. As in the male, this engorgement is caused by parasympathetic stimulation of the blood vessels supplying these tissues. The nipples of the breasts may also become erect. Mucus secretions are provided both by Bartholin's glands that are adjacent to the labia minora and by the vaginal epithelium.

These secretions facilitate entry of the penis into the vagina and ensure that intercourse is associated with a pleasurable massaging sensation rather than dry frictional irritation, which can inhibit both ejaculation and the female climax. The tactile stimulation of the female genitalia associated with sexual intercourse, coupled with psychological stimuli, will normally trigger an orgasm. This is associated with rhythmic contraction of the perineal muscles, dilation of the cervical canal, and increased motility of the uterus and possibly the Fallopian tubes. These contractions may help to transport sperm towards the site of fertilization in the Fallopian tube, although female orgasm is not a requirement for fertilization. In both males and females, orgasm is followed by a feeling of warm relaxation known as *resolution*.

Contraception

The prevention of unwanted pregnancies (contraception) is now a major issue in most societies as human fertility leads to inexorable population growth. The principal means of preventing conception are summarized in Box 21.1 and rely on either

BOX 21.1 AN OUTLINE OF CONTRACEPTIVE METHODS

Method	Effectiveness (estimated as % of couples remaining childless after 1 year)	Comments
Oral contraceptives (the birth control pill)	99.5	Contraceptive preparations are of two types: a) Combination—consisting of estrogens and progesterone in sufficiently high concentrations to exert powerful negative feedback on gonadotropin output. They therefore mimic the luteal phase of the menstrual cycle and prevent ovulation. b) The mini-pill—this contains only progesterone and acts by modifying the secretions and environment of the reproductive tract. Side-effects include minor symptoms of early pregnancy—nausea, breast tenderness, fluid retention, hypertension, and, in rare cases, thromboses (mainly in smokers over 35 years of age).
Intrauterine device (IUD)	98.5	Probably works by creating an environment within the uterus that is hostile to fertilization or implantation. May cause uterine bleeding. Increased risk of inflammatory disease in the pelvic region.
Condom (sheath)	96	A barrier method which is more effective if used with a spermicide. Its chief advantage is that it protects against AIDS and other sexually transmitted diseases. It may also protect against cervical cancer. Its disadvantage is that it may split in use.
Diaphragm	98 when used with spermicide	An alternative barrier method. Needs to be inserted before intercourse. It fits over the cervix and blocks the entrance to the uterus. Its use carries a small risk of infection.
Rhythm method	Highly variable; dependent on regularity of cycle and accuracy with which mid cycle is calculated	Couple must refrain from intercourse during the fertile period of the cycle (i.e. 2 or 3 days on either side of ovulation). The time of ovulation must be calculated on the basis of the previous cycle, assuming that ovulation occurred on day 14 before menstruation. Other indicators such as the change in body temperature at mid-cycle and the constitution of cervical mucus may also be used.
Sterilization (vasectomy in males and ligation of the Fallopian tubes in females).	c. 100	Requires surgery and is not always reversible. Risks are similar to those of other minor surgical procedures.

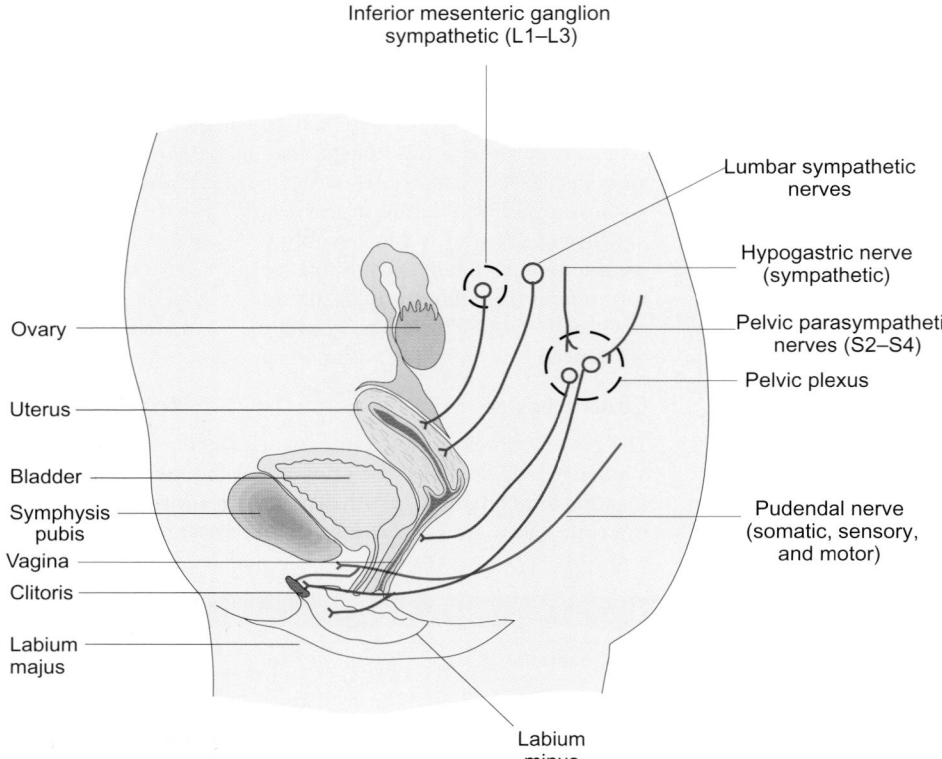

Inferior mesenteric ganglion
sympathetic (L1–L3)

Lumbar sympathetic
nerves

Hypogastric nerve
(sympathetic)

Pelvic parasympathetic
nerves (S2–S4)

Pelvic plexus

Ovary

Uterus

Bladder

Symphysis
pubis

Vagina

Clitoris

Labium
majus

Pudendal nerve
(somatic, sensory,
and motor)

Labium
minus

Fig. 21.2 A schematic drawing to show the nerve supply to the female sex organs.

preventing contact between sperm and ovum by a physical barrier (e.g. a condom) or preventing ovulation.

21.3 Fertilization and the implantation of the embryo

Around 3 ml of seminal fluid are released in each ejaculation and this will normally contain about 200 million sperm. Sperm deposited in the vagina during intercourse swim through the cervical mucus and the uterus to the Fallopian tube which is the site of fertilization. Both the male and female gametes have a limited period of viability within the female reproductive tract. Sperm are thought to retain their fertility for up to 48 hours, while ova remain viable for only around 12–24 hours after ovulation. Therefore there is a relatively short time during each menstrual cycle in which intercourse must take place if pregnancy is to be achieved.

Before a sperm can fertilize an ovum it must undergo a process known as *capacitation*. (Sperm that are intended for *in vitro* fertilization are unable to fertilize an egg unless they are first washed free of seminal fluid.) Capacitation occurs naturally within the female reproductive tract and depends on the high concentration of bicarbonate ions in the fluid of the uterus and Fallopian tubes. The biochemical processes involved in capacitation are complex and not fully understood but involve activation of intracellular signaling pathways. One early event is activation of a soluble adenylyl cyclase in the sperm head by bicarbonate ions. The subsequent rise in intracellular cyclic AMP and activation of protein kinase A then initiates a series of protein phosphorylation reactions. Other changes involve activation of

receptor tyrosine kinases in the sperm head by a number of factors including progesterone. The process of capacitation takes around 5 hours in humans, during which time the composition of the plasma membrane of the sperm head is modified. Should an activated sperm meet a viable egg in the Fallopian tube, the sperm first swims through the follicular cells that surround the egg and then undergoes the *acrosome reaction* as it binds to the zona pellucida. The acrosome reaction is a calcium dependent process in which the acrosomal vesicle fuses with the surface membrane of the head of the sperm (see Fig. 20.5). The resulting exocytosis releases hydrolytic enzymes that, together with the force generated by the sperm tail, aid the passage of the sperm through the zona pellucida. The acrosome reacted sperm is now capable of fusing with the plasma membrane of the egg to complete the first stage of fertilization.

The fertilized egg must now complete several important tasks in order to ensure its continued successful development.

1. It must complete its second meiotic division to avoid triploidy now that it has fused with a sperm (remember that the secondary oocyte arrested in the second meiotic metaphase —see p. 441). This division is normally completed within 2 or 3 hours of fertilization and the second polar body is extruded.

2. It must avoid fusing with any further sperm (polyspermy). To avoid polyspermy, a newly fertilized egg undergoes the so-called 'cortical reaction'. Much of our information concerning this reaction has come from experiments using mouse eggs and the extent to which it is applicable to humans is unclear. A simple diagram of the likely key events is shown in Figure 21.3. After fertilization, the sperm head enters the

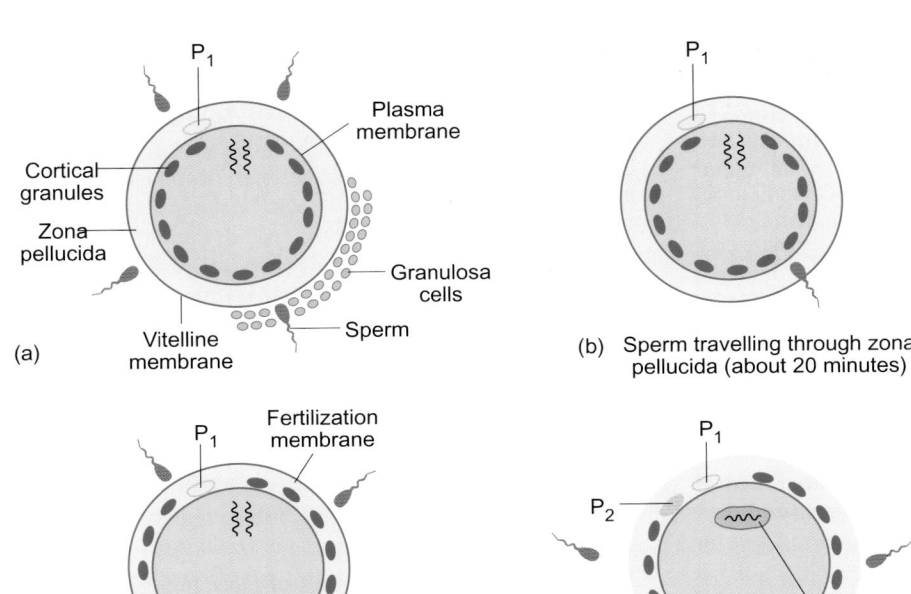

(a)

(b) Sperm travelling through zona pelucida (about 20 minutes)

(c)

(d)

Fig. 21.3 Diagrammatic representation of the changes that occur during fertilization. For a sperm to fuse with an egg it must first penetrate the layer of granulosa cells and the zona pellucida as shown in (a) and (b). Fusion with the egg leads to the formation of the fertilization membrane (c) and this is followed by the second meiotic division and the extrusion of the second polar body (d). P_1 and P_2 are the first and second polar bodies.

oocyte. In doing so it introduces a form of phospholipase C (known as PLC zeta) that is able to catalyze the formation of IP_3 from membrane lipids. As the IP_3 concentration rises, it triggers the release of calcium from internal stores within the oocyte. As a result, the free calcium levels within the cytoplasm of the oocyte rise. This rise triggers the cortical reaction during which granules derived from the Golgi apparatus fuse with the oocyte membrane, releasing their contents into the perivittelline space by exocytosis (see Chapter 4). These contents form the 'fertilization membrane' which appears to consist mainly of enzymes that act to prevent penetration of the egg by any further sperm.

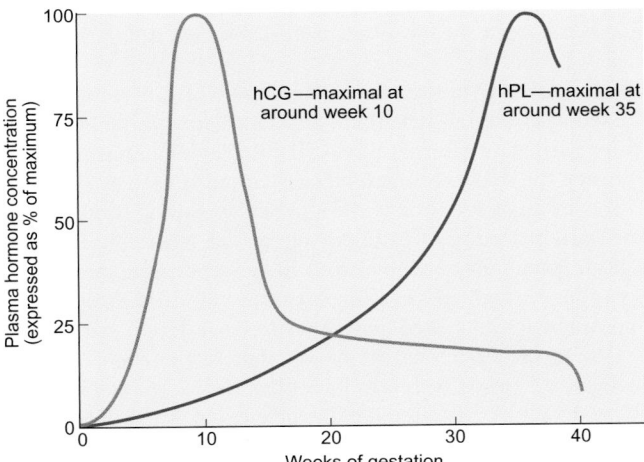

Fig. 21.4 The changes in the plasma concentrations of human chorionic gonadotrophin (hCG) and human placental lactogen (hPL) that occur during gestation.

3. Finally, the fertilized ovum initiates changes that prevent regression of the corpus luteum and shedding of the endometrium. The maintenance of the structure and function of the corpus luteum beyond the end of the cycle is dependent on the secretion of a glycoprotein hormone known as *human chorionic gonadotrophin* (hCG). This hormone is a potent luteotrophic agent that is secreted by the zygote. By its luteotrophic action, hCG is believed to maintain the corpus luteum beyond its normal lifespan so that it will continue to secrete the progesterone that is required for the continuation of pregnancy.

hCG is structurally very similar to anterior pituitary LH but has a longer half-life. It appears in the maternal circulation within a few days of fertilization and can be detected in the urine by about 2 weeks after ovulation. Levels of hCG then continue to increase steadily up until weeks 8–10 of gestation before falling rather sharply over the next few weeks. This profile of secretion is illustrated in Fig. 21.4. After 6 or 8 weeks, the placenta is well established (see below) and is able to synthesize and secrete sufficient progesterone to maintain the remaining gestation period. The pregnancy is then said to be autonomous, and the fall in hCG secretion seen after about 8 weeks probably reflects this diminishing requirement for the hormones of the corpus luteum.

Clinically, hCG is a very important hormone, chiefly because of its early appearance in the maternal body fluids following fertilization. hCG can be detected in the maternal plasma as early as 7 days after ovulation. Its presence in the urine 2 weeks or so after ovulation is used as a reliable and simple test for pregnancy; indeed, it is so simple that it can be carried out, using a kit, by a woman herself at home. More sophisticated assay techniques can also be used by clinicians to gain information about the pregnancy. For example, levels of hormone above the nor-

Summary

1. Sexual intercourse involves penile erection, penetration of the vagina by the penis, and ejaculation of about 3 ml of seminal fluid containing around 200 million sperm. Both male and female sexual responses are mediated by sacral reflexes involving the autonomic nervous system.

2. A sperm is only capable of fertilizing an egg if it first undergoes capacitation followed by the acrosome reaction. Fertilization can occur if an activated sperm meets a viable ovum in the Fallopian tube.

3. The first stage of fertilization occurs when an activated sperm fuses with the oocyte. The newly fertilized egg then completes its second meiotic division, and undergoes the cortical reaction to create a fertilization membrane, which prevents further sperm from fusing with it.

4. The newly fertilized egg (the zygote) secretes a powerful luteotrophic hormone hCG, which prolongs the secretory life of the corpus luteum. This ensures that progesterone continues to be secreted and the specialized endometrial layers of the uterus are maintained until the pregnancy can be supported by progesterone of placental origin. This occurs at around 6–8 weeks of gestation.

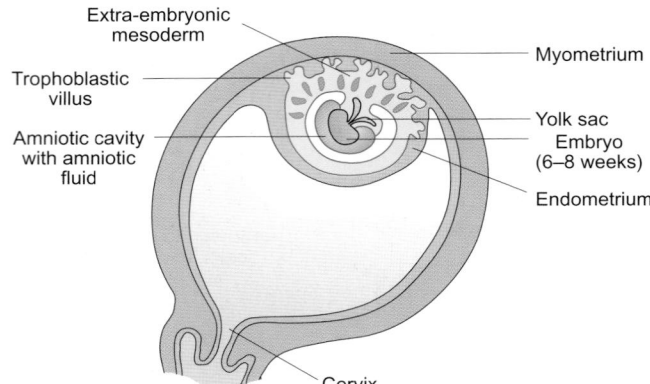

Fig. 21.5 The placenta in relation to adjacent structures during early pregnancy. NB. For clarity, the placenta and embryo are shown disproportionately large in relation to the uterus.

mal range, suggest the presence of twins, a situation that may be confirmed later by ultrasound scans. In women who have suffered habitual miscarriages because of insensitivity of the corpus luteum to hCG, it may be useful to be able to detect the presence of an embryo within a few days of fertilization so that progesterone can be administered exogenously to prevent loss of the pregnancy when the corpus luteum regresses.

21.4 The formation of the placenta

All organisms throughout their embryonic development need a large and continuous supply of nutrients. They must also be able to respire and to dispose of the waste products of their metabolism. Most mammalian species accomplish this crucial task by the process of *placentation*. A specialized organ, the placenta, is developed which brings the blood supply of the fetus into close proximity with that of its mother, thereby permitting the exchange of substances between the two circulations. In this way the placenta can perform those functions carried out by the lungs, GI tract, and kidneys in the adult. Indeed, the placenta is the only source of nourishment, gas exchange, and waste disposal available to the fetus. Therefore it is crucial to the success of a pregnancy that the placenta should develop and function efficiently. Furthermore, placental growth, particularly in the first months of gestation, must keep pace with the requirements of the growing fetus. Without an adequate surface area for transplacental exchange, the growth of the fetus will be impaired and its life may be threatened. The association between maternal and fetal circulations established by the placenta allows for prolonged development within the uterus and, as a result, the delivery of a highly developed baby.

Although the anatomical details of placental formation are beyond the scope of this book, it is important to understand how the placenta is adapted structurally for carrying out its role as an organ of exchange. Figure 21.5 illustrates the gross spatial relationships between the fetal and maternal tissues. Essentially, the developing placenta is an association between the uterine endometrium and the embryonic membranes derived from a layer of cells known as the *trophoblast*.

How is this interface created? Figure 21.6 shows an enlarged view of a section of human placenta soon after the start of embryonic implantation. This is known as the *stem-villus* stage of development because the fetal tissue grows up into the maternal endometrial tissue in the form of finger-like projections or *villi*. These are formed from a membrane called the *chorion* which is derived from trophoblastic tissue and mesoderm that lies outside the developing embryo (the *extra-embryonic mesoderm*). Blood vessels form within this outgrowth to give rise to the fetal component of the placental circulation, which will later become the umbilical vessels in the umbilical cord. As the trophoblastic tissue invades the endometrium it secretes digestive enzymes that break down the spiral arteries. As a result of this erosion, blood spills out of the maternal vessels to create blood-filled spaces between the chorionic villi. These are called the intervillous blood spaces. At the same time, the villi themselves become cored with mesoderm (Fig. 21.6) which becomes vascularized with fetal vessels that carry fetal blood into close proximity with the maternal blood spaces. The essential interface between the maternal and fetal circulations is then established.

This invasive behavior of the trophoblast during implantation and early placentation is highly aggressive and reminiscent of that of a malignant neoplasm. It has indeed been shown that occasionally fragments of tissue break away from the trophoblast and can be found lodged in distant maternal tissues, particularly the lung, rather in the same way that malignancies tend to metastasize. However, regression of the ectopic trophoblastic tissue generally occurs quite rapidly following delivery of the baby.

Invasion of the endometrium and villus formation take place during the first month after conception. This establishes proximity between maternal and fetal blood. In the succeeding two months, the highly vascularized villi become much more highly

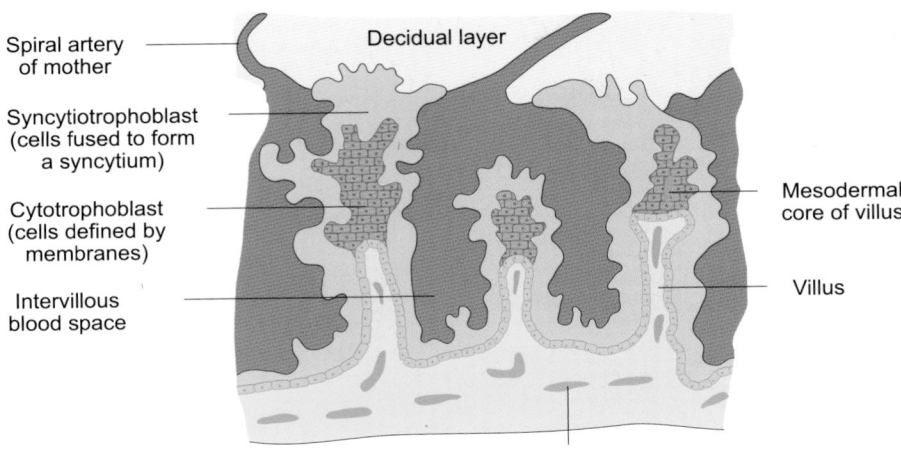

Fig. 21.6 The stem-villus stage of placental development at around week 3 of gestation.

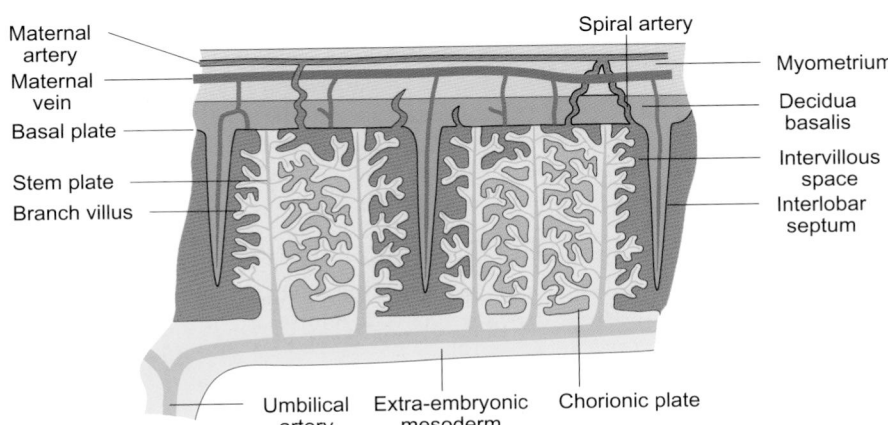

Fig. 21.7 The definitive placenta at the end of the first trimester.

branched, thereby increasing the surface area of the fetal capillaries available for transplacental exchange of nutrients and waste products. By the end of the first trimester (3 months) of gestation the placenta is known as the *definitive placenta* because, although it will increase in size, it will show few further changes to its basic structure for the remainder of the pregnancy.

A diagrammatic and highly simplified representation of the placenta at 3 months is shown in Fig. 21.7. Trophoblastic cells at the outer margins of the villi fuse together to form a syncytium (the *syncytiotrophoblast*), while those further from the outer margins retain their individual membranes (the *cytotrophoblast*). There is also elaborate branching of the primary villi. The erosion of the maternal spiral arteries is now so extensive that their blood is simply discharged into the intervillous spaces. The fetal blood supply runs in the capillaries within the branched villi, and the arrangement of the placental blood flow is such that the fetal capillaries essentially dip into the maternal blood spaces so that they are virtually surrounded by maternal blood. This arrangement of the two circulations within the placenta is termed a *dialysis pattern*, and it allows movement of solutes in either direction according to the concentration gradient over the entire surface of the fetal placental capillaries. This arrangement is shown schematically in Fig. 21.8.

21.5 The placenta as an organ of exchange between mother and fetus

The rate and extent of diffusion of a substance across any cellular barrier depends upon a variety of factors. In addition to the chemical characteristics of the substance itself, these include:

◆ the nature and thickness of the barrier to diffusion;

◆ the surface area available for exchange;

◆ the concentration gradient of the substance.

Each of these factors will now be examined with reference to the human placenta in order to understand more about the ways in which it is adapted to carry out its functions as an organ of exchange.

The barrier to diffusion in this case is the so-called 'placental barrier' between the maternal and fetal blood. What does this consist of? For a substance to diffuse from the maternal blood space to the fetal capillary blood (or vice versa), it must cross the syncytiotrophoblast and the fetal capillary endothelial layer. The latter consists merely of a single layer of cells on a basement membrane and therefore is very thin, but the syncytiotrophoblast contains more layers of cells with no paracellular pathways to act as shortcuts for diffusion and therefore has a relatively low permeability. Overall, then, the placental barrier is rather impermeable.

However, this low solute permeability is largely offset by the enormous surface area available for placental exchange created by the extensive branching of the fetal capillaries within the villi and the dialysis arrangement of the two circulations.

Within the placenta, the concentration gradient of any substance between the maternal and fetal blood will be influenced by the blood supply to the maternal and fetal circulations, particularly the relative blood flow rates on either side. The rate of entry of blood to the maternal intervillous spaces is rather difficult to measure since the uterine arteries supply both the uterus and the placenta, making it hard to differentiate between blood destined for the intervillous spaces and that supplying the uterus. Total blood flow in the uterine artery at full term is 600–1000 ml min^{-1}, measured just prior to delivery by Cesarean section. About half of this (or 10 per cent of the total maternal cardiac output at full term) is thought to perfuse the maternal side of the placenta. The maternal blood space has a total volume of about 250 ml so that the blood on the maternal side of the placenta is exchanged roughly twice each minute. The blood entering the intervillous spaces is at a relatively high pressure (about 13 kPa (100 mmHg)), as it is discharged from the eroded spiral arteries, thus ensuring a fair degree of turbulence and good mixing.

In a full-term fetus, perfusion of the fetal capillaries in the placental villi is estimated at around 360 ml min^{-1}, roughly half the fetal cardiac output. The total volume of blood in the capillaries is 45 ml or so, so that the blood in the fetal placental compartment is exchanged about eight times a minute. To summarize, the maternal blood spaces present a large volume of well-mixed blood with a moderate turnover, while the fetal capillaries have a much smaller volume but a greater turnover.

This pattern of circulation emphasizes the dialysis nature of the placental blood flow. It optimizes the conditions for passive exchange of solutes by maximizing concentration gradients. Consider the diffusion of a solute from maternal to fetal blood. As the fetal blood flow is very high, solute diffusing into the fetal blood from the maternal side will be removed from the placenta rapidly, keeping its concentration low in the fetal capillaries. However, the maternal blood has a much larger volume, so that, despite its lower flow, it will not readily become depleted of solute. Through this arrangement, the concentration gradient for the solute between the maternal and fetal blood is maintained and ensures efficient diffusion.

The dialysis arrangement of the placental blood flow also helps to optimize conditions for the removal of waste products from the fetal circulation by maintaining a steep concentration gradient for the waste substance across the placental barrier. The rapid turnover of blood on the fetal side will ensure a constant delivery of waste to the placenta, while the relatively large volume of blood in the intervillous spaces will keep the concentration of waste product low on the maternal side.

Towards the end of pregnancy, the exchange capacity of the placenta tends to diminish. This is chiefly due to changes in the perfusion of the organ. Maternal blood flow may be somewhat reduced as the spiral arteries become progressively occluded

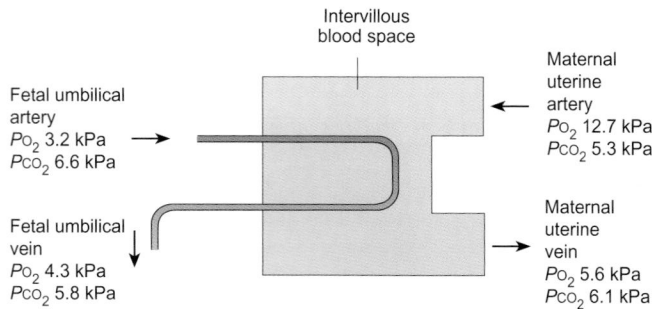

Fig. 21.8 The dialysis organization of the placenta with typical blood gas values. Note that the maternal and fetal circulations are separate and that the blood leaving the placenta via the umbilical vein is not fully equilibrated with the maternal blood.

during pregnancy. At the same time, the fetal capillaries tend to become blocked with small clots and other debris. This leads to a progressive decline in placental perfusion towards term. As parts of the chorionic villi become poorly perfused, they can no longer participate efficiently in exchange and the effective surface area for diffusion is reduced. As a result of this declining efficiency, the placenta is said to become 'senescent' near to term and is less and less able to meet the demands of the fetus. This may be one of the many factors involved in the triggering of parturition (see Section 21.7).

Gas exchange across the placenta occurs by diffusion

Oxygen diffuses passively from the maternal to the fetal side of the placenta. Carbon dioxide diffuses in the opposite direction. How efficient is the placenta as an organ of gas exchange? To answer this question, it is necessary to consider the gas tensions on both the maternal and fetal sides of the placenta. Figure 21.8 shows the values for oxygen and carbon dioxide tensions in each compartment. The normal arteriovenous differences in partial pressures seem to prevail on the maternal side with a P_{aO_2} of around 12.6 kPa (95 mmHg) in the uterine artery falling to around 5.6 kPa (42 mmHg) in the uterine venous blood. The maternal P_{CO_2} rises from 5–5.6 kPa (38–40 mmHg) in the arterial blood to 6.1 kPa (46 mmHg) on the venous side. Since the maternal blood space is large and the blood well mixed, equilibrium values of gas partial pressures are quickly reached and it is therefore reasonable to assume that the blood of the intervillous spaces has a P_{O_2} of 5.6 kPa (42 mmHg) and a P_{CO_2} of 6.1 kPa (46 mmHg).

On the fetal side, the blood in the umbilical artery traveling to the placenta from the fetus is both highly deoxygenated and hypercapnic, having a P_{O_2} of around 3.2 kPa (24 mmHg) and a P_{CO_2} of about 6.6 kPa (50 mmHg). Blood returning from the placenta to the fetus in the umbilical vein has a P_{CO_2} of 5.8 kPa (44 mmHg) and a P_{O_2} of about 4.25 kPa (32 mmHg). The fetal umbilical venous blood is not in equilibrium with the maternal blood. In this respect, the placenta differs from the lung, in which complete equilibration is normally achieved between the pulmonary blood and the alveolar air. There are two important reasons for this failure to reach equilibrium. First, not all the maternal blood is in direct contact with the villi (which is the area of gas exchange) so that 'shunts' exist which are analogous with a ventilation–perfusion mismatch in the lungs (see Chapter 16). Secondly, the placental tissue itself, which is highly metabolically active, uses around 20 per cent of the oxygen in the maternal blood before it has a chance to reach the fetal capillaries. Therefore the placenta is a less efficient organ of gas exchange than the lung. However, it is able to satisfy the oxygen demand of the fetus because of a variety of specific adaptations which ensure that the transfer of oxygen to the fetal tissues is maximized. These are discussed in Chapter 22.

In addition to supplying oxygen to the fetus, the placenta must, in its capacity as the fetal organ of gas exchange, remove the carbon dioxide produced by fetal metabolism. As for oxygen, the passive movement of carbon dioxide depends on blood flow and diffusion gradients, but the placental barrier is more permeable to carbon dioxide than it is to oxygen and exchange is more or less complete.

Placental exchange of glucose and amino acids is carrier mediated

The placenta is the sole source for the fetus of the nutrients essential for its growth. The most important of these is glucose. Because glucose is a polar molecule, and so rather lipid insoluble, it cannot rely solely upon passive diffusion across the lipid-rich placental barrier. Instead, it moves from the maternal to the fetal side by facilitated diffusion, a mechanism which is discussed elsewhere (Chapter 4). The process is mediated by a carrier that is located in the membrane of the cells of the syncytiotrophoblast. Unless the carrier becomes saturated, fetal levels of glucose will be in equilibrium with those of the mother. While this would normally be desirable, there are circumstances in which it is not. Consider the case of poorly controlled maternal diabetes. Here, maternal glucose levels may be abnormally high and, because of facilitated diffusion across the placenta, fetal levels will also be high. This can lead to overnourishment and obesity of the baby—indeed, the babies of diabetic mothers are often larger than normal for their gestational age (see Section 21.8).

Amino acids are vital to the fetus during its development in the uterus. They are needed to support the high rate of protein synthesis that occurs during gestation. The concentrations of most amino acids are higher in the fetal plasma than in the maternal plasma. This suggests that amino acids are actively transported across the placenta. Recent studies using the techniques of molecular biology have shown that specific transporters exist in the human placenta for all the essential amino acids.

Lipids are of great importance in fetal development. While phospholipids do not pass readily across the placental barrier, free fatty acids are able to do so and it is in this form that the fetus receives most of its lipid. Phospholipid in the maternal blood is hydrolyzed by enzymes on the placental surface to form free fatty acids, which diffuse passively down their concentration gradients to the fetal blood. They then pass to the fetal liver where they undergo reconjugation to form new phospholipid.

The excretion of fetal waste products by placental exchange

Like carbon dioxide, which diffuses from the fetal to the maternal blood across the placenta, other fetal waste products are removed in a similar fashion and are then excreted together with those of the mother herself. One of the most important of these metabolic waste products is urea, the nitrogenous waste product of protein metabolism. Although much of the metabolism of the fetus is concerned with the synthesis of new structural protein, there is also a fair amount of tissue destruction throughout gestation as fetal tissues are remodeled. Indeed, of the total nitrogen that enters the fetus in the form of amino acids transported across the placenta from the maternal blood, about 40 per cent ends up in the form of urea which must be disposed of. Excretion occurs by the passive diffusion of urea down its the concentration gradient between the fetal and maternal blood.

Another fetal waste product of considerable clinical significance is bilirubin, which is produced by the breakdown of hemoglobin. In adults, bilirubin is conjugated by hepatic enzymes to form bilirubin glucuronide. This is water soluble and therefore can be excreted without difficulty. However, the fetal liver is relatively immature and does not possess sufficient amounts of the necessary conjugation enzymes. During fetal life, there is significant destruction of red blood cells and the bilirubin produced crosses the placenta. It is then conjugated by maternal liver

Summary

1. The placenta forms an interface between the maternal and fetal circulations. At implantation, the trophoblastic tissue of the fertilized egg invades the endometrial tissue of the uterus by means of chorionic villi containing the fetal capillaries. As a result of this invasive behavior, the spiral arteries of the uterus are eroded and spill their blood into the spaces between adjacent chorionic villi. In this way a dialysis pattern of blood flow is set up within the placenta such that fetal capillaries essentially dip into maternal blood spaces.

2. During fetal life, the placenta carries out the functions normally performed in the adult by the lungs, kidneys, and GI tract.

3. Many substances, including oxygen, carbon dioxide, and essential nutrients, cross the placenta by means of either passive diffusion or carrier-mediated transport. Although the placental barrier itself is relatively impermeable to polar molecules, the surface area available for exchange is immense due to the considerable branching of the chorionic villi. The dialysis pattern of the fetal and maternal blood supplies within the placenta optimizes concentration gradients for solutes, thereby ensuring efficient exchange.

4. Oxygen diffuses passively from maternal to fetal blood, although full equilibration does not occur. Carbon dioxide diffuses in the opposite direction, normally to complete equilibration. Glucose and amino acids move across the placenta from the maternal to the fetal plasma by carrier-mediated transport, while free fatty acids diffuse passively across the lipid-rich placental barrier. Fetal waste products such as urea and bilirubin diffuse from fetal to maternal plasma down their concentration gradients.

enzymes before being excreted in the bile. If it is not disposed of but allowed to build up in the fetal blood, bilirubin can cross the blood–brain barrier and cause severe brain damage. The basal ganglia are the most commonly affected brain regions, giving rise to the condition known as *kernicterus* in which there may be permanent impairment of motor function.

21.6 The placenta as an endocrine organ

In addition to its crucial transporting role described above, the placenta is also an extremely important endocrine organ. At full term it normally weighs around 650 g and, at this time, it is the largest endocrine organ in the body of a pregnant woman. More significant, pehaps, is its versatility. The placenta secretes a wide variety of different hormones, both peptide and steroid, which are important in the maintenance of pregnancy and for the preparation of the mother's body for parturition and lactation.

The major peptide hormones secreted by the placenta are:

◆ human chorionic gonadotrophin (hCG)

◆ human placental lactogen (hPL).

The major placental steroids are:

◆ estrogens

◆ progesterone.

The physiological role of each of these hormones will be considered in turn.

Human chorionic gonadotrophin

This hormone was discussed in some detail in Section 21.3. It is secreted from a very early stage of pregnancy by the trophoblastic tissue of the embryo, and it is believed that it provides the signal that enables the mother's body to recognize the existence of a fertilized egg. It is a powerfully luteotrophic hormone which prolongs the life of the corpus luteum beyond the normal 12–14 days. As a result, progesterone secretion continues so that shedding of the uterine endometrium is prevented and spontaneous contractile activity of the myometrium is inhibited. Luteal progesterone is required for about the first 6–8 weeks of pregnancy and loss of the ovaries during this period will result in miscarriage. After this time, the placenta takes over as the main source of progesterone and the pregnancy is said to become autonomous. Indeed, levels of hCG output decline sharply after about 10 weeks although the hormone continues to be produced by the placenta for the remainder of the pregnancy (Fig. 21.4).

In addition to its luteotrophic role, a number of other actions have been attributed to hCG. It has been suggested that hCG may exert a direct effect on the maternal hypothalamus to inhibit the synthesis of FSH and LH. If so, this might contribute to the suppression of ovulation during pregnancy. hCG is also thought to play a role in preventing rejection of the fetus by the mother.

hCG exerts a stimulatory effect on the Leydig cells of the testes in male fetuses and is believed to play a part in the differentiation of the male reproductive tract. Plasma levels of hCG are high during the period of Wolffian duct development and the start of differentiation of the external genitalia (see Chapter 22, p. 000).

Human placental lactogen

The pattern of secretion of human placental lactogen (hPL), sometimes known as human chorionic somatomammotrophin (hCS), is shown in Fig. 21.4. It can be seen that this hormone appears in the maternal circulation at the time that hCG levels are beginning to fall, around 8 weeks of gestation. The concentration of hPL in the maternal plasma then continues to rise during the pregnancy, reaching a peak around week 35. This hormone is secreted by the syncytiotrophoblastic tissue of the placenta and, unlike hCG which appears equally in both the fetal and maternal circulations, hPL is secreted preferentially into the maternal blood.

Like hCG, hPL is structurally and functionally related to other peptide hormones. It has a high degree of homology with both growth hormone and prolactin, two hormones secreted by the anterior pituitary. Like these hormones, hPL can stimulate both somatic growth and milk secretion, though only weakly. Its principal action is to encourage the proliferation of breast tissue during pregnancy in preparation for lactation following delivery.

In addition to its mammotrophic action, hPL exerts some important metabolic effects. These are mainly concerned with adjusting the maternal plasma levels of certain metabolites in order to favor fetal uptake via the placenta without undue depletion of the maternal blood. For example, maternal plasma glucose levels tend to rise under the influence of hPL as a result of the inhibition of glucose uptake into cells (the so-called anti-insulin or diabetogenic effect). Gluconeogenesis (the synthesis of glucose from amino acids) also appears to be suppressed by hPL, leading to an increase in maternal plasma amino acid levels, while increased lipolysis causes an increase in plasma free fatty acids. By these actions, it is thought that hPL counteracts the fall in metabolites that might otherwise occur as a result of fetal uptake and ensures the maintenance of favorable gradients of important nutrients for placental transport.

The monitoring of hPL levels during pregnancy can have clinical importance. hPL can be measured accurately and simply by a variety of techniques, including radio-immunoassay, and its concentration in the maternal plasma provides a valuable indication of placental sufficiency. Although it is not unknown for pregnancies to proceed successfully in the absence of hPL (in the case of specific genetic deficiencies for example), in general, falling levels of this hormone are indicative of placental insufficiency which may put the fetus at risk.

Other placental polypeptide hormones

In addition to the two major placental peptide hormones described above, a large number of other proteins are produced by the placenta. New agents continue to be discovered but in the majority of cases no specific actions have yet been ascribed to them. Among the more important of the placental peptides are a chorionic FSH, and a chorionic thyrotrophin. Clinically, measurement of these hormones may be of value in assessing possible risks to the fetus from placental insufficiency.

The placenta secretes large amounts of progesterone and estrogens

It should be clear from Chapter 20, Section 20.8, that for a pregnancy to proceed successfully to term, adequate amounts of

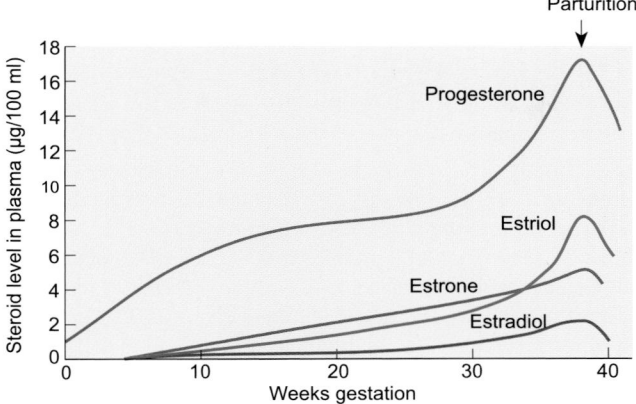

Fig. 21.9 The plasma levels of various steroid hormones during pregnancy. Note that progesterone secretion dominates the period of gestation, falling only after parturition.

the steroid hormone progesterone are crucial. There is no recorded case of a pregnancy continuing normally in the face of insufficient progesterone secretion. During the early weeks after conception, this progesterone is supplied by the corpus luteum, rescued from its declining phase by hCG. By about 8 weeks, however, the placenta is becoming well established and starts to produce large amounts of progesterone. For the remainder of gestation placental progesterone output is extremely high. The pattern of progesterone secretion by the placenta is illustrated in Fig. 21.9. During late gestation it is produced at a rate of 250–350 mg a day. This compares with about 20 mg a day during the luteal phase of the menstrual cycle. The placental tissue is capable of synthesizing progesterone without the need for precursors from elsewhere, and so levels of this hormone during pregnancy are determined solely by the synthetic and secretory capacity of the placenta itself. Consequently, plasma progesterone provides another valuable clinical index of placental performance.

Why is progesterone so important during pregnancy? A number of different functions have been suggested, but the most important of these appear to be the maintenance of a quiescent myometrium and the prevention of endometrial shedding prior to placentation. Progesterone also plays a role in the stimulation of breast development, the suppression of ovulation, and the inhibition of immunorejection of the embryo.

In addition to its progesterone production, the placenta secretes large amounts of a number of estrogenic hormones including estrone, estriol and estradiol-17β. The patterns of secretion of these hormones are illustrated in Fig. 21.9. By comparing these values with those seen in non-pregnant women it is possible to obtain some idea of the scale of estrogen secretion during pregnancy. In the non-pregnant state, the principal estrogen (secreted by the developing follicle) is estradiol-17β. In pregnancy, the levels of this hormone rise to about 100 times the non-pregnant values. Circulating levels of the other estrogens of pregnancy are even higher.

What is the role of estrogens in pregnancy? This question remains largely unanswered. Although it is usual for high levels of estrogens to be present throughout gestation, pregnancies

Summary

1. The placenta secretes a wide variety of peptide and steroid hormones.

2. The major peptide hormones are human chorionic gonadotrophin (hCG) and human placental lactogen (hPL).

3. hCG is a potent luteotrophic agent whose central function seems to be to prevent regression of the corpus luteum to ensure the continued secretion of progesterone during the early weeks of pregnancy.

4. hPL is secreted from around week 10 of gestation. It contributes to the proliferative changes seen in the mammary tissue in preparation for lactation and exerts important metabolic effects in the mother. It stimulates an increase in the maternal plasma levels of glucose, amino acids, and free fatty acids, and these actions ensure that placental transport of essential metabolites from mother to fetus is optimized.

5. Steroid hormones are produced in huge amounts by the placenta throughout gestation. Progesterone is essential for successful pregnancy, and the placenta takes over from the corpus luteum as the major source of this steroid at around week 10. Progesterone maintains the endometrium and reduces myometrial excitability as well as stimulating mammary development in readiness for lactation.

6. A variety of estrogenic hormones are secreted by the placenta, using precursors of both maternal and fetal origin. Their precise role in pregnancy is unclear though they appear to prepare the body for labor and lactation.

can continue successfully even when estrogen levels are rather low. However, they do seem to have a role in preparing the body for giving birth and for lactation. They seem to bring about relaxation of the symphysis pubis and to act alongside hPL to stimulate proliferation of the mammary tissue. They may also play a part in the initiation of parturition—this will be discussed in more detail later.

21.7 The infant is delivered around 38 weeks after conception—what triggers parturition?

In humans, as in all mammals, the length of the gestation period is remarkably constant under normal conditions—40 weeks after the start of the last menstrual period or 38 weeks after conception. This constancy suggests that there is a well-coordinated trigger for the onset of *parturition* (the process of expulsion of the fetus). What brings about the conversion from the stable maintenance of pregnancy with minimal myometrial activity to one in which the cervix dilates and the myometrium contracts efficiently to expel the fetus?

What are the critical signals that bring pregnancy to an end? Over the last few years, considerable research effort has gone into trying to shed more light upon this problem. Most of this work has been carried out on small laboratory animals, such as the rat, and larger domestic animals, particularly sheep. It is now apparent that there is no single trigger for the initiation of labor but rather a combination of different factors both physical and endocrine.

The physical factors that seem most likely to contribute to the onset of parturition are stretching of the myometrium and placental insufficiency. The uterine musculature or myometrium is progressively stretched as the fetus grows during gestation. As it stretches, it becomes thinner and its excitability increases. Once a certain level of excitability is reached, spontaneous contractions occur which tend to squeeze the contents of the uterus down towards the cervix. Small areas of myometrium act as pacemaker cells to initiate action potentials that are then conducted throughout the myometrium, which behaves as a syncytium.

The second physical factor that may play a part in the process of parturition is the increasing inability of the placenta to meet the ever-growing nutritional demands of the fetus. Growth of the fetus far outstrips that of the placenta after the first trimester of pregnancy, and the fetal capillaries tend to become clogged with clots and other debris as the pregnancy nears term. As a result, the placenta becomes less efficient as an organ of exchange and, although it is difficult to verify experimentally, this decline may contribute to the onset of labor. The stages of labor are described in Box 21.2

The fetus may control the timing of its own birth

Whether or not the physical changes described above form a truly significant part of the trigger for labor is open to question. What is clear, however, is that a number of hormonal factors also have a role to play, at least in non-human species. A number of observations have led to the belief that the endocrine system of the fetus itself plays a key role in triggering its own delivery. For example:

- anencephalic fetuses (i.e. fetuses in whom the brain is absent or severely damaged) are frequently born postmature, i.e. after the normal gestation time;

- in sheep, an infusion of cortisol or ACTH to the fetus brings about premature delivery.

Such findings have led to the idea that maturation of the fetal adrenal cortex is in some way responsible for triggering the onset of parturition. Using a sensitive assay for cortisol, it has been shown in sheep that fetal plasma cortisol rises 15–20 days before full term. This rise correlates well with an increase in fetal adrenal enzyme activity and in the abundance of ACTH receptors on the adrenal cortical cells. As might be expected, fetal ACTH levels appear to rise as term approaches and the paraventricular nucleus of the hypothalamus shows an increased content of both corticotropin-releasing hormone and arginine vasopressin, the hormones which act as releasing factors for ACTH (see Chapter 12). The mechanisms that underpin this series of changes in the fetus at the end of gestation are unclear at present.

At the end of gestation the uterus is released from the 'progesterone block' which has dominated pregnancy

The experimental findings discussed in the previous section suggest very strongly that fetal cortisol plays a significant role in the onset of labor at least in some species. It has been recognized for some time that other hormones also show marked changes in their pattern of secretion as pregnancy proceeds. Throughout most of the gestation period, placental progesterone secretion outstrips that of estrogens (although these are also secreted in very large amounts—see Fig. 21.9). This is, of course, vital for the success of the pregnancy since progesterone acts as a myometrial relaxant that prevents premature expulsion of the fetus from the uterus. It also reduces the sensitivity of the uterine smooth muscle to other agents that increase uterine excitability. Estrogens, on the other hand, increase myometrial excitability

---

BOX 21.2 THE STAGES OF LABOR

Labor (parturition) is the process by which a pregnant woman expels the fetus at the end of gestation. It normally begins between 250 and 285 days after presumed ovulation, although during the last month of gestation the woman often experiences irregular uterine contractions. These are believed to occur in response to a gradual increase in sensitivity of the uterine smooth muscle to spasmogenic agents such as oxytocin. The process of labor takes place in three stages.

- Stage 1. Uterine contractions start to occur regularly, at intervals of 20–30 min to begin with, then more frequently until they are occurring every 2–3 min. The purpose of these contractions is to dilate the cervix to around 10 cm to allow the baby to move into the vagina (birth canal). Although the average time for this stage is about 15 h for a first baby, full cervical dilation may be achieved much more quickly in subsequent deliveries. The amniotic sac that surrounds the baby in the uterus may rupture at any time during labor but this often occurs at the onset of labor, a process sometimes described as the 'waters breaking'. The

amniotic membranes may also be ruptured artificially to induce labor. Occasionally a 'show', consisting of blood and mucus from the cervix (the 'mucus plug') is passed from the vagina at the onset of labor.

- Stage 2. The baby is delivered in this stage. Using the Valsalva maneuver to raise the intra-abdominal pressure during uterine contractions, the woman pushes to expel the baby through the cervix and vagina. Delivery is normally head first, although other presentations occasionally occur (e.g. breech delivery, in which the baby is delivered feet first). This stage may take anything from a few minutes to several hours. In some cases forceps or suction are required to facilitate expulsion of the infant.

- Stage 3. About 30 min after the birth of the baby, the placenta, fetal membranes, and any remaining amniotic fluid (the 'afterbirth') are expelled from the uterus in response to further uterine contractions. Their expulsion also helps to seal the blood vessels ruptured by separation of the placenta from the wall of the uterus.

and enhance its sensitivity to other substances that may cause increased contractile activity. An estrogen-primed uterus is highly sensitive to agents such as histamine, oxytocin, acetylcholine, and prostaglandins, reflecting alterations in resting membrane potential of the smooth muscle cells.

Fetal cortisol may initiate the switch from progesterone to estrogen dominance

It is known that cortisol stimulates the conversion of progesterone to estrogens in the placenta. The following is a hypothetical sequence of events that could lead up to the onset of parturition. Fig. 21.10 shows a diagrammatic representation of the scheme.

- Fetal cortisol stimulates the conversion of progesterone to estrogens in the placenta and so redirects placental steroidogenesis in favor of the myometrial stimulant.

- Estrogens in turn stimulate the production of prostaglandin $F_{2\alpha}$ ($PGF_{2\alpha}$) by the placenta, which may help to enhance rhythmical contractions of the uterus.

- Contractions, in their turn, stimulate the secretion of oxytocin from the posterior pituitary via the reflex described in Chapter 12, Section 12.2. Oxytocin increases the excitability of the musculature during labor itself.

It must be emphasized that this scheme is purely speculative —the nature of the interactions between the neural and hormonal changes that occur as pregnancy nears full term is far from clear even in well-studied species like the sheep. Human parturition is even less well understood as obvious ethical difficulties surround research using human subjects.

Evidence from non-human primates suggests that there is a rise in fetal ACTH secretion just prior to delivery and that maternal estrogen levels rise close to full term. Furthermore, removal of the fetus, but not the placenta, some days before the expected

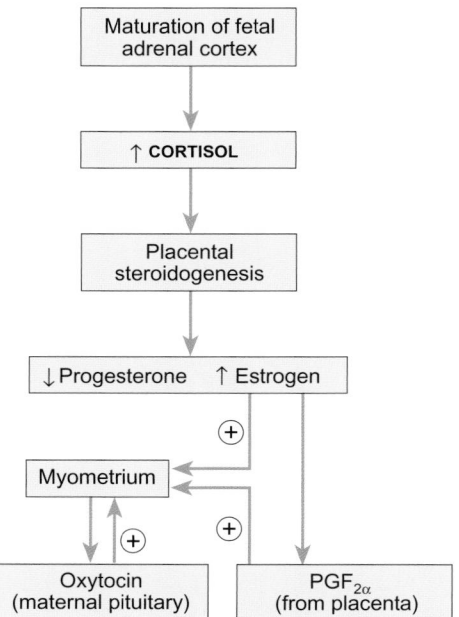

Fig. 21.10 Some endocrine factors involved in the initiation of parturition. ⊕ indicates a stimulatory action.

Summary

1. Parturition is a complex process involving both the maternal and fetal nervous and endocrine systems.

2. The nature of the trigger for parturition is still poorly understood but it is widely believed that the fetus plays a part in determining the time of its own birth.

3. Fetal cortisol appears to initiate a switch in the placenta away from progesterone synthesis to allow estrogenic hormones to dominate the hormonal profile in the last days of pregnancy. Estrogens and other agents such as $PGF_{2\alpha}$ and oxytocin may then increase the contractility of the myometrium still further to bring about delivery of the infant.

date of delivery delays expulsion of the placenta for up to 50 days after full term. This is highly suggestive of a role for the fetus itself in controlling the onset of labor.

21.8 Changes in maternal physiology during gestation

Throughout gestation, the growing fetus makes considerable metabolic demands upon its mother, effectively plundering maternal resources to ensure its survival and successful development. Consequently, the anatomy, physiology, and metabolism of a woman's body undergo a number of significant alterations during the months of pregnancy. These create favorable conditions in which the fetus can grow. They also prepare the mother's reproductive tract and mammary glands for the delivery and subsequent nourishment of the baby.

There is considerable individual variation in the exact nature and extent of the physiological and anatomical changes shown by the mother's body during pregnancy. Typically, however, in the first trimester (3 months) of pregnancy, the changes that occur are designed to prepare the body for the additional metabolic demands of the later stages of gestation. The latter part of pregnancy represents a state of 'accelerated starvation' for the mother's body in which nutrients and amino acids are conserved for use by the fetus. Parallel changes take place throughout pregnancy in all the major systems of the body and most revert to normal after delivery of the infant. Many of these, particularly towards the end of pregnancy, occur because of the marked changes in spatial relationships of the maternal abdominal organs caused by the increasing amount of space occupied by the fetus. In particular, the organs of the cardiovascular, respiratory, renal, and GI systems are progressively displaced or compressed by the enlarging uterus. Furthermore, the mechanical demands placed upon the structures of the musculo-skeletal system are significantly altered during pregnancy. Figure 21.11 illustrates the growth of the fetus and the space occupied by the uterus within the abdomen during gestation.

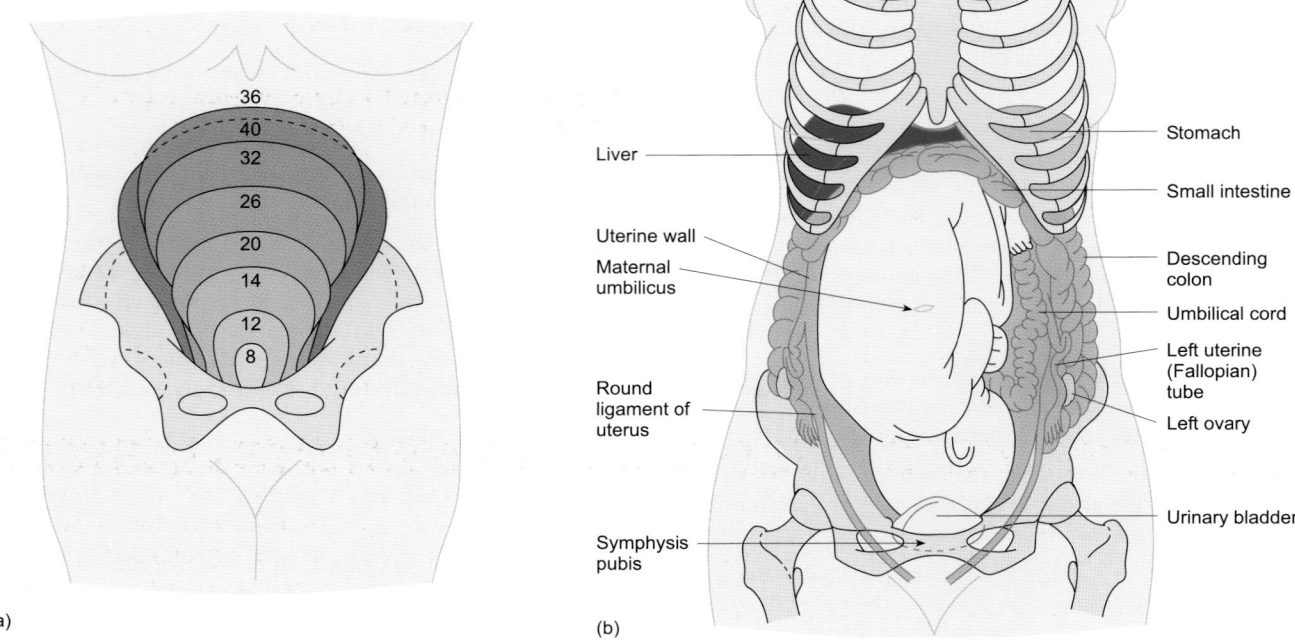

Fig. 21.11 (a) The height of the fundus of the uterus throughout gestation. Heights vary considerably between women but a general rule is that at 20 weeks gestation the fundus is usually around the height of the umbilicus. (b) The position and space occupied by the fetus at full term. Note the displacement of the GI structures and the pressure on the bladder and diaphragm.

Cardiovascular changes during pregnancy

Cardiac output meets the requirements of the placenta

During pregnancy, blood flow through the uterine artery increases about six-fold (Fig. 21.12). There is also a general increase in the maternal metabolic rate (see below). As a consequence of these changes, maternal cardiac output increases by 30–50 per cent between weeks 6 and 28 of gestation. Typically, this represents an increase from around 4.5 to 6.0 l min^{-1}. Cardiac output is increased as a result of changes in both heart rate (from 70 to 80 or 90 b.p.m.) and stroke volume, which increases by around 10 per cent. Despite the increase in uterine blood flow (Fig. 21.12), there may be a slight fall in resting car-

diac output at the very end of pregnancy because of partial obstruction of the inferior vena cava by the uterus. Cardiac output increases further during labor itself but then drops rapidly post-partum so that pre-pregnancy values are reached by about 6 weeks after delivery. The increased cardiac output during gestation supplies the increasing demands of the uterus and the placenta for nutrients. It also assures the removal of waste products (see Section 21.5). Functional murmurs are heard more frequently in pregnancy because the circulation is more dynamic.

Maternal blood changes in volume and composition

The changes in plasma volume and red blood cell mass occurring during pregnancy are illustrated in Fig. 21.13. Plasma vol-

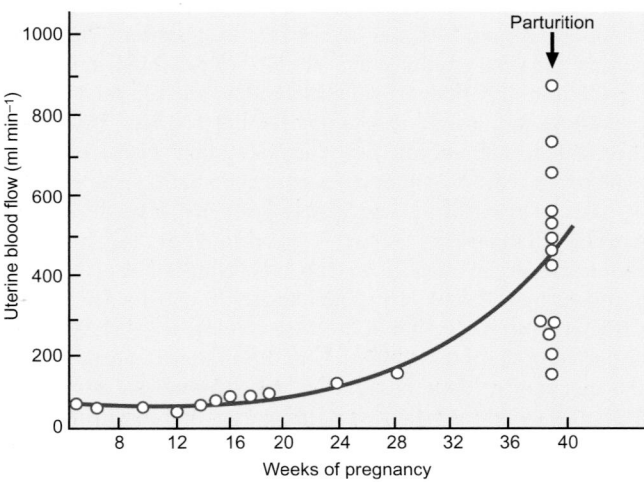

Fig. 21.12 Uterine blood flow at various stages of pregnancy.

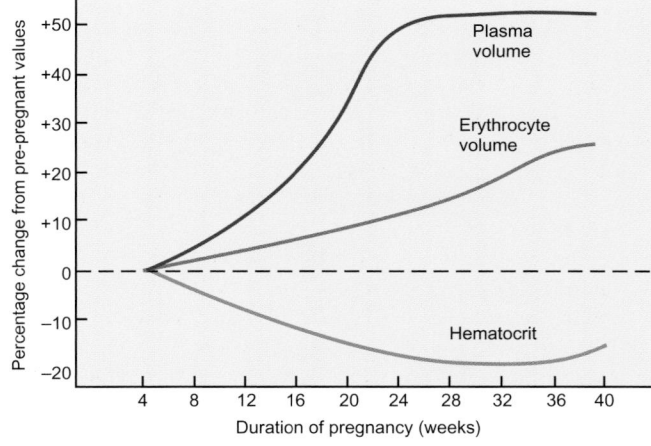

Fig. 21.13 Changes in plasma volume, erythrocyte volume, and hematocrit during pregnancy.

ume increases by about 50 per cent so that the total circulating blood volume is raised. However, red blood cell mass normally rises by only about 30 per cent and, as a result, there is often a fall in the hemoglobin content of the maternal blood during pregnancy. Published figures vary, but a typical drop in hemoglobin content might be from 13.3 to 12.1 g dl⁻¹ blood. Although some degree of dilutional anemia is normal during pregnancy, iron supplements may be required if hemoglobin levels fall significantly (see also Section 21.9). Other changes in the composition of the maternal blood include a small increase in the white blood cell count and a fall in the platelet count (although leukocyte and platelet numbers generally remain within normal limits—see Chapter 13, Table 13.2). However, despite the fall in platelets, pregnancy is sometimes regarded as a 'hypercoagulable state'

because levels of fibrinogen and Factors VII–X (see p. 239) are often increased. Consequently, there is an increased risk of deep vein thrombosis.

Maternal arterial blood pressure changes are related to circulating volume and peripheral resistance

Maternal systemic arterial blood pressure shows some interesting changes throughout gestation that are related to changes in both blood volume and peripheral vascular resistance. Despite the increased cardiac output described above, arterial blood pressure often falls from the pre-pregnant value during the second trimester because the placental circulation is expanding and peripheral vascular resistance falls. A typical fall in mid-pregnancy would be about 0.66–1.32 kPa (5–10 mmHg) for systolic pressure and 1.32–1.98 kPa (10–15 mmHg) for diastolic pressure.

BOX 21.3 PRE-ECLAMPSIA, ECLAMPSIA, AND HELLP SYNDROME—SERIOUS DISORDERS OF PREGNANCY

Hypertensive disease in pregnancy is a major cause of perinatal morbidity and mortality. It displays a wide spectrum of severity from mild pre-eclampsia with subtle non-specific symptoms through to full-blown eclampsia with life-threatening complications. A third related condition of pregnancy is the so-called HELLP syndrome in which there are damaging alterations in liver function and in certain properties of blood cells.

Pre-eclampsia (PE) occurs in around 3 per cent of pregnancies. It is characterized by high blood pressure, the appearance of protein in the urine (proteinuria), and often edema. In the majority of cases it begins after week 20 of gestation and is most common in the final trimester. Mild pre-eclampsia is defined as an arterial blood pressure (BP) of 18.6/12 kPa (140/90 mmHg) or an increase of 4 kPa (30 mmHg) in systolic BP and of 2 kPa (15 mmHg) in diastolic BP compared with the pre-pregnancy values. The pre-eclampsia is classified as severe if an arterial BP of 21.3/14.6 kPa (160/110 mmHg) is maintained even during bed rest, and if proteinuria reaches a value of 5 g/24 hours.

Although pre-eclampsia has been the subject of extensive research, the underlying pathophysiology of the disease remains poorly understood and its precise causes unknown. While both maternal cardiac output and peripheral resistance are increased in pre-eclampsia (both changes that could contribute to hypertension), the placenta is believed to have the primary role since delivery of this organ results in reversal of the signs and symptoms of pre-eclampsia. The disease also occurs more frequently with twin pregnancies and hydatidiform disease in which there is an increased amount of trophoblastic tissue. Microscopic examination of the trophoblastic tissue in patients with pre-eclampsia reveals some interesting changes. In normal pregnancies, the uterine spiral arteries are invaded by trophoblast at weeks 14–16 of gestation. As a result of this invasion, the arteries are converted into low-resistance high-volume vessels that do not respond to vasoconstrictors. These changes maximize perfusion of the maternal side of the placenta. In pre-eclampsia, trophoblastic invasion of the spiral arteries appears to be incomplete and the vessels remain responsive to vasoconstrictors. Some studies have also shown evidence of endothelial dysfunction in women with pre-eclampsia. There is a reduction in the secretion of endothelial vasodilator substances but an

increase in the concentration of the potent vasoconstrictor endothelin. These changes may further exacerbate the hypertension and hasten the progression of damage to the capillary beds of the placenta and elsewhere in the maternal circulation. The net effect of all these changes is likely to be a reduction in placental blood flow. Indeed, pre-eclampsia is frequently associated with intra-uterine growth retardation of the fetus, presumably as a result of the reduction in nutrient supply.

Although pre-eclampsia can affect any pregnancy, there is an increased risk in women at the extremes of age for pregnancy, in those who are pregnant for the first time, and in women who have had normal pregnancies but who are now pregnant with the child of a new partner. These findings suggest that genetic or immunological factors may have a role in the development of the disease.

In a small proportion of cases, pre-eclampsia progresses to the potentially fatal condition known as eclampsia in which extreme hypertension results in changes in intracranial pressure, seizures, and coma. There is a significant risk to the mother of cerebral hemorrhage, renal failure, and *abruptio placentae* (in which the placenta starts to detach from the wall of the uterus, causing massive hemorrhage). Eclampsia occurs in roughly 0.05 per cent of all deliveries, and in such cases the fetal mortality rate is between 13 and 30 per cent while the maternal mortality rate is from 8 to 36 per cent depending upon the speed with which a diagnosis is made and the facilities available for treatment and immediate delivery of the infant.

Occasionally, pre-eclampsia may progress to another illness known as the HELLP syndrome in which there is hemolysis (H), elevated liver function tests (EL) and low platelets (less than 100 × 10⁹/l⁻¹) (LP). This condition, which is characterized by nausea, malaise, and abdominal pain, affects up to 10 per cent of women with severe pre-eclampsia and 30–50 per cent of those who have progressed to eclampsia. In a few patients, however, the HELLP syndrome presents without any prior hypertension or proteinuria. Although the symptoms can be vague and relatively mild, the underlying hepatocellular necrosis and bleeding problems associated with thrombocytopenia are potentially life threatening even in those women who do not have severe hypertension.

As a result, there will be a small increase in the pulse pressure (the difference between the systolic and diastolic values). Arterial pressure may rise a little towards term, and blood pressure is usually monitored very carefully during the final trimester since any significant increase, particularly if seen in conjunction with edema or proteinuria (urinary excretion of proteins), may indicate a risk of pre-eclampsia, a potentially life-threatening condition for both mother and fetus (see Box 21.3).

The maternal respiratory system is influenced by increasing fetal size

There are normally relatively few significant changes to respiratory function during pregnancy, but those changes which do occur can be attributed to either positional changes associated with increasing uterine size or the effects of placental progesterone. A number of respiratory parameters remain largely unchanged in a pregnant woman. These include vital capacity, respiratory rate and inspiratory reserve volume.

Furthermore, the partial pressure of oxygen in the arterial blood remains unaffected by pregnancy. However, small reductions in the following respiratory volumes are commonly observed close to the end of pregnancy, probably because of the obstruction caused by the uterus as it pushes up under the diaphragm:

◆ functional residual capacity (20 per cent)

◆ expiratory reserve volume (20 per cent)

◆ residual volume (20 per cent)

◆ total lung capacity (5 per cent).

In contrast, tidal volume may increase by as much as 40 per cent by the end of pregnancy and there may be a slight increase in thoracic circumference. Mild dyspnea is commonly seen during exertion, particularly in late pregnancy, and may be associated with a very slight respiratory alkalosis (plasma pH rising to about 7.44).

Progesterone causes hyperemia and edema of the tissues of the respiratory tract which, in some women, leads to nasal stuffiness, obstruction of the Eustachian tube, and changes in the quality of the voice.

Maternal renal function and body fluid balance reflect cardiovascular changes

Many of the major changes in renal function and body fluid homeostasis occurring during pregnancy can be explained in terms of the cardiovascular changes discussed earlier. As cardiac output increases, there is a rise in renal blood flow and, as a result of this, an increase in glomerular filtration rate (GFR) of between 30 and 50 per cent. Peak GFR values are normally recorded at around weeks 16–24 of gestation. As renal function increases, the concentrations of both urea and creatinine in the maternal plasma fall from their pre-pregnancy values of 4–7 mmol l^{-1} and 100 μmol l^{-1} to around 3.6 mmol l^{-1} and 60 μmol l^{-1}, respectively. Near to term, the expanding uterus starts to exert pressure on the inferior vena cava and renal blood flow may fall slightly. This effect is known as positional stasis. It may lead to a slight drop in GFR in late gestation. It can be seen from Fig. 21.11(b) that the abdominal organs are displaced and

compressed by the fetus as it enlarges, and pressure on the ureters may cause urine to back up rather than pass freely to the bladder. This effect, coupled with the direct smooth muscle relaxing effects of progesterone, may cause dilatation of the ureters.

Many pregnant women experience an increase in the frequency of micturition. There are several reasons for this. In the first and second trimesters of pregnancy, the increased urinary flow rate probably reflects increased renal function. In later weeks, as the fetus increases in both size and activity, there is likely to be increased pressure on the bladder causing urgency and discomfort.

Many of the hormones that act on the renal tubules alter their output during pregnancy

In response to increased estrogenic stimulation in pregnancy, there is an increase in the secretion of renin from granular cells of the afferent arteriole (see Chapter 17, Fig. 17.3). This in turn elicits a rise in plasma angiotensin levels (Chapter 17, Fig. 17.19). This might be expected to produce an increase in arterial blood pressure, but there is also a reduction in the sensitivity of blood vessels to these hormones which seems to cancel out any hypertensive effect. However, the increased concentration of angiotensin will enhance the secretion of aldosterone from the adrenal cortex. By the end of pregnancy, aldosterone secretion may be six to eight times higher than in the non-pregnant state. Aldosterone stimulates the reabsorption of salt and water from the distal tubular fluid and more than balances the increase in filtered water and salt resulting from the increased GFR.

Progesterone, which is secreted by the placenta in very large amounts, is both natriuretic and potassium sparing in its effect on the renal tubules. This is in direct contrast to the effects of aldosterone, which favor potassium excretion and sodium retention. Consequently, the two steroid hormones largely offset one another's actions during pregnancy but, if anything, there is a small degree of both potassium and sodium retention. This may contribute to the expansion of total body fluid volume that occurs during gestation.

Total body water increases by 6–8 liters during pregnancy

The increased volume of water in the body of a pregnant woman contributes significantly to the overall weight gain normally seen (see Section 21.9). Extracellular fluid volume is increased by about 3 liters at term, about half of which is in the plasma and the rest is in the interstitial fluid. In some women, there is significant edema in pregnancy because of this expansion of the extracellular fluid compartment. Provided that it is not associated with hypertension or proteinuria, a small degree of edema is considered a normal consequence of pregnancy.

In addition to the expansion of the extracellular compartment, plasma osmolality may fall during gestation by as much as 10 mOsm kg^{-1}. This fall may be accounted for, at least in part, by the drop in urea and creatinine levels described earlier. In the non-pregnant state, such a fall would normally be countered by a reduction in the secretion of ADH so that increased amounts of dilute urine would be excreted. This does not appear to occur during pregnancy, possibly because the hypothalamic osmoreceptors are reset at a lower value throughout gestation.

Many pregnant women experience an increased thirst. This may be the result of increased circulating levels of angiotensin, a hormone known to stimulate thirst (see p. 550). Increased fluid intake may contribute to the increased extracellular volume of pregnancy.

Minor changes to gastrointestinal function are common features of pregnancy

Nausea and vomiting are experienced by the majority of women, particularly in the first 12–14 weeks of pregnancy. After this time, symptoms usually become less severe and often disappear completely, but in rare cases severe problems may persist throughout gestation (a condition known as *hyperemesis gravidarum*). The precise cause of sickness in early pregnancy is not clear, but it seems to parallel the rising level of hCG secretion by the syncytiotrophoblast in the first 10–12 weeks after fertilization (see Section 21.6).

Constipation is another common feature of pregnancy and, while it may occur at any stage, is most frequently experienced during late pregnancy as the enlarging uterus presses against the rectum and lower colon. Certain other factors may also contribute to constipation during gestation. These include relaxation of the smooth muscle of the colon under the influence of placental progesterone, and a reduction in the water content of the stool brought about by increased water absorption from the colon in response to increased concentrations of angiotensin II and aldosterone (see above).

Many aspects of GI function are reduced or slowed during pregnancy, principally as a result of relaxation of the smooth muscle of the gut wall in response to progesterone. For example, gastric emptying occurs more slowly and there is an overall decrease in GI motility. Relaxation of the lower esophageal sphincter, particularly at night in late pregnancy, may result in acidic gastric contents being regurgitated into the esophagus, causing belching and heartburn. Furthermore, the pressure of the fetus on the diaphragm and intra-abdominal organs, particularly when the mother is lying down, add to the discomfort and the likelihood of reflux.

The secretion of gastric acid decreases and peptic ulcers may show an improvement during pregnancy. Gallstones, however, are more common because of the increasingly sluggish flow of bile caused by smooth muscle relaxation. Increased levels of plasma progesterone may lead to swelling and edema of the gums, which tend to become spongy and bleed more easily.

The functions of most endocrine glands are altered during pregnancy

The specific actions of placental progesterone, estrogens, hCG and hPL are described in Section 21.6. However, pregnancy also alters the functions of most of the other endocrine tissues throughout the body. These changes occur partly through the actions of the placental hormones themselves, and partly because many hormones (particularly steroids) circulate in the blood in combination with plasma proteins whose levels are often altered in pregnancy.

The placenta secretes a variety of hormonal factors whose actions are similar to those of the anterior pituitary trophic hormones. One of these, which is similar to anterior pituitary

thyroid-stimulating hormone (TSH) (see p. 206) increases thyroid function and stimulates the secretion of thyroxine. Occasionally this may lead to symptoms reminiscent of hyperthyroidism—tachycardia, palpitations, excessive sweating and anxiety—although in most women levels of thyroxine-binding globulins in the plasma also increase so that free plasma thyroxine levels remain unchanged.

The placenta also secretes a substance similar to ACTH which stimulates the output of adrenal cortical hormones, and a melanocyte-stimulating hormone (MSH), which causes increased pigmentation of the skin. The latter probably accounts for the relatively common so-called mask of pregnancy (melasma), in which a blotchy brownish pigment appears over the forehead and cheeks. Pigmentation of the areolae of the nipples also increases, and there is sometimes a line of darkness down the midline of the lower abdomen. Furthermore, MRI scans show that the maternal pituitary gland itself increases in size during pregnancy. The enlarged gland secretes additional amounts of both ACTH and MSH whose effects are added to those of the placental factors. In rare instances, enlargement of the pituitary during pregnancy may be sufficient to cause some visual disturbances as a result of increased pressure on the optic chiasm which lies very close to it.

Secretion by the anterior pituitary of both growth hormone (GH) and the gonadotrophins FSH and LH is reduced during pregnancy. This reduction is mediated by the negative feedback effects of hPL (a powerfully somatotropic hormone) on GH and of the placental sex steroids on FSH and LH. In contrast, prolactin secretion is enhanced, showing an eightfold increase by the end of gestation. This hormone is partly responsible for the preparatory changes that take place within the mammary glands in readiness for milk synthesis and secretion. Its presence also seems to be required for the expression within the glands of the effects of estrogens and progesterone (see Section 21.10).

Relaxin, a large polypeptide hormone, is secreted by the corpus luteum and the decidual tissue of the uterus during pregnancy. It appears to ripen the cervix and to cause softening and relaxation of the pelvic ligaments and the symphysis pubis of the pelvic bones to accommodate the expanding uterus.

Parathyroid hormone secretion increases during pregnancy

The fetus represents a significant drain on maternal stores of calcium, largely because of the demands of its developing skeleton. As a result of this, maternal plasma calcium is at risk of being reduced. To prevent any such fall, there is normally an increase in the rate of secretion of parathyroid hormone during gestation. This mild degree of hyperparathyroidism augments plasma levels of 1,25-dihydroxycholecalciferol (calcitriol—see pp. 218–220), which in turn stimulates the intestinal absorption of calcium. These changes are especially important during the final weeks of pregnancy.

Gestational diabetes is a relatively common condition of late pregnancy

In early pregnancy, tissues typically show an increased sensitivity to insulin. Consequently plasma glucose may fall slightly. Later in gestation, insulin sensitivity falls and plasma glucose may rise. These changes reflect the initial increase in the requirement of the maternal tissues for glucose and the later needs of the developing fetal tissues. Glycosuria is often seen in pregnancy and may be explained by the combination of raised plasma glucose, increased GFR and a reduction in the tubular reabsorption of glucose.

Gestational diabetes occurs in between 1 and 3 per cent of pregnancies. In this condition, carbohydrate intolerance of varying severity develops during the pregnancy and often (though not invariably) resolves once the baby has been delivered. The women who develop gestational diabetes are of several kinds. While some are obese, hyperinsulinemic, and insulin resistant, others are relatively thin but insulin deficient. In either case,

they are unable to respond effectively to the metabolic stress of pregnancy. Furthermore, hPL may increase the risk of developing gestational diabetes in susceptible women. The fetus of a diabetic mother with hyperglycemia will also be hyperglycemic since glucose is in equilibrium between the fetal and maternal sides of the placental circulation. Not surprisingly, in many cases the fetus has a higher than normal weight for date (which can pose difficulties during labor and delivery). Also associated with maternal gestational diabetes is an increased risk of fetal respiratory distress due to retarded lung maturation and surfactant production. The incidence of fetal abnormalities is also increased. Early recognition of gestational diabetes and good maternal plasma glucose control are vital if such problems are to be minimized. Glucose tolerance tests are routine in many antenatal clinics.

Summary

1. Throughout pregnancy, changes take place in all the major systems of the mother's body. Many of these are adapted to create favorable conditions for fetal development and to prepare the body for delivery and subsequent nourishment of the baby. Others occur as the inevitable consequence of the progressive changes in spatial relationships of the maternal abdominal organs produced by the growing fetus.

2. Many aspects of the maternal cardiovascular system alter throughout the course of gestation. Both heart rate and stroke volume are increased. This results in a rise in cardiac output that satisfies the increasing demands of the uterus and placenta for nutrients and waste disposal. Arterial blood pressure often falls in mid-pregnancy as peripheral resistance decreases, but may increase slightly towards term. Total maternal plasma volume increases and, although red cell mass also rises, there is often a dilutional anemia with a typical fall in hemoglobin content of around 8–10 per cent.

3. Changes in maternal respiratory function are largely connected with the positional changes associated with increasing uterine size and compression of the diaphragm.

4. Changes in maternal renal function may be explained by the alterations in cardiovascular function. Glomerular filtration is increased by up to 50 per cent and there are small changes in plasma urea and creatinine concentrations. Output of both renin and aldosterone is stimulated during gestation. Total body water increases by 6–8 liters during pregnancy and plasma osmolality falls by about 10 mOsm kg^{-1}.

5. Minor changes in GI function such as nausea, vomiting, and constipation are often seen in pregnancy. Relaxation of GI smooth muscle in response to progesterone results in a reduction of both motility and secretory activity.

6. Pregnancy alters the function of most endocrine glands. The placenta secretes a number of hormones, which have a similar action to those of the anterior pituitary. Changes in the insulin sensitivity of maternal tissues take place during gestation, with an increase in sensitivity in the early weeks and a fall later on. Gestational diabetes occurs in 1–3 per cent of pregnancies.

21.9 Nutritional requirements of pregnancy

Adequate nutrition both prior to conception and during pregnancy itself is essential in order to ensure that optimal intra-uterine conditions are maintained for fetal development, while at the same time ensuring that the health of the mother is not compromised by the additional metabolic and nutritional demands of the growing fetus. Poor maternal nutrition is associated with an increased risk of low birth weight, fetal abnormalities, and neonatal mortality.

Weight gain during pregnancy is normally between 7 and 14 kg, averaging around 0.4 kg per week during the second and third trimesters. Twin pregnancies may result in a total weight gain of between 16 and 20 kg. Although women with a high body mass index (BMI > 25—see Chapter 19, p. 426) are generally advised to gain less weight than those who are underweight (BMI < 20) before becoming pregnant, even obese women will need to gain a minimum amount of weight (about 6 kg) in order to optimize the chances of delivering an infant of healthy birth weight. However, excessive maternal weight gain is undesirable. It is associated with prolonged gestation (late delivery), problems during labor, and an increased risk that delivery will need to be performed by Cesarean section. Table 21.1 shows figures for the distribution of weight gained during a typical pregnancy in which maternal weight increases by around 12 kg.

Of the total weight gain, the fetus contributes around 3.5 kg, the placenta and amniotic fluid around 1.5 kg, and the breasts 0.5–1 kg. Increases in maternal fat and extracellular fluid account for the remaining weight gained.

TABLE 21.1 The distribution of maternal weight gain at 40 weeks gestation

	Weight (kg)
Fetus	3.3–3.5
Placenta	0.65
Additional blood volume	1.3
Amniotic fluid	0.8
Weight gain of uterus and breasts	1.3
Additional fluid retention and fat deposits	4.2–6.0
Total	11.5–13.5

The daily calorific requirement increases by 15 per cent in pregnancy

In order to supply the demands of the fetus and to maintain adequate maternal nutrition, an increased daily energy intake of 1050–1250 kJ (250–300 kcal) is required. This represents an increase of about 15 per cent in energy requirements. In very late pregnancy this need may be somewhat reduced because of a reduction in maternal energy expenditure. Many women experience an increase in appetite during pregnancy. Contributory factors could include the presence of high circulating concentrations of progesterone and the fall in plasma glucose concentration, which is characteristic of early pregnancy (see above). There may also be odd cravings, particularly for highly flavored foods, perhaps related to dulled taste sensation, as well as aversions to certain foods and flavors. However, true pica (craving for inappropriate foods such as coal) is rare in pregnancy.

During early pregnancy the mother's body prepares for the metabolic demands to come

The metabolism of the mother alters in a variety of ways during early pregnancy to ensure that reserves are available to meet the requirements of both mother and fetus during later gestation. During the first half of a typical pregnancy about 3 kg of maternal fat is laid down and will provide a store of energy for the final trimester during which fetal growth is particularly rapid.

In addition, maternal tissues become more sensitive to insulin and, because of these changes, carbohydrate loads are readily assimilated. As discussed earlier, a small decrease in plasma glucose is often seen after about week 6 of gestation. At the same time, protein synthesis increases, the net effect of which is to stimulate growth of the uterus, breasts, and essential musculature of the mother.

During later pregnancy, the metabolism of the mother adapts to accommodate the growing fetal demands

As the fetal requirements for oxygen and nutrients rise and the sheer physical size of the fetus increases during the second half of gestation, the metabolism of the mother enters a state similar to that occurring during starvation. Glucose is the chief metabolic fuel for the developing fetus which, at full term, uses up to 25 g of glucose each day. In late pregnancy, maternal tissues show a fall in insulin sensitivity. This ensures that the maternal plasma glucose concentration is adequate to maintain a constant supply of glucose both for transfer across the placenta and for the needs of the mother's central nervous system. At the same time, mobilization of lipids is facilitated, possibly by hPL, to ensure an alternative source of energy for the mother. Increased plasma levels of triglycerides are seen as the hepatic synthesis of very-low-density lipoprotein increases under the influence of placental estrogens. Some of these triglycerides are stored by the mammary tissue in preparation for milk synthesis during lactation.

During late gestation, the mother's body requires a small quantity of additional protein (no more than 6–10 g per day) mainly to provide the substrate for the synthesis of fetal protein. By the time of its birth, the fetus will normally have accumulated 400–500 g of protein. In most cases, a normal diet is sufficient to supply the additional protein requirements of pregnancy.

The maternal requirement for some micronutrients increases during pregnancy

The need for certain vitamins and minerals increases by a small amount during pregnancy, both to satisfy the demands of the fetus and to prepare the mother's body for lactation. A normal mixed diet will usually provide adequate concentrations of these micronutrients, but in some women (for example those following a strict vegetarian or vegan diet) nutrient supplements may be necessary. An example of such a nutrient is vitamin B_{12}, which is found only in animal products. Table 21.2 shows the recommended daily nutrient intakes during pregnancy. It should be noted that recommendations often vary widely between different countries.

Folic acid is now recognized as a very important dietary requirement of pregnancy and is particularly important in the first weeks after conception. It is needed for normal cell proliferation, and folate deficiency has been linked to an increased risk of neural tube defects. These are abnormalities in the closure of the neural tube, which normally occurs between weeks 3 and 6 of gestation. Neural tube defects include anencephaly (in which the cerebral hemispheres fail to develop normally) and spina bifida in which there is defective fusion of the vertebral arches, most often in the lumbar region. Women are encouraged to supplement their folic acid intake both in the weeks prior to conception and throughout pregnancy by eating leafy green vegetables, wheat grains, and legumes.

A vitamin A intake of around 750 µg per day is recommended for adult females. In contrast to most vitamins, the requirement for vitamin A remains unchanged in pregnancy. Indeed, there is some concern that a high daily intake of this vitamin may be

TABLE 21.2 Recommended daily protein and micronutrient intake in pregnancy

Nutrient	USA	Canada	UK	Australia
Protein (g)	60	75	51	51
Vitamin A (µg)	800	800	700	750
Vitamin D (µg)	10	5	10	Not set
Vitamin E (mg)	10	8	Not set	7
Vitamin C (mg)	70	40	50	60
Thiamin (mg)	1.5	0.9	0.9	1.0
Riboflavin (mg)	1.6	1.3	1.4	1.5
Niacin (mg) (nicotinic acid)	17	16	13	15
Folate (µg)	400	385	300	400
Vitamin B_{12} (µg)	2.2	1.2	1.5	3.0
Calcium (g)	1.2	1.2	0.7	1.1
Magnesium (g)	0.32	0.25	0.27	0.30
Iron (mg)	30	13	14.8	22–36
Zinc (mg)	15	15	7	16
Iodine (µg)	175	185	140	150

teratogenic, leading to abnormalities of the fetal nervous system and heart. This effect may be linked with the role of vitamin A in cell differentiation. However, a normal intake of vitamin A during pregnancy has been shown to be important in reducing the likelihood of transmission of the human immunodeficiency virus (HIV) from mother to fetus.

Calcium, iron, and zinc are the most important minerals of pregnancy

During the third trimester of gestation, the fetus needs 0.3 g of calcium a day to allow calcification of the skeleton. The infant is normally born with about 28 g of calcium, which must come from the maternal reserves. During pregnancy, maternal calcium absorption by the small intestine increases under the influence of raised levels of calcitriol, while urinary calcium loss is reduced by the increased secretion of parathyroid hormone (see pp. 218–221). Nevertheless, to ensure that the increased calcium requirement is met, a pregnant woman will need to increase her daily intake of calcium by about 70 per cent, i.e. to about 1200 mg per day. If dietary calcium is not increased during pregnancy, the fetus will draw on calcium stored in the maternal skeleton. This may increase the mother's susceptibility to osteoporosis in later life, particularly if repeated in several pregnancies.

There is much debate surrounding the requirement for iron during pregnancy. Some studies have indicated that iron deficiency may be associated with low birth weight and premature delivery. Others indicate that the fetal demand for iron can be met by the maternal stores provided that the diet contains sufficient amounts of meat and vitamin C (which enhances the absorption of iron in the gut) and that the mother was not iron deficient prior to her pregnancy. The fetus and placenta use about 300 mg of iron during a typical pregnancy, while the increased red blood cell mass of the mother requires a further 500 mg. The average Western diet contains 10–15 mg of iron per day, of which about 10 per cent is absorbed in the non-pregnant state. The efficiency of absorption appears to rise during pregnancy, reaching around 66 per cent by week 36 of gestation. This increase contributes to the ability of the maternal iron stores to supply the needs of the fetus. However, if absorption of dietary iron falls short of the demands of pregnancy, maternal iron stores will become depleted and iron deficiency anemia may occur. This can be avoided by regular monitoring of the maternal hemoglobin content, so that iron supplements can be given if necessary.

The normal requirement for dietary zinc in a non-pregnant woman is around 7 mg per day. Zinc is a constituent of a large number of enzymes including carbonic anhydrase, reverse transcriptase and DNA/RNA polymerase. Therefore it plays an important part in a number of metabolic processes including the synthesis of proteins and nucleic acids. Zinc is also essential for the synthesis and activity of insulin. Not surprisingly, therefore, there is a small increase in the requirement for this trace element during pregnancy. In most women, a normal mixed diet containing meat and fish will provide sufficient zinc to support the feto-placental unit, but strict vegans may be unable to absorb enough zinc from their food. Women taking iron supplements may also become deficient in zinc because iron-to-zinc

Summary

1. Weight gain during pregnancy is typically 7–14 kg and is associated with an increase in daily calorific requirement of 1050–1260 kJ (250–300 kcal) per day. Adequate maternal nutrition during pregnancy is required to ensure the continued health of the mother, to provide for the demands of lactation, and to minimize the risks of fetal abnormalities and low birth weight.

2. In early pregnancy, lipogenesis is favored and the insulin sensitivity of tissues is increased. Maternal adipose tissue is deposited. Protein synthesis also increases so that growth of breast and uterine tissues occurs.

3. In late gestation, the maternal metabolism mimics that seen during starvation. Insulin sensitivity falls to maintain an adequate concentration of plasma glucose for placental transfer and lipids are mobilized as an alternative energy source for the mother.

4. The requirement for certain vitamins is enhanced, while there is an increased utilization of calcium, iron, and zinc. In most cases, a normal mixed diet will be sufficient to meet the demand for extra micronutrients, although occasionally supplements, particularly vitamin B_{12} and iron, are needed.

ratios greater than 3:1 have been shown to interfere with the intestinal absorption of zinc.

21.10 Lactation—the synthesis and secretion of milk after delivery

While the fetus develops within its mother's uterus, it receives all the nutrients it requires via the placenta. However, once it has been delivered, the baby needs a regular and plentiful supply of milk. Although in certain human communities, particularly in Western society, bottle-feeding with powdered 'formula milk' offers a suitable alternative, in many parts of the world the mother's breast is the only source of nourishment for the newborn infant. Here, as in all other mammalian species, lactation—the synthesis and secretion of milk following delivery—is as vital to the process of reproduction as gamete fertilization and fetal development. The formation and synthesis of milk is also called *galactopoiesis*.

Most placental mammals are born at a relatively advanced stage of development. This is the result of the relatively long gestation period, which is made possible by the direct link between the maternal and fetal circulations established in the first weeks of pregnancy. Consequently, the infant makes considerable nutritional demands upon its mother. The human infant is no exception, and to ensure that sufficient milk of adequate calorific value is produced from the very start of lactation, preparatory changes must occur within the mammary glands during pregnancy. These changes are regulated by hormones from the placenta and the pituitary and adrenal glands. They will be discussed in later sections. First, however, it will be helpful to consider the growth and development of the breasts prior to pregnancy.

The non-pregnant mammary gland is incapable of lactation

Until puberty, the immature breast consists almost entirely of ducts known as *lactiferous ducts*. Around puberty, gonadotrophins begin to be secreted in larger amounts from the anterior pituitary and, under their influence (see Chapter 20), the ovaries start to increase their production of estrogenic hormones. These steroids initiate further breast development—in particular, the ducts begin to sprout and to become more highly branched. Once menstruation has commenced, progesterone, secreted during the luteal phase of each cycle, stimulates the formation of small spherical masses of granular cells at the end of each duct. These are known as immature *alveoli* and are the cells which will later develop into the milk-secreting alveoli of the lactating gland in the event of a successful pregnancy. Once the mammary glands have been exposed successively to estrogens and progesterone during the follicular and luteal phases of many cycles, the non-pregnant breast is said to be fully developed.

Throughout this adolescent period, there is deposition of fat and connective tissue, which, together with the ductal growth, brings about a considerable increase in breast size and results in

a gland that is highly developed even though no pregnancy has yet occurred. This is in marked contrast to most other mammals, including non-human primates, in which very little mammary growth is seen at all until mid or late pregnancy. Figures 21.14 and 21.15 show the basic organization of the human mammary gland. Figure 21.15 also illustrates the way in which the ducts separate the gland into lobes. In each gland there are between 15 and 20 lobes separated by fat. Each lobe consists of clusters of granular cells at the ends of the lactiferous ducts. The ducts dilate near the areola (the area of brownish pigment surrounding the nipple) to form lactiferous sinuses, each of which runs up into the nipple and opens onto its surface. Dotted around the areola are small sebaceous glands called Montgomery glands.

Although the mammary gland is fully developed by the end of puberty, small changes do take place during each menstrual cycle as first estrogens and then progesterone influence the mammary tissue. Occasionally there is some degree of secretory activity during the luteal phase of the cycle, and there is often an increase in both size and weight of the gland throughout the premenstrual period caused by the retention of fluid. However, despite its comparatively advanced stage of development, the

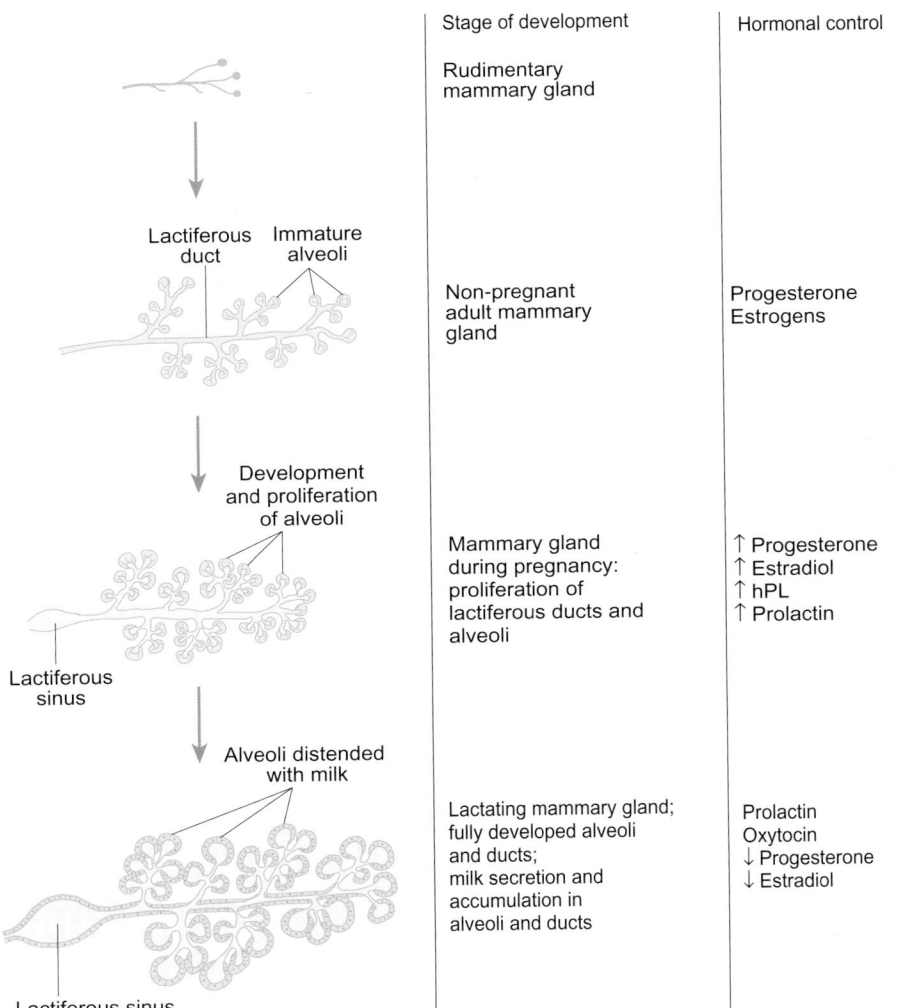

Stage of development	Hormonal control
Rudimentary mammary gland	
Non-pregnant adult mammary gland	Progesterone Estrogens
Mammary gland during pregnancy: proliferation of lactiferous ducts and alveoli	↑ Progesterone ↑ Estradiol ↑ hPL ↑ Prolactin
Lactating mammary gland; fully developed alveoli and ducts; milk secretion and accumulation in alveoli and ducts	Prolactin Oxytocin ↓ Progesterone ↓ Estradiol

Labels on figure: Lactiferous duct; Immature alveoli; Development and proliferation of alveoli; Lactiferous sinus; Alveoli distended with milk; Lactiferous sinus filled with milk

Fig. 21.14 The development of the mammary gland. During pregnancy the gland matures under the influence of a number of hormones. Lactation does not commence until the circulating levels of progesterone and estrogens decline.

breast is not capable of large-scale milk production (lactogenesis) at this stage. Before that can happen, further development, particularly of the alveoli, must take place. This occurs during pregnancy under the influence of a variety of hormones.

The development of the mammary gland during pregnancy

Most of the growth and structural changes that are essential for successful lactation take place during the first 4 months or so of pregnancy. By mid-term, the mammary gland is fully developed for milk secretion. The lobular ductal–alveolar system that was laid down during adolescence undergoes hypertrophy. The ducts proliferate further and the alveoli mature—the balls of granular cells become hollowed out so that alveolar cells surround a central lumen, which is drained by a branch of one of the lactiferous ducts (Fig. 21.14). The hormones thought to be responsible for these changes are the placental steroids estradiol and progesterone and the placental peptide hormone hPL (see Section 21.6). Progesterone in particular seems to be required for the alveolar changes characteristic of early pregnancy. In addition to the placental hormones, pituitary growth hormone and prolactin may also be mammotropic, although their contribution at this stage is not clear. Additional adipose tissue is also deposited between the lobules of the gland during early pregnancy, adding further to the size and weight of the breast. Figure 21.15 shows the appearance of the breast tissue during pregnancy.

The alveoli are the primary sites of milk production

The mature alveoli develop under the influence of placental progesterone, prolactin, and hPL. Figure 21.16 is a highly simplified diagram showing the basic organization of a mature alveolus. The alveolar wall is formed by a single layer of epithelial cells whose shape can vary from low cuboidal to tall columnar

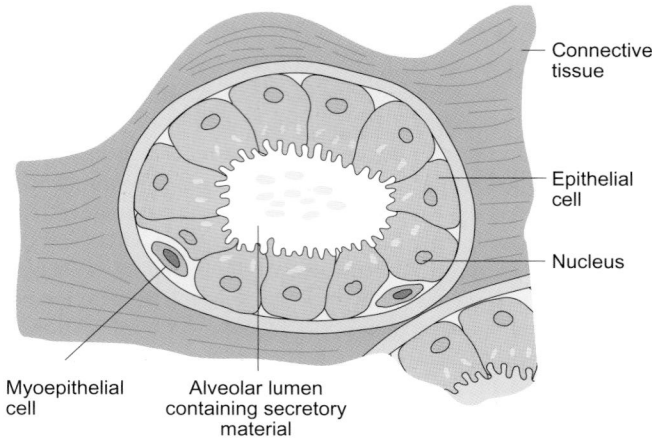

Fig. 21.16 Cross-section of a mature (lactiferous) alveolus.

depending on the amount of secretory material filling the central lumen. During pregnancy, but before the onset of full-scale lactation, the epithelial cells are columnar in appearance, while after delivery, once milk production is under way, the cells are usually squashed flat by material within the lumen. These epithelial cells are the cells that synthesize and secrete the constituents of milk, and they show all the classical characteristics of secretory cells. They possess microvilli on their luminal surfaces and their cytoplasm is rich in mitochondria, Golgi membranes, rough endoplasmic reticulum, secretory granules, and lipid droplets. Adjacent alveolar cells are connected by junctional complexes near to the luminal surfaces, and between the basement membrane and the secretory alveolar cells are specialized cells called myoepithelial cells. As their name suggests, these cells are contractile, and they are important for moving milk into the lactiferous ducts prior to ejection from the nipple when the baby suckles.

Lactation is triggered by the fall in steroid secretion that follows delivery

Although the breast is fully developed for milk production by the middle of pregnancy, no significant lactogenesis takes place

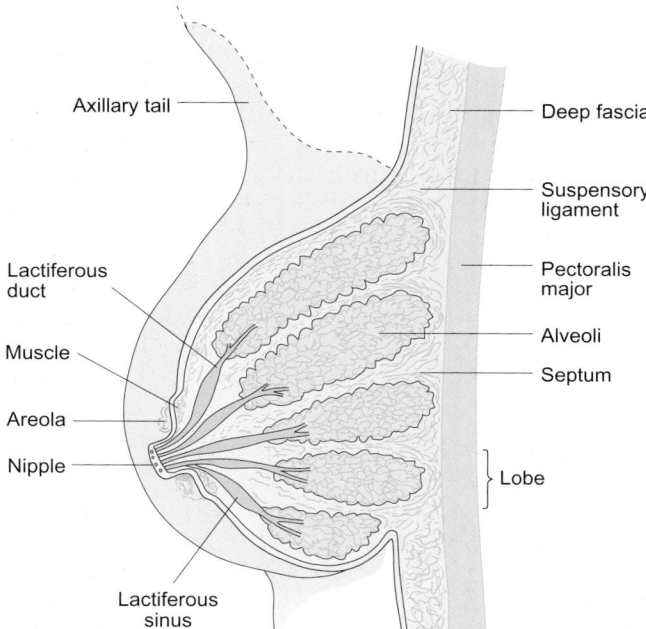

Fig. 21.15 Sectional view of the mammary gland during pregnancy. Note the development of the alveoli.

Summary

1. Lactation is the synthesis and secretion of milk by the mammary glands.

2. The prepubertal mammary gland is composed largely of lactiferous ducts. The ovarian steroid hormones secreted from puberty are responsible for subsequent development of the gland. In particular, progesterone stimulates development of alveoli at the ends of the ducts. These are spherical masses of cells that will produce milk.

3. Under the influence of placental hormones (estrogens, progesterone, and hPL), the alveoli mature during pregnancy and the breast acquires the potential for milk secretion.

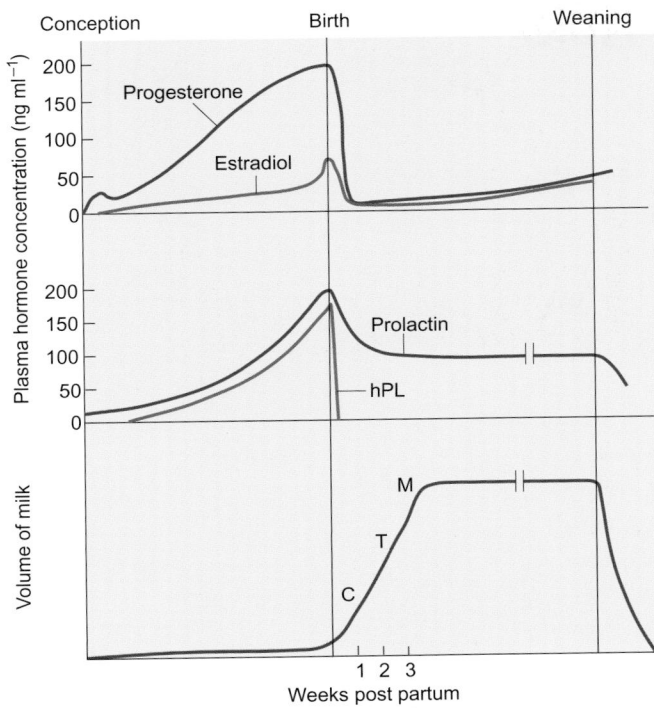

Fig. 21.17 Changes in the patterns of secretion of hormones before and after birth in relation to their role in the control of lactation. When milk secretion is first initiated, colostrum (C) is secreted which gradually undergoes a transition (T) until mature milk is produced (M). Milk is not secreted until the level of steroid hormones falls following delivery, despite the high level of prolactin that prevails in the latter stages of pregnancy.

TABLE 21.3 The composition of human breast milk			
	Colostrum	**Transitional milk**	**Mature milk**
Total fats (g l⁻¹)	30	35	45
Total protein (g l⁻¹)	23	16	11
Lactose (g l⁻¹)	57	64	71
Total solids (g l⁻¹)	128	133	130
Calorific value (MJ l⁻¹)	2.81	3.08	3.13

Note that these values are approximate as the composition changes both during a single feed and during the course of the day. In general, the fat content rises from the beginning to the end of a feed.

a sticky yellowish fluid which, while relatively low in fats, lactose, and some B vitamins, is rich in protein, minerals, and vitamins A, D, E, and K. It also contains significant quantities of immunoglobulins (IgAs) that provide the newborn infant with some resistance to infection. During the second and third weeks after birth, the composition of the fluid secreted gradually changes (Table 21.3). Although the proportion of immunoglobulins and other proteins decreases, the milk becomes much richer in fats and sugars. Its calorific value increases as a result. At this time the fluid is known as 'transitional milk', but by the time the baby is 3 weeks old the milk has attained its 'mature' composition—it is high in fats, sugars, and essential amino acids; it is iso-osmotic with plasma, and it has a calorific value of about 3.1 MJ l⁻¹ (75 kcal per 100 ml).

Breast milk synthesis is hormonally controlled

Most of the fat in human milk consists of medium-chain (10–12 carbons) fatty acids. These are thought to be synthesized *de novo* within the alveoli of the mammary tissue under the control of the enzymes fatty-acid synthetase and medium-chain acylthioester hydrolase. Prolactin and insulin are thought to regulate fatty-acid synthesis by the alveoli, whose epithelial cell membranes are especially rich in prolactin receptors. Prolactin is also believed to stimulate the secretion of lipid from the cells into the central lumen of the alveolus. The mechanism by which this secretion takes place is rather interesting. The lipid is manufactured in the endoplasmic reticulum of the alveolar cell and leaves it in the form of lipid droplets that migrate towards the luminal surface of the cell, increasing in size as they do so. Once a droplet reaches the luminal surface, it pushes out against the cell surface membrane causing a bulge. The area behind the lipid droplet gradually thins and eventually the membrane 'pinches off' so that the membrane-bound droplet is released into the lumen.

The major milk proteins are casein, α-lactalbumin, and lactoglobulin

The three major milk proteins have both nutritional and immunological significance, while α-lactalbumin has an additional and specific role in the synthesis of milk sugar. The basic processes of milk protein secretion and synthesis are similar to those occurring in other protein-secreting tissues such as the pancreas and liver. Amino acids, the precursors of protein synthesis, are supplied to the mammary tissue by the maternal

until the infant has been delivered. The endocrine changes that follow delivery are necessary to activate the prepared gland and trigger the synthesis and secretion of milk.

It is well established that the primary lactogenic hormone is prolactin, secreted by the anterior pituitary, and that high levels of prolactin must be maintained for sustained lactation. It is also known that this hormone is secreted in large amounts throughout gestation, so why does lactation not occur during this time? The most widely accepted explanation is that the high circulating levels of the placental steroids (estrogens and progesterone) exert a direct inhibitory effect on the secretory activity of mammary tissue. Therefore the gland remains unresponsive to high levels of prolactin until the baby has been born. Figure 21.17 shows that, while estrogen and progesterone levels drop dramatically following delivery, prolactin levels remain high and are then able to initiate milk production by the fully prepared breast.

The composition of human milk changes gradually during the first weeks after delivery

The composition of the milk produced by the mammary gland varies with the time that has elapsed since parturition. So-called 'mature milk' is not secreted until about 2 or 3 weeks postpartum. Prior to this time, fluids of varying composition are produced. For the first week or so after delivery, a fluid called *colostrum* is secreted at a rate of around 40 ml a day. Colostrum is

circulation and pass from the blood into the alveolar cells via specific carrier systems. The milk proteins are synthesized in the usual way (Chapter 3) by the endoplasmic reticulum and Golgi membranes, and are then packaged into vesicles which bud off from the Golgi apparatus into the cytoplasm of the alveolar cell. These vesicles or granules move to the luminal surface, possibly by the action of microtubules, and release their contents into the alveolar lumen by the process of exocytosis. The vesicle membrane fuses with the plasma membrane, allowing release of the vesicular contents without the cell cytoplasm being exposed to the extracellular fluid. This release of protein vesicles, like the budding off of the lipid droplets (see above), is controlled by prolactin. It is important to realize that, while the release of fat results in a loss of cell membrane as the droplet pinches off, the exocytotic release of protein granules adds to the cell membrane by fusion of the granules with the plasma membrane.

Lactose is the most abundant milk sugar

Lactose is synthesized within the Golgi apparatus of the alveolar epithelial cells. Its synthesis is dependent on the prior production of α-lactalbumin, which is made in the endoplasmic reticulum and passed to the Golgi. Once there, it combines with galactosyltransferase, an enzyme present within the Golgi membranes, and this enzyme system metabolizes blood glucose to form lactose which is packaged together with the proteins, in granules that bud off from the Golgi and undergo exocytotic release into the alveolar lumen, as described earlier. The enzyme system comprising α-lactalbumin and galactosyltransferase is stimulated by prolactin but inhibited by the high levels of progesterone circulating throughout gestation.

Human breast milk contains more than 50 different oligosaccharides, most of which are synthesized from lactose. In addition to providing the source of many of the other milk sugars, lactose also promotes the growth of intestinal flora which is very important to the newborn infant. Furthermore, galactose, one of the digestion products of lactose, is an essential component of the myelin which surrounds many nerve fibers (see Chapter 6).

Nutritional requirements of lactating women

During lactation, a woman needs to take sufficient nutrients to provide for her own bodily requirements together with those needed by the infant for its growth and development. Obviously, the extra requirement will depend on the amount of milk she is producing. As a guide, a baby weighing 5–6 kg will drink about 750 ml of breast milk each day. This volume of mature milk has an energy equivalent of about 2.6 MJ (630 kcal). Although a proportion of the additional energy requirement will come from mobilized maternal fat stores, it is usually recommended that a lactating mother's energy intake should increase by around 2 MJ (400–500 kcal) per day. Table 21.4 illustrates some other important nutritional requirements of lactation. Of particular significance is the need for an adequate intake of calcium and phosphate. The normal requirement for a woman of childbearing age is about 800 mg a day for both calcium and phosphate. An extra 400 mg a day of both minerals is normally sufficient to match the quantities secreted in the milk.

TABLE 21.4 Recommended daily intake of protein and micronutrients during lactation

Nutrient	USA	UK
Protein (g)	65	56
Vitamin A (mg)	1.30	0.95
Vitamin D (μg)	10	10
Vitamin E (mg)	12	10
Vitamin C (mg)	95	70
Vitamin B$_1$ (Thiamin) (mg)	1.6	1.0
Vitamin B$_2$ Riboflavin (mg)	1.8	1.6
Vitamin B$_3$ (niacin or nicotinamide) (mg)	20	15
Vitamin B$_{12}$ (μg)	2.6	2.0
Folate (μg)	280	260
Calcium (g)	1.2	1.25
Magnesium (mg)	355	320
Iron (mg)	15	15
Zinc (mg)	19	13
Iodine (μg)	200	140

Summary

1. During gestation progesterone and estrogens inhibit the lactogenic action of prolactin, but after delivery this inhibitory influence is lost and lactation commences.

2. The composition of milk changes during the first weeks after parturition. Colostrum is secreted in the first few days after delivery. This is rich in proteins, minerals, and immunoglobulins, but low in fats and sugars. Gradually, the composition of the milk changes and by 3 weeks postpartum mature milk is produced. This is rich in fats, proteins, and sugars.

3. Lactose is the major milk sugar, while casein, lactoglobulin, and α-lactalbumin are the chief milk proteins.

After delivery, milk production is maintained by regular suckling

Lactation is initiated by the precipitous drop in steroid levels that occurs following removal of the placenta at the time of delivery. Why does the breast then continue to secrete milk for as long as the baby requires it after this time? What is the hormonal basis underlying the maintenance of lactation? It is known that lactation will continue normally in women who have undergone removal of their ovaries but not in those who have damaged or absent pituitaries. The critical hormone for continued milk secretion appears to be prolactin. Levels of this hormone must remain high for efficient lactogenesis, and the only way that this can be ensured is through regular suckling by the infant. Indeed, the suckling stimulus is the single most important factor in the maintenance of established lactation—in the absence of suckling, milk production ceases after 2 or 3 weeks. Suckling, or more correctly, nipple stimulation, induces the release of prolactin from the anterior pituitary gland via a neuroendocrine reflex arc in which the afferent limb is neural and the efferent limb is endocrine. Nerve impulses set up by the mechanical stimulation of the baby suckling at the breast pass via the spinal cord and brainstem to the hypothalamus. The result would seem to be a fall in the output of prolactin inhibitory hormone (which is now known to be dopamine) from the hypothalamic neurons and a subsequent increase in the secretion of prolactin—remember that prolactin release is usually suppressed by prolactin inhibitory hormone (PIH) (see Chapter 12, Table 12.3). The prolactin then stimulates the synthesis and secretion of milk.

Prolactin output is a direct consequence of nipple stimulation

Denervation of the nipple abolishes the release of prolactin in response to suckling. It has also been shown that the amount of prolactin released depends directly on the strength and duration of the suckling stimulus. If both breasts are suckled together, for example during the feeding of twins, more prolactin is released than when a single infant is suckled. In turn, more milk is produced, from which it appears that milk production is determined by the levels of circulating prolactin.

To summarise, once lactation is initiated by the removal of inhibitory steroidal influences, milk production is ensured by the release of bursts of prolactin occurring each time the infant suckles—the baby makes sure of its next meal while enjoying its current one.

Milk ejection is a direct response to the suckling stimulus

The constituents of breast milk are produced by the alveolar epithelial cells and secreted into the alveolar lumen under the influence of prolactin. However, for this to be of any value to the baby, the milk must be moved from the lumen to the nipple. This is the process of milk let-down and subsequent ejection, and is another example of a neuroendocrine reflex occurring as a direct response to the suckling stimulus. The hormone responsible for the let-down and ejection of milk is *oxytocin*, a peptide hormone synthesized within the hypothalamus, and stored and secreted by the posterior pituitary gland (see pp. 196–199). When the baby suckles, afferent impulses are initiated in the nipple and areola and travel to the hypothalamus, in particular the paraventricular and supraoptic nuclei. In response to this stimulation, the synthesis and secretion of oxytocin are enhanced. Once oxytocin is released into the general circulation, it reaches the breast where it stimulates contraction of the myoepithelial cells that lie within the alveolar basement membrane (Fig. 21.16). When these cells contract, the contents of the alveolar lumen are squeezed out into the lactiferous ducts. As the ducts and sinuses fill with milk, the intramammary pressure rises. When it reaches a high enough level milk is actually ejected from the nipple to the suckling baby.

Figure 21.18 illustrates in a highly simplified diagram the reflex control of both prolactin and oxytocin release during suckling. The output of both hormones rises in synchrony with the episodes of suckling. While suckling seems to be the only effective stimulus for prolactin release under normal circum-

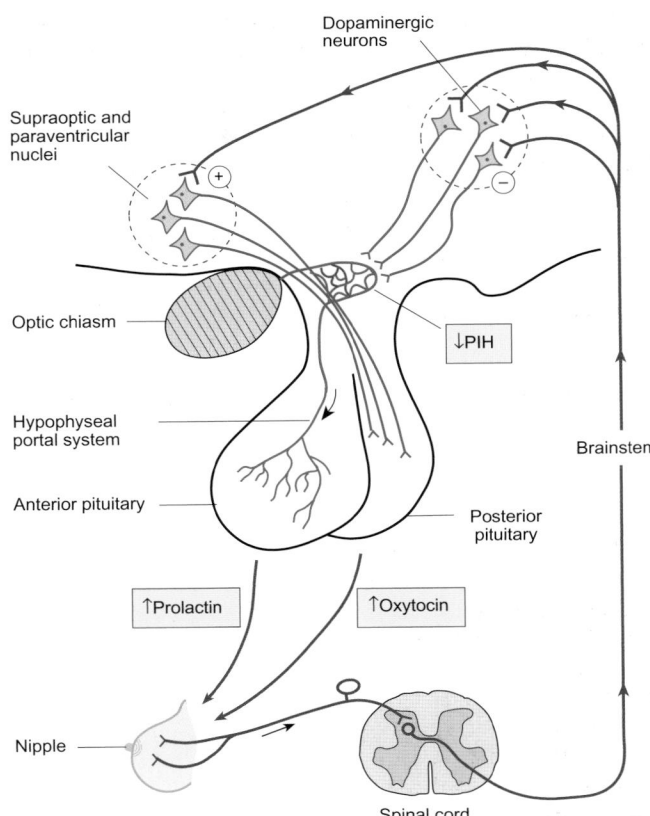

Fig. 21.18 Summary of the principal neuroendocrine pathways responsible for the reflex release of prolactin and oxytocin during suckling.

stances, this is not the case for oxytocin whose secretion may be enhanced by a number of other stimuli. For example, uterine contractions, and mechanical stimulation of the cervix and vagina can initiate oxytocin release (e.g. during parturition—see Section 21.7). The milk-ejection reflex is readily conditioned. In cows, the rattling of the milking equipment may be sufficient to start the release of milk, while in humans the cry of a hungry baby may induce the secretion of oxytocin. By the same token, the milk-ejection reflex seems particularly susceptible to inhibition by stress, both physical and psychological, a response that may be mediated by catecholamines.

After weaning, cessation of suckling suppresses milk production

Lactation normally ceases within 2 or 3 weeks of weaning the baby onto a bottle or solid foods. This is entirely due to the loss of the suckling stimulus. In the absence of mechanical stimulation of the nipple, prolactin secretion declines and lactogenesis gradually slows down. Although milk production itself stops relatively quickly, complete involution of the mammary gland takes about 3 months. At first, milk accumulates in the alveoli and small lactiferous ducts causing distension and mechanical atrophy of the epithelial structures. The alveolar cells are ruptured and hollow spaces form within the mammary tissue. The distension also causes compression of the capillary network supplying the alveoli and, as a result of the reduced perfusion, the alveolar cells become hypoxic and lack nutrients. This in turn depresses milk production. Desquamated alveolar cells and glandular debris are phagocytosed and the alveoli disappear almost completely. Consequently, the ductal system starts to dominate and the involuted alveolar epithelial cells revert to the granular non-secretory type characteristic of the non-pregnant state. All these changes occur quite naturally as a direct result of removing the suckling stimulus at the time of weaning.

It is occasionally necessary to suppress lactation artificially and rather more quickly than would occur naturally. The human mammary gland is fully prepared for lactation by the fourth month of pregnancy (see above). This means that in the event of a miscarriage or abortion after this time, milk production will commence because of the decline in steroid secretion following removal of the placenta. Under these circumstances, it is clearly desirable to inhibit lactation as rapidly as possible. Years ago this would have been assisted by the application of ice packs and tight bandages to the breasts. Since the discovery that PIH is the neurotransmitter dopamine, pharmacological suppression of lactation has been possible through the administration of dopamine agonists such as bromocriptine.

Recommended reading

Alberts, B., Johnson, A., Lewis, J., Raff, M., Roberts, K., and Walter, P. (2002). *Molecular biology of the cell* (4th edn), Chapter 20. Garland, New York.

Case, R.M., and Waterhouse, J.M. (eds.) (1994). *Human physiology: age, stress and the environment*. Oxford Science Publications, Oxford.

Ferin, M., Jewckwicz, R., and Warres, M. (1993). *The menstrual cycle*. Oxford University Press, Oxford.

Griffin, N.E., and Ojeda, S.R. (2000). *Textbook of endocrine physiology* (4th edn). Oxford University Press, Oxford.

Johnson, M.H., and Everitt, B.J. (1999). *Essential reproduction* (5th edn). Blackwell Scientific, Oxford.

Summary

1. After delivery, milk production is maintained by regular suckling. The hormone responsible is prolactin, secreted in direct response to nipple stimulation.

2. Milk let-down and ejection are the processes by which milk is moved from the alveoli of the mammary gland to the nipple. This occurs in response to oxytocin secreted by the posterior pituitary during suckling.

3. Once the baby is weaned and the suckling stimulus is lost, prolactin output falls. Lactation slows and ceases altogether 2 or 3 weeks after the cessation of suckling.

Multiple choice questions

Each statement is either true or false. Answers are given below.

1. a. Erection is a sympathetic reflex.
    b. Erection is due to pooling of blood in the erectile tissues of the penis.
    c. During ejaculation the sperm are mixed with secretions from the prostate and seminal vesicles.
    d. Fertilization normally occurs in the uterus.
    e. Fertilization may occur 3 or more days after ovulation.

2. a. A normal sperm count is about 20 million.
    b. Immediately after ejaculation, the sperm are capable of fertilizing an ovum.
    c. The fertilized egg is also known as a zygote.
    d. Following fertilization the egg completes its second meiotic division.
    e. The fertilized egg secretes hCG to maintain the corpus luteum.

3. a. The placenta nourishes the fetus by establishing a dialysis pattern for exchange between the maternal and fetal circulations.
    b. The placental barrier is very thin and freely permeable to glucose and amino acids.
    c. The P_{O_2} of the blood of the umbilical vein is the same as that in the maternal arterial blood.
    d. The placenta secretes about 10 times as much progesterone as the corpus luteum.
    e. The estrogens secreted by the placenta tend to increase uterine excitability during late pregnancy.

4. a. Thyroid function is increased during pregnancy.
    b. Placental progesterone secretion is highest in the first 8 weeks of gestation.
    c. Blood flow to the kidneys is increased in pregnancy.
    d. Maternal plasma urea concentration decreases during pregnancy.
    e. Maternal respiratory rate is increased during pregnancy.

5. a. Suckling is the single most important stimulus for milk secretion and ejection.
    b. Prolactin stimulates milk ejection.
    c. Oxytocin is secreted in response to suckling.
    d. Prolactin stimulates the synthesis and secretion of all the major constituents of milk.
    e. Lactation can be suppressed by dopamine agonists.

6. a. Mature milk has a higher protein content than colostrum.
    b. Colostrum has a higher calorific value than mature milk
    c. During pregnancy, estrogens stimulate ductal development while progesterone stimulates development of the alveoli.
    d. Following a miscarriage at 5 months gestation, lactation will commence.
    e. Placental steroid secretion inhibits lactation during pregnancy.

Answers to multiple choice questions

1. Erection is initiated by activity in the parasympathetic nerves arising from S2–S4, which cause the vessels of the pudendal artery to dilate. Fertilization normally occurs in one of the Fallopian tubes. Since the unfertilized egg is believed to be viable for only 12–24 hours after ovulation, fertilization must take place within this period.
   a. False
   b. True
   c. True
   d. False
   e. False

2. A normal sperm count is about 200 million. Values as low as 20 million are associated with male infertility. Sperm must undergo capacitation before they can fertilize an egg. This normally takes place in the female reproductive tract.
   a. False
   b. False
   c. True
   d. True
   e. True

3. The placental barrier is relatively thick and impermeable. This prevents fetal proteins passing into the maternal circulation and eliciting an immune response. While blood gases cross the placental barrier by passive diffusion, glucose and amino acids pass from mother to fetus by carrier-mediated facilitated diffusion. The P_{O_2} of the blood in the umbilical vein is about 4.25 kPa (c. 32 mmHg), which is much lower than that in the uterine artery where P_{O_2} is about 12.6 kPa (95 mmHg).
   a. True
   b. False
   c. False
   d. True
   e. True

4. Thyroid function is often enhanced in pregnancy by the actions of a placental TSH-like hormone. Placental secretion of progesterone peaks at around week 36. Blood flow to the kidneys is increased because of the increased cardiac output of pregnancy. Plasma urea falls because GFR is raised. Respiratory rate is normally unchanged during gestation.
   a. True
   b. False
   c. True
   d. True
   e. False

5. Oxytocin is the hormone responsible for milk ejection. Prolactin stimulates the synthesis of milk. The prolactin inhibitory hormone (PIH) is dopamine, so that dopamine agonists will suppress prolactin secretion and milk production.
   a. True
   b. False
   c. True
   d. True
   e. True

6. Colostrum has a higher protein content than mature milk, but a lower fat and carbohydrate content. Therefore it has a lower calorific value than mature milk. The mammary gland is prepared for lactation early in pregnancy so that loss of placental steroid hormones will initiate lactation.
   a. False
   b. False
   c. True
   d. True
   e. True

Fetal and neonatal physiology

After reading this chapter you should understand:

- The differences in the organization of the fetal and adult cardiovascular systems
- The carriage of oxygen in the fetal blood
- The changes that take place following birth
- The first breath and the cardiovascular changes that follow the first breaths
- The role of surfactant in lung inflation
- The factors responsible for closure of the foramen ovale, ductus arteriosus and ductus venosus
- Respiration in the neonate
- The differences between the fetal, neonatal, and adult gut, kidneys, and adrenal glands
- The mechanisms underlying temperature regulation in the neonate
- The differentiation of the male and female reproductive tracts

22.1 Introduction

The fetus is totally dependent upon the placenta for gas exchange, nutrition and waste disposal while it remains within the uterus of its mother. It is adapted in a number of important ways to life within a fluid-filled bag. This chapter will consider some of the more important aspects of the physiology of the fetus itself, as well as the physiological changes that take place at, or soon after, birth to enable the baby to make a successful transition from its uterine existence to a semi-independent air-breathing life. The changes occurring around the time of birth within the cardiovascular and pulmonary systems are of paramount importance to the survival of the infant. However it is also important to remember that a number of other organs are functioning throughout fetal life and that these too must adapt at birth to the different requirements of extra-uterine life. Of these, the adrenal glands, kidneys, thermoregulatory tissues and GI tract will be considered briefly. The chapter will end with a simple account of the differentiation of the male and female fetal sexual organs.

22.2 The fetal circulation is arranged to make the best use of a poor oxygen supply

The fetal circulation differs from that of the adult in a number of important ways. The pattern of circulation is adapted for placental rather than pulmonary gas exchange, and organs which are virtually non-functional, such as the lungs, gut, and liver, are largely bypassed.

The fetal heartbeat is detectable at 4–5 weeks of gestation and by week 11 the cardiovascular system is fully developed in miniature. Figure 22.1 shows a simplified plan of the organization of the fully developed fetal cardiovascular system. It illustrates the three important shunts that differentiate the fetal from the adult circulations.

1. The *foramen ovale* is a gap formed by incomplete fusion of the septum between the left and right atria. As shown in Figs 22.1 and 22.2, the foramen ovale provides a direct path for blood to pass between the inferior vena cava and the left atrium.

2. The *ductus arteriosus* forms a direct link between the pulmonary artery and the aorta.

3. The *ductus venosus* links the umbilical vein with the inferior vena cava.

These shunts, which normally close at birth, enable the two sides of the fetal heart to work in parallel, with mixing of the right ventricular and left ventricular outputs. This arrangement differs from that of the adult in which the pulmonary and systemic circulations are perfused entirely separately (see Fig. 22.4 below).

In order to understand how the parallel organization of the fetal circulation operates, it will be helpful to refer to Fig. 22.1 while considering the route taken by the blood as it flows round the fetal cardiovascular system.

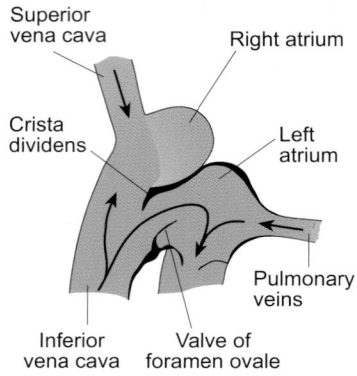

Fig. 22.2 The pattern of blood flow in the right atrium. Note how the crista dividens diverts the flow of oxygenated blood in the inferior vena cava through the foramen ovale to the left atrium.

Venous return to the right side of the heart consists of deoxygenated blood carried by the superior vena cava together with oxygenated blood from the placenta carried by the inferior vena cava. About 80 per cent of the blood in the umbilical veins bypasses the immature fetal liver and passes directly to the inferior vena cava via the ductus venosus, while the remaining 20 per cent travels to the liver via the hepatic portal vein. Therefore, oxygenated blood in the umbilical vein is mixed with the deoxygenated blood returning from the lower parts of the body in the inferior vena cava. If this blood were also to be mixed with the deoxygenated blood traveling in the superior vena cava from the upper parts of the body, as it is in the adult, oxygen saturation of the blood from the placenta would be reduced still further. In the fetus, complete mixing of inferior and superior vena caval blood in the heart is avoided as a result of the anatomical position and mode of operation of the *crista dividens* and the foramen ovale. These allow most of the blood

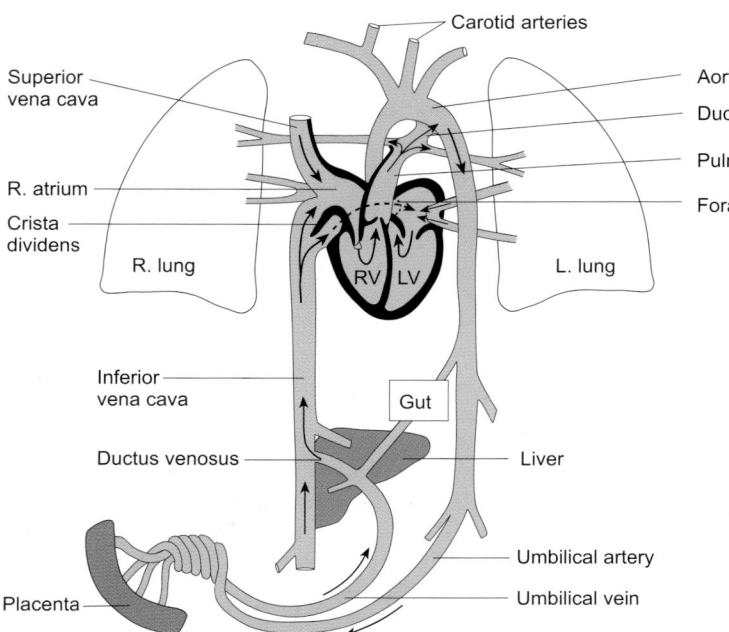

Fig. 22.1 The organization of the fetal cardiovascular system. Note that the structure of the heart has been distorted to show the operation of the crista dividens. The broken arrow indicates the direct flow of blood from the inferior vena cava through the foramen ovale to the left atrium.

from the inferior vena cava to pass directly to the left side of the heart instead of to the right as it would in the adult. This is shown in Figs 22.1 and 22.2. The remainder continues upwards to enter the right atrium. No blood from the superior vena cava normally passes the crista dividens (the upper part of the incomplete septum dividing the two sides of the heart). Instead, all of it enters the right atrium.

Blood from the left ventricle is pumped into the ascending aorta from which the carotid arteries supplying the brain branch off (Fig. 22.1). Blood from the right ventricle enters the pulmonary artery. However, there is a direct link between the pulmonary artery and the descending aorta—the ductus arteriosus—so that when blood reaches the point at which the ductus arteriosus branches off from the pulmonary artery, it may take one of two routes. It may travel to the fetal lungs in the pulmonary arteries, or it may bypass the lungs and travel directly to the descending aorta via the ductus. In practice, only around 20 per cent of the blood from the right ventricle perfuses the lungs because of the high vascular resistance of the pulmonary circulation (see below). The rest flows through the ductus arteriosus.

An important consequence of this arrangement of the fetal circulation is that the blood in the ascending aorta has a higher oxygen content than that in the descending aorta because of the operation of the foramen ovale. This means that the blood supplying the brain is comparatively well oxygenated.

The control of the fetal circulation

By week 11 of gestation, when the fetal cardiovascular system has been laid down, the fetal heart is beating at around 160 b.p.m., a very high rate compared with the average resting heart rate of the adult, which is about 70 b.p.m. Later, during the third trimester (the final 3 months) of gestation, the autonomic nervous system becomes functional and the parasympathetic innervation of the heart is established. At this stage the fetal heart rate slows to around 140 b.p.m. This gradual development of the autonomic control of the cardiovascular system is also evident in the changes in blood pressure that occur during fetal life. Pressure is comparatively low—around 9/6 kPa (c. 70/45 mmHg)—in the early months of gestation when there is very little peripheral vascular tone and therefore a low total peripheral resistance. Blood pressure gradually rises as autonomic activity becomes established and vascular tone is increased. At the same time, aortic and carotid baroreceptors begin to function. This increase in blood pressure and drop in heart rate continues after delivery, until by the age of about 7 years, values similar to those of the adult are achieved.

The fetus depends upon the placenta for gas exchange as its lungs are collapsed and the alveoli filled with fluid

The role of the placenta in the transport of oxygen and carbon dioxide between the maternal and fetal blood was described in some detail in Chapter 21 (Sections 21.4 and 21.5). As far as gas exchange is concerned, the fetal lungs are non-functional and the alveoli are almost collapsed and filled with fluid. This fluid is secreted by type I alveolar cells (epithelial cells which overlie the pulmonary capillaries) and its composition differs from that of the amniotic fluid. It first appears around mid-gestation and by full term the lungs contain a total of about 40 ml of this fluid.

Because the alveoli are collapsed, their capillaries are tortuous and offer a high resistance to blood flow. Consequently, the lungs are relatively poorly perfused.

Breathing movements develop before birth

Although the fetal lungs do not participate in gas exchange, ventilatory movements are seen during gestation. Ultrasound scans have revealed that fetal breathing movements begin at around weeks 10 of gestation. They remain shallow and irregular up until around week 34 of gestation, after which they start to display a more rhythmical pattern with periods of activity interspersed with periods when movements are absent. Occasionally, gasping movements are seen, especially if the fetus experiences hypercapnia—for example, as a result of placental insufficiency or compression of the umbilical cord. This response suggests that chemoreceptors (see Chapter 16) are functional during the latter part of gestation. It is now believed that fetal breathing movements are important in the preparation of the respiratory system for its postnatal function of gas exchange.

Fetal blood has a higher affinity for oxygen than adult blood

There are species variations in the magnitude of the fetal–adult difference in hemoglobin oxygen affinity. Most of the research in this area has been done using sheep in which the difference is rather large. In humans, the differences are thought to be smaller, though still significant.

In the sheep the partial pressure of oxygen in the blood traveling from the placenta to the fetus in the umbilical veins is around 5 kPa (35–40 mmHg), much lower than that of normal adult arterial blood (around 12.6 kPa (95 mmHg)—see Chapter 21, Fig. 21.8). Moreover, Pa_{O_2} in the fetal aorta is only about 3.0 kPa (c. 22 mm Hg) (Fig. 22.3) as the umbilical blood is mixed with blood returning from the lower part of the body before reaching the left heart. Such a low value for Pa_{O_2} would be expected to result in a very low value for blood oxygen satura-

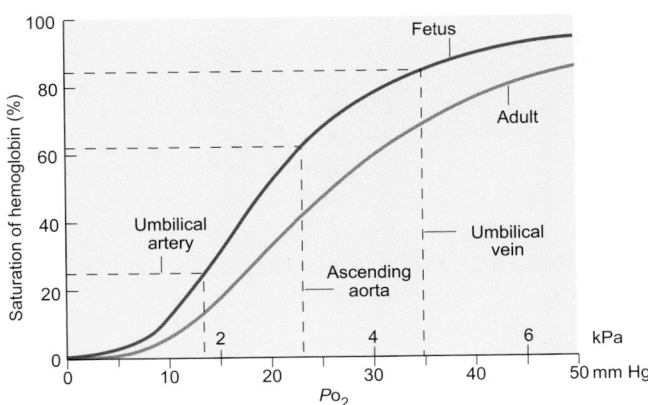

Fig. 22.3 The oxygen dissociation curves for adult and fetal sheep hemoglobin. Note that the dissociation curve for the fetus is displaced to the left, indicating a higher degree of saturation of fetal hemoglobin for a given partial pressure of oxygen. The approximate values for the partial pressure and percentage saturation in the ascending aorta of the fetus and in the umbilical artery and vein of the sheep are indicated. The difference in oxygen carriage between adult and fetal blood is believed to be slightly smaller in man.

Summary

1. The fetus is dependent on the placenta for gas exchange, waste disposal, and nutrition.

2. The fetal circulation is arranged so that the two sides of the heart work essentially in parallel and the three fetal shunts permit the blood to bypass those organs with little or no function.

3. The fetal heart rate is high and the blood pressure is low.

4. The fetal lungs are fluid filled and virtually collapsed. The pulmonary vascular resistance is elevated and so pulmonary blood flow is only 20 per cent of the right ventricular output; the other 80 per cent passes through the ductus arteriosus.

5. Despite its low Pa_{O_2}, the oxygen content of fetal blood is about 16 ml dl⁻¹. This capacity is a result of a high hemoglobin concentration and the high affinity of fetal hemoglobin for oxygen.

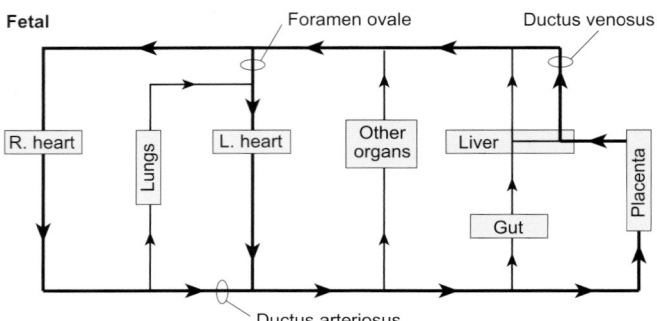

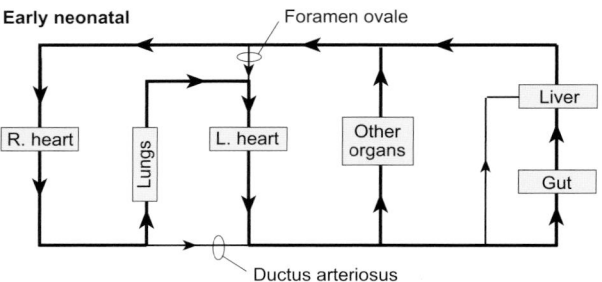

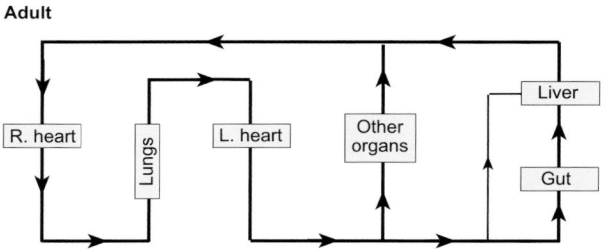

Fig. 22.4 Changes in the pattern of the circulation following birth.

tion and oxygen content in the fetus. In fact, even at this low Pa_{O_2}, the arterial blood in the fetus is about 60 per cent saturated compared with 98 per cent or so in the adult arteries. The oxygen content of the arterial blood is about 16 ml dl⁻¹. Two important factors are responsible for this state of affairs.

1. Fetal blood has a higher hemoglobin concentration than adult blood (around 20 g dl⁻¹ compared with around 15 g dl⁻¹). This increases the oxygen-carrying capacity of fetal blood.

2. Fetal hemoglobin (Hb-F), which has a different structure from that of the adult, has a higher affinity for oxygen than adult hemoglobin. The oxygen dissociation curve for fetal blood is shifted to the left in relation to that of the adult (Fig. 22.3). As a result of this, even at the low P_{O_2} values experienced by the fetus, fetal blood has a relatively high oxygen saturation and total oxygen content. Furthermore, the portion of the curve over which the fetus normally operates is very steep. Consequently, even though there is a relatively small difference between arterial and systemic venous blood P_{O_2}, there is a large difference in oxygen saturation and oxygen is readily off-loaded to the tissues.

22.3 Respiratory and cardiovascular changes at birth

The changes that occur at birth can be summarized as follows:

1. The initiation of the first breath.

2. The expansion of the lungs followed by a reduction in the pulmonary vascular resistance which results in a striking rise in pulmonary blood flow.

3. The ductus arteriosus gradually closes so that all the right ventricular output eventually passes through the pulmonary circulation.

4. The increase in pulmonary blood flow into the left atrium leads to the closure of the foramen ovale.

5. The ductus venosus closes so that all the blood in the portal vein passes through the liver.

The changes in the pattern of the circulation following birth are summarized in Fig. 22.4.

The first breath is probably triggered by cooling and hypercapnia

Once the infant has been delivered, it is cut off from the placenta that has acted as its site of gas exchange for the previous nine months. If the cord is not clamped surgically in the delivery room, the umbilical vessels quickly shut down of their own accord. Therefore, despite the fact that both the fetus and the neonate are able to tolerate degrees of hypercapnia and hypoxia that would probably kill an adult, the infant has to start breathing independently if it is to survive for more than about 10 minutes. What mechanisms are responsible for triggering the first breath?

1. The infant experiences a drop in ambient temperature following delivery.

2. There is an increase in P_{CO_2} following delivery. During labor and delivery, particularly if this has been long and difficult, the placental blood supply may be partially or wholly occluded for short periods of time. Certainly, once the baby has been born the placental blood supply will be lost. Consequently, the baby will experience a considerable degree of hypercapnia. Indeed, measurements of the P_{CO_2} of scalp blood sampled during and just after delivery have revealed a marked increase which may well provide the neonate with an important stimulus to gasp for air. This reflex is known to be functional in the fetus whose breathing movements become more pronounced during hypercapnia (see above).

3. Once it has been born, the neonate is subjected to a barrage of sensory information from which it has been insulated while in the womb. It is known that tactile and painful stimuli can stimulate breathing movements even in the fetus.

It is not known which, if any, of these factors is responsible for the initiation of ventilation following delivery—perhaps a combination of both physical and chemical stimuli is required.

Lung inflation is facilitated by surfactant

The lungs have remained collapsed and fluid filled throughout embryonic life. In order to inflate its lungs for the first time, the newborn baby must overcome the enormous surface tension forces at the gas–liquid interface of the alveoli. Around week 20 of gestation, type II epithelial cells start to appear in the developing alveoli, and 8–10 weeks later, under the control of fetal cortisol, these cells start to secrete phospholipid surfactants, the most important of which are lecithin and sphingomyelin. Surfactant molecules are responsible for reducing the surface tension forces that oppose lung inflation. This reduction makes it possible for the newborn infant to inflate its lungs. Nevertheless, a considerable ventilatory effort is still required in order to expand the lungs for the first time.

It can be seen from Figure 22.5 that an enormous negative pressure must be generated to make the initial inspiration possible. The figure illustrates the pressure–volume relationships for

the neonatal lung operating during the first and subsequent breaths. It also shows that an equally large positive pressure must be generated to bring about expiration because lung compliance is still rather low. A great mechanical effort is demanded from the baby. The diaphragm contracts strongly and the ribs and sternum, which at this stage are very flexible, become slightly concave during the initial breaths. After the first breath, the volume of the lungs does not return to zero after expiration but a little air remains to form the beginnings of the residual volume, which persists throughout life. Subsequent breaths are achieved with much smaller pressure changes and consequently require much less mechanical effort, indicating that lung compliance has increased. (See Chapter 16 for more details of the mechanics of breathing.)

Once the fetal shunts are closed after delivery and the lungs are expanded, the pulmonary vascular resistance is decreased and the pulmonary blood flow is greatly increased (see Section 22.4). As a result, the fluid that filled the alveoli during fetal life is quickly reabsorbed into the pulmonary capillary blood which has a higher oncotic pressure than the alveolar fluid.

What happens if surfactant is inadequate?

The presence of sufficient quantities of surfactant is crucial to the initiation of ventilation following delivery and, even then, the first breath requires a considerable mechanical effort on the part of the baby (Fig. 22.5). Imagine, therefore, the problems faced by babies born prematurely, i.e. before adequate surfactant secretion has been established. If a baby is born before weeks 28–30 of gestation, it will almost certainly have difficulty overcoming the surface tension forces opposing ventilation and is very likely to show respiratory distress. If such infants are to have a chance of survival, they must be ventilated artificially until their lungs are sufficiently well developed to permit independent respiration.

How does neonatal respiration differ from that of the adult?

Once the first few, rather difficult, breaths have been accomplished, respiration settles down into the 'neonatal' pattern—a

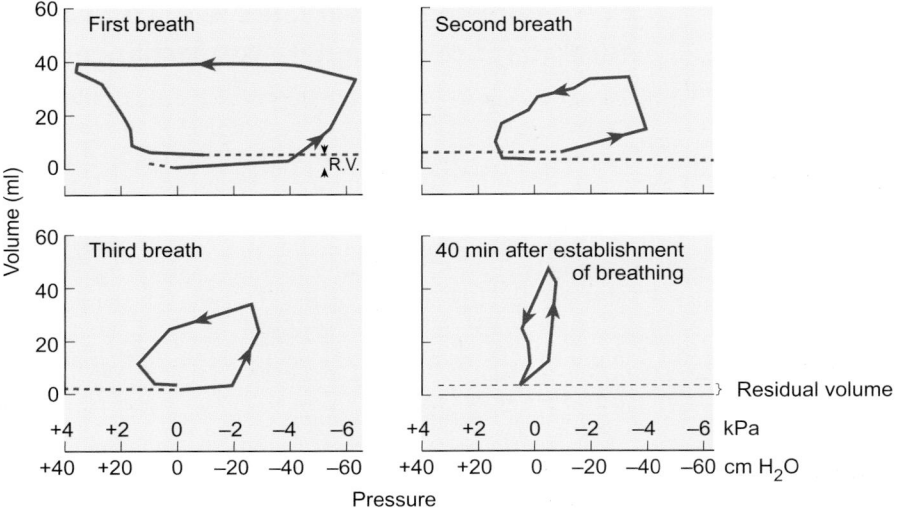

Fig. 22.5 The lung pressure–volume relationships during the first, second, and third breaths, together with that for breathing about 40 minutes after the first breath. Note that the residual volume (RV) is established with the first breath and that the compliance increases with subsequent breaths (i.e. the pressure change required for a given change in volume falls after the first breath).

TABLE 22.1 Comparison of respiratory variables between the neonate and the adult

Variable	Neonate	Adult
Body weight (kg)	3.3	70
Ventilation rate (breaths min⁻¹)	20–50	12–15
Minute volume (ml)	c. 500	c. 6500
Tidal volume (ml)	18	500
Vital capacity (ml)	120	4500
Surface area for gas exchange in lungs (m^2)	3	60
Compliance		
($l\ kPa^{-1}$)	0.051	1.7
($ml\ cmH_2O^{-1}$)	5	165
Bronchiole diameter (mm)	0.1	0.2
Oxygen diffusion capacity		
($ml\ s^{-1}\ kPa^{-1}$)	0.6	6
($ml\ min^{-1}\ mmHg^{-1}$)	2.5	25
Energy expended in breathing (% of total O_2 consumption)	6	2

Summary

1. After delivery, the baby is cut off from its placental blood supply and must begin to use its lungs. Physical factors such as temperature changes may stimulate breathing, but a likely trigger for the first breath is hypercapnia following cord compression during delivery.

2. To inflate its lungs for the first time, the baby must overcome the enormous surface tension forces at the gas–liquid interface in the alveoli. Surfactant plays a vital role in reducing these forces, but a massive effort is still required by the baby to generate the intrathoracic pressures needed to open the lungs.

3. Babies born before adequate levels of surfactant have been produced (around week 28) are at risk of showing respiratory distress.

4. Neonatal respiration differs in certain respects from that of the adult. Ventilatory rate is higher but more erratic, airway resistance is higher, and the work of breathing is greater in the neonate.

5. The ventilatory response to hypercapnia is well developed in both the fetus and neonate, with central medullary chemoreceptor activity present from about mid-gestation. In the fetus the peripheral chemoreceptors have a very low level of activity but start to respond to reductions in oxygen tension following delivery.

somewhat erratic rhythm with certain characteristics that differ significantly from those of a more mature child or adult. There are, understandably, considerable difficulties associated with the study of lung function in very small babies but some information has been obtained, much of which is contained in Table 22.1. The ventilatory rate of a newborn infant (i.e. less than about 1 month old) is rather high and extremely variable. Neonatal breathing often resembles the fetal pattern of respiratory movements with episodes of shallow breathing (or even apnea) interspersed with periods of normal respiration.

Because of its high rate of ventilation, the infant's minute volume is relatively high in relation to its body weight. As Table 22.1 shows, the neonatal respiratory system has a rather low compliance, which indicates high airway resistance when compared with that of an adult. A number of factors contribute to this resistance: first, during the early weeks of its life a baby tends to breathe mostly through its nose; secondly, the bronchioles are very narrow; and thirdly, lung compliance is itself still rather low. These characteristics mean that the energy expenditure of breathing in the neonate is high.

Regulation of ventilation in the neonate

Regulation of ventilation during the first weeks of life represents a transitional stage somewhere between that of the fetus and that of the adult. Central medullary chemoreceptor activity seems to be present from about the time of the onset of fetal breathing movements. The evidence for this is that fetal hypercapnia seems to initiate 'gasping' movements (see above). The peripheral chemoreceptors appear to be desensitized or switched off in the fetus, probably because the partial pressure of oxygen in the fetal blood is always low. They show slight tonic activity in the full-term fetus at the normal Pa_{O_2} of 3.0 kPa (c. 23 mmHg). After birth, however, the hypoxic sensitivity gradu-

ally changes to that of the adult. The mechanism for this change in sensitivity is still unclear. This means that, in the neonate, the same tonic activity is now present at much higher levels of Pa_{O_2} and a reduction in the Pa_{O_2} below that level causes the expected stimulation of the chemoreceptors. The ventilatory response to hypercapnia is very marked in the newborn baby—the addition of only 2 per cent carbon dioxide to the inspired air produces an increase in the minute volume of around 80 per cent.

22.4 Following delivery, the fetal circulation must adapt to pulmonary gas exchange

The anatomical arrangement of the fetal circulation differs from that of an adult in a number of ways (Fig. 22.4). In essence, the two sides of the circulation work in parallel, with the three fetal shunts permitting blood to bypass those organs with little or no function. Such an arrangement is well adapted to gas exchange via the placenta, but would be quite inappropriate once the baby has begun to breathe for itself. Following delivery, the placental blood supply is lost and the lungs become the sole source of oxygen. As the infant takes its first breaths of air, the fetal circulation must start to adapt to the adult pattern so that blood no longer bypasses the pulmonary circulation. To achieve this it is essential that the three fetal shunts close.

Probably the most important step in the initiation of shunt closure is the increase in pulmonary perfusion which accompanies the establishment of ventilation. During fetal life only about 20 per cent of the cardiac output enters the pulmonary circulation because resistance to blood flow through the vessels is high in the collapsed lungs. After the first breath pulmonary blood flow is dramatically increased. Two key factors are involved:

◆ Inflation of the alveoli reduces the tortuosity of the pulmonary capillaries, thereby reducing their resistance to blood flow.

◆ As a result of breathing air, there is a considerable increase in the partial pressure of oxygen in the blood perfusing the lungs. In response to this rise, there is vasodilatation of the pulmonary vessels and a corresponding fall in resistance. Both these changes stimulate a large increase in the volume of blood circulating in the pulmonary vessels.

At the same time, the fetus is separated from its placental blood supply. Following delivery, the umbilical cord is normally clamped, but even if this is not done, the umbilical vessels appear to shut down spontaneously as a result of vasoconstriction in response to the raised systemic P_{O_2}. (Note that this is the opposite reaction to that of the pulmonary arterioles, which dilate in response to a raised P_{O_2}—see also Chapter 16.)

How do the changes in the pattern of blood flow following delivery bring about closure of the fetal shunts?

Consider first the foramen ovale, i.e. the shunt between the left and right atria. During fetal life, right atrial pressure is similar to, or just exceeds, left atrial pressure because of the relatively high pulmonary resistance and the relatively low systemic resistance. After birth and the onset of independent breathing, increased pulmonary perfusion results in an increase in venous return to the left atrium. At the same time, loss of the umbilical

blood supply reduces the venous return in the inferior vena cava to the right atrium. Consequently, left atrial pressure rises above right atrial pressure. The foramen ovale consists of two unfused septa. When right atrial pressure exceeds left atrial pressure, the septa part and the shunt is open. Once the pressures are reversed, the septa will be forced against each other and the shunt will be closed. At first, closure is purely physiological, but within a few days the septa fuse permanently and closure is anatomically complete. Figure 22.6 shows how the pressure changes bring about the closure of the foramen ovale.

Consider next the closure of the ductus arteriosus, the fetal shunt between the aorta and the pulmonary artery. This is an extremely wide channel, almost as large in diameter as the aorta itself (see Fig. 22.1.), and the mechanisms by which closure is effected are not established beyond doubt. The ductus receives little or no innervation, so the most likely cause of closure following delivery and the onset of breathing is constriction in response to blood-borne factors. The smooth muscle of the ductus arteriosus, like that of umbilical vessels, is thought to constrict in response to the substantial rise in P_{O_2} seen after the first breaths. Permanent closure of the shunt occurs within 10 days or so as a result of fibrosis within the lumen of the vessel.

The third shunt is the ductus venosus, the channel which bypasses the liver and carries about 80 per cent of the blood in the umbilical veins directly to the inferior vena cava during fetal life. Once again, the exact mechanisms of closure are poorly understood, but it is believed to occur as a result of constriction of the umbilical vessels following delivery.

In essence, the parallel arrangement of the two sides of the heart characteristic of the fetus is converted to a serial arrangement soon after birth and the commencement of pulmonary gas exchange (Fig. 22.4). At the same time, the relative workloads of the two sides of the heart are altered. Because the resistance to flow in the vascular bed of the lung is only about 12 per cent of that of the systemic circulation, the workload of the right side of the heart is considerably less than that of the left. As a consequence of this, there is an accelerated growth of the more heavily loaded left ventricle, which eventually develops a mass of muscle about three times that of the right.

Occasionally, the fetal shunts fail to close

While there have been no reports of the ductus venosus failing to close during the first days of life, persistent fetal connections in the form of a patent foramen ovale or ductus arteriosus are seen. Indeed, each accounts for probably about 15–20 per cent of congenital heart defects. During very early neonatal life it is not unusual to see intermittent flow through the fetal shunts, but if they remain open for a prolonged period after birth, circulatory function is impaired and surgical intervention will be required to correct the defect. For example, if the foramen ovale remains patent, the volume of blood ejected by the right ventricle is often increased and there is likely to be persistent admixture of oxygenated and deoxygenated blood (the blue baby syndrome). Where the ductus arteriosus remains open, 50 per cent or more of the left ventricular stroke volume can be diverted into the pulmonary circulation, resulting in pulmonary hypertension and heart failure. This condition can be treated surgically by tying the ductus arteriosus.

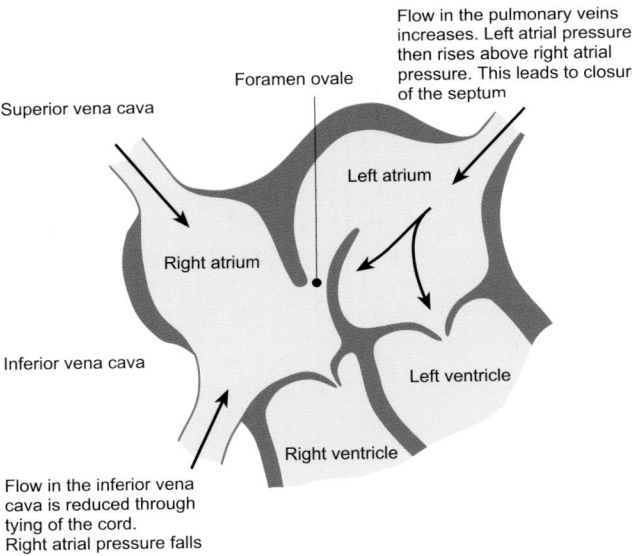

Flow in the pulmonary veins increases. Left atrial pressure then rises above right atrial pressure. This leads to closure of the septum

Foramen ovale

Superior vena cava

Left atrium

Right atrium

Inferior vena cava

Left ventricle

Right ventricle

Flow in the inferior vena cava is reduced through tying of the cord. Right atrial pressure falls

Fig. 22.6 The changes in the pressures of the right and left atria that lead to closure of the foramen ovale. While the lungs are non-functional with respect to gas exchange the pressure in the right atrium is greater than that in the left and blood passes through the foramen ovale. Following the first breath, the pressure in the left atrium becomes greater than that in the right and this leads to the closure of the foramen ovale.

Summary

1. After delivery, the circulation of the infant must adapt to pulmonary gas exchange.

2. The parallel arrangement of the two sides of the heart seen during fetal life converts to a serial arrangement because the three fetal shunts close.

3. Closure of the shunts depends upon ventilation itself. As the lungs expand, pulmonary perfusion increases while at the same time the umbilical vessels close. This causes left atrial pressure to rise above right atrial pressure with subsequent closure of the septa that form the foramen ovale. The ductus venosus and the ductus arteriosus are thought to close as a result of vasoconstriction in response to a rise in arterial $P\text{O}_2$.

22.5 The fetal adrenal glands and kidneys

The fetal adrenal gland secretes large quantities of cortisol during development

The adrenal glands are vital endocrine organs in the adult. They consist of two distinct regions, the cortex, which synthesizes and secretes a variety of steroids (see Chapter 12), and the medulla, which produces the catecholamines epinephrine and norepinephrine. During fetal life, the adrenal glands appear to be, if anything, more important as they play a key role in the development of many of the fetal organ systems and are important in the initiation of parturition (see Chapter 21, Section 21.7).

In relation to the overall body size of the fetus, the fetal adrenal gland is much larger than that of the adult. Furthermore, it is organized in a different way. Unlike the adult gland which consists of a medulla and a zoned cortex, the fetal adrenal is divided into three areas with differing characteristics. These are a small region of medullary tissue derived from embryonic neural crest cells, a small zoned cortex—the so-called definitive cortex—which resembles that of the adult, and a third very large region, the *fetal zone*. The relative sizes of these areas are shown in Figure 22.7.

The functions of the different regions of the fetal adrenal gland are not fully established, but the medullary tissue is capable of secreting catecholamines. This occurs particularly in response to hypoxic stress and, immediately after delivery, to cold stress. The fetal zone seems to be very important in producing the precursors required for the placental synthesis of estrogens (see Chapter 21, p. 460) and its large size probably reflects the enormous output of these steroids during gestation. The 'definitive cortex' seems to carry out little in the way of steroid synthesis itself during fetal life but it does perform one very important task—it converts progesterone to cortisol, especially during the last 3 months of pregnancy.

Fetal cortisol has a number of crucial functions:

- It is linked to the production of surfactant by type 2 alveolar cells.

- It accelerates functional differentiation of the liver and induces the enzymes that are involved in glycogen synthesis.

- It plays an important role in the triggering of parturition.

After delivery, the fetal zone of the adrenal gland regresses rapidly, while the definitive zoned cortex grows quickly to establish the adult organizational pattern of the adult.

Renal function and fluid balance in the fetus and the neonate

Although the placenta is the major organ of homeostasis and excretion of metabolic waste products during gestation, the fetal kidneys do play a role in the regulation of fluid balance and the control of fetal arterial blood pressure. In addition, amniotic fluid volume is regulated chiefly by the formation of fetal urine. The human fetus begins to produce urine at about week 8 of gestation. Its volume increases progressively throughout gestation and is roughly equivalent to the volume of amniotic fluid swallowed by the fetus (around 28 ml h^{-1} in late gestation). Fetal urine is usually hypotonic with respect to the plasma. Indeed, the ability of the kidney to concentrate the urine is not fully developed until after birth when the organ matures, the loops of Henle increase in length, and sensitivity of the tubules to ADH increases.

In the adult, virtually all the filtered sodium is reabsorbed by the renal tubules. In the fetus, sodium reabsorption is comparatively low (85–95 per cent of the filtered load). The exact reasons for this difference are not clear but may be partly explained by a low tubular sensitivity to aldosterone. Fetal glucose reabsorption is thought to occur by sodium-dependent transport as in the adult. Furthermore, the tubular maximum for reabsorption (when corrected for the lower GFR) is higher than that of the

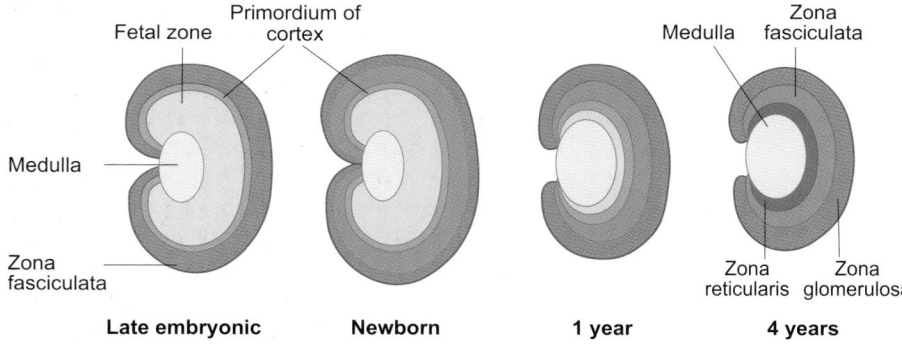

Fig. 22.7 The changes in the adrenal gland during development. Note the regression of the fetal cortex following birth.

adult, as is the renal plasma threshold for glucose.

The fetal kidney plays a part in the regulation of acid–base balance during gestation. Between 80 and 100 per cent of the filtered bicarbonate is reabsorbed by the tubules. The fetal response to metabolic acidosis is relatively poor, but in severe acidosis there is an increase in hydrogen ion excretion.

Renal changes occurring at or soon after birth

Although the changes in kidney function that accompany birth are less dramatic than those in the respiratory and cardiovascular systems, they are just as important. Once the placenta is lost, the kidneys of the newborn infant become solely responsible for maintaining fluid balance and disposing of waste products.

GFR and urine output increase gradually over the first weeks of life although adult levels (relative to body surface area) are not reached for 2 or 3 years. Tubular function is difficult to assess in neonates although it is thought that, while glucose and phosphate are reabsorbed efficiently, bicarbonate and amino acids are reabsorbed less well. Babies cannot concentrate their urine to the degree seen in adults. Possible reasons for this include immaturity of the tubules, shorter loops of Henle, lower sensitivity to ADH, and a low plasma concentration of urea. The reason for this lack of urea is that nearly all the amino acids derived from the protein in a baby's diet are used in the formation of new tissue—very little is metabolized in the liver to form urea.

Newborn babies are at risk from dehydration

The inability of young babies to concentrate their urine efficiently means that they can quickly become dehydrated, particularly during episodes of diarrhea or vomiting. It is essential that fluids are replenished by mouth and, if this is not possible, intravenous fluid replacement may be needed (see also Chapter 28, Section 28.4).

22.6 Temperature regulation in the newborn infant

The fetus has no problems with temperature regulation as it is surrounded by amniotic fluid which is at body temperature. The mother is responsible for generating and dissipating heat. At delivery, the newborn infant has to make a rapid adjustment from the warm moist constant environment of its mother's uterus to an outside world in which the temperature is much lower and heat is readily lost by radiant, convective, and evaporative routes. The neonate has a high surface area to volume ratio, which means that heat is readily lost from the skin's surface, its cardiac output is high in relation to its surface area, and its layer of insulating fat is comparatively thin. These factors combine to cause the core temperature of the baby to drop to around 35°C during the first hours of its life.

Babies generate large quantities of heat through the metabolism of brown adipose tissue

Normally, when a critical temperature difference of 1.5 °C is reached between the skin and the environment, thermogenesis begins and oxygen consumption increases in order to restore the body temperature to normal. While thermoregulatory mechanisms are only partly functional at birth, newborns are capable of maintaining their body temperature above ambient temperature. They respond to a lowered ambient temperature by increased muscular movement, although this is limited. They respond by shivering only in a very minor way. These responses cannot account for all the heat generated in response to cold. The extra heat is generated by non-shivering thermogenesis via the metabolism of *brown adipose tissue* or *brown fat*, which is abundant in the infant. It is situated between the scapulae, at the nape of the neck, in the axillae, between the trachea and the esophagus, and in large amounts around the kidneys and adrenal glands. In all, the neonate possesses about 200 g of brown fat which represents a relatively high proportion of the total body mass (see Fig. 26.6).

The brown fat is well vascularized and exhibits unique metabolic properties which are triggered either by increased plasma levels of circulating catecholamines or by norepinephrine released by sympathetic nerve endings. Cold stress results in an increase in sympathetic nerve activity and an increased secretion of epinephrine and norepinephrine by the adrenal medulla. These hormones stimulate the metabolism of brown fat cells by interacting with β-adrenoceptors on the cell surface to activate a lipase (hormone-sensitive lipase (HSL)) which then releases glycerol and free fatty acids from cellular stores of triglycerides (see Figure 22.8). Most of these free fatty acids are resynthesized directly into triglycerides by the incorporation of α-glycerophosphate so that the brown fat stores are not unduly depleted. The inner mitochondrial membrane of brown fat cells contains a protein that uncouples oxidation from ATP generation and heat is generated instead of the energy being stored as ATP for subsequent use during cellular metabolism. Furthermore, the free

Summary

1. The fetal adrenal gland consists of a medullary region, a small zoned 'definitive' cortex, and a larger fetal zone.

2. The medulla secretes catecholamines, the fetal zone provides the precursors for the synthesis of estrogens by the placenta, and the definitive zone converts progesterone to cortisol.

3. Fetal cortisol has several key functions. It stimulates surfactant production, accelerates maturation of the liver, and has a role in the initiation of parturition.

4. The fetal kidneys play a part in fluid and acid–base balance. While urine is produced from about week 8 of gestation, the fetal kidneys are unable to concentrate it effectively. Therefore, the urine they produce is hypotonic.

5. Glucose reabsorption is comparable to that of adults relative to GFR, but sodium reabsorption is comparatively low.

6. At birth, the kidneys assume sole responsibility for fluid balance and waste disposal. GFR and urine output increase gradually, as does the ability to concentrate urine.

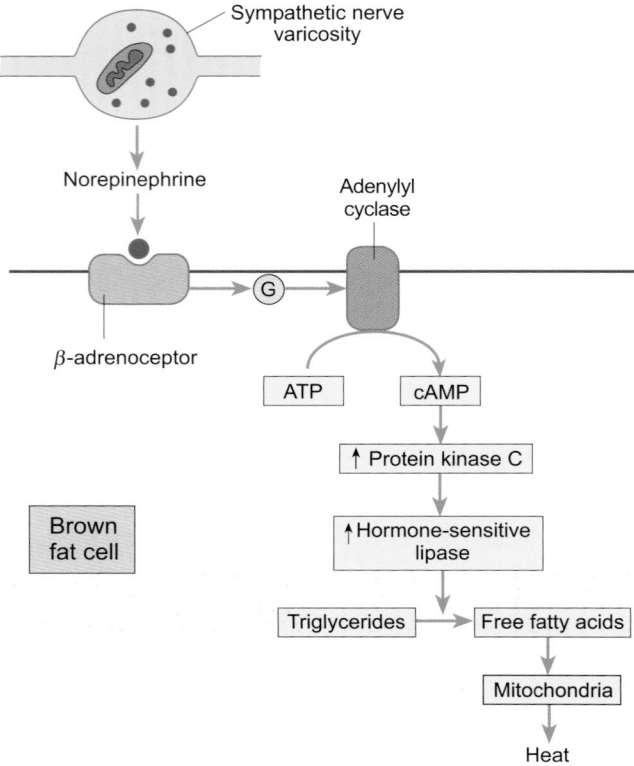

Fig. 22.8 The metabolism of a brown fat cell. Activation of β-adrenoceptors on the cell surface leads to a signal cascade that results in an increased breakdown of triglycerides. These are metabolized in the mitochondria to generate heat (see text for further details).

fatty acids and glycerides which are not immediately resynthesized become available for oxidation by the usual biochemical pathway to provide still more heat energy. Since the tissue is well supplied with blood, the heat that is generated by this pathway is quickly carried to the rest of the body and in this way the brown fat acts as a rather effective source of heat for the newborn baby.

Summary

1. A newborn infant can lose heat very rapidly. Its high surface area to volume ratio, relatively high cardiac output, and lack of insulating fat combine to cause a drop in core temperature after birth.

2. Babies can generate large quantities of heat through the metabolism of brown fat, a well-vascularized tissue situated around the kidneys, at the nape of the neck, between the scapulae and in the axillae. The metabolism of brown fat is stimulated by catecholamines released in response to cold stress.

3. Premature infants have even greater difficulty in maintaining their body temperature and frequently need to be kept in a thermally controlled incubator.

Premature infants have special thermoregulatory problems

Premature babies have even greater difficulty in maintaining their body temperature than normal infants born at full term. Their surface area to volume ratio is even higher, allowing more rapid heat loss, their insulating fat layer is even thinner, and their brown fat stores are less well developed. For this reason it is almost always necessary to keep premature babies in a thermally controlled environment—an incubator—until their thermoregulatory mechanisms develop sufficiently to permit independent temperature regulation.

22.7 The gastrointestinal tract of the fetus and the neonate

The role of the placenta in delivering essential nutrients to the fetus is described in Chapter 21 (see pp. 457–460). The fetus obtains glucose, amino acids, and fatty acids from its mother. Towards the end of gestation, glycogen is stored in the muscles and liver of the fetus, while deposits of both brown and white fat are laid down. These stores will be crucial to the survival of the infant immediately after its birth.

The gut of the fetus is relatively immature, with limited movements and secretion of digestive enzymes. Some salivary and pancreatic secretion commences during the second half of gestation. Gastric glands appear at around the same time, although they do not appear to be secretory since the gastric contents are neutral at birth. Most of the major GI hormones are secreted during fetal life, although at a low level. Motilin is especially low, possibly accounting for the low level of gut motility when compared with that of the adult.

The fetus passes little, if any, feces while it remains in the uterus. The contents of the large intestine accumulate as *meconium*, a sticky greenish-black substance. Meconium does not normally enter the amniotic fluid, although if the fetus becomes distressed, for example during a prolonged or difficult labor, motilin levels rise, gut motility increases, and meconium is passed. Meconium-stained amniotic fluid is recognized as a sign of fetal distress and can cause damage to the lungs if it is aspirated.

At birth placental nutrients are lost but oral feeding is not yet established

With clamping of the umbilical cord soon after delivery, intravenous nutrition of the baby ceases. However, it will be several days before oral feeding is fully established. During this time the neonate must rely on the stores of fat and carbohydrate laid down during late gestation. This accounts for the typical weight loss seen in the first few days of life. A baby will normally regain its birth weight within 7–10 days. Most importantly, glycogen is broken down to glucose under the influence of catecholamines secreted by the adrenal medulla. Premature or low-birth-weight babies may experience problems because of inadequate stores and may require intravenous nutrition.

As milk feeds are established, the chief metabolic substrate switches from glucose to fat

As the first milk feeds are ingested by the baby, its gut increases rapidly in size to accommodate the relatively large volume of

Summary

1. The fetal gut is relatively immature. The fetus is nourished solely by the placenta and the chief metabolic substrate is glucose. There is a limited degree of motility and secretory activity.

2. The contents of the fetal large intestine accumulate as meconium, which may be passed into the amniotic fluid during fetal distress.

3. After birth, but before the full establishment of oral feeding, the baby relies largely on stores of fat and carbohydrate laid down during late gestation.

4. With the establishment of milk feeds, the major metabolic substrate switches to fats. Digestive juice secretion and motility increase.

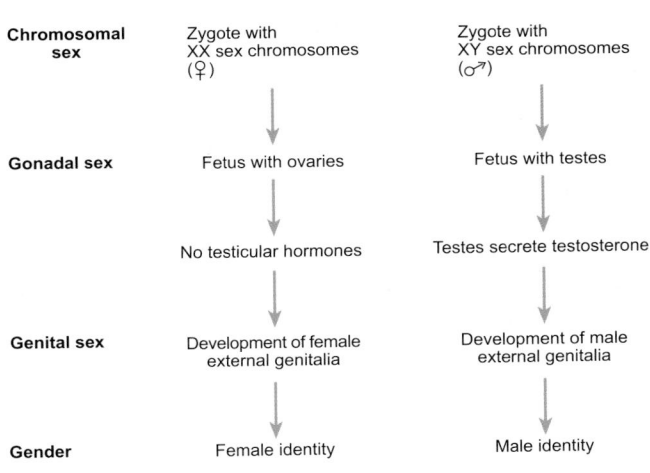

Fig. 22.9 The sequence of prenatal development of gender including differentiation of the appropriate gonads and genitalia.

fluid it must now handle. At the same time, secretion of digestive juices is stimulated and motility increases. Mature human milk is rich in fat (see Chapter 21, p. 473) and this becomes the major metabolic substrate. Lactose, the chief carbohydrate of milk, is hydrolyzed by lactase, an enzyme located in the small intestinal brush border. A specific lack of this enzyme, or a more generalized reduction in pancreatic enzymes, as occurs in cystic fibrosis for example, will result in a substantial reduction in digestion and absorption. The tarry meconium that was present in the fetal large intestine is usually passed during the first few days, after which the semi-liquid stools change to green and then yellowish-brown in color. Bowel movements are generally frequent in young babies, although there is also great variability —there may be as many as 12 stools a day or as few as one every 3 or 4 days.

22.8 Development of the male and female reproductive tissues

Humans have 23 pairs of chromosomes, one pair of which are the sex chromosomes. In the female both of these are X chromosomes and all her ova will carry a single X chromosome. However, in the male, the sex chromosomes consist of one X and one Y. Therefore sperm may carry either an X or a Y chromosome. In humans, the female is said to be the homogametic sex (XX), while the male is the heterogametic sex (XY). Thus it follows that if an ovum is fertilized by a sperm carrying an X chromosome, the resulting baby will be a girl, while fertilization by a sperm carrying a Y chromosome will produce a boy (Fig. 22.9). Studies of patients with a range of chromosomal abnormalities have revealed that the presence of a Y chromosome is the critical determinant of 'maleness', at least as far as gonadal development in the embryo is concerned (Box 22.1). If a Y chromosome is present, male gonads (testes) will develop, but in the absence of a Y chromosome female gonads (ovaries) will form. Recently it has been shown that only a small part of the Y chromosome is actually required for the determination of 'maleness'. This is the so-called sex-determining region of the Y chromosome, the *SRY* gene, in whose presence testes develop. Indeed, studies using mice have shown that the *SRY* gene can induce maleness in XX

BOX 22.1 ABNORMALITIES OF SEXUAL DIFFERENTIATION

Sex is determined genetically in humans. The normal male chromosomal karyotype is XY while that of the female is XX. As described in the main text, in the presence of a Y chromosome testes develop, while in its absence ovaries are formed. Subsequent differentiation of the male and female genitalia depends upon the existence of either functional ovaries or testes. Genetic errors can result in anatomical aberrations and distortion of sexual differentiation. A few of the more widely occurring abnormalities of this kind are described below;

1. Turner's syndrome (karyotype XO). Here there is only a single sex chromosome and, in the absence of either a second X chromosome to stimulate normal ovarian development or a Y chromosome to stimulate testicular formation, the gonad remains as a primitive streak. In the absence of functional testes, the external genitalia develop as the female type.

2. Klinefelter's syndrome (karyotype XXY). In this condition, the internal and external genitalia develop as male because a Y chromosome is present, but the ability of the testes to carry out spermatogenesis is severely impaired by the presence of an additional X chromosome. Females who carry additional X chromosomes (e.g. karyotype XXX or XXXX) may also have a shortened or impaired reproductive life because of damage to germ cell function, although the mechanism for this is not understood.

3. Certain individuals with a normal XY (male) karyotype lack the capacity to respond to androgens owing to a receptor deficiency. Such individuals will develop testes but show no growth or development of the Wolffian ducts nor masculinization of the external genitalia.

4. Certain enzymatic deficiencies in otherwise normal XX individuals can result in overproduction of androgens during fetal life. In such cases there may be mild or severe masculinization of the external genitalia, despite the presence of normal ovaries.

individuals otherwise lacking in all other genes normally carried by the Y chromosome.

In the early embryo, the gonads of males and females are indistinguishable

For the first 5 or 6 weeks of fetal life the gonads of males and females develop identically. They are made up of two different types of tissue, *somatic mesenchymal tissue*, which forms the matrix of the organ, and the *primordial germ cells*, which form the gametes. Ridges of mesenchymal tissue (the primitive sex cords) develop on either side of the dorsal aorta between weeks 3 and 4 of gestation. The primordial germ cells originate outside these ridges but migrate via the developing hindgut, gut mesentery,

and the region of the kidneys to lie between and within the sex cords by the sixth week of fetal life. At the same time the population of germ cells is expanding by mitosis.

The development of testes depends on the presence of a Y chromosome

Up until around week 6 of gestation, the gonads of males and females are indistinguishable and are said to be 'indifferent'. After completion of the migration of the germ cells to the primitive sex cords, divergence of the gonads resulting from Y-chromosome determination of 'maleness' starts to become apparent. The primitive sex cords of the male embryo undergo considerable proliferation to make contact with ingrowing mesonephric

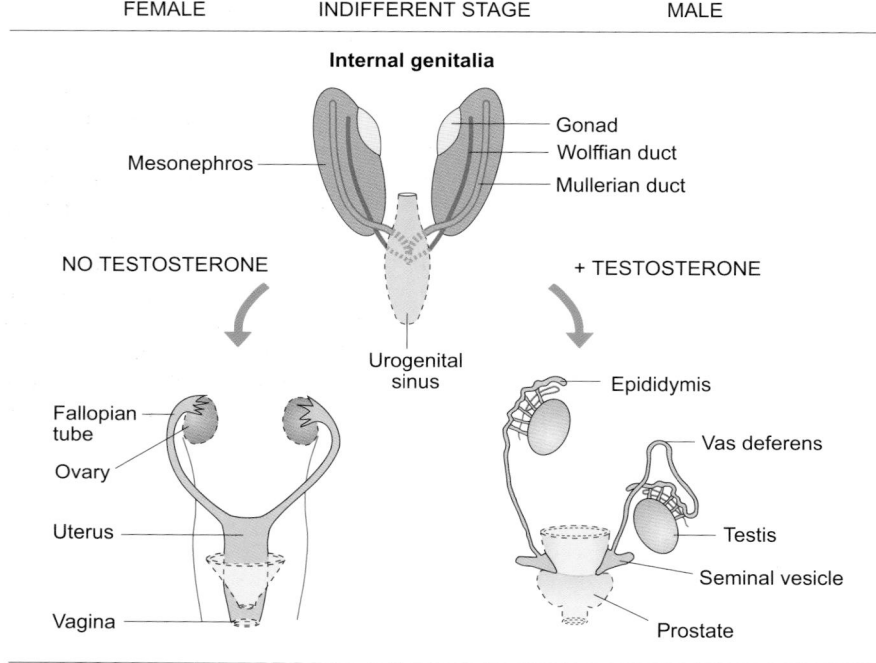

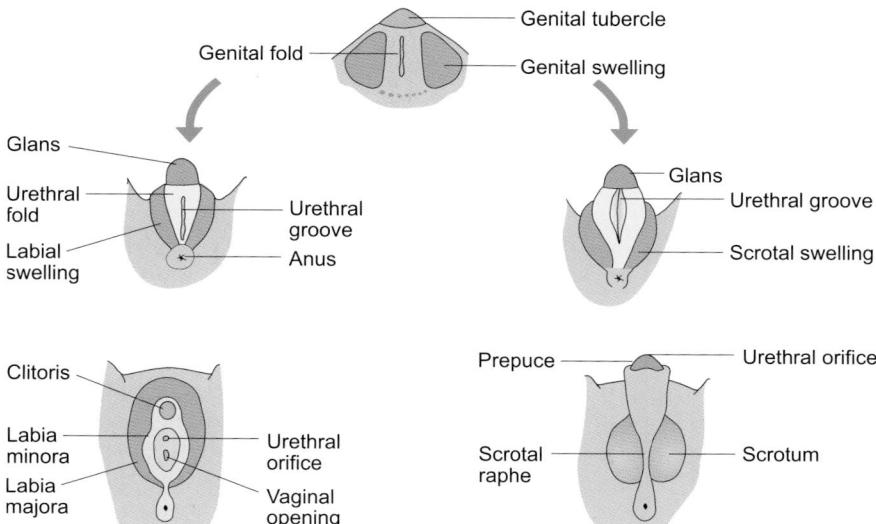

Fig. 22.10 The role of the sex hormones in the development of the internal and external genitalia. In the upper panel one of the testes is shown in the process of descent.

tissue and form a structured organ surrounded by a fibrous layer—the tunica albuginea. The cells of the sex cords, incorporating primordial germ cells, secrete a basement membrane. They are then known as the seminiferous cords and will give rise to the seminiferous tubules of the fully developed testis. Within these cords, the primordial germ cells will give rise to spermatozoa while the mesenchymal cord cells will form the Sertoli cells. The specific endocrine Leydig cells form as clusters within the stromal mesenchymal tissue lying between the cords.

The presence of a Y chromosome within the mesodermal cells of the genital ridge initiates the conversion of an indifferent gonad into a testis. In the absence of a Y chromosome, the changes in gonadal organization described above do not occur—the developing female gonad appears to remain indifferent. The primordial germ cells continue to proliferate mitotically and the primitive sex cords disappear. A second set of cords arises in the cortical region of the gonad and breaks up into clusters of cells that surround the germ cells. In this way, the primitive follicles that characterize the ovary are laid down—the germ cells forming the oocytes and the cord cells forming the granulosa cells of the follicles. Between the follicles, groups of interstitial cells are laid down.

To summarize, during the early development of the fetal gonads, activity of a small part of the Y chromosome appears to play an essential role in triggering the divergence of the primitive sex organs. If the SRY gene is present, the indifferent gonad is converted to a testis with seminiferous cords, primordial germ cells which will form sperm, and tissue that will give rise to the Sertoli and Leydig cells. In the absence of a Y chromosome the

indifferent organ forms an ovary containing a population of primordial follicles.

Subsequent development of the male and female genitalia depends on the hormones secreted by the gonads

Once the fetal gonads are established, the role of the sex chromosomes in the determination of sex is largely complete. Subsequent steps in the development of the male and female genital organs seem to be determined by the nature of the gonads themselves. This is particularly so in the case of the male in whom the fetal testes secrete two hormones which appear to play a key role in differentiation of the male genitalia. These are testosterone from the Leydig tissue and a substance known as Müllerian inhibiting hormone (MIH) from the Sertoli cells. In their absence, i.e. when ovaries are present, female genitalia are formed (Fig. 22.9 and 22.10).

The fetus possesses two primordial internal genital tissues, the Wolffian duct which forms male organs and the Müllerian duct which gives rise to female parts. In a female fetus in whom ovaries have developed, the male (Wolffian) duct disappears (possibly as a consequence of the lack of testosterone) and the Müllerian ducts go on to develop into the Fallopian tubes, uterus, cervix, and upper vagina. In a male fetus, testosterone seems to stimulate development of the Wolffian ducts to give rise to the epididymis, seminal vesicles, and vas deferens. At the same time, the female Mullerian ducts regress under the influence of MIH secreted by the Sertoli cells. As with the divergence of the fetal sex organs, the male pattern of differentiation must

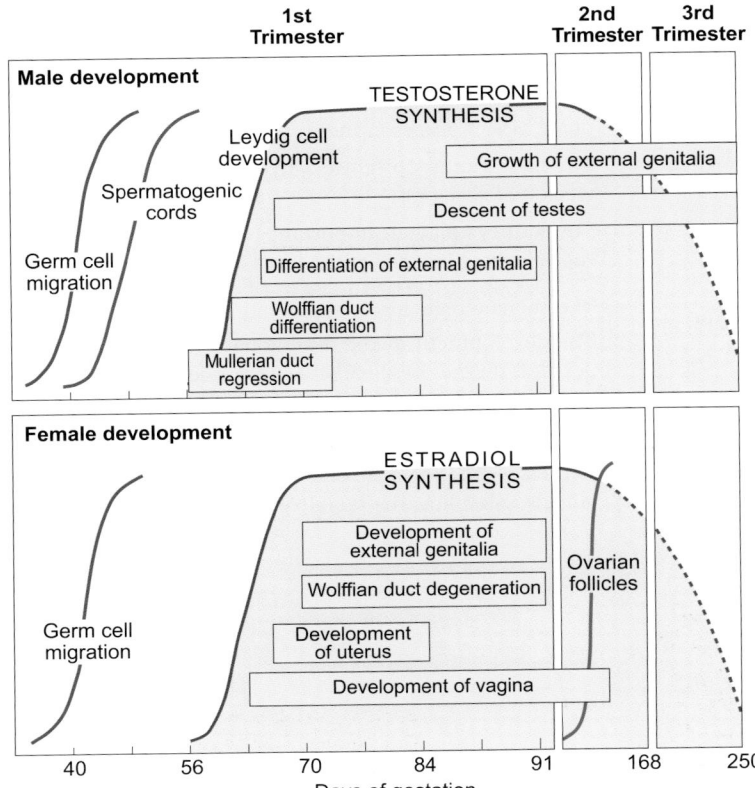

Fig. 22.11 The timing of the prenatal sexual differentiation of the internal and external genitalia of the human fetus.

Summary

1. Humans have 23 pairs of chromosomes, one pair of which are the sex chromosomes. The female (homogametic sex) has two X chromosomes, while the male (heterogametic sex) has one X and one Y chromosome.

2. In the presence of a Y chromosome the indifferent gonads of the fetus develop as testes, but in its absence ovaries develop. Subsequent steps in the development of the male and female genital organs seem to depend on the gonads themselves.

3. Androgens from the fetal testes play a particularly important role in stimulating the development of the internal male genitalia from the Wolffian ducts. In the presence of ovaries, the Müllerian ducts develop into the Fallopian tubes, uterus, cervix, and upper vagina.

be actively induced. In the absence of intervention, the female pattern develops inherently.

Fetal androgens also play a part in the development of the male external genitalia. They bring about fusion of the urethral folds to enclose the urethral tube and fusion of the genital swellings to form the scrotum. There is also enlargement of the genital tubercle to form the penis. In the female, the urethral folds and genital swellings remain separate to form the labia while the genital tubercle forms the small clitoris. These stages of development are represented diagrammatically in Figs 22.10 and 22.11.

Recommended reading

Griffin, N.E., and Ojeda, S.R. (2000). *Textbook of endocrine physiology* (4th edn). Oxford University Press, Oxford.

Johnson, M.H., and Everitt, B.J. (1999). *Essential reproduction* (5th edn), Chapter 11. Blackwell Scientific, Oxford.

Thorburn, G.D., and Harding, R. (1994). *Textbook of fetal physiology*. Oxford Medical Publications, Oxford.

Multiple choice questions

Each statement is either true or false. Answers are given below.

1. a. Blood returning to the fetus via the umbilical vein is fully saturated with oxygen.
    b. Fetal hemoglobin has a higher affinity for oxygen than adult hemoglobin.
    c. Fetal blood has a higher hemoglobin content than adult blood.
    d. Blood perfusing the brain of a fetus has the same P_{aO_2} as that of blood in the descending aorta.

2. a. The fetal heart rate is nearly double that of a healthy adult.
    b. All the blood in the umbilical vein enters the right atrium.
    c. The ductus arteriosus carries blood from the pulmonary artery to the descending aorta.
    d. The three fetal shunts normally close within a few days of birth.
    e. Fetal blood pressure is similar to that of an adult.

3. a. The fetus performs breathing movements *in utero*.
    b. The first breath is achieved by large changes in the intrathoracic pressure.
    c. Lung compliance in the newborn is much lower than that of an adult.
    d. Lack of surfactant in the neonate may cause respiratory distress.
    e. The peripheral chemoreceptors are active in the fetus.
    f. Immediately after birth, all the output of the right ventricle passes through the lungs.

4. a. Fetal cortisol stimulates the production of pulmonary surfactant by alveolar type II cells.
    b. The fetal zone of the adrenal gland synthesizes large quantities of progesterone.
    c. The fetal kidneys produce a hypotonic urine after about week 8 of gestation.
    d. The kidneys play an important role in the regulation of the acid–base balance of the fetus.

5. a. The neonate regulates its temperature mainly by shivering.
    b. The development of the fetal gonads into the male type depends on the presence of testosterone.
    c. The sex of an individual is determined by a single gene on the Y chromosome.
    d. In the absence of a Y chromosome the development of the gonads will follow the female pattern.

Answers to multiple choice questions

1. The blood in the umbilical vein is about 80 per cent saturated with oxygen, but as fetal blood has a higher hemoglobin content and as fetal hemoglobin has a higher affinity for oxygen than adult hemoglobin, the oxygen content of blood in the umbilical vein is about 16 ml dl⁻¹. As the blood perfusing the brain is supplied via the ascending aorta and as the ductus arteriosus supplies deoxygenated blood to the descending aorta, the blood perfusing the brain has a higher P_{O_2} than that of the descending aorta.

   a. False

   b. True

   c. True

   d. False

2. Because of the operation of the crista dividens, most of the blood returning in the umbilical vein passes directly to the left atrium via the foramen ovale. Fetal blood pressure is normally low (c. 70/45 mmHg).

   a. True

   b. False

   c. True

   d. True

   e. False

3. To inflate the lungs for the first time, the neonate generates a very large negative intrathoracic pressure (about 10–15 times the pressure needed for a normal inspiration in a healthy adult). To exhale the first breath requires a large positive intrathoracic pressure. The fact that very large pressure changes are required during the establishment of breathing demonstrates that the lung compliance is very low compared with that of the adult. The central chemoreceptors are active but not the peripheral chemoreceptors, which become progressively more sensitive in the weeks following birth. Although the proportion of the right ventricular output passing through the lungs greatly increases after the first breath, it is some days before closure of the ductus arteriosus results in all the blood from the right ventricle passing through the lungs.

   a. True

   b. True

   c. True

   d. True

   e. False

   f. False

4. The fetal zone of the adrenal gland synthesizes large amounts of estrogen precursors which are converted to estrogenic hormones by the placenta.

   a. True

   b. False

   c. True

   d. True

5. The neonate is unable to generate very much heat by shivering; instead it utilizes non-shivering thermogenesis. Much of this additional heat is generated by the metabolism of brown adipose tissue.

   a. False

   b. True

   c. True

   d. True

The control of growth

After reading this chapter you should understand:

♦ Patterns of growth before and after birth

♦ The physiology of bone and the growth of long bones during childhood

♦ The importance of growth hormone and the consequences of over- and underproduction of growth hormone in childhood and adulthood

♦ The role of other hormones in growth and the physiology of the skeleton

♦ The adolescent growth spurt and the influence of the sex steroids

♦ The factors which govern the overall size of tissues and organs

♦ The processes involved in the transformation of normal into malignant cells

23.1 Introduction

All biological tissues are made up of cells. Life begins as a single cell, the fertilized egg, from which all the diverse cell types of the body arise within a few weeks. Very early in development, cells begin to specialize and develop into particular types—liver cells, nerve cells, epithelial cells, muscle cells, and so on. Each cell type has its appropriate place within the organism. This development of specific and distinctive features is known as *differentiation*. Differentiated cells maintain their specialized character and pass it on to their progeny through the process of mitosis (see Chapter 3).

Overall growth of the body involves an increase in size and weight of the body tissues with the deposition of additional protein, and is thus a measurable quantitative change. In contrast, development occurs through a series of coordinated qualitative changes that affect the complexity and function of body tissues. Developmental change is most rapid while an individual is young. Growth and development are complex processes that are influenced by a number of different factors, both genetic and environmental. It is believed that genetic factors set both the basic guidelines for the overall height that may be achieved (as indicated by the correlation of adult height between parents and children) and the pattern and timing of growth spurts.

The major influence superimposed upon the genetic make-up of an individual is probably nutritional, although illness, trauma, and other socio-economic factors such as smoking can also modify the processes involved in growth. A child who has a diet that is inadequate with regard to either its quality or quantity will be unlikely to achieve his or her full genetic potential in terms of adult height. Indeed, improved nutrition is cited as one of the most important factors in the increase in average height that has been noted in Western societies over the last century.

Growth occurs at the level of individual cells, in populations of cells (the tissues and organs), and at the level of the whole body. The underlying processes are regulated by a number of different hormones including growth hormone, thyroid hormones, and the sex steroids. The general properties of these hormones are discussed in Chapter 12 and only those aspects of endocrine activity which relate specifically to growth will be considered in this chapter. Overall body growth will be considered first and given the greatest emphasis. A detailed discussion of the many factors believed to be responsible for the development and maintenance of appropriately sized populations of differentiated cells is beyond the scope of this book. Nevertheless, the importance of tissue growth and the factors that control the overall size of cell populations cannot be ignored, and a brief overview, including abnormalities of cell and tissue growth, will be given at the end of the chapter.

Patterns of growth during fetal life

The period of prenatal growth is of great importance to an individual's future well-being. The development of sensitive ultrasound techniques has meant that it is now possible to monitor fetal size throughout pregnancy. Measurements of abdominal circumference, femur length, and biparietal diameter (the distance across the head measured from one ear to the other) are commonly taken to assess the increasing size of the fetus. Figure 23.1 shows that the rate of increase in body length is at its greatest at around weeks 16–20 of gestation. Prior to this, particularly during the so-called embryonic period (the first 8 weeks after fertilization), growth velocity is lower, and there is a greater emphasis on the differentiation of body parts, such as the head, arms, and legs, and of cells into specialized tissues such as muscle and nerve. Each region is molded into a definite shape by the processes of cell migration and differential growth rates (morphogenesis).

Until weeks 26–28 of gestation, the increase in fetal weight is due largely to the accumulation of protein, as the major cells of the body are multiplying and enlarging. During the last 10 weeks or so, the fetus starts to accumulate a considerable amount (up to 400 g) of fat that is distributed both subcutaneously and deep within the body. Peak velocity of fetal weight gain occurs at around week 34 of gestation. The rate of increase then declines towards the time of delivery. The exact mechanism for this slowing is not clear, but it seems likely that the placental blood supply is less and less able to meet the ever-increasing nutritional demands of the fetus (see Chapter 21).

A large number of factors may influence the rate of fetal growth, but their relative importance remains unclear. Genetic, endocrine, and environmental factors are likely to be as important in fetal life as they are in postnatal development, with the

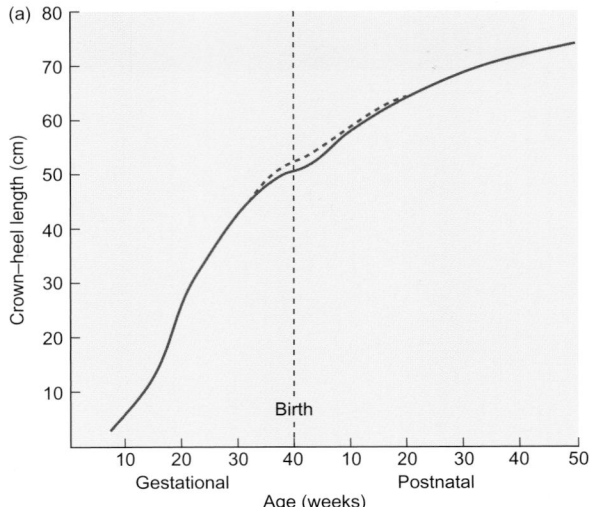

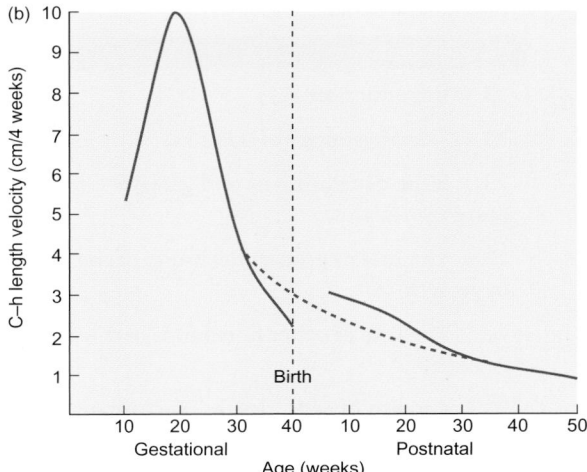

Fig. 23.1 Fetal and early postnatal growth in boys: (a) The increase in length; (b) The rate of growth or growth velocity. In each case, the dotted lines represent the theoretical values expected in the absence of any uterine restrictions. In reality, growth commonly slows towards the end of gestation as the placenta becomes less able to meet the ever-growing demands of the fetus. 'Catch-up' growth is seen following delivery if the child is adequately nourished.

genetic constitution setting the upper limits of fetal size and the level of nourishment provided by the placenta determining to what extent the genetic potential is achieved. In turn, placental efficiency will be affected by numerous maternal influences such as smoking, medication, alcohol consumption, and nutritional status.

Patterns of growth and development during childhood and adolescence

The rapid rate of growth seen in fetal life continues into the postnatal period but declines significantly through early childhood. There is further deceleration prior to the growth spurt of puberty. This pattern is illustrated in Fig. 23.2, which shows the oldest known longitudinal record of growth, carried out during the years 1759–1777.

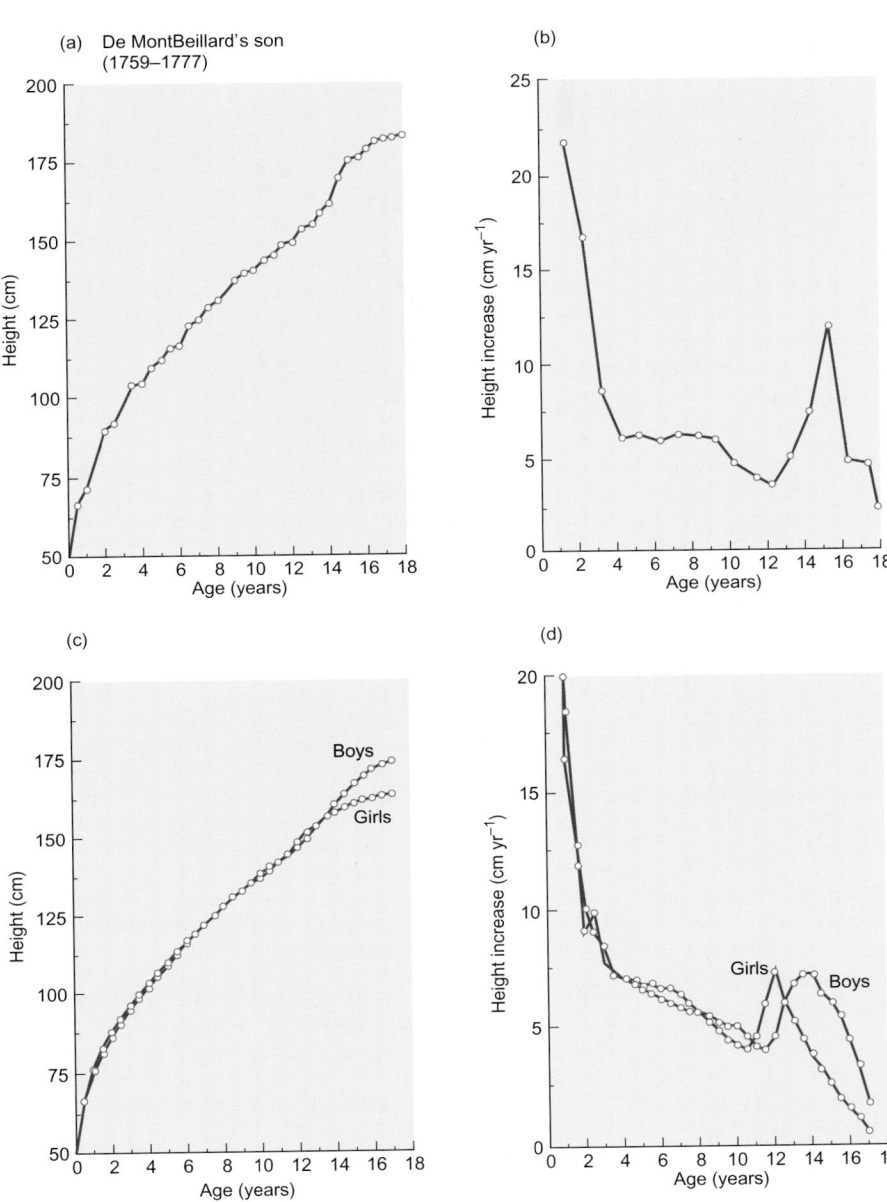

Fig. 23.2 (a), (b) The growth record of an individual child as recorded by his father in the eighteenth century. (c), (d) A comparison of average growth velocity curves for boys and girls from early childhood to adulthood. Note the timing of the adolescent growth spurts in boys and girls and the difference in final heights achieved.

The age at which the adolescent growth spurt takes place varies considerably between individuals. It occurs on average between 10.5 and 13 years in girls and between 12.5 and 15 years in boys. In general, the earlier the growth spurt occurs, the shorter will be the final stature. During this period, there is considerable variation in both stature and development between individuals of the same chronological age. The endocrine changes that accompany and contribute to the adolescent growth spurt will be considered further in Section 23.5.

Most body measurements follow approximately the growth curves described for height. The skeleton and muscles grow in this manner, as do many internal organs such as the liver, spleen and kidneys. However, certain tissues do not conform to this pattern and vary in their rate and timing of growth (Fig. 23.3). Examples include the reproductive organs (which show a significant growth spurt during puberty), the brain and skull, and

the lymphoid tissue. The brain, together with the skull, eyes, and ears, develops earlier than any other part of the body and thus has a characteristic postnatal curve. The lymphoid tissue also shows a characteristic pattern of growth. It reaches its maximum mass before adolescence and then, probably under the influence of the sex hormones, declines to its adult value. In particular, the thymus gland, a well-developed structure in children that plays a major role in the early development of the immune system, atrophies after puberty. It is no more than a residual nodule of tissue in adults.

Growth, even of the skeleton, does not cease entirely at the end of the adolescent period. Although there is no further increase in the length of the limb bones, the vertebral column continues to grow until the age of about 30 by the addition of bone to the upper and lower surfaces of the vertebrae. This gives rise to an additional height increase of 3–5 mm in the post-

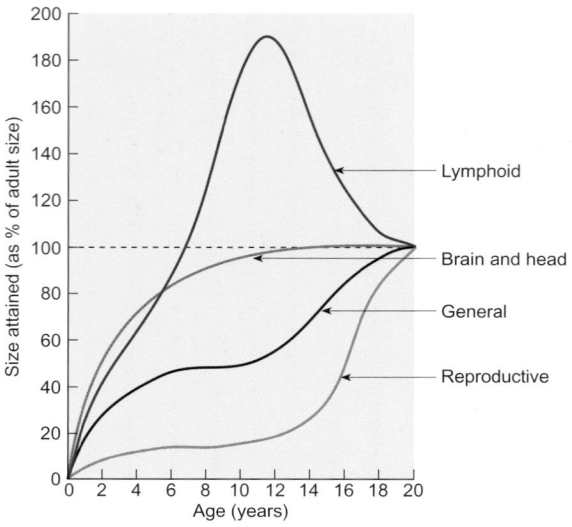

Fig. 23.3 Growth curves for individual body tissues.

Summary

1. Growth occurs at the level of the cells, tissues, and the whole body. Normal growth is influenced by many factors—genetic and environmental. A number of hormones are involved in the regulation of growth and development including growth hormone, thyroid hormones, and sex steroids.

2. Fetal growth is at its greatest between weeks 16 and 20 of gestation as new protein is laid down through the process of cell division. Before this time the emphasis is on morphogenesis—the differentiation and specialization of cells into tissues and organs. Fat is deposited predominantly during the last 10 weeks of gestation.

3. Growth velocity progressively declines until puberty when there is a growth spurt. The adult height is achieved by the end of puberty and growth in stature ceases. Individual tissues such as the brain and lymphoid tissue show characteristic growth patterns.

adolescent period. However, for practical purposes it can be considered that the average boy stops growing at around 17.5 years of age and the average girl at around 15.5 years of age with a 2-year variability range on either side.

23.2 The physiology of bone

Bone is a specialized form of connective tissue that is made durable by the deposition of mineral within its infrastructure. In an adult, skeletal bone forms one of the largest masses of tissue, weighing 10–12 kg. Far from being the inert supporting structure its outward appearance might suggest, bone is a dynamic tissue with a high rate of metabolic activity which is continuously undergoing complex structural alterations under the

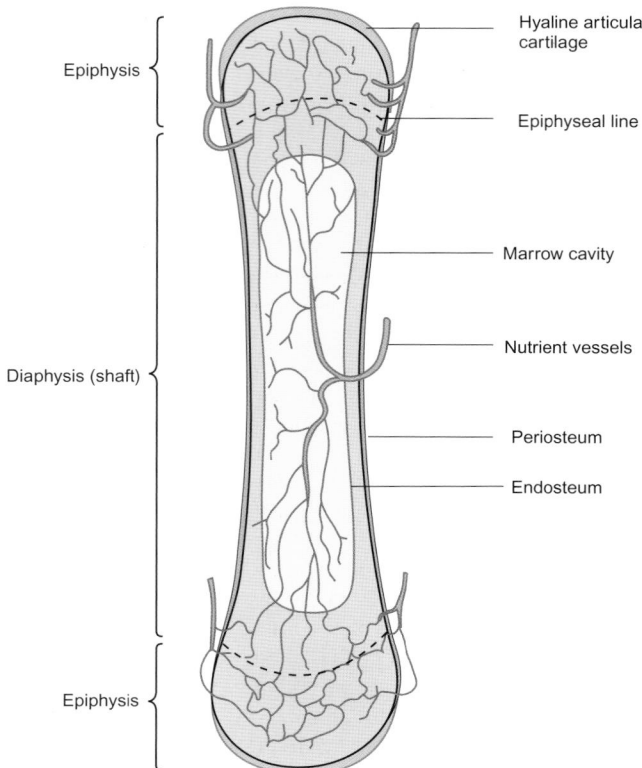

Fig. 23.4 The structural components of a typical long bone.

influence of mechanical stresses and a variety of hormones. Four main functions are ascribed to bone:

(1) to provide protection and structural support for the body and an attachment for muscles, tendons, and ligaments;

(2) to allow movement by means of articulations (joints);

(3) the homeostasis of mineral (calcium and phosphate);

(4) to form blood cells from hematopoietic tissue in the red bone marrow, which is found particularly in the short, flat, and irregular bones.

Three major tissue components are found in bone. About 30 per cent of total skeletal mass is made up of *osteoid*, an organic matrix consisting largely of collagen together with hyaluronic acid, chondroitin sulfate, and a vitamin-K-dependent protein called osteocalcin which is an important calcium-binding molecule. The remainder consists of a *mineral matrix* of calcium phosphate (hydroxyapatite) crystals and *bone cells* including osteoblasts (bone-forming cells), osteoclasts (bone-resorbing cells), osteocytes (mature bone cells), and fibroblasts.

The anatomical features of a typical long bone are illustrated in Fig. 23.4. The central shaft is called the *diaphysis* while the regions at either end of the bone are the *epiphyses*. Between the diaphysis and epiphysis is a region of bone known as the *epiphyseal plate* or growth plate. Adjacent to this is the growing end of the diaphysis, known as the *metaphysis*. During growth, this region is made of cartilage, but once growth is completed following puberty, the plate becomes fully calcified and remains as the *epiphyseal line*. Growth in length occurs by deposition of new cartilage at the metaphysis and its subsequent mineralization. The process by which bone becomes mineralized is not fully understood. Calcium phosphate appears to become oriented along the collagen molecules of the organic matrix. Surface ions of the crystals are hydrated, forming a layer through which exchange of substances with the extracellular medium can occur. The adult skeleton contains between 1 and 2 kg of calcium (about 99 per cent of the body total) and between 0.5 and 0.75 kg of phosphorus (about 88 per cent of the body total).

The surfaces of the bones are covered by *periosteum*, which consists of an outer layer of tough fibrous connective tissue and an inner layer of osteogenic ('bone-forming') tissue. A central space runs through the center of bones. This is the *marrow* (or medullary) space, which is lined with osteogenic tissue (the *endosteum*). The marrow spaces of the long bones contain mainly fatty yellow marrow that is not involved in hematopoiesis under normal circumstances. Red marrow containing hematopoietic tissue is found within the small, flat, and irregular bones of the skeleton, such as the sternum, ilium, and vertebrae. It is here that blood cell production is carried out.

Long bones are supplied by the nutrient artery, the periosteal arteries, and the metaphyseal and epiphyseal arteries. The nutrient artery branches from a systemic artery and pierces the diaphysis before giving rise to ascending and descending medullary arteries within the marrow cavity. In turn, these give rise to arteries supplying the endosteum and diaphysis. The periosteal blood supply takes the form of a capillary network, while the metaphyseal and epiphyseal vessels branch off from the nutrient artery. At rest, the arterial flow rate to the skeleton is around 12 per cent of the total cardiac output (2–3 ml per 100 mg tissue per minute). The mechanisms that control skeletal circulation are poorly understood, but it is known that blood flow is significantly increased during inflammation and infection and following fracture (see below). The blood flow to the red bone marrow is increased during chronic hypoxia when red blood cell production is enhanced in response to erythropoietin secreted by the kidney.

Bone is not uniformly solid but contains spaces that provide channels for blood vessels and also reduce the weight of the skeleton. Bone can be classified as either *compact* (*dense*) or *spongy* (*trabecular, cancellous*) according to the size and distribution of the spaces. Compact bone forms the outer regions of all bones, the diaphysis of long bones, and the outer and inner regions of flat bones. It contains few spaces and provides protection and support especially for the long bones in which it helps to reduce the stress of weight bearing. The functional units of compact bone are the *Haversian systems* or *osteons*. These consist of a central canal, which contains blood vessels, lymphatics, and nerves, surrounded by concentric rings of hard intercellular substance (lamellae) between which are spaces (lacunae) containing osteocytes (mature bone cells) (Fig. 23.5). Radiating from the lacunae are tiny canals (canaliculi) that connect with adjacent lacunae to form a branching network through which nutrients and waste products can be transported to and from the osteocytes. In contrast, spongy bone contains no true osteons but consists of an irregular lattice of thin plates or spicules of bone (trabeculae) between which are large spaces filled with bone marrow. Lacunae containing osteocytes lie within the trabeculae. The

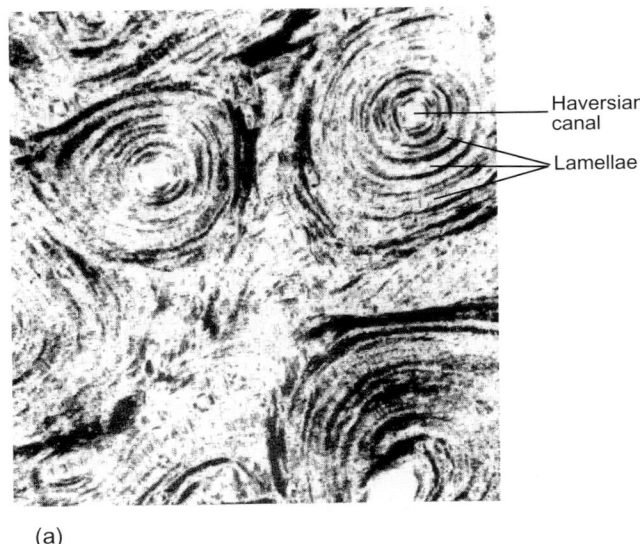

(a)

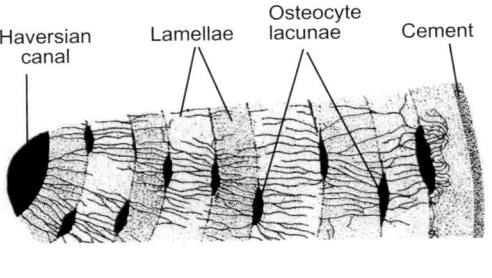

(b)

Fig. 23.5 The structure of compact bone. Note the lamellar organization of the Haversian systems and the lacunae which imprison the osteocytes.

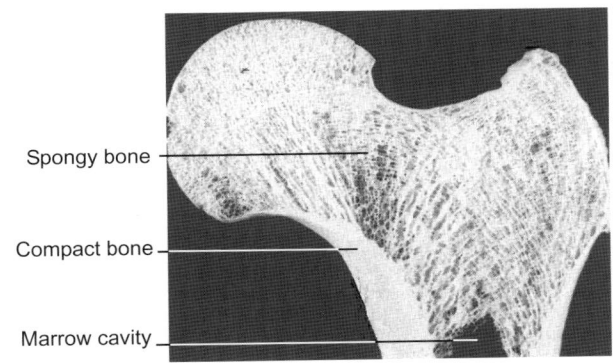

Fig. 23.6 A section through the head of the femur showing regions of dense and spongy bone.

osteocytes are nourished directly by blood circulating through the marrow cavities from blood vessels penetrating to the spongy bone from the periosteum. Spongy bone makes up most of the mass of short, flat, and irregular bones and is present within the epiphyses of long bones and at the growth plates. Figure 23.6 illustrates the different organization of dense and spongy bone.

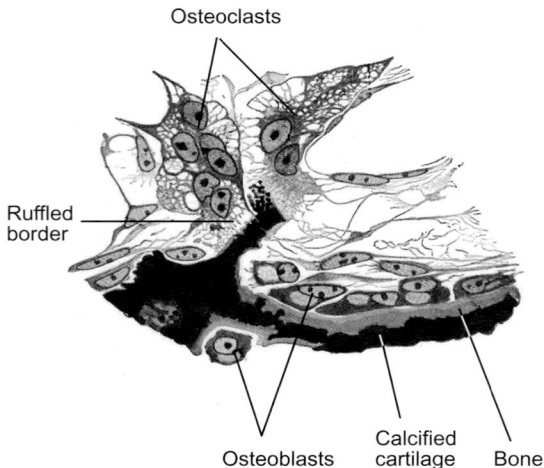

Osteoclasts

Ruffled
border

Osteoblasts   Calcified   Bone
cartilage

Fig. 23.7 A section through the marrow cavity of bone showing calcified cartilage (here stained dark blue), osteoblasts, and two large multinucleate osteoclasts. Note the ruffled border next to the calcified cartilage.

The bone cells

Three major cell types are recognized in histological sections of bone. These are *osteoblasts*, *osteocytes*, and *osteoclasts*. Their general appearance is shown in Fig. 23.7. The first two types originate from progenitor cells within the osteogenic tissue of the bone. Osteoclasts are believed to differentiate separately from mononuclear phagocytic cells.

Osteoblasts are present on the surfaces of all bones and line the internal marrow cavities. They contain numerous mitochondria and an extensive Golgi apparatus associated with rapid protein synthesis. They secrete the constituents of the organic matrix of bone including collagen, proteoglycans, and glycoproteins. They are also important in the process of mineralization (calcification) of this matrix. Osteoblasts possess specific receptors for parathyroid hormone and calcitriol (see below).

Osteocytes are mature bone cells derived from osteoblasts that have become trapped in lacunae (small spaces) within the matrix that they have secreted. As described above, adjacent osteocytes are linked by fine cytoplasmic processes that pass through tiny canals (canaliculi) between lacunae (Fig. 23.7). This arrangement permits the exchange of calcium from the interior to the exterior of bones and thence into the extracellular fluid. This transfer is known as osteocytic osteolysis and can be used to remove calcium from the most recently formed mineral crystals when plasma calcium levels fall (for more details regarding whole-body calcium balance, see Chapter 12).

Osteoclasts are giant multinucleated cells that are believed to arise from the fusion of several precursor cells and therefore contain numerous mitochondria and lysosomes. They are highly mobile cells that are responsible for the resorption of bone during growth and skeletal remodeling. They are abundant at or near the surfaces of bone undergoing erosion. At their site of contact with the bone is a highly folded 'ruffled border' of microvilli that infiltrates the disintegrating bone surface. Bone dissolution is brought about by the actions of collagenase, lysosomal enzymes, and acid phosphatase. Calcium, phosphate, and

the constituents of the bone matrix are released into the extracellular fluid as bone mass is reduced. The activity of the osteoclasts appears to be controlled by a number of hormones, notably parathyroid hormone, calcitonin, thyroxine, estrogens, and the metabolites of vitamin D (see Chapter 12 for further details).

23.3 Bone development and growth (osteogenesis)

At week 6 of gestation the fetal skeleton is constructed entirely of fibrous membranes and hyaline cartilage. From this time, bone tissue begins to develop and eventually replaces most of the existing structures. Although this process of ossification begins early in fetal life, it is not complete until the third decade of adult life. The bones of the cranium, lower jaw, scapula, pelvis, and the clavicles develop from fibrous membranes by a process called *intramembranous ossification*. In this process, new bone is formed on the surface of existing bone. The bones of the rest of the skeleton grow in length as hyaline cartilage templates are replaced by bone (a process known as *endochondral ossification*).

Growth of bone length

A long bone such as the radius in the forearm is laid down first as a cartilage model. At the center of this model, the so-called primary center of ossification, the cartilage cells break down and bone appears. This process begins early in fetal life and, shortly before birth, secondary centers of ossification have also developed, predominantly at the ends of the bone or epiphyses. Smaller bones such as the carpals and tarsals of the hands and feet develop from a single ossification center. The areas of cartilage between the diaphysis and the epiphyses are known as the growth plates. In the part of the growth plate immediately under the epiphysis is a layer of stem cells or chondroblasts. These give rise to clones of cells (chondrocytes) arranged in columns extending inwards from the epiphysis towards the diaphysis.

Several zones can be distinguished within the columns of chondrocytes. The outer zone is one of proliferation in which

Summary

1. Growth of the long bones is responsible for the increase in height seen during childhood and adolescence. Growth occurs through the proliferation and hypertrophy of cartilage cells at the growth plates of the long bones followed by their subsequent mineralization.

2. Bone is a dynamic tissue that continually undergoes remodeling and renewal even during adulthood. It consists of osteoid, an organic matrix, strengthened by the deposition of complex crystals of calcium and phosphate (hydroxyapatite).

3. There are three types of bone cells: osteoblasts, which secrete the organic matrix, osteocytes (mature bone cells), and osteoclasts, which are responsible for the resorption of old bone during growth and remodeling of the skeleton.

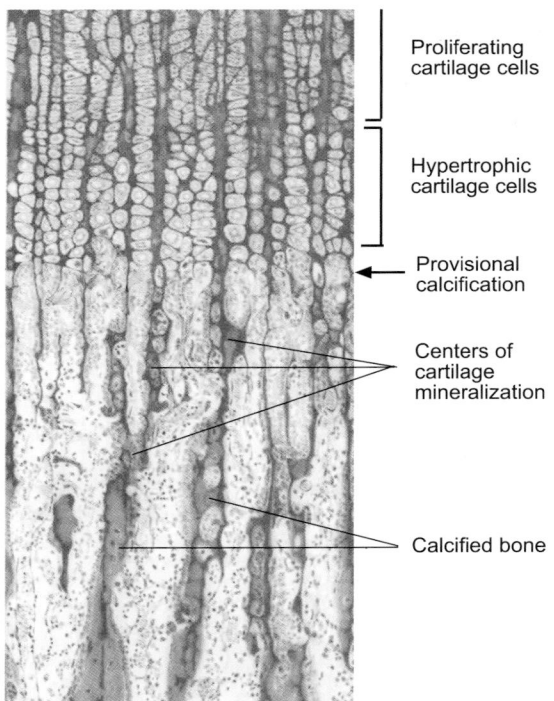

Proliferating
cartilage cells

Hypertrophic
cartilage cells

Provisional
calcification

Centers of
cartilage
mineralization

Calcified bone

Fig. 23.8 Decalcified bone showing the process of bone growth and ossification in a typical long bone. New cells formed in the proliferative region move down to the hypertrophic region to add to bone accumulating on the top of the diaphysis.

the cells are dividing rapidly. Beneath this are layers in which the cells mature, enlarge, and eventually degenerate as shown in Fig. 23.8. The innermost layer of cells is the region of calcification. Here, the osteogenic cells differentiate into osteoblasts and lay down bone. In the radiographs of growing hands illustrated in Fig. 23.9 the regions at which rapid calcification are occurring are clearly visible as areas of high density.

Thus, cartilage is produced at one end of the epiphyseal plate, while at the other end it is degenerating. Therefore growth in length is dependent upon the proliferation of new cartilage cells. In humans, it takes around 20 days for a cartilage cell to complete the journey from the start of proliferation to degeneration. Clearly, the bone marrow cavity must also increase in size as the bone grows, and to ensure this, osteoclasts erode bone within the diaphysis.

At the end of the growth period, the growth plate thins as it is gradually replaced by bone until it is eliminated altogether and the epiphysis and diaphysis are unified, a process known as *synostosis*. Following this 'fusion' of the epiphyseal plate no further increase in bone length is possible at this site. Although growth in length of most bones is complete by the age of 20, the clavicles do not ossify completely until the third decade of life. The dates of ossification are fairly constant between individuals but different between bones. This fact is exploited in forensic science to determine the age of a body according to which bones have, and which have not, ossified.

Growth of bone diameter

The growth in width of long bones is achieved by *appositional* bone growth in which osteoblasts beneath the periosteum of the bone form new osteons on the external surface of the bone.

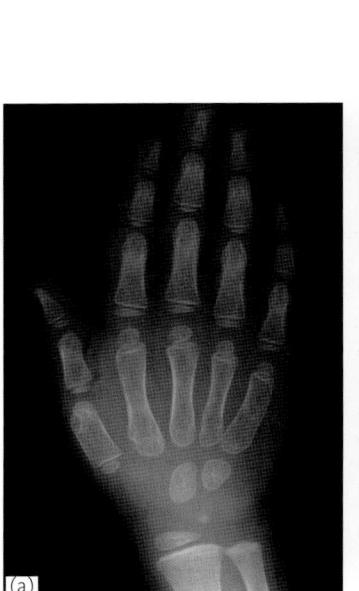

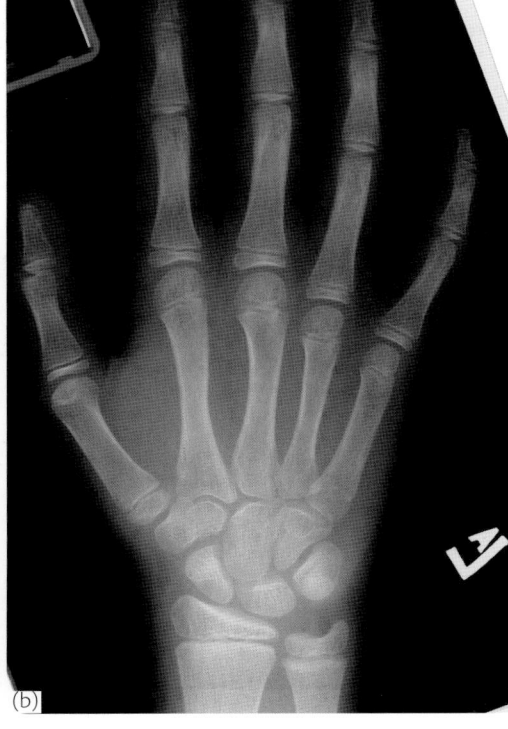

(a)

(b)

Fig. 23.9 Radiographs of the hands of two children aged (a) 2 years and (b) 11 years. Note the increase in the number of ossified carpal bones in the wrist of the 11-year-old.

BOX 23.1 BONE HEALING FOLLOWING A FRACTURE

When bone is fractured its original structure and strength are restored quite rapidly through the formation of new bone tissue. Provided that the edges of the fractured bone are repositioned and the bone is immobilized by splinting, repair will normally occur with no deformity of the skeleton. There are three stages in the repair of a fractured bone. The first stage occurs during the first 4 or 5 days after injury and involves the removal of debris resulting from the tissue damage. This includes bone and other tissue fragments as well as blood clots formed by bleeding between the bone ends and into surrounding muscle when the periosteum is damaged. Phagocytic cells such as macrophages clear the area and granulation tissue forms. This is a loosely gelled protein-rich exudate that forms at any site of tissue damage and which later becomes fibrosed and organized into scar tissue. As it revascularizes from undamaged capillaries in adjacent tissue, it takes on a pink granular appearance. Osteoblasts within the endos-

teum and periosteum migrate to the site of damage to initiate the second stage of healing (Fig. 1(a)). During this stage, which normally lasts for the next 3 weeks or so, osteoid is secreted by the osteoblasts into the granulation tissue to form a mass between the fractured bone to bridge the gap. This tissue mass is also known as *soft callus* (Fig. 1(b)). The soft callus gradually becomes ossified to form a region of woven bone (similar to cancellous bone), also called *hard callus* (Fig. 1(c)). At this stage of healing there is normally some degree of local swelling at the site of the fracture caused by the hard callus deposit. During the final stage in the process of healing the mass of hard callus is restructured to restore the original architecture of the bone. This stage may take place over many months and involves the actions of both osteoblasts and osteoclasts. During this time, the periosteum also re-forms and the bone is able to tolerate normal loads and stresses. The radiographs shown in Fig. 2 illustrate the main stages of healing following a fracture.

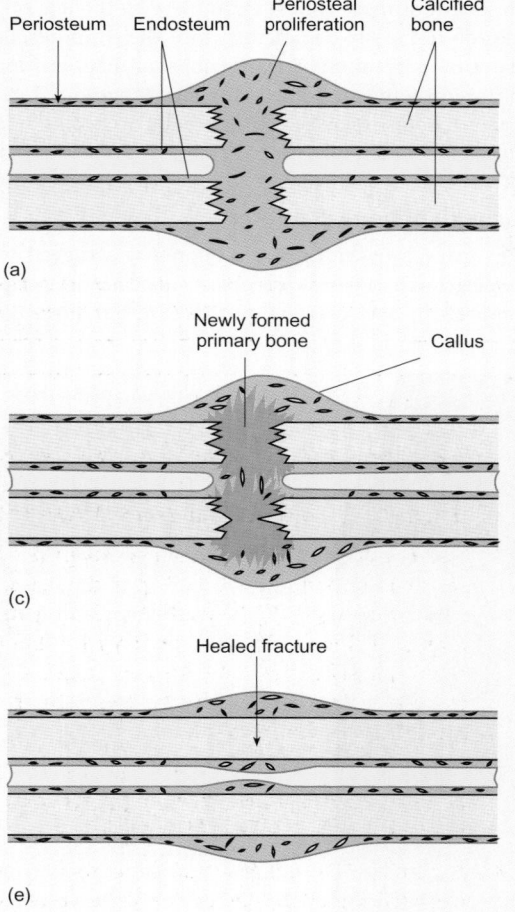

Fig. 1. The principal stages of bone repair following a fracture.

BOX 23.1 *(cont.)*

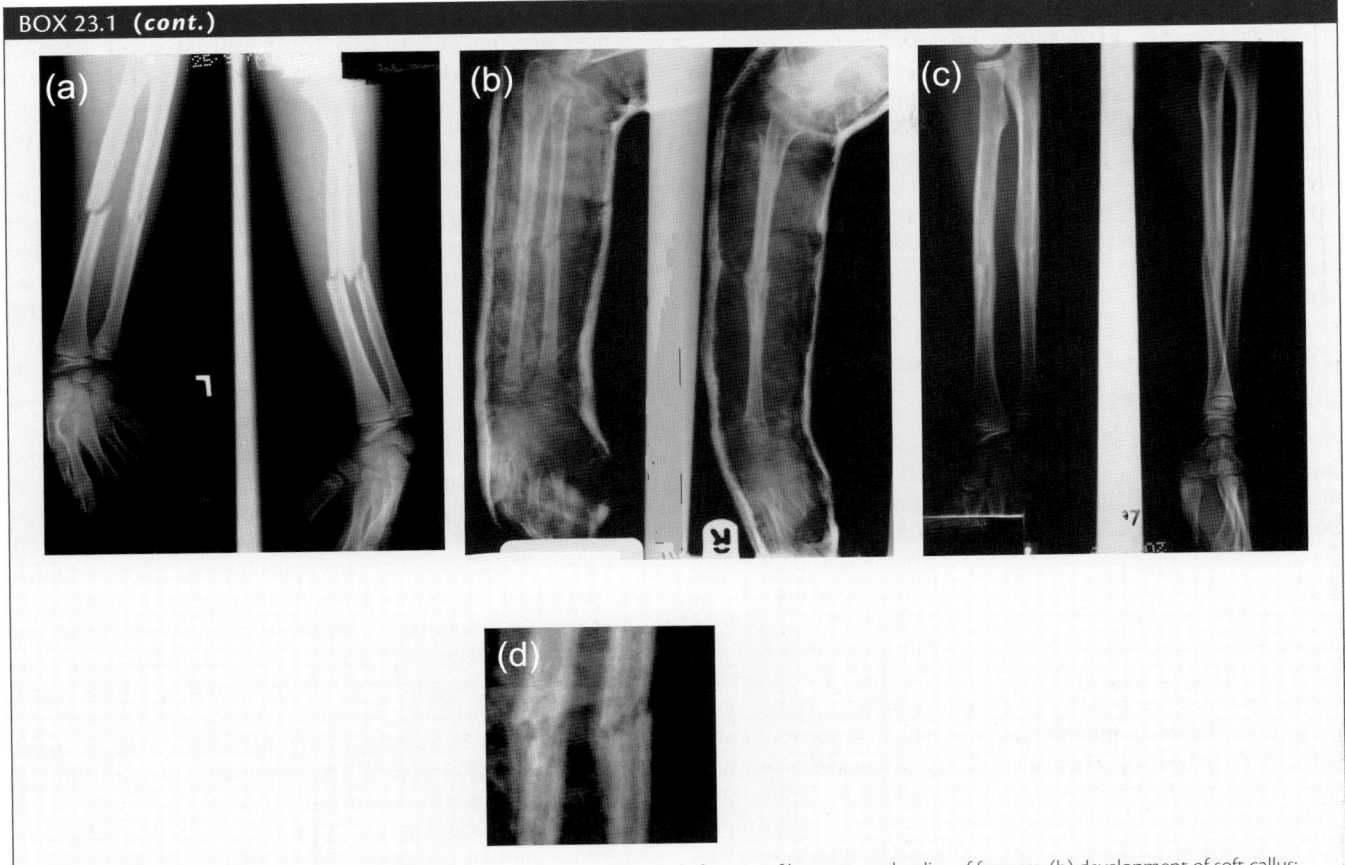

Fig. 2. Radiographs showing the stages of repair following a fracture: (a) simple fracture of humerus and radius of forearm; (b) development of soft callus; (c) final healing of fracture. (d) shows the soft callus in more detail. Note that the radiograph shown in (b) was taken though the immobilizing plaster cast.

Thus the bone becomes thicker and stronger. Rapid ossification of this new tissue takes place to keep pace with the growth in length of the bone. This process is similar to the mechanism by which the flat bones grow.

Remodeling of bone

Even after growth has ended, the skeleton is in a continuous state of remodeling as it is renewed and revitalized at the tissue level. Large volumes of bone are removed and replaced, and bone architecture continually changes as 5–7 per cent of bone mass is recycled each week. Furthermore, following a break to a bone, self-repair takes place remarkably quickly (Box 23.1). Remodeling allows bone to adapt to external stresses, adjusting its formation to increase strength when necessary. Remodeling occurs in cycles of activity in which resorption precedes formation. First, bone is eroded by the osteoclasts. This erosion is followed by a period of intense osteoblastic activity in which new bone is laid down to replace that which has been resorbed.

In general terms, bone is deposited in proportion to the load it must bear. Therefore it follows that in an immobilized person

Summary

1. During fetal life, the skeleton is laid down as a cartilage 'model' which subsequently becomes ossified. Ossification is not complete until the third decade of life.

2. The areas of cartilage between the diaphysis and epiphyses of long bones are called growth plates. Stem cells (chondroblasts) lying beneath the plates give rise to proliferating clones of cells (chondrocytes) that extend into the diaphysis of the bone. As they progress, they mature and later degenerate. Calcification occurs within the innermost layer of cells.

3. After puberty, the growth plate thins as it is replaced by bone and the diaphysis and epiphysis fuse. No further increase in length is possible after fusion.

4. Around 5–7 per cent of total bone mass is recycled each week during adulthood as the skeleton responds to the changing demands placed upon it.

23 THE CONTROL OF GROWTH

bone mass is rapidly (though reversibly) lost—a process known as *disuse osteoporosis*. Astronauts experiencing prolonged periods of weightlessness in space have been shown to lose up to 20 per cent of their bone mass in the absence of properly planned exercise programs. Similarly, appropriate exercise during childhood and adolescence is thought to enhance the development of bone and result in a stronger healthier skeleton in adult life, a factor that may be particularly important in females. However, the exact mechanisms that control the rate of deposition and loss of bone in response to mechanical requirements remain largely unknown.

23.4 The role of growth hormone in the control of growth

Growth is the result of the multiple interactions of circulating hormones, tissue responsiveness, and the supply of nutrients and energy for growing tissues. Many hormones are known to be involved in the regulation of growth at different stages of life. Nevertheless, growth hormone is the hormone that undoubtedly exerts a dominant effect on normally coordinated growth.

The nature and secretion of growth hormone (GH) has been discussed fully in Chapter 12, but a brief summary may be helpful here. GH is a polypeptide (M_r 22 kDa) derived from the pituitary somatotrophs. It bears a marked structural similarity to prolactin and human placental lactogen. The secretion of GH is controlled by hypothalamic releasing hormones. Growth hormone releasing hormone (GHRH) stimulates the output of GH while somatostatin inhibits it. GH shows a marked irregular pulsatile pattern of release which is influenced by a number of physiological stimuli. For example, stress and exercise both stimulate GH secretion, and there is a significant increase in the rate of secretion during slow-wave (deep) sleep, particularly in children. Both the pulsatile character and the sleep-induced patterns of release are lost in patients suffering from hypo- or hypersecretion of GH (Fig. 23.10).

Other hormones and products of metabolism also influence the rate of GH secretion. For example, estrogens increase the sensitivity of the pituitary to GHRH, an effect that contributes to the earlier growth spurt seen in adolescent girls compared with boys. GH secretion is decreased by the adrenal glucocorticoid hormones and stimulated by insulin. Oral glucose depresses GH release, while secretion is promoted by low levels of plasma glucose.

In common with most endocrine systems, the secretion of GH is under negative feedback control. This is probably mediated both by GH itself (chiefly at the level of the hypothalamus) and by the insulin-like growth factors (IGFs) (see below) that are thought to act at both pituitary and hypothalamic levels. GH interacts with its target cells at the plasma membrane where it binds to surface receptors. Synthesis of these receptors requires the presence of GH itself, while an excess of GH causes downregulation of the receptors. The mechanisms of signal transduction have now been clarified. GH activates membrane-bound tyrosine kinases which phosphorylate a group of proteins that activate gene transcription.

The actions of GH can be divided into metabolic and growth-promoting effects. The metabolic actions of GH tend to oppose those of insulin and are largely direct in nature. GH exerts its direct actions on a variety of target tissues, principally the liver, muscle and adipocytes. It depresses glucose metabolism (to spare glucose for use by the central nervous system in times of fasting or starvation). Furthermore, GH stimulates lipolysis, which increases the availability of fatty acids for oxidation, and facilitates the uptake of amino acids into cells for protein synthesis.

The growth-promoting actions of GH embrace both direct and indirect effects. GH seems to exert a direct stimulatory effect on chondrocytes, increasing the rate of differentiation of these cells and therefore of cartilage formation. Many of the direct metabolic actions of GH, such as the increase in uptake of amino acids and the rate of protein synthesis, will also contribute to the overall processes of growth and repair.

The indirect actions of growth hormone are mediated by a family of peptide hormone intermediaries called *insulin-like growth factors* (IGFs) formerly known as somatomedins. They have a molecular weight of around 7000 and are structurally related to proinsulin, the precursor of insulin. The IGFs are synthesized in direct response to GH, chiefly by the liver but also by other tissues including cartilage and adipose tissue. Plasma IGF-1 is increased by the administration of GH, with a time lag of 12–18 hours, and is reduced in individuals who lack GH. IGFs have plasma half-lives in excess of that of GH because they are carried in the blood bound to several proteins. The blood level of IGF-1 is low in infancy, rises gradually until puberty, and then increases more swiftly to reach a peak which coincides with the peak height increase (Fig. 23.2), after which it falls to its adult (and prepubertal) value.

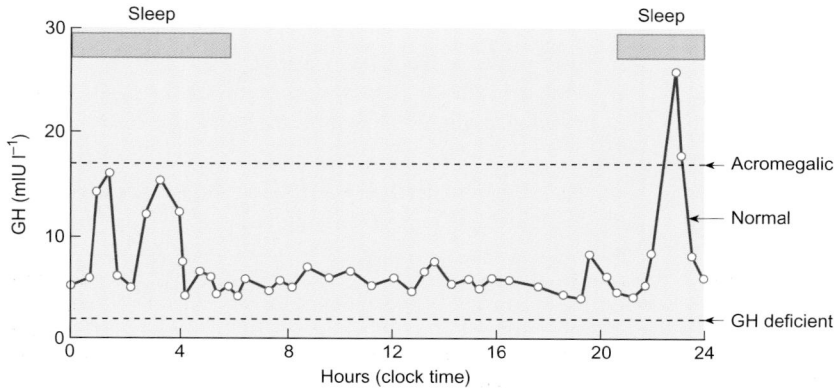

Fig. 23.10 A comparison of the pattern of secretion of GH in normal individuals and in patients suffering from acromegaly (GH excess in adulthood) and GH deficiency. Note that pulsatile secretion is lost in each of the abnormal states. In acromegaly, there is a sustained high circulating level of GH.

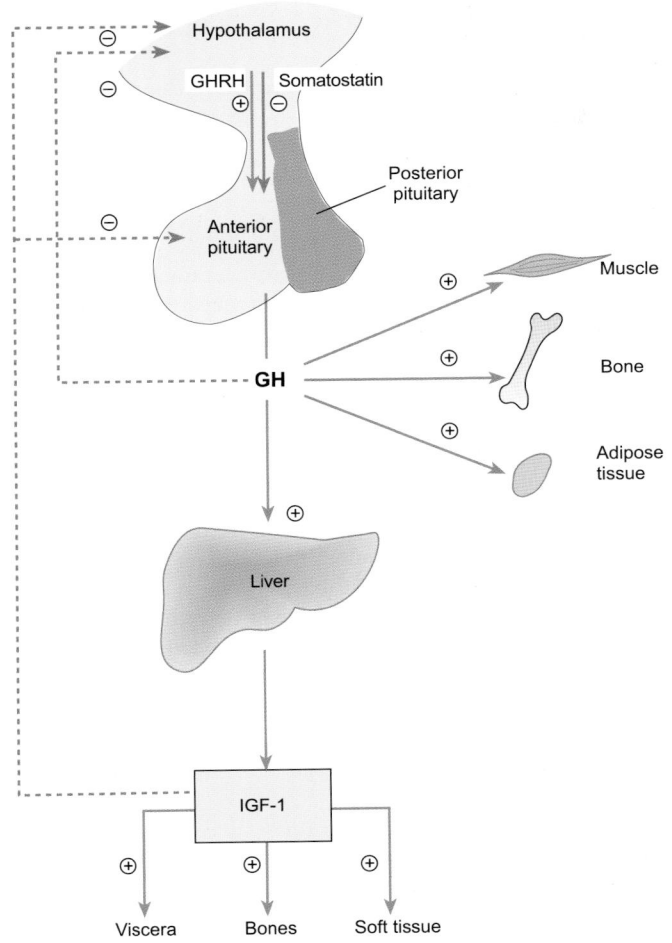

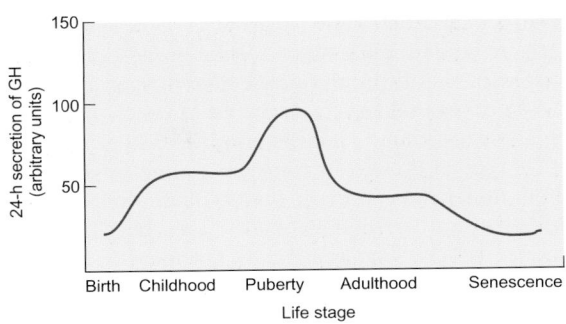

Fig. 23.12 The lifetime pattern of GH secretion.

Fig. 23.11 An overview of the regulation and effects of GH on its target tissues. ⊕ indicates a stimulatory action ⊖ an inhibitory one.

the control of growth and GH receptors do not appear until the final 2 months of gestation. The growth factors IGF-1 and IGF-2 appear to play a dominant role in fetal growth.

Following delivery, and in the early part of childhood, GH secretion increases considerably, and during this phase, overall growth and increase in stature seems to depend almost entirely on the actions of GH itself and of IGF-1. At puberty, there is a further significant rise in GH secretion (probably associated with an increase in the output of sex steroids) with a parallel increase in IGF-1 output. This promotes the further growth of the long bones and contributes to the adolescent growth spurt.

During the final phases of puberty the sex steroids cause the epiphyses to fuse, and during subsequent adult life no further increase in stature occurs. However, GH, may still play a part in the remodeling of bone and in the repair and maintenance of cartilage.

GH deficiency

As the preceding discussion suggests, GH is needed for normal growth between birth and adulthood. Individuals who lack

The actions of the IGFs, as their name suggests, tend to be insulin-like in character and account principally for the growth-promoting effects of GH. They act on cartilage, muscle, fat cells, fibroblasts, and tumor cells. More specifically related to bone growth is the action of IGFs (particularly IGF-1 and IGF-2) in stimulating the clonal expansion of chondrocytes and the formation and maturation of osteoblasts in the growth plates of the long bones. All aspects of the functions of the chondrocytes are stimulated, including the incorporation of the amino acid proline into collagen and its subsequent conversion to hydroxyproline. Furthermore, GH (via IGFs) stimulates the incorporation of sulfate into chondroitin. Chondroitin sulfate and collagen together form the tough inorganic matrix of cartilage. Growth of soft tissue and the viscera is also attributed to the indirect actions of GH via the IGFs.

A summary of the direct and indirect actions of GH, and of the factors regulating GH output, is shown in Fig. 23.11.

The importance of GH in growth at different stages of life

Figure 23.12 illustrates the pattern of GH secretion throughout life. During the fetal period, GH itself is of little importance in

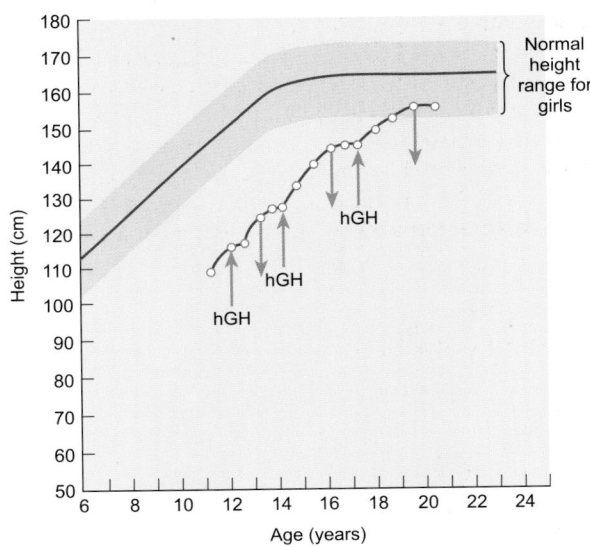

Fig. 23.13 The pattern of growth in a girl with isolated pituitary GH deficiency treated with three periods of exogenous GH administration. Note the 'catch-up' growth seen during the treatment periods. The downward-pointing arrows represent the end of each period of treatment. The range of heights for normal girls is indicated by the shaded area.

GH (so-called pituitary dwarfs) grow to a height of around 120–130 cm while remaining of normal proportions. This is in contrast with the disproportionate growth seen in achondroplasia, the congenital type of dwarfism in which growth of the bones is impaired due to defects in other local growth factors. A further type of growth impairment caused by defective GH receptors rather than a lack of the hormone itself is known as Laron dwarfism. These individuals have the same physical appearance as those who lack growth hormone.

GH-deficient children can be treated by injections of human GH. After treatment, they usually achieve significant catch-up growth and reach an acceptable adult height (Fig. 23.13). Unlike other hormones such as insulin and ACTH, growth hormone is species specific, i.e. animal GH is without effect in humans. From 1958 until 1985, the GH administered to patients was extracted from the pituitary glands of human cadavers at post-mortem. Unfortunately a few of the children treated in this way have since become ill or died from the degenerative brain disease Creutzfeld–Jakob Disease (CJD). In recent years recombinant DNA technology has developed, and now human GH can be manufactured and used to treat GH deficiency without risk of CJD.

Finally, short stature may be caused by a failure to produce the IGFs in response to GH rather than a simple lack of GH. In conditions of this kind, GH treatment will be of no value but such children can be treated with recombinant IGF-1.

GH excess

Although hypersecretion of GH may occur at any stage of life, the incidence of pituitary gigantism resulting from an excess of GH in childhood is extremely rare. Tumors of the pituitary gland or overgrowth of the GH-producing cells can occasionally cause vastly excessive (though proportionate) growth. A further condi-

Summary

1. The anterior pituitary peptide growth hormone (GH) has a dominant effect on postnatal growth. GH secretion rises during early infancy and shows a further peak during puberty. It is under hypothalamic control and shows pulsatile release. Moreover, GH secretion is stimulated by exercise, stress, fasting, and during non-REM sleep.

2. GH exerts both metabolic and growth-promoting effects. It directly stimulates the formation of cartilage by chondrocytes and enhances the rate of uptake and incorporation of amino acids into protein.

3. GH exerts a number of indirect effects through insulin-like growth factors (IGFs). IGFs stimulate the clonal expansion of chondrocytes and the formation of the organic bone matrix.

4. GH deficiency during childhood results in pituitary dwarfism. It can be treated by injections of genetically engineered human GH. Pituitary gigantism, which is very rare, results from hypersecretion of GH before the age of puberty. Hypersecretion of GH in adulthood is more common, resulting in acromegaly in which there are metabolic disturbances, thickening of bones, and overgrowth of soft tissues.

tion characterized by extreme tallness is cerebral gigantism (Sotos' syndrome) which seems to be caused by an over-reaction to GH by its target tissues rather than an excess of GH itself. This is extremely rare. Figure 23.14 illustrates the extremes of height that may be caused by hyper- and hyposecretion of GH.

In adulthood, GH-secreting pituitary tumors are much more common and result in a condition called *acromegaly*. After epiphyseal fusion, no further increase in bone length can occur. Instead, in response to the raised levels of plasma GH, the bones start to thicken. This gives the patient a characteristic appearance, with large hands and feet. There is also coarsening of the facial features as illustrated in Fig. 12.11. Furthermore, there is overgrowth of soft tissues and the patient may become diabetic because of the anti-insulin-like metabolic effects of GH. Such individuals are at increased risk from cardiovascular disease and neoplasias, owing to the effects of GH on the heart and colon.

23.5 The role of other hormones in the process of growth

Although growth hormone undoubtedly plays a pivotal role in the process of physical growth, many other hormones are also important. Indeed, the number of hormones involved in the normal growth and development of an individual is indicated by the range of abnormalities of hormone secretion that can result in disturbed growth and abnormal development. Hormones of particular significance include thyroxine and the sex steroids. A number of other hormones, including insulin, the metabolites of vitamin D, parathyroid hormone, calcitonin, and cortisol, may indirectly influence growth and development through their

Fig. 23.14 A pituitary giant and a pituitary dwarf standing next to the well-known British television personality David Frost.

general metabolic actions or their actions on the physiology of bone. The growth-promoting actions of most of these have been discussed in the relevant sections of Chapter 12 and so only a brief summary will be given here.

Thyroid hormone

Thyroxine is necessary for normal growth from early fetal life onwards and for normal physiological function in both children and adults. Its secretion begins at weeks 15–20 of gestation and it seems to be essential for protein synthesis in the brain of the fetus and very young children. It is also required for the normal development of nerve cells. As the brain matures, this action assumes less importance. Children born with thyroid hormone deficiency will be mentally handicapped unless treated quickly—a condition known as cretinism (see also Chapter 12).

Children who develop thyroid hormone deficiency at a later stage have increasingly slowed bodily growth and delayed skeletal and dental maturity, but do not suffer obvious brain damage. Catch-up growth is achieved rapidly following treatment with exogenous thyroxine. Thyroid hormones appear to play a permissive rather than a direct role in growth, allowing cells (including the somatotrophs of the anterior pituitary) to function normally.

Corticosteroids

If present in excess of normal concentrations, hormones of the adrenal cortex, principally cortisol, appear to have an inhibitory action on growth. Such a situation may develop pathologically, for example in Cushing's syndrome or following therapeutic administration of steroids to treat asthma, rheumatoid arthritis, kidney disease, or severe eczema. In such cases, the rate at which the skeleton matures is increased so that the potential for further growth is reduced.

Insulin

Insulin is produced by the islets of Langerhans in the pancreas. It has no particular significance as far as growth is concerned except that it must be secreted in normal concentrations for normal growth to take place. The plasma level of insulin, both in the fasting state and following a meal, rises during puberty and falls back again at the end of puberty. Even small imbalances of plasma insulin and glucose levels can result in stunting and retardation of growth. However, diabetic children whose disease is well controlled by injected insulin and a suitable diet will grow normally.

Vitamin D metabolites and parathyroid hormone

The hormones that regulate plasma mineral levels have indirect effects on growth through their actions on the development and maintenance of the skeleton. Of particular importance are the metabolites of vitamin D (see Chapter 12, p. 218). Calcitriol (1,25-dihydroxycholecalciferol) stimulates the intestinal uptake of calcium, thereby helping to maintain normal plasma levels of calcium. Calcitriol may also have a direct effect on bone to stimulate mineralization.

Vitamin D deficiency causes the disorder of skeletal development known as *rickets* in children and *osteomalacia* in adults. Both conditions are characterized by failure of the matrix of bone

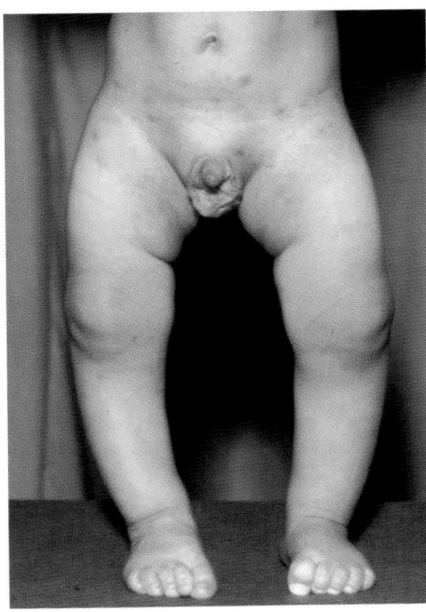

Fig. 23.15 Typical curvature of the legs in a child suffering from rickets.

(osteoid) to calcify. In children whose bones are still growing there is a reduction in the rate of remodeling, which results in swelling of the growth regions of the bones, lack of ossification, and a thickened growth plate of cartilage which is soft and weak. The weight-bearing bones bend, leading to bow legs or knock-knees as shown in Fig. 23.15. In osteomalacia, layers of osteoid are produced which eventually cover practically the entire surface of the skeleton. The main feature of the condition is pain, and bones may show partial fractures.

Parathyroid hormone (PTH) is important in whole-body calcium and phosphate homeostasis. Normal secretion of this hormone is needed for normal bone formation. PTH is believed to bind to osteoblasts (possibly under the permissive influence of calcitriol) and to stimulate their activity. Calcitonin, secreted by parafollicular cells of the thyroid gland, is hypocalcemic in its action, encouraging the binding of calcium to bone. Although its importance in adults is questioned, it is possible that calcitonin contributes to the growth or preservation of the skeleton during childhood and possibly throughout pregnancy through an inhibition of osteoclast activity.

Sex steroids and the adolescent growth spurt

The growth velocity curves shown in Fig. 23.2 illustrate the timing of the growth spurt that is evident in both girls and boys at puberty. The growth spurt can be divided into three stages. These are the age at 'take-off' (i.e. the age at which growth velocity begins to increase), the period of peak height velocity, and the time during which growth velocity declines and finally ceases at epiphyseal fusion. In general, boys begin their growth spurt 2 years later than girls. Therefore boys are taller at the time of 'take-off' and reach their peak height velocity 2 years later. During the growth spurt, boys increase their height by an average of 28 cm and girls by 25 cm. The average 10 cm difference in height between boys and girls is due more to the height

Summary

1. Many hormones in addition to GH are involved in the regulation of growth. Thyroxine is required for growth from the early fetal period onwards and plays an important part in maturation of the CNS. Excessive secretion (or therapeutic administration) of corticosteroids can inhibit normal growth and maturation of the skeleton in children. Small imbalances in insulin secretion and plasma glucose seem also to interfere with normal development.

2. Calcitriol (an active metabolite of vitamin D) stimulates the intestinal uptake of calcium and seems to be essential for normal growth, calcification and remodeling of the skeleton. Parathyroid hormone stimulates the activity of the osteoblasts and is important for the normal growth of bone.

3. The co-operative action of the gonadal steroids and pituitary GH underlie the growth spurt of puberty. Both estradiol-17β and testosterone stimulate the secretion of GH which in turn increases the rate of long-bone growth. The male and female sex steroids also exert specific effects that give rise to the secondary sex characteristics and the differences in musculo-skeletal morphology between men and women.

difference at 'take-off' than to the height gained during the spurt.

Virtually every aspect of muscular and skeletal growth is altered during puberty, and sex differences are seen (e.g. in shoulder growth) which result in accentuation of sexual dimorphism (the differences between men and women) in adulthood.

The hormonal mechanisms that underlie the growth spurt of puberty involve the cooperative actions of pituitary growth hormone and the gonadal steroids. At puberty, estradiol-17β from the ovaries and testosterone from the testicular Leydig cells are secreted in increasing amounts under the influence of pituitary gonadotrophins. These steroids stimulate the secretion of GH, which in turn stimulates growth of the long bones resulting in an increase in height. Estradiol-17β is also responsible for the development of the breasts, uterus, and vagina, and for the growth of parts of the pelvis. Testosterone stimulates the development of male secondary sexual characteristics and has a direct action on the bones and muscles, which accounts for the differences in lean body mass and skeletal morphology seen between men and women. The increased secretion of sex steroids at puberty is important in triggering the process of epiphyseal fusion, limiting long-bone growth at the end of puberty.

23.6 Growth of cells, tissues, and organs

All biological tissues are made up of cells, which continually renew their constituents through metabolism (see Chapter 3). In terms of overall growth characteristics, however, tissues can be divided into three categories. In the first are nerve and muscle, which manufacture few, if any, new cells once the period of growth is over. Once formed, the cells in these tissues last for most or all of the individual's life. In the second category are tissues such as skin, blood, and the GI epithelium whose cells are continually dying and being replaced by new cells. Tissues such as these have a special germinative zone (e.g. hematopoietic tissue in red bone marrow) wherein new cells are born. In the third category, cells are relatively long lived and stable, but new cells can be generated if the tissue is damaged or when increased activity is required of it. This group of tissues with significant powers of regeneration includes parts of the liver and kidneys and most glands.

An organ may enlarge in three ways:

(1) the number of its constituent cells increases (hyperplasia);

(2) the size of its constituent cells increases (hypertrophy);

(3) The amount of substance between the cells increases.

In non-regenerating tissue, growth occurs in three phases. First, the tissue increases its size through cell division and an increase in cell numbers. Further details of the stages of mitotic division are given in Chapter 3, pp. 22–24. During the second phase, the rate of cell division falls but the cells increase in size as proteins continue to be synthesized and enter the cytoplasm. In the third phase, cell division stops almost completely and the tissue expands only by increasing cell size. The age at which the cells stop dividing depends upon the individual tissue or organ. The neurons of the CNS are the first cells to stop dividing, at around 18 weeks of gestation in the case of the cerebral cortex.

During early development, the overall number of cells in the body is increasing. In general, more cells than are needed are produced, and the excess is eliminated by pre-programmed cell death known as *apoptosis*. Once adult size is reached, cell division is important mainly for wound repair and the replacement of short-lived cells. During young adulthood, cell numbers remain fairly constant. However, local changes in the rate of cell division are seen, for example in anemia, when the bone marrow undergoes hyperplasia, or accelerated growth, so that red blood cells are produced at an increased rate. In contrast, atrophy (loss of tissue mass) can result from the loss of normal stimulation. Muscles that lose their nerve supply will atrophy, while loss of TSH, which normally exerts a trophic effect on the thyroid gland, will similarly lead to atrophy of the thyroid tissue and a reduction in thyroid hormone output.

Alterations in cell differentiation: carcinogenesis

The body consists of cells that are organized into populations that form the tissues and organs. Cells reproduce by cell division and are programmed to die. The balance between cell proliferation and cell death within a tissue determines its overall size.

Under normal circumstances, it seems that differentiated cells can continually sense their environment and adjust their rate of proliferation to suit the prevailing requirements. For example, liver cells increase their rate of proliferation in response to loss of liver tissue caused by alcohol. However, when cells fail to obey the normal rules governing their proliferation and multiply excessively, an abnormal mass of rapidly dividing cells is formed. This is called a *neoplasm* (new formation) and the process is called *neoplasia*. Neoplasms are composed of two types of tissue: parenchymal tissue which represents the functional component of the organ from which it is derived, and stroma, or supporting tissue, consisting of blood vessels, connective tissue, and lymph structures.

Neoplasms are classified as benign or malignant according to their growth characteristics. Benign neoplasms are well-defined local structures that usually grow slowly and do not *metastasize* (spread to distant sites to seed secondary tumors). Malignant neoplasms, however, are poorly differentiated, grow rapidly, and readily metastasize via the blood or lymph. Cancer cells consume large amounts of nutrients, thus depriving other cells of necessary metabolic fuels. This leads to the characteristic weight loss and tissue wasting which often contribute to the death of cancer patients. Cancers can arise from almost any cell type except neurons, but the most common cancers originate in the skin, lung, colon, breast, prostate gland, and urinary bladder. About 20 per cent of all inhabitants of the prosperous countries of the world die of cancer.

Summary

1. Tissues are either non-regenerating (such as nerve) or regenerating. The latter include those that are in a continual state of renewal (skin, blood cells, etc.) and those, like liver, which can regenerate in response to tissue loss or damage.

2. Organs increase in size by cell division, by an increase in cell size, and by an increase in volume of intercellular material. In non-regenerating tissues, cell division stops when the tissue has reached an appropriate size.

3. Under normal conditions, differentiated cells adjust their rate of proliferation to requirements in response to a variety of signals. Failure to do so results in the formation of a benign or malignant neoplasm. Malignant neoplasms are poorly differentiated, grow rapidly, and metastasize.

4. A cell may be transformed into a cancer cell when its DNA undergoes mutation and the expression of certain genes is altered. Specific cancer-promoting genes and tumor-suppressor genes have been discovered.

What are the factors that cause transformation of a normal cell into a cancer cell?

It is well known that certain physical and chemical factors, including irradiation, tobacco tars, and saccharine can act as carcinogens. They do so by causing mutations—changes in the DNA that alter the expression of certain genes. Cancer-causing genes (*oncogenes*) have been detected in certain rapidly spreading tumors, and proto-oncogenes (benign forms of oncogenes) have been discovered in normal cells. Proto-oncogenes code for the proteins that are essential for cell division, growth, and cellular adhesion, and it is believed that they may be converted to oncogenes when fragile sites within them are exposed to and damaged by carcinogens. As a result, dormant genes may be switched on that allow cells to become invasive and to metastasize. These capabilities are possessed by embryonic cells and cancer cells but not by differentiated adult cells.

Recently, tumor-suppressor genes (anti-oncogenes) have been discovered. They seem to protect cells against cancer by influencing processes that inactivate carcinogens, aid in the repair of DNA, or enhance the ability of the immune system to destroy cancer cells.

Recommended reading

Alberts, B., Johnson, A., Lewis, J., Raff, M., Roberts, K., and Walters, P. (2002). *Molecular biology of the cell* (4th edn), Chapters 17, 21, 22, and 23. Garland, New York.

Campbell, E.J.M., Dickinson, C.J., Slater, J.D.H., Edwards, C.R.W., and Sikora, E.K. (eds.) (1984). *Clinical physiology* (5th edn), Chapter 9. Blackwell Scientific, Oxford.

Laycock, J.F., and Wise, P.H. (1996). *Essential endocrinology* (3rd edn), Chapter 13. Oxford University Press, Oxford.

Luskey, K.L. (2000). Growth and development. In *Textbook of endocrine physiology* (4th edn) (ed. J.E. Griffin and S.R. Ojeda), Chapter 11. Oxford University Press, Oxford.

Tanner, J.M. (1989). *Foetus into man* (2nd edn). Castlemead, London.

Multiple choice questions

Each statement is either true or false. Answers are given below.

1. a. The adolescent growth spurt occurs, on average, earlier in boys than in girls.
   b. Long bones increase in length as new cartilage is deposited at the epiphyseal plate.
   c. Bone mass starts to fall after puberty.
   d. Lymphoid tissue mass is proportionately greater at age 6 years than at age 20 years.
   e. The adolescent growth spurt is absent in pituitary dwarfism.

2. a. The shaft of a long bone is called the diaphysis.
   b. The organic matrix of bone is secreted by chondroblasts.
   c. Dividing cartilage cells are situated in the marrow cavity of long bones.
   d. Bone remodeling is a function of the activity of osteoblasts and osteoclasts.
   e. The long bones develop from fibrous connective tissue.
   f. Most bones are formed by endochondral ossification of a hyaline cartilage model.
   g. The principal inorganic components of bone are calcium and phosphates.

3. a. Bone growth during fetal life is controlled primarily by thyroid hormone.

   b. Growth hormone secretion is stimulated by IGF-1.

   c. An excess of vitamin D causes rickets.

   d. Parathyroid hormone is required for normal osteoblastic activity.

   e. The sex steroids are chiefly responsible for closure of the epiphyses.

4. a. Liver cells respond to loss of liver tissue by increasing their rate of division.

   b. Nerve is a regenerative tissue.

   c. Oncogenes protect normal cells from transformation into cancer cells.

   d. Cells from a malignant neoplasm are less well differentiated than normal cells.

Answers to multiple choice questions

1. The adolescent growth spurt occurs on average about 2 years earlier in girls than boys. It is not seen in GH-deficient pituitary dwarves unless they are treated with exogenous hormone. Bones grow by the addition of new cartilage at the epiphyses, which later becomes ossified. Bone mass remains virtually stable after puberty, beginning to fall from around the age of 40 years. Lymphoid tissue has a greater total mass in children, falling to adult levels after puberty.

   a. False

   b. True

   c. False

   d. True

   e. True

2. The organic matrix of bone (osteoid) is secreted by osteoblasts. The dividing cartilage cells are situated in the inner regions of the epiphyseal junction. Bone remodeling involves the resorption of old bone by osteoclasts and the accretion of new bone, which is dependent on osteoblastic activity. The flat bones (e.g. the skull) develop from fibrous connective tissue. The long bones, like most of the skeleton, develop by endochondral ossification of a cartilage model. Complex crystals of calcium phosphate (hydroxyapatite) are the principal inorganic components of bone.

   a. True

   b. False

   c. False

   d. True

   e. False

   f. True

   g. True

3. During fetal life, the principal hormonal regulator of skeletal growth and development is thought to be thyroxine. Growth hormone stimulates the synthesis and secretion of the IGFs. Rickets is caused by vitamin D deficiency. Normal circulating levels of parathyroid hormone are required for osteoblastic activity and the secretion of osteoid. Estrogens and androgens, secreted in large amounts at puberty, bring about closure of the epiphyses.

   a. True

   b. False

   c. False

   d. True

   e. True

4. Liver cells increase their mitotic activity following loss or damage to liver tissue. In this way large amounts of liver tissue may be regenerated. Nervous tissue does not regenerate significantly. Nerve cells stop dividing during infancy. Oncogenes are thought to be responsible for the genetic transformations leading to certain cancers. Malignant cancer cells show a loss of differentiation.

   a. True

   b. False

   c. False

   d. True

CHAPTER 24

Energy balance and the control of metabolic rate

After reading this chapter you should understand:

- The general concepts of metabolism, metabolic rate, and basal metabolic rate
- That cellular reactions produce heat
- The concept of the energy equivalent of oxygen for carbohydrates, fats, and proteins
- The relationship between oxygen consumption and carbon dioxide production—the respiratory quotient
- The energy requirements of different tasks
- The actions of circulating hormones on the metabolic rate

24.1 Introduction

The vital processes of the body involve a multitude of chemical reactions. To gain some idea of the scale of these reactions, consider the fact that an average human being uses about 360 liters of oxygen each day to 'burn' several hundred grams of carbohydrates and fats and that about 7500 kJ of heat are generated. This is roughly equivalent to a steady heat output of 90 W. The chemical processes of the body that are responsible for generating this heat constitute the *metabolism* of the body. Metabolism can be divided into two categories: anabolic and catabolic. *Catabolic metabolism* involves the breakdown of large complex molecules to smaller simpler ones. This process is accompanied by the liberation of energy. The oxidative metabolism of glucose to carbon dioxide and water is a typical example. *Anabolic metabolism* involves the synthesis of complex molecules from simpler ones, such as the synthesis of proteins from amino acids.

24.2 The chemical processes of the body produce heat

In cells, anabolic and catabolic reactions take place side by side. The structural components of the cell are continually being broken down and replaced. Some of the energy released by the catabolic processes is harnessed to drive the energy-requiring anabolic processes of the cell and the important cellular activ-

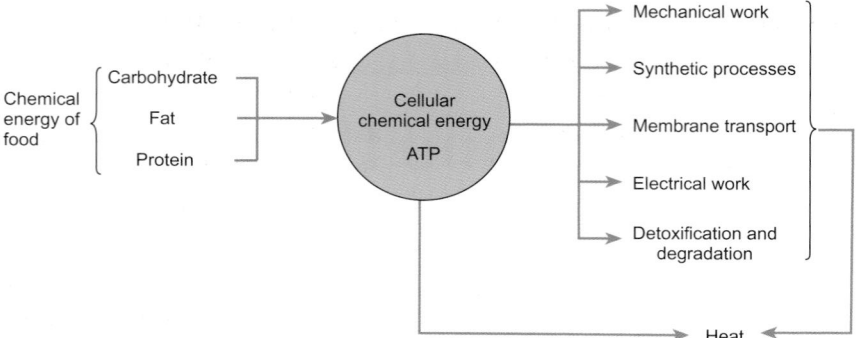

Fig. 24.1 An overview of the biological transformation of various forms of energy.

ities such as muscle contraction, membrane ion pumps, secretory processes, etc. The rest is 'lost' as heat. However, this heat is of vital importance in the maintenance and regulation of body temperature (see Chapter 26).

ATP plays a crucial role in all these processes. It is manufactured in large amounts in the mitochondria during the oxidative metabolism of glucose, fats, and proteins as described in Chapter 3. Its function is to transfer, in the form of phosphate bonds, the energy liberated by catabolism to the anabolic reactions of the cells or to provide energy for the active processes carried out by the cells as shown in Fig. 24.1. Examples of the crucial role of ATP in active transport and in the contraction of skeletal muscle are given in Chapters 4 and 7.

Virtually all the work performed by the body eventually ends up as heat. This is strikingly evident during exercise, where heat production rises with the amount of work done. Only the work done outside the body, such as that involved in lifting a heavy object, is not directly converted to heat. To take another example, the heart does mechanical work in pumping blood against the pressures in the aorta and pulmonary artery. However, the kinetic energy imparted to the blood is converted to heat in overcoming friction as it passes through the circulatory system. Even in anabolic reactions, such as protein synthesis, the chemical energy stored within a protein is lost as heat when that protein is eventually broken down.

24.3 Energy balance

The unit of energy used in the study of energy metabolism is the kilojoule (kJ). The joule is defined as the amount of work done when a force of one newton moves through a distance of one meter in the direction of the force. The rate at which work is done (power) is measured in watts where one watt is equivalent to one joule of work per second. In the earlier literature, the unit of energy is given as kilocalories (or calories). In terms of heat, one kilocalorie is the amount of heat needed to raise the temperature of one kilogram of water by 1°C. One kilocalorie is equivalent to 1000 calories or 4187 joules (4.187 kJ).

According to the first law of thermodynamics, energy can neither be created nor destroyed. Applied to the human body, this means that the total amount of energy taken in by the body must be accounted for by the energy put out by the body. Thus

$$\text{energy input} = \text{energy output}.$$

This can be expressed as

$$\text{chemical energy of food} = \text{heat energy} + \text{work energy} \pm \text{stored chemical energy.} \quad (24.1)$$

When the amount of energy ingested as food is sufficient to balance the amount of energy put out in the forms of heat and work, the chemical energy of the body remains constant. In reality, this is rarely the case. More often there is a small imbalance such that energy is either stored within the body or depleted by the catabolism of carbohydrates, fats, and, in more prolonged fasting, proteins. Even during the course of a day, there are small fluctuations in the chemical energy stores of a normal individual. At night, the glycogen stores of the liver are slowly consumed and are restored following the first meal of the next day. During growth, there is substantial gain in body mass due to the synthesis of new proteins. In later life, as activity declines, there is a tendency for excess food intake to result in the deposition of fat as discussed in Chapter 19.

To study the energy balance of a person under the conditions normally encountered in life, three of the variables in equation (24.1) would have to be measured so that the fourth can be calculated. This can be done in a whole-body calorimeter, which measures the total heat output of a subject. This technique is impractical for the clinical assessment of patients. Instead, the measurement of energy balance can be simplified by eliminating some of the variables in equation (24.1). If measurements are carried out on an individual in the postabsorptive state of metabolism (8–12 hours after a meal—see Chapter 27), the input of chemical energy from food can be taken as zero. If the person is at complete rest, energy output in the form of external work can also be disregarded. Under these conditions, all the chemical energy derived from the metabolic pool is used to maintain vital bodily functions and is ultimately completely converted to heat. Equation (24.1) now simplifies to

$$\text{loss of stored chemical energy} = \text{heat energy.} \quad (24.2)$$

The rate at which chemical energy is expended by the body is known as the *metabolic rate*. As equation (24.2) shows, the rate at which heat is produced by the body is equal to the rate at which chemical energy is expended. Consequently, the heat production is a direct measure of metabolic rate.

Although the metabolic rate will vary with activity, under the resting conditions described above, the rate of heat production is said to represent the *basal metabolic rate* (BMR), which is defined as the energy requirement of the fasting body *during complete rest* in a thermoneutral environment (*c*. 20 °C). The BMR is an index of metabolism under standardized conditions. It does not represent the minimum metabolic rate. This may be achieved during sleep. As it varies with body size, it is usual to relate the BMR to body surface area or to lean body mass. Abnormal values of BMR often result from disturbances of hormonal function, some of which will be discussed in Section 24.7.

24.4 How much heat is liberated by metabolism?

The answer to this question depends upon the nature and proportions of the foods being metabolized. If known amounts of pure fat, carbohydrate, or protein are burned in a calorimeter together with 1 liter of oxygen, different amounts of heat energy are liberated from each foodstuff. For example, pure carbohydrate will liberate 20.93 kJ (5 kcal) of energy per liter of oxygen, whereas pure fat will liberate 19.68 kJ (4.7 kcal) per liter of oxygen, and protein will liberate about 18.84 kJ (4.5 kcal) per liter of oxygen. These values represent the *energy equivalent of oxygen* for the complete oxidation of carbohydrate, fat, and protein.

Another way of looking at the energy value of different foods is to consider how much energy is liberated by the complete oxidation of a standard amount. Thus the complete oxidation of 1 g of carbohydrate or protein will liberate 17.16 kJ (4.1 kcal), while 1 g of fat will liberate 38.94 kJ (9.3 kcal). The body's metabolism can be considered in much the same way, as carbohydrates, fats, and proteins liberate energy as they react with oxygen.

The measurement of oxygen consumption

In order to measure the metabolic rate of an individual, it is necessary to know, or calculate, two values:

(1) the rate at which the individual is using up oxygen—his or her *oxygen consumption* (normally expressed in l min⁻¹);

(2) the energy equivalent of oxygen for the food being metabolized (expressed in kJ per liter of oxygen).

By multiplying these two values by 60 (the number of minutes in an hour), a figure for that individual's metabolic rate is obtained in kJ h⁻¹. (If the values are divided by 60, the number of seconds in a minute, the energy consumption can be expressed in kW (kJ s⁻¹).) A relatively simple way of measuring the rate of oxygen consumption of an individual is by means of a spirometer (see Chapter 16). The spirometer bell is filled with oxygen from which the subject inspires. Expired air passes back to the bell via a canister of soda lime, which removes all the carbon dioxide from it. As the subject consumes oxygen, individual respiratory cycles are recorded and the volume of the oxygen in the bell (represented by the height of the bell) gradually falls (Fig. 24.2). The spirometer is calibrated in terms of volume per unit distance fall of the bell. Therefore the difference between the height of the bell at the start and finish of the experimental period is a measure of the volume of oxygen consumed during that time (Fig. 24.3). Modern methods rely on measuring the airflow and difference in the oxygen and carbon dioxide content between the inspired and expired air. As this can be measured for each breath, it is possible to obtain continuous measure-

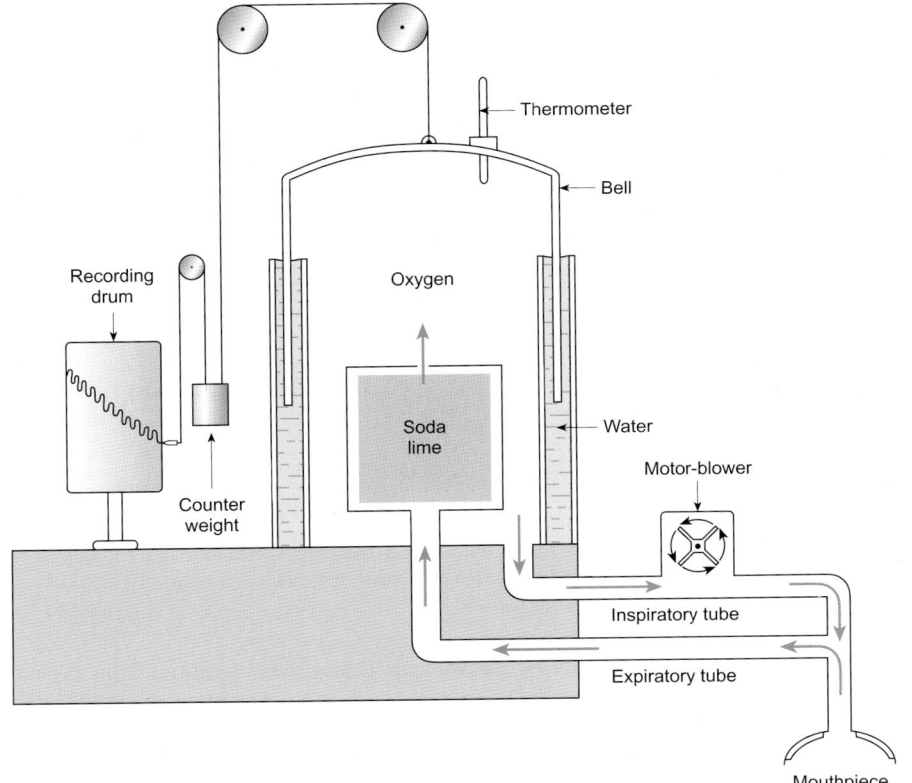

Fig. 24.2 A schematic diagram of a spirometer adapted to measure oxygen consumption.

Thermometer

Bell

Recording drum

Oxygen

Soda lime

Water

Motor-blower

Counter weight

Inspiratory tube

Expiratory tube

Mouthpiece

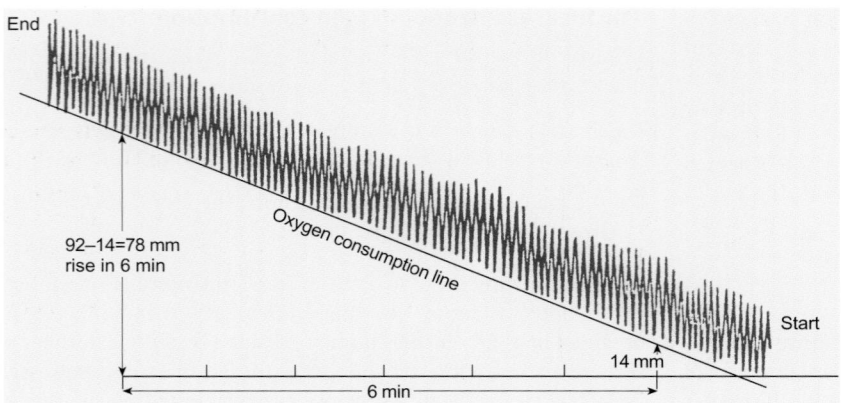

Fig. 24.3 A graphical record illustrating the measurement of oxygen consumption. The record is read from right to left and can be obtained from apparatus of the kind shown in Fig. 24.2. Each millimeter fall of the spirometer bell (and corresponding rise in the experimental record) represents a known volume of oxygen used. The slope of the trace can then be used to calculate the oxygen consumption.

ments of a subject's work rate while performing a variety of tasks.

The energy equivalent of oxygen

As explained above, the energy equivalent of oxygen will vary according to the relative amounts of protein, fat, and carbohydrate being utilized at the time of metabolic rate determination. These amounts cannot be determined directly. Therefore, the energy equivalent of oxygen has to be calculated indirectly from the respiratory quotient (RQ) or respiratory exchange ratio. The RQ is the ratio of the volume of carbon dioxide produced by the lungs to the volume of oxygen absorbed from the lungs in 1 min (see Chapter 16):

$$RQ = \frac{\text{volume of } CO_2 \text{ produced per minute}}{\text{volume of oxygen consumed per minute}}. \quad (24.3)$$

RQ depends upon the nature of the foodstuffs undergoing metabolism. If pure carbohydrate is oxidized, an RQ of 1.0 is obtained, while the specific oxidation of fat gives an RQ of 0.7 and that of protein an RQ of about 0.8. In reality, the body uses a variable mixture of all three metabolic fuels (although protein is usually a minor energy source). Thus the RQ will be a mean value weighted towards the principal fuel. The RQ of a subject on an ordinary mixed diet will be around 0.85, while that of a fasting individual is about 0.82 as relatively more fat (or even protein) is metabolized.

Once the RQ value for the subject under investigation has been obtained, it is possible to compare this with standard values relating RQ to energy equivalent of oxygen (Fig. 24.4). The metabolic rate can then be calculated by multiplying oxygen consumption by the energy equivalent of oxygen obtained from the tables. This simple indirect method is subject to a number of assumptions and inaccuracies and can only give information about the total amount of heat being generated by the body. Nevertheless, such information can be very useful as an aid to diagnosing conditions such as thyroid disorders.

Although the RQ is a useful index of the type of fuel being oxidized by the body at a particular time, it is not reliable during changes in the acid–base status of an individual. In heavy exercise in particular, the rise in blood lactate (and consequent fall in arterial pH) favors the formation of carbonic acid, with the result that more carbon dioxide is lost from the body than is pro-

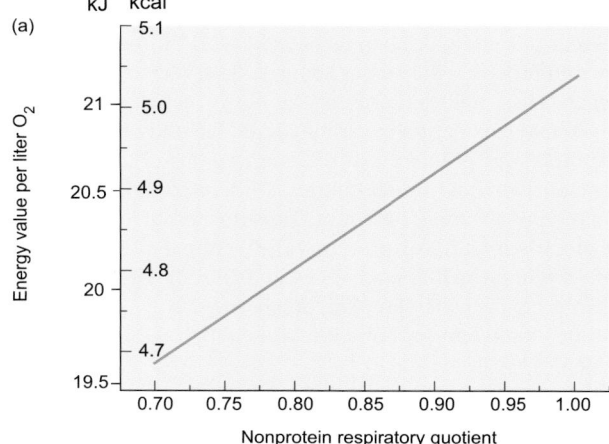

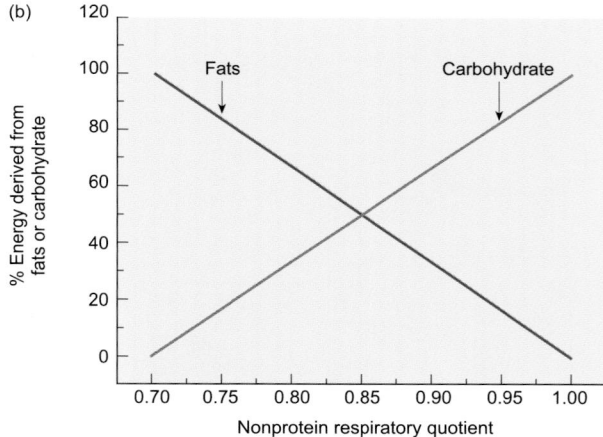

Fig. 24.4 (a) The relationship between the non-protein respiratory quotient and the energy equivalent of oxygen, and (b) the proportion of energy derived from carbohydrates and fats.

duced during metabolism. Under these conditions the RQ may exceed unity. Following a period of exercise, much of the lactate that was produced is resynthesized into glucose and glycogen and blood pH rises. During this phase, the RQ may be less than normal as bicarbonate is re-formed.

24.5 Basal metabolic rate and the factors that affect it

In adults, BMR amounts to an average daily expenditure of 84–105 kJ (20–25 kcal) per kilogram body weight. This requires the consumption of some 200–250 ml of oxygen each minute. About 20 per cent of the BMR is accounted for by the CNS, 25 per cent by the liver, 20–30 per cent by the skeletal muscle mass, and about 16 per cent by the heart and kidneys.

The BMR depends on many different factors. Based on evidence obtained from identical twins and families, it is believed that the BMR is determined genetically, at least in part. It is also affected by a number of physiological variables including body weight, body surface area, lean body mass, age, and the sex of the individual. Although there is no simple, direct relationship between BMR and body size, it is customary to relate metabolism to the surface area (kilojoules or kilocalories per square meter per hour) since heat is produced in proportion to surface area or the lean body mass. (Muscle has a much higher metabolic rate than adipose tissue.) A standard nomogram is available for computing body surface area from the height and weight of an individual.

Any comparison of the BMR of different individuals must take into account the factors that might alter it. As a result of numerous measurements of BMR in a wide range of subjects, tables are now available which give the expected normal range of BMR values for people of different body size, age, and sex. This information is valuable in clinical situations where it is necessary to determine whether, and by how much, the BMR of a patient lies outside the normal range. BMR is increased during fever and is affected by changes in the circulating levels of thyroid hormones (see below).

Figure 24.5 illustrates the effects of age on the normal metabolic rates of both males and females. The BMR of women is 6–10 per cent lower than that of men of comparable size and age, although it shows a significant increase during pregnancy because of the additional metabolic activity of the fetus. For adult males between 20 and 60 years of age, the normal BMR lies in the range 142–168 kJ m^{-2} h^{-1} (34–40 kcal m^{-2} h^{-1}). For non-pregnant women the range is 134–146 kJ m^{-2} h^{-1} (32–35 kcal m^{-2} h^{-1}).

The metabolic rate of a young child, in relation to its size, is almost twice that of an elderly person. The high BMR of childhood results from high rates of cellular reactions, rapid synthesis of cellular materials, and growth, all of which require moderate quantities of energy. Part of the decline in BMR seen in old age is due to the fall in lean body mass with age. The percentage of total body weight represented by fat more than doubles for men between the ages of 20 and 55 years.

24.6 Physiological factors that affect metabolic rate

The *metabolic rate* is the rate at which chemical energy is expended by the body. It varies from minute to minute depending on the nature of the activities being undertaken by an indi-

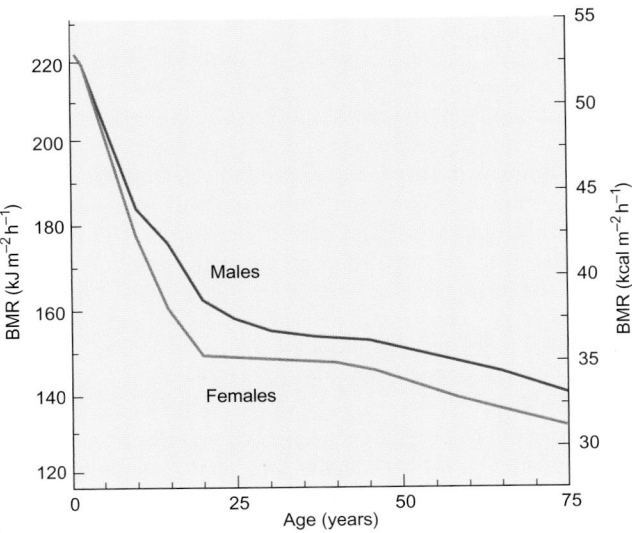

Fig. 24.5 The average age-related changes in BMR for males and females corrected for body size.

BOX 24.1 DAILY ENERGY BALANCE CALCULATION

A worker spends about 8 hours a day working with an average energy consumption of 700 kJ h^{-1}. He spends another 8 hours sleeping or lying down (276 kJ h^{-1}). He walks to and from the office at a brisk pace (1250 kJ h^{-1}), which takes 30 minutes a day. In his remaining time, he spends 7 hours sitting and attending to personal needs (dressing, eating, etc.) at an average energy consumption of 376 kJ h^{-1} and exercises by jogging for 30 minutes each evening at an average energy consumption of 3000 kJ h^{-1}. His total energy requirement each day can be calculated as follows.

Activity	Time spent (h)	Rate of energy consumption (kJ h^{-1})	Total energy requirement (kJ)
Sleeping and lying	8	276	2 208
Sitting, etc.	7	376	2 632
Walking	0.5	1250	625
Jogging	0.5	3000	1 500
Working	8	700	5 600
Total	24	—	12 565

His total daily energy consumption is 12.56 MJ (3000 kcal), which must be met by the diet. For optimal nutrition, about 60 per cent of the energy value of food should be in the form of carbohydrate, 30 per cent as fats, and the remainder as protein. Since each gram of carbohydrate or protein yields approximately 17 kJ of energy and each gram of fat yields approximately 39 kJ, his recommended dietary intake should be:

carbohydrates $(12\,560 \times 0.6)/17 = 443$ g
fats $(12\,560 \times 0.3)/9.3 = 97$ g
protein $(12\,560 \times 0.1)/4.1 = 74$ g.

vidual. It is important to distinguish between the metabolic rate and the BMR, which refers to the metabolic activity of a subject during complete rest. The BMR provides a standard way of comparing the metabolic rate of different subjects, which is of particular value in clinical situations. However, there are very few times during everyday life when body metabolism is occurring this slowly, as any activity that alters the chemical activity of cells will alter the metabolic rate of the body. In this section, some of the physiological factors that alter the metabolic rate are discussed.

The energy requirements of different tasks and different occupations have been investigated very thoroughly. People involved in sedentary work have a modest energy requirement—about 8.25 MJ (*c.* 2000 kcal) per day—while manual workers in heavy industries such as mining may have an energy requirement three times greater than this (see Box 24.1).

Exercise

Any degree of exercise, from that required simply to carry out the normal daily activities of life such as walking, eating, dressing, etc. to the severe exercise involved in manual labor, produces an increase in metabolic rate. Walking slowly (4.2 k.p.h.) uses about three times as many kilojoules per hour as lying in bed asleep, while maximal muscular exercise can, in short bursts, increase the metabolic rate by around 20-fold. Table 24.1 shows some examples of the energy requirements of particular forms of muscular activity.

Ingestion of food

After eating a protein-rich meal, there is an increase in the rate of heat production by the body, i.e. an increase in metabolic rate. Although part of this rise may be due to the cellular processes involved in the digestion and storage of foods, it is believed to result mainly from specific effects exerted by certain amino acids derived from the protein in the meal. This is known as the *specific dynamic action of protein*. Heat production also increases following the ingestion of a meal rich in carbohydrates or fats, but

to a lesser extent. It is thought that the liver is the major site of the extra heat production because of its importance in the intermediary metabolism of absorbed foods (see Chapters 18 and 27).

Fever

Whatever its cause, fever elicits an increase in metabolic rate. This is due simply to the fact that all chemical reactions, including those taking place within cells, proceed more rapidly as temperature increases. For every 0.5°C rise in body temperature, the BMR increases by about 7 per cent. Thus the metabolic rate of a person with a fever of 39°C will be nearly a third higher than normal.

Fasting and malnutrition

The basal daily energy requirement of a man with a surface area of 1.8 m² is about 7–7.5 MJ (1700–1800 kcal). This will normally be supplied by the daily ingestion of foods. When little or no food is being eaten, the component tissues of the body are used for the production of energy. The hepatic glycogen stores are consumed first, followed by the mobilization of stored fats and the conversion of amino acids to glucose (gluconeogenesis—see Chapter 3, Section 3.5). Later, tissue protein is broken down to provide amino acids for glucose production and creatine for skeletal muscle activity. The fasting person is said to be in a state of catabolism, with stores of fat, carbohydrate, and protein all diminishing.

Prolonged starvation decreases the basal metabolic rate, often by as much as 20–30 per cent, as the availability of necessary food substances in cells is reduced. This decrease serves to limit the drain on available resources.

Sleep

There is a fall in metabolic rate of between 10 and 15 per cent during sleep as a result of a reduction in the level of muscular tone and a fall in activity of the sympathetic nervous system.

24.7 The actions of hormones on energy metabolism

A variety of hormones are able to alter the way in which fats, carbohydrates, and proteins are utilized by the cells and tissues of the body and therefore may alter the metabolic rate. The principal hormones in this category are the catecholamines (from both the adrenal medulla and the sympathetic nervous system), the thyroid hormones, and growth hormone. The male sex steroids also exert a small effect on metabolic rate. Full details of the actions of all these hormones are given in the appropriate chapters of this book (particularly Chapters 12 and 27), and so only a brief description will be given here.

Epinephrine and norepinephrine stimulate the metabolic rates of virtually all the tissues of the body. Their major effect is to increase the rate of glycogenolysis within cells. Maximal sympathetic stimulation is thought to increase metabolic rate by between 25 and 100 per cent.

The thyroid hormones exert a powerful effect on the metabolic rate. Loss of thyroidal activity gives rise to a condition known as myxedema (see Chapter 12) in which metabolic rate

TABLE 24.1 Approximate energy consumption of an adult male engaged in various physical activities	
Activity	**Energy consumed** W (kcal min⁻¹)
Sleeping	77 (1.1)
Sitting	105 (1.5)
Standing	174 (2.5)
Slow walking (3.2 k.p.h)	195 (2.8)
Brisk walking (6.4 k.p.h)	363 (5.2)
Cycling (16 k.p.h)	433 (6.2)
Jogging (10 k.p.h)	712 (10.2)
Jogging (14 k.p.h)	963 (13.8)
Swimming (fast crawl)	872 (12.5)
Cross-country skiing	1033 (14.8)

1 watt = 1 Joule s⁻¹

Summary

1. The chemical processes of the body constitute metabolism. Metabolism gives rise to heat and provides energy in the form of ATP. Energy is measured in kilojoules (kJ) or kilocalories (kcal), where 1 kcal equals 1000 cal or 4187 J. The rate at which chemical energy is expended by the body is called the metabolic rate. During the postabsorptive period, under conditions of complete rest, this represents the basal metabolic rate (BMR). BMR is partly determined genetically. It also varies with the age, sex, and size of the individual.

2. The body metabolizes variable quantities of fats, carbohydrates and sometimes proteins to produce energy. Metabolic rate can be calculated from oxygen consumption and the respiratory quotient (RQ), which is related to the energy equivalent of oxygen for the foods being metabolized. Metabolic rate is increased by any form of exercise, the ingestion of food, and fever. It is reduced in malnutrition and during sleep.

3. A variety of hormones can modify the metabolic rate. Catecholamines and thyroid hormones are potent stimulators of metabolism, and growth hormone and the sex steroids exert a mild stimulatory effect.

may fall to half its normal value. In hyperthyroid individuals, metabolic rate is correspondingly enhanced. Because of these effects, hypothyroidism produces a reduced tolerance to cold conditions, whereas increased thyroid hormone secretion often causes the patient to feel hot. Thyroid hormones raise metabolic rate because they stimulate many of the chemical reactions taking place within cells.

Growth hormone and the male and female gonadal steroids influence metabolic rate to a minor degree, all having a mild stimulatory effect. Male sex steroids are more potent than female hormones in this respect and probably account for the higher BMR seen in males.

Recommended reading

Astrand, P.-O., Rodahl, K., Dahl, H., and Stromme, S. (2003). *Textbook of work physiology* (4th edn), Chapters 12 and 17. Human Kinetics, Champaign, IL.

The physiology of exercise

After reading this chapter you should understand:

◆ The energy requirements of exercising muscle and how they are met

◆ The adjustments to the circulation that accompany exercise

◆ The relationship between muscle work and ventilation

◆ The effects of training on performance

25.1 Introduction

At rest, the skeletal muscles have relatively low metabolic needs. In adults, although the skeletal muscles account for about 40 per cent of total body weight they utilize only about 20–30 per cent of the oxygen taken up by the body. During exercise, the muscles perform work and their metabolic requirements increase. The oxygen consumption of the skeletal muscle mass may rise from about 75 ml min^{-1} to as much as 3000 ml min^{-1} in severe exercise—a 40-fold increase. In addition, glucose and fats are mobilized from body stores for oxidation to yield the ATP required for muscle contraction. To meet these metabolic needs, there are major adjustments of the cardiovascular, respiratory, and endocrine systems. Consequently, the whole body responds to the stress of exercise.

This chapter will be concerned with four main issues.

1. How much energy is expended in the performance of a given task?

2. What is the source of that energy?

3. How are the cardiovascular and respiratory systems adjusted to meet the demands of exercise?

4. How far can performance be improved by training?

The principal units of measurement in exercise physiology are joules, the unit of physical work. *Power* is the ability to work, which is expressed as work per unit time (joules per second or watts) (see Chapter 7, Box 7.1). Since the energy for muscular work is supplied by the oxidation of foodstuffs, the amount of work carried out in the performance of different tasks can be assessed by measuring the increase in oxygen uptake over that for the resting body. The greater the oxygen uptake, the greater is the amount of work performed. The metabolic rate can be obtained by multiplying the oxygen consumption (in l min^{-1}) by the energy equivalent of oxygen for the food being metabolized.

Categories of work and exercise

The intensity of exercise obviously varies from the most mild, for example the exercise undertaken during a gentle stroll, to the very severe, such as that encountered in sprinting and other athletic pursuits. It is a matter of everyday experience that the ability to sustain physical activity depends on its intensity.

Light and moderate work is that which requires an average oxygen uptake of less than four times the resting oxygen uptake (which is about 250 ml min^{-1}). This corresponds to an oxygen consumption between about 300 ml min^{-1} and 1 l min^{-1}. This includes most everyday tasks—dressing, washing, walking, etc. Clearly, this type of work can be carried out for many hours without fatigue.

Heavy work requires an oxygen consumption between four and eight times the resting value, which corresponds to an oxygen consumption of 1–2 l min^{-1}. This category includes most of the laborious jobs in heavy industry such as building and mining. While these levels of energy expenditure can be sustained by very fit individuals for an average eight-hour day, individuals who are not physically fit cannot sustain such levels of activity without periods of rest.

Severe work includes work rates that require an oxygen consumption in excess of 2 l min^{-1}. Higher levels of oxygen consumption can be attained for short periods (as in competitive athletics). Such levels of work cannot be sustained even by fit individuals for long periods.

In the course of daily life, exercise takes two forms—dynamic and static. In dynamic exercise, such as walking, there is a rhythmical movement of the limbs with flexing of the joints and alternating periods of contraction and relaxation of the skeletal muscles. In static exercise, such as lifting, specific muscles are maintained in an isometric contraction for a period of time and the muscles perform no external work. Nevertheless, the body responds to both kinds of exercise by adjustments to the cardiovascular and respiratory systems.

25.2 Metabolism in exercise

Sources of energy in exercise

The energy for muscle contraction is provided by ATP, which is rapidly replenished by the transfer of a phosphate group from creatine phosphate as discussed in Chapter 7. ATP can be produced by oxidative metabolism via the Krebs cycle and the electron transport chain or by the anaerobic breakdown of glucose via glycolysis (see Chapter 3). However, anaerobic metabolism is much less efficient at generating ATP than oxidative metabolism and results in the production of large quantities of lactic acid.

The proportion of aerobic and anaerobic metabolism varies with the severity and duration of the exercise. The relative contributions of aerobic and anaerobic metabolism during maximal exercise lasting for up to an hour are shown in Fig. 25.1. Note the

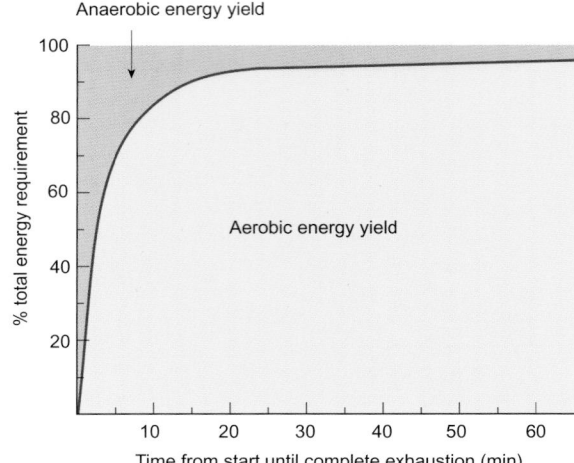

Fig. 25.1 The relative contribution of anaerobic and aerobic metabolism to energy consumption during maximal efforts of various durations. Note that during short bursts of high-intensity exercise the anaerobic energy yield is very high compared with its contribution during prolonged periods of exercise.

increasing proportion of energy contributed by aerobic metabolism with longer periods of exercise. In sustainable exercise almost all the energy required is derived from aerobic metabolism, but in short periods of intense exercise anaerobic metabolism (including the breakdown of ATP and creatine phosphate) may account for more than half the total energy expended. The ATP and creatine phosphate utilized during the anaerobic phase are subsequently replenished by aerobic metabolism.

Carbohydrates provide most of the energy for muscular exercise

The energy requirements of muscular exercise are met mainly by the oxidation of carbohydrates, with a smaller contribution from the oxidation of fats. In people who are well nourished, proteins are not used as a significant source of fuel in exercise. The proportion of carbohydrates and fats used can be assessed by measuring the respiratory quotient (see Chapter 24). This proportion depends on the type of exercise (whether it is continuous or intermittent), the diet, and the physical fitness and state of health of the individual. Nevertheless, with increasing severity of exercise, the oxidation of carbohydrates provides an increasing share of the energy needs.

During exercise, plasma glucose falls very little unless the exercise is both severe and prolonged. Even after 3 hours of continuous exercise at half the maximal rate, the plasma glucose level falls by less than 10 per cent. The glucose utilized during exercise is derived from the glycogen stored in the skeletal muscles and the liver. The breakdown of glycogen into glucose (glycogenolysis) is stimulated by a rise in circulating epinephrine in both the liver and muscle. In prolonged exercise, the glycogen reserves become depleted and glucose is generated in the liver from non-carbohydrate precursors (gluconeogenesis—see Chapter 3). This process is stimulated by the increased circulating levels of cortisol, epinephrine and growth hormone which occur during exercise. Additionally, the lipolytic actions of these hormones mobilize free fatty acids for oxidation.

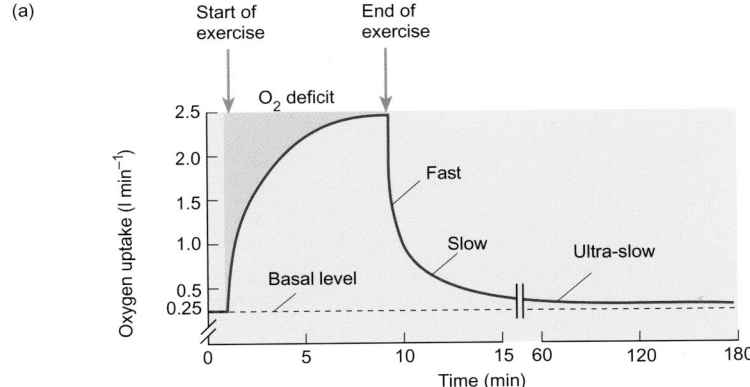

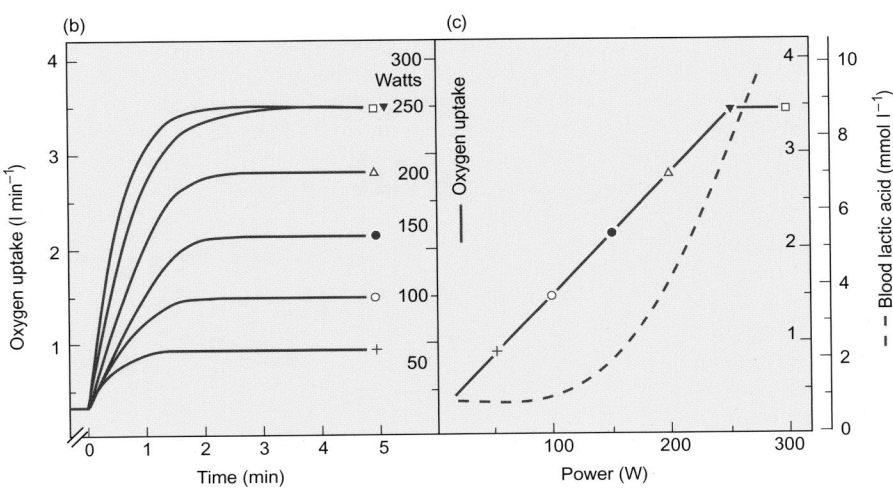

Fig. 25.2 Oxygen consumption during and after a period of exercise. (a) The increase in oxygen uptake that follows the start of a period of exercise and the phases of recovery after exercise ceases. (b), (c) The relationship between oxygen uptake and energy expenditure. The highest levels of exercise are associated with large increases in blood lactate concentrations.

Oxygen consumption rises in proportion to the work done

As soon as exercise begins, the muscles begin to expend energy in proportion to the work done. However, oxygen consumption does not rise immediately to match the energy requirements. Instead, it rises progressively over several minutes until it matches the needs of the exercising muscles. As the work continues, the oxygen uptake remains at a level appropriate to the severity of the exercise. Thus, at the commencement of exercise the body builds up an oxygen deficit ('oxygen debt') (Fig. 25.2).

In the steady state the oxygen consumption is proportional to the work done until the work rate approaches the maximum capacity. In the example shown in Fig. 25.2(c), the maximal oxygen uptake is 3.5 l min^{-1} which provides the *maximal aerobic power*. The further increase in work done (from 250 to 300 W) is not accompanied by additional oxygen uptake and blood lactate rises steeply. Blood lactate levels begin to rise above 1–2 mmol l^{-1} (the normal plasma level) when the oxygen consumption exceeds about 2 l min^{-1}. This is known as the *anaerobic threshold*.

At the end of a period of exercise, the oxygen consumption declines rapidly but does not reach normal resting levels for up to 60 minutes. The first phase of the decline in oxygen consump-

Summary

1. The energy for muscle contraction is provided by the breakdown of ATP. The ATP levels in exercising muscle are maintained by the transfer of a phosphate group from creatine phosphate.

2. Skeletal muscle contains large amounts of glycogen, which is broken down to glucose during periods of exercise. The glucose can be metabolized either anaerobically via the glycolytic chain to lactate or aerobically via the tricarboxylic acid cycle to generate ATP. Aerobic metabolism is much more efficient in generating ATP than anaerobic metabolism.

3. At the onset of a period of aerobic exercise, oxygen consumption rises exponentially to its steady state value. During maintained exercise, oxygen consumption is in direct proportion to the work rate. At the end of a bout of exercise, oxygen consumption falls quickly but may not reach resting values for some time.

tion is very fast, with a half-time of about 30 seconds. During this period, ATP and creatine phosphate are resynthesized via oxidative pathways. This is followed by a slower decline that has a half-time of about 15 minutes. Excess lactate is resynthesized into glucose and glycogen during this period. After severe and sustained exercise, oxygen consumption remains elevated for several hours, perhaps due to stimulation of metabolism as a result of the heat generated during the period of exercise.

25.3 Cardiovascular and respiratory adjustments during exercise

When exercise is undertaken by a healthy person, the circulatory and respiratory systems are adjusted to meet the increased metabolic demands. Cardiac output rises as a result of an increase in both heart rate and stroke volume, and the proportion of the cardiac output distributed to the active muscles increases (Fig. 25.3). There is a rise in oxygen uptake that is related to the amount of work done. This is achieved by an increase in ventilation and greater extraction of oxygen from the circulating blood. In this section, the detailed mechanisms by which these changes are accomplished are considered.

Cardiovascular changes in exercise

At rest, cardiac output is about 5 l min⁻¹. Of this only 15–20 per cent is distributed to the skeletal muscles (i.e. 0.75–1 l min⁻¹). In heavy exercise, cardiac output may rise to 25 l min⁻¹ of which approximately 80 per cent is distributed to the exercising muscles (20 l min⁻¹). In contrast, the blood flow to the brain remains essentially constant, while that to the splanchnic and renal circulations declines. The blood flow in the splanchnic bed falls from 1–1.2 l min⁻¹ at rest to about 0.75 l min⁻¹ in exercise, while renal blood flow declines from about 1 l min⁻¹ to less than half this amount (Fig. 25.3).

Effects on heart rate and stroke volume.

At rest, the heart rate is kept low by the activity in the vagus nerves and most blood vessels are partially constricted by activity in the sympathetic nerves (see Chapter 15). As exercise begins, vagal activity declines and sympathetic activity increases. This results in an increase in heart rate and mobilization of blood from the large veins. The increased sympathetic activity has a positive inotropic effect that leads to an increase in stroke volume. Stroke volume would increase even in the absence of an inotropic response because the augmented venous return increases left ventricular end-diastolic pressure and stroke volume by Starling's law. The positive inotropic response allows an increase in stroke work at the same filling pressure (see Chapter 15, Fig. 15.16). As a result, the end-diastolic volume of the heart does not increase with exercise. The increased stroke volume is achieved by a more complete emptying of the ventricles so that end-systolic volume falls. The increases in heart rate and stroke volume combine to increase cardiac output. At moderate levels of exercise (up to about 40 per cent of maximal oxygen uptake), both heart rate and stroke volume increase in proportion to the work done. Above this level, stroke volume does not increase and any additional increase in cardiac output is due to an increase in heart rate (Fig. 25.4).

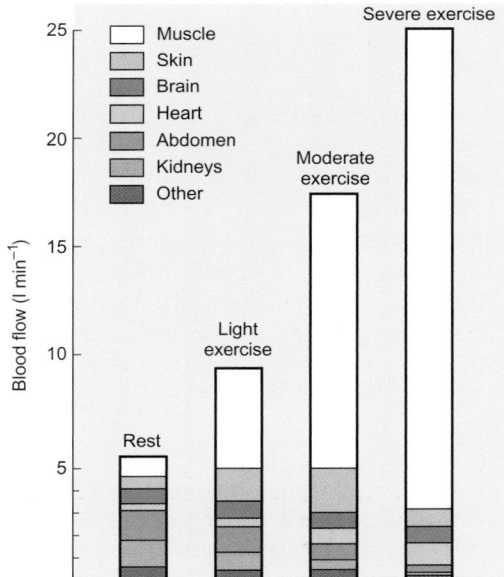

Fig. 25.3 Estimated values of the distribution of cardiac output to various tissues at rest and during light, moderate, and severe exercise.

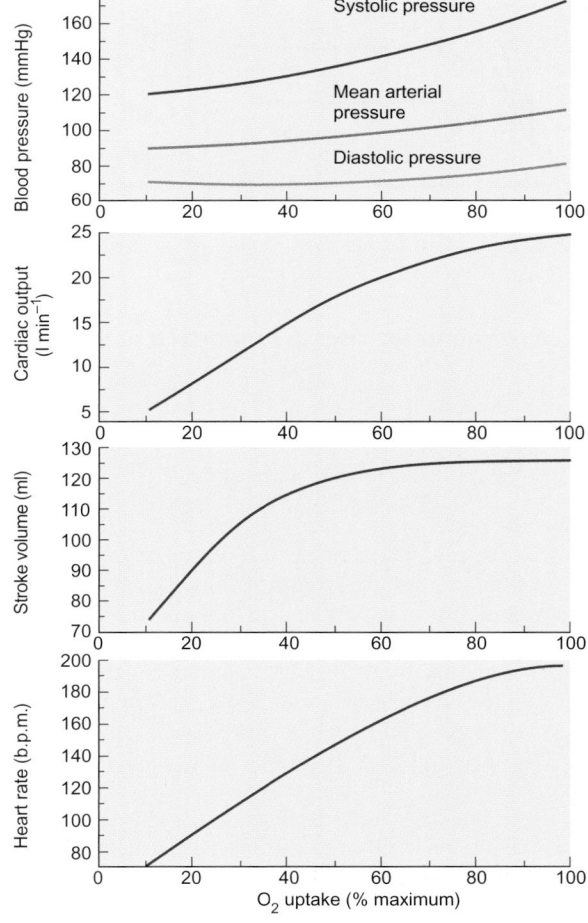

Fig. 25.4 Typical changes in the principal cardiovascular variables with increasing oxygen uptake (taken as a measure of the intensity of exercise).

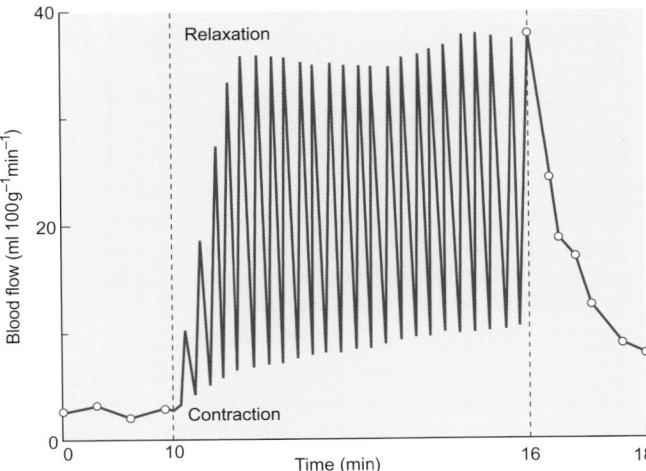

Fig. 25.5 A diagrammatic representation of the changes in blood flow through the calf muscle during strong rhythmic contractions. Note the large increase in blood flow (functional hyperemia) during the period of exercise and its rapid decline during the recovery phase. During the periods of contraction, the blood within the muscles is pumped towards the heart and inflow of blood to the muscles is greatly reduced.

Effects on regional blood flow

The increase in sympathetic activity that both precedes and accompanies a period of exercise results in vasoconstriction in most vascular beds, partly as a result of the direct effects of the sympathetic nerves and partly as a result of the rise in circulating catecholamines secreted by the adrenal medulla.

At rest, the caliber of the arterioles of skeletal muscle is mainly regulated by the activity of their sympathetic nerves and by myogenic activity. Consequently, there is a high level of resting smooth muscle tone and the arterioles are held in a state of partial constriction. Blood flow is low. During exercise, there is a marked vasodilatation of the arterioles of the active muscle which is mainly due to the increased production of metabolites and represents an example of functional hyperemia (see Chapter 15). In addition, there is a vasodilatation in response to circulating epinephrine secreted by the adrenal medulla. Overall, the arteriolar dilatation results in a greatly increased perfusion of the capillary beds within the active muscle.

Exercising muscle has a problem similar to that of the myocardium with respect to its perfusion. A strongly contracting calf muscle, for example, squeezes its blood vessels with each contraction thereby reducing the amount of blood that can flow to it. Therefore in continuous rhythmical exercise such as walking, there will be regular surges in blood flow. During the phase of muscular contraction, the blood within the muscles is pumped towards the heart to augment the venous return. This forward propulsion of blood is aided by the valves of the limb veins. The periodic changes in blood flow in an exercising muscle during such exercise are shown schematically in Fig. 25.5.

Blood pressure during exercise

During exercise, the increased force of ventricular contraction causes an increase in systolic pressure, which becomes more marked as exercise intensifies. In dynamic exercise, the diastolic pressure remains relatively stable and may even decline as the peripheral resistance falls owing to the dilatation of the arterioles in the skeletal muscles. Consequently, under these conditions the rise in mean arterial pressure is modest and there may even be a slight decline.

In static exercise, however, the contraction of the muscles compresses the blood vessels and reduces blood flow. There is a marked pressor response with an abrupt increase in heart rate. Peripheral resistance, diastolic pressure, and mean arterial pressure all rise. When large groups of muscles are engaged, as in weight lifting, the systolic pressure can rise briefly to 40 kPa (300 mmHg) and the diastolic pressure may reach 20 kPa (150 mmHg).

During exercise, large molecules (notably glycogen and creatine phosphate) break down into smaller ones so that the osmotic pressure within the exercising muscle increases. In prolonged exercise, this results in a movement of fluid from the plasma into the exercising muscle cells and interstitial space. This phenomenon is known as *hemoconcentration*. Consequently, during heavy exercise the hematocrit and the oxygen-carrying capacity of the blood are increased. Nevertheless, the limited time for gas exchange in the lungs restricts the rise in the oxygen content of the arterial blood to a few per cent. The viscosity of the blood is increased at the same time.

Ventilation increases in proportion to the work done

At rest, pulmonary ventilation is about 8 l min^{-1} but in heavy exercise it may increase to 100 l min^{-1} or more. At the start of exercise, pulmonary ventilation increases immediately and continues to increase progressively until a new steady state is reached that is appropriate to the work being done. At the end of exercise ventilation rapidly falls, although it may not reach normal resting values for up to an hour if the period of exercise has been intense.

At moderate work rates the steady state ventilation is directly proportional to the work done as measured by the oxygen con-

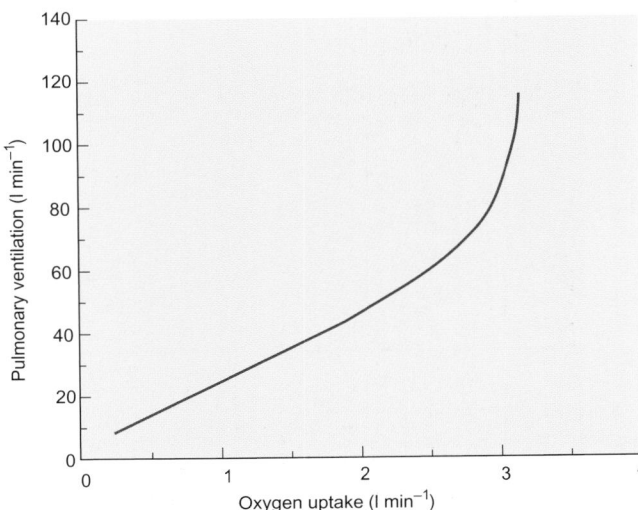

Fig. 25.6 The relationship between pulmonary ventilation and oxygen uptake during exercise. Note the steep rise in ventilation as oxygen uptake approaches its maximum.

sumption. However, during very severe exercise the increase in ventilation is disproportionately large in relation to the oxygen uptake and this may be one limiting factor in the capacity for exercise (Fig. 25.6).

Blood gases in exercise

At rest, the oxygen content of the arterial blood is 19.8 ml dl^{-1} and the hemoglobin is about 97 per cent saturated. The mixed venous blood is approximately 75 per cent saturated and has an oxygen content of 15.2 ml dl^{-1}, so that approximately 4.6 ml of oxygen are extracted from each deciliter of blood as it passes through the tissues.

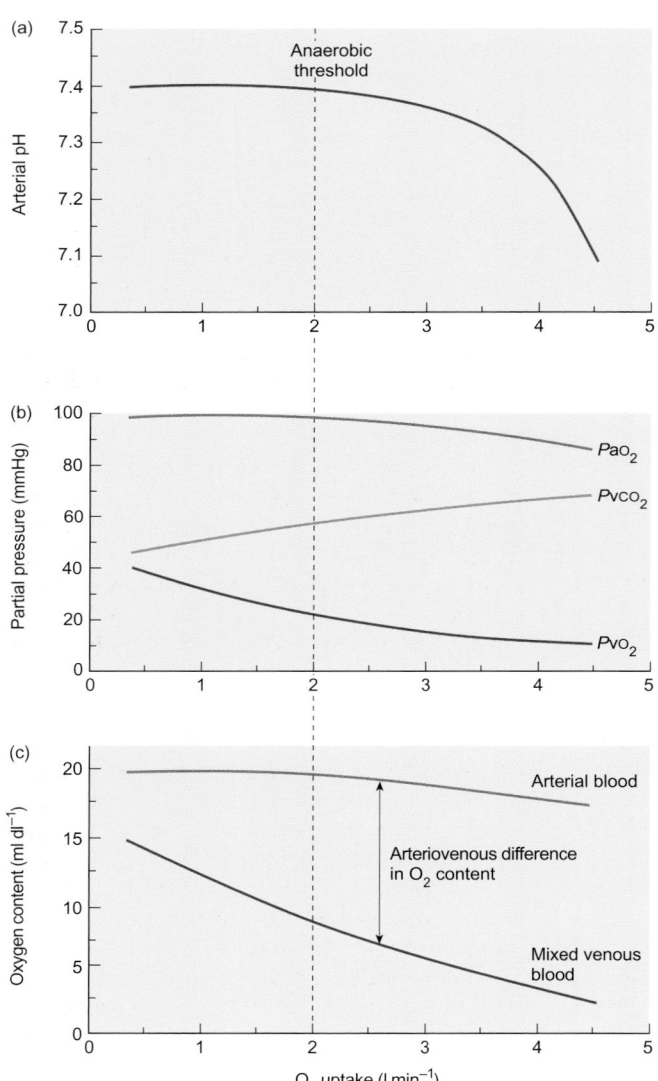

Fig. 25.7 The changes in blood gases and arterial pH during exercise. a) Arterial pH falls as the intensity of exercise increases. (b) As the oxygen consumption rises, the partial pressure of oxygen in the mixed venous (P_{VO_2}) blood falls while the P_{VCO_2} rises. At high levels of exercise the partial pressure of oxygen in the arterial blood declines slightly. (c) The oxygen content of arterial blood falls slightly but the arteriovenous difference in oxygen content increases markedly.

Summary

1. In exercise the cardiac output increases in proportion to the metabolic demand. The increase is due to both an increase in heart rate and an increase in stroke volume.

2. Blood flow is redistributed from the splanchnic circulation to the exercising muscles. Systolic blood pressure rises, but diastolic pressure is stable and may even fall due to the reduction in peripheral resistance that follows vasodilatation of the skeletal muscle bed. As a result, mean arterial pressure rises only slightly.

3. In mild and moderate exercise, pulmonary ventilation increases in direct proportion to the work done. As exercise becomes more strenuous, the additional increase in ventilation becomes disproportionately large.

4. At workloads below the anaerobic threshold, the P_{O_2} and P_{CO_2} of the arterial blood do not change significantly. However, in the venous blood, there is a fall in P_{O_2} and a rise in P_{CO_2}. Thus the oxygen requirement of the exercising muscles is met by an increase in cardiac output, an increased blood flow caused by a vasodilatation of the arterioles and an increased extraction of oxygen from the circulating blood.

At work loads below the anaerobic threshold, the Pa_{O_2}, Pa_{CO_2}, and pH of the arterial blood remain relatively constant, which raises the intriguing question of the mechanism by which the pulmonary ventilation is increased. Nevertheless, the P_{O_2} of the venous blood draining the active muscles and that of the mixed venous blood declines progressively as the intensity of the exercise increases. At the same time, the mixed venous P_{CO_2} rises from its normal value of 46 mmHg. The rise in P_{CO_2} and the associated fall in pH favor the delivery of oxygen to the active tissues (the Bohr effect). At workloads above the anaerobic threshold, there is a gradual reduction in the P_{O_2} and pH of the arterial blood. Overall, the amount of oxygen extracted from the blood increases with the intensity of exercise as shown in Fig. 25.7.

25.4 How are cardiac output and ventilation matched to the metabolic demands of exercise?

During exercise, both cardiac output and pulmonary ventilation are adjusted precisely to meet the metabolic demands until fatigue sets in. How is this remarkable feat achieved? This is the central problem of exercise physiology, which is still not completely resolved. Two principal factors seem to be involved: commands from the brain—the *central command*—and reflexes elicited in response to the exercise itself. In considering how these different processes interact, it is convenient to divide a period of exercise into three phases:

- *Phase 1*, in which exercise starts, ventilation increases and the partial pressures of oxygen and carbon dioxide in the mixed venous blood begin to change.

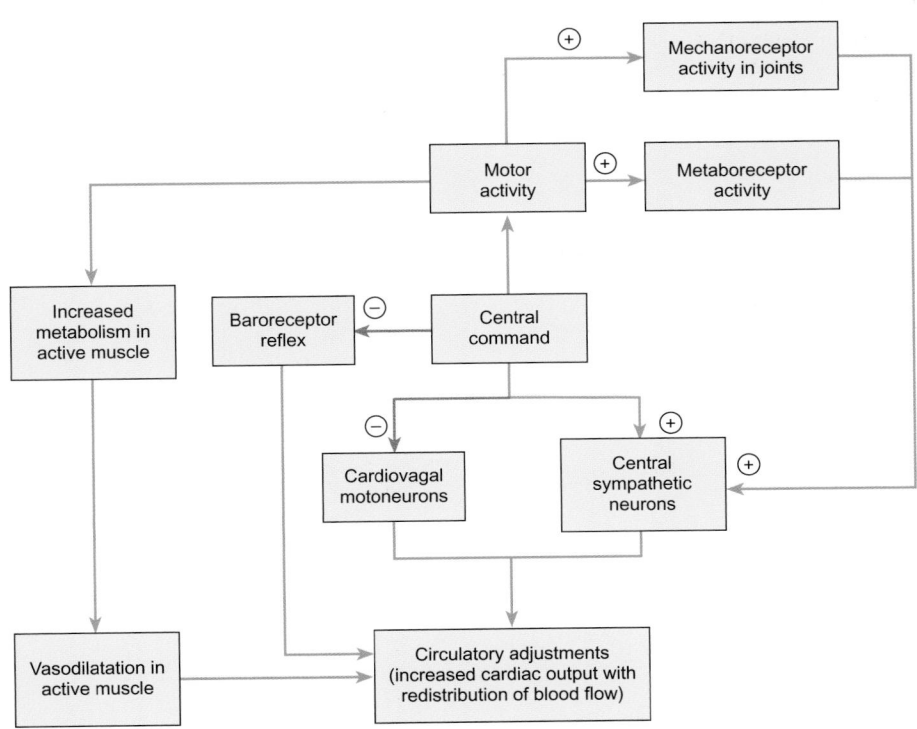

Fig. 25.8 The factors that regulate the cardiovascular response to exercise: ⊕ an increase in activity or an activation; ⊖ a decrease in sensitivity or an inhibition.

- *Phase 2*, during which ventilation, cardiac output and the partial pressures of the respiratory gases in the mixed venous blood approach their steady state values.

- *Phase 3*, during which the steady state levels of ventilation, cardiac output, Pa_{O_2}, Pa_{CO_2}, and arterial pH are maintained. This phase is not reached until the exercise approaches its maximal sustainable capacity. In severe exercise, pH continues to fall as lactate accumulates.

Regulation of the circulation

During phase 1, or even before exercise begins, the heart rate increases and there is an increased force of contraction. These changes are due to an inhibition of parasympathetic activity and an increase in sympathetic activity (including the secretion of epinephrine from the adrenal glands). The increased sympathetic activity causes vasoconstriction in most vascular beds. In some animals, there is a vasodilatation in the skeletal muscles due to activity in sympathetic cholinergic vasodilator fibers, although this probably does not occur in humans. These changes are believed to be due to signals from the higher levels of the brain to those regions of the brainstem concerned with regulation of the cardiovascular system. This central command also acts to reduce the sensitivity of the baroreceptor reflex.

During phases 2 and 3, the central command is reinforced by reflexes which are triggered by increased activity in the afferent nerves in the exercising limbs (Fig. 25.8). In animal experiments, passive movements of the hindlimbs elicit an increase in heart rate and blood pressure. This increase can be blocked by cutting the nerves from the joints, so it appears that afferent activity arising in the joints contributes to the maintenance of the cardiovascular response to exercise. In addition, metaboreceptors in the exercising muscles respond to the fall in extracellular pH

and rise in extracellular potassium and reinforce the cardiovascular response. These reflexes are probably responsible for the matching of cardiac output to the metabolic requirements of the exercise.

The metabolites released by the active muscles and the increased circulating levels of epinephrine cause a vasodilatation in the arterioles that augments local blood flow. As exercise continues, body temperature begins to rise and this is sensed by the hypothalamic thermoreceptors. The change in the activity of these receptors elicits a reflex vasodilatation in the skin vessels to aid in the dissipation of the heat generated by the active muscles (Fig. 25.3).

Regulation of pulmonary ventilation during exercise

The increase in ventilation during exercise is believed to be due to both neural and humoral factors. The neural mechanisms activate the respiratory muscles, but the fine tuning of ventilation to match the oxygen utilization appears to be accomplished by various chemical agents. Ventilation increases as soon as exercise begins. This can only be explained by a central command that is associated with the initiation of motor activity from the premotor area of the cerebral cortex.

During phase 2, pulmonary ventilation rises exponentially and this is believed to be due to changes in the composition of the arterial blood acting on the peripheral chemoreceptors. This would explain the delay in the response; before they could influence ventilation, the metabolites would first need to build up in the exercising muscle and then be transported to the peripheral chemoreceptors. Moreover, it is known that patients with denervated carotid bodies have a slower ventilatory response to exercise than normal healthy subjects. In addition, there is an increased neural input arising from afferent activity in the joints.

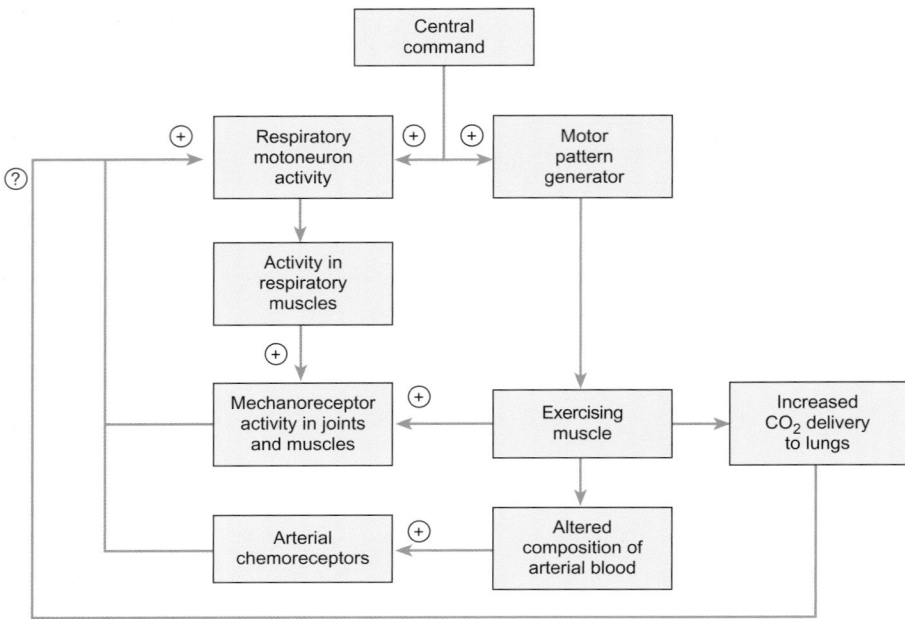

Fig. 25.9 The neural and chemical factors involved in the regulation of breathing during exercise: ⊕ an increase in activity or an activation; ⊙ a possible but unconfirmed interaction.

What chemical stimuli regulate pulmonary ventilation? The arterial blood gases and pH change very little during this phase. Therefore it is evident that the changes in the partial pressures of the respiratory gases in the arterial blood are quite inadequate by themselves to explain the close matching between oxygen uptake and ventilation. What other chemical signals might be involved? During exercise, the plasma potassium concentration is elevated in both the arterial and venous blood. Indeed, in humans, the increase in ventilation during exercise is correlated with the rise in plasma potassium concentration. In anesthetized animals, similar concentrations of potassium elicit a powerful ventilatory response from the carotid bodies. Therefore it is possible that the plasma potassium provides an additional stimulus to the peripheral chemoreceptors.

In phase 3 of exercise, ventilation is in a steady state and matched to the metabolic requirements. As indicated in Fig. 25.9, both chemical and neural stimuli are likely to be involved in maintaining the respiratory effort. The rise in body temperature may also contribute to the respiratory drive. The $P_{a_{CO_2}}$ appears to be more closely regulated than the $P_{a_{O_2}}$, and carbon dioxide elimination is closely matched to carbon dioxide production by the exercising muscles. How this is achieved is not known, but there is evidence that the relationship between ventilation and $P_{a_{CO_2}}$ is reset so that there is a higher ventilation for a given $P_{a_{CO_2}}$. The involvement of neural mechanisms in phases 2 and 3 of exercise is indicated by the matching of the respiratory rhythm to that of the exercise. The afferent barrage from the muscle spindles and the mechanoreceptors in the muscles and joints also contributes to the activation of the respiratory motoneurons.

Summary

1. During exercise, both cardiac output and pulmonary ventilation are adjusted precisely to meet the metabolic demands.

2. The cardiovascular response to exercise is initiated by signals from the higher levels of the brain which act to inhibit parasympathetic activity and increase sympathetic activity. As a result, heart rate and stroke volume increase and blood flow is preferentially distributed to the exercising muscles.

3. Afferent signals arising from the active joints and muscle metaboreceptors activate cardiovascular reflexes that act to maintain the cardiovascular response at a level appropriate to the intensity of the exercise. The associated rise in body temperature initiates a reflex vasodilatation of the skin vessels that promotes heat loss.

4. The ventilatory response to exercise is initiated by signals from the higher levels of the brain. During exercise, the neural activation is supported by signals arising in the muscle spindles and the mechanoreceptors in the muscles and joints. As the arterial partial pressures of the respiratory gases and arterial pH hardly change except in severe exercise, the chemical regulation appears to be dependent on other humoral factors, most notably the plasma potassium concentration.

25.5 Effects of training

The performance of exercise can be improved by training. This requires the regular undertaking of physical exercise that is of an appropriate intensity, duration and frequency. The intensity must increase as performance improves to achieve optimal results. However, the load must be related to the fitness and strength of the individual. Frequent regular exercise of an appropriate kind is important if the improvement in performance is to be maintained. Regular training with strenuous exercise will lower the resting heart rate and increase the size of the heart and the thickness of the ventricular wall. As the end-diastolic volume and stroke volume increase the resting cardiac output is maintained, despite the fall in resting heart rate. Maximal cardiac output increases from 20–25 l min^{-1} in untrained people to values that may exceed 35 l min^{-1}. The cardiovascular changes increase the maximal oxygen uptake and the capacity for physical work is thereby increased.

In addition to its effects on the cardiovascular system, training affects the bone, the connective tissues, and the muscle mass of the limbs involved in the type of exercise being undertaken. In response to the stress of exercise, the bone becomes remodeled so that the stressed areas have a greater degree of mineralization and therefore a higher strength. The importance of exercise in the maintenance of bone structure is shown by the effects of prolonged immobility where the bone becomes progressively demineralized and more brittle. During training, the cartilage of the joints becomes thicker so that it is more compliant. Its contact area increases and, since pressure is force per unit area, the pressures generated within the joints are smaller for a given task.

Within the muscle mass, both connective tissue and muscle mass hypertrophy; the diameter of the muscle fibers increases although their numbers remain the same. The increased mass of the tendons and connective tissue of the muscles improves their ability to transmit the force generated by the muscle fibers to the skeleton. In the early stages of training, the force generated by the muscles improves without hypertrophy of the muscle fibers themselves. This is believed to reflect an improvement in the recruitment of motor units during the performance of a specific type of exercise (i.e. the CNS learns to perform the task more efficiently). Further training leads to hypertrophy of the muscle fibers. In dynamic exercise (e.g. running) there is a progressive increase in the capillary density of the muscle with the duration of training, and a consequent improvement in the transfer of oxygen from the blood to the tissues.

Factors that limit the performance of exercise

Different individuals have different capacities for physical work. The capacity for work for any individual depends on their size, age, sex, genetic make-up, and general state of health. In addition, the performance of strenuous exercise is greatly influenced by the level of motivation.

The ability to undertake muscular exercise ultimately depends on the ability of the muscles to generate sufficient ATP to sustain their contractile activity. In part, the metabolic needs of exercising muscles are met by their glycogen reserves and, for slow-twitch muscle, by the small reserve of oxygen stored in myoglobin. Ultimately, however, the energy requirements of the muscles are dependent on the delivery of glucose and oxygen by the circulation. At low work rates, the energy requirements are met chiefly by aerobic metabolism and exercise can be maintained for considerable periods of time. At high work rates, anaerobic metabolism contributes an increasing fraction of the energy requirements and fatigue sets in relatively quickly (see Fig. 25.1). The fatigue associated with intense exercise is due, at least in part, to the accumulation of lactate.

Exercise is beneficial

Studies of a number of different groups of people have shown that a state of maintained physical fitness reduces the likelihood of coronary heart disease. In an extensive study of office workers, it was found that those who undertook some regular and vigorous exercise had an incidence of coronary heart disease that was less than half that of those who did not. The fit group also had a lower death rate from coronary heart disease. The benefits of exercise were found to extend to all groups including the obese and smokers.

Recommended reading

Astrand, P.-O., Rodahl, K., Dahl, H., and Stromme, S. (2003). *Textbook of work physiology* (4th edn). Human Kinetics, Champaign, IL.

Coote, J.H. (1995), in *Cardiovascular regulation* (ed. D. Jordan, and J. Marshall), Chapter 6. Portland Press, London.

Dejours, P. (1966). *Respiration*, Chapter 8. Oxford University Press, New York.

Levick, J.R. (2003). *An introduction to cardiovascular physiology* (4th edn), Chapter 17. Hodder Arnold, London.

The regulation of body temperature

After reading this chapter you should understand:

- The importance of thermoregulation
- The concept of core and shell temperatures and the ways in which heat may be lost from the body surface to the environment
- The importance of the cutaneous circulation as a means of regulating heat exchange between the core and the surface of the body
- The responses of the body to cold, including behavioural changes, shivering, vasoconstriction, and non-shivering thermogenesis
- The responses of the body to heat including behavioural changes, vasodilatation, and sweating
- The consequences of hypothermia, hyperthermia, and fever

26.1 Introduction

People maintain a normal body temperature of about 37°C despite wide variations in both their metabolic activity and the temperature of their environment (Fig. 26.1). Different regions of the body have different temperatures at rest. The brain and organs within the thoracic and abdominal cavities have the highest temperature. This is known as the *core temperature*. Under most conditions the body surface, the skin, has the lowest temperature—the *shell temperature*. It is the core temperature that is precisely regulated. In contrast, the shell temperature may vary substantially depending on the temperature of the surroundings (Fig. 26.1). As we shall see later on, the blood acts as the vehicle of heat exchange between the core of the body and its shell. For clinical purposes core temperature is usually recorded from the outer ear or mouth although it can also be measured in the axilla and rectum. When temperature is measured in the mouth 95 per cent of people will have a temperature within the range 36.3–37.1 °C.

A healthy individual's body temperature fluctuates approximately 1 °C in 24 hours, with a low occurring in the early morning (around 4 a.m.) and a high in the late afternoon or early

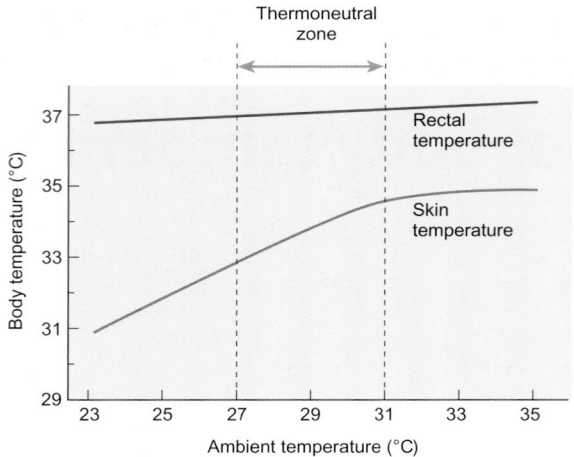

Fig. 26.1 The effects of different ambient temperatures on rectal temperature (core body temperature) and skin temperature in unclothed healthy adult male subjects.

evening (Fig. 26.2). In women of child-bearing age, there is an additional increase in body temperature of about 0.5 °C following ovulation which persists until steroid levels fall just prior to the onset of menstruation (see p. 443). In the event of pregnancy, this elevation in temperature is sustained until delivery. Only during heavy exercise or in fever (see below) does body temperature exceed 37.5 °C for any significant period of time.

Why is body temperature regulated so precisely? With each rise of 1 °C the rate of chemical reactions increases by around 10 per cent, so that as temperature rises the rate of enzymatic activity within cells is accelerated. However, most enzymes have an optimum temperature above which their activity declines as they begin to denature (degrade). The denaturation of enzymes and other cellular proteins accelerates with temperatures above about 42 °C, with the result that cell damage will occur if body temperature rises to this level. Exposure to temperatures in excess of this will result in cell death. Indeed, if core temperat-

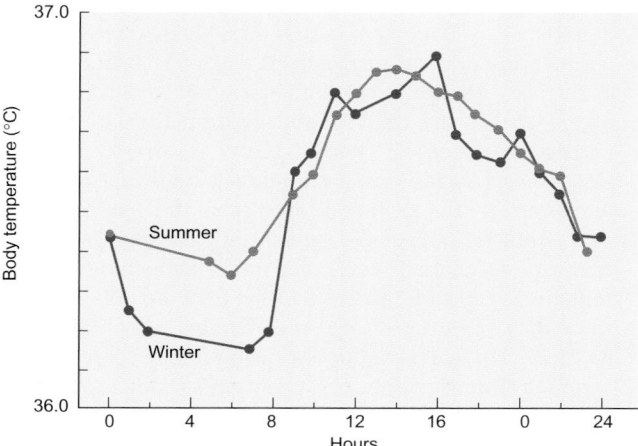

Fig. 26.2 The circadian rhythm of core body temperature. The lowest temperature occurs in the early morning and the highest in mid to late afternoon.

ure exceeds 41 °C most adults will go into convulsions, and a temperature of 43 °C appears to represent the absolute limit for life. In contrast, most cells can withstand marked reductions in temperature, although they function more slowly as their temperature falls. Nevertheless, when core temperature falls below about 33 °C, temperature regulation is impaired and consciousness is lost. Unless body temperature is restored to normal, death will ensue. Therefore, there is a relatively narrow range of temperatures within which the body can function normally.

Thus it is necessary to control core body temperature. This process is known as *thermoregulation* and is one of the important homeostatic mechanisms of the body.

26.2 Heat exchange between the body surface and the environment

Almost all the cellular processes of the body will ultimately result in the production of heat. The more metabolically active a tissue is, the more heat it produces. The organs that produce the most heat are the brain, the skeletal muscle, and the visceral organs such as liver and kidneys (see Chapter 24). To maintain normal body temperature, heat loss to the environment must be balanced by heat generated through metabolism. This is the situation when body temperature is stable. In a cold environment heat is lost continually from the body surface and this must be minimized by appropriate physiological adjustments. In addition, the lost heat must be balanced by some form of heat production. In a hot environment it is possible for the body to gain heat, and the mechanisms governing heat loss need to be activated to prevent a rise in body temperature.

The role of the cutaneous circulation in thermoregulation

The range of environmental temperatures over which it is easy for the body to maintain its core temperature is known as the *thermoneutral zone*. It is between 27 and 31 °C for a naked individual, but can extend well below 27 °C when appropriate clothing is worn. In the thermoneutral zone, skin temperature is about 33 °C. For ambient temperatures within the thermoneutral zone, thermoregulation is achieved solely by alterations in the blood flow to the skin.

The skin and the subcutaneous tissues provide a layer of insulation between the body and the external environment. The skin is well endowed with blood vessels that are organized to provide an efficient means of regulating heat exchange between the body and the environment (see p. 304). Arterioles supplying the skin break up into dense capillary networks beneath the epidermis. These networks drain into a superficial venous plexus, which is able to accommodate a large volume of blood. In addition, in the most exposed areas of the body, such as the fingers, toes, ears, and nose, arteriovenous anastomoses are present which can open and close according to thermoregulatory requirements. When they are open, the blood flow through the venous plexuses is augmented. The presence of a large volume of warm blood close to the skin surface increases the loss of heat from the body surface. Conversely, when the arteriovenous anastomoses are closed, blood flow through the superficial venous plexus is reduced and heat is conserved.

Mechanisms of heat exchange

Heat may be lost from the body by:

- radiation
- conduction
- convection
- evaporation of sweat.

Radiation

This is the loss of heat in the form of infrared rays (thermal energy). If a human sits naked at normal room temperature, radiation will account for around 60 per cent of the total heat loss. The body can also gain heat by radiation, as it does during sunbathing for example.

Conduction and convection

Conduction is the transfer of heat between objects that are in direct contact with one another. A very small amount of heat is lost from the body by conduction to objects (e.g. warming a chair by sitting in it). Much more significant heat loss occurs from the body surface to the air. First the heat is conducted to the air immediately outside the skin and then the warmed air is removed by convection. As the warmed air is replaced by new cool air, further heat loss can occur. This form of heat loss becomes especially noticeable when there is a cool breeze blowing, as the air close to the body surface is being replaced by cold air more rapidly (the *wind chill* effect). Together, conduction and convection to the air account for between 15 and 20 per cent of heat loss to the environment. Light clothing reduces loss by this route to about half that from a naked body, while arctic clothing can reduce it much further.

Because the specific heat of water is several thousand times greater than that of air, water in close contact with skin is able to absorb much more heat than air. Furthermore, the conductivity of heat through water is greater than that through air. Consequently, the body loses heat very rapidly indeed when immersed in water, even at moderate temperatures. A naked person will become hypothermic (see below) after only 1.5–2 hours in water at 15 °C, while in arctic waters (c. 5 °C) hypothermia occurs within 20 minutes.

Evaporation

Water evaporates when its molecules absorb heat from their environment and become energetic enough to escape as gas (water vapour). Approximately 2.4 MJ (570 kcal) of heat are lost for each litre of water that evaporates from the body surface. Thus when water evaporates continuously from the lungs, from the mucosa of the mouth, and through the skin, there is a basal level of 'insensible heat loss'. Evaporative heat loss becomes obvious when body temperature rises and sweating provides increased amounts of water for vaporization. The insensible water loss is about 600 ml per day, amounting to a loss of about 1.4 MJ (340 kcal) of heat per day. During vigorous muscular activity such as cross-country running, sweat may be produced at a rate of 1–2 l h^{-1}, thus removing about 2.4–4.8 MJ (570–1140 kcal) of heat from the body per hour. In some industrial and athletic activities sweat production may reach 6 litres or more an hour for short period. It is vitally important that this water loss is replaced if dehydration is to be avoided (see Chapter 28).

26.3 The role of the hypothalamus in the regulation of body temperature

Body temperature is chiefly regulated by neurons that lie within the hypothalamus, which is a part of the diencephalon. Of particular importance is the preoptic anterior hypothalamic area, although neurons within the septal area of the hypothalamus and the reticular substance of the midbrain also seem to play a role. Together these brain areas control body temperature almost entirely by nervous feedback mechanisms.

The hypothalamus receives afferent input from both peripheral thermoreceptors located in the skin (see Chapter 8) and central thermoreceptors located in the body core, including some within the anterior portion of the hypothalamus itself. The preoptic area of the hypothalamus contains some neurons whose rate of discharge increases markedly in response to a rise in core temperature while others respond to a fall in core temperature by reducing their rates of discharge. Receptors located chiefly in the skin and the spinal cord provide additional information concerning body temperature. Figure 26.3 shows the patterns of static discharge of cold and warm cutaneous thermoreceptors as a function of skin temperature. The cutaneous temperature receptors are probably bare nerve endings. Their afferents are unmyelinated C fibres (warm receptors) or small-diameter myelinated Aδ fibres (cold receptors). In a thermoneutral environment, both cold and warm cutaneous thermoreceptors show a moderate level of activity.

Body temperature is controlled around a set-point

The critical level at which the thermoregulatory mechanisms of the body try to maintain core temperature is known as the *set-point* of the system. While it is not clear how this is determined physiologically, the set-point seems to be under the control of the hypothalamus which behaves as a kind of thermostat maintaining a balance between heat loss and heat production. The set-point is normally close to 37 °C; above this temperature, mechanisms promoting heat loss come into operation, while below it heat-conserving and heat-generating mechanisms are initiated. However, the set-point can be altered as a result of

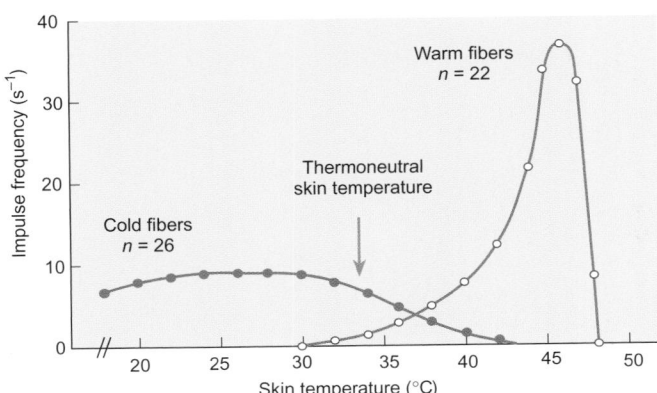

Fig. 26.3 The average response of cold-sensing and warm-sensing cutaneous thermoreceptors. Note that at a skin temperature of 33 °C, both sets of receptors are active. This is the skin temperature in a thermoneutral environment.

signals arriving from the peripheral thermoreceptors or in response to fever-producing agents (pyrogens). The hypothalamus exerts its thermoregulatory actions on the vasculature of the skin, sweat glands, and adipose tissue, mainly via the autonomic nervous system.

26.4 Thermoregulatory responses to cold

When the body's core temperature starts to fall, two kinds of homeostatic responses are initiated. Some are designed to increase the rate at which the body generates heat, while others serve to reduce the rate at which heat is lost from the body surface. Furthermore, people will initiate appropriate behavioural responses to cold including seeking a warmer environment, putting on additional clothing, turning the heating up, eating, drinking warm fluids, and so on. The physiological responses to low temperature include cutaneous vasoconstriction, shivering, and non-shivering thermogenesis.

Cutaneous vasoconstriction

Skin has an extremely wide range of blood flow. When the ambient temperature falls within the thermoneutral zone, the blood flow to the skin is around 0.15 l min^{-1} kg^{-1}. At very low temperatures it may fall as low as 0.01 l min^{-1} kg^{-1}. The skin vessels, particularly the arteriovenous anastomoses, are richly supplied by sympathetic adrenergic fibers. Vasoconstriction in response to cold results from an increase in sympathetic activity and seems to be mediated chiefly by the action of norepinephrine at α-adrenoceptors. This vasoconstriction enhances the insulating properties of the skin and reduces the blood flow to the superficial venous plexuses with the result that less heat is lost from the skin surface. The increase in sympathetic outflow to the cutaneous vessels is believed to be initiated by neurons in the posterior hypothalamus.

Paradoxically, during long periods of cold exposure the cutaneous circulation of the extremities will often show intermittent periods of vasodilatation. This is known as the 'hunting reaction' (Fig. 26.4) and may represent a safety mechanism designed to

reduce the risk of ischemic tissue damage (frost bite) in extreme cold. The underlying physiological mechanism is unclear but may result from a temporary loss of sensitivity to norepinephrine.

In up to 5 per cent of the population, mainly young women, the peripheral arterioles constrict excessively in response to cold. The fingers in particular may appear dead white and feel numb. In extreme cases, there may be local ischemia and tissue damage. This disorder is called *Raynaud's disease*. Its physiological basis is unknown.

Increased heat production from shivering

When peripheral vasoconstriction is inadequate to prevent heat loss, metabolic heat production is increased by voluntary muscle contraction or shivering. It is well known that in cold weather there is a tendency for voluntary skeletal muscle activity to be increased. Foot-stamping, hand-rubbing, and a faster walking speed are all examples of such behaviour. When muscle contracts, ATP is hydrolyzed and heat is produced (see Chapter 7). This additional heat production will contribute to the restoration of a normal core temperature.

Shivering is a specialized form of muscular activity in which the muscles themselves perform no external work and virtually all the energy of contraction is converted directly to heat. It is predominantly an involuntary activity consisting of the contraction and relaxation of small antagonistic muscle groups. Prior to the onset of overt shivering, there is an overall increase in the degree of muscle tone. Shivering begins in the extensor muscles and the proximal muscles of the trunk and upper limbs. The muscles contract in response to signals from the somatic motor neurons. These signals, which are believed to arise in the hypothalamus, are sustained by the response to afferent input from proprioceptors and stretch receptors in the joints and muscles.

Although shivering produces a considerable amount of additional heat, it is insufficient on its own to maintain body temperature for long if ambient temperature is very low. It also places a substantial burden on energy reserves and is exhausting for the individual.

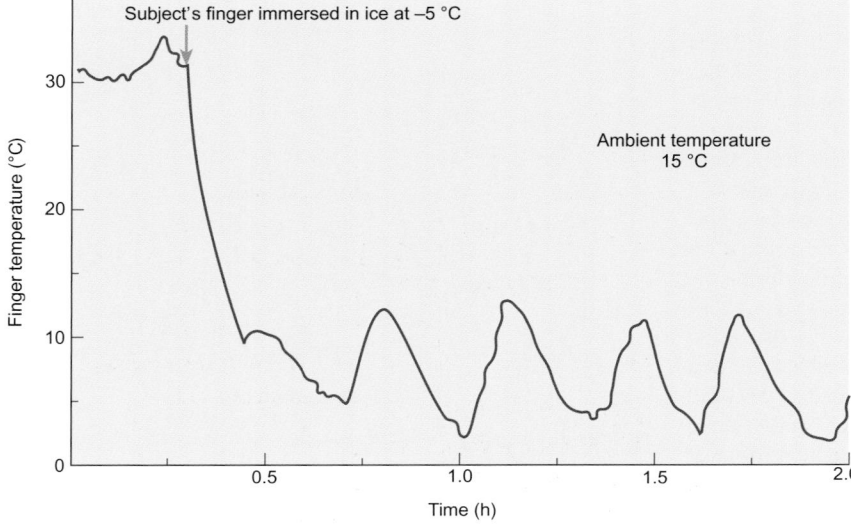

Fig. 26.4 The 'hunting reaction' in a human finger immersed in ice. Periodic vasodilatation is thought to delay the onset of tissue damage.

Non-shivering thermogenesis

This is the generation of heat through processes other than muscle contraction and includes the actions of a variety of calorigenic hormones and the stimulation of brown fat metabolism. The catecholamines epinephrine and norepinephrine stimulate thermogenesis in both skeletal muscle and brown fat. Thyroid hormones appear to increase the oxygen consumption of most cells and thereby cause an increase in basal metabolic rate during prolonged exposure to cold. Other calorigenic hormones include glucocorticoids, insulin, and glucagon. Although brown fat thermogenesis is of importance in neonates (see Chapter 22 and below), its significance in the overall process of thermoregulation in adults is uncertain.

26.5 Thermoregulatory responses to heat

When the core temperature of the body starts to rise, physiological responses are initiated which increase the rate at which heat is lost from the body surface. A number of behavioural modifications are also seen, including a reduction in activity, the shedding of clothing, drinking cold fluids, seeking a cooler environment, and so on. The major physiological responses in humans are cutaneous vasodilatation and sweating.

Cutaneous vasodilatation

When the core temperature rises above its normal value, the blood vessels of the skin dilate and cutaneous blood flow may reach $2 \, l \, min^{-1} \, kg^{-1}$. This dilatation is mediated by the autonomic nervous system, mainly through a reduction in vasomotor tone. As the vasculature swells with warm blood, heat is lost more readily from the body surface by radiation, conduction, and convection as described earlier.

Enhanced sweating

When heat production exceeds heat loss, body temperature rises, sweating is initiated, and large quantities of sweat may be produced. As long as the air is dry, the evaporation of sweat is an efficient means of losing heat from the body. However, when the relative humidity is high, evaporation occurs much more slowly and sweating is a less effective mechanism of core cooling.

There are two kinds of sweat gland, apocrine and eccrine. The *apocrine glands* are found in the skin of the axilla, the areola, and the anal regions. They secrete a viscous milky secretion that has much to do with individual body odour but nothing to do with thermoregulation.

Sweat is secreted by the *eccrine glands*. Humans possess around 2.5 million of these, about half of which are situated in the dermis of the back and chest. In cold conditions less than 500 ml of sweat is produced a day, but in very hot weather (particularly during heavy exercise) this can rise to between 1.5 and $6 \, l \, h^{-1}$. The sweating mechanism is initiated at an ambient temperature of 30–32 °C in a resting individual. The eccrine sweat glands are simple tubular structures. A coiled portion deep within the dermis secretes a fluid called the precursor fluid. This is similar in composition to the plasma. Reabsorption of some of the constituents of this fluid takes place in the duct of the gland, which opens onto the surface of the skin. The final composition of the sweat depends upon the rate at which it is being produced. At low sweat rates much of the sodium and chloride in the precursor fluid is reabsorbed and the sweat is very dilute. However, at higher sweat rates there is less time for reabsorption as the sweat flows along the duct, and consequently more sodium and chloride is lost from the body. Appreciable amounts of urea, lactic acid, and potassium ions are also present in sweat.

Sympathetic fibers, most of which are cholinergic, innervate the sweat glands. Stimulation of these fibers leads to an increase in the rate of sweat production, which is mediated by muscarinic receptors as it is blocked by atropine (see Chapter 10). Circulating catecholamines from the adrenal medulla are also thought to increase the rate of sweating.

Although sweating provides an effective means of heat loss from the body, it also represents a potentially dangerous loss of water and sodium chloride. Although the secretion of both ADH and aldosterone (see Chapter 28) is increased during heavy sweating, it is important that the lost fluid and salt is replaced quickly.

Humans can adapt (acclimatize) to hot climates relatively quickly (within a few weeks). The most important adaptation is in the rate of sweating, which may double, and in the lowering of the threshold temperature at which the sweating mechanism is initiated. Furthermore, the sodium chloride content of sweat also appears to fall after prolonged exposure to heat, possibly as a result of increased aldosterone secretion.

26.6 Disorders of thermoregulation

Hypothermia

From the above discussion it is evident that the body employs a variety of strategies to prevent a fall in core temperature during exposure to a cold environment. Nevertheless, in extreme conditions such compensatory mechanisms may prove inadequate and, under such conditions, hypothermia occurs (Fig. 26.5). This is defined as a core temperature below 35°C. Below this temperature the muscles become weaker and both voluntary movement and shivering will be reduced. With the loss of these heat-generating mechanisms, the core temperature may start to fall quite rapidly.

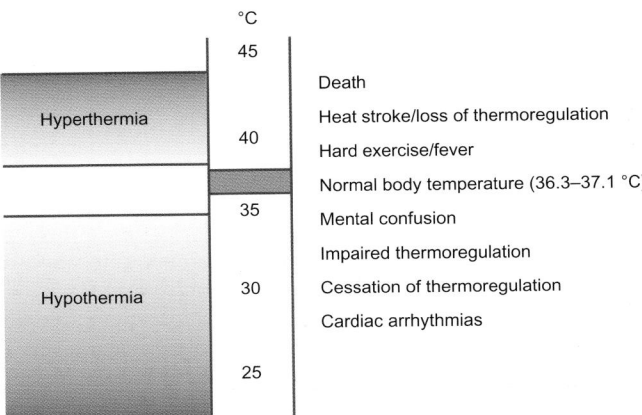

Fig. 26.5 A summary of the effects of extreme core temperatures.

As the core temperature falls to 34 °C, mental confusion is seen, with a loss of consciousness soon following. When the temperature falls below 28 °C, serious cardiovascular changes may occur including a fall in heart rate and arrhythmias leading to ventricular fibrillation, which is fatal. However, complete recovery from even extreme hypothermia is possible, provided that the subject is warmed slowly, preferably 'from the inside out'. This is because the reduced temperature of the body tissues, especially the brain, considerably reduces their metabolic requirements and enables them to be sustained by a severely restricted blood supply. In such cases it is important that efforts to resuscitate the subject should not be discontinued prematurely.

It is inadvisable to warm the surface of a hypothermic patient too rapidly, i.e. by hot blankets or vigorous rubbing, since the increased blood flow to the periphery may compromise blood flow to the body's vital organs and lead to further problems. Extracorporeal warming of the patient's blood by dialysis or peritoneal lavage with warmed fluids are the preferred means of restoring body temperature.

Hypothermia in the elderly

The ability to regulate core body temperature declines with advancing age. There is a reduction in the awareness of temperature changes and impairment of thermoregulatory responses. While the young can detect a skin temperature change of 0.8 °C, many elderly people are unable to discriminate a difference in temperature of 2.5 °C—indeed, some cannot discriminate a difference of 5.0 °C. In these people, shivering in response to moderate cooling is reduced and the change in metabolic rate in response to cold is also small compared with that shown by young adults. In addition to these age-related changes, elderly people are often relatively immobile, suffering in many cases from orthopedic and rheumatic problems, Parkinson's disease, or cerebrovascular disease, which may make it difficult for them to increase their voluntary muscle activity. Therefore, old people must live in surroundings where the temperature is maintained at a minimum of 20 °C.

The clinical use of hypothermia

In open-heart surgery, during which the heart must be stopped, it has now become routine to reduce the body temperature of patients in order to minimize the metabolic requirements of the tissues. This allows more time for the surgical procedures to be completed without the risk of hypoxic tissue damage.

Hyperthermia

Although short-term increases in body temperature to as much as 43 °C can be tolerated without permanent harm, prolonged exposure to temperatures above 40 °C or so may result in the serious condition of *heat stroke* in which there is a loss of thermoregulation. Sweating is reduced and core temperature starts to soar. The skin feels hot and dry, respiration becomes weak, blood pressure falls, and reflexes are sluggish or absent. Initially there is a loss of energy and irritability, followed by cerebral edema, convulsions, and neural damage as the core temperature exceeds 42 °C. Death usually follows unless rapid cooling is achieved (see Fig. 26.5).

Heat exhaustion is another potential consequence of hyperthermia and may be the result of either dehydration (a loss of body fluid) or salt deficiency (excessive loss of salt through sweating, which is not replaced in the diet). Dehydration is characterized by fatigue and dizziness in the early stages, progressing to intracellular dehydration and cellular damage. Salt deficiency produces a fall in tissue osmolality, which causes muscle cramps.

26.7 Special thermoregulatory problems of the newborn

The mechanisms of heat loss from the body were discussed in Section 26.2. When the surface area to volume ratio is high, as in small babies, heat loss occurs more readily. This is exacerbated by the relatively thin layer of insulating fat forming the shell of the baby's body, which renders vasoconstriction rather ineffective at minimizing heat loss. Furthermore, the shivering mechanism of newborn infants seems poorly developed so that shivering only occurs in response to extreme cold. Newborn babies, therefore, are at greater risk of hypothermia than older children and adults. Indeed, they require an ambient temperature of 32–34 °C to maintain their core temperature without increasing their metabolic rate.

However, the capacity of the newborn infant to generate heat by non-shivering (metabolic) pathways is four to five times greater (per unit body weight) than that of an adult. In the neonate the proportion of brown adipose tissue is very high relative to body weight (Fig. 26.6). Brown fat metabolism generates large amounts of heat in response to catecholamines from the adrenal medulla and the sympathetic nervous system.

Premature babies are even more susceptible to heat loss. They have an even larger surface area to volume ratio, an even thinner insulating layer, and incompletely developed brown fat reserves. Such infants need to be kept in incubators until they are sufficiently mature to regulate their own core temperature.

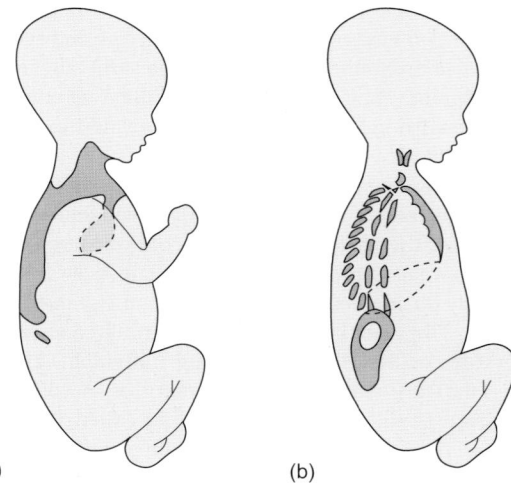

(a)   (b)

Fig. 26.6 The distribution of brown fat in a neonate showing (a) the interscapular pad and (b) the thoracic and abdominal depots.

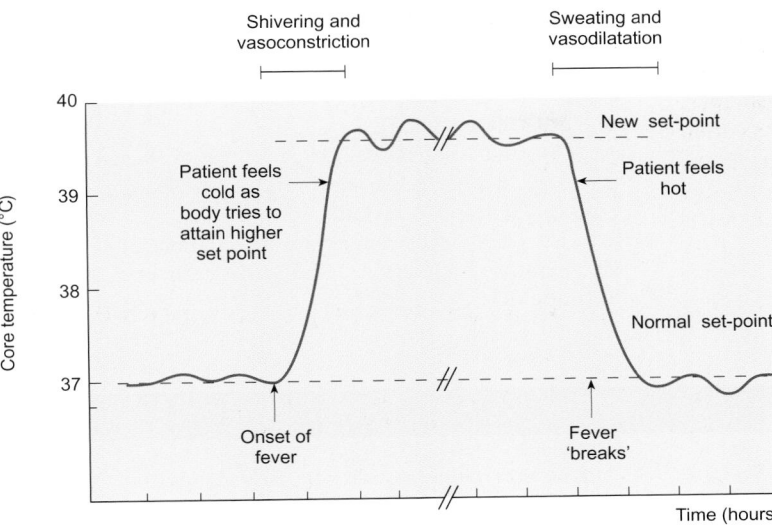

Fig. 26.7 The time course of a typical febrile episode.

26.8 Fever

Fever (or pyrexia) is the most common disorder of thermo-regulation and represents an elevation of normal body temperature which is not related to work or exposure to hyperthermic conditions. It is most often associated with infectious disease, although it can also arise as a consequence of certain neurological conditions or as a result of dehydration. The chemical substances that cause fever are known as pyrogens. They may be produced by bacteria themselves (endotoxins) or they may be endogenous pyrogens produced by the immune system (monocytes, macrophages, and astrocytes in the brain) in response to infection, for example interleukin 6 (IL-6).

The development of fever seems to involve a shift in the set-point around which core temperature is regulated. The exact mechanism whereby this occurs is unclear but may involve an alteration in the firing rates of preoptic neurons in the hypothalamus. As a result, heat is conserved inappropriately through the thermoregulatory mechanisms of the body. The patient may perceive himself to be cold and may show peripheral vasoconstriction and shivering despite a raised core temperature. When the febrile agent is no longer active or present, the set-point returns to normal and the patient perceives himself to be hot. Peripheral vasodilatation and sweating occur and body temperature returns to normal. Figure 26.7 shows the time course of a typical febrile episode.

Although fever is considered to form part of the body's defense against infectious organisms, prolonged increases in core temperature are believed to be potentially harmful because

Summary

1. To ensure optimal conditions for enzyme activity, humans maintain core body temperature between 36 and 38 °C. Heat is generated within the body by metabolic reactions and is lost from the body surface to the environment by radiation, conduction, and convection. Heat is also lost by evaporation. For effective thermoregulation, heat loss must be balanced by heat gain.

2. The cutaneous circulation plays an important role in thermoregulation. Vasodilatation of the skin vessels is an important heat loss mechanism, while vasoconstriction of the skin vessels reduces heat loss.

3. The hypothalamus receives input from thermoreceptors in the skin and the body core and acts as a thermostat to initiate appropriate mechanisms to conserve or lose heat. In this way the core temperature is kept at the set-point value of around 37 °C.

4. Physiological responses which act to conserve heat during exposure to cold include cutaneous vasoconstriction, shivering, and non-shivering thermogenesis. Physiological responses to overheating include cutaneous vasodilatation and sweating, which brings about heat loss through evaporation.

5. Hypothermia occurs when core temperature falls below 35 °C. As temperature drops, heat-conserving mechanisms start to fail; there is mental confusion and cardiovascular complications, followed by a loss of consciousness. Newborn infants and the elderly are particularly at risk of hypothermia.

6. Hyperthermia (a core temperature in excess of 40 °C) may have grave consequences. As the body's heat loss mechanisms fail, cerebral edema develops and later there is irreversible neuronal damage.

7. Fever is an elevation of body temperature usually associated with the presence of infectious agents. In response to pyrogens, the body conserves heat inappropriately. Febrile convulsions may occur in young children.

of the risk of neurological damage. Therefore, it is customary to treat the febrile state with antipyretic agents such as aspirin and other non-steroidal anti-inflammatory drugs.

In children under the age of 5 years, fever can cause convulsions (presumably due to the relative immaturity of their nervous systems). While febrile convulsions rarely have any long-term consequences, they are distressing for both child and parents, and the appropriate course of action is to cool any small child with fever quickly and effectively.

Further reading

Astrand, P.-O., Rodahl, K., Dahl, H., and Stromme, S. (2003). *Textbook of work physiology* (4th edn), Chapter 13. Human Kinetics, Champaign, IL.

Case, R.M., and Waterhouse, J.M. (eds.) (1994). *Human physiology: age, stress, and the environment* (2nd edn), Chapter 6. Oxford Science Publications, Oxford.

Hardy, J.D. (1980). Body temperature regulation. In *Medical physiology* (14th edn) (ed. V.B. Mountcastle), Chapter 59. Mosby, St Louis, MO.

Multiple choice questions

The following statements are either true or false. Answers are given below.

1. a. In a temperate climate under normal conditions, the greatest loss of body heat occurs through radiation.
   b. Heat is lost less readily from the body surface during immersion in water than in air.
   c. The skin temperature is regulated more closely than the core temperature.
   d. The hypothalamus is the centre for integrating thermal information.
   e. Vasodilatation of cutaneous vessels promotes heat loss from the body.

2. a. Non-shivering thermogenesis is more efficient in neonates than in adults.
   b. Heat loss through sweating is controlled by the parasympathetic nervous system.
   c. The skin temperature is generally higher than the temperature within the abdominal cavity.
   d. Profuse sweating can lead to heat exhaustion.
   e. The efficacy of sweating as a mechanism of heat loss is increased in humid conditions.
   f. If environmental cold is prolonged, the thyroid gland secretes more thyroxine.

3. A man moves from a temperate to a tropical area to live.
   a. During the first few weeks in the tropical area his sweat rate will rise.
   b. The salt content of his sweat will fall.
   c. His rate of urine production will fall.
   d. The rate at which aldosterone is secreted from his adrenal cortex will fall.

Answers to multiple choice questions

1. In a temperate climate more than half the heat loss from the body occurs by radiation. Water has a greater thermal conductivity than air, allowing heat to be lost more readily from a body immersed in it. Core temperature (surrounding the internal organs) is closely regulated. The skin temperature can fluctuate by 20 °C on either side of the normal core value (37 °C or so). The hypothalamus acts as the 'thermostat' of the body. It receives input from thermoreceptors and initiates appropriate heat-conserving or heat-loss mechanisms. Increased blood flow to the skin brings warm blood closer to the surface, allowing more heat to be lost by radiation, conduction, and convection.
   a. True
   b. False
   c. False
   d. True
   e. True

2. Neonates can raise their metabolic rates more effectively than adults, largely through the metabolism of brown fat. Sweat rates are controlled by the sympathetic nervous system. Skin temperature is usually lower than core temperature. During profuse sweating, fluid and salts are lost from the body in large amounts. This can lead to a variety of symptoms collectively known as heat exhaustion. In humid conditions (when the atmosphere is moist) sweat evaporates more slowly and therefore is a less effective means of heat loss. Thyroid hormone stimulates oxidative metabolism and generates heat. Its secretion is enhanced during prolonged exposure to cold.
   a. True
   b. False
   c. False
   d. True
   e. False
   f. True

3. Sweat rates rise progressively during the initial weeks of acclimatization to high temperatures. Aldosterone secretion is increased. This leads to increased reabsorption of salt from the sweat and a fall in its salt content. As the sweat rate increases, the body will lose water. ADH secretion will be increased in response to this loss and urine output will fall.
   a. True
   b. True
   c. True
   d. False

The regulation of plasma glucose

After reading this chapter you should understand:

- The mechanisms involved in the maintenance of plasma glucose levels including glycogen storage, glycogenolysis, and gluconeogenesis
- The reciprocal actions of the pancreatic hormones insulin and glucagon on plasma glucose regulation
- The actions of other hormones concerned with the regulation of plasma glucose
- The mechanisms of glucose regulation that occur during the absorptive and postabsorptive states
- The consequences of hyperglycemia, with particular reference to diabetes mellitus
- The consequences of hypoglycemia

27.1 Introduction

To carry out their normal metabolism, the cells of the body must have continuous access to glucose, the major fuel used to produce cellular energy (ATP). Indeed, certain tissues, notably the CNS, the retina, and the germinal epithelium, rely almost entirely on glucose metabolism for the generation of ATP. The nervous system alone requires around 110 g of glucose each day to meet its metabolic needs. Therefore it is vital that the concentration gradient for glucose between the blood and the extracellular environment of the brain cells is maintained. Blood glucose levels are normally maintained between 70 and 140 mg dl^{-1} (4–8 mmol l^{-1}) despite wide fluctuations in dietary intake. A variety of hormones contribute to this regulation, which takes place continuously over a time-scale of minutes.

27.2 Organs involved in the handling of glucose

Following their absorption from the GI tract, the nutrient sugars including glucose are transported directly to the liver through the portal vein. The liver is able both to store glucose as glycogen and to synthesize it from non-carbohydrate precursors.

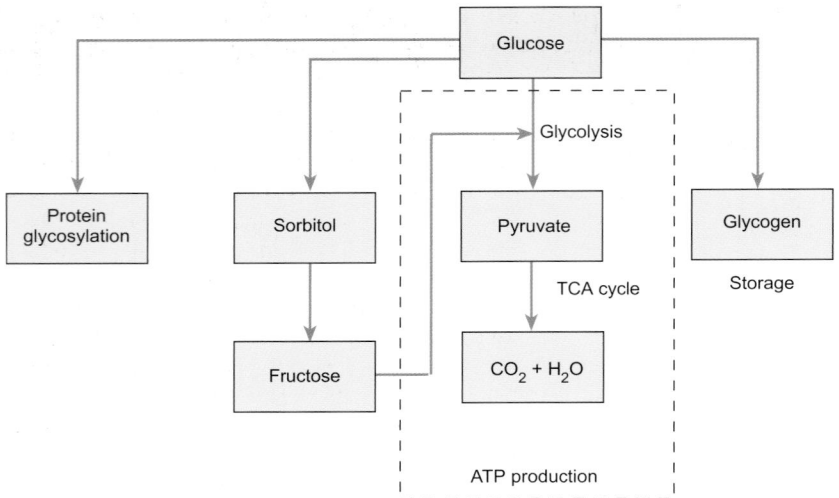

Fig. 27.1 The principal pathways involved in glucose metabolism.

When plasma glucose is raised, the liver removes glucose from the blood and stores it for future use. When plasma glucose falls, the liver releases glucose into the circulation. Therefore the liver is able to act as a buffer system to regulate blood sugar levels.

Excess glucose can be stored as glycogen or as fat

Glucose and other nutrient sugars arriving at the liver may be broken down by *glycolysis* to intermediates, which may be used as energy sources, or converted to glycogen (*glycogenesis*) which forms the liver's store of carbohydrate. The relative amounts utilized and stored will depend on the current plasma level of glucose and the glucose requirements of the tissues at the time. Smaller amounts of glucose are also stored as glycogen in other tissues, particularly the kidneys, skeletal muscle, skin, and certain glands. When the body's glycogen stores are fully saturated, any excess plasma glucose is converted to fatty acids and stored in adipocytes (fat cells) in the form of triglycerides. In cells that do not manufacture glycogen, glucose is metabolized by enzymes involved in the glycolytic pathway. Pyruvate enters the tricarboxylic acid cycle and is metabolized to produce ATP.

Liberation of stored glucose by glycogenolysis

All cells that store glycogen are capable of utilizing it for their own metabolism, but the cells of the liver and kidneys also release glucose into the general circulation so that it can be used by other cells as well. When plasma glucose falls during a period of fasting, the glycogen in these cells is broken down by phosphorylase enzymes to glucose, which is then released into the blood to provide a source of energy for the glucose-dependent tissues.

Glucose can be synthesized from non-carbohydrate precursors

Cells of the liver and kidney are able to synthesize glucose from non-carbohydrate precursors such as glycerol, lactate, and certain amino acids (a process known as *gluconeogenesis* see p. 31). However, the kidneys only become a significant source of plasma glucose in times of starvation. Under most conditions, the liver is the chief source of glucose for the general circulating pool and plays a crucial role in supplying the brain with the glucose on which it depends. The principal pathways involved in glucose metabolism are illustrated in Fig. 27.1.

Figure 27.1 also illustrates two pathways of glucose metabolism that may become significant when plasma glucose is elevated. These are the protein glycosylation pathway, in which certain proteins, especially hemoglobin, become glycosylated, and the polyol pathway where glucose is converted to sorbitol, which is then oxidized slowly to fructose. This occurs chiefly in the retina, lens, Schwann cells, kidneys, and aorta.

Many endocrine systems can influence the metabolic pathways in order to ensure that plasma glucose levels are maintained within the normal range. In view of its importance in cerebral metabolism, it is vital that glucose levels should not be allowed to fall too low and all but one of the hormones concerned with glucose metabolism act to raise plasma glucose. These so-called *glucogenic* hormones include glucagon, the catecholamines (epinephrine and norepinephrine), the glucocorticoids (chiefly cortisol), the thyroid hormones, and certain anterior pituitary hormones, notably growth hormone. In contrast, there is only one hormone whose action elicits a decrease in plasma glucose—pancreatic insulin. Nevertheless, this hormone plays a key role both in the overall homeostasis of plasma glucose and in regulating glucose uptake by non-neural tissues.

27.3 Insulin and glucagon provide short-term regulation of plasma glucose levels

Dispersed throughout the pancreas and occupying about 2 per cent of its total mass are small clumps of endocrine tissue known as the *islets of Langerhans*. These secrete insulin and glucagon, which are peptide hormones vital to the regulation of plasma glucose levels. Three main cell types have been identified within the islets. These are the α, β and δ cells. Glucagon and insulin are synthesized, stored and secreted by the α and β cells respectively, while the δ cells secrete somatostatin (which acts to inhibit the secretion of both hormones, possibly by a paracrine action). The islets are highly vascular, and are innervated by the sympathetic and parasympathetic branches of the autonomic nervous system.

Insulin

Insulin is a small (51 amino acid) protein hormone derived from a larger single-chain precursor called proinsulin, which is synthesized in the rough endoplasmic reticulum. Most of this precursor molecule is converted to insulin in the Golgi apparatus as a result of proteolytic enzyme cleavage to form a structure in which two amino acid chains are joined by two disulphide (S–S) bridges. The hormone is then stored within granules in the β cells until it is secreted into the blood stream by exocytosis in response to an appropriate stimulus.

Much of the circulating insulin is loosely bound to a β globulin, but the half-life of insulin within the plasma is short (around 5 minutes) because it is avidly taken up by tissues, particularly the liver, kidneys, muscle, and adipose tissues. Very little insulin appears in the urine.

Control of insulin secretion

Although a large number of factors can alter the output of insulin from the pancreatic β cells, glucose is the stimulant of greatest importance in humans. Glucose is believed to act directly on the islet cells to stimulate the secretion of insulin. At low plasma glucose levels (0–3 mmol l⁻¹) the β cells have a resting potential of about –60 mV. As plasma glucose is raised above 3 mmol l⁻¹ there is a progressive depolarization of the β cells. When this depolarization reaches threshold (around –55 mV), the β cells generate action potentials superimposed on slow waves. The action potentials open voltage-gated calcium channels, allowing calcium ions

to enter the cells and trigger the release of insulin by exocytosis.

During fasting, when plasma glucose is relatively low (around 3–4 mmol l⁻¹), insulin is secreted at a very low rate and is barely detectable in the blood. Following a meal, insulin secretion increases as plasma glucose rises (Fig. 27.2). After a typical meal, plasma insulin rises three- to tenfold, usually peaking 30–60 minutes after eating begins. The secretion of insulin in response to a glucose load is normally biphasic in nature. The early rise in the rate of secretion (phase 1) reflects the release of available insulin, while the later rise (phase 2) is believed to depend on the synthesis of new insulin in response to the glucose load. Plasma glucose reaches a maximum about an hour after ingestion of a meal and then declines until it is less than the normal fasting level before returning to normal. The close link between insulin secretion and the plasma glucose concentration prevents the latter from reaching excessively high values.

The major targets for insulin action are the liver, the adipose tissue, and the muscle mass. Insulin binds to glycoprotein receptors on the surfaces of cells and this leads to the insertion of preformed glucose carriers into the plasma membrane from intracellular vesicles. The overall result is an increase in the rate of uptake of glucose by the insulin-responsive cells. Some of the glucose is used for metabolism while the remainder enters anabolic pathways, leading to the conversion of glucose to glycogen and fat (see above). As glucose enters the cells, plasma levels fall once more and, after a short lag, insulin secretion is inhibited and returns to basal levels.

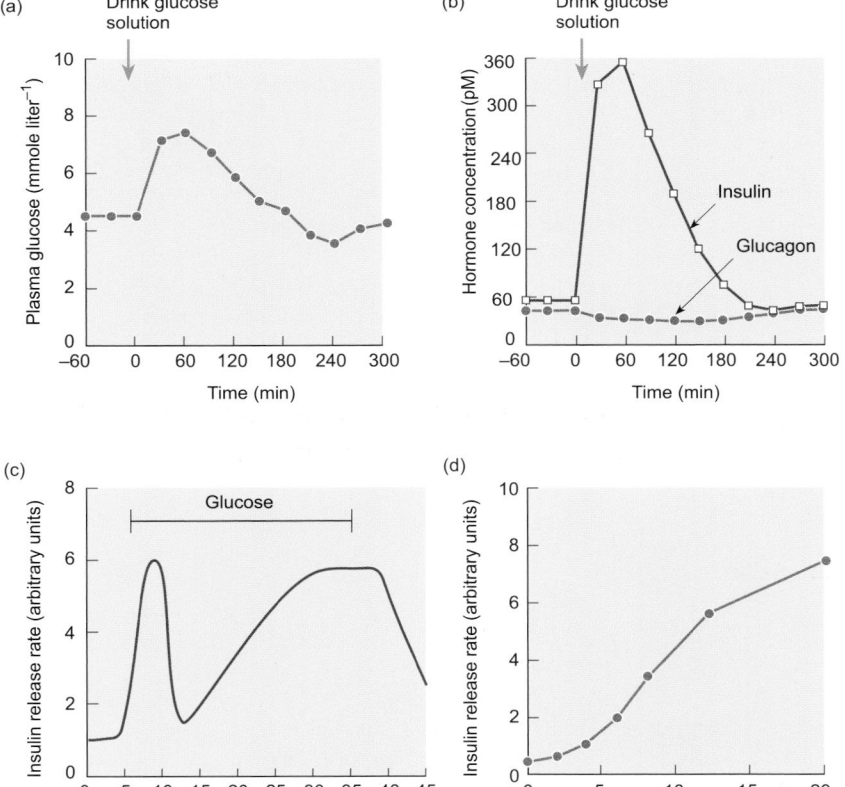

Fig. 27.2 The relationship between plasma glucose and insulin secretion. (a), (b) Changes in plasma glucose, insulin, and glucagon following a carbohydrate-rich meal. (c) Variation of insulin secretion (expressed in arbitrary units) with time following a glucose load. (d) The rate of insulin secretion as a function of the plasma glucose concentration.

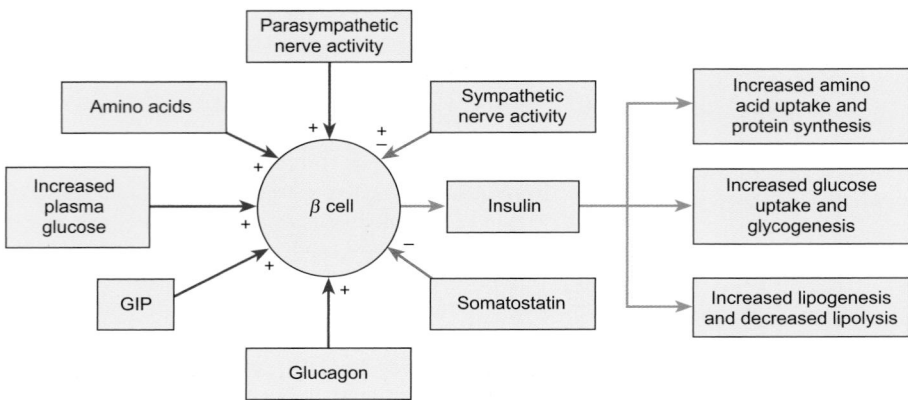

Fig. 27.3 The major factors involved in the regulation of insulin secretion by the β cells of the pancreatic islets and the principal actions of insulin: + enhancement of secretion; – inhibition of secretion.

The consumption of glucose by the CNS is independent of insulin. However, certain areas of the brain, such as the hypothalamus, are responsive to insulin and it is possible that it plays a role in the control of appetite. Neurons within the so-called 'satiety center' of the hypothalamus increase their rates of discharge in response to insulin, before showing a decline in firing as the plasma glucose level falls in response to insulin.

When glucose is given orally, a greater insulin response is elicited than when it is infused intravenously. This is believed to be the result of the actions of certain other GI hormones, notably gastric inhibitory polypeptide (GIP), gastrin, secretin, and cholecystokinin (CCK), which seem to enhance the β-cell response to glucose. They are probably important in preventing a large rise in plasma glucose immediately following the absorption of a carbohydrate-rich meal.

The autonomic nervous inputs to the pancreatic β cells also seem to play a role in the regulation of insulin release. The major effect of sympathetic stimulation (and circulating catecholamines) is a reduction in insulin release (seen for instance during stress), while parasympathetic stimulation enhances the rate of insulin secretion.

Certain amino acids (particularly leucine) are able to stimulate the secretion of insulin. Amino acid uptake into cells is stimulated as a result of the increased insulin release and there is a net stimulation of protein synthesis. The major actions of insulin and the factors that control its secretion are summarized diagrammatically in Fig. 27.3.

Glucagon

Although there are a number of hyperglycemic hormones (see later), the most potent is glucagon, many of whose actions directly oppose those of insulin. Glucagon is a 29 amino acid single-chain polypeptide hormone synthesized by the α cells of the pancreatic islets and released from them by exocytosis. Like insulin, it has a short half-life in the plasma (around 6 minutes).

The regulation of glucagon secretion

Glucagon is secreted in response to glucose deficiency and acts to raise circulating plasma glucose levels. Insulin appears to modulate the efficacy of glucose as a stimulus for glucagon release—glucagon secretion is stimulated much more by hypoglycemia if insulin is absent. Indeed, insulin also appears to inhibit glucagon secretion directly (Fig. 27.2 (b)). Glucagon secretion is also powerfully stimulated by certain amino acids (especially arginine and alanine), by both parasympathetic and sympathetic stimulation, and by the GI hormone CCK. Somatostatin inhibits glucagon release. The factors influencing the secretion of glucagon are summarized diagrammatically in Fig. 27.4.

The hyperglycemic actions of glucagon

Glucagon exerts effects on carbohydrate, fat, and protein metabolism. Its actions raise plasma levels of glucose. It also acts to maintain adequate circulating levels of other energy substrates during periods of fasting. The major target of glucagon is the liver. Hepatic glycogenolysis is stimulated and glycogen synthesis

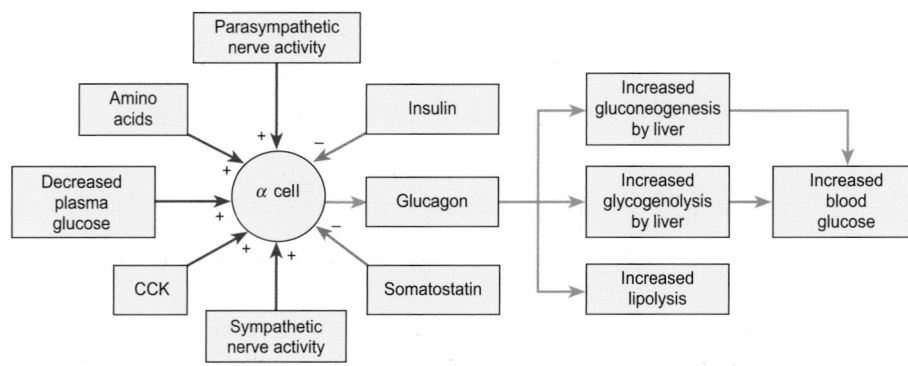

Fig. 27.4 The factors controlling the secretion of glucagon and the principal actions of this hormone.

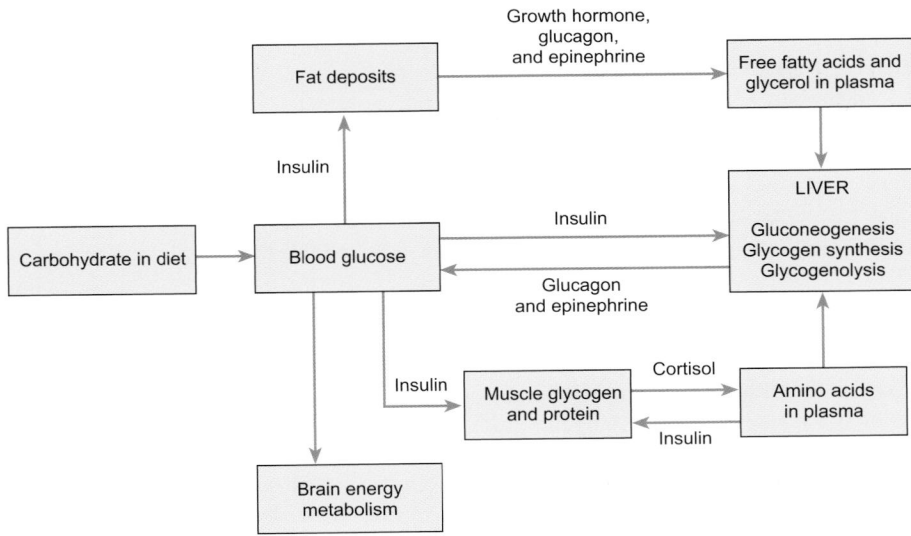

Fig. 27.5 An overview of the hormonal regulation of plasma glucose concentration. During the absorptive state insulin promotes the uptake of glucose by the liver, and by muscle and adipose tissue. In the postabsorptive state, glucose levels are maintained by glycogenolysis in the liver (which is stimulated by glucagon and epinephrine) and by gluconeogenesis, which is regulated by cortisol and glucagon. Lipolysis makes fatty acids available for oxidation, and this process is promoted by growth hormone, glucagon, and epinephrine.

is inhibited in response to glucagon. Hepatic uptake of certain amino acids and gluconeogenesis (the synthesis of new glucose from amino acids) are also enhanced, contributing further to the overall increase in plasma glucose. This hormone has also been shown to exert a significant lipolytic effect, mobilizing fatty acids and glycerol from adipose tissue. This provides a ready supply of metabolic substrates, thus enabling glucose to be spared for use by the brain, as well as providing glycerol which can act as a precursor for glucose in the hepatic gluconeogenic pathway. The actions of glucagon are thought to be mediated by the second-messenger cyclic AMP whose intracellular concentration is increased in response to the binding of glucagon to the plasma membrane receptors of its target cells. The major actions of both insulin and glucagon are summarized in Fig. 27.5.

27.4 Other hormones involved in the regulation of plasma glucose

Insulin and glucagon play a pivotal role in the fine regulation of plasma glucose levels—indeed, insulin is the only hormone capable of lowering plasma glucose and glucagon is the most important hyperglycemic hormone. Nevertheless, a number of other hormones also contribute to the maintenance of a stable blood glucose, as well as mobilizing glucose when necessary. These include the adrenal corticosteroids, growth hormone (GH), the catecholamines, and the thyroid hormones. The actions of all these regulators are described in detail elsewhere (particularly in Chapter 12) and so their effects on plasma glucose levels will be discussed briefly here.

The role of the glucocorticoid hormones in plasma glucose regulation

The adrenal cortical hormones are steroids. They are able to cross the plasma membrane of their target cells and bind to their intracellular receptors. The hormone–receptor complex then binds reversibly with DNA and functions as a gene activator, stimulating the production of appropriate mRNA as described in Chapter 5. The mRNA then leaves the nucleus to promote the synthesis of specific proteins by the ribosomes. Hormones that exert their effects in this way tend to have a rather prolonged time course of action with a lag between stimulus and effect. Therefore the glucocorticoids, of which cortisol is the most important, do not play a role in minute-to-minute regulation of plasma glucose levels in the same way as insulin and glucagon, but exert a longer-term influence, particularly in pathological conditions of steroid excess.

The primary role of the glucocorticoids in carbohydrate metabolism seems to be to maintain the reserves of glycogen in the liver (and to a lesser extent the heart and skeletal muscle). In so doing they have an overall hyperglycemic effect and are secreted in response to a fall in plasma glucose. In addition to this, they also exert an anti-insulin action in peripheral tissues, particularly adipose and lymphoid tissue, where they inhibit glucose uptake. At the same time, they promote the release of free fatty acids by lipolysis, thereby increasing the supply of these substrates and reducing glucose utilization by many tissues, so sparing glucose for use by the brain.

The glucocorticoids promote glycogen production by stimulating gluconeogenesis. They do this both by enhancing amino acid release from skeletal muscle protein and by inducing the synthesis of specific glycogenic proteins by the liver (Fig. 27.5).

The role of growth hormone in plasma glucose regulation

The actions of GH on glucose metabolism become significant in times of fasting or starvation. Its overall effect is as an 'anti-insulin' glucose-sparing agent whose role is to depress glucose uptake by muscle tissue in particular, to stimulate hepatic glycogenolysis, and to increase the breakdown of adipose tissue. These actions provide fatty acids for metabolism while increasing the availability of glucose for use by the brain.

The role of thyroid hormones in plasma glucose regulation

As explained in Chapter 12, the effects of the thyroid hormones on carbohydrate metabolism are rather complex and depend upon the levels of circulating hormone. All aspects of glucose metabolism seem to be enhanced, including glycogenolysis, gluconeogenesis, and absorption from the GI tract. These actions will result in an increase in plasma glucose. At the same time, however, thyroid hormones increase the rate of glucose uptake by cells (particularly muscle) and enhance the rate of insulin-dependent glycogenesis, effects which will tend to reduce plasma glucose. In general, it appears that low concentrations of thyroid hormones are anabolic and tend to reduce plasma glucose, while higher doses are catabolic and hyperglycemic.

Thyroid hormones are released in response to stresses of all kinds, including hypoglycemia and starvation. Under such conditions, they are of importance, together with other hormones (particularly the catecholamines), in ensuring that glucose reserves are mobilized and glucose-dependent tissues receive an adequate supply.

The role of catecholamines in the regulation of plasma glucose

When plasma glucose falls below 4 mmol l^{-1}, there is an increase in the secretion of catecholamines from the adrenal medulla and an increased output of norepinephrine from sympathetic nerve terminals. Epinephrine and norepinephrine work alongside the glucocorticoids, GH, and glucagon to maintain plasma glucose levels during the period of hypoglycemia and spare the available glucose for use by the brain. Indeed, catecholamines become very important in this regard if plasma glucose falls to very low levels or if glucagon secretion is impaired.

Epinephrine increases plasma glucose in two ways: directly by stimulating the mobilization of glucose from hepatic glycogen (see Chapter 5), and indirectly by enhancing the rate of glucagon secretion and by inhibiting the secretion of insulin. High concentrations of norepinephrine also inhibit insulin secretion (although low concentrations stimulate it), and so the net effect of generalized sympathetic stimulation is a stimulation of the secretion of hyperglycemic hormones and a depression of insulin secretion. The catecholamines exert a potent lipolytic action in adipose tissue, stimulating lipases to break down fats to liberate free fatty acids and glycerol. These metabolites may then be utilized in preference to glucose by many tissues. These changes result in an elevation of plasma glucose (Fig. 27.5).

27.5 Plasma glucose regulation following a meal

Having discussed the various hormones which play a role in the regulation and cellular handling of glucose within the body, it may be helpful to examine the changes in plasma glucose that follow the ingestion of a glucose-rich meal and during fasting. Immediately prior to a meal, plasma glucose is likely to be relatively low (around 3–4 mmol l^{-1}) and plasma concentrations of insulin will be low. However, levels of glucagon and the other hyperglycemic hormones will be elevated as the body attempts to conserve its glucose and maintain the supply of glucose to the brain. If a glucose-rich meal is then eaten, glucose is absorbed from the GI tract and a glucose load is presented to the body as plasma levels rise (see Fig. 27.2).

During the 90 minutes or so following a meal the body is said to be in the *absorptive state of metabolism* and plasma glucose may rise to 7 or 8 mmol l^{-1}. However, the exact rise will depend on the form of the ingested carbohydrate, the relative proportions of the other constituents of the meal (fats, proteins, etc.), the time of day, and so on. About 2 hours after the meal, plasma glucose will start to fall once more as glucose is utilized by cells or converted to its stored forms. As a rule, roughly 35 per cent of the glucose ingested will be oxidized, chiefly by the brain and muscle tissue, while the remaining 65 per cent will be stored. Glucose is stored principally in the form of glycogen, mostly in the liver but also in renal, muscle, and other tissues. If glycogen stores become saturated, the additional glucose is converted into fatty acids and stored in fat cells as triglyceride.

The pancreatic hormones, insulin in particular, are the dominant regulators of the events underlying the changes in plasma glucose concentration in the period following a meal. Insulin is secreted in response to the initial rise in plasma glucose (see Fig. 27.2). It brings about a fall in plasma glucose by the actions described earlier, notably an increase in cellular uptake of glucose, stimulation of glycogen production, and a reduction in gluconeogenesis. There is also decreased glycogenolysis, lipolysis, and ketogenesis.

Within 3–4 hours of the completion of a meal, we see the onset of processes whereby the body defends itself against hypoglycemia. These are known as the *counter-regulatory effects* and, at this time, the body is said to have entered the *postabsorptive state of metabolism*. By now, plasma glucose will have fallen back to around 4 mmol l^{-1}, a little below the normal fasting level. Insulin secretion will be low and there will be stimulation of the secretion of the hyperglycemic hormones. Glucagon is the most important of these and it acts largely on the liver to stimulate the breakdown of glycogen and the release of glucose into the bloodstream. The glucose stored as glycogen in other tissues such as muscle cannot be released into the plasma but is metabolized directly by those tissues, thereby helping to conserve plasma glucose. At the same time, in response to adrenal glucocorticoids, amino acids are mobilized from muscle and used by the liver to create new glucose (gluconeogenesis).

The catecholamines, GH, and glucagon all stimulate lipolysis in adipose cells. The free fatty acids released can be used by the liver and muscle as a metabolic substrate in preference to glucose. Ketone bodies, produced in the liver by the metabolism of fatty acids, provide an energy source for muscle and, in longer periods of fasting, for the brain. As a result of the counter-regulatory mechanisms described above and summarized in Fig. 27.5, plasma glucose rarely falls below 3 mmol l^{-1} even during quite prolonged periods of fasting.

27.6 Lack of pancreatic insulin results in diabetes mellitus

Diabetes mellitus is a condition that occurs when inadequate uptake of glucose by the cells of the body causes high levels of glucose in the blood (*hyperglycemia*). It is a relatively common disorder, affecting around 1–2 per cent of the population. A diagnosis of diabetes is usually made in patients with a fasting plasma glucose level of 8 mmol l^{-1} or more. This diagnosis may be con-

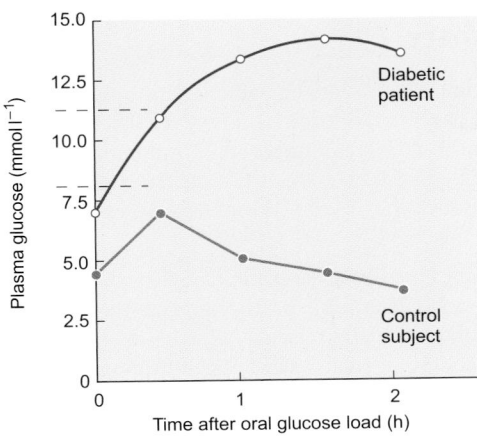

Fig. 27.6 Changes in plasma glucose following a glucose load in a diabetic patient compared with those seen in a normal subject. The glucose load was administered at time zero. The dotted lines show the range of plasma glucose levels which prompt a diagnosis of diabetes.

firmed by the *oral glucose tolerance test* in which a patient is given 300 ml of a 25 per cent solution of glucose to drink (corresponding to an intake of 75 g of glucose). Blood samples are taken before ingestion of the glucose and at 30 minute intervals thereafter and plasma glucose is measured. In a normal healthy individual blood glucose returns to normal within 2 hours of ingestion of a normal glucose load, while in diabetic individuals plasma glucose exceeds 11 mmol l^{-1} 2 hours after a meal (Fig. 27.6).

Diabetes mellitus may be caused by a failure of the pancreas to produce sufficient insulin or to insulin resistance due to a lack of normal insulin receptors. Failure of insulin secretion by the pancreas normally develops before the age of 40 and is due to autoimmune destruction of the β cells by islet cell antibodies, possibly triggered by chemicals or viruses. This form of the disease (type I diabetes) is treated by administering insulin either by multiple daily subcutaneous injections or by continuous subcutaneous infusion. For this reason, it is called *insulin-dependent diabetes*. In older patients, a non-insulin-dependent form of diabetes (type II diabetes) may develop. In these patients insulin is secreted in response to hyperglycemia but there is a loss of sensitivity to the hormone. Type II diabetes can be treated by careful management of the diet.

The classical symptoms of diabetes mellitus are the passage of an increased volume of urine (polyuria), intense thirst (polydipsia) and, in many cases, weight loss. The characteristic metabolic changes occurring in uncontrolled diabetes include a rise in blood glucose, an increase in glycogen breakdown, an increase in the rate of gluconeogenesis, increased fatty acid oxidation, and increased production of ketones (which can lead to the severe medical problem of ketoacidosis). The excessive plasma glucose levels lead to saturation of the renal glucose carriers so that glucose appears in the urine (glycosuria). Glycosuria induces an osmotic diuresis by reducing the amount of water reabsorbed in the distal part of the nephron (see Chapter 17). This loss of water induces thirst. Weight loss is due to the increased rate of breakdown of fats and protein. These changes are summarized in Fig. 27.7. Another feature of diabetes is an increased rate of secretion of glucagon, since the hyperglycemic suppression of glucagon secretion requires insulin. This will tend to exacerbate the effects of the insulin deficiency.

A number of chronic problems are associated with diabetes and are the result of prolonged exposure to hyperglycemia. Tissues such as muscle and fat, which depend on insulin for glucose uptake and utilization, remain relatively unaffected by high levels of plasma glucose. However, in non-insulin-dependent tissues, glucose uptake depends only on the concentration gradient for glucose between the extracellular and intracellular fluid. In hyperglycemia, this will tend to drive large amounts of glucose into the cells, resulting in over-utilization of glucose. Over long periods of time, this over-utilization can cause complications. The most common chronic problems associated with both types of diabetes are changes to the lens of the eye, degenerative

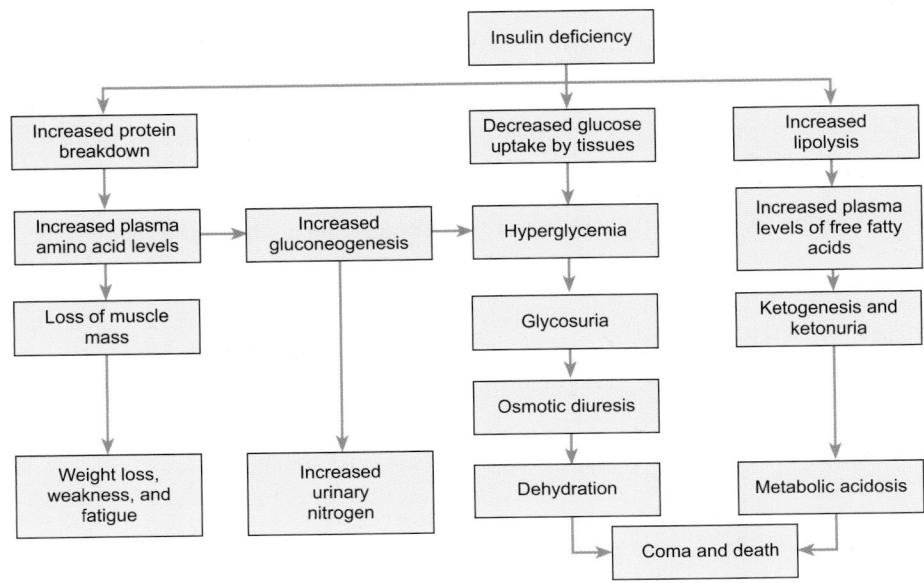

Fig. 27.7 The acute and chronic consequences of insulin deficiency.

changes in the retina, and peripheral nerves (*retinopathy* and *peripheral neuropathy*), thickening of the filtration membrane of the nephron (*nephropathy*), peripheral vascular lesions, and chronic skin infections. These pathological changes may be the result of increased protein glycosylation or they may result from the accumulation of sorbitol in cells, as described in Section 27.2 and shown in Fig. 27.1.

In view of the problems associated with chronically elevated blood glucose, it is important to monitor long-term glycemic control in diabetic individuals. This has been made possible by measurements of the blood concentration of *glycosylated hemoglobin* (HbA$_1$). Hemoglobin (Hb) is glycosylated by a non-enzymatic process at a rate proportional to the prevailing level of blood glucose, so that, as long as the lifespan of a person's red blood cells is normal, the level of glycosylated Hb will provide an indication of the mean blood glucose concentration over the preceding 60 days or so (half the average lifespan of a red blood cell, see Chapter 13). Although total HbA$_1$ is sometimes measured clinically, the value used most widely in the monitoring of diabetics is that of HbA$_{1c}$, which is produced by glycosylation of the N-terminal valine of the β-chain of the Hb molecule (see Chapter 13). Normoglycemic individuals with normal insulin responses will show a HbA$_{1c}$ value in the range of 3.5–5.5 per cent. Diabetics generally aim for values below 7 per cent as this is evidence of good long-term control of blood glucose. The figures presented in Table 27.1 show how the mean blood glucose concentration influences the measured level of HbA$_{1c}$.

TABLE 27.1 The relationship between plasma glucose concentration and glycosylation of hemoglobin

Blood glucose (mmol l^{-1})	HbA$_{1c}$ (% of total Hb)
18	13
13	10
10	8
5	5

27.7 Consequences of hypoglycemia

Hypoglycemia is defined as a blood glucose level below 2.5 mmol l^{-1}. It may arise from a number of causes, including overdosage of insulin in diabetics, an insulin-secreting tumor of the islet cells, lack of GH or cortisol, severe liver disease, very severe exercise, or inherited defects of gluconeogenic enzymes. The consequences of hypoglycemia fall into two categories: those associated with activation of the autonomic nervous system, and those caused by altered cerebral function.

A hypoglycemic episode is often heralded by a feeling of hunger mediated by the parasympathetic nervous system. Later, symptoms of sympathetic activation are seen. These include tachycardia, sweating, pallor, and anxiety. Because the brain relies on glucose as its prime energy source, hypoglycemia also elicits symptoms related to altered cerebral function. Headache, difficulty with problem solving, confusion, irritability, convulsions, and eventually coma occur. As soon as these symptoms appear, the treatment is to ingest carbohydrate (e.g. as glucose tablets). If plasma glucose falls below 1 mmol l^{-1}, there may be irreversible neuronal damage and the patient may die.

Summary

1. Glucose is an essential metabolic substrate, particularly for the CNS. Plasma glucose is closely regulated between 4 and 8 mmol l^{-1} by a number of hormones. These include insulin, glucagon, glucocorticoids, catecholamines, growth hormone, and thyroid hormones. All except insulin are hyperglycemic in their actions.

2. Plasma glucose is determined by the balance between intestinal absorption, storage and synthesis by the liver, and uptake by other tissues. Glucose may be stored in the form of glycogen and synthesized from non-carbohydrate precursors (gluconeogenesis). The latter takes place chiefly in the liver.

3. Insulin is synthesized by the β cells of the pancreatic islets. It is secreted chiefly in response to a rise in plasma glucose and amino acids. It stimulates the uptake and utilization of glucose by cells (particularly liver, adipose, and skeletal muscle), thereby causing a fall in plasma glucose. It stimulates both glycogen and protein synthesis. A lack of insulin or of insulin receptors gives rise to diabetes mellitus in which there is hyperglycemia leading to polyuria, excessive thirst, and weight loss.

4. Glucagon is the most potent hyperglycemic hormone. It is secreted by the α cells of the pancreatic islets and acts to promote the release of glucose into the blood. In effect, it acts in antagonism to insulin to provide short-term regulation of plasma glucose. Hypoglycemia is the principal stimulus for the secretion of glucagon, which then stimulates glycogenolysis, lipolysis, and gluconeogenesis.

5. During more prolonged periods of fasting, other hyperglycemic hormones such as cortisol, catecholamines, growth hormone, and thyroid hormones play a more significant role. Their effects are geared to the maintenance of glycogen stores, the stimulation of gluconeogenesis, and the mobilization of fatty acids and proteins to provide other metabolic substrates for those tissues able to use them. In this way glucose is spared for use by the tissues reliant upon it, notably the brain.

6. In the absorptive phase of metabolism (just after a meal), plasma glucose is high, insulin levels are high, and the hyperglycemic hormones are inhibited. During this time, glycogen stores are replenished and glucose uptake and utilization by cells is high. In the postabsorptive state (from 3 hours or so after a meal until the next intake of food), counter-regulatory mechanisms are initiated. Levels of the hyperglycemic hormones are enhanced, and there is an increase in the mobilization of fats (and, in prolonged fasting, proteins) so that glucose is spared for use by the central nervous system.

Further reading

Bell, J.I., and Hockaday, T.D.R. (2000). Diabetes mellitus. In *Concise Oxford textbook of medicine* (ed. J.G.G. Ledingham and D.A. Warrell), Chapter 6.13. Oxford University Press, Oxford.

Brook, C.G.D., and Marshall, N. (2001). *Essential endocrinology* (4th edn), Chapter 8. Blackwell Science, Oxford.

Campbell, E.J.M., Dickinson, C.J., Slater, J.D.H., Edwards, C.R.W., and Sikora, E.K. (eds.) (1984). *Clinical physiology* (5th edn), Chapter 14. Blackwell Scientific, Oxford.

Genuth, S.M. (1998). Hormones of the pancreatic islets. In *Physiology* (4th edn) (ed. R.M. Berne and M.N. Levy), Chapter 47. Mosby Yearbook, St Louis, MO.

Laycock, J., and Wise, P. (1997). *Essential endocrinology* (3rd edn), Chapter 11. Oxford Medical Publications, Oxford.

Porth, C.M., and Hurwitz, L.S. (1994). Diabetes mellitus. In *Pathophysiology: concepts of altered health states* (ed. C.M. Porth). J.B. Lippincott, Philadelphia, PA.

Wilding, J., and Williams, G. (2003). Diabetes mellitus and lipid metabolism. In *Textbook of medicine* (4th edn) (ed. R.L. Souhami and J. Moxham), Chapter 18. Churchill-Livingstone, Edinburgh.

Multiple choice questions

Each statement is either true or false. Answers are given below.

1. Concerning some general principals of glucose metabolism:
   a. Gluconeogenesis is the formation of glucose from glycogen.
   b. Glycogenolysis and gluconeogenesis are functions of the liver.
   c. Excess glucose may be stored as glycogen or fat.
   d. Gluconeogenesis is stimulated when plasma glucose is low.
   e. Glycogenesis is stimulated when cellular ATP reserves are low.

2. Concerning the absorptive and postabsorptive states:
   a. During the absorptive state, glucose is the major energy source.
   b. Events of the absorptive state are controlled by insulin.
   c. In the postabsorptive state glucagon secretion is inhibited.
   d. In the postabsorptive state glycogen and fat reserves are mobilized.
   e. Hypoglycemia inhibits the secretion of growth hormone.

3. A person has been on hunger strike for a week. Compared with normal he has:
   a. Increased release of fatty acids from adipose tissue.

   b. Ketosis and ketonuria.
   c. Elevated plasma glucose.
   d. Increased activity of hepatic glycogen synthetase.
   e. Increased plasma catecholamine levels.

4. Shortly after a carbohydrate-rich meal, the following metabolic processes are likely to be enhanced:
   a. glycogenesis
   b. glycogenolysis
   c. cortisol secretion.

   Shortly after waking up, the following processes are likely to be enhanced:
   d. growth hormone secretion.
   e. gluconeogenesis.
   f. lipolysis.

5. Diabetes mellitus:
   a. May be caused by a reduction in functional insulin receptors.
   b. Is characterized by a fall in urine output.
   c. Is associated with an increase in lipolysis.
   d. Leads to hypoglycemia.
   e. Is a common symptom of acromegaly.

Answers

1. When plasma glucose levels are high and cellular ATP reserves are adequate, glucose is stored in the form of glycogen and triglyceride. When plasma glucose levels fall, glycogen is broken down (glycogenolysis) to liberate glucose and glucose is synthesized from non-carbohydrate precursors (gluconeogenesis). The liver carries out both these processes.
   a. False
   b. True
   c. True

   d. True
   e. False

2. In the absorptive state (after a meal), plasma glucose is high, insulin secretion is stimulated, and glucose entry and utilization by cells is enhanced. In the postabsorptive state, plasma glucose falls, output of the hyperglycemic hormones (including GH and glucagon) is stimulated, and reserves of fat and glycogen are mobilized.
   a. True
   b. True

c. False

d. True

e. False

3. During fasting, plasma glucose is likely to be low. Levels of all the hyperglycemic hormones (including the catecholamines) will be increased. Glycogen synthesis will be inhibited, but there will be an increase in lipolysis leading to the liberation of fatty acids from adipose tissue. In prolonged starvation, excessive mobilization of fats will lead to the production of ketones in the liver. Although some of these are metabolized by the tissues of the CNS, excess ketones will appear in the plasma and urine.

a. True

b. True

c. False

d. False

e. True

4. After eating, plasma glucose is elevated and so glycogen synthesis is favored. The secretion of hyperglycemic hormones such as cortisol is depressed. In fasting, hyperglycemic hormones, particularly GH, are secreted. There is enhanced synthesis of glucose from non-carbohydrate sources and an increase in the utilization of fats.

a. True

b. False

c. False

d. True

e. True

f. True

5. Diabetes is the result of either a lack of pancreatic insulin or the absence of normal insulin receptors. It is characterized by raised plasma glucose, which leads to glycosuria and an osmotic diuresis. As glucose cannot be utilized effectively by many cells, fats are broken down to create alternative sources of energy. Acromegaly is caused by the excessive secretion of GH in adulthood. Because of the anti-insulin action of GH, acromegaly can result in diabetes.

a. True

b. False

c. True

d. False

e. True

The regulation of body fluid volume

After reading this chapter you should understand:

♦ How body water is distributed between the various body fluid compartments

♦ How the fluid volumes can be measured

♦ The mechanisms involved in maintaining the fluid balance between body fluid compartments

♦ How total body fluid is sensed and regulated

♦ The importance of sodium in the determination of body fluid volume

♦ How water intake is regulated by thirst

♦ Some common disorders of fluid balance and the physiological principles that are the basis of their treatment: dehydration, hemorrhage, and edema

28.1 Introduction

In healthy individuals, the volume and osmolality of the tissue fluids are maintained within closely defined limits. This chapter is mainly concerned with the mechanisms that regulate the quantity of water that is present in the body and maintain its distribution between the different body compartments. The detailed mechanisms by which the osmolality of the body fluids is regulated have already been discussed in Chapter 17, pp. 367–371.

Water is the principal constituent of the human body and is essential for life. The proportion of total body weight contributed by water varies with the age and sex of an individual. In both men and women, the water content of the lean body mass (i.e. the non-adipose tissues) is about 73 per cent. However, adipose tissue contains about 10 per cent water. For this reason the proportion of body weight contributed by water varies both between the sexes and with age. Since newborn infants possess little body fat, water accounts for nearly 75 per cent of their body weight. As adipose tissue forms and other tissues develop, this proportion declines so that, by the end of the first year, body water accounts for around 65 per cent of body weight. By the third decade of life, water makes up about 60 per cent of the

TABLE 28.1 Approximate distribution of body water (% total body weight)

	Adult males	Adult females	Neonates
Total body water	60	50	75
Intracellular water	40	30	40
Extracellular water	20	20	35
Plasma	4	4	5
Interstitial fluid[a]	16	16	30

[a] The interstitial water includes the lymph and the transcellular fluid.

total body weight of normal healthy adult males. However, as women of the same age have more adipose tissue than men, their body water accounts for a smaller proportion of body weight—about 51 per cent. By the seventh decade of life, body water accounts for about 50 per cent of body weight in males and about 45 per cent in females.

Broadly speaking, body water is distributed between the intracellular fluid (ICF) and the extracellular fluid (ECF) as discussed in Chapter 2, p. 6. The ECF can be further subdivided into the interstitial fluid, the plasma, and the transcellular fluid (see Fig. 2.1). The small contribution from the lymph is included in the interstitial fluid. Table 28.1 gives the normal distribution of water between the major body compartments for males, females, and neonates.

The space between the cells (the interstitium) consists of connective tissue, chiefly collagen, hyaluronate, and proteoglycan filaments together with an ultrafiltrate of plasma. The water of the interstitial fluid hydrates the proteoglycan filaments to form a gel (much like a thin jelly) and in normal tissues there is very little free liquid. This important adaptation prevents the extracellular fluid flowing to the lower regions of the body under the influence of gravity. Nevertheless, while exchange of water and solutes between the cells and the tissue fluid occurs mainly through diffusion, there is a bulk flow of isotonic fluid between the capillaries and the interstitium which is returned to the blood via the lymphatics (see Chapter 15, pp. 293–298, for further details). When the lymphatic drainage is obstructed, the tissues swell (edema—see below) and free fluid is found within the interstitial space.

28.2 The distribution of body water between compartments

The amount of water in the main fluid compartments can be determined by the dilution of specific markers. For a marker to permit the accurate measurement of the volume of a particular compartment, it must be evenly distributed throughout that compartment and it should be physiologically inert (i.e. it should not be metabolized or alter any physiological variable). In practice, it is necessary to correct for the loss of the markers in the urine. Fortunately, it is not difficult to make the appropriate corrections.

The *plasma volume* can be estimated from the dilution of the dye Evans blue (see Box 28.1) which does not readily pass across the capillary endothelium into the interstitial space. Radiolabeled albumin has also been used to measure plasma volume. Since the amount of marker injected is known, it is a simple matter to calculate the volume in which it has been diluted (the principle is explained in Box 28.1).

To determine the *total body water* a known amount of tritiated water (^3H_2O) or deuterium oxide (^2H_2O) is injected and sufficient time allowed for the label to distribute throughout the body. A sample of blood is then taken and the concentration of label measured. Measurement of the *extracellular fluid volume* requires a substance that passes freely between the circulation and the interstitial fluid but does not enter the cells. These requirements are met by inulin and mannitol, although other markers have been used. The volume of the intracellular fluid is simply the difference between the total body water and the volume of the

---

BOX 28.1 THE USE OF DILUTION METHODS TO ESTIMATE THE VOLUME OF FLUID COMPARTMENTS

Evans blue does not enter the red cells and is largely retained within the circulation. Therefore this dye is useful for estimating the plasma volume. As an example, assume that a patient with a body weight of 70 kg was injected with 10 ml of a 1 per cent (w/v) solution of the dye. Further assume that a sample of blood was taken after 10 minutes, and the plasma was found to contain 0.037 mg ml^{-1} of dye. What is the plasma volume?

Since

$$\text{concentration} = \frac{\text{amount of dye}}{\text{volume}},$$

$$\text{volume} = \frac{\text{amount of dye}}{\text{concentration}}.$$

The total amount of dye injected was 0.1 g (or 100 mg) and the concentration in the plasma 10 minutes after injection was 0.037 mg ml^{-1}. Therefore

$$\text{plasma volume} = \frac{100}{0.037} = 2702 \text{ ml.}$$

Note that this calculation assumes (i) that the dye is evenly distributed and (ii) that all of the dye remains in the circulation. In practice, some dye is lost from the circulation and corrections for the lost dye need to be applied to improve the accuracy of the estimate. Similar limitations apply to estimates of the ECF using inulin (and other markers) and to estimates of total body water. After allowing sufficient time for equilibration, the volume of a fluid compartment is given to a first approximation by

$$\text{volume} = \frac{\text{amount of marker infused} - \text{amount excreted}}{\text{concentration in plasma}}$$

extracellular fluid. Thus

$$\text{total body water} = \text{ECF} + \text{ICF}$$

and

$$\text{ECF} = \text{plasma} + \text{interstitial water.}$$

Osmosis and hydrostatic pressure determine the distribution of water between the fluid compartments of the body

With the exception of the apical membranes of the cells of the distal nephron (see Chapter 17), cell membranes are very permeable to water and this permits its free movement from one body compartment to another. Two forces govern this movement: osmosis and hydrostatic pressure. The hydrostatic pressure is derived from both the pumping action of the heart and the influence of gravity.

The movement of fluid between the plasma and the interstitial fluid is determined by the net filtration pressure and the capillary permeability. The net filtration pressure is determined chiefly by the difference between the capillary pressure and the plasma oncotic pressure (Starling forces—see Chapter 15, Box 15.8). Thus, when the hydrostatic pressure of the capillaries exceeds the plasma oncotic pressure, the hydrodynamic forces favor fluid movement from the capillaries to the interstitial space. When the oncotic pressure of the plasma exceeds the hydrostatic pressure, the hydrodynamic forces favor the absorption of fluid from the interstitium into the capillaries. Any excess fluid in the tissues is drained by the lymphatic system and re-enters the blood via the subclavian vein as described in Chapter 15, p. 297.

Exchange of water between the cells and the interstitial fluid is governed by osmotic forces. To illustrate how osmotic pressure regulates the movement of water between the intracellular and extracellular compartments, consider what happens when a 70 kg man drinks a liter of water. As the water is absorbed, the osmolality of the ECF falls. If there was no exchange of water between the ICF and the ECF, the volume of the ECF (20 per cent

of 70 kg) would increase from 14 to 15 liters and its osmolality would fall from 285 to 270 mOsm kg^{-1}, assuming that no solutes entered or left the ECF. This would result in a 15 mOsm kg^{-1} gradient in favor of water movement into the cells or of solute movement out of the cells. Since cell membranes are much more permeable to water than they are to ions and other small solutes, water moves from the ECF into the ICF and total body water increases from 42 to 43 liters, limiting the fall in osmolality to 2 mOsm kg^{-1} (rather than 15 mOsm kg^{-1}). Similar considerations govern the movement of water from the ICF to the ECF in response to water loss (e.g. during sweating). It should now be clear that the intracellular and extracellular compartments are normally in osmotic equilibrium and any departure from this situation will be transitory.

28.3 Body fluid osmolality and volume are regulated independently

If a person drinks 1 liter of water, their urine output will increase rapidly (a diuresis). Urine output peaks within an hour of drinking and returns to normal about an hour later, by which time the excess water has been eliminated (Fig. 28.1). If the same person were to drink a liter of isotonic saline, there would be no diuresis and only a very small rise in urine output. In this case, it takes many hours for the body to eliminate the excess fluid. In both situations, there is an initial increase in total body water but when there is a pure water load, the osmolality of the body fluids falls. To restore the normal osmolality of the plasma, only the excess water needs to be excreted. In contrast, when a liter of isotonic saline is drunk, the osmolality of the tissue fluids is unchanged but the total volume of the ECF is increased. To restore the normal ECF volume, the body must eliminate both the excess salt and the excess water. This simple experiment illustrates an important principle: the osmolality and volume of the body fluids are regulated by separate mechanisms.

Summary

1. Body water is distributed between the intracellular fluid and the extracellular fluid. The extracellular fluid is further subdivided into the plasma and the interstitial fluid.

2. Total body water can be estimated by measuring the dilution of radioactive water (^3H_2O), while the volumes of the various fluid compartments can be estimated using markers specific for each body compartment. Thus, inulin can be used to estimate the extracellular fluid volume and Evans blue can be used to estimate the plasma volume.

3. The volume of the intracellular fluid is given by the difference between total body water and the volume of the extracellular fluid.

4. The movement of fluid between the plasma and the interstitial fluid is governed by Starling forces, and the exchange of water between the interstitial fluid and the intracellular fluid is governed by osmotic forces.

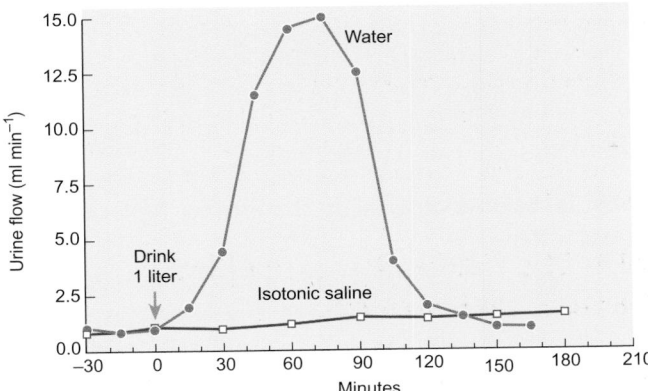

Fig. 28.1 The effect of drinking 1 liter of water or 1 liter of isotonic saline on the urine flow rate of a normal subject in water balance. Drinking occurred at the arrow. The urine flow rate increased from about 1 ml min^{-1} to nearly 15 ml min^{-1} following the intake of a liter of water. Moreover, the excess water was excreted in about 2 hours, by which time the urine flow rate had returned to normal levels. In contrast, drinking 1 liter of isotonic saline had little effect on urine flow rate.

Body water balance is maintained by the activity of osmoreceptors in the hypothalamus

Water requirements depend on body size, specifically on body surface area as this determines the extent of water loss via the skin and lungs. This loss, together with the water loss in the urine and feces, is replaced by water in the diet and by that generated during metabolism. For a normal adult male with a body weight of 70 kg, a typical balance sheet could be:

	Water gains (ml (24 h)$^{-1}$)		Water losses (ml (24 h)$^{-1}$)
Water intake (drinking)	1600	Urine	1500
Water content of food	500	Skin	500
Water generated by metabolism	500	Lungs	500
		Feces	100
Total	2600	Total	2600

The water loss from the lungs and skin obviously depends on prevailing conditions. In temperate climates water is lost from the lungs and skin without sweating (the *insensible water loss*). This loss cannot be reduced, and so any restriction of water intake must be balanced by a decline in urine output. As the urine osmolality cannot be greater than about 1250 mOsm kg^{-1} and the quantity of solids excreted in the urine each day is between 50 and 70 g (chiefly as sodium chloride and urea), the minimum volume of urine required for excretion is about 700 ml per day. To balance this and the other losses, a minimum fluid intake of about 1.75 liters each day is required to maintain water balance. As noted above, this figure is for a 70 kg adult in a temperate environment. In hot environments there is additional water loss via sweating which may amount to several liters a day and which must be matched by an appropriate intake of water.

The full extent of water exchange within the body is not revealed by these considerations. In addition to the obvious water exchanges that occur between the body and the environment, the GI tract secretes and reabsorbs some 7–8 liters of fluid each day and the kidneys form about 180 liters of filtrate a day, of which about 178.5 liters are normally reabsorbed. Therefore any reduction in fluid reabsorption arising from a disturbance of GI or renal function will have dramatic consequences for water balance. This is discussed further below.

Thirst is the physiological mechanism for replacement of lost water

Loss of water from the body is known as *dehydration* and results when water intake is not sufficient to balance water loss. To maintain the osmolality of the body fluids, it is essential that the water lost from the lungs together with that lost in the urine, sweat, and feces is replaced. Two sources provide water: oxidative metabolism of fats and carbohydrates, and water intake via the diet. In humans, the generation of water during metabolism is not sufficient to meet the needs of the body and drinking is essential for the maintenance of water balance. The stimulus for water intake is *thirst*, which can be defined as the appetite for water.

What factors stimulate drinking?

The state of water balance is monitored by the osmoreceptors of the anterior hypothalamus (see Chapter 12, p. 198). These receptors regulate the amount of ADH secreted by the posterior pituitary, increasing ADH secretion in dehydration and decreasing it during a water load. When the osmolality of the body fluids rises by about 4 mOsm kg^{-1} the desire for water is stimulated.

Drinking behavior can be triggered by electrically stimulating the preoptic area of the hypothalamus or by injecting hypertonic solutions into the same brain region. This suggests that the osmoreceptors that stimulate ADH secretion also play an important part in the regulation of water intake. Diarrhea and hemorrhage both cause a loss of isotonic fluid which results in a reduction in the circulating volume and this also stimulates thirst, probably as a result of increased plasma levels of angiotensin II (see below). Thus, an increase in the osmolality of the ECF or a decrease in the circulating volume will lead to increased thirst and to an increase in water intake by drinking.

Drinking provides relief from thirst long before the GI tract has been able to absorb the ingested water. Nevertheless, if water intake is not sufficient to satisfy the body's needs, drinking is resumed after a period of 15–20 minutes. This regulation of water intake is probably mediated by stretch receptors in the stomach wall, as inflation of a balloon in the stomach can inhibit drinking behavior. This ability of the body to determine water intake by the degree of distension of the stomach avoids excessive intake of water and the consequent dilution of the body fluids.

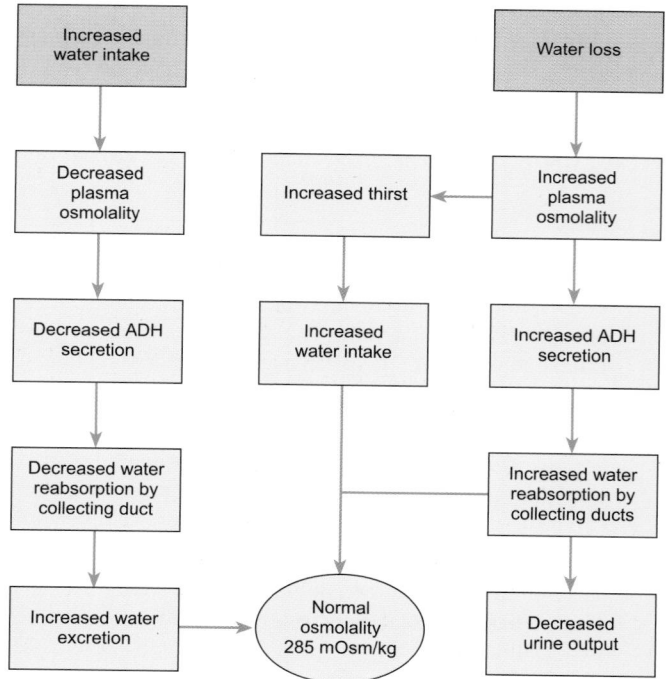

Fig. 28.2 The principal mechanisms by which the osmolality of the body fluids is restored following a water load (left-hand side) or pure water loss (right-hand side).

During dehydration, water is initially lost from the extracellular compartment but, since there is an osmotic equilibrium between the extracellular and intracellular compartments, this loss of water will ultimately result in cellular dehydration. The increase in plasma osmolality during dehydration is detected by the osmoreceptors of the hypothalamus which, in turn, stimulate ADH secretion from the posterior pituitary. ADH then acts on the distal nephron to increase water reabsorption as described in Chapter 17, p. 369. There is a reduction in urine flow rate and an increase in urine osmolality. As a result, body water is conserved. Restoration of water balance requires an increase in water intake and the high plasma osmolality stimulates thirst. The increased water intake restores plasma osmolality to normal. These processes are summarized in Fig. 28.2.

A water load causes a fall in plasma osmolality. This is detected by the osmoreceptors, which inhibit the secretion of ADH from the posterior pituitary. As a result, water reabsorption in the distal nephron declines and the urine flow rate increases while urine osmolality falls. The net effect is an increase in solute-free water excretion and the restoration of plasma osmolality to its normal value as illustrated in Fig. 28.2.

Alterations to the effective circulating volume regulate sodium balance via changes in the activity of the renal sympathetic nerves

As it is the most abundant ion in the extracellular fluid, sodium is the principal determinant of plasma osmolality and, since the osmolality of the plasma is closely regulated, total body sodium is also the principal determinant of body fluid volume. Moreover, as the equilibrium between the extracellular fluid and plasma volume is determined by Starling forces, any change in the total body sodium will affect the volume of both the ECF and the plasma.

In healthy people, the *effective circulating volume* (ECV) (the degree of 'fullness' of the circulatory system) is essentially constant and the sodium chloride and water losses are balanced by dietary intake. While the regulation of osmolality is relatively well understood (see above), little is known about the factors that control salt intake. However, it is known that animals will seek out salt when it is deficient in their diet. Moreover, patients with Addison's disease (in which the adrenal cortex is no longer capable of secreting aldosterone) crave salty foods. Therefore, there is an appetite for salt, which is regulated according to need.

The loss of body sodium is mainly governed by the kidneys which can regulate the amount of sodium they excrete over a wide range. This is achieved by regulating the filtered load (glomerulotubular balance—see Chapter 17, p. 359) and by adjusting the amount of sodium absorbed by the distal nephron. Sodium uptake by the distal nephron is regulated by the plasma levels of renin, which influence the circulating levels of aldosterone via the formation of angiotensin II (see Chapter 17, p. 364). When the ECV is changed, various mechanisms operate to adjust the whole-body sodium content so that the normal situation is restored.

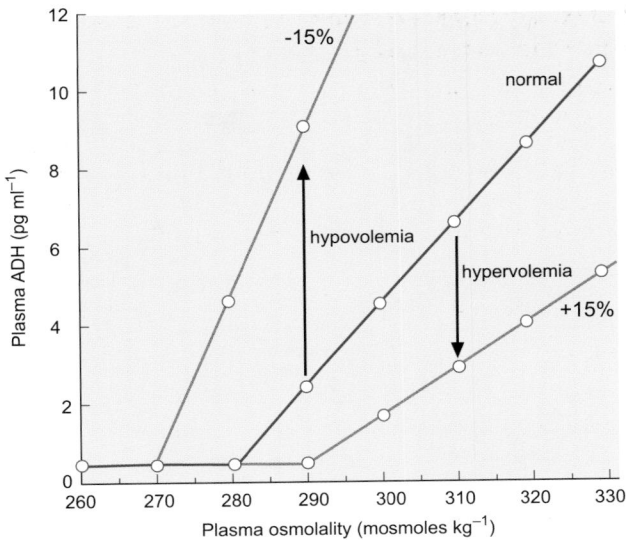

Fig. 28.3 The influence of changes in the ECV on the secretion of ADH. Note that the slope of the relationship between ADH secretion and plasma osmolality is altered during deviations from normovolemia. If body fluid falls by 15% (hypovolemia), both the secretion of ADH and the sensitivity to changes in plasma osmolality are increased (the slope of the relationship is steeper). If body fluid is increased by 15% (hypervolemia), the reverse applies; ADH secretion is decreased and is less sensitive to changes in plasma osmolality.

Where are the ECV receptors and how do they act to promote alterations to sodium balance?

The cardiac output, the vascular tone, and the ECV determine both the systemic blood pressure and the end-diastolic pressure. Systemic blood pressure is regulated by the baroreceptor reflex, whereby a reduction in pressure leads to an increase in cardiac output and total peripheral resistance as discussed earlier (see Chapter 15, pp. 299–301). The end-diastolic pressure is sensed by low-pressure receptors in the great veins, the pulmonary circulation, and the cardiac atria (also known as the central volume receptors). When the ECV is diminished, central venous pressure falls and this results in a decline in the afferent activity of the central volume receptors. This fall elicits a sympathetic reflex which results in a peripheral vasoconstriction. Atrial myocytes also play a role in regulating the ECV. In response to increased end-diastolic pressure, these cells secrete a hormone called *atrial natriuretic peptide* (ANP) which acts on the renal tubules to promote sodium excretion.

In addition to its regulation by plasma osmolality, the secretion of vasopressin (ADH) by the posterior pituitary gland is also modulated by the arterial baroreceptors and the central volume receptors. As shown in Fig. 28.3, a decrease in ECV (hypovolemia) leads to an increase in the secretion of vasopressin, while an increase in ECV (hypervolemia) leads to a fall in the secretion of vasopressin.

When the ECV falls

There is an increase in the activity of the renal sympathetic nerves which acts to promote sodium reabsorption and thus restore ECV. The principal mechanisms involved are illustrated in Fig. 28.4.

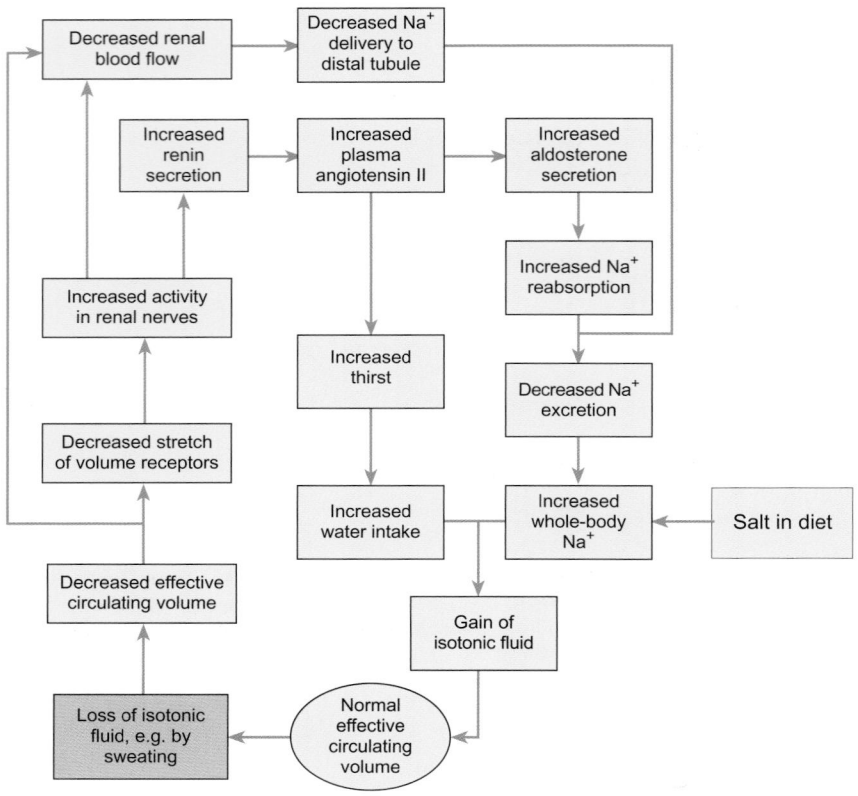

Fig. 28.4 The principal mechanisms responsible for restoring the ECV following loss of isotonic fluid (e.g. as a result of heavy sweating or diarrhea).

1. In response to the reduced filling pressures of the right side of the heart, there is vasoconstriction of the arterioles of the muscle and splanchnic vascular beds. This is a reflex response that is mediated via the sympathetic nerves. This response is triggered by the unloading of the stretch receptors of the atria and great veins (the central volume receptors) and occurs before there is any significant change in arterial blood pressure. The afferent and efferent arterioles of the nephrons become constricted and this reduces the glomerular filtration rate. Consequently, the filtered load of sodium is reduced and a higher proportion of the filtered sodium is reabsorbed (see Chapter 17, p. 359).

2. The increased sympathetic activity leads to an increase in renin secretion which, in turn, leads to an increase in the amount of aldosterone in the circulation. As described in Chapter 17, aldosterone increases the ability of the distal nephron to reabsorb sodium. In addition, the increased activity of the renal sympathetic nerves directly stimulates the uptake of sodium by the cells of the proximal tubule. The secretion of ADH is increased, promoting the reabsorption of water in the distal nephron.

3. Finally, as these changes serve to increase sodium reabsorption by the renal tubules, there will be an increase in plasma osmolality which, together with the increased plasma levels of angiotensin II, will stimulate thirst. The increased water intake together with the increased retention of dietary sodium leads to an increase in ECV and the restoration of body fluid volume.

When the ECV rises above normal

Full correction requires elimination of both the excess sodium and the excess water. This is achieved by the following processes (Fig. 28.5).

1. An increase in ECV activates both the arterial baroreceptors and the central volume receptors. This triggers the baroreceptor reflex and the activity of the renal sympathetic nerves is decreased. The afferent and efferent arterioles dilate, and both renal blood flow and GFR increase. This increases the filtered load of sodium and the delivery of sodium to the distal tubule. Renin secretion is inhibited and the concentration of aldosterone in the plasma falls. As a result, less of the filtered sodium is reabsorbed. In addition, the secretion of ADH is decreased (Fig. 28.3). These changes promote the excretion of salt and water by the kidneys.

2. The increased stretch of the atrial myocytes triggers the secretion of ANP which acts to increase sodium excretion. ANP causes dilatation of the renal afferent and efferent arterioles and thus augments both renal blood flow and GFR. ANP has actions that antagonize those of the renin–angiotensin system. It inhibits aldosterone formation and directly inhibits sodium chloride uptake by the distal tubule and collecting duct. It also inhibits both ADH secretion and its action on the distal nephron, so promoting the loss of water. Overall, these effects lead to a loss of sodium chloride and water and re-establishment of the normal ECV and body fluid volumes.

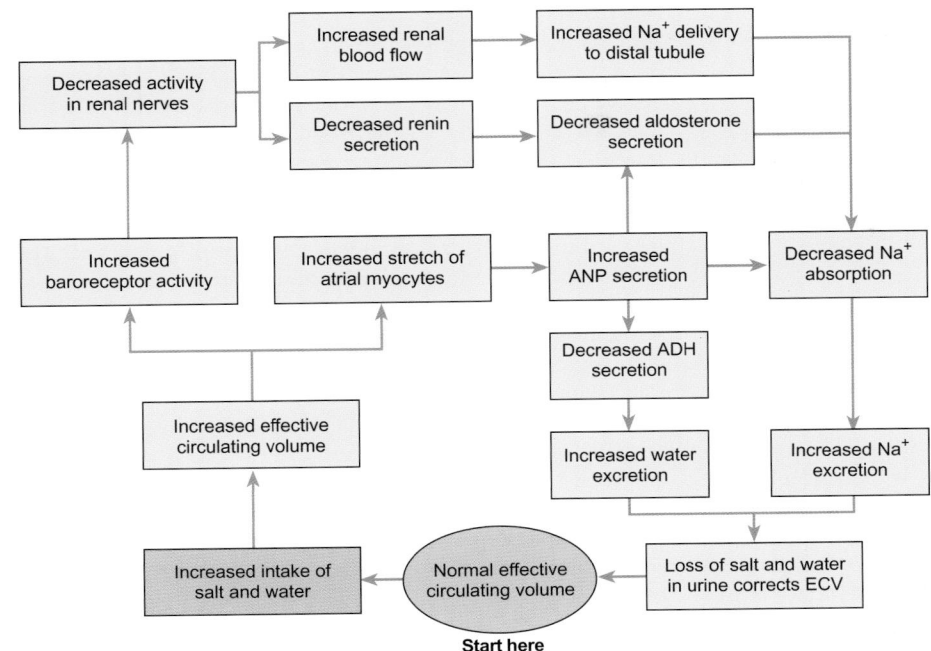

Fig. 28.5 The principal mechanisms responsible for restoring the ECV after its expansion by the intake of salt and water. ANP, atrial natriuretic peptide.

Summary

1. The osmolality of the body fluids is regulated by the activity of the osmoreceptors in the hypothalamus, which control the secretion of ADH from the posterior pituitary. Extracellular fluid volume is sensed via the effective circulating volume and is regulated by adjustment of sodium intake and excretion. Thus, osmolality and fluid volume are regulated independently of each other.

2. During a water load, plasma ADH levels fall and there is less water reabsorption by the distal nephron. As a result, the excess solute-free water is excreted.

3. When the body is dehydrated, the osmolality of the tissue fluids rises and ADH secretion is increased. This increases water reabsorption by the distal tubule and collecting ducts, thus conserving body water. In addition, the increase in plasma osmolality stimulates thirst.

4. Total body water is determined mainly by total body sodium. In response to a fall in the ECV the kidneys increase sodium reabsorption via activation of the renin–angiotensin system. As angiotensin II stimulates thirst, water intake is also increased and the ECV is restored.

5. In response to an increase in the ECV, excretion of sodium by the kidneys is increased. This is accomplished by an increase in the filtered load of sodium and a decrease in tubular absorption. An osmotic equivalent of water is also excreted and this results in the restoration of the normal ECV.

28.4 Dehydration and disorders of water balance

Dehydration has a number of causes:

- excessive water loss from the lungs and sweat glands;
- excessive urine production;
- excessive loss of fluids from the GI tract either by persistent vomiting or as a result of chronic diarrhea;
- inadequate intake of fluid and electrolytes.

In most of these situations *both water and electrolytes* are lost. For example, after heavy exercise significant quantities of water and salt are lost in the sweat and need to be replaced. (Sweat contains 30–90 mmol sodium chloride per liter.) If the lost water is replaced by drinking but the lost salt is not replaced, there is a deficiency of sodium chloride and the osmolality of the tissue fluids falls. The fall in tissue fluid osmolality results in muscle cramps (known as heat cramp), which can be relieved by drinking a weak saline solution (0.5 per cent sodium chloride) or by drinking water and taking salt tablets.

Pure water loss can occur when drinking water is unavailable or in situations where the kidneys are unable to reabsorb water from the distal tubule and collecting ducts. The classical example of this is diabetes insipidus where ADH secretion by the posterior pituitary is insufficient (see Chapter 12, p. 198). Despite their deficiency, patients with diabetes insipidus remain in good health provided that enough water is available for them to drink. However, if they are rendered unconscious, their situation rapidly becomes perilous. In the normal course of events, pure water loss causes a rise in plasma osmolality and an increased thirst.

When water intake is persistently less than water loss, there is a progressive dehydration of the tissues. When more than 6–10 per cent of body water has been lost, the plasma volume falls and circulatory failure commences. The poor circulation causes a failure of urine production and a metabolic acidosis develops. In addition, during severe dehydration the lack of water leads to a reduction in evaporative heat loss and fever may occur. The fever may be associated with drowsiness and delirium and, unless the lost fluid is replaced, these signs will be followed by coma and eventually death.

Overhydration of the tissues (water intoxication) is much less common. Nevertheless, when it occurs, a new osmotic equilibrium between the plasma and the tissues is established and the cells swell. When the brain cells swell, the intracranial pressure increases and brain function becomes impaired. The clinical signs include nausea, headache, fits, and coma. The most common cause is acute oliguric renal failure where the ingested water cannot be excreted. Inappropriate vasopressin (ADH) secretion either from the posterior pituitary (perhaps as a result of a head injury) or from tumors can also lead to water retention. Failure of the anterior pituitary gland to secrete ACTH also leads to water intoxication as the excretion of water depends on normal circulating levels of glucocorticoids. The exact mechanism of this effect is not known but may reflect a role of glucocorticoids in regulating the secretion of vasopressin.

Oral rehydration therapy

In vomiting and diarrhea there is a large loss of water and electrolytes as the fluids of the GI tract are isotonic with the blood. This results in both dehydration and changes in acid–base balance (see Chapter 29 for a more detailed discussion of acid–base balance). As the average person secretes about 7 liters of fluid into the GI tract each day, persistent diarrhea can lead to severe dehydration even in mild gastroenteritis. In cholera, the fluid loss is greater than with other causes of diarrhea as the causative organism *Vibrio cholerae* stimulates intestinal secretions through the action of its toxin on cyclic AMP production. Consequently, fluid loss in the stools during an attack of cholera may be 10 liters or more each day. Clearly, unless the effects of the diarrhea are rapidly countered, death will inevitably occur. Indeed, in many poor countries dehydration caused by fluid loss in the stools is a common cause of death, particularly in children.

Drinking water will not be sufficient by itself to combat the loss of fluid, as the lost fluid will have been isotonic with the plasma and both sodium chloride and water will be required to rehydrate the tissues. While intravenous infusions of isotonic sodium chloride (0.9 per cent NaCl) could be used to restore body fluid volume, this approach needs suitable resources. Nevertheless, it may be the only route of administration in cases of persistent vomiting and diarrhea. The use of simple oral rehydration fluids has been found to provide a very effective alternative therapy. In diarrhea the intestinal absorption of glucose is unimpaired despite the high fluid loss in the stools. Oral rehydration is achieved by giving the patients a solution of salt and sugar to drink. The sugar (as glucose) is absorbed across the intestinal wall by cotransport with sodium, and water follows osmotically. The solution must not be significantly hypertonic, otherwise water loss will be enhanced. The use of sucrose or starch has the advantage that these sugars are readily available. Moreover, as they are broken down to glucose in the intestine before being absorbed, the amount of glucose available for absorption (together with sodium and water) can be increased without making the rehydration fluid hypertonic.

28.5 Hemorrhage

If there is an acute loss of blood, the cardiovascular system rapidly adjusts to maintain blood pressure and to preserve blood flow to vital organs (Fig. 28.6). Nevertheless, the response of the body to blood loss depends on the amount lost. In conscious peo-

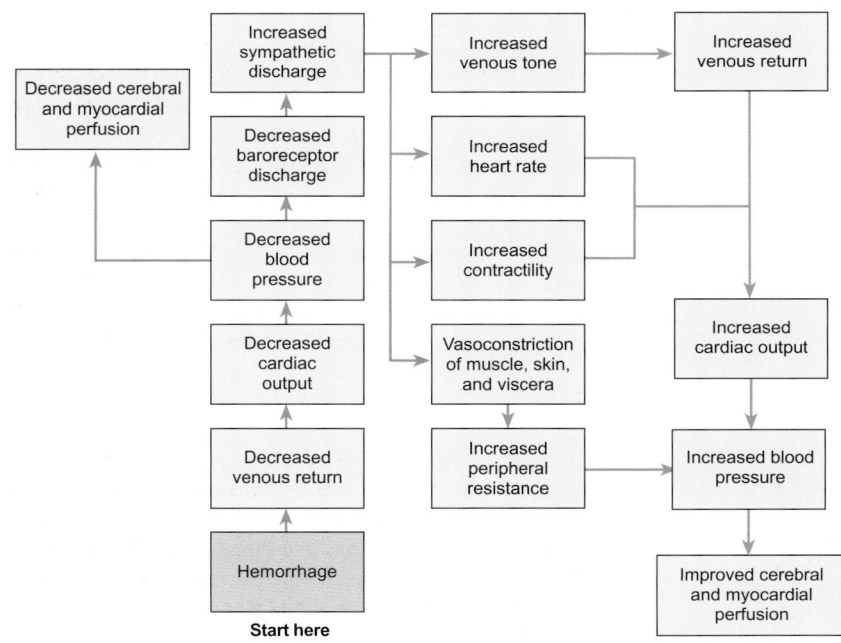

Start here

Fig. 28.6 The hemodynamic changes that occur during moderate hemorrhage. Although cardiac output falls following blood loss, the compensatory mechanisms indicated act to restore the blood pressure so that perfusion of the brain and the heart muscle is not compromised. Note, however, that these cardiovascular adjustments result in a marked decrease in blood flow to other vascular beds, particularly the renal and splanchnic circulations.

ple, significant cardiovascular adjustments begin to occur as blood loss exceeds about 5 per cent (*c.* 250 ml). Initially the loss of blood leads to a diminished venous return and a reduced stimulation of the low-pressure receptors. This results in a vasoconstriction of the cutaneous, muscle, and splanchnic vessels, which occurs before there is any significant change in either mean blood pressure or pulse pressure.

As blood loss increases, venous return falls further and there is a fall in both cardiac output and blood pressure. The fall in blood pressure activates the arterial baroreceptor reflex so that there is an increase in heart rate and in arteriolar tone. Together, these changes restore the blood pressure. The increase in sym-

pathetic drive also increases venous tone. This results in a mobilization of the blood from the capacitance vessels to the distribution vessels and so helps to prevent the cardiac output falling further than it otherwise would. These adjustments occur rapidly following blood loss.

If blood loss continues, the cardiac output falls further and the activity of the sympathetic nerves becomes more intense. The heart rate continues to rise and the peripheral vasoconstriction increasingly diverts blood away from the skin, muscles, and viscera towards the brain and heart. As plasma levels of angiotensin II rise during the early stages of hemorrhage, this hormone may contribute to the early vasoconstriction. Despite the fall in cardiac output, these changes help to maintain blood pressure and they are the principal mechanisms by which the cardiovascular system adjusts to mild hemorrhage (loss of less than 10 per cent of the blood volume).

Further loss of blood intensifies the activation of the sympathetic nerves but venous return is no longer sufficient to maintain blood pressure, which begins to fall. During this phase, catecholamine secretion by the adrenal medulla increases, as does the secretion of ADH by the posterior pituitary. These hormones further intensify the vasoconstriction. As a consequence of these adjustments, the pulse is weak and rapid, the skin is cold and clammy, and the mouth becomes dry as salivary secretion stops.

When blood loss reaches about 20 per cent the patient faints (known as *syncope*), there is a fall in heart rate, and blood pressure falls precipitously. Note that in this stage of severe hemorrhage the patient has a *bradycardia* despite the extensive blood loss (Fig. 28.7). The bradycardia is the result of a vagal reflex, but the decline in peripheral resistance appears to result from inhibition of the central sympathetic drive. Catecholamine secretion from the adrenal medulla, ADH secretion, and angiotensin II all continue to increase. Curiously, cardiac output does not necessarily fall further and may even rise a little. Further blood loss results in intense sympathetic activation with pronounced vasoconstriction and tachycardia.

This description of the cardiovascular responses to hemorrhage assumes that there are no other factors influencing the autonomic responses. In practice, hemorrhage is often accompanied by traumatic tissue injury and the sympathetic response to the trauma tends to mask the bradycardia. The reflex bradycardia is also reduced by administration of general anesthetics and may be complicated by drug therapy.

Following extensive blood loss, there is an urgent need to restore blood volume. Clinically, this can be achieved by blood transfusion or by infusion of fluids that expand the plasma volume, for example normal saline or a plasma substitute (see below).

The physiological processes responsible for restoring the plasma volume are as follows.

◆ The peripheral vasoconstriction lowers capillary pressure and the oncotic pressure of the plasma draws fluid from the extracellular space to the circulation. In addition, there is reabsorption of fluid from the GI tract. The increase in ADH secretion and diminished renal blood flow cause a marked fall in urine output (oliguria). These are the first steps towards restoration of blood volume.

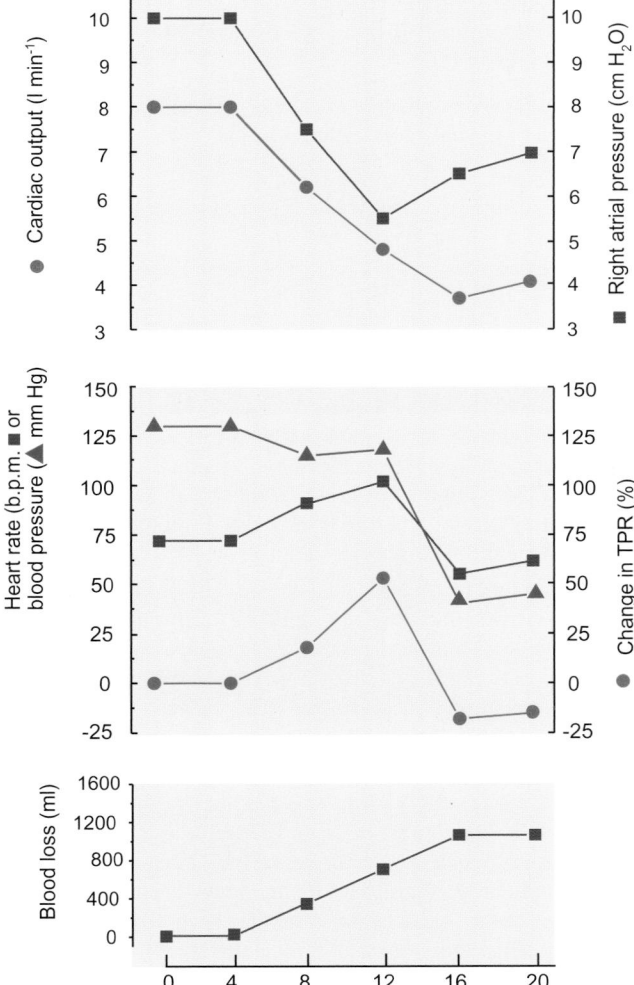

Fig. 28.7 The cardiovascular changes that occur during acute blood loss. The bottom panel shows the amount of blood lost by a human subject. The center panel shows the changes in heart rate, systolic blood pressure, and total peripheral resistance, and the top panel shows the changes in cardiac output and right atrial pressure. Note that cardiac output and right atrial pressure (equivalent to central venous pressure) progressively fell during the hemorrhage. The heart rate, systolic blood pressure, and total peripheral resistance showed a biphasic response, rising during the early phase of blood loss and abruptly falling after 1080 ml of blood had been lost. At this point, the subject fainted.

- The fall in renal blood flow increases renin secretion and thus increases the level of circulating angiotensin II, which increases the secretion of aldosterone by the adrenal cortex. The aldosterone increases sodium reabsorption by the renal tubules and colon. These events lead to the retention of sodium.

- Loss of ECV also leads to intense thirst (probably as a result of the increased formation of angiotensin II) and so increases water intake.

Taken together, these processes bring about a relatively rapid restoration of the ECV and permit adequate perfusion of the tissues. The secretion of both angiotensin and ADH aids spontaneous recovery from hemorrhage by minimizing water loss and increasing sodium retention by the kidneys. If antagonists of either angiotensin or ADH are administered, recovery is delayed.

Over the ensuing several days, the lost plasma proteins are replaced by the liver but the absorption of fluid from the interstitial space leads to a further reduction in hemoglobin concentration (known as hemodilution). In the first 2 weeks after blood loss, the reticulocyte count is elevated (revealing a rapid increase in red cell production). Thereafter, the red cell count increases slowly, taking some weeks to return to normal. The increase in red cell production is stimulated by the hormone erythropoietin, which is released from the kidney in response to tissue hypoxia.

Circulatory shock

Circulatory shock exists when the circulation fails to provide adequate perfusion of the tissues. Once it has reached a critical stage, circulatory shock becomes irreversible. The failure of the circulation leads to the accumulation of metabolites and toxins within the vascular beds. These then cause more peripheral vasodilatation and a further fall in perfusion pressure until complete circulatory failure occurs and the patient dies. There are four main causes:

1. *Hemorrhagic* or *hypovolemic shock* in which the circulating volume is inadequate to sustain the blood pressure.

2. *Distributive shock* due to excessive vasodilatation. In this case, the capacity of the circulation exceeds the ability of the heart to pump sufficient blood to maintain an adequate blood pressure.

3. *Obstructive shock* which is caused by restriction of the venous return or of ventricular filling.

4. *Cardiogenic shock* in which the ability of the heart to pump blood is impaired. This is one aspect of heart failure which is discussed in Chapter 31, pp. 600–603.

BOX 28.2 CLASSIFICATION OF HYPOVOLEMIC SHOCK

The clinical response to hemorrhage depends on the extent of blood loss. However, the assessment of blood loss is not straightforward. Measurements of blood hemoglobin or the hematocrit are unreliable indicators of the extent of blood loss in acute hemorrhage. They do not change until there has been a significant redistribution of fluid from the interstitial space, which can take between 24 and 36 hours. No single measure gives a reliable indication of the extent of blood loss, but a combination of signs provides a means of classifying the degree of hypovolemia into four categories:

Class I Blood loss is less than 15 per cent of blood volume (i.e. less than 750 ml for an adult). There are no signs of shock. The blood pressure and pulse pressure are normal but there may be a slight tachycardia. The capillary refill time is normal (<2 seconds). (This is assessed by pressing the subject's skin for 5 seconds with sufficient pressure to blanch the skin, releasing the pressure, and measuring the time taken for the color of the skin to return to normal.) Respiration is normal and the subject is mentally alert. Normal healthy adults can sustain this level of blood loss without ill effect. In blood donation, 450 ml of blood (a blood unit) are routinely taken.

Class II When blood loss is between 15 and 30 per cent of blood volume, the clinical signs of shock become evident. The pulse is rapid (the heart rate is greater than 100 b.p.m.), and because of the tachycardia the diastolic pressure is increased, although systolic blood pressure is normal. Consequently, the pulse pressure is reduced. Capillary refill time is slower than normal (i.e. >2 seconds). The subject is pale, thirsty, anxious,

and breathing rapidly (>20 breaths min⁻¹). This extent of blood loss is rarely life threatening. Nevertheless, if blood loss is greater than about 20 per cent (1 liter for an adult, and proportionately less for a child), transfusion of blood or a plasma replacement should be considered.

Class III When blood loss is 30–40 per cent of blood volume, the pulse is weak and usually rapid. Be aware, however, that there may be a bradycardia (see text). Systolic blood pressure is very low and capillary refill time is slow. Breathing is fast (>20 breaths min⁻¹) and shallow. The subject is confused, lethargic, pale, and has clammy skin. In white subjects, the skin may be cyanosed (blue). Urine output is low (10–20 ml h⁻¹).

Class IV If blood loss is greater than 40 per cent of blood volume, systolic blood pressure is very low, there is tachycardia, and the skin is pale and cold. Breathing is fast and shallow. Urine output is less than 10 ml h⁻¹ and may fail altogether. The patient is drowsy and confused, and may become unconscious. Following this degree of blood loss, the shock may quickly become irreversible and urgent measures are required to restore the circulating volume in order to prevent circulatory collapse.

If severe (i.e. class IV) hypovolemic shock is suspected, the condition of the patient can be assessed by measuring the central venous pressure and its response to a rapid infusion of 100–200 ml of saline. If the patient is hypovolemic, there will be little change in central venous pressure but there will be an improvement in cardiovascular performance shown by a fall in heart rate and a rise in blood pressure.

Hypovolemic shock

The normal circulatory adjustments that are made in response to hemorrhage have already been described. The clinical management should be to eliminate the cause of blood loss and replace the lost blood by transfusion where this is possible. A basic classification of hypovolemic shock is described in Box 28.2. In some situations (e.g. road accidents and battlefield injuries), immediate blood transfusion is not possible. In these cases, intravenous infusion of plasma or plasma substitutes (such as 0.9 per cent saline solution containing a colloid such as polygelatin or a high-molecular-weight dextran) will provide fluid to expand the circulating volume and thus help to maintain blood pressure and adequate tissue perfusion.

Burn injuries to the skin are another potential cause of hypovolemia. These lead to a loss of a protein-rich fluid from the damaged area that is equivalent to a loss of plasma. The hematocrit rises as the ECV falls and, in cases of severe burns, the fall in ECV may be sufficient to produce hypovolemic shock which can be treated by infusing plasma.

Distributive shock

This occurs when the blood volume is normal but the volume of the circulatory system has become increased as a result of a generalized vasodilatation. The cardiac output is then insufficient to maintain the blood pressure and adequate tissue perfusion. This type of shock may result from a powerful emotional experience (extreme grief or fear) resulting in a powerful inhibition of sympathetic activity. This is called *neurogenic shock*. Distributive shock may also occur when a person is exposed to an antigen to which they have become previously sensitized. The resulting reaction releases histamine in large quantities and causes a profound peripheral vasodilatation. There is an increase in capillary permeability and fluid loss to the interstitial space. This is called *anaphylactic shock*.

Severe hypotension analogous to distributive shock can also arise following administration of local anesthetics during epidural anesthesia and spinal anesthesia. The activity in the sympathetic nerves is blocked and the blood vessels dilate. The hypotension can be countered by administration of the β-agonist ephedrine.

In severe bacterial infection, the release of toxins into the blood causes both vasodilatation and an increase in capillary permeability. This type of distributive shock is known as *septic shock*.

As with hypovolemic shock, the treatment of distributive shock should be aimed at eliminating the source of the vasodilatation and restoring the circulating volume.

Obstructive shock

This arises when the venous return is inadequate or the heart is unable to fill adequately. Restricted venous return may occur as a direct consequence of a large pulmonary embolism, limiting the flow of blood in the pulmonary circulation. If the pericardium becomes inflamed (*pericarditis*), fluid may collect in the pericardial sac (a pericardial effusion) and the pericardium becomes stretched. As the fluid accumulates, the pressure outside the heart rises, preventing the proper filling of the ventricles during diastole. The result is a fall in cardiac output. This condition is known as *cardiac tamponade* and is associated with a raised jugular pressure (see Chapter 15, p. 275).

Summary

1. The response of the body to hemorrhage depends on the extent of blood loss. In mild and moderate hemorrhage, the blood pressure is maintained by increased levels of sympathetic activity. Heart rate, peripheral resistance, and venous tone are all increased. Consequently, cardiac output is sufficient to maintain blood pressure. When blood loss exceeds 20–30 per cent, there is a fall in heart rate and blood pressure, and the patient faints. Blood flow to the brain and heart falls, with potentially disastrous consequences.

2. After hemorrhage, the blood volume is progressively restored by a number of adjustments. The vasoconstriction tips the balance of the Starling forces in favor of fluid uptake from the interstitium to the plasma, and the increase in circulating levels of angiotensin II and aldosterone promote both retention of sodium and an increased thirst. These changes act to restore the ECV. Over the ensuing days, the lost plasma proteins are replaced by the liver. Restoration of the red cell count may take several weeks.

3. If the circulating volume is insufficient to provide adequate perfusion of the tissues, circulatory shock develops. Unless this is treated promptly, a vicious downward spiral occurs in which local metabolites accumulate and further vasodilatation occurs, thus accentuating the mismatch between the ECV and the capacity of the circulation. The principal types of circulatory shock are hypovolemic shock, distributive shock, obstructive shock and cardiogenic shock.

28.6 Edema

Edema is the abnormal accumulation of fluid in the interstitial space. It arises when alterations to the Starling forces occur as a result of various pathologies. As described earlier (Chapter 15) fluid moves from the plasma to the interstitium when the capillary hydrostatic pressure exceeds the sum of the plasma oncotic pressure and the hydrostatic pressure within the tissues. Fluid moves from the tissues to the plasma when the sum of the plasma oncotic pressure and the tissue hydrostatic pressure exceeds the hydrostatic pressure in the capillaries. Under normal conditions about 8 liters of fluid per day pass from the circulation into the interstitium. About half of this is reabsorbed by the circulation either in the tissues or in the lymph nodes and the remainder is returned to the circulation as lymph via the thoracic duct into the subclavian vein on the left side.

The hydrostatic pressure in the capillaries is normally closely regulated by the tone of the afferent arterioles. However, the average capillary pressure also depends on the venous pressure and the ratio of pre- to postcapillary resistance. A small increase in venous pressure has a disproportionately large effect on the capillary pressure and the absorption of tissue fluid (see Box 28.3). Thus, when venous pressure is elevated as a result of a venous thrombosis or chronic right-sided heart failure, the aver-

BOX 28.3 STARLING FORCES AND EDEMA

The direction of fluid movement between a capillary and the surrounding interstitial fluid depends on four factors (Starling forces—see Chapter 15, Box 15.8):
(1) the capillary pressure (P_c).
(2) the interstitial pressure, (P_i).
(3) the oncotic pressure exerted by the plasma proteins (π_p).
(4) the oncotic pressure of the proteins present within the interstitial fluid (π_i).

   The algebraic sum P_f of the various pressures is called the net filtration pressure. It is given by

$$P_f = (P_c - P_i) - (\pi_p - \pi_i) \qquad (1)$$

The capillary pressure depends on the pressure in the arteries, the venous pressure, and ratio of pre- to postcapillary resistance (R_a/R_v). This exact relationship can be determined as follows. The blood flow from an artery to the midpoint of a capillary depends on the pressure difference and the precapillary resistance (see Chapter 15, Box 15.4):

$$\text{blood flow} = \frac{P_a - P_c}{R_a}.$$

Similarly, the blood flow from the midpoint of a capillary into the vein is

$$\text{blood flow} = \frac{P_c - P_v}{R_v}$$

Since virtually all the blood entering the capillary leaves via the veins (only a very small proportion is returned via the lymphatic drainage),

$$\frac{P_a - P_c}{R_a} = \frac{P_c - P_v}{R_v}$$

which can be rearranged to give the capillary pressure:

$$P_c = \frac{P_a + P_v \dfrac{R_a}{R_v}}{1 + \dfrac{R_a}{R_v}} \qquad (2)$$

Normally the precapillary resistance R_a is about four times the postcapillary resistance R_v and the capillary pressure is determined mainly by the arteriolar resistance. An important consequence of this relationship is that a small increase in venous pressure has a disproportionately large effect on the capillary pressure (see below). Thus, when venous pressure is raised, average capillary pressure P_c rises and filtration is favored over absorption. Fluid accumulates in the tissues, giving rise to edema. However, the swelling of the tissues raises the interstitial pressure and this reduces the net filtration pressure. A new balance becomes established and further accumulation of fluid in the tissues is prevented.

A fall in the oncotic pressure of the plasma raises the net filtration pressure and fluid may accumulate in the tissues. In practice, edema does not develop until the protein content of the plasma falls below about 30 g l^{-1}. This may arise because of liver failure, proteinuria, malabsorption, or malnutrition. If the colloid osmotic pressure of the proteins in the interstitium is increased (e.g. due to an increase in capillary permeability following an inflammatory reaction), this also has the effect of increasing the net filtration pressure as the presence of additional protein in the interstitial fluid diminishes the osmotic force drawing fluid from the interstitium to the plasma.

Worked example

Assume that the arterial pressure is 13.3 kPa (100 mmHg), the venous pressure is 0.67 kPa (5 mmHg), and the precapillary resistance is four times that of the postcapillary resistance. Inserting these values into equation (2) gives the capillary pressure.

$$P_c = \frac{13.3 + (0.67 \times 4)}{1 + 4} = 3.2 \text{ kPa (24 mmHg).}$$

If venous pressure increases to 1.33 kPa (10 mmHg), the capillary pressure increases to

$$P_c = \frac{13.3 + (1.33 \times 4)}{1 + 4} = 3.73 \text{ kPa (28 mmHg).}$$

Hence the doubling of venous pressure results in a rise of capillary pressure of 0.53 kPa (c.4 mmHg). This entire rise contributes to the net filtration pressure, which therefore rises proportionately more than the venous pressure, increasing the rate of fluid filtration. If the lymphatics become overloaded, the affected region will develop edema.

---

age capillary pressure is increased and more fluid passes from the plasma to the tissues. The resulting fall in plasma volume is reflected in a fall in the ECV and this in turn leads to the retention of sodium and water by the mechanisms discussed earlier. Thus a situation exists in which fluid can progressively accumulate in the tissues. As the fluid accumulates, the hydrostatic pressure in the tissues increases and opposes further accumulation of fluid in the interstitial space. A new equilibrium is established between the plasma and the volume of the interstitial fluid.

Since about half the fluid passing from the capillaries to the interstitial space is returned to the circulation via the lymphatic drainage (see Chapter 15, pp. 297–298), any obstruction of the flow of lymph will lead to fluid accumulation in the affected region. In the industrialized countries of Western Europe and North America lymphatic insufficiency is relatively rare, but is seen when the lymph nodes have been damaged during radical surgery (as in the example shown in Fig. 28.8) or where cancerous growths have invaded the lymph glands (lymphomas). In Third World countries, edema resulting from obstruction of the lymphatic circulation is commonly the result of the invasion of the lymph nodes by parasitic nematode worms (filariasis). This results in obstruction of lymph flow from the limbs and scrotum that is manifest by a gross edema (lymphedema) known as elephantiasis.

Edema also occurs when the plasma oncotic pressure is low. In this situation, the net filtration pressure rises and fluid accumulates in the tissues. This can occur during nephritis when the glomerular capillaries become abnormally permeable to albu-

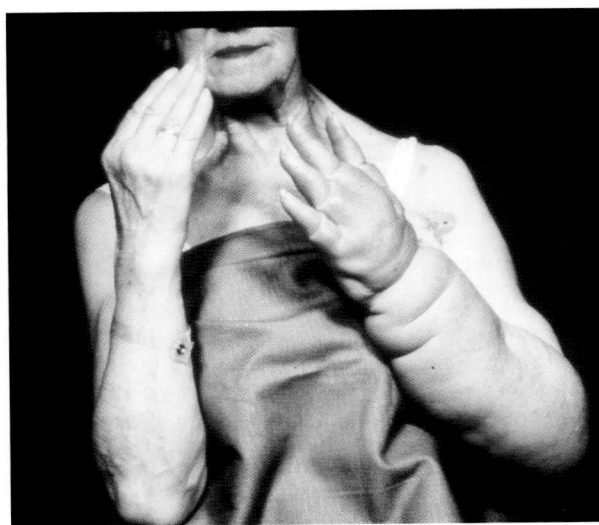

Fig. 28.8 Severe lymphedema in the left arm following radical surgery to treat breast cancer.

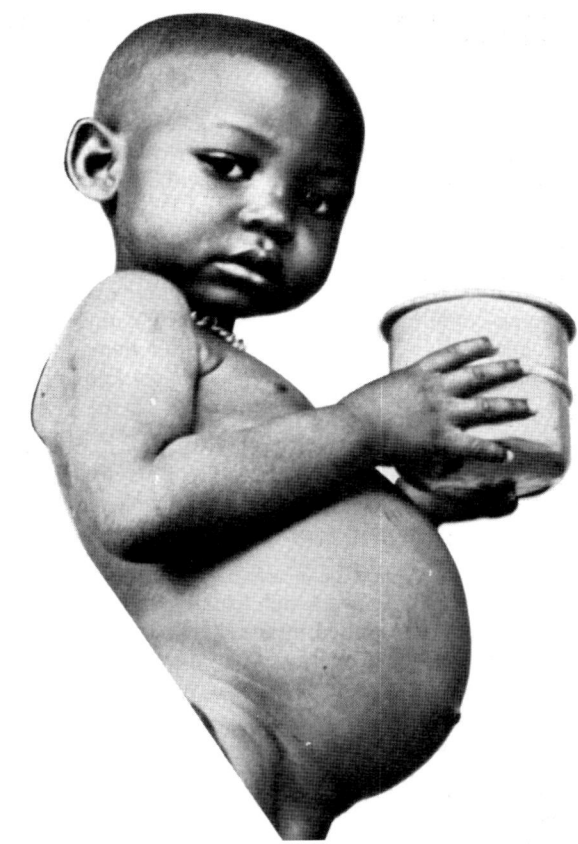

Fig. 28.9 The appearance of a child with kwashiorkor. Note the widespread edema, particularly the swollen abdomen.

min and other plasma proteins, so that significant quantities of protein are lost in the urine. It may also arise when the liver is unable to synthesize adequate quantities of the plasma proteins. A similar situation arises during severe malnutrition, when the diet may be rich in carbohydrate but contains little or no protein. This gives rise to a disease common in children in the poorer parts of the world known as *kwashiorkor*. A typical example is shown in Fig. 28.9.

The release of certain local mediators such as histamine causes the capillaries to become permeable to albumin and other plasma proteins. This results in an increase in the net filtration pressure of the capillaries and fluid accumulates in the affected area. The local swelling that accompanies insect bites arises in this way.

Systemic edema first appears in the lower parts of the body (the *dependent regions* of the body), particularly in the ankles as the venous pressure in the legs is elevated during prolonged periods of standing. Edema in the ankles can be distinguished from tissue fat by applying firm pressure to the affected area with a finger or thumb for a short period. If edema is present, the pressure will have forced fluid from the area and a depression in the skin will remain for some time after the pressure has been removed (*pitting edema*) (Fig. 28.10). If the swelling is due

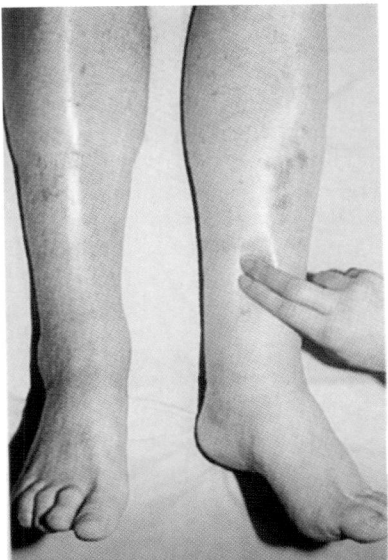
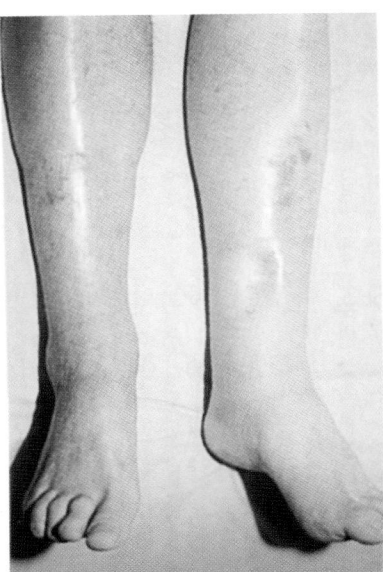

Fig. 28.10 An example of pitting edema.

simply to tissue fat, the skin springs back as soon as the pressure is removed.

In normal subjects, hydrostatic edema can arise when the leg muscles are relatively inactive and the muscle pumps contribute little to the venous return from the lower body. There is venous pooling and swelling of the ankles. This situation is exacerbated in persons with varicose veins, where the walls of the veins have become stretched rendering their valves incompetent. The accumulation of excess fluid is readily reversed by a short period of rest in a horizontal position or by mild exercise.

The fluids of the serosal spaces are separated from the extracellular fluid by an epithelial layer. These include the fluids of the pericardial, pleural, and peritoneal spaces. These fluids are essentially ultrafiltrates of plasma and their formation is governed by Starling forces. Normally the amount of fluid in these spaces is relatively small as the plasma oncotic pressure exceeds the hydrostatic pressure in the capillaries, but in certain disease states abnormal accumulations of fluid occur. For example, the volume of fluid between the visceral and parietal pleural membranes is normally only about 10 ml, but when fluid formation exceeds reabsorption, fluid accumulates between the pleural membranes in a process known as *pleural effusion*.

An accumulation of excessive amounts of fluid in the peritoneal cavity known as *ascites* can arise when there is a rise in pressure within the hepatic venous circulation, when there is obstruction of hepatic lymph flow, or when plasma albumin is abnormally low (as in kwashiorkor—see above). It also occurs during right-sided heart failure when the pressure within the systemic veins rises.

Treatment of edema by diuretics

From the previous section, it is clear that edema can arise as a result of various pathologies. Effective treatment requires identification and elimination of the underlying cause. Nevertheless, it may be desirable to eliminate the edema and, in many cases, this can be achieved by treating the patient with drugs that promote the loss of *both sodium and water* in the urine. Since this results in an increase in urine output known as a diuresis, these drugs are called *diuretics*. They are classified according to their modes of action. Diuretics may act indirectly by exerting an osmotic pressure that is sufficient to inhibit the reabsorption of water and sodium chloride from the renal tubules or they may act directly by inhibiting active transport in various parts of the nephron. Note that lymphedema cannot be treated in this way.

Osmotic diuretics such as the sugar mannitol are filtered at the glomerulus but they are not transported by the cells of the proximal tubule. In consequence, as other substances are transported and the proportion of the original filtered volume falls, these substances accumulate and exert sufficient osmotic pressure to inhibit tubular reabsorption of water. Since absorption by the proximal tubule is iso-osmotic, a decrease in fluid reabsorption allows more sodium to reach the distal nephron so that an increase in sodium excretion results. Therefore, there is a loss of both sodium and water. Nevertheless, as the capacity of the distal nephron to transport sodium is increased when the sodium load is increased (glomerulotubular feedback—see Chapter 17, p. 359), the increase in water excretion is not accompanied by an equivalent increase in sodium excretion. Consequently, the osmotic diuretics are more effective in increasing water excretion than they are in increasing sodium excretion.

Diuretics that act by inhibiting active transport are exemplified by *loop diuretics* such as frusemide. These compounds inhibit the cotransport of sodium, potassium and chloride by the ascending thick limb of the loop of Henle (see Chapter 17, p. 363). They appear to act only from the luminal side of the tubule. Inhibition of this transport decreases the ability of the nephron to concentrate urine. The effects of the loop diuretics are thus twofold: an increase in sodium excretion by inhibiting sodium chloride transport, and an increase in water loss through impairment of the countercurrent mechanism. They are the most potent diuretics in current clinical use and produce a pronounced increase in sodium excretion (known as a *natriuresis*).

One of the consequences of inhibiting the cotransport of sodium, potassium, and chloride in the ascending thick limb is an increase in potassium excretion. Unless this is carefully monitored, potassium balance will be disturbed and cardiac arrhythmias may result. To avoid this, a group of potassium-sparing diuretics has been developed. These drugs, of which amiloride is an example, act on the distal tubule, connecting tubules and collecting ducts to inhibit both sodium absorption and potassium excretion. The diuretic spironolactone exerts its effect on the distal tubule by antagonizing the sodium-retaining action of aldosterone.

Summary

1. Edema occurs when there is an abnormal accumulation of fluid in the tissues. There are a number of causes. Alterations to the balance of the Starling forces such as an increase in capillary or central venous pressure will give rise to edema, as will a reduction in plasma oncotic pressure. In addition, edema can arise from lymphatic obstruction

2. Several forms of edema can be treated by administration of drugs known as diuretics that promote the excretion of both water and sodium. An increase in water excretion alone would not be sufficient to eliminate the accumulated fluid. Diuretics can either act directly by inhibiting sodium transport by the nephron (e.g. loop diuretics such as frusemide) or indirectly by modifying the filtrate (osmotic diuretics such as mannitol). Lymphedema cannot be treated in this way.

Further reading

Physiology of body fluids

Koppen, B.M., and Stanton, B.A. (2001). *Renal physiology* (3rd edn), Chapters 1, 6–8, 12. Mosby Year Book, St Louis, MO.

Levick, J.R. (2003). *An introduction to cardiovascular physiology* (4th edn), pp. 318–23. Hodder Arnold, London.

Pharmacology

Rang, H.P., Dale, M.M., Ritter, J.M., and Moore, P. (2003). *Pharmacology* (5th edn), Chapter 23. Churchill-Livingstone, Edinburgh.

Clinical physiology and medicine

Campbell, E.J.M., Dickinson, C.J., Slater, J.D.H., Edwards, C.R.W., and Sikora, E.K. (eds.) (1984). *Clinical physiology* (5th edn), Chapter 1. Blackwell Scientific, Oxford.

Kumar, P., and Clark, M. (eds.) (1998). *Clinical medicine* (4th edn), Chapters 10 and 13. W.B. Saunders, London.

Quantitative problems

Answers are given below.

1. A patient with a body weight of 65 kg was injected with 10 ml of a 1 per cent (w/v) solution of Evans blue. After 10 minutes, the blood was sampled and found to contain 0.037 mg ml^{-1} of the dye. (a) What is the plasma volume? (b) If the hematocrit is 45 per cent, what is the blood volume? (c) Are these values within the normal range?

2. A patient was given an intravenous infusion containing 10 g ^{14}C-labeled inulin and 10 ml of ^3H_2O. After 90 minutes, the plasma concentration of inulin was 0.3 mg ml^{-1} and that of ^3H_2O was equivalent to 0.18 μl per ml of plasma. Over the same period, 5.2 g of inulin and 2.26 ml of ^3H_2O were excreted in the urine. Calculate (a) the total body water, (b) the extracellular volume, and (c) the intracellular volume. (d) Assuming that the patient is a normal adult male, what is his approximate body weight?

3. A miner with a body weight of 75 kg loses 4 liters of sweat during his day's work. (a) If the sweat contained 50 mmol l^{-1} of sodium chloride, what would be the osmolality of the body fluids after he finished work? (b) If he replaces the lost fluid by drinking pure water, what would be the osmolality of the tissue fluid? (The initial osmolality of the body fluids can be taken as equivalent to 290 mOsm kg^{-1} of sodium chloride and initially total body water accounted for 60 per cent of his body weight.)

Answers to quantitative problems

1. a. 2.70 liters
   b. 4.91 liters
   c. Yes

2. a. 43 liters
   b. 16 liters
   c. 27 liters
   d. 72 kg

3. a. 308 mOsm kg^{-1}
   b. 281 mOsm kg^{-1}

Acid–base balance

After reading this chapter you should understand:

♦ The physical chemistry of acids, bases and hydrogen ions

♦ The pH scale of hydrogen ion concentration

♦ The role of the physiological buffers in maintaining the pH of body fluids

♦ The physicochemical factors that determine hydrogen ion concentration in physiological solutions

♦ The principal physiological mechanisms that regulate the pH of the body fluids

♦ The common disorders of acid–base metabolism

♦ The compensatory mechanisms that are used by the body to minimize the effects of acid–base disorders

29.1 Introduction

Although the body continually produces carbon dioxide and non-volatile acids as a result of metabolic activity, the blood hydrogen ion concentration $[H^+]$ is normally maintained within the relatively narrow range of 40–45 nmol (40–45×10^{-9} mol) of free hydrogen ions per liter. This corresponds to a blood pH between 7.35 and 7.4, and the extreme limits that are generally held to be compatible with life range from pH 6.8 to pH 7.7. This regulation is achieved in two ways: hydrogen ions are absorbed by other molecules in a process known as *buffering* and acid products are subsequently eliminated from the body via the lungs and kidneys. *The concept of acid–base balance refers to the processes that maintain the hydrogen ion concentration of the body fluids within its normal limits.* An understanding of these processes provides the basis for a rational approach to the clinical treatment of acid–base disorders.

The importance of the hydrogen ion concentration in the regulation of body function may not be as readily appreciated as that of the concentration of other ions such as sodium or potassium, which are present in far higher concentrations and play an important role in cellular physiology. Nevertheless, changes in the hydrogen ion concentration of body fluids can have profound consequences for cellular physiology as the hydrogen ions can bind to charged groups on proteins (e.g. carboxyl, phosphate, and imidazole groups). When hydrogen ions have been bound or lost from a protein its net ionic charge will change, and this in turn can lead to altered function. To take a dramatic example, the activity of the enzyme phosphofructokinase (a key regulatory enzyme in the glycolytic pathway) increases nearly 20-fold as the hydrogen ion concentration decreases from about 80×10^{-9} to 60×10^{-9} mol l^{-1} (corresponding to a rise of about 0.1 pH unit from pH 7.1 to pH 7.2). The functions of many other proteins are also affected by the prevailing hydrogen ion concentration, although most are not as exquisitely sensitive as phosphofructokinase.

This chapter is organized so that the basic physical chemistry of acids and bases is discussed first, followed by the principles of hydrogen ion buffering. *These topics are included for completeness and may be omitted on a first reading.* The physiological mechanisms of pH regulation are then discussed, and the chapter ends with a detailed discussion of acid–base disturbances and their clinical assessment.

29.2 The physical chemistry of acid–base balance

Hydrogen ions in solution and the pH scale

Historically, the concentration of hydrogen ions in the body fluids has been given in terms of pH units. The pH of a solution is the logarithm to the base 10 of the reciprocal of the hydrogen ion concentration:

$$pH = \log_{10}\left(\frac{1}{[H^+]}\right). \qquad (29.1a)$$

However, since

$$\log_{10}\left(\frac{1}{[H^+]}\right) = -\log_{10}[H^+]$$
$$pH = -\log_{10}[H^+]. \qquad (29.1b)$$

Therefore a common alternative definition is, that the pH of a solution is the negative logarithm of the hydrogen ion concentration. A change of one pH unit corresponds to a tenfold change in hydrogen ion concentration (because $\log_{10}10 = 1$). While the pH notation is convenient for expressing a wide concentration range, it is somewhat confusing as *a decrease in pH reflects an increase in hydrogen ion concentration and vice versa.*

To calculate the pH of a solution of known hydrogen ion concentration first take the logarithm of the hydrogen ion concentration [H$^+$] (as moles per liter) and then change the sign. Use the following relationship to convert from pH to free [H$^+$]:

$$[H^+] = 10^{-pH}$$

Thus if the pH of a blood sample is 7.4 (a normal value), [H$^+$] is $10^{-7.4}$ mol l^{-1} or 39.8×10^{-9} mol l^{-1}. If a urine sample has a pH of 5, [H$^+$] is 10^{-5} mol l^{-1}. Note that the 2.4 unit difference in pH between these two samples corresponds to a 250-fold difference in hydrogen ion concentration.

A neutral solution is one in which the concentrations of hydrogen and hydroxyl ions are equal. When pure water dissociates, both ions are produced in equal quantities:

$$H_2O \rightleftharpoons [H^+] + [OH^-].$$

Applying the law of mass action,

$$K_w = \frac{[H^+][OH^-]}{[H_2O]} \qquad (29.2)$$

where K_w is the dissociation constant for water. However, since undissociated water is present in much higher concentration (about 55.5 mol l^{-1}) than either [H$^+$] or [OH$^-$] (which are present at 100×10^{-9} mol l^{-1}), the ionization of water has almost no effect on the concentration of non-ionized water. Therefore equation (2) can be simplified to

$$K'_w = [H^+][OH^-] \qquad (29.3)$$

where K'_w has a value of 10^{-14} (mol l^{-1})^2 at 25 °C.
   Since [H$^+$] = [OH$^-$],

$$[H^+] = \sqrt{K'_w}$$

so that for pure water

$$[H^+] = 10^{-7} \text{ mol l}^{-1}$$

and the pH is

$$pH = -\log_{10}[10^{-7}]$$
$$= 7.0.$$

Thus at 25 °C a neutral solution has a pH of 7.0. The value of K'_w is dependent on temperature and, at body temperature, it is $10^{-13.6}$ (mol l^{-1})^2 so that at body temperature neutral pH is 6.8. *Acid solutions have a pH value less than the value for neutrality, i.e. less than 7 at 25 °C or less than 6.8 at 37°C, while alkaline solutions have a pH value greater than neutrality.*

Weak acids and weak bases are only partially ionized in aqueous solution

In physiological terms, an acid is a substance that generates hydrogen ions in solution and a base is a substance that absorbs hydrogen ions. Acids and bases are further classified as weak or strong according to how completely they dissociate in solution. Strong acids such as hydrochloric acid (HCl) and bases such as sodium hydroxide (NaOH) are completely dissociated in solution, while weak acids such as dihydrogen phosphate (H$_2$PO$^-_4$) and weak bases such as ammonium hydroxide (NH$_4$OH) are only partly dissociated. Thus, when sodium hydroxide is added to water only sodium and hydroxyl ions are present. There are no neutral sodium hydroxide molecules in the solution. Similarly, when hydrochloric acid is added to water, only hydrogen ions and chloride ions are present. In contrast, the dissociation of a weak acid can be represented by the following equation:

$$HA \rightleftharpoons H^+ + A^-$$

Note that when a weak acid dissociates it generates an anion [A$^-$], which is known as the *conjugate base* of that acid.

Similarly, the dissociation of a weak base can be represented as follows:

$$BH^+ \rightleftharpoons B + H^+$$

The acid dissociation constant K_a is defined as

$$K_a = \frac{[H^+][A^-]}{[HA]} \qquad (29.4)$$

TABLE 29.1 The pK_a values of some common weak acids and bases

Name	Acid	Conjugate base	pK_a
Acetoacetic acid	CH_3COCH_2COOH	$CH_3COCH_2COO^-$	3.6
Lactic acid	$CH_3CH(OH)COOH$	$CH_3CH(OH)COO^-$	3.86
Acetic acid	CH_3COOH	CH_3COO^-	4.75
Carbonic acid	H_2CO_3	HCO_3^-	6.1
Dihydrogen phosphate	$H_2PO_4^-$	HPO_4^{2-}	6.8
Imidazole			6.95
Ammonium	NH_4^+	NH_3	9.25

The table is arranged in order of acid strength so that the strongest acids, which have the smallest pKa values, are at the top of the table. Note that while ammonium is a very weak acid (it has a large pKa value), its conjugate base (ammonia) is a moderately strong base.

and, by analogy with pH, the pK_a value for an acid or base is defined as

$$pK_a = \log_{10} \frac{1}{K_a} = -\log_{10} K_a \qquad (29.5)$$

Acids have pK_a values less than neutrality (i.e. less than 7) and bases have pK_a values greater than neutrality (Table 29.1).

Equation (29.4) can be rewritten as

$$\frac{1}{[H^+]} = \frac{1}{K_a} \frac{[A^-]}{[HA]}. \qquad (29.6)$$

Summary

1. The acidity of a solution is determined by its hydrogen ion concentration. The greater the hydrogen ion concentration, the more acidic is the solution.

2. The degree of acidity is often expressed using the pH scale. Pure water is neutral in acid–base terms and has a pH of 7 at 25°C. Acid solutions have pH values below 7 and alkaline solutions have pH values above 7.

3. An acid generates hydrogen ions in solution and a base absorbs hydrogen ions.

4. Strong acids and bases in aqueous solutions dissociate completely into their constituent ions. Weak acids and bases are only partially dissociated, and the degree of dissociation depends on the hydrogen ion concentration. The ratio of dissociated to undissociated weak acid can be calculated from the Henderson–Hasselbalch equation.

By taking logarithms the following equation can be derived:

$$pH = pK_a + \log_{10} \frac{[A^-]}{[HA]} \qquad (29.7)$$

which can be written as

$$pH = pK_a + \log_{10} \frac{[base]}{[acid]} \qquad (29.8)$$

This important relationship is known as the *Henderson–Hasselbalch equation*. It shows that for any weak acid (or weak base) the ratio of base [A$^-$] to undissociated acid [HA] is determined by the pH of the solution. When the concentrations of acid and conjugate base are equal, the pH is equal to the pK_a (as [base]/[acid] = 1 and $\log_{10}1 = 0$). Moreover, where the pK_a is known, knowledge of any two of the variables will define the third.

29.3 What factors determine the pH of an aqueous solution?

Before proceeding to discuss the detailed mechanisms that regulate the hydrogen ion concentration in plasma it is important to establish the factors that determine the pH of aqueous solutions (see Box 29.1 for further details). At constant temperature, the hydrogen ion concentration of any physiological solution is determined by the following.

1. The difference between the total concentration of fully dissociated cations ('strong' cations, e.g. Na$^+$) and that of the fully dissociated anions ('strong' anions, e.g. Cl$^-$). This difference indicates whether there is an excess of strong base or strong acid.

2. The quantity and pK_a values of the weak acids that are present (e.g. phosphate ions and the ionizable groups on proteins).

3. The partial pressure of carbon dioxide.

In all cases of acid–base disturbance, it is a change in one or more of these factors that is the underlying cause.

When other factors such as the P_{CO_2} are kept constant, changes in the difference between the sum of all the fully dissociated cations and the sum of all the fully dissociated anions will alter the hydrogen ion concentration of an aqueous solution. For example, in diarrhea large quantities of sodium are lost in the stools. As a result, the concentration of fully dissociated anions in the plasma (mainly chloride) is increased relative to that of the fully dissociated cations (chiefly sodium) and the plasma becomes more acid, giving rise to a metabolic acidosis.

If the difference in the concentrations of the fully dissociated ions is kept constant and the concentration of weak acids is increased, as in uncontrolled diabetes mellitus, the plasma again becomes more acid.

Finally, if the P_{CO_2} is increased because of poor alveolar ventilation, the plasma hydrogen ion concentration will increase, leading to a respiratory acidosis. Conversely, if the alveolar ventilation is increased, the Pa_{CO_2} will fall and so will plasma hydrogen ion concentration (respiratory alkalosis).

BOX 29.1   WHAT DETERMINES THE PH OF AN AQUEOUS SOLUTION?

In a physiological solution, the free H^+ concentration is determined by a set of chemical equilibria that must be simultaneously satisfied. These are as follows.

1. The dissociation of water:

$$H_2O \rightleftharpoons [H^+] + [OH^-].$$

The amount of free H^+ and OH^- is governed by the equilibrium constant, so that

$$K'_w = [H^+][OH^-].$$

2. The dissociation of the weak acids present. These reactions can be represented by the equation

$$HA \rightleftharpoons H^+ + A^-.$$

The total amount of weak acid present $[T_a]$ is given by

$$[T_a] = [HA] + [A^-]$$

and the degree of dissociation is determined by the equilibrium constant K_a:

$$K_a = \frac{[H^+][A^-]}{[HA]}.$$

In physiological solutions the most important weak acid is carbonic acid, derived from dissolved carbon dioxide, and the amount of carbonic acid formed is directly related to the P_{CO_2}.

3. Finally, as electroneutrality must be maintained, the concentration of all cations must be equal to that of all anions. As fully dissociated cations and anions do not liberate or bind hydrogen ions it is convenient to define the difference between the total concentration of all the fully dissociated cations ('strong cations') and the total concentration of all the fully dissociated anions ('strong anions') as the strong ion differences with units of moles per liter:

S = (sum of all 'strong' cations – sum of all 'strong' anions)

and

$$S + [H^+] = [OH^-] + [A^-].$$

This relationship can be expressed as an equation with only $[H^+]$ as a variable. Since

$$[OH^-] = \frac{K'_w}{[H^+]}$$

and

$$[A^-] = \frac{K_a[T_a]}{([H^+] + K_a)},$$

the free H^+ concentration is given by the solution of the following equation:

$$[H^+] - \frac{K'_w}{[H^+]} - \frac{K_a T_a}{([H^+] + K_a)} + S = 0.$$

While only a single weak acid has been considered in this simplified analysis, the basic principle can be extended to include any number of weak acids or bases. In all cases, the basic relationship holds: at constant temperature the hydrogen ion concentration of any aqueous solution is determined by the difference in concentration of the fully dissociated anions and cations and by the amounts and acid dissociation constants of the weak acids and weak bases that are present.

While it is clear that adding a weak acid or a weak base to a solution will alter the hydrogen ion concentration of that solution, the role of the strong ions in determining the hydrogen ion concentration may need a little further explanation. Consider the effect of adding NaOH to pure water. The OH^- ions react with H^+ ions to form water until the H^+ and the added Na^+ ions exactly balance the free OH^- ions. In effect, the Na^+ ions have replaced the H^+ ions as the principal carriers of positive charge. Note that the solution is alkaline as the concentration of hydroxyl ions exceeds that of hydrogen ions. If HCl is subsequently added, the H^+ from the HCl reacts with OH^- to form water and the negative charge on the Cl^- ions replaces that of the OH^- ions. When $[Na^+]$ exceeds $[Cl^-]$ the solution will be alkaline, when $[Na^+] = [Cl^-]$ the solution will be neutral (as $[H^+]$ must also equal $[OH^-]$ to maintain electroneutrality), and when $[Na^+]$ is less than $[Cl^-]$ the solution will be acid.

The hydrogen ion buffers of the body fluids

From the previous section, it should be evident that, when an acid or a base is added to an aqueous solution, the magnitude of the change in hydrogen ion concentration (or the pH change) will depend on the nature and quantities of other ionizable substances present. When the change in hydrogen ion concentration is less than the quantity of acid or base added, the solution is said to be buffered and the substances responsible for this effect are called buffers (see Box 29.2 for a fuller explanation of how buffers work).

The buffers that have the greatest quantitative importance in whole blood are the bicarbonate of the plasma and the hemoglobin of the red cells (Table 29.2). The buffering of the extracellular fluid is principally due to bicarbonate and phosphate, and intracellular buffering is provided by proteins, organic and inorganic phosphates, and bicarbonate.

The effectiveness of any buffer depends on its concentration and its pK_a value. In general, the closer the pK_a of a buffer is to the pH of the plasma, the more effective the buffer. For phosphate the buffer reactions can be summarized as follows:

$$H_2PO_4^- \rightleftharpoons H^+ + HPO_4^{2-}.$$

In this reaction the total amount of phosphate ($[HPO_4^{2-}]$ + $[H_2PO_4^-]$) is unchanged. Only the degree of ionization alters as the hydrogen ion concentration varies. Thus the effect of adding or removing hydrogen ions is to alter the ratio $[H_2PO_4^-] / [HPO_4^{2-}]$

Similarly, the buffer reaction for hemoglobin can be represented by the following reaction:

$$HHb^+ \rightleftharpoons H^+ + Hb.$$

There are two successive reactions for the bicarbonate buffer system: first the dissolution of CO_2 in water to form carbonic acid

$$H_2O + CO_2 \rightleftharpoons H_2CO_3$$

followed by the ionization of carbonic acid

$$H_2CO_3 \rightleftharpoons H^+ + HCO_3^-.$$

Since the concentration of H_2CO_3 depends on the partial pressure of CO_2,

$$[H_2CO_3] = \alpha . P_{CO_2}$$

where is α the solubility coefficient for CO_2 in plasma (Henry's law—see Chapter 16, p. 314). The Henderson–Hasselbalch equation for the bicarbonate buffer system can be written as

$$pH = 6.1 + \log_{10} \frac{[HCO_3^-]}{[\alpha P_{CO_2}]}$$

where 6.1 is the pK_a for the dissociation of H_2CO_3 in plasma and $\alpha = 0.225$ mmol l^{-1} kPa^{-1} (0.03 mmol l^{-1} $mmHg^{-1}$). When P_{CO_2} is 5.3 kPa (40 mmHg) there is $5.3 \times 0.225 = 1.2$ mmol CO_2 per liter of plasma.

Since carbon dioxide can be eliminated from solution, a bicarbonate buffer has very different pH buffering properties to phosphate where the total amount of buffer is unchanged. This difference has important consequences for the efficiency of the

---

BOX 29.2 HOW BUFFERS WORK: THE BUFFER ACTION OF INORGANIC PHOSPHATE AND BICARBONATE

If 0.1 mmol of a strong acid such as HCl were to be added to a liter of pure water, the total hydrogen ion concentration [H$^+$] would be 0.1001 mmol l^{-1} (the amount added plus the amount due to the dissociation of water) and the pH of the solution would be

$$pH = -\log_{10}[0.0001001] = 4.$$

However, if the same quantity of HCl were to be added to a mixture of 0.5 mmol l^{-1} NaH_2PO_4 and 0.8 mmol l^{-1} Na_2HPO_4 the pH change would be far smaller. This can be shown by the following calculation:

From the Henderson–Hasselbalch equation the pH of the phosphate solution is

$$pH = 6.8 + \log \frac{[HPO_4^{2-}]}{[H_2PO_4^-]}$$

$$= 6.8 + \log \frac{[0.8]}{[0.5]}$$

$$= 6.8 + 0.2 = 7.0$$

(where 6.8 is the pK_a for phosphate). This pH corresponds to a free hydrogen ion concentration of 100 nmol l^{-1}, which is the same as that of pure water at 25 °C. When 0.1 mmol of HCl is added to the mixture, the hydrogen ion reacts with the HPO_4^{2-} ion to form $H_2PO_4^-$:

$$H^+ + HPO_4^{2-} \rightleftharpoons H_2PO_4^-$$

If we assume that the reaction is complete, $[HPO_4^{2-}]$ has been reduced from 0.8 to 0.7 mmol l^{-1} and $[H_2PO_4^-]$ has been increased from 0.5 to 0.6 mmol l^{-1}. Entering these values into the Henderson–Hasselbalch equation gives

$$pH = 6.8 + \log \frac{0.7}{0.6}$$

$$= 6.8 + 0.067 = 6.87$$

The pH of the solution is now 6.87 so that the presence of the phosphate buffer has reduced the pH change from 3 units to just 0.13 units. As all the reactants remain in the solution, this is called a closed buffer system.

For comparison, consider how the CO_2–HCO_3^- buffer system would react to the addition of 0.1 mmol of acid to a liter of solution. For direct comparison with the previous calculation for phosphate, let us assume that bicarbonate was 1.2 mmol l^{-1} and CO_2 0.15 mmol l^{-1}. From the Henderson–Hasselbalch equa-

tion, the pH of this solution can be calculated to be 7.0. Addition of 0.1 mmol of acid would reduce bicarbonate to 1.1 mmol l^{-1} and increase the CO_2 to 0.26 mmol l^{-1} and the pH after addition of acid would be 6.73, a fall of 0.27 pH units. Thus mole for mole bicarbonate is a less effective buffer than phosphate at neutral pH.

In normal plasma the P_{CO_2} of arterial blood is 5.3 kPa (40 mmHg) which is equivalent to 1.2 mmol l^{-1} of carbonic acid, and plasma bicarbonate is about 24 mmol l^{-1} so that the pH of arterial blood calculated from the Henderson–Hasselbalch equation is

$$pH = 6.1 + \log_{10} \frac{24}{1.2} = 7.4$$

If 0.1 mmol of a strong acid were to be added to a liter of plasma, how large would the pH change be? Assuming that all the H$^+$ reacts with HCO_3^-, the concentration of HCO_3^- after addition of acid will be 23.9 mmol l^{-1} and the CO_2 would rise to 1.3 mmol l^{-1}. The net effect would be to change the pH from 7.4 to 7.36—a fall of only 0.04 pH units. The pH fall is far less than in the previous example, as the amount of HCO_3^- present in the solution is greater (24 mmol l^{-1} compared with 1.2 mmol l^{-1}).

If excess CO_2 is removed, the efficiency of bicarbonate as a buffer is dramatically enhanced. In the first example the CO_2 would be kept at 0.15 mmol l^{-1} and the bicarbonate would fall from 1.2 to 1.1 mmol l^{-1}. The fall in pH of the solution would be limited to 6.965—a fall of just 0.035 of a pH unit. In normal plasma, the bicarbonate would fall to 23.9 mmol l^{-1} and the fall in pH would be only 0.001 of a pH unit! (Use the Henderson–Hasselbalch equation to show this for yourself.) In these two situations, the bicarbonate is acting as an open buffer system.

To summarize, the effectiveness of any buffer depends on its concentration and its pK_a value. For a closed buffer system where all the reactants remain in solution (e.g. phosphate) a buffer is at its most effective when the pH is within one pH unit of the pK_a for the buffer system. When CO_2–HCO_3^- buffering occurs in an open system and the P_{CO_2} is maintained constant, the buffering power of the CO_2–HCO_3^- system is greatly increased. Consequently, despite operating far from its pK_a the CO_2–HCO_3^- system is a very effective buffer because of its high concentration in plasma and because blood P_{CO_2} is maintained relatively constant by adjustment of the alveolar ventilation.

CO_2–HCO_3^- buffering system in the body as a whole. The amount of carbon dioxide in the plasma can be regulated by the alveolar ventilation, thereby regulating the pH of the plasma to a significant degree. As a result, the buffering power of the CO_2–HCO_3^- system is far greater at normal plasma pH than its relatively low pK_a (6.1) would suggest (see Box 29.2).

Buffers in combination: the isohydric principle

When a buffer reacts with hydrogen ions, the change in hydrogen ion concentration affects all the other buffer reactions in the same body compartment. This is known as the *isohydric principle*. Thus knowledge of the status of one physiological buffer will serve to define the changes in [H+] that affect all the other buffer systems in that compartment. In clinical practice, the state of acid–base balance (sometimes called the acid–base status) of patients is usually assessed by measurement of arterial pH, P_{CO_2} and [HCO_3^-], and this approach will be adopted here.

Hydrogen ion buffering provides short-term stabilization of plasma pH

In blood, the principal buffers are inorganic phosphate (HPO_4^{2-} and $H_2PO_4^-$), bicarbonate, plasma protein, and the hemoglobin of the red cells. As Table 29.2 shows, the bicarbonate and hemoglobin are quantitatively the most important.

The large contribution that bicarbonate makes towards the total buffer capacity of whole blood depends on the amount present (24–26 mmol l^{-1}) and on the fact that, when it absorbs hydrogen ions, bicarbonate becomes converted to carbonic acid which then dissociates to form carbon dioxide and water. Since the excess carbon dioxide will stimulate the central chemoreceptors, alveolar ventilation will increase and the excess carbon dioxide generated will be excreted via the lungs. Consequently, the partial pressure of carbon dioxide in the blood is maintained relatively constant. The amount of acid buffered in this way is equal to the amount of bicarbonate that is lost.

If acid is slowly infused into a vein, the fall in blood pH is remarkably small, far smaller than can be accounted for by the blood buffers alone. This difference is accounted for by the buffers of the interstitial fluid and those within the cells. The total buffering capacity of the body is thus made up of three components:

TABLE 29.2 The amount of hydrogen ion absorbed by the principal buffers of the blood as blood pH falls from 7.4 to 7.0 following the addition of acid

Buffer	Amount per liter of blood	H+ absorbed (mmol l^{-1})
Bicarbonate	25 mmol	18[a]
Phosphate	1.25 mmol	0.3
Plasma protein	40 g	1.7
Hemoglobin	150 g	8
Total		28

[a]This assumes that the bicarbonate buffer operates in an open system so that the CO_2 generated can be eliminated (see Box 29.2).

Summary

1. At constant temperature, the concentration of hydrogen ions [H+] in the plasma is determined by three factors: the difference between the total concentration of fully dissociated cations and that of the fully dissociated anions, the quantity and the pK_a values of the weak acids that are present, and the partial pressure of carbon dioxide. A change in any one of these will result in a change in plasma [H+].

2. The weak acids and bases of the plasma are able to absorb some of the hydrogen ions that are formed during metabolism. Therefore they provide the first line of defense against changes in plasma [H+]. This is known as buffering. Of the buffers present in the plasma and interstitial fluid, the CO_2–HCO_3^- buffer system is quantitatively the most important—partly because it is the most plentiful buffer and partly because the P_{CO_2} can be regulated by the respiratory system.

- blood buffers—chiefly hemoglobin and bicarbonate;
- buffers in the interstitial fluid—chiefly bicarbonate;
- buffers in the cells—chiefly organic phosphates (e.g. ATP) and proteins.

Experimental evidence suggests that following infusion of acid, about half of the added hydrogen ions are buffered by the phosphates and protein within the cells.

While this whole-body buffering is capable of absorbing large quantities of hydrogen ions relative to the total amount of free hydrogen ion present in the body, it offers only a temporary defense against metabolic acid production. *Maintenance of blood pH within the normal range ultimately depends on the elimination of excess acid from the body.*

29.4 How the body regulates plasma pH

In the course of a day, an average healthy person eating a typical Western diet produces between 12 and 15 moles of carbon dioxide and excretes about 50 mmol of acid in the urine. From the previous section, it will be clear that the dissolved carbon dioxide of the body fluids leads to the formation of hydrogen ions in solution. Since it can be excreted via the lungs as a gas, carbon dioxide is often referred to as *volatile acid*. The acid that is excreted in the urine is chiefly sulfate derived from the metabolism of sulfur-containing amino acids (cystine and methionine). This is known as *non-volatile acid* and must be excreted in the urine.

Under normal conditions no organic acids appear in the urine, but in severe uncontrolled diabetic ketosis, large quantities of β-hydroxybutyric acid and acetoacetic acid are produced each day and a significant portion of these acids may be eliminated in the urine as 'ketone bodies'.

Respiratory regulation of carbon dioxide

The concentration of carbon dioxide in the alveolar gas is governed by both the rate of carbon dioxide production by the body and the rate of alveolar ventilation. In healthy people, if the arterial P_{CO_2} rises because of an increase in carbon dioxide production ($\dot{V}_{CO_2}$) or of insufficient alveolar ventilation ($\dot{V}_A$), the chemoreceptors of the medulla and carotid bodies will detect the change and increase alveolar ventilation. The increase in ventilation then tends to bring the $P_{A}CO_2$ back to normal.

Over a period of time, the $P_{A}CO_2$ will be directly related to the amount of carbon dioxide produced by the body and inversely related to the alveolar ventilation. This relationship can be written as follows:

$$P_{A}CO_2 = k \frac{\dot{V}_{CO_2}}{\dot{V}_A}$$

Since the P_{CO_2} of alveolar air and arterial blood in healthy people is the same, a maintained doubling of the alveolar ventilation will halve the P_{CO_2} in the blood, provided that the rate of carbon dioxide production is unchanged. The relationship between $P_{A}CO_2$ and alveolar ventilation for a normal adult is shown in Fig. 29.1(a).

A fall in plasma pH stimulates the rate of respiration

Altering plasma $[H^+]$ by injection of a weak solution of acid or base changes alveolar ventilation. Rapid injection of an acid

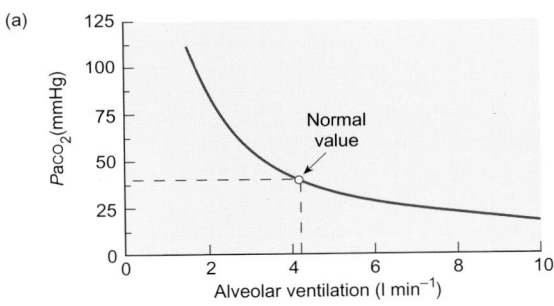

(a)

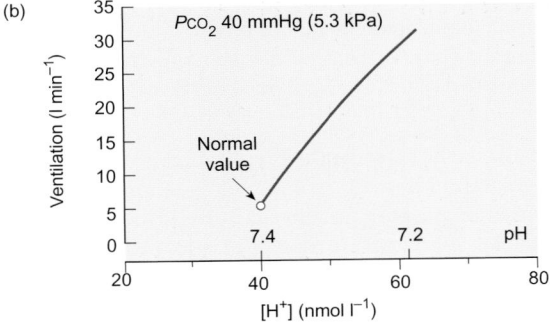

(b)

Fig. 29.1 The relationship between P_{CO_2}, pH, and alveolar ventilation. (a) The effect of changes in alveolar ventilation on the P_{CO_2} of arterial blood. Note that as ventilation is increased P_{CO_2} falls, and vice versa. This relationship assumes a constant rate of CO_2 production. (b) The effect of changes in plasma hydrogen ion concentration $[H^+]$ on the rate of ventilation. In this experiment, the plasma P_{CO_2} was kept constant by allowing the subject to inhale gas mixtures containing varying amounts of CO_2. An increase in plasma $[H^+]$ (fall in pH) stimulates ventilation independently of any change in P_{CO_2} and this increase in ventilation will lead to a fall in P_{CO_2} (see (a)) which will tend to offset the rise in $[H^+]$.

solution into a vein causes an increase in the frequency and depth of respiration. Conversely, rapid injection of an alkaline solution depresses ventilation (hypoventilation) or briefly stops it (apnea). By themselves, these observations do not prove that plasma hydrogen ion concentration directly affects respiration, as an increase of hydrogen ion concentration (lowering of pH) by injection of acid will result in an increase in P_{CO_2} and a fall in plasma bicarbonate. Conversely, a fall in plasma hydrogen ion concentration (an increase in plasma pH) caused by injection of base will result in a fall in P_{CO_2} and a rise in plasma bicarbonate. These changes are a consequence of the chemical reactions that govern the formation of carbonic acid (see Section 29.3).

The role of hydrogen ions as an independent stimulus to ventilation can be examined by the continuous infusion of an acid solution. Initially, ventilation will increase and the P_{CO_2} of the plasma will tend to fall for the reasons discussed above. This fall in P_{CO_2} can be offset by allowing the subject to breathe a gas mixture containing carbon dioxide. Under these conditions, alveolar ventilation is found to increase progressively as the hydrogen ion concentration rises (i.e. as pH falls) (Fig. 29.1 (b)). As this increase in ventilation occurs without any change in P_{CO_2}, the hydrogen ion concentration of the plasma must be capable of stimulating respiration independently of changes in P_{CO_2}. Therefore, although ventilation increases in response to both to a rise in plasma hydrogen ion concentration (a fall in plasma pH) and a rise in arterial P_{CO_2}, the two stimuli act independently.

When a subject is breathing air, the change in ventilation that follows a change in plasma hydrogen ion concentration will itself alter the P_{CO_2} of the plasma over a period of time. The hyperventilation that follows an increase in plasma hydrogen ion concentration will lead to a fall in plasma P_{CO_2}, whereas the apnea or hypoventilation that follows a fall in plasma hydrogen ion concentration will lead to retention of carbon dioxide and a rise in plasma P_{CO_2}. These changes to P_{CO_2} tend to bring plasma hydrogen ion concentration within the normal range. Final correction occurs when the excess acid or base has been removed from the body by the action of the kidneys.

The kidney excretes about 50 mmol of non-volatile acid each day in the urine as ammonium or in combination with phosphate

In addition to its elimination of carbon dioxide, the body needs to excrete about 50 mmol of non-volatile acid per day to maintain plasma pH within the normal range. The mechanisms responsible for the secretion of hydrogen ions into the proximal and distal tubules have already been discussed in Chapter 17, (p. 360 and p. 366). As the average volume of urine produced each day is 1–1.5 liters, and as the pH of the urine is usually between 5 and 6, less than 0.05 per cent of this acid is excreted as free hydrogen ion. Far larger quantities of hydrogen ions are eliminated in combination with the urinary buffers, of which the most important is phosphate.

A normal person eating a typical Western diet excretes about 30 mmol of phosphate a day. If the urine pH is 5, almost all of this will be in the $H_2PO_4^-$ form (compared with only 20 per cent in the plasma). Thus about 80 per cent of the excreted phosphate, or 24 mmol a day, is available to buffer hydrogen ions in the urine. This accounts for about half of the hydrogen ions derived from non-volatile acids. The remainder is excreted as ammonium (NH_4^+) which is derived from the metabolism of glutamine, a process known as *ammoniagenesis*.

The total amount of non-volatile acid that is excreted is the sum of that buffered by the urinary buffers (principally phosphate) and the amount of ammonium in the urine. It can be measured by titrating the urine with base until the urine pH is the same as that of the plasma (the *titratable acid*) and by separately measuring the total amount of ammonium excreted. The total acid excretion is the sum of the two figures.

The formation of ammonium and acid excretion

The generation of ammonium ions (NH_4^+) from glutamine occurs primarily in the proximal tubules of the kidneys. The reaction occurs in two stages, first the enzyme glutaminase deaminates glutamine to form NH_4^+ and glutamate. Then the glutamate that is formed by this reaction is deaminated under the influence of the enzyme glutamate dehydrogenase to form NH_4^+ and α-ketoglutarate. The sequence of reactions is shown in Fig. 29.2.

The ammonium ions produced by these reactions are secreted into the lumen in exchange for sodium ions. The subsequent metabolism of the α-ketoglutarate via the Krebs cycle and the NADH via the electron transport chain consumes two protons. In this way, the secretion of ammonium ions leads to a loss of hydrogen ions from the body. It is important to note that the key to the success of this process is the spatial separation of the NH_4^+ from the α-ketoglutarate. Chemically, the formation and excretion of ammonium is equivalent to the titration of hydrogen

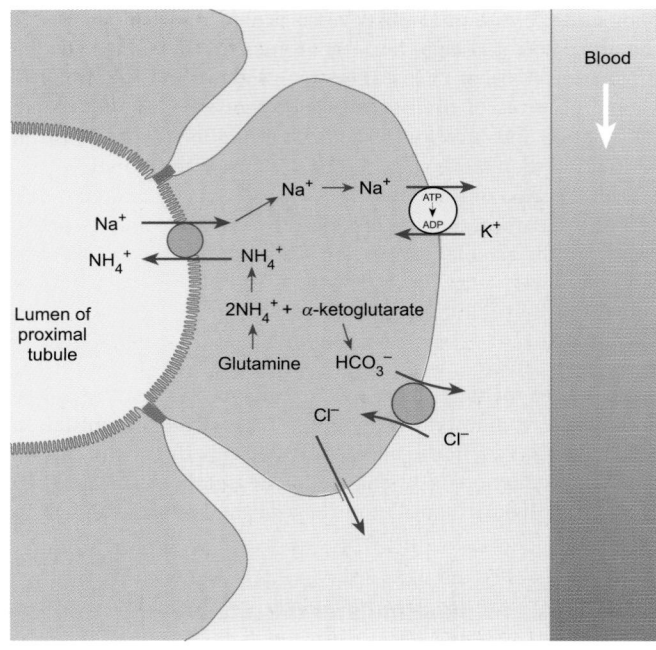

Fig. 29.3 The mechanism by which cells of the proximal tubule generate and secrete NH_4^+ into the tubular lumen. NH_4^+ is generated in the tubular cells by deamination of glutamine. It is secreted into the tubular lumen in exchange for Na^+ and the Na^+ entering the cell is removed by the Na^+–K^+-ATPase located on the basolateral surface. The metabolism of α-ketoglutarate generates HCO_3^-, which passes into the interstitial fluid in exchange for extracellular Cl^-. Excess Cl^- leaves the cell via Cl^- channels located on the basolateral surface.

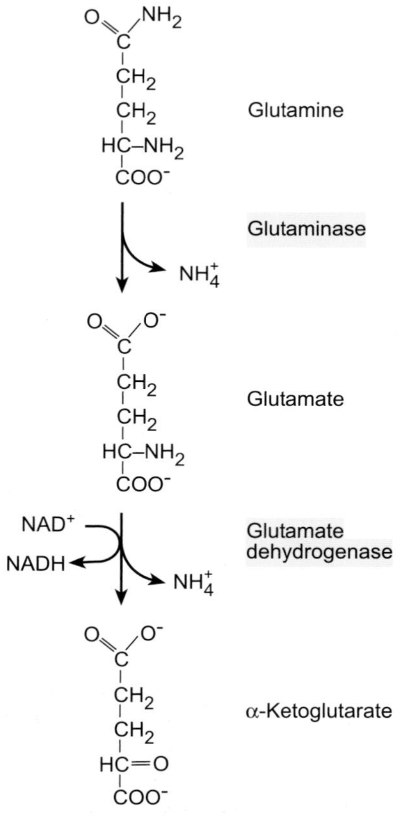

Fig. 29.2 Generation of ammonium ions from glutamine.

Summary

1. Each day a normal person produces about 15 mol of carbon dioxide and 50 mmol of non-volatile acid (chiefly sulfate). To maintain plasma hydrogen ion concentration within normal limits these metabolic products must be eliminated from the body. This occurs via two routes: carbon dioxide is excreted via the lungs, and non-volatile acid is excreted via the kidneys.

2. The frequency and depth of respiration is stimulated by an increase in plasma P_{CO_2} and hydrogen ion concentration so that an increase in either factor leads to an increase in alveolar ventilation. A rise in plasma hydrogen ion concentration leads to an increased loss of carbon dioxide, and this mechanism provides a rapid means of adjusting plasma hydrogen ion concentration.

3. The excretion of non-volatile acid in the urine depends on the ability of phosphate to buffer free hydrogen ions and on the ability of the kidneys to generate NH_4^+. Under normal circumstances about half of the non-volatile acid is excreted as NH_4^+ salts but, if non-volatile acid production increases, the amount of NH_4^+ in the urine increases proportionately.

ions derived from non-volatile acids by the weak base ammonia. The non-volatile acids that are formed by the metabolism of amino acids are therefore excreted as their ammonium salts. The cellular mechanism by which ammonium ions are excreted is shown in Fig. 29.3.

29.5 Primary disturbances in acid–base balance

When the pH of the arterial blood is less than 7.35 (i.e. when plasma hydrogen ion concentration is greater than 45×10^{-9} mol l^{-1}), it is regarded as being acid with respect to normal. Patients with such blood are said to have *acidemia* (literally acid in the blood). The processes responsible for this increase in plasma hydrogen ion concentration can be attributed to increased amounts of non-volatile acid (*metabolic acidosis*) or to a failure to remove carbon dioxide from the blood (*respiratory acidosis*). Conversely, when the pH of the arterial blood is greater than 7.45 (i.e. when plasma hydrogen ion concentration is less than 35×10^{-9} mol l^{-1}), it is regarded as being alkaline with respect to normal. Patients with such a blood pH are said to have *alkalemia*. The underlying causes define the kind of *alkalosis* responsible for the condition (*respiratory* or *metabolic alkalosis*).

Since the pH of body fluids is determined by the difference in the sum of the concentrations of strong cations and strong anions and by the quantity and pK_a values of the weak acids that are present (including carbon dioxide), any derangement in acid–base balance must arise from a change in one or more of these factors. Some of the common causes of acid–base imbalance are given in Table 29.3.

Respiratory acidosis is the result of an increase in the P_{CO_2} of the plasma due to inadequate ventilation or to the presence of significant amounts of carbon dioxide in the inspired air. The rise in plasma P_{CO_2} results in an increase in the formation of carbonic acid, which dissociates giving rise to H^+ and HCO_3^-. The increase in hydrogen ion concentration is directly related to the P_{CO_2} as discussed in Section 29.4 and the plasma bicarbonate increases in proportion to the fall in plasma pH. These relationships are indicated in Fig. 29.4 by the line linking point A to the normal value (P_{aCO_2} 40 mmHg, pH 7.4, and bicarbonate = 24 mmol l^{-1}).

Respiratory alkalosis is the result of a fall in plasma P_{CO_2} due to an increase in alveolar ventilation. The fall in P_{CO_2} shifts the $[CO_2]$–$[HCO_3^-]$ equilibrium and leads to a decreased concentration of carbonic acid and so to a rise in plasma pH and a fall in plasma bicarbonate. The magnitude of the pH change is directly related to the increase in ventilation (See Fig. 29.1). This condition is a common feature of life at high altitude where the fall in atmospheric P_{O_2} and hence in arterial P_{O_2} stimulates respiration (see p. 579). The changes in plasma pH and bicarbonate as P_{aCO_2} falls are shown in Fig. 29.4 by the line linking the normal value to point B.

In *metabolic acidosis* the fall in plasma pH is accompanied by a fall in plasma bicarbonate. There are many causes (Table 29.3) including an increase in metabolically derived acids, a loss of base ($NaHCO_3$) from the gut during diarrhea and a failure of the renal tubules to excrete hydrogen ions. The pH and bicarbonate changes in metabolic acidosis are indicated in Fig. 29.4 by the line linking the normal value to point C.

TABLE 29.3	Some causes of acid–base disturbance
Respiratory acidosis (alveolar hypoventilation)	Impaired ventilation due to airway obstruction
	Impaired alveolar gas exchange
	Decreased respiratory drive
	Inhalation of CO_2
Respiratory alkalosis (alveolar hyperventilation)	Hypoxia (e.g. while living at high altitude)
	Increased respiratory drive due to cerebrovascular disease
	Hepatic failure
	Effects of drugs and poisons
Metabolic acidosis	Endogenous acid loading (e.g. diabetic ketoacidosis)
	Loss of base from the gut (e.g. diarrhea)
	Impaired acid secretion by the renal tubules (renal tubular acidosis)
	Exogenous acid loading (e.g. methanol ingestion)
Metabolic alkalosis	Loss of gastric juice (e.g. by vomiting)
	Excessive base ingestion
	Aldosterone excess

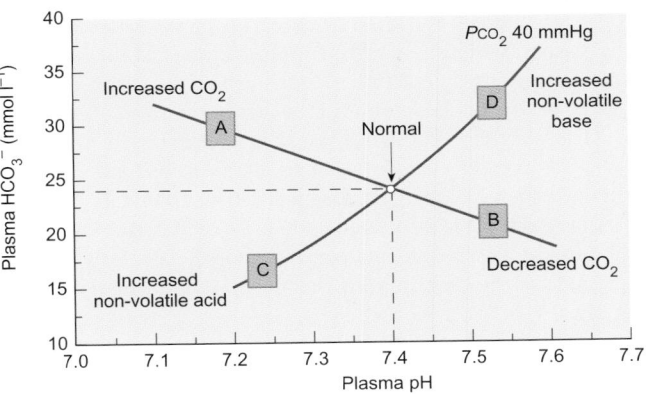

Fig. 29.4 The pH–[HCO_3^-] diagram for blood plasma. The line AB shows the plasma pH as P_{CO_2} is varied. Values below the line reflect increased metabolic acid, while values above it reflect excess base. If ventilation is less than normal, pH falls and [HCO_3^-] rises because CO_2 accumulates (normal value to point A). This is respiratory acidosis. In hyperventilation plasma P_{CO_2} and [HCO_3^-] fall while pH rises (normal value to point B). This is respiratory alkalosis. The line CD shows the change in pH that occurs at a constant P_{aCO_2} of 40 mmHg as non-volatile acid is gained in metabolic acidosis (normal to point C) or lost in metabolic alkalosis (normal value to point D).

When diabetes mellitus is inadequately controlled, energy metabolism shifts from carbohydrates to fats and the amounts of β-hydroxybutyric acid and acetoacetic acid in the plasma increase. As a result, there is a fall in plasma pH. These changes are known as *ketoacidosis*. The hydrogen ions react with plasma bicarbonate to form carbonic acid and the carbon dioxide liberated is excreted via the lungs.

The kidney excretes acid in three ways:

- by Na⁺–H⁺ exchange across the apical membrane of the cells of the proximal tubule;

- by the secretion of NH_4^+ into the proximal tubule;

- by direct proton secretion via the H⁺-ATPase of the intercalated cells of the distal tubule and collecting ducts.

As a consequence of this acid secretion, an equimolar amount of bicarbonate is generated and this 'new' bicarbonate replaces that lost by the reaction of plasma bicarbonate with metabolic acid during buffering. Clearly, if plasma pH is to be maintained constant, the amount of acid secretion must be sufficient to match the generation of non-volatile acids. If this is not the case, both plasma pH and plasma bicarbonate will fall and a type of metabolic acidosis known as renal tubular acidosis will result.

The bony skeleton contains a vast number of mineral crystallites bound together by cells, collagen, and ground substance rich in mucopolysaccharides. The mineral crystallites consist of calcium phosphate and calcium carbonate, and their surface is negatively charged. Normally these charges are neutralized largely by sodium and potassium ions, but when plasma pH falls these ions are displaced by protons so that the mineral phase of the skeleton provides an additional source of extracellular buffering. This additional buffering is bought at a price—the ions displaced (sodium, potassium, and calcium) are excreted in the urine and there can be a significant loss of calcium during chronic metabolic acidosis, leading to a slow dissolution of the bone.

Metabolic alkalosis is caused by an excess of non-volatile base in the plasma, which may arise from a number of factors (Table 29.3). Commonly, metabolic alkalosis arises as a result of vomiting the gastric contents and can be attributed to the loss of HCl which is a strong acid. Since the P_{CO_2} is unchanged, the fall in hydrogen ion concentration that results from this loss is accompanied by an increase in plasma bicarbonate. In metabolic alkalosis, the rise in pH is associated with a rise in plasma bicarbonate. The pH and bicarbonate changes in metabolic alkalosis are summarized by the line linking the normal value to point D in Fig. 29.4.

Summary

1. Normal arterial blood pH is generally taken as 7.4. When it is less than 7.35 there is acidemia and an underlying acidosis. When it is greater than 7.45 there is alkalemia and an underlying alkalosis.

2. If the deviation in plasma pH results from a change in alveolar ventilation, the disorder is of respiratory origin (respiratory acidosis or respiratory alkalosis).

3. All other disorders are classified as metabolic irrespective of their underlying cause. Thus a metabolic acidosis develops if the production of non-volatile acids exceeds their rate of excretion via the kidneys, or if there is a loss of non-volatile base from the gut. Conversely, loss of acid from the stomach or ingestion of non-volatile base gives rise to a metabolic alkalosis.

29.6 Disorders of acid–base balance are compensated by respiratory and renal mechanisms

When acid–base balance has become disturbed, various mechanisms operate to bring plasma pH closer to the normal range in a process called *compensation*. The mechanisms that act to restore plasma pH can be grouped under two headings:

1. respiratory compensation, which is fast but not very sensitive (pH adjustment occurs in minutes but the changes in P_{CO_2} offset the original stimulus);

2. renal compensation, which is sensitive but slow (pH adjustment takes hours to days).

While compensatory mechanisms operate to minimize the change in plasma pH, complete restoration of acid–base balance (i.e. correction) requires treatment or elimination of the underlying cause.

Compensation of chronic respiratory acidosis and alkalosis can only occur by renal means, as the primary deficit is due to a change in alveolar ventilation

While respiratory acidosis and alkalosis can be produced voluntarily by breath-holding or hyperventilation, this discussion is concerned with persistent alterations to ventilation arising from disease or from adaptation to a particular environment. The effects of short-term voluntary changes to ventilation are readily reversed by resumption of normal patterns of breathing.

In *chronic (i.e. long-term) respiratory acidosis* the plasma P_{CO_2} is elevated as the alveolar ventilation is insufficient to eliminate all the carbon dioxide generated during metabolism. This leads to a fall in plasma pH and an increased hydrogen ion secretion into the proximal tubule and collecting ducts. Two mechanisms are responsible.

1. In the proximal tubule the increased hydrogen ion secretion occurs via Na⁺–H⁺ exchange (see Chapter 17, p. 360). This process leads to increased reabsorption of the filtered bicarbonate. Normally the kidneys filter about 3 mmol of bicarbonate each minute and the rate of hydrogen ion secretion by the proximal tubule is sufficiently high to permit the absorption of about 85 per cent of the filtered load—the remainder is normally reabsorbed in the thick ascending loop and collecting duct. In respiratory acidosis the rate of hydrogen ion secretion by the proximal tubule is increased and up to 98 per cent of the filtered load of bicarbonate can be reabsorbed in this part of the nephron. To preserve the electroneutrality of the plasma, the increase in bicarbonate reabsorption is associated with an increase in Cl⁻ excretion.

2. Excretion of excess hydrogen ions is performed by the intercalated cells of the distal tubule and collecting ducts where acid secretion occurs via a H⁺-ATPase. This secretion of acid differs from that occurring in the proximal tubule as it can take place against a steep pH gradient. Moreover, as the protons are derived from carbonic acid, the secretion of acid leads to the generation of bicarbonate. This bicarbonate is reabsorbed across the basolateral membrane (see Fig. 17.22)

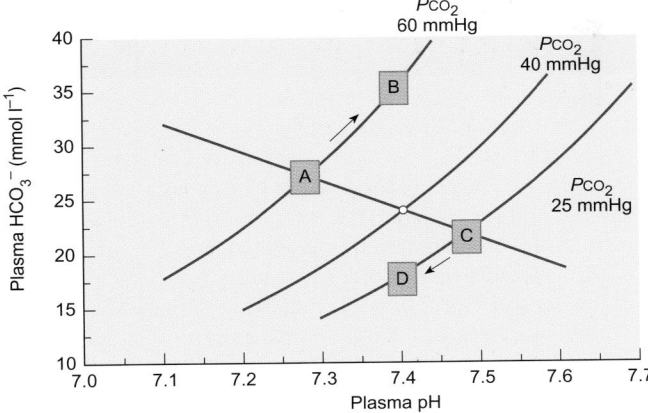

Fig. 29.5 The changes in plasma [HCO$_3^-$] that occur during *renal compensation* in a patient with respiratory acidosis (points A to B) and in one with respiratory alkalosis (points C to D). The closed circle represents normal values. See text for further information.

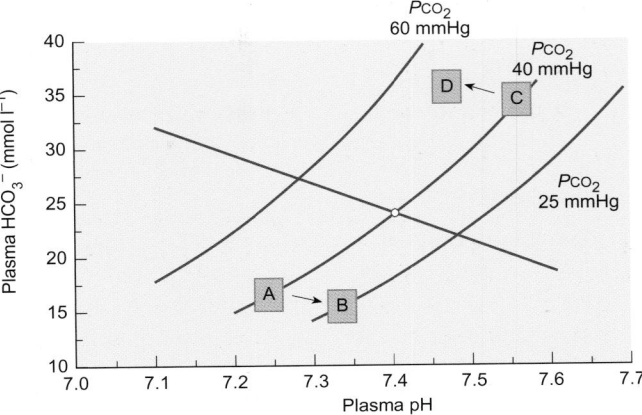

Fig. 29.6 The changes in plasma [HCO$_3^-$] during metabolic acidosis and metabolic alkalosis and the effects of *respiratory compensation*. In metabolic acidosis (point A) the PCO$_2$ is initially normal (40 mmHg) but the low plasma pH stimulates respiration and the PCO$_2$ falls, thereby shifting the plasma pH closer to normal (point B). In metabolic alkalosis, the loss of fixed acid initially shifts the pH in the alkaline direction even though PCO$_2$ remains normal (point C). The increase in pH tends to depress respiration and leads to CO$_2$ retention and a fall in plasma pH (point D). As in the previous figures, the closed circle represents normal values.

and represents new bicarbonate. As the line joining points A and B in Fig. 29.5 shows, the plasma bicarbonate increases progressively as the pH returns to normal.

Overall, the higher the PCO$_2$ the greater is the secretion of hydrogen ions and the larger the quantity of bicarbonate generated and reabsorbed. Thus in chronic respiratory acidosis increased renal hydrogen ion secretion leads to an increase in plasma bicarbonate. This, in turn, helps to limit the fall in plasma pH as the bicarbonate/PCO$_2$ ratio increases towards normal.

In *chronic respiratory alkalosis* the situation is reversed. As the line joining points C and D in Fig. 29.5 shows, if the kidneys are to restore plasma pH to normal they must excrete bicarbonate. This is accomplished by a reduction in hydrogen ion secretion in the proximal tubule so that less of the filtered load of bicarbonate is reabsorbed. As the PCO$_2$ of the plasma is low, less carbonic acid is formed within the tubular cells so that fewer hydrogen ions are secreted and less of the filtered bicarbonate is reabsorbed. In the distal tubule, the same considerations apply: hydrogen ion secretion by the H$^+$-ATPase of the intercalated cells is reduced because the PCO$_2$ is lower than normal. Therefore a smaller proportion of the HCO$_3^-$ reaching the distal nephron will be reabsorbed. Consequently, during the early stages of compensation, the kidneys excrete bicarbonate and the urine is relatively alkaline. In the long term plasma bicarbonate falls and bicarbonate excretion is reduced. Indeed, renal compensation for respiratory hyperventilation is sufficiently powerful to enable healthy people living at high altitude to have a normal plasma pH.

In metabolic acidosis and metabolic alkalosis, the changes in plasma pH are first minimized by respiratory compensation; fine adjustment occurs over a longer period by altering the amount of H$^+$ or HCO$_3^-$ excreted by the kidneys

Metabolic acidosis is usually due to one of three factors:

1. an increase in the production of metabolic acid;

2. a loss of base from the lower gut;

3. a reduced ability to excrete acid (in renal tubular acidosis).

Therefore the compensatory mechanisms employed will vary according to the underlying cause. In the short term, the decrease in blood pH stimulates respiration and this increases the loss of carbon dioxide from the lungs. The resulting fall in PCO$_2$ causes the pH of the plasma to rise towards normal. This restoration of pH is relatively rapid (it occurs within minutes) but as the pH approaches 7.3 the stimulus to the central and peripheral chemoreceptors becomes less and the hyperventilation declines. Consequently, respiratory compensation will bring plasma pH within about 0.1 of a pH unit of its normal range, but this is achieved at the cost of a fall in plasma bicarbonate as shown by the change from point A to point B in Fig. 29.6.

Secondly, the fall in plasma pH results in the secretion of a more acid urine. The effectiveness of this mechanism in eliminating the excess hydrogen ions is limited by the availability of the urinary buffers. However, a low plasma pH also stimulates ammoniagenesis and the kidneys excrete more NH$_4^+$. This process helps to eliminate excess acid by the mechanism shown in Fig. 29.3. This stage of compensation can bring pH back to normal but, as it takes time to filter plasma and excrete excess acid, hours or days are required for further compensation to occur. If metabolic acid production exceeds the ability of the kidneys to excrete NH$_4^+$, full compensation will not occur until the underlying cause is treated.

In acidosis due to loss of base from the gut, there is a fall in the filtered load of Na$^+$ which stimulates aldosterone production. This leads to an increase in Na$^+$ reabsorption by the mechanism discussed in Chapter 17 (Section 17.7).

Metabolic alkalosis is commonly caused by vomiting gastric juice or by ingestion of alkali. Initially respiratory compensation

Summary

1. Following a disturbance of acid–base balance, compensatory mechanisms come into play to bring plasma pH within the normal range. Full correction requires treatment or elimination of the underlying cause and restoration of the plasma [HCO$_3^-$] to normal levels.

2. Respiratory disorders are compensated by renal adjustments of plasma [HCO$_3^-$], which may take days to complete.

3. Metabolic disorders are initially compensated by alterations to the rate of alveolar ventilation (respiratory compensation), but this is always insufficient to restore plasma pH to the normal range. Full compensation and correction occurs via renal mechanisms.

occurs as the high plasma pH depresses respiration. In consequence, the P_{aCO_2} rises and the plasma pH tends to fall towards normal as shown by the shift from point C to point D in Fig. 29.6. Nevertheless, as plasma pH approaches the normal range the fall in pH offsets the depression of ventilation so that respiratory compensation is only partial. Final correction of metabolic alkalosis due to loss of gastric juice requires excretion of HCO$_3^-$ and retention of Cl$^-$ by the proximal tubule. Alkalosis due to ingestion of base is corrected by renal excretion of the excess base.

29.7 Clinical evaluation of the acid–base status of a patient using the pH–[HCO₃⁻] diagram

The state of acid–base balance in any patient can be deduced from the pH–HCO$_3^-$ diagram discussed in the previous two sections of this chapter. By measuring the arterial pH, bicarbonate, and P_{aCO_2} an unambiguous interpretation can be made (see Table 29.4). There are a number of different ways of representing the relationship between plasma pH and plasma bicarbonate. In this chapter the pH–bicarbonate diagram (sometimes called a Davenport diagram) has provided the basis for this discussion, but an alternative representation is shown in Box 29.3. The pH–bicarbonate diagram can be divided into six zones as shown in Fig. 29.7. The numbers plotted on the graph show the pH and [HCO$_3^-$] values for various types of disorder. The primary uncompensated changes are shown by the solid lines (see also

TABLE 29.4 The direction of the changes in plasma pH, P_{CO_2}, and bicarbonate that characterize the primary metabolic and respiratory disturbances of acid–base balance

Condition	Plasma pH	Plasma P_{CO_2}	Plasma [HCO$_3^-$]
Respiratory acidosis	Decreased	Increased	Increased
Metabolic acidosis	Decreased	Normal	Decreased
Respiratory alkalosis	Increased	Decreased	Decreased
Metabolic alkalosis	Increased	Normal	Increased

BOX 29.3 THE BICARBONATE–P_{CO_2} DIAGRAM AND ITS USE IN ASSESSING ACID–BASE STATUS

The Henderson–Hasselbalch equation can be represented in a number of different ways, each of which gives a slightly different emphasis to the basic relationship. In the diagram shown here, plasma bicarbonate is plotted on a logarithmic abscissa and P_{CO_2} is plotted as the ordinate on a linear scale. Plasma pH is represented by a diagonal axis. The advantage of this representation is that respiratory changes are shown in the vertical direction while metabolic changes are shown in the horizontal direction. The range of normal values is shown by the shaded ellipse in the center of the figure.

In respiratory acidosis, P_{CO_2} rises and pH falls as shown for example (a). Compensation requires the excretion of acid as

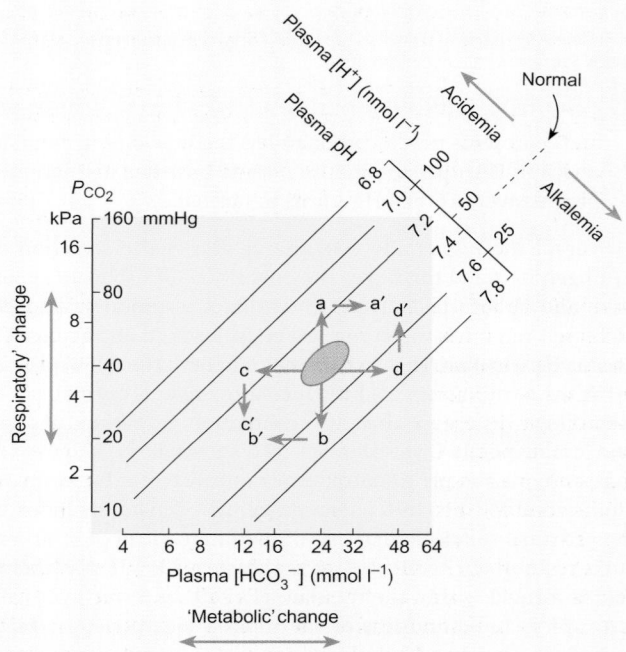

indicated by the horizontal line a-a'. Conversely, in respiratory alkalosis, P_{CO_2} falls and pH rises as shown for example (b). Compensation requires the excretion of excess bicarbonate as shown by the line b-b' and pH falls.

In metabolic acidosis, the fall in pH results in a decrease in plasma bicarbonate as shown in example (c). Short-term compensation occurs as hyperventilation lowers the P_{CO_2} as shown by line c-c'. Conversely, in metabolic alkalosis there is a rise in pH as shown in example (d). Short-term compensation occurs as respiration is inhibited and P_{CO_2} rises as shown by the line d-d'.

Fig. 29.4) and the vertical dotted line shows complete compensation (i.e. restoration of plasma pH to the normal range). As discussed earlier, complete correction requires that the plasma [HCO$_3^-$] is returned to normal.

Determination of the *anion gap* provides a quick way of assessing metabolic disorders of acid–base balance. Since blood pH is

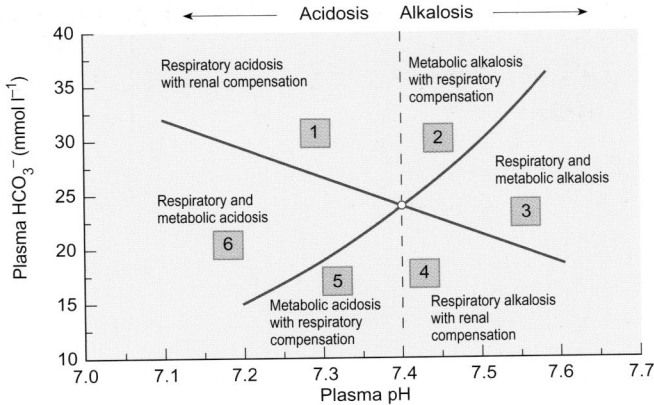

Fig. 29.7 The pH–[HCO$_3^-$] diagram for blood plasma can be used to determine the acid–base status of a patient. The white circle represents normal values. The diagram is labeled to show the underlying deficits for any given pair of pH–[HCO$_3^-$] values. The numbers denote values that occur in six different conditions.

1. [HCO$_3^-$] 31 mmol l^{-1}, pH 7.30: partially compensated respiratory acidosis.
2. [HCO$_3^-$] 30 mmol l^{-1}, pH 7.45: partially compensated metabolic alkalosis.
3. [HCO$_3^-$] 24 mmol l^{-1}, pH 7.55: metabolic and respiratory alkalosis.
4. [HCO$_3^-$] 17 mmol l^{-1}, pH 7.42: compensated respiratory alkalosis.
5. [HCO$_3^-$] 16 mmol l^{-1}, pH 7.31: partially compensated metabolic acidosis.
6. [HCO$_3^-$] 20 mmol l^{-1}, pH 7.18: mixed metabolic and respiratory acidosis.

normally about 7.4, the fully dissociated cations (principally sodium) are present in greater quantities than the fully dissociated anions (principally chloride). The difference in charge is made up by bicarbonate, phosphate, and organic anions such as lactate. The difference between the total cation content and the total amount of chloride plus bicarbonate is known as the 'anion gap' and is an indirect measure of the amounts of phosphate, sulfate, and organic acids present in plasma. Normally the anion gap is about 15 mmol l^{-1}.

If the anion gap increases, this will be the result of either an increase in the quantity of anions (other than Cl$^-$ and HCO$_3^-$) or a loss of strong cations (principally Na$^+$). The plasma pH will be lower than normal. This situation typifies metabolic acidosis. Conversely, if the gap narrows, this will be the result of ingestion of base or the loss of strong anion (e.g. loss of Cl$^-$ from the stomach as a result of vomiting) and the plasma pH will be higher than normal. This situation typifies metabolic alkalosis.

Another common way of determining the extent of metabolic disturbance of acid–base balance is to measure the *base excess* or *base deficit*. The base excess is measured by titrating the blood or plasma to pH 7.4 with a strong acid or base while the P_{CO_2} is kept constant at 40 mmHg. If strong acid (e.g. HCl) needs to be added to bring the pH to 7.4, there is a base excess, and if strong alkali (e.g. NaOH) is required, there is a base deficit. By definition, in normal people the base excess is zero, but deviations of ± 2.5 mmol l^{-1} are considered to lie within the normal range.

The effect of changes to the base excess can be understood by reference to Fig. 29.8. The line AB in Fig. 29.8 is called the blood buffer line. It represents the titration of the blood buffers with carbon dioxide (see also Fig. 29.4). If there is a base excess or base deficit, there is a parallel displacement of the buffer line. When there is a base excess, the line is displaced upwards as shown in

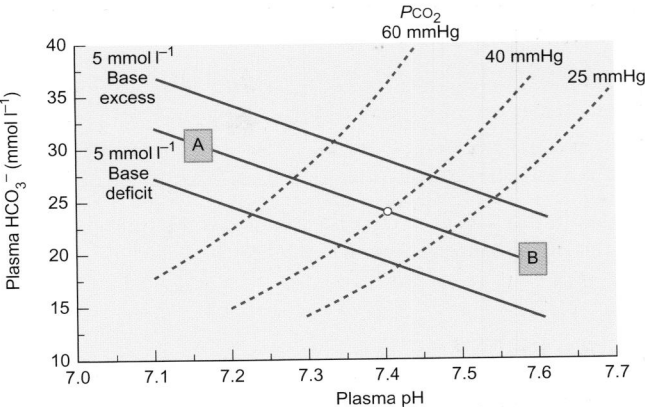

Fig. 29.8 The effect of a base excess or a base deficit on the pH–[HCO$_3^-$] diagram for blood plasma. The line AB is the normal buffer line and reflects the change in plasma pH as P_{CO_2} is varied. When there is a base deficit (as in metabolic acidosis), the buffer line is displaced downwards and results in a lower plasma pH for a given P_{CO_2}. When there is a base excess (as in metabolic alkalosis), the buffer line is displaced upwards. The plasma pH will be higher than normal for a given P_{CO_2}.

Fig. 29.8 and the blood pH becomes more alkaline for any given P_{CO_2}. Conversely, if there is a base deficit the buffer line is displaced downwards and the blood pH will be relatively more acid for any given P_{CO_2}.

Recommended reading

Campbell, E.J., Dickenson, C.J., Slater, J.D.H., Edwards, C.R.W., and Sikora, E.K. (eds.) (1984). *Clinical physiology* (5th edn), Chapter 5. Blackwell Scientific, Oxford.

Holmes, O. (1993). *Human acid–base physiology*. Chapman & Hall, London.

Lowenstein, J. (1993). *Acid and basics: a guide to understanding acid–base disorders*. Oxford University Press, New York.

Quantitative problems

Answers are given below.

1. Convert the following pH values into free H$^+$ concentrations:
    a. 5.6
    b. 6.8
    c. 7.5
    d. 8.2

2. Convert the following H$^+$ concentrations into pH units:
    a. 100×10^{-3} M
    b. 1×10^{-3} M
    c. 5×10^{-6} M
    d. 2×10^{-7} M
    e. 3×10^{-8} M

3. Use the Henderson–Hasselbalch equation to calculate the pH of the following solutions (the solubility coefficient of CO$_2$ is 0.03 mmol l^{-1} mmHg^{-1} and the pK_a for carbonic acid is 6.1).

Pco₂ (mmHg)	[HCO₃⁻] (mmol l⁻¹)
38	15
45	26
55	32
60	20

4. Assuming that the P_{CO_2} of a solution is 40 mmHg calculate the HCO_3^- concentration when its pH is:

   a. 6.8

   b. 7.6.

5. What is the acid–base status of the following patients:

	pH	Paco₂ (mmHg)	[HCO₃⁻] (mmol l⁻¹)
a	7.5	50	35
b	7.6	18	20
c	7.2	60	22
d	7.4	58	33
e	7.54	30	23
f	7.4	25	16
g	7.25	40	16

6. In a man undergoing surgery, it was necessary to aspirate the contents of the upper GI tract. After surgery, the following values were obtained from an arterial blood sample: pH 7.55, P_{CO_2} = 52 mmHg and $[HCO_3^-]$ = 40 mmol l⁻¹. What is the underlying disorder and what remedy would restore normal acid–base balance?

7. A patient was admitted in a diabetic coma. Analysis of the arterial blood gave the following values: P_{CO_2} = 16 mmHg, $[HCO_3^-]$ = 5 mmol l⁻¹, and pH 7.1. What is the underlying acid–base disorder? Why is the P_{CO_2} so low? What measures would you take to restore the acid–base variable to normal?

8. A patient with pulmonary edema was admitted to hospital. Analysis of the arterial blood revealed that the P_{CO_2} was 85 mmHg, $[HCO_3^-]$ was 22 mmol l⁻¹ and the pH was 7.04. Following mechanical ventilation the P_{CO_2} was reduced to 60 mmHg, $[HCO_3^-]$ was 20 mmol l⁻¹ and plasma pH was 7.15. What was the original acid–base disorder? What further measures could be employed to restore the acid–base variables to normal?

Answers to quantitative problems

1. a. 2.5×10^{-6} mol l⁻¹

   b. 1.58×10^{-7} mol l⁻¹

   c. 3.16×10^{-8} mol l⁻¹

   d. 6.31×10^{-9} mol l⁻¹

2. a. 1.00

   b. 3.00

   c. 5.30

   d. 6.70

   e. 7.52

3. a. 7.22

   b. 7.38

   c. 7.39

   d. 7.15

4. a. 6.01 mmol l⁻¹

   b. 37.9 mmol l⁻¹

5. a. Metabolic alkalosis with respiratory compensation

   b. Uncompensated respiratory alkalosis

   c. Mixed respiratory and metabolic acidosis

   d. Fully compensated respiratory acidosis

   e. Mixed metabolic and respiratory alkalosis

   f. Fully compensated respiratory alkalosis

   g. Uncompensated metabolic acidosis

   (These answers can be best understood by reference to Fig. 29.7)

6. The aspiration of the GI contents has caused a metabolic alkalosis as a result of loss of gastric acid. The situation can be reversed by the slow infusion of a quantity of dilute HCl.

7. In this case, both the plasma pH and bicarbonate are low so there is a metabolic acidosis. The P_{CO_2} is also low because the very low plasma pH has stimulated ventilation excessively. In this case, the first priority is to stabilize blood pH to reverse the coma. After checking that blood glucose is high, insulin can be injected intravenously to correct the hyperglycemia and restore the glucose supply to the tissues. Infusion of normal saline will restore fluid and electrolyte balance (as dehydration generally accompanies the development of the ketoacidosis). The normal values for plasma pH and bicarbonate will be re-established over time.

8. The pulmonary edema prevented proper gas exchange and led to a respiratory acidosis. In addition, there was a metabolic acidosis (see Fig. 29.7). The persistent low pH following ventilation could be corrected by an infusion of sodium bicarbonate, but full correction would require identification of the cause of the underlying metabolic acidosis followed by appropriate corrective treatment.

The physiology of high altitude and diving

After reading this chapter you should understand:

♦ The physiological changes that occur following ascent to high altitude

♦ The physiological problems associated with high environmental pressures

♦ Breath-hold diving and the diving response

30.1 Introduction

In earlier chapters, the physiology of the various body systems and the ways in which they interact have been discussed. This chapter is concerned with the physiology of the respiratory system under stress. It will discuss the adaptations that occur when a person moves to high altitude and the problems associated with high ambient atmospheric pressures such as those experienced by divers.

30.2 The physiology of high altitude

The barometric pressure falls progressively with altitude and, since the fraction of oxygen in the air (20.9 per cent) does not change, the partial pressure of oxygen in the inspired air will also progressively fall with increasing altitude (Fig. 30.1). Consequently, the oxygen partial pressure and content of the blood will decline and this will ultimately limit the capacity of the body to perform work.

The hypoxia of altitude can be divided into *acute hypoxia*, which is experienced by subjects who have been exposed to high altitude for a few minutes or hours, and *chronic hypoxia*, which is experienced by people living for long periods at high altitude or by mountaineers who have become acclimatized to high altitude. The principal features of each will be discussed in turn.

Acute hypoxia

Although the partial pressure of oxygen in the inspired air falls with altitude, the arterial P_{O_2} is a relatively weak stimulus to breathing. Not until the alveolar partial pressure of oxygen falls to about 8 kPa (60 mmHg) does the rate and depth of breathing

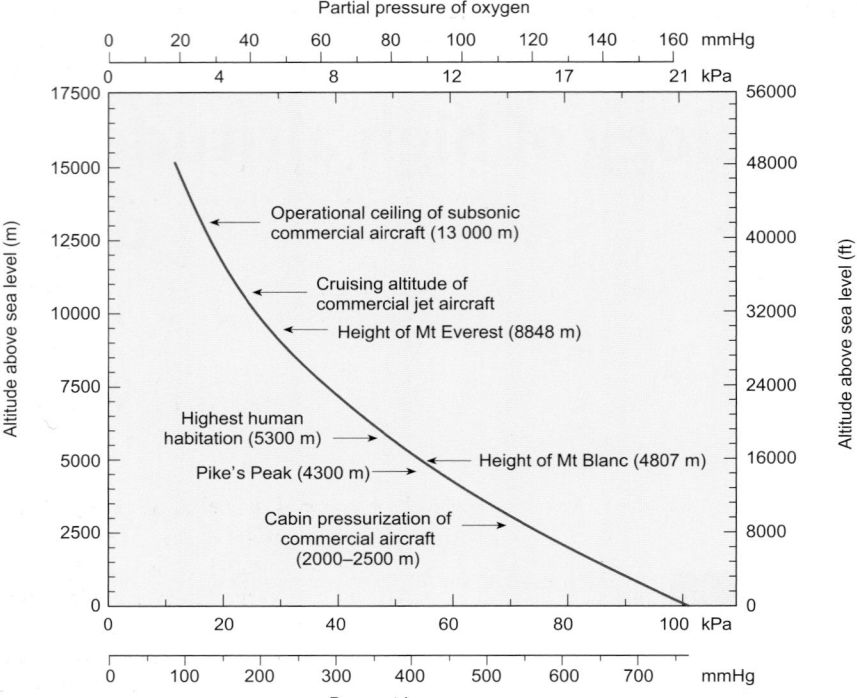

Fig. 30.1 The change in barometric pressure and the partial pressure of oxygen with altitude.

TABLE 30.1   The effect of altitude on the partial pressures of carbon dioxide and oxygen in the alveoli

Location	Altitude (m)	Barometric pressure (kPa (mm Hg))	P_{ACO_2} (kPa (mm Hg))	P_{AO_2} (kPa (mm Hg))
Sea level	0	101 (760)	5.3 (40)	13.3 (100)
Colorado Springs	1800	82 (620)	4.8 (36)	10.5 (79)
Pike's Peak Colorado	4300	61 (460)	3.7 (28)	7 (53)
North Col of Everest	6400	45 (355)	2.5 (19)	5.1 (39)
Summit of Everest	9100	32 (240)	2.0 (15)	3.2 (24)

increase substantially. Below this level, the respiratory minute volume increases progressively with declining P_{O_2} as discussed in Chapter 16, (p. 337). The stimulus for this increase in ventilation comes from the carotid bodies, which sense the fall in arterial P_{O_2}. As ventilation increases, carbon dioxide is lost from the lungs faster than it is produced by the body. Consequently, alveolar P_{CO_2} falls (Table 30.1). This fall in alveolar P_{CO_2} leads to a diminished respiratory drive from the central chemoreceptors that tends to offset the respiratory drive due to the hypoxic stimulation of the carotid bodies (see Chapter 16, Fig. 16.27). In addition, the fall in alveolar P_{CO_2} results in a respiratory alkalosis (see Chapter 29).

While the carotid bodies have little influence on the circulation at normal blood gas tensions, during hypoxia they elicit a reflex vasoconstriction in the resistance vessels and the large veins of the splanchnic circulation. In hypoxia, there is also an increase in heart rate and cardiac output. As a result of these

changes, blood is diverted from the skin and splanchnic circulation to increase the proportion of the available oxygen for use by the brain and exercising muscles.

Severe acute hypoxia

Rapid ascent in an unpressurized aircraft or balloon is associated with a rapid change in the alveolar P_{O_2} and severe acute hypoxia can be experienced by aviators and balloonists. When people are exposed to low P_{O_2}, they first experience physical weakness, which progresses to full paralysis of the limbs when the P_{IO_2} falls below about 8 kPa (60 mmHg). This corresponds to an altitude of 6000 m (about 19 600 ft). As P_{O_2} falls further, there is loss of consciousness and death rapidly ensues. Even at more moderate altitudes, there are psychological changes that often result in poor judgment and elementary mistakes. For this reason, commercial aircraft maintain a cabin pressure equivalent to 2000–2500 m (6500–8200 ft.)

The effect of reduced P_{IO_2} on the oxygen content of the blood

The stimulation of ventilation by low P_{O_2} results in a fall of the alveolar P_{CO_2}. This increases the ability of hemoglobin to bind oxygen because of the reversal of the Bohr shift. (As P_{ACO_2} falls and pH rises, the position of the oxyhemoglobin dissociation curve shifts to the left—see Chapter 13, Fig. 13.7). Consequently, the hemoglobin is more saturated at a given P_{AO_2} than it would be if the P_{ACO_2} had remained at 5.3 kPa (40 mmHg), the value it would normally be at sea level. Despite this, above about 3000 m (c. 10 000 ft), the oxygen content of the blood is significantly less than it would be at sea level. To meet the demands of the tissues, more oxygen is extracted from the blood and this reduces the P_{O_2} of the venous blood.

Mountain sickness and other effects of high altitude

When people initially experience the hypoxia of altitude, they often develop an illness known as *mountain sickness*. The typical symptoms are headache, nausea, giddiness, gastrointestinal disturbances, lassitude and psychological disturbances. Severe dyspnea occurs in response to mild exercise, as the circulation is unable to supply the tissues with adequate amounts of oxygen.

Sleep apnea, which is usually associated with periodic breathing, occurs at altitudes above 4000 m (c. 13 000 ft). The primary cause of the periodic breathing is the hypoxemia due to the low P_{IO_2}. At sea level, respiration is principally driven by the response of the central chemoreceptors to arterial P_{CO_2}. The increase in ventilation that occurs following ascent to high altitude reduces the P_{ACO_2} and the ventilatory drive from the central chemoreceptors falls. As a result, breathing is not adequately stimulated by the P_{CO_2} and ventilation becomes periodic as described in Chapter 16, p. 339. The periods of apnea exacerbate the hypoxemia that is already present as a result of the low inspired P_{O_2}.

Pulmonary edema can occur when unacclimatized climbers reach altitudes in excess of 3000 m (c. 10 000 ft). The cause of this edema is not clear and it is likely that several factors are involved. Perhaps the most important factor is pulmonary hypertension resulting from the hypoxic vasoconstriction of the pulmonary vessels (Chapter 16, p. 329). The pulmonary hypertension is increased during exercise. As a result, the balance of the Starling forces in the pulmonary capillaries favors the movement of fluid into the interstitial space and, eventually, into the alveoli.

There is wide variation in the susceptibility of different people to mountain sickness: some are affected at relatively low altitude (c. 2500 m or 8200 ft) while others show few signs. Very few subjects can venture to altitudes in excess of 6000 m (19 600 ft) without experiencing severe symptoms. The causes of mountain sickness are not entirely clear but include hypoxia and dehydration. In severe cases, administration of oxygen and descent to lower altitude are the primary steps in treatment.

Acclimatization to chronic hypoxia

Once at altitude, the body adapts to the new circumstances. Indeed, many people live permanently at high altitude and their physiology has become so well adapted that they are able to perform the everyday tasks of life as easily as those who live near sea level. Even for subjects who have newly arrived at high altitude, the process of acclimatization begins almost immediately. The principal changes that occur during acclimatization to high altitude are as follows:

The respiratory minute volume increases

This occurs despite the low P_{ACO_2}. The respiratory alkalosis associated with the hyperventilation is gradually compensated by the excretion of excess bicarbonate by the kidneys. Similar compensation occurs in the CSF and this is believed to restore the respiratory drive provided by the central chemoreceptors in response to carbon dioxide. As a result, the minute volume increases rapidly at first and then more slowly to reach its final level. The adjustment of CSF pH occurs within about 24 hours. The pH of the plasma is restored to normal levels by renal excretion of excess bicarbonate within a week.

There is an increase in the number of red cells and hemoglobin

Following ascent to a high altitude, the red cell count increases significantly. This trend continues for several weeks (Fig. 30.2) and red cell counts as high as 8×10^{12} have been recorded after long periods of acclimatization (the normal red cell count is 4–5×10^{12}). The increased hematopoiesis is stimulated by the hormone erythropoietin, which is secreted by the kidney in response to the low blood P_{O_2} (see Chapter 13, p. 230). The increased hemoglobin content of the blood leads to an increase in its oxygen-carrying capacity. In addition, the low P_{O_2} leads to an increased production of 2,3-bisphosphoglycerate by the red cells which enhances the rightward shift in the oxygen dissociation curve (the Bohr shift) and so facilitates the release of oxygen in the tissues.

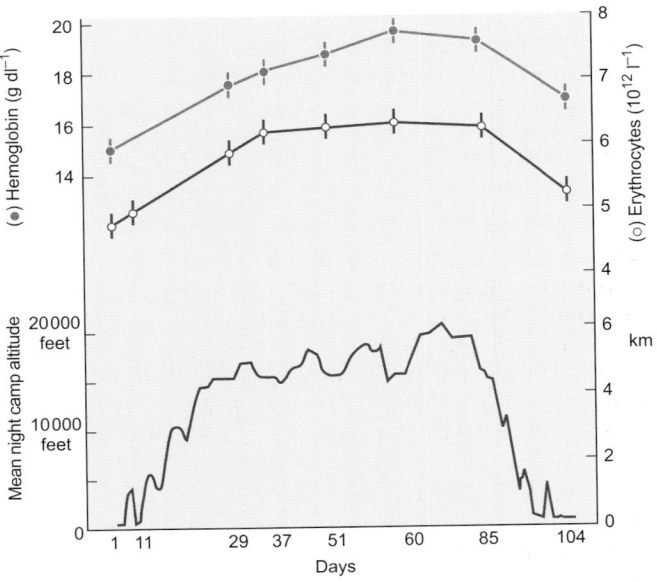

Fig. 30.2 The change in red cell count with time spent at altitude. The data were collected during a period of 62 days spent continuously above 4600 m (15 000 ft). Note that the red cell count and blood hemoglobin increased over several weeks spent at altitude. Recovery on descent to sea level was much more rapid. The symbols and vertical bars indicate mean ± standard error of the mean.

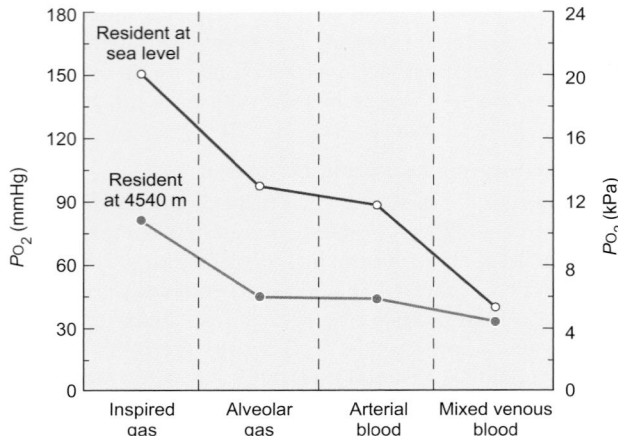

Fig. 30.3 The partial pressure of oxygen in the inspired air, alveoli, arterial blood, and mixed venous blood for subjects acclimatized to sea level and 4540 m (14 755 ft).

There is an increase in cardiac output

The rise in heart rate seen in the early stages of acute hypoxia is maintained during acclimatization and gives rise to an increased cardiac output. This results in greater blood flow through the tissues, thus improving their oxygenation. The enhanced oxygen delivery to the tissues is so efficient that the P_{O_2} of mixed venous blood of people acclimatized to very high altitude is only slightly less than that of those who live at sea level (Fig. 30.3). During a prolonged sojourn at high altitude, the cardiac output slowly declines to near normal values.

There is an increase in the vascularization of the tissues

The diameter of the capillaries is increased and they become more tortuous in their course through the tissues. Blood volume is also increased. These changes greatly enhance the ability of the blood to supply the tissues with the oxygen that they require for metabolism. The high red cell count will increase blood viscosity and this, coupled with the increase in cardiac output, might be expected to result in an increase in arterial blood pressure. In fact, people who live at high altitudes do not have elevated blood pressure compared with people living at sea

Summary

1. In acute hypoxia, the fall in inspired P_{O_2} leads to an increase in ventilation and in cardiac output. The increased ventilation leads to a respiratory alkalosis. These physiological changes may be accompanied by the symptoms of mountain sickness.

2. Following acclimatization to chronic hypoxia, there is a maintained increased in ventilation, an increased vascularization of the tissues, and an increased oxygen-carrying capacity of the blood. The respiratory alkalosis resulting from the hyperventilation is compensated by renal excretion of bicarbonate. The cardiac output slowly returns to near-normal values.

level. The increased vascularization leads to a fall in total peripheral resistance and it is this change that accounts for the normal value of arterial blood pressure.

30.3 The effects of high environmental pressure

Exposure to high atmospheric pressures occurs during diving. The atmospheric pressure at sea level is 1 atm but, for every 10 m (33 ft) of descent into the sea, it rises by 1 atm. Thus the total pressure at a depth of 99 ft is 1 + 3 = 4 atms. Increased environmental pressures are also experienced by engineering workers when they are employed in tunneling, as the air must be maintained under pressure to prevent water seeping into the workings. In both cases, the increased air pressure increases the amount of gas dissolved in the blood and tissues in accordance with Henry's law (see Chapter 16, p. 314). Since the alveolar P_{CO_2} remains almost constant under these circumstances, the increases in the partial pressures and amounts of dissolved oxygen and nitrogen are of prime concern.

Oxygen toxicity

While it is essential for life, oxygen is a very reactive gas. Therefore it is perhaps not surprising that certain hazards are associated with breathing pure oxygen. Breathing oxygen up to a partial pressure of 60 kPa (60 per cent in inspired air at normal pressure) is perfectly safe for an adult even for long periods. However, breathing oxygen at normal atmospheric pressure (101 kPa or 760 mmHg) for more than 8 hours leads to signs of pharyngitis, tracheitis, and cough. Subsequently there are signs of pulmonary congestion and sluggish mental activity. Breathing oxygen at pressures above about 1.7–2 atm (170–200 kPa) leads to overt signs of oxygen toxicity. These signs include nausea, dizziness, feelings of intoxication, tremor, and even convulsions or syncope. For this reason, the partial pressure of oxygen must be carefully controlled in deep diving. It follows that the safe use of 100 per cent oxygen in diving is extremely limited (Fig. 30.4).

In moderate hyperoxia (increased alveolar P_{O_2}), only the alveoli, pulmonary vessels, and systemic arteries experience significantly elevated P_{O_2} values (100–150 kPa). Once the blood reaches the tissues, P_{O_2} rapidly falls. Therefore it is not surprising that the lungs are the most vulnerable to the effects of prolonged hyperoxia.

Premature and newborn babies are particularly sensitive to the toxic effects of increased partial pressures of oxygen. The oxygen pressure to which newborn babies are exposed must not exceed 40 kPa (c. 300 mmHg or 40 per cent of normal atmospheric pressure). There is a risk of permanent blindness if these values are exceeded. This arises because of intense constriction of the immature retinal vessels resulting in retinal ischemia, which eventually leads to secondary pathological changes in the retina.

Problems associated with breathing compressed air during diving

Inhalation of compressed air at high pressure can be dangerous. Although nitrogen is less soluble in water than either oxygen or

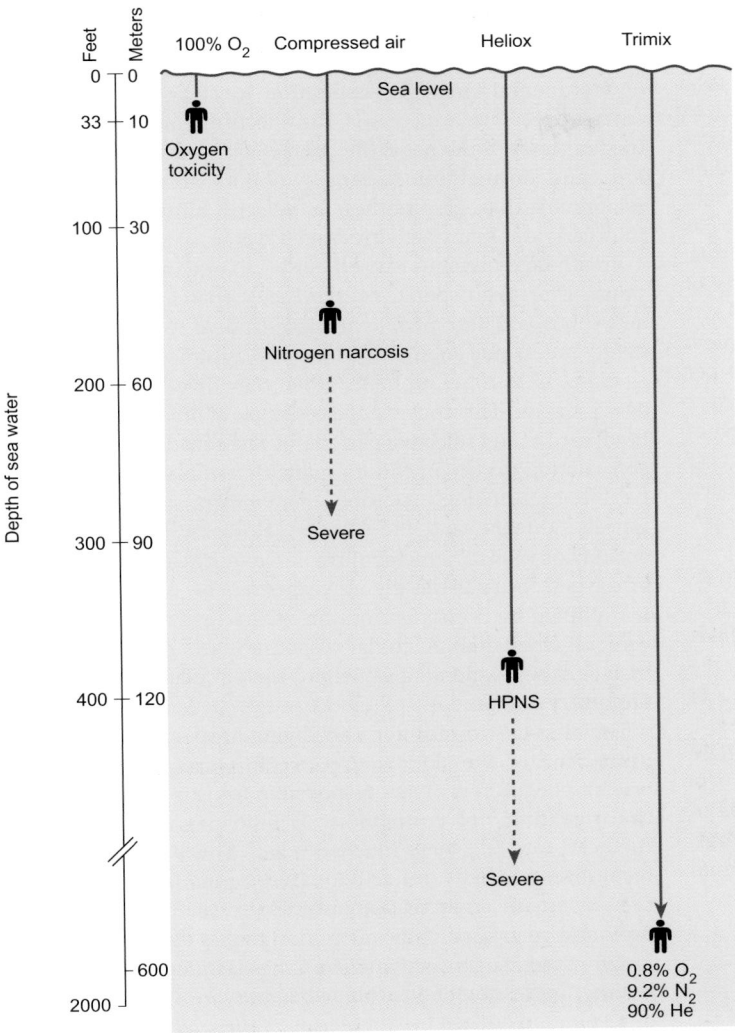

Fig. 30.4 The use of various gases in diving. Oxygen toxicity limits the breathing of 100 per cent oxygen to dives of only 33 ft (10 m) of seawater. If air (*c*. 20 per cent oxygen, 80 per cent nitrogen) is supplied under pressure, depths of about 150 ft (*c*. 50 m) of seawater are possible before there is a risk of nitrogen narcosis. (Note that with compressed air this sets in before oxygen toxicity becomes important). Greater depths are possible with helium–oxygen mixtures (heliox) and helium–oxygen–nitrogen mixtures (trimix). See text for further information.

carbon dioxide, it is about five times more soluble in fat than it is in blood. Therefore prolonged exposure to compressed air will lead to an accumulation of nitrogen in the tissues. Unless suitable precautions are taken when a diver returns to normal ambient pressure, this dissolved gas will come out of solution and form bubbles in the tissues, causing traumatic damage and intense pain. This disorder, *decompression illness*, was once common in divers and caisson workers and is known as the 'bends' or 'chokes'. It can be avoided by slow ascent according to specific diving schedules. If decompression illness becomes evident on completing a dive, it should be treated by immediate recompression in a pressure chamber followed by slower decompression according to specific decompression tables.

Nitrogen narcosis

Breathing a gas mixture containing nitrogen at high pressure has a second danger—that of narcosis. Breathing air at a pressure greater than about 5 atms (which would correspond to a depth of about 40 m) brings about the early signs of anesthesia.

There is a sense of euphoria ('rapture of the deep'), mental confusion, and a lack of proper motor coordination. These symptoms become more severe with depth. For these reasons the breathing of air during diving is limited to depths of less than 50 m (see Fig. 30.4).

The use of helium–oxygen mixtures in diving

Nitrogen narcosis can be overcome by substituting helium for the nitrogen of the inhaled gas. Helium avoids the problem of narcosis and is less soluble in the tissues, so that decompression from a given depth can be somewhat faster. For these reasons mixtures of oxygen and helium (known as heliox) are widely used in commercial diving (Fig. 30.4). Helium has the disadvantage that is a very efficient conductor of heat, so that precautions must be taken against hypothermia. It has the additional disadvantage that its low density leads to a rise in the pitch of the voice. As a result, speech becomes less intelligible and this can make communication between the diver and the surface difficult.

The high-pressure nervous syndrome

Exposure to very high pressures is a hazard even with helium–oxygen mixtures. To avoid oxygen toxicity the fraction of oxygen in the inspired air is kept around 50 kPa (0.5 atm) as explained above. In addition, there are direct effects of pressure on performance. Pressures above about 1200 kPa (corresponding to a depth of about 130 m or 400 ft) cause characteristic changes known as the high-pressure nervous syndrome (HPNS). There is tremor of the hands and arms, dizziness, and nausea. Intellectual performance is much less affected. The severity of the symptoms appears to relate to the rate of compression.

The excitatory effects of HPNS can be countered by a slow descent and by adding a small quantity of nitrogen (5–10 per cent) to the helium–oxygen mixture. This gas mixture (known as trimix) was developed to exploit the known depressant effect of nitrogen on the nervous system to offset the excitatory effects of high pressure. Dives in excess of 650 m (c. 2000 feet) have been made using this mixture (Fig. 30.4).

Barotrauma

This is the most common occupational disease of divers. It is caused by the contraction or expansion of gas spaces in the body that are for one reason or another unable to equilibrate with the ambient air. Air trapped in the sinuses and even the cavities of the teeth may cause pain as it is compressed and decompressed. If the Eustachian tube is blocked, the resulting pressure changes in the middle ear cause pain and may lead to rupture of the tympanic membrane. When a diver ascends, he must exhale to ensure that the pulmonary gases do not overdistend the lungs. Gases trapped in the intestinal tract may cause abdominal discomfort during decompression.

Summary

1. At elevated ambient pressures, the quantity of dissolved gas is directly related to the pressure. Thus when divers or tunneling engineers work in an atmosphere of compressed air, the quantity of nitrogen dissolved in the tissues increases. Rapid decompression can lead to bubble formation and consequential tissue damage. This can be avoided by progressive decompression according to specified tables.

2. A second problem of breathing air at elevated pressure is nitrogen narcosis. Breathing pure oxygen at pressures greater than 1 atm leads to signs of oxygen toxicity so that it cannot be used as a safe alternative in diving. Instead, the problems associated with breathing air at high pressure are largely overcome by breathing mixtures of helium and oxygen. Even so, it is important that the partial pressure of oxygen is kept below 1 atm to avoid the problems of oxygen toxicity (usually the P_{O_2} employed is 0.5 atm or 50 kPa).

30.4 Breath-hold diving

In breath-hold diving, the body is subjected to a number of environmental changes. Respiration must be suspended, the ambient pressure is increased, the effects of gravity are offset by the buoyancy of the body, the contact with water increases heat loss, and normal somatosensory inputs are impaired. It is remarkable that, despite these handicaps, humans can successfully carry out many activities under water.

Breath-hold diving is carried out by a large number of professional divers who operate off the Pacific coasts collecting pearls and sponges. These people are able to dive to depths of 20 m (66 ft) and to stay submerged for up to a minute. They can dive as many as 20 times an hour. When they dive, they exhibit the *diving response*. This consists of a cessation of breathing, profound bradycardia, and selective peripheral vasoconstriction. The vasoconstriction occurs in those organs that can survive for short periods by utilizing anaerobic metabolism, such as the skin, muscle, kidneys, and the GI tract. However, the brain relies entirely on oxidative metabolism and requires a constant supply of oxygen. During the diving response, the oxygen supply is maintained by the redistribution of the cardiac output to the cerebral circulation. A similar response is seen in aquatic mammals (e.g. seals and whales) where the circulatory changes are often very dramatic.

The initial stimulus for the diving response seems to be immersion of the skin of the face in water (especially cold water). The afferent fibers responsible for this response run in the trigeminal nerve. Immediately following face immersion, the heart rate falls by as much as a half. This bradycardia is the result of increased vagal activity. Despite the fall in heart rate, there is usually an increase in arterial pressure resulting from a profound peripheral vasoconstriction that is due to an increase in sympathetic nerve activity. The apnea is induced partly by voluntary suppression of breathing and partly by a reflex inhibition of respiration elicited by stimulation of the trigeminal receptors.

During prolonged breath-holding, both hypoxia and hypercapnia occur, and P_{aO_2} levels as low as 4–4.6 kPa (30–35 mmHg) have been recorded in synchronized swimmers at the end of their performance. The fall in P_{aO_2} and rise in P_{aCO_2} is known as *asphyxia*. These changes in P_{aO_2} and P_{aCO_2} stimulate the carotid body chemoreceptors. This chemoreceptor activity does not, at least initially, lead to stimulation of breathing, but it contributes to the bradycardia and peripheral vasoconstriction of the diving response. Under these conditions, the chemoreceptor stimulus to breathing is powerfully inhibited by the activation of the trigeminal receptors. Eventually, however, the hypoxia and hypercapnia become so intense that the trigeminal inhibition is overcome and the desire to breathe becomes compelling.

During the descent stage of a breath-hold dive, the lung volume will decrease and the alveolar gases will be compressed. This raises their partial pressure and facilitates their transport into the blood. Note that even for a dive of only 10 m (33 ft), the total gas pressure will approximately double. This will increase the partial pressure of both oxygen and carbon dioxide. However, the increased pressure will not increase the oxygen content of the blood very significantly. The increased arterial P_{CO_2} that occurs during the dive acts to stimulate the desire to breathe (the 'break-point') and serves to limit the duration of the dive.

Before diving, many people hyperventilate. This has comparatively little effect on the alveolar P_{O_2} but greatly reduces the alveolar P_{CO_2}. As a result, the respiratory drive from the central chemoreceptors is much reduced and this increases the time to the 'break-point' where the desire to breathe becomes overwhelming. This practice is potentially dangerous. If the arterial P_{CO_2} does not rise sufficiently to stimulate breathing before cerebral hypoxia occurs, consciousness may be lost, possibly with fatal consequences. Remember that during the diving response, the respiratory drive from the peripheral chemoreceptors (which sense Pa_{O_2}) is powerfully inhibited.

Summary

During breath-hold diving, there is a reflex slowing of the heart, peripheral vasoconstriction, and apnea. The duration of the dive is limited by the break-point which is determined largely by the arterial P_{CO_2}.

Further reading

Dejours, P. (1966). *Respiration*, Chapter 9. Oxford University Press, New York.

Hlastala, M.P., and Berger, A.J. (2001). *Physiology of respiration* (2nd edn), Chapters 12 and 13. Oxford University Press, New York.

Levitzky, M.G. (2003). *Pulmonary physiology* (6th edn), Chapter 11. McGraw-Hill, New York.

Lumb, A.B. (2000). *Nunn's applied respiratory physiology* (5th edn), Chapter 15 and 16. Butterworth Heinemann, Oxford.

Minors, D.S. (1994). In *Human physiology: age, stress, and the environment* (ed. R.M. Case and J.M. Waterhouse), Chapters 7 and 8. Oxford Science Publications. Oxford.

Slonim, N.B., and Hamilton, L.H. (1987). *Respiratory physiology*, Chapter 17. Mosby, St. Louis, MO.

Clinical physiology

After reading this chapter you should understand:

- The physiological changes associated with age
- Hypertension
- Causes of abnormal heart sounds
- How to interpret changes in the ECG
- The pathophysiology of heart failure
- Renal failure
- Liver failure
- Multiple organ dysfunction syndrome

31.1 Introduction

Earlier chapters of this book discussed the physiology of the various body systems. The interactions between the systems under various circumstances (e.g. exercise, high altitude) were subsequently considered. This chapter is concerned with the physiological changes that occur throughout life and with the pathophysiological changes that are responsible for some common clinical conditions. The physiological principles that underpin the treatment of these conditions are also briefly discussed where appropriate.

31.2 The physiological changes associated with aging

Gerontology (the scientific study of aging and the problems associated with elderly people) is becoming an increasingly important area of investigation. Although the maximum lifespan of about 110 years has remained largely unchanged during the last century, in the industrialized countries of the West more and more people are reaching old age. Figure 31.1 shows that the increased life expectancy attained in the United Kingdom during the twentieth century is primarily due to a marked fall in infant mortality. Nevertheless, as the figure shows, all age groups had a significantly lower mortality in 1975 compared with 1901. Current life expectancy in the United Kingdom and the United

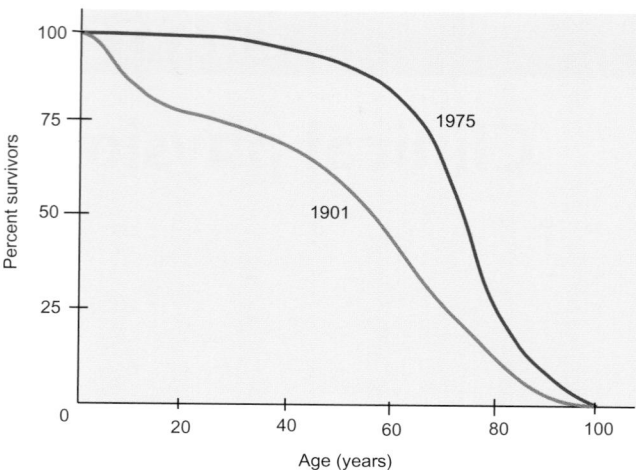

Fig. 31.1 The changes in survival rates for individuals of different ages in the United Kingdom in 1901 and 1975. Note that the maximum age attained is very similar but that between 1901 and 1975 there was a marked drop in infant mortality and a significant decrease in mortality at other ages.

States is about 75 years for males and 80 years for females. At the age of 75 women outnumber men by 2 to 1, while at 100 years of age the ratio of women to men has risen to 6 to 1. The vast majority of older people lead healthy independent lives. Even so, their ability to respond to physiological stresses gradually becomes compromised and this section is concerned with the underlying changes in their physiology.

Aging is the sum of all the changes that take place in the body with the passage of time that ultimately lead to functional impairment and death. Aging might be defined as an increased vulnerability to internal and external stress—a gradual inability to respond to physiological challenges. Those alterations in homeostatic function that occur as a result of aging have been discussed where appropriate throughout this text. For example,

---

BOX 31.1   DEFINITIONS OF OLD AGE

'To me, old age is fifteen years older than I am' (Bernard Baruch (1870–1965), quoted in an interview when he was 85 years old). Definitions of what constitutes old age are naturally somewhat arbitrary, even if we cannot accept Baruch's definition. Hippocrates, the Greek physician regarded as the father of medicine, divided old age into five categories. Up to 70 years, he described as the springtime of old age, 70–75 years as green old age, and 75–80 as real old age. The years 80–90 he considered ultimate old age and over 90 he regarded as senility. The modern World Health Organization definitions of life's stages are no less arbitrary but provide a basis for modern work. They are given below.

45–59 years	Middle age
60–74 years	Elderly
75–89	Old
90+	Very old

---

the reduced ability of the elderly to regulate core temperature is discussed in Chapter 26. Definitions of aging have changed over the centuries and Box 31.1 gives the current World Health Organization definitions compared to those given by the father of Western medicine, Hippocrates.

Although the lifespan of some animals appears to be determined by a single gene (e.g. certain strains of mice), it is more likely that multiple factors are involved in determining the lifespan of humans. Nonetheless, aging is the result both of intrinsic factors, which are in part genetically determined, and of environmental factors, which may reflect the employment or lifestyle of an individual. In practice, it is often difficult to separate these two influences.

There have been many theories of aging some cellular, some relating to whole body changes. Examples include:

- accumulation of toxic metabolites;
- exhaustion of irreplaceable substances (e.g. loss of elastic tissue);
- endocrine changes;
- accumulation of errors in genetic material, caused for example by free-radical damage, the switching on of lethal genes, or a reduction in immunological surveillance.

All are hypothetical and it is possible that some, or all, may play a part in the progressive alterations of function that are characteristic of old age.

Old age is associated with disease but does not cause it

When considering age-related changes to anatomy, biochemistry, and physiology, several immediate and obvious questions arise. First, do specific changes happen to everyone? If so, is the rate of change the same in everyone? Undoubtedly, things happen at different times and to different extents in different people. Some people considered elderly are fitter than many of those less than half their age. Related to this is a second question: clearly an 80-year-old does not exhibit the same anatomical and physiological characteristics that he or she did at the age of 20, but are the changes those of aging or pathology? It is always difficult to differentiate between a degenerative change that leads to dysfunction and a change that is to be expected as a normal consequence of aging, but it is important to realize that many of the changes that accompany aging are normal and not pathological. They are simply the expected consequences of changes in physiology and anatomy that occur as a result of progression through life. Old age itself does not cause disease.

All organ systems are affected in some way by the aging process

A small selection of the changes associated with the aging process will be considered briefly here but, for fuller discussion, a specialist text on geriatric medicine should be consulted. Perhaps the most obvious changes at the whole-body level are those that involve appearance and gait. The skin loses its elasticity over time, giving rise to wrinkling. The skin also thins, increasing the risk of pressure sores in immobile individuals. Loss of teeth is still common, despite improvements in dental care, and results in restructuring of the mandible and changes to the facial appearance. There is loss of muscle power and mass, stiffening of the joints, and loss of motor nerve fibers. These changes result in the characteristic changes in gait and balance seen in the elderly. With advancing age, the gait tends to be slow, often on a widened base, and shuffling is common. The standing posture is less erect due to loss of muscle power and the ability to maintain posture is reduced, particularly when the eyes are closed. These are the natural consequences of degenerative changes to the vestibular apparatus, cerebellum, skeletal musculature, and proprioceptors.

The skeleton shows significant changes during life. While osteoporosis is associated mainly with postmenopausal women, all aging individuals show changes to bone tissue. Gradually bone mass is lost as bone erodes without equivalent periosteal deposition. Furthermore, there is an age-related decrease in circulating levels of hydroxylated vitamin D_3 (25-hydroxycholecalciferol), leading to impaired intestinal absorption of calcium and an increasing reliance on the skeletal mineral reserves for maintaining adequate plasma calcium levels. There is enlargement of the Haversian canals and the developing spaces fill with adipose or fibrous tissue. As a result, the bone weakens and becomes liable to fracture even under relatively light loads. This helps to explain the high proportion of elderly orthopedic patients.

Salivary secretion is often reduced in the elderly. This can cause dysphagia which, together with a reduction in the sensitivity of the taste buds, may lead to a loss of appetite and enjoyment of certain foods. Over time, the resulting alterations in diet may contribute to malnutrition. Age-related changes are also seen throughout the GI tract. For example, small intestinal villi shorten, thereby reducing the surface area available for absorption, and the motility of the colon and rectum is reduced. The latter may lead to constipation, which can in turn exacerbate problems of fecal incontinence.

Cardiovascular changes associated with aging

There are subtle changes to cardiovascular function throughout the normal lifespan. These appear to be related to a gradual decline in the ability of the body to exert control over the heart and blood vessels as well as to specific alterations of cardiac and vascular properties. Although resting heart rate alters little throughout life, there is a gradual reduction in both the intrinsic heart rate (in the absence of any autonomic stimulation) and the maximum heart rate that can be achieved. At the age of 45 years, maximum achievable heart rate is about 94 per cent of that at 25 years, while at 85 years this figure has fallen to 80 per cent. These changes are caused by a loss of pacemaker cells in the sinoatrial node as they become replaced by fibrous tissue. The ability of the heart to respond to challenges such as heavy exercise is thus limited. While resting stroke volume does not appear to change with age, the ability of the heart to increase stroke volume is reduced. This effect, coupled with the reduced maximal heart rate, brings about a fall in the maximum cardiac output achievable. Valvular thickening and calcification may further reduce the efficiency of the heart.

Figure 31.2 illustrates typical age-related changes in systemic arterial blood pressure. Between the ages of about 40 and

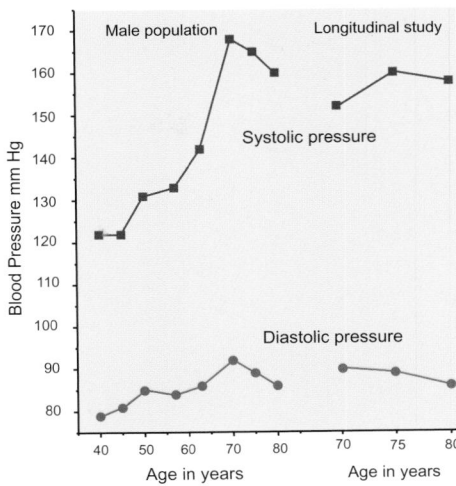

Fig. 31.2 The progressive increase in blood pressure with age. On the left, the figure shows the trend for a cross-section of the male population. On the right, the changes for a group of male subjects at three different ages (a longitudinal study) are shown. The trend in women is similar.

70 years systolic pressure rises from an average of 16 kPa (120 mmHg) to an average of around 22.5 kPa (170 mmHg), while diastolic pressure increases from about 10.7 kPa (80 mmHg) to around 12.5 kPa (90–95 mmHg). The increase in systolic pressure is probably due to decreased aortic compliance and a loss of elasticity, while the increase in diastolic pressure may reflect an increased total peripheral resistance, probably caused by narrowing of the blood vessels. After the age of 75 or so, blood pressure begins to fall again, perhaps because of a gradual reduction in the strength of contraction of the myocardium. Elderly people also show a reduced baroreceptor response and this frequently results in postural hypotension on moving from a supine to an upright position. Old people may feel dizzy for a while after standing up, and many falls may be attributed to this delay in restoration of normal blood pressure.

Changes in respiratory function

Although some older people suffer from chronic respiratory illnesses as a result of smoking or industrial disease, alveolar function is relatively unchanged with age. The most noticeable changes to the respiratory system include a gradual reduction in lung compliance associated with a loss of elasticity and a reduction in the strength of the muscles of the ribcage. Consequently, there is an increase in the work of breathing, a fall in vital capacity, and an increase in residual volume. Closing volume increases with age and may encroach upon the normal tidal volume. Usable lung capacity typically falls to around 82 per cent of its maximum value by the age of 45, to 62 per cent at the age of 65, and to around 50 per cent by the age of 85. Over time, some alveoli are replaced by fibrous tissue and gas exchange is reduced, so that arterial P_{O_2} declines with age (Fig. 31.3). There may be a small rise in arterial P_{CO_2}, particularly during exercise. Overall, the respiratory changes seen in the elderly tend to limit the ability to increase ventilation and oxygen delivery to tissues during periods of increased demand such as exercise.

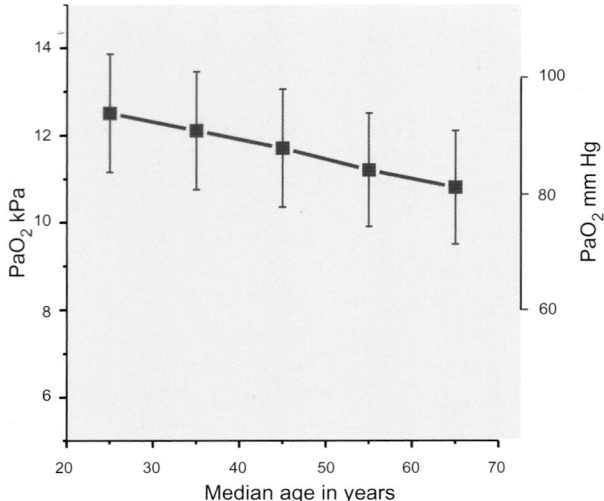

Fig. 31.3 The fall in arterial P_{O_2} with age. Note that not all healthy young subjects achieve the theoretical value of 13.3 kPa (100 mmHg). However, even when P_{O_2} has fallen to around 10.6 kPa (80 mmHg), there is adequate oxygenation of the arterial blood.

Changes in renal function with age

The renal system shows significant changes with age and these may have important consequences for the ability of the kidneys to excrete the metabolites of drugs. When the vessels supplying the nephrons become atherosclerotic and narrow, parts of the kidney may experience a fall in blood flow or even complete ischemia. This results in a corresponding fall in GFR which, by the age of 80, may be as little as 50 per cent of its value at the age of 30. Even at 45 years of age, GFR is less than 90 per cent of its value in young adulthood. The chief consequence of these changes is a reduced ability to respond to homeostatic challenges (e.g. a sodium load or sodium depletion). It may also compromise the ability to respond to an acid or alkaline load so that acid–base balance may be disturbed. Drug dose regimes and timings may need to be altered to compensate for the reduction in the rate of renal clearance of drugs or their metabolites. Figure 31.4 illustrates the effects of aging on the plasma half-life of diazepam (an anti-anxiety and sedative drug).

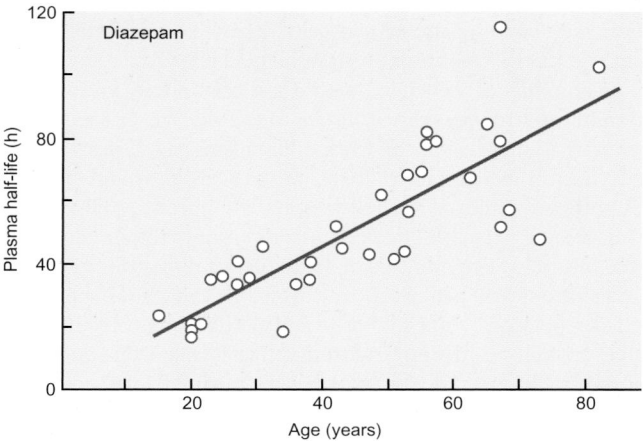

Fig. 31.4 The increase in the half-life of diazepam in the plasma with age.

A common problem seen in the elderly is urinary incontinence. There are a variety of possible reasons for this, but the most common is that the muscles that control the release of urine from the bladder are weakened. In men, frequent urination with poor flow may result from an enlarged prostate, a condition seen in 75 per cent of men over 55.

Changes in the nervous system

The aging nervous system shows a wide array of changes. Some of these are obvious, such as a loss of acuity of the special sense organs. Examples include the gradual loss of accommodation for near vision, loss of high frequencies in hearing, loss of taste cells, and a fall in olfactory sensitivity. The use of glasses and hearing aids can offset these changes to some extent. Other alterations in nervous function include changes in autonomic function (manifested, for example, in poor thermoregulatory capacity), reduced motor control, loss of short-term memory (due to degenerative changes in the hippocampal area), and alterations in sleep patterns and circadian rhythms.

In summary, the healthy elderly person may be seen to represent one extreme of the spectrum of the normal anatomical, biochemical, and physiological parameters that constitute the human body. Although susceptibility to disease increases with the passage of time, the elderly person is neither abnormal nor unhealthy simply because he or she is old. Nevertheless, it is important to realize that the approach to treating disease in elderly people may need to be different from that employed in a younger individual, since complications are more likely and recovery may well be slower. These alterations reflect the impairment of homeostatic mechanisms and the loss of 'physiological reserve' which are the inevitable consequences of growing old.

31.3 Hypertension

Hypertension, or high blood pressure, is probably the most common of all cardiovascular disorders and accounts for a large number of deaths each year, mainly through vascular complications that lead to stroke, coronary heart disease and chronic renal failure. The disease is characterized by an arterial blood pressure which is persistently higher than normal (generally taken to be in excess of 18.7 kPa (140 mmHg) for systolic pressure and 12 kPa (90 mmHg) for diastolic pressure in the young adult). However, it should be remembered that the diastolic pressure is the chief determinant of mean blood pressure. Accordingly, the diastolic pressure is a particularly important indicator of hypertension. As mentioned earlier, systolic blood pressure normally rises with age and an approximate value of 100 + age in years can be considered to be a 'normal' systolic blood pressure for adults (in mmHg). High blood pressure is commonly divided into the categories of primary (or essential) hypertension and secondary hypertension.

Hypertension can be caused by either an increase in cardiac output or an increase in total peripheral vascular resistance. In practice, cardiac output is generally relatively normal and the raised blood pressure is due almost entirely to an increase in vascular resistance. The arterioles are particularly affected, with veins, capillaries, and pulmonary vessels usually remaining normal.

Primary hypertension

About 80 per cent of all patients with high blood pressure fall into this group. Primary hypertension is further subdivided into a number of categories that are classified according to the severity of the condition as shown in Table 31.1. The most significant types are:

- borderline hypertension with diastolic blood pressures in the range 12–12.6 kPa (90–95 mmHg);
- mild hypertension with diastolic pressures in the range 12.6–13.3 kPa (95–100 mmHg);
- moderate hypertension with diastolic pressures in the range from 13.3–15.2 kPa (100–114 mmHg);
- severe hypertension where the diastolic pressure exceeds 15.2 kPa (114 mmHg).

Systolic pressures less than 18.6 kPa (140 mmHg) together with a normal diastolic pressure would be considered normal. If systolic pressure lies between 18.7 and 21.2 kPa (140 and 159 mmHg), the hypertension is borderline provided that diastolic pressure is normal. A systolic pressure in excess of 21.3 kPa (160 mmHg) is considered hypertensive.

Nevertheless, it is important to remember that one elevated blood pressure reading should not constitute a diagnosis of hypertension. Such a diagnosis should be based on repeated observations that give consistent readings. Indeed, the method of carrying out continuous 24-hour monitoring of blood pressure on an ambulatory patient is now considered to be of more value in determining the true extent of hypertension. Some people show an automatic elevation in blood pressure when they are confronted by a doctor ('white coat hypertension') while others do so in response to the blood pressure cuff itself ('cuff responders'). These patients are found to have normal blood pressure when it is monitored over a 24-hour period.

All cases of primary hypertension have a chronically elevated arterial blood pressure without evidence of other disease. The causes of this type of hypertension remain unclear. The patients show no evidence of an increase in renin levels and possess active baroreceptor reflexes, although the sensitivity of these reflexes is often reduced. While the cause or causes of primary hypertension are largely unknown, several risk factors have been implicated as contributing to its development. These include advancing age, family history, and obesity. Other factors may also contribute to the development of high blood pressure. These include high salt intake, excessive alcohol consumption, stress, and the use of certain types of oral contraceptives.

Secondary hypertension

About 20 per cent of patients suffering from high blood pressure are classified as having secondary hypertension, i.e. hypertension resulting from another condition. Such conditions include pregnancy, endocrine disorders, renal disease, vascular disorders, and certain types of brain lesions.

Pregnancy-induced hypertension

About 10 per cent of all pregnancies are accompanied by hypertension. It normally occurs after week 20 of gestation and is often associated with the appearance of protein in the urine and edema. Women showing all of these symptoms are said to be suffering from *pre-eclampsia*. Eclampsia is an exaggerated form of pre-eclampsia which has progressed to include convulsions and sometimes coma (see Chapter 21, Box 21.2). It is difficult to define the causes of hypertension in pregnancy. It has been suggested that elevated levels of estrogens, progesterone, prolactin, and ADH may result in an increased vascular responsiveness to angiotensin II.

Endocrine causes of hypertension

A variety of different hormonal disturbances can result in a raised blood pressure. The most common (although still comparatively rare) disorders are pheochromocytoma, Conn's syndrome and Cushing's syndrome.

A pheochromocytoma is a tumor of the adrenal medulla which, like normal chromaffin tissue, produces the catecholamines epinephrine and norepinephrine. Hypertension results from the large circulating quantities of these hormones. In many cases, release of the catecholamines occurs in spurts rather than continuously, giving rise to episodes of extreme hypertension, tachycardia, sweating and anxiety. A diagnosis is often made by monitoring the excretion of the catecholamine metabolite vanillylmandelic acid (VMA) in the urine over several days. Treatment is usually by surgical removal of the adrenal tumor.

Elevated levels of adrenocorticosteroid hormones can also produce secondary hypertension. Excess production of aldosterone by a tumor of the adrenal cortex (Conn's syndrome) leads to the excessive reabsorption of sodium in exchange for potassium in the distal tubules of the kidneys. Water and chloride ions accompany the reabsorbed sodium and the extracellular fluid volume is expanded. In turn, this leads to an expanded plasma volume and an increase in arterial blood pressure.

Excessive secretion of glucocorticoid hormones from the adrenal cortex (Cushing's syndrome) can also lead to hypertension, since high concentrations of these hormones have aldosterone-like effects and can promote the retention of salt and water.

TABLE 31.1 Blood pressure and hypertension

Diastolic pressure (kPa (mmHg))	Classification
<11.3 (<85)	Normal
<12 (<90)	High normal
12–12.6 (90–95)	Borderline hypertensive
12.6–13.3 (95–100)	Mild hypertensive
13.3–15.2 (100–114)	Moderate hypertensive
>15.2 (>114)	Severe hypertensive

Systolic pressure (kPa (mmHg))	Classification (assuming normal diastolic pressure)
<18.6 (<140)	Normal
18.6–21.2 (140–159)	Borderline hypertensive
>21.3 (>160)	Hypertensive

Renal causes of hypertension

The kidneys play an important role in the long-term regulation of arterial blood pressure by regulating the volume of the extracellular fluid (Chapter 28). Therefore it is not surprising that renal disease accounts for more cases of secondary hypertension than any other disorder. Any condition that restricts the blood supply to the kidneys can produce a rise in blood pressure. A reduction in renal blood flow activates the renin–angiotensin system leading to vasoconstriction and the retention of sodium and water. As a result, the extracellular volume is expanded and hypertension results.

Vascular causes of hypertension

Vascular pathology tends to produce or exacerbate high blood pressure. Arteriosclerosis can bring about narrowing of the vessels, which leads to an increase in total peripheral resistance and a rise in arterial pressure. Coarctation of the aorta (narrowing of the aorta as it leaves the heart) can also give rise to an increase in systolic blood pressure and blood flow, particularly to the upper part of the body. In polycythemia (a condition in which the red blood cell count is elevated), blood pressure may rise because of the increase in blood viscosity.

Neural causes of hypertension

While hypertension can be caused by some abnormality in the parts of the brain responsible for the control of blood pressure, most cases of neurogenic hypertension are relatively short-lived responses to a reduction in cerebral blood flow. A raised intracranial pressure will tend to compress the vessels supplying the brain and will result in a reduction in cerebral perfusion because the brain is located within the rigid confines of the skull with no room for expansion. In response, there is a rise in arterial pressure brought about by a reflex vasoconstriction in the peripheral vessels, which acts to restore cerebral blood flow (Cushing's reflex, see p. 306). However, if the blood supply to the brainstem becomes inadequate, vasoconstrictor tone will be lost and blood pressure will fall.

The effects of high blood pressure

The complications and mortality associated with both essential and secondary hypertension can be explained in terms of increased wear and tear on the heart and the blood vessels (Fig. 31.5).

Cardiac changes in hypertension

When the arterial blood pressure increases, the heart must work harder to eject the same stroke volume (see Chapter 15, p. 276). As a result of the Starling mechanism, the increase in left ventricular stroke work brings about a rise in left ventricular end-diastolic pressure. At the same time, the heart muscle hypertrophies to compensate for its increased work load. The effect of this is to shift the ventricular function curve upwards so that the ventricle is able to achieve a greater stroke work at the normal filling pressure. As a result of these changes, the increase in ventricular end-diastolic filling pressure which would arise from the operation of the Starling mechanism is minimized.

However, the hypertrophy of the cardiac muscle that occurs in response to the increased demands placed upon the heart in hypertension can only partially compensate for the increased cardiac work. Diffusion of oxygen from the capillaries to the enlarged cardiac muscle fibers becomes less efficient. Gradually, the left ventricular function curve becomes depressed and, with prolonged and severe hypertension, left ventricular failure will eventually occur.

Vascular effects of hypertension

Hypertension is often caused by an increase in total peripheral resistance. This increase in vascular resistance affects the vasculature of virtually every organ, with small arteries and arterioles the most affected. Narrowing of these vessels is seen in hypertensive patients. Initially this is caused by the myogenic response of the vascular smooth muscle to the increased degree of stretching. This reflex narrowing is responsive to vasodilator drugs. However, when hypertension has been present for some

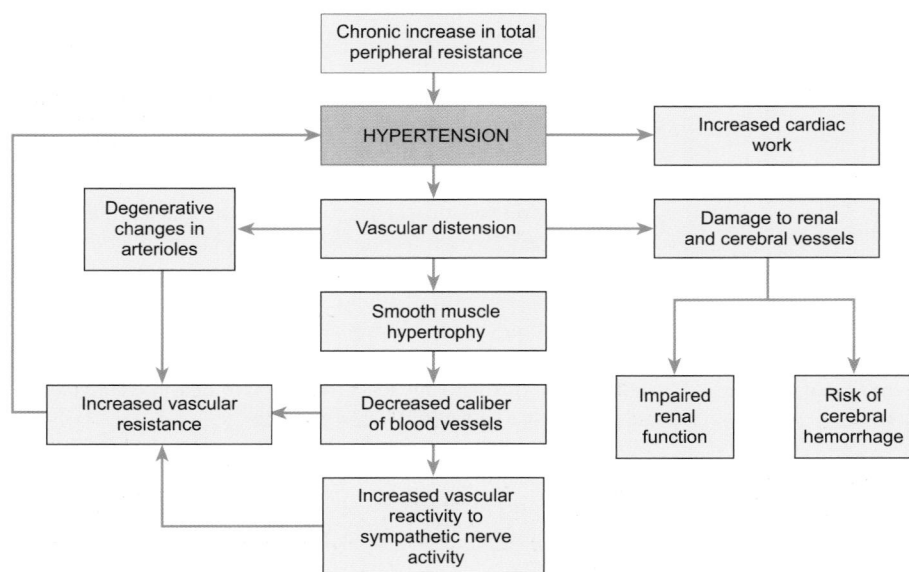

Fig. 31.5 The relationships between various cardiovascular variables during hypertension.

time, the smooth muscle of the tunica media of the vessel starts to hypertrophy in response to the chronically raised pressure. Hypertrophy of the walls of vessels causes their lumens to narrow and since this kind of narrowing is organic in nature, it cannot be fully reversed even during maximal vasodilatation. Furthermore, the vascular response of hypertensive patients to sympathetic stimulation is more pronounced than that of normal subjects. This is because the resistance of a vessel varies inversely as the fourth power of its radius (see Chapter 15, p. 284) so that a given level of sympathetic activity will have a relatively larger effect on the already constricted hypertensive vessel than on a normal one.

A second feature of the vasculature in severe chronic hypertension is *rarefaction*. This is a reduction in the number of vessels present per unit volume of tissue and has been observed in both the retina and intestine of hypertensives.

It might be expected that the baroreceptor reflex would act to restore blood pressure to normal in cases of hypertension. However, it should be remembered, that the baroreflex is a dynamic and rapidly adapting reflex (see Chapter 15, pp. 299–301). It continues to operate in patients with raised blood pressure, but appears to be reset to operate around a higher set-point. In severe cases, the sensitivity of the baroreflex may be reduced by stiffening of the arterial wall.

The treatment of hypertension

The aim of treatment in hypertension is to reduce arterial blood pressure to around 140/85 mmHg (18.7/11.3 kPa). If left untreated for any length of time, there is a high risk that the patient will suffer heart failure, renal failure or cerebral complications such as strokes. A variety of drug treatments are employed. Diuretics are given to reduce the extracellular fluid volume, while peripheral vasodilators are given to lower peripheral resistance. Drugs such as propranolol (a β-blocker) may be administered to reduce the cardiac output, while angiotensin and aldosterone levels may be lowered by drugs such as captopril, which block the angiotensin-converting enzyme (ACE inhibitors).

31.4 Heart sounds and murmurs

Normal heart sounds are described in Chapter 15 (p. 276). They are caused by the closure of the heart valves which sets up oscillations in the wall of the heart, the heart valves themselves and in the blood within the heart. The heart sounds can be heard clearly at the chest wall using a stethoscope. They can also be recorded with a microphone applied to the chest wall (a phonocardiogram). The first two heart sounds (S1 and S2) are caused by the closure of valves as the sudden tension in the valve cusps sets up vibrations in the heart wall and eddy currents in the blood. The first heart sound is caused by closure of the tricuspid and mitral valves at the start of ventricular systole, while the second sound occurs as the aortic and pulmonary valves close at the end of ventricular systole. The second sound is often split into aortic (A_2) and pulmonary (P_2) components during inspiration, the aortic component fractionally preceding the pulmonary component. This occurs normally in healthy young people. As the sounds made by each of the four heart valves radiate outwards in the direction of the blood flow, each is most clearly audible in a particular region of the chest (see Fig. 31.6 and accompanying legend).

Two further heart sounds may be present during a normal cardiac cycle, although they are much quieter and more difficult to hear than S1 and S2. The third sound (S3) occurs as blood rushes into the relaxed ventricles in early diastole. It is heard best at the apex and is often heard in children and young adults, and during pregnancy. The fourth sound (S4) is due to turbulence set up by atrial systole during the final stage of ventricular filling. It may be a common finding for an elderly individual, but in younger people it indicates an abnormal increase in the stiffness of the ventricular wall. When either S3 or S4 occurs it gives rise to a triple rhythm that can sound like the hooves of a galloping horse. For that reason they are called 'gallop' sounds—S3 is referred to as a ventricular gallop and S4 as an atrial gallop. When S3 is audible the rhythm set up is said to sound like KEN-TU-CKY (the sequence is S1—S2–S3) while S4 sounds like TE-NNE-SSEE (the sequence is S4–S1—S2).

Summary

1. Hypertension exists when the arterial blood pressure is persistently higher than normal. It is classified as either *primary hypertension*, in which there is a chronic elevation of blood pressure without evidence of other disease, or *secondary hypertension*, which is hypertension resulting from another disease process. Hypertension is highly undesirable as it increases the work of the heart and the stresses on the vasculature.

2. Hypertension leads to narrowing of the vessels because of the myogenic response of the vascular smooth muscle. This exacerbates the initial hypertension and leads to further hypertrophy of the smooth muscle in the walls of the resistance vessels. There is a further narrowing of the vessels and an additional rise in blood pressure. Breaking this vicious cycle is an important consideration in the treatment of hypertension.

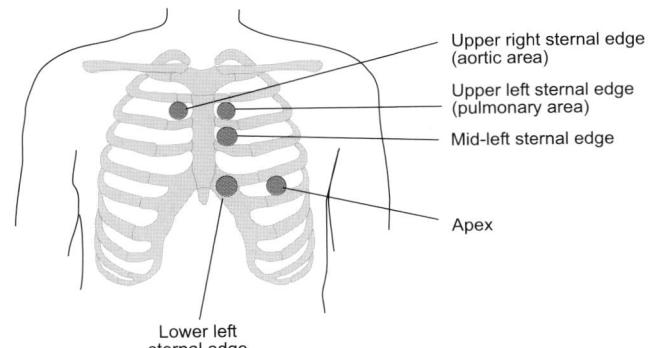

Fig. 31.6 A diagram of the chest to show the key auscultation points for listening to the heart sounds. The aortic sounds are best heard in the second intercostal space on the right side; the pulmonary sounds are best heard in the second intercostal space on the left side. Aortic and pulmonary regurgitation are best heard in the third intercostal space on the left side (mid-left sternal edge). Tricuspid regurgitation is best heard at the lower left sternal edge while mitral regurgitation, mitral valve prolapse, and mitral valve stenosis are best heard in the mid-clavicular line of the fifth intercostal space (apex).

Heart sounds occurring at other times are known as murmurs. Some of these are normal and are referred to as 'innocent' or 'functional' murmurs as they arise from physiological rather than pathological causes. Such additional heart sounds may be audible during pregnancy, during heavy exercise, and in young children. In such cases, the sounds are the result of an increased rate of blood flow through the normal heart and, as for S3 discussed earlier, they are not associated with any underlying cardiac pathology. Often, however, murmurs are the result of abnormalities of the valves, alterations in the diameter of the aortic or pulmonary arterial roots, or septal defects (perforations in the septum that separates the right and left sides of the heart). An ability to assess the significance of such sounds and to differentiate between benign and clinically significant murmurs is an important aspect of diagnostic cardiology. Murmurs sometimes give rise to vibrations that can be felt during palpation of the chest. These are known as precordial thrills. Careful palpation and auscultation of the chest is often sufficient to permit a diagnosis of the cause of murmurs but confirmation and assessment of the severity of the abnormality usually rely on echocardiography and other imaging methods.

Abnormalities of the valves fall into 2 categories:

- Valvular incompetence in which a valve fails to close completely and is therefore leaky. This allows regurgitation of blood through the valve. This regurgitation sets up eddy currents and turbulence in the blood.

- Valvular stenosis in which a valve is narrowed so that a high pressure gradient is needed to push blood through it. As a result, a high pressure stream of turbulent blood is created.

These abnormalities may be congenital or arise in later life as a result of disease and all four valves may exhibit either pathology. The most common problems, however, are acquired lesions of the valves of the left side of the heart which include aortic stenosis, mitral stenosis, aortic regurgitation, and mitral regurgitation. Tricuspid and pulmonary valve disease is relatively uncommon.

Murmurs are described in a variety of ways according either to the type of sound they create or to their timing within the cardiac cycle. For example the murmur may be of a blowing or swishing quality or it may be low-pitched and rumbling. It is generally true that volume of the sound reflects the severity of the underlying disease, and murmurs are often graded in terms of their volume.

Murmurs may be heard during either the systolic or diastolic phase of the cardiac cycle. Systolic murmurs are further subdivided into three categories:

- Ejection murmurs which are heard at the start of ventricular systole.

- Late systolic murmurs which are audible towards the end of systole.

- Pansystolic murmurs which are present throughout the whole of the systolic phase.

Diastolic murmurs are usually quieter than systolic murmurs but always indicate pathology. They are most often audible during either the early or middle part of diastole but may be heard throughout this phase of the cardiac cycle. Some examples of the most common pathologies that give rise to murmurs are

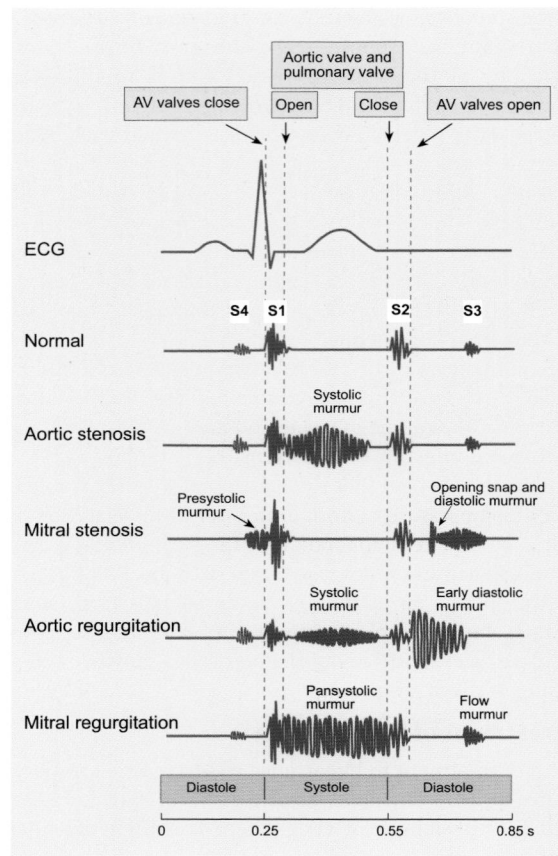

Fig. 31.7 Phonocardiograms for various conditions including the normal heart sounds.

described below and a diagrammatic representation of the phonocardiograms showing the magnitude and timing of these murmurs is illustrated in Fig. 31.7.

Aortic stenosis

This condition is most commonly caused by inflammatory damage (e.g. rheumatic heart disease), congenital malformation or degenerative thickening and calcification of the valve leaflets. In each case there is narrowing of the orifice of the valve and increased pressure in the left ventricle which creates a jet of turbulent blood as it enters the root of the aorta. The murmur associated with this condition is often very loud and, although it may persist throughout systole, it begins at the start of this phase (an ejection murmur) first becoming louder, then quieter (crescendo-decrescendo).

Mitral stenosis

Narrowing of the mitral valve is most often the result of thickening and scarring following inflammation. This impedes the flow of blood from the left atrium to the left ventricle thus creating turbulent blood flow that can produce low-frequency rumbling sounds in mid- to late-diastole as the valve cusps flutter in the turbulent stream. Murmurs of this kind can be mistaken for a third heart sound and are easy to miss. As the cusps of the mitral valve are more rigid than normal, they open rather suddenly. This gives rise to an opening snap. There is also a presystolic murmur produced by the turbulent flow of blood into the left ventricle following atrial contraction.

Aortic valve incompetence

Aortic valve incompetence ('leakiness') causes the ejected blood to flow back into the ventricle (regurgitation). It may be caused by many conditions including hypertension, inflammation of the valve, atherosclerosis, and connective tissue disorders such as Marfan syndrome. Aortic regurgitation causes a 'blowing' murmur during diastole as blood flows backwards from the aorta into the left ventricle creating turbulence. The murmur is usually loudest in early diastole and then fades towards the end of the diastolic phase. As the stroke volume is increased there may also be a softer systolic murmur. (The stroke volume increases because blood enters the left ventricle both from the pulmonary circulation via the left atrium and from the aorta by regurgitation through the aortic valve.)

Mitral valve incompetence

Incomplete closure of the mitral valve causes a typical pansystolic murmur as regurgitation occurs into the left atrium throughout ventricular systole. This murmur is heard most loudly over the mitral area and obscures both S1 and S2. Incompetence of the tricuspid valve produces a similar murmur. The regurgitated volume plus the normal venous return from the pulmonary circulation passing through the mitral valve ori-

fice can increase blood flow sufficiently to cause turbulence during the rapid filling phase. This can give rise to a flow murmur that occurs around the same time as S3. Prolapse of the mitral valve (MVP or floppy valve syndrome) may give rise to a murmur that is heard towards the end of systole (a late systolic murmur). MVP is a very common cardiac condition which is often asymptomatic and harmless.

A variety of other pathologies can give rise to abnormal heart sounds but a detailed account of these is beyond the scope of this book. Further information can be found in specialist texts of cardiology.

31.5 Clinical aspects of electrocardiography

In this section, some common abnormalities in the ECG and their underlying causes are discussed. The ECG provides a non-invasive technique for monitoring the spread of excitation throughout the heart. Normally, a 12 lead ECG record is used with three limb leads (leads I, II and III), three augmented limb leads (aVL, aVR, and aVF) and six chest leads (V_1–V_6), see Chapter 15 pp. 267–273. The three limb leads and three augmented limb leads each provide information about the activity of the heart viewed from a specific point on the frontal (vertical) plane while each of the six chest leads provide information about the activity of the heart from a specific point in the transverse plane as shown in Fig. 31.8. In interpreting the ECG it is essential to keep in mind that the *polarity* of the signal in any lead at a given time *reflects the average direction of current flow with respect to that lead*, while the *amplitude* of the signal *reflects the mass of cardiac muscle involved*.

The ECG can be used to gain information about the following aspects of cardiac function:

- The electrical rhythm of the heart.
- The size of the muscle mass of the individual chambers of the heart.
- Disorders in the origin or in the conduction of the wave of excitation throughout the heart.
- The site of any abnormal pacemaker activity.
- Abnormal cardiac excitability caused by altered plasma electrolytes.
- The metabolic state and viability of the myocardium.

Before discussing the origins of abnormal patterns of ECG activity, it is helpful to keep in mind the pattern of excitation of the normal heart. Briefly, as discussed on pp. 266–267, the heart beat is initiated by the pacemaker activity of the sinoatrial or SA node, the excitation spreads across the atria from right to left. This activity is recorded as the P wave of the ECG. It then passes to the ventricles via the atrioventricular or AV node where conduction is delayed by around 0.1s to permit the atria to complete their contraction before the ventricles begin theirs. From the AV node, the excitation spreads via the bundle of His and its left and right branches to excite the ventricles. The spread of excitation through the ventricles is seen in the ECG as the QRS complex while the repolarization of the ventricles is recorded as the T wave. The standard intervals for reporting the ECG and their normal values are given in Table 31.2.

Summary

1. The first two normal heart sounds (S1 and S2) arise from vibrations set up by closure of the valves. The vibrations occur in the valve cusps themselves, in the heart wall and as eddy currents in the blood. The third and fourth heart sounds are much softer and arise from rapid blood flow into the ventricles in diastole (S3) and from rapid blood flow into the ventricles during atrial systole (S4).

2. In various heart diseases, the flow of blood may become very turbulent and the resulting vibrations give rise to additional heart sounds that may mask the normal heart sounds. These are known as murmurs and commonly arise as a result of defects in the heart valves.

3. A valve may fail to close completely and this allows regurgitation of blood. This is known as incompetence or insufficiency. This regurgitation sets up eddy currents and turbulence in the blood. A valve may be narrowed (stenosis) so that an abnormally high pressure is needed to push blood through it. As a result, a high pressure jet of turbulent blood is created.

4. The most common problems are acquired lesions of the valves of the left side of the heart which include aortic stenosis, mitral stenosis, aortic regurgitation, and mitral regurgitation. In aortic stenosis there is a prominent systolic murmur. In contrast, in mitral stenosis, there is a diastolic murmur. In aortic regurgitation the main feature is an early diastolic murmur, while in mitral regurgitation, there is a pansystolic murmur.

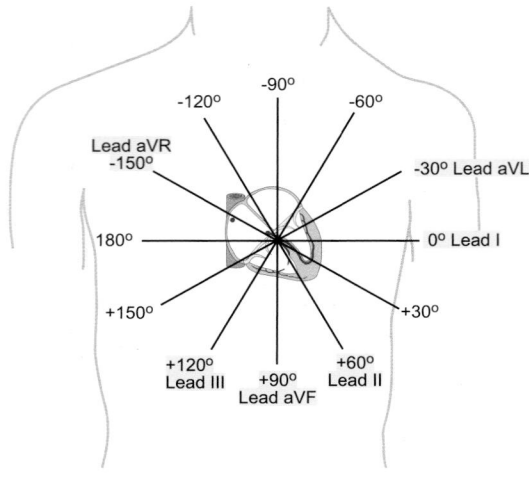

a) The limb leads record the ECG from the frontal plane.

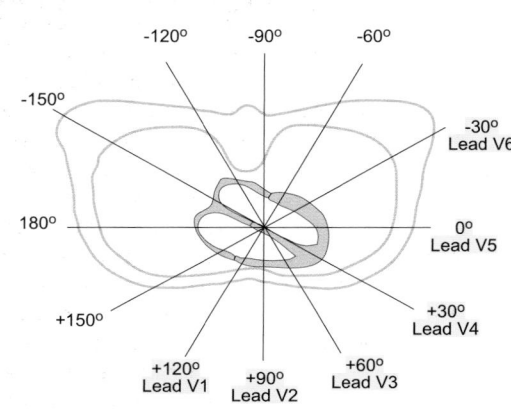

b) The chest leads record the ECG in the transverse plane.

Fig. 31.8 Each ECG lead records the activity of the heart from a different perspective and the complete array provides detailed information about the origin and spread of excitation within the heart. (a) The angles formed by the limb leads and augmented limb leads gives rise to a hexaxial reference system for the frontal plane. (b) The chest leads record the electrical activity of the heart in the transverse plane.

Measurement	Duration	Comments
TABLE 31.2 Standard ECG intervals		
P wave	<0.12 s	Normally smooth and rounded. Entirely positive or entirely negative in all leads except V1.
PR interval	0.12–0.21s	Interval is shorter in children and in adolescents. Varies with heart rate.
QRS width	0.07–0.11s	Tends to be slightly shorter in females. Amplitude and polarity of R wave in different leads used to determine main cardiac axis.
QT interval	0.3–0.4s (QT$_c$ 0.35–0.43s)	Varies with heart rate and often expressed as the QT$_c$ interval which is QT divided by the square root of the R-R interval.
R-R interval	c.0.75–0.85s	The heart rate can be quickly calculated as follows: b.p.m. = 60 ÷ R-R interval

attacks, hypothermia, and hypothyroidism. A rapid heart beat (>100 b.p.m.) is known as *sinus tachycardia* and is associated with exercise, stress, hemorrhage, and hyperthyroidism. The definitions of the terms bradycardia and tachycardia are given on p. 279 (Chapter 15).

If the rhythmical activity of the SA node activity is much slower than normal, the fundamental rhythm may be taken over by another part of the heart. These rhythms are known as *escape rhythms* and are named after their site of origin. Those that originate in the atria are called atrial rhythms, those that originate in or close to the AV node are junctional rhythms and those that originate in the ventricles are ventricular rhythms.

Sick sinus syndrome

In some people, particularly the elderly, the SA node fails to excite the atria in a regular manner, resulting in a slow resting heart rate that does not increase appropriately with exercise. This is called *sick sinus syndrome* and has many causes. It may be drug induced, it may reflect abnormally intense vagal activity, but a non-specific, scar-like degeneration of the pacemaker region following ischemia of the muscle supplied by the sinus-nodal artery accounts for around one third of all cases. Although the most common manifestation of sick sinus syndrome is brady-cardia, it may also appear as alternating fast and slow rhythms or as a tachycardia that originates in the atria. Arrhythmias caused by failure of the SA node to excite the heart can often be treated by the implantation of an artifical pacemaker.

Ventricular hypertrophy and the electrical axis of the heart

In humans, the heart lies mainly in the lower part of the left side of the chest with the right ventricle in front of the left ventricle (see Fig. 15.3). The ventricles normally point obliquely down-wards away from the atria which lie close to the mid-line. During the cardiac cycle the pattern of current flow resulting from depolarization and repolarization of the heart muscle con-tinually changes, and, as mentioned above, each ECG lead

Each action potential originating in the SA node initiates one beat of the heart. This is known as *sinus rhythm* and is the normal state of affairs. Any deviation from the normal sinus rhythm is known as an *arrhythmia*. While some arrhythmias are of no clin-ical significance, others may reflect serious and life threatening disorders of the myocardium. Indeed, many arrhythmias arise as a consequence of ischemic heart disease.

The heart rate commonly varies with the respiratory cycle in healthy young people (sinus arrhythmia, see p. 302). In highly trained athletes the heart rate is usually slower than normal (though stroke volume is higher). This is known as *sinus brady-cardia* (< 60 b.p.m.). Sinus bradycardia is also seen in fainting

detects the average current flowing towards or away from it at any instant of time (for more detail see Box 15.1 pp. 272–273). The electrical axis of the heart (the principal cardiac axis) refers to the direction of the largest electrical dipole recorded from the frontal plane and can be assessed by comparing the amplitude of the R wave in the three limb leads or by calculating the resultant vector of the activity recorded by lead I and lead aVF. These two leads record the electrical activity in the horizontal and vertical planes respectively—see Box 31.2. The cardiac axis depends on body size and shape but normally lies between –30° and +100° to the horizontal. In right axis deviation the cardiac axis lies between +100° and +150°, while in left axis deviation the cardiac axis lies between –30° and –90° (Fig. 31.9). The determination of the cardiac axis is helpful in the diagnosis of a number of conditions including right ventricular hypertrophy, conduction defects, and pulmonary embolus.

Atrial enlargement is associated with changes to the P wave. In right atrial enlargement (which may be the result of pulmonary hypertension or of stenosis of the tricuspid valve) the P wave is larger than normal, because the right atrium has an

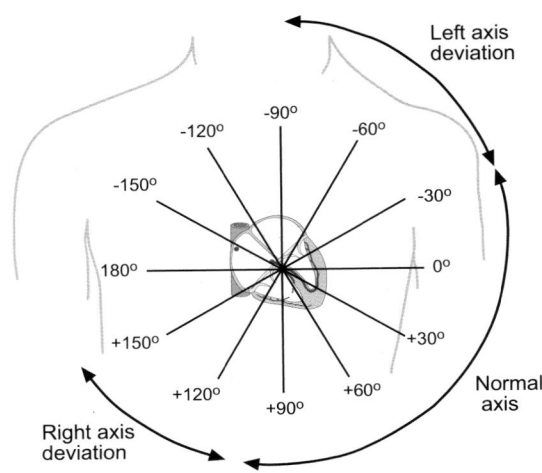

Fig. 31.9 The mean QRS axis of the heart. Note the wide range for normal hearts. The region between −90° and +150° is familiarly known as North West territory or no man's land. The cardiac axis may lie in this region in emphysema, hyperkalemia, and during ventricular tachycardia.

BOX 31.2 DETERMINATION OF THE CARDIAC AXIS

The direction of the largest electrical dipole recorded by the ECG during the cardiac cycle is called the mean cardiac axis. Together with the cardiac rhythm, conduction intervals, and the description of the QRS complex, ST segment, and T waves, it is one of the reported ECG variables. As mentioned in the main text, the normal range for the cardiac axis is wide (–30° to +100°). Right axis deviation is normal in children and tall thin adults but it occurs in various pathologies: it occurs in right ventricular hypertrophy (which may be caused by pulmonary hypertension or pulmonary stenosis); it also occurs in pulmonary embolism and in left posterior hemiblock. Left axis deviation occurs in left anterior hemiblock but *not* in left ventricular hypertrophy. (Left ventricular hypertrophy is indicated when at least one of the R waves seen in the left chest leads (V4-V6) is abnormally large (>2.7 mV) and one of the S waves seen in the right chest leads exceeds 3mV.) A quick assessment of the cardiac axis can be made by examining the polarity of the R wave in the limb leads as shown in Fig. 1.

A more accurate assessment is made by vector analysis of the R wave. Traditionally this has been based on Einthoven's triangle. To compute the cardiac axis, the three limb leads are represented by the three sides of an equilateral triangle as shown in Fig. 2(a). Each side is then bisected by a perpendicular line and the amplitude of the R wave represented as an arrow on the sides of the triangle for each lead. Positive potentials are plotted from the zero point (where the perpendicular line intersects the side of the triangle) with the head of the arrow towards the positive terminal of the lead and negative values are plotted in the other direction. (NB the same scale must be used for each lead.) Lines parallel to the original perpendiculars are then drawn from the head of each arrow. An arrow drawn from the centre of the triangle to the point where these lines intersect gives the angle of the cardiac axis. A slightly simpler alternative is to plot the amplitude of the R wave in lead I as a horizontal vector and the amplitude of the R wave in lead aVF as a vertical

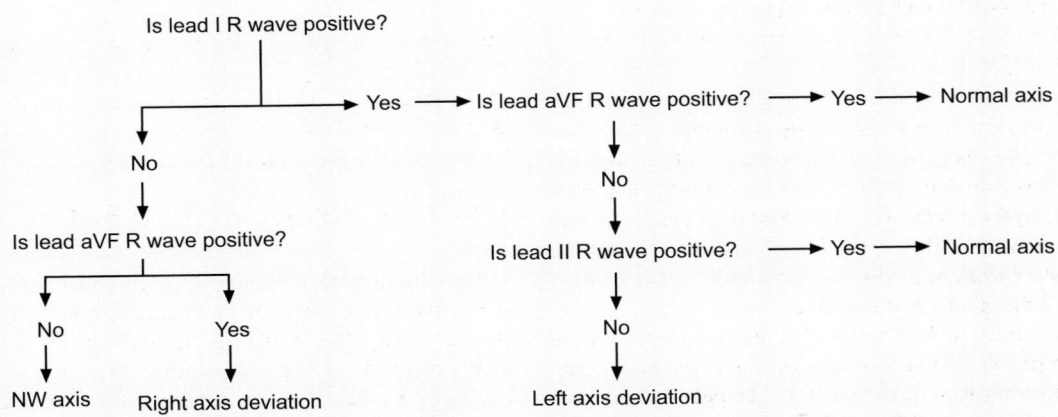

Fig. 1 A simple flow diagram to show how the cardiac axis can be quickly estimated by inspection of the polarity of the R wave recorded in leads I, aVF, and II.

BOX 31.2 DETERMINATION OF THE CARDIAC AXIS
Continued

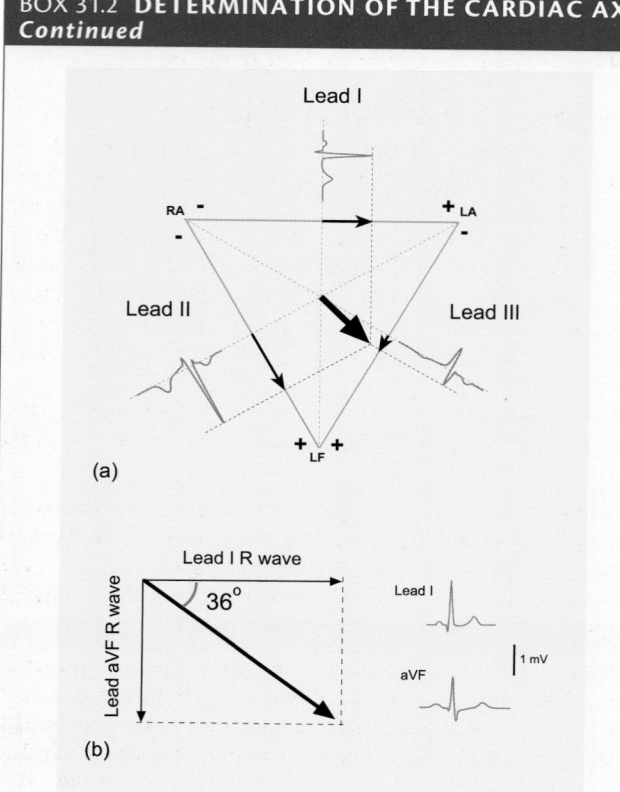

(a)

(b)

Fig. 2 The electrical axis of the heart may be precisely determined from vector diagrams of the amplitude of the R wave recorded in the standard limb leads (a) and from its amplitude recorded lead I and lead aVF (b). (A vector is a variable that has both magnitude and direction. Vectors are represented graphically by lines in which the magnitude of the variable is indicated by the length of the line and its direction is indicated by the angle plotted on suitable coordinates. They can be added by means of vector diagrams as represented here by the Einthoven triangle (a) or rectangle (b)).

vector as shown in Fig. 2(b). The diagonal of the completed rectangle gives the main cardiac axis. This method exploits the fact that lead I records the ECG in the horizontal plane while lead aVF records the ECG in the vertical plane (see Fig. 31.8 of the main text).

increased muscle mass (atrial hypertrophy). This alters the appearance of the P wave which has a peaked appearance. This arises because the right atrium depolarizes before the left atrium. In left atrial enlargement (usually caused by mitral stenosis) the P wave is broadened and may have a double peak (a bifid P wave). The appearance of the P wave in left atrial enlargement can be attributed to the later depolarization of the larger muscle mass of the left atrium.

Defects of conduction

Defects in the conduction of the electrical activity can, in principle, occur at any point in the pathway between the SA node and the ventricular muscle. These problems are known as *heart*

block. Problems with conduction from the atria to the ventricles are known as atrio-ventricular block and are classified according to their severity as first, second, or third degree heart block. Impairment of conduction through the bundle of His is known as *bundle block* and may affect the main bundle or any of its branches (branch bundle block).

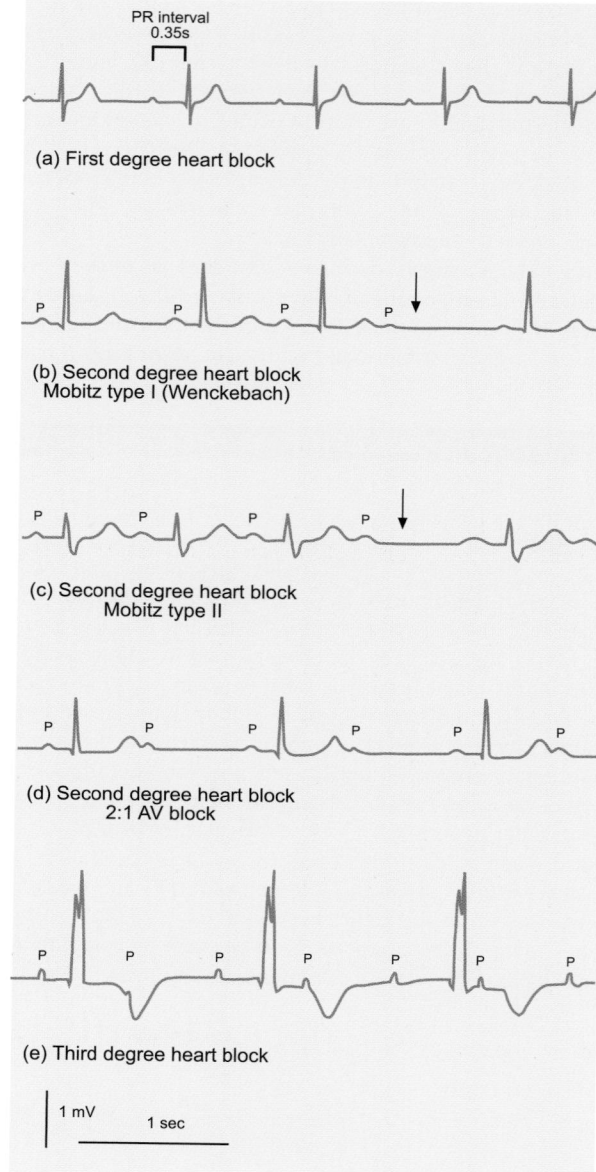

(a) First degree heart block

(b) Second degree heart block
Mobitz type I (Wenckebach)

(c) Second degree heart block
Mobitz type II

(d) Second degree heart block
2:1 AV block

(e) Third degree heart block

Fig. 31.10 The ECG seen in various types of heart block. (a) First degree heart block; note the long PR interval. (b) Second degree heart block showing the Mobitz type I or Wenckebach pattern in which the PR interval progressively lengthens until one beat is missed (arrow). The cycle then repeats itself. (c) Second degree heart block of the Mobitz type II pattern. Here the PR interval is constant but occasionally a beat is missed (arrow). This pattern reflects a failure of the conducting system. (d) Second degree heart block in which every second P wave elicits an R wave—this is 2:1 block. Other regular patterns may also be seen, such as 3:1 or 4:1 block. (e) Third degree heart block—here there is complete dissociation of the QRS complex from the P wave. Moreover, the QRS complex is broader than normal and has an abnormal appearance.

In *first degree heart block* each wave of depolarization originating in the SA node is conducted to the ventricles but there is an abnormally long PR interval (>0.2s), indicating that somewhere in the pathway a delay has occurred (Fig. 31.10(a)). This is usually at the AV node. First degree heart block is not usually a problem but may be a sign of some other disease process (e.g. coronary artery disease or electrolyte disturbance).

In *second degree heart block* there is an intermittent failure of excitation to pass to the ventricles. There are three main types of second degree heart block:

- Mobitz type I, in which the PR interval becomes progressively longer until one P wave fails to elicit a QRS complex (and thus a ventricular contraction) as shown in Fig. 31.10(b). The pattern is then repeated. This type of block is also known as the Wenckebach phenomenon or Wenckebach block.

- Mobitz type II, in which most P waves elicit a ventricular contraction but occasionally a P wave fails to be conducted to the ventricles so it is not followed by a QRS complex (Fig. 31.10(c)). In this type of block there is no progressive prolongation of the PR interval.

- 2:1 or 3:1 block occurs when every second or third P wave elicits a QRS complex (Fig. 31.10(d)).

In *third degree heart block* the wave of excitation originating in the SA node fails to excite the ventricles (atrioventricular block). The ventricles then show a slow intrinsic rhythm of their own (sometimes called an escape rhythm) with abnormally shaped QRS complexes that are not associated with P waves, see Fig. 31.10(e). While the shape of the abnormal QRS complexes will depend on the site at which the pacemaker activity originates, the QRS complexes will be broader than normal unless a nodal rhythm is established. The broadening of the QRS complex arises because the conduction of the wave of excitation through the cardiac muscle is significantly slower than that through the Purkinje fibers.

In some situations, the excitation reaches the AV node and passes through the bundle of His only to be delayed or fail to pass from one or other branch to the ventricular muscle. This is known as *bundle branch block*. The QRS complex is broadened (>0.12s) as the wave of excitation will spread more slowly across the ventricles than it normally would via the Purkinje fiber system (Fig. 31.11). On the left side, the bundle of His branches to form an anterior fascicle and a posterior fascicle. Block of the

conduction through the left anterior bundle will cause the main cardiac axis to deviate towards the left as the wave of excitation will spread from the septum and base of the left ventricle.

Arrhythmias of atrial or ventricular origin

As mentioned on p. 594, abnormal rhythms may begin in the atria—away from the SA node, they may begin at the AV node (nodal or junctional rhythms) or they may begin in the ventricles. Arrhythmias originating in the atria or nodal region are often called *supraventricular rhythms*. When the cycle of excitation begins at a site remote from the SA node, the resulting rhythm is slower than normal (bradycardia). Sometimes the atria or ventricles contract earlier than expected. This is known as an *extrasystole* (an ectopic beat).

For both escape rhythms and extrasystoles, the appearance of the ECG trace will depend on the origin of the abnormal excitation (Fig. 31.12):

- If the excitation has its origin in the atria, the P wave will be abnormal (as the excitation begins away from the SA node) but the QRS complex will be normal in appearance as the ventricles will be excited via the bundle of His and the Purkinje fibers in the normal way.

- If the excitation begins in or near the AV node (a junctional rhythm), there may be no P wave as the atria may not be excited or their depolarization may be masked by the QRS complex, which will be normal in appearance. Alternatively, a P wave may be seen in the ECG trace but it occurs later than normal either just before or just after the QRS complex.

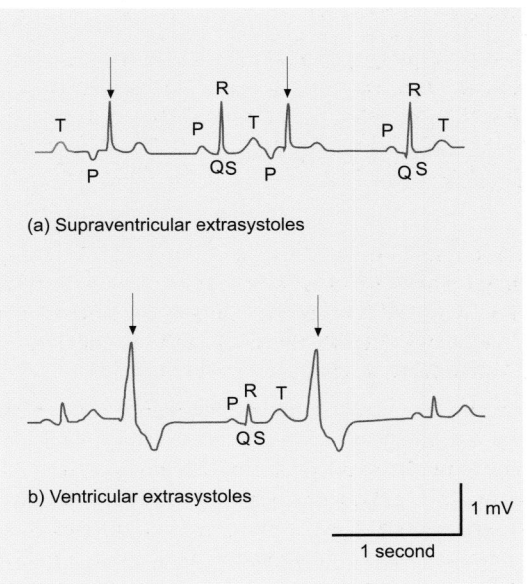

(a) Supraventricular extrasystoles

b) Ventricular extrasystoles

1 mV

1 second

Fig. 31.12 Examples of atrial and ventricular extrasystoles recorded via lead II. (a) Two supraventricular extrasystoles (arrows) with preceding inverted P waves. The normal appearance of the QRS complex should be compared with the normal pattern of excitation seen on the right of this trace. The inverted P wave and short PR interval preceding the first and third R waves indicates that the ectopic focus is located near the base of one of the atria or in the nodal tissue. (b) Two ventricular extrasystoles (arrows). Note the broad QRS complexes and their abnormal appearance compared to the normal ECG pattern seen in the middle of this trace.

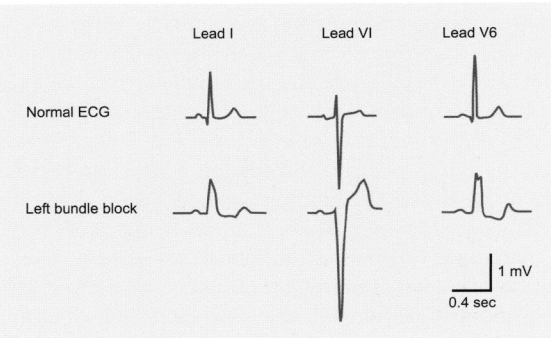

Fig. 31.11 The appearance of the ECG in left bundle block. Note the wide QRS complex and the notched peak in lead V6.

- If the excitation begins somewhere in the ventricles, the QRS complex will be abnormally wide and distorted in appearance as the wave of excitation will be conducted from the point of excitation via the cardiac muscle fibers, rather than by the conducting system of the heart. There will be no consistent relation between the P waves and the QRS complex. The appearance of the ECG trace in this situation is similar to that seen for a ventricular escape rhythm as seen in third degree heart block.

The normal pattern of excitation in the heart relies on the refractory period of the cardiac cells to ensure that the conduction proceeds in an orderly way: from the SA node to the atria, then to the AV node and bundle of His. Finally, the excitation reaches the Purkinje cells which excite the ventricular myocytes. Retrograde excitation is normally prevented by the long refractory period of the cardiac cells. In some circumstances this inherent regulation breaks down and premature excitation can occur via *re-entry circuits*. These circuits give rise to premature beats and tachycardias, which may be so rapid that they compromise the effective filling of the heart. The site of the abnormal excitability can be deduced using the analysis discussed above.

- When the atria depolarize more frequently than about 220 times a minute, *atrial flutter* is present. The ECG baseline is not flat but often shows a continuous regular sawtooth pattern (Fig. 31.13b). As the AV node cannot be activated more than about 200 times a minute, the ventricles are activated at a lower rate. In some people the ventricles are activated in a completely regular manner with, for example, a 2:1 or 4:1 ratio of P waves to QRS complexes. In others, the ventricles may be activated in a very irregular manner.

- If the atria are activated more than about 350 times a minute, they do not contract in a coordinated way. The individual muscle fiber bundles contract asynchronously. The ECG trace shows no P waves, only an irregular baseline (Fig. 31.13c) and the ventricles show an irregular rhythm. This is known as *atrial fibrillation*.

- If the site of abnormal activity is in the wall of one of the ventricles, the rhythm is known as *ventricular tachycardia* and, for the reasons discussed above, the QRS complex will be wide and abnormal in appearance as seen in Fig. 31.13d.

- When the ventricular muscle fibers fail to contract in a concerted way, the heart is in a state of *ventricular fibrillation*, and is unable to pump the blood around the body.

As the quantity of blood forced from the atria into the ventricles contributes only about 20% of the end diastolic volume, neither atrial flutter nor atrial fibrillation is immediately life-threatening. In contrast, ventricular fibrillation requires urgent action to restore normal function by means of a defibrillator, which uses a strong electrical shock to terminate the fibrillation and reset a normal heart beat.

Wolf-Parkinson-White syndrome

An example of a re-entry circuit is provided by the Wolf-Parkinson-White syndrome. Affected individuals have an access-

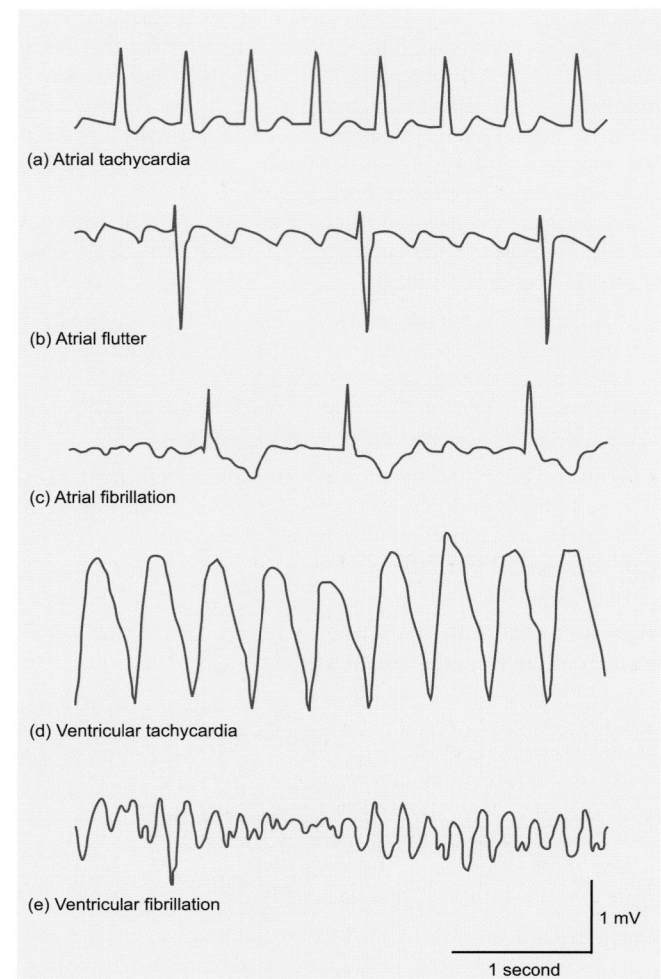

Fig. 31.13 Abnormal cardiac rhythms. (a) Atrial tachycardia. (b) Atrial flutter—note the sawtooth pattern of the P waves and the broad QRS complexes. The ventricular rate is 45 b.p.m. (c) Atrial fibrillation—note the irregular baseline and the highly abnormal QRST pattern. (d) Ventricular tachycardia—a regular but highly abnormal pattern of excitation is seen as the point of excitation lies in the ventricles themselves. The ventricular rate is 136 b.p.m. (e) Ventricular fibrillation—note the highly irregular pattern with no identifiable waves present.

ory pathway connecting atrium and ventricle, usually on the left side of the heart. This is known as the bundle of Kent. Unlike the excitation via the AV node and the bundle of His, the excitation via the bundle of Kent is not delayed. Consequently, the left ventricle is excited earlier than in normal individuals. As a result, the PR interval is reduced and there is a small upstroke on the rising phase of the QRS complex called a *delta wave* reflecting this early excitation (Fig. 31.14). The vast majority of people with this anatomical feature have no symptoms of heart disease but, in a few cases, an abnormal circuit of excitation is set up in which impulses travel from the ventricle to the atrium via the bundle of Kent. This results in the early excitation of the AV node and a sustained ventricular tachycardia.

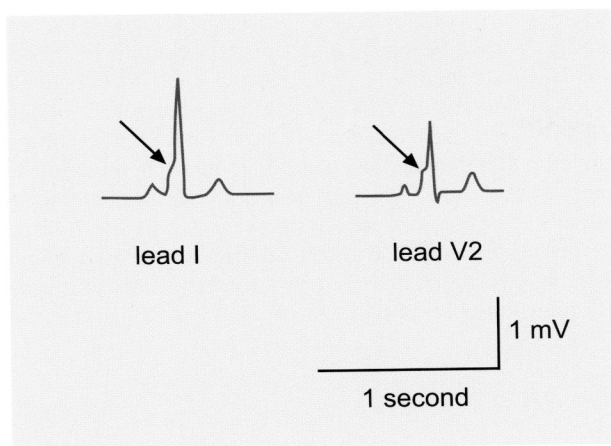

Fig. 31.14 Delta waves (arrows) recorded from an individual with Wolf-Parkinson-White syndrome.

Electrolyte imbalance and the ECG

As for other cells, the resting potential of cardiac myocytes depends on the distribution of potassium ions across the plasma membrane. Consequently, changes to plasma potassium concentrations will alter the excitability of the cardiac myocytes and these changes are reflected in characteristic changes in the ECG. If plasma potassium levels are low (*hypokalemia*), the T wave is flattened and a U wave may appear. If plasma potassium is elevated (*hyperkalemia*), the T wave has a greater amplitude and width than normal, with a peaked appearance (Fig. 31.15).

The plateau phase of the cardiac action potential is maintained by an influx of calcium through voltage-gated calcium

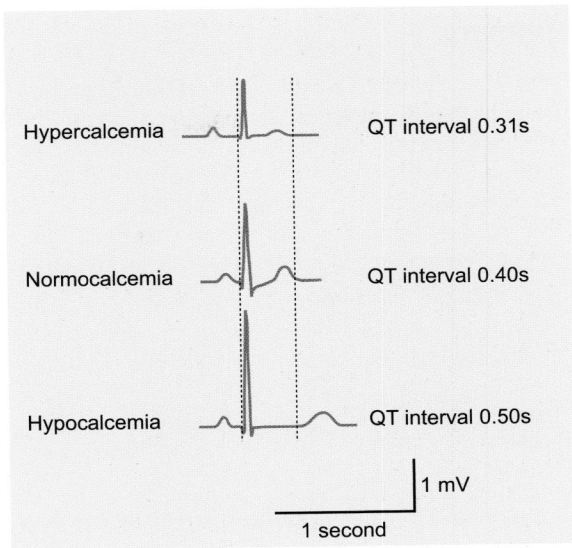

Fig. 31.16 The changes in the ECG seen in hypercalcemia and hypocalcemia compared to the ECG in normal plasma calcium. Note the changed QT intervals.

channels. This calcium entry both triggers contraction of the cardiac myocytes and determines the rate at which they repolarize. If plasma calcium levels are lower than normal, the cardiac action potential is prolonged and this is reflected in the ECG trace as a long QT interval. If plasma calcium levels are higher than normal, the QT interval is shorter than normal (Fig. 31.16) and the heart is prone to arrhythmias.

ECG changes in ischemia and infarction

Cardiac ischemia refers to a situation in which the blood supply is inadequate to meet the metabolic requirements of the myocardium. If this situation is prolonged, the affected tissue becomes damaged and may die. This will result in necrosis of

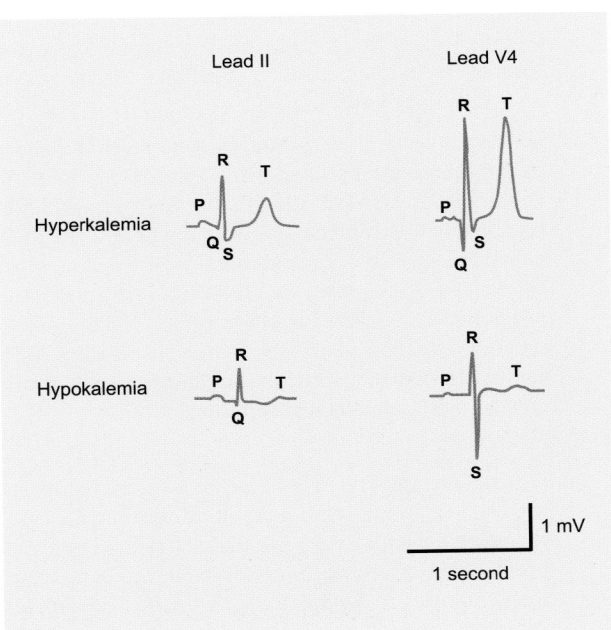

Fig. 31.15 The appearance of the ECG in hyperkalemia and hypokalemia. Note the peaked T wave in hyperkalemia and the low amplitude T wave in hypokalemia.

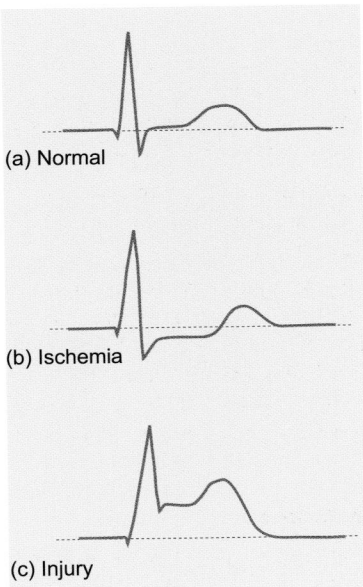

Fig. 31.17 Characteristic changes in the QRST waves seen in ischemia (depressed ST segment) and infarct (elevated ST segment).

Summary

1. The ECG provides a non-invasive tool for examining the state of the myocardium. The polarity and amplitude of the components of the ECG are determined by the pattern of excitation and size of muscle mass, respectively. Consequently, careful analysis of the ECG as it appears in the standard leads allows the site and nature of abnormal activity within the myocardium to be determined.

2. The ECG provides information regarding the electrical rhythm of the heart, the size of the muscle mass of the individual chambers of the heart and the conduction of the wave of excitation throughout the heart. It can also provide information regarding the site of any abnormal pacemaker activity, plasma electrolyte imbalance, and the site of any damage to the myocardium.

3. The mean QRS axis of the heart normally lies between −30° and +100° to the horizontal. In right axis deviation the cardiac axis lies between +100° and +150°, while in left axis deviation the cardiac axis lies between −30° and −90°.

4. In first degree heart block, the PR interval is prolonged. The pattern in second degree heart block depends on the type of block: in the Mobitz type I (or Wenckebach) block, the PR interval progressively lengthens until a beat is missed. The cycle then repeats itself. In the Mobitz type II block, the PR interval is constant but a beat is occasionally missed. Second degree heart block may also be manifested as 2:1 or 3:1 block in which every second or third P wave elicits an R wave. In third degree heart block there is complete dissociation of the QRS complex from the P wave.

5. In atrial flutter the P waves have a very high frequency (>220 b.p.m.) with a sawtooth pattern. In atrial fibrillation the baseline is irregular and the QRST is highly abnormal in appearance as the ventricles are not activated via the normal conducting system. In ventricular tachycardia a regular but highly abnormal pattern of excitation is seen as the point of excitation lies in the ventricles themselves. In ventricular fibrillation the ECG shows a highly irregular pattern with no identifiable waves present.

6. In electrolyte imbalance, the ECG shows characteristic changes: In hyperkalemia, the T wave is accentuated; in hypokalemia it is flattened. In hypercalcemia the QT interval is shortened and in hypocalcemia it is prolonged.

7. The ST segment lies below baseline in ischemia. Immediately following an infarct, the ST segment is elev-

the affected area (a *cardiac infarct*). Since the spread of excitation throughout the myocardium will be affected in both these conditions, the ECG will show characteristic changes. When ischemia affects the inner part of the wall of the left ventricle, the affected myocytes may be unable to maintain the prolonged action potential characteristic of normal cells, they begin to repolarize early, and the T wave is inverted in many of the leads. More generally, during ischemia the ST segment recorded in long axis leads such as lead II is negative (i.e. it lies below the isoelectric line, Fig. 31.17b). If the ischemia progresses to cause a cardiac infarct, the ECG shows characteristic changes immediately after the damage has been sustained. The most obvious change is that the ST segment does not return to baseline as it normally does, but is elevated as shown in Fig. 31.17c. A detailed interpretation of the ECG records during cardiac ischemia and following myocardial infarction is beyond the scope of this book but can be found in specialist texts of electrocardiography.

31.6 Heart failure

Heart failure occurs when the heart is unable to pump sufficient blood at normal filling pressures to meet the metabolic demands of the body. Failure may involve either of the two ventricles alone or both the left and right ventricles simultaneously. People with heart failure are unable to exercise normally and complain of excessive fatigue because their cardiac output does not increase in proportion to the work performed as it would in a normal healthy person.

The causes of heart failure fall into two main categories:

1. **Cardiac**

   (a) *Impairment of myocardial contractility*. Loss of healthy muscle mass can occur if the myocardium experiences hypoxia or ischaemia because of the loss of the normal blood supply (e.g. following a coronary thrombosis). The loss of healthy muscle mass is known as *myocardial infarction*. Loss of myocardium resulting from other forms of cardiac myopathy also results in a decreased contractility. Inflammation of the heart muscle is known as *myocarditis* and this too impairs the contractile performance of the heart. In all such situations, cardiac performance falls in proportion to the extent of the pathology.

   (b) *Impairment of the filling or emptying of the heart*. Impairment of the filling or emptying of the heart is seen in various conditions. In atrial fibrillation, the ventricles will not be completely filled and cardiac output will fall. The narrow valve opening in mitral stenosis will prevent normal filling of the left ventricle. In pericarditis and cardiac tamponade the stiffening of the pericardial sac prevents normal expansion of the ventricles during filling. Emptying of the ventricles is compromised in aortic stenosis and in mitral regurgitation.

2. **Extracardiac**

   (a) *Increased preload*. The preload is the tension that exists in the walls of the heart as a result of diastolic filling. It is therefore determined by the end-diastolic pressure. The preload may become excessively elevated because of renal failure. (In which there is an increase in blood volume as a result of sodium and water retention.) It may also increase due to valvular regurgitation or myocardial infarction as discussed above. The heart performs more work when it pumps a given volume of blood from distended ventricles (see Chapter 15 p. 276). As a result, a situation can develop where the coronary arteries are unable to supply sufficient oxygen to meet the requirements of the heart (myocardial ischemia) and the initial heart failure is exacerbated.

(b) *Increased afterload*. The *afterload* is the pressure in the aorta during the period that the aortic valve is open. It is the load that the heart must overcome in order to pump the blood from the left ventricle into the aorta. Whenever the afterload is increased, the work of the heart is also increased for any given stroke volume. As with the effects of an increased preload, a situation can arise in which the coronary arteries are unable to supply sufficient oxygen to meet the requirements of the heart. The afterload is elevated in patients with hypertension and in those with aortic stenosis (narrowing of the aortic valves).

Acute heart failure

Consider the situation when a blood clot occludes one of the coronary arteries supplying the left heart. The loss of the blood supply prevents the affected region of the myocardium from

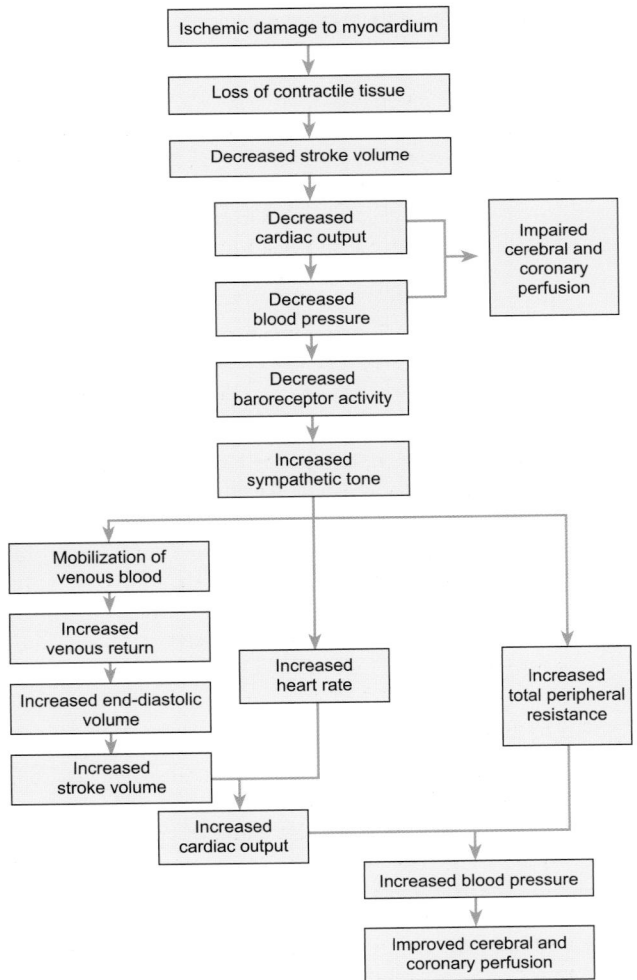

Fig. 31.18 The compensatory changes in the circulation that follow occlusion of one of the left coronary arteries.

contracting normally so that the left ventricle does not pump the blood it contains as efficiently as it should (left-sided heart failure). The first effects are a fall in cardiac output and arterial blood pressure. These changes are followed by a variety of compensatory mechanisms that act to restore the cardiac output as far as possible (see Figure 31.18).

The fall in arterial blood pressure will result in unloading of the arterial baroreceptors, causing a reflex increase in sympathetic stimulation of the heart and the blood vessels. The secretion of epinephrine and norepinephrine by the adrenal medulla is also increased. At the same time, vagal tone will be inhibited. Consequently, the heart rate will increase. This effect is supported by an increase in the contractility of the part of the myocardium that is still being normally perfused with blood. As a result of these two mechanisms, cardiac output increases although it may still be well below normal. The increased vascular sympathetic tone leads to an increased peripheral resistance that acts to support the blood pressure and maintain cerebral perfusion.

These changes occur within about 30 seconds of the occurrence of the coronary thrombosis and give rise to the characteristic symptoms of a heart attack: tachycardia, severe pain in the chest which may radiate to the left arm (*angina pectoris*), pallor (resulting from vasoconstriction of the skin vessels), and sweating ('cold sweat').

The venoconstriction that occurs in response to the increased sympathetic tone leads to mobilization of the blood in the veins and an increase in the venous return to the right heart which, being unaffected, will increase its stroke volume and its output to the pulmonary circulation. In the normal course of events, this would lead to a rapid increase in the output of the left ventricle by the Frank–Starling mechanism. In the present situation, the function of the left ventricle has been impaired by loss of part of the active muscle mass so that it is unable to respond normally to the increased venous return. Consequently, both the end-diastolic volume and the end-diastolic pressure of the left ventricle increase (increased preload). Unless the damage to the myocardium is so extensive that the ventricle cannot respond (in which case death will rapidly follow), the increased preload will eventually result in an increase in left ventricular output so that it matches that of the right ventricle.

Chronic compensated heart failure

The previous sections have discussed the basic mechanisms that operate to restore cardiac output following a coronary thrombosis. In chronic heart failure, these same mechanisms act to compensate for the poor cardiac output that results from ischemic damage to the myocardium, valvular incompetence, or chronic hypertension. This can be a relatively stable condition in which the patient may scarcely be aware of the situation until he or she undertakes exercise, when dyspnea and fatigue set in rapidly. Nevertheless, in all cases of heart failure there is an increase in the end-diastolic pressure required to achieve a given stroke volume, as shown in the ventricular function curves of Fig. 31.19.

In heart failure, the cardiac output is diverted away from the skin and visceral organs in favor of the heart, brain, and skeletal muscle. The reduction in renal blood flow has profound conse-

(a)

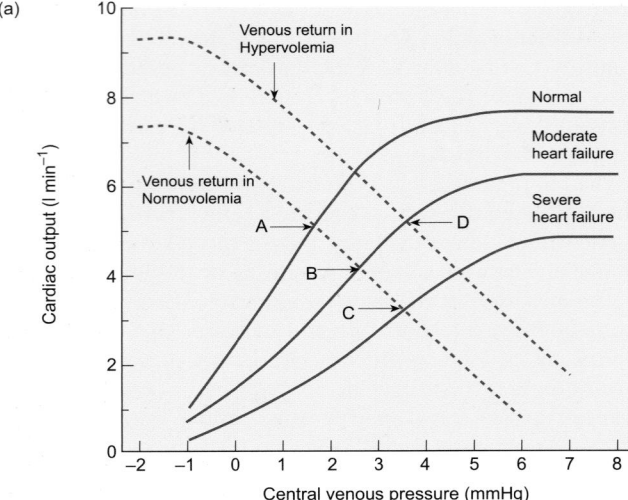

(b)

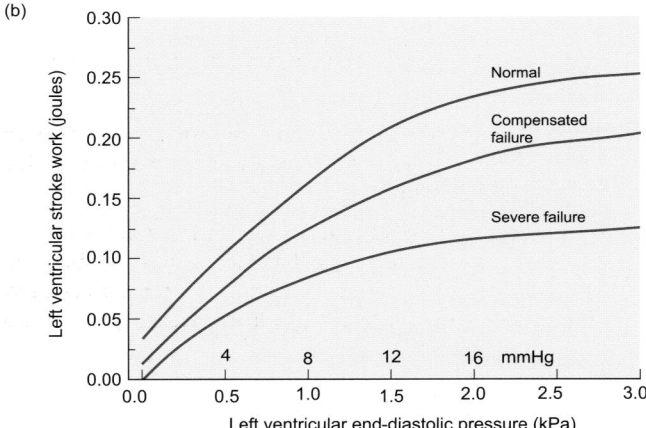

Fig. 31.19 (a) The relationship between central venous pressure and cardiac output for the normal heart, during moderate heart failure, and during severe heart failure. In normal volemia a cardiac output of 5 l min^{-1} is attained when the central venous pressure is less than 0.25 kPa (2 mmHg) (point A). In heart failure, a higher central venous pressure is required to maintain a smaller cardiac output as shown by points B and C. In chronic heart failure, there is hypervolemia and the normal resting cardiac output is attained only with a central venous pressure of 0.5 kPa (4 mmHg) as shown by point D. (b) The relationship between left ventricular stroke work and left ventricular end-diastolic pressure for a normal heart during acute failure, and in compensated failure. (The end-diastolic pressure is proportional to the end-diastolic volume.) Note that in heart failure the function curve is flattened so that a higher end-diastolic pressure is required for a given stroke work. Moreover, the capacity of the heart to perform work becomes severely limited.

quences: following the vasoconstriction of the afferent arterioles, there is an increase in the secretion of renin leading to elevated levels of angiotensin II (see p. 364). This hormone has two important effects: it is a powerful vasoconstrictor and it stimulates the secretion of aldosterone from the adrenal cortex. The low GFR caused by the constriction of the afferent arterioles, coupled with the increased sodium retention caused by the elevated levels of aldosterone, leads to increased fluid retention and expansion of the plasma volume. Similar compensatory mechanisms operate to restore cardiac output following severe

hemorrhage (see Chapter 28), but in this case end-diastolic pressure is not elevated.

Edema in cardiac failure

In chronic heart failure, the plasma volume is expanded because of fluid retention. This lowers the oncotic pressure and increases the capillary pressure so that the Starling forces increasingly favor filtration. As a result, fluid accumulates in the tissues and edema results. The edema may be more evident in the limbs or the lungs depending upon which side of the heart is most affected.

In right-sided heart failure, the increased capillary pressure leads to edema in the periphery. This is most noticeable in the ankles. In acute left-sided heart failure (the most common form of ischemic heart disease), there is a raised pressure in the left atrium and pulmonary vessels. The raised pressure in the pulmonary veins causes them to become engorged and there is increased backpressure on the pulmonary capillaries. The increased pressure in the pulmonary capillaries leads to greater transfer of fluid into the pulmonary interstitium and ultimately to pulmonary edema. In this situation, the diffusing capacity of the lungs is impaired, further compounding the effects of low cardiac output. A vicious cycle develops, leading to death (Fig. 31.20).

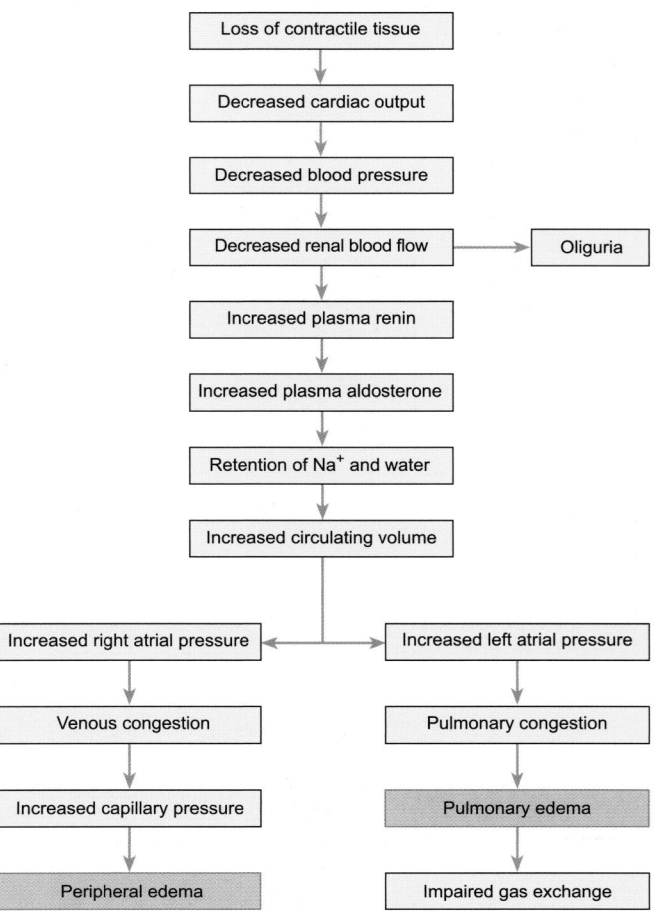

Fig. 31.20 The factors that lead to edema during chronic heart failure.

Principles of treatment

Since the main problem in cardiac failure is the inability of the heart to pump sufficient blood to meet the needs of the circulation, the first requirement is to minimize the demands on the circulation (see Box 31.3 for the classification of the severity of heart failure). This can be achieved by rest. Following a mild coronary thrombosis, the physiological compensatory mechanisms discussed earlier help to maintain an adequate cardiac output and blood pressure. Over the ensuing months the remaining heart muscle hypertrophies and there is an increase in vascularization in those areas of the myocardium that became hypoxic as a result of the occlusion of part of the blood supply. These adaptations improve the contractility of the heart and eventually a normal pattern of life can be resumed.

In severe heart failure the situation is more complicated and treatment has three aims.

1. To reduce the work of the heart—this can be achieved by rest.

Summary

1. Heart failure occurs when the heart is unable to pump sufficient blood at normal filling pressures to meet the metabolic demands of the body. Failure may involve either of the two ventricles alone or both the left and right ventricles simultaneously.

2. Heart failure may result from ischemic damage to the myocardium resulting in loss of contractility, incompetence or narrowing of the heart valves resulting in increased preload, or excessive peripheral resistance resulting in increased afterload.

3. When heart failure occurs, there is increased sympathetic activity—increased heart rate, increased contractility of the unaffected myocardium, and vasoconstriction. These changes help to compensate for the reduced cardiac output. The increased venous return leads to a greater end-diastolic volume and restoration of the stroke volume by the Frank–Starling mechanism.

2. To reduce the circulating volume and the resultant cardiac dilatation—this can be achieved by administration of diuretics such as frusemide.

3. To improve myocardial contractility—this is often attempted by administration of drugs that have a positive inotropic effect on the myocardium. Examples are the cardiac glycosides (e.g. digoxin and ouabain) and β_1-adrenoceptor agonists such as dobutamine.

31.7 Renal failure

Renal failure occurs when the function of the kidneys is depressed to such an extent that they are unable to maintain the composition of the plasma within normal limits. Almost any factor that seriously impairs the function of the kidneys can cause renal failure. Broadly speaking, renal failure may develop rapidly (*acute renal failure*) or it may develop over a considerable period because of a significant loss of functional tissue (*chronic renal failure*). Acute renal failure may be caused by glomerulonephritis (inflammation of the glomerulus resulting from immune reactions to infection), renal ischemia, and nephrotoxic poisons such as the mercuric ion (Hg^{2+}). Chronic renal failure occurs when the proportion of damaged nephrons is so high that the kidneys are unable to perform their normal functions. This generally occurs when the GFR falls below about 25 ml min^{-1}. (It is normally about 120 ml min^{-1}—see Chapter 17).

The first sign of acute renal failure is a marked reduction in urine output. In glomerulonephritis, the inflammation of the glomeruli leads to their occlusion and a decrease in the GFR. As a result, the normal transport mechanisms of the proximal convoluted tubule are able to absorb a higher proportion of the filtered Na^+ and this results in a low Na^+ concentration in the fluid reaching the macula densa. Consequently, the secretion of renin increases and plasma angiotensin II levels become elevated. In turn, this leads to increased secretion of aldosterone

---

BOX 31.3 CLASSIFICATION OF HEART FAILURE

To guide the treatment of patients with a history of cardiac disease, heart failure can be classified according to the remaining functional capacity. This idea was first proposed by the New York Heart Association in 1928 and has been revised on several occasions. The main features of the most recent classification are summarized below:

◆ Class I. Patients who have had a history of heart disease but have recovered sufficiently to undertake normal physical activity that does not cause undue fatigue, palpitations, breathlessness (dyspnea), or chest pain.

◆ Class II. Patients who are comfortable at rest but in whom the ability to undertake physical activity is slightly limited. Ordinary physical activity results in fatigue, palpitation, breathlessness, or chest pain.

◆ Class III. Patients who are comfortable at rest but are very limited in the amount of physical activity they can undertake. Minor physical activity results in fatigue, palpitation, breathlessness, or chest pain.

◆ Class IV. Patients who cannot undertake any physical activity without discomfort. Symptoms of heart failure or chest pain may be present even at rest. Discomfort increases with physical activity.

Objective assessments:

A. No objective evidence of cardiovascular disease.
B. Objective evidence of cardiovascular disease.
C. Objective evidence of moderately severe cardiovascular disease.
D. Objective evidence of severe cardiovascular disease.

These two classifications are combined to provide a concise clinical description. For example, a patient who is comfortable at rest but is quickly fatigued when exercising normally and has a severe obstruction of a major coronary artery is classified as Functional Capacity II, Objective Assessment D.

from the adrenal cortex and a greater Na$^+$ absorption by the distal tubule. Since the absorbed Na$^+$ will be accompanied by its isotonic equivalent of water, there is fluid retention and edema develops. This may be followed by hypertension, which persists until the condition subsides.

Apart from glomerulonephritis, the most common cause of acute renal failure is a sudden reduction in cardiac output resulting from heart failure or hemorrhage. This elicits the normal circulatory adjustments discussed earlier. There is a pronounced vasoconstriction of the afferent arterioles resulting from increased sympathetic activity in the renal nerves and increased plasma levels of epinephrine and vasopressin (ADH). This vasoconstriction overrides the normal autoregulatory response of the renal circulation and the GFR falls sharply. Since the afferent arterioles are constricted, the pressure in the peritubular capillaries is reduced and there is increased absorption of the tubular fluid (which is mainly isotonic NaCl). This leads to a reduction in concentration of Na$^+$ in the fluid reaching the macula densa of the distal tubule. In response to this there is an increased secretion of renin, leading to increased plasma levels of angiotensin II and further vasoconstriction.

The increased plasma levels of angiotensin II lead to increased secretion of aldosterone from the adrenal cortex and further Na$^+$ absorption in the distal tubule. The decline in the concentration of Na$^+$ in the fluid of the distal tubule tends to decrease the secretion of H$^+$ and K$^+$ into the tubular fluid which, if not corrected, will lead to acidosis and hyperkalemia.

The increased plasma levels of ADH promote further water absorption in the collecting tubule. As a result, the kidneys produce a small volume of concentrated urine that is low in Na$^+$. As discussed in Chapter 28, this renal response will tend to correct the fall in the effective circulating volume. Prompt restoration of the effective circulating volume at this stage will restore the normal function of the kidneys. If the circulatory insufficiency is not corrected, the severe vasoconstriction will lead to tissue ischemia, cell death, and tubular necrosis.

Chronic renal failure results when more than three-quarters of the functional renal tissue is lost. As a result, the GFR falls

substantially and the concentration of urea in the blood rises (uremia). The impaired tubular function leads to failure of normal ionic regulation, acidosis, and the accumulation of metabolites. Unless corrective measures are applied, the accumulation of metabolites (particularly nitrogenous metabolites from protein catabolism) and the disturbance of normal ionic balance lead to CNS depression, coma, and eventually death. In relatively mild chronic renal failure, dietary control of protein intake plus supplementation of the diet by administration of vitamins and essential amino acids may be adequate to maintain the patient in a reasonable state of health. If the GFR falls below about 5 ml min^{-1}, dialysis becomes necessary. A more or less normal lifestyle is possible following renal transplantation.

31.8 Liver failure

The liver is the largest and, in function, one of the most versatile organs in the body. Despite its great importance in metabolism, hepatic disease only rarely leads to serious illness. This fortunate circumstance arises because the liver has a great reserve capacity and a remarkable ability to regenerate. Liver failure is diagnosed when there is evidence of jaundice, fluid accumulation in the peritoneal cavity (ascites), failure of blood clotting, and marked psychological changes.

The many functions of the liver are summarized in Table 31.3. Its principal role in digestion is the secretion of bile. Bile salts are essential for the digestion and absorption of fat and fat-

Summary

1. Renal failure occurs when the kidneys are unable to regulate the composition of the plasma. It may be acute in onset (as in glomerulonephritis and hemorrhage) or it may develop over time (chronic renal failure). In either case, there is a marked reduction in the production of urine.

2. In acute renal failure due to glomerulonephritis there is activation of the renin–angiotensin system leading to fluid retention. During hemorrhage, there is also a pronounced vasoconstriction due to the secretion of vasopressin.

3. Chronic renal failure occurs when more than three-quarters of the functional renal tissue is lost. This leads to a progressive failure of normal ionic regulation and an increased accumulation of urea and other metabolites in the plasma.

TABLE 31.3   The normal functions of the liver

Function	Comments
Synthesis and secretion of bile	Bile is important for fat absorption and for the excretion of bile pigments, which are derived from the breakdown of hemoglobin
Carbohydrate metabolism	The liver stores the body's reserve of carbohydrate in the form of glycogen. It is also able to form glucose by gluconeogenesis
Fat metabolism	Absorption of fats and fat soluble vitamins (see above)
	Storage of fat-soluble vitamins
	Synthesis of lipoproteins
Protein metabolism	The liver is the major source of plasma proteins including albumin and the clotting factors
Detoxification	Inactivation of hormones
	Conjugation of various drugs and toxins for excretion
	Conversion of ammonia to urea
Iron storage	Essential for erythropoiesis
Storage of vitamin B$_{12}$	Required for normal erythropoiesis
General	As the liver is situated between the gut and the general circulation, it can protect the body by inactivating toxic materials absorbed from the gut

soluble vitamins (see Chapter 18). The liver plays a vital role in the maintenance of normal plasma glucose levels (see Chapter 27); it stores glycogen when plasma glucose levels are high, and releases glucose to the blood by glycogenolysis when plasma glucose is low. It can also synthesize glucose from non-carbohydrate precursors (gluconeogenesis).

The liver plays an important part in the formation and destruction of red blood cells (see Chapter 13, pp. 230–232). It removes the bilirubin formed by the breakdown of red cells and excretes it in the form of bile pigments. The liver is also concerned with the destruction or modification of toxic substances prior to their removal by the kidneys. Drugs and other chemicals, as well as certain normal physiological substances, are 'detoxified' by the liver cells in this way. For example, barbiturates are completely destroyed by the liver, while other agents are conjugated with glucuronic acid, glycine, acetate, etc. to render them suitable for excretion. Steroid hormones, including estrogens, progesterone, testosterone, glucocorticoids, and aldosterone, are also metabolized by the liver.

The hepatocytes manufacture the majority of the plasma proteins including clotting factors such as prothrombin, fibrinogen, and Factors V, VII, IX, and X. The liver is also involved in the catabolism of proteins and converts the ammonia formed by deamination of excess amino acids to urea.

It will be clear from the above brief description of the numerous functions subserved by the liver that the consequences of hepatic failure or insufficiency will be widespread and extremely grave. The liver is subject to most of the disease processes that can affect other body structures including inflammation, vascular disorders, metabolic disease, toxic injury and neoplasms. Two of the most common hepatic diseases are hepatitis, characterized by inflammation of the liver, and cirrhosis, characterized by fibrosis and a loss of normal structure and function. Alcoholic cirrhosis is the most common form of the disease. The manifestations of liver failure reflect its various functions. For the purposes of discussion, the consequences of liver failure can be divided into digestive effects, effects on the blood, and effects concerning the excretion of toxic substances.

The effects of liver failure

In liver failure, severe hypoglycemia develops rapidly because of the central role played by the liver in the maintenance of plasma glucose. However, galactose levels tend to rise because galactose is no longer converted to glucose by the liver. Bile production will slow or cease, and this will impair the digestion and absorption of fats. Since the ammonia formed by deamination of amino acids is only converted to urea in the liver, the level of ammonia in the plasma will rise while the concentration of urea will fall.

The protein profile of the blood will change radically following liver failure. The production of plasma protein by the liver will be impaired, leading to a loss of albumin and certain globulins. If the total plasma protein level falls significantly there will be a reduction in the plasma oncotic pressure, which may lead to generalized edema or fluid accumulation in the peritoneal cavity (ascites). At the same time, many of the enzymes contained within the hepatocytes will enter the blood after leaking out of the damaged cells. These enzymes include the amino-

Summary

1. During liver failure, many physiological processes are disrupted. In acute hepatic failure there is hypoglycemia, disordered lipid metabolism, and decreased protein synthesis. The diminished synthesis of albumin leads to the formation of ascites and peripheral edema. The loss of those clotting factors that are synthesized in the liver leads to disorders of blood clotting.

2. The liver plays a central role in the destruction of red blood cells and iron metabolism. When liver failure occurs, there is an increased level of bilirubin in the plasma, leading to jaundice.

3. Liver failure leads to failure of detoxification mechanisms, which prolongs the action of the steroid hormones giving rise to the symptoms of endocrine disease.

transferases (or transaminases), alkaline phosphatase, and lactate dehydrogenase which catalyses the conversion of pyruvic acid to lactic acid. The levels of these enzymes in the plasma may be used clinically to provide an index of the extent of liver damage.

The liver plays a vital role in the excretion of bilirubin derived from the breakdown of red blood cells. In liver failure, this pathway is blocked, and unbound unconjugated bilirubin accumulates in the plasma. Gradually the patient will develop jaundice, a characteristic yellowing of the skin resulting from the presence of high levels of bilirubin. In severe cases, damage to the basal ganglia of the brain may occur, a condition known as kernicterus or brain jaundice.

In liver disease, the production of clotting factors will fall, leading to an impairment of blood clotting which may result in spontaneous bleeding from the skin and mucous membranes.

Finally, a failing liver is less able to carry out detoxification of poisonous chemicals. For this reason, drug actions are likely to be prolonged in patients in liver failure. Disturbances in gonadal function may also arise from the impairment of sex steroid hormone metabolism. There may also be signs of Cushing's syndrome, sodium retention with subsequent edema, and hypokalemia arising from a reduction in the rate of metabolism of gluco- and mineralocorticoids.

The treatment of liver failure

The treatment of liver failure is, of necessity, complex and must address the many physiological processes throughout the body that are disturbed. If the initial problem was alcoholic cirrhosis, the first step in treatment will be to eliminate alcohol intake. Sufficient carbohydrate and calories will be required to prevent protein breakdown, fluid and electrolyte imbalances must be corrected, and protein intake may be limited to inhibit the production of ammonia. Liver transplantation is now a realistic form of treatment for patients in the final stages of liver failure.

31.9 Multiple organ dysfunction syndrome

Multiple organ dysfunction syndrome (MODS) is the progressive dysfunction of two or more organ systems so that the body is unable to maintain normal homeostasis. Although sepsis and septic shock are the most common causes of MODS, it may also be triggered by many other conditions including burns, adult respiratory distress syndrome (ARDS, see Chapter 16 p. 342), acute pancreatitis, and major surgery. In many cases, MODS progresses to multiple organ failure and death. If two organ systems have failed, mortality is 45–55 per cent while if three or more systems are affected, mortality rises to 80 per cent. If multiple organ failure persists for more than four days, death is virtually inevitable.

Although MODS may arise as a direct result of injury to individual organs (primary MODS), more often it develops some time after the initial trauma, as a result of a massive inflammatory reaction (secondary MODS). The mechanisms thought to be responsible for secondary MODS are summarized below.

Damage to an organ such as the lungs can cause macrophages to release large quantities of inflammatory mediators such as IL-1 (interleukin-1), IL-6 (interleukin-6), and TNF-α (tumor necrosis factor α). These inflammatory mediators have a number of important actions on the vascular endothelium. First, IL-1 and TNF-α induce the synthesis of the powerful chemotatic agent IL-8, which attracts neutrophils to the site of synthesis where they are induced to congregate by the expression of cell adhesion proteins known as *selectins* by the endothelial cells. Normally, this is a local response that permits neutrophils to migrate into infected or damaged tissues in order to engulf invading organisms or cell debris (see Chapter 14 p. 251). However, in severe inflammation, the endothelial cells themselves are attacked by the activated neutrophils and the 'respiratory burst' associated with neutrophil activation (see Chapter 14) produces many highly toxic oxygen free radicals (reactive oxygen species that are oxygen atoms with an unpaired electron in their outer electron ring) that damage the vascular endothelium. The reactive oxygen species attack DNA and cause peroxidation of membrane lipids. Both lead to necrosis. Furthermore, the neutrophils release proteases that directly damage the endothelium and other cell membranes.

In MODS the coagulation pathway may become activated in many parts of the circulation as part of the systemic inflammatory response. Circulating IL-1 and TNF-α activate phospholipase A2 on the endothelial cell membranes. This leads to the release of arachidonic acid from the membrane phospholipids and the synthesis of thromboxane A_2. Platelet activation factor is also released by the endothelial cells and the platelets aggregate. These changes initiate the blood clotting cascade (see p. 239). The clotting cascade is also activated when Factor XII comes into direct contact with the tissues following the breakdown of the capillary walls that follows the damage caused by the activated neutrophils. As a result, microvascular thrombosis occurs throughout the body, resulting in impaired circulation in the microvasculature and tissue ischemia. Furthermore, activated Factor XII forms kallikrein from prekallikrein at the sites of inflammation and kallikrein releases the potent vasodilator bradykinin from an α-globulin known as kininogen. In addition, kallikrein activates the complement cascade and can convert plasminogen into plasmin (see p. 241). Components of the complement cascade, in particular C3a and C5a (the *anaphylatoxins*),

stimulate the release of histamine from mast cells. The released histamine produces vasodilatation and increases vascular permeability. As a result of all these changes, there is increased capillary permeability, loss of fluid and protein from the circulation, and organ damage.

The gastrointestinal mucosa is especially vulnerable to the inflammatory mediators released by neutrophils and macrophages and, as the gut barrier is breached, bacteria gain access to the circulation from the intestine. Because of this, systemic spread of infection becomes more likely.

An endocrine stress response accompanies the systemic inflammatory response as the body strives to maintain homeostasis. There is an increase in catecholamine secretion which contributes to tachycardia, increased metabolic rate, and oxygen consumption. The secretion of many other hormones such as cortisol, GH, insulin, glucagon, and vasopressin is also enhanced. However, as the stress persists and homeostatic mechanisms begin to fail, the endocrine changes become excessive and injurious, particularly the vasoconstrictor actions of vasopressin on the gastrointestinal tract and the kidneys.

The clinical manifestations of MODS and the order in which they generally appear are:

◆ Fever, tachycardia, tachypnea, dyspnea, and a generalized increase in metabolism are seen within hours of the physical injury or onset of infection.

◆ Adult respiratory distress syndrome (ARDS) develops within a few days of the onset of the first symptoms.

◆ Signs of gastrointestinal, hepatic, and renal failure may develop within 7–10 days. There is some evidence that bacteria cross the damaged gut wall. The hypermetabolic state intensifies. Liver failure (see p. 604) may occur and is characterized by raised blood levels of liver enzymes, decreased serum transferrin, and jaundice. As discussed on p. 603, renal failure is characterized by increased blood levels of urea and creatinine, with oliguria (falling urinary flow rate) that later progresses to anuria (failure of urine production). Signs of gastrointestinal failure include abdominal distension, ascites (see p. 560), bleeding, acute gastritis, mucosal ulceration, and decreased bowel sounds.

◆ All of the above become more severe with time and cardiovascular dysfunction begins to develop with decreased systemic vascular resistance, decreased right atrial pressure, and increased heart rate.

◆ Unless ameliorated by drug treatments, later signs include CNS changes (with convulsions and eventual loss of consciousness), disseminated intravascular coagulation, leucopenia, thrombocytopenia, and profound hypotension.

The changes in MODS may evolve slowly and progressively or may occur suddenly. Dysfunction frequently progresses to failure. The patient's prognosis is related to the original cause of the organ failure. Mortality from trauma induced MODS is significantly lower than that which is secondary to sepsis or pancreatitis. Nevertheless, the eventual outcome is highly dependent upon the facilities available for treatment of the multiple pathological changes that occur in this condition.

Further reading

Campbell, E.J.M., Dickinson, C.J., Slater, J.D.H., Edwards, C.R.W, and Sikora, E.K. (eds.) (1984). *Clinical physiology* (5th edn). Blackwell Scientific, Oxford.

Case, R.M., and Waterhouse, J.M. (eds.) (1994). *Human physiology: age, stress and the environment*, Chapter 4. Oxford Science Publications, Oxford.

Levick, J.R. (2003). *An introduction to cardiovascular physiology* (5th edn), Chapters 2 and 18. Hodder Arnold, London.

Pharmacology of the heart and circulation:

Grahame-Smith, D.G., and Aronson, J.K. (2002). *Clinical pharmacology and drug therapy* (3rd edn), Chapter 23. Oxford University Press, Oxford.

Rang H.P., Dale, M.M., Ritter, J.M., and Moore. P. (2003). *Pharmacology* (5th edn), Chapters 17 and 18. Churchill-Livingstone, Edinburgh.

Medicine

Brown, E.M., Collis, W., Leung, T., and Salmon, A.P. (2002). *Heart Sounds Made Easy*. Churchill-Livingstone, Edinburgh.

Hampton, J.R. (1998). *The ECG made easy* (5th edn). Churchill-Livingstone, Edinburgh.

Ledingham J.G.G., and Warrell, D.A. (2000). *Concise Oxford textbook of medicine*, Chapters 2.2, 2.31–2.34, 5.37, 12.28–12.29, and 16.4. Oxford University Press, Oxford.

Kumar, P., and Clark, M. (eds.) (1998). *Clinical medicine* (4th edn), Chapter 13. Saunders, Edinburgh.

Wagner, G.S. (2001). *Marriott's practical electrocardiography* (10th edn). Lippincott-Williams & Wilkins, Philadelphia.

SI units

A system of units based on the meter, kilogram, and second has now been adopted internationally. This system is known as the 'Système International des Unités' or SI system of units. There are seven basic units:

Physical quantity	Name of unit	Standard symbol
Mass	kilogram	kg
Length	meter	m
Time	second	s
Electric current	ampere	A
Temperature	degree Kelvin	K
Light intensity	candela	cd
Amount of substance	mole	mol

All other units are derived from these base units. The principal derived SI units are:

Physical quantity	Unit	Standard symbol	Definition
Electrical potential	volt	V	$J\,A^{-1}\,s^{-1}$
Energy	joule	J	$kg\,m^2\,s^{-2}$
Force	newton	N	$J\,m^{-1}$
Frequency	hertz	Hz	s^{-1}
Power	watt	W	$J\,s^{-1}$
Pressure	pascal	Pa	$N\,m^{-2}$
Volume	liter	l (or dm^3)	$10^{-3}\,m^3$

Each unit can be expressed as a multiple of ten or as a decimal fraction. The most important of these in physiology are:

Multiple		Name	Symbol
1000	$= 10^3$	kilo	k
0.1	$= 10^{-1}$	deci	d
0.001	$= 10^{-3}$	milli	m
0.000001	$= 10^{-6}$	micro	μ
0.000000001	$= 10^{-9}$	nano	n
0.000000000001	$= 10^{-12}$	pico	p
0.000000000000001	$= 10^{-15}$	femto	f

Under the SI system the standard volume for expressing concentrations is the liter. Thus a plasma protein concentration of 7g per 100 ml should be expressed as 70 g l^{-1}, although 7 g dl^{-1} (7 grams per deciliter) is equally correct. Where the molecular weight of a constituent of one of the body fluids is known, its concentration should be expressed as its molar concentration (moles per liter). Thus the plasma sodium concentration should be expressed as 0.14 mol l^{-1} or 140 mmol l^{-1}. The same rule applies for expressing cell counts in blood, so that a red cell count of 5×10^6 cells per microliter on the old system is now expressed as 5×10^{12} cells per liter.

The unit of pressure in the SI system is the pascal, which is 1 newton per square meter (N m^{-2}) as pressure is force per unit area. The conventional unit of pressure is millimeters of mercury (mmHg) which is still widely used. To convert from mmHg to pascals multiply by 133.325. For kilopascals multiply by 0.133325. A pressure of 7.5 mmHg is equivalent to 7.5×0.133325 kPa $= 0.9999375$ kPa. Thus, to a good approximation:

7.5 mmHg	= 1 kPa
15 mmHg	= 2 kPa
40 mmHg	= 5.3 kPa
60 mmHg	= 8 kPa
75 mmHg	= 10 kPa
100 mmHg	= 13.3 kPa
150 mmHg	= 20 kPa
760 mmHg	= 101 kPa

The unit of temperature is K (degrees Kelvin) but °C (degrees Celsius) is still commonly used. To convert from degrees Celsius to degrees Kelvin add 273.15. Thus

$$37\,°C = 37 + 273.15$$
$$= 310.15\,K.$$

The calorie is not an SI unit as the joule is used as the unit for energy. Heat is merely one form of energy. To convert from calories to joules multiply by 4.185. For example, the energy equivalent of 100 g of bread is 240 calories or 240×4.185 J (= 1004 J).

Resuscitation following cardiac or respiratory arrest

This appendix describes the basic life support measures that can be taken immediately, without equipment, to revive someone who has collapsed or drowned. The appearance of a person who has collapsed will depend on the immediate cause, but it may be very blue (cyanosis resulting from asphyxia), gray (primary cardiac arrest), or very pale (hypovolemia and hypothermia). First, establish that resuscitation is necessary by feeling for a pulse either in the neck or in the groin and by observing the chest and abdomen for respiratory movements. Only when it is established that these are absent will it be necessary or desirable to begin artificial ventilation or cardiac massage.

Basic resuscitation has three components: ABC—airway, breathing, and circulation. The aim is to maintain a flow of oxygenated blood to the vital organs until the victim's own breathing and circulation can be re-established.

Step 1—Airway

If the airway is obstructed, air cannot enter the lungs and asphyxia will result. Therefore the first step is to clear the airway. The most common cause of airway obstruction is the tongue falling back to block the oropharynx. Other common causes are the inhalation of regurgitated gastric contents or choking on food. Lay the victim on one side, prize the jaws apart, and clear the mouth with the index finger. Remove any false teeth. Lie the victim on his or her back and tilt the head backwards as this pulls the tongue away from the back of the pharynx (see figure). This may be sufficient to permit the resumption of normal breathing.

In conscious people who are choking, it may be possible to dislodge any impacted material by blows to the back. The procedure is as follows. Lay the victim on his or her side or, in the case of a small child, along the rescuer's thigh. Then, with the flat of the hand, thump the middle of the back four to six

times as the victim attempts to breathe out. The increase in intrathoracic pressure will help to expel any material lodged in the airway.

An alternative method that can be applied to adults is known as the Heimlich maneuver. The victim should be held firmly around the upper abdomen just below the rib cage. Then a series of sharp upward thrusts should be delivered to dislodge the obstructing material.

Step 2—Breathing

If respiration does not commence spontaneously, it will be necessary to begin artificial respiration. The recommended method is mouth-to-mouth respiration as shown in the figure. The rescuer should pinch the nose with one hand, take a good breath, seal his or her mouth over that of the victim, and exhale, watching to see if the victim's chest rises and then falls as the rescuer takes another good breath. If there are no chest movements, check the airway again. It may be necessary to tilt the head back further, but extreme care should be taken if a neck injury is suspected.

Step 3—Circulation

Circulation of the blood is promoted by external cardiac massage. This is performed by placing the heel of one hand over the lower two-thirds of the sternum with the other hand on top as shown in the figure. The chest is then compressed by 4–5 cm and released at approximately the same rate. About 15 compressions should be given at a rate of about 80 per minute followed by two further breaths by the mouth-to-mouth method described above. This should be followed by a further 15 compressions and the cycle repeated until the patient is able to breathe spontaneously.

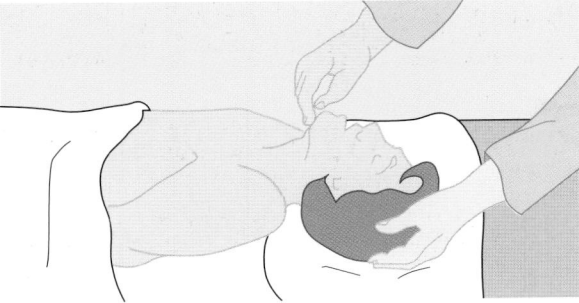

Step 1 Clear airway

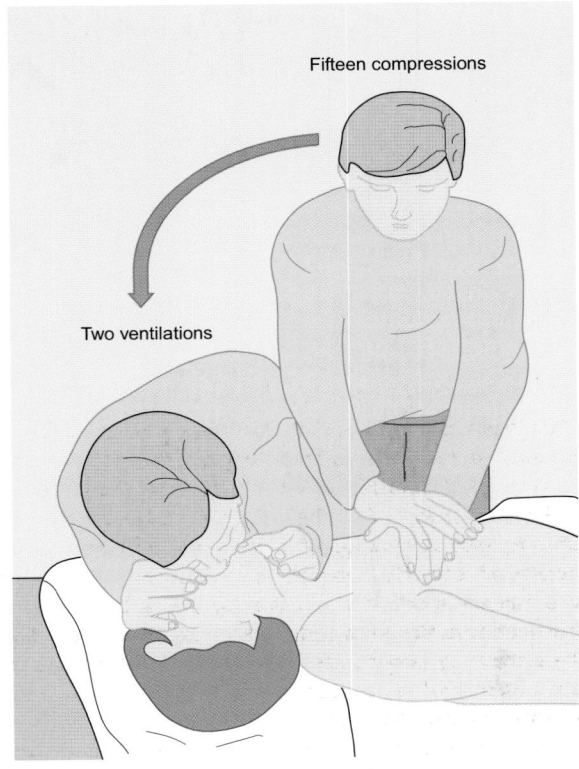

Fifteen compressions

Two ventilations

Step 3 One-rescuer CPR sequence

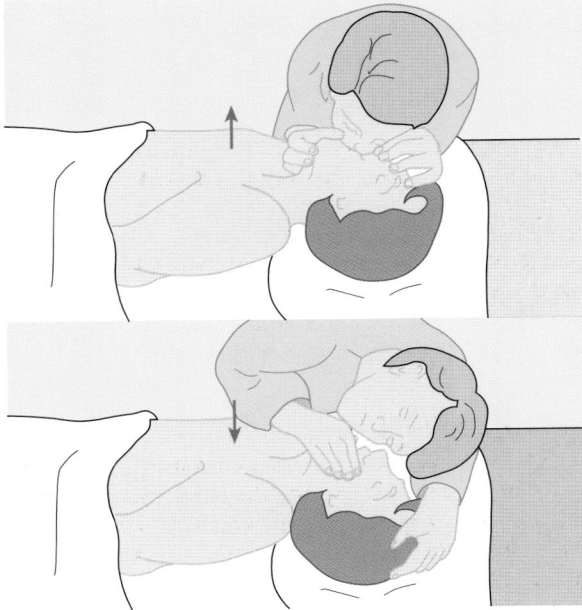

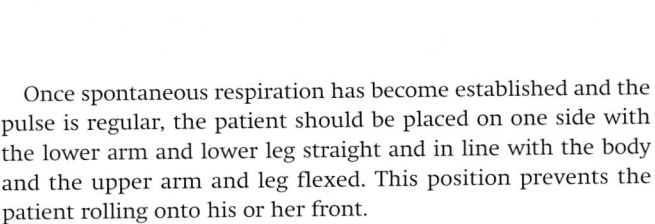

Step 2 Mouth-to-mouth ventilation

Once spontaneous respiration has become established and the pulse is regular, the patient should be placed on one side with the lower arm and lower leg straight and in line with the body and the upper arm and leg flexed. This position prevents the patient rolling onto his or her front.

Further reading

Baskett, P.J.F. (1993). *Resuscitation handbook* (2nd edn). Wolfe, London.

Index

electrocardiogram 267–73
heartbeat 266–7
hemodynamics: blood flow and
        pressure 283–9
microcirculation and tissue fluid
        exchange 293–9
skeletal muscle, circulation of 304–5
skin, circulation of 304
gross anatomy of 263–4
rate 306, 464, 522
sounds 591–3
heart failure 600–3
acute 601
cardiac 600
chronic compensated 601–2
classification 603
edema 602
extracardiac 600–1
treatment 603
heartbeat 266–7
heartburn 112, 387, 467
heat:
cramp 553
exhaustion 534
production (calorigenesis) 206, 511–12, 516
stroke 534
thermoregulatory responses 533
see also body temperature
height 324, 425, 495, 496
Heimlich maneuver 610
Helicobacter pylori 390, 392
helicotrema 131
heliox 581
helium dilution method 319
helium-oxygen mixtures 581, 582
HELLP syndrome 465
helper cells 254
hematocrit 228, 285, 523
hematopoiesis 230–1, 579
heme group 233
hemiballismus 161
hemidesmosomes 26
hemoconcentration 523
hemodilution 556
hemodynamics: blood flow and pressure 283–9
arterioles and vascular resistance 287
blood viscosity and blood flow 284–5
capillary pressure 287
measurement 285–6
normal arterial pressure 286–7
skeletal muscle pump 287–9
systemic arteries 285
vascular resistance 284
venous pressure 287
hemoglobin 228, 230, 231, 237, 341
acid–base balance 566, 568
altitude, high 579
and carbon monoxide binding 235
exercise 524
fertilization and pregnancy 459, 465, 470
fetal and neonatal physiology 482
and high altitude 579
kidneys and internal environment
        regulation 348, 354, 355
and oxygen 233–4, 340
pH, PCO$_2$, 2,3-BPG, and temperature 234–5
plasma glucose regulation 538
see also anemia
hemolysis 243

hemolytic anemia 423
hemolytic disease of newborn 244
hemolytic jaundice 230
hemophilia 15, 240, 242–3
hemorrhage 232, 281, 293, 305, 352, 360, 550,
        554–7
hemorrhoids 413
hemostasis 239–43
blood coagulation 239–40
and calcium 241
clot dissolution 241
clot formation 242
clot retraction 240–1
clotting and endogenous anticoagulants 241
clotting factors, impaired synthesis of 243
failure of normal clotting mechanisms 242
hemophilias 242–3
platelets, role of 239
vascular disorders 243
vasoconstriction 239
Henderson–Hasselbalch equation 565, 567, 574
Henry's law 314, 567, 580
heparin 241
hepatic arterioles 404
hepatic artery 382, 404
hepatic bile 404, 408
hepatic duct 403
hepatic jaundice 407
hepatic vein 382, 404
hepatitis 407, 605
hepatocellular necrosis 465
hepatocytes 403, 404, 405, 406, 605
hepcidin 232
Hering, E. 126–7
Hering-Breuer lung inflation reflex 333–4
Hess, R. 181
heterochromatin 21
heterotrimeric G proteins 54–5
hexosaminidase A 22
hexose 9
hiatus hernia 387
high blood pressure *see* hypertension
high-pressure nervous syndrome (HPNS) 582
higher nervous function *see* cerebral
        hemispheres
hilus 348–9
hippocampus 185, 186
Hippocrates 586
Hirschsprung's disease 380
hirsutism 215
histamine 50, 339
digestive system 381, 391
distributive shock 557, 559
inflammation and immunity 252
multiple organ dysfunction syndrome 606
sensory systems 113
histones 21
Hodgkin, A.L. 68
homeostasis 3–4
homologs 24, 25
horizontal canal 138
hormonal regulation 189–224
adrenal glands 210–17
of blood vessels 292–3
of blood volume 301
calcium balance 218
digestive system 401
female reproductive system 443–5
fertilization and pregnancy 460

gastrointestinal tract 222, 381
growth hormone 199–203
of heart rate 279–80
hypocalcemia and tetany 218
and lactation 473–4
phosphate, whole-body handling of 218
plasma calcium and phosphate 218–22
thyroid gland 203–10
see also pituitary gland and hypothalamus
hormone replacement therapy 210, 447
hormone-dependent cyclical changes 444–5
hormones 3, 50, 52, 290, 516–17
chemical nature, carriage in blood, and
        modes of action 191
measurement of levels in body fluids 191
patterns of secretion—circadian rhythms
        and feedback control 191–2
secretion of 50–1
see also specific hormones
hormone-sensitive lipase (HSL) 487
host cells 248, 249
'housekeeper' contractions 398
human chorionic gonadotropin (hCG) 455,
        460, 461, 467
human chorionic somatomammotropin *see*
        human placental lactogen
Human Genome Project 15
human immunodeficiency virus 258, 470
human placental lactogen (hPL) 460, 467, 468,
        469, 472
human split-brain operation 175–7
hunger 425
'hunting reaction' 532
Huntingdon's chorea 161, 183
Huxley, A.F. 68
hydatidiform disease 465
hydrocephalus 307
hydrochloric acid (HCl) 388, 389, 391
hydrogen/hydrogen ions 5, 37–8
acid–base balance 563, 564, 566, 567, 569,
        570, 571, 572–4
-ATPase 572–3
blood, properties of 227, 236, 237
buffers of body fluids 566–8
digestive system 388, 389, 391, 400, 409–10,
        413
fetal and neonatal physiology 487
heart and circulation 303, 305
intercalated cells 366
-K+-ATPase 389
kidneys and internal environment
        regulation 354, 360, 361, 366
plasma membrane 37
respiratory system 336
sensory systems 139
hydrogen sulphide 413, 415
hydrophobic molecules 34
hydrophobic signaling 53
hydrostatic pressure 288, 329, 355, 356, 549,
        557
hydroxybutyrate 30, 568, 571
hydroxyl ions 389, 564, 566
hydroxyproline 505
5-hydroxytryptamine *see* serotonin
5-hydroxytryptophan 181
hyperbaric oxygen therapy 342
hyperbilirubinemia 406
hypercalcemia 220, 221, 423
hypercapnia 301, 305, 329, 336, 337, 341, 582